The Flavonoids

ADVANCES IN RESEARCH
SINCE 1980

The Flavonoids

ADVANCES IN RESEARCH SINCE 1980

Edited by

J.B. HARBORNE

London
CHAPMAN AND HALL
New York

First published in 1988 by
Chapman and Hall Ltd
11 New Fetter Lane, London EC4P 4EE
Published in the USA by
Chapman and Hall
29 West 35th Street, New York NY 10001

Printed in Great Britain by
J.W. Arrowsmith Ltd., Bristol

ISBN 0 412 28770 6

British Library Cataloguing in Publication Data

The flavonoids: advances in research.
Since 1980
1. Flavonoids
I. Harborne, J.B.
547.7 QP925.F5

ISBN 0-412-28770-6

Library of Congress Cataloging in Publication Data

The Flavonoids: advances in research since 1980.

1. Flavonoids. I. Harborne, J.B. (Jeffrey B.)
QK898.F5F554 1988 582'.019'218 88-11879
ISBN 0-412-28770-6

Contents

Contributors

B.A. Bohm
Department of Botany, University of British Columbia, Vancouver, Canada

R. Brouillard
Institut de Chimie, Université Louis Pasteur, Strasbourg, France

J. Chopin
Laboratoire de Chimie Biologie, Université Claude Bernard, Villeurbanne, France

G. Dellamonica
Laboratoire de Chimie Biologie, Université Claude Bernard, Villeurbanne, France

P.M. Dewick
Department of Pharmacy, University of Nottingham, UK

D.M.X. Donnelly
Department of Chemistry, University College, Dublin, Eire

G. Forkmann
Biologisches Institut II, Tubingen, West Germany

H. Geiger
Windhalmweg 14, D 7000 Stuttgart 70, West Germany

D.E. Giannasi
Department of Botany, University of Georgia, Athens, USA

R.J. Grayer
Plant Science Laboratories, University of Reading, UK

J.B. Harborne
Plant Science Laboratories, University of Reading, UK

W. Heller
Institut für biochemisches Pflanzenpathologie, Neuherberg, West Germany

M. Jay
Institut für Botanik, Technische Hochschule, Darmstadt, West Germany

K.R. Markham
Chemistry Division, DSIR, Petone, New Zealand

G.J. Niemann
Botanisch Laboratoire, Rijksuniversiteit, Utrecht, The Netherlands

L.J. Porter
Chemistry Division, DSIR, Petone, New Zealand

C.J. Quinn
School of Botany, University of New South Wales, Australia

M. Helen Sheridan
Department of Chemistry, University College, Dublin, Eire

E. Wollenweber
Institut für Botanik, Technische Hochschule, Darmstadt, West Germany

Preface

The major purpose of this third volume in *The Flavonoids* series is to provide a detailed review of progress in the field during the five years, 1981–1985 inclusive. It thus continues the comprehensive coverage of the literature on these fascinating and important plant pigments which began in 1975 with the publication of *The Flavonoids* and which was followed in 1982 with *The Flavonoids: Advances in Research.* As with the two previous volumes, this one is entirely self-contained and where necessary tabular data and references from earlier volumes are included and expanded here. A unique feature is the complete listing in the Appendix of all known flavonoids, which now number over 4 000 structures; in this list, structures newly reported during the period 1981–1985 are so indicated.

The first ten chapters of this book provide a critical review of the new substances that have been discovered among each of the main classes of flavonoid during the period under review. Again, the number of new isoflavonoids reported outweighs that of other classes and a hundred pages are needed to describe all the novel findings. Neoflavonoids, which were omitted in the first supplement, have been included again and a special chapter on miscellaneous flavonoids has been introduced to cope with those structures (e.g. homoisoflavonoids) which do not fit in easily anywhere else. Although there have been advances in flavonoid methodology, these have not been as spectacular as in earlier years. Hence, literature reports on new chromatographic and spectral procedures are included here in the individual chapters under the different flavonoid classes.

Major developments have taken place in flavonoid biosynthesis with the description of many new enzymes of the pathway and a separate chapter by W. Heller and G. Forkmann covering the 1981–1985 literature is therefore provided. Significant advances have also occurred in our understanding of the way that anthocyanins provide *in vivo* colour in flower petals and R. Brouillard reviews these advances in the last chapter. The dominant themes of the remaining four chapters in the second half of the book are the natural distribution and the evolution of the flavonoids. The first really critical and comprehensive listing of flavonoid occurrences in algae, bryophytes and ferns is provided by K.R. Markham in his chapter on their distribution in lower plants, while G.J. Niemann likewise provides a detailed account of their presence in gymnosperms. The most abundant and prolific sources of flavonoids are the flowering plants and so much is known here that separate books would be required to provide comprehensive coverage of flavonoid occurrences in the angiosperms. Here, we have two contrasting chapters outlining the relationship between flavonoid patterns and plant evolution, first in the dicotyledons and then in the monocotyledons.

Inevitably, some areas of recent flavonoid research have had to be omitted in order to keep the book within the limit set by the publishers. For example, flavonoids have recently been shown to have a signalling function in nitrogen fixation through their ability to regulate the nodulating genes of the *Rhizobium* bacterium. Other recent studies have shown that flavone glycosides are involved as oviposition stimulants to swallowtail butterflies. I hope to review these and other developments in the ecology, physiology and biochemistry of flavonoids in a third supplement. In the meantime I would welcome comments, criticisms and suggestions from readers about this series. As editor, I am most grateful to all the contributors who have once again carried out their taxing assignments on time. Because of other commitments, my co-editor, T.J. Mabry was unable to contribute to this volume but he hopes to take part in future developments. I would personally like to thank my two co-workers at Reading, who have joined me in writing three of the chapters, and also the many students and visiting scientists who have worked with me on flavonoid projects in recent times. Finally, I am most grateful to the publishers for their continued support and interest in this endeavour.

December, 1987 JEFFREY B. HARBORNE
Reading

1

The anthocyanins

JEFFREY B. HARBORNE and RENEE J. GRAYER

1.1 INTRODUCTION

Anthocyanin pigmentation is almost universal in the flowering plants and provides scarlet to blue colours in flowers, fruits, leaves and storage organs. It continues to provide a challenge to plant biochemists because of the intricate chemical variation and the complexity of biosynthesis, metabolism and regulation. The two most important recent advances in the structural characterization of anthocyanin pigments have been the application of high performance liquid chromatography (HPLC) and of fast atom bombardment mass spectrometry (FAB-MS) to their analyses. Both these procedures have proved of value in studying zwitterionic anthocyanins, a relatively new class of acylated anthocyanin recently recognized to be widespread in the plant kingdom (Harborne and Boardley, 1985). These anthocyanins, which are acylated through sugar by such acids as malonic, are labile in solution and when such pigments are isolated using solvents containing mineral acid, they are rapidly degraded to the corresponding unacylated glycoside. They are, however, stable in the solid state and molecular ions can be obtained by means of FAB-MS (Saito *et al.*, 1983). Work has continued on anthocyanins substituted by hydroxycinnamic acids and extensive proton NMR studies have allowed the characterization of the complex pigment of *Ipomoea purpurea*, a peonidin glycoside which contains six glucose and three caffeyl substituents (Goto, 1984). Studies of the blue pigment of *Commelina communis* have finally shown, after some earlier controversy, that this anthocyanin probably occurs *in vivo* as a dimagnesium complex in which six anthocyanin and six flavone molecules are linked either covalently or by hydrogen bonding (Goto *et al.*, 1986).

Our knowledge of the natural distribution of anthocyanins has been enhanced during the period 1981–1985 by surveys in families such as Araceae, Bromeliaceae, Commelinaceae and Polygonaceae (See Section 1.4). Only one new anthocyanidin has been reported, 6-hydroxycyanidin, but a variety of new glycosides have been described, including several with sugars unusually substituted in the B-ring. Such substitution can cause a hypsochromic shift in flower colour and a bright scarlet pigment from bromeliads has been recognized as cyanidin 3, 5, 3′-triglucoside (Saito and Harborne, 1983).

An important discovery about the biosynthetic pathway to anthocyanins has been the recognition of flavan-3, 4-diols (leucoanthocyanidins) as the immediate precursors of the flavylium cations (see Chapter 11). Genetic studies have also enhanced our knowledge of biosynthesis: the structure of the gene for chalcone synthase in *Antirrhinum* has been elucidated (Sommer and Saedler, 1986) and the *c* locus in *Zea mays*, which regulates anthocyanin synthesis, has been cloned by transposon tagging (Paz-Ares *et al.*, 1986).

In this review of anthocyanin research of the period 1981–1985, advances in analytical procedures will be described first. New aglycones, glycosides and acylated glycosides will then be enumerated. This will be followed by a review of the new reports on the natural distribution of these pigments in plants. This review updates three earlier listings (Harborne, 1967; Timberlake and Bridle, 1975; Hrazdina, 1982). These four listings together provide a complete record of anthocyanin occurrences in plants to the generic, and in most cases, to the specific level. A book on anthocyanins as food colours was published by Markakis (1982), and the industrial and medical applications of anthocyanin research will be considered in the final section. The role of anthocyanins in

The Flavonoids. Edited by J.B. Harborne
Published in 1988 by Chapman and Hall Ltd,
11 New Fetter Lane, London EC4P 4EE.

flower coloration will not be mentioned here, since this is the subject of a separate review later in this volume (see Chapter 16).

1.2 ANALYTICAL PROCEDURES

1.2.1 New chromatographic techniques

The HPLC of anthocyanins, which was pioneered during the 1970 s (see Hrazdina, 1982), has now become routine in most laboratories. It has the advantage over other chromatographic procedures of sensitivity, rapidity and easy quantification but the apparatus is expensive to set up and maintain. When coupled with a diode array detector, HPLC provides an almost ideal procedure for accurately analysing, quantitatively and qualitatively, the complex mixtures of pigments present in cultivated flowers and fruits.

Asen, who first applied the HPLC of anthocyanins to cultivar identification in the case of poinsettias, has described its use in the case of geraniums (Asen and Griesbach, 1983) and of *Gerbera* flowers (Asen, 1984). Akavia *et al.* (1981) have used HPLC to separate and quantify the nine anthocyanins variously present in *Gladiolus* cultivars. Van Sumere and van de Casteele (1985) have similarly applied the HPLC of anthocyanins to the differentiation of *Rhododendron* cultivars.

HPLC is also useful in many other ways. It can be applied to the detection of intermediates in the partial hydrolysis of anthocyanins with more than one sugar residue and may reveal compounds which are not apparent when PC or TLC is used (Strack *et al.*, 1980). Its greater sensitivity means that it is the method of choice when only small amounts of fresh plant tissue are available for anthocyanin analysis. Schram *et al.* (1983) were able to detect cyanidin 3-glucoside and 3-diglucoside in a rare flower mutant of *Petunia hybrida*, when only 9 mg wet weight of flower were available. Again in plant tissue culture, HPLC has proved useful for comparing pigmentation in callus or suspension culture with the pigments of the parent plant. Such analyses, in the case of *Petunia*, have shown mainly malvidin glycoside with traces of petunidin glycoside in the intact plant, with these pigment concentrations being reversed in tissue culture (Colijn *et al.*, 1981).

Typical HPLC retention times for some of the more common anthocyanins are shown in Table 1.1. From these data, it appears that glycosylation increases mobility on a reversed phase column, with 3, 5-diglycosides moving off the column faster than 3-diglycosides. Increasing hydroxylation of the anthocyanidin improves the mobility, and *O*-methylation reverses this trend so that malvidin glycosides are generally eluted after the other anthocyanins. Acylation with either aromatic acids (e.g. *p*-coumaric) or aliphatic acids (e.g. malonic) increases the retention time. Some typical values for acylated anthocyanins and their unacylated analogues are shown in Table 1.2.

TLC continues to be widely employed for anthocyanin separation, because it is a highly effective, convenient and inexpensive technique. Andersen and Francis (1985) have shown that it is possible to analyse anthocyanins and anthocyanidins simultaneously on cellulose layers, if the solvent system conc. HCl–formic acid–water (24.9:23.7:51.4) is used.

For preparative work, droplet counter current chromatography (DCCC) has been used with the anthocyanins of

Table 1.1 Retention times (t_R) of anthocyanins on a Lichrosorb RP-18 column eluted with a water–acetic acid–acetonitrile gradient

Pigment		t_R *(min)*★
3-Rhamnosylglucoside-5-glucoside	Dp[†]	15.4
	Cy	19.1
	Pt	22.0
	Pg	22.2
	Pn	25.8
	Mv	28.4
3, 5-Diglucoside	Dp	18.2
	Cy	21.7
	Pt	25.0
	Pn	25.0
	Pg	25.3
	Mv	32.1
3-Rhamnosylglucoside	Dp	27.0
	Cy	31.1
	Pt	33.6
	Pg	34.8
	Pn	39.0
	Mv	42.8

[†]For key to anthocyanidin abbreviations, see Table 1.3.
★Data from Akavia *et al.* (1981).

Table 1.2 Effect of acylation on anthocyanin HPLC retention times

Pigment	RR_t
Pg 3, 5-diglucoside	1.00★
Pg 3-(6″-malonylglucoside)-5-glucoside	1.87
Pg 3, 5-di(malonylglucoside)	2.67
Cy 3-glucoside	1.00★
Cy 3-malonylglucoside	2.14
Cy 3-dimalonylglucoside	2.96
Mv 3-rhamnosylglucoside-5-glucoside	1.00[†]
Mv 3-(*p*-coumarylrhamnosylglucoside)-5-glucoside	4.00

★On a Spherisorb-hexyl column at 35 °C eluted with a 0.6% aq. $HClO_4$–MeOH gradient (Takeda *et al.*, 1986a).
[†]On a Lichrosorb 10 RP-18 column eluted with a MeOH–HCO_2H–H_2O gradient (Schram *et al.*, 1983).

blackcurrants and the solvent system: butan-1-ol–acetic acid–water (4:1:5) (Francis and Andersen, 1984). However, HPLC can also be used for semi-preparative separations, with a wider column than in analytical work, and this probably has the edge over DCCC. A technique for the automated HPLC separation of anthocyanins in blackberry and cranberry on the preparative scale has been described by Hicks *et al.* (1985).

1.2.2 Spectral methods

The first application of FAB-MS to anthocyanin identification was that of Saito *et al.* (1983) on the known pigment violanin from *Viola* and on a novel pigment platyconin from *Platycodon grandiflorum*. Since then the method has been widely used for the characterization of acylated anthocyanins, especially zwitterionic pigments carrying malonyl substitution (Bridle *et al.*, 1984; Takeda *et al.*, 1986a) (see Section 1.3.4). Proton and ^{13}C NMR spectroscopy also continue to be applied with success to anthocyanin structural elucidation, together with the newer ^{1}H–^{1}H correlated spectroscopy (COSY) (Tamura *et al.*, 1983; Bridle *et al.*, 1984; Kondo *et al.*, 1985). Further details of the application of spectral techniques to anthocyanin studies are discussed by R. Brouillard in Chapter 16.

1.3 CHEMISTRY

1.3.1 New anthocyanidins

To the 16 known anthocyanidins, one new structure has to be added, namely 6-hydroxycyanidin (Table 1.3). This has been found in the red flowers of *Alstroemeria* (Alstroemeriaceae) where it occurs as the 3-glucoside and 3-rutinoside (Saito *et al.*, 1985b). Its structure was established by spectral measurements and comparison with the literature data. The related 6-hydroxypelargonidin (aurantinidin) is known (Clevenger, 1964; Jurd and Harborne, 1968) but the delphinidin analogue has yet to be described.

Table 1.3 Known anthocyanidins

Name	*Abbreviation*	*Structure*
Apigeninidin	Ap	5, 7, 4′-TriOH
Luteolinidin	Lt	5, 7, 3′, 4′-TetraOH
Tricetinidin	Tr	5, 7, 3′, 4′, 5′-PentaOH
Pelargonidin	Pg	3, 5, 7, 4′-TetraOH
Aurantinidin	Au	3, 5, 6, 7, 4′-PentaOH
Cyanidin	Cy	3, 5, 7, 3′, 4′-PentaOH
5-Methylcyanidin	5MCy	5-Methyl ether
Peonidin	Pn	3′-Methyl ether
Rosinidin	Rs	7, 3′-Dimethyl ether
6-Hydroxycyanidin	6OHCy	3, 5, 6, 7, 3′, 4′-HexaOH
Delphinidin	Dp	3, 5, 7, 3′, 4′, 5′-HexaOH
Petunidin	Pt	3′-Methyl ether
Malvidin	Mv	3′, 5′-Dimethyl ether
Pulchellidin	Pl	5-Methyl ether
Europinidin	Eu	5, 3′-Dimethyl ether
Capensinidin	Cp	5, 3′, 5′-Trimethyl ether
Hirsutidin	Hs	7, 3′, 5′-Trimethyl ether

The first report of 5-methylcyanidin as a new anthocyanidin in *Egeria densa* (Elodeaceae) (Momose *et al.*, 1977) was inadvertently omitted from the text of the last review (Hrazdina, 1982) but is now included in Table 1.3. Europinidin, which was found in petals of *Plumbago europaea* (Harborne, 1966), was not analysed at the time because of shortage of material. Its structure as the 5, 3′-dimethyl ether of delphinidin has now been confirmed by FAB-MS, which gave the required molecular ion $[M+H]^+$, at 331 (Harborne and Self, unpublished results).

Several other anthocyanidins have been partly described (i.e. carajurin from *Arrabidaea*, carexidin from *Carex*, columnidin from *Columnea*, purpurinidin from *Salix* and margicassidin from *Cassia*) (cf. Timberlake and Bridle, 1975) but these still await complete characterization. X-ray crystallography of pelargonidin bromide monohydrate has shown that this anthocyanidin is nearly planar in the solid state (Saito and Ueno, 1985). The phenyl ring makes a dihedral angle of only 3.8° with the benzopyrylium moiety. A new total synthesis of apigeninidin and luteolinidin has been described (Sweeny and Iacobucci, 1981). It involves the oxidative decarboxylation of the corresponding 4-carboxyflav-2-enes with lead tetra-acetate and the yields are much higher than in the traditional method starting from 2-*O*-benzoylphloroglucinaldehyde.

1.3.2 New glycosides

The most unusual feature of the new anthocyanidin glycosides reported over the last five years (Table 1.4) is the regular presence, in nine pigments, of B-ring hydroxyl groups which are linked to glucose. Such compounds are not completely new; for example, an acylated derivative of delphinidin 3-rutinoside-5, 3′, 5′-triglucoside was reported in *Lobelia* by Yoshitama in 1977. However, such substances have been reported rarely; it is now apparent that they may occur in a range of plants and have a variety of substitution patterns. Major sources are members of the Bromeliaceae (Saito and Harborne, 1983) but they have also been found in plants of the Commelinaceae, Compositae, Gentianaceae, Leguminosae, Liliaceae and Lobeliaceae.

Substitution of a B-ring hydroxyl by sugar causes a hypsochromic shift in the visible spectrum which makes it easy to recognize such glycosides. For example, while cyanidin 3, 5-diglucoside has a visible maximum at 526 nm in methanolic HCl, the 3, 3′-diglucoside has a maximum at 519 nm and the 3, 5, 3′-triglucoside at 518 nm. This has an

Table 1.4 New glycosides of anthocyanidins

Pigment	*Source*	*References*
Cy 3-rhamnosylarabinoside	*Cissus sicyoides*	Toledo *et al.* (1983)
Cy 3,3′-diglucoside Cy 3,5,3′-triglucoside Cy 3-rutinoside-3′-glucoside Cy 3-rutinoside-5,3′-diglucoside	From Bromeliaceae in leaf, sepal, bract or petal	Saito & Harborne (1983)
Cy 3,7,3′-triglucoside*	*Senecio cruentus* flowers	Yoshitama (1981)
Cy 3-glucuronosylglucoside*	*Helenium autumnale* flower rays	Takeda *et al.* (1986a)
6OHCy 3-glucoside 6OHCy 3-rutinoside	*Alstroemeria* cvs	Saito *et al.* (1985b)
Pn 3-arabinoside-5-glucoside	*Polygonum longisetum* sepal	Yoshitama *et al.* (1984)
Dp 3,7-diglucoside	*Aristotelia chilensis* fruit	Diaz *et al.* (1984)
Dp 3-rhamnosylglucoside-7-xyloside	*Olea europaea* fruit	Tanchev *et al.* (1980a,b)
Dp 3,5,3′-triglucoside*	*Gentiana makinoi* petals	Goto *et al.* (1982).
Dp 3,7,3′-triglucoside†	*Puya* petals	Scogin (1985)
Dp 3,3′,5′-triglucoside*	*Clitoria ternatea* petals	Saito *et al.* (1985a)
Pt 3-(2^G-glucosylrutinoside)-5′-glucoside‡	*Ophiopogon jaburan* seed coat	Yoshitama (1984)
Mv 3-xyloside-5-glucoside*	*Tibouchina granulosa* petals	Francis *et al.* (1982)
Mv 3,7-diglucoside	*Aristotelia chilensis* fruit	Diaz *et al.* (1984)

*Only occurring in acylated form (see Tables 1.5 or 1.6 for details).
†Previously known in acylated form; now reported as the simple glycoside.
‡First reported in 1976–1982 period, but omitted from the 1982 list.

effect *in vivo* on flower colour, and, in the Bromeliaceae, these unusual cyanidin glycosides are found in bracts or petals which are scarlet rather than crimson in colour (Saito and Harborne, 1983). Pigments such as cyanidin 3,5,3′-triglucoside can be characterized by partial hydrolysis, since they produce all the expected intermediates before complete hydrolysis to the aglycone. Their structures can be confirmed by FAB-MS, when intense molecular ions are obtainable, with fragment ions caused by sequential loss of the different glucose moieties.

The only new monosaccharide reported since 1982 in association with anthocyanidins is glucuronic acid. Cyanidin 3-glucuronosylglucoside has been found in malonated form (see Section 1.3.4) in flowers of *Helenium autumnale* (Takeda *et al.*, 1986a). As expected, it is zwitterionic in character and is highly resistant to acid hydrolysis.

In addition to the glycosides listed in Table 1.4, there have been several tentative reports of 'new' glycosides which need further investigation. Bobbio *et al.* (1983) provisionally detected 3-maltosides in the seed and skin of *Cyphomandra betacea* in spite of an earlier finding of 3-rutinosides in the same plant cultivated in New Zealand (Wrolstad and Heatherbell, 1974). Again, Tsukui *et al.* 1983 describe an acylated cyanidin 3-glucosylfructoside-5-xyloside as a tuber skin pigment of the sweet potato *Ipomoea batatas*. This is at variance with earlier identifications of acylated 3-sophoroside-5-glucosides in *Ipomoea* (see Hrazdina, 1982, and Table 1.8). Since fructose is so rarely found as a glycosidic component of flavonoids, this work needs confirmation before it can be accepted.

1.3.3 Anthocyanins with aromatic acylation

An example of the structural complexity that can be present in anthocyanins acylated with hydroxycinnamic acids is provided by the pigment 'heavenly blue anthocyanin' (HBA) from the flowers of *Ipomoea purpurea*. This is now recognized as having structure (*1.1*) and is probably the largest anthocyanin so far reported with a molecular weight of 1759. Its structure was established by extended proton NMR studies on the original pigment and on two partly deacylated derivatives (Goto, 1984). It is a peonidin 3-sophoroside-5-glucoside substituted through sugar with three caffeylglucose residues. Two other similar pigments have also recently been characterized: gentiodelphin (*1.2*) from *Gentiana makinoi* and platyconin (*1.3*) from *Platycodon grandiflorum*. Other related pigments that still await full characterization are present in flowers of members of the Commelinaceae (Stirton and Harborne, 1980), other than in the genus *Commelina*.

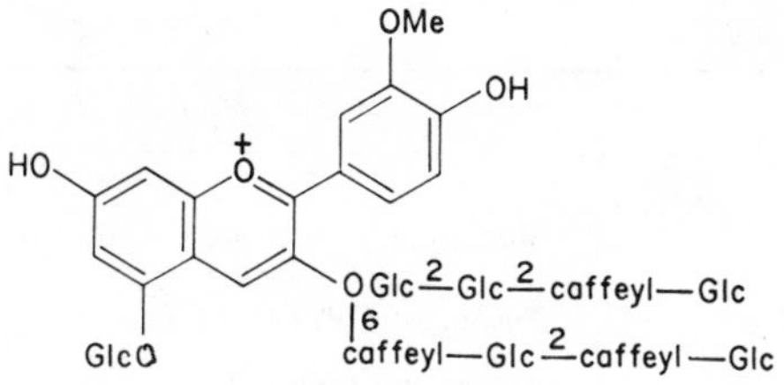

Heavenly blue anthocyanin (*1.1*)

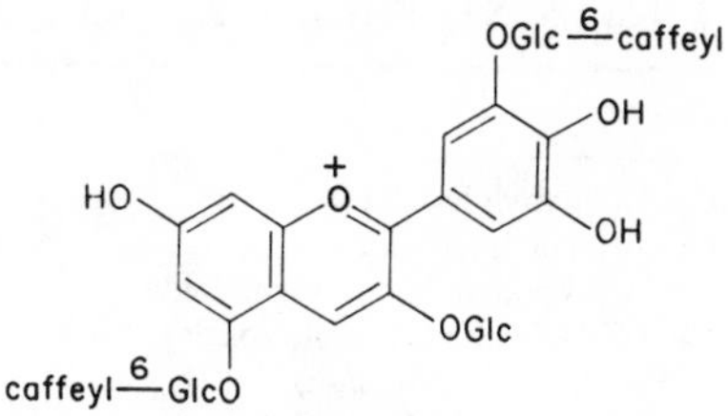

Gentiodelphin (*1.2*)

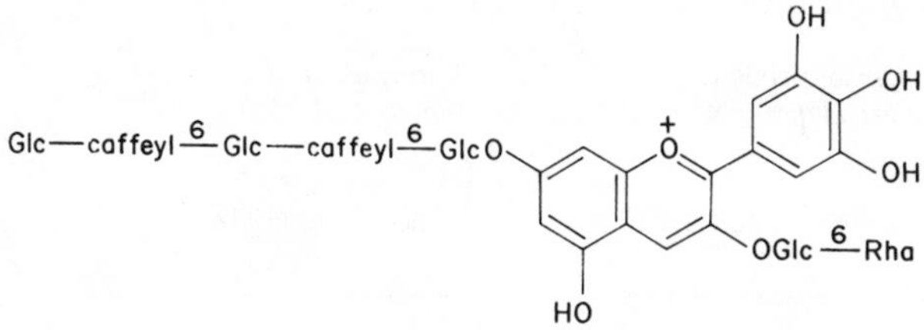

Platyconin (*1.3*)

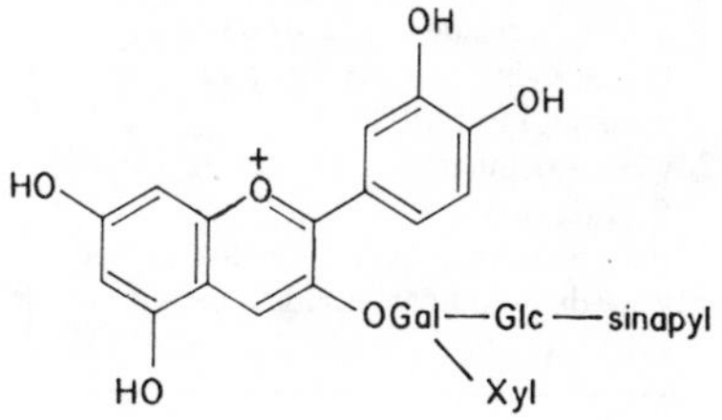

Carrot anthocyanin (*1.4*)

Cyanidin and delphinidin 3, 7, 3′-triglucosides are present variously trisubstituted with caffeic and/or ferulic acid residues but the positions of attachment have yet to be determined.

A list of recently characterized anthocyanins of this general type are shown in Table 1.5. Pigments which have both aromatic and aliphatic acyl substituents are discussed in the next section. Apart from pigments (*1.1*)–(*1.3*), most of the anthocyanins in the table are still incompletely characterized in as much as the position of acylation on the sugar is often unknown. One of these acylated pigments deserves further comment, namely cyanidin 3-(sinapylxylosylglucosylgalactoside) (*1.4*), one of the constituents of *Daucus carota* (Harborne, 1976). In 1982, Hemingson and Collins described four simple glycosides of cyanidin from tissue cultures of the same plant. Since none of these pigments occurs in the intact carrot, their presence in tissue culture was unexpected. Re-examination of carrot tissue culture showed, however, that pigment (*1.4*) was the only major constituent (Harborne *et al.*, 1983). The erroneous report of Hemingson and Collins arose mainly from the fact that these authors inadvertently partly hydrolysed the pigment during purification. Recent FAB-MS studies on (*1.4*) (Harborne and Self, unpublished results) have indicated that it has a branched trisaccharide with the sinapyl residue located on the glucose residue, but further work is needed to determine the exact modes of linkage. A report by Canbas (1985) based only on spectral measurements of malvidin and peonidin 3-glucosides in the purple–black carrot root can be dismissed in view of the earlier detailed characterization of acylated cyanidin glycosides from the same source (Harborne, 1976).

It is remarkable how many reports appear in the anthocyanin literature, especially on food plants, where the investigators appear to completely overlook previous studies on the same or on a closely related species. If the results agree with the previous work, as in the case of Lu (1985) reinvestigating radish pigments, there is no harm done. However, if the results are completely at variance with earlier investigations, then there can be considerable confusion. For example, anthocyanins with aromatic acylation have recently been described in raspberries (Joo and Park, 1983) and in elderberries (Shin and Ahn, 1980) in spite of the fact that earlier investigators only found simple glycosides in these fruits (cf. Timberlake and Bridle, 1975; Hrazdina, 1982). Since acyl groups are labile and can be lost during isolation and purification (see next section), it is always possible that acylated pigments may have been overlooked during the earlier work. However, in these two cases, the new data are ambiguous and it is our view that further work is needed before it can be accepted that acylated pigments occur in these two fruits.

A report of vanillic acid as an acyl group in pigments of *Helianthus annuus* achenes (Vaccari *et al.*, 1981) is based on incorrect spectral evidence: a peak at 240 nm in one of the pigments is assumed to be due to the vanillyl substituent in spite of the fact that vanillic acid actually has a maximum at 258 nm. Again, further investigation is necessary before this phenolic acid is added to the list of known acylating groups.

1.3.4 Zwitterionic anthocyanins

Until recently, acylated anthocyanins were known to be substituted by hydroxycinnamic acids (*p*-coumaric,

Table 1.5 New anthocyanins with aromatic acid acylation

Pigment	*Source*	*Reference*
Cy 3-(sinapylglucoside)		
Cy 3-(ferulylglucoside)	*Citrus sinensis*	Maccarone *et al.* (1985)
Cy 3-(*p*-coumarylferuloylglucoside)		
Cy 3-(6″-*p*-coumarylsophoroside)	*Epimedium grandiflorum* leaf	Yoshitama (1984)
Cy 3-(caffeylsophoroside)	*Cynara scolymus* bract	Aubert and Foury (1982)
Cy 3-(dicaffeylsophoroside)		
Cy 3-(sinapylxylosylglucosylgalactoside) (*1.4*)	*Daucus carota* tissue culture	Harborne *et al.* (1983)
Cy 3, 7, 3′-tri (caffeylglucoside)	*Tradescantia* spp.	Stirton and Harborne (1980)
Pn 3-sophoroside-5-glucoside tri(caffeylglucose) ester (*1.1*)	*Ipomoea purpurea* petals	Goto (1984)
Gentiodelphin (*1.2*)	*Gentiana makinoi* petals	Goto *et al.* (1983a)
Mv 3-(*p*-coumarylsambubioside)-5-glucoside	*Tibouchina grandiflora* petals	Bobbio *et al.* (1983)
Mv 3-(*p*-coumarylxyloside)-5-glucoside		
Mv 3-(di-*p*-coumarylxyloside)-5-glucoside	*Tibouchina granulosa* petals	Francis *et al.* (1982)

Table 1.6 Known anthocyanins with aliphatic dicarboxylic acids as acylating groups

Pigment	*Source*	*Reference*
PELARGONIDIN GLYCOSIDES		
3-(6″-Malonylglucoside)	*Callistephus chinensis*	Takeda *et al.* (1986a)
3-(6″-Malylglucoside)	*Dianthus caryophyllus*	Terahara *et al.* (1986); Terahara and Yamaguchi (1986)
3-Malonylsophoroside	*Papaver nudicaule*	Cornuz *et al.* (1981)
3-(6″-Malonylglucoside)-5-glucoside	*Dahlia variabilis*	Takeda *et al.* (1986a)
3, 5-Di(malonylglucoside)	*Dahlia variabilis*	
3-(6″-*p*-Coumaroylglucoside)-5-(4‴, 6‴-dimalonylglucoside) (*1.6*)	*Monarda didyma*	Kondo *et al.* (1985)
CYANIDIN GLYCOSIDES		
3-(6″-Malonylglucoside)	*Cichorium intybus*	Bridle *et al.* (1984)
3-(6″-Malylglucoside)	*Dianthus deltoides*	Terahara *et al.* (1986)
3-(6″-Oxalylglucoside)	*Ophrys*	Strack *et al.* (1986)
3-Dimalonylglucoside	*Coleostephus myconis*	
3-Malonylglucuronosylglucoside	*Helenium* cv. Bruno	Takeda *et al.* (1986a)
3-(6″-Malonylglucoside)-5-glucoside	*Dahlia variabilis*	
3, 5-Di(malonylglucoside)	*Dahlia variabilis*	
3-(6″-Succinylglucoside)-5-glucoside	*Centaurea cyanus*	Tamura *et al.* (1983)
3-(*p*-Coumarylglucoside)-5-malonylglucoside	*Stachys* sp.	Takeda *et al.* (1986b)
Rubrocinerarin*	*Senecio cruentus*	Goto *et al.* (1984)
PEONIDIN GLYCOSIDE		
3, 5-(Malonyl-*p*-coumaryldiglucoside)*	*Plectranthus argenteus*	Harborne and Boardley (1985)
DELPHINIDIN GLYCOSIDES		
3, 5-Di(malonylglucoside)	*Cichorium intybus*	Takeda *et al.* (1986a)
3-(*p*-Coumarylglucoside)-5-malonylglucoside (*1.5*)	*Commelina communis*	Goto *et al.* (1983b)
Cinerarin* (*1.7*)	*Senecio cruentus*	Goto *et al.* (1984)
Ternatins A-F*	*Clitoria ternatea*	Saito *et al.* (1985a)

*Complex pigments including both malonic acid and aromatic acid substituents.

caffeic, ferulic or sinapic), by *p*-hydroxybenzoic acid or by acetic acid. However, it is now apparent from extensive electrophoretic surveys (Harborne and Boardley, 1985; Harborne, 1986) and from detailed investigations of individual pigments (e.g. Cornuz *et al.*, 1981) that anthocyanins are also acylated in nature with aliphatic dicarboxylic acids, such as malonic, malic, oxalic and succinic. Such acylation renders the cationic anthocyanin a zwitterion, which means that it is possible to distinguish these pigments from other anthocyanins by paper electrophoresis in a weakly acidic buffer.

Another significant feature of this type of acylation is the instability of the acyl link *in vitro*, particularly when compared with aromatic acid acylation. If malonated anthocyanins are extracted by standard procedures using methanolic HCl, there is intermediate methyl ester formation at the free carboxyl group, but the main reaction is loss of the malonyl group within a short time. By contrast, hydroxycinnamyl residues are relatively unaffected by such treatment. Successful extraction of zwitterionic anthocyanins depends on the substitution of mineral acid by weaker acids such as acetic, formic or perchloric, and the careful monitoring of the acidity of extracts during the processes of purification. [Exceptionally, a stronger acid (3% aqueous trifluoroacetic) was used by Kondo *et al.* (1985) to isolate malonated pigments from *Monarda didyma*.]

Because of this unusual lability, a variety of pigments now known to be zwitterionic were previously reported to be unacylated. In particular, the petals of the blue cornflower, *Centaurea cyanus*, the classic source of cyanidin 3, 5-diglucoside, have now been shown to contain the 3-(6″-succinylglucoside)-5-glucoside using milder methods of extraction and purification (Takeda and Tominaga, 1983; Tamura *et al.*, 1983). Subsequent reinvestigation of other members of the Compositae has shown that most flower pigments are zwitterionic, with malonated anthocyanins occurring in species of *Callistephus*, *Cichorium*, *Coleostephus*, *Dahlia* and *Helenium* (Takeda *et al.*, 1986a). Zwitterionic anthocyanins are also distinguished from other anthocyanins by their chromatographic properties and by their high retention times on HPLC (see Table 1.2). FAB-MS and NMR spectroscopy have proved invaluable for their characterization.

Structures of the zwitterionic anthocyanins known to date are listed in Table 1.6. Most are malonic acid derivatives but anthocyanins substituted by malic, oxalic and succinic acids have also been described. Commonly, a single malonic acid is present substituted at the 6-position of the 3-sugar moiety, as in cyanidin 3-(6″-malonylglucoside) from leaves of *Cichorium intybus* (Bridle *et al.*, 1984). Dimalonates are known, with the second malonic acid substituting at a 5-glucose (as in the 3, 5-dimalonylglucosides in *Dahlia*) or with disubstitution at the same glucose residue (as in the 3-dimalonylglucoside of cyanidin in *Coleostephus*) (Takeda *et al.*, 1986a).

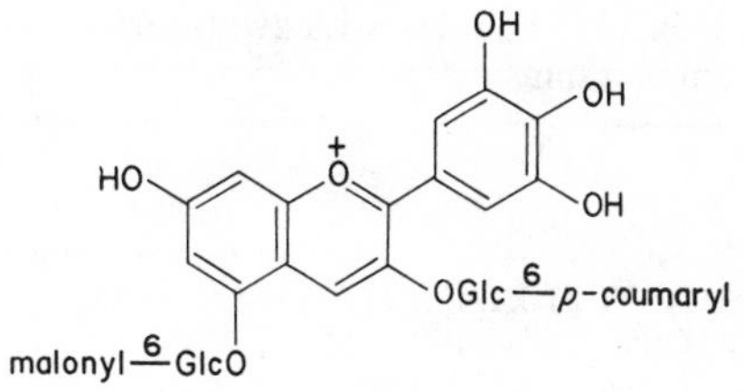

Malonylawobanin (*1.5*)

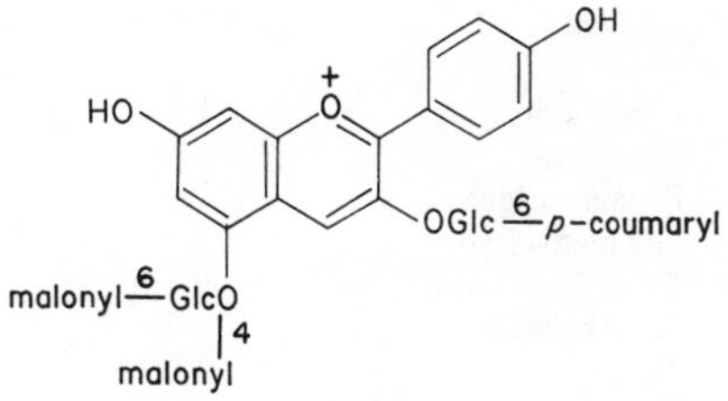

Monardaein (*1.6*)

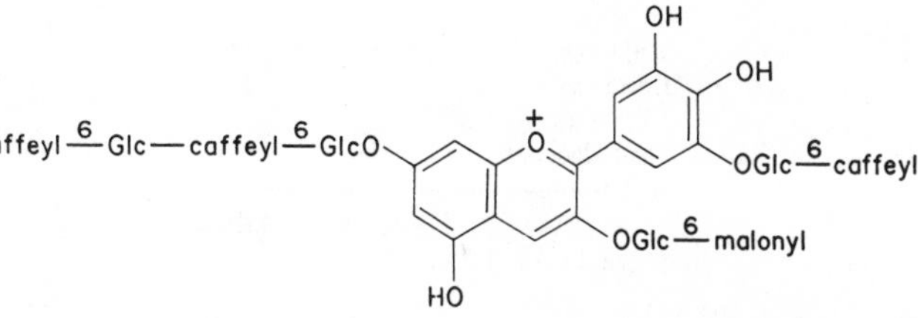

Cinerarin (*1.7*)

Pigments with both malonic acid and aromatic acid substitution have been encountered. Typical is malonylawobanin (*1.5*) with *p*-coumaryl and malonyl residues, which occurs in the blue flowers of *Commelina communis* (Goto *et al.*, 1983b) and in the bluebell, *Hyacinthoides non-scripta* (Takeda *et al.*, 1986b). Another example is the pigment of *Monarda didyma*, monardaein (*1.6*), which has been the subject of several earlier investigations but is now known to be pelargonidin 3, 5-diglucoside with *p*-coumaric acid acylating the 3-sugar and two malonyl residues at the 5-sugar (Kondo *et al.*, 1985). The most complex structure of this type to date is cinerarin (*1.7*) from *Senecio cruentus* (Goto *et al.*, 1984) which is substituted by caffeic and malonic acids. The six pigments, ternatins A–F, from flowers of *Clitoria ternatea* are also known to have both *p*-coumaric and malonic acid residues linked to delphinidin 3, 3′, 5′-triglucoside, but the precise modes of linkage are still not known (Saito *et al.*, 1985a).

Table 1.7 Families with zwitterionic anthocyanins

Family	*Species frequency*
DICOTYLEDONS	
Caryophyllaceae*	5/12
Compositae*	35/39
Convolvulaceae	1/3
Cruciferae*	5/6
Euphorbiaceae	1/1
Gesneriaceae	1/5
Labiatae*	17/17
Leguminosae*	3/14
Lobeliaceae	1/1
Malvaceae	2/4
Papaveraceae*	1/4
Polemoniaceae	2/3
Ranunculaceae	9/10
Scrophulariaceae	10/14
Solanaceae	1/7
Thymeliaceae	2/2
Verbenaceae*	1/1
MONOCOTYLEDONS	
Alliaceae	4/4
Commelinaceae*	1/6
Gramineae*	2/2
Iridaceae	1/17
Lemnaceae*	1/1
Liliaceae*	3/11
Orchidaceae*	3/3

*Pigments identified in members of these families; see Tables 1.7 and 1.9.

Many of the pigments listed in Table 1.6 as being zwitterionic were originally thought to be unacylated. Structural revision of further known anthocyanins will probably be necessitated by the discovery that malonylation is a regular feature of anthocyanin chemistry. Since zwitterionic anthocyanins are relatively widespread, having already been recorded in 24 plant families (Table 1.7) (Harborne, 1986), many more structures of this type remain to be elucidated. Organic acids other than the four so far identified will almost certainly be encountered in future investigations.

1.4 DISTRIBUTION

1.4.1 Cellular localization

The detection of discrete red 'spots' within the vacuole of higher plant species has led to the proposal that membrane-bound bodies containing these pigments, called 'anthocyanoplasts', are present. These have been observed transiently in hypocotyls of red cabbage (Pecket and Small, 1980) and of radishes (Yasuda and Shinoda, 1985). They are only formed when pigment synthesis is in operation and they disappear as the whole vacuole eventually becomes red in colour. This appears to be a general phenomenon in that these organelles have been detected in 70 species representing 33 families. Ultrastructural studies in red cabbage cells indicate that these anthocyanoplasts are bound by a single tripartite membrane 10 nm thick (Small and Pecket, 1982). These latter authors have suggested that the later enzymes of anthocyanin biosynthesis might actually be attached to the inner membrane, but Neumann (1983) has argued convincingly that the existence of functional organelles within the vacuole is very unlikely. In any case, the pH optima of the known anthocyanin-synthesizing enzymes are basic, so it is difficult to see how they could operate in what is presumably an acidic environment (cf. Hrazdina and Wagner, 1985, but see also Merlin *et al.*, 1985).

A survey of petals of 201 species from 60 families has shown that anthocyanins are specifically and exclusively located in the epidermal cells in almost all cases. Very occasionally pigment also occurs in the mesophyll, but the cases where pigment is confined to the mesophyll are very few. This happens only in members of the Boraginaceae and is correlated with morphological differences in the mesophyll cells (Kay *et al.*, 1981). What happens in the case of leaf anthocyanins is less clear. Investigation of the *Argenteum* mutant of the garden pea *Pisum sativum* by Hrazdina *et al.* (1982) has indicated that anthocyanin is only present in epidermal cells, yet a similar study of the leaves of rye, *Secale cereale*, has shown that anthocyanin, while absent from the epidermis, is restricted to the mesophyll (see McClure, 1986). Clearly, it will be of interest to examine the localization of leaf anthocyanin in a range of other species, to see whether epidermal or mesophyll deposition is favoured in most cases.

1.4.2 Taxonomic patterns

The results of anthocyanin identifications carried out during 1981–1985 inclusive (a few 1986 references are also included) are collected in Table 1.8. This is an alphabetical listing by species, genus and family. These data can be usefully compared with previous listings (Harborne, 1967; Timberlake and Bridle, 1975; Hrazdina, 1982). Such comparisons indicate that glycosidic and acylating patterns are generally consistent both at the generic and familial levels. However, it is still difficult to use anthocyanins as taxonomic characters on any scale because of the relatively limited sampling at the species level within those families that have been investigated. Additionally, many families have yet to be studied at all. Within the monocotyledons, our information is rather better than in the dicotyledons and a discussion of anthocyanin patterns in these families is possible, as will be found later in this volume in Chapter 15. Here it is only appropriate to comment briefly on the patterns in those families which have been surveyed in depth during the last five years.

Table 1.8 Occurrence of anthocyanins in plant classes, families, genera and species

Class, family, genus, species	*Organ examined*	*Pigments present*	*Reference*
ANGIOSPERMAE			
MONOCOTYLEDONEAE			
ALSTROEMERIACEAE			
Alstroemeria cultivars	Flower	6-OH Cy 3-rutinoside, 3-glucoside, Cy 3-rutinoside	Saito *et al.* (1985b)
ARACEAE			
Alocasia 6 spp.	Leaf, petiole	Cy 3-rutinoside	Williams *et al.* (1981)
Amorphophallus 2 spp.	Spathe	Cy 3-rutinoside	Williams *et al.* (1981)
Anchomanes abbreviatus	Fruit	Cy, Pg 3-gentiobiosides, Pg 3-glucoside	Williams *et al.* (1981)
Anthurium 17 spp.	Fruit, spadix, spathe, petiole	Cy, Pg 3-rutinosides, Cy 3-glucoside	Williams *et al.* (1981)
Arisaema sp.	Petiole	Cy 3-rutinoside, 3-glucoside	Williams *et al.* (1981)
Arum maculatum	Spadix, spathe	Cy 3-rutinoside, 3-glucoside	Williams *et al.* (1981)
Asterostigma riedelianum	Stem	Cy 3-rutinoside, 3-glucoside	Williams *et al.* (1981)
Caladium bicolor	Leaf	Cy 3-rutinoside	Williams *et al.* (1981)
Dracontium 2 spp.	Petiole, spadix, spathe	Cy, Pg 3-rutinosides	Williams *et al.* (1981)
Dracunculus canariensis	Spathe	Cy 3-rutinoside	Williams *et al.* (1981)
Helicodiceros muscivorus	Spathe	Cy 3-rutinoside	Williams *et al.* (1981)
Homalomena rubescens	Leaf, petiole	Cy 3-glucoside	Williams *et al.* (1981)
Philodendron 10 spp.	Leaf, petiole, stem, spadix, spathe	Cy 3-rutinoside, 3-glucoside	Williams *et al.* (1981)
Pinellia tripartita	Stem	Cy 3-rutinoside	Williams *et al.* (1981)
Rhektophyllum mirabile	Petiole	Cy 3-glucoside, 3-gentiobioside	Williams *et al.* (1981)
Schismatoglottis concinna var. *immaculata*	Leaf, petiole	Dp 3-rutinoside	Williams *et al.* (1981)
Stylochiton spp.	Stem	Cy 3-rutinoside	Williams *et al.* (1981)
Typhonium giraldii	Stem	Cy 3-rutinoside	Williams *et al.* (1981)
Typhonodorum lindleyanum	Leaf	Cy 3-rutinoside, 3-glucoside	Williams *et al.* (1981)
Xanthosoma 3 spp.	Stem, petiole, leaf	Cy, Pg 3-rutinosides	Williams *et al.* (1981)
Xenophya lauterbachiana	Leaf, stem	Cy 3-rutinoside	Williams *et al.* (1981)
BROMELIACEAE			
Aechmea 4 spp.	Bract, leaf, flower, calyx, fruit	Cy 3, 5, 3′-triglucoside, Cy, Mv, Pn 3, 5-diglucosides	
Ananas comosus	Leaf	Cy 3, 5, 3′-triglucoside, Cy, Pn 3, 5-diglucosides	
Billbergia cv. *fantasia*	Leaf	Cy 3, 5, 3′-triglucoside, 3, 5-diglucoside	
Billbergia buchholtzii	Bract	Pg, Cy 3, 5-diglucosides, Pg 3-rutinoside-5-glucoside, Cy 3, 5, 3′-triglucoside, 3-rutinoside-5, 3′-diglucoside?	
	Flower	Mv 3, 5-diglucoside	

(*Contd.*)

Table 1.8 (*Contd.*)

Class, family, genus, species	*Organ examined*	*Pigments present*	*Reference*
Canistrum cyathiforme	Leaf	Cy 3, 5, 3′-triglucoside, 3, 5-diglucoside	
Cryptanthus 5 spp.	Leaf	Cy 3, 5, 3′-triglucoside, Cy, Mv, Pn 3, 5-diglucosides	
Dyckia niederleinii	Leaf	Cy 3-glucoside	Saito and Harborne (1983)
Fascicularia kirchoffiana	Leaf	Cy 3, 5, 3′-triglucoside	
Guzmania zahnii	Leaf	Cy, Pn 3, 5-diglucosides	
Neoregelia 4 spp.	Leaf	Cy, Mv, Pn 3, 5-diglucosides, Cy 3-glucoside, 3, 5, 3′-triglucoside	
Nidularium 2 spp.	Leaf	Cy 3, 5, 3′-triglucoside, Cy, Mv, Pn 3, 5-diglucosides, Cy 3-rutinoside-3′-glucoside, 3, 3′-diglucoside?	
	Flower	Mv 3, 5-diglucoside, Dp, Pt glycosides	
Pitcairnia 3 spp.	Leaf	Cy 3, 5, 3′-triglucoside, 3, 5-diglucoside, 3-glucoside, Mv, Pn 3, 5-diglucosides	
	Flower	Pg 3-rutinoside	
Portea petropolitana	Bract	Cy, Pn 3, 5-diglucosides	
	Flower	Cy, Mv 3, 5-diglucosides	
Puya spp.	Leaf	Cy 3, 5-diglucoside, 3-glucoside	
	Corolla	Dp 3, 7, 3′-triglucoside, Dp, Cy 3, 5-diglucosides, Dp, Cy 3-glucosides	Scogin (1985)
	Calyx	Cy 3, 5-diglucoside, 3-glucoside, traces of Dp 3, 7, 3′-triglucoside	
Streptocalyx poeppigii	Bract, leaf	Cy 3, 5, 3′-triglucoside, 3, 5-diglucoside	
Tillandsia 3 spp.	Bract, leaf	Cy, Mv, Pg, Pn 3, 5-diglucosides	Saito and Harborne (1983)
	Flower	Mv 3, 5-diglucoside	
Vriesea 2 spp.	Leaf	Mv, Pn 3, 5-diglucosides, Cy, Pn 3-glucosides	
COMMELINACEAE			
Commelina communis	Flower	Dp 3-(6″-*p*-coumarylglucoside)-5-(6‴-malonylglucoside) (= malonylawobanin, *1.5*)	Goto *et al.* (1983b)
Commelina 7 spp.	Flower	Dp 3-(*p*-coumarylglucoside)-5-glucoside	
Callisia, Campelia, Cyanotis, Dichorisandra, Gibasis, Rhoeo,	Leaf, flower, stem	tricaffeyl Cy 3, 7, 3′-triglucoside	Stirton and Harborne (1980)
Tradescantia, Tripogandra, Zanonia and *Zebrina* spp.	Flower	tricaffeyl Dp 3, 7, 3′-triglucoside	

GRAMINEAE			
Zizania aquatica	Leaf sheath, staminate florets	Cy 3-glucoside, 3-rhamnosylglucoside	Gutek *et al.* (1981)
IRIDACEAE			
Gladiolus cultivars	Flower	Cy, Dp, Mv, Pg, Pn, Pt 3-rhamnosylglucoside-5-glucosides, 3, 5-diglucosides, 3-glucosides, 3-rhamnosylglucosides	Akavia *et al.* (1981)
LILIACEAE			
Hyacinthoides non-scripta	Flower	Dp 3-(*p*-coumarylglucoside)-5-(malonylglucoside)	Takeda *et al.* (1986b)
Liriope muscari	Fruit pulp	Dp, Mv, Pt-arabinosides, Dp-xyloside	Hruska *et al.* (1982)
Ophiopogon jaburan	Seed coat	Pt 3-(2^G-glucosylrutinoside)-5′-glucoside	Ishikura and Yoshitama (1984)
Scilla pensylvanica	Flower	Dp 3-(*p*-coumarylglucoside)-5-(malonylglucoside)	Takeda *et al.* (1986b)
ORCHIDACEAE			
Anacamptis, Dactylorhiza, Epipactis, Gymnadenia, Neottianthe, Nigritella, Ophrys, Orchis, Trannsteinera spp.	Flower	Cy 3-(6″-oxalylglucoside)	Strack *et al.* (1986)
DICOTYLEDONEAE			
ACANTHACEAE			
Strobilanthus dyeriana	Callus tissue	Cy, Pn 3, 5-diglucosides	Smith *et al.* (1981)
ANACARDIACEAE			
Pistacia vera	Nut kernel skin	Cy 3-galactoside	Miniati (1981)
ARALIACEAE			
Panax ginseng	Fruit	Pg 3-glucoside	Park-Lee and Park (1980)
P. quinquefolium	Fruit	Pg 3-glucoside	
ARISTOLOCHIACEAE			
Asarum canadense	Flower	Cy glycoside, unacylated	Wilson and Brown (1981)
Hexastylis heterophylla	Flower	Cy, Mv glycosides, acylated	
H. virginica	Flower	Cy, Mv glycosides, acylated	
BERBERIDACEAE			
Berberis coreana	Fruit	Cy, Pn, Pt 3-glucosides	
B. crataegina	Fruit	Dp, Mv, Pn, Pt 3-glucosides	Vereskovskii and Shapiro (1985 a,b)
B. cretica	Fruit	Dp, Mv, Pn, Pt 3-glucosides	Cubukcu and Dortunc (1982)
B. darwinii	Fruit	Dp, Mv, Pn, Pt 3-glucosides, Pt 3-gentiobioside, Dp, Pt 3-rutinosides	Medrano *et al.* (1985)
B. integerrima	Fruit	Cy, Dp, Pn, Pt 3-glucosides, Cy 3-rutinoside	Vereskovskii and Shapiro (1985b)
B. sieboldii	Fruit	Cy, Pn, Pt 3-glucosides	
B. sphaerocarpa	Fruit	Cy, Dp, Pn, Pt 3-glucosides, Cy 3-rutinoside	
B. vulgaris	Fruit	Cy, Pn, Pt 3-glucosides	

(*Contd.*)

Table 1.8 (*Contd.*)

Class, family, genus, species	*Organ examined*	*Pigments present*	*Reference*
Epimedium 5 spp.	Stem, leaf	Cy 3-(*p*-coumarylsophoroside)	Yoshitama (1984)
	Flower, stem, leaf	Dp 3-(*p*-coumarylsophoroside)-5-glucoside	
BURSERACEAE			
Commiphora mukul	Flower	Pg 3, 5-diglucoside	Kakrani (1981)
CAMPANULACEAE			
Platycodon grandiflorum	Flower	Dp 3-rutinoside-7-(di-(caffeylglucosyl))glucoside (= platyconin, *1.3*)	Goto *et al.* (1983a)
CAPRIFOLIACEAE			
Sambucus ebulus	Fruit	Cy 3, 5-diglucoside, 3-glucoside	Novruzov and Aslanov (1984)
S. nigra	Fruit	Cy 3-glucoside, 3-sambubioside, 3-sambubioside-5-glucoside, 3, 5-diglucoside, 3-xylosylglucoside, 3-rhamnosylglucoside	Broennum-Hansen and Hansen (1983); Pogorzelski (1983)
CARYOPHYLLACEAE			
Agrostemma githago	Flower	Cy 3-glucoside, 3, 5-diglucoside, triglucoside	Ferry and Darbour (1980)
Dianthus caryophyllus cvs.	Flower	Pg, Cy 3-(6″-malylglucoside)	Terahara *et al.* (1986)
D. deltoides	Flower	Cy 3-(6″-malylglucoside)	
COMPOSITAE			
Aster cultivars	Flower	Cy, Dp, Pg 3-glucosides, Cy, Dp, Pg 3, 5-diglucosides	Inazu and Onodera (1980)
Callistephus chinensis cv.	Flower	Pg 3-(6″-malonylglucoside)	Takeda *et al.* (1986a)
Centaurea cyanus	Flower	Cy 3-(6″-succinylglucoside)-5-glucoside	Takeda and Tominaga (1983); Tamura *et al.* (1983)
Centaurea 7 spp.	Flower	Cy 3-(6″-succinylglucoside)-5-glucoside	Sulyok and Laszlo-Bencsik (1985)
Cichorium intybus cv.	Flower	Dp 3, 5-di(malonylglucoside)	Takeda *et al.* (1986a)
	Leaf	Cy 3-(6″-malonylglucoside)	Bridle *et al.* (1984)
Coleostephus myconis	Stem	Cy 3-dimalonylglucoside	Takeda *et al.* (1986a)
Cynara scolymus	Bract, leaf	Cy 3-sophoroside, 3-glucoside, 3-(caffeylsophoroside), 3-(caffeylglucoside), 3-(dicaffeylsophoroside)	Aubert and Foury (1982)
	Flower	Cy 3, 5-diglucoside, 3-(caffeylsophoroside)-5-glucoside	
Dahlia cultivars	Flower	Pg, Cy 3-(6″-malonylglucoside)-5-glucoside, Pg, Cy 3, 5-di(malonylglucoside)	Takeda *et al.* (1986a)
Gerbera jamesonii cvs.	Flower	Cy, Pg 3-malonylglucosides, Cy, Pg 3-glucosides	Asen (1984)
Helenium autumnale cv.	Flower	Cy 3-malonylglucuronosylglucoside	Takeda *et al.* (1986a)
Helichrysum sanguineum	Capitulum	Pg, Pn 3-glucosides	Mericli *et al.* (1984)
Senecio cruentus	Flower	Rubrocinerarin	Yoshitama (1981)
		Cinerarin (*1.7*)	Goto *et al.* (1984)

CONVOLVULACEAE			
Ipomoea purpurea	Flower	Heavenly blue anthocyanin (*1.1*)	Goto *et al.* (1981a); Goto *et al.* (1981b); Goto (1984)
CRUCIFERAE			
Raphanus raphanistrum	Root	Pg 3-(caffeylsophoroside)-5-glucoside, Pg 3-(ferulylsophoroside)-5-glucoside	Lu (1985)
DIOSCOREACEAE			
Dioscorea alata	Tuber	Cy 3-glucoside	Ozo *et al.* (1984)
ELAEOCARPACEAE			
Aristotelia chilensis	Fruit	Cy, Mv, Pt 3,5-diglucosides, Dp, Mv 3,7-diglucosides, Dp 3-(*p*-coumarylglucoside), 3-glucoside, *p*-coumaryl Mv 3,5-diglucoside, Pt 3-glucoside	Diaz *et al.* (1984)
EMPETRACEAE			
Empetrum nigrum ssp. *hermaphroditicum*	Fruit	Cy, Dp, Mv, Pn, Pt 3-galactosides, 3-arabinosides, Dp, Pt 3-glucosides	Karppa *et al.* (1984)
ERICACEAE			
Arbutus unedo	Fruit	Cy 3-glucoside	Proliac and Raynaud (1981)
Vaccinium arboreum	Fruit	Cy, Dp, Mv, Pn, Pt 3-arabinosides, 3-galactosides, Dp, Pn, Pt 3-glucosides	Ballinger *et al.* (1982)
V. corymbosum	Fruit	Cy, Dp, Mv, Pt 3-galactosides, Cy 3-glucoside, Dp, Mv, Pt 3-arabinosides	Vereskovskii and Shapiro (1985)
V. stamineum	Fruit	Cy 3-galactoside, 3-arabinoside, 3-glucoside	Ballinger *et al.* (1981)
V. vitis-idaea	Fruit	Cy 3-galactoside, 3-arabinoside, Cy, Dp 3-glucosides	Andersen (1985)
EUPHORBIACEAE			
Euphorbia minuta	Whole plant	Cy 3-galactoside	Del. V. Galarza *et al.* (1983)
E. serpens microphylla	Whole plant	Cy, Dp 3-glucosides	Del. V. Galarza *et al.* (1983)
E. tirucalli	Root	Cy 3,5-diglucoside, Dp 3-rhamnosylglucoside	Baslas and Gupta (1983)
GENTIANACEAE			
Gentiana makinoi	Flower	Dp 3-glucoside-5,3′-di(6″-caffeylglucoside) (= gentiodelphin, *1.2*)	Goto *et al.* (1982)
GERANIACEAE			
Pelargonium × *hortorum* cultivars	Flower	Cy, Dp, Mv, Pg, Pn, Pt 3,5-diglucosides	Asen and Griesbach (1983)
GESNERIACEAE			
Saintpaulia ionantha cvs.	Flower	Mv, Pg, Pn 3-rutinoside-5-glucosides	Khokhar *et al.* (1982)
GUTTIFERAE			
Garcinia indica	Fruit	Cy 3-glucoside, 3-sambubioside	Krishnamurthy *et al.* (1982)
LABIATAE			
Monarda didyma	Flower	Pg 3-(6″-*p*-coumarylglucoside)-5-(4‴,6‴-dimalonylglucoside) (= monardaein, *1.6*)	Kondo *et al.* (1985)
Perilla frutescens	Leaf, seed	Cy 3,5-diglucoside, 3-(*p*-coumarylglucoside)-5-glucoside	Ishikura (1981a)

(*Contd.*)

Table 1.8 (*Contd.*)

Class, family, genus, species	*Organ examined*	*Pigments present*	*Reference*
Prunella vulgaris	Flower	Hs, Mv, Pn 3, 5-diglucosides	Saxena and Archana (1984)
Salvia spp.	Flower	Cy, Dp, Pg 3-rhamnosides	Haque *et al.* (1981)
Stachys sp.	Flower	Cy 3-(*p*-Coumarylglucoside)-5-(malonylglucoside)	Takeda *et al.* (1986b)
LAURACEAE			
Persea americana	Fruit	Cy 3-galactoside, 3, 5-diglucoside *p*-coumarate	Prabha *et al.* (1980)
LEGUMINOSAE			
Clitoria ternatea	Flower	Dp 3, 3′, 5′-triglucosides acylated with *p*-coumaric and malonic acids (= ternatins A–F)	Saito *et al.* (1985a)
Hedysarum coronarium	Flower	Pn 3-glucoside, Mv, Pn 3, 5-diglucosides	Chriki and Harborne (1983)
Phaseolus 5 spp.	Seedling	Mv glycosides	Nozzolillo and McNeil (1985)
Vigna 3 spp.	Seedling	Dp and Cy glycosides	Nozzolillo and McNeil (1985)
V. mungo	Hypocotyl	Cy, Dp 3-glucosides	Ishikura *et al.* (1981)
	Seed coat	Dp 3-glucoside	Ishikura *et al.* (1981)
V. radiata	Hypocotyl	Dp 3-(*p*-coumarylglucoside), 3-glucoside	Ishikura *et al.* (1981)
	Seed coat	Dp 3-glucoside	Ishikura *et al.* (1981)
MALVACEAE			
Alcea rosea var. *nigra*	Flower	Dp, Mv, Pt 3, 5-diglucosides, Dp, Mv, Pt 3-glucosides	Kohlmunzer *et al.* (1983)
Hibiscus sabdariffa var. *sabdariffa*	Calyx	Cy 3-sambubioside, 3-glucoside, Dp glycosides	Khafaga and Koch (1980)
MELASTOMATACEAE			
Tibouchina grandiflora	Flower	Pn 3-sophoroside, 3-sambubioside, Mv 3, 5-diglucoside, 3-(*p*-coumaryl-sambubioside)-5-glucoside	Bobbio *et al.* (1985)
T. granulosa	Flower	Mv 3-(di-*p*-coumarylxyloside)-5-glucoside, 3-(*p*-coumarylxyloside)-5-glucoside	Francis *et al.* (1982)
OLEACEAE			
Olea europaea cultivars	Fruit	Dp 3-rhamnosylglucoside-7-xyloside, Cy 3-rutinoside	Tanchev *et al.* (1980a,b)
ONAGRACEAE			
Clarkia 2 spp.	Flower	Cy, Dp, Mv 3-glucosides, 3, 5-diglucosides	Dorn and Bloom (1984)
PAPAVERACEAE			
Papaver nudicaule	Flower	Pg 3-(6″-malonylsophoroside)	Cornuz *et al.* (1981)
POLEMONIACEAE			
Collomia 14 spp.	Flower	Cy and/or Dp 3-(*p*-coumarylglucoside)-5-glucosides	Wilken *et al.* (1982)

POLYGONACEAE			
Antigonon leptopus	Flower	Mv, Pg 3, 5-diglucosides	Tiwari and Minocha (1980)
Fagopyrum sagittatum	Hypocotyl	Cy 3-glucoside, 3-galactoside, 3-rhamnosylgalactoside (?)	Inouye *et al.* (1982)
Polygonum 23 spp.	Stem, sepal, seedling, leaf	Cy 3-glucoside, 3-galactoside, 3-arabinoside, 3-rutinoside, 3-arabinosylglucoside, Cy, Pn 3-arabinoside-5-glucosides, Mv 3, 5-diglucoside	Yoshitama *et al.* (1984)
Rheum 2 spp.	Petiole	Cy 3-glucoside, 3-rutinoside	
Rumex 5 spp.	Sepal, leaf, stem	Cy 3-glucoside, 3-rutinoside	
PRIMULACEAE			
Anagallis arvensis f. *coerulea*	Flower	Mv 3-rhamnoside	Ishikura (1981b)
Lysimachia nummularia	Flower stem	Cy, Dp, Pn glycosides	Prum *et al.* (1983)
PUNICACEAE			
Punica granatum	Seed coat	Cy, Dp 3-glucosides, 3, 5-diglucosides	Santagi *et al.* (1984)
RANUNCULACEAE			
Aquilegia 15 spp.	Flower	Acylated Cy, Dp, Pg 3-glucosides, 3, 5-diglucosides	Taylor (1984)
A. alpina	Flower	Dp, Pg 3-xylosylglucoside-5-glucosides	
A. formosa	Flower	Dp, Pg 3-xylosylglucoside-5-glucosides	
ROSACEAE			
Amelanchier 4 spp.	Fruit	Cy, Pg 3, 5-diglucosides, 3-glucosides, Cy 3-galactoside	Vereskovskii *et al.* (1982)
Cotoneaster 53 taxa	Fruit	Cy, Pg 3-galactosides	Kruegel and Krainhoefner (1985)
Potentilla atrosanguinea	Flower	Cy 3-rutinoside	Harborne and Nash (1984)
P. nepalensis	Flower	Cy 3-glucoside	
Prunus salicina cultivars	Fruit	Cy 3-glucoside, 3-rutinoside	Tsuji *et al.* (1981) Itoo *et al.* (1982)
Rosa platyacantha	Fruit	Cy 3-glucoside, 3, 5-diglucoside	Beisekova *et al.* (1984)
Rubus spp. (raspberry cultivars)	Fruit	Cy 3-glucoside, 3-sophoroside, 3-rutinoside, 3-glucosylrutinoside, 3-sambubioside, 3-xylosylrutinoside	Jennings and Carmichael (1980)
RUBIACEAE			
Warszewiczia coccinea	Aerial parts	Cy 3-glucoside	Mohammed and Seaford (1981)
RUTACEAE			
Citrus sinensis cultivars	Fruit	Cy, Dp 3-glucosides, Cy, Dp, Pn 3, 5-diglucosides, Cy 3-(acetylglucoside), 3-(ferulylglucoside), 3-(*p*-coumarylferulylglucoside), 3-(sinapylglucoside), Pn 3-(*p*-coumarylglucoside)	Maccarone *et al.* (1983) Maccarone *et al.* (1985)
SAXIFRAGACEAE			
Ribes aureum	Fruit	Cy, Dp 3-glucosides, Cy 3-rutinoside	Medrano *et al.* (1985)
R. magellanicum	Fruit	Cy, Dp 3-glucosides, Cy 3-rutinoside	
Saxifraga 4 spp.	Axial bulbil	Cy, Dp, Mv, Pn, Pt 3-glucosides	Andersen and Oevstedal (1983)
	Flower	Cy, Dp, Pt 3-glucosides	
S. hirculus	Whole plant	Mv, Dp or Pt 3-glucosides	Miller and Bohm (1980)

(*Contd.*)

Table 1.8 (*Contd.*)

Class, family, genus, species	Organ examined	Pigments present	Reference
SCHISANDRACEAE			
Schisandra chinensis	Flower?	Pn 3-glucoside	Yang *et al.* (1982)
SCROPHULARIACEAE			
Digitalis purpurea mutant	Flower	Cy, Pn 3-glucosides	Brown (1980)
SOLANACEAE			
Petunia hybrida mutants	Flower	Cy, Dp 3-glucosides, 3-rhamnosylglucosides, Cy 3-diglucoside, Mv 3-rhamnosylglucoside-5-glucoside, *p*-coumaryl Cy, Dp, Mv, Pn, Pt 3-rhamnosylglucoside-5-glucosides	Schram *et al.* (1983)
Solanum scabrum spp. *scabrum*	Fruit, stem, leaf	Mv, Pt glycosides	Gbile and Adesina (1985)
S. scabrum spp. *nigericum*	Fruit	Mv, Pt glycosides	Gbile and Adesina (1985)
THEACEAE			
Camellia japonica cvs.	Flower	Cy 3-glucoside	Sakata *et al.* (1981)
TILIACEAE			
Grewia subinaeqalis	Fruit	Cy, Dp 3-glucosides	Khurdiya and Anand (1981)
UMBELLIFERAE			
Daucus carota	Cell culture	Cy 3-(sinapylxylosylglucosylgalactoside) (*1.4*)	Harborne et al. (1983)
VITACEAE			
Cissus sicyoides	Fruit	Cy 3-rhamnosylarabinoside, Dp 3-rutinoside, 3-rhamnoside	Toledo *et al.* (1983)
Vitis sp.	Callus cell culture	Cy, Pn 3-glucosides, Pn 3, 5-diglucoside	Yamakawa *et al.* (1983)

In Araceae of the monocotyledons, a survey of 59 representative species showed that the anthocyanin pattern is simple: 3-glucosides and 3-rutinosides are present and there is an absence of methylated pigments. This is to be expected in a family where fly pollination occurs and in which spathe and spadix may be dull in colour. Pelargonidin glycosides are uncommon and delphinidin is rare. Exceptionally, cyanidin 3-gentiobioside was found to replace cyanidin 3-rutinoside in two genera, *Anchomanes* and *Rhektophyllum*. These are both in the same subfamily and there are cytological and morphological grounds for including them in the same tribe (Williams *et al.*, 1981).

In Bromeliaceae, by contrast with Araceae, the anthocyanins are diverse. Cyanidin 3, 5, 3′-triglucoside is present characteristically in many species and genera (Table 1.8) (Saito and Harborne, 1983). *Puya* is exceptional in the family in having delphinidin 3, 7, 3′-triglucoside instead of cyanidin 3, 5, 3′-triglucoside (Scogin, 1985). This pattern links it with the Commelinaceae, where cyanidin and delphinidin 3, 7, 3′-triglucosides are widespread, albeit in acylated form (Stirton and Harborne, 1980). In Commelinaceae, *Commelina* is unusual in that the acylated delphinidin 3, 5-diglucoside, malonylawobanin, is the characteristic petal pigment.

In Orchidaceae, the recent discovery of cyanidin 3-oxalylglucoside as a taxonomic marker for certain European species (see Table 1.8) is of considerable interest, but further surveys are needed to see how widespread this novel pigment is (Strack *et al.*, 1986). Considering the enormous biological and popular interest in orchid flowers, it is surprising how little is known of the anthocyanins in the family and further studies are bound to be rewarding.

In Berberidaceae of the dicotyledons, there are at least two distinct patterns. *Berberis* fruit pigmentation is based on the 3-glucosides of cyanidin, peonidin, delphinidin, petunidin and malvidin, whereas in *Epimedium*, stem, leaf and flower contain delphinidin 3-(*p*-coumarylsophoroside)-5-glucoside (Yoshitama, 1984). Two patterns are also apparent in the Compositae, this time based on different acylating acids. In members of the Cynareae, pigments are based on succinic acids and cyanidin 3-(6′-succinylglucoside)-5-glucoside is widespread (Sulyok and Raszlo-Bencsik, 1985, and unpublished results). By contrast, all members of other tribes that have been examined contain anthocyanins with malonyl substitution (see Table 1.8).

In Polygonaceae, a survey of 33 species and three genera indicated a simple pattern (3-glucoside or 3-galactoside of cyanidin) in most species (Yoshitama *et al.*, 1984). Methylated pigments are uncommon, but interestingly there is a change in glycosylation pattern (to 3, 5-disubstitution) when these are present. Thus, malvin is confined to species of section *Echinocaulon* of genus *Polygonum* and peonidin 3-arabinoside-5-glucoside to the same section and the related section *Persicaria*.

1.5 APPLICATIONS

The most important industrial application of plant anthocyanin research is in the food industry, and especially in the wine trade. Anthocyanin analyses have proved of value in differentiating red wines, in determining which grape varieties were used in manufacture and for recognizing when exogenous colouring matters have been added to wines. The anthocyanins of grapes and the derived red wines have been investigated in great detail and the transformations that occur during wine storage and ageing have been carefully monitored. An excellent review of this field has been provided by Ribereau-Gayon (1982). Since that review was written, attention has been given to the anthocyanin content of ruby ports and of the grapes from which they are derived. It has been found that the cultivar used for port production can be identified by measuring through HPLC the ratio of malvidin 3-(acetylglucoside) to total malvidin glycosides in the wine (Bakker and Timberlake, 1985).

The other major interest of the food industry in anthocyanins is in their use as natural colorants to replace synthetic red dyes. A problem with the anthocyanins for colouring food and drinks is their instability in solution to light and pH changes, and especially their bleaching by sulphur dioxide which is often used as a preservative. This instability can be avoided by reacting anthocyanins with carbonyl compounds, such as acetaldehyde, to stabilize them. Alternatively, it is possible to employ acylated anthocyanins, which are more stable to light than simple glycosides (Timberlake and Henry, 1986). The flower pigments of *Clitorea ternatea* (Leguminosae), which have been utilized in Malaysia to colour rice cakes blue, promise to be of value because of their exceptional stability in solution (Saito *et al.*, 1985a). Commercial interest in elaborating anthocyanins in plant tissue culture has also been based on their potential application as colorants in the food industry and here, the acylated cyanidin triglycoside of carrot cells (see Section 1.3.3) is advantageous because of the high yields that can be obtained in suspension cultures.

There has also been current medicinal interest in anthocyanins, following the recognition that pigment extracts are more effective than *O*-(β-hydroxyethyl)rutin in decreasing capillary permeability and fragility and in their anti-inflammatory and anti-oedematic activities (Wagner, 1985). Anthocyanins thus appear to have replaced rutin and its derivatives in the treatment of illnesses involving tissue inflammation or capillary fragility. The crude anthocyanin extracts of *Rubus occidentalis, Sambucus nigra* and *Vaccinium myrtillus* have all been applied, but the *Vaccinium* extracts have become the most popular. All these fruits contain mixtures of simple glycosides and there does not seem to be any particular specificity in the type of anthocyanidins present or their sugar substituents. The key feature they share in common is the high anthocyanin concentrations; bilberry preparations with as much as 27% anthocyanins are now on the market. As a result of this medicinal application, many papers have appeared describing the effects of *Vaccinium* anthocyanins on isolated arterial tissues (e.g. Bettini *et al.*, 1985).

One other biological activity can now be ascribed to anthocyanins; inhibition of larval growth in insects. Because of their ready metabolism *in vivo*, anthocyanins have usually been thought of as being dietarily beneficial in humans (Pierpoint, 1986) and toxicological effects have rarely been considered in any animals. Yet, Hedin and his co-workers (1983) have observed that the cotton leaf anthocyanin, cyanidin 3-glucoside, is particularly inhibitory to larval growth in the tobacco budworm *Heliothis viriscens*. It is even more inhibitory to this moth than the well-known cotton toxin, gossypol. Since anthocyanin is present in the leaf of many cotton cultivars, it could be a factor in limiting insect predation on this crop. Indeed, the levels of anthocyanin in leaves of different cultivars are more closely correlated with insect resistance than are the levels of condensed tannin, also present in the leaf.

The reasons for such dramatic effects on insect growth are not yet clear, but presumably the anthocyanin interacts in some way with the insects' ability to absorb its nutritional requirements from the host plant. It appears to be rather specific, since the related *Heliothis zea* which feeds on tomatoes, is known to be quite unaffected by the tomato leaf anthocyanin, petanin (Isman and Duffey, 1982). Nevertheless, it would be interesting to know whether or not cultivars rich in anthocyanin, in other crop plants, were less attacked by insects as a result of an increase in pigmentation.

One other interaction between anthocyanins and insects has been proposed in the early entomological literature – pigment sequestration from the diet and subsequent contribution to insect colour (Fox and Vevers, 1960) – but this has never been substantiated by modern experiments. Barbier (1984) isolated a water-soluble green pigment with pH indicator properties from the moth *Euchloron megaera*, which he suggested might be anthocyanin-derived. A specimen of this pigment was examined in this laboratory, through the courtesy of Dr Barbier, but it showed no characteristic anthocyanin properties and appeared from our observations to be quinone-based.

REFERENCES

Akavia, N., Strack, D. and Cohen A. (1981), *Z. Naturforsch.* **36c**, 378.

Andersen, O.M. (1985), *J. Food Sci.* **50**, 1230.

Andersen, O.M. and Francis, G.W. (1985), *J. Chromatogr.* **318**, 450.

Andersen, O.M. and Oevstedal, D.O. (1983), *Biochem. Syst. Ecol.* **11**, 239.

Asen, S. (1984), *Phytochemistry* **23**, 2523.

Asen, S. and Griesbach, R. (1983), *J. Am. Soc. Hortic. Sci.* **108**, 845.

Aubert, S. and Foury, C. (1982), *Chem. Abstr.* **97**, 195777.

Bakker, J. and Timberlake, C.F. (1985), *J. Sci. Food Agric.* **36**, 1315 and 1325.

Ballinger, W.E., Maness, E.P. and Ballington, J.R. (1981), *Sci. Hortic.* **15**, 173.

Ballinger, W.E., Maness, E.P. and Ballington, J.R. (1982), *Can. J. Plant Sci.* **62**, 683.

Barbier, M. (1984), *J. Chem. Ecol.* **10**, 1109.

Baslas, R.K. and Gupta, N.C. (1983), *Herba Hung.* **22**, 35.

Beisekova, K.D., Mukhametgaliev, A.G. and Bikbulatova, T.N. (1984), *Zdravookhr. Kav.* **1984**, 64.

Bettini, V., Fiori, A., Martino, R., Mayellaro, F. and Ton, P. (1985), *Fitoterapia* **56**, 67.

Bobbio, F.O., Bobbio, P.A. and Rodriguez-Amaya, D.B. (1983), *Food Chem.* **12**, 189.

Bobbio, F.O., Bobbio, P.A. and Degaspari, C.H. (1985), *Food Chem.* **18**, 153.

Bridle, P., Loeffler, R.S.T., Timberlake, C.F. and Self, R. (1984), *Phytochemistry* **23**, 2968.

Broennum-Hansen, K. and Hansen, S.H. (1983). *J. Chromatogr.* **262**, 385.

Brown, A. (1980), *Bull. Liaison – Groupe Polyphenols* **1980**, 250.

Canbas, A. (1985), *Doga Bilim Derg.*, Seri Oz, **9**, 394.

Chriki, A. and Harborne, J.B. (1983), *Phytochemistry* **22**, 2322.

Clevenger, S. (1964), *Can J. Biochem.* **42**, 154.

Colijn, C.M., Jonsson, L.M.V., Schram, A.W. and Kool, A.J. (1981), *Protoplasma* **107**, 63.

Cornuz, G., Wyler, H. and Lauterwein, J. (1981), *Phytochemistry* **20**, 1461.

Cubukcu, B. and Dortunc, T. (1982), *Doga*, Seri C, **6**, 11.

Del V. Galarza, S., Cabreya, J.L. and Juliani, H.R. (1983), *An. Asoc. Quim. Argent.* **71**, 505.

Diaz, L.S., Rosende, C.G. and Antunez, M.I. (1984), *Rev. Agroquim. Tecnol. Aliment.* **24**, 538.

Dorn, P.S. and Bloom, W.L. (1984), *Biochem. Syst. Ecol.* **12**, 311.

Ferry, S. and Darbour, N. (1980), *Plant. Med. Phytother.* **14**, 148.

Fox, H.M. and Vevers, G. (1960). In *The Nature of Animal Colours*, Sidgwick and Jackson, London, pp. 166–168.

Francis, F.J., Draetta, I., Baldini, V. and Iaderoza, M. (1982), *J. Am. Soc. Hortic. Sci.* **107**, 789.

Francis, G.W. and Andersen, O.M. (1984), *J. Chromatogr.* **283**, 445.

Gbile, Z.O. and Adesina, S.K. (1985), *Fitoterapia* **56**, 11.

Goto, T. (1984), *Proc. 5th Asian Symp. Med. Plants Spices* (Seoul, Korea), pp. 593–604.

Goto, T., Kondo, T., Imagawa, H. and Miura, I. (1981a), *Tetrahedron Lett.* **22**, 3213.

Goto, T., Kondo, T., Imagawa, H., Takase, S., Atobe, M. and Miura, I. (1981b), *Chem. Lett.* **1981**, 883.

Goto, T., Kondo, T., Tamura, H., Imagawa, H., Iino, A. and Takeda, K. (1982), *Tetrahedron Lett.* **23**, 3695.

Goto, T., Kondo, T., Tamura, H., Kawahori, K. and Hattori, H. (1983a), *Tetrahedron Lett.* **24**, 2181.

Goto, T., Kondo, T., Tamura, H. and Takase, S. (1983b), *Tetrahedron Lett.* **24**, 4863.

Goto, T., Kondo, T., Kawai, T. and Tamura, H. (1984), *Tetrahedron Lett.* **25**, 6021.

Goto, T., Tamura, H., Kawai, T., Hoshino, T., Harado, N. and Kondo, T. (1986), *Ann N.Y. Acad. Sci.* **471**, 155.

Gutek, L.H., Woods, D.L. and Clark, K.W. (1981), *Crop Sci.* **21**, 79.

Haque, M.S., Ghoshal, D.N. and Ghoshal, K.K. (1981), *Proc. Ind. Natl. Acad. Sci.* **B47**, 204.

Harborne, J.B. (1966), *Phytochemistry* **5**, 589.

Harborne, J.B. (1967), *Comparative Biochemistry of the Flavonoids*, Academic Press, London.

Harborne, J.B. (1976), *Biochem. Syst. Ecol.* **4**, 31.

Harborne, J.B. (1986), *Phytochemistry* **25**, 1887.

Harborne, J.B. and Boardley, M. (1985), *Z. Naturforsch.* **40c**, 305.

Harborne, J.B. and Nash, R.J. (1984), *Biochem. Syst. Ecol.* 12, 315.

Harborne, J.B., Mayer, A.M. and Bar-Nun, N. (1983), *Z. Naturforsch.* **38c**, 1055.

Hedin, P.A., Jenkins, J.N., Collum, D.H., White, W.H. and Parrott, W.L. (1983). In *Plant Resistance to Insects* (ed. P.A. Hedin), American Chemical Society, Washington DC, pp. 347–367.

Hemingson, J.C. and Collins, R.P. (1982), *J. Nat. Prod.* (*Lloydia*) **45**, 385.

Hicks, K.B., Sondey, S.M., Hargrave, D., Sapers, G.M. and Bilyk, A. (1985), *L.C. Mag.* **3**, 981 and 984.

Hrazdina, G. (1982). In *The Flavonoids – Advances in Research* (eds J.B. Harborne and T.J. Mabry), Chapman and Hall, London and New York, pp. 135–188.

Hrazdina, G. and Wagner, G.J. (1985), *Ann. Proc. Phytochem. Soc. Eur.* **25**, 119.

Hrazdina, G., Marx, G.A. and Hoch, H.C. (1982), *Plant Physiol.* **70**, 745.

Hruska, A.F., Dirr, M.A. and Pokorny, F.A. (1982), *J. Am. Soc. Hortic. Sci.* **107**, 468.

Inazu, K. and Onodera, T. (1980), *Tamagawa Daiguku Nogakubu Kenkyu Hokoku* **1980**, 56.

Inouye, K., Hosoyama, Y. and Shimadate, T. (1982), *Chem. Abstr.* **97**, 88690.

Ishikura, N. (1981a), *Agric. Biol. Chem.* **45**, 1855.

Ishikura, N. (1981b), *Z. Pflanzenphysiol.* **103**, 469.

Ishikura, N. and Yoshitama, K. (1984), *J. Plant Physiol.* **115**, 171.

Ishikura, N., Iwata, M. and Miyazaki, S. (1981), *Bot. Mag. Tokyo* **94**, 197.

Isman, M.B. and Duffy, S.S. (1982), *Ent. Exp. Appl.* **31**, 370.

Itoo, S., Matsuo, T., Noguchi, K. and Kodama, K. (1982), *Kogoshima Daigaku Nogakubu Gakujutsu Hokoku* **1982**, 35.

Jennings, D.L. and Carmichael, E. (1980), *New Phytol.* **84**, 505.

Joo, K.J. and Park, J.M. (1983), *Han'guk Yongyang Sikyong Hakhoechi* **12**, 31.

Jurd, L. and Harborne, J.B. (1968), *Phytochemistry* **7**, 1209.

Kakrani, H.K. (1981), *Fitoterapia* **52**, 221.

Karppa, J., Kallio, H., Peltonen, I. and Linko, R. (1984), *J. Food. Sci.* **49**, 634.

Kay, Q.O.N., Daoud, H.S. and Stirton, C.H. (1981), *Bot. J. Linn. Soc.* **83**, 57.

Khafaga, E.-S.R. and Koch, H. (1980), *Angew. Bot.* **54**, 295.

Khokhar, J.A., Humphries, J.M., Short, K.C. and Grout, B.W.W. (1982), *Hortic. Sci.* **17**, 810.

Khurdiya, D.S. and Anand, J.C. (1981), *J. Food. Sci. Technol.* **18**, 112.

Kohlmunzer, S., Konska, G. and Wiatr, E. (1983), *Herba Hung.* **22**, 13.

Kondo, T., Nakane, Y., Tamura, H., Goto, T. and Eugster, C.H. (1985), *Tetrahedron Lett.* **26**, 5879.

Krishnamurthy, N., Lewis, Y.S. and Ravindranath, B. (1982), *J. Food Sci. Technol.* **19**, 97.

Kruegel, T. and Krainhoefner, A. (1985), *Chem. Abstr.* **103**, 51155.

Lu, X. (1985), *Shipin Yu Fajiao Gongye* **1985**, 19.

Maccarone, E., Maccarone, A., Perrini, G. and Rapisarda, P. (1983), *Ann. Chim.* (*Rome*) **73**, 533.

Maccarone, E., Maccarone, A. and Rapisarda, P. (1985), *Ann. Chim.* (*Rome*) **75**, 79.

Markakis, P. (ed.) (1982). *Anthocyanins as Food Colors*. Academic Press, New York.

McClure, J.W. (1986). In *Plant Flavonoids in Biology and Medicine* (eds V. Cody, E. Middleton and J.B. Harborne), Alan Liss, New York, pp. 77–86.

Medrano, M.A., Tomas, M.A. and Frontera, M.A. (1985), *Rev. Latinoam. Quim.* **16**, 84.

Mericli, A.H., Cubukcu, B. and Dortunc, T. (1984), *Fitoterapia* **55**, 112.

Merlin, J.C., Statona, A. and Brouillard, R. (1985), *Phytochemistry* **24**, 1575.

Miller, J.M. and Bohm, B.A. (1980), *Biochem. Syst. Ecol.* **8**, 279.

Miniati, E. (1981), *Fitoterapia* **52**, 267.

Mohammed, M.J. and Seaforth, C.E. (1981), *Rev. Latinoam. Quim.* **12**, 72.

Momose, T., Abe, K. and Yoshitama, K. (1977), *Phytochemistry* **16**, 1321.

Neumann, D. (1983), *Biochem. Physiol. Pflanzen* **178**, 405.

Novruzov, E.N. and Aslanov, S.M. (1984), *Dokl. Akad. Nauk Az. Az. SSR* **40**, 61.

Nozzolillo, C. and McNeil, J. (1985), *Can. J. Bot.* **63**, 1066.

Ozo, O.N., Caygill, J.C. and Coursey, D.G. (1984), *Phytochemistry* **23**, 329.

Park-Lee, Q. and Park, H. (1980), *Hanguk Nonghwa Hakhoe Chi* **23**, 242.

Paz-Ares, J., Wienand, U., Peterson, P.A. and Saedler, H. (1986), *EMBO J.* **5**, 829.

Pecket, R.C. and Small, C.J. (1980), *Phytochemistry* **19**, 2571.

Pierpoint, W.S. (1986). In *Plant Flavonoids in Biology and Medicine* (eds V. Cody, E. Middleton and J.B. Harborne), Alan Liss, New York, pp. 125–140.

Pogorzelski, E. (1983), *Przem. Spozyw.* **37**, 167.

Prabha, T.N., Ravindranath, B. and Patwardhan, M.V. (1980), *J. Food Sci. Technol.* **17**, 241.

Proliac, A. and Raynaud, J. (1981), *Plant Med. Phytother.* **15**, 109.

Prum, N., Prum, A. and Raynaud, J. (1983), *Die Pharmazie* **38**, 494.

Ribereau-Gayon, P. (1982). In *Anthocyanins as Food Colors* (ed. P. Markakis), Academic Press, New York, pp. 209–244.

Saito, N. and Harborne, J.B. (1983), *Phytochemistry* **22**, 1735.

Saito, N. and Ueno, K. (1985), *Heterocycles* **23**, 2709.

Saito, N., Timberlake, C.F., Tucknott, O.G. and Lewis, I.A.S. (1983), *Phytochemistry* **22**, 1007.

Saito, N., Abe, K., Honda, T., Timberlake, C.F. and Bridle, P. (1985a), *Phytochemistry* **24**, 1583.

Saito, N., Yokoi, M., Yamaji, M. and Honda, T. (1985b), *Phytochemistry* **24**, 2125.

Sakata, Y., Nagayoshi, S. and Arisumi, K. (1981), *Mem. Fac. Agric., Kagoshima Univ.* **17**, 79.

Santagi, N.F., Duro, R. and Duro, F. (1984), *Riv. Merceol.* **23**, 247.

Saxena, V.K. and Archana, S. (1984), *Acta Cienc. Indica.* [*Ser.*] *Chem.* **10**, 37.
Schram, A.W., Jonsson, L.M.W. and De Vlaming, P. (1983), *Z. Naturforsch.* **38c**, 342.
Scogin, R. (1985), *Biochem. Syst. Ecol.* **13**, 387.
Shin, M.S. and Ahn, S.Y. (1980), *Hanguk Sikp'um Kwahakhoe Chi* **12**, 305.
Small, C.J. and Pecket, R.C. (1982), *Planta* **154**, 97.
Smith, S.L., Slywka, G.W. and Krueger, R.J. (1981), *J. Nat. Prod. (Lloydia)* **44**, 609.
Sommer, H. and Saedler, H. (1986), *Mol. Gen. Genet.* **202**, 429.
Stirton, J.Z. and Harborne, J.B. (1980), *Biochem. Syst. Ecol.* **8**, 285.
Strack, D., Akavia, N. and Reznik, H. (1980), *Z. Naturforsch.* **35c**, 533.
Strack, D., Busch, E., Wray, V., Grotjahn, L. and Klein, E. (1986), *Z. Naturforsch.* **41c**, 707.
Sulyok, G. and Laszlo-Bencsik, A. (1985), *Phytochemistry* **24**, 1121.
Sweeney, J.G. and Iacobucci, G.A. (1981), *Tetrahedron* **37**, 1481.
Takeda, K. and Tominaga, S. (1983), *Bot. Mag. Tokyo* **96**, 359.
Takeda, K., Harborne, J.B. and Self, R. (1986a), *Phytochemistry* **25**, 1337.
Takeda, K., Harborne, J.B. and Self, R. (1986b), *Phytochemistry* **25**, 2191.
Tamura, H., Kondo, T., Kato, Y. and Goto, T. (1983), *Tetrahedron Lett.* **24**, 5749.
Tanchev, S.S., Ontcheva, Y., Genov, N. and Codounis, M. (1980a), *Bull. Liaison – Groupe Polyphenols* **1980**, 114.
Tanchev, S.S., Iencheva, N., Genov, N. and Codounis, M. (1980b), *Georgike Eureuna* **4**, 5.
Taylor, R.J. (1984), *Bull. Torrey Bot. Club* **111**, 462.
Terahara, N. and Yamaguchi, M. (1986), *Phytochemistry* **25**, 2906.
Terahara, N., Yamaguchi, M., Takeda, K., Harborne, J.B. and Self, R. (1986), *Phytochemistry* **25**, 1715.
Timberlake, C.F. and Bridle, P. (1975). In *The Flavonoids* (eds J.B. Harborne, T.J. Mabry and H. Mabry), Chapman and Hall, London, pp. 214–266.
Timberlake, C.F. and Henry, B.S. (1986), *Endeavour (New Ser.)* **10**, 31.
Tiwari, K.P. and Minocha, P.K. (1980), *Vijana Parishad Anusandhan Patrika* **23**, 305.
Toledo, M.C.F., Reyes, F.G.R., Iaderoza, M., Francis, F.J. and Draetta, I.S. (1983), *J. Food. Sci.* **48**, 1368.
Tsuji, M., Harakawa, M. and Komiyama, Y. (1981), *Nippon Shokuhin Kogyo Gakkaishi* **28**, 517.
Tsukui, A., Kuwano, K. and Mitamura, T. (1983), *Kaseigaku Zasshi* **34**, 153.
Vaccari, A., Pifferi, P.G. and Zaccherini, G. (1981), *J. Food Sci.* **47**, 40.
Van Sumere, C.F. and Van de Casteele, K. (1985), *Ann. Proc. Phytochemistry. Soc. Eur.* **25**, 17.
Vereskovskii, V.V. and Shapiro, D.K. (1985a), *Khim. Prir. Soedin.* **1985**, 569.
Vereskovskii, V.V. and Shapiro, D.K. (1985b), *Khim. Prir. Soedin.* **1985**, 570.
Vereskovskii, V.V., Shapiro, D.K. and Narizhnaya, T.I. (1982), *Khim. Prir. Soedin.* **1982**, 522.
Wagner, H. (1985), *Ann. Proc. Phytochem. Soc. Eur.* **25**, 409.
Wilken, D.H., Smith, D.M., Harborne, J.B. and Glennie, C.W. (1982), *Biochem. Syst. Ecol.* **10**, 239.
Williams, C.A., Harborne, J.B. and Mayo, S.J. (1981), *Phytochemistry* **20**, 217.
Wilson, J.E. and Brown, L.C. (1981), *Castanea* **46**, 36.
Wrolstad, R.E. and Heatherbell, D.A. (1974), *J. Sci. Food Agric.* **25**, 1221.
Yamakawa, T., Ishida, K., Kato, S., Kodama, T. and Minoda, Y. (1983), *Agric. Biol. Chem. Tokyo* **47**, 997.
Yang, H.C., Lee, J.M. and Song, K.B. (1982), *Hanguk Nonghwa Hakhoe Chi* **25**, 35.
Yasuda, H. and Shinoda, H. (1985), *Cytologia* **50**, 397.
Yoshitama, K. (1977), *Phytochemistry* **16**, 1857.
Yoshitama, K. (1981), *Phytochemistry* **20**, 186.
Yoshitama, K. (1984), *Bot. Mag. Tokyo* **97**, 429.
Yoshitama, K., Hisada, M. and Ishikura, N. (1984), *Bot. Mag. Tokyo* **97**, 31.

2

Flavans and proanthocyanidins

LAWRENCE J. PORTER

2.1 INTRODUCTION

Haslam's previous accounts of this area of flavonoid chemistry (Haslam, 1975, 1982a) eloquently summarized the historical developments in the chemistry and biochemistry of this important and complex group of plant phenolics. Also covered were their relationship with condensed tannins and aspects of their characteristic chemical and spectroscopic properties and metabolism.

The past five years has witnessed spectacular advances in all aspects of our knowledge of the proanthocyanidins, perhaps the most dramatic feature of which is an increase by an order of magnitude in the number of known compounds in this class. In a review of the metabolites of gallic acid, Haslam (1982b) suggested that these and other phenolics, including the proanthocyanidins, were better named 'complex polyphenols', by analogy with carbohydrates. This proposal was timely, as a feature of proanthocyanidin chemistry revealed over the past 5 years has been the high incidence of compounds probably best called 'mixed metabolites'. This is typified by a number of structures where a flavan-3-ol (most often catechin) is bound to a variety of other secondary plant metabolites: examples include alkaloids, cinnamic acids, chalcones and ellagitannins.

This makes the task of adequately reviewing recent advances particularly difficult because of the sheer volume of new data. I have therefore departed considerably from the format used previously and made my primary aim construction of an accurate and hopefully comprehensive register of the known proanthocyanidins and their plant sources. The purpose of this is twofold: firstly, to bring the structure of the chapter more into line with those of other flavonoid classes, and consequently to make future reviews of the field simpler, and secondly to present the results of a most complex and scattered literature in a readily comprehensible form.

In this context it must be pointed out that Haslam's (1975, 1982a) previous reviews were confined to compounds based on the flavan-3-ol skeleton (often called catechins) while ignoring a large body of compounds with an unsubstituted C-ring (the flavans). As these clearly belong to the proanthocyanidin class of flavonoid (the flavans arising from a double reduction of a flavanone rather than a dihydroflavonol), I have decided to include these as well, particularly as a number of proanthocyanidins with an unsubstituted C-ring are now known. Furthermore, a number of naturally occurring flavan-4-ols are now known (Table 2.3) which clearly also belong to this class. Also covered are the peltogynoid flavans as these have not been reviewed previously.

2.2 NOMENCLATURE

As the number and complexity of flavanoid and proanthocyanidin structures increase, so does the need for a straightforward system of nomenclature for this group of plant metabolites. Although this issue has been addressed in each of the previous reviews (Haslam 1975, 1982a) the spectacular increase in the number of compounds with novel structure demands nomenclatural reappraisal.

Following the earlier suggestions of Weinges and Freudenberg, Haslam (1982a) defined leucoanthocyanidins as monomeric proanthocyanidins and condensed proanthocyanidins as flavan-3-ol oligomers. It is

The Flavonoids. Edited by J.B. Harborne
Published in 1988 by Chapman and Hall Ltd,
11 New Fetter Lane, London EC4P 4EE.

Table 2.1 Proanthocyanidin nomenclature: types of proanthocyanidin and suggested names for the (2*R*, 3*S*) monomer units

Proanthocyanidin class	*Monomer unit*	*Substitution pattern* 3	5	7	8	3′	4′	5′
Proapigeninidin	Apigeniflavan (*2.3*)	H	OH	OH	H	H	OH	H
Proluteolinidin	Luteoliflavan (*2.4*)	H	OH	OH	H	OH	OH	H
Protricetinidin	Tricetiflavan (*2.5*)	H	OH	OH	H	OH	OH	OH
Propelargonidin	Afzelechin (*2.6*)	OH	OH	OH	H	H	OH	H
Procyanidin	Catechin (*2.7*)	OH	OH	OH	H	OH	OH	H
Prodelphinidin	Gallocatechin (*2.8*)	OH	OH	OH	H	OH	OH	OH
Proguibourtinidin	Guibourtinidol (*2.9*)	OH	H	OH	H	H	OH	H
Profisetinidin	Fisetinidol (*2.10*)	OH	H	OH	H	OH	OH	H
Prorobinetinidin	Robinetinidol (*2.11*)	OH	H	OH	H	OH	OH	OH
Proteracacinidin	Oritin (*2.12*)	OH	H	OH	OH	H	OH	H
Promelacacinidin	Prosopin (*2.13*)	OH	H	OH	OH	OH	OH	H

convenient to retain this terminology but with the following modifications. The term leucoanthocyanidins should include all monomeric flavanoids which produce anthocyanidins by cleavage of a C—O bond on heating with mineral acid, including not only flavan-3,4-diols, but also flavan-4-ols, and unusual metabolites such as cyanomaclurin (Table 2.5). In the case of the proanthocyanidins, it is suggested that 'condensed' should be dispensed with in view of the above, and this group should be redefined to include all compounds which produce anthocyanidins by cleavage of a C–C bond.

A system of nomenclature for naming proanthocyanidins was introduced by Hemingway *et al.* (1982) and outlined by Haslam (1982a). Subsequent developments have shown the merit of this system, especially as it relates to the naming of oligoproanthocyanidins. However, it is evident that this system must now be extended to encompass the more complex structures of this type and a greater variety of flavanoid monomer (configurational base) units.

Briefly, the original form of this system (Hemingway *et al.*, 1982) was inspired by the realization that the previous trivial system of nomenclature used to distinguish procyanidin dimers and trimers (Weinges *et al.*, 1968b; Thompson *et al.*, 1972) was unsuitable for naming oligomers containing monomer units with differing oxidation patterns. Moreover, use of a fully systematic system based on IUPAC rules for absolute stereochemistry and ring substitution patterns is excessively cumbersome and potentially misleading.

The system is as follows: proanthocyanidins are named in a similar way to polysaccharides (see Polysaccharide nomenclature, 1980) where C-4 of the flavan monomer unit is equivalent (in the nomenclatural sense) to C-1 of a monosaccharide in an oligo- or poly-saccharide chain. The interflavanoid linkage is indicated in the same way as polysaccharides, the bond and its direction being contained in brackets (4→). The configuration of the interflavanoid bond at C-4 is indicated by the $\alpha\beta$ nomenclature (IUPAC, 1979) within the above brackets. The flavanoid monomer units are defined in terms of the trivial names of monomeric flavan-3-ols, the names catechin, epicatechin, etc. being reserved for those units with a 2*R*-configuration, whereas those with a 2*S*-configuration are distinguished by the enantio prefix. Typical examples are the dimer structure (*2.1*) which is named epicatechin-(4β→8)-catechin and the dimer (*2.2*) named *ent*-epicatechin-(4α→8)-epicatechin.

This system is extended so that it is as generally applicable as possible. This firstly requires that an agreed system of nomenclature is adopted to name all monomer units likely to be encountered in proanthocyanidins. These are listed in Table 2.1.

The names for the monomer units (*2.6*)–(*2.11*) are those already established for those flavan-3-ols with (2*R*, 3*S*) absolute stereochemistry with the particular A- and B-ring hydroxylation patterns listed in Table 2.1. New names are prosopin (*2.13*) after its isolation from *Prosopis glandulosa* (Roux, 1986) and oritin (*2.12*) which is named from the fact that the first flavan-3,4-diols isolated with this phenolic hydroxylation pattern were obtained from *Acacia orites* (misnamed *A. intertexta*, Clark-Lewis and Dainis, 1967). The names for (*2.3*)–(*2.5*) are also new and stress their relationship to flavan (i.e. they lack a 3-hydroxy group) and also relates them to the corresponding anthocyanidin. The corresponding (2*R*, 3*R*) or 3α-hydroxy isomers of the units (*2.6*)–(*2.13*) are distinguished by adding 'epi' to the beginning of each monomer name, i.e. epiafzelechin, epiquibourtinidol, epipubeschin, etc. This does not arise for (*2.3*)–(*2.5*) as they lack a 3-substituent. The (2*R*), (2*S*, 3*R*) or (2*S*, 3*S*) isomers are indicated by adding *ent* to the beginning of the appropriate monomer name (see Section F: IUPAC, 1979), i.e. *ent*-luteoliflavan, *ent*-epigallocatechin, etc. Many examples of the use of this nomenclature follow.

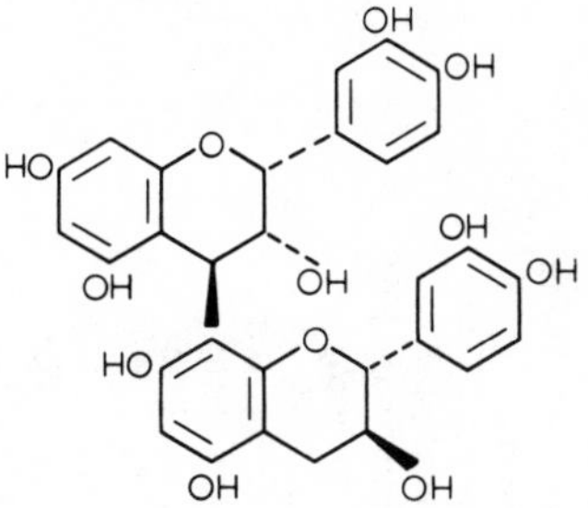

(*2.1*)

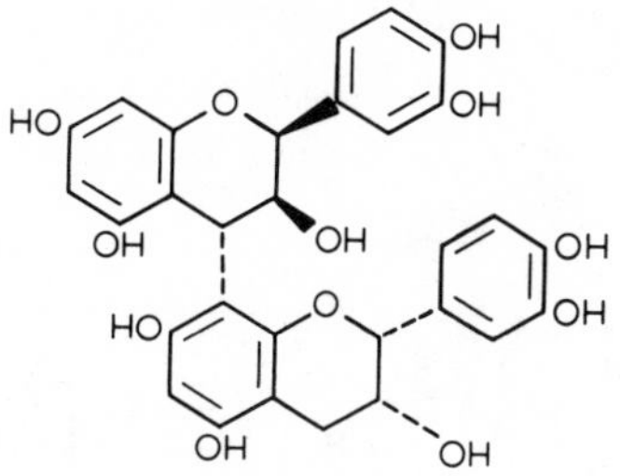

(*2.2*)

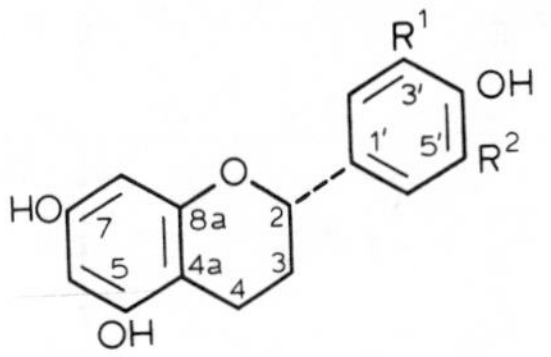

(*2.3*) $R^1 = R^2 = H$

(*2.4*) $R^1 = OH, R^2 = H$

(*2.5*) $R^1 = R^2 = OH$

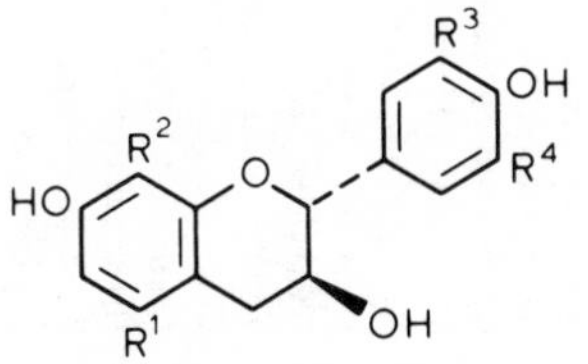

(*2.6*) $R^1 = OH, R^2 = R^3 = R^4 = H$

(*2.7*) $R^1 = R^3 = OH, R^2 = R^4 = H$

(*2.8*) $R^1 = R^3 = R^4 = OH, R^2 = H$

(*2.9*) $R^1 - R^4 = H$

(*2.10*) $R^1 = R^2 = R^4 = H, R^3 = OH$

(*2.11*) $R^1 = R^2 = H, R^3 = R^4 = OH$

(*2.12*) $R^1 = R^3 = R^4 = H, R^2 = OH$

(*2.13*) $R^1 = R^4 = H, R^2 = R^3 = OH$

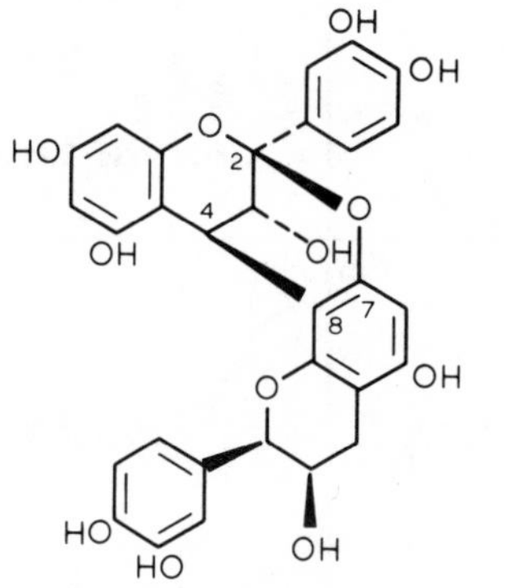

(*2.14*)

Proanthocyanidin A2

A considerable number of doubly linked (so called A-type) proanthocyanidins are now known, often this type of linkage co-occurring with the above single linkages in the same molecule. Typical is proanthocyanidin A2 (*2.14*) where two epicatechin units are linked through a normal $4\beta \rightarrow 8$ linkage and also through C-2 to *O*-7 of the adjacent epicatechin unit. The naming of such compounds is readily accommodated by the proposed system by including both types of linkage within the brackets, as follows: epicatechin-($2\beta \rightarrow 7, 4\beta \rightarrow 8$)-epicatechin. As in the case of interglycosidic linkages, there is no need to nominate *O* in the $2\beta \rightarrow 7$ linkage as this is obvious from the epicatechin substitution pattern.

The system may also be used to name the leucoanthocyanidins. The flavan-3, 4-diols are currently named by a confusing system of trivial names, some actually having more than one. It is proposed, for example, that (+)-mollisacacidin is called fisetinidol-4α-ol and (−)-melacacidin is called epiprosopin-4α-ol.

A convenient shorthand method for distinguishing the units of proanthocyanidin oligomers is also needed. The units in an oligomer chain may be readily distinguished by the degree of substitution on the A- and C-rings (Porter, 1984). A proposed system for doing this is described in Scheme 2.1.

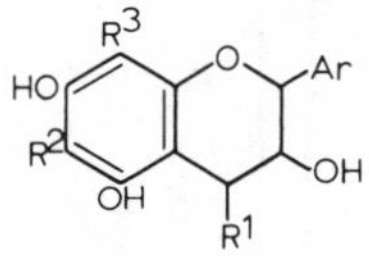

Scheme 2.1 Method for naming flavanoid units in proanthocyanidin oligomers. R^1 = flavanoid; $R^2 = R^3 = H$; T unit. $R^1 = R^2$ or R^3 = flavanoid; R^2 or $R^3 = H$; M unit. $R^1 = R^2 = R^3$ = flavanoid; J unit. $R^1 = R^2$ or $R^3 = H$; R^2 or R^3 = flavanoid; B unit.

Table 2.2 The natural flavans

Class	*Compound (trivial name)*	$[\alpha]_D$ *(solvent)*	*Source (reference)*
(1) Hydroxylated	7-OH	−30.5(m)	*Narcissus pseudonarcissus* bulb (Coxon *et al.*, 1980)
	4′,7-diOH	−30.5(m)	*Narcissus pseudonarcissus* bulb (Coxon *et al.*, 1980)
			Broussonetia papyrifera shoots (Takasugi *et al.*, 1984)
(2) *O*-Methylated and/or methylenated	4′-OH-7-OMe	−15.6(e)	*Stypandra grandis* root (Cooke & Down, 1971)
	7-OH-5-OMe	−6.4(c)	*Dracaena draco* resin (Cardillo *et al.*, 1971)
	7-OH-4′-OMe (broussin)	−17.4(c)	*Broussonetia papyrifera* shoot (Takasugi *et al.*, 1980)
	5,7-diOMe (tephrowatsin E)		*Tephrosia watsoniana* leaf (Gomez *et al.*, 1985)
	4′,5,7-triOMe	0(c)	*Xanthorrhea preissii* resin (Birch & Salahuddin, 1964)
	4′,7-diOH-3′-OMe	−6.9(m)	*Iryanthera elliptica* wood (Braz Filho *et al.*, 1980a)
			Dracaena draco resin (Camarda *et al.*, 1983)
			Zephyranthes flava bulbs (Ghosal *et al.*, 1985b)
	7-OH-3′,4′-methylenedioxy	−14.2	*Zephyranthes flava* bulbs (Ghosal *et al.*, 1985b)
	3′,4′-diOH-5,7-diOMe	0	*Iryanthera coriacea* wood (Franca *et al.*, 1974)
	2′-OH-7-OMe-4′,5′-methylenedioxy	0	*Iryanthera juruensis* wood (Franca *et al.*, 1974)
			Zephyranthes flava bulb (Ghosal *et al.*, 1985)
(3) *C*-Methylated	7-OH-5-OMe-6-Me	−9.2(c)	*Dracaena draco* resin (Cardillo *et al.*, 1971)
	4′-OH-7-OMe-8-Me	−22.4(c)	*Dianella revoluta* root (Cooke & Downe, 1971)
	4′,5-diOH-7-OMe-8-Me	−9.4(m)	*Draaena draco* resin (Camarda *et al.*, 1983)
	4′,7-diOH-8-Me	−36.4(c)	*Narcissus pseudonarcissus* bulb (Coxon *et al.*, 1980)
			Dracaena draco resin (Camarda *et al.*, 1983)
	3′,7-diOH-4′-OMe-8-Me	−31.0(c)	*Lycoris radiata* bulb (Numata *et al.*, 1983)
	4′,7-diOH-3′-OMe-8-Me	−45.8(m)	*Dracaena draco* resin (Camarda *et al.*, 1983)
	2′,7-diOH-4′,5′-methylenedioxy-6,8-diMe		*Iryanthera laevis* wood (Braz Filho *et al.*, 1980b)
	2′,7-diOH-4′,5′-methylenedioxy-5,8-diMe		*Iryanthera laevis* wood (Braz Filho *et al.*, 1980b)
	2′,5-diOH-4′,5′-methylenedioxy-6,8-diMe		*Iryanthera laevis* wood (Braz Filho *et al.*, 1980b)
(4) Prenylated	5,7-diOMe-8-pr*	−79.5(c)	*Tephrosia watsoniana* leaf (Gomez *et al.*, 1985)
			Tephrosia madrensis leaf (Gomez *et al.*, 1983)
	3′,4′,7-triOH-2′,5′-di-pr (kazinol A)	−11(c)	*Broussonetia papyrifera* cortex (Ikuta *et al.*, 1985)
	kazinol B (*2.15*)	−20(c)	*Broussonetia papyrifera* cortex (Ikuta *et al.*, 1985)
	nitenin (*2.16*)		*Tephrosia nitens* leaf (Gomez *et al.*, 1984)
			Tephrosia watsoniana leaf (Gomez *et al.*, 1985)
(5) *O*-Glycosides	5-OH-7-OGlc (koaburanin)	−64.2(e)	*Enkianthus nudipes* leaf (Ogawa *et al.*, 1970)
	7,4′-diOMe-5-OGlc (dichotosin)	−34.5(m)	*Hoppea dichotoma* root (Ghosal *et al.*, 1985a)
	4′,7-diOH-5-OXyl	−31.8(e)	*Buckleya lanceolata* leaf (Sashida *et al.*, 1976)
	3′,7-diOH-4′-OMe-5′-OGlu (auriculoside)	−77.0(m)	*Acacia auriculiformis* leaf (Sahai *et al.*, 1980)
	3′,4′,7-triOMe-5-OGlu (dichotosinin)		*Hoppea dichotoma* root (Ghosal *et al.*, 1985a)
	7-OH-3′,4′-diOMe-5′-OGlu (diffutin)	−46.3(m)	*Canscora diffusa* leaf (Ghosal *et al.*, 1983)

(6) Biflavanoids	Daphnodorin A (*2.17*)	− 63.2(d)	*Daphne odora* root (Baba *et al.*, 1985)
	Daphnodorin C (*2.18*)	− 263(d)	*Daphne odora* root (Baba *et al.*, 1985)
	Compound (*2.19*)	− 60(c)	*Daemonorops draco* resin (Merlini & Nasini, 1976)
	Dracorubin (*2.20*)	− 103(c)	*D. draco* resin (Cardillo *et al.*, 1971)
	Nordracorubin (*2.21*)	− 77.5(m)	*D. draco* resin (Cardillo *et al.*, 1971)
	Compound (*2.22*)		
	Compound (*2.23*)		
	Compound (*2.24*)		*D. draco* resin (Camarda *et al.*, 1981)
	Compound (*2.25*)		
	Compound (*2.26*)	+ 105(e)	*Xanthorrhoea preissii* resin (Birch *et al.*, 1967)
	Compound (*2.27*)	+ 244(e)	*Xanthorrhoea preissii* resin (Birch *et al.*, 1967)
	Compound (*2.28*)		*D. draco* resin (Camarda *et al.*, 1981)

*pr = 3, 3-dimethylallyl.

Key to Tables 2.2 to 2.9: The reference in the 'Source' column refers to the isolation of the compound from that particular plant source. It does not necessarily refer to the work involving the structural elucidation of the compound, or its spectroscopic characteristics, including the specific rotation. The latter are the best value available of $[\alpha]_{589}$ or $[\alpha]_{578}$ for that compound (or a derivative if this is not available) in the opinion of the author. Key to the abbreviations used for solvents in the specific rotations: a = acetone; aw = acetone/water (usually 1:1, v/v); c = chloroform; d = dioxane; e = ethanol; ea = ethyl acetate; m = methanol; mw = methanol/water (1:1, v/v); w = water.

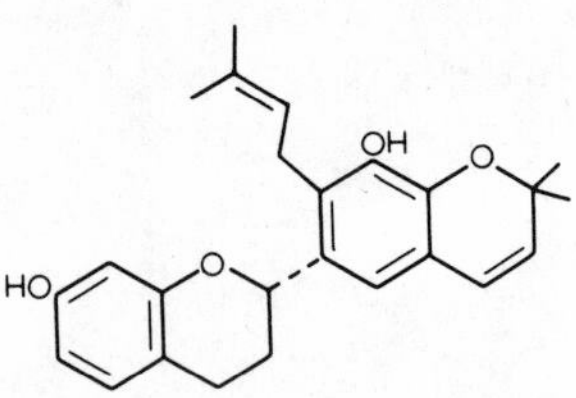

(*2.15*) Kazinol B

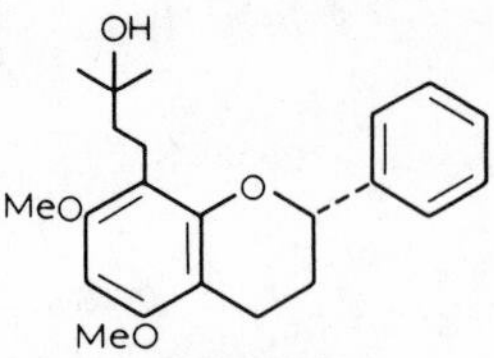

(*2.16*) Nitenin

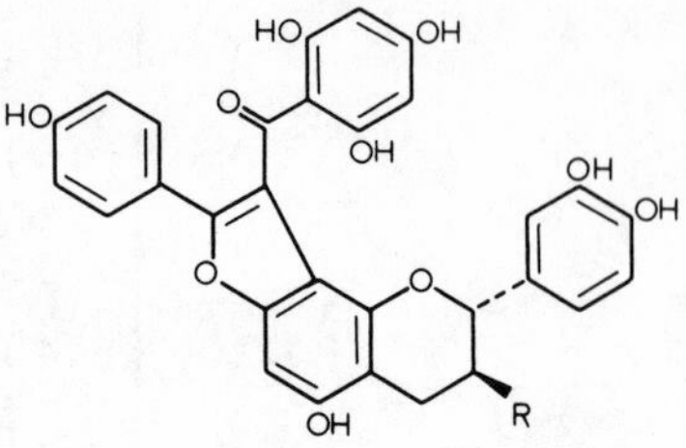

(*2.17*) R = H; daphnodorin A
(*2.30*) R = OH; daphnodorin B

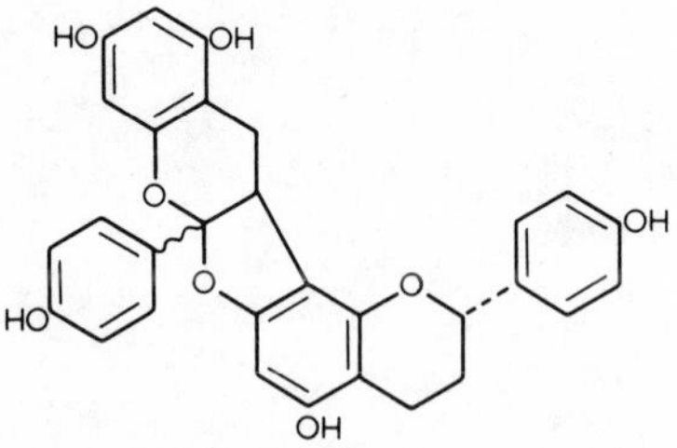

(*2.18*) daphnodorin C

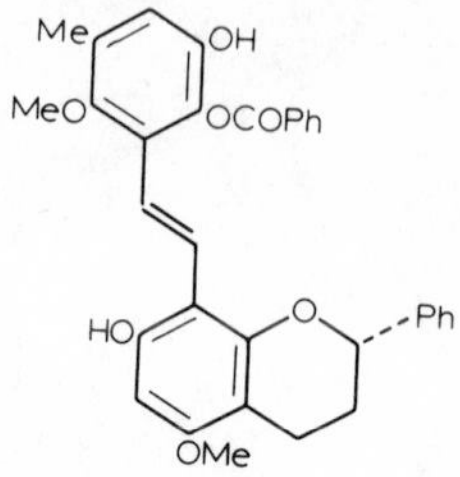

(*2.19*)

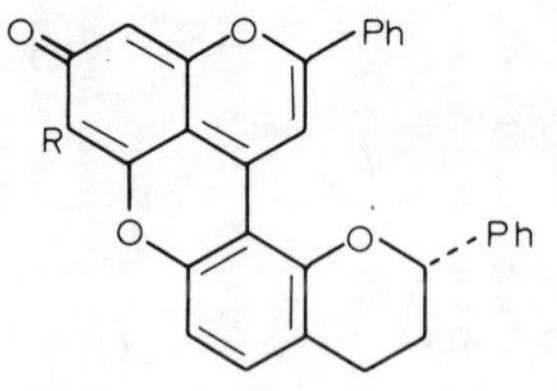

(*2.20*) R = Me; dracorubin
(*2.21*) R = H; nordracorubin

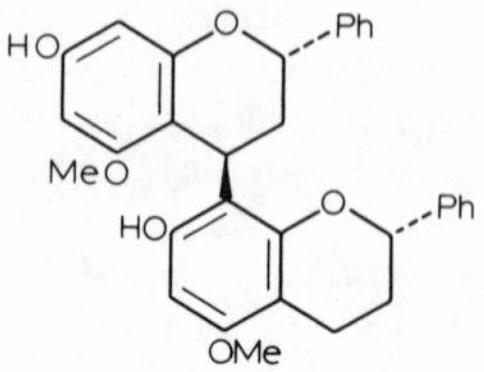

(*2.22*)

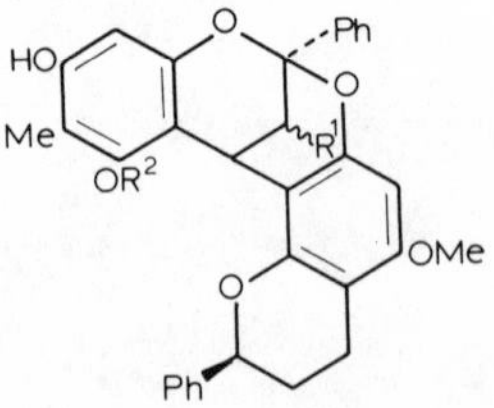

(*2.23*) $R^1 = OH$; $R^2 = Me$
(*2.24*) $R^1 = H$; $R^2 = Me$
(*2.25*) $R^1 = R^2 = H$

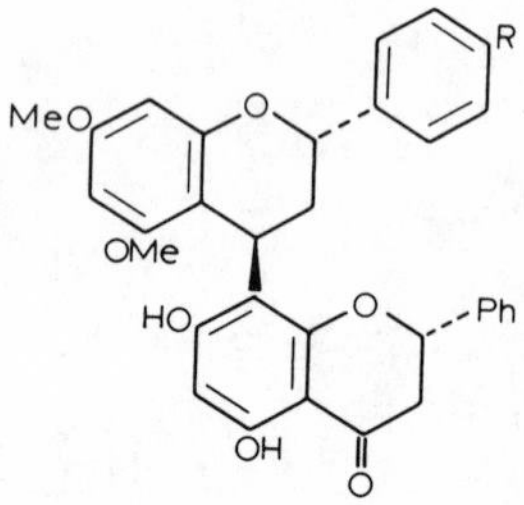

(*2.26*) R = H
(*2.27*) R = OH

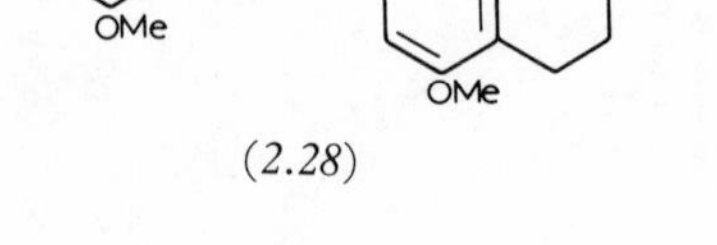

(*2.28*)

2.3 STRUCTURE AND DISTRIBUTION

A comprehensive list of the naturally occurring compounds in the flavan, flavan-3-ol, leucoanthocyanidin and proanthocyanidin (oligomeric and polymeric) classes are listed in Tables 2.2–2.9, together with their plant sources. All compounds are named according to the system outlined in Table 2.1, or consistent with well-established principles for naming flavonoids. Where difficulties arise because of unwieldy nomenclature, or the use of a trivial name, the structure is also given in the text. In addition, a specific rotation (generally at 589 nm) is given for the natural product, where available, or a simple derivative, also where available. These data are included as virtually all the flavonoids in these classes are optically active, and a specific rotation remains the sole readily available physical constant for these compounds. In contrast to other flavonoids, the ultraviolet spectrum is generally uninformative. As pointed out earlier, there have been a considerable number of new natural products isolated in these classes over the 1981–85 period. The more interesting of these will be highlighted as each class is considered in turn.

2.3.1 Flavans

These were not considered earlier although a list of biflavans may be found in Geiger and Quinn (1982). A very useful review of the natural flavans has appeared (Saini and Ghosal, 1984) and the reader is referred to this article for details of the spectroscopy of flavans. A number of flavans additional to this review appear in Table 2.2.

There is little doubt that the flavans arise from a double reduction of a flavanone in the same way as proposed for flavan-3-ols (Section 2.3.2). It is common for flavans to co-occur with the flavanone of identical substitution pattern, in spite of the most unusual patterns of substitution possessed by many of them. All the flavans whose ORD or CD spectra have been studied have the (2*S*) absolute configuration, as would be expected from the flavanone origin. A flavan glucoside, koaburanin, from the leaves of *Enkianthus nudipes* (Table 2.2) was considered by the authors to possess a (2*R*) configuration (Ogawa *et al.*, 1970). However, perusal of their CD data revealed that it is, in fact, correlated with epicatechin and therefore of (2*S*) absolute stereochemistry. Many of the natural flavans are listed as racemic because of nil rotation at 589 nm. This assumption is dangerous in view of the low specific rotation of flavans and the stereochemistry can only be judged from full CD data on the compound in question.

Many natural flavans are lipid-soluble and appear to be leaf-surface constituents. A number are phytoalexins – such as those from *Broussonetia papyrifera* and *Narcissus pseudonarcissus* (Table 2.2). The flavan from *Lycoris radiata* bulbs was found to be an anti-feedant for the larvae of the yellow butterfly *Eurema hecabe mandarina* (Numata *et al.*, 1983).

The first natural flavan isolated was from the resin of the Australian 'blackboy' grass (*Xanthorrhea preissii*) (Birch and co-workers, 1964, 1967), and later a pair of flavans linked through C-4 to pinocembrin were isolated (Table 2.2).

Subsequently other red resins from *Dracaena draco* and *Daemonorops draco* (the so-called 'Dragons blood' resins) also yielded a great variety of mono- and dimeric flavans (see Table 2.2). The sensitivity of these flavans to oxidation to form stable quinone-methides (such as dracorubin) largely accounts for the intense red colours of these resins.

2.3.2 Flavan-3-ols (Table 2.3)

This is by far the largest class of monomeric flavans, and two compounds, catechin and epicatechin, are among the commonest flavonoids known, sharing a distribution almost as widespread as quercetin (the flavonol of equivalent oxidation pattern) in the Dicotyledoneae. The 3, 4, 5-trihydroxy B-ring flavan-3-ols, gallocatechin and epigallocatechin, also have an extremely widespread distribution (paralleling myricetin) especially in more primitive plants (the Coniferae being outstanding).

Ent-catechin and *ent*-epicatechin are relatively rare, the latter being widespread in the Palmae and also being isolated recently from *Polygonum multiflorum* and *Uncaria gambir*. Freudenberg and Purrman reported *ent*-epicatechin from the latter source in 1923, but this occurrence was later suggested to be an artefact of the non-enzymic epimerization of catechin, now disproved. *Ent*-catechin has only very recently been isolated from *Rhaphiolepsis umbellata* and *Polygonum multiflorum* (as the 7-glucoside and 3-gallate respectively). *Ent*-catechin and *ent*-epicatechin usually co-occur with epicatechin and catechin respectively, and it seems plausible that the *ent* isomers result from the action of a C-2 epimerase enzyme (see Ellis *et al.*, 1983).

Reports of the occurrence of resorcinol A-ring flavan-3-ols (i.e. fisetinidol, etc.) have been confined to the Leguminosae and Anacardiaceae. However, fisetinidol has recently been isolated from the heartwood of two *Virola* (Myristicaceae) species (Kijjoa *et al.*, 1981).

Two new hydroxyflavan-3-ols were reported in 1983. Prosopin (so named, Roux, 1986) was isolated from *Prosopis glandulosa* (mesquite) and shown to have (2*R*, 3*S*) absolute stereochemistry by CD correlation with catechin (Roux, 1986). This compound was previously reported from *Piptadenia macrocarpa*, but insufficient data were presented to confirm its absolute configuration (Miyauchi *et al.*, 1976). Both these are leguminous plants.

The second example is the reported isolation of (2*R*, 3*R*)-3, 5, 7, 3′, 5′-pentahydroxyflavan from *Humboltia laurifolia* (Leguminosae). Structural elucidation involved high-resolution mass spectroscopy and ^{1}H NMR of the tetra-*O*-methyl ether. The absolute configuration was

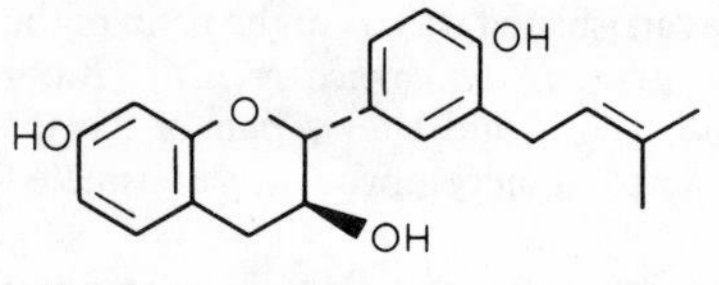

(2.29) Broussinol

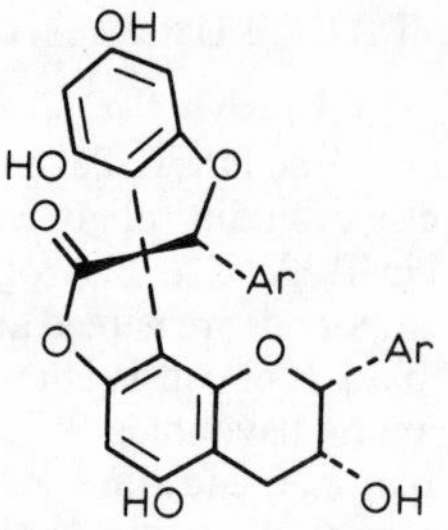

(2.31) Ar = p-C_6H_4-OH; larixinol

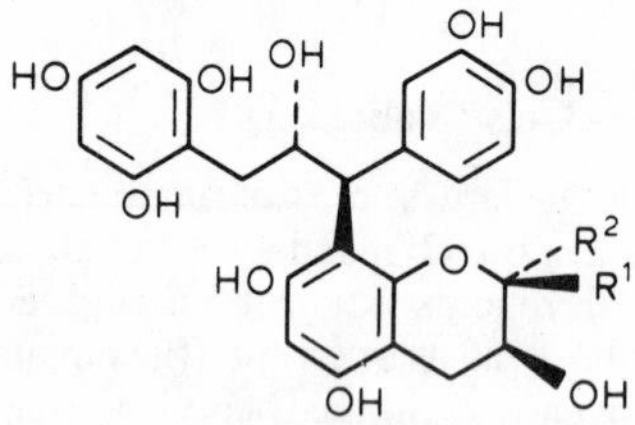

(2.32) R^1 = H; R^2 = 3, 4-dihydroxyphenyl; gambiriin A1
(2.33) R^1 = 3, 4-dihydroxyphenyl; R^2 = H; gambiriin A2

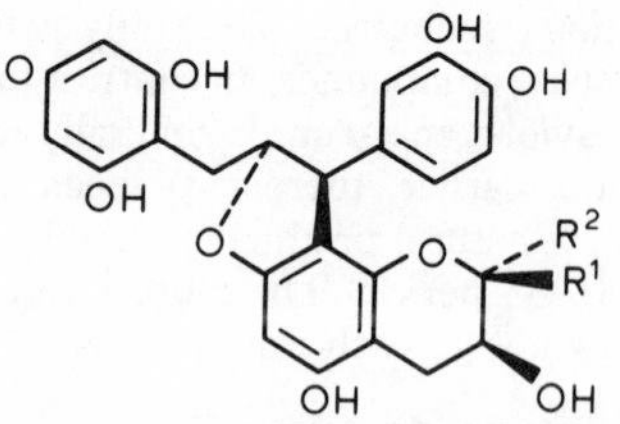

(2.34) R^1 = H; R^2 = 3, 4-dihydroxyphenyl; gambiriin B1
(2.35) R = 3, 4-dihydroxyphenyl; R = H; gambiriin B2

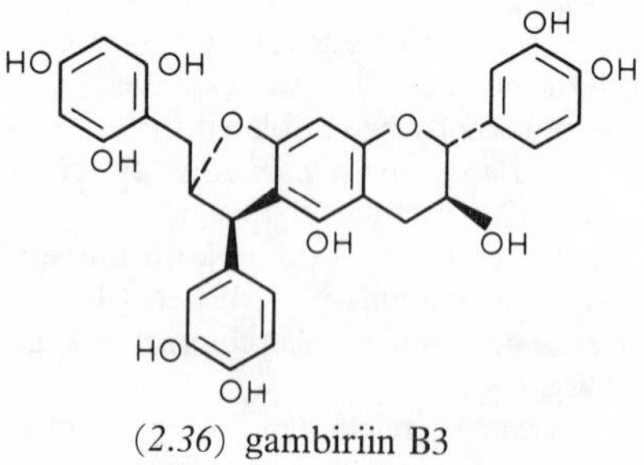

(2.36) gambiriin B3

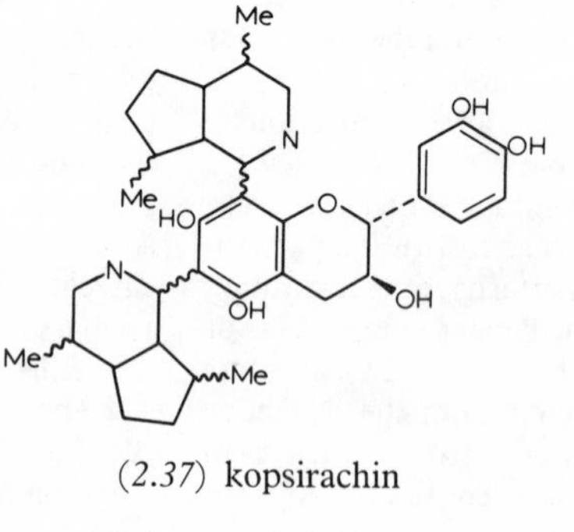

(2.37) kopsirachin

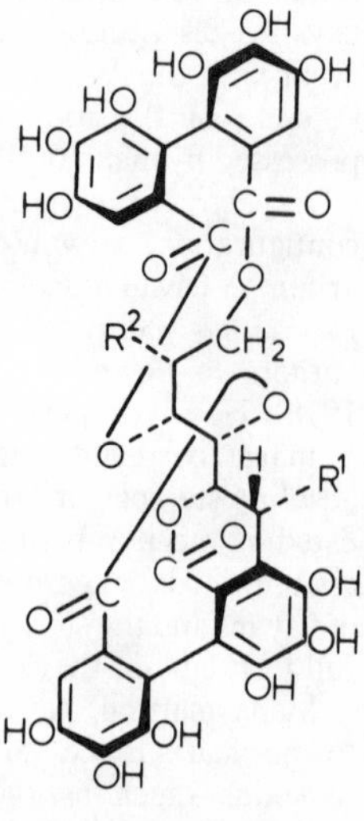

(2.38) R^1=8-catechin, R^2=galloyl; Stenophyllanin A
(2.39) R^1=6-catechin, R^2=galloyl; Stenophyllanin B
(2.40) R^1=8-catechin, R^2=H; Stenophyllanin C

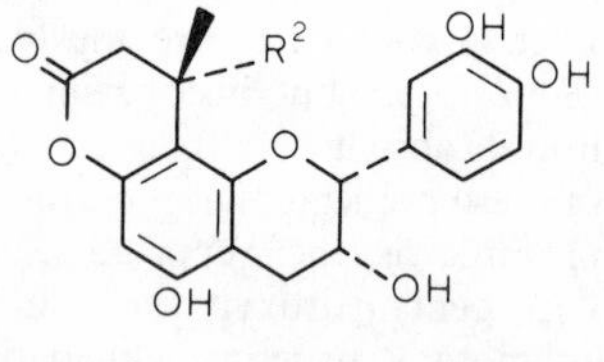

(2.41) R^1=H, R^2=3,4-dihydroxyphenyl; Cinchonain 1a
(2.42) R^1=3,4-dihydroxyphenyl; R^2=H; Cinchonain 1b

Table 2.3 The naturally occurring flavan-3-ols and their natural derivatives and related compounds

Class and compound	$[\alpha]_D$ *(solvent)*	*Source (reference)*
(1) Hydroxylated A- and B-rings		
Afzelechin	+ 20.6(aw)	*Eucalyptus calophylla* wood (Hillis & Carle, 1960)
		Nothofagus fusca heartwood (Hillis & Inoue, 1967)
		Saxifraga ligulata root (Poce Tucci *et al.*, 1969)
		Desmoncus polycanthus leaf (Marletti *et al.*, 1976)
		Juniperus communis fruit (Friedrich & Engelstowe, 1978)
		Prunus persica cv. Hakuto roots (Ohigashi *et al.*, 1982)
		Kandelia candel bark (Hsu *et al.*, 1985a)
Epiafzelechin	− 58.9(e)	*Afzelia* sp. heartwood (King *et al.*, 1955)
		Larix gmelini bark (Shen *et al.*, 1985)
		Larix sibirica bark (Pashinana *et al.*, 1970)
		Juniperus communis fruit (Friedrich & Engelstowe, 1978)
		Cassia sieberiana root (Waterman & Faulkner, 1979)
		Ephedra sp. root (Hikino *et al.*, 1982a)
ent-Epiafzelechin	+ 66(e)*	*Livinstona chinensis* leaf (Delle Monache *et al.*, 1972)
		Crataeva religiosa leaf (as 5-Glu, Sethi *et al.*, 1984)
Catechin	+ 17(aw)	Widespread
ent-Catechin	− 7.9(a)[†]	*Polygonum multiflorum* root (as 3-gallate, Nonaka *et al.*, 1982b)
		Rhaphiolepsis umbellata bark (as 7-Glu, Nonaka *et al.*, 1983a)
Gallocatechin	+ 14.7(aw)	Widespread
Epigallocatechin	− 60(e)	Widespread
Epicatechin	− 68(e)	Widespread
ent-Epicatechin	+ 52(a)	Palmae fruit and leaf (Delle Monache *et al.*, 1972)
		Uncaria gambir leaf (Nonaka & Nishioka, 1980)
		Polygonum multiflorum root (Nonaka *et al.*, 1982b)
Fisetinidol	− 8.6(aw)	*Acacia mearnsii* heartwood (Roux & Paulus, 1961)
		Colophospermum mopane heartwood (Drewes & Roux, 1966a)
		Virola minutifolia, *V. elongata* wood (Kijjoa *et al.*, 1981)
ent-Fisetinidol	+ 8.0(aw)	*Afzelia xylocarpa* wood (Dean *et al.*, 1965)
ent-Epifisetinidol	+ 82(aw)	*Colophospermum mopane* heartwood (Drewes & Roux, 1966a)
Robinetinidol	− 10.7(aw)	*Acacia mearnsii* bark (Roux & Maihs, 1960)
(2*R*, 3*R*)-3, 3′, 5′, 7-Pentahydroxyflavan[‡]	− 39.7(m)	*Humboltia laurifolia* bark (Samaraweera *et al.*, 1983)
Prosopin	n.d.	*Piptadenia macrocarpa* heartwood (Miyauchi *et al.*, 1976)
		Prosopis glandulosa heartwood (Jacobs *et al.*, 1983)
(2) *O*-Methylated		
4′, 5-DiOMe-afzelechin	n.d.	*Lagascea rigida* leaf (Bohlmann & Jakupovic, 1978)
4′-OMe-Catechin	+ 6.7(a)	*Cinnamomum cassia* bark (Morimoto *et al.*, 1985b)
4′, 7-DiOMe-catechin	− 3.8(m)	*C. cassia* bark (Morimoto *et al.*, 1985b)
3′, 5, 7-TriOMe-Catechin		*Viguiera quinqueradiata* leaf (Delgado *et al.*, 1984)
4′, 5, 7-TriOMe-catechin	− 8.2(a)	*C. cassia* bark (Morimoto *et al.*, 1985b)
3′-OMe-Epicatechin	− 56.1(a)	*C. cassia, C. obtusifolium* bark (Morimoto *et al.*, 1985b)
		Lindera umbellata twig (Morimoto *et al.*, 1985b)
		Symplocus uniflora bark (as 7-Glu, Tschesche *et al.*, 1980)
3′, 4′-DiOMe-epicatechin	− 59.3(a)	*L. umbellata* twig (Morimoto *et al.*, 1985b)
3′, 5-DiOMe-epicatechin	− 66.4(a)	*C. cassia, C. obtusifolium* bark (Morimoto *et al.*, 1985b)
		L. umbellata twig (Morimoto *et al.*, 1985b)
3′, 5, 7-TriOMe-epicatechin	− 61.2(a)	*C. cassia, C. obtusifolium* bark (Morimoto *et al.*, 1985b)
		L. umbellata twig (Morimoto *et al.*, 1985b)
3′, 4′-Methylenedioxy-5, 7-diOMe-epicatechin	0(c)	*C. cassia* bark (Miyamura *et al.*, 1983)

(*Contd.*)

Table 2.3 (*Contd.*)

Class and compound	$[\alpha]_D$ (*solvent*)	Source (*reference*)
4′-OMe-Epigallocatechin (ourateacatechin)	−62(e)	*Ouratea* sp. root bark (Delle Monache *et al.*, 1967)
		Elaeodendron balae root bark (Weeratunga *et al.*, 1985)
		Kokoona zeylanica root bark (Kamal *et al.*, 1982)
		Prionostema aspera root bark (Delle Monache *et al.*, 1976)
		Maytenus rigida root bark (Delle Monache *et al.*, 1976)
(3) Simple esters of flavan-3-ols		
ent-Epicatechin 3-*O*-δ-(3,4-diOH-phenyl)-β-OH-pentanoate (phylloflavan)	−7(m)	*Phyllocladus alpinus* twig (Foo *et al.*, 1985)
Catechin 3-*O*-(1-OH-6-oxo-2-cyclo-hexene-1-carboxylate)	−102(a)	*Salix sieboldiana* bark (Hsu *et al.*, 1985c)
Catechin 3-*O*-(1,6-diOH-2-cyclo-hexene-1-carboxylate)	−87(a)	*Salix sieboldiana* bark (Hsu *et al.*, 1985c)
Epigallocatechin 3-*O*-*p*-coumarate	−158(a)	*Thea sinensis* leaf (Nonaka *et al.*, 1983b)
Catechin 3-*O*-GA[§]	+56(e)	*Bergenia crassifolia, B. cordifolia* roots (Haslam, 1969)
Catechin 7-*O*-GA	+39(a)	*Sanguisorba officinalis* root (Tanaka *et al.*, 1983)
ent-Catechin 3-*O*-GA	−52(e)	*Polygonum multiflorum* root (Nonaka *et al.*, 1982b)
Epicatechin 3-*O*-GA	−190(e)	*Thea sinensis* leaf (Bradfield & Penney, 1948)
		Vitis cv fruit (Weinges & Piretti, 1971)
		Acacia pycnantha bark (Roux *et al.*, 1961)
		Hamamelis virginiana bark (Friedrich & Krueger, 1974)
		Davidsonia pruriens leaf (Wilkins & Bohm, 1977)
		Rheum rhizoma root (Nonaka *et al.*, 1981)
		Myrica rubra bark (Nonaka *et al.*, 1983d)
Epicatechin 3-*O*-(3-OMe)-GA	−168(e)	*Thea sinensis* leaf (Saijo, 1982)
Epicatechin 3,5-di-*O*-GA	−9(e)	*T. sinensis* leaf (Coxon *et al.*, 1972)
Epigallocatechin 3-*O*-GA	−179(e)	*T. sinensis* leaf (Bradfield & Penney, 1948)
		Vitis cv. fruit (Iankov, 1966)
		Acacia pycnantha bark (Roux *et al.*, 1961)
		Hamamelis virginiana bark (Friedrich & Krueger, 1974)
		Davidsonia pruriens leaf (Wilkins & Bohm, 1977)
Epigallocatechin 3-*O*-(3-OMe)-GA	−162(e)	*T. sinensis* leaf (Saijo, 1982)
Epigallocatechin 7-*O*-GA	−24(e)	*Acacia nilotica* bark & fruit (Ayoub, 1985)
Epigallocatechin 3,5-di-*O*-GA	−13(e)	*T. sinensis* leaf (Coxon *et al.*, 1972)
Epigallocatechin 5,7-di-*O*-GA	−18(e)	*Acacia nilotica* bark & fruit (Ayoub, 1985)
(4) *C*- and *O*-Glycosides of flavan-3-ols		
Epiafzelechin 5-*O*-β-D-Glc*p*	−38(m)	*Crataeva religiosa* bark (Sethi *et al.*, 1984)
Catechin 5-*O*-β-D-Glc*p*	−29(m)	*Rheum rhizoma* root (Nonaka *et al.*, 1983a)
Catechin 7-*O*-α-L-Ara*f*	−122(m)	*Polypodium vulgare* frond (Uvarova *et al.*, 1967)
Catechin 7-*O*-Apioside		*Polypodium vulgare* frond (Karl *et al.*, 1982)
Catechin 7-*O*-β-D-Xyl*p*	−34(w)	*Ulmus americana* bark (Doskotch *et al.*, 1973)
Catechin 5-*O*-β-D-(2″-*O*-ferulyl-6″-*O*-*p*-coumaryl)-Glc*p*	−52(a)	*Cinnamomum obtusifolium* bark (Morimoto *et al.*, 1986b)
ent-Catechin 7-*O*-β-D-Glc*p*	−21(m)	*Rhaphiolepsis umbellata* bark (Nonaka *et al.*, 1983a)
Epicatechin 3-*O*-β-D-Glc*p*	−72(m)	*Cinnamomum cassia* bark (Morimoto *et al.*, 1986a)
Epicatechin 3-*O*-β-D-All*p*	−30(m)	*Davallia divaricata* fronds (Murakami *et al.*, 1985)
Epicatechin 3-*O*-β-D-(2″-*trans*-cinnamyl)-All*p*	−92(m)	*Davallia divaricata* fronds (Murakami *et al.*, 1985)
Epicatechin 3-*O*-β-D-(3″-trans-cinnamyl)-Allp	−75(m)	*Davallia divaricata* fronds (Murakami *et al.*, 1985)
3′-OMe-Epicatechin 7-*O*-β-D-Glc*p* (symplocoside)	−77(m)	*Symplocus uniflora* bark (Tschesche *et al.*, 1980)
Epicatechin 6-*C*-D-Glc*p*	+8.8(m)	*Cinnamomum cassia* bark (Morimoto *et al.*, 1986a)
Epicatechin 8-*C*-β-D-Glc*p*	−38(a)	*Cinnamomum cassia* bark (Morimoto *et al.*, 1986a)
(5) A- or B-ring *C*-substituted flavan-3-ols		
Broussinol (*2.29*)	−22(e)	*Broussonetia papyrifera* shoot (Takasugi *et al.*, 1984)
Daphnodorin B (*2.30*)	−108(d)	*Daphne odora* root (Baba *et al.*, 1985)

Table 2.3 *(Contd.)*

Class and compound	$[\alpha]_D$ *(solvent)*	*Source (reference)*
Larixinol (*2.31*)	−151(a)	*Larix gmelini* bark (Shen *et al.*, 1985)
Catechin 6-carboxylic acid		*Acacia luederitzii* heartwood (Ferreira *et al.*, 1985)
Gambiriin A1 (*2.32*)	−14(a)	*Uncaria gambir* leaf (Nonaka & Nishioka, 1980)
		Sanguisorba officinalis roots (Tanaka *et al.*, 1983)
Gambiriin A2 (*2.33*)	+67(a)	*Uncaria gambir* leaf (Nonaka & Nishioka, 1980)
Gambiriin B1 (*2.34*)	−45(a)	
Gambiriin B2 (*2.35*)		
Gambiriin B3 (*2.36*)	−20(m)	
		Sanguisorba officinalis roots (Tanaka *et al.*, 1983)
Kopsirachin (*2.37*)	+66(c)	*Kopsia dasyrachis* leaf (Homberger & Hesse, 1984)
Stenophyllanin A (*2.38*)	+94(m)	*Quercus stenophylla* bark (Nonaka *et al.*, 1985)
Stenophyllanin B (*2.39*)	+48(m)	*Quercus stenophylla* bark (Nonaka *et al.*, 1985)
Stenophyllanin C (*2.40*)	+81(m)	*Quercus stenophylla* bark (Nonaka *et al.*, 1985)
Cinchonain 1a (*2.41*)	−214(a)	*Cinchona succirubra* bark (Nonaka & Nishioka, 1982)
		Kandelia candel bark (Hsu *et al.*, 1985a)
		Uncaria rhynchophylla (Nonaka & Nishioka, 1982)
		Polygonum bistorta (Nonaka & Nishioka, 1982)
		Raphiolepsis umbellata (Nonaka & Nishioka, 1982)
Cinchonain 1b (*2.42*)	+12(a)	*Cinchona succirubra* bark (Nonaka & Nishioka, 1982)
		Kandelia candel bark (Hsu *et al.*, 1985a)
Cinchonain 1c (*2.43*)	−25(a)	*Cinchona succirubra* bark (Nonaka & Nishioka, 1982)
Cinchonain 1d (*2.44*)	+29(a)	*Cinchona succirubra* bark (Nonaka & Nishioka, 1982)
(6) Biflavan-3-ols		
Prosopin-(5→5)-prosopin		*Prosopis glandulosa* heartwood (Jacobs *et al.*, 1983)
Prosopin-(5→6)-prosopin		*Prosopis glandulosa* heartwood (Jacobs *et al.*, 1983)
Epigallocatechin-(2′→2′)-epigallocatechin (*2.45*)	−20(a)	Tannase product from theasinensins A and B
Theasinensin A (*2.46*)	−227(a)	*Thea sinensis* leaf (Nonaka *et al.*, 1983b)
Theasinensin B (*2.47*)	−147(a)	*Thea sinensis* leaf (Nonaka *et al.*, 1983b)

*Specific rotation of the trimethylether.
†Enzymic hydrolysis product.
‡But see the discussion.
§GA = gallate or galloyl.
For definitions and explanation see footnote to Table 2.2.

(*2.43*) R^1 = 3,4-dihydroxyphenyl, R^2 = H; Cinchonain 1c
(*2.44*) R^1 = H, R^2 = 3,4-dihydroxyphenyl; Cinchonain 1d

(*2.45*) $R^1 = R^2$ = H
(*2.46*) $R^1 = R^2$ = gallate; Theasinensin A
(*2.47*) R^1 = gallate; R^2 = H; Theasinensin B

Table 2.4 The natural leucoanthocyanidins: flavan-3, 4-diols and their derivatives

Structural type and compound (trivial name)	$[\alpha]_D$ *(solvent)*	*Source (reference)*
(1) Leucopelargonidin		
Afzelechin-4β-ol		Enzymic product from aromadendrin (Heller *et al.*, 1985b)
(2) Leucocyanidin		
Catechin-4β-ol	+34(c)*	Enzymic product from taxifolin (Stafford *et al.*, 1985)
(3) Leucoguibourtinidin		
Epiguibourtinidol-4β-ol	+12(aw)	*Guibourtia coleosperma* heartwood (Roux & de Bruyn, 1963)
Guibourtinidol-4α-ol	−18(a)[†]	*Guibourtia coleosperma* heartwood (Saayman & Roux, 1965) *Acacia cultriformis* heartwood (du Preez & Roux, 1970) *Acacia luederitzii* heartwood (Ferreira *et al.*, 1985)
Guibourtinidol-4β-ol	+115(a)[†]	*Guibourtia coleosperma* heartwood (Saayman & Roux, 1965) *Acacia luederitzii* heartwood (Ferreira *et al.*, 1985)
(4) Leucofisetinidin		
Fisetinidol-4α-ol (mollisacacidin)	+31(aw)	*Gleditsia japonica* leaf (Clark-Lewis & Mitsuno, 1958) *Acacia* spp. heartwood – Section Bipinnatae, Subsection Botryocephalae, Pulchae and Gummiferae; Section Phyllodinae, Subsections Brunioideae and Uninerves – only some species (Tindale & Roux, 1974)
Fisetinidol-4β-ol	+43(a)*	*Guibourtia coleosperma* heartwood (Drewes & Roux, 1965) *Colophospermum mopane* heartwood (Drewes & Roux, 1966a) *Neorautanenia amboensis* root (Oberholzer *et al.*, 1980) *Acacia* spp. heartwood-same distribution as fisetinidol-4α-ol.
ent-Fisetinidol-4β-ol	−32(aw)	*Schinopsis lorentzii* heartwood (Drewes & Roux, 1964)
Epifisetinidol-4α-ol	−29(aw)	*Guibourtia coleosperma* heartwood (Drewes & Roux, 1965)
Epifisetinidol-4β-ol	−18(e)[‡]	*Guibourtia coleosperma* heartwood (Drewes & Roux, 1965)
ent-epifisetinidol-4β-ol	+49(aw)	*Acacia mearnsii* heartwood (Drewes & Ilsley, 1969)
(5) Leucorobinetinidin		
Robinetinidol-4α-ol	+34(aw)	*Robinia pseudacacia* heartwood (Roux & Paulus, 1962)
(6) Leucoteracacinidin		
Oritin-4β-ol	−7.0(e)*	*Acacia auriculiformis* heartwood (Drewes & Roux, 1966b)
Epioritin-4α-ol (teracacidin)	−80(aw)	*Acacia* spp. heartwood – scattered distribution in Subsections Pungentes, Uninerves, Plurinerves, and Juliflorae (Tindale & Roux, 1974)
Epioritin-4β-ol (isoteracacidin)	−40(e)*	*Acacia* spp. heartwood – same distribution as epioritin-4α-ol
(7) Leucomelacacinidin		
Prosopin-4β-ol	−6.0(e)*	*Acacia nigrescens* heartwood (Fourie *et al.*, 1972) *Prosopis glandulosa* heartwood (Jacobs *et al.*, 1985)
Epiprosopin-4α-ol (melacacidin)	−85(aw)	*Acacia* spp. heartwood – widely distributed in Subsections Pungentes, Calamiformes, Uninerves (except Racemosae), Plurinerves, and Juliflorae (Tindale & Roux, 1974)
Epiprosopin-4β-ol	−32(e)	*Acacia* spp. heartwood – distribution as for epiprosopin-4α-ol
(8) Miscellaneous		
Cyanomaclurin (*2.48*)	+192(w)	*Artocarpus integrifolia* heartwood (Appel & Robinson, 1935)

Table 2.4 (*Contd.*)

Structural type and compound (trivial name)	$[\alpha]_D$ *(solvent)*	*Source (reference)*
(9) Alkylated flavanoids		
4α-OEt-Epiprosopin	−48(e)	*Acacia melanoxylon* heartwood (Foo & Wong, 1986)
4β-OEt-Epiprosopin	−75(e)	*Acacia melanoxylon* heartwood (Foo & Wong, 1986)
8-OMe-Oritin-4α-ol	−8.8(a)[§]	*Acacia cultriformis* heartwood (du Preez & Roux, 1970)
8-OMe-Prosopin-4β-ol		*Acacia cultriformis* heartwood (du Preez & Roux, 1970)
		Acacia saxatilis heartwood (Fourie *et al.*, 1974)
8-OMe-Prosopin-4α-ol		*Acacia saxatilis* heartwood (Fourie *et al.*, 1974)
3, 8-diOMe-Prosopin-4β-ol		*Acacia saxatilis* heartwood (Fourie *et al.*, 1974)
3′, 4′, 7-TriOMe-fisetinidol-4β-ol	+52(m)	*Neorautanenia amboensis* root (Oberholzer *et al.*, 1980)
3, 3′, 4, 4′, 7-PentaOMe-fisetinidol-4β-ol	−132(c)	*Neorautanenia amboensis* root (Oberholzer *et al.*, 1980)
5-OMe-8-pr-Afzelechin-4β-ol[‖]	+38(c)	*Marshallia grandiflora* roots (Bohlmann *et al.*, 1979)
		Marshallia obovata root (Bohlmann *et al.*, 1980)
4, 5-DiOMe-8-pr-afzelechin-4β-ol[‖]		*Marshallia grandiflora* root (Bohlmann *et al.*, 1979)
		Marshallia obovata root (Bohlmann *et al.*, 1980)
3, 5-DiOMe-8-pr-afzelechin-4β-ol[‖]		*Marshallia obovata* root (Bohlmann *et al.*, 1980)
Compound (*2.49*)		*Marshallia obovata* root (Bohlmann *et al.*, 1980)
Compound (*2.50*)	−11(c)	*Derris araripensis* root (Nascimento & Mors, 1981)

For definitions and explanation, see footnote to Table 2.2.
[*]Phenolic permethyl ether
[†]4′, 7-DiOMe-3, 4-diOAc.
[‡]4-OEt.
[§]Peracetate.
[‖]pr = 3, 3-dimethylallyl.

inferred from the sign and magnitude of the specific rotation being similar to epicatechin. In the author's opinion this report must be treated with caution as all the spectroscopic and physical data presented are very similar to epicatechin. The proposal of a 3′, 5′-dihydroxy B-ring was based on observing *meta* couplings for 2′-H and 6′-H with 4′-H in the tetramethyl ether. However, comparison of these data with those for epicatechin tetramethyl ether shows close similarity with the reported chemical shifts and couplings of the B-ring protons. The assigned structure must be confirmed by ^{13}C NMR (which would provide unequivocal evidence) before final acceptance.

Eleven *O*-methylated flavan-3-ols are now known (Table 2.3). The original example, ourateacatechin, has now been isolated from the root bark of further species in the Celastraceae. Most of the other examples are catechin or epicatechin derivatives. 3′-*O*-Methylepicatechin, earlier isolated as its 7-*O*-glucoside (symplocoside) from *Symplocus uniflora*, has now been isolated from the bark of *Cinnamomum* species and *Lindera umbellata*.

Three novel 3-*O*-esters of flavan-3-ols were reported in 1985. The twigs of *Phyllocladus alpinus* yielded phylloflavan, a 3-*O*-ester of *ent*-epicatechin with δ-(3, 4-dihydroxyphenyl)-β-hydroxypentanoic acid, probably formed by acylation of caffeic acid (Foo *et al.*, 1985). The bark of *Salix sieboldiana* yielded esters of 1, 6-dihydroxy-2-cyclohexene-1-carboxylic acid and 1-hydroxy-6-oxo-2-cyclohexene-1-carboxylic acid, which appear to be biosynthetic precursors of saligenin, a ubiquitous constituent of the Salicaceae in the form of the glycosides salicin and populin (Hsu *et al.*, 1985c). The ester of the keto acid was also incorporated into a series of procyanidin oligomers isolated from the same source (see Table 2.7, and Hsu *et al.*, 1985c).

The simple cinnamic acid ester epigallocatechin 3-*O*-*p*-coumarate was isolated from green tea leaves (Nonaka *et al.*, 1983b) where it accompanies a host of gallate esters of epicatechin, epigallocatechin and associated proanthocyanidins (Table 2.7). While a number of these were known previously, Saijo (1982) isolated a new pair of 3-*O*-esters of epicatechin and epigallocatechin in which the 3-hydroxy group of gallic acid is methylated. The most interesting green tea metabolites reported, however, are the pair of dimeric epigallocatechin gallates, theasinensins A and B, in which the flavan-3-ol moieties are linked by a biphenyl 2′ → 2′ bond [structures (*2.46*) and (*2.47*)]. These compounds are the metabolic equivalent of the ellagitannins, and are also the previously unrecognized natural precursors of the theaflavins, the benztropolone pigments of black tea (Haslam, 1975), although these may also be formed by oxidative coupling during fermentation.

The theasinensins together with a pair of prosopin

Table 2.5 Leucoanthocyanidins: natural flavan-4-ols

Compound	$[\alpha]_D$ *(solvent)*	*Source (reference)*
4′,5,7-TriOMe-2,4-*trans*-flavan-4-ol		*Dahlia tenuicaulis* leaf (Lam & Wrang, 1975)
Tephrowatsin A (*2.51*)	−43.3(c)	*Tephrosia watsoniana* leaf (Gomez *et al.*, 1985)
4′,5-DiOH-7-OMe-8-pr-2,4-*trans*-flavan-4-ol*		*Marshallia obovata* roots (Bohlmann *et al.*, 1980)
Erubin A (*2.52*)	+88(m)	*Glaphyropteridopsis erubescens* fronds (Tanaka *et al.*, 1984)
Erubin B (*2.53*)	+10(m)	*Glaphyropteridopsis erubescens* fronds (Tanaka *et al.*, 1984)
Triphyllin A (*2.54*)	+17(m)	*Pronephrium triphyllum* fronds (Tanaka *et al.*, 1985)
Triphyllin B (*2.55*)	+18(m)	*Pronephrium triphyllum* fronds (Tanaka *et al.*, 1985)

For definitions and explanation, see footnote to Table 2.2.
*pr = 3,3-dimethylallyl.

Table 2.6 Leucoanthocyanidins: peltogynoid flavans

Compound (structure)	$[\alpha]_D$ *(solvent)*	*Plant source (reference)*
Pubeschin (*2.56*)	+188(c)*	*Peltogyne pubescens*, *P. venosa* heartwood (Malan & Roux, 1974)
Peltagynol (*2.57*)	+273(ea)	*Peltogyne porphyrocardia* heartwood (Robinson & Robinson, 1935) *P. pubescens*, *P. venosa* heartwood (Malan & Roux, 1974) *P. confertifolia*, *P. catingue* wood (de Almeida *et al.*, 1974) *Acacia peuce* heartwood (Brandt & Roux, 1979) *Acacia carnei* heartwood (Brandt *et al.*, 1981a) *Acacia crombei* heartwood (Brandt *et al.*, 1981b) *Acacia fasciculifera* heartwood (van Heerden *et al.*, 1981) *Trachylobium verrucosum* heartwood (Ferreira *et al.*, 1974) *Colophospermum mopane* heartwood (Drewes & Roux, 1966a)
Peltagynol B (*2.58*)	+221(c)†	accompanies peltagynol
Mopanol (*2.59*)	+209(ea + e)	*Colophospermum mopane* heartwood (Drewes & Roux, 1966a) *Peltogyne confertifolia, P. catingue* wood (de Almeida *et al.*, 1974) *P. porphyrocardia, P. confertifolia, P. venosa* heartwood (Drewes & Roux, 1967)
Mopanol B (*2.60*)	+221(c)†	accompanies mopanol
ent-12a,6a-*cis*-Peltagynol B (*2.61*)	−220(c)†	*Acacia peuce* heartwood (Brandt & Roux, 1979)
1-*O*-OMe-Peltagynol		*Peltogyne paniculata, P. confertifolia* wood (de Almeida *et al.*, 1974)
7-OMe-2,3-O,O-Methylenemopanol		
3-OMe-4-OH-5,5-DiMe-peltagynol	+209(a)	*Elaeodendron balae* root bark (Weeratunga *et al.*, 1985)

For definitions and explanation, see footnote on Table 2.2.
*Tri-*O*-methyl ether derivative.
†Peracetate derivative.

Table 2.7 The natural proanthocyanidins

Structural type, chain length and proanthocyanidin class *Compound (trivial name)*	$[\alpha]_D$ *(solvent)*	*Plant source (reference)*
(A) UNSUBSTITUTED FLAVANOID UNITS		
(1) **Dimers**		
(a) *Propelargonidin*		
Afzelechin-(4α→8)-afzelechin	–227(a)	*Kandelia candel* bark (Hsu *et al.*, 1985a)
Afzelechin-(4α→8)-catechin	–190(a)	*Kandelia candel* bark (Hsu *et al.*, 1985a)
Afzelechin-(4α→8)-epicatechin	–164(a)	*Kandelia candel* bark (Hsu *et al.*, 1985a)
Epiafzelechin-(4β→8)-catechin (gambiriin C)	+6.0(a)	*Uncaria gambir* leaf (Nonaka & Nishioka, 1980)
Epiafzelechin-(4β→8)-4′-OMe-epigallocatechin	+54(a)	*Ouratea* sp. root bark (Delle Monache *et al.*, 1967) *Prionostema aspera* root bark (Delle Monache *et al.*, 1976) *Maytenus rigida* root bark (Delle Monache *et al.*, 1976) *Elaeodendron balae* root bark (Weeratunga *et al.*, 1985)
(b) *Procyanidin*		
Catechin-(4α→6)-catechin (procyanidin B6)	–130(e)	Widespread
Catechin-(4α→8)-catechin (procyanidin B3)	–235(mw)	Widespread
Catechin-(4α→6)-epicatechin (procyanidin B8)	–160(m)	Widespread
Catechin-(4α→8)-epicatechin (procyanidin B4)	–194(e)	Widespread
Epicatechin-(4β→6)-catechin (procyanidin B7)	+142(w)	Widespread
Epicatechin-(4β→8)-catechin (procyanidin B1)	+104(w)	Widespread
Epicatechin-(4β→6)-epicatechin (procyanidin B5)	+102(w)	Widespread
Epicatechin-(4β→8)-epicatechin (procyanidin B2)	+26(w)	Widespread
ent-Epicatechin-(4α→8)-*ent*-epicatechin	–24(mw)	*Chamaerops humulis* fruits (Delle Monache *et al.*, 1971)
Catechin-(4α→8)-epiafzelechin	–180(a)[*]	*Wisteria sinensis* fruit (Weinges *et al.*, 1968a)
Epicatechin-(4β→8)-epiafzelechin		*Saraca asoca* bark (Middelkoop & Labadie, 1985)
Epicatechin-(4β→8)-*ent*-epicatechin	+187(mw)	*Chamaerops humulis* fruit (Foo & Porter, 1983)
(c) *Prodelphinidin*		
Gallocatechin-(4α→8)-epigallocatechin	–77(c)[*]	*Ribes sanguineum* leaf (Foo & Porter, 1978)
Gallocatechin-(4α→8)-catechin	–69(c)[*]	*Salix caprea* catkins (Foo & Porter, 1978) *Hordeum vulgare* seed coat (Mulkay *et al.*, 1981) *Myrica nagi* stem bark[†] (Krishnamurthy & Seshadri, 1966)
Epigallocatechin-(4β→8)-catechin	+22(e)	*Pinus sylvestris* male flower (Gupta & Haslam, 1981)
(d) *Proguibourtinidin*		
Guibourtinidol-(4α→8)-epiafzelechin	–71	*Cassia fistula* sapwood (Patil & Deshpande, 1982)
Guibourtinidol-(4α→6)-catechin	–88(a)[‡]	*Acacia luederitzii* heartwood (Ferreira *et al.*, 1985)

(*Contd.*)

Table 2.7 (*Contd.*)

Structural type, chain length and proanthocyanidin class *Compound (trivial name)*	$[\alpha]_D$ *(solvent)*	*Plant source (reference)*
Guibourtinidol-($4\alpha \rightarrow 8$)-catechin	− 120(a)‡	*Julbernadia globiflora* heartwood (Pelter *et al.*, 1969) *Acacia luederitzii* heartwood (Ferreira *et al.*, 1985)
Guibourtinidol-($4\alpha \rightarrow 6$)-epicatechin		*Acacia luederitzii* heartwood (Ferreira *et al.*, 1985)
Guibourtinidol-($4\alpha \rightarrow 8$)-epicatechin	− 114(a)‡	*Julbernadia globiflora* heartwood (Pelter *et al.*, 1969) *Acacia luederitzii* heartwood (Ferreira *et al.* 1985)
(e) *Profisetinidin*		
Fisetinidol-($4\alpha \rightarrow 6$)-fisetinidol		*Colophospermum mopane* heartwood (Botha *et al.*, 1981a)
Fisetinidol-($4\beta \rightarrow 6$)-fisetinidol		*Colophospermum mopane* heartwood (Botha *et al.*, 1981a)
Fisetinidol-($4\alpha \rightarrow 6$)-fisetinidol-4α-ol		*Acacia mearnsii* heartwood (Viviers *et al.*, 1982)
Fisetinidol-($4\alpha \rightarrow 6$)-fisetinidol-4β-ol		*Acacia mearnsii* heartwood (Viviers *et al.*, 1982)
Fisetinidol-($4\beta \rightarrow 6$)-fisetinidol-4α-ol		*Acacia mearnsii* heartwood (Viviers *et al.*, 1982) *Acacia fasciculifera* heartwood (van Heerden *et al.*, 1981)
Fisetinidol-($4\beta \rightarrow 6$)-fisetinidol-4β-ol		*Acacia mearnsii* heartwood (Viviers *et al.*, 1982) *Acacia fasciculifera* heartwood (van Heerden *et al.*, 1981)
Fisetinidol-($4\alpha \rightarrow 8$)-catechin		*Acacia mearnsii* bark (Botha *et al.*, 1981a)
Fisetinidol-($4\beta \rightarrow 8$)-catechin		*Acacia mearnsii* bark (Botha *et al.*, 1981a)
ent-Fisetinidol-($4\beta \rightarrow 6$)-catechin *ent*-Fisetinidol-($4\beta \rightarrow 8$)-catechin *ent*-Fisetinidol-($4\alpha \rightarrow 6$)-catechin *ent*-Fisetinidol-($4\alpha \rightarrow 8$)-catechin		} *Schinopsis balansae*, *S. lorentzii*, *Rhus lancea*, and *R. leptodictya* heartwood (Viviers *et al.*, 1983b)
ent-Fisetinidol-($4\beta \rightarrow 8$)-epicatechin		*Rhus leptodictya* heartwood (Viviers *et al.*, 1983b)
(f) *Prorobinetinidin*		
Robinetinidol-($4\alpha \rightarrow 8$)-catechin		*Acacia mearnsii* bark (Botha *et al.*, 1981a)
Robinetinidol-($4\alpha \rightarrow 8$)-gallocatechin		*Acacia mearnsii* bark (Botha *et al.*, 1981a)
(g) *Promelacacinidin*		
Prosopin-($4\alpha \rightarrow 6$)-prosopin		*Prosopis glandulosa* heartwood (Jacobs *et al.*, 1983)
Epiprosopin-($4\alpha \rightarrow 6$)-epiprosopin-4α-ol	+ 26(e)	*Acacia melanoxylon* heartwood (Foo, 1986a)
(2) Trimers		
(a) *Procyanidin*		
[Catechin-($4\alpha \rightarrow 8$)]$_2$-catechin		*Hordeum vulgare* seed coat (Outtrup & Schaumberg, 1981) *Salix caprea* male catkins (Thompson *et al.*, 1972) *Fragaria* cv *annanasa* fruit (Thompson *et al.*, 1972) *Pinus radiata* bark (Porter, 1974) *Cryptomeria japonica* bark (Samejima & Yoshimoto, 1979)
[Catechin-($4\alpha \rightarrow 8$)]$_2$-epicatechin		*Cryptomeria japonica* bark (Samejima & Yoshimoto, 1979)
[Epicatechin-($4\beta \rightarrow 8$)]$_2$-epicatechin	+ 102(w)	Widespread, but specific isolations include: *Dioscorea cirrhosa* root (Hsu *et al.*, 1985b) *Kandelia candel* bark (Hsu *et al.*, 1985a) *Thea sinensis* leaf (Nonaka *et al.*, 1984) *Cinchona succirubra* bark (Nonaka *et al.*, 1982a) *Aesculus hippocastanum* fruit (Hemingway *et al.*, 1982) *Crataegus oxyacantha* fruit (Kolodziej *et al.*, 1984) *Nelia meyeri* leaf (Kolodziej, 1984)

Epicatechin-(4$\beta \rightarrow$6)-epicatechin- (4$\beta \rightarrow$8)-epicatechin	+138(a)	*Kandelia candel* bark (Hsu *et al.*, 1985a)
[Epicatechin-(4$\beta \rightarrow$6)]$_2$ epicatechin	+128(a)	*Kandelia candel* bark (Hsu *et al.*, 1985a)
[Epicatechin-(4$\beta \rightarrow$8)]$_2$ catechin	+74(w)	*Pinus taeda* phloem (Hemingway *et al.*, 1982) *Areca catechu* seed (Nonaka *et al.*, 1981a)
Epicatechin-(4$\beta \rightarrow$8)-epicatechin- (4$\beta \rightarrow$6)-catechin	+207(w)	*Pinus taeda* phloem (Hemingway *et al.*, 1982) *Areca catechu* seed (Nonaka *et al.*, 1981a)
Epicatechin-(4$\beta \rightarrow$6)-epicatechin- (4$\beta \rightarrow$8)-catechin	+138(a)	*Pinus taeda* phloem (Hemingway *et al.*, 1982) *Dioscorea cirrhosa* tuber (Hsu *et al.*, 1985b) *Kandelia candel* bark (Hsu *et al.*, 1985a)
Catechin-(4$\alpha \rightarrow$6)-epicatechin (4$\beta \rightarrow$8)-epicatechin		*Rhaphiolepsis umbellata* (Hsu *et al.*, 1985b) *Dioscorea cirrhosa* tuber (Hsu *et al.*, 1985b)
Epicatechin-(4$\beta \rightarrow$8)-catechin- (4$\alpha \rightarrow$8)-catechin	−97(a)	*Dioscorea cirrhosa* tuber (Hsu *et al.*, 1985b)
Epicatechin-(4$\beta \rightarrow$8)-catechin- (4$\alpha \rightarrow$8)-epicatechin	−98(m)	*Dioscorea cirrhosa* tuber (Hsu *et al.*, 1985b)
(b) *Prodelphinidin*		
[Gallocatechin-(4$\alpha \rightarrow$8)]$_2$ catechin		*Hordeum vulgare* seed coat (Outtrup & Schaumberg, 1981)
(c) *Procyanidin and prodelphinidin*		
Catechin-(4$\alpha \rightarrow$8)-gallocatechin- (4$\alpha \rightarrow$8)-catechin		*Hordeum vulgare* seed coat (Outtrup & Schaumberg, 1981)
Gallocatechin-(4$\alpha \rightarrow$8)-catechin- (4$\alpha \rightarrow$8)-catechin		*Hordeum vulgare* seed coat (Outtrup & Schaumberg, 1981)
(d) *Profisetinidin*		
Fisetinidol-(4$\alpha \rightarrow$8)-catechin- (6$\rightarrow$4α)-fisetinidol	−134(a)[‡]	*Acacia mearnsii* heartwood (Botha *et al.*, 1982)
Fisetinidol-(4$\alpha \rightarrow$8)-catechin- (6$\rightarrow$4β)-fisetinidol		*Colophospermum mopane* heartwood (Botha *et al.*, 1982)
Fisetinidol-(4$\beta \rightarrow$8)-catechin- (6$\rightarrow$4α)-fisetinidol		*Acacia mearnsii* heartwood (Botha *et al.*, 1982)
Fisetinidol-(4$\beta \rightarrow$8)-catechin- (6$\rightarrow$4β)-fisetinidol		*Acacia mearnsii* heartwood *Colophospermum mopane* heartwood *Acacia mearnsii* heartwood (Botha *et al.*, 1982)
ent-Fisetinidol-(4$\beta \rightarrow$8)-catechin- (6$\rightarrow$4β)-*ent*-fisetinidol *ent*-Fisetinidol-(4$\beta \rightarrow$8)-catechin- (6$\rightarrow$4α)-*ent*-fisetinidol *ent*-Fisetinidol-(4$\alpha \rightarrow$8)-catechin- (6$\rightarrow$4β)-*ent*-fisetinidol *ent*-Fisetinidol-(4$\alpha \rightarrow$8)-catechin- (6$\rightarrow$4α)-*ent*-fisetinidol		*Schinopsis balansae*, *S. lorentzii*, *Rhus leptodictya*, *R. lancea*, heartwood (Viviers *et al.*, 1983b)
(e) *Prorobinetinidin*		
Robinetinidol-(4$\alpha \rightarrow$8)-catechin- (6$\rightarrow$4β)-robinetinidol Robinetinidol-(4$\beta \rightarrow$8)-catechin- (6$\rightarrow$4β)-robinetinidol Robinetinidol-(4$\alpha \rightarrow$8)-gallocatechin- (6$\rightarrow$4α)-robinetinidol Robinetinidol-(4$\alpha \rightarrow$8)-gallocatechin- (6$\rightarrow$4β)-robinetinidol		*Acacia mearnsii* bark (Viviers *et al.*, 1983a)

(*Contd.*)

Table 2.7 (*Contd.*)

Structural type, chain length and proanthocyanidin class *Compound (trivial name)*	$[\alpha]_D$ *(solvent)*	*Plant source (reference)*
(3) **Tetramers**		
(a) *Procyanidin*		
[Epicatechin-(4β→8)]$_3$-epicatechin	+89(a)	*Dioscorea cirrhosa* tuber (Hsu *et al.*, 1985b) *Crataegus oxyacantha* fruit (Kolodziej *et al.*, 1984) *Cinnamomum cassia* bark (Morimoto *et al.*, 1986a)
[Epicatechin-(4β→8)]$_3$-catechin	+99(a)	*Areca catechu* seed (Nonaka *et al.*, 1981a)
[Epicatechin-(4β→8)]$_2$-epicatechin-(4β→6)-catechin	+135(a)	*Areca catechu* seed (Nonaka *et al.*, 1981a)
Catechin-(4α→6)-[epicatechin-(4β→8)]$_2$-epicatechin	−2.2(a)	*Dioscorea cirrhosa* tuber (Hsu *et al.*, 1985b)
(b) *Profisetinidin*		
Fisetinidol-(4β→6)-fisetinidol-(4β→8)-catechin-(6→4β)-fisetinidol		*Acacia mearnsii* heartwood (Young *et al.*, 1985a)
Fisetinidol-(4β→6)-fisetinidol-(4β→8)-catechin-(6→4α)-fisetinidol		
Fisetinidol-(4α→6)-fisetinidol-(4α→8)-catechin-(6→4α)-fisetinidol		
Fisetinidol-(4α→6)-fisetinidol-(4α→8)-catechin-(6→4β)-fisetinidol		
ent-Fisetinidol-(4β→6)-*ent*-fisetinidol-(4β→8)-catechin-(6→4β)-*ent*-fisetinidol		*Rhus lancea* heartwood (Young *et al.*, 1985b)
ent-Fisetinidol-(4β→6)-*ent*-fisetinidol-(4β→8)-catechin-(6→4α)-*ent*-fisetinidol		
ent-Fisetinidol-(4α→6)-*ent*-fisetinidol-(4α→8)-catechin-(6→4β)-*ent*-fisetinidol		
ent-Fisetinidol-(4α→6)-*ent*-fisetinidol-(4α→8)-catechin-(6→4α)-*ent*-fisetinidol		
(4) **Pentamer**		
(a) *Procyanidin*		
[Epicatechin-(4β→8)]$_4$-epicatechin	+102(a)	*Cinnamomum cassia* bark (Morimoto *et al.*, 1986a)
(5) **Hexamer**		
(a) *Procyanidin*		
[Epicatechin-(4β→8)]$_5$-epicatechin	+112(a)	*Cinnamomum cassia* bark (Morimoto *et al.*, 1986a)
(B) SIMPLE ESTERS		
(i) *Gallate esters*		
(1) **Dimers**		
(a) *Procyanidin*		
3-*O*-GA-catechin-(4α→8)-catechin	−170(a)	*Sanguisorba officinalis* root (Tanaka *et al.*, 1983)
Catechin-(4α→8)-epicatechin- 3-*O*-GA	−256(a)	*Thea sinensis* leaf (Nonaka *et al.*, 1983b)
Epicatechin-(4β→8)-epicatechin-3-*O*-GA	−46(a)	*Thea sinensis* leaf (Nonaka *et al.*, 1983b) *Vitis* cv 'Siebel' fruit (Czochanska *et al.*, 1979)

3-*O*-GA-epicatechin-(4β→8)-epicatechin-3-*O*-GA	−95(a)	*Thea sinensis* leaf (Nonaka *et al.*, 1983b) *Rhei rhizoma* root (Nonaka *et al.*, 1981b) *Polygonum multiflorum* root (Nonaka *et al.*, 1982b)
3-*O*-GA-epicatechin-(4β→8)-catechin	−21(a)	*Rhei rhizoma* root (Nonaka *et al.*, 1981b) *Polygonum multiflorum* root (Nonaka *et al.*, 1982b)
(b) *Prodelphinidin*		
Epigallocatechin-(4β→8)-epicatechin-3-*O*-GA	−72(a)	*Thea sinensis* leaf (Nonaka *et al.*, 1984)
Epigallocatechin-(4β→8)-epigallocatechin-3-*O*-GA	−53(a) +73(m)	*Thea sinensis* leaf (Nonaka *et al.*, 1983b) *Myrica rubra* bark (Nonaka *et al.*, 1983c)
3-*O*-GA-epigallocatechin-(4β→6)-epigallocatechin-3-*O*-GA	+57(a)	*Myrica rubra* bark (Nonaka *et al.*, 1983c)
3-*O*-GA-epigallocatechin-(4β→8)-epigallocatechin-3-*O*-GA	−61(a)	*Myrica rubra* bark (Nonaka *et al.*, 1983c)
3-*O*-GA-epigallocatechin-(4β→8)-gallocatechin-3-*O*-GA	+27(a)	*Myrica rubra* bark (Nonaka *et al.*, 1983c)
(ii) *Cyclohexene carboxylates*		
(1) Dimers		
(a) *Procyanidin*		
Epicatechin-(4β→8)-catechin-3-*O*-(1-OH-6-oxo-2-cyclohexene-1-carboxylate)	−63(a)	
Catechin-(4α→8)-catechin-3-*O*-(1-OH-6-oxo-2-cyclohexene-1-carboxylate)	−292(a)	*Salix sieboldiana* bark (Hsu *et al.*, 1985c)
(2) Trimer		
(a) *Procyanidin*		
Epicatechin-(4β→8)-catechin-(4α→8)-catechin-3-*O*-(1-OH-6-oxo-2-cyclohexene-1-carboxylate)	−160(a)	
(C) *C*- AND *O*-GLUCOSIDES		
(1) Dimer		
(a) *Proluteolinidin*		
5-*O*-Glc-luteoliflavan-(4β→8)-eriodictyol	−99(m)	*Sorghum vulgare* cv 'Szegeditoerpe' (Gujer *et al.*, 1986)
5-*O*-Glc-luteoliflavan-(4β→8)-eriodictyol-3-*O*-Glc	−124(m)	*Sorghum vulgare* cf 'Szegeditoerpe' (Gujer *et al.*, 1986)
(b) *Procyanidin*		
6-*C*-β-D-Glc*p*-epicatechin-(4β→8)-epicatechin	+12(a)	*Cinnamomum cassia* bark (Morimoto *et al.*, 1986b)
8-*C*-β-D-Glc*p*-epicatechin-(4β→8)-epicatechin	+18(a)	*Cinnamomum cassia* bark (Morimoto *et al.*, 1986b)
(2) Trimer		
(a) *Proluteolinidin*		
[5-*O*-Glc-luteoliflavan-(4β→8)]$_2$-eriodictyol	−110(m)	*Sorghum vulgare* cv 'Szegeditoerpe' (Gujer *et al.*, 1986)
[5-*O*-Glc-luteoliflavan-(4β→8)]$_2$-eriodictyol-5-*O*-Glc	−134(m)	*Sorghum vulgare* cv 'Szegeditoerpe' (Gujer *et al.*, 1986)
(D) *C*-SUBSTITUTED FLAVONOID UNITS		
(1) Dimer		
(a) *Proguibourtinidin*		
Guibourtinidol-(4α→8)-epicatechin-6-carboxylic acid	−126(a)‡	
Guibourtinidol-(4α→6)-epicatechin-8-carboxylic acid	n.d.	*Acacia luederitzii* heartwood (Ferreira *et al.*, 1985)
Guibourtinidol-(4α→8)-catechin-6-carboxylic acid	−114(a)‡	
Guibourtinidol-(4α→6)-catechin-8-carboxylic acid	−114(a)‡	

(*Contd.*)

Table 2.7 (*Contd.*)

Structural type, chain length and proanthocyanidin class *Compound (trivial name)*	$[\alpha]_D$ *(solvent)*	*Plant source (reference)*
(b) *Procyanidin*		
Cinchonain-1a-(4β→8)-epicatechin (cinchonain IIa)	−48(a)	*Cinchona succirubra* bark (Nonaka *et al.*, 1982a) *Kandelia candel* bark (Hsu *et al.*, 1985a)
Cinchonain-1b-(4β→8)-epicatechin (cinchonain IIb)	+135(a)	*Cinchona succirubra* bark (Nonaka *et al.*, 1982a)
Cinchonain-1a-(4β→8)-catechin (kandelin A-1)	−58(a)	*Kandelia candel* bark (Hsu *et al.*, 1985a)
Cinchonain-1b-(4β→8)-catechin (kandelin A-2)	+9.3(a)	
(2) **Trimer**		
(a) *Procyanidin*		
Cinchonain-1a-(4β→8)-epicatechin-(4β→8)-epicatechin (kandelin B-1)	+36(a)	
Cinchonain-1b-(4β→8)-epicatechin-(4β→8)-epicatechin (kandelin B-2)	+164(a)	
Cinchonain-1b-(4β→8)-epicatechin-(4β→8)-catechin (kandelin B-4)	+198(a)	
Epicatechin-(4β→6)-cinchonain-1a-(4β→8)-epicatechin (kandelin B-3)	+65(a)	
(E) MISCELLANEOUS		
(1) **Monomer**		
(a) *Flavan*		
4′-OH-7-OMe-4-(4-OH-Styryl)flavan	+59(m)*	*Xanthorrhoea* sp gum (Camarda *et al.*, 1977)
4′-OH-5, 7-DiOMe-4-(4-OH-Styryl)flavan	+78(m)*	*Xanthorrhoea* sp gum (Camarda *et al.*, 1977)
(b) *Procyanidin*		
Dryopterin (*2.62*)	−51(m)	*Dryopteris filix-mas* frond (Karl *et al.*, 1981)
Epicatechin-(4β→2)-phloroglucinol	+122(w)	*Nelia meyeri* leaf (Kolodziej, 1983)
(c) *Proguibourtinidin*		
Guibourtinidol-(4α→2)-3, 5, 3′, 4′-tetrahydroxystilbene		*Guibourtia coleospermum* heartwood (Steynberg *et al.*, 1983)
Epiguibourtinidol-(4β→2)-3, 5, 3′, 4′-tetrahydroxystilbene (*2.63*)		
(2) **Dimer**		
(a) *Proguibourtinidin*		
Guibourtinidol-(4α→2)-3, 5, 3′, 4′-tetrahydroxystilbene-(6→4β)-epiguibourtinidol		
Epiguibourtinidol-(4β→2)-3, 5, 3′, 4′-tetrahydroxystilbene-(6→4β)-epiguibourtinidol		
(b) *Isoflavanoid proanthocyanidins*		
Compound (*2.64*)		*Dalbergia nitidula* heartwood (Bezuidenhoudt *et al.*, 1984)
Compound (*2.65*)	−131(m)	*Dalbergia odorifera* heartwood (Yahura *et al.*, 1985)
Compound (*2.66*)	−67(m)	
Compound (*2.67*)	−149(m)	
Compound (*2.68*)	−111(m)	

For definitions and explanation, see footnote to Table 2.2.
*Peracetate.
†Probable structure.
‡Phenolic permethyl ether 3-*O*-acetate.

Table 2.8 Proanthocyanidins containing a double interflavanoid (A-type) linkage

Chain length, proanthocyanidin class, and compound (trivial name)	$[\alpha]_D$ *(solvent)*	*Plant source (reference)*
(1) Dimer		
(a) Proapigeninidin		
ent-Apigeniflavan-(2α→7, 4α→8)-epiafzelechin(mahuannin D)	−104(m)	*Ephedra* sp. roots (Kasahara & Hikino, 1983)
(b) Propelargonidin		
ent-Epiafzelechin-(2α→7, 4α→8)-afzelechin	−103(m)	*Prunus persica* cv. 'Hakuto' root (Ohigashi *et al.*, 1982)
ent-Epiafzelechin-(2α→7, 4α→8)-catechin	−53(c)*	*Prunus persica* cv. 'Hakuto' root (Ohigashi *et al.*, 1982)
Epiafzelechin-(2β→7, 4β→8)-epiafzelechin (mahuannin B)		*Ephedra* sp. root (Hikino *et al.*, 1982a)
Epiafzelechin-(2β→7, 4β→6)-epiafzelechin (mahuannin C)		*Ephedra* sp. root (Kasahara *et al.*, 1983)
ent-Epiafzelechin-(2α→7, 4α→8)-epiafzelechin (mahuannin A)		*Ephedra* sp. root (Hikino *et al.*, 1982a)
ent-Epiafzelechin-(2α→7, 4α→8)-kaempferol (ephedrannin A)	−357(m)	*Ephedra* sp. root (Hikino *et al.*, 1982b)
(c) Procyanidin		
Epicatechin-(2β→7, 4β→8)-catechin (proanthocyanidin A1)	+61(aw)	*Cola accuminata* nut (Weinges *et al.*, 1968a)
		Vaccinium vitis-idaea berry (Weinges *et al.*, 1968a)
		Aesculus hippocastanum fruit (Thompson *et al.*, 1972)
		Persea gratissima fruit (Thompson *et al.*, 1972)
Epicatechin-(2β→7, 4β→8)-epicatechin (proanthocyanidin A2)	+64(a)	*Aesculus hippocastanum* fruit (Mayer *et al.*, 1966)
		Vaccinium vitis-idaea fruit (Weinges *et al.*, 1968a)
		Malus sylvestris fruit (Thompson *et al.*, 1972)
		Persea gratissima fruit (Thompson *et al.*, 1972)
		Cinnamomum spp bark (Nonaka *et al.*, 1983c)
(2) Trimer		
(a) Procyanidin		
Epicatechin-(2β→7, 4β→8)-epicatechin-(4β→8)-epicatechin	+70(a)	*Cinnamomum zeylanicum* bark (Nonaka *et al.*, 1983c)
		C. sieboldii root bark (Otsuka *et al.*, 1982; Morimoto *et al.*, 1985a)
		Lindera umbellata stem (Ezaki *et al.*, 1985)
		Kandelia candel bark (Hsu *et al.*, 1985a)
Epicatechin-(2β→7, 4β→8)-epicatechin-(4β→8)-catechin	+102(a)	*Cinnamomum sieboldii* bark (Morimoto *et al.*, 1985a)
		Linderae umbellata stem (Ezaki *et al.*, 1985)
(3) Tetramer		
(a) Procyanidin		
Epicatechin-(4β→8)-epicatechin-(2β→7, 4β→8)-epicatechin-(4β→8)-epicatechin	+51(a)	*Cinnamomum zeylanicum* bark (Nonaka *et al.*, 1983c)
		C. sieboldii bark (Morimoto *et al.*, 1985a)
		Lindera umbellata stem (Ezaki *et al.*, 1985)
Epicatechin-(4β→6)-epicatechin-(2β→7, 4β→8)-epicatechin-(4β→8)-epicatechin	+93(a)	*C. zeylanicum* bark (Nonaka *et al.*, 1983c)
		C. sieboldii bark (Morimoto *et al.*, 1985a)
Epicatechin-(4β→8)-epicatechin-(2β→7, 4β→8)-epicatechin-(4β→8)-catechin	+86(a)	*C. sieboldii* bark (Morimoto *et al.*, 1985a)
		L. umbellata stem (Ezaki *et al.*, 1985)
Epicatechin-(4β→6)-epicatechin-(2β→7, 4β→8)-epicatechin-(4β→8)-catechin	+111(a)	*C. sieboldii* bark (Morimoto *et al.*, 1985a)
(4) Pentamer		
(a) Procyanidin		
[Epicatechin-(4β→8)]$_2$-epicatechin-(2β→7, 4β→8)-epicatechin-(4β→8)-epicatechin	+71(a)	*C. zeylanicum* bark (Nonaka *et al.*, 1983c)
		C. sieboldii bark (Morimoto *et al.*, 1985a)
[Epicatechin-(4β→8)]$_2$-epicatechin-(2β→7, 4β→8)-epicatechin-(4β→8)-catechin	+116(a)	*C. sieboldii* bark (Morimoto *et al.*, 1985a)

For definitions and explanation, see footnote to Table 2.2. *Phenolic methyl ether derivative.

Table 2.9 Structure and distribution of proanthocyanidin polymers

(A) Class, (B) Group, (C) Order, (D) Family	Species	Organ	PC:PD[a]	X_{cis}[b]	$[\alpha]_{578}$[c]
(A) Filices					
(D) Dennstaedtiaceae	*Pteridium aquilinum*	Frond	80:20[d]	0.95	+150
(D) Cyatheaceae	*Cyathea dealbata*	Frond	60:40	0.96	+155
(D) Dicksoniaceae	*Dicksonia squarrosa*	Frond	80:20	0.94	+146
(A) Coniferae					
(C) Coniferales					
(D) Pinaceae	*Pinus brutia*	Bark	100:0[e]	0.6	−19
	Pinus contorta	Bark[f]	69:31	0.44	−102
	Pinus elliottii	Bark[g]	37:63	0.85	+102
	Pinus radiata	Bark[g]	52:48	0.82	+86
		Bark[h]	85:15	0.45	−114
		Bark[f]	90:10	0.25	−130
		Needle	20:80	0.90	+125
		Winter bud	43:57	0.88	+105
		Male cone	44:56	0.79	+59
	Pinus palustris	Bark[g]	100:0	0.90	+125
		Bark[h]	80:20	0.75	+49
		Bark[f]	86:14	0.74	+46
	Pinus patula	Bark[h]	65:35	0.75	+49
		Bark[f]	90:10	0.45	−96
	Pinus sylvestris	Bark[h]	100:0	0.88	+105
		Bark[f]	100:0	0.67	0[i]
		Needle	59:41	0.78	+65
	Picea abies	Bark	100:0[e]	0.7	+38
	Pseudotsuga taxifolia	Bark	100:0	0.73	+54
	Tsuga heterophylla	Bark	100:0	0.59	−30
(D) Araucariaceae	*Agathis australis*	Leaf	100:0	0.89	+121
(D) Cupressaceae	*Chamaecyparis nootkatensis*	Cone	79:21	0.83	+77
(D) Podocarpaceae	*Podocarpus totara*	Leaf	100:0	0.77	+60
(A) Monocotyledoneae					
(B) Arecidae					
(C) Arecales					
(D) Palmae	*Phoenix canariensis*	Frond	100:0	1.0	−91[j]
	Rhopalostylis sapida	Fruit	100:0	1.0	−77[j]
	Cocos nucifera	Fruit	100:0	1.0	−75[j]
(C) Pandanales					
(D) Pandanaceae	*Freycinetia baueriana* ssp. *banksii*	Frond	42:58	0.90	+125
(C) Arales					
(D) Araceae	*Lemna minor*	Leaf	100:0	0.72	+38
	Zantedeschia aethiopica[k]	Fruit	44:0[l]	1.0	+164
	Zantedeschia aethiopica	Fruit	15:85	0.25	−195
	Zantedeschia rehmannii	Fruit	100:0	0.87	+110
(B) Alismatidae					
(C) Najadales					
(D) Aponogetonaceae	*Aponogeton distachyus*	Leaf	54:0[l]	0.80	+74
(D) Potomogetonaceae	*Syringodium filiforme*	Leaf	100:0	1.0	+164
(B) Commelinidae					
(C) Restionales					
(D) Restionaceae	*Leptocarpus similis*	Inflorescence	100:0	0.74	+45
(C) Poales					
(D) Graminaceae	*Hordeum vulgare*	Ear	40:60	0.12	−259
	Sorghum vulgare	Seed coat	92:8	0.77	+76
(C) Juncales					
(D) Juncaceae	*Juncus bufonius*	Inflorescence	61:39	0.94	+142
(C) Cyperales	*Cyperus eragrostis*	Inflorescence	100:0	0.88	+116

Table 2.9 (*Contd.*)

(A) Class, (B) Group, (C) Order, (D) Family	*Species*	*Organ*	*PC:PD*[a]	X_{cis}[b]	$[\alpha]_{578}$[c]
(D) Cyperaceae					
(C) Typhales					
(D) Typhaceae	*Typha orientalis*	Inflorescence	83:0[l]	1.0	−113[j]
(C) Zingiberales					
(D) Musaceae	*Musa sapientum*	Fruit skin	78:22	1.0	−69[j]
(D) Strelitziaceae	*Strelitzia reginae*	Leaf	75:0[l]	1.0	−81[j]
(D) Zingiberaceae	*Hedychium flavescens*	Leaf	67:33	0.80	+77
	Hedychium gardnerianum	Leaf	53:47	0.80	+76
(D) Cannaceae	*Canna indica*	Leaf	76:24	0.70	+27
(D) Marantaceae	*Ctenanthe oppenheimiana* cv. *tricolor*	Leaf	67:33	0.77	+59
(B) Lilliidae					
(C) Liliales					
(D) Pontederiaceae	*Eichhornia crassipes*	Leaf	56:19[l]	0.40	−120
		Root	61:0[l]	0.50	−73
(D) Iridaceae	*Gladiolus* cv.	Leaf	43:57	0.16	−238
	Iris germanica	Fruit	89:11	0.65	0[i]
	Iris pseudacorus	Fruit	20:80	0.40	−125
		Leaf	100:0	0.65	0[i]
	Watsonia pyramidata	Leaf	0:100	0.05	−295
	Watsonia ardernii	Leaf	0:100	0.01	−315
(D) Liliaceae	*Astelia fragrans*	Inflorescence	93:7	0.45	−98
(D) Agavaceae	*Phormium cookianum*	Leaf	100:0	0.78	+66
(D) Haemodoraceae	*Anigozanthos flavidus*	Leaf	20:0[l]	1.0	−115[j]
(D) Smilacaceae	*Ripogonum scandens*	Leaf	100:0	1.0	−73[j]
(A) Dicotyledoneae					
(B) Magnoliidae					
(C) Laurales					
(D) Lauraceae	*Persea americana*	Fruit skin	100:0	0.94	+138
(B) Hamamelidae					
(C) Fagales					
(D) Betulaceae	*Betula alba*	Catkin	85:15	0.83	+84
(D) Fagaceae	*Quercus robur*	Leaf	67:33	—[m]	0
(C) Casuarinales					
(D) Casuarinaceae	*Casuarina stricta*	Leaf	70:30	—[m]	+64
(B) Dilleniidae					
(C) Dilleniales					
(D) Actinidaceae	*Actinidia chinensis*	Leaf	88:12	1.0	+161
(C) Malvales					
(D) Sterculiaceae	*Theobroma cacao*	Bean	100:0	0.96	+155
(D) Malvaceae	*Gossypium hirsutum*	Flower bud	40:60	0.78	+65
(C) Salicales					
(D) Salicaceae	*Salix caprea*	Catkin	90:10	0.38	−130
		Leaf	56:44	0.32	−162
	Salix fragilis	Leaf	90:10	0.88	+111
(C) Ericales					
(D) Ericaceae	*Rhododendron* cv.	Leaf	73:27	—[m]	+102
	Vaccinium corymbosum	Fruit unripe	100:0	0.94	+144
		Fruit ripe	100:0	0.95	+150
	Vaccinium oxycoccus	Fruit unripe	78:22	0.87	+111
(C) Ebenales					
(D) Ebenaceae	*Diospyros kaki*	Fruit unripe	59:41	0.94	+144
(B) Rosidae					
(C) Rosales					
(D) Rosaceae	*Chaenomeles sinensis*	Fruit unripe	100:0	0.94	+144
	Chaenomeles speciosa	Fruit unripe	100:0	0.91	+128

(*Contd.*)

Table 2.9 (*Contd.*)

(A) Class, (B) Group, (C) Order, (D) Family	*Species*	*Organ*	*PC:PD*[a]	X_{cis}[b]	$[\alpha]_{578}$[c]
	Cotoneaster serotina	Fruit unripe	93:7	0.90	+125
	Crataegus oxyacantha	Fruit ripe	100:0	0.94	+146
	Cydonia oblonga	Fruit unripe	100:0	0.95	+151
	Fragaria ananassa cv Redgauntlet	Fruit unripe	100:0	—[m]	+34
		Fruit ripe	100:0	—[m]	+36
	Malus pumila cv. Granny Smith	Fruit ripe	100:0	0.93	+139
	Photinia glabrescens cv *rubra* x *P. serrulata* (Red robin)	Leaf	100:0	0.97	+161
	Prunus domestica	Leaf	84:16	0.62	−11
	Rosa cv.	Hip unripe	90:10	—[m]	+23
(D) Saxifragaceae	*Ribes alpinum*	Leaf	0:100	0.20	−220
	Ribes glutinosum cv *albidum*	Leaf	28:72	0.45	−97
	Ribes grossularia	Fruit unripe	63:37	0.77	+61
		Leaf	20:80	0.80	+78
	Ribes menziesii	Leaf	27:73	—[m]	+28
	Ribes nigrum	Fruit unripe	40:60	0.75	+52
		Leaf	4:96	0.13	−258
	Ribes rubrum	Fruit unripe	24:76	0.79	+70
		Leaf	8:92	0.92	+137
	Ribes sanguineum	Fruit ripe	12:88	0.72	+38
		Leaf	10:90	0.12	−260
	Ribes speciosum	Leaf	66:34	—[m]	+44
(C) Fabales					
(D) Leguminosae	*Acacia pravissima*	Leaf	46:54	—[m]	+26
	Coronilla varia	Leaf	29:71	0.95	+148
	Lathyrus latifolius	Leaf	10:90	0.60	−18
	Lotus corniculatus	Root	78:22	0.87	+109
		Leaf	73:27	0.94	+145
	Lotus pedunculatus	Root	23:77	0.73	+43
		Leaf (January)	38:62	0.75	+52
		Leaf (March)	18:82	0.80	+76
	Lotus tenuis	Root	80:20	0.84	+94
	Onobrychis viciifolia	Leaf	23:77	0.91	+130
	Robinia fertilis	Leaf	71:29	0.55	−49
	Trifolium affine	Leaf	63:37	0.81	+81
	Trifolium arvense	Leaf	47:53	0.91	+132
	Trifolium repens	Flower	0:100	0.66	+6.0
	Vicia hirsuta	Leaf	100:0	0.90	+126
	Vicia sativa	Leaf	100:0	1.0	+169
	Virgilia divaricata	Leaf	83:17	—[m]	−142
	Wisteria sinensis	Leaf	64:36	0.82	+85
(C) Myrtales					
(D) Rhizophoraceae	*Rhizophora mangle* cv *samoensis*	Leaf	90:10	0.79	+69
(D) Myrtaceae	*Eucalyptus diversicolor*	Bark	100:0	0.74	+44
	Eucalyptus globulus	Leaf	50:50	—[m]	+112
	Feijoa sellowiana	Fruit	70:30	0.85	+101
(C) Proteales					
(D) Proteaceae	*Grevillea robusta*	Leaf	39:61	0.72	+35
	Grevillea rosmarinifolia	Leaf	18:82	0.90	+123
(C) Rhamnales					
(D) Vitaceae	*Vitis vinifera* cv Beaujolais	Fruit	87:13	0.92[n]	+134
	Vitis vinifera cv Siebel	Fruit	90:10	0.95[n]	+150
	Vitis vinifera cv Albany Surprise	Leaf	68:32	0.97[n]	+159

Table 2.9 *(Contd.)*

(A) Class, (B) Group, (C) Order, (D) Family	*Species*	*Organ*	*PC:PD*[a]	X_{cis}[b]	$[\alpha]_{578}$[c]
(C) Sapindales					
(D) Hippocastanaceae	*Aesculus* x *carnea*	Fruit unripe	100:0	0.93	+141
	Aesculus hippocastanum	Fruit unripe	100:0	1.0	+162
(D) Aceraceae	*Acer rubrum*	Bark	100:0	1.0[n]	—[o]

[a]Procyanidin: prodelphinidin.
[b]X_{cis} = mole fraction of 2,3-*cis*-(epicatechin-4 or epigallocatechin-4) proanthocyanidin units. This may be calculated from ^{13}C NMR spectra, or from the specific rotation at 578 nm using the formula: $X_{cis} = ([\alpha]_{578} + 320)/494$.
[c]Solvents used for the specific rotation measurements were water or methanol:water, 1:1, v/v;
[d]B unit was epiafzelechin;
[e]Procyanidin *O*-glucosides.
[f]Outer bark.
[g]Inner bark.
[h]Middle bark.
[i]Interpretation checked by ^{13}C NMR, i.e. the units were not racemic.
[j]Contains enantiomeric 2,3-*cis*-proanthocyanidin units.
[k]Variegated variety.
[l]Balance are propelargonidin units.
[m]Cannot be estimated as *O*-gallate units or hydrolysable tannins are present.
[n]Contains small amount of *O*-gallate units.
[o]Not recorded.

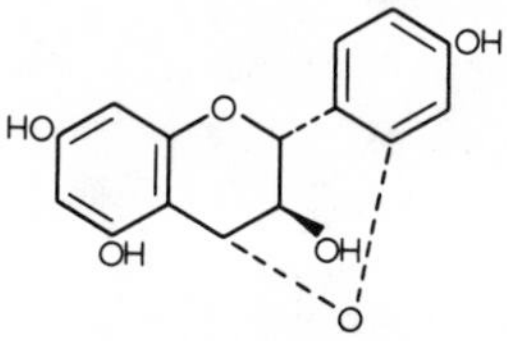

(2.48) Cyanomaclurin

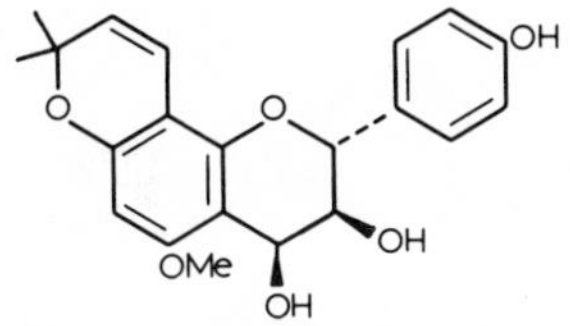

(2.49)

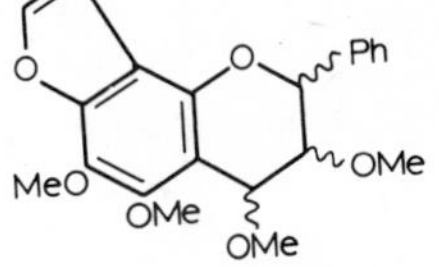

(2.50)

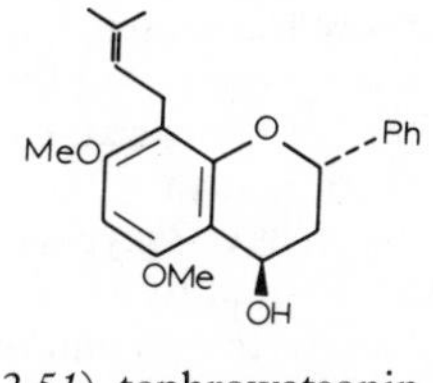

(*2.51*) tephrowatsonin A

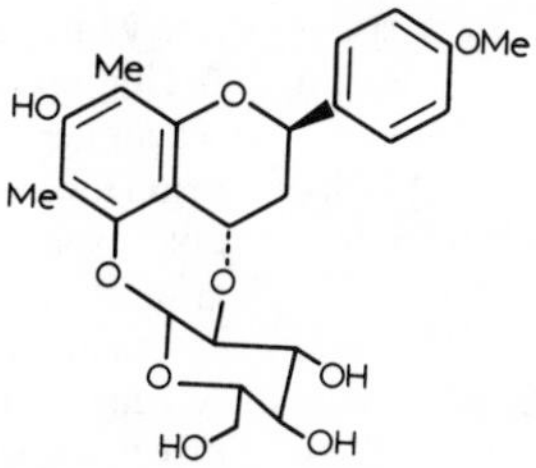

(*2.52*) erubin A

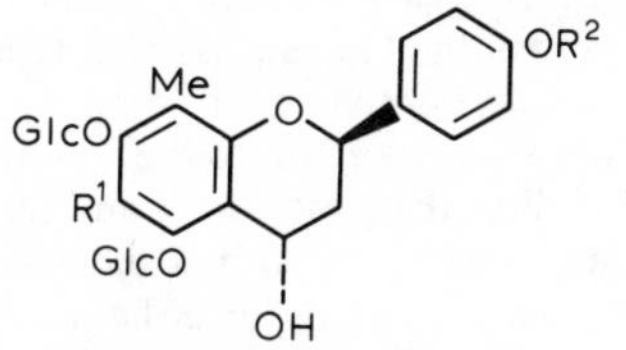

(*2.53*) $R^1 = R^2 = Me$; erubin B
(*2.54*) $R^1 = CH_2OH$; $R^2 = Me$; triphyllin A
(*2.55*) $R^1 = CH_2OH$; $R^2 = H$; triphyllin B

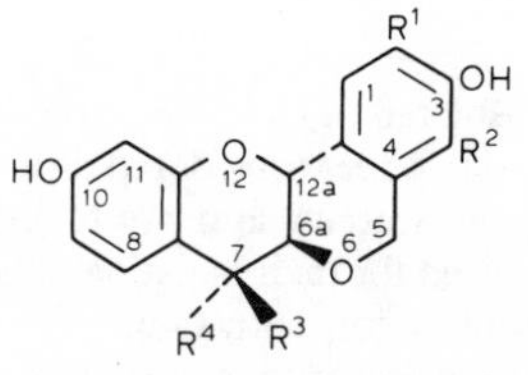

(2.56) $R^1=OH, R^2=R^3=R^4=H$; Pubeschin
(2.57) $R^1=R^4=OH, R^2=R^3=H$; Peltagynol
(2.58) $R^1=R^3=OH, R^2=R^4=H$; Peltagynol B
(2.59) $R^1=R^3=H, R^2=R^4=OH$; Mopanol
(2.60) $R^1=R^4=H, R^2=R^3=OH$; Mopanol B

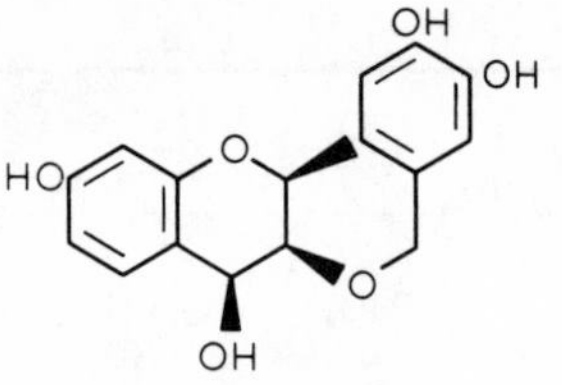

(2.61)

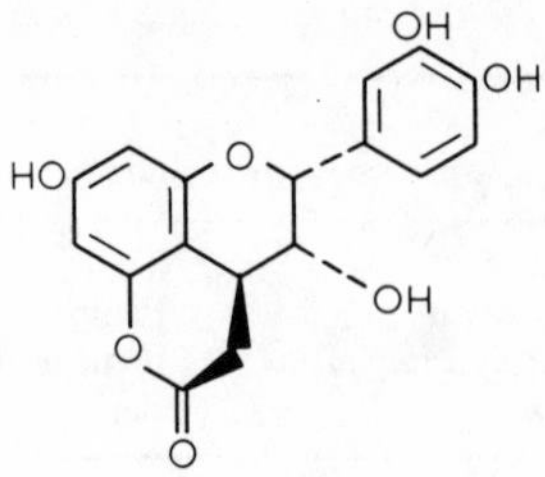

(2.62) Dryopterin

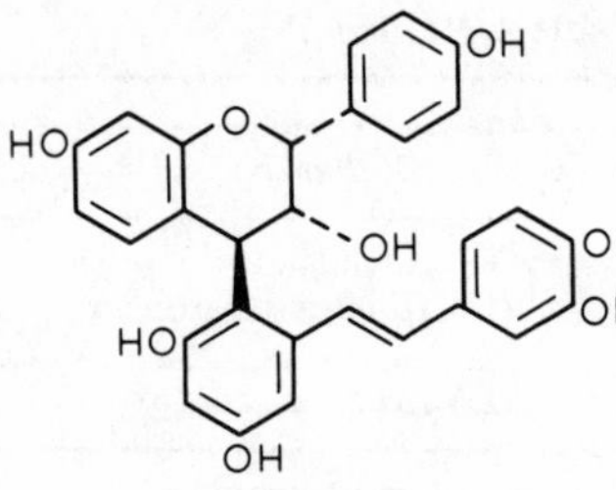

(2.63)

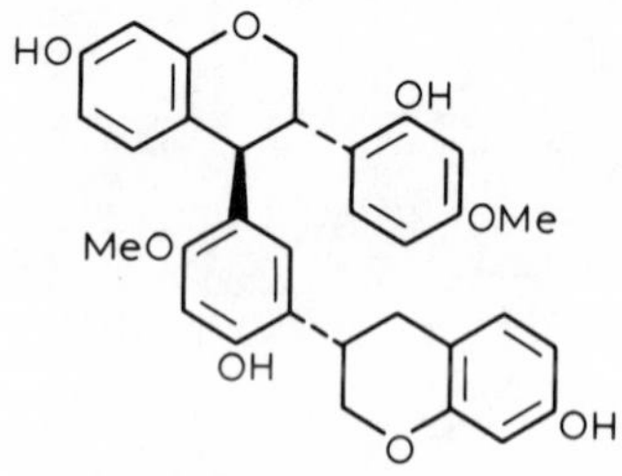

(2.64)

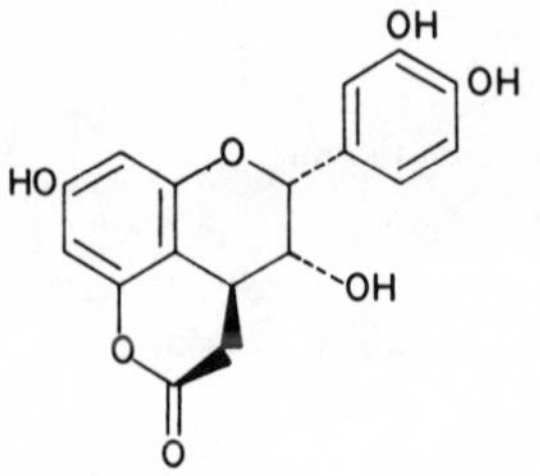

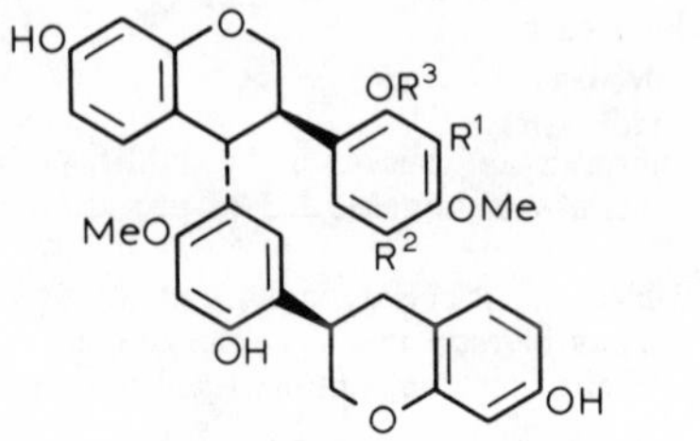

(2.65) $R^1=OH, R^2=R^3=H$

(2.66) $R^1=R^2=R^3=H$; Enantiomer of (2.64)

(2.67) $R^1=R^3=H, R^2=OMe$

(2.68) $R^1=OH, R^2=H, R^3=Me$

dimers also isolated in 1983 by Jacobs *et al.* represent the first authenticated examples of biflavonoids of the flavan-3-ol class, although Ahn and Gstirner (1973) had previously reported catechin and gallocatechin $6' \rightarrow 6$ (or 8) biflavonoids from oak bark on the basis of mass spectrometric data. Both the theasinensins and the prosopin dimers possess characteristically intense CD bands expected for biaryl compounds (see Nonaka *et al.*, 1983b, and Jacobs *et al.*, 1983). The chirality of the biaryl unit is (*S*), that shown, the same as that for ellagitannins (Haslam, 1982b).

A considerable number of flavan-3-ol glycosides have been isolated in addition to those reported in 1982. They can be no longer considered rare, and part of their previous apparent scarcity appears to lie in incorrect techniques being used for their isolation. They are sometimes present in very low concentrations compared with other proanthocyanidin constituents such as the *C*-glucosides from *Cinnamomum cassia* bark (Morimoto *et al.*, 1986a). Examples of 3-, 5- and 7-*O*- and 6- and 8-*C*-glycosides are now known, which covers the range of glycosides likely to be encountered, in view of the rarity of B-ring glycosides in other flavonoid classes.

As illustrated by the selective ^{13}C NMR data in Table 2.10, the three main types of glycosylation, i.e. 3-*O* versus 5-*O* or 7-*O* versus 6-*C* or 8-*C* may be readily distinguished. As anticipated, the C-3 experiences considerable downfield, and C-2 and C-4 upfield shifts on glycosylation, whereas the heterocyclic carbons are relatively unaffected compared with the parent compound, on A-ring glycosylation. *C*-Glycosylation may be readily recognized by the downfield position of the sugar C-1 resonance (δ 77 for glucose) compared with that for *O*-glycosylation (δ 102 for glucose). C-6 versus C-8 glycosylation may be distinguished by the position of the A-ring resonances, but 5- or 7-glycosylation leads to very similar A-ring chemical shifts (Table 2.10). These may only be distinguished by methylation/hydrolysis studies (Nonaka *et al.*, 1983a).

A feature of the past five years has been the isolation of a number of novel *C*-substituted flavan-3-ols. These range from broussinol, a simple isoprenylated flavan-3-ol phytoalexin from *Broussonetia papyrifera*, to the complex and closely related metabolites daphnodorin B and larixinol. Daphnodorin B is apparently formed by condensation of afzelechin with chalconaringenin to form the novel furan (*2.30*). Larixinol is thought to be formed by a similar attack of C-3 on C-8, this time by naringenin, to form the novel spiro-compound (*2.31*). The absolute stereochemistry was established by X-ray crystallography in both cases (Baba

Table 2.10 Some useful ^{13}C NMR chemical shifts for structural fragments of proanthocyanidins and flavan-3-ols and their derivatives (all shifts in ppm relative to TMS)

(A) Characteristic signals for common A-ring oxidation patterns

Ring	Carbon number *4a*	*5*	*6*	*7*	*8*	*8a*
Resorcinol (7-OH)	117	130	110	158	103	156
Phloroglucinol (5,7-diOH)	102	157	97	157	96	156
Pyrogallol (7,8-diOH)	115	123	110	144	132	146

(B) Characteristic signals for common B-ring oxidation patterns

Ring	Carbon number *1′*	*2′*	*3′*	*4′*	*5′*	*6′*
4′-Hydroxy	131	129	116	157	116	129
3′,4′-Dihydroxy	132	115	145	145	116	120
3′,4′,5′-Trihydroxy	131	108	146	133	146	108

(C) Effect of glycosylation on A- and C-ring carbons

Compound	*Solvent*	Carbon number *2*	*3*	*4*	*4a*	*6*	*8*
Catechin	DMSO-d_6	81.2	66.6	28.1	99.4	95.5	94.4
Catechin 5-*O*-Glc*p*	DMSO-d_6	80.9	65.9	27.3	100.3	96.0	94.9
Catechin 7-*O*-Glc*p*	DMSO-d_6	80.8	65.8	27.3	100.5	96.1	94.8
Epicatechin	acetone-d_6	79.4	66.9	29.1	99.7	96.2	95.7
Epicatechin 3-*O*-Glc*p*	acetone-d_6	78.2	73.3	24.2	—	96.3	95.4
Epicatechin 6-*C*-Glc*p*	acetone-d_6	79.3	66.3	28.1	—	105.0	96.5
Epicatechin 8-*C*-Glc*p*	acetone-d_6	79.2	66.7	29.2	—	97.2	103.8

(D) Signals characteristic of common conjugates

Conjugate				Carbon number 1	2,6	3,5	4	CO	
O-Gallate				122	110	145	138	165	
	1	2	3	4	5	6	α	β	CO
Cinchonain	134	115	145	145	116	118	38	34	169

et al., 1985; Shen *et al.*, 1985).

Catechin 6-carboxylic acid is a minor metabolite in the heartwood of *Acacia luederitzii* (Ferreira *et al.*, 1985). It may be readily synthesized by carboxylation of catechin with aqueous $NaHCO_3/CO_2$. The same compound is incorporated in a series of proguibourtinidin dimers (Table 2.7) in the same plant, together with their normal catechin and epicatechin analogues.

A series of catechin or *ent*-epicatechin metabolites, the gambiriins, have been isolated from *Uncaria gambir* leaves and twigs (gambir tannin) by Nonaka and Nishioka (1980). These involve condensation between the flavan-3-ols and C-3 of the diarylpropan-2-ol analogues of catechin. The latter are relatively well-known natural products (Kijjoa *et al.*, 1981). The gambiriins are in fact identical with the products of acid self-condensation of catechin which were extensively investigated by Freudenberg and Mayer and their colleagues (see Weinges *et al.*, 1969, and references therein). However, it would seem implausible that compounds isolated from fresh plant tissue could arise this way. Subsequently, gambiriins A1 and B3 have been isolated from *Sanguisorba officinalis* (Tanaka *et al.*, 1983).

Continuing the theme of 'mixed metabolites' formed between catechin and other natural products, the past five years has also seen the first examples of conjugates with an alkaloid and a hydrolysable tannin. Homberger and Hesse

(1984) isolated kopsirachin from *Kopsia dasyrachis* (Orobanchaceae) and found it to consist of a molecule of catechin condensed at both C-6 and C-8 with alkaloids of the skytanthine type. The absolute configuration of the alkaloid moieties has yet to be established. The second examples of novel catechin metabolites were isolated from the bark of *Quercus stenophylla*. This species was earlier found to contain the ellagitannins casuarinin, casuariin, castalagin and castalin. Three catechin-containing metabolites were isolated, stenophyllanins A–C, which had a close structural relationship with casuarinin and casuariin. This was confirmed by a facile synthesis of stenophyllanin A by acid-catalysed condensation of casuarinin with catechin, where the latter is linked through C-8 to the benzylic C-1 of the open-chain *C*-glucosyl moiety of casuarinin (Nonaka *et al.*, 1985). Stenophyllanin B is the minor C-6-linked isomer, and stenophyllanin C is the casuariin analogue of stenophyllanin A.

The remaining group of flavan-3-ol metabolites are the cinchonains, which are derived from epicatechin and caffeic acid and first isolated from *Cinchona succirubra* (Nonaka and Nishioka, 1982). The formation of cinchonain 1a may be rationalized by 7-*O*-ester formation between caffeic acid and epicatechin, followed by nucleophilic attack of C-8 of epicatechin on the β-face of C-3 of the caffeyl group. Conversely, cinchonain 1b is formed by α-face attack. Nonaka and Nishioka (1982) were able to distinguish the two isomers from the anisotropic effects on the *C*-ring protons being consistent with the aromatic ring of the caffeyl moiety being on the *same* side as 2-H for cinchonain 1b. Cinchonains 1c and 1d are the 5-*O*-ester C-6-linked analogues of cinchonains 1b and 1a respectively.

Nonaka and Nishioka (1982) achieved a low yield of cinchonains 1a and 1d by reaction of epicatechin with caffeic acid in dioxan solution at 100°C using *p*-toluenesulphonic acid as catalyst. However, in the plant cinchonain 1b predominates over 1a by a 2:1 molar ratio. The cinchonains are most interesting metabolites and illustrate well the difficulties of natural products' nomenclature. They are obviously lactones, and this structural fragment may qualify the cinchonains as flavonolignans. Equally well they may be considered to be neoflavonoids!

A final comment is that all of the *C*-substituted metabolites: the cinchonains, gambiriins, stenophyllanins, kopsirachin, larixinol, etc. all reflect the high reactivity of C-6 or C-8 of catechin (or epicatechin) towards electrophilic substitution.

2.3.3 Leucoanthocyanidins: flavan-3, 4-diols

Haslam (1975, 1982a) has previously discussed the structural elucidation and reactions of flavan-3, 4-diols, and these will not be discussed further. Table 2.4 gives a list of the currently known flavan-3, 4-diols and their occurrence. Few new sources have been added in recent years. Afzelechin- and catechin-4β-ol have been isolated as enzymic reduction products from the corresponding dihydroflavonols as discussed in Section 2.7. A new source of guibourtinidol-4α- and -4β-ol has been reported to be the heartwood *Acacia luederitzii* (Ferreira *et al.*, 1985), and prosopin-4β-ol has been isolated for the first time from a source outside *Acacia* from *Prosopis glandulosa* (mesquite, Jacobs *et al.*, 1983).

Foo and Wong (1986) have reinvestigated the flavonoids of *Acacia melanoxylon* heartwood – the source of the original flavan-3, 4-diol, melacacidin. They isolated the 4-*O*-ethyl derivatives of melacacidin and isomelacacidin. The latter was originally thought to be an artefact (Clark-Lewis and Mortimer, 1960) of the isolation method, as ethanol reacts very readily under mildly acidic conditions with melacacidin to produce this product. However, the isolation conditions used by Foo and Wong (1986) were performed in the absence of ethanol or ether, and thus the 4-ethoxy compounds cannot be artefacts.

A wide variety of methylated and *C*-alkylated flavan-3, 4-diols have been isolated from a variety of sources. The methylated products from *Neorautanenia amboensis* accompanied the parent phenol (Oberholzer *et al.*, 1980). Bohlmann and co-workers (1979, 1980) have isolated a number of methylated/prenylated flavan-3, 4-diols based on afzelechin-4β-ol from the lipid fraction from root extracts of two *Marshallia* (Compositae) species.

2.3.4 Leucoanthocyanidins: flavan-4-ols

These may be chemically (and probably also biosynthetically) derived by a single reduction step from a flavanone. To the author's knowledge only a handful are known to exist (Table 2.5). The first to be isolated was by Lam and Wrang (1975) from *Dahlia tenuicaulis* leaf and it remained until the 1980s before further examples were isolated from *Tephrosia watsoniana* and *Marshallia obovata*. Tanaka *et al.* (1984, 1985) have also isolated four flavan-4-ol glycosides from ferns.

If the aglycones are considered first, all were found to possess the 2, 4-*trans* stereochemistry, using Clark-Lewis (1968) ^{1}H NMR analysis for heterocyclic ring proton couplings in the C-ring. Most flavan-4-ols were found to co-exist with the corresponding flavanone, and therefore probably all possess the (2*S*, 4*R*) absolute stereochemistry. The negative rotation of tephrowatsin A may be significant as all the (2*S*)-flavans also have a negative rotation (Table 2.2).

In contrast, the flavan-4-ol glycosides from the ferns, erubin A and B and triphyllin A and B were shown to possess the (2*R*, 4*S*) stereochemistry by a clever use of n.O.e. measurements (nuclear Overhauser enhancement) on the doubly linked (C-1″ and C-2″) glucoside, erubin A (*2.52*). The fact that the erubins and triphyllins all possess positive specific rotations is also probably significant. In both cases the normal (2*S*)-flavanone co-occurred with the flavan-4-ols.

2.3.5 Leucoanthocyanidins: peltogynoid flavans (Table 2.6)

The peltogynoids differ from other flavonoids by the presence of a fourth ring formed by a one-carbon bridge between the 3-oxy function and C-2′ or -6′ of the B-ring. The accepted (Clark-Lewis and Mahandru, 1972) numbering system for peltogynoids also differs from that of the other flavonoids and is given in structure (*2.56*). Peltogynol (*2.57*) along with cyanomaclurin was the first leucoanthocyanidin to be isolated pure. It was isolated by Robinson and Robinson (1935) from the heartwood of the tropical leguminous hardwood *Peltogyne porphyrocardia* and its structure was established by Chan *et al.* (1958). Subsequently it has been isolated from the heartwoods of a number of other legumes in the genera *Acacia*, *Colophospermum*, *Peltogyne* and *Trachylobium*.

Peltogynol is always accompanied by its 7β-ol isomer peltogynol B, and the corresponding pair of regioisomers mopanol and mopanol B. The peltogynols and mopanols differ only by the position of ring closure on the flavonoid B-ring, and this led van der Merwe *et al.* (1972) to suggest that they were formed by oxidative cyclization from (2*R*, 3*R*)-3′, 4′, 7-trihydroxy-3-methoxyflavanone (3-*O*-methyl-2, 3-*trans*-fustin). This compound co-occurs with mopanol and peltogynol in the heartwood of *Trachylobium verrucosum*.

The 6a, 12a-*cis* isomer (*2.61*) of peltogynol B has been isolated, in low concentration, from the heartwood of *Acacia peuce* (Brandt and Roux, 1979) and this, presumably, is formed from 3-*O*-methyl-2, 3-*cis*-fustin (which was also present in *T. verrucosum*; van der Merwe *et al.*, 1972). The only non-leguminous source of a peltogynoid was recently reported from the root bark of *Elaeodendron balae* (Celastraceae). The structure is unusual because the methylene bridge of peltogynol is methyl disubstituted.

Solvolysis studies showed that peltogynol could be condensed with phloroglucinol or catechin, but under much more drastic conditions than those used for leucocyanidin or mollisacacidin (van Heerden *et al.*, 1981). This may mean that peltogynoids are unlikely to occur as oligomeric proanthocyanidins.

2.3.6 Proanthocyanidin oligomers (B-type, Table 2.7)

The past five years have seen very large additions of the number and type of natural proanthocyanidins – particularly higher oligomers and various naturally occurring ester derivatives. Propelargonidins are relatively rare, the sole previous example being a dimer isolated from *Ouratea* sp. root bark and other species of the Celastraceae. Subsequently a number of other dimers have been isolated from *Kandelia candel* bark and *Uncaria gambir* leaf. The dimeric procyanidins and prodelphinidins have been considered in detail earlier (Haslam, 1982a), the only further recent example being epicatechin-(4β→8)-epiafzelechin from *Saraca asoca* bark.

A considerable number of procyanidin trimers are now known. Those isolated from *Kandelia candel* bark (Hsu *et al.*, 1985a) are of particular interest as they call in to question our current concept of procyanidin biosynthesis. Arguments based on the products of 'biomimetic' syntheses of procyanidin dimers and trimers indicate that there will be a preponderance of 4→8 linkages over 4→6 linkages (Hemingway *et al.*, 1982), probably of the order of 3–4:1. Yet three of the four trimers isolated from *Kandelia candel* contain 4→6 linkages (see Table 2.7) and these products may be rationalized on the basis that addition of an epicatechin-4 unit at C-6 of the T-unit of a dimer is a biosynthetically favoured process in this plant.

Also of particular interest is the fact that a number of trimers have been isolated consisting of procyanidin units with different structures. For example, epicatechin-(4β→8)-catechin-(4α→8)-epicatechin and epicatechin-(4β→8)-catechin-(4α→8)-catechin were isolated from yam tubers (Hsu *et al.*, 1985b), and catechin-(4α→8)-gallocatechin-(4α→8)-catechin and gallocatechin-(4α→8)-catechin-(4α→8)-catechin were isolated from barley (Outtrup and Schaumberg, 1981). Such patterns of mixed stereochemistry and/or B-ring oxidation pattern were known to be common in proanthocyanidin polymers (Czochanska *et al.*, 1980), and the above results favour the view that this will be due to irregular polymer chains, rather than mixtures of regular (isotactic) polymer chains.

Examples of tetra-, penta- and hexa-meric procyanidins are now known. The isolation of a series of oligomers of [epicatechin-(4β→8)]*n* epicatechin from *Cinnamomum cassia* bark, where $n = 1$–5, is consistent with the finding that proanthocyanidins are polydisperse (Section 2.9).

A series of guibourtinidol–catechin or –epicatechin dimers were isolated from *Acacia luederitzii* (Ferreira *et al.*, 1985), two of which had been isolated previously from *Julbernadia globiflora* (Pelter *et al.*, 1969). Together with these dimers were a duplicate, and unique series of dimers in which catechin or epicatechin were carboxylated at C-6 or C-8 (Ferreira *et al.*, 1985). It is unknown whether the latter are products of biosynthesis or *post mortem*.

Roux and Ferreira's group have mounted an extensive reinvestigation of the 5-deoxyproanthocyanidins using an approach based on synthesis of the natural products (as their phenolic permethyl ether 3-*O*-acetates) and spectroscopic characterization using ^{1}H NMR (medium and high-field with variable temperature measurements) and CD.

Synthesis of a series of profisetinidin and prorobinetinidin oligomers by 'biomimetic' syntheses under acid-catalysed conditions was facilitated by the fact that the interflavanoid bond, once formed, is stable to acid cleavage under mild conditions in 5-deoxyproanthocyanidins. This has allowed elucidation of the structure of a considerable

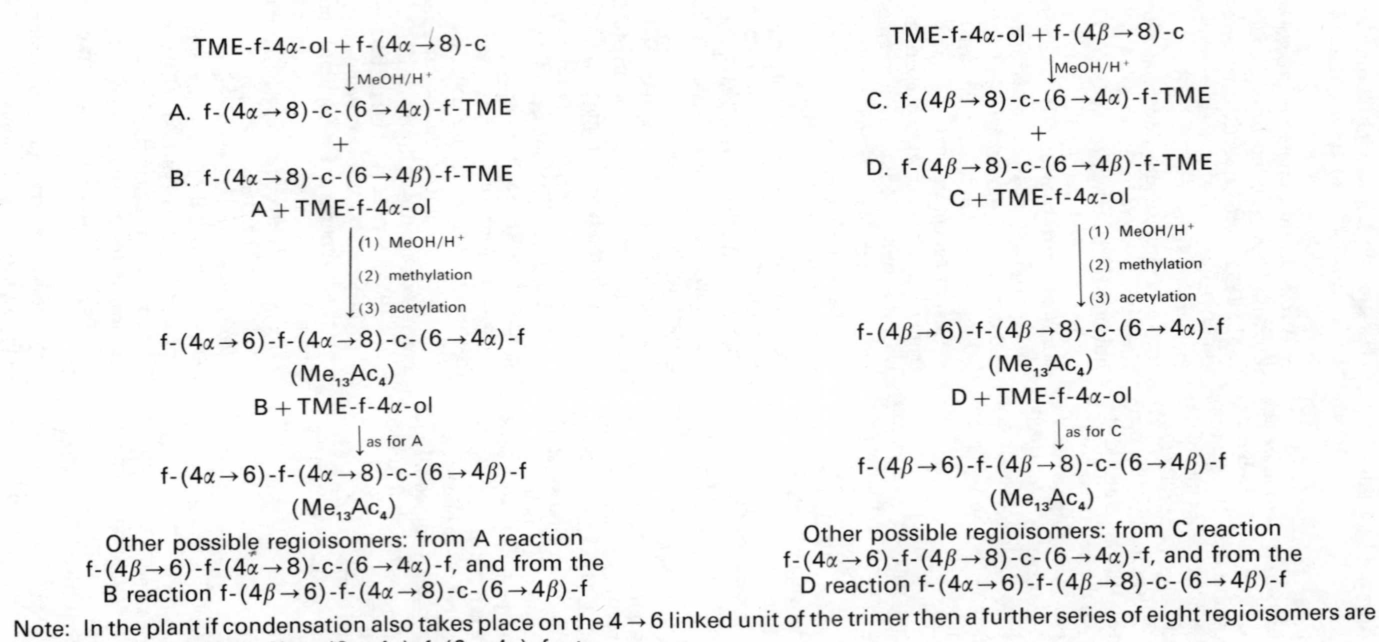

Note: In the plant if condensation also takes place on the 4 → 6 linked unit of the trimer then a further series of eight regioisomers are possible: f-(4α → 8)-c-(6 → 4α)-f-(6 → 4α)-f, etc.

Scheme 2.2 Syntheses of four profisetinidin regioisomers as their tridecamethyl ether tetra-acetate derivatives.
Note: f = fisetinidol; c = catechin; TME = tri-*O*-methyl ether.

number of proanthocyanidins (Table 2.7) including those from the commercially important sources of condensed tannins, black wattle (mimosa, *Acacia mearnsii*) bark and quebracho (*Schinopsis* spp.) wood.

The approach used is illustrated in Scheme 2.2. Attack of the carbocation produced by solvolysis of fisetinidol-4α-ol on catechin is reasonably regiospecific (ratio of 4→8:4→6 > 7:1, Botha *et al.*, 1981a) but stereoselectivity is not high (4α:4β = 1.5–2:1) in contrast with the procyanidins (Botha *et al.*, 1981a). Further reaction of fisetinidol-4α-ol with the dimers results in exclusive attack on the A-ring of catechin, because of its much greater nucleophilicity than the resorcinol A-ring, to produce the four isomeric 'bent' trimers (Scheme 2.2). The process was extended by repeating the trimer synthesis, but this time with the phenolic trimethyl ether of fisetinidol-4α-ol to produce the same four (but partly methylated) trimers, and these were further reacted with the trimethyl ether to produce the four tetrameric methyl ether acetates (after methylation and acetylation) as shown in Scheme 2.2. The second condensation must, of course, occur exclusively on the sole phenolic A-ring site available.

The final condensations occurred with a high degree of stereospecificity, attributed to asymmetric induction, and the four isomers synthesized were reportedly the same as the four naturally occurring tetramers (out of eight possible isomers, Scheme 2.2) linked to the 4→8 bonded fisetinidol unit (Young *et al.*, 1985a). Why, in the biosynthesis, the second condensation should not take place on the 4→6 bonded unit (to produce, theoretically, a further eight regioisomers) was not discussed, nor how such products might be distinguished from those synthesized. In view of this there must be some reservations as to the identity of the tetrameric profisetinidins reported from *Acacia mearnsii* and *Rhus lancea* (Table 2.7).

Ten dimeric 3-*O*-gallate esters of procyanidin and prodelphinidin are listed in Table 2.7, examples including esterification of either, or both, the T and B units. While most are linked to 2,3-*cis* units, there are two examples of esterification of a 2,3-*trans* unit, so that the galloyltransferase enzyme is not specific for the former stereochemistry. Green tea leaves and *Myrica rubra* bark are very rich sources of gallate esters. Esterification is confined to the B unit (catechin) of the di- and tri-meric procyanidin cyclohexene carboxylate 3-*O*-esters from *Salix sieboldiana*. Esterification therefore appears to be specific for the flavan-3-ol unit in this case.

C-Glucosylation is apparently confined to the T-unit of the dimeric procyanidins from *Cinnamomum cassia* bark and are accompanied by the equivalent epicatechin *C*-glucosides (Table 2.4). These observations imply that *C*-glucosylation probably takes place at an earlier stage in the biosynthesis, most probably at the flavanone/chalcone stage (Wallace and Grisebach, 1973).

The nature of the proluteolinidin pigments in sorghum has been a problem of long-standing. Bate-Smith (1969) made the initial observations and proposed that it was due to the presence of a flavan-4-ol (luteoforol). The origin of luteolinidin has now been shown to be due to a series of glucosylated dimers and trimers based on luteoliflavan and eriodictyol (Table 2.7). The structures were established by hydrolysis to luteolinidin, glucose and (2*S*)-eriodictyol; location of the bond as 4→8 in the dimers was achieved by bromination to yield 6-bromoeriodictyol, and location of glucosylation at C-5 by Hakomori permethylation and hydrolysis. The relative and absolute configuration of the proluteolinidin units and the interflavanoid bonds remain to be established by NMR and CD, although it may be argued that the configuration of the flavan unit will be 2α-phenyl-4β-flavanyl on biosynthetic and chemical grounds.

A series of procyanidin dimers and trimers with cinchonain-1a or -1b constituting the T unit, and one example with 1a being the M unit of a trimer, have been isolated from *Cinchona succirubra* and *Kandelia candel* bark. The fact that caffeic acid is conjugated with both the T and B units implies that the condensation probably takes place earlier in the biosynthesis at the dihydroflavonol level at least.

Among miscellaneous proanthocyanidins is the novel metabolite, apparently resulting from condensation between epicatechin-4-ol and acetyl-CoA, dryopterin (*2.62*), which has been isolated from the fronds of *Dryopteris filix-mas* (Karl *et al.*, 1981). The authors did not assign a relative, or absolute stereochemistry, but the ^{1}H NMR data are consistent with a 4β-*C*-acyl derivative of epicatechin, as shown. Confirmation of the absolute stereochemistry will require CD or conversion to a suitable model compound such as epicatechin-(4β→2)-phloroglucinol, which Kolodziej (1983) has shown to be a natural product in *Nelia meyeri*. This compound had been synthesized earlier (Fletcher *et al.*, 1977).

Steynberg *et al.* (1983) have isolated a series of novel proguibourtinidins from the heartwood of *Guibourtia coleosperma*. These represent a further example of a 'mixed metabolite' – this time involving a flavan-3-ol and a stilbene. Here the stilbene occupies the same position as the B unit in normal proanthocyanidin dimers. *G. coleosperma* apparently does not possess significant amounts of a flavan-3-ol, as the stilbenes are the major proanthocyanidin metabolites (Steynberg *et al.*, 1983).

The remaining series are the first examples of isoflavanoid proanthocyanidins. The *Dalbergia odorifera* dimers are associated with (3*R*)-vestitol (Yahara *et al.*, 1985), whereas those from *Dalbergia nitidula* are enantiomeric, being associated with (3*S*)-vestitol (Bezuidenhoudt *et al.*, 1984). The structures of the *D. odorifera* series were established by ^{1}H and ^{13}C NMR, and CD measurements, and confirmed by X-ray crystallography of compound (*2.65*). Bezuidenhoudt *et al.* (1984) showed that condensation of (3*S*)-vestitol with the pterocarpan, (6a*S*, 11a*S*)-medicarpin, under acid or photolysis conditions opened the benzofuran ring of the latter to produce an intermediate 4-carbocation (quinone methide) to produce the natural dimer (*2.64*) together with its 4$\beta \rightarrow$6 isomer and trimeric products.

2.3.7 Proanthocyanidin oligomers (A-type, Table 2.8)

Prior to 1982 only two double linked (A-type) proanthocyanidin dimers and a series of trimers of unresolved structure were known to exist (Haslam, 1975). Of these, only the structure of proanthocyanidin A2 was known with any certainty, and this was thought to be structurally correlated with epicatechin-(4$\beta \rightarrow$8)-epicatechin (Jacques *et al.*, 1977). Subsequently the structure of proanthocyanidin A-2 has been confirmed by X-ray crystallography (Otsuka *et al.*, 1982), and shown to be epicatechin-(2$\beta \rightarrow$7, 4$\beta \rightarrow$8)-epicatechin (*2.14*). The structure of proanthocyanidin A1 may be assigned as epicatechin-(2$\beta \rightarrow$7, 4$\beta \rightarrow$8)-catechin on the basis that its spectroscopic properties match those of A2 closely, and it produces a low yield of catechin on acid hydrolysis.

Of outstanding interest are a series of five A-type dimers isolated by Hikino and co-workers from *Ephedra* roots (Hikino *et al.*, 1982a and b). Their structures were established by ^{1}H and ^{13}C NMR, using n.O.e. and C–H connectivity methods, and their absolute stereochemistry was established by the sign of the low-wavelength couplet in the CD spectrum (Barrett *et al.*, 1979). This work broke much new ground and provided the first authenticated examples of a proanthocyanidin with a flavan C-ring (mahuannin D), with a flavonol (kaempferol) B unit (ephedrannin A), and A-type dimers with a 4α configuration (mahuannins A and D, and ephedrannin A) and with *ent*-flavan-3-ol units (mahuannin A and ephedrannin A). A further pair of growth-inhibiting propelargonidins were isolated from the roots of *Prunus persica* and were originally assigned the structures afzelechin-(2$\beta \rightarrow$7, 4$\beta \rightarrow$8)-afzelechin and -catechin (Ohigashi *et al.*, 1982) on the basis of NMR measurements and the determination of the absolute configuration at C-3 of the T unit as (*S*) by the Horeau Brooks method. These structures were subsequently revised to *ent*-epiafzelechin-(2$\alpha \rightarrow$7, 4$\alpha \rightarrow$8)-afzelechin and -catechin when it was shown that they possessed a negative low-wavelength couplet in the CD spectrum (Ohigashi and Porter, unpublished results) and therefore must possess the 4α configuration (Barrett *et al.*, 1979). The structures are therefore similar to mahuannin A.

The biosynthesis of the *ent* compounds are presumably from the dimer afzelechin-(4$\alpha \rightarrow$8)-epiafzelechin (in the case of mahuannin A), inversion of the B-ring occurring because formation of the C-2 ether linkage is directed from the α-face (see Section 2.7). A series of procyanidin oligomers containing both A- and B-type linkages have been isolated from a variety of sources (Table 2.8), *Cinnamomum zeylanicum* (cinnamon) and *C. sieboldii* being especially rich sources. Elucidation of these structures largely rested on the products of hydrolysis of phenylmethanethiol (Nonaka *et al.*, 1983c, and Section 2.5). Of special interest is the fact that the corresponding dimer, proanthocyanidin A2, is not always associated with the higher oligomers in a plant. Therefore the distribution of A-type proanthocyanidins may be very much wider than previously suspected, and surveys based on the detection of proanthocyanidin A1 or A2 alone (Bate-Smith, 1975) may be misleading.

2.3.8 Proanthocyanidins: polymers

The greater proportion of proanthocyanidins in plant tissue is often in the form of higher oligomers and polymers. The average chain length for such proanthocyanidins may vary widely. It may be as low as two to three in *Rubus* or barley, to 20–30, or greater, in other plant tissues (Porter, 1984).

The utility of chiroptical and ^{13}C NMR methods for the study of the proanthocyanidin polymers (PAP) was discussed previously by Haslam (1982a). Further, and considerable, progress has been made since that time. For example gel-permeation chromatography (Williams *et al.*, 1983, and references therein) and measurements based on vapour pressure osmometry and low-angle laser light scattering (Porter, 1984) have clearly demonstrated the polydisperse nature of PAP and the great variability in the shape and width of their molecular weight distributions. Measurements based on ^{13}C NMR signal intensities and a theoretical appraisal of the dispersivities of higher-molecular-weight procyanidin and prodelphinidin polymers have indicated that they are often branched (i.e. contain J units see Scheme 2.1), rather than being linear polymers (Mattice and Porter, 1984).

Surveys of the structure of the PAP of edible fruits (Foo and Porter, 1981), barley and sorghum (Brandon *et al.*, 1982), fodder legumes (Foo *et al.*, 1982), monocotyledonous plants (Ellis *et al.*, 1983) and other plant sources (Foo and Porter, 1980) have been carried out, using the methodology described by Czochanska *et al.* (1980). These data for 129 PAP are summarized in Table 2.9, together with information on the PAP from a number of other plant sources, collected at our laboratory over the past several years. ^{13}C NMR spectra have been run on $>90\%$ of the PAP samples reported. This information serves to give some idea of the distribution of PAP structures in the plant kingdom.

The following inferences may be drawn from the data in Table 2.9: (1) PAP containing predominantly epicatechin units (i.e. procyanidin 2, 3-*cis* units) have by far the widest distribution. (2) PAP containing PD (prodelphinidin) units are also very common, but the PD monomers predominate over PC (procyanidin) monomers in $<18\%$ of the polymers isolated. (3) PAP containing propelargonidin units are exceedingly rare. Only seven PAP from monocotyledonous plants contained these units, and none from dicotyledonous sources (see also Table 2.7). (4) The occurrence of PAP containing a mixture of epicatechin-4 and *ent*-epicatechin-4 units is a feature of many monocotyledonous plants (Ellis *et al.*, 1983). The mole ratio of normal to *ent* units was found to be consistently 1:3, and these units could be estimated spectroscopically (Ellis *et al.*, 1983), or by acid hydrolysis of a PAP in the presence of epicatechin, and chromatographic separation of the diastereoisomeric procyanidins epicatechin-(4$\beta \rightarrow$8)-epicatechin and *ent*-epicatechin-(4$\alpha \rightarrow$8)-epicatechin (Foo and Porter, 1983). (5) The structure of PAP may vary considerably from one plant organ to another. *Pinus radiata* is an excellent example; PD units predominate in the needles, whereas PC units predominate in the outer bark. The monomer unit stereochemistry also varies considerably (Table 2.9). (6) Examples of *O*-glucosyl (Porter *et al.*, 1985a) and *O*-galloyl PAP may also be found in Table 2.9. In some examples gallotannins (e.g. *Quercus robur*) or ellagitannins (*Fragaria ananassa*) are co-isolated with the PAP.

It may be concluded that the composition of PAP may vary very widely – even in the gross structural features summarized in Table 2.9. Enormous gaps exist in our knowledge of the chemosystematics of PAP – especially about their occurrence in the ferns and their allies, and in the majority of dicotyledonous families.

2.4 METHODS OF ISOLATION AND PURIFICATION

Much of the emphasis in contemporary natural products chemistry lies in elucidation of the biological activity of plant metabolites, rather than with the intrinsic interest of the structures themselves. The raison d'être for the study of the proanthocyanidins is to develop a knowledge of their role in plant biochemistry and how this is related to their structure and chemistry. In this context it may be noted that the impetus behind the outstanding contributions of Nishioka and Nonaka's group has been investigation of the pharmacological properties of proanthocyanidins (Nishioka, 1983).

These requirements necessitate the development of strategies to separate and purify proanthocyanidins in an unmodified state. Thompson *et al.* (1972) earlier showed the usefulness of Sephadex LH-20, using largely alcoholic solvents, to separate mono-, di- and tri-meric flavan-3-ols from a variety of plant sources. More recently Nonaka and Nishioka's group has greatly extended the range of separations achieved on LH-20 by using aqueous methanol or ethanol. These separations are used in concert with high-porosity polystyrene gel (Mitsubishi Chemicals MCI gel CHP 20P) using similar solvent systems. A wide variety of proanthocyanidins may be thus purified by juggling between LH-20 and CHP 20P and different solvent mixtures or solvent programmes (see for example, Morimoto *et al.*, 1985a).

More recently Nonaka and Nishioka's group have added column chromatography on reverse-phase C-8, C-18 or CN supports to the above chromatographic systems, once again using aqueous methanol or ethanol solvents (see Hsu *et al.*, 1985b). A further useful chromatographic medium for separating proanthocyanidin oligomers was introduced by Derdelinckx and Jerumanis (1984), the Toyo Soda product TSK HW-40(S), which may be used with organic solvents for molecular size separations in the 10^2–10^4 daltons range. The above workers separated malt and hop proanthocyanidins on TSK gel, and subsequently Delcour *et al.* (1985) used the same medium to separate procyanidin oligomers preparatively.

HPLC on reverse-phase columns has been used extensively during the 1980s to achieve separation of proanthocyanidins, both analytically and preparatively. The analytical potential of HPLC has been explored most thoroughly by Lea (1978, 1980, 1982) in studies of the procyanidins of cider. High-resolution separations may be achieved by solvent programming and dilute acid (usually acetic) is normally required as a component of the solvent (to repress the ionization of the phenolic hydroxy groups) to obtain satisfactory peak shapes. The most commonly used support is C-18, with C-8 and CN columns also being used.

Preparative applications of HPLC include the separation of procyanidin trimers by Hemingway *et al.* (1982) on nitrile columns, while Gujer *et al.* (1986) obtained pure samples of proluteolinidin glucoside and procyanidin oligomers from C-8 columns. It should be noted that the acid present in a preparative HPLC solvent requires that the samples must be carefully evaporated to avoid disproportionation. Freeze-drying is obligatory in such circumstances, and its use is preferable for column chromatography also, where aqueous solvents are used.

Some representative examples of the chromatographic behaviour of flavan-3-ols on reverse-phase HPLC are given in Table 2.11. While the factors which determine the

Table 2.11 Order of elution of flavan-3-ols from an HPLC C-18 column (solvent MeOH–1% HOAc, 20:80, v/v)

Compound	*Relative retention time (catechin standard)*
Catechin	1.00
Epicatechin	1.83
Epicatechin-(4$\beta \rightarrow$8)-catechin	0.65
Epicatechin-(4$\beta \rightarrow$6)-catechin	2.55
[Epicatechin-(4$\beta \rightarrow$8)]$_2$catechin	0.55
Epicatechin-(4$\beta \rightarrow$8)-epicatechin	1.18
[Epicatechin-(4$\beta \rightarrow$8)]$_2$epicatechin	1.90
Catechin-(4$\alpha \rightarrow$8)-catechin	0.60
Epicatechin-(4$\beta \rightarrow$2)-phloroglucinol	0.57
Epigallocatechin-(4$\beta \rightarrow$2)-phloroglucinol	0.38

order of elution of proanthocyanidin oligomers are complex, three trends are readily apparent. (1) Oligomers with a catechin B unit are eluted before those of similar constitution, but with an epicatechin B unit. This is probably due to the fact that the strongest binding to the reverse-phase is through the C-4 methylene of the B unit, and this is more 'exposed' in the case of epicatechin, because of the axial disposition of the 3-hydroxy-group. (2) Dimers with a 4→8 linkage are eluted before their 4→6 isomers. (3) Prodelphinidins are eluted before procyanidins of equivalent constitution.

2.5 STRUCTURAL ELUCIDATION

A key feature of the structure of proanthocyanidin oligomers is the position and absolute stereochemistry of interflavanoid bonds. While the latter may readily be determined by NMR and CD spectroscopy (Haslam, 1982a), similar information regarding their position (4→6 or 4→8 for procyanidins) is rather more difficult to obtain.

This has prompted a number of studies of the partial bromination and subsequent cleavage of proanthocyanidin oligomers to determine the position of the interflavanoid linkage on the A-ring. Several studies have shown that catechin (Hundt and Roux, 1981; McGraw and Hemingway, 1982; Kiehlmann *et al.*, 1983) or epicatechin (Nonaka *et al.*, 1983c; Kolodziej *et al.*, 1984) may be brominated with perbromopyridinium hydrobromide at C-6 and/or C-8 to produce a mixture of regioisomers, which may be readily separated. Kiehlmann *et al.* (1983) have also shown that 6, 8-dibromocatechin is selectively debrominated with Na_2SO_3/Na_2CO_3 to produce the 6-isomer. Subsequent derivatization and ^1H NMR studies have shown that the H-8 proton consistently resonates at a significantly lower field than the H-6 isomer for the tetramethyl ether derivatives, but the relative chemical shift positions are equivocal for the phenols or the penta-acetates.

The above chemical shift difference was used to distinguish regioisomers of epicatechin oligomers, isolated from *Crataegus oxyacantha* by ^1H NMR of their phenolic methyl ether 3-*O*-acetates (Kolodziej *et al.*, 1984), measurements being carried out in $CDCl_3$ and at elevated temperature to overcome the effects of rotational isomerism (Haslam, 1982a).

The bromination/cleavage approach was successfully applied by Nonaka *et al.* (1983c) to elucidation of the structure of A-type procyanidin oligomers from *Cinnamomum zeylanicum* bark. For example, the trimer epicatechin-(2β→7, 4β→8)-epicatechin-(4β→8)- epicatechin was brominated with perbromopyridinium hydrobromide in acetonitrile to yield a tetrabromo-trimer (i.e. substitution of all available A-ring carbons). Cleavage in refluxing 0.2 M HCl in ethanol yielded 6-bromoepicatechin, and a mixture of brominated anthocyanidins, thus establishing the B unit interflavanoid linkage as 4→8. This approach, unfortunately, appears to have limited usefulness. Attempts to cleave the tetrabromo-trimer with phenylmethanethiol, and so obtain brominated T and M fragments, failed because of debromination (Nonaka *et al.*, 1983c).

Cleavage of procyanidin or prodelphinidin oligomers may be readily achieved by reaction with phenylmethanethiol in the presence of acetic acid in refluxing ethanol (Thompson *et al.*, 1972; Haslam 1975, 1982a). This method has the advantage of producing oligomeric fragments from which may be deduced both the position and stereochemistry of the interflavanoid bonds. For example, Hemingway *et al.* (1982) were able to establish the structure of three regioisomers of $(\text{epicatechin-4})_2$ catechin from the products of partial thiolysis (Scheme 2.3). The method requires that the structures of the dimeric fragments have been independently established.

The method has been used extensively by Nonaka and Nishioka's group to investigate a wide range of proanthocyanidin oligomers (Table 2.7), the most dramatic example perhaps being establishment of the structure of a series of $[\text{epicatechin-}(4\beta\rightarrow8)]_n$ epicatechin ($n = 1$–5) homologues, isolated from *Cinnamomum cassia* (Morimoto *et al.*, 1986a). This required separation of the products of partial cleavage and subsequent desulphurization of the 4-thiobenzyl ethers with Raney nickel (Thompson *et al.*, 1972) and correlation of the subsequently formed oligomers with those of previously determined structure. In this way the complete series of oligomers was characterized and an HPLC method developed to analyse the thiolysis products.

Other advantages of the thiolysis method are the retention of the ester and lactone moieties of *O*-gallates and cinchonains respectively. For instance, Nonaka *et al.* (1984) cleaved the prodelphinidin epigallocatechin-(4β→8)-epicatechin-3-*O*-gallate to obtain epigallocatechin-4β-thiobenzyl ether and epicatechin-3-*O*-gallate. Similarly cleavage of the dimer cinchonain-1a-(4β→8)-epicatechin yielded cinchonain-1a-4β-thiobenzyl ether and epicatechin (Nonaka *et al.*, 1982a). Some caution must be exercized in the reaction times used in the thiolysis reaction. Prolonged heating causes significant epimerization and disproportionation of products. Heating for moderate periods, 6 h or so, is sufficient to yield useful concentrations of oligomeric products.

^{13}C NMR spectroscopy has proved to be the most generally useful technique for the study of flavans and proanthocyanidins, especially as the phenols usually provide good-quality spectra, whereas their ^1H NMR spectra are considerably broadened due to proton-exchange processes. The ^{13}C NMR spectra of oligomers and polymers immediately yield such information as the A- or B-ring substitution pattern, the relative stereochemistry of the C-ring, and in favourable cases, the position of interflavanoid linkages. Extensive compilations of spectra, especially for procyanidins, have appeared (Porter *et al.*, 1982; Foo and Porter, 1983). To illustrate the usefulness of

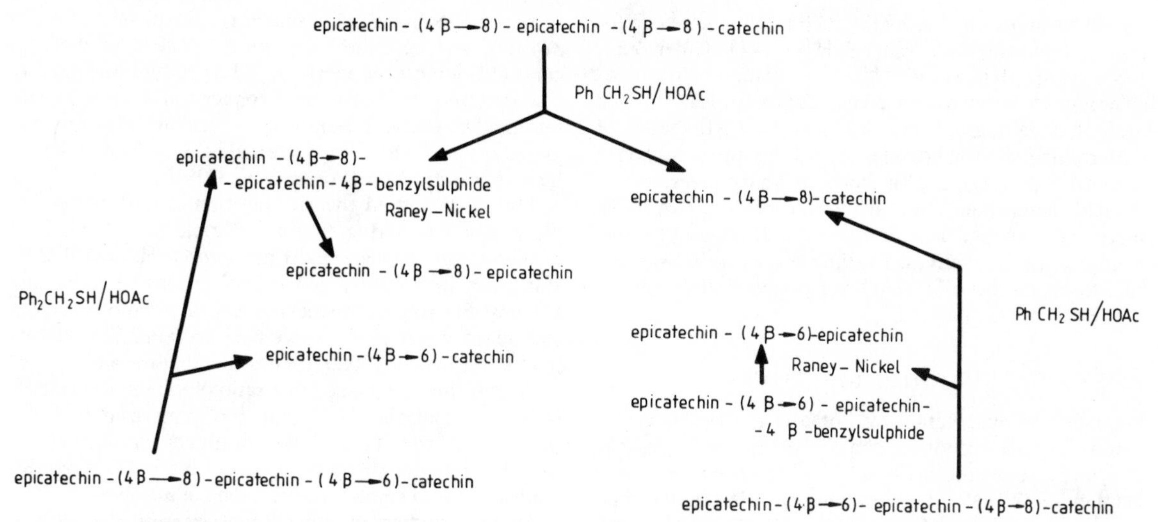

Scheme 2.3 Partial cleavage of three regioisomers of (epicatechin-4)$_2$ catechin with phenylmethanethiol and acetic acid.

^{13}C NMR the patterns of signals characteristic of various structural features of flavan-3-ols and proanthocyanidins are given in Table 2.10.

^{13}C NMR is extremely useful for giving immediate indications of the presence of non-flavanoid metabolites in proanthocyanidins. Thus the presence of *O*-gallates, *O*-glycosides, or cinchonain-type functionality may be deduced. In particular, glycosides are indicated by extra signals in the C-ring region (δ 60–80), while esters (gallates) or cinchonains are indicated by resonances in the carbonyl region (δ 160–170). Resonances typical of these features are summarized in Table 2.10, while many specific examples may be found in Nonaka and Nishioka's work.

Successful application of ^{1}H NMR spectroscopy to proanthocyanidin structural problems requires suitable derivatization (see above). Roux and Ferreira's group have made extensive use of this approach and have elucidated the structure of a wide variety of proanthocyanidins as the phenolic permethyl ether 3-*O*-acetates. This often requires measurement of the spectra at high temperature (for example Delcour *et al.*, 1983) to overcome the effects of rotational isomerism which leads to signal multiplicities (Fletcher *et al.*, 1977). Roux and co-workers have also used high-field ^{1}H NMR (at 300 or 500 MHz) to resolve the complex pattern of heterocyclic and aromatic protons in proanthocyanidin oligomers (see, for example, Young *et al.*, 1985a).

Application of CD spectroscopy has been central to our ability to assign with confidence the absolute stereochemistry of interflavanoid linkages. The CD bands of proanthocyanidins are much more intense than those of their constituent flavan units because of the close proximity of the A-ring chromophores (Snatzke's second chiral sphere, Snatzke *et al.*, 1973) in contrast to the more remote placement of the A- and B-ring chromophores in monomeric flavans (Snatzke's third chiral sphere). Two regions of the CD spectrum have been used for stereochemical assignments. Barrett *et al.* (1979) showed that the CD spectra of procyanidin oligomers possess a short-wavelength couplet at low wavelength, centred near 200 nm. A positive couplet was correlated with a 4β, and a negative couplet with a 4α, interflavanoid bond configuration.

Similarly Roux and co-workers (Botha *et al.*, 1981b) in a detailed study of a wide variety of flavan-3-ol-4-aryl chromophores showed that the absolute configuration of the interflavanoid bond could be correlated with the sign of the CD band near 230 nm (probably an ^1L_a transition), a positive sign being correlated with a 4β, and negative with a 4α, configuration, *regardless* of the configuration of the rest of the molecule. It should be noted that the 230 nm band is often merged with the intense high-wavelength component of the couplet discussed earlier. These rules have found wide application in a variety of proanthocyanidin structural problems and many of the oligomers listed in Tables 2.7 and 2.8 had their absolute stereochemistry assigned by one rule or the other.

In common with most other areas of natural products chemistry, this field has already derived considerable benefits from the application of fast atom bombardment (FAB) MS to the determination of structure and molecular weight. The question of the molecular weight of oligomeric proanthocyanidins has always posed problems and it is fortunate that these polyphenols are particularly well suited to the FAB method. The highest molecular weight so far obtained using the FAB method has been for procyanidin pentamers where $MH^+ = 1441$ (Gujer *et al.*, 1986) or $[M + K]^+ = 1481$ (Morimoto *et al.*, 1986a) molecular ions were recorded. Gujer *et al.* (1986) also observed useful hexose fragmentation losses in the FAB spectra of the proluteolinidin glucosides from sorghum. A standard glycerol matrix was used in both the above cases.

Field desorption mass spectrometry has also been successfully applied in a number of studies on phenolic methyl ether derivatives. Morimoto *et al.* (1986a) were able to obtain the MH^+ ion for a procyanidin hexamer, m/z 2067, by this method.

2.6 SYNTHESIS AND REACTIONS

Following an observation by Botha *et al.* (1981a) that a flavan-3,4-diol product could be trapped as the 3′,4,4′,5,7-pentamethyl ether from the products of reduction of (2*R*,3*R*)-dihydroquercetin, Porter and Foo (1982) achieved a synthesis of catechin-4α-ol using the same route (Scheme 2.4). Isolation of the phenolic flavan-3,4-diol was dependent on careful control of the acid concentration during work up. The product was initially characterized by NMR and conversion to the known tetramethyl ether. Subsequently catechin-4α-ol was obtained as a crystalline dihydrate and its structure confirmed by X-ray diffraction (Porter *et al.*, 1985b).

This work settled the old question of whether or not phloroglucinol A-ring flavan-3,4-diols (leucocyanidins) are sufficiently stable to exist in solution (Haslam, 1982a) and, more importantly, led several groups to investigate the possible role of leucocyanidins in proanthocyanidin and anthocyanin biosynthesis (see Section 2.7). Subsequently Kristiansen (1986) established that the 4α- and 4β-ols of catechin are formed in a ratio of 95:5 in the $NaBH_4$ reduction, and the 4α-ol may be epimerized to a 4:6 mixture of the 4α and 4β products in acetic acid (Scheme 2.4), the 4β-ol being the (expected) thermodynamically more stable product (Baig *et al.*, 1969).

Another finding of outstanding interest was that of

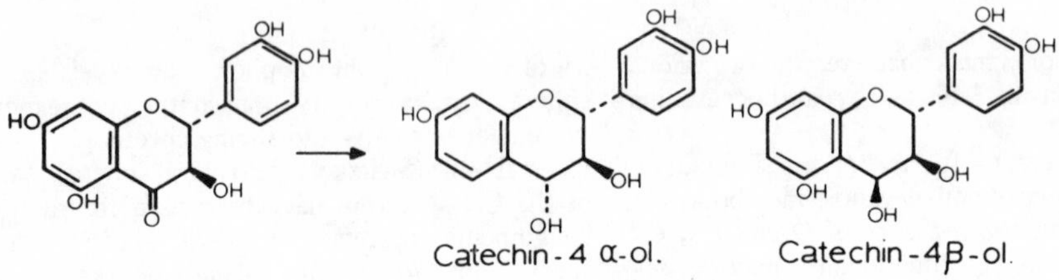

Ratio of products	α		β
1) enzymic reduction	0	:	100
2) $NaBH_4$ reduction	90	:	10
3) acid epimerization	40	:	60

Scheme 2.4 Products produced by chemical or enzymic reduction of (2*R*, 3*R*)-dihydroquercetin, or epimerization of the diols.

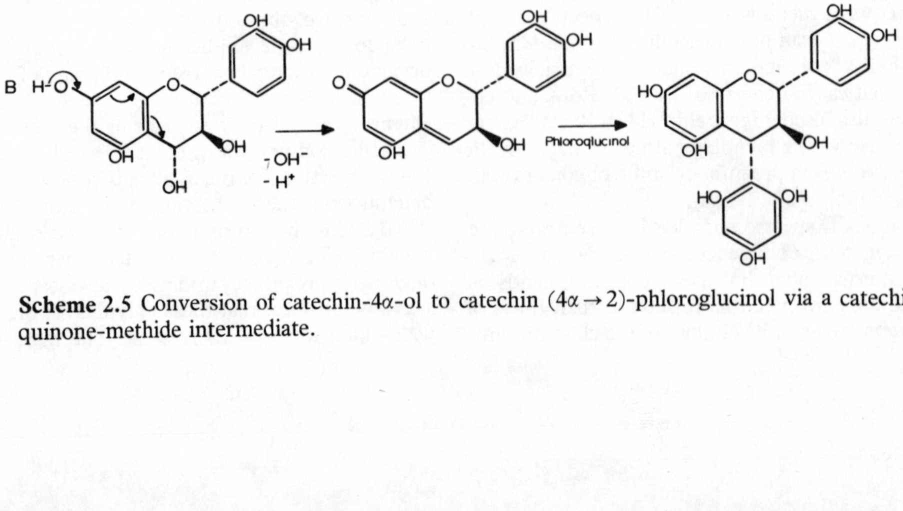

Scheme 2.5 Conversion of catechin-4α-ol to catechin (4α→2)-phloroglucinol via a catechin quinone-methide intermediate.

Hemingway and Foo (1983) who observed that reaction of catechin-4α-ol or epicatechin-4β-thiobenzyl ether at pH 9 in the presence of excess phloroglucinol led to formation of high yields of the 4α and 4β *C*-phloroglucinol addition products respectively. This could only be interpreted in terms of formation of a quinone methide intermediate (Scheme 2.5, a *p*-quinone methide is shown, but an *o*-quinone methide is also possible) which is formed by deprotonation of the 7- (or 5-) hydroxy group and elimination of the group at C-4 (by either a concerted or stepwise mechanism). The quinone methide then reacts with phloroglucinol to form the final product.

Further studies by Brown and co-workers (Attwood *et al.*, 1984) soon showed that a quinone methide intermediate was also formed under mildly acidic conditions, rather than a carbocation (see Haslam, 1982a). Reaction of the tetrahydropyranoxy derivative of 7-hydroxyflavan-4α-ol with HCl in aqueous dioxan resulted in the slow formation of an absorption band at 305 nm (the position expected for a quinone methide), and subsequent addition of benzenesulphinic acid caused gradual disappearance of the band and formation of the 4β-benzenesulphinate. 7-Hydroxyflavan-4β-ol was regenerated by hydrolysis of the sulphinate with alkali (Attwood *et al.*, 1984). The above studies established that quinone methides are key intermediates in proanthocyanidin transformations, although they will probably be in equilibrium with the carbocation in strongly acidic solutions.

The ability to carry out condensations in mildly alkaline solutions presents considerable advantages for the synthesis of phloroglucinol A-ring proanthocyanidins. Under these conditions (pH 9) the interflavanoid bond is stable and it allowed Foo and Hemingway (1984) to synthesize the first example of a 'branched' trimer in the procyanidin series by condensation of the epicatechin quinone methide with epicatechin-(4β→6)-catechin to form epicatechin-(4β→6)-catechin-(8→4β)-epicatechin with high regiospecificity.

The interflavanoid bond of procyanidins is extremely labile in acid solutions and two groups have measured its rate of cleavage in dimers (Hemingway *et al.*, 1983; Beart *et al.*, 1985). These experiments showed that a 4β→6 interflavanoid bond is cleaved at a considerably slower rate than a 4β→8 bond in epicatechin procyanidins. Further, that the rate-determining step is protonation of the B unit A-ring and that the reaction is specific acid catalysed which is first-order in hydrogen ion concentration (Beart *et al.*, 1985). These studies show that the rate of interflavanoid bond cleavage will be significant at ambient temperatures and at the natural pH (3–4) of procyanidins in aqueous solution. This implies that some disproportionation may occur in procyanidin oligomers during isolation and purification.

The above findings also imply that a low degree of regiospecificity will be achieved in procyanidin syntheses in acid solutions. The products will be determined by the relative rates of formation/cleavage of the interflavanoid bonds and the stoichiometry of the reactants. For instance, Beart *et al.* (1985) noted that an equilibrium was reached after 3–4 days when epicatechin-(4β→8)-epicatechin was hydrolysed with acid, corresponding to 80% decomposition, and that this equilibrium could be displaced towards the dimer by addition of equimolar epicatechin.

In contrast, the interflavanoid bond in resorcinol A-ring proanthocyanidins is relatively stable in acid solutions. For example, the dimer guibourtinidol-(4α→8)-epiafzelechin was stable to acetic acid and phenylmethanethiol in refluxing ethanol, and heating to 140°C under pressure was required to achieve cleavage (Patil and Deshpande, 1982). Similarly treatment of the dimer fisetinidol-(4α→8)-catechin with strong acid resulted not in disproportionation, but opening of the C-ring and reclosure on the 7-hydroxyl group of catechin to form compound (*2.69*) as the predominant product (Steenkamp *et al.*, 1985).

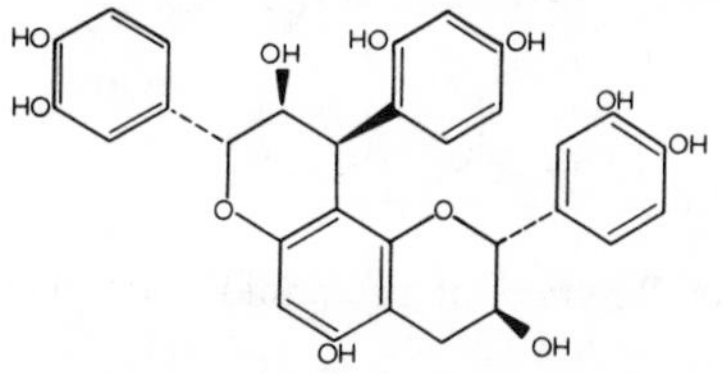

(*2.69*)

Our understanding of the reactions of proanthocyanidins in strongly acidic or basic solutions has also advanced considerably. A traditional method for determining the presence of proanthocyanidins is their conversion to anthocyanidins in BuOH/HCl. In a detailed study of this reaction Porter *et al.* (1986) showed that the hydrolysis is an autoxidation in which the yield of anthocyanidin is critically dependent on trace impurities. Reproducible yields of anthocyanidins may be achieved if iron(III) salts are added to the reaction medium. The reaction is also catalysed by quinones. It is likely that the rate-determining step in the above reaction is initial protonation, as the yield of cyanidin was higher from a 4→8 than a 4→6 linked dimer. However, in the strongly acidic and dehydrating conditions used for this reaction the formation of products is irreversible, so that the actual yield of anthocyanidin is dependent only on its relative rate of formation compared with other products.

Foo *et al.* (1983) investigated the products of the reaction of an epicatechin-4 procyanidin (from loblolly pine bark) with sodium bisulphite (a reaction with considerable commercial importance). Heating at 100°C and pH 5.5 for 20 h resulted in the formation of sodium epicatechin-4β-sulphonate in 20% yield, together with other C(4)-sulphonated oligomeric products. The fact that this reaction caused cleavage (depolymerization) of procyanidins was most significant, as it had previously been

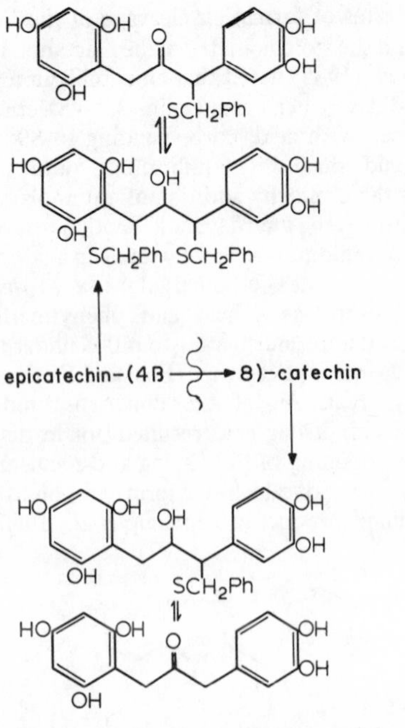

Scheme 2.6 Reaction of procyanidin B1 with phenylmethanethiol at pH 12.

maintained that the predominant products of this reaction were C(2)-sulphonates (Sears, 1972).

Further studies by Hemingway and Laks (1985) investigated the reaction between epicatechin-(4$\beta \rightarrow$8)-catechin and phenylmethanethiol at pH 12 and ambient temperatures. Four products were formed (Scheme 2.6) whose formation may be rationalized on the basis of an initial cleavage of the interflavanoid bond to form epicatechin-4β-benzylsulphide and catechin, followed by base-catalysed ring opening and further attack of phenylmethanethiol to form the two initial alcohols. These react further by base-catalysed tautomerism, (Scheme 2.6) to form the final pair of ketones.

The base-catalysed epimerization of catechin and epicatechin was used by Foo and Porter (1983) to synthesize *ent*-epicatechin and *ent*-catechin respectively. It has been established that catechin or *ent*-catechin are the thermodynamically favoured products of the reaction (if carried out under nitrogen), the ratio of catechin: epicatechin being 3:1 (Kiatgrajai *et al.*, 1982). Catechinic acid is rapidly formed in an irreversible reaction in the presence of oxygen (Sears *et al.*, 1974; Kiatgrajai *et al.*, 1982). The epimerization has long been considered to proceed via an ionic mechanism involving initial deprotonation of the 4′-hydroxyl group (Kiatgrajai *et al.*, 1982). Kennedy *et al.* (1984) have subsequently shown that the epimerization requires traces of oxygen as well as base and therefore must proceed instead by a radical-anion mechanism.

2.7 BIOSYNTHESIS

Debate (Stafford, 1983; Platt *et al.*, 1984) has continued over the final steps in the biosynthesis of proanthocyanidins. The issue at point is whether or not a flavan-3,4-diol is the *direct* precursor of flavan-3-ols and proanthocyanidins, or if in fact a flav-3-en-3-ol is the key intermediate (Haslam, 1982a). Support for the latter view was apparently obtained by Platt *et al.* (1984) who showed that doubly labelled β-phenylalanine suffered a 40–60% loss of the pro-*R* or -*S* label when incorporated into C-3 of catechin or epicatechin. Hemingway and Laks (1985) later suggested that these observations could be explained by tautomerism (enzyme-mediated?) of an intermediate quinone methide.

In support of the former proposal was the independent demonstration by three groups (Kristiansen, 1984, 1986; Stafford, *et al.*, 1985; Heller *et al.*, 1985b) that crude enzyme preparations from Douglas fir tissue cultures, barley, and *Matthiola incana* flowers respectively, all converted (2*R*, 3*R*)-dihydroquercetin, stereospecifically, to catechin-4β-ol, and further to catechin, through NADPH-mediated reductions. Stafford and Lester (1985) have also shown that (2*R*, 3*R*)-dihydromyricetin may be converted to gallocatechin-4β-ol and gallocatechin by the Douglas fir enzyme complex.

A major factor which led Haslam *et al.* (1977) to propose the flav-3-en-3-ol intermediate was the route by which epicatechin units are formed, and why often the stereochemistry of the B unit differs from that of the T and M units in procyanidins. One argument against a dihydroflavonol being the direct precursor of epicatechin units was the fact that only the thermodynamically favoured 2, 3-*trans*-dihydroflavonols had been observed in nature. Stafford (1983) countered this by proposing that 2, 3-*cis*-dihydroflavonols probably did exist, but were stabilized as a glycoside, or on an enzyme surface, immediately prior to reduction.

Support for this view is in fact available. Sakurai *et al.* (1982) isolated (2*R*, 3*S*)-dihydroquercetin 3-*O*-β-D-glucopyranoside from *Taxillus kaempferi*, confirming that phloroglucinol A-ring 2, 3-*cis*-dihydroflavonols do occur naturally, if suitably protected, as was previously indicated by the epimerization studies of astilbin by Gaffield *et al.* (1975). More direct evidence has been obtained by Foo (1986b) who isolated the first 2, 3-*cis*-dihydroflavonol from the heartwood of *Acacia melanoxylon*, (2*R*, 3*S*)-3, 3′, 4′, 7, 8-pentahydroxyflavanone, where it is accompanied by two flavan-3, 4-diols with identical oxidation pattern and C-ring stereochemistry, melacacidin and isomelacacidin (see also Table 2.4). Earlier the corresponding *racemic* 2, 3-*trans*-dihydroflavonol had been isolated from this source, but Foo (1986b) established that this was an artefact of the isolation method.

The above observations and the enzymic reduction experiments would appear to clinch the case for flavan-3, 4-diols being the *direct* precursor of the flavan-3-ols and proanthocyanidins. The contrasting stereochemistry of T and M units versus B units in some procyanidins (or even the contrasting structure of T and B units in many dimers, see Table 2.7) may be explained by the T and B units being synthesized in distinct metabolic pools. Stafford and Lester (1984) have shown from competitive labelling experiments, for example, that catechin-4 and epicatechin-4 units are synthesized in different metabolic pools in Douglas fir needles.

Heller *et al.* (1985a, b) have shown that catechin-4β-ol is the likely precursor of cyanidin 3-*O*-glucoside in *Matthiola incana* flowers. It is thought that the interconversion probably takes place via an enzyme-mediated hydroxylation at C-2 of the flavan-3, 4-diol, in very much the same way as the autoxidation at C-2 in the *chemical* conversion of procyanidins to cyanidin (Porter *et al.*, 1986).

These data suggest a possible biosynthetic route from B-type to A-type proanthocyanidins. As discussed previously proanthocyanidin A2 and procyanidin B2 have a close structural relationship, differing only by an ether linkage between C-2 of the T unit and O-7 of the B unit. Scheme 2.7 illustrates a route for the conversion of the singly to the doubly linked dimer. The B unit is ideally placed for this reaction as it is orthogonal to the T unit C-ring plane and O-7 is almost directly over C-2 in one of the preferred (Haslam, 1982a) conformations of the dimer. It is also anticipated that the C-2 epimerase enzyme postulated by Ellis *et al.* (1983) to account for the formation of *ent*-epicatechin-4 units in the procyanidins of monocotyledonous plants would also be of this type.

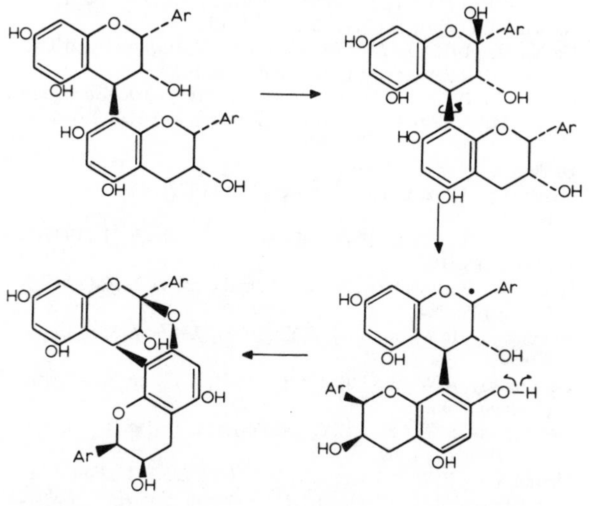

Scheme 2.7 Proposed biosynthetic pathway for the conversion of B-type to A-type procyanidins. Ar = 3, 4-dihydroxyphenyl.

REFERENCES

Ahn, B. and Gstirner, F. (1983), *Arch. Pharm. (Weinheim)* **306**, 338, 353.

Appel, H. and Robinson, R. (1935), *J. Chem. Soc.*, 752.

Attwood, M.R., Brown, B.R., Lisseter, S.G., Torrero, C.L. and Weaver, P.M. (1984), *J. Chem. Soc., Chem. Commun.*, 177.

Ayoub, S.M.H. (1985), *Int. J. Crude Drug Res.* **23**, 87.

Baba, K., Takeuchi, K., Hamasaki, F. and Kozawa, M. (1985), *Chem. Pharm. Bull.* **33**, 416.

Baba, K., Takeuchi, K., Doi, M., Inoue, M. and Kozawa, M. (1986), *Chem. Pharm. Bull.* **34**, 1540.

Baig, M.I., Clark-Lewis, J.W. and Thompson, M.J. (1969), *Aust. J. Chem.* **22**, 2645.

Barrett, M.W., Klyne, W., Scopes, P.M., Fletcher, A.C., Porter, L.J. and Haslam, E. (1979), *J. Chem. Soc., Perkin 1*, 2375.

Bate-Smith, E. C. (1969), *J. Food Sci.* **34**, 203.

Bate-Smith, E.C. (1975), *Phytochemistry* **14**, 1107.

Beart, J.E., Lilley, T.H. and Haslam, E. (1985), *J. Chem. Soc., Perkin 2*, 1439.

Bezuidenhoudt, B.C.B., Brandt, E.V. and Roux, D.G. (1984), *J. Chem. Soc., Perkin 1*, 2767.

Birch, A.J. and Salahuddin, M. (1964), *Tetrahedron Lett.*, 2211.

Birch, A.J., Dahl, C.J. and Pelter, A. (1967), *Tetrahedron Lett.*, 481.

Bohlmann, F. and Jakupovic, J. (1978), *Phytochemistry* **17**, 1677.

Bohlmann, F., Zdero, C., King, R.M. and Robinson, H. (1979), *Phytochemistry* **18**, 1246.

Bohlmann, F., Jakupovic, J., King, R.M. and Robinson, H. (1980), *Phytochemistry* **19**, 1815.

Botha, J.J., Ferreira, D. and Roux, D.G. (1981a), *J. Chem. Soc., Perkin 1*, 1235.

Botha, J.J., Young, D.A., Ferreira, D. and Roux, D.G. (1981b), *J. Chem. Soc., Perkin 1*, 1213.

Botha, J.J., Viviers, P.M., Young, D.A., du Preez, I.C., Ferreira, D., Roux, D.G. and Hull, W.E. (1982), *J. Chem. Soc., Perkin 1*, 527.

Bradfield, A.E. and Penney, M. (1948), *J. Chem. Soc.*, 2249.

Brandon, M.J., Foo, L.Y., Porter, L.J. and Meredith, P. (1982), *Phytochemistry* **21**, 2953.

Brandt, E.V. and Roux, D.G. (1979), *J. Chem. Soc., Perkin 1*, 777.

Brandt, E.V., Ferreira, D. and Roux, D.G. (1981a), *J. Chem. Soc., Perkin 1*, 514.

Brandt, E.V., Ferreira, D. and Roux, D.G. (1981b), *J. Chem. Soc., Perkin 1*, 1879.

Braz Filho, R., Diaz Diaz, P.P. and Gottlieb, O.R. (1980a), *Phytochemistry* **19**, 455.

Braz Filho, R., Da Silva, M.S. and Gottlieb, O.R. (1980b), *Phytochemistry* **19**, 1195.

Camarda, L., Merlini, L. and Nasini, G. (1977), *Aust. J. Chem.* **30**, 873.

Camarda, L., Merlini, L. and Nasini, G. (1981), *Proc. Int. Bioflavonoid Symposium*, Munich, pp. 311–320.

Camarda, L., Merlini, L. and Nasini, G. (1983), *Heterocycles* **20**, 39.

Cardillo, G., Merlini, L., Nasini, G. and Salvadori, P. (1971), *J. Chem. Soc., (C)*, 3967.

Chan, W.R., Forsyth, W.G.C. and Hassal, C.H. (1958), *J. Chem. Soc.*, 3174.

Clark-Lewis, J.W. (1968), *Aust. J. Chem.* **21**, 2059.

Clark-Lewis, J.W. and Dainis, I. (1967), *Aust. J. Chem.* **17**, 1170.

Clark-Lewis, J.W. and Mahandru, M.M (1972). In *Some Recent*

Developments in the Chemistry of Natural Products, Prentice-Hall, Englewood Cliffs, New Jersey, pp. 3–11.
Clark-Lewis, J.W. and Mitsuno, M. (1958), *J. Chem. Soc.*, 1724.
Clark-Lewis, J.W. and Mortimer, P.I. (1960), *J. Chem. Soc.*, 4106.
Cooke, R.G. and Down, J.G. (1971), *Aust. J. Chem.* **24**, 1257.
Coxon, D.T., Holmes, A., Ollis, W.D., Vora, V.C., Grant, M.S. and Tee, J.L. (1972), *Tetrahedron* **28**, 2819.
Coxon, D.T., O'Neill, T.M., Mansfield, J.W. and Porter, A.E.A. (1980), *Phytochemistry* **19**, 889.
Czochanska, Z., Foo, L.Y. and Porter, L.J. (1979), *Phytochemistry* **18**, 1819.
Czochanska, Z., Foo. L.Y., Newman, R.H., and Porter, L.J. (1980), *J. Chem. Soc., Perkin 1*, 2278.
de Almeida, M.E., Gottlieb, O.R., de Sousa, J.R. and Teixeira, M.A. (1974), *Phytochemistry* **13**, 1225.
Dean, F.M., Mongkolsuk, S. and Podimuang, V. (1965), *J. Chem. Soc.*, 828.
Delcour, J.A., Ferreira, D. and Roux, D.G. (1983), *J. Chem. Soc., Perkin 1*, 1711.
Delcour, J.A., Serneels, E.J., Ferreira, D. and Roux, D.G. (1985), *J. Chem. Soc., Perkin 1*, 669.
Delgado, G., Alvarez, L. and Romode Vivar, A. (1984), *Phytochemistry* **23**, 675.
Delle Monache, F., D'Albuqueque, I.L., Ferrari, F. and Marini-Bettolo, G.B. (1967), *Tetrahedron Lett.*, 4211.
Delle Monache, F., Ferrari, F. and Marini-Bettolo, G.B. (1971), *Gazz. Chim. Ital.* **101**, 387.
Delle Monache, F., Ferrari, F., Poce-Tucci, A. and Marini-Bettolo, G.B. (1972), *Phytochemistry* **11**, 2333.
Delle Monache, F., Pomponi, M., Marini-Bettolo, G.B. D'Albuquerque, I.L. and de Lima, O.G., (1976), *Phytochemistry* **15**, 573.
Derdelinckx, G. and Jerumanis, J. (1984), *J. Chromatogr.* **285**, 231.
do Nascimento, M.C. and Mors, W.B. (1981), *Phytochemistry* **20**, 147.
Doskotch, R.W., Mikhail, A.A. and Chatterji, S.K. (1973), *Phytochemistry* **12**, 1153.
Drewes, S.E. and Ilsley, A.H. (1969), *Phytochemistry* **8**, 1039.
Drewes, S.E. and Roux, D.G. (1964), *Biochem. J.* **90**, 343.
Drewes, S.E. and Roux, D.G. (1965), *Biochem. J.* **96**, 681.
Drewes, S.E. and Roux, D.G. (1966a), *J. Chem. Soc. (C)*, 1644.
Drewes, S.E. and Roux, D.G. (1966b), *Biochem. J.* **98**, 493.
Drewes, S.E. and Roux, D.G. (1967), *J. Chem. Soc. (C)*, 1407.
du Preez, I.C. and Roux, D.G. (1970), *J. Chem. Soc. (C)*, 1800.
Ellis, C.J., Foo, L.Y. and Porter, L.J. (1983), *Phytochemistry* **22**, 483.
Ezaki, N., Kato, M., Takizawa, N., Morimoto, S., Nonaka, G. and Nishioka, I. (1985), *Plant. Med.*, 34.
Ferreira, D., van der Merwe, J.P. and Roux, D.G. (1974), *J. Chem. Soc., Perkin 1*, 1492.
Ferreira, D., du Preez, I.C., Wijamaalen, J.C. and Roux, D.G. (1985), *Phytochemistry* **24**, 2415.
Fletcher, A.C., Porter, L.J., Haslam, E. and Gupta, R.K. (1977), *J. Chem. Soc., Perkin 2*, 1628.
Foo, L.Y. (1986a), *J. Chem. Soc., Chem. Commun.*, 236.
Foo, L.Y. (1986b), *J. Chem. Soc., Chem. Commun.*, 675.
Foo, L.Y. and Hemingway, R.W. (1984), *J. Chem. Soc., Chem. Commun.*, 85.
Foo, L.Y. and Porter, L.J. (1978), *J. Chem. Soc., Perkin 1*, 1186.
Foo, L.Y. and Porter, L.J. (1980), *Phytochemistry* **19**, 1747.
Foo, L.Y. and Porter, L.J. (1981), *J. Sci. Food Agric.* **32**, 711.
Foo, L.Y. and Porter, L.J. (1983), *J. Chem. Soc., Perkin 1*, 1535.
Foo, L.Y. and Wong, H. (1986), *Phytochemistry* **25**, 1961.
Foo, L.Y., Jones, W.T., Porter, L.J. and Williams, V.M. (1982), *Phytochemistry* **21**, 93.
Foo, L.Y., McGraw, G.W. and Hemingway, R.W. (1983), *J. Chem. Soc., Chem. Commun.*, 672.
Foo, L.Y., Hrstich, L.N. and Vilain, C. (1985), *Phytochemistry* **24**, 1495.
Fourie, T.G., du Preez, I.C. and Roux, D.G. (1972), *Phytochemistry* **11**, 1763.
Fourie, T.G., Ferreira, D. and Roux, D.G. (1974), *Phytochemistry* **13**, 2573.
Franca, N.C., Diaz Diaz, P.P., Gottlieb, O.R. and Rosa, B.P. (1974), *Phytochemistry* **13**, 1631.
Freudenberg, K. and Purrmann, L. (1923), *Chem. Ber.* **56**, 1185.
Friedrich, H. and Engelstowe, R. (1978), *Plant. Med.* **33**, 251.
Friedrich, H. and Krueger, N. (1974), *Plant. Med.* **25**, 138.
Gaffield, W., Waiss, A.C. and Tominaga, T. (1975), *J. Org. Chem.* **40**, 1057.
Geiger, H. and Quinn, C. (1982). In *The Flavonoids – Advances in Research* (eds J.B. Harborne and T.J. Mabry), Chapman and Hall, London, p. 505.
Ghosal, S., Saini, K.S. and Sinha, B.N. (1983), *J. Chem. Res.* (S), 330; (M), 2601.
Ghosal, S., Jaiswal, D.K., Singh, S.K. and Srivastava, R.S. (1985a), *Phytochemistry* **24**, 831.
Ghosal, S., Singh, S.K. and Srivastava, R.S. (1985b), *Phytochemistry* **24**, 151.
Gomez, F., Quijano, L., Garcia, G., Calderon, J.S. and Rios, T. (1983), *Phytochemistry* **22**, 1305.
Gomez, F., Calderon, J., Quijano, L., Cruz, O. and Rios, T. (1984), *Chem. Ind.*, 632.
Gomez, F., Quijano, L., Calderon, J.S., Rodriquez, C. and Rios, T. (1985), *Phytochemistry* **24**, 1057.
Gujer, R., Magnolato, D. and Self, R. (1986), *Phytochemistry* **25**, 1431.
Gupta, R.K. and Haslam, E. (1981), *J. Chem. Soc., Perkin 1*, 1148.
Haslam, E. (1969), *J. Chem. Soc., (C)*, 1824.
Haslam, E. (1975). In *The Flavonoids* (eds J.B. Harborne and T.J. Mabry), Chapman and Hall, London, p. 505.
Haslam, E. (1982a). In *The Flavonoids – Advances in Research* (eds J.B. Harborne and T.J. Mabry), Chapman and Hall, London, p. 417.
Haslam, E. (1982b), *Fortschr. Chem. Org. Naturst.* **41**, 1.
Haslam, E., Opie, C.T. and Porter, L.J. (1977), *Phytochemistry* **16**, 99.
Heller, W., Britsch, L., Forkmann, G. and Grisebach, H. (1985a), *Planta* **163**, 191.
Heller, W., Forkman, G., Britsch, L. and Grisebach, H. (1985b), *Planta* **165**, 284.
Hemingway, R.W. and Foo, L.Y. (1983), *J. Chem. Soc., Chem. Commun.*, 1035.
Hemingway, R.W. and Laks, P.E. (1985), *J. Chem. Soc., Chem. Commun.*, 746.
Hemingway, R.W., Foo, L.Y. and Porter, L.J. (1982), *J. Chem. Soc., Perkin 1*, 1209.
Hemingway, R.W., McGraw, G.W., Karchesy, J.J., Foo, L.Y. and Porter, L.J. (1983), *J. Appl. Polym. Sci: Appl. Polym. Symp.* **37**, 967.
Hikino, H., Shimoyama, N., Kasahara, Y., Takahashi, M. and Konno, C. (1982a), *Heterocycles* **19**, 1381.

Hikino, H., Takahashi, M. and Konno, C. (1982b), *Tetrahedron Lett.* **23**, 673.
Hillis, W.E. and Carle, A. (1960), *Aust. J. Chem.* **13**, 390.
Hillis, W.E. and Inoue, T. (1967), *Phytochemistry* **6**, 59.
Homberger, K. and Hesse, M. (1984), *Helv. Chim. Acta* **67**, 237.
Hsu, F., Nonaka, G. and Nishioka, I. (1985a), *Chem. Pharm. Bull.* **33**, 3142.
Hsu, F., Nonaka, G. and Nishioka, I. (1985b), *Chem Pharm. Bull*, **33**, 3293.
Hsu, F., Nonaka, G. and Nishioka, I. (1985c), *Phytochemistry* **23**, 2089.
Hundt, H.K.L. and Roux, D.G. (1981), *J. Chem. Soc., Perkin 1*, 1227.
Iankov, I.A. (1966), *Bull. Offic. Intern. Vigne. Vin* **40**, 1145.
Ikuta, J., Hano, Y. and Nomura, T. (1985), *Heterocycles* **23**, 2835.
IUPAC (1979), *Nomenclature of Organic Chemistry*, Pergamon, Oxford.
Jacobs, E., Ferreira, D. and Roux, D.G. (1983), *Tetrahedron Lett.* **24**, 4627.
Jacques, D., Opie, C.T., Porter, L.J. and Haslam, E. (1977), *J. Chem. Soc., Perkin 1*, 1637.
Kamal, G.M., Gunaherath, B., Gunatilaka, A.A.L., Sultanbawa, M.U.S. and Balasubramanian, S. (1982), *J. Nat. Prod.* **45**, 140.
Karl, C., Pedersen, P.A. and Muller, G. (1981), *Z. Naturforsch.* **36c**, 607.
Karl, C., Muller, G. and Pedersen, P.A. (1982), *Z. Naturforsch.* **37c**, 148.
Kasahara, Y. and Hikino, H. (1983), *Heterocycles* **20**, 1953.
Kasahara, Y., Shimoyama, N., Konno, C. and Hikino, H. (1983), *Heterocycles* **20**, 1741.
Kennedy, J.A., Munro, M.H.G., Powell, H.K.J., Porter, L.J. and Foo, L.Y. (1984), *Aust. J. Chem.* **37**, 885.
Kiatgrajai, P., Wellons, J.D., Gollob, L. and White, J.D. (1982), *J. Org. Chem.* **47**, 2910.
Kiehlmann, E., van der Merwe, P.J. and Hundt, H.K.L. (1983), *Org. Prep. Proc. Int.* **15**, 341.
Kijjoa, A., Giesbrecht, A.M., Gottlieb, O.R. and Gottlieb, H. (1981), *Phytochemistry* **20**, 1385.
King, F.E., Clark-Lewis, J.W. and Forbes, W.F. (1955), *J. Chem. Soc.*, 2948.
Kolodziej, H. (1983), *Tetrahedron Lett.* **24**, 1825.
Kolodziej, H. (1984), *Phytochemistry* **23**, 1745.
Kolodziej, H. Ferreira, D. and Roux, D.G. (1984), *J. Chem. Soc., Perkin 1*, 343.
Krishnamurthy, V. and Seshadri, T.R. (1966), *Tetrahedron* **22**, 2367.
Kristiansen, K.N. (1984), *Carlsberg Res. Commun.* **49**, 503.
Kristiansen, K.N. (1986), *Carlsberg Res. Commun.* **51**, 51.
Lam, J. and Wrang, P. (1975), *Phytochemistry* **14**, 1621.
Lea, A.G.H. (1978), *J. Sci. Food Agric.* **29**, 471.
Lea, A.G.H. (1980), *J. Chromatogr.* **194**, 62.
Lea, A.G.H. (1982), *J. Chromatogr.* **238**, 253.
McGraw, G.W. and Hemingway, R.W. (1982), *J. Chem. Soc., Perkin 1*, 973.
Malan, E. and Roux, D.G. (1974), *Phytochemistry* **13**, 1575.
Marletti, F., Delle Monache, F., Marini-Bettolo, G.B. and D'Albuquerque, I.L. (1976), *Phytochemistry* **15**, 443.
Mattice, W.L. and Porter, L.J. (1984), *Phytochemistry* **23**, 1309.
Mayer, W., Goll, L., von Arndt, E.M. and Mannschreck, A. (1966), *Tetrahedron Lett.*, 429.
Merlini, L. and Nasini, G. (1976), *J. Chem. Soc., Perkin 1*, 1570.
Middelkoop, T.B. and Labadie, R.P. (1985), *Z. Naturforsch.* **40b**, 855.
Miyauchi, Y., Yoshimoto, T. and Minami, K. (1976), *Mokuzai Gakkaishi*, **22**, 47.
Miyamura, M., Nohara, T., Tomimatsu, T. and Nishioka, I. (1983), *Phytochemistry* **22**, 215.
Morimoto, S., Nonaka, G. and Nishioka, I. (1985a), *Chem. Pharm. Bull.* **33**, 4338.
Morimoto, S., Nonaka, G., Nishioka, I., Ezaki, N. and Takizawa, N. (1985b), *Chem. Pharm. Bull.* **33**, 2281.
Morimoto, S., Nonaka, G. and Nishioka, I. (1986a), *Chem. Pharm. Bull.* **34**, 633.
Morimoto, S., Nonaka, G. and Nishioka, I. (1986b), *Chem. Pharm. Bull.* **34**, 643.
Mulkay, P., Touillaux, R. and Jerumanis, J. (1981), *J. Chromatogr.* **208**, 419.
Murakami, T., Wada, H., Tanaka, N., Kuraishi, T., Saiki, Y. and Chen, C. (1985), *Yakugaku Zasshi* **105**, 649.
Nishioka, I. (1983), *Yakugaku Zasshi* **103**, 125.
Nonaka, G. and Nishioka, I. (1980), *Chem. Pharm. Bull.* **28**, 3145.
Nonaka, G. and Nishioka, I. (1982), *Chem. Pharm. Bull.* **30**, 4268.
Nonaka, G., Hsu, F. and Nishioka, I. (1981a), *J. Chem. Soc., Chem. Commun.*, 781.
Nonaka, G., Nishioka, I., Nagasawa, T. and Oura, H. (1981b), *Chem. Pharm. Bull.* **29**, 2862.
Nonaka, G., Kawahara, O. and Nishioka, I. (1982a), *Chem. Pharm. Bull.* **30**, 4277.
Nonaka, G., Miwa, N. and Nishioka, I. (1982b), *Phytochemistry* **21**, 429.
Nonaka, G., Ezaki, E., Hayashi, K. and Nishioka, I. (1983a), *Phytochemistry* **22**, 1659.
Nonaka, G., Kawahara, O. and Nishioka, I. (1983b), *Chem. Pharm. Bull.* **31**, 3906.
Nonaka, G., Morimoto, S. and Nishioka, I. (1983c), *J. Chem. Soc., Perkin 1*, 2139.
Nonaka, G., Muta, M. and Nishioka, I. (1983d), *Phytochemistry* **22**, 237.
Nonaka, G., Sakai, R. and Nishioka, I. (1984), *Phytochemistry* **23**, 1753.
Nonaka, G., Nishimura, H. and Nishioka, I. (1985), *J. Chem. Soc., Perkin 1*, 163.
Numata, A., Takemura, T., Ohbayashi, H., Katsuno, T., Yamamoto, K., Sato, K. and Kobayashi, S. (1983), *Chem. Pharm. Bull.* **31**, 2146.
Oberholzer, M.E., Rall, G.J.H. and Roux, D.G. (1980), *Phytochemistry* **19**, 2503.
Ogawa, M., Hisada, S. and Inagaki, I. (1970), *Yakugaku Zasshi* **90**, 1081.
Ohigashi, H., Minami, S., Fukui, H., Koshimizu, K., Mizutani, F., Sugiura, A. and Tomana, T. (1982), *Agric. Biol. Chem.* **46**, 2555.
Otsuka, H., Fujioka, S., Komiya, T., Mizuta, E. and Takumoto, M. (1982), *Yakugaku Zasshi* **102**, 162.
Outtrup, H. and Schaumberg, K. (1981), *Carlsberg Res. Commun.* **46**, 43.
Pashinana, L.T., Chumbalov, T.K. and Leiman, Z.A. (1970), *Khim. Prirod. Soed.* **6**, 478.
Patil, A.D. and Deshpande, V.H. (1982), *Indian J. Chem.* **21B**, 626.
Pelter, A., Amenechi, P.I., Warren, R. and Harper, S.H. (1969), *J. Chem. Soc. (C)*, 2572.
Platt, R.V., Opie, C.T. and Haslam, E. (1984), *Phytochemistry* **23**, 2211.

Poce Tucci, A., Delle Monache, F. and Marini-Bettolo, G.B. (1969), *Ann. Ist. Sanita* **5**, 555.
Polysaccharide nomenclature, 1980 (1982), *J. Biol. Chem.* **257**, 3352.
Porter, L.J. (1974), *N.Z. J. Sci.* **17**, 213.
Porter, L.J. (1984), *Rev. Latinomer. Quim.* **15**, 43.
Porter, L.J. and Foo, L.Y. (1982), *Phytochemistry* **21**, 2947.
Porter, L.J., Newman, R.H., Foo, L.Y., Wong, H. and Hemingway, R.W. (1982), *J. Chem. Soc., Perkin 1*, 1217.
Porter, L.J., Foo, L.Y. and Furneaux, R.H. (1985a), *Phytochemistry* **24**, 567.
Porter, L.J., Wong, R.Y. and Chan, B.G. (1985b), *J. Chem. Soc., Perkin 1*, 1413.
Porter, L.J., Hrstich, L.N. and Chan, B.G. (1986), *Phytochemistry* **25**, 223.
Robinson, G.M. and Robinson, R. (1935), *J. Chem. Soc.*, 744.
Roux, D.G. (1986), Private communication.
Roux, D.G. and de Bruyn, G.C. (1963), *Biochem. J.* **87**, 439.
Roux, D.G. and Maihs, E.A. (1960), *Biochem. J.* **74**, 44.
Roux, D.G. and Paulus, E. (1961), *Biochem. J.* **78**, 120.
Roux, D.G. and Paulus, E. (1962), *Biochem. J.* **82**, 324.
Roux, D.G., Maihs, E.A. and Paulus, E. (1961), *Biochem. J.* **78**, 834.
Saayman, H.M. and Roux, D.G. (1965), *Biochem. J.* **96**, 36.
Sahai, R., Agarawal, S.K. and Rastogi, R.P. (1980), *Phytochemistry* **19**, 1560.
Saijo, R. (1982), *Agric. Biol. Chem.* **46**, 1969.
Saini, K.S. and Ghosal, S. (1984), *Phytochemistry* **23**, 2415.
Sakurai, A., Okada, K. and Okumura, Y. (1982), *Bull. Chem. Soc. (Japan)* **55**, 3051.
Samaraweera, U., Sotheeswaran, S. and Sultanbawa, M.U. (1983), *Phytochemistry* **22**, 565.
Samejima, M. and Yoshimoto, T. (1979), *Mokuzai Gakkaishi* **25**, 671.
Sashida, Y., Yamamoto, T., Koike, C. and Shimomura, H. (1976), *Phytochemistry* **15**, 1185.
Sears, K.D. (1972), *J. Org. Chem.* **37**, 3546.
Sears, K.D., Casebier, R.L., Hergert, H.L., Stout, G.H. and McCandlish, L.E. (1974), *J. Org. Chem.* **39**, 3244.
Sethi, V.K., Taneja, S.C., Dhar, K.L. and Atal, C.K. (1984), *Phytochemistry* **23**, 2402.
Shen, Z., Falshaw, C.P., Haslam, E. and Begley, M.J. (1985), *J. Chem. Soc., Chem. Commun.*, 1135.
Snatzke, G., Kajtar, M. and Snatzke, F. (1973). In *Fundamental Aspects and Recent Developments in Optical Rotatory Dispersion and Circular Dichroism* (eds F. Ciardelli and P. Salvadori) Heyden, London, p. 148.
Stafford, H.A. (1983), *Phytochemistry* **22**, 2643.
Stafford, H.A. and Lester, H.H. (1984), *Plant Physiol.* **76**, 184.
Stafford, H.A. and Lester, H.H. (1985), *Plant Physiol.* **78**, 791.
Stafford, H.A., Lester, H.H. and Porter, L.J. (1985), *Phytochemistry* **24**, 333.
Steenkamp, J.A., Steynberg, J.P., Brandt, E.V., Ferreira, D. and Roux, D.G. (1985), *J. Chem. Soc., Chem. Commun.*, 1678.
Steynberg, J.P., Ferreira, D. and Roux, D.G. (1983), *Tetrahedron Lett.* **24**, 4147.
Takasugi, M., Kumagai, Y., Nagao, S., Masamune, T., Shirata, A. and Takahashi, K. (1980), *Chem. Lett.*, 1459.
Takasugi, M., Niino, N., Nagao, S., Anetai, M., Masamune, T., Shirata, A. and Takahashi, K. (1984), *Chem. Lett.*, 689.
Tanaka, T., Nonaka, G. and Nishioka, I. (1983), *Phytochemistry* **22**, 2575.
Tanaka, N., Sada, T., Murakami, T., Saiki, Y. and Chen, C. (1984), *Chem. Pharm. Bull.* **32**, 490.
Tanaka, N., Murakami, T., Wada, H., Gutierrez, A.B., Saiki, Y. and Chen, C. (1985), *Chem. Pharm. Bull.* **33**, 5231.
Thompson, R.S., Jacques, D., Haslam, E. and Tanner, R.J.N. (1972), *J. Chem. Soc., Perkin 1*, 1387.
Tindale, M.D. and Roux, D.G. (1974), *Phytochemistry* **13**, 829.
Tschesche, R., Braun, T.M. and von Sassen, W. (1980), *Phytochemistry* **19**, 1825.
Uvarova, N.J., Jizba, J. and Herout, V. (1967), *Coll. Czech. Chem. Commun.* **32**, 3075.
van der Merwe, J.P., Ferreira, D., Brandt, E.V. and Roux, D.G. (1972), *J. Chem. Soc., Chem. Commun.*, 521.
van Heerden, F.R., Brandt, E.V., Ferreira, D. and Roux, D.G. (1981), *J. Chem. Soc., Perkin 1*, 2483.
Viviers, P.M., Young, D.A., Botha, J.J., Ferreira, D. and Roux, D.G. (1982), *J. Chem. Soc., Perkin 1*, 535.
Viviers, P.M., Botha, J.J., Ferreira, D., Roux, D.G. and Saayman, H.M. (1983a), *J. Chem. Soc., Perkin 1*, 17.
Viviers, P.M., Kolodziej, H., Young, D.A., Ferreira, D. and Roux, D.G. (1983b), *J. Chem. Soc., Perkin 1*, 2555.
Wallace, J.W. and Grisebach, H. (1973), *Biochim. Biophys. Acta* **304**, 837.
Waterman, P.G. and Faulkner, D.F. (1979), *Plant. Med.* **37**, 178.
Weeratunga, G., Bohlin, L., Verpoorte, R. and Kumar, V. (1985), *Phytochemistry* **24**, 2093.
Weinges, K., Goritz, K. and Nader, F. (1968a), *Annalen* **715**, 164.
Weinges, K., Kaltenhauser, W., Marx, H., Nader, E., Nader, F., Perner, J. and Seiler, D. (1968b), *Annalen* **711**, 184.
Weinges, K., Bahr, W., Ebert, W., Goritz, K. and Marx, H. (1969), *Fortschr. Chem. Org. Naturst.* **27**, 158.
Weinges, K. and Piretti, M.V. (1971), *Annalen* **748**, 218.
Wilkins, C.K. and Bohm, B.A. (1977), *Phytochemistry* **16**, 144.
Williams, V.M., Porter, L.J. and Hemingway, R.W. (1983), *Phytochemistry* **22**, 569.
Yahara, S., Saijo, R., Nohara, T., Konishi, R., Yamahara, J., Kawasaki, T. and Miyahara, K. (1985), *Chem. Pharm. Bull.* **33**, 5130.
Young, D.A., Ferreira, D., Roux, D.G. and Hull, W.E. (1985a), *J. Chem. Soc., Perkin 1*, 2529.
Young, D.A., Kolodziej, H., Ferreira, D. and Roux, D.G. (1985b), *J. Chem. Soc., Perkin 1*, 2537.

3

C-glycosylflavonoids

J. CHOPIN and G. DELLAMONICA

3.1 INTRODUCTION

The last five years have seen a steady growth of the number of *C*-glycosylflavonoids known to occur in plants, which now amounts to a little more than three hundred. New natural *C*-glucosyl-, *C*-galactosyl-, *C*-xylosyl-, *C*-arabinosyl- and *C*-rhamnosyl-flavonoids have been reported to occur either as free compounds or as *O*-glycosides or *O*-acyl derivatives. The *C*-β-L-arabinopyranosyl residue (Fig. 3.1) has been identified in the 'neo' isomer of schaftoside as the result of the characterization of the four 6-*C*-α and β-L-arabinopyranosyl and furanosylacacetins produced by the acid isomerization of synthetic 6-*C*-α-L-arabinopyranosylacacetin (Besson and Chopin, 1983; Besson *et al.*, 1984). Several new types of *C*-glycosylflavonoids have been described: *C*-glycosyl-α- and β-hydroxydihydrochalcones, *C*-glycosylquinochalcone, *C*-glycosylflavanol, *C*-glycosylprocyanidin, di-*C*-glycosylflavanone, di-*C*-glycosylquinochalcone. In di-*C*-glycosylflavones, an unusual type of substitution has been reported with the isolation of 3, 6- and 3, 8-di-*C*-glucosylflavones from *Citrus* peels. In *C*-glycosylflavonoid *O*-glycosides, two sugars have been found *O*-linked for the first time: D-mannose and apiofuranose. In *O*-acyl derivatives, the new 2, 4, 5-trihydroxycinnamoyl group has been identified.

The natural sources of *C*-glycosylflavonoids reported in the last five years (or omitted in the preceding volume) are listed in Table 3.1. The number of species containing *C*-glycosylflavonoids is given for each genus, except in the case of the large family surveys that have been carried out in the Araceae, Cyperaceae, Iridaceae, Palmae, Oleaceae (Harborne and co-workers) and Commelinaceae (Swain and co-workers). Continued interest has been given to Hepaticae by Markham, Mues, Zinsmeister and co-workers, to pteridophytes by Wallace and Markham, to Gnetopsida by Wallace, to Podocarpaceae by Markham, to Gentianaceae by Hostettmann, Molho and Chulia, to Labiatae by Husain and Markham, to Polemoniaceae by Smith. The following genera have been intensively studied: *Radula* (Mues), *Oryza* (Boyer), *Hoya* (Niemann), *Silene* (van Brederode), *Achillea* and *Leucocyclus* (Valant-Vetschera), *Lupinus* (Bohm, Williams), *Prosopis* (Juliani), *Rynchosia* (Adinarayana), *Mollugo* (Nair), *Passiflora* (Mabry), and *Crataegus* (Nikolov).

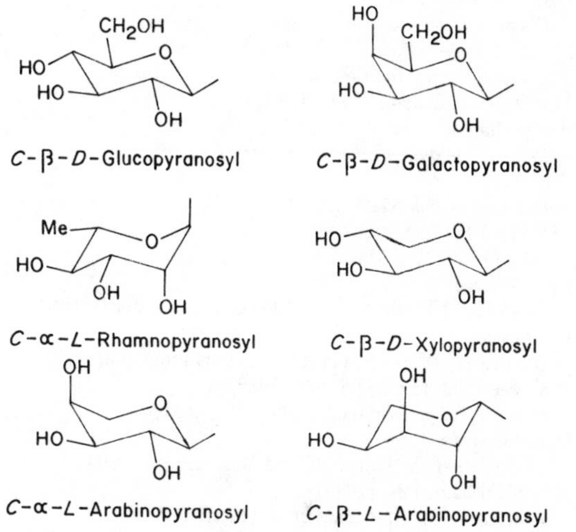

At the species level, *Avena sativa* and *Secale cereale* (Weissenböck), *Hordeum vulgare and Fagopyrum*

Fig. 3.1 *C*-Glycosyl residues in new natural *C*-glycosylflavonoids.

The Flavonoids. Edited by J.B. Harborne
Published in 1988 by Chapman and Hall Ltd,
11 New Fetter Lane, London EC4P 4EE.

Table 3.1 Natural sources of *C*-glycosylflavonoids*

CHLOROPHYTA
Chaetophoraceae, Chaetosiphonaceae O'Kelly (1982)
BRYOPHYTA
Apometzgeria pubescens Theodor *et al.* (1981a)
Blepharostoma trichophyllum Mues (1982b)
Conocephalum conicum Porter (1981)
Frullania (2) Mues *et al.* (1984)
Metzgeria (3) Theodor *et al.* (1981b, 1983); Markham *et al.* (1982)
Plagiomnium affine Freitag *et al.* (1986)
Porella (7) Mues (1982b)
Radula (10) Markham and Mues (1984); Mues (1984)
Trichocolea (4) Mues (1982a, b)

PTERIDOPHYTA
Angiopteridaceae
Angiopteris (4) Wallace *et al.* (1981)
Daennstaedtiaceae
Daennstaedtia scandens Tanaka *et al.* (1980)
Lycopodiaceae
Lycopodium cernuum Markham *et al.* (1983)
Marsileaceae Wallace *et al.* (1984)
Polypodiaceae
Polypodium vulgare Karl *et al.* (1982)
Psilotaceae
Psilotum nudum Markham (1984)
SPERMAPHYTA
GYMNOSPERMAE
Ephedraceae
Ephedra (2) Wallace *et al.* (1982); Nawwar *et al.* (1984)
Gnetaceae
Gnetum (3) Wallace *et al.* (1982); Ouabonzi *et al.* (1983)
Pinaceae Lebreton and Sartre (1983)
Larix leptolepis Niemann (1980)
Podocarpaceae
Dacrycarpus (3) Markham and Whitehouse (1984); Markham *et al.* (1985)
Podocarpus (5) Markham *et al.* (1984, 1985)
Welwitschiaceae
Welwitschia mirabilis Wallace *et al.* (1982)
ANGIOSPERMAE
MONOCOTYLEDONAE
Araceae
(82% of 144 species from 58 genera) Williams *et al.* (1981)
Arisarum vulgare Pagani (1982b)
Arum orientale Koleva *et al.* (1982, 1984)
Bromeliaceae
Tillandsia utriculata Ulubelen and Mabry (1982)
Cyperaceae
Cyperus (5% of 92 species) Harborne *et al.* (1982)
(4% of 160 species from 35 genera outside *Cyperus*) Harborne *et al.* (1985)
Kyllinga brevifolia Huang *et al.* (1980)
Trichophorum cespitosum Salmenkallio *et al.* (1982)
Gramineae
Aeluropus lagopoides Abdalla and Mansour (1984)
Arrhenaterum (5) Ismaili (1982)
Avena sativa Proksch *et al.* (1981); Knogge and Weissenböck (1984, 1986)
Cymbopogon citratus Gunasingh *et al.* (1981b)
Dactylis glomerata Jay *et al.* (1984b); Ardouin (1985)
Hordeum vulgare Laanest and Vainjärv (1980); Laanest (1981a,b); Margna *et al.* (1981)
Oryza (23) Boyet (1985)
Oryza sativa Besson *et al.* (1985); Kim *et al.* (1985)
Poa annua Rofi and Pomilio (1986)
Saccharum (sugar cane) Mabry *et al.* (1984) Ulubelen *et al.* (1985)
Secale cereale Strack *et al.* (1982); Dellamonica *et al.* (1983); Schulz and Weissenböck (1986)
Setaria (2) Gluchoff-Fiasson (1986)
Stenotaphrum secundatum Ferreres *et al.* (1984)
Triticum aestivum Wagner *et al.* (1980); Neuman *et al.* (1983)
Iridaceae (66% of 255 species from 57 genera) Williams *et al.* (1986)
Gemmingia chinensis Shirane *et al.* (1982)
Iris (4) Hirose *et al.* (1981); Fujita and Inoue (1982); Pryakhina and Blinova (1984); Pryakhina *et al.* (1984); Kitanov and Pryakhina (1985).
Liliaceae
Ornithogalum gussonei Bandyukova (1979)
Palmae
Cocosoideae (77% of 52 species) Williams *et al.* (1983b)
Attalea (11) Williams *et al.* (1983b, 1985)
Orbigyna (3) Williams *et al.* (1983b, 1985)
Plectocomia (14) Madulid (1980)
Rhapis (2) Hirai *et al.* (1984)
Scheelea (7) Williams *et al.* (1983b, 1985)
DICOTYLEDONAE
Acanthaceae
Siphonoglossa sessilis Hilsenbeck and Mabry (1983)
Yeatesia viridiflora Hilsenbeck *et al.* (1984)
Aizoaceae
Gliscrothamnus ulei Richardson (1981)
Apocynaceae
Trachelospermum jasminoides Sakushima *et al.* (1982)
Asclepiadaceae
Hoya (13) Baas *et al.* (1981); Niemann *et al.* (1981)
Marsdenia cundurango Koch and Steinegger (1982)
Boraginaceae
Arnebia hispidissima Prabhakar *et al.* (1981)
Caryophyllaceae
Cerastium arvense Dubois *et al.* (1982a, b, 1983, 1984, 1985)
Dianthus (9) Boguslavskaya *et al.* (1983a,b)
Lychnis (2) Ferry and Darbour (1980); Balabanova-Radonova *et al.* (1981, 1982);
Malachium aquaticum Darmograi (1980b)
Minuartia (9) Wolf *et al.* (1979); Darmograi (1980b)
Sagina japonica Zhuang (1983)
Silene pratensis (Silene alba, Melandrium album) Niemann *et al.* (1980, 1983); van Brederode and Kamps-Heinsbroek (1981a,b); Niemann (1981, 1982, 1984a,b); van Brederode *et al.* (1982a,b); Mastenbroek *et al.* (1982); van Brederode and Kooten (1983); van Brederode and Mastenbroek (1983); van Brederode and Steyns (1983); van Genderen *et al.* (1983a,b); Mastenbroek *et al.* (1983a,b,c); Steyns *et al.* (1983, 1984); van

Table 3.1 (*Contd.*)

Brederode and Steyns (1985); Steyns and van Brederode (1986)
Silene spp. (6) van Brederode and Mastenbroek (1983); Mastenbroek *et al.* (1983b).
Pseudostellaria palibiniana Yang and Kim (1983)
Spergularia (2) Darmograi (1980a); Bouillant *et al.* (1984)
Stellaria (2) Yasukawa *et al.* (1982); Bouillant *et al.* (1984); Boguslavskaya *et al.* (1985)

Chloranthaceae
Ascarina lucida Soltis and Bohm (1982)

Clusiaceae (Guttiferae)
Garcinia kola Olaniyi and Phillips (1979)
Hypericum (3) Kitanov *et al.* (1979); Calie *et al.* (1983)
Ochrocarpus longifolius Prabakhar *et al.* (1981); Roy *et al.* (1983)

Combretaceae
Combretum micranthum Bassene *et al.* (1985)

Commelinaceae (78% of 152 species) Del Pero Martinez and Swain (1985)

Compositae
Achillea Valant-Vetschera (1982a,b, 1984, 1985a,b,c)
Artemisia (3) Liu *et al.* (1982); Saleh *et al.* (1985)
Carlina (3) Dombris and Raynaud (1980, 1981a,b); Raynaud *et al.* (1981); Proliac and Raynaud (1983)
Carthamus tinctorius Takahashi *et al.* (1982, 1984)
Centaurea (4) Kamanzi *et al.* (1983a,b); Oksuz *et al.* (1984); Gonnet (1986)
Coreopsis Crawford and Smith (1983)
Gutierrezia (2) Fang *et al.* (1986a,b)
Haplopappus (2) Ates *et al.* (1982)
Launaea asplenifolia Gupta *et al.* (1985)
Leucanthemum vulgare Sagareishvili *et al.* (1982)
Leucocyclus Valant-Vetschera (1982b)
Notobasis syriaca Mericli and Dellamonica (1983)
Tithonia calva La Duke (1982)
Vernonia (9) Gunasingh *et al.* (1981a); King and Jones (1982)

Cruciferae
Isatis tinctoria Raynaud and Prum (1980)
Lepidium ruderale Kurkin *et al.* (1981)

Cucurbitaceae
Cayaponia tayuya Bauer *et al.* (1985)

Degeneriaceae
Degeneria vitiensis Young and Sterner (1981)

Dilleniaceae
Dillenia (1)
Doliocarpus (7) Gurni and Kubitzki (1981)

Dipsacaceae
Cephalaria (7) Aliev and Movsumov (1981)
Scabiosa (10) Aliev and Movsumov (1981)

Euphorbiaceae
Euphorbia (2) Del Galarza *et al.* (1983)

Gentianaceae
Enicostemma hyssopifolium Ghosal and Jaiswal (1980)
Frasera tetrapetala Agata *et al.* (1984)
Gentiana Nikolaeva *et al.* (1980); Hostettmann-Kaldas *et al.* (1981); Luong Minh Duc *et al.* (1981); Massias *et al.* (1981, 1982) Chulia (1984); Oyuungerel *et al.* (1984); Chulia and Mariotte (1985)
Lomatogonium carinthiacum Schaufelberger and Hostettmann (1984)
Swertia (3) Kubota *et al.* (1983); Khetwal and Verma (1984); Snyder *et al.* (1984)

Globulariaceae
Globularia alypum Hassine *et al.* (1982)

Idiospermaceae
Idiospermum australiense Young and Sterner (1981)

Labiatae
Hyssopus officinalis Husain and Markham (1981)
Ocimum sanctum Nair *et al.* (1982)
Origanum (5) Husain and Markham (1981); Husain *et al.* (1982)
Perilla frutescens Aritomi (1982)
Phlomis (2) El-Negoumy *et al.* (1986)
Salvia triloba Abdalla *et al.* (1983)
Scutellaria baicalensis Takagi *et al.* (1981)
Thymbra capitata Barberan *et al.* (1986)
Thymus (15) Husain and Markham (1981); Barberan *et al.* (1985)

Lauraceae
Cinnamomum cassia Morimoto *et al.* (1986a, b)

Leguminosae
Abrus precatorius Bhardwaj *et al.* (1980)
Acacia (4) Lorente *et al.* (1982, 1983); Suarez *et al.* (1982)
Amorpha fruticosa Rozsa *et al.* (1982)
Carmichaelia Purdie (1984)
Cassia sophora Tiwari and Bajpai (1981)
Cladrastis shikokiana Ohashi *et al.* (1980)
Coronilla Kovalev and Komissarenko (1983)
Crotalaria retusa Srinivasan and Subramanian (1983)
Dalbergia nitidula Van Heerden *et al.* (1980)
Desmodium (3) Adinarayana and Syamasundar (1982); Naderuzzaman (1983); Sreenivasan and Sankarasubramanian (1984)
Eysenhardtia polystachya Beltrami *et al.* (1982)
Genista patula Ozimina (1981)
Glycine Jay *et al.* (1984a); Vaughan and Hymowitz (1984)
Glycyrrhiza (3) Afchar *et al.* (1984a,b); Yahara and Nishioka (1984)
Heylandia latebrosa Sebastian *et al.* (1984)
Lathyrus pratensis Ismaili *et al.* (1981)
Lespedeza capitata Linard *et al.* (1982)
Lupinus (88% of 65 species) Nicholls and Bohm (1983); (55% of 54 specimens of 11 species) Williams *et al.* (1983a)
Ononis spinosa Baztan *et al.* (1981)
Prosopis (7) Gitelli *et al.* (1981); Chiale *et al.* (1984)
Psophocarpus tetragonolobus Lachman *et al.* (1982)
Pterocarpus marsupium Adinarayana *et al.* (1982); Bezuidenhoudt *et al.* (1986)
Pueraria Fang *et al.* (1983)
Rynchosia (7) Adinarayana and Ramachandraiah (1985a,b); Adinarayana *et al.* (1985)
Trigonella (2) Bandyukova *et al.* (1985)
Vigna (2) Ishikura *et al.* (1981)

Loranthaceae
Phoradendron tomentosum Dossaji *et al.* (1983)

(*Contd.*)

Table 3.1 (*Contd.*)

Lythraceae	Rhamnaceae
Ammania coccinea Graham *et al.* (1980)	*Zizyphus* Woo *et al.* (1980a,b); Okamura *et al.* (1981); Shin *et al.* (1982)
Martyniaceae	Rosaceae
Martynia proboscidata Pagani (1982a)	*Crataegus* Pangarova *et al.* (1980); Nikolov *et al.* (1981, 1982a,b); Simova and Pangarova (1983); Ficarra *et al.* (1984a,b); Hiermann and Katnig (1984); Kashnikova *et al.* (1984)
Molluginaceae	Rutaceae
Mollugo (4) Chopin *et al.* (1982); Singh *et al.* (1982); Nair and Gunasegaran (1983a,b); Chopin *et al.* (1984)	*Almeidea guyanensis* Wirasutisna *et al.* (1986)
Moraceae	*Citrus* (3) Kumamoto *et al.* (1984, 1985a); Matsubara *et al.* (1984, 1985a,b)
Ficus carica Siewek *et al.* (1985)	*Fortunella japonica* Kumamoto *et al.* (1985b)
Ochnaceae	*Monnieria trifolia* Keita *et al.* (1985)
Ochna squarrosa Mohammad *et al.* (1982)	*Vepris heterophylla* Gomes *et al.* (1983)
Oleaceae	Sterculiaceae
Ligustrum sempervirens (1 of 97 species) Harborne and Green (1980)	*Kleinhovia hospita* Ramesh and Subramanian (1984)
Olea europaea Touati (1985)	Theaceae
Phillyrea angustifolia Touati (1985)	*Camellia sinensis* Chkhikvishvili *et al.* (1985)
Onagraceae	Thymeleaceae
Gaura triangulata Hilsenbeck *et al.* (1984)	*Daphne* (2) Cabrera and Garcia-Granados (1981); Ulubelen *et al.* (1982a)
Ludwigia (5) Huang (1985)	*Thymelaea tartonraira* Garcia-Granados and Saenz de Buruaga (1980)
Papilionaceae	*Wikstroemia chamaedaphne* Qin *et al.* (1981, 1982)
Retama raetam Abdalla and Saleh (1983)	Ulmaceae
Passifloraceae	*Trema cannabina* Higa *et al.* (1983)
Adenia mannii Ulubelen *et al.* (1982c)	Umbelliferae
Passiflora (46) Escobar (1980); Ulubelen *et al.* (1981, 1982b,c,d, 1984); McCormick and Mabry (1981, 1982, 1983); Ayanoglu *et al.* (1982); Escobar *et al.* (1983); Ulubelen and Mabry (1983); Oga *et al.* (1984)	*Daucus littoralis* Saleh *et al.* (1983)
Polemoniaceae	Verbenaceae
Leptodactylon (3) Smith *et al.* (1982)	*Premna tomentosa* Jyotsna *et al.* (1984)
Linanthus (16) Smith *et al.* (1982)	*Vitex peduncularis* Sahu *et al.* (1984)
Polygonaceae	Vitaceae
Fagopyrum esculentum Margna and Margna (1980, 1982); Margna and Vainjärv (1981, 1983, 1985); Margna *et al.* (1983, 1985)	*Cissus rheifolia* Saifah *et al.* (1983)
Ranunculaceae	Winteraceae
Adonis (2) Komissarenko *et al.* (1981); Evdokimov (1979)	*Drimys winteri* Williams and Harvey (1982)
Trollius chinensis Kang *et al.* (1984)	*Tasmannia piperita* Williams and Harvey (1982)

*The number of species in which C-glycosylflavonoids have been found is enclosed in parentheses.

esculentum (Margna) have been the subject of metabolic studies. *Cerastium arvense* has been thoroughly analysed by Dubois and *Silene pratensis* has been extensively studied by van Brederode and co-workers, who have shown that there is a remarkable correlation between the glycosylation pattern of isovitexin and seed morphology. Two further isovitexin glycosylating loci controlling respectively 7-*O*-galactosylation and 2″-*O*-arabinosylation have been added to the three loci already known to control the binding of glucose and xylose to the 7-hydroxyl and the binding of arabinose, rhamnose and glucose to the 2″-hydroxyl group. Expression and regulation of these genes in the leaves as well as in petals and their biochemical and morphological consequences have been studied, leading to the identification of a new locus controlling 7-*O*-galactosylation in the stem, leaves and petals and of a new allele controlling 2″-*O*-xylosylation. Recent serological studies suggest the presence of common epitopes on isovitexin glycosyltransferases which differ in substrate specificity and are controlled by different and differently regulated genes. A critical review on the accumulation of *C*-glycosylflavones in plant tissues has been written by Valant-Vetschera (1985b). Vitexin, isovitexin, orientin, isoorientin and their respective 7-*O*-glucosides were identified in the wings and body of the butterfly *Melanargia galathea* (Wilson, 1985).

C-Glycosylflavones have been isolated as probing stimulants of planthoppers feeding on rice plants (Kim *et al.*, 1985) and vicenin 2 has been identified as an oviposition stimulant of *Papilio xuthus*, a *Citrus*-feeding swallowtail butterfly (Ohsugi *et al.*, 1986). By contrast, vitexin has been shown to deter the feeding of *Schizaphis graminum* and *Myzus persicae* on wheat (Dreyer and Jones, 1981). *C*-Glycosylflavones also inhibit the *in vitro* growth of *Phytophthora parasitica* (Ravisé and Chopin, 1981). Inhibitory effects of *C*-glycosylflavones on c-AMP phosphodiesterase activity of rat heart have been studied by Petkov *et al.* (1981). Hypotensive effects in animal tissues have been reported for vitexin (Prabhakar *et al.*, 1981) and *C*-glycosylflavonoids from *Citrus* fruit peels (Matsubara *et al.*, 1984), and sedative effects have been noted for spinosin from *Zizyphus* seeds (Woo *et al.*, 1980b).

The chemistry, occurrence and biosynthesis of *C*-glycosyl compounds in plants have been reviewed by Franz and Grün (1983). Apart from the low-yield *C*-glycosylation of 5,7-dihydroxyflavones and 5,7-dihydroxy-8-*C*-glycosylflavones still used for preparing specific *C*-glycosides, the only total synthesis described during the period under review is that of the unknown 7,4′-di-*O*-methylisobayin by Tschesche and Widera (1982). Chemical methods such as ferric chloride oxidation have still been used occasionally in the identification of the *C*-glycosyl residue, but in most cases ^{13}C NMR spectroscopy has been invaluable, when sufficient amounts of substance were available, for the simultaneous identification and location of the *C*-glycosyl, *O*-glycosyl and *O*-acyl residues. In other cases permethylation (or perdeuteriomethylation) and mass spectrometry of permethyl (or perdeuteriomethyl) ethers have been widely used. New applications of this technique have provided the means of differentiating furanosyl and pyranosyl isomers of 6-*C*-arabinosylflavones and naturally occurring 6-*C*-glycosyl-8-*C*-diglycosyl- and 6-*C*-diglycosyl-8-*C*-glycosyl-flavones. The effect of conformation and configuration of arabinose residues on the circular dichroism of *C*-glycosylflavones has been analysed by Gaffield *et al.* (1984). The difficult problem of differentiating *C*-glucosyl from *C*-galactosyl and *C*-xylosyl from *C*-arabinosyl residues in *C*-glycosylflavones, which required permethylation and thin layer chromatography of the permethyl derivatives, has now found a simpler solution in the use of reversed-phase high-performance liquid chromatography of the free compounds with isocratic elution (Lardy *et al.*, 1984).

3.2 NATURALLY OCCURRING *C*-GLYCOSYLFLAVONOIDS

Natural *C*-glycosylflavonoids can be divided into two groups: the mono- and di-*C*-glycosylflavonoids which are resistant to acid hydrolysis and their hydrolysable derivatives (*O*-glycosides and *O*-acyl derivatives). Mono-*C*-glycosylflavonoids are listed in Table 3.2, di-*C*-glycosylflavonoids in Table 3.3, *C*-glycosylflavonoid *O*-glycosides in Table 3.4 and *O*-acyl-*C*-glycosylflavonoids in Table 3.5. New structures are indicated with an asterisk.

3.2.1 Mono-*C*-glycosylflavonoids

Five new types of mono-*C*-glycosylflavonoids, *C*-glycosyl-α-hydroxydihydrochalcone, *C*-glycosyl-β-hydroxydihydrochalcone, *C*-glycosylquinochalcone, *C*-glycosylflavanol and *C*-glycosylprocyanidin, have been added to the previously eight known types, namely *C*-glycosylflavone, *C*-glycosylflavonol, *C*-glycosylflavanone, *C*-glycosylflavanonol, *C*-glycosylchalcone, *C*-glycosyldihydrochalcone, *C*-glycosylisoflavone and *C*-glycosylisoflavanone. However, *C*-glycosylflavones still remain by far the most important group and most *C*-glycosylflavonoids have a phloroglucinol-derived A ring.

(a) *Mono-C-glycosylflavones*

Two Wessely–Moser isomers of previously known natural *C*-glycosylflavones have been isolated: isoembigenin (7,4′-di-*O*-methylvitexin) from *Siphonoglossa sessilis* by Hilsenbeck and Mabry (1983) and 7,3′-di-*O*-methylorientin from sugar cane by Mabry *et al.* (1984). The first natural *C*-rhamnosylflavone, isofurcatain (6-*C*-α-L-rhamnopyranosylapigenin), has now been found as the 7-*O*-glucoside in *Metzgeria furcata* by Markham *et al.* (1982). Its structure was established by ^{1}H and ^{13}CNMR spectroscopy and mass spectrometry of the permethyl derivative, and confirmed by direct comparison with the synthetic compound.

A new substitution type in the flavone series has been reported by Yasukawa *et al.* (1982): 6-*C*-galactopyranosylisoscutellarein from the roots of *Stellaria dichotoma*. Its structure was deduced from UV and ^{1}H NMR spectroscopy and the nature of the sugar determined from ferric chloride oxidation, giving galactose and lyxose. 6-*C*-Xylosylapigenin, already known as the *O*-rhamnoside, has been found in the free state in *Cerastium arvense* and named cerarvensin by Dubois *et al.*, (1982a).

(b) *Mono-C-glycosylflavonols and flavanonols*

The previously mentioned isolation of 6-*C*-glucosylkaempferol, 6-*C*-glucosylquercetin, 6-*C*-glucosyldihydrokaempferol and 6-*C*-glucosyldihydroquercetin from *Zelkowa serrata* by Hayashi (1980) has not yet been published in detail. A new *C*-glycosylflavonol has been isolated from *Cassia sophera* by Tiwari and Bajpai (1981) and assigned the structure 3,5,3′,4′,5′-pentahydroxy-7-methoxyflavone-8-*C*-L-rhamnopyranoside (8-*C*-rhamnosyleuropetin) on the basis of ferric chloride oxidation, giving rhamnose and (surprisingly) the flavonol aglycone. Permanganate oxidation of the latter gave gallic acid and alkali degradation gave phloroglucinol monomethyl ether, in agreement with the ^{1}H NMR spectra of the glycoside and its acetyl derivative. The position of the sugar residue was deduced from the presence of one acetyl signal at δ 1.73 in the latter and from the mass fragmentation pattern of the glycoside. Moreover the pyranose form of the rhamnose was confirmed by the consumption of two moles of periodate and liberation of one mole of formic acid per mole of glycoside.

Table 3.2 Naturally occurring mono-*C*-glycosylflavonoids

Compounds	*Sources*	*References*
C-GLYCOSYLFLAVONES		
Bayin (7, 4′-diOH-8-Glc)	*Cladrastis shikokiana* (Leg.) leaf *Globularia alypum* (Glo.) root	Ohashi *et al.* (1980) Hassine *et al.* (1982)
Isovitexin (saponaretin) (5, 7, 4′-triOH 6-Glc)	Many sources	Horowitz and Gentili (1964) (structure)
Vitexin (5, 7, 4′-triOH 8-Glc)	Many sources	Horowitz and Gentili (1964) (structure)
8-*C*-β-D-Galactopyranosylapigenin (5, 7, 4′-triOH 8-Gal)	*Carlina acanthifolia* (Comp.) leaf	Proliac and Raynaud (1983)
*Cerarvensin (5, 7, 4′-tri-OH 6-Xyl)	*Cerastium arvense* (Car.) whole plant	Dubois *et al.* (1982a)
Isomollupentin (5, 7, 4′-triOH 6-Ara)	*Passiflora platyloba* (Pas.) leaf *Cerastium arvense* (Car.) fresh leaf and flower	Ayanoglu *et al.* (1982) Dubois *et al.* (1985)
Mollupentin (5, 7, 4′-triOH 8-Ara)	*Mollugo pentaphylla* (Mol.) fresh aerial part	Chopin *et al.* (1979)
*Isofurcatain (5, 7, 4′-triOH 6-Rha)	*Metzgeria furcata* (Hep.) (as 7-*O*-Glc)	Markham *et al.* (1982)
3′-Deoxyderhamnosylmaysin [5, 7, 4′-triOH 6-(6-deoxy-*xylo*-hexos-4-ulosyl)]	*Zea mays* (Gra.) leaf	Elliger *et al.* (1980)
Swertisin (5, 4′-diOH 7-OMe 6-Glc)	Many sources	Komatsu and Tomimori (1966) (structure)
Isoswertisin (5, 4′-diOH 7-OMe 8-Glc)	*Passiflora sexflora* (Pas.) leaf *Gnetum sp.* (Gne.) leaf *Swertia paniculata* (Gen.) stem *Almeidea guyanensis* (Rut.) bark of stem and root	McCormick and Mabry (1982) Ouabonzi *et al.* (1983) Khetwal and Verma (1984) Wirasutisna *et al.* (1986)
Isomolludistin (5, 4′-diOH 7-OMe 6-Ara)	*Asterostigma riedelianum* (Ara.) leaf (as 2″-*O*-Glc)	Markham and Williams (1980)
Molludistin (5, 4′-diOH 7-OMe 8-Ara)	*Ocimum sanctum* (Lab.) leaf	Nair *et al.* (1982)
Isocytisoside (5, 7-diOH 4′-OMe 6-Glc)	*Fortunella japonica* (Rut.) (as 2″-*O*-Rha) peeling	Kumamoto *et al.* (1985b)
Cytisoside (5, 7-diOH 4′-OMe 8-Glc)	*Lupinus* sp. (Leg.) leaf *Trema cannabina* (Ulm.) (as 7-*O*-Glc) leaf *Fortunella japonica* (Rut.) (as 2″-*O*-Rha) peeling	Nicholls and Bohm (1983) Higa *et al.* (1983) Kumamoto *et al.* (1985b)
Embigenin (5-OH 7, 4′-diOMe 6-Glc)	*Siphonoglossa sessilis* (Aca.) stem and leaf (as 2″-*O*-Glc and 2″-*O*-Rha)	Hilsenbeck and Mabry (1983)
*Isoembigenin (5-OH 7, 4′-diOMe 8-Glc)	*Siphonoglossa sessilis* (Aca.) stem and leaf	Hilsenbeck and Mabry (1983)
7, 4′-Di-*O*-methylisomollupentin (5-OH 7, 4′-diOMe 6-Ara)	*Asterostigma riedelianum* (Ara.) leaf	Markham and Williams (1980)
Isoorientin (5, 7, 3′, 4′-tetraOH 6-Glc)	Many sources	Koeppen (1964) (structure)
Orientin (5, 7, 3′, 4′-tetraOH 8-Glc)	Many sources	Koeppen (1964) (structure)
6-*C*-β-D-Xylopyranosyl-luteolin (5, 7, 3′, 4′-tetraOH 6-Xyl)	*Phlox drummondii* (Pol.) (as 2″-*O*-Rha) flower	Mabry *et al.* (1971)
Derhamnosylmaysin [5, 7, 3′, 4′-tetraOH 6-(6-deoxy-*xylo*-hexos-4-ulosyl)]	*Zea mays* (Gra.) leaf (as 2″-*O*-Rha) corn silk	Elliger *et al.* (1980)
Swertiajaponin (5, 3′, 4′-triOH 7-OMe 6-Glc)	*Iris nertshinskia* (Iri.) fresh petal *Passiflora* sp. (Pas.) leaf *Achillea* sp. (Comp.) leaf	Hirose *et al.* (1981) McCormick and Mabry (1982, 1983) Valant-Vetschera (1982a)

Table 3.2 (*Contd.*)

Compounds	*Sources*	*References*
Isoswertiajaponin (5, 3′, 4′-triOH 7-OMe 8-Glc)	*Passiflora sexflora* (Pas.) leaf	McCormick and Mabry (1982)
	Iris florentina (Iri.) fresh leaf	Fujita and Inoue (1982)
Isoscoparin (5, 7, 4′-triOH 3′-OMe 6-Glc)	*Isatis tinctoria* (Cru.) seed	Raynaud and Prum (1980)
	Silene alba (Car.) green part	van Brederode and Kamps-Heinsbroek (1981b)
	Lychnis coronaria (Car.) leaf	Balabanova-Radonova *et al.* (1982)
	Dianthus sp. (Car.) aerial part	Boguslavskaya *et al.* (1983a,b)
	Passiflora palmeri (Pas.) dried leaf	Ulubelen *et al.* (1984)
	Gentiana sp. (Gen.) leaf	Chulia (1984)
Scoparin (5, 7, 4′-triOH 3′-OMe 8-Glc)	*Passiflora coactilis* (Pas.) leaf	Escobar *et al.* (1983)
		Chopin *et al.* (1968) (structure)
3′-*O*-Methylderhamnosylmaysin [5, 7, 4′-triOH 3′-OMe 6-(6-deoxy-*xylo*-hexos-4-ulosyl)]	*Zea mays* (Gra.) leaf	Elliger *et al.* (1980)
6-*C*-*β*-D-Glucopyranosyldiosmetin (5, 7, 3′-triOH 4′-OMe 6-Glc)	*Citrus limon* (Rut.) peeling	Kumamoto *et al.* (1985a)
8-*C*-*β*-D-Glucopyranosyldiosmetin (5, 7, 3′-triOH 4′-OMe 8-Glc)	*Lupinus sp.* (Leg.) leaf	Nicholls and Bohm (1983)
	Passiflora coactilis (Pas.) leaf	Escobar *et al.* (1983)
	Citrus limon (Rut.) peeling	Kumamoto *et al.* (1985a)
7, 3′-Di-*O*-methylisoorientin (5, 4′-diOH 7, 3′-diOMe 6-Glc)	*Achillea cretica* (Comp.) leaf	Valant *et al.* (1980)
	Saccharum (Gra.) mill sirup	Mabry *et al.* (1984)
*7, 3′-Di-*O*-methylorientin (5, 4′-diOH 7, 3′-diOMe 8-Glc)	*Saccharum* (Gra.) mill sirup	Mabry *et al.* (1984)
7, 3′, 4′-Tri-*O*-methylisoorientin (5-OH 7, 3′, 4′-triOMe 6-Glc)	*Linum maritimum* (Lin.) (as 2″-*O*-Rha) aerial part	Wagner *et al.* (1972)
*6-*C*-Galactosylisoscutellarein (5, 7, 8, 4′-tetraOH 6-Gal)	*Stellaria dichotoma* (Car.) root	Yasukawa *et al.* (1982)
Isoaffinetin (5, 7, 3′, 4′, 5′-pentaOH 6-Glc)	*Trichomanes venosum* (Hym.)	Markham and Wallace (1980)
	Radula lindenbergiana (Hep.)	Mues (1984)
Affinetin (5, 7, 3′, 4′, 5′-pentaOH 8-Glc)	*Trichomanes venosum* (Hym.)	Markham and Wallace (1980)
Isopyrenin (5, 7, 4′-triOH 3′, 5′-diOMe 6-Glc)	*Gentiana argentea* (Gen.) leaf	Chulia (1984)
6-*C*-Glucosyl-5, 7-dihydroxy-8, 3′, 4′, 5′-tetramethoxyflavone	*Vitex negundo* (Ver.) (as 5-*O*-Rha) bark	Subramanian and Misra (1979)
C-GLYCOSYLFLAVONOLS		
*8-*C*-Glucosyl-5-deoxykaempferol (3, 7, 4′-triOH 8-Glc)		
*8-*C*-Glucosylfisetin (3, 7, 3′, 4′-tetraOH 8-Glc)	*Pterocarpus marsupium* (Leg.) heartwood	Bezuidenhoudt *et al.* (1986)
6-*C*-Glucosylkaempferol (3, 5, 7, 4′-tetraOH 6-Glc)	*Zelkowa serrata* (Ulm.)	Hayashi (1980)
Keyakinin (3, 5, 4′-triOH 7-OMe 6-Glc)	*Zelkowa serrata* (Ulm.) wood	Hillis and Horn (1966) (structure)
6-*C*-Glucosylquercetin (3, 5, 7, 3′, 4′-pentaOH 6-Glc)	*Zelkowa serrata* (Ulm.)	Hayashi (1980)
Keyakinin B (3, 5, 3′, 4′-tetraOH 7-OMe 6-Glc)	*Zelkowa serrata* (Ulm.) wood	Hillis and Horn (1966)
*8-*C*-Rhamnosyleuropetin (3, 5, 3′, 4′, 5′-pentaOH 7-OMe 8-Rha)	*Cassia sophera* (Leg.) leaf	Tiwari and Bajpai (1981)
C-GLYCOSYLFLAVANONES		
Aervanone (7, 4′-diOH 8-Gal)	*Aerva persica* (Ama.) root	Garg *et al.* (1980)
Hemiphloin (5, 7, 4′-triOH 6-Glc)	*Acacia retinoide* (Leg.) flower	Lorente *et al.* (1982)
Isohemiphloin (5, 7, 4′-triOH 8-Glc)	*Acacia retinoide* (Leg.) flower	Lorente *et al.* (1983)
C-GLYCOSYLFLAVANONOLS		
6-*C*-Glucosyldihydrokaempferol (3, 5, 7, 4′-tetraOH 6-Glc)	*Zelkowa serrata* (Ulm.)	Hayashi (1980)

(*Contd.*)

Table 3.2 (*Contd.*)

Compounds	*Sources*	*References*
Keyakinol (3, 5, 4′-triOH 7-OMe 6-Glc)	*Zelkowa serrata* (Ulm.) wood	Funaoka and Tanaka (1957)
6-*C*-Glucosyldihydroquercetin (3, 5, 7, 3′, 4′-pentaOH 6-Glc)	*Zelkowa serrata* (Ulm.)	Hayashi (1980)
C-GLYCOSYLCHALCONES		
3′-*C*-Glucosylisoliquiritigenin (2′, 4′, 4-triOH 3′-Glc)	*Cladrastis shikokiana* (Leg.) leaf	Ohashi *et al.* (1980)
C-GLYCOSYLDIHYDROCHALCONES		
Nothofagin (2′, 4′, 6′, 4-tetraOH *C*-Gly)	*Nothofagus fusca* (Fag.) heartwood	Hillis and Inoue (1967)
Konnanin (2′, 4′, 6′, 3, 4-pentaOH *C*-Gly)	*Nothofagus fusca* (Fag.) heartwood	Hillis and Inoue (1967)
Aspalathin (2′, 4′, 6′, 3, 4-pentaOH 3′-Glc)	*Aspalathus linearis (acuminatus)* (Leg.) leaf	Koeppen and Roux (1965)
C-GLYCOSYL-α-HYDROXY DIHYDROCHALCONES		
*Coatline A (α, 2′, 4′, 4-tetraOH 3′-Glc)	*Eysenhardtia polystachya* (Leg.) wood	Beltrami *et al.* (1982)
*Coatline B (α, 2′, 4′, 3, 4-pentaOH 3′-Glc)	*Eysenhardtia polystachya* (Leg.) wood	Beltrami *et al.* (1982)
C-GLYCOSYL-β-HYDROXY DIHYDROCHALCONES		
*Pterosupin (β, 2′, 4′, 4-tetraOH 3′-Glc)	*Pterocarpus marsupium* (Leg.) root	Adinarayana *et al.* (1982)
C-GLYCOSYLISOFLAVONES		
Puerarin (7, 4′-diOH 8-Glc)	*Pueraria lobata* (Leg.) root	Fang *et al.* (1983)
8-*C*-β-D-Glucopyranosylgenistein (5, 7, 4′-triOH 8-Glc)	*Dalbergia nitidula* (Leg.) bark	van Heerden *et al.* (1980)
8-*C*-Glucosylprunetin (5, 4′-diOH 7-OMe 8-Glc)	*Dalbergia paniculata* (Leg.) bark	Parthasarathy *et al.* (1974, 1976)
Volubiline (5-OH 7, 4′-diOMe 8-Rha)	*Dalbergia volubilis* (Leg.) flower	Chawla *et al.* (1974)
Isovolubiline (5-OH 7, 4′-diOMe 6-Rha)	*Dalbergia volubilis* (Leg.) flower	Chawla *et al.* (1975)
8-*C*-Glucosylorobol (5, 7, 3′, 4′-tetraOH 8-Glc)	*Dalbergia nitidula* (Leg.) bark	van Heerden *et al.* (1980)
Dalpanitin (5, 7, 4′-triOH 3′-OMe 8-Glc)	*Dalbergia paniculata* (Leg.) seed	Adinarayana and Rao (1972)
Volubilinin (5, 7-diOH 6, 4′-diOMe 8-Glc)	*Dalbergia volubilis* (Leg.) flower	Chawla *et al.* (1976)
C-GLYCOSYLISOFLAVANONES		
Dalpanin	*Dalbergia paniculata* (Leg.) flower	Adinarayana and Rao (1975)

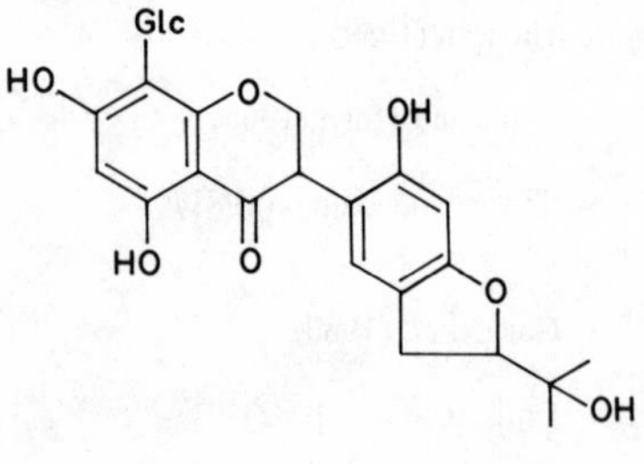

Table 3.2 (*Contd.*)

Compounds	*Sources*	*References*
C-GLYCOSYLFLAVANOLS		
*6-*C*-Glucosyl-(−)-epicatechin (3,5,7,3′,4′-pentaOH 6-Glc) *8-*C*-Glucosyl-(−)-epicatechin (3,5,7,3′,4′-pentaOH 8-Glc)	*Cinnamomum cassia* (Lau.) bark	Morimoto *et al.* (1986a)
C-GLYCOSYLPROANTHOCYANIDINS		
*6-*C*-Glucosylprocyanidin B2 (*3.4*) *8-*C*-Glucosylprocyanidin B2 (*3.5*)	*Cinnamomum cassia* (Lau.) bark	Morimoto *et al.* (1986b)
C-GLYCOSYLQUINOCHALCONE		
*Carthamin (*3.6*)	*Carthamus tinctorius* (Comp.) petal	Takahashi *et al.* (1982)

Quite recently, the first natural 5-deoxy-*C*-glycosylflavonols have been found in *Pterocarpus marsupium* heartwood by Bezuidenhoudt *et al.* (1987). 8-*C*-Glucosyl-5-deoxykaempferol and 8-*C*-glucosylfisetin were isolated, after methylation and acetylation, and their structures were established by high-temperature (180°C) ^{1}H and ^{13}C NMR spectroscopy of the methyl ether acetates.

(c) *Mono-C-glycosyl-α-hydroxydihydrochalcones*

Palodulcins A and B, isolated from *Eysenhardtia polystachya* by Vita-Finzi *et al.* (1980) and first considered to be *C*-glycosylflavanones, have not been listed in Table 3.2 since their structures have been revised and their names changed to coatlines A and B by Beltrami *et al.* (1982). These latter authors showed them to be 3′-*C*-β-glucopyranosyl-α, 2′, 4′, 4-tetrahydroxydihydrochalcone (*3.1*) and 3′-*C*-β-glucopyranosyl-α, 2′, 4′, 3, 4-pentahydroxydihydrochalcone (*3.2*) respectively on the basis of UV, IR, ^{1}H and ^{13}C NMR spectra and of FD-MS (Field Desorption-Mass Spectrometry) of the free compounds and their acetates. The presence of a methylene linked to an aromatic ring was proved by the n.O.e (nuclear Overhauser enhancement) effect observed for two of the aromatic protons by irradiation of the CH_2 group and by the presence of appropriate aromatic fragments in the EI-MS (Electron Impact-Mass Spectrometry) of non-acetylcoatline B. Coatline A and B are the first naturally occurring *C*-glycosyl-α-hydroxydihydrochalcones to be discovered.

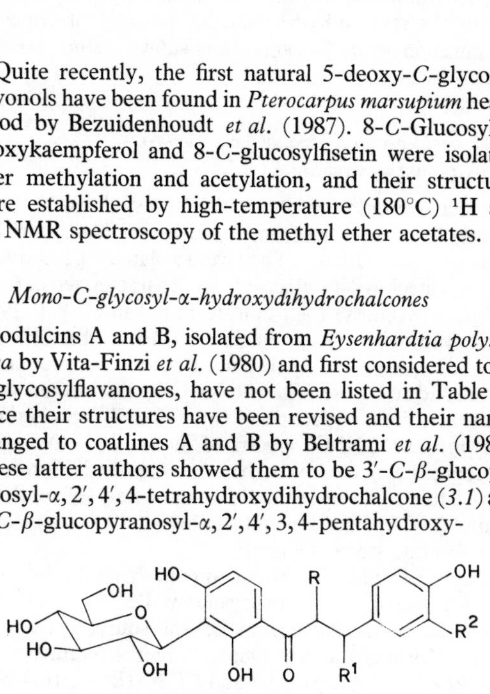

(*3.1*) R = OH, R^1 = R^2 = H Coatline A

(*3.2*) R = OH, R^1 = H, R^2 = OH Coatline B

(*3.3*) R = H, R^1 = OH, R^2 = H Pterosupin

(d) *Mono-C-glycosyl-β-hydroxydihydrochalcone*

Almost simultaneous with the preceding paper was the report by Adinarayana *et al.* (1982) of the isolation and structure elucidation of pterosupin from *Pterocarpus marsupium*, the first naturally occurring *C*-glycosyl-β-hydroxydihydrochalcone to be found. The 3′-β-D-glucopyranosyl-2′, 4, 4′, β-tetrahydroxydihydrochalcone (*3.3*) structure was proposed on the basis of UV, IR, ^{1}H and ^{13}C NMR spectra. The α-hydroxydihydrochalcone structure was excluded from consideration by the similarity of the shifts of α-CH_2 and β-CHOH signals to those of flavanone C–3 and C–2 respectively and from the comparison of the ^{1}H NMR data (for the ABMX system–CH_2–CHOH-) with those of gliricidol (considered to be 2′, 4′, 3, 5, β-pentahydroxy-4-methoxydihydrochalcone) (Manners and Jurd, 1979) and nubigenol (considered to be 2′,4′,6′,4, α-pentahydroxydihydrochalcone) (Bhakuni *et al.*, 1973). The pterosupin data were in good agreement with those for gliricidol, but not with those reported for nubigenol. Thus pterosupin and coatline A are closely related compounds since they possess the same *C*-glycosyldihydrochalcone skeleton, the only difference being the position of substitution of the aliphatic hydroxyl.

Indeed the reported ^{1}H and ^{13}C NMR data for pterosupin (in DMSO-d_6) and coatline A (in CD_3OD) are strikingly similar and the difference mentioned above between pterosupin and nubigenol may be ascribed to the use of pyridine-d_5 as the solvent in the latter case. Therefore the use of shift values in ^{1}H and ^{13}C NMR spectra does not allow a clear-cut distinction between α- and β-hydroxydihydrochalcones. By contrast, the MS data reported for pterosupin were not used in the structure determination. They showed, as usual in the MS of *C*-

glycosylflavonoids, an extremely low molecular peak (*m/z* 436) followed by two peaks resulting from the loss of one and two molecules of water. The base peak was found at *m/z* 107. Unfortunately, the only EI-MS data reported for coatlines are those of nonacetyl coatline B, in which the most important peak above *m/z* 100 is one at *m/z* 122 resulting from the highly favoured rupture of the α–β bond. In the case of coatline A, this rupture should have given a *p*-hydroxybenzyl ion with *m/z* 107 as at least a very important peak.

Turning now to the MS data reported for nubigenol, the molecular peak (*m/z* 290) was very weak and the following peak (*m/z* 272) resulted from the loss of one molecule of water. Unfortunately no peak lower than *m/z* 123 was reported and the relative intensities were not given. However, one peak at *m/z* 165 may be ascribed to the loss of 107 mass units from dehydrated nubigenol and there is a parallel in the MS of pterosupin, in which a loss of 106 mass units is observed between the two dehydration peaks *m/z* 418 and 400 and the two following peaks *m/z* 312 and 294.

Finally, in the MS data reported for gliricidol, the molecular peak (*m/z* 320, 2%) was low and no dehydration peak was mentioned. However, the main peaks were *m/z* 153 (96%), corresponding to the 3,5-dihydroxy-4-methoxybenzyl ion and *m/z* 137 (99%) corresponding to the 2,4-dihydroxyphenylketone ion. These ions find a parallel in the MS of pterosupin with the ions: *p*-hydroxybenzyl (*m/z* 107, 100%) and 2,4-dihydroxy-3-glucosylphenylketone-H_2O (*m/z* 281, 60%).

As the ions HO^+=CH—Ar and Ar′—C(OH)=CH^+ expected from a β-hydroxy structure Ar′–CO–CH_2–CHOH–Ar were not found in the mass spectra of pterosupin and gliricidol, we conclude that both compounds might be α-hydroxydihydrochalcones since the α-hydroxy structure was conclusively proved only in the case of coatline B.

This conclusion was supported by the MS data reported for a novel α-hydroxydihydrochalcone isolated from the heartwood of *Pterocarpus angolensis* by Bezuidenhoudt *et al.* (1981), the α,2′-dihydroxy-4,4′-dimethoxydihydrochalcone structure of which has been confirmed by synthesis. The molecular peak (*m/z* 302, 6%) was followed by an important dehydration peak (*m/z* 284, 75%) and the main peaks were *m/z* 151(85%) corresponding to a 2-hydroxy-4-methoxyphenylketo ion and *m/z* 121 (100%) corresponding to a 4-methoxybenzyl ion. Except for the dehydration peak, these results are quite similar to those reported for gliricidol. In the same paper, Bezuidenhoudt *et al.* (1981) prepared the α-hydroxy and β-hydroxy 2′,4′-dimethoxy-4-hydroxydihydrochalcones by photochemical methods. Unfortunately, the mass spectra were limited to the molecular peak but the slight differences observed in signals of the methylene protons in the ^{1}H NMR spectra favour the α-hydroxydihydrochalcone structures of pterosupin and gliricidol. Indeed the most striking difference observed between the α- and β-hydroxy compounds was the spontaneous dehydration of the latter to the corresponding chalcone by addition of acid. Pterosupin, being resistant to acid treatment, must therefore belong to the α-hydroxydihydrochalcone group.

Further support for this conclusion came quite recently from the isolation of coatline A from *Pterocarpus marsupium* heartwood by Bezuidenhoudt *et al.* (1986). The compound could be isolated only after acetylation of the mixture and its α-hydroxydihydrochalcone structure was unambiguously defined by spin–spin decoupling of the doublet at δ7.16 (H2 and 6 of the B ring) leading to selective sharpening of the non-equivalent methylene resonances at δ3.15 and 2.97 in the ^{1}H NMR spectrum of the acetate. Although pterosupin has been isolated from *Pterocarpus marsupium* rootwood, it now seems probable that pterosupin and coatline A are both α-hydroxy derivatives of the same dihydrochalcone, possibly of opposite configuration at the C-α since they show rotatory powers of opposite sign.

(e) *Mono-C-glycosylflavanols and mono-C-glycosylproanthocyanidins*

The first examples of natural *C*-glycosylflavanols have recently been reported from *Cinnamomum cassia* bark by Morimoto *et al.* (1986a). They were isolated together with (−)-epicatechin 3-*O*-glucoside and characterized as 6-*C*-β-D-glucopyranosyl-(−)-epicatechin and 8-*C*-β-D-glucopyranosyl-(−)-epicatechin by FAB-MS, ^{1}H and ^{13}C NMR spectra and ferric chloride oxidation producing glucose and arabinose. The C-6 and C-8 substitutions were determined by comparison of the C-6 and C-8 chemical shifts in the ^{13}C NMR spectra of the methyl ethers, the differences being too low in the free compounds. Synthesis of the latter could be realized by acid-catalysed condensation of (−)-epicatechin and D-glucose, thus establishing their absolute stereochemistry.

Closely related to the above compounds are the 6-*C*- and 8-*C*-β-D-glucopyranosylprocyanidins B2 (*3.4*) and (*3.5*) which have been isolated from the same source by the same group (Morimoto *et al.*, 1986b). Their structures were established by FAB-MS, ^{1}H and ^{13}C NMR spectroscopy,

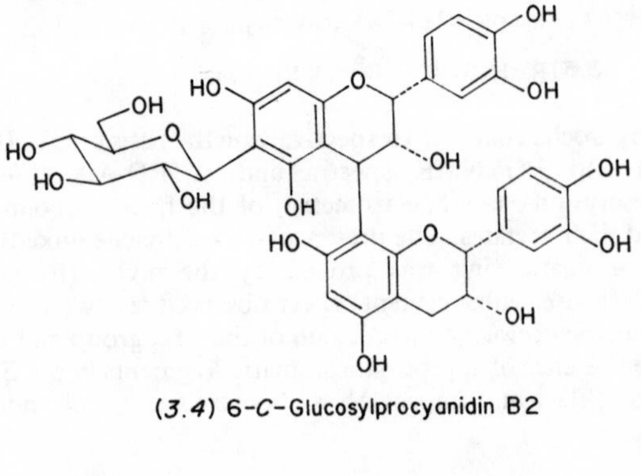

(*3.4*) 6-*C*-Glucosylprocyanidin B2

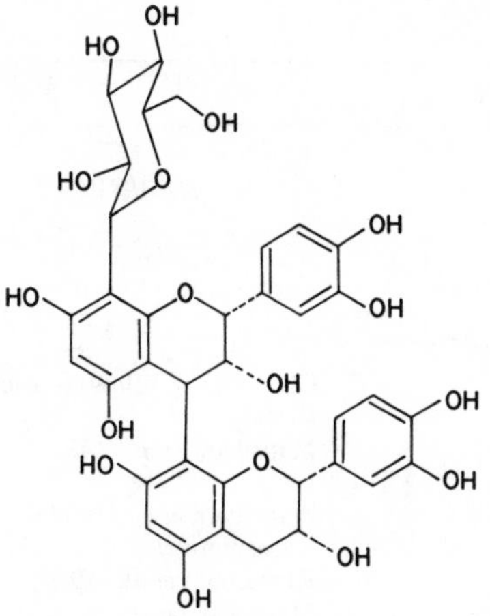

(*3.5*) 8-*C*-Glucosylprocyanidin B2

and thiolytic degradation to (−)-epicatechin and a thioether characterized by ^{1}H NMR analysis and desulphurization to (−)-epicatechin 6-*C*- and 8-*C*-β-D-glucopyranoside respectively. The location of the interflavanoid linkage was determined on the basis of the ^{1}H NMR chemical shifts for H-2'' and H-3'' which were in good agreement or identical with those of procyanidin B2.

(f) *Mono-C-glycosylquinochalcone*

Carthamin, the red pigment isolated from the flower petals of *Carthamus tinctorius*, has been assigned the quinochalcone *C*-glycoside structure (*3.6*) by Takahashi *et al.* (1982) on the basis of FAB-MS, UV, ^{1}H and

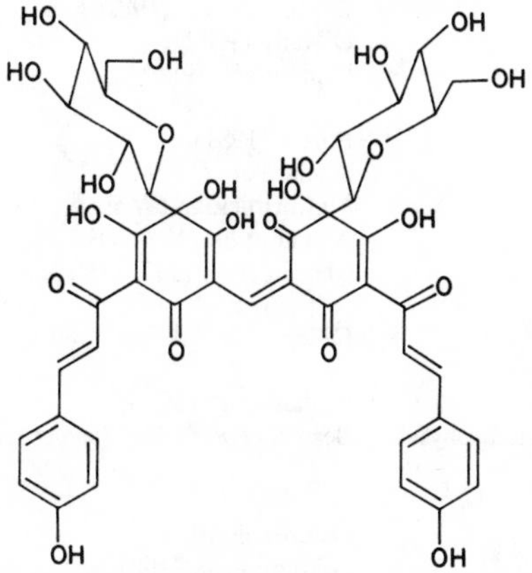

(*3.6*) Carthamin

^{13}C NMR spectra. Each quinochalcone unit being linked to one *C*-glycosyl residue, it can be considered as a mono-*C*-glycosylquinochalcone derivative.

(g) *Mono-C-glycosylbiflavonoid* (?)

Occidentoside, isolated from the nut shells of *Anacardium occidentale*, was considered to be tetrahydrohinokiflavone-*C*-glucoside (*3.7*) by Murthy *et al.* (1981) in a paper

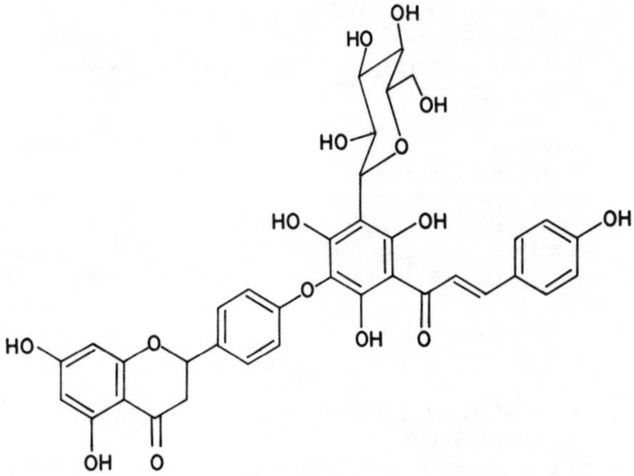

(*3.7*) Occidentoside

published without the prior consent of the senior authors (R.S. Ward, personal communication). The compound was reported to give a dimethyl ether and a monomethyl ether on methylation with diazomethane and a hexa-acetate on acetylation. No signal at $\delta < 2$ characteristic of *C*-glycosylflavonoid acetates was observed in the ^{1}H NMR spectrum of the latter, and most spectral data are in agreement with the structure of prunin (naringenin 7-*O*-glucoside) 6''-*O*-*p*-coumarate previously reported from the same source. Because of these discrepancies, occidentoside is not included in Table 3.2.

3.2.2 Di-*C*-glycosylflavonoids

Except for one di-*C*-glucosylflavanone, one di-*C*-glucosylisoflavone and two di-*C*-glucosylquinochalcones, all new compounds of this group are di-*C*-glycosylflavones. Most of them are 6,8-di-*C*-glycosylflavones, but an unusual type of substitution has recently been reported with the isolation of 3,6- and 3,8-di-*C*-glucosylflavones from hot water extracts of *Citrus* peels. The existence of such compounds is not quite unexpected, however, since 3,6- and 3,8-prenylated flavones are known to occur in nature, but their biosynthesis requires *C*-glycosylation of a dibenzoylmethane structure. It must be recalled here that in *Spirodela polyrhiza* *C*-glycosylation takes place at the flavanone/chalcone level, since labelled naringenin was incorporated into vitexin and orientin (Wallace and Grisebach, 1973), and not at the flavone level, since labelled apigenin or luteolin were not incorporated (Wallace *et al.*, 1969). More recently, Britsch *et al.* (1981) have postulated from studies with an enzyme

Table 3.3 Naturally occurring di-*C*-glycosylflavonoids

Compounds	*Sources*	*References*
DI-*C*-GLYCOSYLFLAVONES		
*6-*C*-Glucopyranosyl-8-*C*-arabinopy-ranosylchrysin (5, 7-diOH 6-Glc-8-Ara)	*Scutellaria baicalensis* (Lab.) root	Takagi *et al.* (1981)
*6-*C*-Arabinopyranosyl-8-*C*-glucopy-ranosylchrysin (5, 7-diOH 6-Ara-8-Glc)	*Scutellaria baicalensis* (Lab.) root	Takagi *et al.* (1981)
Vicenin-2	Many sources	Chopin *et al.* (1969) (structure)
(5, 7, 4′-triOH 6, 8-diGlc)		
*3, 6-Di-*C*-Glucosylapigenin	*Citrus unshiu* (Rut.) peelings	Matsubara *et al.* (1985a)
(5, 7, 4′-triOH 3, 6-diGlc)		
*3, 8-Di-*C*-Glucosylapigenin	*Citrus sudachi* (Rut.) peelings	Matsubara *et al.* (1985b)
(5, 7, 4′-triOH 3, 8-diGlc)	*Citrus junos* (Rut.) peelings	Kumamoto *et al.* (1985c)
	Citrus sinensis (Rut.) peelings	Kumamoto *et al.* (1986)
6, 8-Di-*C*-hexosylapigenin	*Plagiochila asplenioides* (Hep.) gametophyte, sporophyte	Mues and Zinsmeister (1976)
	Hedwigia ciliata (Bry.)	Österdahl (1979a, b)
*6, 8-Di-*C*-Galactopyranosylapigenin	*Stellaria dichotoma* (Car.) root	Yasukawa *et al.* (1982)
(5, 7, 4′-triOH 6, 8-diGal)		
*6-*C*-Glucopyranosyl-8-*C*-galactopy-ranosyl apigenin (5, 7, 4′-triOH 6-Glc-8-Gal)	*Cerastium arvense* (Car.) aerial part	Dubois *et al.* (1984)
Vicenin-3	*Angiopteris* sp. (Ang.)	Wallace *et al.* (1981)
(5, 7, 4′-triOH 6-Glc-8-Xyl)	*Ephedra antisyphilitica* (Eph.)	Wallace *et al.* (1982)
	Leucocyclus formosus (Comp.) leaf	Valant-Vetschera (1982b)
	Gnetum sp. (Gne.) leaf and stem	Ouabonzi *et al.* (1983)
	Monnieria trifolia (Rut.) leaf	Keita *et al.* (1985)
Vicenin-1	*Cladrastis shikokiana* (Leg.) leaf	Ohashi *et al.* (1980)
(5, 7, 4′-triOH 6-Xyl-8-Glc)	*Angiopteris evecta* (Ang.)	Wallace *et al.* (1981)
	Ephedra antisyphilitica (Eph.)	Wallace *et al.* (1982)
	Gnetum buchholzianum (Gne.)leaf and stem	Ouabonzi *et al.* (1983)
Violanthin	*Conocephalum conicum* (Bry.)	Porter (1981)
(5, 7, 4′-triOH 6-Glc-8-Rha)	*Angiopteris* sp. (Ang.)	Wallace *et el.* (1981)
	Adenia mannii (Pas.) leaf	Ulubelen *et al.* (1982c)
	Chlorophyta chloroplasts	O'Kelly (1982)
	Gutierrezia microcephala (Comp.) aerial part	Fang *et al.* (1986b)
Isoviolanthin	*Conocephalum conicum* (Bry.)	Porter (1981)
(5, 7, 4′-triOH 6-Rha-8-Glc)	*Angiopteris evecta* (Ang.)	Wallace *et al.* (1981)
	Passiflora sexflora (Pas.) leaf	McCormick and Mabry (1982)
	Glycyrrhiza sp. (Leg.) root	Afchar *et al.* (1984a, b)
Schaftoside	*Silene schafta* (Car.) stem and leaf	Chopin *et al.* (1974) (structure)
(5, 7, 4′-triOH 6-Glc-8-Ara)	Many sources	
Isoschaftoside	*Flourensia cernua* (Comp.) leaf	Dillon *et al.* (1976)
(5, 7, 4′-triOH 6-Ara-8-Glc)	Many sources	
Neoschaftoside	*Catananche cerulea* (Comp.) leaf	Proliac *et al.* (1973)
(5, 7, 4′-triOH 6-Glc-8-Ara)	*Radula complanata* (Bry.)gametophyte, sporophyte	Besson *et al.* (1984) (structure)
	Many sources	
Neoisoschaftoside	*Mnium undulatum* (Bry.)	Österdahl (1979a, c)
(5, 7, 4′-triOH 6-Ara-8-Glc)	*Crataegus monogyna* (Ros.) leaf	Nikolov *et al.* (1981)
	Trichophorum cespitosum (Cyp.) stem	Salmenkallio *et al.* (1982)
	Gemmingia chinensis (Iri.) leaf	Shirane *et al.* (1982)
	Saccharum (Gra.) sugarcane sirup	Ulubelen *et al.* (1985)
Isocorymboside	*Polygonatum multiflorum* (Lil.) fresh leaf	Chopin *et al.* (1977a)
(5, 7, 4′-triOH 6-Gal-8-Ara)	*Cerastium arvense* (Car.) fresh leaf, whole plant	Dubois *et al.* (1982a, b)

Table 3.3 (*Contd.*)

Compounds	*Sources*	*References*
Corymboside (5, 7, 4′-triOH 6-Ara-8-Gal)	*Carlina corymbosa* (Comp.) root	Besson *et al.* (1979)
	Triticum aestivum (Gra.) (as acyl derivs.) seedling	Wagner *et al.* (1980)
6, 8-Di-*C*-arabinosylapigenin (5, 7, 4′-triOH 6, 8-diAra)	*Angiopteris* sp. (Ang.)	Wallace *et al.* (1981)
	Hoya sp. (Asc.) fresh leaf	Niemann *et al.* (1981)
*6-*C*-β-D-Xylopyranosyl-8-*C*-α-L-arabinopyranosylapigenin (5, 7, 4′-triOH 6-Xyl-8-Ara)	*Mollugo pentaphylla* (Mol.) fresh aerial part	Chopin *et al.* (1982)
	Cerastium arvense (Car.) fresh leaf, whole plant	Dubois *et al.* (1982a, b)
*6-*C*-Arabinosyl-8-*C*-xylosylapigenin (5, 7, 4′-triOH 6-Ara-8-Xyl)	*Mollugo pentaphylla* (Mol.) fresh aerial part	Chopin *et al* (1982)
*6-*C*-Glucosyl-8-*C*-galactosylgenkwanin (5, 4′-diOH 7-OMe 6-Glc-8-Gal)	*Glycine max* (Leg.) root	Jay *et al.* (1984a)
*6-*C*-Glucosyl-8-*C*-arabinosylgenkwanin (5, 4′-diOH 7-OMe 6-Glc-8-Ara)	*Almeidea guyanensis* (Rut.) bark	Wirasutisna *et al.* (1986)
Almeidein (5, 4′-diOH 7-OMe 6, 8-diAra)	*Almeidea guyanensis* (Rut.) bark	Jay *et al.* (1979)
		Wirasutisna *et al.* (1986)
*3, 6-Di-*C*-glucosylacacetin (5, 7-diOH 4′-OMe 3, 6-diGlc)	*Fortunella japonica* (Rut.) peelings	Kumamoto *et al.* (1985b)
6-*C*-Pentosyl-8-*C*-hexosylacacetin (5, 7-diOH 4′-OMe 6-Pen-8-Hex)	*Trigonella corniculata* (Leg.)	Bouillant *et al.* (1975) (MS)
7, 4′-Di-*O*-methyl-6, 8-di-*C*-arabinopyranosylapigenin	*Asterostigma riedelianum* (Ara.) leaf	Markham and Williams (1980)
Lucenin-2 (5, 7, 3′, 4′-tetraOH 6, 8-diGlc)	Many sources	Osterdahl (1978) (structure)
Lucenin-3 (5, 7, 3′, 4′-tetraOH 6-Glc-8-Xyl)	*Ephedra antisyphilitica* (Eph.)	Wallace *et al.* (1982)
	Ephedra alata (Eph.) whole plant	Nawwar *et al.* (1984)
Lucenin-1 (5, 7, 3′, 4′-tetraOH 6-Xyl-8-Glc)	*Ephedra antisyphilitica* (Eph.)	Wallace *et al.* (1982)
Carlinoside (5, 7, 3′, 4′-tetraOH 6-Glc-8-Ara)	*Lespedeza capitata* (Leg.) leaf	Linard *et al.* (1982)
	Passiflora sexflora (Pas.) leaf	McCormick and Mabry (1982)
	Blepharostoma trichophyllum (Hep.)	Mues (1982b)
	Radula complanata (Hep.)	Mues (1984)
	Glycine max (Leg.) root	Jay *et al.* (1984a)
	Oryza sativa (Gra.) leaf	Besson *et al.* (1985) (structure)
*Isocarlinoside (5, 7, 3′, 4′-tetraOH 6-Ara-8-Glc)	*Lespedeza capitata* (Leg.) leaf	Linard *et al.* (1982)
	Blepharostoma trichophyllum (Hep.)	Mues (1982b)
	Glycine max (Leg.) root	Jay *et al.* (1984a)
Neocarlinoside (5, 7, 3′, 4′-tetraOH 6-Glc-8-Ara)	*Lespedeza capitata* (Leg.) leaf	Linard *et al.* (1982)
	Radula complanata (Hep.)	Mues (1984)
	Oryza sativa (Gra.) leaf	Besson *et al.* (1985) (structure)
	Saccharum (Gra.) sugarcane syrup	Ulubelen *et al.* (1985)
6, 8-Di-*C*-pentosyl-luteolin (5, 7, 3′, 4′-tetraOH 6, 8-diPen)	*Takakia lepidozioides* (Hep.)	Markham and Porter (1979)
6, 8-Di-*C*-glucosylchrysoeriol (stellarin-2) (5, 7, 4′-triOH 3′-OMe 6, 8-diGlc)	*Conocephalum conicum* (Bry.)	Porter (1981)
	Trichocolea tomentella (Hep.) gametophyte	Mues (1982a)
	Trichophorum cespitosum (Cyp.) stem	Salmenkallio *et al.* (1982)
	Welwitschia mirabilis (Web.)	Wallace *et al.* (1982)
	Aeluropus lagopoides (Gra.) leaf, stem	Abdalla and Mansour (1984)
*6-*C*-Glucosyl-8-*C*-arabinosylchrysoeriol (5, 7, 4′-triOH 3′-OMe 6-Glc-8-Ara)	*Trichophorum cespitosum* (Cyp.) stem	Salmenkallio *et al.* (1982)
*6-*C*-Arabinosyl-8-*C*-glucosylchrysoeriol (5, 7, 4′-triOH 3′-OMe 6-Ara-8-Glc)	*Trichophorum cespitosum* (Cyp.) stem	Salmenkallio *et al.* (1982)

(*Contd.*)

Table 3.3 (*Contd.*)

Compounds	*Sources*	*References*
6-*C*-Arabinosyl-8-*C*-hexosyl-chrysoeriol (5,7,4′-triOH 3′-OMe 6-Ara-8-Hex)	*Mnium undulatum* (Bry.)	Österdahl (1979a, c)
*6,8-Di-*C*-glucosyldiosmetin (5,7,3′-triOH 4′-OMe 6,8-diGlc)	*Citrus limon* (Rut.) peelings	Kumamoto *et al.* (1985a)
*3, 8-Di-*C*-glucosyldiosmetin (5,7,3′-triOH 4′-OMe 3,8-diGlc)	*Citrus sudachi* (Rut.) peelings	Matsubara *et al.* (1985b)
	Citrus sinensis (Rut.) peelings	Kumamoto *et al.* (1986)
6,8-Di-*C*-glucosyltricetin (5,7,3′,4′5′-pentaOH 6,8-diGlc)	*Plagiochila asplenioides* (Hep.) gametophyte sporophyte	Mues and Zinsmeister (1976)
	Apometzgeria pubescens (Hep.) gametophyte	Theodor *et al.* (1980)
	Metzgeria furcata (Hep.)	Theodor *et al.* (1983)
	Radula spp. (Hep.)	Mues (1984)
6,8-Di-*C*-hexosyltricetin	*Plagiochila asplenioides* (Hep.) gametophyte sporophyte	Mues and Zinsmeister (1976)
*6-*C*-Glucosyl-8-*C*-arabinosyl-tricetin (5,7,3′,4′,5′-pentaOH 6-Glc-8-Ara)	*Metzgeria furcata* (Hep.)	Theodor *et al.* (1983)
	Radula complanata (Hep.)	Markham and Mues (1984)
6-*C*-Hexosyl-8-*C*-pentosyl-tricetins	*Plagiochila asplenioides* (Hep.) gametophyte sporophyte	Mues and Zinsmeister (1976)
	Takakia lepidozioides (Hep.)	Markham and Porter (1979)
6-*C*-Arabinosyl-8-*C*-glucosyltricetin (5,7,3′,4′,5′-pentaOH 6-Ara-8-Glc)	*Apometzgeria pubescens* (Hep.) gametophyte	Theodor *et al.* (1980, 1981a)
	Metzgeria furcata (Hep.)	Theodor *et al.* (1983)
	Radula complanata (Hep.)	Mues (1984)
	Takakia lepidozioides (Hep.)	Markham and Porter (1979)
6,8-Di-*C*-pentosyltricetin (5,7,3′,4′,5-pentaOH 6,8-diPen)		
6-*C*-Hexosyl-8-*C*-pentosyl-3′-*O*-methyltricetin (5, 7, 4′, 5′-tetraOH 3′-OMe 6-Hex-8-Pen)	*Plagiochila asplenioides* (Hep.) gametophyte sporophyte	Mues and Zinsmeister (1976)
6,8-Di-*C*-glucosyltricin (5,7,4′-triOH 3′,5′-diOMe 6,8-diGlc)	*Apometzgeria pubescens* (Hep.) gametophyte	Theodor *et al.* (1980, 1981a)
	Metzgeria conjugata (Hep.) gametophyte	Theodor *et al.* (1981b)
	Frullania sp. (Hep.)	Mues *et al.* (1984)
6-*C*-Glucosyl-8-*C*-arabinosyltricin (5,7,4′-triOH 3′,5′-diOMe 6-Glc-8-Ara)	*Apometzgeria pubescens* (Hep.) gametophyte	Theodor *et al.* (1980, 1981a)
	Metzgeria conjugata (Hep.) gametophyte	Theodor *et al.* (1981b)
6-*C*-Arabinosyl-8-*C*-glucosyltricin (5,7,4′-triOH 3′,5′-diOMe 6-Ara-8-Glc)	*Apometzgeria pubescens* (Hep.) gametophyte	Theodor *et al.* (1980, 1981a)
	Metzgeria conjugata (Hep.) gametophyte	Theodor *et al.* (1981b)
6-*C*-Xylosyl-8-*C*-hexosyltricin (5,7,4′-triOH 3′,5′-diOMe 6-Xyl-8-Hex)	*Metzgeria leptoneura* (Hep.) gametophyte	Theodor *et al.* (1981b)
6,8-Di-*C*-arabinosyltricin (5,7,4′-triOH 3′,5′-diOMe 6,8-diAra)	*Apometzgeria pubescens* (Hep.) gametophyte	Theodor *et al.* (1980, 1981a)
6-*C*-Hexosyl-8-*C*-pentosylapometzgerin (5, 7, 5′-triOH 3′, 4′-diOMe 6-Hex-8-Pen)	*Apometzgeria pubescens* (Hep.) gametophyte	Theodor *et al.* (1980, 1981a)
6, 8-Di-*C*-Arabinosylapometzgerin (5, 7, 5′-triOH 3′, 4′-diOMe 6, 8-diAra)	*Apometzgeria pubescens* (Hep.) gametophyte	Theodor *et al.* (1980, 1981a)
DI-*C*-GLYCOSYLFLAVANONES		
*6,8-Di-*C*-glucosylnaringenin (5,7,4′-triOH 6,8-diGlc)	*Zizyphus jujuba* (Rha) fructus	Okamura *et al.* (1981)

Table 3.3 (*Contd.*)

Compounds	*Sources*	*References*
DI-*C*-GLYCOSYLDIHYDROCHALCONES		
3′,5′-Di-*C*-glycosylphloretin (2′,4′,6′,4-tetraOH 3′,5′-diGly)	*Hymenophyton leptopodum* (Hep.)	Markham *et al.* (1976)
DI-*C*-GLYCOSYLISOFLAVONES		
Paniculatin (5,7,4′-triOH 6,8-diGlc)	*Dalbergia paniculata* (Leg.) bark	Narayanan and Seshadri (1971)
* 6,8-Di-*C*-glucosylorobol (5,7,3′,4′-tetraOH 6,8-diGlc)	*Dalbergia nitidula* (Leg.) bark	van Heerden *et al.* (1980)
DI-*C*-GLYCOSYLQUINOCHALCONES		
*Safflor yellow A (3.8)	*Carthamus tinctorius* (Com.) petal	Takahashi *et al.* (1982)
*Safflor yellow B (3.9)	*Carthamus tinctorius* (Com.) petal	Takahashi *et al.* (1984)

system from cell cultures of parsley (*Petroselinum hortense*) that synthesis of apigenin from naringenin proceeds via 2-hydroxynaringenin, a 2-hydroxyflavanone, in rapid equilibrium with the corresponding 2,6-dihydroxydibenzoylmethane as shown by Chadenson *et al.* (1972) and Hauteville *et al.* (1973a, b). It seems likely that 6- or 8-*C*-glycosylation occurs at the flavanone/chalcone level as a first step, followed by 2-hydroxylation of the 6- or 8-*C*-glycosylflavanone and *C*-glycosylation of the corresponding 2,6-dihydroxydibenzoylmethane, leading to the dehydration of the 3,6- or 3,8-di-*C*-glycosyl-2-hydroxyflavanone to the corresponding flavone.

6-*C*-β-D-Glucopyranosyl-8-*C*-α-L-arabinopyranosylchrysin and 6-*C*-α-L-arabinopyranosyl-8-*C*-β-D-glucopyranosylchrysin are the first natural examples of *C*-glycosylchrysins to be found. They have been isolated from *Scutellaria baicalensis* roots by Takagi *et al.* (1981) and fully characterized by UV, IR, ^{1}H and ^{13}C NMR spectra, mass spectrometry of permethyl ethers, Wessely-Moser isomerization and $FeCl_3$ oxidation.

6-*C*-β-D-Glucopyranosyl-8-*C*-α-L-arabinopyranosylchrysoeriol and 6-*C*-α-L-arabinopyranosyl-8-*C*-β-D-glucopyranosylchrysoeriol have been found in *Trichophorum cespitosum* by Salmenkallio *et al.* (1982). They were characterized by UV spectrometry and TLC as chrysoeriol *C*-glycosides, by mass spectrometry of permethyl ethers as *C*-hexoside-*C*-pentoside and *C*-arabinopyranoside-*C*-hexoside and by co-TLC of the permethyl ethers with permethyl derivatives of carlinoside and isocarlinoside, the corresponding di-*C*-glycosylluteolins of known structure.

Isocarlinoside (6-*C*-α-L-arabinopyranosyl-8-*C*-β-D-glucopyranosyl-luteolin) has now been found in nature along with the already known carlinoside and neocarlinoside in *Lespedeza capitata* by Linard *et al.* (1982). From TLC and UV data and mass spectrometry of the permethyl ether its structure was deduced as 6-*C*-arabinopyranosyl-8-*C*-hexosyl-luteolin, the glucopyranosyl nature of the hexosyl residue resulting from Wessely–Moser isomerization to carlinoside and neocarlinoside. The structure of carlinoside (6-*C*-β-D-glucopyranosyl-8-*C*-α-L-arabinopyranosyl-luteolin), first proposed on the basis of the mass spectrum of its permethyl ether and later supported by its Wessely–Moser isomerization to isocarlinoside and neocarlinoside, has now been confirmed by ^{1}H and ^{13}C NMR spectra of the same compound isolated from *Oryza sativa* (Besson *et al.*, 1985).

Neocarlinoside was first isolated from *Oryza sativa* by Fukami's group in 1978 along with carlinoside, schaftoside and neochaftoside. Its name was derived by E. Besson from its Wessely–Moser isomerization to carlinoside and isocarlinoside and from the identity of the sugar moiety from the ^{1}H NMR spectra of perdeuteriomethylneocarlinoside and perdeuteriomethylneoschaftoside. This showed that neocarlinoside has the structure 6-*C*-β-D-glucopyranosyl-8-*C*-β-L-arabinopyranosyl–luteolin, now confirmed by comparison of ^{13}C NMR and CD spectra (Besson *et al.*, 1985), since the 6-*C*-β-D-glucopyranosyl-8-*C*-β-L-arabinopyranosylapigenin structure of neoschaftoside had been deduced from the study of synthetic 6-*C*-α and β-L-arabinopyranosyl and furanosylacacetins (Besson and Chopin, 1983). This structure has now been confirmed by ^{13}C NMR spectrometry of neoschaftoside isolated from *Radula complanata* by Mues (Besson *et al.*, 1984).

The 6-*C*-β-L-arabinopyranosyl-8-*C*-β-D-glucopyranosylapigenin structure of neoisoschaftoside, first isolated from *Mnium undulatum* by Österdahl

(1979a, c) and now found in several higher plants (see Table 3.3), is supported by the fragmentation pattern observed in the mass spectrum of its permethyl derivative, which is the same as that of permethylisoschaftoside, since permethyl 6-*C*-arabinofuranosylapigenins can be differentiated from permethyl 6-*C*-arabinopyranosylapigenins by their mass spectra (Besson and Chopin, 1983). Moreover the ^{1}H NMR spectrum (in DMSO-d_6) of neoisoschaftoside isolated from *Gemmingia chinensis* by Shirane *et al.* (1982) shows the same signals as neoschaftoside for the anomeric sugar proton of the arabinosyl residue (Yagishita, 1983).

6-*C*-β-D-Glucopyranosyl-8-*C*-α-L-arabinopyranosyltricetin was first isolated from *Metzgeria furcata* by Theodor *et al.* (1983) and its structure deduced from chromatographic and UV data, mass spectrometry of the permethyl and perdeuteriomethyl derivatives and co-chromatography with PM (permethylated) 6-*C*-β-D-glucopyranosyl-8-*C*-α-L-arabinopyranosyltricin. The same compound was later isolated from *Radula complanata* and its structure confirmed by ^{1}H and ^{13}C NMR spectroscopy by Markham and Mues (1984). It co-occurs in *Rudula complanata* with the corresponding Wessely–Moser and 'neo' isomers, together with schaftoside and neoschaftoside, carlinoside and neocarlinoside (Mues, 1984).

6-*C*-β-D-Glucopyranosyl-8-*C*-α-L-arabinopyranosylgenkwanin has recently been found in *Almeidea guyanensis* by Wirasutisna *et al.* (1986) and its structure derived from chromatographic and UV data, ^{1}H NMR spectra, mass spectrometry of the permethyl derivative and co-chromatography with permethylschaftoside. Three new di-*C*-glucosylflavones containing *C*-galactosyl residues have been found in nature. 6, 8-Di-*C*-galactopyranosylapigenin was isolated from *Stellaria dichotoma* roots by Yasukawa *et al.* (1982) and its structure deduced from UV, IR and ^{1}H NMR spectra of the free compound and its acetate, and ferric chloride oxidation giving galactose and lyxose. 6-*C*-β-D-Glucopyranosyl-8-*C*-β-D-galactopyranosylapigenin, the first natural asymmetrical di-*C*-hexosylflavone of known structure, was isolated from *Cerastium arvense* by Dubois *et al.* (1984). It was shown to be a di-*C*-hexosylapigenin by MS of its permethyl derivative and to be a *C*-β-D-glucopyranosyl-*C*-β-D-galactopyranosylapigenin by ^{13}C NMR spectroscopy. The position of the hexosyl residues was established by comparison with the 6-*C*-β-D-galactopyranosyl-8-*C*-β-D-glucopyranosylapigenin obtained from 6-*C*-glycosylation of vitexin with acetobromogalactose. Direct comparison of the permethyl derivatives of the above compounds with that of a di-*C*-hexosylgenkwanin isolated from *Glycine max* roots by Jay *et al.* (1984a) indicated that the latter is 6-*C*-glucosyl-8-*C*-galactosylgenkwanin. The asymmetrical nature of a 6, 8-di-*C*-pentosylapigenin isolated from *Mollugo pentaphylla* by Chopin *et al.* (1979) has been demonstrated by Chopin *et al.* (1982). Comparison of the ^{1}H NMR spectrum of the perdeuteriomethylated compound with those of the perdeuteriomethylated derivatives of molludistin (8-*C*-α-L-arabinopyranosyl-7-*O*-methylapigenin), vitexin (8-*C*-β-D-glucopyranosylapigenin) and vicenin-1 (6-*C*-β-D-xylopyranosyl-8-*C*-β-D-glucopyranosylapigenin) suggested that xylose is attached at C-6 and arabinose is attached at C-8, in agreement with the CD curve being similar to that of schaftoside. This was finally proved by the MS of the permethyl derivative of the monoisopropylidene ketal of the compound, which showed that the isopropylidene group is not attached to the 6-*C*-glycosyl residue. The presence of β-D-xylopyranosyl and α-L-arabinopyranosyl residues was later confirmed by ^{13}C NMR spectroscopy (K. R. Markham, unpublished), in agreement with the proposed 6-*C*-β-D-xylopyranosyl-8-*C*-α-L-arabinopyranosylapigenin structure. It was also shown that the other 6, 8-di-*C*-pentosylapigenin co-occurring in *Mollugo pentaphylla* is probably the Wessely–Moser isomer, i.e. 6-*C*-arabinosyl-8-*C*-xylosylapigenin.

As mentioned above, a series of unusual 3, 6- and 3, 8-di-*C*-glycosylflavones was recently isolated from hot water extracts of *Citrus* peelings by a group of Japanese workers interested in hypotensive substances. 3, 6-Di-*C*-glucosylapigenin was found in *Citrus unshiu* peelings (Matsubara *et al.*, 1985a) and its structure was established by FAB–MS, UV, ^{1}H and ^{13}C NMR spectroscopy: H-8 was found at δ 6.49, the anomeric sugar protons at 4.67 and 4.84 ($J = 10$ Hz), C-8 at 95.3, C-6 at 102.0 and C-3 at 108.3 in DMSO-d_6. 3, 8-Di-*C*-glucosylapigenin and 3, 8-di-*C*-glucosyldiosmetin were isolated from *Citrus sudachi* peelings (Matsubara *et al.*, 1985b) and their structures were similarly established by FAB–MS, UV, ^{1}H and ^{13}C NMR spectroscopy: H-6 was found at δ 6.22 and 6.26, the anomeric sugar protons at 4.52 and 4.77 ($J = 10$ Hz) C-6 at 100.3, C-8 at 104.7 and C-3 at 109.4 in DMSO-d_6. 3, 8-Di-*C*-glucosylapigenin was also isolated from *Citrus junos* peelings (Kumamoto *et al.*, 1985c). 3, 6-Di-*C*-glucosylacacetin was found in *Fortunella japonica* peelings (Kumamoto *et al.*, 1985b) along with 6, 8-di-*C*-glucosylapigenin, and its structure was established by FAB–MS, UV and ^{1}H NMR spectroscopy. Similarly, the structure of 6, 8-di-*C*-glucosyldiosmetin isolated from lemon peelings (Kumamoto *et al.*, 1985a) was only based upon UV and ^{1}H NMR spectra. Later 3, 8-di-*C*-glucosylapigenin and 3, 8-di-*C*-glucosyldiosmetin were also isolated from *Citrus sinensis* peelings (Kumamoto *et al.*, 1986).

6, 8-Di-*C*-glucosylnaringenin (a mixture of (2*R*) and (2*S*) forms) isolated from *Zizyphus jujuba* by Okamura *et al.* (1981) is the first example of a natural di-*C*-glycosylflavanone. Its structure was determined on the basis of UV, IR, ^{1}H and ^{13}C NMR spectra and ferric chloride oxidation to glucose and arabinose. Acetylation gave two acetates corresponding to (2*R*) and (2*S*) configurations, from their CD spectra.

A new 6,8-di-*C*-glycosylisoflavone, 6,8-di-*C*-β-D-glucopyranosylorobol, has been found in *Dalbergia nitidula* bark by van Heerden *et al.* (1980), together with already known paniculatin (6,8-di-*C*-β-D-glucopyranosylgenistein), 8-*C*-β-D-glucopyranosylgenistein and 8-*C*-β-D-glucopyranosylorobol. The latter gave glucose on ferric chloride oxidation and all compounds readily yielded the aglycones on treatment with hydriodic acid/phenol. All structures were established by ^{13}C NMR spectroscopy.

Two yellow pigments isolated from the flower petals of *Carthamus tinctorius*, safflor yellow A and B, have been assigned the quinochalcone *C*-glycoside structures (*3.8*) and (*3.9*) respectively on the basis of FAB–MS, UV, ^{1}H and ^{13}C NMR spectra (Takahashi *et al.*, 1982, 1984). With each quinochalcone unit being linked to two *C*-glycosyl residues, they can be considered to be di-*C*-glycosylquinochalcone derivatives.

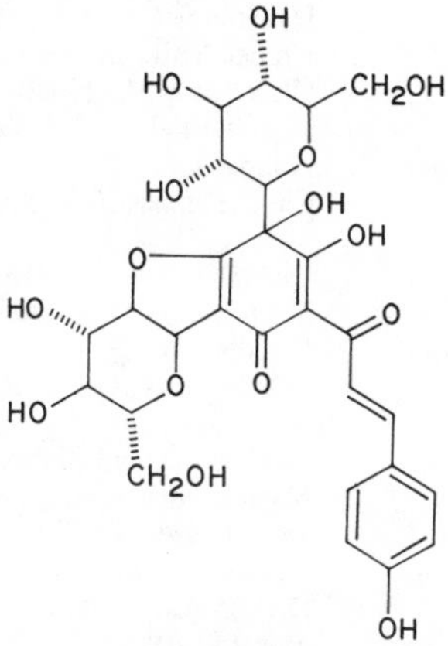

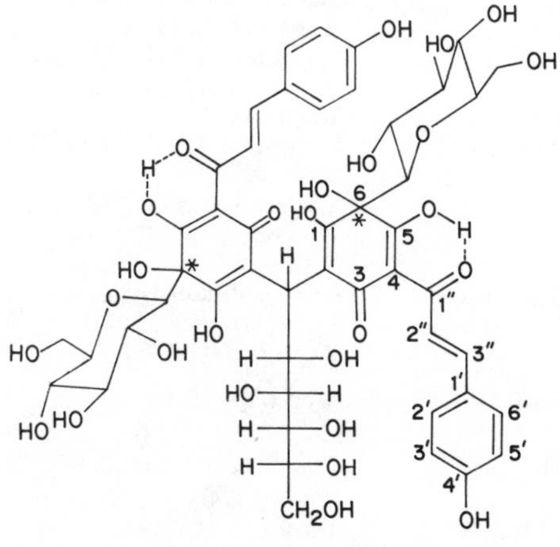

(*3.8*) Safflor yellow A

(*3.9*) Safflor yellow B

3.2.3 *C*-Glycosylflavonoid *O*-glycosides

C-Glycosylflavones are often found in nature as *O*-glycosides in which the hydrolysable sugar is linked to a phenolic hydroxyl group (X or X′-*O*-glycosides) or to a hydroxyl group of the *C*-glycosyl residue (X″-*O*-glycosides of 6- or 8-*C*-glycosylflavones; X″-*O*-glycosides of 6,8-di-*C*-glycosylflavones if linked to a hydroxyl group of the 6-*C*-glycosyl residue; X‴-*O*-glycosides of 6,8-di-*C*-glycosylflavones if linked to a hydroxyl group of the 8-*C*-glycosyl residue). The only interest in this rather confusing nomenclature is to clearly distinguish the hydrolysable sugar moiety from the unhydrolysable sugar, i.e. the *C*-glycosylflavonoid remaining after acid hydrolysis of the compound under study. Glycosylation of a phenolic hydroxyl group is easily detected and usually located by changes in UV spectral shifts after hydrolysis. No change occurs with X″- or X‴-*O*-glycosides which show the same UV spectrum as the corresponding *C*-glycosylflavones. ^{13}C NMR spectroscopy is the only general method for the location of the sugar linked to a *C*-glycosyl residue, but it does not distinguish X″- from X‴-*O*-glycosides of 6,8-di-*C*-glycosylflavones. This distinction is now possible from the fragmentation pattern observed in the mass spectra of the permethyl derivatives (see Section 3.4). On the other hand, mass spectrometry of permethyl ethers can locate the sugar only in the case of X″-*O*-glycosides of 6-*C*-glycosylflavones or 6,8-di-*C*-glycosylflavones. Except for one isoflavone *O*-glycoside, all new compounds in Table 3.4 are *C*-glycosylflavone *O*-glycosides. Most of the new X″-*O*-glycosides are 2″- and only a few are 6″-*O*-glycosides, two of them being 6″-*O*-glycosides of 6,8-di-*C*-glycosylflavones. No natural 3″- or 4″-*O*-glycoside has yet been described. D-Mannose and apiofuranose have been found for the first time in iso-orientin 2″-*O*-mannoside and puerarin 6″-*O*-apiofuranoside.

Isovitexin 7-*O*-rhamnoside has been isolated from *Yeatesia viridiflora* by Hilsenbeck *et al.* (1984b) and its structure deduced from UV and ^{1}H NMR data. Isovitexin 4′-*O*-arabinoside has been identified by UV data in *Leptodactylon* species by Smith *et al.* (1982). The structure of isovitexin 2″-*O*-galactoside isolated from *Secale cereale* was established by mass spectrometry of the permethyl ether (Dellamonica *et al.*, 1983). 6-*C*-Glucosylapigenin and 6-*C*-arabinosylapigenin 7-*O*-rhamnosylglucosides were isolated from *Passiflora platyloba* and their structures deduced from mass spectrometry of the permethyl ethers (Ayanoglu *et al.*, 1982). Isovitexin 7,2″-di-*O*-galactoside isolated from *Gentiana depressa* by Chulia (1984) has been identified by mass spectrometry of the permethyl ether. Isovitexin 7-*O*-galactoside 2″-*O*-arabinoside and isovitexin 7-*O*-arabinoside 2″-*O*-glucoside have been found respectively in *Silene pratensis* (Steyns *et al.*, 1983) and *Silene*

Table 3.4 Naturally occurring *C*-glycosylflavonoid *O*-glycosides

Compounds	*Sources*	*References*
ISOVITEXIN		
7-*O*-Glucoside (saponarin)	Many sources	
7-*O*-Galactoside (neosaponarin)	*Silene pratensis* (Car.) leaf	Steyns *et al.* (1983)
*7-*O*-Rhamnoside	*Yeatesia viridiflora* (Aca.) stem leaf	Hilsenbeck *et al.* (1984b)
7-*O*-Xyloside	*Silene dioica* (Car.) petal	Mastenbroek *et al.* (1983c)
4′-*O*-Glucoside (isosaponarin)	*Cerastium arvense* (Car.) leaf, flower	Dubois *et al.* (1985)
*4′-*O*-Arabinoside	*Leptodactylon* sp. (Pol.) leaf	Smith *et al.* (1982)
2″-*O*-Glucoside	*Oxalis acetosella* (Oxa.) aerial part	Tschesche and Struckmeyer (1976) (structure)
	Vepris heterophylla (Rut.) leaf	Gomes *et al.* (1983)
	Silene sp. (Car.) petal, leaf	Mastenbroek *et al.* (1983c)
	Cerastium arvense (Car.) leaf, flower	Dubois *et al.* (1985)
*2″-*O*-Galactoside	*Secale cereale* (Gra.) leaf	Dellamonica *et al.* (1983)
2″-*O*-Xyloside	*Passiflora serratifolia* (Pas.) leaf	Ulubelen and Mabry (1980)
	Passiflora sp. (Pas.) aerial part	Ulubelen *et al.* (1982b, d)
	Silene dioica (Car.) petal	Mastenbroek *et al.* (1983c)
	Cerastium arvense (Car.) leaf, flower	Dubois *et al.* (1985)
2″-*O*-Arabinoside	*Avena sativa* (Gra.) leaf, stem, inflorescence	Chopin *et al.* (1977b) (structure)
	Secale cereale (Gra.) leaf	Dellamonica *et al.* (1983)
	Silene dioica (Car.) petal	Mastenbroek *et al.* (1983c)
	Cerastium arvense (Car.) leaf, flower	Dubois *et al.* (1985)
2″-*O*-Rhamnoside	*Crataegus pentagyna* (Ros.) leaf and flower	Nikolov *et al.* (1976, 1982a, b)
	Passiflora biflora (Pas.) leaf	McCormick and Mabry (1983)
	Silene sp. (Car.) petal, leaf	Mastenbroek *et al.* (1983c)
6″-*O*-Arabinoside	*Swertia perennis* (Gen.) stem, leaf	Snyder *et al.* (1984)
*7-*O*-Rhamnosylglucoside	*Passiflora platyloba* (Pas.) leaf	Ayanoglu *et al.* (1982)
7, 2″-Di-*O*-glucoside	*Cerastium arvense* (Car.) leaf	Dubois *et al.* (1982a)
	Silene sp. (Car.) petal, leaf	Mastenbroek *et al.* (1983c)
*7, 2″-Di-*O*-galactoside	*Gentiana depressa* (Gen.) leaf	Chulia (1984)
7-*O*-Glucoside 2″-*O*-arabinoside	*Cerastium arvense* (Car.) fresh leaf and flower	Dubois *et al.* (1983c)
7-*O*-Glucoside 2″-*O*-rhamnoside	*Silene* sp. (Car.) petal, leaf	Mastenbroek *et al.* (1983c)
7-*O*-Galactoside 2″-*O*-glucoside	*Melandrium album* (Car.)	Wagner *et al.* (1979)
	Silene pratensis (Car.) cotyledon	Steyns *et al.* (1984)
*7-*O*-Galactoside 2″-*O*-arabinoside	*Silene pratensis* (Car.) leaf	Steyns *et al.* (1983)
7-*O*-Galactoside 2″-*O*-rhamnoside	*Melandrium album* (Car.)	Wagner *et al.* (1979)
	Silene pratensis (Car.) cotyledon	Steyns *et al.* (1984)
7-*O*-Xyloside 2″-*O*-glucoside }		
7-*O*-Xyloside 2″-*O*-arabinoside }	*Silene dioica* (Car.) petal	Mastenbroek *et al.* (1983c)
7-*O*-Xyloside 2″-*O*-rhamnoside }		
*7-*O*-Arabinoside 2″-*O*-glucoside }		
4′, 2″-Di-*O*-glucoside	*Gentiana asclepiadea* (Gen.) leaf	Goetz and Jacot-Guillarmod (1977a)
4′-*O*-Glucoside 2″-arabinoside	*Vaccaria segetalis* (Car.)	Baeva *et al.* (1974)
*X″-Arabinosylglucoside }	*Minuartia rossii* (Car.) leaf	Wolf *et al.* (1979)
*X″-Diglucoside }		
VITEXIN		
7-*O*-Glucoside	*Mollugo oppositifolia* (Mol.) fresh aerial part	Chopin *et al.* (1984)
4′-*O*-Glucoside	*Passiflora coactilis* (Pas.) leaf	Escobar *et al.* (1983)
*4′-*O*-Galactoside	*Crotalaria retusa* (Leg.) seeds	Srinivasan and Subramanian (1983)
2″-*O*-Glucoside	*Desmodium triflorum* (Leg.)	Adinarayana and Syamasundar (1982)

Table 3.4 (*Contd.*)

Compounds	*Sources*	*References*
	Mollugo cerviana (Mol.) whole plant	Nair and Gunasegaran (1983b)
	Vepris heterophylla (Rut.) leaf	Gomes *et al.* (1983)
	Silene pratensis (Car.) petal, leaf	Mastenbroek *et al.* (1983c)
	Passiflora palmeri (Pas.) dried leaf	Ulubelen *et al.* (1984)
	Gnetum buchholzianum (Gne.) leaf, stem	Ouabonzi *et al.* (1983)
	Podocarpus sp. (Pod.) foliage	Markham *et al.* (1984, 1985)
*X″-*O*-Glucoside	*Linanthus* sp. (Pol.) leaf	Smith *et al.* (1982)
2″-*O*-Xyloside	*Passiflora* sp. (Pas.) leaf	Ulubelen *et al.* (1981, 1982b, c, d) Ulubelen and Mabry (1983)
	Silene alba (Car.) petal	van Brederode and Kamps-Heinsbroek (1981a)
	Gnetum buchholzianum (Gne) leaf, stem	Ouabonzi *et al.* (1983)
	Desmodium triflorum (Leg.) aerial part	Sreenivasan and Sankara-subramanian (1984)
6″-*O*-Xyloside	*Gypsophila paniculata* (Car.) leaf	Darmograi *et al.* (1968)
X″-*O*-Arabinoside	*Trichomanes venosum* (Hym.)	Markham and Wallace (1980)
2″-*O*-Rhamnoside	*Crataegus monogyna* (Ros.)	Nikolov *et al.* (1976, 1982a, b)
(ex 4′-*O*-rhamnoside)	*Crataegus oxyacantha* (Ros.) leaf, stem, inflorescence	Chopin *et al.* (1977b) (structure)
	Passiflora palmeri (Pas.) dried leaf	Ulubelen *et al.* (1984)
	Glycine max (Leg.) root	Jay *et al.* (1984a)
	Dacrycarpus dacrydioides (Pod.) foliage	Markham and Whitehouse (1984)
	Fortunella japonica (Rut.) peelings	Kumamoto *et al.* (1985b)
	Podocarpus acutifolius (Pod.) foliage	Markham *et al.* (1985)
6″-*O*-Rhamnoside (?)	*Larix sibirica* (Pin.) needles	Medvedeva *et al.* (1974)
7-*O*-Rutinoside	*Phoenix* sp. (Pal.) leaf	Williams *et al.* (1973)
2″-*O*-Sophoroside	*Polygonatum multiflorum* (Lil.) leaf	Morita *et al.* (1976)
6″-*O*-Gentiobioside (marginatoside)	*Piper marginatum* (Pip.) leaf	Tillequin *et al.* (1978)
*4′-*O*-Glucoside 2″-*O*-rhamnoside	*Passiflora coactilis* (Pas.) leaf	Escobar *et al.* (1983)
6-*C*-XYLOSYLAPIGENIN (cerarvensin)		
*7-*O*-Glucoside	*Cerastium arvense* (Car.) fresh leaf	Dubois *et al.* (1982a)
2″-*O*-Rhamnoside	*Phlox drummondii* (Pol.) flower	Bouillant *et al.* (1978) (structure)
6-*C*-ARABINOSYLAPIGENIN (isomollupentin)		
*7-*O*-Glucoside	*Cerastium arvense* (Car.) leaf, flower	Dubois *et al.* (1985)
*4′-*O*-Glucoside	*Cerastium arvense* (Car.) leaf, flower	Dubois *et al.* (1985)
*2″-*O*-Glucoside	*Cerastium arvense* (Car.) leaf, flower	Dubois *et al.* (1985)
*7-*O*-Rhamnosylglucoside	*Passiflora platyloba* (Pas.) leaf	Ayanoglu *et al.* (1982)
7,2″-Di-*O*-glucoside	*Cerastium arvense* (Car.) fresh (leaf and flower)	Dubois *et al.* (1983)
*7-*O*-Glucoside 2″-*O*-xyloside	*Cerastium arvense* (Car.) fresh (leaf and flower)	Dubois *et al.* (1983)
*7-*O*-Glucoside 2″-*O*-arabinoside	*Cerastium arvense* (Car.) fresh (leaf and flower)	Dubois *et al.* (1983)
6-*C*-RHAMNOSYLAPIGENIN (isofurcatain)		
*7-*O*-Glucoside	*Metzgeria furcata* (Hep.)	Markham *et al.* (1982)
3′-DEOXYDERHAMNOSYLMAYSIN		
2″-*O*-Rhamnoside (3′-deoxymaysin)	*Zea mays* (Gra.) corn silk	Elliger *et al.* (1980)
SWERTISIN		
*5-*O*-Glucoside	*Enicostemma hyssopifolium* (Gen.) whole plant	Ghosal and Jaiswal (1980)
4′-*O*-Glucoside	*Commelina* sp. (Comm.) flower	Komatsu *et al.* (1968)
*4′-*O*-Rhamnoside	*Passiflora biflora* (Pas.) leaf	McCormick and Mabry (1983)

(*Contd.*)

Table 3.4 (*Contd.*)

Compounds	*Sources*	*References*
2″-*O*-Glucoside (spinosin)	*Zizyphus vulgaris* (Rha) leaf	Woo *et al.* (1980b); Shin *et al.* (1982)
(flavoayamenin)	*Iris nertshinskia* (Iri.) fresh petal	Hirose *et al.* (1981)
	Cayaponia tayuya (Cuc.) root	Bauer *et al.* (1985)
*2″-*O*-Rhamnoside	*Gemmingia chinensis* leaf	Shirane *et al.* (1982)
X″-*O*-Xyloside	*Iris tingitana* (Iri.)	Asen *et al.* (1970)
ISOSWERTISIN		
*5-*O*-Glucoside	*Enicostemma hyssopifolium* (Gen.) whole plant	Ghosal and Jaiswal (1980)
4′-*O*-Glucoside	*Triticum aestivum* (Gra.) leaf	Julian *et al.* (1971)
*2″-*O*-Glucoside	*Gnetum* sp. (Gne.) leaf, stem	Ouabonzi *et al.* (1983)
*2″-*O*-Xyloside	*Gnetum* sp. (Gne.) leaf, stem	Ouabonzi *et al.* (1983)
2″-*O*-Rhamnoside	*Gnetum* sp. (Gne.) leaf, stem	Ouabonzi *et al.* (1983)
	Avena sativa (Gra.) leaf	Knogge and Weissenböck (1984)
ISOMOLLUDISTIN		
2″-*O*-Glucoside	*Asterostigma riedelianum* (Ara.) leaf	Markham and Williams (1980)
MOLLUDISTIN		
2″-*O*-Glucoside	*Almeidea guyanensis* (Rut.) bark of stem	Jay *et al.* (1979)
*2″-*O*-Xyloside	*Almeidea guyanensis* (Rut.) bark of stem and root	Wirasutisna *et al.* (1986)
2″-*O*-Rhamnoside	*Mollugo distica* (Mol.) aerial part	Chopin *et al.* (1978)
ISOCYTISOSIDE		
7-*O*-Glucoside	*Gentiana pyrenaica* (Gen.) aerial part	Marston *et al.* (1976)
2″-*O*-Glucoside	*Securigera coronilla* (Leg.) leaf	Jay *et al.* (1980)
2″-*O*-Rhamnoside	*Fortunella japonica* (Rut.) peelings	Kumamoto *et al.* (1985b)
CYTISOSIDE		
7-*O*-Glucoside	*Trema cannabina* (Ulm.) leaf	Higa *et al.* (1983)
2″-*O*-Rhamnoside	*Fortunella japonica* (Rut.) peelings	Kumamoto *et al.* (1985b)
EMBIGENIN		
*2″-*O*-Glucoside (embinoidin)	*Siphonoglossa sessilis* (Aca.) stem, leaf	Hilsenbeck and Mabry (1983)
2″-*O*-Rhamnoside (embinin)	*Siphonoglossa sessilis* (Aca.) stem, leaf	Hilsenbeck and Mabry (1983)
	Iris ensata (Iri.)	Kitanov and Pryakhina (1985)
*X″-*O*-Glucoside	*Linanthus* sp. (Pol.) leaf	Smith *et al.* (1982)
7, 4′-DI-*O*-METHYLISOMOLLUPENTIN		
2″-*O*-Glucoside	*Asterostigma riedelianum* (Ara.) leaf	Markham and Williams (1980)
ISO-ORIENTIN		
7-*O*-Glucoside (lutonarin)	Many sources	
3′-*O*-Glucoside	*Gentiana pedicellata* (Gen.) leaf	Chulia (1984)
3′-*O*-Glucuronide	*Rhynchospora eximia* (Cyp.) leaf	Williams and Harborne (1977a)
4′-*O*-Glucoside	*Gentiana* sp. (Gen.) aerial part	Hostettmann-Kaldas *et al.* (1981)
	Passiflora pavonis (Pas.) leaf	McCormick and Mabry (1981)
	Gentiana algida (Gen.) aerial part	Oyuungerel *et al.* (1984)
2″-*O*-Mannoside	*Poa annua* (Gra.)	Rofi and Pomilio (1986)
2″-*O*-Glucoside	*Gentiana depressa* (Gen.) leaf	Chulia (1984)
	Silene pratensis (Car.) leaf	Mastenbroek *et al.* (1983c)
	Oryza sativa (Gra.) whole plant	Besson *et al.* (1985)
2″-*O*-Xyloside (?)	*Desmodium canadense* (Leg.)	Chernobrovaya (1973)
2″-*O*-Arabinoside	*Leucocyclus formosus* (Comp.) leaf	Valant-Vetschera (1982b)
2″-*O*-β-L-Arabinofuranoside	*Trichomanes venosum* (Hym.)	Markham and Wallace (1980)
2″-*O*-Rhamnoside	*Crataegus pentagyna* (Ros.) leaf and flower	Nikolov *et al.* (1976, 1982a, b)
	Passiflora biflora (Pas.) leaf	McCormick and Mabry (1983)
	Silene alba (Car.) green part	van Brederode and Kamps-Heinsbroek (1981b)

Table 3.4 (*Contd.*)

Compounds	*Sources*	*References*
*6″-*O*-Glucoside	*Gentiana pedicellata* (Gen.) leaf	Chulia and Mariotte (1985)
6″-*O*-Arabinoside	*Swertia perennis* (Gen.) leaf, stem	Snyder *et al.* (1984)
7-*O*-Rutinoside	*Triticum aestivum* (Gra.) leaf	Julian *et al.* (1971)
*3′-*O*-Sophoroside	*Plagiomnium affine* gametophyte	Freitag *et al.* (1986)
*3′-*O*-Neohesperidoside	*Plagiomnium affine* gametophyte	Freitag *et al.* (1986)
*3′,6″-Di-*O*-glucoside	*Gentiana pedicellata* (Gen.) leaf	Chulia and Mariotte (1985)
4′,2″-Di-*O*-glucoside	*Gentiana asclepiadea* (Gen.) leaf	Goetz and Jacot-Guillarmod (1977a)
ORIENTIN		
7-*O*-Glucoside (bisulphate)	*Phoenix canariensis* (Pal.) female flower	Harborne *et al.* (1974)
7-*O*-Rhamnoside	*Linum usitatissimum* (Lin.)	Ibrahim and Shaw (1970)
4′-*O*-Glucoside	*Passiflora coactilis* (Pas.) leaf	Escobar *et al.* (1983)
	Genista patula (Leg.) aerial part	Ozimina (1981)
2″-*O*-Glucoside	*Silene pratensis* (Car.) leaf	Mastenbroek *et al.* (1983c)
	Mollugo cerviana (Mol.) whole plant	Nair and Gunasegaran (1983b)
	Podocarpus sp. (Pod.) foliage	Markham *et al.* (1984, 1985)
	Stenotaphrum secundatum (Gra.) aerial part	Ferreres *et al.* (1984)
*X″-*O*-Glucoside	*Linanthus* sp. (Pol.) leaf	Smith *et al.* (1982)
2″-*O*-Xyloside	*Adonis leiosepala* (Ran.)	Evdokimov (1979)
	Adonis flammens (Ran.)	Komissarenko *et al.* (1981)
	Silene pratensis (Car.) leaf	Mastenbroek *et al.* (1983c)
	Stenotaphrum secundatum (Gra.) aerial part	Ferreres *et al.* (1984)
2″-*O*-β-L-Arabinofuranoside	*Trichomanes venosum* (Hym.)	Markham and Wallace (1980)
2″-*O*-Rhamnoside	*Crataegus pentagyna* (Ros.) leaf and flower	Nikolov *et al.* (1976, 1982a, b)
	Passiflora coactilis (Pas.) leaf	Escobar *et al.* (1983)
	Dacrycarpus dacrydioides (Pod.) foliage	Markham and Whitehouse (1984)
	Fortunella japonica (Rut.) peelings	Kumamoto *et al.* (1985b)
	Podocarpus acutifolius (Pod.) foliage	Markham *et al.* (1985)
*4′-*O*-Glucoside 2″-*O*-rhamnoside	*Passiflora coactilis* (Pas.) leaf	Escobar *et al.* (1983)
*X″-*O*-Rutinoside	*Linanthus* sp. (Pol.) leaf	Smith et *al.* (1982)
6-*C*-XYLOSYL-LUTEOLIN		
2″-*O*-Rhamnoside	*Phlox drummondii* (Pol.) flower	Bouillant *et al.* (1978) (structure)
SWERTIAJAPONIN		
3′-*O*-Glucoside	*Phragmites australis* (Gra.) flower	Nawwar *et al.* (1980)
*4′-*O*-Rhamnoside	*Passiflora biflora* (Pas.) leaf	McCormick and Mabry (1983)
*2″-*O*-Glucoside (luteoayamenin)	*Iris nertshinskia* (Iri.) fresh petal	Hirose *et al.* (1981)
2″-*O*-Rhamnoside	*Securigera coronilla* (Leg.) leaf	Jay *et al.* (1980)
3′-*O*-Gentiobioside	*Phragmites australis* (Gra.) flower	Nawwar *et al.* (1980)
ISOSCOPARIN		
7-*O*-Glucoside	*Gentiana pyrenaica* (Gen.) aerial part	Marston *et al.* (1976)
	Silene pratensis (Car.) leaf	Mastenbroek *et al.* (1983c)
*2″-*O*-Glucoside	*Oryza sativa* (Gra.) whole plant	Besson *et al.* (1985) (structure)
	Silene pratensis (Car.) leaf	Mastenbroek *et al.* (1983c)
*2″-*O*-Rhamnoside	*Silene alba* (Car.) green part	van Brederode and Kamps-Heinsbroek (1981b)
SCOPARIN		
*2″-*O*-Rhamnoside	*Passiflora coactilis* (Pas.) leaf	Escobar *et al.* (1983)
X″-*O*-Rhamnosylglucoside	*Crocus reticulatus* (Iri.) leaf	Sergeyeva (1977)
Episcoparin 7-*O*-glucoside (knautoside)	*Knautia montana* (Dip.) flower	Zemtsova and Bandyukova (1974)
	Scabiosa sp., *Cephalaria* sp. (Dip.)	Aliev and Movsumov (1981)

(*Contd.*)

Table 3.4 (*Contd.*)

Compounds	*Sources*	*References*
8-*C*-GLUCOSYLDIOSMETIN		
*2″-*O*-Rhamnoside	*Fortunella japonica* (Rut.) peelings	Kumamoto *et al.* (1985b)
7, 3′, 4′-TRI-*O*-METHYLISO-ORIENTIN		
2″-*O*-Rhamnoside (linoside B)	*Linum maritimum* (Lin.) aerial part	Chari *et al.* (1978)
ISOPYRENIN		
7-*O*-Glucoside	*Gentiana pyrenaica* (Gen.) aerial part	Marston *et al.* (1976)
6-*C*-GLUCOSYL-5, 7-DIHYDROXY -8, 3′, 4′, 5′-TETRAMETHOXYFLAVONE		
5-*O*-Rhamnoside	*Vitex negundo* (Ver.) bark	Subramanian and Misra (1979)
PUERARIN		
O-Xyloside	*Pueraria thunbergiana* (Leg.) root	Murakami *et al.* (1960)
*6″-*O*-*β*-Apiofuranoside (mirificin)	*Pueraria mirifica* (Leg.) root	Ingham *et al.* (1986)
VICENIN-2		
*6″-*O*-Glucoside	*Stellaria holostea* (Car.) whole plant	Bouillant *et al.* (1984)
SCHAFTOSIDE		
*6″-*O*-Glucoside	*Stellaria holostea* (Car.) whole plant	Bouillant *et al.* (1984)
6, 8-DI-*C*-GLUCOSYLTRICETIN		
*X‴-*O*-Rhamnoside	*Radula complanata* (Hep.)	Mues (1984)

dioica (Mastenbroek *et al.*, 1983b). An isovitexin X″-*O*-arabinosylglucoside and an isovitexin X″-*O*-diglucoside have been identified by UV data in *Minuartia rossii* (Wolf *et al.*, 1979). The structure of vitexin 4′-*O*-galactoside from *Crotalaria retusa* seeds (Srinivasan and Subramanian, 1983) was deduced from UV data. Vitexin 4′-*O*-glucoside-2″-*O*-rhamnoside, orientin 4′-*O*-glucoside-2″-*O*-rhamnoside and scoparin 2″-*O*-rhamnoside have been isolated from *Passiflora coactilis* by Escobar *et al.* (1983) and their structures deduced from UV, ^{1}H and ^{13}C NMR spectroscopy and enzymic partial hydrolysis. *Cerastium arvense* was found to be a rich source of new 6-*C*-pentosylapigenin *O*-glycosides: 6-*C*-xylosylapigenin 7-*O*-glucoside (Dubois *et al.*, 1982a), 6-*C*-arabinosylapigenin 7-*O*-glucoside, 4′-*O*-glucoside, 2″-*O*-glucoside (Dubois *et al.*, 1985), 7,2″-di-*O*-glucoside, 7-*O*-glucoside 2″-*O*-xyloside and 7-*O*-glucoside 2″-*O*-arabinoside (Dubois *et al.*, 1983), which have been identified by mass spectrometry of the permethyl ethers. Isofurcatain (6-*C*-rhamnosylapigenin 7-*O*-glucoside) isolated from *Metzgeria furcata* has been fully characterized by UV, ^{1}H and ^{13}C NMR and mass spectrometry of the permethyl ether (Markham *et al.*, 1982).

Swertisin and isoswertisin 5-*O*-glucosides have been identified from *Enicostemma hyssopifolium* by UV data (Ghosal and Jaiswal, 1980), swertisin and swertiajaponin 4′-*O*-rhamnosides from *Passiflora biflora* by UV, ^{1}H NMR and mass spectrometry of permethyl ethers (McCormick and Mabry, 1983), swertisin 2″-*O*-rhamnoside from *Gemmingia chinensis* by UV, ^{1}H and ^{13}C NMR spectra and mass spectrometry of permethyl ether (Shirane *et al.*, 1982), isoswertisin 2″-*O*-glucoside and 2″-*O*-xyloside from *Gnetum* species by mass spectrometry of permethyl ethers (Ouabonzi *et al.*, 1983). 8-*C*-Arabinosylacacetin 2″-*O*-xyloside has been characterized from *Almeidea guyanensis* by mass spectrometry of the permethyl ether (Wirasutisna *et al.*, 1986), embigenin 2″-*O*-glucoside (embinoidin) from *Siphonoglossa sessilis* by mass spectrometry of the permethyl ether (Hilsenbeck and Mabry, 1983) and an embigenin X″-*O*-glucoside from *Linanthus* species by UV data (Smith *et al.*, 1982). Iso-orientin 2″-*O*-α-D-mannopyranoside (gordin) isolated from *Poa annua* by Rofi and Pomilio (1986) is the first example of an *O*-mannoside among the flavonoids. Its structure has been fully established by UV, ^{1}H and ^{13}C NMR spectra, mass spectrometry of the permethyl ether and enzymic hydrolysis. Iso-orientin 6″-*O*-glucoside and 3′,6″-di-*O*-glucoside have been characterized from *Gentiana pedicellata* by UV, ^{1}H and ^{13}C NMR spectra and mass spectrometry of the permethyl ethers (Chulia and Mariotte, 1985). Iso-orientin 3′-*O*-sophoroside and 3′-*O*-neohesperidoside have been isolated from *Plagiomnium affine* by Freitag *et al.* (1986) and identified by UV, ^{1}H and ^{13}C NMR spectra and mass spectrometry of the permethyl ethers.

An orientin X″-*O*-glucoside and an orientin X″-*O*-rutinoside from *Linanthus* species have been identified by UV data (Smith *et al.*, 1982). Swertiajaponin 2″-*O*-glucoside (luteoayamenin) from *Iris nertshinskia* has been characterized by UV, ^{1}H and ^{13}C NMR spectra (Hirose *et al.*, 1981). Isoscoparin 2″-*O*-glucoside was found in *Silene pratensis* (Mastenbroek *et al.*, 1983c) and fully characterized in *Oryza sativa* (Besson *et al.*, 1985). Isoscoparin 2″-*O*-rhamnoside from *Silene alba* has been identified by UV data and the 2″-substitution was arrived at on the basis of genetic experiments (van Brederode and Kamps-Heinsbroek, 1981b). 8-*C*-Glucosyldiosmetin 2″-*O*-rhamnoside isolated from *Fortunella japonica* by Kumamoto *et al.* (1985b) has been fully characterized by UV, ^{1}H and ^{13}C NMR spectra.

Mirificin, the first *O*-apioside found among *C*-glycosylflavonoids, is puerarin 6″-*O*-β-apiofuranoside isolated from *Pueraria mirifica*. Its structure has been thoroughly established by UV, ^{1}H and ^{13}C NMR spectroscopy (Ingham *et al.*, 1986). 6″-*O*-Glucosides of vicenin 2 and schaftoside isolated from *Stellaria holostea* by Bouillant *et al.* (1984) and an X‴-*O*-rhamnoside of 6, 8-di-*C*-glucosyltricetin isolated from *Radula complanata* by Mues (1984) have been characterized by mass spectrometry of the permethyl ethers.

3.2.4 Acyl *C*-glycosylflavonoids

Fifteen new *O*-acylated derivatives of mono- and di-*C*-glycosylflavones, *C*-glycosylflavone and di-*C*-glycosylflavone *O*-glycosides have been reported during the last five years. Acetyl, *p*-coumaryl, ferulyl, sinapyl and 2, 4, 5-trihydroxycinnamyl groups are linked to sugar hydroxyl groups and, in most cases, ^{13}C NMR spectroscopy has been used to locate the acyl group.

An isovitexin 7-*O*-ferulylglucoside isolated from *Silene pratensis* by Niemann (1981, 1982) was characterized by UV spectrometry. 2″-*O*-Acetylvitexin isolated from

Table 3.5 Naturally occurring acyl-C-glycosylflavonoids

Compounds	*Sources*	*References*
ISOVITEXIN		
★7-*O*-Ferulylglucoside	*Silene pratensis* (Car.)	Niemann (1981, 1982)
2″-*O*-(*E*)-Ferulyl	*Gentiana punctata* (Gen.) leaf	Luong and Jacot-Guillarmod (1977)
2″-*O*-(*E*)-Ferulyl 4′-*O*-glucoside }	*Cerastium arvense* (Car.) leaf, flower	Dubois *et al.* (1985)
X″-*O*-(*E*)-Caffeyl 2″-*O*-glucoside }		
7-Sulphate	*Cucumis melo* (Cuc.) leaf	Monties *et al.* (1976)
	Palmae leaf, flower	Williams *et al.* (1973); Harborne *et al.* (1974)
VITEXIN		
★2″-*O*-Acetyl	*Crataegus sanguinea* (Ros.) flower	Kashnikova *et al.* (1984)
2″-*O*-*p*-Hydroxybenzoyl	*Vitex lucens* (Ver.) wood	Horowitz and Gentili (1966)
2″-*O*-*p*-Coumaryl	*Trigonella foenum-graecum* (Leg.) seed	Sood *et al.* (1976)
★2″-*O*-*p*-Coumaryl 7-*O*-glucoside	*Mollugo oppositifolia* (Mol.) fresh aerial part	Chopin *et al.* (1984)
4‴-*O*-Acetyl 2″-*O*-rhamnoside	*Crataegus monogyna* (Ros.) leaf and flower	Nikolov *et al.* (1976, 1982a, b)
7-Sulphate }	Palmae leaf, flower	Williams *et al.* (1973)
7-*O*-Rutinosidesulphate }		Harborne *et al.* (1974)
8-*C*-GALACTOSYLAPIGENIN		
6″-*O*-Acetyl	*Briza media* (Gra.) leaf	Chari *et al.* (1980) (structure)
SWERTISIN		
★6‴-*O*-*p*-Coumaryl 2″-*O*-glucoside }		
★6‴-*O*-Ferulyl 2″-*O*-glucoside }	*Zizyphus jujuba* (Rha.) seed	Woo *et al.* (1980a)
★6‴-*O*-Sinapyl 2″-*O*-glucoside }		
ISOSWERTISIN		
2″-*O*-Acetyl	*Brackenridgea zanguebarica* (Och.) leaf	Bombardelli *et al.* (1974)
CYTISOSIDE		
O-Acetyl-7-glucoside (tremasperin)	*Trema aspera* (Ulm.) leaf	Oelrichs *et al.* (1968)
	Trema cannabina (Ulm.) leaf	Higa *et al.* (1983)
EMBIGENIN		
★2‴-*O*-Acetyl 2″-*O*-rhamnoside	*Iris lactea* (Iri.) aerial part	Pryakhina *et al.* (1984)
★X″, 2‴-Di-*O*-acetyl 2″-*O*-rhamnoside		

Table 3.5 (*Contd.*)

Compounds	*Sources*	*References*
7, 4′-DI-*O*-METHYLISOMOLLUPENTIN		
2″-*O*-Caffeylglucoside	*Asterostigma riedelianum* (Ara.) leaf	Markham and Williams (1980)
ISO-ORIENTIN		
X″-*O*-Acetyl	*Crataegus monogyna* (Ros.)	Nikolov *et al.* (1982b)
2″-*O*-*p*-Hydroxybenzoyl	*Gentiana asclepiadea* (Gen.) leaf	Goetz and Jacot-Guillarmod (1978)
2″-*O*-(*E*)-*p*-Coumaryl } 2″-*O*-(*E*)-Caffeyl }	*Gentiana X marcailhouana* (Gen.) leaf	Luong *et al.* (1980)
2″-*O*-(*E*)-Ferulyl		
2″-*O*-*p*-Hydroxybenzoyl 4′-*O*-glucoside	*Gentiana asclepiadea* (Gen.) leaf	Goetz and Jacot-Guillarmod (1978)
2″-*O*-(*E*)-Caffeyl 4′-*O*-glucoside	*Gentiana punctata* (Gen.) leaf	Luong and Jacot-Guillarmod (1977)
2″-*p*-*O*-Glucosyl-(*E*)-caffeyl 4′-*O*-glucoside	*Gentiana burseri* (Gen.) leaf	Jacot-Guillarmod *et al.* (1975)
2″-*O*-(*E*)-Ferulyl 4′-*O*-glucoside	*Cerastium arvense* (Car.) leaf, flower	Dubois *et al.* (1985)
2″-*O*-(*E*)-Caffeylglucoside	*Cucumis melo* (Cuc.) leaf	Monties *et al.* (1976)
*2″-*O*-[4-*O*-Glucosyl-2, 4, 5-trihydroxy-(*E*)-cinnamyl] 4′-*O*-glucoside	*Gentiana X marcailhouana* (Gen.) leaf	Luong *et al.* (1981)
7-Sulphate	Palmae leaf, flower	Williams *et al.* (1973), Harborne *et al.* (1974)
ORIENTIN		
*2″-*O*-Acetyl	*Hypericum hirsutum* (Clu.) aerial part	Kitanov *et al.* (1979)
7-Sulphate } 7-*O*-Glucosidesulphate }	Palmae leaf, flower	Williams *et al.* (1973); Harborne *et al.* (1974)
ISOSCOPARIN		
*6‴-*O*-*p*-Coumaryl 2″-*O*-glucoside } *6‴-*O*-Ferulyl 2″-*O*-glucoside }	*Oryza sativa* (Gra.) whole plant	Besson *et al.* (1985)
SCOPARIN		
6″-*O*-Acetyl	*Sarothamnus scoparius* (Leg.) twig	Brum-Bousquet *et al.* (1977)
7, 3′, 4′-TRI-*O*-METHYLISO-ORIENTIN		
6″-*O*-Acetyl 2″-*O*-rhamnoside (linoside A)	*Linum maritimum* (Lin.) aerial part	Chari *et al.* (1978) (structure)
VICENIN-2		
*X‴-*O*-Diferulylglucoside	*Spergularia rubra* (Car.) whole plant	Bouillant *et al.* (1984)
VICENIN-1		
Sinapyl	*Triticum* (wheat germ) (Gra.)	Galle (1974)
ISOSCHAFTOSIDE		
*2‴-*O*-Ferulyl	*Metzgeria* sp. (Hep.)	Theodor *et al.* (1981b, 1983)
Sinapyl	*Triticum* (wheat germ) (Gra.)	Galle (1974)
CORYMBOSIDE		
*X‴-*O*-Ferulyl } *X‴-*O*-Sinapyl }	*Triticum aestivum* (Gra.) seedling	Wagner *et al.* (1980)
PUERARIN		
4′, 6″-Di-*O*-acetyl	*Pueraria tuberosa* (Leg.) root	Bhutani *et al.* (1969)

Crataegus sanguinea by Kashnikova *et al.* (1984) and 2″-*O*-acetylorientin isolated from *Hypericum hirsutum* by Kitanov *et al.* (1979) have been identified by the high-field shift of the acetyl group. 2″-*O*-*p*-Coumarylvitexin 7-*O*-glucoside has been isolated from *Mollugo oppositifolia* by Chopin *et al.* (1984) and its structure established by ^{13}C NMR spectroscopy.

Similarly the structures of 6‴-*O*-*p*-coumaryl-, 6‴-*O*-ferulyl- and 6‴-*O*-sinapyl-swertisin 2″-*O*-glucoside isolated from *Zizyphus jujuba* seed by Woo *et al.* (1980a) and of 6‴-*O*-*p*-coumaryl- and 6‴-*O*-ferulyl-isoscoparin 2″-*O*-glucoside isolated from *Oryza sativa* by Besson *et al.* (1985) were deduced from their ^{13}C NMR spectra. The positions of the acetyl groups in 2‴*O*-acetyl and X″, 2‴-di-*O*-acetylembinin isolated from *Iris lactea* by Pryakhina *et al.* (1984) were determined only on the basis of their chemical shifts in the ^{1}H NMR spectra.

The new acyl group, 2, 4, 5-trihydroxycinnamyl, has been found as the 4-*O*-glucoside in iso-orientin 2″-*O*-[4-*O*-*β*-D-glucosyl-2, 4, 5-trihydroxy-(*E*)cinnamyl] 4′-*O*-glucoside isolated from *Gentiana marcailhouana* by Luong *et al.* (1981). Its structure was demonstrated by enzymic, chemical and spectroscopic (UV, ^{1}H NMR of the acetate) procedures.

X‴-*O*-ferulyl and X‴-*O*-sinapyl corymbosides have been found in *Triticum aestivum* seedlings by Wagner *et al.* (1980) and characterized by mass spectrometry of permethyl and permethyl isopropylidene derivatives and ^{13}C NMR spectrometry. The structure of 2‴-*O*-ferulylisoschaftoside from *Metzgeria* species first deduced from mass spectrometry of the permethyl ether has now been confirmed by ^{13}C NMR spectrometry (Theodor *et al.*, 1981b, 1983). An X‴-*O*-diferulylglucoside of vicenin-2 isolated from *Spergularia rubra* was characterized by mass spectrometry of the permethyl derivative (Bouillant *et al.*, 1984).

3.3 SYNTHESIS OF *C*-GLYCOSYLFLAVONOIDS

The high-yield *C*-glucosylation of 1, 3, 5-trimethoxybenzene with tetra-acetyl-α-D-glucosyl bromide described by Eade and Pham (1979) in their synthesis of 5, 7, 4′-tri-*O*-methylvitexin has not yet been followed by the synthesis of other 5, 7, 4′-tri-*O*-methyl-8-*C*-glycosylapigenins. However, the reaction of 1, 3, 5-trimethoxybenzene with tetra-acetyl-α-D-galactopyranosyl bromide, triacetyl-α-D-xylopyranosyl bromide, triacetyl-*β*-L-arabinopyranosyl bromide and triacetyl-α-L-rhamnopyranosyl bromide was successfully used by Chari (unpublished) to synthesize the corresponding 1-(2′, 4′, 6′-trimethoxyphenyl)-1, 5-anhydroalditols for ^{13}C NMR spectroscopy (Markham and Chari, 1982).

A synthesis of 7, 4′-di-*O*-methylisobayin (6-*C*-*β*-D-glucopyranosyl-7, 4′-dimethoxyflavone) has been described by Tschesche and Widera (1982) (Fig. 3.2). The reaction between 2, 4-dimethoxyphenylmagnesium bromide (*3.10*) and 2, 3, 4, 6-tetra-*O*-benzylglucopyranosyl chloride (*3.11*) gave 2′, 3′, 4′, 6′-tetra-*O*-benzyl-*β*-D-glucopyranosyl-2, 4-dimethoxybenzene

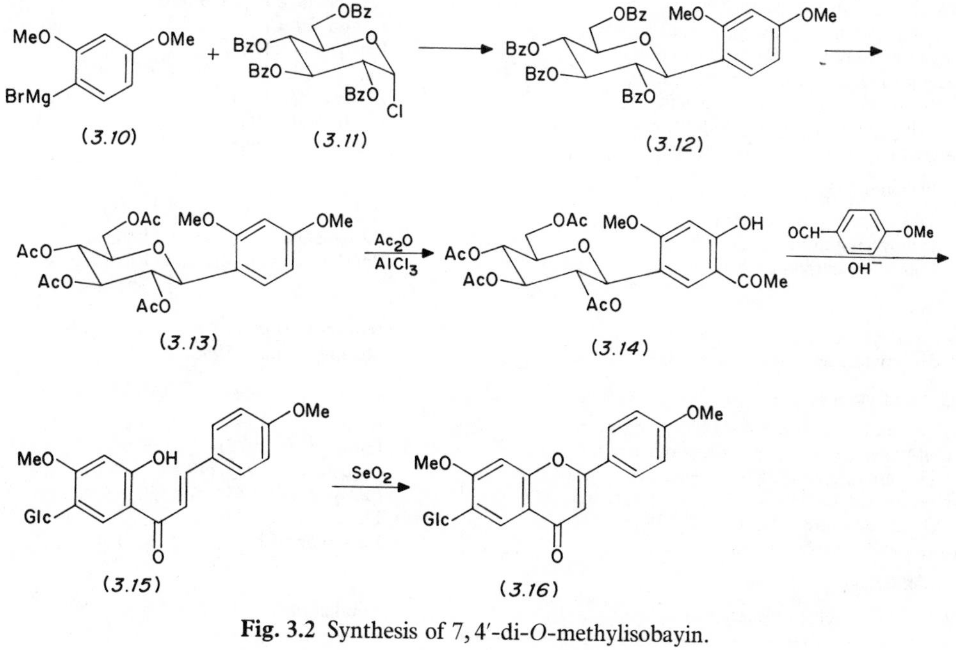

Fig. 3.2 Synthesis of 7, 4′-di-*O*-methylisobayin.

(*3.12*), converted to the tetra-acetate (*3.13*) after debenzylation. Reaction of (*3.13*) with acetic anhydride and anhydrous aluminium chloride led to 5-β-D-glucopyranosyl-2-hydroxy-4-methoxyacetophenone tetra-acetate (*3.14*). Condensation of the latter with 4-methoxybenzaldehyde in alkaline medium gave 5′-β-D-glucopyranosyl-2′-hydroxy-4,4′-dimethoxychalcone (*3.15*) from which reaction with selenium dioxide gave 7,4′, di-*O*-methylisobayin (*3.16*). It must be mentioned here that isobayin has not been found in nature and cannot be obtained from natural bayin by Wessely–Moser isomerization since the necessary 5-hydroxy group is lacking.

In spite of the low yields of 6-*C*-glycosylflavones obtained in the reaction of acetobromo sugars with 5,7-dihydroxyflavones, this method has been used in the authors' laboratory to prepare some standard compounds for mass spectral studies or to solve some specific problems. For example, 6-*C*-α-L-arabinofuranosylacacetin and 6-*C*-α-L-arabinopyranosylacacetin were prepared from acacetin and the corresponding acetobromo-L-arabinoses in order to study their acid isomerization products and to compare their chromatographic and spectral properties (Besson and Chopin, 1983). 6-*C*-α-D-Arabinopyranosylapigenin was similarly prepared from apigenin and the corresponding acetobromo-D-arabinose for circular dichroism studies (Dubois, unpublished). *C*-Glycosylation of 5,7-dihydroxy-3′,4′5′-trimethoxyflavone has been achieved with acetobromo-α-D-glucose, acetobromo-α-D-galactose, acetobromo-α-D-xylose, acetobromo-β-L-arabinose and acetobromo-α-L-rhamnose following the discovery of natural *C*-glycosyltricetins and *C*-glycosyltricins. The respective 6-*C*-glycosides and 6,8-di-*C*-glycosides (except for galactose) were obtained and permethylated for comparison with the permethyl derivatives of some natural compounds (Lardy *et al.*, 1983). 6-*C*-Galactosylation of vitexin with acetobromo-α-D-galactose led to 6-*C*-β-D-galactopyranosyl-8-*C*-β-D-glucopyranosylapigenin, the Wessely–Moser isomer of a new natural compound isolated from *Cerastium arvense* (Dubois *et al.*, 1984). 6-*C*-Glycosylation of cytisoside (8-*C*-glucosylacacetin) with the easily accessible acetobromocellobiose gave 6-*C*-cellobiosyl-8-*C*-glucosylacacetin, used for mass spectrometry of permethyl ethers (Bouillant *et al.*, 1984). 6-*C*-Glycosylation of cytisoside with acetobromo-α-L-arabinofuranose led to 6-*C*-β-L-arabinofuranosyl-8-*C*-β-D-glucopyranosylacacetin, used for ^{1}H NMR and mass spectrometry of the perdeuteriomethyl derivate (Besson and Chopin, unpublished). 6-*C*-β-D-Galactopyranosyl-, 6-*C*-β-D-xylopyranosyl- and 6-*C*-α-L-arabinopyranosylquercetin have been synthesized by Rasolojaona and Mastagli (1985) by *C*-glycosylation of quercetin with the

Table 3.6 New synthetic *C*-glycosylflavonoids from flavonoid *C*-glycosylation

Compound	Reference
6-*C*-Glycosylflavones	
6-*C*-α-D-Arabinopyranosylapigenin	Dubois, unpublished
6-*C*-α-L-Arabinopyranosylacacetin	Besson and Chopin (1983)
6-*C*-α-L-Arabinofuranosylacacetin	Besson and Chopin (1983)
6-*C*-β-D-Glucopyranosyl-4′-*O*-methyltricin	Lardy *et al.* (1983)
6-*C*-β-D-Galactopyranosyl-4′-*O*-methyltricin	Lardy *et al.* (1983)
6-*C*-β-D-Xylopyranosyl-4′-*O*-methyltricin	Lardy *et al.* (1983)
6-*C*-α-L-Arabinopyranosyl-4′-*O*-methyltricin	Lardy *et al.* (1983)
6-*C*-α-L-Rhamnopyranosyl-4′-*O*-methyltricin	Lardy *et al.* (1983)
6-*C*-Glycosylflavonols	
6-*C*-β-D-Galactopyranosylquercetin	Rasolojaona and Mastagli (1985)
6-*C*-β-D-Xylopyranosylquercetin	Rasolojaona and Mastagli (1985)
6-*C*-α-L-Arabinopyranosylquercetin	Rasolojaona and Mastagli (1985)
C-Glycosylflavanols	
6-*C*-β-D-Glucopyranosyl-(−)-epicatechin	Morimoto *et al.* (1986a)
8-*C*-β-D-Glucopyranosyl-(−)-epicatechin	Morimoto *et al.* (1986a)
6,8-di-*C*-glycosylflavones	
6,8-Di-*C*-β-D-glucopyranosyl-4′-*O*-methyltricin	Lardy *et al.* (1983)
6,8-Di-*C*-β-D-xylopyranosyl-4′-*O*-methyltricin	Lardy *et al.* (1983)
6,8-Di-*C*-α-L-arabinopyranosyl-4′-*O*-methyltricin	Lardy *et al.* (1983)
6,8-Di-*C*-α-L-rhamnopyranosyl-4′-*O*-methyltricin	Lardy *et al.* (1983)
6-*C*-β-D-Galactopyranosylvitexin	Dubois *et al.* (1984)
6-*C*-β-L-Arabinofuranosylcytisoside	Besson and Chopin, unpublished
6-*C*-Diglycosyl-8-*C*-glycosylflavone	
6-*C*-Cellobiosyl-8-*C*-glucosylacacetin	Bouillant *et al.* (1984)

appropriate acetobromo sugars. Acid-catalysed condensation of (−)-epicatechin and D-glucose (Morimoto *et al.*, 1986a) gave a mixture of 6-*C*- and 8-*C*-β-D-glucopyranosyl-(−)epicatechins. These new synthetic compounds are listed in Table 3.6.

3.4 IDENTIFICATION OF *C*-GLYCOSYLFLAVONOIDS

3.4.1 Chemical methods

Acid isomerization is still widely used for the characterization of new *C*-glycosylflavones. It can give complex mixtures in the case of *C*-arabinosylflavones such as 6,8-di-*C*-arabinosylapigenin due to the fact that both Wessely–Moser and sugar ring isomerizations occur. Besson and Chopin (1983) have shown that sugar ring isomerization of the *C*-arabinosyl moiety is faster than Wessely–Moser isomerization of the flavone moiety when *O*-methyl groups are present in the 7- or 4′-positions. Thus the four 6-*C*-α and β-L-arabinopyranosyl and furanosylacacetins were isolated via the acid isomerization of synthetic 6-*C*-α-L-arabinopyranosylacacetin, whereas natural 8-*C*-α-L-arabinopyranosylgenkwanin gave a mixture of the two 8-*C*-α-pyranosyl and furanosyl isomers as the main products. However, it must be mentioned here that no natural *C*-arabinofuranosylflavonoid has yet been isolated, since neoschaftoside, neoisoschaftoside and neocarlinoside have been shown to contain a *C*-β-L-arabinopyranosyl moiety. The use of trifluoroacetic acid may prevent Wessely–Moser isomerization (Markham, 1982). Treatment with hydriodic acid/phenol readily yielded the corresponding isoflavones from 8-*C*-glucosylgenistein, 8-*C*-glucosylorobol, 6,8-di-*C*-glucosylgenistein and 6,8-di-*C*-glucosylorobol (van Heerden *et al.*, 1980) but destroyed the hydroxydihydrochalcone from coatline A (Bezuidenhoudt *et al.*, 1986).

Ferric chloride oxidation has been used to identify the *C*-galactosyl residue in 6-*C*-galactopyranosylisoscutellarein and 6,8-di-*C*-galactopyranosylapigenin (Yasukawa *et al.*, 1982) and the *C*-rhamnosyl residue in 8-*C*-rhamnosyleuropetin (Tiwari and Bajpai, 1981) without ^{13}C NMR confirmation. Permethylation has been commonly used for structure determination by mass spectrometry and for thin layer chromatographic comparison with standard compounds.

Isopropylidenation of 6-*C*-β-xylopyranosyl-8-*C*-α-L-arabinopyranosylapigenin was used by Chopin *et al.* (1982) to establish the position of the reactive *C*-arabinopyranosyl moiety by mass spectrometry of the permethyl derivative. Tritylation of the 6-CH_2OH of the 8-*C*-glucopyranosyl residue in vicenin-l and of the 6-CH_2OH of the 6-*C*-glucopyranosyl residue in schaftoside with triphenylmethyl chloride was used by Bouillant *et al.* (1984) in the structural determination of 6-*C*-diglycosyl-8-*C*-glycosylflavones and 6-*C*-glycosyl-8-*C*-diglycosylflavones by mass spectrometry of their permethyl ethers.

3.4.2 Physical methods

(a) ^1H *NMR spectrometry*

Proton NMR spectrometry is still widely used to assign the 6- or 8-position to the glycosyl residue in mono-*C*-glycosylflavones and in 3,6- or 3,8-di-*C*-glycosylflavones. The method depends on measuring the chemical shift differences exhibited by H-3, H-6 and H-8 protons in 5,7-dihydroxyflavones. The high-field shift of the 2″-*O*-acetyl signal in the spectrum of acetylated derivatives has been observed for all new types of *C*-glucosylflavonoids, but it must be recalled that the presence of acetyl signals around δ 1.80 has been reported in the spectrum of peracetyl 6-*C*-glycosylflavone 2″-*O*-glycosides and cannot therefore be used as diagnostic for a 2″-*O*-acetyl group where 6-*C*-diglycosylflavones are concerned. The β-glucopyranosyl structure of the *C*-glycosyl residue could be deduced from 300 mHz spectra of coatline B (Beltrami *et al.*, 1982) and coatline A acetates (Bezuidenhoudt *et al.*, 1986) and from 400 mHz spectra of carthamin and safflor yellow A (with 2D *J*-resolution) (Takahashi *et al.*, 1982) and 500 mHz spectra of safflor yellow B acetate (Takahashi *et al.*, 1984). Perdeuteriomethyl ethers have been useful for the separation of anomeric protons from other sugar protons in the identification of α- and β-L-arabino-pyranosyl and -furanosyl isomers of *C*-arabinosylflavones (Besson and Chopin, 1983) and in the identification of the *C*-glycosyl residues in 6-*C*-β-D-xylopyranosyl-8-*C*-α-L-arabinopyranosylapigenin (Chopin *et al.*, 1982). A very high temperature (180°C) was required to induce free rotation of the sugar and B ring in the analysis of 300 mHz spectra of the methyl ether acetates of 8-*C*-glucosyl-5-deoxykaempferol and 8-*C*-glucosylfisetin (Bezuidenhoudt *et al.*, 1986).

(b) ^{13}C *NMR spectroscopy*

As expected, ^{13}C NMR spectroscopy remains by far the most important procedure for the structural analysis of *C*-glycosylflavonoids and Markham and Wong have contributed greatly to the structure determination (or confirmation) of a large number of new natural substances during the last five years. The main interest derives from the immediate distinction between the signals given by any flavonoid *O*-glycoside and by the corresponding *C*-glycoside and from the fact that the common sugars found in plants, either *O*- or *C*-linked give different patterns of signals. Moreover, characteristic shifts are induced by *O*-glycosylation of a flavonoid hydroxyl or a sugar hydroxyl group, by *C*-glycosylation of a flavonoid or by acylation of a sugar. It follows that, if sufficient amounts of substance are available, ^{13}C NMR spectroscopy can be used for the simultaneous identification of the *C*-glycosyl, *O*-glycosyl and *O*-acyl residues and of their respective sites of linkage. The only restriction is that the sugar signals remain the same in 6-*C*-glycosyl- and 8-*C*-glycosyl-flavones. When the two sugar residues are different in 6,8-di-*C*-glycosylflavones, it is not possible to determine their positions by ^{13}C NMR.

It must be noted that a downfield shift of about 3 ppm may be observed for the signals of C-1 and C-2 in the *C*-β-D-glucopyranosyl residue of epicatechin and procyanidin B2 *C*-glucosides, when compared with *C*-glucosylflavonoids bearing a conjugated carbonyl group.

(c) *Mass spectrometry*

The main advance in this field during the last five years has been the generalized use of fast atom bombardment (FAB-MS) and field-desorption (FD-MS) mass spectrometry in the determination of the molecular weight without prior derivatization. This is especially important in the case of alkali-labile compounds, such as acyl derivatives, which are often deacylated during permethylation. However, the usual electron impact mass spectrometry (EI-MS) of permethyl (PM) or perdeuteriomethyl (PDM) ethers still remains the best source of structural information and is more generally important when the available amounts of substance preclude the use of ^{13}C NMR spectroscopy. During the last five years, it has been successfully applied in many laboratories to the structural study of *C*-glycosylflavones, 6,8-di-*C*-glycosylflavones and 6-*C*-glycosylflavone *O*-glycosides, and the derived conclusions (see Chopin *et al.*, 1982) have been confirmed by ^{13}C NMR spectroscopy when both techniques were used.

Only once has the use of the relative intensities led to an incorrect conclusion. In the MS of 6-*C*-β-D-xylopyranosyl-8-*C*-α-L-arabinopyranosylapigenin isolated from *Mollugo pentaphylla*, the relative intensities of $[M-119]^+$, $[M-131]^+$ and $[M-145]^+$ ions were in the order $[M-131]^+ > [M-119]^+ > [M-145]^+$. Such an order had always been observed in the MS of PM 6-*C*-arabinosyl-8-*C*-hexosylflavones and in the MS of PM 6,8-di-*C*-arabinosylflavones. It was therefore postulated as characteristic of a 6-*C*-arabinosyl residue in 6,8-di-*C*-pentosylflavones in general. It has now been shown that this is not valid for *C*-arabinosyl-*C*-xylosylflavones and this was proved by the use of the MS of the PM isopropylidene derivative, which showed that the 6-*C*-glycosyl residue did not bear the isopropylidene group (Chopin *et al.*, 1982).

Superimposable peaks between m/z 510 and 297 (flavone ion) have been observed in the EI-MS of PM 7-*O*-glucosylvitexin and PM 7-*O*-glucosylisovitexin. This shows that, after loss of the 7-*O*-hexosyl residue, the fragmentation of the *C*-glycosyl residue is the same in both cases. However, the unambiguous determination of the position of the *C*-glucosyl residue remains possible from the base peak (for m/z values higher than 250) being $[M-31]^+$ for the 6-*C*- and $[M]^+$ for the 8-*C*-glycosylflavone derivative. Moreover, the rupture of the 7-*O*-glycosidic bond takes place without hydrogen transfer to the oxygen atom in the case of PM 7-*O*-glycosyl-6-*C*-glycosylflavones, whereas this transfer does take place in the case of PM 7-*O*-glycosyl-8-*C*-glycosylflavones (Chopin *et al.*, 1984).

Another interesting application of EI-MS of PM ethers has been in the field of 6-*C*-arabinosylflavones, since PM 6-*C*-arabinofuranosylflavones show the same fragmentation pattern as their 6-*C*-arabinopyranosyl isomers, but characteristic differences are observed in the relative intensities of some ions and in the metastable peaks (Besson and Chopin, 1983). This may be useful for the detection of as yet unknown 6-*C*-arabinofuranosylflavones in nature.

EI-MS of PM ethers has also been successfully extended to natural 6,8-di-*C*-glycosylflavones *O*-glycosylated on one sugar, which thus bear one *C*-monoglycosyl and one *C*-diglycosyl residue. In these cases, ^{13}C NMR spectroscopy can indicate the nature of the monoglycosyl and diglycosyl residues, but not their 6- or 8-position, because the signals of the *C*-glycosyl residue remain about the same in both cases. Fortunately, as in the case of PM 6,8-di-*C*-glycosylflavones themselves, the fragmentation pattern is characteristic of the 6-*C*-glycosyl residue, thus allowing easy separation of 6-*C*-monosaccharide residues (as found in X‴-*O*-glycosides, i.e. 6-*C*-glycosyl-8-*C*-diglycosylflavones) from 6-*C*-disaccharide residues (as found in X″-*O*-glycosides, i.e. 6-*C*-diglycosyl-8-*C*-glycosylflavones). Moreover, in the latter case, the position of the *O*-glycosidic bond can be deduced from the fragmentation patterns of the PM ether and its acid hydrolysis product, as in the case of PM 6-*C*-diglycosylflavones, since these fragmentation patterns are not affected by the presence of a 8-*C*-glycosyl residue (Bouillant *et al.*, 1984).

(d) *Circular dichroism*

The effect of conformation and configuration of arabinose residues on the circular dichroism (CD) of *C*-glycosylflavones has been analysed by Gaffield *et al.* (1984). Examination of a variety of arabinose-containing *C*-glycosylflavones has shown that the sign and intensity of the CD band at 250–275 nm (charge-transfer band) reflect not only the point of attachment of the sugar to the flavone, but also depend upon the absolute and anomeric configuration, ring size and ring conformation in addition to the preferred rotameric conformation of the sugar about the C—aryl—C—1″ bond. A change in stereochemistry of arabinose from the α- to β-anomer resulted in sign inversion of the 250–275 nm CD band for 6-*C*-L-arabinosylflavones. Furthermore, 6-*C*-α-D-arabinopyranosylapigenin exhibits an oppositely signed charge transfer CD band in comparison with 6-*C*-α-L-arabinopyranosylapigenin.

3.4.3 Chromatographic methods

In chemotaxonomic studies, the identification of *C*-glycosides is usually based on paper (PC) or thin-layer (TLC) co-chromatography with standard compounds. However, in the commonly used solvent systems it is not possible to distinguish *C*-galactosides from *C*-glucosides and *C*-arabinosides from *C*-xylosides. This probably explains why 8-*C*-galactosylapigenin has been so rarely identified, since permethylation and TLC of the permethyl ether is necessary to distinguish it from vitexin. Because the last five years have seen a rapid expansion of high-performance liquid chromatography (HPLC) in the field of flavonoid analysis, it was of interest to explore the possibilities of this technique to solve the problem, without derivatization. Lardy *et al.* (1984) have shown that reversed-phase HPLC using Lichrosorb RP-18 and simple isocratic eluting systems (methanol–water–acetic acid or acetonitrile–water) is able to separate *C*-galactosides from *C*-glucosides, *C*-arabinosides from *C*-xylosides and *C*-arabinofuranosides from *C*-arabinopyranosides. A very useful application is the clean separation of vicenins-1, -2 and -3 whereas their permethyl ethers are difficult to separate on TLC. Schaftoside, isoschaftoside and neoschaftoside are as well separated on HPLC as their permethyl ethers are on TLC. Isocorymboside is better separated from schaftoside than corymboside from isoschaftoside. Vicenin-3 is well separated from the former two and vicenin-1 from the latter two compounds. It may be hoped that this work will lead to a better recognition of the importance of *C*-galactosylflavones in plants.

On the other hand, HPLC has been mainly used for quantitative analysis of plant extracts of pharmacological interest: *Zizyphus* seeds (Shin *et al.*, 1982), *Crataegus oxyacantha* (Ficarra *et al.*, 1984a, b), *Combretum micranthum* (Bassene *et al.*, 1985), for chemotaxonomic comparison in *Dactylis glomerata* (Jay *et al.*, 1984b) and for flavonoid accumulation studies in *Avena sativa* and *Secale cereale* by Weissenböck and co-workers, and in *Silene pratensis* by Niemann *et al.* (1983). Post-column derivatization was applied by Hostettmann *et al.* (1984) to the analysis of crude extracts of various *Gentiana* species by HPLC coupled with UV-visible spectroscopy. A remarkable demonstration of the possibilities of reversed-phase HPLC on a semi-preparative scale has been furnished by Dubois with the isolation of about thirty apigenin- and luteolin-derived *C*-glycosylflavonoids from *Cerastium arvense*. However, most of the new *C*-glycosylflavonoids have been isolated by the usual chromatographic techniques. Rotation locular counter-current chromatography has been successfully applied by Snyder *et al.* (1984) to the isolation of *C*-glycosylflavonoids from *Swertia perennis*. The intense blue colour given by iodine/potassium iodide spray reagent is specific for isovitexin 7-*O*-glucoside among the many closely related *C*-glucosylflavones present in *Silene*, thus providing a rapid screening method in population-genetic studies (van Genderen *et al.*, 1983a).

ACKNOWLEDGEMENTS

We are grateful to Professors P. Lebreton, E. Stanislas, G. Weissenböck and H.D. Zinsmeister and Drs J. van Brederode, K. Gluchoff-Fiasson, J.F. Gonnet, M. Jay, K.R. Markham, R. Mues, N. Nikolov and K. Valant-Vetschera for supplying papers and information prior to publication.

REFERENCES

Abdalla, M.F. and Mansour, R.M.A. (1984), *J. Nat. Prod.* **47**, 184.

Abdalla, M.F. and Saleh, N.A.M. (1983), *J. Nat. Prod.* **46**, 755.

Abdalla, M.F., Saleh, N.A.M., Garb, S., Abu-Eyta, A.M. and El-Said, H. (1983), *Phytochemistry* **22**, 2057.

Adinarayana, D. and Ramachandraiah, P. (1985a), *J. Nat. Prod.* **48**, 156.

Adinarayana, D. and Ramachandraiah, P. (1985b), *Indian. J. Chem.* **24B**, 453.

Adinarayana, D. and Rao, J.R. (1972), *Tetrahedron* **28**, 5377.

Adinarayana, D. and Rao, J.R. (1975), *Proc. Indian Acad. Sci.* **81A**, 23.

Adinarayana, D. and Syamasundar, K.V. (1982), *Curr. Sci.* **51**, 936.

Adinarayana, D., Syamasundar, K.V., Seligmann, O. and Wagner, H. (1982), *Z. Naturforsch.* **37C**, 145.

Adinarayana, D., Ramachandraiah, P. and Rao, K.N. (1985), *Experientia* **41**, 251.

Afchar, D., Cave, A. and Vaquette, J. (1984a), *Plant. Med. Phytother.* **18**, 55.

Afchar, D., Cave, A., Guinaudeau, H. and Vaquette, J. (1984b), *Plant. Med. Phytother.* **18**, 170.

Agata, I., Nakaya, Y., Nishibe, S., Hisada, S. and Kimura, K. (1984), *Yakugaku Zasshi* **104**, 418.

Aliev, A.M. and Movsumov, I.S. (1981), *Rastit. Resur.* **17**, 602.

Ardouin, P. (1985), Thèse de Doctorat de 3ème cycle, Université Claude Bernard, Lyon N° 1 652.

Aritomi, M. (1982), *Kaseigaku Zasshi* **33**, 353.

Asen, S., Stewart, R.N., Norris, K.H. and Massie, D.R. (1970), *Phytochemistry* **9**, 619.

Ates, N., Ulubelen, A., Clark, W.D., Brown, G.K., Mabry, T.J., Dellamonica, G. and Chopin, J. (1982), *J. Nat. Prod.* **45**, 189.

Ayanoglu, E., Ulubelen, A., Mabry, T.J., Dellamonica, G. and Chopin, J. (1982), *Phytochemistry* **21**, 799.

Baas, W.J., Warnaar, F. and Niemann, G.J. (1981), *Acta Bot. Neerl.* **30**, 257.

Baeva, R.T., Karryev, M.O., Litvinenko, V.I. and Abubakirov, N.K. (1974), *Khim. Prir. Soedin.* 171.

Balabanova-Radonova, E., Georgieva, I. and Mondeshka, D. (1981), *Dokl. Bolg. Akad. Nauk.* **34**, 1517.

Balabanova-Radonova, E., Georgieva, I. and Mondeshka, D. (1982), *Dokl. Bolg. Akad. Nauk.* **35**, 463.

Bandyukova, V.A. (1979), *Khim. Prir. Soedin.* 724.

Bandyukova, V.A., Khalmatov, K.K. and Yunusova, K.K. (1985), *Khim. Prir. Soedin.* 562.

Barberan, F.A.T., Hernandez, L., Ferreres, F. and Tomas, F. (1985), *Plant. Med.* **47**, 452.

Barberan, F.A.T., Hernandez, L. and Tomas, F. (1986), *Phytochemistry* **25**, 561.

Bassene, E., Laurence, A., Olschwang, D. and Pousset, J.L. (1985), *J. Chromatogr.* **346**, 428.

Bauer, R. Berganza, L.H., Seligmann, O. and Wagner, H. (1985), *Phytochemistry* **24**, 1587.

Baztan, J.M., Rebuelta, M. and Vivas, J.M. (1981), *An. R. Acad. Farm.* **47**, 303.

Beltrami, E., Bernardi, M. de, Fronza, G., Mellerio, G., Vidari, G. and Vita-Finzi, P. (1982), *Phytochemistry* **21**, 2931.

Besson, E. and Chopin, J. (1983), *Phytochemistry* **22**, 2051.

Besson, E., Dombris, A., Raynaud, J. and Chopin, J. (1979), *Phytochemistry* **18**, 1899.

Besson, E., Chopin, J., Markham, K.R., Mues, R., Wong, H. and Bouillant, M.L. (1984), *Phytochemistry* **23**, 159.

Besson, E., Dellamonica, G., Chopin, J., Markham, K.R., Kim M., Koh, H.S. and Fukami, H. (1985), *Phytochemistry* **24**, 1061.

Bezuidenhoudt, B.C.B., Brandt, E.V. and Roux, D.G. (1981), *J. Chem. Soc. Perkin I* 263.

Bezuidenhoudt, B.C.B., Brandt, E.V. and Ferreira, D. (1987), *Phytochemistry* **26**, 531.

Bhakuni, D., Bittner, M., Silva, M. and Sammes, P.G. (1973), *Phytochemistry* **12**, 2777.

Bhardwaj, D.K., Bisht, M.S. and Mehta, C.K. (1980), *Phytochemistry* **19**, 2040.

Bhutani, S.P., Chibber, S.S. and Seshadri, T.R. (1969), *Indian. J. Chem.* 7, 210.

Boguslavskaya, L.I., Demyanenko, S.I. and Salam, J.H. (1983a), *Khim. Prir. Soedin.* 386.

Boguslavskaya, L.I., Demyanenko, S.I., Salam, J.H. and Soboleva, V.A. (1983b), *Khim. Prir. Soedin.* 783.

Boguslavskaya, L.I., Tikhonov, A.I., Pashnev, P.D., Bekkari, J. and Sklyar, V.I. (1985), *Khim. Prir. Soedin.* 410.

Bombardelli, E., Bonati, A., Gabetta, B. and Mustich, G. (1974), *Phytochemistry* **13**, 295.

Bouillant, M.L., Favre-Bonvin, J. and Chopin, J. (1975), *Phytochemistry* **14**, 2267.

Bouillant, M.L., Besset, A., Favre-Bouvin, J. and Chopin, J. (1978), *Phytochemistry* **17**, 527.

Bouillant, M.L., Ferreres, F., Favre-Bonvin, J., Chopin, J., Zoll, A. and Mathieu, G. (1984), *Phytochemistry* **23**, 2653.

Boyet, C. (1985) Thèse de Doctorat de 3ème cycle, Université Claude Bernard, Lyon N° 1675.

Britsch, L., Heller, W. and Grisebach, H. (1981), *Z. Naturforsch.* **36C**, 742.

Brum-Bousquet, M., Tillequin, F. and Paris, R.R. (1977), *Lloydia* **40**, 591.

Cabrera, E. and Garcia-Granados, A. (1981), *An Quim. Ser. C* 77, 31.

Calie, P.J., Schilling, E.E. and Webb, D.H. (1983), *Biochem. Syst. Ecol.* **11**, 107.

Chadenson, M., Hauteville, M. and Chopin, J. (1972), *J. Chem. Soc. Chem. Commun.* 107.

Chari, V.M., Wagner, H., Schilling, G. and Nesmelyi, A. (1978), *11th IUPAC International Symposium: Chemistry of Natural Products*, Symposium papers, vol. 2, p. 279.

Chari, V.M., Harborne, J.B. and Williams, C.A. (1980), *Phytochemistry* **19**, 983.

Chawla, H., Chibber, S.S. and Seshadri, T.R. (1974), *Phytochemistry* **13**, 2301.

Chawla, H., Chibber, S.S. and Seshadri, T.R. (1975), *Indian J. Chem.* **13**, 444.

Chawla, H., Chibber, S.S. and Seshadri, T.R. (1976), *Phytochemistry* **15**, 235.

Chernobrovaya, N.V. (1973), *Khim. Prir. Soedin.* 801.

Chiale, C.A., Cabrera, J.L. and Juliani, H.R. (1984), *An. Asoc. Quim. Argent.* 72, 501.

Chkhikvishvili, I.D., Kurkin, V.A. and Zaprometov, M.N. (1985), *Khim. Prir. Soedin.* 118.

Chopin, J., Durix, A. and Bouillant, M.L. (1968), *C.R. Acad. Sci., Ser. C.* **266**, 1334.

Chopin, J., Roux, B., Bouillant, M.L., Durix, A., d'Arcy, A., Mabry, T.J. and Yoshioka, H. (1969), *C.R. Acad. Sci., Ser. C* **268**, 980.

Chopin, J., Bouillant, M.L., Wagner, H. and Galle, K. (1974), *Phytochemistry* **13**, 2583.

Chopin, J., Dellamonica, G., Besson, E., Skrzypczakova, L., Budzianowski, J. and Mabry, T.J. (1977a), *Phytochemistry* **16**, 1999.

Chopin, J., Dellamonica, G., Bouillant, M.L., Besset, A., Popovici, G. and Weissenböck, G. (1977b), *Phytochemistry* **16**, 2041.

Chopin, J., Bouillant, M.L., Nair, A.G.R., Ramesh, P. and Mabry, T.J. (1978), *Phytochemistry* **17**, 299.

Chopin, J., Besson, E. and Nair, A.G.R. (1979), *Phytochemistry* **18**, 2059.

Chopin, J., Besson, E., Dellamonica, G. and Nair, A.G.R. (1982), *Phytochemistry* **21**, 2367.

Chopin, J., Dellamonica, G., Markham, K.R., Nair, A.G.R. and Gunasegaran, R. (1984), *Phytochemistry* **23**, 2106.

Chulia, A.J. (1984), Thèse de Doctorat ès Sciences pharmaceutiques, Université de Grenoble 1.

Chulia, A.J. and Mariotte, A.M. (1985), *J. Nat. Prod.* **48**, 480.

Crawford, D.J. and Smith, E.B. (1983), *Bot. Gaz. (Chicago)* **144**, 577.

Darmograi, V.N. (1980a), *Khim. Prir. Soedin.* 722.

Darmograi, V.N. (1980b), *Khim. Prir. Soedin.* 838.

Darmograi, V.N., Litvinenko, V.I. and Krivenchuk, P.E. (1968), *Khim. Prir. Soedin.* 248.

Del Galarza, V.S., Cabreya, J.A. and Juliani, H.R. (1983), *An Asoc. Quim. Argent.* **71**, 505.

Dellamonica, G., Meurer, B., Strack, D. Weissenböck, G. and Chopin, J. (1983), *Phytochemistry* **22**, 2627.

Del Pero Martinez, M.A. and Swain, T. (1985), *Biochem. Syst. Ecol.* **13**, 391.

Dillon, M.O., Mabry, T.J., Besson, E., Bouillant, M.L. and Chopin, J. (1976), *Phytochemistry* **15**, 1085.

Dombris, A. and Raynaud, J. (1980), *Plant. Med. Phytother.* **14**, 69.

Dombris, A. and Raynaud, J. (1981a), *Pharmazie* **36**, 384.

Dombris, A. and Raynaud, J. (1981b), *Plant. Med. Phytother.* **15**, 21.

Dossaji, S.F., Becker, H. and Exner, J. (1983), *Phytochemistry* **22**, 311.

Dreyer, D.L. and Jones, K.C. (1981), *Phytochemistry* **20**, 2489.

Dubois, M.A., Zoll, A., Bouillant, M.L. and Chopin, J. (1982a), *Phytochemistry* **21**, 1141.

Dubois, M.A., Zoll. A., Bouillant, M.L. and Delaveau, P. (1982b), *Plant. Med.* **46**, 56.

Dubois, M.A., Zoll, A. and Chopin, J. (1983), *Phytochemistry* **22**, 2879.

Dubois, M.A., Zoll, A., Markham, K.R., Bouillant, M.L., Dellamonica, G. and Chopin, J. (1984), *Phytochemistry* **23**, 706.

Dubois, M.A., Zoll, A., and Chopin, J. (1985), *Phytochemistry* **24**, 1077.

Eade, R.A. and Pham, H.P. (1979), *Aust. J. Chem.* **32**, 2483.
Elliger, C.A., Chan, B.G., Waiss, A.C., Lundin, R.E. and Haddon, W.F. (1980), *Phytochemistry* **19**, 293.
El-Negoumy, S.I., Abdalla, M.F. and Saleh, N.A.M. (1986), *Phytochemistry* **25**, 772.
Escobar, L.K. (1980), Ph.D. dissertation, University of Texas at Austin.
Escobar, L.K., Liu, Y.L. and Mabry, T.J. (1983), *Phytochemistry* **22**, 796.
Evdokimov, P.K. (1979), *Khim. Prir. Soedin.* 736.
Fang, N., Mabry, T.J. and Le-Van, N. (1986a), *Phytochemistry* **25**, 235.
Fang, N., Leidig, M. and Mabry, T.J. (1986b), *Phytochemistry* **25**, 927.
Fang, Q., Wu, P. and Yang, L. (1983), *Yaoxue Xuebao* **18**, 695.
Ferreres, F., Lorente, F.T. and Guirado, A. (1984), *An. Quim. Ser. C* **80**, 198.
Ferry, S. and Darbour, N. (1980), *Plant. Med. Phytother.* **14**, 148.
Ficarra, P., Tommasini, A., Pasquale, A. de, Fenech, C.G. and Iauk, L. (1984b), *Farm. Ed. Prat.* **39**, 148.
Ficarra, P., Ficarra, R., Tommasini, A., Pasquale, A., de, Fenech, C.G. and Iauk, L. (1984a), *Farm. Ed. Prat.* **39**, 342.
Franz, G. and Grün, M. (1983), *Plant. Med.* **47**, 131.
Freitag, P., Mues, R., Brill-Fess, C., Stoll, M., Zinsmeister H.D. and Markham, K.R. (1986), *Phytochemistry* **25**, 669.
Fujita, M. and Inoue, T. (1982), *Chem. Pharm. Bull.* **30**, 2342.
Funaoka, K. and Tanaka, M. (1957), *Mokuzai Gakkaishi* **3**, 144.
Gaffield, W., Besson, E. and Chopin, J. (1984), *Phytochemistry* **23**, 1317.
Galle, K. (1974), Dissertation, Universität München.
Garcia-Granados, A. and Saenz de Buruaga, J.M. (1980), *An. Quim. Ser. C* **76**, 96.
Garg, S.P., Bhushan, R. and Kapoor, R.C. (1980), *Phytochemistry* **19**, 1265.
Ghosal, S. and Jaiswal, D.K. (1980), *J. Pharm. Sci.* **69**, 53.
Gitelli, A.M., Gianinetto, I.B., Cabrera, J.L. and Juliani, H.R. (1981), *An. Asoc. Quim. Argent.* **69**, 33.
Gluchoff-Fiasson, K. (1986), Private communication.
Goetz, M. and Jacot-Guillarmod, A. (1977), *Helv. Chim. Acta* **60**, 1322.
Goetz, M. and Jacot-Guillarmod, A. (1978), *Helv. Chim. Acta* **61**, 1373.
Gomes, E., Dellamonica, G., Gleye, J., Moulis, C., Chopin, J. and Stanislas, E. (1983), *Phytochemistry* **22**, 2628.
Gonnet, J.F. (1986), Private communication.
Graham, S.A., Timmermann, B.N. and Mabry, T.J. (1980), *J. Nat. Prod.* **43**, 644.
Gunasingh, C., Gnanaraj, B. and Nagarajan, S. (1981a), *Indian J. Pharm. Sci.* **43**, 114.
Gunasingh, C., Gnanaraj, B. and Nagarajan, S. (1981b), *Indian J. Pharm. Sci.* **43**, 115.
Gupta, D.R., Dhiman, R.P. and Bahar, A. (1985), *Pharmazie* **40**, 273.
Gurni, A.A. and Kubitzki, K. (1981), *Biochem. Syst. Ecol.* **9**, 109.
Harborne, J.B. and Green, P.S. (1980), *Bot. J. Linn. Soc.* **81**, 155.
Harborne, J.B., Williams, C.A., Greenham, J. and Moyna, P. (1974), *Phytochemistry* **13**, 1557.
Harborne, J.B., Williams, C.A. and Wilson, K.L. (1982), *Phytochemistry* **21**, 2491.
Harborne, J.B., Williams, C.A. and Wilson, K.L. (1985), *Phytochemistry* **24**, 751.
Hassine, B.B., Bui, A.M., Mighri, Z. and Cave, A. (1982), *Plant. Med. Phytother.* **16**, 197.
Hauteville, M., Chadenson, M. and Chopin, J. (1973a), *Bull Soc. Chim. France*, 1781.
Hauteville, M., Chadenson, M. and Chopin, J. (1973b), *Bull. Soc. Chim. France*, 1784.
Hayashi, Y. (1980), Private communication.
Hiermann, A. and Katnig, T. (1984), *Sci. Pharm.* **52**, 30.
Higa, M., Miyagi, Y., Yogi, S. and Hokama, K. (1983), *Bull. Coll. Sci. Univ. Ryukyus* **35**, 53.
Hillis, W.E. and Horn, D.H.S. (1966), *Aust. J. Chem.* **19**, 705.
Hillis, W.E. and Inoue, T. (1967), *Phytochemistry* **6**, 59.
Hilsenbeck, R.A. and Mabry, T.J. (1983), *Phytochemistry* **22**, 2215.
Hilsenbeck, R.A., Levin, D.A., Mabry, T.J. and Raven, P.H. (1984a), *Phytochemistry* **23**, 1077.
Hilsenbeck, R.A., Wright, S.J. and Mabry, T.J. (1984b), *J. Nat. Prod.* **47**, 312.
Hirai, Y., Sanada, S., Ida, Y. and Shoji, Y. (1984), *Chem. Pharm. Bull.* **32**, 4003.
Hirose, R., Kazuta, Y., Koga, D., Ide, A. and Yagishita, K. (1981), *Agric. Biol. Chem.* **45**, 551.
Horowitz, R.M. and Gentili, B. (1964), *Chem. Ind.*, 498.
Horowitz, R.M. and Gentili, B. (1966), *Chem. Ind.*, 625.
Hostettmann, K., Domon, B., Schaufelberger, D. and Hostettmann, M. (1984), *J. Chromatogr.* **283**, 137.
Hostettmann-Kaldas, M., Hostettmann, K. and Sticher, O. (1981), *Phytochemistry* **20**, 443.
Huang, S. (1985), *Shih Ta Hsueh Pao (Taipei)* **30**, 547.
Huang, Y.Z., Chang, Y.S., Sun, Y.C., Fan, P.T., Hu, C.P. and Chou, P.N. (1980), *Chung Tsao Yao* **11**, 342, 349.
Husain, S.Z. and Markham, K.R. (1981), *Phytochemistry* **20**, 1171.
Husain, S.Z., Heywood, V.H. and Markham, K.R. (1982), *World Crops: Prod. Util. Descr.* 7, 141.
Ibrahim, R.K. and Shaw, M. (1970), *Phytochemistry* **9**, 1855.
Ingham, J.L., Markham, K.R., Dziedzic, S.Z. and Pope, G.S. (1986), *Phytochemistry* **25**, 1772.
Ishikura, N., Iwata, M. and Miyazaki, S. (1981), *Bot. Mag.* **94**, 197.
Ismaili, A., Reynaud, J., Jay, M. and Raynaud, J. (1981) *Plant. Med. Phytother.* **15**, 155.
Ismaili, A. (1982), Thèse de Doctorat de 3ème cycle, Université Claude Bernard, Lyon, N° 1135.
Jacot-Guillarmod, A., Luong Minh Duc and Hostettmann, K. (1975), *Helv. Chim. Acta* **58**, 1477.
Jay, M., Gleye, J., Bouillant, M.L., Stanislas, E. and Moretti, C. (1979), *Phytochemistry* **18**, 184.
Jay, M., Voirin, B., Hasan, A., Gonnet, J.F. and Viricel, M.R. (1980), *Biochem. Syst. Ecol.* **8**, 127.
Jay, M., Lameta-d'Arcy, A. and Viricel, M.R. (1984a), *Phytochemistry* **23**, 1153.
Jay, M., Plenet, D., Ardouin, P., Lumaret, R. and Jacquard, P. (1984b), *Biochem. Syst. Ecol.* **12**, 193.
Julian, E.A., Johnson, G., Johnson, D.K. and Donnelly, B.J. (1971), *Phytochemistry* **10**, 3185.
Jyotsna, D., Sarma, P.N., Srimannarayana, G. and Rao, A.V. Subba (1984), *Curr. Sci.* **53**, 573.
Kamanzi, K., Raynaud, J. and Voirin, B. (1983a), *Pharmazie* **38**, 494.
Kamanzi, K., Raynaud, J. and Voirin, B. (1983b), *Plant. Med. Phytother.* **17**, 47.
Kang, S., Yu, Y. and Wang, P. (1984), *Zhongcaoyao* **15**, 247.

Karl, C., Mueller, G. and Pedersen, P.A. (1982), *Z, Naturforsch.* **37C**, 148.
Kashnikova, M.V., Sheichenko, V.I., Glyzin, V.I. and Samylina, I.A. (1984), *Khim. Prir. Soedin.* 108.
Keita, A., Gleye, J., Stanislas, E. and Fouraste, I. (1985), *J. Nat. Prod.* **48**, 675.
Khetwal, K.S. and Verma, D.L. (1984), *Indian J. Pharm. Sci.* **46**, 25.
Kim, M., Koh, H.S. and Fukami, H. (1985), *J. Chem. Ecol.* **1**, 441.
King, B.L. and Jones, S.B. (1982), *Bull. Torrey Bot. Club* **109**, 279.
Kitanov, G., Blinova, K.F. and Akhtardzhiev, K. (1979), *Khim. Prir. Soedin.* **154**, 231.
Kitanov, G. and Pryakhina, N.I. (1985), *Farmatsiya (Sofia)* **35**, 10.
Knogge, W. and Weissenböck, G. (1984), *Eur. J. Biochem.* **140**, 113.
Knogge, W. and Weissenböck, G. (1986), *Planta* **167**, 196.
Koch, H. and Steinegger, E. (1982), *Pharm. Acta Helv.* **57**, 211.
Koeppen, B.H. (1964), *Z. Naturforsch.* **19B**, 173.
Koeppen, B.H. and Roux, D.G. (1965), *Tetrahedron Lett.*, 3497.
Koleva, M., Nikolova, V. and Akhtardzhiev, K. (1982), *Probl. Farm.* **10**, 53.
Koleva, M., Nikolova, V. and Akhtardzhiev, K. (1984), *Probl. Farm.* **12**, 35.
Komatsu, M. and Tomimori, T. (1966), *Tetrahedron Lett.*, 1611.
Komatsu, M., Tomimori, T., Takeda, K. and Hayashi, K. (1968), *Chem. Pharm. Bull.* **16**, 1413.
Komissarenko, N.F., Stupakova, E.P. and Pakalu, D.A. (1981), *Khim. Prir. Soedin.*, 249.
Kovalev, V.N. and Komissarenko, A.N. (1983), *Khim. Prir. Soedin.* 235.
Kubota, M., Hattori, M. and Namba, T. (1983), *Shoyakugaku Zasshi* **37**, 229.
Kumamoto, H., Matsubara, Y., Iizuka, Y., Murakami, T., Okamoto, K., Miyake, H. and Yokoi, K. (1984), *Nippon Nogeikagaku Kaishi* **58**, 137.
Kumamoto, H., Matsubara, Y., Iizuka, Y., Okamoto, K. and Yokoi, K. (1985a), *Nippon Nogeikagaku Kaishi* **59**, 677.
Kumamoto, H., Matsubara, Y., Iizuka, Y., Okamoto, K. and Yokoi, K. (1985b), *Agric. Biol. Chem.* **49**, 2613.
Kumamoto, H., Matsubara, Y., Iizuka, Y., Okamoto, K. and Yokoi, K. (1985c), *Nippon Nogeikagaku Kaishi* **59**, 683.
Kumamoto, H., Matsubara, Y., Iizuka, Y., Okamoto, K. and Yokoi, K. (1986), *Agric. Biol. Chem.* **50**, 781.
Kurkin, V.A., Zapesochnaya, G.G. and Krivenchuk, P.E. (1981), *Khim. Prir. Soedin.*, 661.
Laanest, L. (1981a), *Fiziol. Rast.* (*Moscow*) **28**, 103.
Laanest, L. (1981b), *Eesti NSV Tead. Akad. Toim. Biol.* **30**, 207.
Laanest, L. and Margna, U. (1985), *Plant Sci.* **41**, 19.
Laanest, L. and Vainjärv, T. (1980), *Regul. Rosta Pitan. Rast.*, 106.
Lachman, J., Rehakova V., Hubacek, J. and Pivec, V. (1982), *Sci. Agric. Bohemoslov.* **14**, 265.
La Duke, J.C. (1982), *Am. J. Bot.* **69**, 784.
Lardy, C., Bouillant, M.L. and Chopin, J. (1983), *Phytochemistry* **22**, 2571.
Lardy, C., Bouillant, M.L. and Chopin, J. (1984), *J. Chromatogr.* **291**, 307.
Lebreton, P. and Sartre, J. (1983), *Canad. J. For. Res.* **13**, 145.
Linard, A., Delaveau, P., Paris, R.R., Dellamonica, G. and Chopin, J. (1982), *Phytochemistry* **21**, 797.
Liu, Y.L., Mabry, T.J., Lardy, C. and Chopin, J. (1982), *Rev. Latinoamer. Quim.* **13**, 56.
Lorente, F.T., Ferreres, F. and Barberan, F.T. (1982), *Phytochemistry* **21**, 1461.
Lorente, F.T., Ferreres, F. and Barberan, F.T. (1983), *An. Quim. Ser. C.* **79**, 456.
Luong Minh Duc and Jacot-Guillarmod, A. (1977), *Helv. Chim. Acta* **60**, 2099.
Luong Minh Duc, Fombasso, P. and Jacot-Guillarmod, A. (1980), *Helv. Chim. Acta* **63**, 244.
Luong Minh Duc, Saeby, J., Fombasso, P. and Jacot-Guillarmod, A. (1981), *Helv. Chim. Acta* **64**, 2741.
Mabry, T.J., Yoshioka, H., Sutherland, S., Woodland, S., Rahman, W., Ilyas, M., Usmani, J.N., Hameed, N., Chopin, J. and Bouillant, M.L. (1971), *Phytochemistry* **10**, 677.
Mabry, T.J., Liu, Y.L., Pearce, J., Dellamonica, G., Chopin, J., Markham, K.R., Paton, N.H. and Smith, P. (1984), *J. Nat. Prod.* **47**, 127.
Madulid, D.A. (1980), *Kalikasan Philipp. J. Biol.* **9**, 69.
Manners, G.D. and Jurd, L. (1979), *Phytochemistry* **18**, 1037.
Margna, U. and Margna, E. (1980), *Regul. Rosta Pitan. Rast.*, 100.
Margna, U. and Margna, E. (1982), *Fiziol. Rast.* (*Moscow*) **29**, 293.
Margna, U. and Vainjärv, T. (1981), *Biochem. Physiol. Pflanz.* **176**, 44.
Margna, U. and Vainjärv, T. (1983), *Z. Naturforsch.* **38c**, 711.
Margna, U. and Vainjärv, T. (1985), *Biochem. Physiol. Pflanz.* **180**, 291.
Margna, U., Margna, E., Taht, R. and Orav, I. (1981), *Biochem. Physiol. Pflanz.* **176**, 191.
Margna, U., Laanest, L., Margna, E. and Vainjärv, T. (1983), *Regul. Rosta Metab. Rast.*, 169.
Margna, U., Laanest, L., Margna, E. and Vainjärv, T. (1985), *Z. Naturforsch.* **40C**, 154.
Markham, K.R. (1982), *Techniques of Flavonoid Identification*, Academic Press, London, p. 53.
Markham, K.R. (1984), *Phytochemistry* **23**, 2053.
Markham, K.R. and Chari, V.M. (1982). In *The Flavonoids – Advances in Research* (eds J.B. Harborne and T.J. Mabry), Chapman and Hall, London, p. 19.
Markham, K.R. and Mues, R. (1984), *Z. Naturforsch.* **39C**, 309.
Markham, K.R. and Porter, L.J. (1979), *Phytochemistry* **18**, 611.
Markham, K.R. and Wallace, J.W. (1980), *Phytochemistry* **19**, 415.
Markham, K.R. and Williams, C.A. (1980), *Phytochemistry* **19**, 2789.
Markham, K.R. and Whitehouse, L.A. (1984), *Phytochemistry* **23**, 1931.
Markham, K.R., Porter, L.J., Campbell, E.O., Chopin, J. and Bouillant, M.L. (1976), *Phytochemistry* **15**, 1517.
Markham, K.R., Theodor, R., Mues, R. and Zinsmeister, H.D. (1982), *Z. Naturforsch.* **37C**, 562.
Markham, K.R., Moore, N.A. and Given, D.R. (1983), *N.Z.J. Bot.* **21**, 113.
Markham, K.R., Webby, R.F. and Vilain, C. (1984), *Phytochemistry* **23**, 2049.
Markham, K.R., Webby, R.F., Whitehouse, L.A., Molloy, B.P.J., Vilain, C. and Mues, R. (1985), *N.Z.J. Bot.* **23**, 1.
Marston, A., Hostettmann, K. and Jacot-Guillarmod, A. (1976), *Helv. Chim. Acta* **59**, 2596.

Massias, M., Carbonnier, J. and Molho, D. (1981), *Phytochemistry* **20**, 1577.
Massias, M., Carbonnier, J. and Molho, D. (1982), *Biochem. Syst. Ecol.* **10**, 319.
Mastenbroek, O., Maas, J.W., van Brederode, J., Niemann, G.J. and van Nigtevecht, G. (1982), *Genetica* **59**, 139.
Mastenbroek, O., Hogeweg, P., van Brederode, J. and van Nigtevecht, G. (1983a), *Biochem. Syst. Ecol.* **11**, 91.
Mastenbroek, O., Knorr, J.J., Kamps-Heinsbroek, R., Maas, J.W., Steyns, J.M. and van Brederode J. (1983b), *Z. Naturforsch.* **38C**, 894.
Mastenbroek, O., Prentice, H.C., Kamps-Heinsbroek, R., van Brederode, J., Niemann, G.J. and van Nigtevecht, G. (1983c), *Plant Syst. Evol.* **141**, 257.
Matsubara, Y., Kumamoto, H., Yonemoto, H., Iizuka, Y., Murakami, T., Okamoto, K., Miyake, H. and Yokoi, K. (1984), *Kinki Daigaku Igaku Zasshi* **9**, 61.
Matsubara, Y., Kumamoto, H., Iizuka, Y., Murakami, T., Okamoto, K., Miyake, H. and Yokoi, K. (1985a), *Agric. Biol. Chem.* **49**, 909.
Matsubara, Y., Kumamoto, H., Yonemoto, H., Iizuka, Y., Okamoto, K. and Yokoi, K. (1985b), *Nippon Nogeikagaku Kaishi* **59**, 405.
McCormick, S. and Mabry, T.J. (1981), *J. Nat. Prod.* **44**, 623.
McCormick, S. and Mabry, T.J. (1982), *J. Nat. Prod.* **45**, 782.
McCormick, S. and Mabry, T.J. (1983), *Phytochemistry* **22**, 798.
Medvedeva, S.A., Tijukavkina, N.A. and Ivanova, S.Z. (1974), *Khim. Drev.* **15**, 144.
Mericli, A.H. and Dellamonica, G. (1983), *Fitorerapia* **54**, 35.
Mohammad, F., Taufeeq, H.M., Ilyas, M., Rahman, W. and Chopin, J. (1982), *Indian J. Chem.* **21B**, 167.
Monties, B., Bouillant, M.L. and Chopin, J. (1976), *Phytochemistry* **15**, 1053.
Morimoto, S., Nonaka, G. and Nishioka, I. (1986a), *Chem. Pharm. Bull.* **34**, 633.
Morimoto, S., Nonaka, G. and Nishioka, I. (1986b), *Chem. Pharm. Bull.* **34**, 643.
Morita, N., Arisawa, M. and Yoshikawa, A. (1976), *J. Pharm. Soc. Jpn.* **96**, 1180.
Mues, R. (1982a), *J. Hattori Bot. Lab.* **51**, 61.
Mues, R. (1982b), *J. Hattori Bot. Lab.* **53**, 271.
Mues, R. (1984), *Proc. Third Meet Bryol. Central and East Europe, Praha 1982*, Universita Karlova, Praha, p. 37.
Mues, R. and Zinsmeister, H.D. (1976), *Phytochemistry* **15**, 1757.
Mues, R., Hattori, S., Asakawa, Y. and Grolle, R. (1984), *J. Hattori Bot. Lab.* **56**, 227.
Murakami, T., Nishikawa, Y. and Ando, T. (1960), *Chem. Pharm. Bull.* **8**, 688.
Murthy, S.S.N., Rao, N.S.P., Anjaneyulu, A.S.R., Row, L.R., Pelter, A. and Ward, R.S. (1981), *Curr. Sci.* **50**, 227.
Naderuzzaman, A.T.M. (1983), *Bangladesh J. Bot.* **12**, 27.
Nair, A.G.R. and Gunasegaran, R. (1983a), *Curr. Sci.* **52**, 248.
Nair, A.G.R. and Gunasegaran, R. (1983b), *J. Indian Chem. Soc.* **60**, 307.
Nair, A.G.R., Gunasegaran, R. and Joshi, B.S. (1982), *Indian J. Chem.* **21B**, 979.
Narayanan, V. and Seshadri, T.R. (1971), *Indian J. Chem.* **9**, 14.
Nawwar, M.A.M., El-Sissi, H.I. and Barakat, H.H. (1980), *Phytochemistry* **19**, 1854.
Nawwar, M.A.M., El-Sissi, H.I. and Barakat, H.H. (1984), *Phytochemistry* **23**, 2937.
Neuman, P.R., Waines, J.G., Hilu, K.W. and Barnhart, D. (1983), *Genetics* **103**, 313.
Nicholls, K.W. and Bohm, B.A. (1983), *Canad. J. Bot.* **61**, 708.
Niemann, G.J. (1980), *Canad. J. Bot.* **58**, 2313.
Niemann, G.J. (1981), *Acta Bot. Neerl.* **30**, 475.
Niemann, G.J. (1982), *Rev. Latinoamer. Quim.* **13**, 74.
Niemann, G.J. (1984a), *Acta Bot. Neerl.* **33**, 123.
Niemann, G.J. (1984b), *J. Plant Physiol.* **115**, 311.
Niemann, G.J., van Nigtevecht, G. and van Brederode, J. (1980), *Plant Physiol.* **65**, 99.
Niemann, G.J., Dellamonica, G. and Chopin, J. (1981), *Z. Naturforsch.* **36C**, 1084.
Niemann, G.J., Koerselman-Kooy, J.W., Steyns, J.M. and van Brederode, J. (1983), *Z. Pflanzenphysiol.* **109**, 105.
Nikolaeva, G.G., Glyzin, V.I., Krivut, B.A., Silla, A. and Patudin, A.V. (1980), *Khim. Prir. Soedin.*, 833.
Nikolov, N., Horowitz, R.M. and Gentili, B. (1976), *International Congress for Research on Medicinal Plants*, Munich, Abstracts of papers, Section A.
Nikolov, N., Dellamonica, G. and Chopin, J. (1981), *Phytochemistry* **20**, 2780.
Nikolov, N., Seligmann, O., Wagner, H., Horowitz, R.M. and Gentili, B. (1982a), *Plant. Med.* **44**, 50.
Nikolov, N., Wagner, H., Chopin, J., Dellamonica, G., Chari, V.M. and Seligmann, O. (1982b). In *Flavonoids and Bioflavonoids 1981* (eds L. Farkas, M. Gabor, F. Kallay and H. Wagner), Akademiai Kiado, Budapest, p. 325.
Oelrichs, P., Marshall, J.T.B. and Williams, D.H. (1968), *J. Chem. Soc.*, 941.
Oga, S., de Freitas, P.C.D., Gomes da Silva, A.C. and Hanada, S. (1984), *Plant. Med.*, 303.
Ohashi, H., Goto, M., Fukuda, J. and Imamura, H. (1980), *Bull. Fac. Agric. Gifu Univ.* **43**, 83.
Ohsugi T., Nishida, R. and Fukami, H. (1986), *Agric. Biol. Chem.* (In press).
Okamura, N., Yagi, A. and Nishioka, I. (1981), *Chem. Pharm. Bull.* **29**, 3507.
O'Kelly, C.J. (1982), *Bot. Mar.* **25**, 133.
Oksuz, S., Ayyildiz H. and Johansson, C. (1984), *J. Nat. Prod.* **47**, 902.
Olaniyi, A.A. and Phillips, A.A. (1979), *Niger. J. Pharm.* **10**, 208.
Österdahl, B.G. (1978), *Acta Chem. Scand.* **32**, 93.
Österdahl, B.G. (1979a), Doctoral dissertation, University of Uppsala, N. 524.
Österdahl, B.G. (1979b), *Acta Chem. Scand.* **33B**, 119.
Österdahl, B.G. (1979c), *Acta Chem. Scand.* **33B**, 400.
Ouabonzi, A., Bouillant, M.L. and Chopin, J. (1983), *Phytochemistry* **22**, 2632.
Oyuungerel, Z., Komissarenko, N.F., Batyuk, V.S. and Lamzhav, A. (1984), *Khim. Farm. Zh.* **18**, 967.
Ozimina, I.I. (1981), *Khim. Prir. Soedin.* **242.**
Pagani, F. (1982a), *Boll. Chim. Farm.* **121**, 178.
Pagani, F. (1982b), *Boll. Chim. Farm.* **121**, 460.
Pangarova, T., Pavlova, A. and Dimov, N. (1980), *Pharmazie* **35**, 501.
Parthasarathy, M.R., Seshadri, T.R. and Varma, R.S. (1974), *Curr. Sci.* **43**, 74.
Parthasarathy, M.R., Seshadri, T.R. and Varma, R.S. (1976), *Phytochemistry* **15**, 1025.
Petkov, E., Nikolov, N. and Uzunov, P. (1981), *Plant. Med.* **43**, 183.
Porter, L.J. (1981), *Taxon* **30**, 739.

Prabhakar, M.C., Bano, H., Kumar, I., Shamsi, M.A. and Khan, M.S.Y. (1981), *Plant. Med.* **43**, 396.

Proksch, M., Strack, D. and Weissenböck, G. (1981), *Z. Naturforsch.* **36C**, 222.

Proliac, A. and Raynaud, J. (1983), *Helv. Chim. Acta* **66**, 2412.

Proliac, A., Raynaud, J., Combier, H., Bouillant, M.L. and Chopin, J. (1973), *C.R. Acad. Sci. Ser. D* **277**, 2813.

Pryakhina, N.I. and Blinova, K.F. (1984), *Khim. Prir. Soedin.*, 109.

Pryakhina, N.I., Sheichenko, V.L. and Blinova, K.F. (1984), *Khim. Prir. Soedin.*, 589.

Purdie, A.W. (1984), *N.Z.J. Bot.* **22**, 7.

Qin, Y., Shi, J., Zhang, W. and Zhang, G. (1981), *Yaoxue Tangbao* **16**, 756.

Qin, Y.Q., Shi, J.Z., Zhang, W.P. and Zhang, Q.Q. (1982), *Zhivu Xuebao* **24**, 558.

Ramesh, P. and Subramanian, S.S. (1984), *Arogya (Manipal, India)* **10**, 76.

Rasolojaona, L. and Mastagli, P. (1985), *Carbohydr. Res.* **143**, 246.

Ravisé, A. and Chopin, J. (1981), *Phytopathol. Z.* **100**, 257.

Raynaud, J. and Prum, N. (1980), *Pharmazie* **35**, 712.

Raynaud, J., Dombris, A. and Prum, N. (1981), *Pharmazie* **36**, 444.

Richardson, P.M. (1981), *Plant Syst. Evol.* **138**, 227.

Rofi, R.D. and Pomilio, A.B. (1987), *Phytochemistry* **26**, 859.

Roy, S.K., Qasim, M.A., Kamil, M., Ilyas, M. and Rahman, W. (1983), *Indian J. Chem.* **22B**, 609.

Rozsa, Z., Hohmann, J., Czako, M., Mester, I., Reisch, J. and Szendrei, K. (1982). In *Flavonoids and Bioflavonoids 1981* (eds L. Farkas, F. Kallay, M. Gabor and H. Wagner), Akademiai Kiado, Budapest, p. 245.

Sagareishvili, T.G., Alaniya, M.D., Kikoladze, V.S. and Kemertelidze, E.P. (1982), *Khim. Prir. Soedin.*, 442.

Sahu, N.P., Roy, S.K. and Mahato, S.B. (1984), *Plant. Med.* **50**, 527.

Saifah, E., Kelley, C.J. and Leary, J.O. (1983), *J. Nat. Prod.* **46**, 353.

Sakushima, A., Hisada, S., Agata, I. and Nishibe, S. (1982), *Shoyakugaku Zasshi* **36**, 82.

Saleh, N.A.M., El-Negoumi, S.I., El-Hadidi, M.N. and Hosni, H.A. (1983), *Phytochemistry* **22**, 1417.

Saleh, N.A.M., El-Negoumi, S.I., Abdalla, M.F., Abouzaid, M.M., Dellamonica, G. and Chopin, J. (1985), *Phytochemistry* **24**, 201.

Salmenkallio, S., McCormick, S., Mabry, T.J., Dellamonica, G. and Chopin, J. (1982), *Phytochemistry* **21**, 2990.

Schaufelberger, D. and Hostettmann, K. (1984), *Phytochemistry* **23**, 787.

Schulz, M. and Weissenböck, G. (1986), *Z. Naturforsch.* **41C**, 22.

Sebastian, E., Nidiry, J. and Gupta, R.K. (1984), *Indian J. Pharm. Sci.* **46**, 203.

Sergeyeva, N.V. (1977), *Khim. Prir. Soedin.*, 124.

Shin, K.H., Kim, H.Y. and Woo, W.S. (1982), *Plant. Med.* **44**, 94.

Shirane, S., Ohya, S., Matsuo, T., Hirose, R., Koga, D., Ide, A. and Yagishita, K. (1982), *Agric. Biol. Chem.* **46**, 2595.

Siewek, F., Herrmann, K., Grotjahn, L. and Wray, V. (1985), *Z. Naturforsch.* **40C**, 8.

Simova, M. and Pangarova, T. (1983), *Pharmazie* **38**, 791.

Singh, B.P., Singh, R.P. and Jha, O.P. (1982), *Biol. Bull. India* **4**, 157.

Smith, D.M., Glennie, C.W. and Harborne, J.B. (1982), *Biochem. Syst. Ecol.* **10**, 37.

Snyder, J.K., Nakanishi, K., Hostettmann, K. and Hostettmann, M. (1984), *J. Liq. Chromatogr.* **7**, 243.

Soltis, D.E. and Bohm, B.A. (1982), *J. Nat. Prod.* **45**, 415.

Sood, A.R., Boutard, B., Chadenson, M., Chopin, J. and Lebreton, P. (1976), *Phytochemistry* **15**, 351.

Sreenivasan, K.K. and Sankarasubramanian, S. (1984), *Arogya (Manipal, India)* **10**, 156.

Srinivasan, K.K. and Subramanian, S.S. (1983), *Arogya (Manipal, India)* **9**, 89.

Steyns, J.M. and van Brederode, J. (1986), *Z. Naturforsch.* **41C**, 9.

Steyns, J.M., van Nigtevecht, G., Niemann, G.J. and van Brederode, J. (1983), *Z. Naturforsch.* **38C**, 544.

Steyns, J.M., Mastenbroek, O., van Nigtevecht, G. and van Brederode, J. (1984), *Z. Naturforsch.* **39C**, 568.

Strack, D., Meurer, B. and Weissenböck, G. (1982), *Z. Pflanzenphysiol.* **108**, 131.

Suarez, S.S., Cabrera, J.L. and Juliani, H.R. (1982), *An. Asoc. Quim. Argent.* **70**, 647.

Subramanian, P.M. and Misra, G.S. (1979), *J. Nat. Prod.* **42**, 540.

Takagi, S., Yamaki, M. and Inoue, K. (1981), *Phytochemistry* **20**, 2443.

Takahashi, Y., Miyasaka, N., Tasaka, S., Miura, I., Urano, S., Ikura, M., Hikichi, K., Matsumoto, T. and Wada, M. (1982), *Tetrahedron Lett.* **23**, 5163.

Takahashi, Y., Saito, K., Yanagiya, M., Ikura, M., Hikichi, K., Matsumoto, T. and Wada, M. (1984), *Tetrahedron Lett.* **25**, 2471.

Tanaka, N., Nagase, S., Wachi, K., Murakami, T., Saiki, Y. and Chen, C. (1980), *Chem. Pharm. Bull.* **28**, 2843.

Theodor, R., Zinsmeister, H.D., Mues, R. and Markham, K.R. (1980), *Phytochemistry* **19**, 1695.

Theodor, R., Markham, K.R., Mues, R. and Zinsmeister, H.D. (1981a), *Phytochemistry* **20**, 1457.

Theodor, R., Zinsmeister, H.D., Mues, R. and Markham, K.R. (1981b), *Phytochemistry* **20**, 1851.

Theodor, R., Mues, R. and Zinsmeister, H.D. (1983), *Z. Naturforsch.* **38C**, 165.

Tillequin, F., Paris, M., Jacquemin, H. and Paris, R.R. (1978), *Plant. Med.* **33**, 46.

Tiwari, R.D. and Bajpai, M. (1981), *Indian J. Chem.* **20B**, 450.

Touati, D. (1985), Thèse de Doctorat de 3ème cycle, Université Claude Bernard, Lyon N. 1661.

Tschesche, R. and Struckmeyer, K. (1976), *Chem. Ber.* **109**, 2901.

Tschesche, R. and Widera, W. (1982), *Liebigs Ann. Chem.*, 902.

Ulubelen, A. and Mabry, T.J. (1980), *J. Nat. Prod.* **43**, 162.

Ulubelen, A. and Mabry, T.J. (1982), *Rev. Latinoamer. Quim.* **13**, 35.

Ulubelen, A. and Mabry, T.J. (1983), *J. Nat. Prod.* **46**, 597.

Ulubelen, A., Ayyildiz, H. and Mabry, T.J. (1981), *J. Nat. Prod.* **44**, 368.

Ulubelen, A., Bucker, R. and Mabry, T.J. (1982a), *Phytochemistry* **21**, 801.

Ulubelen, A., Kerr, R.R. and Mabry, T.J. (1982b), *Phytochemistry* **21**, 1145.

Ulubelen, A., Oksuz, S., Mabry, T.J., Dellamonica, G. and Chopin, J. (1982c), *J. Nat. Prod.* **45**, 783.

Ulubelen, A., Topcu, G., Mabry, T.J., Dellamonica, G. and Chopin, J. (1982d), *J. Nat. Prod.* **45**, 103.

Ulubelen, A., Mabry, T.J., Dellamonica, G. and Chopin, J. (1984), *J. Nat. Prod.* **47**, 384.

Ulubelen, A. Mabry, T.J., Miski, M., Dellamonica, G., Chopin, J., Paton, N.H. and Smith, P. (1985), *Rev. Latinoamer. Quim.* **16**, 63.
Valant, K., Besson, E. and Chopin, J. (1980), *Phytochemistry* **19**, 156.
Valant-Vetschera, K. (1982a). In *Flavonoids and Bioflavonoids 1981* (eds L. Farkas, F. Kallay, M. Gabor and H. Wagner), Akademiai Kiado, Budapest, p. 213.
Valant-Vetschera, K. (1982b), *Phytochemistry* **21**, 1067.
Valant-Vetschera, K. (1984), *Sci. Pharm.* **52**, 307.
Valant-Vetschera, K. (1985a), *Biochem. Syst. Ecol.* **13**, 15.
Valant-Vetschera, K. (1985b), *Bot. Rev.* **51**, 1.
Valant-Vetschera, K. (1985c), *Biochem. Syst. Ecol.* **13**, 119.
van Brederode, J. and Kamps-Heinsbroek, R. (1981a), *Z. Naturforsch.* **36C**, 484.
van Brederode, J. and Kamps-Heinsbroek, R. (1981b), *Z. Naturforsch.* **36C**, 486.
van Brederode, J. and van Kooten, H. (1983), *Plant Cell Rep.* **2**, 144.
van Brederode, J. and Mastenbroek, O. (1983), *Theor. Appl. Genet.* **64**, 151.
van Brederode, J. and Steyns, J.M. (1983), *Z. Naturforsch.* **38C**, 549.
van Brederode, J. and Steyns, J.M. (1985), *Protoplasma* **128**, 59.
van Brederode, J., van Genderen, H.H. and Berendsen, W. (1982a), *Experientia* **38**, 929.
van Brederode, J., Kamps-Heinsbroek, R. and Mastenbroek, O. (1982b), *Z. Pflanzenphysiol.* **106**, 43.
van Genderen, H.H., van Brederode, J. and Niemann, G.J. (1983a), *J. Chromatogr.* **256**, 151.
van Genderen, H.H., Niemann, G.J. and van Brederode, J. (1983b), *Protoplasma* **118**, 135.
van Heerden, F.R., Brandt, E.V. and Roux, D.G. (1980), *J. Chem. Soc. Perkin Trans.* **1**, 2463.
Vaughan, D.A. and Hymowitz, T. (1984), *Biochem. Syst. Ecol.* **12**, 189.
Vita-Finzi, P., Bernardi, M. de, Fronza, G., Mellerio, G., Servettaz-Grünanger, O., Vidari, G. and Beltrami, E. (1980), *12th IUPAC International Symposium on the Chemistry of Natural Products*, Abstracts, p. 167.
Wagner, H., Budweg, W., Iyengar, M.A., Volk, O. and Sinn, M. (1972), *Z. Naturforsch.* **27B**, 809.
Wagner, H., Obermeier, G., Seligmann, O. and Chari, V.M. (1979), *Phytochemistry* **18**, 907.
Wagner, H., Obermeier, G., Chari, V.M. and Galle, K. (1980), *J. Nat. Prod.* **43**, 583.
Wallace, J.W. and Grisebach, H. (1973), *Biochim. Biophys. Acta* **304**, 837.
Wallace, J.W., Mabry, T.J. and Alston, R.E. (1969), *Phytochemistry* **8**, 93.
Wallace, J.W., Yopp, D.L., Besson, E. and Chopin, J. (1981), *Phytochemistry* **20**, 2701.
Wallace, J.W., Porter, P.L., Besson, E. and Chopin, J. (1982), *Phytochemistry* **21**, 482.
Wallace, J.W., Chapman, N., Sullivan, J.E. and Bhardwaja Trilokin (1984), *Am. J. Bot.* **71**, 660.
Ward, R.S. (1982), Private communication.
Williams, C.A. and Harborne, J.B. (1977a), *Biochem. Syst. Ecol.* **5**, 45.
Williams, C.A. and Harvey, W.J. (1982), *Phytochemistry* **21**, 329.
Williams, C.A., Harborne, J.B. and Clifford, H.T. (1973), *Phytochemistry* **12**, 2417.
Williams, C.A., Harborne, J.B. and Mayo, S.J. (1981), *Phytochemistry* **20**, 217.
Williams, C.A., Demissie, A. and Harborne, J.B. (1983a), *Biochem. Syst. Ecol.* **11**, 221.
Williams, C.A., Harborne, J.B., Clifford, H.T. and Glassman, S.F. (1983b), *Plant Syst. Evol.* **142**, 157.
Williams, C.A., Harborne, J.B. and Glassman, S.F. (1985), *Plant Syst. Evol.* **149**, 233.
Williams, C.A., Harborne, J.B. and Goldblatt, P. (1986), *Phytochemistry* **25**, 2135.
Wilson, A. (1985), *Phytochemistry* **24**, 1685.
Wirasutisna, K.R., Gleye, J., Moulis, C., Stanislas, E. and Moretti, C. (1986), *Phytochemistry* **25**, 558.
Wolf, S.J., Denford, K.E. and Packer, J.G. (1979), *Canad. J. Bot.* **57**, 2374.
Woo, W.S., Kang, S.S., Wagner, H., Seligmann, O. and Chari, V.M. (1980a), *Phytochemistry* **19**, 2791.
Woo, W.S., Shin, K.S. and Kang, S.S. (1980b), *Recent Adv. Nat. Prod. Res., Proc. Int. Symp. 1979*, Seoul National University Press, p. 33.
Yagishita, K. (1983) Private communication.
Yahara, S. and Nishioka, I. (1984), *Phytochemistry* **23**, 2108.
Yang, K. and Kim, T. (1983), *Saengyak Hakhoechi* **14**, 92.
Yasukawa, K., Yamanouchi, S. and Takido, M. (1982), *Yakugaku Zasshi* **102**, 292.
Young, D.A. and Sterner, R.W. (1981), *Biochem. Syst. Ecol.* **9**, 185.
Zemtsova, G.N. and Bandyukova, V.A. (1974), *Khim. Prir. Soedin.*, 107.
Zhuang, L. (1983), *Zhongcaoyao* **14**, 295.

4

Biflavonoids

HANS GEIGER and CHRISTOPHER QUINN

4.1 COMPLETE LIST OF KNOWN BIFLAVONOIDS

The number of new biflavonoids discovered during the last five years is not very large. However, many of them exhibit new structural features. The structures of a few amentoflavone *O*-glycosides have been established and the first biflavone *C*-glycoside has been found. The first two examples of flavone–isoflavone dimers have been isolated; among the luteolin-based dimers two new types of interflavonyl bridgeheads have been discovered, namely the *ortho*- and *para*-positions of the 3′-hydroxyl. These few highlights demonstrate clearly that the field is still expanding. The list of known biflavonoids in Table 4.1 is arranged as in Geiger and Quinn, (1982). The numbering system of the compounds is as follows: the first numeral refers to the type of interflavonyl link, the letter refers to the type of flavonoid moiety (A = biflavones, B = flavanone–flavone, C = biflavanone and D, E, F = others) and the last numeral refers to the individual compound in each group. The names are again those given to them by the authors.

4.2 DETECTION, IDENTIFICATION, ISOLATION AND STRUCTURE DETERMINATION

Considering all the new types of biflavonoids discovered recently, detection and identification of a specific biflavonoid just by paper (PC) or thin layer (TLC) chromatography and the standard UV spectra can no longer be deemed safe. Ideally biflavonoids should always be isolated in a pure state. If, however, as in phytochemical surveys, the amount of plant material is limited, it is a good idea to simplify the chromatograms by permethylation of the biflavonoids (cf. Gadek and Quinn, 1983). Taufeeq *et al.* (1982) have reported the separation of a biflavonoid mixture via their benzyl ethers.

A new TLC system, which yields compact spots and reasonable R_F values with the more highly hydroxylated biflavonoids is polyamide and ethyl acetate–ethylmethylketone–formic acid–water (5:3:1:1) (Becker *et al.*, 1986). HPLC has so far only been used for biflavonoids in a few cases (Briançon-Scheid *et al.*, 1982, 1983). The histochemical detection and localization of biflavonoids by $AlCl_3$-induced fluorescence in microscopic cross-sections has been applied by Gadek *et al.* (1984). No basically new isolation techniques have been reported. Much progress has been made with ^{13}C NMR spectroscopy, which is a powerful method for the elucidation of biflavonoid structures, provided that at least 20 mg, but better 50 mg of pure compound is at hand. When ^{13}C NMR data are available for a given biflavonoid, this is indicated in Table 4.1. There is, however, one pitfall with ^{13}C NMR: the signals of some quaternary carbons may be rather poor due to their long relaxation times and easily overlooked (Markham *et al.*, 1987).

Proton NMR spectroscopy is of course also an essential method, which is complementary to ^{13}C NMR. However most published spectra are those of the trimethylsilyl ethers. Since ^{13}C NMR spectra of biflavonoids are most commonly recorded in DMSO, it is of course convenient to record the proton spectrum in the same solvent. More proton NMR reference spectra recorded in DMSO would therefore be welcome. The enantiomeric purity of atropisomeric biflavones has been determined by use of a chiral NMR shift reagent (Rahman *et al.*, 1982).

Field desorption mass spectrometry (FD-MS) of biflavonoids distinguishes them most clearly from their

The Flavonoids. Edited by J.B. Harborne
Published in 1988 by Chapman and Hall Ltd,
11 New Fetter Lane, London EC4P 4EE.

Table 4.1 Complete list of known biflavonoids

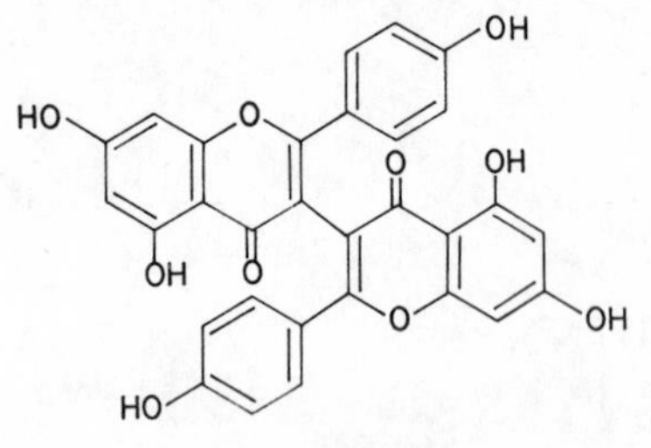

1A.1 3.3″-Biapigenin

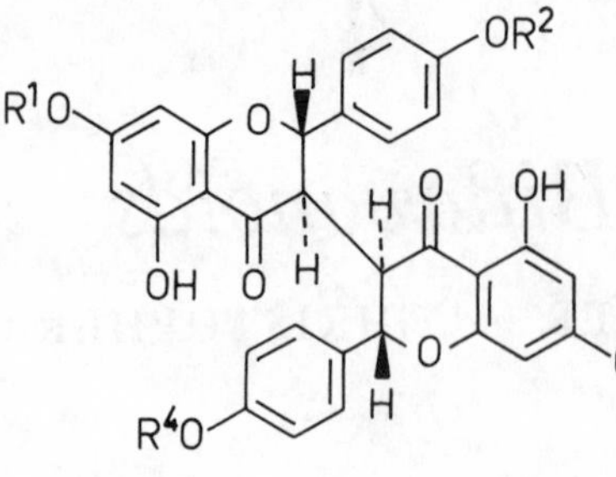

a ‡

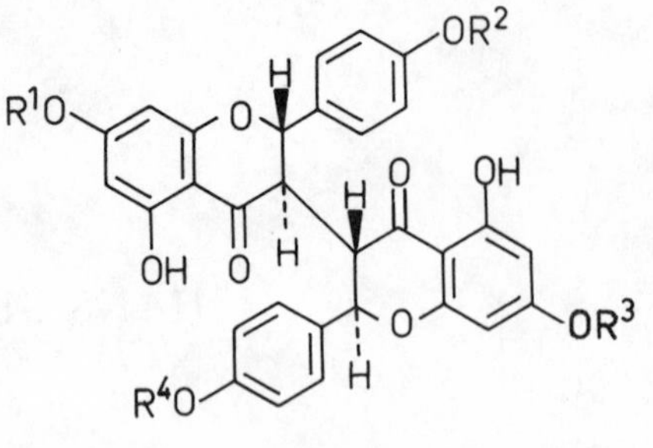

b

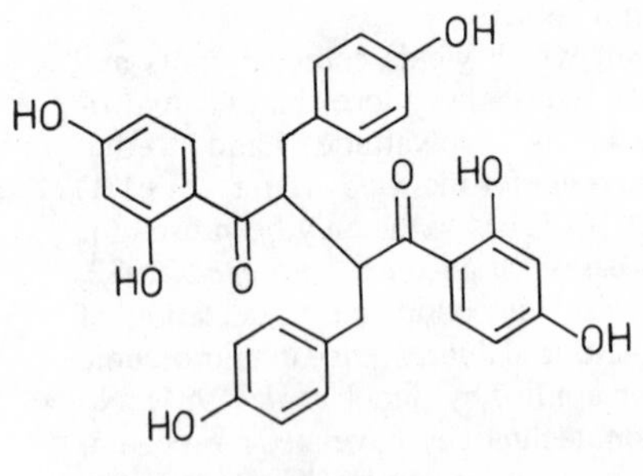

1D.1 Brackenin,
Drewes and Hudson (1983) (^{13}C)

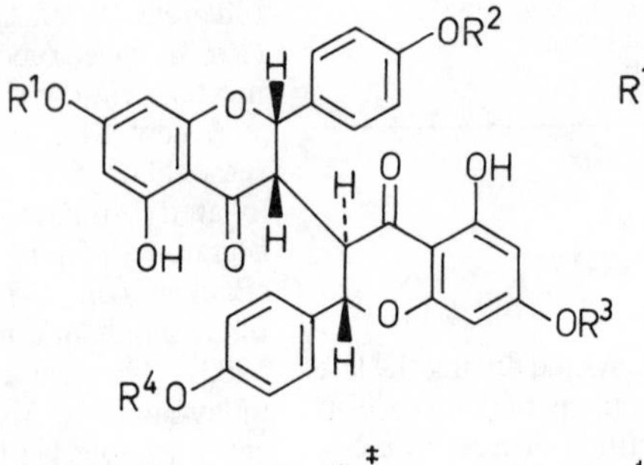

c ‡

1C

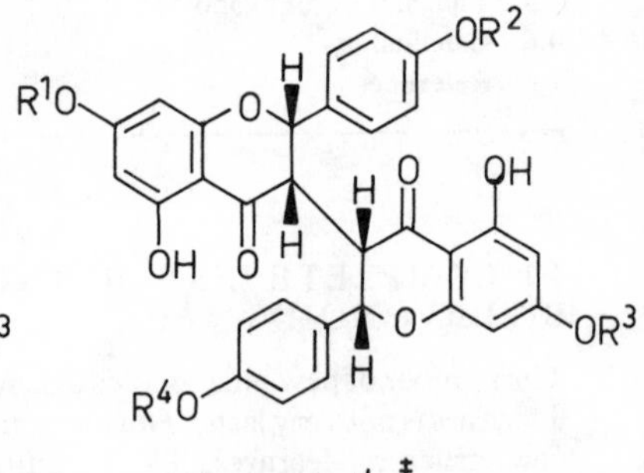

d ‡

‡Relative, not absolute, stereochemistry!

		R^1	R^2	R^3	R^4	*References not in Geiger and Quinn (1982)*
1C.1a	Chamaejasmin	H	H	H	H	Niwa *et al.* (1984a) (stereochem.) Castro and Valverde (1985) (^{13}C)
1C.1b	Isochamaejasmin	H	H	H	H	Niwa *et al.* (1984a) (^{13}C)
1C.1c	Neochamaejasmin B	H	H	H	H	
1C.1d	Neochamaejasmin A	H	H	H	H	Niwa *et al.* (1984b)
1C.2a	7-Methylchamaejasmin	Me	H	H	H	Yang *et al.* (1983, 1984)
1C.3a	Chamaejasmenin A	H	Me	H	Me	Liu *et al.*(1984) (^{13}C)
1C.3d	Chamaejasmenin B	H	Me	H	Me	Liu *et al.*(1984) (^{13}C)
1C.4d	Chamaejasmenin C	Me	Me	H	Me	Liu *et al.*(1984) (^{13}C)
1C.5a	7,7′-Dimethylchamaejasmenin A	Me	Me	Me	Me	Niwa *et al.* (1984b)
1C.5b	7,4′,7″,4‴-Tetramethyl-isochamaejasmin	Me	Me	Me	Me	Niwa *et al.* (1984a) (^{13}C)
1C.5c	7,4′,7″,4‴-tetramethyl-neochamaejasmin B	Me	Me	Me	Me	Niwa *et al.* (1984b)
1C.5d	7,4′,7″,4‴-Tetramethyl-neochamaejasmin A	Me	Me	Me	Me	Niwa *et al.* (1984b)

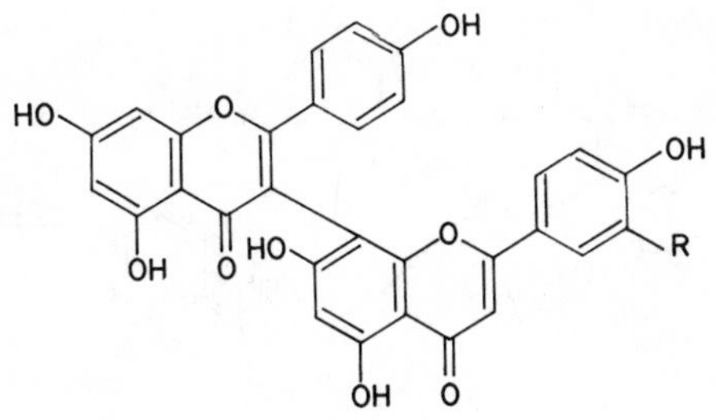

2A

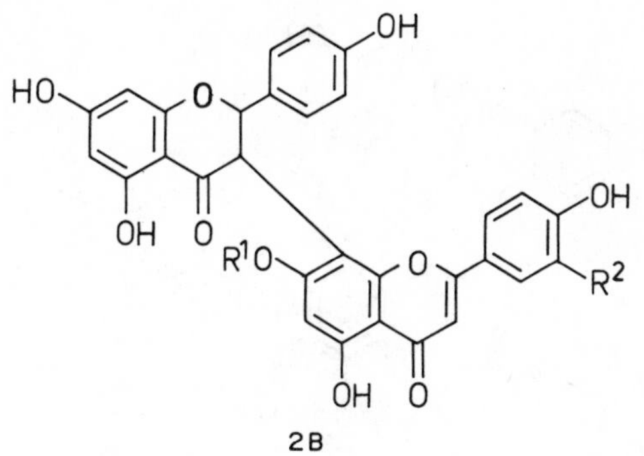

2B

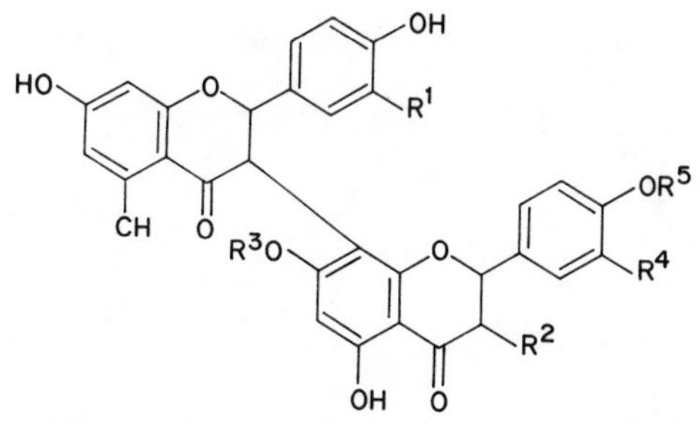

2C

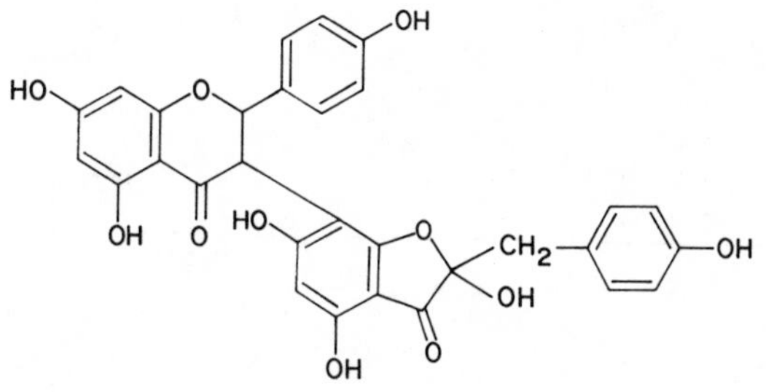

2D

	R					References not in Geiger and Quinn (1982)
2A.1 Bisdehydro-GBla	H					
2A.2 Sahranflavone	OH					
	R[1]	*R*[2]				
2B.1 Volkensiflavone (= BGH-III = talbotiflavone)	H	H				
2B.2 Spicatiside	Glc	H				
2B.3 Fukugetin (= BGH-II = morelloflavone)	H	OH				Duddeck *et al.* (1978) Waterman and Crichton (1980) (^{13}C)
2B.4 3-*O*-methylfugugetin	H	OMe				Waterman and Crichton (1980) (^{13}C)
2B.5 Fukugiside	Glc	OH				
	R[1]	*R*[2]	*R*[3]	*R*[4]	*R*[5]	
2C.1 GBla	H	H	H	H	H	Duddeck *et al.* (1978) (^{13}C)
2C.2 GBla-7″-*O*-glucoside	H	H	Glc	H	H	
2C.3 GBl	H	OH	H	H	H	
2C.4 GB2a	H	H	H	OH	H	Duddeck *et al.* (1978) (^{13}C)
2C.5 GB2	H	OH	H	OH	H	
2C.6 Kolaflavanone	H	OH	H	OH	Me	
2C.7 Xanthochymusside	H	OH	Glc	OH	H	
2C.8 Manniflavanone	OH	OH	H	OH	H	
2D.1 Zeyherin						

(*Contd.*)

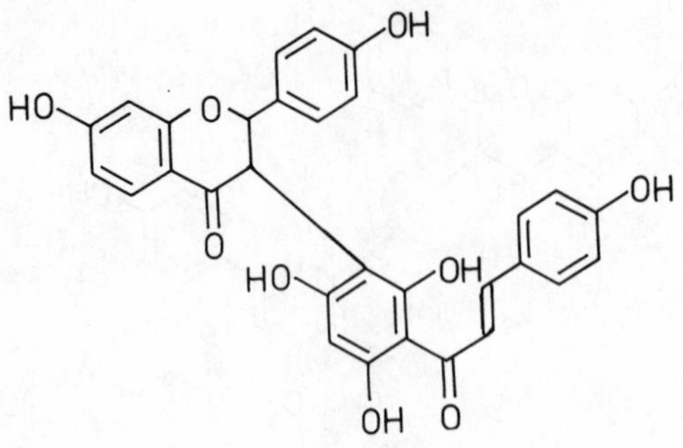

2E.1 'Compound 5' Drewes *et al.* (1984)

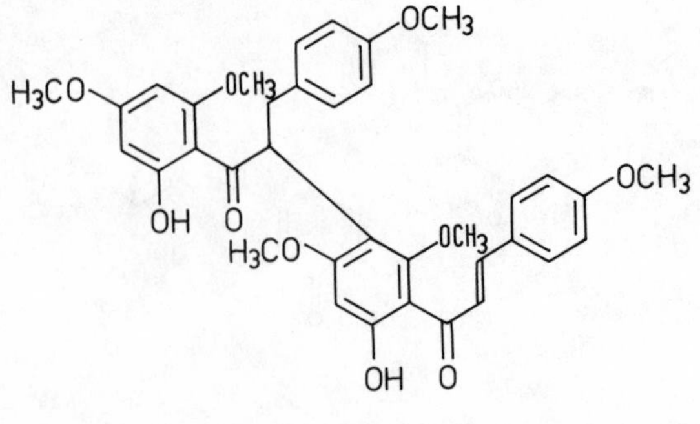

2F.1 'I, II-3'-linked dihydrobichalkone' Manchanda and Khan (1985)

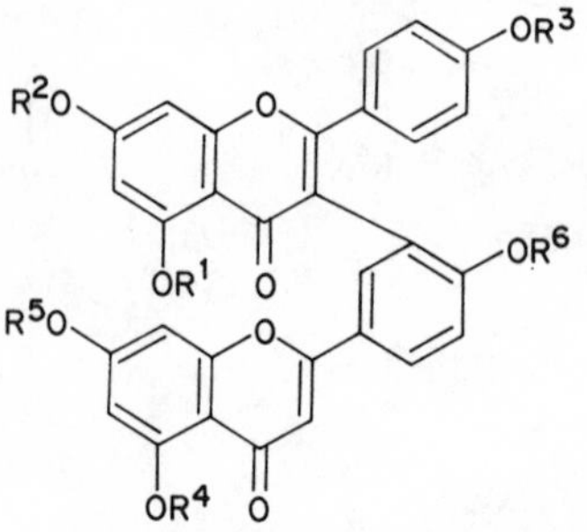

3A

	R^1	R^2	R^3	R^4	R^5	R^6	*References not in Geiger and Quinn (1982)*
3A.1 Taiwaniaflavone	H	H	H	H	H	H	
3A.2 Taiwaniaflavone 7″-methyl ether	H	H	H	H	Me	H	
3A.3 Taiwaniaflavone 4′, 7″-dimethyl ether	H	H	Me	H	Me	H	
3A.4 Taiwaniaflavone hexamethyl ether	Me	Me	Me	Me	Me	Me	Kamil *et al.* (1981) (synth.)

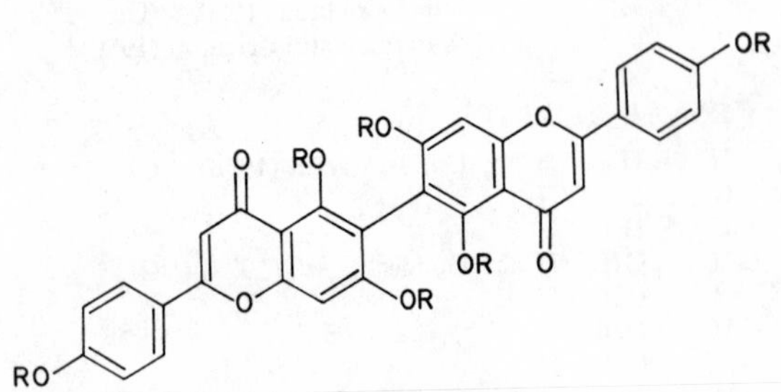

4A

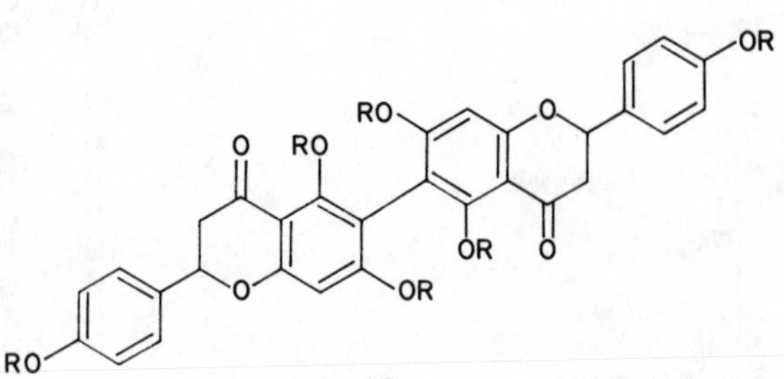

4C

	R	References not in Geiger and Quinn (1982)
4A.1 Didehydrosuccedaneaflavone	H	
4A.2 Didehydrosuccedaneaflavone hexamethyl ether	Me	Parthasarathy *et al.* (1977)
4C.1 Succedaneaflavone	H	
4C.2 Succedaneaflavone hexamethyl ether	Me	

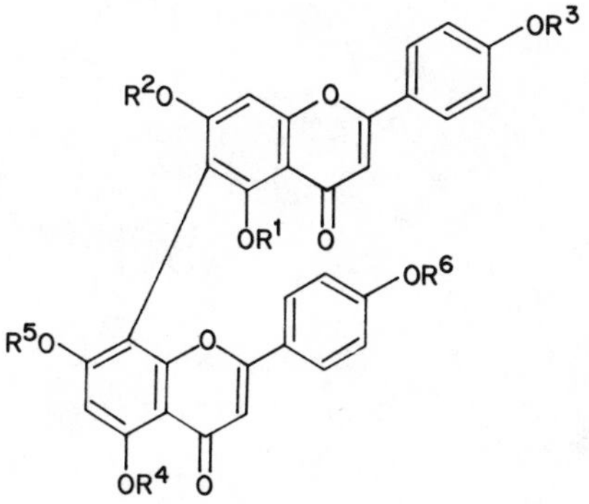

5A

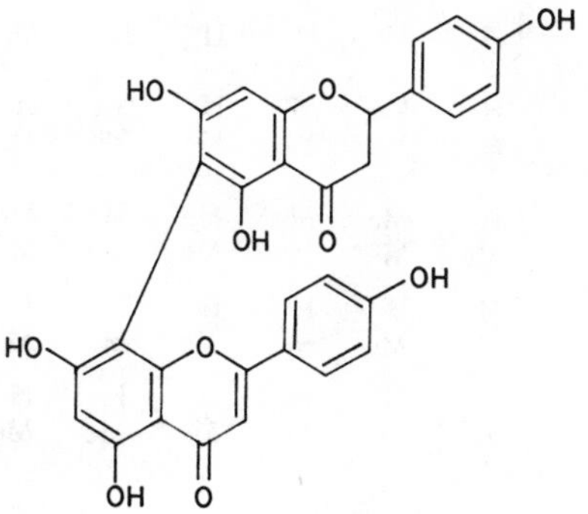

5B

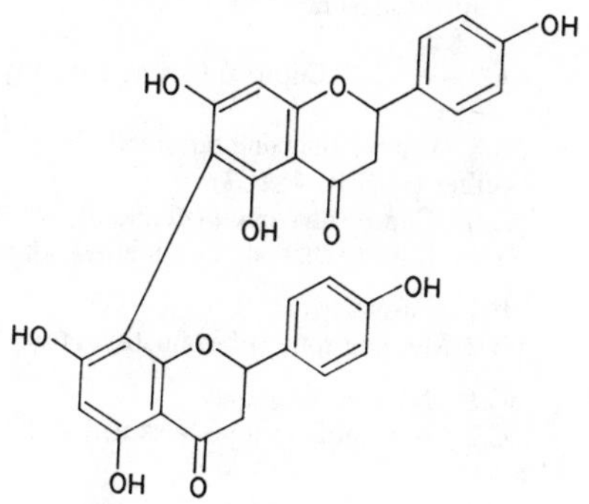

5C

	R^1	R^2	R^3	R^4	R^5	R^6
5A.1 Agathisflavone	H	H	H	H	H	H
5A.2 7-*O*-Methylagathisflavone (=WA-I)	H	Me	H	H	H	H
5A.3 7, 7″-Di-*O*-methylagathisflavone	H	Me	H	H	Me	H
5A.4 7, 4‴-Di-*O*-methylagathisflavone (=WA-VII)	H	Me	H	H	H	Me
5A.5 7, 7″, 4‴-Tri-*O*-methylagathisflavone	H	Me	H	H	Me	Me
5A.6 Agathisflavone tetramethyl ether	H	Me	Me	H	Me	Me
5A.7 Agathisflavone hexamethyl ether	Me	Me	Me	Me	Me	Me
5B.1 Rhusflavone						
5C.1 Rhusflavanone						

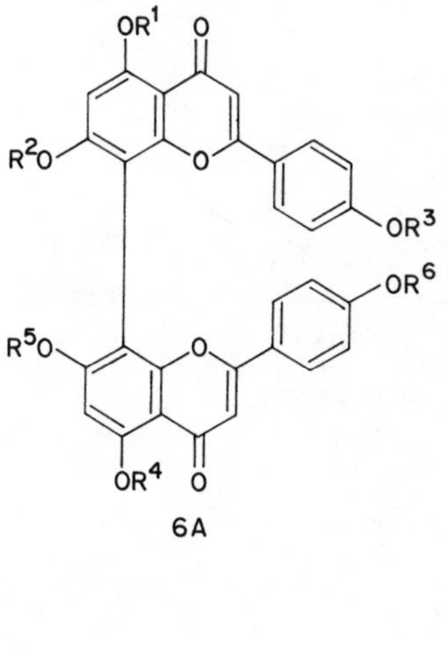

6A

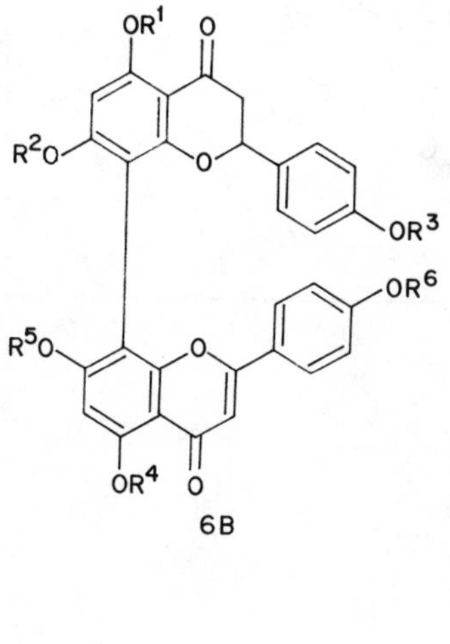

6B

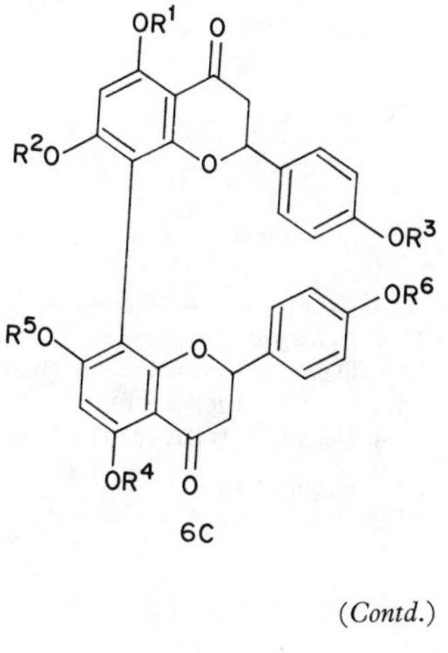

6C

(*Contd.*)

	R^1	R^2	R^3	R^4	R^5	R^6	*References not in Geiger and Quinn (1982)*
6A.1 Cupressuflavone	H	H	H	H	H	H	
6A.2 7-*O*-Methylcupressuflavone	H	Me	H	H	H	H	
6A.3 4′-*O*-Methylcupressuflavone	H	H	Me	H	H	H	
6A.4 5, 5″-Di-*O*-methylcupressuflavone	Me	H	H	Me	H	H	Gopalakrishnan *et al.* (1980)
6A.5 7, 7″-Di-*O*-methylcupressuflavone	H	Me	H	H	Me	H	
6A.6 7, 4′ or 4‴-Di-*O*-methylcupressuflavone	H	Me	Me	H	H	H	
			or				
	H	Me	H	H	H	Me	
6A.7 7, 4′, 7″-Cupressuflavone trimethyl ether	H	Me	Me	H	Me	H	
6A.8 Cupressuflavone tetramethyl ether (=WB1=AC3)	H	Me	Me	H	Me	Me	
6A.9 Cupressuflavone pentamethyl ether	H	Me	Me	Me	Me	Me	
6A.10 Cupressuflavone hexamethyl ether	Me	Me	Me	Me	Me	Me	
6B.1 Mesuaferrone B	H	H	H	H	H	H	
6B.2 Mesuaferrone B hexamethyl ether	Me	Me	Me	Me	Me	Me	
6C.1 Neorhusflavanone	H	H	H	H	H	H	
6C.2 Neorhusflavanone hexamethyl ether	Me	Me	Me	Me	Me	Me	

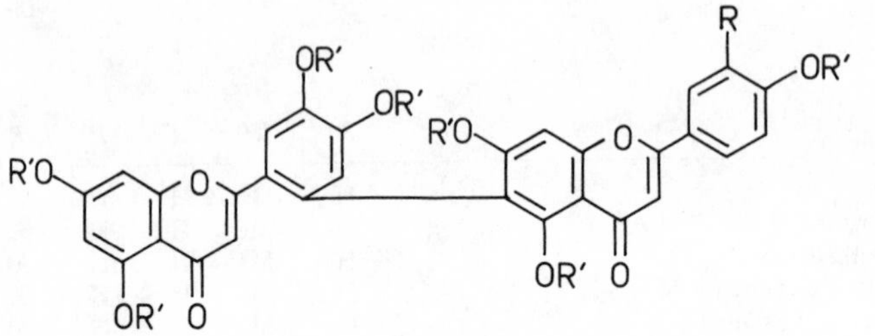

7A

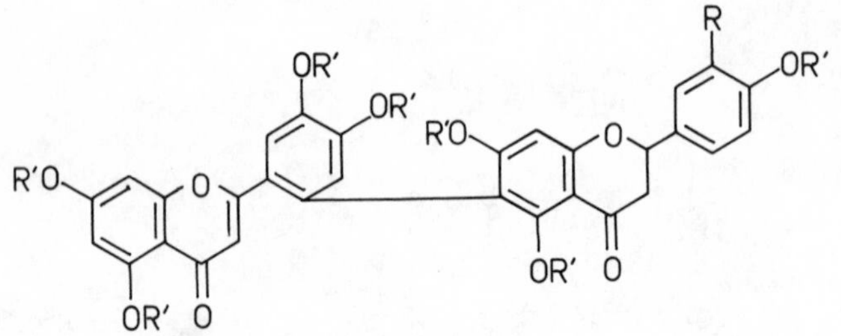

7B

	R	*R′*	*Reference*
7A.1 Dehydrohegoflavone A	H	H	
7A.2 Dehydrohegoflavone A heptamethyl ether	H	Me	
7A.3 Dehydrohegoflavone B	OH	H	Wada *et al.* (1985) (^{13}C)
7A.4 Dehydrohegoflavone B octamethyl ether	OMe	Me	
7B.1 Hegoflavone A	H	H	
7B.2 Hegoflavone B	OH	H	

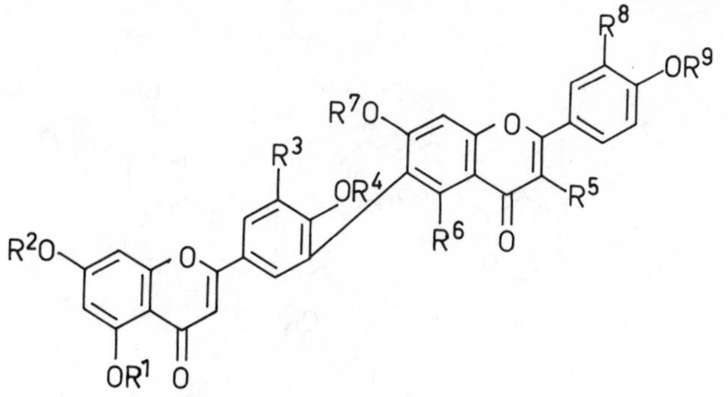

8A

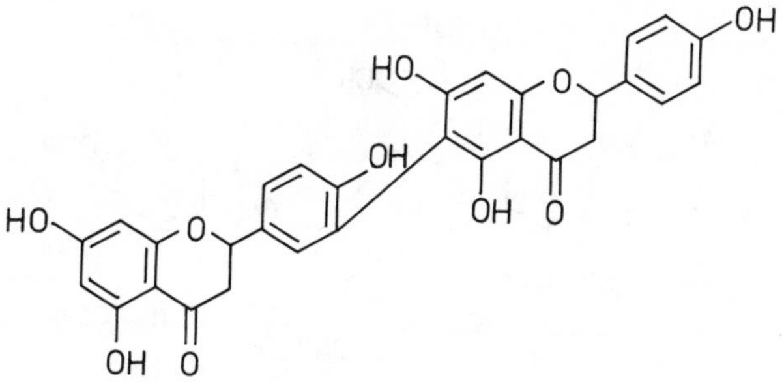

8C.1 Tetrahydrorobustaflavone

	R^1	R^2	R^3	R^4	R^5	R^6	R^7	R^8	R^9	*References not in Geiger and Quinn (1982)*
8A.1 Robustaflavone	H	H	H	H	H	OH	H	H	H	
8A.2 Robustaflavone hexamethyl ether	Me	Me	H	Me	H	OMe	Me	H	Me	
8A.3 Abiesin	H	Me	H	Me	OH	H	H	H	Me	Chatterjee *et al.* (1984) (^{13}C)
8A.4 5′, 3‴-Dihydroxy-robustaflavone	H	H	OH	H	H	OH	H	OH	H	Becker *et al.* (1986) (^{13}C)
8A.5 5′, 3‴-Dihydroxy-robustaflavone octamethyl ether	Me	Me	OMe	Me	H	OMe	Me	OMe	Me	Becker *et al.* (1986) (^{13}C)

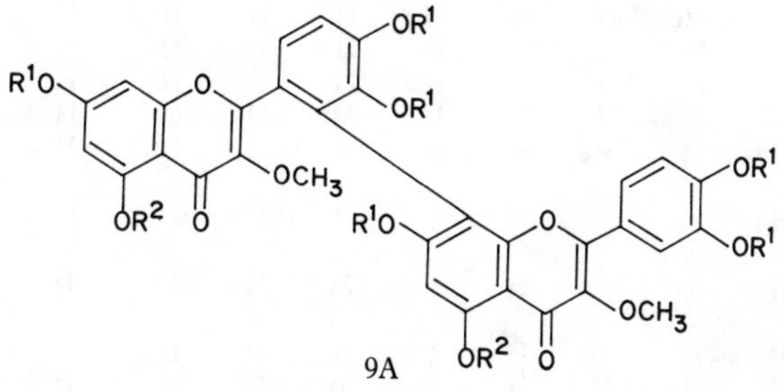

9A

	R^1	R^2	*References*
9A.1 Strychnobiflavone	H	H	
9A.2 Strychnobiflavone hexamethyl ether	Me	H	Nicoletti *et al.* (1984) (^{13}C)
9A.3 Strychnobiflavone octamethyl ether	Me	Me	

(*Contd.*)

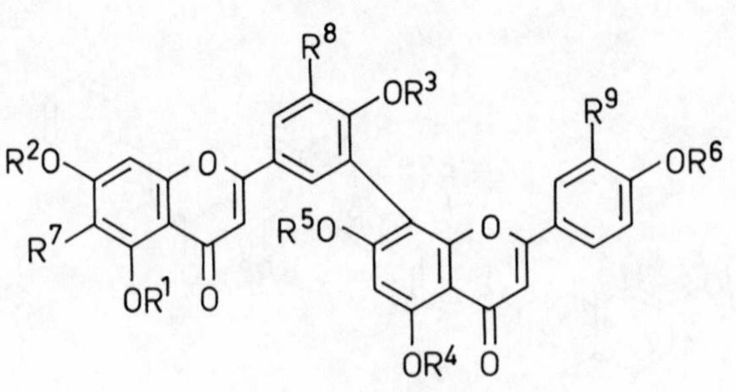

10A

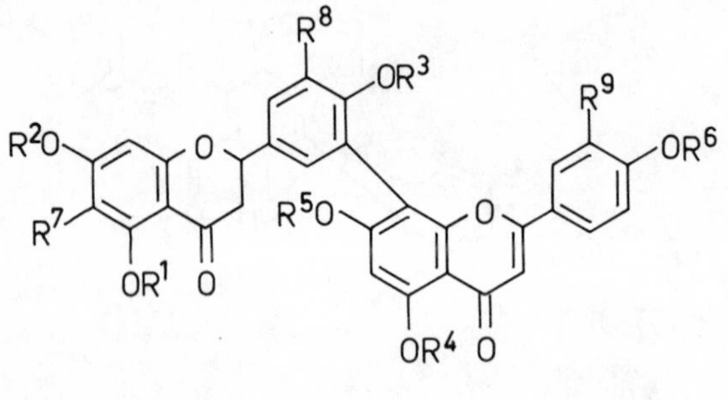

10B

	R^1	R^2	R^3	R^4	R^5	R^6	R^7	R^8	R^9	*References not in Geiger and Quinn (1982)*
10A.1 Amentoflavone	H	H	H	H	H	H	H	H	H	
10A.2 Amentoflavone glucoside Pn II	H	Glc	H	H	H	Glc	H	H	H	
10A.3 Amentoflavone glucoside Pn IV	H	H	Glc	H	H	Glc	H	H	H	Markham (1984)
10A.4 Amentoflavone glucoside Pn V	H	Glc	Glc	H	H	Glc	H	H	H	
10A.5 Sequoiaflavone	H	Me	H	H	H	H	H	H	H	
10A.6 Bilobetin	H	H	Me	H	H	H	H	H	H	
10A.7 Sotetsuflavone	H	H	H	H	Me	H	H	H	H	
10A.8 Podocarpusflavone A	H	H	H	H	H	Me	H	H	H	
10A.9 Ginkgetin	H	Me	Me	H	H	H	H	H	H	
10A.10 7,7″-Di-*O*-methylamentoflavone	H	Me	H	H	Me	H	H	H	H	
10A.11 Podocarpusflavone B (= putraflavone)	H	Me	H	H	H	Me	H	H	H	
10A.12 Amentoflavone 4′, 7″-dimethyl ether	H	H	Me	H	Me	H	H	H	H	
10A.13 Isoginkgetin	H	H	Me	H	H	Me	H	H	H	
10A.14 Amentoflavone 7″, 4‴-dimethyl ether	H	H	H	H	Me	Me	H	H	H	
10A.15 Amentoflavone 7,4′, 7″-tri-*O*-methyl ether	H	Me	Me	H	Me	H	H	H	H	
10A.16 Sciadopitysin	H	Me	Me	H	H	Me	H	H	H	

	R^1	R^2	R^3	R^4	R^5	R^6	R^7	R^8	R^9	*References not in Geiger and Quinn (1982)*
10A.17 Heveaflavone	H	Me	H	H	Me	Me	H	H	H	
10A.18 Kayaflavone	H	H	Me	H	Me	Me	H	H	H	
10A.19 Amentoflavone 7, 7″, 4′, 4‴-tetramethyl ether (= WI3)	H	Me	Me	H	Me	Me	H	H	H	
10A.20 Amentoflavone hexamethyl ether (= Dioonflavone)	Me	Me	Me	Me	Me	Me	H	H	H	
10A.21 5′-Methoxybilobetin	H	H	Me	H	H	H	H	OMe	H	
10A.22 5‴-Hydroxy-amentoflavone	H	H	H	H	H	H	H	H	OH	Ohmoto and Yoshida (1983) (^{13}C)
10A.23 5′, 8″-Biluteolin	H	H	H	H	H	H	H	OH	OH	Österdahl (1983) (^{13}C)
10A.24 7-*O*-Methyl-6-methylamentoflavone	H	Me	H	H	H	H	Me	H	H	
10B.1 2, 3-Dihydro-amentoflavone	H	H	H	H	H	H	H	H	H	
10B.2 2,3-Dihydro-amentoflavone 7″, 4‴-dimethyl ether	H	H	H	H	Me	Me	H	H	H	
10B.3 2,3-Dihydro-sciadipitysin	H	Me	Me	H	H	Me	H	H	H	
10B.4 2, 3-Dihydroamento-flavone (hexamethyl ether)	Me	Me	Me	Me	Me	Me	H	H	H	

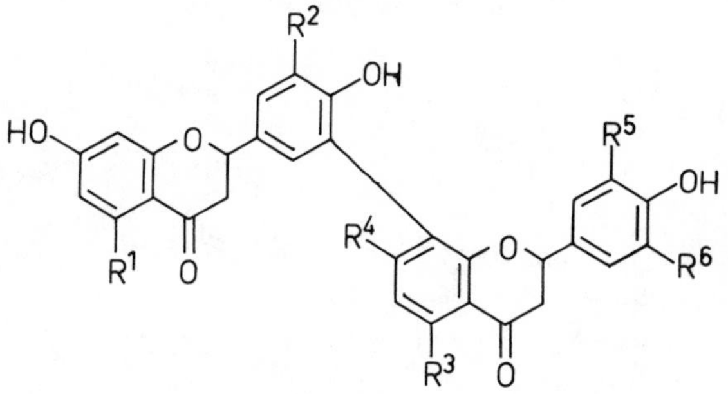

10C

	R^1	R^2	R^3	R^4	R^5	R^6	*References not in Geiger and Quinn (1982)*
10C.1 Biflavanone C	H	H	H	OH	H	H	Murthy (1983a)
10C.2 Tetrahydroamentoflavone	OH	H	OH	OH	H	H	Ahmad *et al.* (1981) (^{13}C)
10C.3 Biflavanone A	OH	H	OH	H	OH	H	Murthy *et al.* (1981a)
10C.4 Semecarpuflavanone	H	H	H	OH	OH	OH	Murthy (1983b)
10C.5 Jeediflavanone	OH	H	OH	OH	OH	H	Murthy (1985a)
10C.6 Galluflavanone	H	OH	H	OH	OH	OH	Murthy (1985b)
10C.7 II-7-*O*-Methyltetra hydro-amentoflavone	OH	H	OH	OMe	H	H	Kamil *et al.* (1986)

(Contd.)

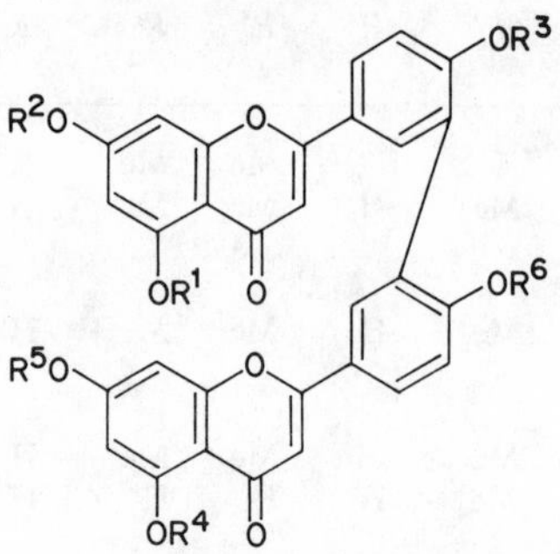

11A.1 ($R^1 = R^2 = R^4 = R^5 = R^6 = Me$)3′, 3‴-biapigenin-5, 5″, 7, 7″, 4″, 4‴-hexamethyl ether

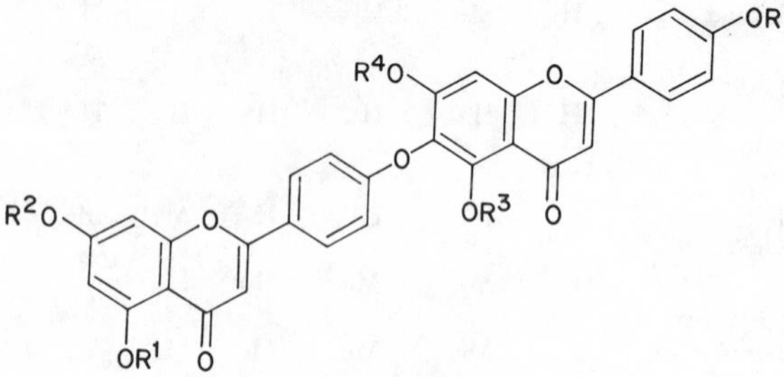

12 A

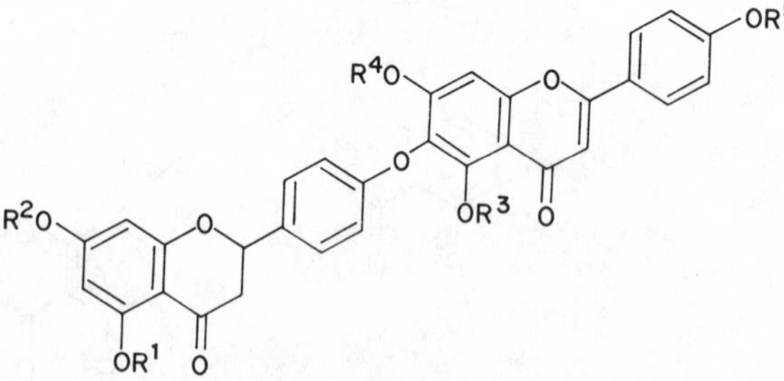

12 B

	R^1	R^2	R^3	R^4	R^5
12A.1 Hinokiflavone	H	H	H	H	H
12A.2 Neocryptomerin	H	Me	H	H	H
12A.3 Isocryptomerin	H	H	H	Me	H
12A.4 Cryptomerin A	H	H	H	H	Me
12A.5 Chamaecyparin	H	Me	H	Me	H
12A.6 Hinokiflavone 7, 4‴-dimethyl ether	H	Me	H	H	Me
12A.7 Cryptomerin B	H	H	H	Me	Me
12A.8 Hinokiflavone 7, 7″, 4‴-trimethyl ether	H	Me	H	Me	Me
12A.9 Hinokiflavone pentamethyl ether	Me	Me	Me	Me	Me
12B.1 2, 3-Dihydrohinokiflavone	H	H	H	H	H
12B.2 2, 3-Dihydrohinokiflavone pentamethyl ether	Me	Me	Me	Me	Me

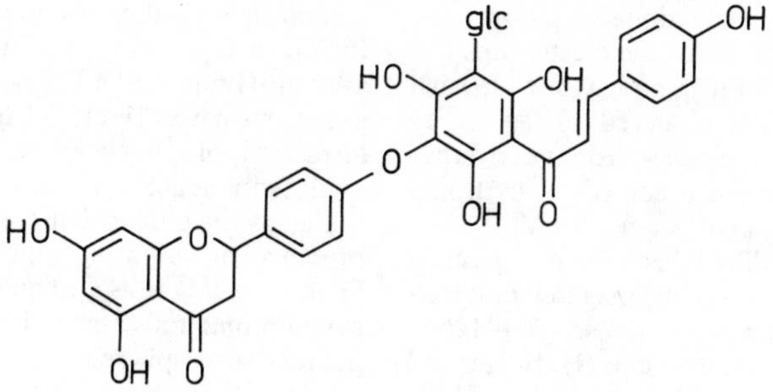

12C.1 Occidentoside Murthy *et al.* (1982)

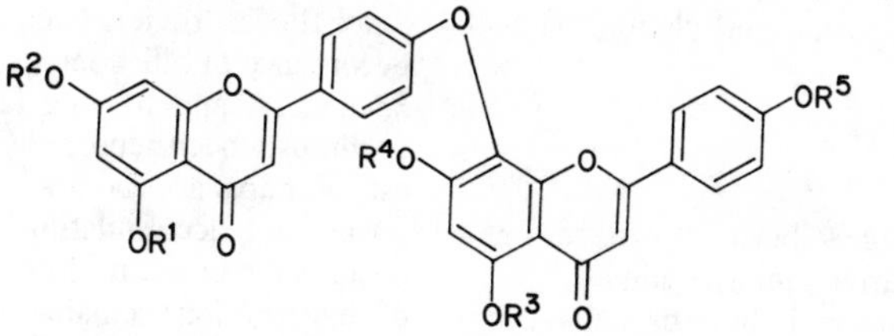

13A.1 ($R^1 = R^2 = R^3 = R^4 = R^5 = Me$)4‴, 5, 5″, 7, 7″-pentamethoxy-4′, 8″-biflavonyl ether

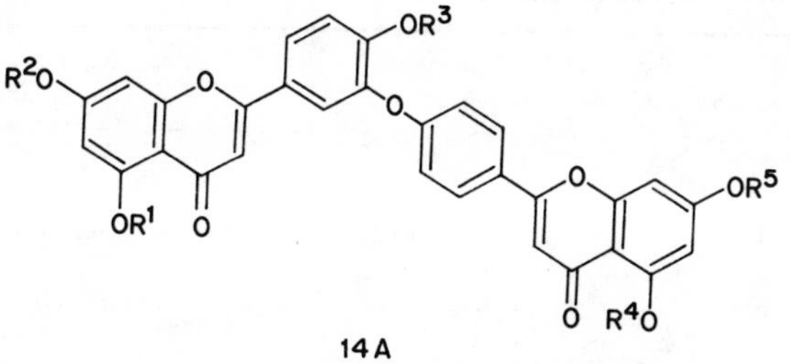

14 A

	R^1	R^2	R^3	R^4	R^5
14A.1 Ochnaflavone (=OSI)	H	H	H	H	H
14A.2 Ochnaflavone 4′-methyl ether (=OSII)	H	Me	H	H	H
14A.3 Ochnaflavone 7, 4′-dimethyl ether (=OSIII)	H	Me	Me	H	H
14A.4 Ochnaflavone 4′, 7, 7″-trimethyl ether (=OSIV)	H	Me	Me	H	Me
14A.5 Ochnaflavone pentamethyl ether	Me	Me	Me	Me	Me

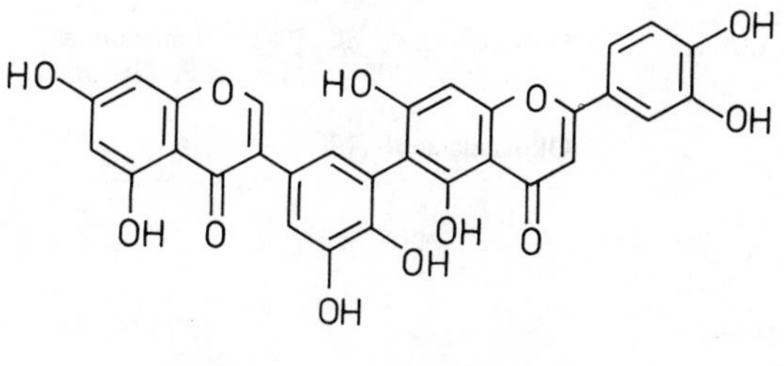

15A.1 Heterobryoflavone
Geiger *et al.* (1986) (^{13}C)

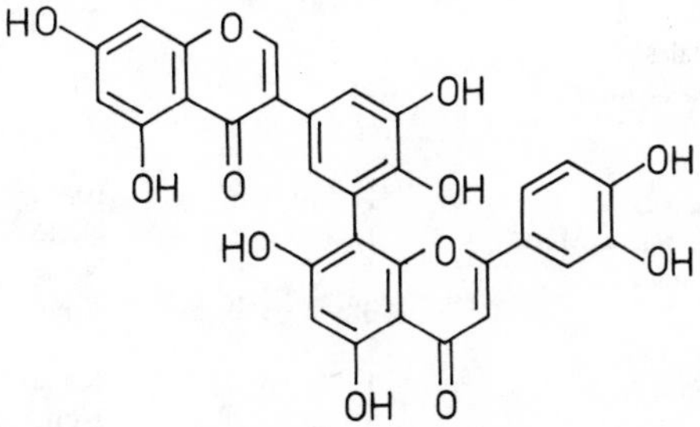

16A.1 Bryoflavone
Geiger *et al.* (1986) (^{13}C)

monomeric counterparts. This technique yields only the molecular ion (observed as $M^{\cdot+}$ or as the cluster ions $[M+H]^+$, $[M+Na]^+$ or $[M+K]^+$) and a few thermal fragments. The fragmentations observed so far are as follows: biflavonoids in which the *ortho*-positions on either side of the interflavonyl link are occupied by hydroxyl groups, may lose water and show a $[M-18]^+$ fragment or the corresponding cluster ions; garcinia biflavonoids (see Table 4.1, 2B and 2C) may lose phloroglucinol and thus give rise to a very intense $[M-126]^+$ fragment or the corresponding cluster ions (H. Geiger and G. Schwinger, unpublished). With biflavonoid-*O*-glycosides, the thermal elimination of sugars has to be considered (Geiger and Schwinger, 1980). Partial methyl ethers may undergo thermal transmethylation (Geiger *et al.*, 1984).

4.3 SYNTHESIS

Silica-bound ferric chloride has been shown to be a versatile reagent for the oxidative dimerization of phloracetophenone, resacetophenone and their partial methyl ethers, thus yielding useful synthons for the synthesis of [6, 6′], [6, 8′]- and [8, 8′]-linked biflavonoids (Parthasarathy and Gupta 1984). The same reagent upon reaction with a chalcone gave a [Iα, II-3′]-linked dihydrobichalcone, the formation of which must be interpreted as an acid catalysed reaction (Manchanda and Khan, 1985).

Finally the photochemical dimerization of flavone in the presence of sulphite, which leads to 2, 2″-biflavanone (Yokoe *et al.*, 1979), should be mentioned, since this reaction might also take place in the presence of an SO_2-polluted atmosphere.

4.4 NATURAL OCCURRENCE

Since the last review (Geiger and Quinn, 1982) the known occurrence of biflavonoids has been extended to a further 33 genera, and to nine families in which they had not previously been reported (Tables 4.2-4.5). Of most interest, perhaps, are the first reports of biflavones in the true ferns, and accumulating evidence of their widespread occurrence in the mosses. Otherwise, the general pattern of distribution remains very much the same: almost

Table 4.2 Biflavonoids in bryophytes and pteridophytes

	No. species	*Organ**	*Biflavonoids reported*	*References not in Geiger and Quinn (1975, 1982)*
BRYOPHYTA				
Bryales				
Bryum L.	1		15A.1; 16A.1	Geiger *et al.* (1987)
Dicranum Hedw.	1		10A.23	
Hylocomium B.S.G.	1		8A.4; 10A.23	Becker *et al.* (1986)
TRACHEOPHYTA				
Psilotales				
Psilotum Sw.	3		10A.1, 2, 3, 4, Pn VI	Markham (1984)
Tmesipteris Bernh.	2		10A.1, 2	
Isoetales				
Isoetes L.	2		None	
Lycopodiales				
Huperzia Bernh.	1		None	
Lepidotis P. Beauv.	2		None	
Lycopodium L.	2		None	
Selaginellales				
Selaginella Beauv.	22		8A.1; 10A.1, 19; 12A.1, 3	Chakravarthy *et al.* (1981); Huneck and Khaidav, (1985); Qasim *et al.* (1985b)
Filicales				
Osmunda L.	1	lf[f]	10A.13, t, 16, 19	Okuyama *et al.* (1979)
Stromatopteris Mett.	1	lf	None	
Gleichenia Sm.	1	lf	None	
Anemia Sw.	3	lf	None	
Mohria Sw.	1	lf	None	
Lygodium Sw.	1	lf	None	
Schizaea Sm.	1	lf	None	
Cyathea Sm.	1	lf	7B.1, 2	Wada *et al.* (1985)

See Table 4.1 for biflavonoid numbering, 15A.1 = heterobryoflavone, etc. lf = leaf; lf[f] = fertile leaf; t = incompletely identified trimethyl ether.

*Organ extracted, where specified.

universal in the gymnosperms and two orders of primitive vascular plants, Psilotales and Selaginellales, but with a very disjunct distribution in the angiosperms. Recent work has confirmed many of the previous reports and expanded the number of species studied and the range of biflavonoids detected. Some of the early reports that have not been confirmed by more recent analyses of the same species have been excluded from the tables. Many of the data are still drawn from piecemeal studies of isolated species, so that the inherent problems of comparability of methods and identity of material detract from their value in plant taxonomy. The field is crying out for broad systematic surveys using standard techniques to establish the distribution of the various biflavonoids. Such studies seem likely to be particularly rewarding in the mosses and angiosperms.

4.4.1 Bryophytes

The isolation from *Hylocomium splendens* (Hedw.) B.S.G. of a dihydroxyrobustaflavone (8A.4) and a biluteolin (10A.23; Becker *et al.*, 1986), and bryoflavone (16A.1) and heterobryoflavone (15A.1) from *Bryum capillare* (Hedw.) (Geiger *et al.*, 1986) bring to three the number of moss genera from which biflavonoids have been isolated. These genera are not closely related: *Hylocomium* is pleurocarpous, while *Dicranum* and *Bryum* are acrocarpous, and all three were placed in different suborders by Vitt (1984). This suggests that biflavonoids may prove to be widespread in the mosses; in fact, Becker *et al.* (1986) report that preliminary two-dimensional paper chromatography of extracts of a wide range of mosses indicated this to be the case. There is still no report of biflavonoids in the hepatics, despite many more studies having been made of the flavonoid content of liverworts than of mosses. It is tempting, therefore, to conclude that the presence of biflavonoids will prove to be a further chemical distinction separating the mosses from the liverworts (see Asakawa, 1986). Although the structures so far reported are unique to the mosses, which reinforces the distinction between bryophytes and vascular plants, the presence of biflavonoids in both mosses and the primitive pteridophytes may be viewed as additional evidence of the chemical affinity between them noted by Asakawa (1986). Alternatively, and especially in the absence of biflavonoid structures that are common to both groups, it may simply reflect 'a parallel evolution probably associated with the discouragement of predators', as suggested for the chemical similarity between the liverworts and brown algae (Smith, 1986, p. 86).

4.4.2 Pteridophytes

The Psilotales and Selaginellales are characterized by the presence of amentoflavone (10A.1) as the major biflavonoid. Wallace and Markham (1978) and Markham (1984) have reported a mixture of di- and tri-*O*-glucosides of amentoflavone (10A.2, 3, 4) as a minor component of the biflavonoid fraction in the aerial parts of *Psilotum nudum* and *Tmesipteris* spp. It is possible that such minor amounts of biflavone glycosides may occur more widely, but the major components typically take the form of the aglycone. The hinokiflavone series has now been reported from two further species of *Selaginella* [*S. krausiana* (Kunze) A. Br. (12A.1, 3) and *S. lepidophylla* (12A.1); Qasim *et al.*, 1985b], although recent studies by Chakravarthy *et al.* (1981) and Huneck and Khaidav (1985) of *S. rupestris* Spring and *S. sanguinolenta* (L.) Spring, respectively, appear to confirm that this series is of only sporadic occurrence in the genus. The isolation of robustaflavone (8A.1) from *S. lepidophylla* (Qasim *et al.*, 1985b) is the first record of this compound outside the seed plants.

The Filicales have for some time been thought to be entirely devoid of biflavonoids (Cooper-Driver, 1977), but the isolation of some highly methylated derivatives of amentoflavone (10A.13, 10A.16, 10A.19) from the sporophylls of *Osmunda japonica* (? = *O. regalis* L.; Okuyama *et al.*, 1979) and the unusual 6,6‴-linked hegoflavones (7B.1, 2) from the leaves of *Cyathea spinulosa* Wall. (*ut Alsophila spinulosa*, Wada *et al.*, 1985) shows this not to be the case. Although neither genus was included in Cooper-Driver's original survey of leaf extracts of some of the more primitive ferns, Okuyama *et al.* (1979) report that biflavonoids are absent from the sterile fronds of *Osmunda japonica*; it is not clear whether the leaves analysed by Wada *et al.* (1985) were fertile or sterile, but in view of the isolation of biflavones from the pollen of seed plants (*v.i.*), spore bearing fronds may prove a more fruitful source of biflavonoids in the ferns. The occurrence of the widely occurring amentoflavone series in a primitive member of the Filicales favours the view that the ability to synthesize biflavonoids is plesiomorphic in the ferns, having been inherited from an ancient common ancestor of other biflavonoid containing vascular plants. Further work is needed to establish the distribution of biflavonoids in the pteridophyte orders.

4.4.3 Gymnosperms

With the notable exception of the monotypic Stangeriaceae, the leaves of all Cycadales contain biflavonoids (Table 4.3). The work of Dossaji *et al.* (1975) remains the benchmark survey of the order. These workers concluded that the Cycadaceae were characterized by amentoflavone and hinokiflavone and their derivatives, while the Zamiaceae contained only the former series. This pattern has subsequently been confirmed in three species of *Encephalartos* (Mohammed *et al.*, 1983), but Qasim *et al.* (1983) have reported hinokiflavone (12A.1) and an unidentified monomethyl ether of it in the leaves of *Zamia angustifolia* Jacq., although the hinokiflavone series had not been detected in two previous studies of the species

Table 4.3 Biflavonoids in the Cycadales, Ginkgoales and Taxales

		No. species	Organ*	Biflavonoids reported	References not in Geiger and Quinn (1975, 1982)
CYCADALES					
Cycadaceae					
Cycas L.		9	lf	10A.1, 6, 7, 8, 9, 13; 10B.1; 12A.1; 12B.1	Varshney *et al.* (1973)
	(11)	2	ts	10A.1, 8	Gadek (1982)
Stangeriaceae					
Stangeria T. Moore		1	lf	None	
Zamiaceae					
Tribe Encephalarteae					
Lepidozamia Regel.		2	lf	10A.1, 8, 9, 13, 16, 19	
	(2)	1	ts	(+traces)	Gadek (1982)
Macrozamia Miq.		13	lf	10A.1, 6, 9, 16	
	(13)	3	ts	6A.1, m, d; 10A.1, 6, 14, t	Gadek (1982)
Encephalartos Lehm.		28	lf	10A.1, 6, 9, 13, 16, 19	Mohammed *et al.* (1983)
	(28)	1	ts	(+traces)	Gadek (1982)
Tribe Diooeae					
Dioon Lindl. corr. Miq.		6	lf	10A.1, 5, 6, 9, 16, 19, 20	
Tribe Zameae					
Ceratozamia Brongn.		5	lf	10A.1, 9, 16	
Zamia L.		14	lf	10A.1, 5, 6, 9, 16, 19; 12A.1, m	Qasim *et al.* (1983)
	(15)	1	ts	None	Gadek (1982)
Bowenia Hook.		2	lf	10A.1, 6, 9, 13, 16, 19	
Microcycas (Miq.) A.D.C.		1	lf	10A, 1, 6, 7, 9	
GINKGOALES					
Ginkgo L.		1	lf	10A.1, 6, 9, 13, 16, 21	Briançon-Scheid *et al.* (1982)
TAXALES					
Amentotaxus Pilg.		1		10A.1	
Taxus L.		2	lf	10A.1, 5, 7, 9, 16	Parveen *et al.* (1985)
Torreya Arn.		1	lf	10A.18	
GNETALES					
Gnetum L.		1	lf	None	
Ephedra L.		1	lf	None	
Welwitschia Hook. f.		1	lf	None	

See Table 4.1 for biflavonoid key. lf = leaf; ts = testa; m, d, t = incompletely identified monomethyl, dimethyl and trimethyl ethers respectively; + = trace amounts of permethyl ethers in permethylated leaf extract only. Total number of species analysed is given in parentheses.
*Organ extracted, where specified.

(Handa *et al.*, 1971; Dossaji *et al.*, 1975). Such records do not necessarily negate the chemotaxonomic disjunction detected by Dossaji *et al.* (1975), since the original comparison was based on standard methods, and the Cycadaceae may well be characterized by much greater accumulation of the hinokiflavone series. That Zamiaceae do possess the genes to enable the synthesis of skeletons other than amentoflavone has been demonstrated by a study of the biflavones in the seed coats of selected members of the order (Gadek, 1982): cupressuflavone (6A.1) was detected in three species of *Macrozamia*. This, together with the finding of amentoflavone derivatives alone (10A.1, 8) in the seed coat of the two representatives of the Cycadaceae examined (*Cycas armstrongii* Miq. and *C. kennedyana* F. Muell.), highlights the need to base chemotaxonomic studies on analyses of homologous organs.

The leaves of *Ginkgo biloba* L. have long been known to be rich in biflavones, and recent work has confirmed most of the earlier reports and has added amentoflavone (Briançon-Scheid *et al.*, 1982) to the list of known constituents; all are 3′,8-linked dimers. The early report of sequoiaflavone (10A.5) was not, however, confirmed by these workers and so must be regarded as suspect.

The only additional report for the Taxales (*Taxus wallichiana* Zucc.; Parveen *et al.*, 1985) confirmed the pattern of amentoflavone derivatives alone in the leaves of this order, although the data base is still minimal (Table 4.3).

In the Coniferales the pattern for most families has been little changed by recent work (Table 4.4). *Lepidothamnus* remains the only genus in the Podocarpaceae from which the cupressuflavone skeleton (6A) has been isolated (Geiger and Quinn, 1982). Otherwise, the members of the family are characterized by a leaf biflavonoid fraction consisting of amentoflavone derivatives, sometimes with the addition of small amounts of hinokiflavone and its partial methyl ethers. Analyses of pollen of *P. macrophyllus* (Thunb.) D. Don (Ohmoto and Yoshida, 1980) and the seeds of *P. neriifolius* D. Don ex Lamb. (Cambie and Sidwell, 1983) in each case yielded amentoflavone alone, although hinokiflavone has been reported from the leaves of both species (Rizvi *et al.*, 1974; Miura *et al.*, 1969). In a survey of permethylated leaf extracts in *Podocarpus sens. lat.* (P.A. Gadek & C.J. Quinn, unpublished), both amentoflavone and hinokiflavone permethyl ethers were detected in all five species of *Dacrycarpus* de Laub. examined, and robustaflavone hexamethyl ether was detected in four of them. The hinokiflavone series was absent from *Decussocarpus falcatus* (Thunb.) de Laub. and *D. wallichianus* (Presl.) de Laub., although robustaflavone hexamethyl ether was detected in the latter species. This agrees with previous reports of amentoflavone derivatives alone in two other species (Geiger and Quinn, 1975, 1982). In *Prumnopitys*, amentoflavone hexamethyl ether was detected in *P. taxifolia* (Sol. ex Lamb.) de Laub. and *P. ferruginea* (D. Don) de Laub., while robustaflavone hexamethyl ether was detected only in the latter. The hinokiflavone series was detected in neither species, nor was it detected in *P. andina* (Poep. ex Endl.) de Laub. by Bhakuni *et al.* (1974). However, a re-examination of *P. montana* (Humb. & Bonpl. ex Willd.) de Laub. [*ut Podocarpus taxifolia* Kunth.] (Ahmad *et al.*, 1982b) has confirmed the earlier report of hinokiflavone (Hameed *et al.*, 1973). All three permethyl ethers (8A.2, 10A.20, 12A.9) were detected in *Podocarpus acutifolius* Kirk, *P. hallii* Kirk, *P. nivalis* Hook. f. and *P. totara* D. Don ex Lamb., and both the amentoflavone and hinokiflavone series have been detected in the leaves of four other species of *Podocarpus sens. str.* (Ahmad *et al.*, 1982b; Bhakuni *et al.*, 1974; Miura *et al.*, 1969; Rizvi *et al.*, 1974). Only the amentoflavone series, however, was detected in *P. urbanii* Pilg. (Dasgupta *et al.*, 1981), and the leaves and stems of *P. nubigena* Lindl. ex Paxt. are reported to be devoid of biflavones (Bhakuni *et al.*, 1974). Hence, there is some evidence of chemical differentiation of the segregate genera proposed by de Laubenfels (1969), although the distinctions are not absolute. A more extensive survey using standard techniques may produce more clear-cut evidence of affinities.

Studies on three species of *Cephalotaxus* (Aquil *et al.*, 1981; Ishratullah *et al.*, 1981; Kamil *et al.*, 1982; Ma *et al.*, 1984) have expanded the range of partial methyl ethers of amentoflavone isolated from the leaves of this monogeneric family (Table 4.4), but have not detected any other biflavonoid series.

There is some diversity in the biflavonoid complements reported for the Taxodiaceae. Both amentoflavone and hinokiflavone derivatives have been isolated from the leaves of *Metasequoia glyptostroboides* Hu & Cheng, *Sequoiadendron giganteum* (Lindl.) Buchholz, *Taxodium distichum* (L.) Rich., *T. mucronatum* Tenore, *Taiwania cryptomerioides* Hayata and *Cunninghamia lanceolata* (Lamb.) Hook. f.; the taiwaniaflavone series (3A,1,2,3) and robustaflavone (8A.1) have also been recorded in the last two, respectively (Geiger and Quinn, 1982; Ansari *et al.*, 1985). There are early reports of amentoflavone derivatives alone in *Sequoia sempervirens* (D. Don) Endl. (10A.5), *Cunninghamia konishii* (10A.5), and *Sciadopitys verticillata* (10A.16); on the other hand only hinokiflavone derivatives are known from the leaves of *Cryptomeria japonica* (L.f.) D. Don and *Glyptostrobus pensilis* (Staunt.) K. Koch (Geiger and Quinn, 1975, 1982). Ohmoto and Yoshida (1983), however, have isolated amentoflavone derivatives (10A.1, 22; 10B.1) from the pollen of *Cryptomeria japonica*, and preliminary studies (P.A. Gadek and C.J. Quinn, unpublished) have detected the permethyl ethers of both amentoflavone and hinokiflavone in permethylated leaf extracts of *C. japonica* var. *sinesis* Sieb, & Zucc. and *Glyptostrobus pensilis*, as well as *Athrotaxis cupressoides* D. Don and *A. selaginoides* D. Don; the amentoflavone series alone, however, was detected in *Sciadopitys verticillata*. Hence, while the amentoflavone and hinokiflavone series appear to be of fairly general occurrence in the family, there is still clear evidence of the chemical differentiation of *Taiwania, Sciadopitys* and perhaps also *Sequoia*. This is supported in the case of *Sciadopitys* by an unusual embryogeny, leaf anatomy and chromosome number (Sporne, 1966), and supports the isolated position accorded this genus by Eckenwalder (1976).

A chemotaxonomic survey of the Cupressaceae *sens. str.* has recently been completed (Gadek and Quinn, 1983, 1985; Gadek, 1986), the results of which agree closely with recent independent analyses of individual species: e.g. the presence of cupressuflavone in *Platycladus orientalis* (L.) Franco and its absence from *Chamaecyparis lawsoniana* (A. Murr.) Parl. [*ut Cupressus lawsoniana*] have been confirmed by Khabir *et al.* (1985) and Ahmad *et al.* (1985) respectively, and the latter authors have reported 6A.1, 8A.1, 10A.1, 12A.1 and an unidentified monomethyl ether of 10A from *Fitzroya cupressoides* (Mollina) Johnson [*ut. F. patagonica*] (c.f. 6A.1, 5, 10A.1, 8, 12A.1; Gadek and Quinn, 1985).

Confusion over the identity of material is still the likely cause of some contradictory reports, and the plea for the preservation and citation of proper voucher specimens and

Table 4.4 Distribution of biflavonoids in the Coniferales

		No. Species	*Organ**	*Biflavonoids reported*	*References not in Geiger and Quinn (1975, 1982)*
ARAUCARIACEAE					
Agathis Salisb.		3	lf	5A.1, 2, 3, 4; 6A.1, 2, 5, 7, d, t; 8A.1, m; 10A.1, 5, 6, 9, t; 12A.1, m	Gadek *et al.* (1984)
Araucaria Juss.		5	lf	5A.1, 2, 3, 4, 5; 6A.1, 2, 5, 6, 7, 8; 8A.1; 10A.1, 6, 7, 10, 12, 15, 16, 18, 19; 12A.1, m	
CUPRESSACEAE					
Cupressus L.		8	lf	6A.1, 3; 10A.1, 5, 6, 8; 12A.1, 3	Gadek and Quinn (1985); Lebreton *et al.* (1978); Qasim *et al.* (1985a); Taufeeq *et al.* (1979)
Juniperus L.		1	c	6A.1; 10A.1, m	Lamer-Zarawaska (1980)
	(23)	23	lf	5A.1; 6A.1, 5, ,m; 8A.1, d; 10A.1, 6, 8, 9; 12A.1, 3	De Pascual Teresa *et al.* (1980); De Pascual Teresa and Sanchez Saez. (1973); Gadek and Quinn (1985); Khatoon *et al.* (1985b); Lamer-Zarawaska (1983); Lebreton *et al.* (1978); Roy *et al.* (1984); Sakar and Friedrich (1984)
Tetraclinis (Vahl.) Mast.		1	lf	6A.1; 10A.1, 8; 12A	Gadek and Quinn (1983)
Chamaecyparis Spach p.p.		5	lf	8A; 10A.1, 5, 6, 8, 9, 10, 13, 17; 12A.1, 3, 4, 5	Ahmad *et al.* (1985); Gadek and Quinn (1985)
C. nootkatensis (D. Don) Spach		1	lf	6A.1; 8A.1; 10A.1, 8, 9, 13; 12A.1	Gadek and Quinn (1985)
Fokienia Henry and Thomas		1	lf	8A.1; 10A.1, 8, 9, 13; 12A.1	Gadek and Quinn (1985)
Thuja L.		4	lf	6A.1; 8A; 10A.1, 8, 9, t; 12A.1	Gadek and Quinn (1985)
Thujopsis Sieb. and Zucc.		1	lf	8A.1, m; 10A.1, 6, 8, 9, t; 12A.1, 3,	Gadek and Quinn (1985)
Platycladus Spach		1	lf	6A.1; 8A; 10A.1, 8; 12A.1	Gadek (1986); Gadek and Quinn (1985); Khabir *et al.* (1985)
Calocedrus Kurz. p.p.		2	lf	10A; 12A	Gadek (1986)
C. decurrens (Torr.) Florin		1	lf	3A.1, m; 6A.1; 8A; 10A.1, 8, 14; 12A.1	Gadek and Quinn (1985)
Neocallitropsis Florin		1	lf	3A.1; 10A.1, 8, 14; 12A.1, 3	Gadek and Quinn (1983, 1985)
Widdringtonia Endl.		2	lf	6A.5; 10A.1, 6, 8; 12A.1, 3	Gadek (1986)
Diselma Hook. f.		1	lf	6A.5; 8A; 10A.1, 6, 14; 12A.1, 3	Gadek and Quinn (1983)
Papuacedrus Li		2	lf	10A.1, 8, 14; 12A.1	Gadek and Quinn (1983)
Pilgerodendron Florin		1	lf	10A.1, 8; 12A.1	Gadek and Quinn (1983)
Libocedrus Endl. emend. Li		3	lf	10A.1, 8; 12A	Gadek (1986); Gadek and Quinn (1983)
Austrocedrus Florin and Boutelje		1	lf	10A.1, 8; 12A	Gadek and Quinn (1983)
Actinostrobus Miq. ex Lehm.		2	lf	10A.1, 8	Gadek and Quinn (1983)
Callitris Vent.		10	lf	10A.1, 5, 6, 8, 14; 12A?	Ansari *et al.* (1981); Gadek (1986); Gadek and Quinn (1982, 1983)
Fitzroya Hook. f.		1	lf	6A.1, 5; 8A.1; 10A.1, 8; 12A.1	Gadek and Quinn (1983); Ahmad *et al.* (1985)

TAXODIACEAE					
Athrotaxis D. Don		2	lf	10A; 12A	Gadek and Quinn (unpublished)
Cryptomeria D. Don		1	lf	10A; 12A.1, 4, 7	Gadek and Quinn (unpublished)
	(1)	1	p	10A.1, 22; 10B.1	Ohmoto and Yoshida (1983)
Cunninghamia R. Br. ex Rich.		2	lf	8A.1; 10A.1, 5, 10, 18; 12A.1, 3	Ansari *et al* (1985)
Glyptostrobus Endl.		1	lf	10A; 12A.1, 4, 5, d	Gadek and Quinn (unpublished)
Metasequoia S. Miki		1		10A.1, 7, 14, 16; 10B.2, 3; 12A.1, 3; 12B.1	
Sciadopitys Sieb. and Zucc.		1	lf	10A	Gadek and Quinn (unpublished)
Sequoia Endl.		1		10A.5	
Sequoiadendron Buchholz		1	lf	10A.1, 8; 12A.1	
Taiwania Hayata		1	lf	3A.1, 2, 3; 10A.1, 5, 10; 12A.1, 3	Kamil *et al* (1981)
Taxodium Rich.		2		10A.1, 6, 8, 14, 16; 12A.1, 3, 4, 7	
PODOCARPACEAE					
Podocarpus L'Herit. ex Pers. emend. de Laub.		5	lf	10A.1, 6, 8, 11, 13, 16, 18; 12A.1, 2, ?4	Dasgupta *et al.* (1981); Ahmad *et al.* (1982b)
		1	p	10A.1	Ohmoto and Yoshida (1980)
		1	s	10A.1	Cambie and Sidwell (1983)
	(6)	1	lf	None	
Decussocarpus de Laub.		2	lf	10A.1, 8, 13, 14, 18	Kubo *et al.* (1983)
Prumnopitys Phil.		2	lf	10A.1, 5, 6, 8, d, t, 19; 12A.1	Ahmad *et al.* (1982b)
Dacrydium Lamb. emend. Quinn		3	lf	10A.1, 6, 7, 8, 9, 13, 14, 18, 19; 12A.1	
Falcatifolium de Laub.		1	lf	10A.1, m, 9, t	
Lagarostrobos Quinn		2	lf	10A.1, 5, 7, 14, t, 19; 12A.1	
Halocarpus Quinn		3	lf	10A.1, 5, 7, 9, 13, t; 12A.1	
Lepidothamnus Phil.		3	lf	6A.m, 5, t, 8; 10A.m, 6, 9, t; 12A.1	
CEPHALOTAXACEAE					
Cephalotaxus Seib. and Zucc.		3	lf	10A.1, 5, 9, d, t, 16, 19, 24	Ma *et al.* (1984); Kamil *et al.* (1982); Aqil *et al.* (1981); Ishratullah *et al.* (1981)
PINACEAE					
Abies (Plin. ex Tourn.) Mill.		1	lf	8A.3	Chatterjee *et al.* (1984)
	(6)	5	lf	None	
Cedrus Trew.		1	lf	None	
Keteleeria Carr.		1	lf	None	
Larix Mill.		1	lf	None	
	(2)	1	bk	Larixinol	Shen *et al.* (1985)
Picea A. Dietr.		6	lf	None	
	(7)	1	lf	5A.1; 6A.1; 10A.1	Azam *et al.* (1985)
Pinus L.		24	lf	None	
Pseudolarix Gordon		1	lf	None	
Pseudotsuga Carr.		1	lf	None	
Tsuga Carr.		2	lf	None	

See Table 4.1 for biflavonoid key. lf = leaf (and primary stem); bk = bark; p = pollen; s = seed; c = cone; m, d, t = incompletely identified monomethyl, dimethyl and trimethyl ethers respectively;
*5A, 6A, 10A, etc. indicate the detection of the permethyl ether of the series in a permethylated raw extract when no derivative of that series was isolated from the extract.
Total number of species analysed is given in parentheses.
*organ extracted, where specified.

the use of authorities for binomials must be reiterated (see Geiger and Quinn, 1975: p. 717). For instance, Khan *et al.* (1979) report amentoflavone and an unidentified monomethyl ether, as well as cupressuflavone, robustaflavone and isocryptomerin (12A.3) in the leaf extract of *Cupressus australis* Desf., for which they cite the synonym *Callitris rhomboidea* R. Br. The latter binomial is taxonomically correct for the taxon to which these names refer, but the material was certainly not of that taxon, since separate analyses of *Callitris rhomboidea* (Prasad and Krishnamurty, 1977; P.A. Gadek and C.J. Quinn, unpublished) have revealed only amentoflavone, and this is characteristic of the genus (Gadek and Quinn, 1983); the pattern reported is certainly consistent with *Cupressus* (Table 4.4), and the authors referred to it as a member of that genus throughout their discussion. But the true identity of the material analysed could only be settled by examination of a voucher specimen taken from the original material. Since no voucher was cited, the results must be excluded from any taxonomic considerations.

The amentoflavone series is of universal occurrence in the Cupressaceae as a major component of the leaf biflavonoid fraction, while robustaflavone has again been detected sporadically as a minor component. The distributions of the cupressuflavone and hinokiflavone series in this family display discontinuities that in general correlate well with generic boundaries; indeed the biflavonoid pattern of *Cupressus funebris* Endl., which has been transferred by some authors to *Chamaecyparis* (Franco, 1941; Silba, 1982), provides critical evidence in favour of its retention in the former genus (Gadek and Quinn, 1987). Discontinuities were found, however, within both *Chamaecyparis* and *Calocedrus* (Table 4.4). *Chamaecyparis nootkatensis* (D. Don) Spach is the only member of that genus to contain cupressuflavone. Neither does it conform with the pattern found in *Cupressus*, where the major bands are unmethylated (Gadek and Quinn, 1985). The species is also distinctive in its foliage oils (Erdtman and Norin, 1966; von Rudloff, 1975), and these chemical distinctions are reinforced by differences in wood and leaf anatomy (Bannan, 1952; Bauch *et al.*, 1972; Gadek and Quinn, 1985, and unpublished). Hence the affinities of this species are not in accord with its inclusion in either genus as presently defined. The complex mixture of biflavones in the north American *Calocedrus decurrens* (Torr.) Florin (3A.1; 6A.1; 8A; 10A.1,8,14; 12A.1; Gadek and Quinn, 1985; Gadek, 1986) is markedly different from the two Asian species, which contain only amentoflavone and hinokiflavone derivatives (Gadek, 1986). Again, there appear to be correlated anatomical distinctions suggesting that the present generic boundaries are unnatural.

The distribution of cupressuflavone within the family cuts straight across the presently accepted subfamilial and tribal boundaries (Li, 1953), occurring in all six tribes, inconsistently so in all save the two monogeneric ones (Gadek and Quinn, 1985). There is a particularly striking chemical similarity between *Cupressus*, *Juniperus* and *Tetraclinis*, which together represent three tribes and both subfamilies under Li's classification. Evidence from other characters supports the view that Li's arrangement is artificial (Gadek, 1986), and work underway to extend the data base should lead to a taxonomy more accurately reflecting affinities within the family, especially between the northern and southern hemisphere genera.

The isolation of the taiwaniaflavone series as a major constituent of the leaves of *Calocedrus decurrens* and a minor constituent of *Neocallitropsis pancheri* (Carr.) de Laub, (Gadek and Quinn, 1985) is an interesting parallel with *Taiwania* in the Taxodiaceae. Although the second species, in particular, has been very little studied from any aspect, there seems to be no supporting evidence of a close affinity between the three.

In the past the Pinaceae have been considered to be devoid of biflavonoids, but now Chatterjee *et al.* (1984) have isolated abiesin (8A.3) from a leaf extract of *Abies webbiana* Lindl., and Azam *et al.* (1985) have detected amentoflavone, cupressuflavone and agathisflavone in a leaf extract of *Picea smithiana* (Wall.) Boiss. The only other record in the family is the 'spirobiflavonoid', larixinol (for the formula see Chapter 2) isolated from the bark of *Larix gmelini* (Rupr) Rupr (Shen *et al.*, 1985). These records certainly reinforce the view that the absence of biflavonoids from many members of the Pinaceae is due to a secondary loss of the metabolic pathways that were present in the ancestral conifers. Since few negative reports get into print, it is unclear just how thoroughly this family has been investigated for biflavonoids; it is to be hoped that the above reports will stimulate further interest.

4.4.4 Angiosperms

Biflavonoids have now been recorded in 32 genera from 15 families of flowering plants (Table 4.5) assigned to eight superorders (Dahlgren, 1980), and the range of known structures, many of which appear to be unique to the angiosperms, has been extended considerably. Several reports are for a single species in the family, but past experience suggests that biflavonoids will be found in some related species and genera; the present dearth of reports of biflavonoids in the angiosperms cannot, then, be accepted at face value, and there is little that can be made of them until the data base is much more extensive. The Anacardiaceae and Clusiaceae have received some recent attention. The study of leaf extracts of Anacardiaceae (Wannan *et al.*, 1985) has shown that biflavonoids are absent from nine species; in total, there are positive records for another seven (Table 4.5). Biflavonoids appear to be of more widespread occurrence and of more diverse structure in the fruits (see *Rhodosphaera*, *Rhus* and *Schinus*, Table 4.5), and this conclusion is supported by a broader survey of the family now under way (Quinn and Graham, unpublished). Future workers might be well advised to concentrate on

Table 4.5 Distribution of biflavonoids in the angiosperms

	No. Species	*Organ**	*Biflavonoids reported*	*References not in Geiger and Quinn (1975, 1982)*
MAGNOLIIDAE (DICOTYLEDONAE)				
RANUNCULIFLORAE				
Nandinaceae				
Nandina Thunb.	1		10A.1	
MALVIFLORAE				
Rhamnaceae				
Phyllogeiton (Weberb.) Herzog.	1	w	2D.1	
Euphorbiaceae				
Hevea Aubl.	1	lf	10A.17	
Manihot Mill.	1		10A.1, 8	
Putranjiva Wall.	1		10A.1, 8, 11	
Thymeliaceae				
Stellera L.	1	r	1C.1a, 1b, 1c, 1d, 3a, 3d, 4d	Liu *et al.* (1984); Niwa *et al.* (1984a, b)
				Yang *et al.* (1983, 1984)
THEIFLORAE				
Ochnaceae				
Ochna L.	3	lf	10C.2, 7; 14A.1, 2, 3, 7m	Kamil *et al.* (1983, 1986); Khan *et al.* (1984, 1985)
Brackenridgea A. Gray	1	w	1C.1; 2E.1	Drewes *et al.* (1984)
(1)	1	st & r w	1D.1	Drewes and Hudson (1983)
Clusiaceae				
Clusioideae—tribe Garciniae				
Allanbackia Oliv.	1	w	2B.1, 3	
Garcinia L.	7	lf	2B.1, 3, 4; 2C.1, 3, 4, 5, 8; 10A.1, 8	Hussain and Waterman (1982); Waterman and Hussain (1983)
				Gunatilaka *et al.* (1984)
	21	w	2B.1, 2, 3, 4, 5; 2C.1, 2, 3, 4, 5, 7, 8	Gunatilaka *et al.* (1983); Waterman and Husain (1982, 1983)
				Hussain and Waterman (1982); Waterman and Crichton (1980)
	1	r	2B.1, 3	
	2	f	2B.1, 3; 2C.1, 3, 6	Baslas and Kumar (1981)
(25)	5	sd	2B.3, 4; 2C.1, 3, 5, 6, 8; 10A.1	Waterman and Husain (1982, 1983); Iwu and Igboko (1982); Iwu (1985)
Pentaphalangium Warb.	1	w	2C.1	
Rheedia L.	3	r	2B.1, 3; 2C.1, 4	
(4)	1	f	2B.1, 3; 2C.4	Botta *et al.* (1984)
Calophylloideae—tribe Calophylleae				
Calophyllum L.	2	lf	10A.1	Gunatilaka *et al.* (1984)
Mesua L.	1	st	6B.1; 6C.1	
Ochrocarpos Thou.	1	lf	10A.1, m, d	Roy *et al* (1983)

(*Contd.*)

Table 4.5 (*Contd.*)

		No. Species	Organ*	Biflavonoids reported	References not in Geiger and Quinn (1975, 1982)
ROSIFLORAE					
Casuarinaceae					
Casuarina Adans.		5	st	6A.1 trace; 12A.1	Saleh and El-Lakany (1979)
	(6)	2	st	None	Saleh and El-Lakany (1979)
Fabaceae					
Diphysa Jacq.		1	w	1C.1	Castro and Valverde (1985)
Caesalpinaceae					
Bauhinia L.		1	lf	5A.1, m, d	Sultana *et al.* (1985)
RUTIFLORAE					
Anacardiaceae					
Anacardium L.		1	lf	5A.1	Quinn and Graham (unpublished)
			f	12C.1	Murthy *et al.* (1981b, 1982)
Blepharocarya F. Muell.		1	lf	5A.1	} Wannan *et al.* (1985)
	(2)	1	lf	None	
Rhodosphaera Engl.		1	lf	None	
			f	5C.1?; 10B.1	
Rhus L.		5	lf	5A.1; 5C.1?; 8A.1; 10A.1; 12A.1	Bagchi *et al.* (1985); Kamil *et al.* (1984); Khatoon *et al.* (1985a); Sarada and Adinarayana (1984); Wannan *et al.* (1985)
		1	lf	None	
	(6)	1	f	4C.1; 5A.1; 5B.1; 5C.1; 6C.1; 8A.1; 10A.1; 12A.1	Wannan *et al.* (1985)
Schinus L.		1	lf	None	Wannan *et al.* (1985)
			f	5A.1; 5C.1?; 10A.1; 10B.1	
Semecarpus L. f.		1	lf	10A.1	
	(2)	2	f	8C.1; 10C.1, 4, 5, 6; 12C.1	Ahmad *et al.* (1982a); Murthy (1983b, c, d); Murthy *et al.* (1981a)
Toxicodendron Mill.		1	f	10A.1; 10C.1	
Buchanania Spreng.		2	lf & f	None	} Wannan *et al.* (1985)
Harpephyllum Bernh. ex Krauss		1	lf & f	None	
Burseraceae					
Canarium L.		2	lf	10A.1	Wannan *et al.* (1985); Quinn and Graham, (unpublished)
Garuga Roxb.		1	lf	10A.1	
Meliaceae					
Melia L.		1	lf & f	None	Wannan *et al.* (1985)
Simaroubaceae					
Ailanthus Desf.		2	lf	None	Wannan *et al.* (1985); Quinn and Graham (unpublished)

CORNIFLORAE				
Caprifoliaceae				
Lonicera L.	1	lf	10A.1, m, d, t	Sultana *et al.* (1984)
Viburnum L.	4	lf	10A.1	Beretz *et al.* (1986); Godeau *et al.* (1978); Khan *et al.* (1983);
(5)	1	lf	None	Pelissier *et al.* (1979); Godeau *et al.* (1978)
GENTIANIFLORAE				
Loganiaceae				
Strychnos L.	1	lf	9A.1	Nicoletti *et al.* (1984)
LILIIDAE (MONOCOTYLEDONAE)				
LILIIFLORAE				
Amaryllidaceae				
Lophiola Ker-Gawl.	1	lf	2C.1	
Iridaceae				
Patersonia R. Br.	1	lf	10A.1	Williams and Harborne (1985)

See Table 4.1 for key to biflavonoids, lf = leaf; f = fruit; r = root; sd = seed; st = stems or cladodes; w = wood and/or bark; m, d, t = incompletely identified monomethyl, dimethyl and trimethyl ethers respectively. Total number of species investigated is given in parentheses.
*Organ extracted, where specified.

the fruit as the organ where the ability to synthesize biflavonoids is more likely to be expressed, at least in this family of angiosperms; unfortunately fruits are not so readily available as leaves, either fresh or from herbarium collections. Amentoflavone, and to a lesser extent hinokiflavone, is widespread, particularly in the leaves. Agathisflavone (5A.1) has been recorded in several species of *Rhus* and in species of *Schinus* and *Anacardium*. According to the current taxonomy of the family (Engler, 1883, 1892), these genera are assigned to two of the four tribes. A general revision of generic affinities within the family, however, has shown these tribes to be artificial, and new data on floral and fruit anatomy have led to the recognition of two subfamily taxa, one of which contains all three of the above genera, as well as those in which the related rhusflavanone (5C.1) has been recorded (Wannan, 1986). Agathisflavone has also been recorded in the Australian endemic *Blepharocarya*, which had been variously placed in the Aceraceae, Anacardiaceae and Sapindaceae, and more recently assigned a family of its own, the Blepharocaryaceae. Detailed analyses of the specialized inflorescence cupule and the secretory canals in the primary stem have shown this genus to be a good member of the Anacardiaceae (Wannan *et al.*, 1987), and the anatomy of the pericarp and flower support its placement near the *Rhus* complex (Wannan, 1986). This is an example of the way in which the full import of chemical data may be masked by the poor state of the current taxonomy of a group. Initially the scattered distribution of agathisflavone appeared to be without significance in the two families, but it now appears that it is a good indicator of affinity; this conclusion is supported by work in progress (Quinn and Graham, unpublished). It is desirable, then, that chemical studies go hand in hand with critical analyses of other data sources.

Although biflavones have been isolated from two members of the Burseraceae, they have not yet been detected in any other family of the Rutiflorae, despite a very extensive preliminary survey of leaf extracts (Wannan *et al.*, 1985; Quinn and Graham, unpublished). Dahlgren (1980) separates these two families into different orders, the Sapindales and Rutales respectively, indicating only a distant relationship between them. They are, however, also linked by the presence of vertical intercellular secretory canals in the primary and secondary phloem (Wannan *et al.*, 1985), a feature found in no other family in the superorder. The biflavonoid data are certainly more in line with the conclusions of Cronquist (1981, p. 805) on the close affinities of the Anacardiaceae and Burseraceae, and support his placement of them in the one order (Sapindales).

In the Theiflorae, a picture is emerging of a fairly widespread occurrence in the Clusiaceae (35 species from seven genera), and there are now records in two genera of the Ochnaceae (Table 4.5.) Each family, however, possesses a unique biflavonoid series. The 3, 8″-linked series (2B, 2C) is prominent in the former, while ochnaflavones (14A) are confined to the latter, and while 3′, 8″-linked types are known from each family (10A and 10C, respectively), no compound is common to both. Since the amentoflavone series is clearly plesiomorphic in the vascular plants as a whole (*v.i.*), and is the most widespread series in the angiosperms (Table 4.5), its presence in both families is not a strong indicator of affinity; they may well have independently developed the ability to synthesize biflavonoids (*v.i.*). Waterman and Hussain (1983) surveyed the occurrence of biflavanones (2C) and flavone–flavanone (2B) dimers of the 3, 8″-linked series in *Garcinia* in particular, as well as in the Clusiaceae in general. Where sufficient species had been sampled they found some agreement with the sectional arrangement of Engler (1925): species of section *Xanthochymus* produced mixed dimers, while biflavanones predominated in section *Tagmanthera*, and the two species studied in section *Rheediopsis* had bark devoid of biflavonoids. The authors drew attention, however, to the markedly different biflavonoid contents of different organs of some species: e.g. mixed dimers occur in the wood of both *G. ovalifolia* Oliv. and *G. conrauana* Engl., but biflavanones are produced in the leaves of the former and the seeds of the latter. This again emphasizes the need for comparisons to be based on homologous organs, or, as in the above case, on a survey of a wide range of plant parts. It is suggested (Waterman and Hussain, 1983, 27) that the mixed dimers may arise by dehydration of a 3′-hydroxybiflavanone, and that the former are therefore probably a derived character-state.

Biflavonoids have been reported from two of the five subfamilies of the Clusiaceae recognized by Engler (1925). The 3, 8″-linked series has now been found in all four of the genera examined from the tribe Garciniae of the subfamily Clusiodeae (Table 4.5), but are not known outside the tribe. Amentoflavone has been reported from the leaves of both *Calophyllum* and *Ochrocarpos*, while 8, 8″-linked dimers (6B.1; 6C.1) have been isolated from the leaves of *Mesua*; all three genera belong to the tribe Calophylleae of the subfamily Calophylloideae. It seems that few if any genera from other tribes have been analysed for biflavones. The diversity of structures already detected, and the apparent differences between genera and tribes certainly suggests that the biflavonoids are a very promising taxonomic data source in this group.

Biflavones have now been found in a second family of monocotyledons, the Iridaceae. While this record of amentoflavone in the leaves of *Patersonia glabrata* R. Br. (Williams and Harborne, 1985) seems to be an isolated occurrence in the family, it is of some interest that it is from the tribe Aristeae, which is also remarkable for the presence of a dracaenoid cambium in the stem. The dracaenoid cambium has a curious distribution amongst seven families in the Liliiflorae (Waterhouse, 1987), and it might be worth investigating other dracaenoid taxa for biflavonoids.

4.5 THE ROLE OF BIFLAVONOIDS

The large accumulations of biflavonoids reported in some species and the wide range of tissues in which they occur (leaves, bark, heartwood, roots, pollen, fruits and seeds) have posed the question as to the function(s) they perform. Leaf flavonoids in general have been suggested as acting as UV filters, at least during early colonization of land in the Silurian, and as deterrents to insect predation and fungal invasion (Lowry *et al.*, 1980, 1983; Swain, 1975, p. 1192, 1981). Further evidence of pharmacological effects of biflavonoids includes their ability to inhibit the release of histamines (Amellal *et al.*, 1985), the adhesion of blood platelets (Cazenave *et al.*, 1986) and the action of lens aldose reductase (Shimizu *et al.*, 1984), to block the inflammatory effects of hepatotoxins (Iwu, 1985), and to act as a heart stimulant (Chakravarthy *et al.*, 1981). The variety of effects, and the markedly different activities of various flavonoids has led to the conclusion that a range of targets is involved. While these studies have mainly been *in vitro*, and leave open the question of the ability of biflavonoids to reach these targets when administered *in vivo*, the biflavonoids deposited in plant leaves will clearly come in direct contact with the enzymes secreted by invading pathogens. There is now direct experimental evidence to support this conclusion. Amentoflavone isolated from *Juniperus communis* L. has been shown to have strong anti-fungal properties, causing strong inhibition of growth at concentrations of $1\,\mu g\,ml^{-1}$ when tested against strains of *Aspergillus fumigatus*, *Botrytis cinerea* and especially *Trichoderma glaucum* (Krolicki and Lamer-Zarawska, 1984); its 4‴- and 7″4‴-methyl ethers (10A.8, 14) have been shown to be one of the defence mechanisms against insects browsing on the leaves of *Decussocarpus gracilior* (Pilg.) de Laub., exhibiting a non-toxic growth inhibition on larvae of three species of non-adapted lepidopteran agricultural pests (Kubo *et al.*, 1983, 1984).

The biflavonoid fraction of the broad leaves of *Agathis robusta* (C. Moore ex F. Muell.) F.M. Bail. has been shown by a combination of autofluorescence techniques on fresh leaf sections, and extraction and TLC of epidermal peels, cuticular scrapings and middle-leaf fractions (Gadek *et al.*, 1984) to be confined to the cuticle and cutinized regions of the epidermal cell wall. They also showed this to be a general feature of a range of biflavone-containing gymnosperms, as well as of the aerial stems of *Psilotum nudum* (L.) Griseb. This concentration of biflavonoids in the surface layer of plant leaves fits well with their involvement in the mechanisms by which the plants resist fungal invasion and deter browsing insects. The cellular location of the biflavonoids found in wood, bark, pollen and fruits has yet to be demonstrated, but it seems likely that they are involved in similar defence mechanisms, particularly in the case of wood, roots and bark.

4.6 CONCLUSIONS

Biflavonoids appear to have been a feature of early land plants of both the bryophyte and tracheophyte lines, but it seems probable that they were developed independently. The view that the amentoflavone series is the primitive character-state in the latter line (Geiger and Quinn, 1975, p. 736) is reinforced by its detection in yet another order, the Filicales.

The important role played by biflavonoids in the defence strategies of land plants must have subjected them to intense selection pressures to keep one step ahead of the predators. Hence the diversity of interflavonoid links and hydroxylation and methylation patterns is not surprising. In this respect, it is interesting that it is in the angiosperms, where in most cases this defensive role has been taken over by other metabolites, that the greatest experimentation in biflavonoid structure is found. Although one must remember that evolution produces very different rates of change in different characters of a group, it still appears likely that the ability to synthesize biflavonoids was lost early in the angiosperm line: they are of wide and sporadic occurrence, and are not a feature of those groups of angiosperms that are considered primitive on other grounds. Extant angiosperms containing biflavonoids therefore appear to represent several separate evolutionary reversals in which the ancestral metabolic pathways have been restored (Geiger and Quinn, 1982, p. 530), frequently with some associated novelty, but in a few cases still producing at least some of the primitive amentoflavone series as a retention of the ancestral condition. Hence the amentoflavone series, being plesiomorphic in the vascular plants, cannot be taken as an indicator of affinity within any one class; it is the more unusual interflavonyl links that are the promising character-states for taxonomy.

There is now ample evidence to demonstrate the reliability of biflavones as a taxonomic data source. There is less reason to doubt the worth of discontinuities demonstrated in metabolites involved in defence mechanisms than many of the characters of "classical taxonomy" (e.g., whorls in the androecium, inferior/superior ovary), despite the cavalier dismissal that can still be solicited from some of the more ignorant practitioners. Biflavonoids are, of course, just as open to reversals and homoplasy during the course of evolution as any other character, and so must be considered in the context of all available data in any taxonomic assessment. Indeed, the occurrence of the 8, 8″-linked cupressuflavone series in, *inter alia*, the Araucariaceae and some members of both the Podocarpaceae and Cupressaceae is best explained as the result of three separate developments. But the distribution within the latter family gives a new insight into affinities, and in conjunction with input from other data sources presently being investigated, is helping to produce a classification that better reflects relationships, especially between northern and southern hemisphere members. Again, the occurrence of 6, 8″-linked biflavones in both the Araucariaceae and the Anacardiaceae appears to be the result of homoplasy, but this does not detract from the value of the character-state, when taken together with data on fruit and floral structure, in helping to define a close-knit *Rhus* affinity group within the latter family. As with any data source, studies of chemical characters will prove the most rewarding when combined with a critical reassessment of the existing taxonomy.

REFERENCES

Ahmad, I., Ishratullah, Kh., Ilyas, M., Rahman, W., Seligmann, O. and Wagner, H. (1981), *Phytochemistry* **20**, 1169.

Ahmad, I., Ishratullah, Kh., Ilyas, M. and Rahman, W. (1982a), *Indian J. Chem.*, **21B**, 78.

Ahmad, I., Ishratullah, Kh., Ilyas, M. and Rahman, W. (1982b), *Indian J. Chem.*, **21B**, 898.

Ahmad, I., Kamil, M., Arya, R., Nasim, K.T. and Ilyas, M. (1985), *Indian J. Chem.* **24B**. 321.

Amellal, M., Bronner, C., Briançon-Scheid, F., Haag, M., Anton, R. and Landry, Y. (1985), *Plant. Med.*, **51**, 16.

Ansari, F.R., Ansari, W.H., Rahman, W., Okigawa, M. and Kawano, N. (1981), *Indian J. Chem.*, **20B**, 724.

Ansari, F.R., Ansari, W.H. and Rahman, W. (1985), *J. Indian Chem. Soc.*, **62**, 406.

Aquil, M., Rahman, W., Hasaka, N., Okigawa, M. and Kawano, N. (1981), *J. Chem. Soc., Perkin Trans. 1*, 1389.

Asakawa, Y. (1986), *J. Bryol.* **14**, 59.

Azam, A., Qasim, M.A. and Khan, M.S.Y. (1985), *J. Indian Chem. Soc.* **52**, 788.

Bagchi, A., Sahai, M. and Ray, A.B. (1985), *Plant. Med.*, **51**, 467.

Bannan, M.W. (1952), *Canad. J. Bot.* **30**, 170.

Baslas, R.K. and Kumar, P. (1981), *Acta Cienc. Indica (Ser.) Chem.*, **7**, 31.

Bauch, J., Liese, W. and Schultze, R. (1972), *Wood Sci. Technol.* **6**, 165.

Becker, R., Mues, R., Zinsmeister, H.D. and Geiger, H. (1986), *Z. Naturforsch.* **41c**, 507.

Beretz, A., Briançon-Scheid, F., Stierle, A., Corre, G., Anton, R. and Cazenave, J.P. (1986), *Biochem. Pharmacol.*, **35**, 257.

Bhakuni, D.S., Bittner, M., Sammes, P.G. and Silva, M. (1974), *Rev. Latinoam. Quim.* **5**, 163.

Botta, B., Marquina McQuhae, M., Delle Monache, G., Delle Monache, F. and De Mello, J.F. (1984), *J. Nat. Prod.*, **47**, 1053.

Briançon-Scheid, F., Guth, A. and Anton, R. (1982), *J. Chromatogr.*, **245**, 261.

Briançon-Scheid, F., Lobstein-Guth, A. and Anton, R. (1983), *Plant. Med.*, **49**, 204.

Cambie, R.C. and Sidwell, D.E. (1983), *Fiji Agric. J.*, **45**, 85.

Castro, C.O. and Valverde, R.V. (1985), *Phytochemistry*, **24**, 367.

Cazenave, J., Beretz, A. and Anton, R. (1986). In *7th Hungarian Bioflavonoid Symposium Szeged*, May 1985, Akademiai Kiado, Budapest.

Chakravarthy, B.K., Rao, Y.V., Gambhir, S.S. and Gode, K.D. (1981), *Plant. Med.*, **43**, 64.

Chatterjee, A., Kotoky, J., Das, K.K., Banerji, J. and Chakraborty, T. (1984), *Phytochemistry* **23**, 704.

Cooper-Driver, G. (1977), *Science* **198**, 1260.

Cronquist, A. (1981). *An Integrated System of Classification of Flowering Plants*, Columbia University Press, New York.

Dahlgren, R.M.T. (1980), *Bot. J. Linn. Soc. (London)*, **80**, 91.

Dasgupta, B., Burke, B.A. and Stuart, K.L. (1981), *Phytochemistry* **20**, 153.

De Pascual Teresa, J. and Sanchez Saez (1973), *Am. Quim.* **69**, 941.

De Pascual Teresa, J., Barrero, A.F., Muriel, L., San Feliciano, A. and Grande, M. (1980), *Phytochemistry*, **19**, 1153.

Dossaji, S.F., Mabry, T.J. and Bell, E.A. (1975), *Biochem. Syst. Ecol.* **2**, 171.

Drewes, S.E. and Hudson, N.A. (1983), *Phytochemistry*, **22**, 2823.

Drewes, S.E., Hudson, N.A., Bates, R.B. and Linz, G.S. (1984), *Tetrahedron Lett.* **25**, 105.

Duddeck, H., Snatzke, G. and Yemul, S.S. (1978), *Phytochemistry* **17**, 1369.

Eckenwalder, J.E. (1976), *Madrono* **23**, 237.

Engler, A. (1883), *Monogr. Phanerogamarum* **4**, 171.

Engler, A. (1892). In *Die Natürlichen Pflanzenfamilien* **III(5)**, 138. (eds. A. Engler and K. Prantl), Englemann, Leipzig, Vol. III(5), p. 138.

Engler, A. (1925). In *Die Natürlichen Pflanzenfamilien* (2nd edn) (eds. A. Engler and K. Prantl), Dunker and Humbolt, Berlin, Vol. 21, p. 154.

Erdtman, H. and Norin, T. (1966), *Fortschr. Chem. org. Naturst.* **24**, 206.

Franco, J.D.A. (1941), *Agros* **24**, 93.

Gadek, P.A. (1982), *Phytochemistry* **21**, 889.

Gadek, P.A. (1986), *Chemotaxonomy of the Cupressaceae*, Ph.D. thesis, University of New South Wales.

Gadek, P.A. and Quinn, C.J. (1982), *Phytochemistry* **21**, 248.

Gadek, P.A. and Quinn, C.J. (1983), *Phytochemistry* **22**, 969.

Gadek, P.A. and Quinn, C.J. (1985), *Phytochemistry* **24**, 267.

Gadek, P.A. and Quinn, C.J. (1987) *Phytochemistry* **26**, 2551.

Gadek, P.A., Quinn, C.J. and Ashford, A.E. (1984), *Aust. J. Bot.* **32**, 15.

Geiger, H. and Quinn, C. (1975). In *The Flavonoids* (eds. J.B. Harborne, T.J. Mabry and H. Mabry), Chapman and Hall, London, p. 692.

Geiger, H. and Quinn, C. (1982) In *Flavonoids – Advances in Research* (eds. J.B. Harborne and T.J. Mabry), Chapman and Hall, London, p. 505.

Geiger, H. and Schwinger, G. (1980), *Phytochemistry* **19**, 897.

Geiger, H., Stein, W., Mues, R. and Zinsmeister, H.D. (1987), *Z. Naturforsch.* **42C**, 863.

Geiger, H., Vande Casteele, K. and Van Sumere, C.F. (1984), *Z. Naturforsch*, **39b**, 393.

Godeau, R.P., Pelissier, Y., Sors, C. and Fouraste, I. (1978), *Plant. Med. Phytother.* **12**, 296.

Gopalakrishnan, S., Neelakantan, S. and Raman, P.V. (1980), *Curr. Sci.*, **49**, 19.

Gunatilaka, A.A.L., Sriyani, H.T.B., Sotheeswaran, S. and Waight, E.S. (1983), *Phytochemistry* **22**, 233.

Gunatilaka, A.A.L., De Silva, A.M.Y., Sotheeswaran, S., Balasubramaniam, S. and Wazeer, M.I.M. (1984), *Phytochemistry* **23**, 323.

Hameed, N., Ilyas, M., Rahman, W., Okigawa, M. and Kawano, N. (1973), *Phytochemistry* **12**, 1494.

Handa, B.K., Chexal, K.K., Rahman, W., Okigawa, M. and Kawano, N. (1971), *Phytochemistry* **10**, 436.

Huneck, S. and Khaidav, T. (1985), *Pharmazie* **40**, 431.

Hussain, R.A. and Waterman, P.G. (1982), *Phytochemistry* **21**, 1393.

Ishratullah, K., Rahman, W., Okigawa, M. and Kawano, N. (1981), *Indian J. Chem.* **20B**, 935.

Iwu, M.M. (1985), *Experientia* **41**, 699.

Iwu, M. and Igboko, O. (1982), *J. Nat. Prod.* **45**, 650.

Kamil, M., Ilyas, M., Rahman, W., Hasaka, N., Okigawa, M. and Kawano, N. (1981), *J. Chem. Soc., Perkin Trans. 1*, 553.

Kamil, M., Khan, N.A., Ahmad, I., Ilyas, M. and Rahman, W. (1982), *J. Indian Chem. Soc.* **59**, 1199.

Kamil, M., Khan, N.A., Ilyas, M. and Rahman, W. (1983), *Indian J. Chem.* **22B**, 608.

Kamil, M., Ahmad, I. and Ilyas, M. (1984), *J. Indian Chem. Soc.* **61**, 375.

Kamil, M., Khan, N.A., Sarwar, A.M. and Ilyas, M. (1987), *Phytochemistry*, **26**, 1171.

Khabir, M., Khatoon, F. and Ansari, W.H. (1985), *Curr. Sci.* **54**, 1180.

Khan, N.A., Kamil, M. and Ilyas, M. (1979), *Indian J. Chem.* **17B**, 536.

Khan, N.A., Kamil, M., Ahmad, I. and Ilyas, M. (1983), *J. Sci. Res. (Bhopal, India)* **5**, 27.

Khan, N.A., Siddiqui, N. and Ilyas, M. (1984), *J. Sci. Res. (Bhopal, India)* **6**, 45.

Khan, N.A., Husain, I., Siddiqui, N.U. and Zia-Ul-Hasan (1985), *Orient. J. Chem.* **1**, 109.

Khatoon, F., Khabir, M. and Ansari, W.H. (1985a), *J. Indian Chem. Soc.* **62**, 560.

Khatoon, F., Khabir, M., Taufeeq, H.M. and Ansari, W.H. (1985b), *J. Indian Chem. Soc.* **62**, 410.

Krolicki, Z. and Lamer-Zarawska, E. (1984), *Herba Pol.* **30**, 53.

Kubo, I., Klocke, J.A., Matsumoto, T. and Naoki, H. (1983), *Rev. Latinoam. Quim.* **14**, 59.

Kubo, I., Matsumoto, T. and Klocke, J.A. (1984), *J. Chem. Ecol.* **10**, 547.

Lamer-Zarawska, E. (1980), *Pol. J. Chem.* **54**, 213.

Lamer-Zarawska, E. (1983), *Pr. Nauk. Akad. Med. Wroclawiu* **16**, 3.

de Laubenfels, D.J. (1969), *J. Arnold Arbor.* **50**, 274.

Lebreton, P., Boutard, B. and Sartre, J. (1978), *Bulletin de l'Institut Scientifique, Rabat 1978*, 155.

Li, H. (1953), *J. Arnold Arbor.* **34**, 17.

Liu, G., Tatematsu, H., Kurokawa, M., Niwa, M. and Hirata, Y. (1984), *Chem. Pharm. Bull.* **32**, 362.

Lowry, J.B., Lee, D.W. and Hebant, C. (1980), *Taxon* 29, 183.

Lowry, J.B., Lee, D.W. and Hebant, C. (1983), *Taxon* **32**, 101.

Ma, Zh., He, G. and Yin, W. (1984), *Zhiwu Xuebao* **26**, 416.

Manchanda, V. and Khan, N.U.D. (1985), *Chem. Ind. (London)*, 127.

Markham, K.R., Sheppard, C. and Geiger, H. (1987) *Phytochemistry* **26**, 3335.

Markham, K.R. (1984), *Phytochemistry* **23**, 2053.

Miura, H., Kihara, T. and Kawano, N. (1969), *Chem. Pharm. Bull. (Tokyo)* **17**, 150.

Mohammed, F., Taufeeq, H.M., Ilyas, M. and Rahman, W. (1983), *Indian J. Chem.* **22B**, 184.

Murthy, S.S.N. (1983a), *Acta Cienc. Indica (ser.) Chem.* **9**, 148.

Murthy, S.S.N. (1983b), *Phytochemistry* **22**, 1518.

Murthy, S.S.N. (1983c), *Indian J. Chem.* **22B**, 1167.

Murthy, S.S.N. (1983d), *Phytochemistry* **22**, 2636.

Murthy, S.S.N. (1985a), *Phytochemistry* **24**, 1065.

Murthy, S.S.N. (1985b), *Indian J. Chem.* **24B**, 398.

Murthy, S.S.N., Rao, N.S.P., Anjaneyulu, A.S.R. and Row, L.R. (1981a), *Plant. Med.* **43**, 46.

Murthy, S.S.N., Rao, N.S.P., Anjaneyulu, A.S.R., Row, L.R., Pelter, A. and Ward, R.S. (1981b), *Curr. Sci.* **50**, 227.

Murthy, S.S.N., Anjaneyulu, A.S.R., Row, L.R., Pelter, A. and Ward, R.S. (1982), *Plant. Med.* **45**, 3.

Nicoletti, M., Goulart, M.O.F., De Lima, R., Goulart, A.E., Delle Monache, F. and Marini, B.G.B. (1984), *J. Nat. Prod.* **47**, 953.

Niwa, M., Chen, X.F., Liu, G.Q., Tatematsu, H. and Hirata, Y. (1984a), *Chem. Lett.*, 1587.

Niwa, M., Tatematsu, H., Liu, G.Q. and Hirata, Y. (1984b), *Chem. Lett.*, 539.

Ohmoto, T. and Yoshida, O. (1980), *Chem. Pharm. Bull.* **28**, 1894.

Ohmoto, T. and Yoshida, O. (1983), *Chem. Pharm. Bull.* **31**, 919.

Okuyama, T., Ohta, Y. and Shibata, S. (1979), *Shoyakugaku Zasshi* **33**, 185.

Österdahl, B.G. (1983), *Acta Chem. Scand.* **B37**, 69.

Parthasarathy, M.R. and Gupta, S. (1984), *Indian J. Chem.* **23B**, 227.

Parthasarathy, M.R., Ranganathan, K.R. and Sharma, D.K. (1977), *Indian J. Chem.* **15B**, 942.

Parveen, N., Taufeeq, H.M. and Khan, N.U.D. (1985), *J. Nat. Prod.* **48**, 994.

Pelissier, Y., Godeau, R.P. and Fouraste, I. (1979), *Trav. Soc. Pharm. Montpellier* **39**, 175.

Prasad, J.S. and Krishnamurty, H.G. (1977), *Phytochemistry* **16**, 801.

Qasim, M.A., Roy, S.K., Azam, A., Kamil, M. and Ilyas, M. (1983), *J. Sci. Res. (Bhopal, India)* **5**, 179.

Qasim, M.A., Roy, S.K., Kamil, M. and Ilyas, M. (1985a), *J. Indian Chem. Soc.* **62**, 170.

Qasim, M.A., Roy, S.K., Kamil, M. and Ilyas, M. (1985b), *Indian J. Chem.* **24B**, 220.

Rahman, W., Ilyas, M., Okigawa, M. and Kawano, N. (1982), *Chem. Pharm. Bull.* **30**, 1491.

Rizvi, S.H.M., Rahman, W., Okigawa, M. and Kawano, N. (1974), *Phytochemistry* **13**, 1990.

Roy, S.K., Qasim, M.A., Kamil, M., Ilyas, M. and Rahman, W. (1983), *Indian J. Chem* **22B**, 609.

Roy, S.K., Qasim, M.A., Kamil, M., Ilyas, N., Ilyas, M. and Rahman, W. (1984), *J. Indian Chem. Soc.* **61**, 172.

Sakar, M.K. and Friedrich, H. (1984), *Plant. Med.* **50**, 108.

Saleh, N.A.M. and El-Lakany, M.H. (1979), *Biochem. Syst. Ecol.* **7**, 13.

Sarada, M. and Adinarayana, D. (1984), *J. Indian Chem. Soc.* **61**, 649.

Shen, Z., Falshaw, C. P., Haslam, E. and Begley, M.J. (1985), *J. Chem. Soc., Chem. Commun.*, 1135.

Shimizu, M., Ito, T., Terashima, S., Hayashi, T., Arasawa, M., Morita, N., Kurokawa, S., Ito, K. and Hashimoto, Y. (1984), *Phytochemistry* **23**, 1885.

Silba, J. (1982), *Phytologia* **51**, 157.

Smith, A.E.J. (1986), *J. Bryol.* **14**, 83.

Sporne, K.R. (1966) *The Morphology of Gymnosperms*, Hutchinson, London.

Sultana, S., Kamil, M. and Ilyas, M, (1984), *J. Indian Chem. Soc.* **61**, 730.

Sultana, S., Ilyas, M., Kamil, M. and Shaida, W.A. (1985), *J. Indian Chem. Soc.* **62**, 337.

Swain, T. (1975). In *The Flavonoids* (eds. J.B. Harborne, T.J. Mabry and H. Mabry), Chapman and Hall, London, p. 1096.

Swain, T. (1981), *Taxon* **30**, 471.

Taufeeq, H.M., Mohd, F. and Ilyas, M. (1979), *Indian J. Chem.* **17B**, 535.

Taufeeq, H.M., Khan, N.U., Ilyas, M., Rahman, W. and Kawano, N. (1982), *Proc. Natl. Acad. Sci., India, Sect. A*, **52**, 43.

Varshney, A.K., Mah, T., Khan, N.V., Rahman, W., Hwa, C.W. Okigawa, M. and Kawano, N. (1973), *Indian J. Chem.* **11**, 1209.

Vitt, D.H. (1984). In *New Manual of Bryology* (ed. R.M. Schuster), Hattori Botanical Laboratory, Nichinan, p. 696.

von Rudolff, E. (1975), *Biochem. Syst. Ecol.* **2**, 131.

Wada, H., Satake, T., Murakami, T., Kojima, T., Saiki, Y. and Chen, C.M. (1985), *Chem. Pharm. Bull.* **33**, 4182.

Wallace, J.W. and Markham, K.R. (1978), *Phytochemistry* **17**, 1313.

Wannan, B.S. (1986), *Generic Affinities in the Anacardiaceae and its Allies*. Ph.D. thesis, University of New South Wales.

Wannan, B.S., Waterhouse, J.T., Gadek, P.A. and Quinn, C.J. (1985), *Biochem. Syst. Ecol.* **13**, 105.

Wannan, B.S., Waterhouse, J.T. and Quinn, C.J. (1987), *Bot. J. Linn. Soc. (London)* **95**, 61.

Waterhouse, J.T. (1987), *Proc. Linn. Soc. N.S.W.* (in press).

Waterman, P.G. and Crichton, E.G. (1980), *Phytochemistry* **19**, 2723.

Waterman, P.G. and Hussain, R.A. (1982), *Phytochemistry* **21**, 2099.

Waterman, P.G. and Hussain, R.A. (1983), *Biochem. Syst. Ecol.* **11**, 21.

Williams, Ch.A. and Harborne, J.B. (1985), *Z. Naturforsch.* **40C**, 325.

Yang, W., Xing, Y., Song, M., Xu, J. and Huang, W. (1983), *Lanzhou Daxue Xuebao, Ziran Kexueban*, **19**, 109.

Yang, W., Xing, Y., Song, M., Xu, J. and Huang, W. (1984), *Gaodeng Xuexiao Huaxue Xuebao*, **5**, 671.

Yokoe, I., Taguchi, M., Shirataki, Y. and Komatsu, M. (1979), *J. Chem. Soc., Chem. Commun.*, 333.

5

Isoflavonoids

PAUL M. DEWICK

5.1 INTRODUCTION

The isoflavonoids represent an important and very distinctive subclass of the flavonoids. These compounds are based on a 3-phenylchroman skeleton that is biogenetically derived by an aryl migration mechanism from the 2-phenylchroman skeleton of the flavonoids. In marked contrast to the flavonoids, the isoflavonoids have a very limited distribution in the plant kingdom, and are almost entirely restricted to the subfamily Papilionoideae of the Leguminosae. Even in the subfamilies Caesalpinioideae and Mimosoideae of the Leguminosae, only one or two plants have been reported to contain isoflavonoids. Amongst the dicotyledons, other families known to produce isoflavonoids include the Amaranthaceae, Chenopodiaceae, Compositae, Cruciferae, Menispermaceae, Moraceae, Myristicaceae, Rosaceae, Scrophulariaceae, Stemonaceae and Zingiberaceae, though in many cases, only isolated plants or genera feature in the list. From the monocotyledons, the Iridaceae is the major source, with *Iris* species making a significant contribution to the number of known naturally occurring isoflavonoids. Two gymnosperm genera, *Juniperus* (Cupressaceae) and *Podocarpus* (Podocarpaceae) are also isoflavonoid producers, and a moss (*Bryum capillare*) has recently been shown to contain a range of isoflavone derivatives. Non-plant sources include the recent report of a pterocarpan isolated from a marine coral, and several microbial cultures. However, in most cases, the formation of isoflavonoids in microbial fermentations can be traced back to leguminous material, e.g. soybean meal, as a component of the culture medium.

Despite the restricted distribution of the isoflavonoids in the plant kingdom, the structural variation encountered in the natural examples is surprisingly large. This arises partly from the number and complexity of substituents on the basic 3-phenylchroman system, but also from the different oxidation levels in this skeleton and the presence of extra heterocyclic rings. Subdivision into several sections according to the latter two criteria aids their systematic classification.

This chapter summarizes important developments in isoflavonoid chemistry and biochemistry since the previous review (Dewick, 1982), and covers the period from late 1980 to the end of 1985. Tabulated data include new compounds isolated as well as new sources of known compounds, but the listings are constructed to reflect the research effort during this period and will thus contain a number of reisolations from known sources. Unless otherwise stated, all plants mentioned are members of the Leguminosae. A check list of the known isoflavonoid

The Flavonoids. Edited by J.B. Harborne
Published in 1988 by Chapman and Hall Ltd,
11 New Fetter Lane, London EC4P 4EE.

Table 5.1 HPLC of isoflavonoids

Isoflavonoids separated	*Column*	*Mobile phase*	*References*
Isoflavones	Lichrosorb RP-18	MeOH–[HCO_2H–H_2O (5:95)], gradient	Casteele *et al.* (1982)
Rotenoids in water samples	Zorbax ODS	CH_3CN–H_2O (70:30)	Bushway (1984)
Rotenoid and 12a-hydroxyrotenoid stereoisomers	Ultrasphere ODS	CH_3CN–1% HOAc (1:1)	Abidi (1984)
	Ultrasphere Si	Hexane–[THF–CH_2Cl_2–propan-2-ol (1:1:0.2)], (9:1)	Abidi (1984)
Baptisia australis isoflavones and glycosides	Lichrosorb RP-18	CH_3CN–3% HOAc, gradient	Köster *et al.* (1983a)
Cicer arietinum isoflavones	Spherisorb-5-ODS	MeOH–H_2O, gradient	Dziedzic and Dick (1982)
isoflavones and glycosides, pterocarpans	Lichrosorb RP-18	CH_3CN–3% HOAc, gradient	Köster *et al.* (1983a, b)
Glycine max isoflavone glycosides	Partisil ODS-3	MeOH–H_2O, gradient	Farmakalidis and Murphy (1984)
isoflavones and glycosides	Zorbax ODS25	MeOH–H_2O, gradient	Eldridge (1982a, b)
	TSK-Gel LS-410 (RP)	MeOH–H_2O, gradient	Ohta *et al.* (1980)
isoflavones and glycosides, coumestans	Ultrasphere ODS	MeOH–H_2O, gradient	Murphy (1981)
pterocarpan phytoalexins	Lichrosorb-Si100	Hexane–THF–HOAc, (88:12:0.5)	Masuda *et al.* (1983)
	Ultrasphere ODS	MeOH–H_2O, gradient	Osswald (1985)
Glycyrrhiza sp. isoflavones, coumestans	Sensyu-Pak SSC-ODS-432 (C_{18})	CH_3CN–3% HOAc, gradient	Hiraga *et al.* (1984)
Lonchocarpus sp. rotenoids	Hypersil C_{22}	CH_3CN–H_2O (4:1)	Westwood *et al.* (1981)
Medicago sativa pterocarpan phytoalexins	Hitachigel 3053	CH_3CN–H_2O (40:60 or 50:50)	Masuda *et al.* (1983)
Onosis spinosa isoflavones	μBondapak C_{18}	CH_3CN–H_2O containing phosphoric acid, gradient	Pietta *et al.* (1983)
isoflavones and glycosides	Lichrosorb RP-18	CH_3CN–3% HOAc, gradient	Köster *et al.* (1983a)
Phaseolus vulgaris pterocarpan phytoalexins	Hitachigel 3053	CH_3CN–H_2O (40:60 or 50:50)	Masuda *et al.* (1983)
pterocarpan, isoflavan, isoflavanone, coumestan phytoalexins	Lichrospher Si100	Hexane–$CHCl_3$ → $CHCl_3$–MeOH, gradient	Goosens and Van Laere (1983)
Piscidia erythrina isoflavone, coumaronochromone	μBondapak C_{18}	MeOH–H_2O containing phosphoric acid (1:3)	Pietta and Zio (1983)
Pisum sativum isoflavones, pterocarpans	μPorasil	Hexane–[CH_2Cl_2–EtOH–HOAc(97:3:0:2)], gradient	Carlson and Dolphin (1981)
pterocarpan phytoalexins	Lichrosorb-Si100	Hexane–THF–HOAc (88:12:0.5)	Masuda *et al.* (1983)
	Hitachigel 3053	CH_3CN–H_2O (40:60 or 50:50)	Masuda *et al.* (1983)
Pueraria sp. isoflavone *O*- and *C*-glycosides	Lichrosorb RP-8	CH_3CN–0.1 M KH_2PO_4 containing phosphoric acid (15:85)	Kitada *et al.* (1985)
	IRICA ODS	CH_3CN–0.1 M KH_2PO_4 containing phosphoric acid (15:85)	Kitada *et al.* (1985)
Trifolium pratense isoflavones	Lichrosorb RP-18	CH_3CN–H_2O containing 0.5% phosphoric acid and 0.5% HCO_2H, gradient	Sachse (1984)
Trifolium subterraneum isoflavones	μBondapak C_{18}	MeOH–H_2O (27:73)	Patroni *et al.* (1982)
isoflavones and glycosides	Partisil-10 ODS-2	MeOH–H_2O (2.1:1)	Nicollier and Thompson (1982)
Trifolium spp. isoflavones and glycosides	Lichrosorb RP-18	CH_3CN–3% HOAc, gradient	Köster *et al.* (1983a)
pterocarpan phytoalexins	Hitachigel 3053	CH_3CN–H_2O (40:60 or 50:50)	Masuda *et al.* (1983)

aglycones is given at the end of the book. The check list now includes 629 structures, which compares with 465 of five years ago, and reflects the continued research activity in this field.

The literature reviewed during this five-year period contains some significant highlights. Undoubtedly, a milestone in isoflavonoid research has now been reached with the isolation of the enzyme that catalyses the crucial aryl migration that differentiates isoflavonoids from the flavonoids (Hagmann and Grisebach, 1984). How appropriate it is that this achievement comes from the laboratory of Hans Grisebach, who started to make his immense contribution to the whole field of isoflavonoid and flavonoid biosynthesis a quarter of a century earlier. An essential aid for any serious isoflavonoid researcher must be a copy of John Ingham's review listing naturally occurring isoflavonoids reported during the period 1855–1981 (Ingham, 1983). This covers all the then-known structures (510 aglycones plus 120 glycosides) and details of their plant sources, and is backed by extremely valuable cross-reference lists of molecular weights, trivial names and plant genera, together with the 845 original literature references. A third highlight is the isolation of natural dimeric isoflavonoid structures (Brandt *et al.*, 1982; Yahara *et al.*, 1985). Whilst flavonoid oligomers have been known for some time, these represent the first reports of isoflavonoid analogues, and one may expect others to follow in due course.

5.2 RECENT DEVELOPMENTS IN ISOLATION TECHNIQUES

Chromatographic techniques for the isolation of flavonoids are, of course, equally applicable to the isoflavonoids. This section is not intended as a comprehensive review of techniques, but highlights some of the developments where special effort has been applied to separate or detect groups of isoflavonoids in various plant systems.

The relatively lower incidence of glycosides in the isoflavonoids compared with other flavonoids means that thin-layer chromatography (TLC) offers considerable versatility, and still remains the main technique employed by researchers. Many separations can readily be accomplished by silica gel TLC using single or multiple developments. A detailed analysis of the mobility of isoflavonoids, and the contribution various functional groups make, allows optimum chromatographic conditions for separation of given isoflavones to be defined (Ozimina and Bandyukov, 1983). Separation of the glyceollin isomers from *Glycine max*, previously achieved only with the aid of high-performance liquid chromatography (HPLC), can be obtained by multiple development TLC on formamide-impregnated silica gel (Komives, 1983).

The use of HPLC as a standard chromatographic technique is becoming increasingly common, and specific applications in the isoflavonoid field are given in Table 5.1. The technique is extremely sensitive and lends itself to quantitative analysis, thus providing a convenient means of analysing legume tissues for biologically active isoflavonoids, e.g. oestrogenic isoflavones in clover species (Nicollier and Thompson, 1982; Patroni *et al.*, 1982; Sachse, 1984), or rotenoid insecticides in water samples (Bushway, 1984). By choosing appropriate solvent programming, isoflavonoids of quite different polarities may be separated and identified in a single analysis, the most common example being analysis of isoflavones with their glycosides. Detection via UV spectroscopy is almost always the method of choice, though a circular dichroism (CD) detector has been applied successfully in the HPLC of rotenoids (Westwood *et al.*, 1981), and amperometric detection was found to be more sensitive than UV for isoflavone *O*- and *C*-glycosides (Kitada *et al.*, 1985). In many laboratories, scale-up of analytical HPLC to semi-preparative or preparative scale may be prohibitively expensive. Chromatographic conditions for the larger separations may also need further optimization from those developed for analytical use (Köster *et al.*, 1983a; Farmakalidas and Murphy, 1984).

As an analytical tool, gas–liquid chromatography–mass spectrometry (GLC–MS) has also found a number of applications in the isoflavonoid field. The isoflavonoids are normally derivatized as their trimethylsilyl (TMS) ethers, and are separated using an OV-101 column or similar. The method has been employed successfully to analyse for isoflavones in *Trifolium pratense* (Popravko *et al.*, 1980; Kononenko *et al.*, 1983) and *Glycine max* (Porter *et al.*, 1985), and biosynthetic precursors of phytoalexins in *Trifolium repens* (Woodward, 1981a). Isomeric isoflavones can be differentiated by their GLC retention and MS characteristics (Woodward, 1981b; see Section 5.3), and similar techniques have been applied to a range of isoflavanones (Woodward, 1982; see Section 5.4).

The detection of low levels of the phytoalexin glyceollin I in tissues of soybean (*Glycine max*) has been achieved by two different techniques, laser microprobe mass analysis (LAMMA) and radioimmunoassay (RIA). LAMMA allows the detection of molecules within a biological matrix, requiring a simple mass spectral fragment from the chosen compound that does not coincide with the spectrum of the matrix. For glyceollin I, the $M - \mathrm{OH}$ fragment was employed (Moesta *et al.*, 1982), and this allowed *relative* levels of glyceollin I content to be assayed near the area of infection. RIA gives *absolute* quantitative levels, and the assay for glyceollin I was achieved by raising antibodies in rabbits against a glyceollin I–bovine serum albumin conjugate, and using [^{125}I] glyceollin I as a tracer (Moesta *et al.*, 1983). The assay showed low cross-reactivity for glyceollin II, glyceollin III and related compounds, and minimal reactivity with daidzein and genistein, and was some 1000-fold more sensitive than any other technique available for glyceollin I assay. This RIA was used to study glyceollin I localization in soybean roots (Hahn *et al.*, 1985).

Table 5.2 Isoflavones

Isoflavone	Substituents 5	6	7	8	2′	3′	4′	5′	6′	Plant sources	References
Daidzein			OH				OH			*Cicer*, *Erythrina*, *Genista*, *Glycine*, *Ononis*, *Phaeolus*, *Pueraria*, *Thermopsis*, *Trifolium*. *Podocarpus amarus* (Podocarpaceae) heartwood	Carman *et al.* (1985)
Formononetin			OH				OMe			*Astragalus*, *Baptisia*, *Centrolobium*, *Centrosema*, *Cicer*, *Dalbergia*, *Genista*, *Glycine*, *Glycyrrhiza*, *Hedysarum*, *Ononis*, *Pisum*, *Thermopsis*, *Trifolium*, *Zollernia*	
Isoformononetin			OMe				OH			*Glycine max* leaf*	Ingham *et al.* (1981b)
Di-*O*-methyldaidzein			OMe				OMe			*Amorpha fruticosa* fruit	Somleva and Ognyanov (1985)
Maximaisoflavone J			OR†				OMe			*Tephrosia maxima* root	Murthy and Rao (1985b)
2′-Hydroxydaidzein			OH		OH		OH			*Phaseolus mungo* seedling*	Adesanya *et al.* (1984)
2′-Hydroxyformononetin (xenognosin B)			OH		OH		OMe			*Astragalus* spp. gum	Steffens *et al.* (1982)
										Trifolium repens leaf*	Woodward (1981a)
3′-Methoxydaidzein			OH			OMe	OH			*Dalbergia odorifera* heartwood	Yahara *et al.* (1985)
3′-Hydroxyformononetin (calycosin)			OH			OH	OMe			*Astragalus membranaceus* root	Wang *et al.* (1983)
										A. mongholicus	Lu *et al.* (1984)
										Glycyrrhiza uralensis callus	Kobayashi *et al.* (1985)
										Hedysarum polybotrys root	Miyase *et al.* (1984)
										Thermopsis facacea root	Arisawa *et al.* (1980)
										Trifolium pratense root	Fraishtat *et al.* (1980)
											Popravko *et al.* (1980)
7-Dimethylallyloxy-3′-hydroxy-4′-methoxy-			OR†			OH	OMe			*Calopogonium mucunoides* seed	Pereira *et al.* (1982)
7-Dimethylallyloxy-3′,4′-dimethoxy-			OR†			OMe	OMe			*Calopogonium mucunoides* seed	Pereira *et al.* (1982)
Pseudobaptigenin (ψ-baptigenin)			OH			OCH_2O				*Pisum sativum* seedling*	Carlson and Dolphin (1981)
										Pterocarpus marsupium root	Adinarayana *et al.* (1982)

							Rothia trifoliata whole plant	Rao and Rao (1985)
							Tephrosia maxima root	Murthy and Rao (1985a)
							Trifolium hybridum root	Fraishtat *et al.* (1981)
							T. pratense root	Fraishtat *et al.* (1980)
								Popravko *et al.* (1980)
							T. repens root	Fraishtat *et al.* (1981)
Maximaisoflavone B		OR†		OCH$_2$O			*Tephrosia maxima* root	Rao and Murthy (1985)
7, 2′, 4′-Trihydroxy-3′-methoxy-		OH	OH	OMe	OH		*Zollernia paraensis* wood	Ferrai *et al.* (1984a)
7-Hydroxy-2′, 4′, 5′-trimethoxy-		OH	OMe		OMe	OMe	*Dalbergia monetaria* seed	Abe *et al.* (1985)
							Eysenhardtia polystacha heartwood	Burns *et al.* (1984)
Cuneatin (maximaisoflavone G)		OH	OMe		OCH$_2$O		*Cicer cuneatum* stem	Ingham (1981a)
							Tephrosia maxima root	Rao *et al.* (1984)
7, 2′-Dimethoxy-4′, 5′-methylenedioxy-		OMe	OMe		OCH$_2$O		*Tephrosia maxima* aerial parts, root	Rao *et al.* (1984)
Maximaisoflavone C		OR†	OMe		OCH$_2$O		*Tephrosia maxima* aerial parts, root	Rao *et al.* (1984)
5, 7-Dihydroxy-	OH	OH					*Arachis hypogaea* hydrolysed flour	Daigle *et al.* (1985)
5, 7, 3′-Trihydroxy-	OH	OH		OH			*Lupinus hirsutus* whole plant	Negi *et al.* (1985)
Genistein	OH	OH			OH		*Apios, Baptisia, Bolusanthus, Cicer, Cytisus, Desmodium, Dolichos, Erythrina, Flemingia, Genista, Glycine, Lupinus, Moghania, Ononis, Ougeinia, Phaseolus, Pueraria, Sophora, Thermopsis, Trifolium.*	
							Podocarpus amarus (Podocarpaceae) heartwood	Carman *et al.* (1985)
Biochanin A	OH	OH			OMe		*Bolusanthus, Cicer, Dalbergia, Monopteryx, Ononis, Trifolium.*	
							Podocarpus amarus (Podocarpaceae) heartwood	Carman *et al.* (1985)
Isoprunetin (5-*O*-methylgenistein)	OMe	OH			OH		*Genista januensis* aerial parts	Nakov and Akhtardzhiev (1983)
							G. rumelica aerial parts	Nakov (1983)
							Phaseolus coccineus seedling*	Adesanya *et al.* (1985)
							Trifolium subterraneum leaf	Nicollier and Thompson (1982)

(*Contd.*)

Table 5.2 (*Contd.*)

Isoflavone	*Substituents* 5	6	7	8	2′	3′	4′	5′	6′	*Plant sources*	*References*
Prunetin	OH		OMe				OH			*Iris milesii* (Iridaceae) rhizome	Agarwal *et al.* (1984a)
2′-Hydroxygenistein	OH		OH		OH		OH			*Apios tuberosa* leaf*	Ingham and Mulheirn (1982)
										Desmodium gangeticum leaf*	Ingham and Dewick (1984)
										Dolichos biflorus leaf*	Keen and Ingham (1980)
										Flemingia stricta root	Rao *et al.* (1982)
										Lupinus spp. leaf	Ingham *et al.* (1983)
										Moghania macrophylla wood	Krishnamurty and Prasad (1980)
										Phaseolus aureus seedling*	O'Neill *et al.* (1983)
										P. coccineus seedling*	Adesanya *et al.* (1985)
										P. mungo seedling*	Adesanya *et al.* (1984)
										P. vulgaris cotyledon*	Whitehead *et al.* (1982)
2′-Hydroxyiso-prunetin	OMe		OH		OH		OH			*Phaseolus coccineus* seedling*	Adesanya *et al.* (1985)
Cajanin	OH		OMe		OH		OH			*Cajanus cajan* seed*	Dahiya *et al.* (1984)
										Centrosema pubescens leaf*	Markham and Ingham (1980)
Orobol	OH		OH			OH	OH			*Bolusanthus speciosus* seed	Asres *et al.* (1985)
										Bryum capillare (Bryales) gametophyte	Anhut *et al.* (1984)
										Ougeinia dalbergioides heartwood	Kalidhar and Sharma (1984)
3′-*O*-Methylorobol	OH		OH			OMe	OH			*Bolusanthus speciosus* seed	Asres *et al.* (1985)
										Cytisus scoparius flower	Viscardi *et al.* (1985)
										Lupinus albus leaf	Tahara *et al.* (1984a)
										Wyethia glabra (Compositae) leaf	McCormick *et al.* (1985)
										W. heleniodes aerial parts	Bohlmann *et al.* (1981)
										W. mollis leaf, stem	Waddell *et al.* (1982)
Pratensein	OH		OH			OH	OMe			*Bolusanthus speciosus* seed	Asres *et al.* (1985)
										Bryum capillare (Bryales) gametophyte	Anhut *et al.* (1984)
										Cicer arietinum leaf	Kazakov *et al.* (1980)

									Monopteryx inpae trunk wood	Albuquerque *et al.* (1981)
									Trifolium pratense root	Fraishtat *et al.* (1980)
									T. pratense aerial parts	Sachse (1984)
									T. spp. leaf	Blanco *et al.* (1982)
Santal	OH		OMe			OH	OH		*Wyethia glabra* (Compositae) leaf	McCormick *et al.* (1985)
									W. heleniodes aerial parts	Bohlmann *et al.* (1981)
									W. mollis leaf, stem	Waddell *et al.* (1982)
3′-*O*-Methylpratensein	OH		OH			OMe	OMe		*Bolusanthus speciosus* seed	Asres *et al.* (1985)
5, 3′-Dihydroxy-7, 4′-dimethoxy-	OH		OMe			OH	OMe		*Simsia foetida* (Compositae)	Vivar *et al.* (1982)
Junipegenin A	OH		OH			OH	OMe	OH	*Iris hookeriana* (Iridaceae) rhizome	Shawl *et al.* (1985)
									I. kumaonensis rhizome	Agarwal *et al.* (1984)
7, 2′-Dihydroxy-6-methoxy-		OMe	OH		OH				*Salicornia europaea* (Chenopodiaceae)	Arakawa *et al.* (1982)
2′-Hydroxy-6, 7-methylenedioxy-		OCH_2O			OH				*Salicornia europaea* (Chenopodiaceae)	Arakawa *et al.* (1982)
6, 7, 4′-Trihydroxy-		OH	OH				OH		*Centrosema haitiense* leaf*	Markham and Ingham (1980)
Glycitein		OMe	OH				OH		*Centrosema haitiense* leaf*	Markham and Ingham (1980)
									C. pubescens leaf*	
Afrormosin		OMe	OH				OMe		*Astragalus clusii* root	Marco *et al.* (1983)
									Centrosema haitiense leaf*	Markham and Ingham (1980)
									C. pubescens leaf*	
									Hedysarum polybotrys root	Miyase *et al.* (1984)
									Millettia reticulata stem	Chen *et al.* (1983)
									Pisum sativum seedling*	Carlson and Dolphin (1981)
Hemerocallone		OCH_2O			OMe			OMe	*Hemerocallis minor* (Liliaceae) root	Xiu *et al.* (1982)
Fujikinetin		OMe	OH			OCH_2O			*Tephrosia maxima* aerial parts	Rao *et al.* (1984)
Retusin			OH	OH			OMe		*Maackia amurensis* heartwood	Maksimov *et al.* (1985)
8-*O*-Methylretusin			OH	OMe			OMe		*Millettia reticulata* stem	Chen *et al.* (1983)
									Monopteryx uaucu trunk wood	Albuquerque *et al.* (1981)

(*Contd.*)

Table 5.2 (*Contd.*)

Isoflavone	*Substituents* 5	6	7	8	2′	3′	4′	5′	6′	*Plant sources*	*References*
Maximaisoflavone H			OCH_2O				OMe			*Tephrosia maxima* root	Rao and Murthy (1985)
8-*O*-Methyl-3′-hydroxyretusin			OH	OMe		OH	OMe			*Monopteryx uaucu* trunk wood	Albuquerque *et al.* (1981)
7-Hydroxy-8, 3′, 4′-trimethoxy-			OH	OMe		OMe	OMe			*Monopteryx uaucu* trunk wood	Albuquerque *et al.* (1981)
Maximaisoflavone E			OH	OMe		OCH_2O				*Tephrosia maxima* aerial parts, root	Rao *et al.* (1984)
7-Dimethylallyloxy-8-methoxy-3′, 4′-methylenedioxy-			OR[†]	OMe		OCH_2O				*Calopogonium mucunoides* seed	Pereira *et al.* (1982)
Maximaisoflavone D			OCH_2O			OMe	OMe			*Tephrosia maxima* aerial parts, root	Rao *et al.* (1984)
Maximaisoflavone A			OCH_2O			OCH_2O				*Tephrosia maxima* aerial parts, root	Rao *et al.* (1984)
Maximaisoflavone F			OH	OMe	OMe		OCH_2O			*Tephrosia maxima* root	Rao *et al.* (1984)
7, 8, 2′-Trimethoxy-4′, 5′-methylenedioxy-			OMe	OMe	OMe		OCH_2O			*Tephrosia maxima* aerial parts, root	Rao *et al.* (1984)
5, 7-Dihydroxy-6, 2′-dimethoxy-	OH	OMe	OH		OMe					*Iris spuria* (Iridaceae) rhizome	Shawl *et al.* (1984)
Betavulgarin	OMe	OCH_2O			OH					*Beta vulgaris* (Chenopodiaceae) root*	Babchenko *et al.* (1982)
										B. spp. leaf*	Richardson (1981)
6-Hydroxygenistein	OH	OH	OH				OH			*Centrosema plumieri* leaf*	Markham and Ingham (1980)
Tectorigenin	OH	OMe	OH				OH			*Centrosema haitiense* leaf*	Markham and Ingham (1980)
										C. pubescens leaf*	
										C. virginianum leaf*	
										Iris milesii (Iridaceae) rhizome	Agarwal *et al.* (1984a)
Irisolidone	OH	OMe	OH				OMe			*Iris germanica* (Iridaceae) rhizome	Ali *et al.* (1983)
										Podocarpus amarus (Podocarpaceae) heartwood	Carman *et al.* (1985)
Isoaurmillone	OH	OMe	OH				OR[†]			*Millettia auriculata* pod	Gupta *et al.* (1983)
Irilone	OH	OCH_2O					OH			*Trifolium pratense* root	Popravko *et al.* (1980)
5, 6, 7, 4′-Tetrahydroxy-3′-methoxy-	OH	OH	OH			OMe	OH			*Iris milesii* (Iridaceae) rhizome	Agarwal *et al.* (1984b)
Iristectorigenin A	OH	OMe	OH			OH	OMe			*Iris spuria* (Iridaceae) rhizome	Shawl *et al.* (1984)
										Monopteryx inpae trunk wood	Albuquerque *et al.* (1981)

Junipegenin B (dalspinosin)	OH	OMe	OH			OMe	OMe		*Dalbergia spinosa* root	Dasan *et al.* (1982)
									Juniperus macropoda (Cupressaceae) leaf	Sethi *et al.* (1981)
Dalspinin	OH	OMe	OH			OCH_2O			*Dalbergia spinosa* root	Dasan *et al.* (1982)
Iriskumaonin	OMe	OCH_2O				OH	OMe		*Iris kumaonensis* (Iridaceae) rhizome	Agarwal *et al.* (1984b)
O-Methyliriskumaonin	OMe	OCH_2O				OMe	OMe		*Iris kumaonensis* (Iridaceae) rhizome	Agarwal *et al.* (1984b)
Caviunin	OH	OMe	OH		OMe		OMe	OMe	*Dalbergia spinosa* root	Dasan *et al.* (1982)
Irigenin	OH	OMe	OH			OH	OMe	OMe	*Belamcanda chinensis* (Iridaceae) rhizome	Ho *et al.* (1982)
									Iris germanica (Iridaceae) rhizome	Ali *et al.* (1983)
										El-Moghazy *et al.* (1980)
									I. hookeriana rhizome	Shawl *et al.* (1985)
									I. kumaonensis rhizome	Agarwal *et al.* (1984b)
									I. milesii rhizome	Agarwal *et al.* (1984a)
Junipegenin C	OH	OMe	OH			OMe	OMe	OMe	*Juniperus macropoda* (Cupressaceae) leaf	Sethi *et al.* (1981)
5, 3′-Dihydroxy-4′, 5′-dimethoxy-6, 7-methylenedioxy-	OH	OCH_2O				OH	OMe	OMe	*Iris germanica* (Iridaceae) rhizome	Ali *et al.* (1983)
Irisflorentin	OMe	OCH_2O				OMe	OMe	OMe	*Iris hookeriana* (Iridaceae) rhizome	Shawl *et al.* (1985)
									I. kumaonensis rhizome	Agarwal *et al.* (1984b)
Aurmillone	OH		OH	OMe			OR†		*Millettia auriculata* seed	Raju *et al.* (1981)
5, 7, 3′-Trihydroxy-8, 4′-dimethoxy-	OH		OH	OMe		OH	OMe		*Monopteryx inpae* trunk wood	Albuquerque *et al.* (1981)
5, 7-Dihydroxy-8, 3′, 4′-trimethoxy-	OH		OH	OMe		OMe	OMe		*Monopteryx inpae* trunk wood	Albuquerque *et al.* (1981)
5, 6, 7, 4′-Tetrahydroxy-8-methoxy-	OH	OH	OH	OMe			OH		*Iris milesii* (Iridaceae) rhizome	Agarwal *et al.* (1984a)
Licoricone (*5.2*)									*Glycyrrhiza uralensis* root	Chang *et al.* (1983)
3′-Dimethylallylgenistein (*5.3*)									*G.* sp.	Chang *et al.* (1981)
									Cajanus cajan seed*	Dahiya *et al.* (1984)
Lupinisoflavone C (*5.5*)									*Lupinus albus* root	Tahara *et al.* (1984a)
Licoisoflavone A (*5.4*) (phaseoluteone)									*Lupinus albus* root	Tahara *et al.* (1984a)
									Phaseolus coccineus seeding*	Adesanya *et al.* (1985)
									P. vulgaris cotyledon*	Goosens and Van Laere (1983)
Licoisoflavone B (*5.7*)									*Glycyrrhiza* sp. root	Hiraga *et al.* (1984)
									Lupinus albus root	Ingham *et al.* (1983)
										Tahara *et al.* (1984a)

(*Contd.*)

Table 5.2 (*Contd.*)

Isoflavone	*Substituents* 5	6	7	8	2′	3′	4′	5′	6′	*Plant sources*	*References*
Lupinisoflavone D (*5.6*)										*Lupinus albus* root	Tahara *et al.* (1984a)
2′-Deoxypiscerythrone (*5.8*)										*Piscida erythrina* root bark	Delle Monache *et al.* (1984)
Piscerythrone (*5.9*)										*Piscidia erythrina* root bark	Delle Monache *et al.* (1984)
Piscidone (*5.10*)										*Piscidia erythrina* root bark	Redaelli and Santaniello (1984) Delle Monache *et al.* (1984)
(*5.11*)										*Piscidia erythrina* root bark	Delle Monache *et al.* (1984) Redaelli and Santaniello (1984)
Calopogonium isoflavone B (*5.12*)										*Tephrosia maxima* root	Murthy and Rao (1985b)
Barbigerone (*5.13*)										*Tephrosia barbigera* seed	Vilain (1983)
Jamaicin (*5.14*)										*Piscidia erythrina* root bark	Pietta and Zio (1983) Delle Monache *et al.* (1984)
6-(or 8?)-(1,1-Dimethylallyl)genistein (*5.15*)										*Moghania macrophylla* wood	Krishnamurty and Prasad (1980)
Wighteone (*5.16*)										*Lupinus albus* root *L.* spp. leaf	Tahara *et al.* (1984a) Ingham *et al.* (1983)
Luteone (*5.17*)										*Lupinus albus* root *L.* spp. leaf	Tahara *et al.* (1984a) Ingham *et al.* (1983)
Lupisoflavone (*5.18*)										*Lupinus albus* leaf *L. luteus leaf*	Ingham *et al.* (1983)
Viridiflorin (*5.19*)										*Tephrosia viridiflora* root, aerial parts	Gomez *et al.* (1985)
Alpinumisoflavone (*5.20*)										*Derris* sp. root *Lupinus albus* root *Millettia thonningii* seed	Da Rocha and Zoghbi (1982) Ingham *et al.* (1983) Khalid and Waterman (1983) Olivares *et al.* (1982)
4′-*O*-Methylalpinumisoflavone (*5.21*)										*Derris* sp. root *Millettia thonningii* seed	Da Rocha and Zoghbi (1982) Khalid and Waterman (1983) Olivares *et al.* (1982)

4′-*O*-Dimethylallylalpinum-isoflavone (*5.22*)	*Derris* sp. root	Da Rocha and Zoghbi (1982)
Di-*O*-Methylalpinumiso-flavone (*5.23*)	*Millettia thonningii* seed	Khalid and Waterman (1983)
Parvisoflavone B (*5.24*)		Olivares *et al.* (1982)
(*5.25*)	*Lupinus albus* root	Tahara *et al.* (1984a)
Robustone (*5.26*)	*Millettia thonningii* seed	Olivares *et al.* (1982)
	Millettia thonningii seed	Khalid and Waterman (1983)
Elongatin (*5.27*)	*Tephrosia viridiflora*	Gomez *et al.* (1985)
Dihydroalpinumisoflavone (*5.28*)	*Crotalaria madurensis* leaf, stem	Bhakuni and Chaturevedi (1984)
Lupinisoflavone A (*5.29*)	*Cajanus cajan* seed*	Dahiya *et al.* (1984)
	Lupinus albus root	Tahara *et al.* (1984a)
Erythrinin C (*5.30*)	*Lupinus albus* root	Tahara *et al.* (1985a)
Lupinisoflavone B (*5.31*)	*Lupinus albus* root	Tahara *et al.* (1984a)
2, 3-Dehydrokievitone (*5.32*)	*Phaseolus aureus* seedling*	O'Neill *et al.* (1983)
4′-*O*-Methylderrone (*5.33*)	*Derris robusta* seed	Chibber *et al.* (1981)
	Millettia pachycarpa seed	Singhal *et al.* (1983)
Lupalbigenin (*5.34*)	*Lupinus albus* root	Ingham and Tahara (1985)
		Tahare *et al.* (1984a)
2′-Hydroxylupalbigenin (*5.35*)	*Millettia pulchra* aerial parts	Baruah *et al.* (1984)
2′-Methoxylupalbigenin (*5.36*)	*Lupinus albus* root	Tahara *et al.* (1984a)
	Millettia pulchra aerial parts	Baruah *et al.* (1984)
Lupinisoflavone E (*5.37*)	*Lupinus albus* root	Tahara *et al.* (1984a)
Lupinisoflavone F (*5.38*)	*Lupinus albus* root	Tahara *et al.* (1984a)
Flemiphyllin (*5.39*)	*Flemingia macrophylla* stem	Rao and Srimannarayana (1984)
6, 8-Di-(dimethylallyl)-genistein (*5.40*)	*Euchresta japonica* root	Shirataki *et al.* (1982)
	Millettia pachycarpa seed	Singhal *et al.* (1983)
6, 8-Di-(dimethylallyl)-orobol (*5.41*)	*Millettia pachycarpa* seed	Singhal *et al.* (1983)
6, 8-Di-(dimethylallyl)-pratensein (*5.42*)	*Millettia pachycarpa* seed	Singhal *et al.* (1983)
Warangalone (scandenone) (*5.43*)	*Erythrina senegalensis* stem bark	Fomum *et al.* (1985)
	Euchresta japonica root	Shirataki *et al.* (1982)
Auriculatin (*5.44*)	*Tephrosia elata* root	Lwande *et al.* (1985a)
	Millettia auriculata seed	Raju *et al.* (1981)

(*Contd.*)

Table 5.2 (*Contd.*)

Isoflavone	*Substituents*									*Plant sources*	*References*
	5	6	7	8	2′	3′	4′	5′	6′		
Auriculasin (*5.45*)										*Millettia auriculata* seed	Raju *et al*. (1981)
(*5.46*)										*Millettia pachycarpa* leaf	Singhal *et al*. (1981)
(*5.47*)										*Millettia pachycarpa* leaf	Singhal *et al*. (1981)
Osajin (*5.48*)										*Euchresta japonica* root	Shirataki *et al*. (1982)
										Maclura pomifera (Moraceae) fruit	Mahmoud (1981)
Pomiferin (*5.49*)										*Maclura pomifera* (Moraceae) fruit	Mahmoud (1981)
										Millettia pachycarpa seed	Singhal *et al*. (1983)
(*5.50*)										*Millettia pachycarpa* leaf	Singhal *et al*. (1981)
(*5.51*)										*Millettia pachycarpa* leaf	Singhal *et al*. (1981)

*Plant part was subjected to physiological stress.
†$R = CH_2CH{=}CMe_2$

5.3 ISOFLAVONES

Isoflavones (*5.1*) constitute the largest group of natural isoflavonoid derivatives, with some 234 known aglycones reported. Table 5.2 lists the structures and sources of isoflavones published since 1980. The four simple isoflavones daidzein, formononetin, genistein and biochanin A are extremely common and new sources of these compounds are listed under genus only. In most cases, these compounds co-occur with more complex isoflavones, for which detailed sources and references are given.

With so many known natural isoflavones, covering a wide range of different oxygenation and substitution patterns, those structures that invite comment are the ones that seem unusual from a biogenetic aspect. The isoflavone synthase enzyme recently reported (see Section 5.18) was able to transform the flavanone substrates (2*S*)-naringenin or (2*S*)-liquiritigenin into the 4′-hydroxyisoflavones genistein or daidzein, respectively. These isoflavones may then act as precursors for other isoflavonoids, gradually building up the more complex substitution patterns. It is not surprising, therefore, that almost all the natural isoflavones contain a 4′-oxygen substituent. In the previous review, a total of nine isoflavones lacking 4′-oxygenation were given, some containing 2′-oxygen substituents, others lacking totally any B-ring substitution. Further examples can now be added to this list. 5,7-Dihydroxyisoflavone from the defatted flour of peanut (*Arachis hypogaea*) (Daigle *et al.*, 1985) comes into the unsubstituted B-ring category, though its dimethyl ether has previously been isolated from peanut fruits. The heartwood of *Pterocarpus*

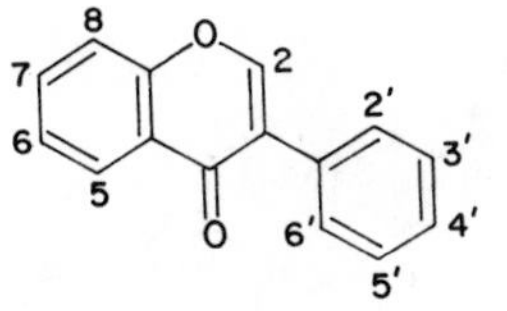

(*5.1*)

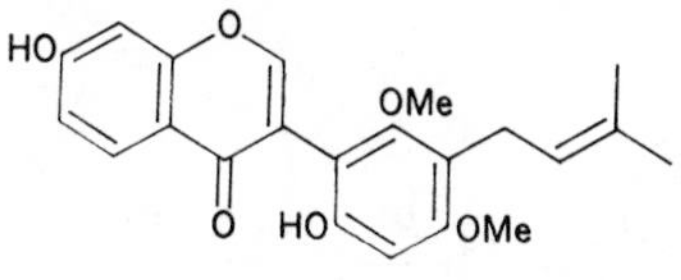

(*5.2*) Licoricone

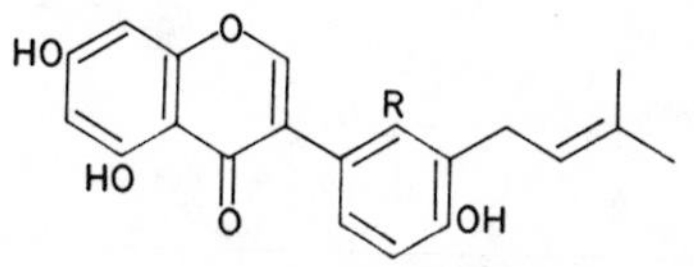

(*5.3*) R=H, 3′-Dimethylallylgenistein

(*5.4*) R=OH, Licoisoflavone A (phaseoluteone)

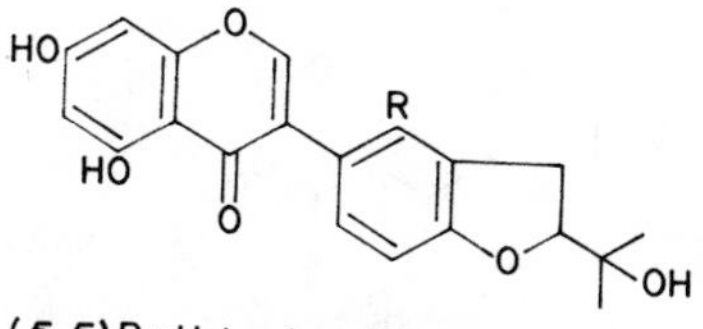

(*5.5*) R=H, Lupinisoflavone C

(*5.6*) R=OH, Lupinisoflavone D

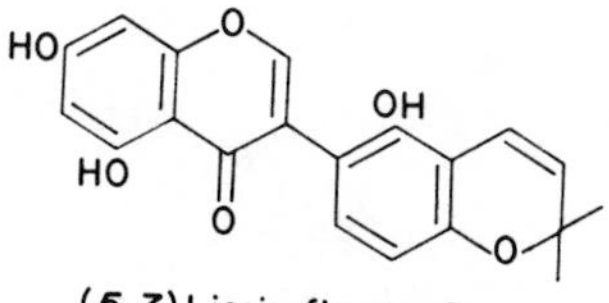

(*5.7*) Licoisoflavone B

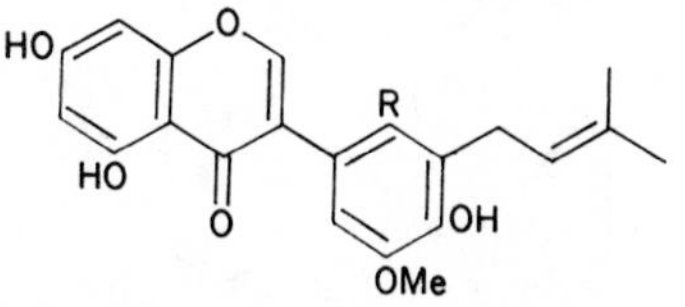

(*5.8*) R=H, 2′-Deoxypiscerythrone

(*5.9*) R=OH, Piscerythrone

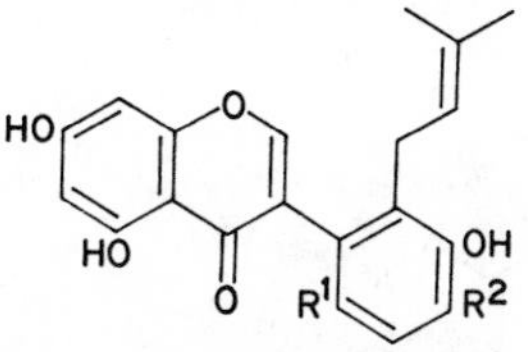

(*5.10*) R^1=OMe, R^2=OH, or R^1=OH, R^2=OMe, Piscidone

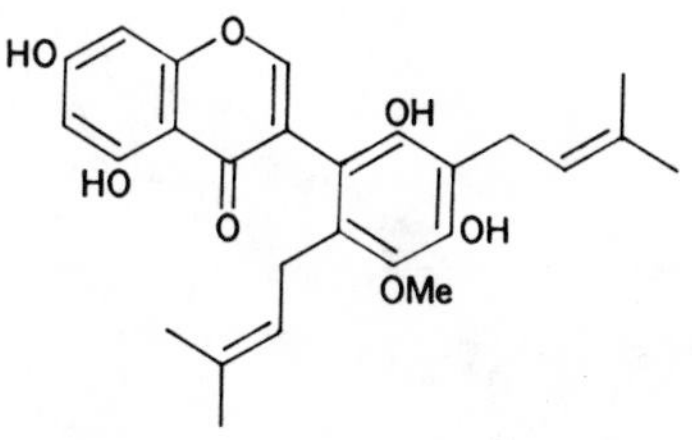

(*5.11*)

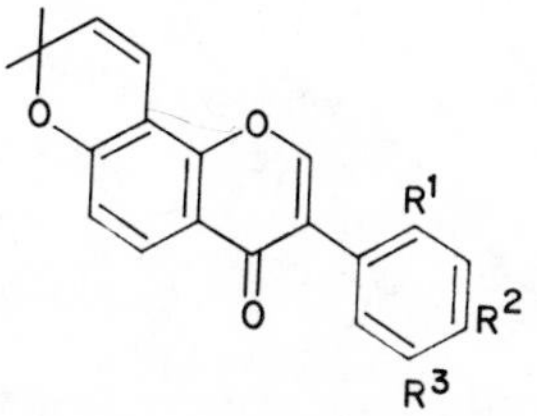

(*5.12*) R^1=H, R^2R^3=OCH_2O, Calopogonium isoflavone B

(*5.13*) R^1=R^2=R^3=OMe, Barbigerone

(*5.14*) R^1=OMe, R^2R^3=OCH_2O, Jamaicin

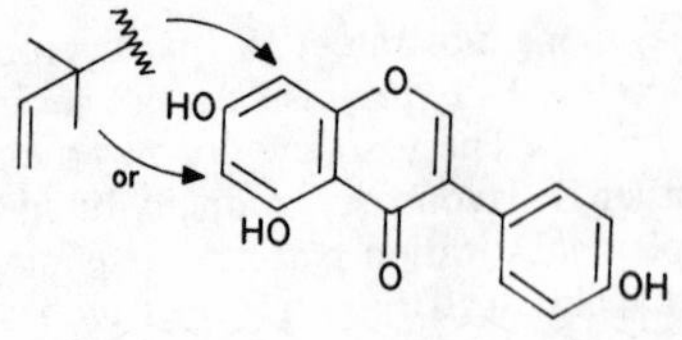

(*5.15*)

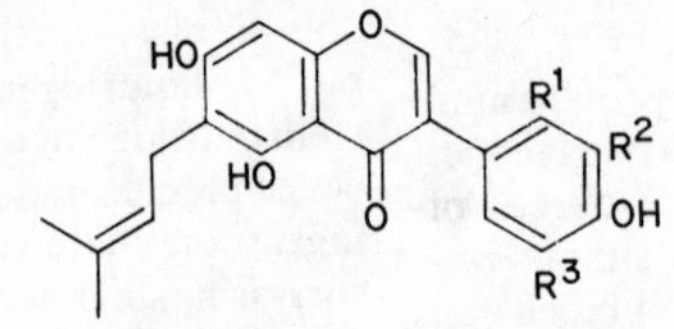

(*5.16*) $R^1=R^2=R^3=H$, Wighteone
(*5.17*) $R^1=OH, R^2=R^3=H$, Luteone
(*5.18*) $R^1=R^3=H, R^2=OMe$, Lupisoflavone
(*5.19*) $R^1=R^3=OMe, R^2=H$, Viridiflorin

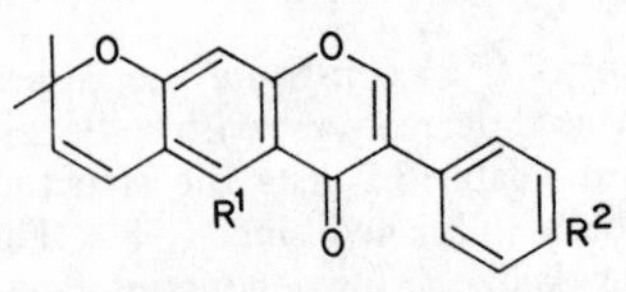

(*5.20*) $R^1=R^2=OH$, Alpinumisoflavone
(*5.21*) $R^1=OH, R^2=OMe$, 4′-*O*-Methylalpinumisoflavone
(*5.22*) $R^1=OH, R^2=OCH_2CH=CMe_2$, 4′-*O*-Dimethylallylalpinumisoflavone
(*5.23*) $R^1=R^2=OMe$, Di-*O*-Methylalpinumisoflavone

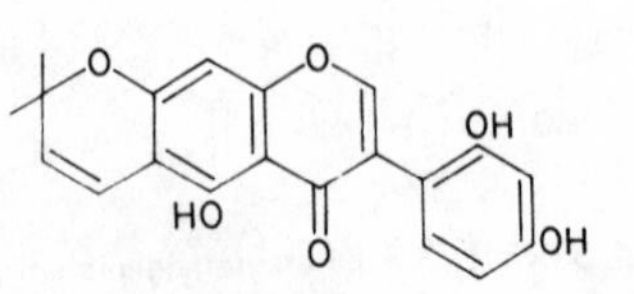

(*5.24*) Parvisoflavone B

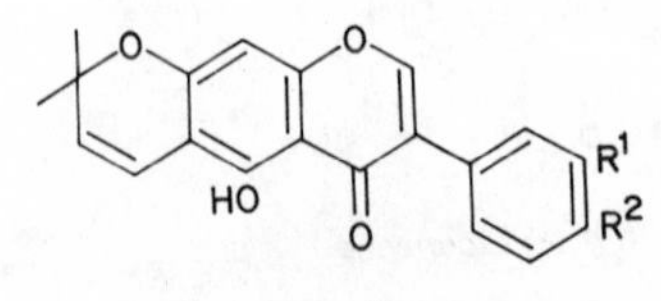

(*5.25*) $R^1=OH, R^2=OMe$
(*5.26*) $R^1R^2=OCH_2O$, Robustone

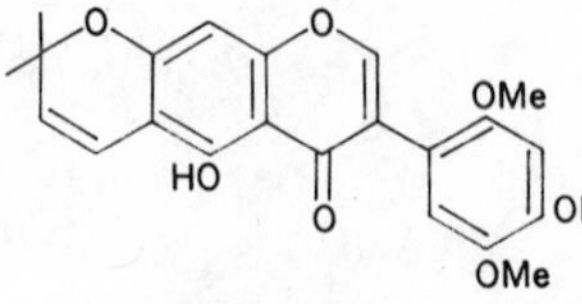

(*5.27*) Elongatin

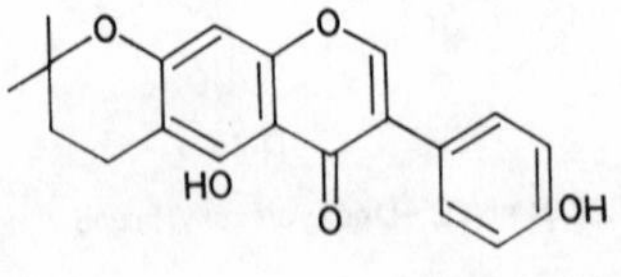

(*5.28*) Dihydroalpinumisoflavone

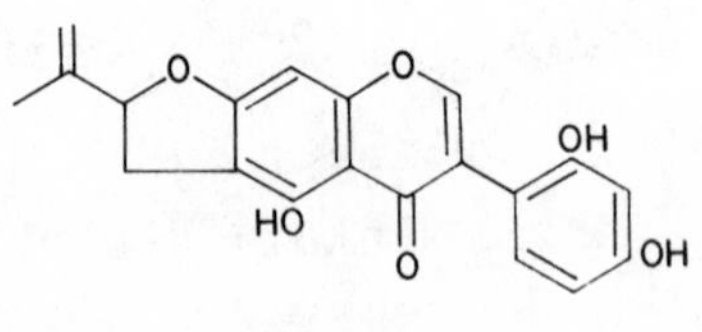

(*5.29*) Lupinisoflavone A

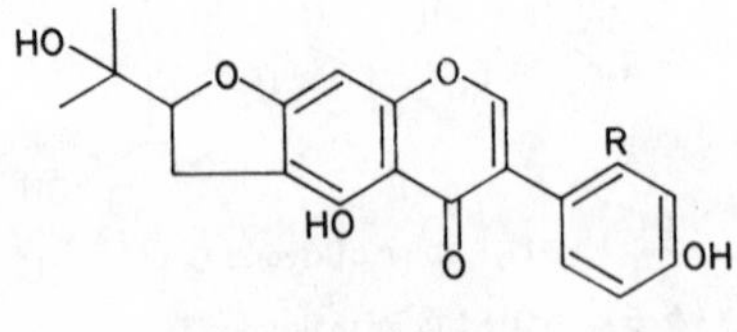

(*5.30*) R=H, Erythrinin C
(*5.31*) R=OH, Lupinisoflavone B

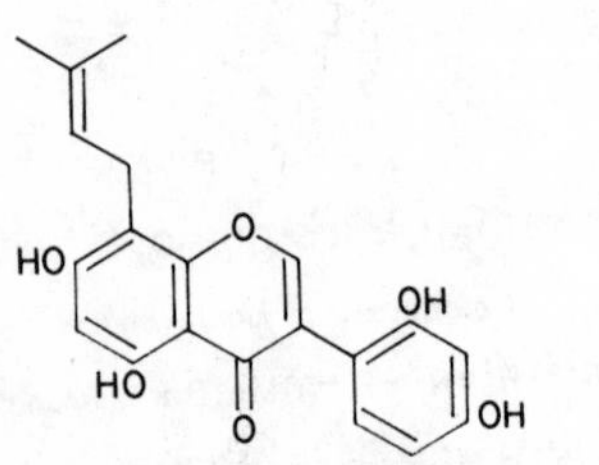

(*5.32*) 2,3-Dehydrokievitone

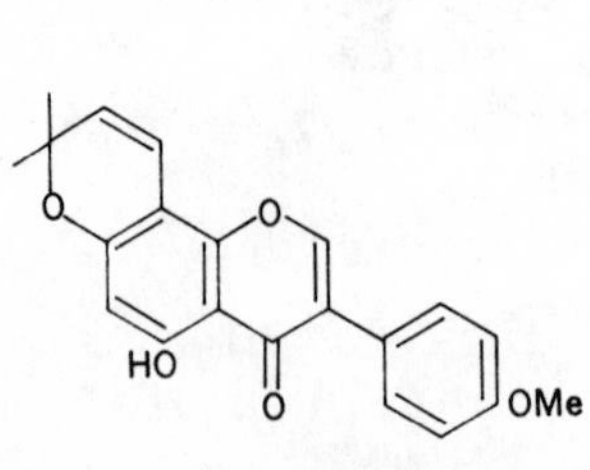

(*5.33*) 4′-*O*-Methylderrone

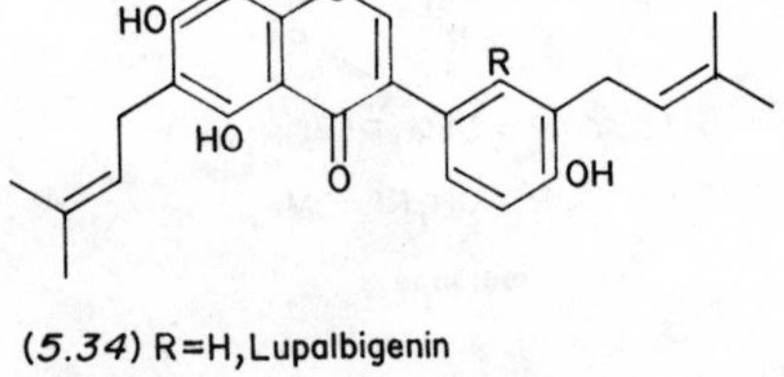

(*5.34*) R=H, Lupalbigenin
(*5.35*) R=OH, 2′-Hydroxylupalbigenin
(*5.36*) R=OMe, 2′-Methoxylupalbigenin

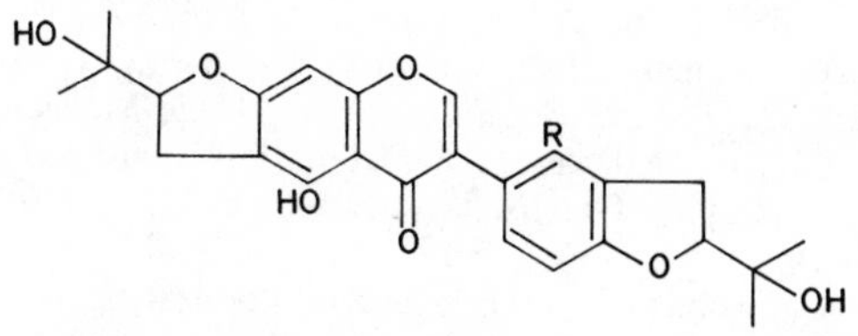

(5.37) R=H, Lupinisoflavone E

(5.38) R=OH, Lupinisoflavone F

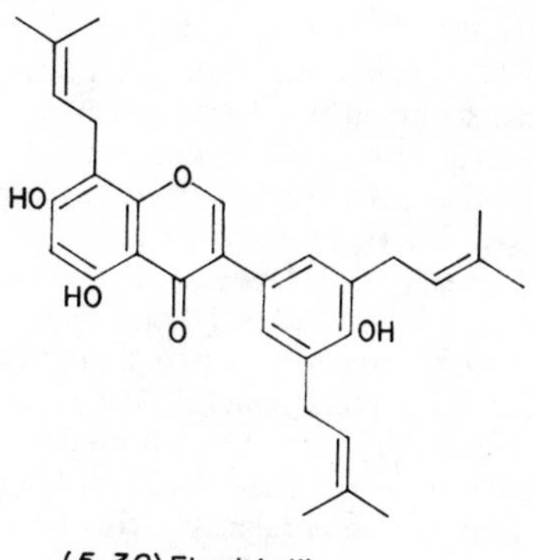

(5.39) Flemiphyllin

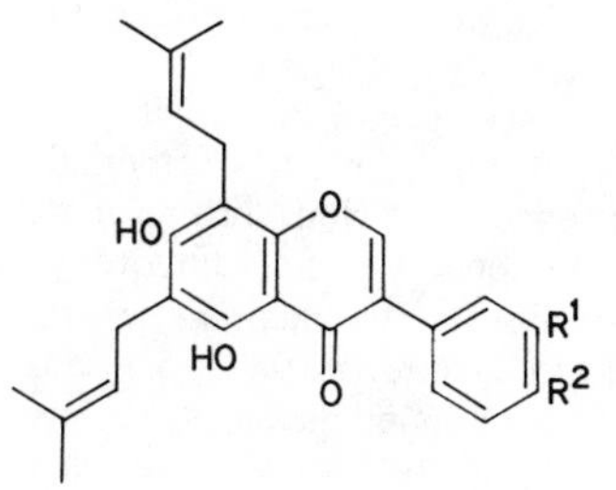

(5.40) R^1=H, R^2=OH, 6,8-Di-(dimethylallyl)genistein

(5.41) R^1=R^2=OH, 6,8-Di-(dimethylallyl)orobol

(5.42) R^1=OH, R^2=OMe, 6,8-Di-(dimethylallyl)pratensein

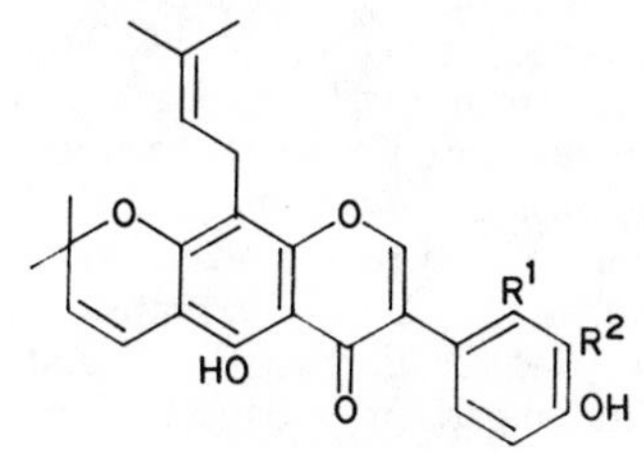

(5.43) R^1=R^2=H, Warangalone (Scandenone)

(5.44) R^1=OH, R^2=H, Auriculatin

(5.45) R^1=H, R^2=OH, Auriculasin

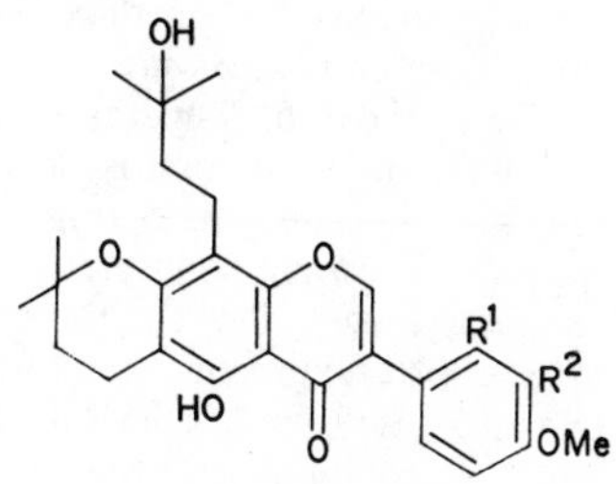

(5.46) R^1=OMe, R^2=H

(5.47) R^1=H, R^2=OMe

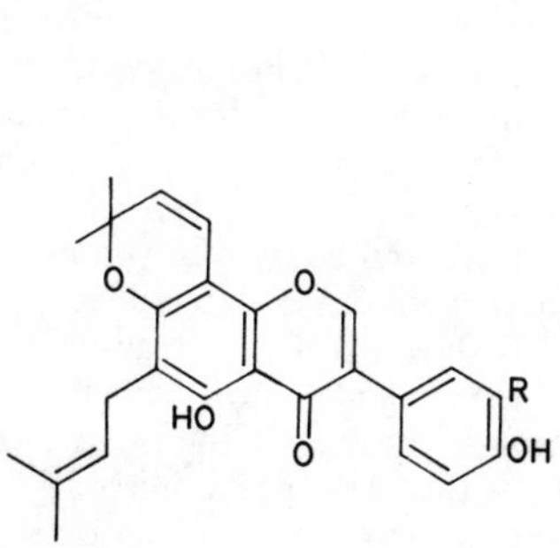

(5.48) R=H, Osajin

(5.49) R=OH, Pomiferin

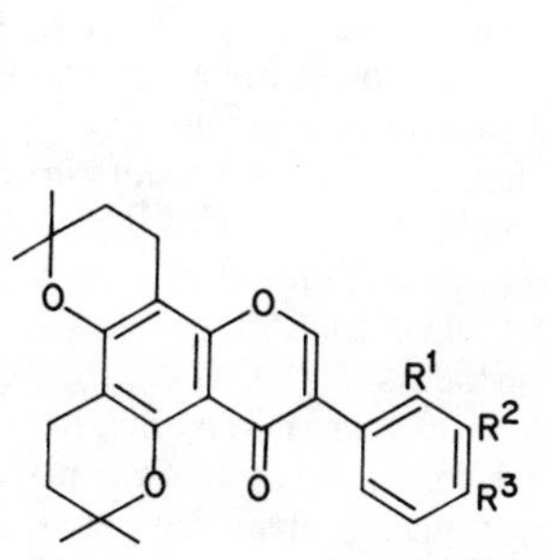

(5.50) R^1=OH, R^2=H, R^3=OMe

(5.51) R^1=H, R^2=OMe, R^3=OH, or R^1=H, R^2=OH, R^3=OMe

marsupium contains 5,7-dihydroxy-6-methoxyisoflavone as its 7-*O*-rhamnoside (Mitra and Joshi, 1983), and again an unsubstituted B-ring is present.

Betavulgarin (2′-hydroxy-5-methoxy-6,7-methylenedioxyisoflavone), originally isolated from fungus-infected leaves of sugar beet (*Beta vulgaris*; Chenopodiaceae), is now shown to be produced as a phytoalexin in other *Beta* species (Richardson, 1981). Glasswort (*Salicornia europaea*) is also a member of the Chenopodiaceae, and this plant has been reported to contain two isoflavones, 7,2′-dihydroxy-6-methoxyisoflavone and 2′-hydroxy-6,7-methylenedioxyisoflavone (Arakawa *et al.*, 1982). A further similarity is that both *Beta vulgaris* and *Salicornia europaea* produce a flavanone having the same substitution pattern as betavulgarin or 2′-hydroxy-6,7-methylenedioxyisoflavone respectively. Rhizomes of *Iris spuria* (Iridaceae) contain a further 2′-oxygenated isoflavone, 5,7-dihydroxy-2′,6-dimethoxyisoflavone (Shawl *et al.*, 1984) which co-occurs with iristectorigenin A, the related 3′-hydroxy-4′-methoxy analogue. Hemerocallone from roots of *Hemerocallis minor* has been identified as 2′,5′-dimethoxy-6, 7-methylenedioxyisoflavone (Xiu *et al.*, 1982) and is the first isoflavonoid reported from the Liliaceae. This unusual B-ring 2′,5′-oxygenation pattern is already known in podospicatin from *Podocarpus spicatus* (Podocarpaceae). Perhaps the most unusual simple isoflavone isolated from the Leguminosae is 5,7,3′-trihydroxyisoflavone from *Lupinus hirsutus* (Negi *et al.*, 1985). There is no precedent for this substitution pattern, but the structure was confirmed by synthesis. The biosynthetic origin of these unusual B-ring substitution patterns has yet to be investigated. It may be that aryl groups without *para*-hydroxylation can be migrated, and it is noteworthy that most of the examples known occur in non-leguminous plants. Alternatively, these *para*-oxygen functions may have been reduced or displaced during biosynthesis. In other fields, there are good precedents for similar removal of phenolic functions.

Structural complexity increases enormously as isoprenyl substituents become incorporated into the isoflavonoid system. Alkylation of phenolic groups giving dimethylallyl ethers is not unknown in the isoflavones, but is rare enough that the report of several new ones is of interest. New examples have been isolated from *Calopogonium mucunoides* (Pereira *et al.*, 1982), *Derris* sp. (Da Rocha and Zoghbi, 1982), *Millettia auriculata* (Gupta *et al.*, 1983) and *Tephrosia maxima* (Murthy and Rao, 1985b). It is more usual to find isoprenyl substituents alkylating the aromatic ring systems at nucleophilic sites generated by neighbouring oxygen functions. 3,3-Dimethylallyl substituents are common, whereas 1,1-dimethylallyl substituents are quite rare. Only two examples of the latter are known, in the 3-hydroxyisoflavanone secondifloran, and in an isoflavone from *Moghania macrophylla* (Krishnamurty and Prasad, 1980). The structure of this compound has yet to be confirmed but is either 6- or 8-(1,1-dimethylallyl)-genistein (*5.15*). The presence of two 3,3-dimethylallyl and three oxygen substituents on the B-ring of an isoflavone (*5.11*) from *Piscidia erythrina* (Delle Monache *et al.*, 1984; Redaelli and Santaniello, 1984) makes this the first example of an isoflavone with a fully substituted B-ring.

The heartwood of *Podocarpus amarus* (Podocarpaceae) is reported to contain a range of simple isoflavones including irisolidone (5,7-dihydroxy-6,4′-dimethoxyisoflavone) which was isolated in a quite remarkable yield of about 5% (Carman *et al.*, 1985). Rough calculations suggest that the average *P. amarus* tree contains approximately 300 kg of irisolidone. Such data must invite speculation about the biological role for this isoflavone.

UV spectra give a very valuable indication of the class of compound to which a particular isoflavonoid belongs (Dewick, 1982), but there is little comprehensive data about the effect of shift reagents and application of this information to assign substitution patterns. Certainly, data for flavonoids are much more comprehensive than for isoflavonoids. In a recent survey (Wolfbeis *et al.*, 1984), UV absorption and fluorescence spectra for twenty isoflavones representing both natural and model compounds have been recorded and the effects of various shift reagents analysed. Methanolic solutions of 5,7-dihydroxyisoflavones have band I absorption maxima around 335 nm, 6,7-dioxygenated isoflavones between 310 and 330 nm, and others (7- or 7,8-oxygenated) below 310 nm. The presence of water can give partial ionization if acidic hydroxyls are present. Addition of sodium acetate gives rise to anion absorption spectra for 7-hydroxyisoflavones, and to partial anion absorption for 6-hydroxyisoflavones, with bathochromic shifts of the order of 13–36 nm and 45–49 nm respectively. 5,7-Dihydroxyisoflavones show hardly any shift, typically 3–6 nm to lower wavelengths. Similar results are obtained with ammonia. Sodium borate will complex with 6,7-dihydroxyisoflavones, giving an additional absorption maximum different from the anion absorption. Aluminium chloride complexes with both 5-hydroxy- and 6,7-dihydroxy-isoflavones, giving bathochromic shifts of about 44 nm and 18 nm respectively. Isoflavones are fluorescent, unless a 5-hydroxy group is present, though in these cases, the aluminium chloride complex gives a green fluorescence.

A mass spectral study of seven isomeric natural dihydroxymonomethoxyisoflavones as their trimethylsilyl ethers has shown that they may be distinguished on the basis of their fragmentation, as well as their GLC mobility (Woodward, 1981b). Of the isoflavones investigated, all had $M^{+\cdot}$ as their base peak, unless a 5- or 2′-OTMS group was present, when $M-15$ was base peak, by loss of a methyl radical from the TMS, giving a stable ion, e.g. (*5.52*). Major fragments were $M-31$ for 2′-methoxyisoflavones, $M-30$ for *ortho*-methoxy-OTMS substitution, and $M-15$ for *ortho*-di(OTMS) substitution. The

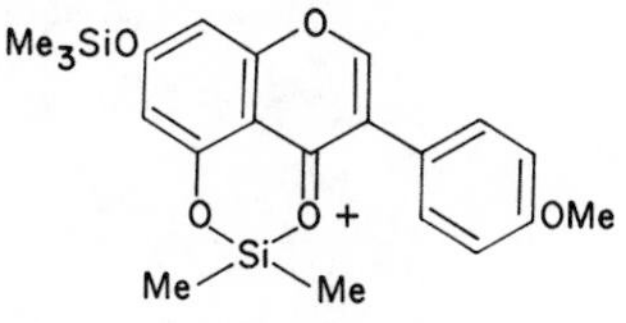

(5.52) 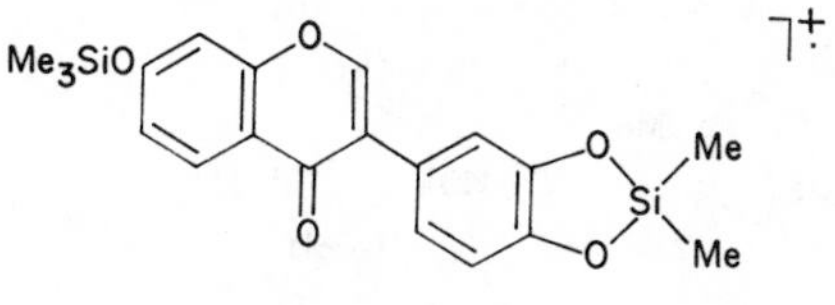 (5.53) 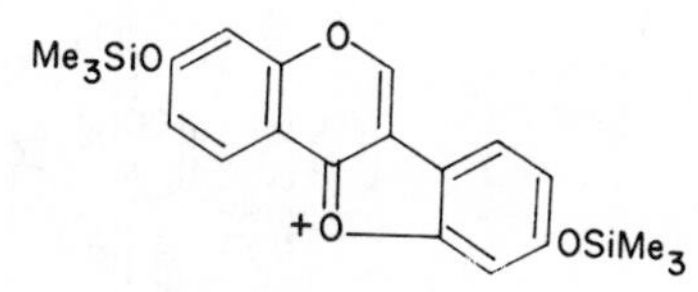 (5.54)

latter two peaks correspond to formation of a cyclic silyl ether, e.g. (*5.53*) with or without loss of the methoxyl methyl respectively, and $M - 31$ to a stable ion, e.g. (*5.54*). Retro-Diels-Alder (RDA) cleavage is much less significant with the TMS ethers than the underivatized compounds.

It is always desirable to confirm the structure of a natural product by total or partial synthesis. This is particularly important if biogenetically unusual structures are proposed, and the synthesis of 2′-hydroxy-6,7-methylenedioxyisoflavone (Arakawa *et al.*, 1982) and 5,7,3′-trihydroxyisoflavone (Negi *et al.*, 1985), unusual isoflavones described above, reinforces the structural analysis. The structure of derrugenin, an isoflavone from *Derris robusta*, has had to be revised to 5,4′-dihydroxy-7,2′,5′-trimethoxyisoflavone as a result of synthetic studies (Tsukayama *et al.*, 1980).

A traditional approach to the synthesis of isoflavones is by ring closure of a suitable C_1 unit on to an appropriate deoxybenzoin. A wide variety of C_1 reagents is now available, and many of these are still routinely employed. Ethoxalyl chloride, triethyl orthoformate, ethyl formate and dimethyl formamide under various conditions all make quite efficient C_1 sources, though protection of functional groups other than the hydroxyl required to make the pyrone ring may be necessary (see Dewick, 1982). Two further reagents have recently been described. 1,3,5-Triazine in the presence of boron trifluoride–etherate, acetic anhydride and acetic acid gives very good yields of isoflavones from deoxybenzoins, is effective with many substitution patterns, and does not require hydroxyls to be protected in either ring (Jha *et al.*, 1981). Acetoformic anhydride (HCO_2Ac) in the presence of either sodium formate or triethylamine has also been used to synthesize 7-hydroxy- and 5,7-dihydroxyisoflavones (Pivovarenko and Khilya, 1985; Pivovarenko *et al.*, 1985).

More frequently, isoflavone syntheses are achieved by oxidative conversion of chalcones. Chalcones are readily obtained by condensation of acetophenones and aromatic aldehydes, and are thus more accessible than deoxybenzoins, particularly if complex substitution patterns are required. The boron trifluoride-catalysed rearrangement of chalcone epoxides is still used, but product yields are often poor. An interesting extension of this type of conversion has been observed (Van der Westhuizen *et al.*, 1980). Thus, reaction of a chalcone epoxide (*5.55*) under ambient conditions with 2,4,6-trihydroxybenzoic acid gave the β-ester (*5.56*) which was then methylated (Fig. 5.1). Acid treatment of this compound (*5.57*) under anhydrous conditions yielded the isoflavone (*5.60*) as the sole product in about 50% yield. The reaction is suggested to proceed via a 1,2-aroyl $O \rightarrow O$ shift giving the α-ester (*5.58*), followed by cyclization and an aryl migration. The unesterified flavanonol corresponding to (*5.59*) is not convertible into the isoflavone under these conditions. Whilst the general applicability of this rearrangement has not been investigated, the mild conditions involved make this approach potentially useful. The biogenetic-type aryl migration observed in this reaction contrasts with the alternative aroyl migration found in the BF_3-catalysed rearrangement of chalcone epoxides.

The thallium nitrate oxidation of 2′-hydroxychalcones in methanol or trimethyl orthoformate is now established as the most satisfactory route to isoflavones. The intermediate acetal arising via an aryl migration mechanism may be transformed into the isoflavone by either acid or base treatment, thus allowing considerable flexibility when acid-sensitive groupings are present. In most cases relating to natural isoflavone synthesis, the migrating aryl group (isoflavone B-ring) is sufficiently activated so that competing rearrangements do not interfere. With moderately activated and deactivated aryls, it is possible to observe rearrangements in which the other aryl (isoflavone A-ring) migrates. A systematic study of substituents and conditions favouring the various processes has been conducted (Taylor *et al.*, 1980). The presence of nitro groups can seriously affect the migratory aptitude of an aryl group (Varma and Varma, 1982a, b), and unwanted nitro by-products may be formed because of nitric acid released during the reaction (Varma, 1982). Nevertheless, in the vast majority of cases, the thallium nitrate oxidation of chalcones gives most satisfactory yields of the required isoflavones.

Problems may arise if chromene rings are present in the starting chalcone. Thallium nitrate reacts with chromene double bonds, often resulting in ring contraction. This limits the reaction's applicability to many natural isoflavones unless adequate steps are taken to protect the chromene double bond. This may conveniently be achieved by carrying out the oxidation on the dihydropyranochalcone, then producing the chromene using DDQ (2,3-dichloro-5,6-dicyanobenzoquinone) oxidation (Jain, A.C. *et al.*, 1985; Suresh *et al.*, 1985; Tsukayama *et al.*, 1984). This is illustrated in Fig. 5.2 for the synthesis

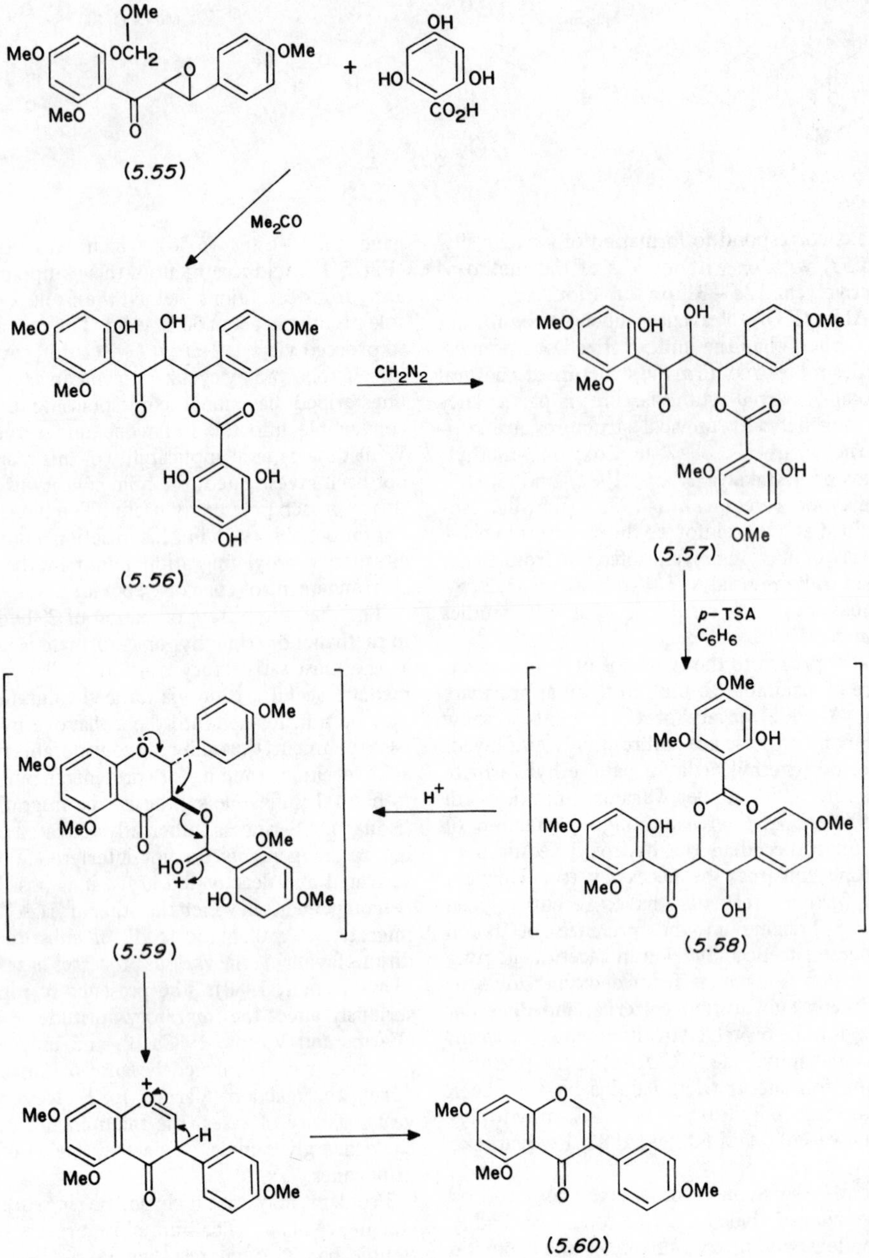

(5.55) + 2,4,6-trihydroxybenzoic acid

Me_2CO

(5.56)

CH_2N_2

(5.57)

p-TSA, C_6H_6

(5.58)

H^+

(5.59)

(5.60)

Fig. 5.1 Synthesis of genistein trimethyl ether.

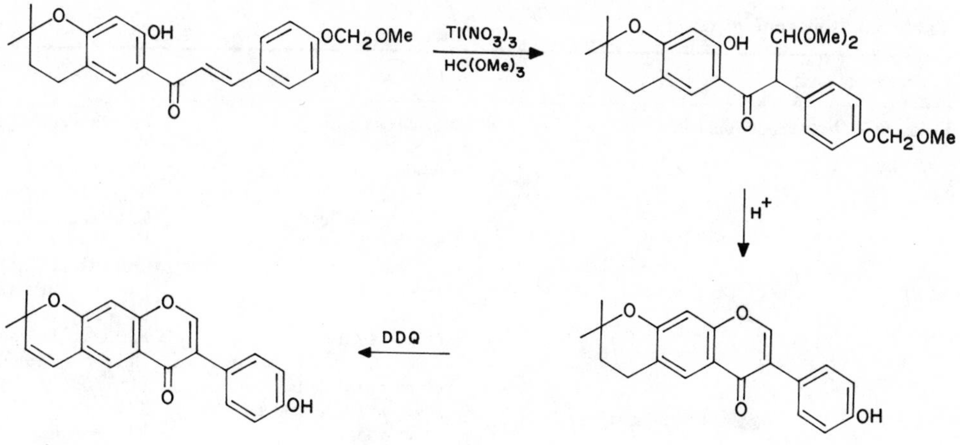

(5.61) Erythrinin A

Fig. 5.2 Synthesis of erythrinin A (*5.61*).

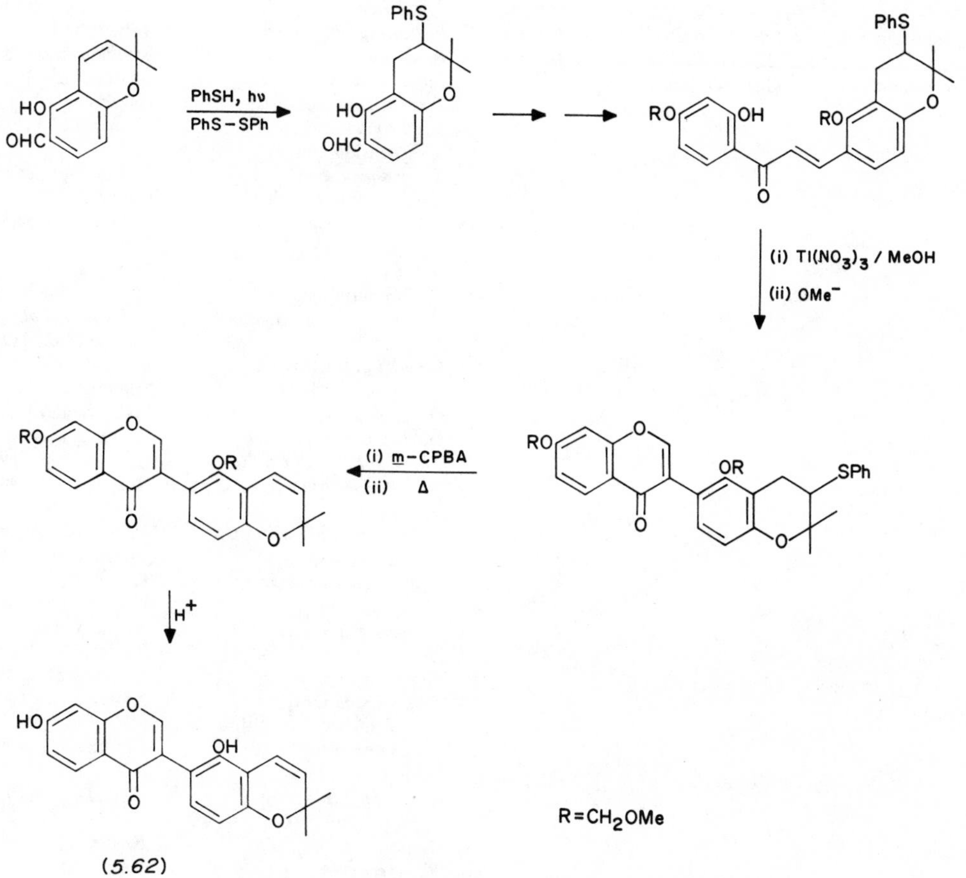

(5.62)

Fig. 5.3 Synthesis of isoflavone (*5.62*).

Table 5.3 Isoflavone glycosides

Isoflavone glycoside	*Plant sources*	*References*
Daidzein 7-*O*-glucoside (daidzin)	*Erythrina crista-galli* bark, trunk wood	Imamura *et al.* (1981)
	Genista rumelica aerial parts	Nakov (1983)
	Glycine max leaf*	Cosio *et al.* (1985)
		Fett (1984)
		Osman and Fett (1983)
	G. max aerial parts	Kovalev and Seraya (1984)
	G. max root	Le-Van (1984)
	G. max seed	Farmakalidas and Murphy (1984)
		Ohta *et al.* (1980)
	G. max seed flour	Eldridge (1982b)
	Pueraria lobata callus	Takeya and Itokawa (1982)
	P. sp. root	Kitada *et al.* (1985)
	Thermopsis fabacea root	Arisawa *et al.* (1980)
	Vigna angularis suspension culture*	Hattori and Ohta (1985)
		Kobayashi and Ohta (1983)
Daidzein 6″-*O*-acetate	*Glycine max* seed	Ohta *et al.* (1980)
Daidzein 7, 4′-di-*O*-glucoside	*Vigna angularis* suspension culture*	Hattori and Ohta (1985)
		Kobayashi and Ohta (1983)
Daidzein 8-*C*-glucoside (puerarin)	*Pueraria lobata* callus	Takeya and Itokawa (1982)
	P. sp. root	Kitada *et al.* (1985)
Formononetin 7-*O*-glucoside (ononin)	*Baptisia australis* root	Köster *et al.* (1983a)
	Cicer arietinum leaf	Kazakov *et al.* (1980)
	C. arietinum root, stem, suspension culture	Köster *et al.* (1983a)
	C. arietinum seed*	Köster *et al.* (1983b)
	Genista patula aerial parts	Ozimina (1981)
	Glycine max leaf*	Cosio *et al.* (1985)
		Fett (1984)
		Osman and Fett (1983)
	G. max aerial parts	Kovalev and Seraya (1984)
	Glycyrrhiza uralensis root	Nakanishi *et al.* (1985)
	Hedysarum polybotrys root	Miyase *et al.* (1984)
	Ononis sp. tincture	Kovalev (1983)
	O. arvensis herb	Spilkova and Hubik (1982)
	O. spinosa herb	Spilkova and Hubik (1982)
	O. spinosa root	Köster *et al.* (1983a)
	Thermopsis fabacea root	Arisawa *et al.* (1980)
	Trifolium incarnatum root, stem	Köster *et al.* (1983a)
	T. polyphyllum whole plant	Luk'yanchikov and Kazakov (1982)
	T. pratense root	Fraishtat *et al.* (1980)
	T. pratense root, stem	Köster *et al.* (1983a)
	T. repens	Luk'yanchikov and Kazakov (1983)
	T. subterraneum leaf	Nicollier and Thompson (1982)
Formononetin 7-*O*-(6″-malonylglucoside)	*Baptisia australis* root	Köster *et al.* (1983a)
	Cicer arietinum root	Jaques *et al.* (1985)
	C. arietinum root, stem, cell suspension culture	Köster *et al.* (1983a)
	C. arietinum seed*	Köster *et al.* (1983b)
	Ononis spinosa root	Köster *et al.* (1983a)
	Trifolium incarnatum root, stem	Köster *et al.* (1983a)
	T. pratense root, stem	Köster *et al.* (1983a)
	T. repens root	Köster *et al.* (1983a)
2′-Hydroxydaidzein 7, 4′-di-*O*-glucoside	*Vigna angularis* suspension culture*	Hattori and Ohta (1985)
		Kobayashi and Ohta (1983)

Table 5.3 (*Contd.*)

Isoflavone glycoside	*Plant sources*	*References*
Calycosin 7-*O*-glucoside	*Astragalus mongholicus*	Lu *et al.* (1984)
Pseudobaptigenin 7-*O*-glucoside (rothindin)	*Rothia trifoliata* whole plant	Rao and Rao (1985)
	Trifolium pratense root	Fraishtat *et al.* (1980)
7-Hydroxy-2′, 4′, 5′-trimethoxyisoflavone 7-*O*-glucoside	*Dalbergia monetaria* seed	Abe *et al.* (1985)
Genistein 7-*O*-glucoside (genistin)	*Erythrina crista-galli* bark, trunk wood	Imamura *et al.* (1981)
	Flemingia stricta root	Rao *et al.* (1982)
	Genista rumelica aerial parts	Nakov (1983)
	Glycine max leaf*	Fett (1984)
		Osman and Fett (1983)
	G. max aerial parts	Kovalev and Seraya (1984)
	G. max root	Le-Van (1984)
	G. max seed	Farmakalidas and Murphy (1984)
		Ohta *et al.* (1980)
	G. max seed flour	Eldridge (1982b)
	G. spp. leaf	Vaughan and Hymowitz (1984)
	Lupinus albus root, leaf	Laman and Oleksyuk (1982)
	L. perennis root, leaf	Laman and Oleksyuk (1982)
	L. polyphyllus leaf	Kolar (1981)
	Thermopsis fabacea root	Arisawa *et al.* (1980)
	Trifolium pratense root	Fraishtat *et al.* (1980)
	T. subterraneum leaf	Nicollier and Thompson (1982)
Genistin 6″-*O*-acetate	*Glycine max* seed	Ohta *et al.* (1980)
Genistein 4′-*O*-glucoside (sophoricoside)	*Lupinus perennis* root	Laman and Oleksyuk (1982)
	Sophora japonica	Ho *et al.* (1982)
Genistein 4′-*O*-neohesperidoside (sophorabioside)	*Sophora japonica*	Ho *et al.* (1982)
Genistein 7, 4′-di-*O*-glucoside	*Lupinus perennis* root, leaf	Laman and Oleksyuk (1982)
Genistein 7, 4′-di-*O*-glucosylapioside (sarothamnoside)	*Cytisus scoparius* (*Sarothamnus scoparius*) seed	Brum-Bousquet *et al.* (1981a, b)
	C. striatus (*S. patens*) seed	Brum-Bousquet *et al.* (1981a, b)
Genistein 8-*C*-glucoside	*Dalbergia nitidula* bark	Van Heerden *et al.* (1980)
	Lupinus luteus flower	Laman and Oleksyuk (1982)
Genistein 6, 8-di-*C*-glucoside	*Dalbergia nitidula* bark	Van Heerden *et al.* (1980)
Biochanin A 7-*O*-glucoside (sissotrin)	*Cicer arietinum* leaf	Kazakov *et al.* (1980)
	C. arietinum root, stem, suspension culture	Köster *et al.* (1983a)
	C. arietinum seed*	Köster *et al.* (1983b)
	Ononis spinosa root	Köster *et al.* (1983a)
	Trifolium incarnatum root	Köster *et al.* (1983a)
	T. pratense root, stem	Köster *et al.* (1983a)
	T. repens	Luk'yanchikov and Kazakov (1983)
	T. subterraneum leaf	Nicollier and Thompson (1982)
Biochanin A 7-*O*-(6″-malonylglucoside)	*Baptisia australis* root	Köster *et al.* (1983a)
	Cicer arietinum root	Jaques *et al.* (1985)
	C. arietinum root, stem, cell suspension culture	Köster *et al.* (1983a)
	C. arietinum seed*	Köster *et al.* (1983b)
	Ononis spinosa root	Köster *et al.* (1983a)
	Trifolium incarnatum root	Köster *et al.* (1983a)
	T. pratense root, stem	Köster *et al.* (1983a)
	T. repens root	Köster *et al.* (1983a)
Biochanin A 7-*O*-xylosylglucoside (kakkanin)	*Pueraria thunbergiana* flower	Kikuchi (1982)
Orobol 7-*O*-glucoside	*Bryum capillare* (Bryales) gametophyte	Anhut *et al.* (1984)
	Lupinus albus root	Laman and Oleksyuk (1982)
	L. perennis root	Laman and Oleksyuk (1982)

(*Contd.*)

Table 5.3 (*Contd.*)

Isoflavone glycoside	*Plant sources*	*References*
Orobol 7-*O*-(6″-malonylglucoside)	*Bryum capillare* (Bryales) gametophyte	Anhut *et al.* (1984)
Orobol 7-*O*-sophoroside (compactin)	*Genista compacta* aerial parts	Ozimina and Bandyukova (1985)
Orobol 8-*C*-glucoside	*Dalbergia nitudula* bark	Van Heerden *et al.* (1980)
Orobol 6, 8-di-*C*-glucoside	*Dalbergia nitudula* bark	Van Heerden *et al.* (1980)
3′-*O*-Methylorobol 7-*O*-glucoside	*Cytisus scoparius* flower	Viscardi *et al.* (1984)
Pratensein 7-*O*-glucoside	*Bryum capillare* (Bryales) gametophyte	Anhut *et al.* (1984)
	Iris germanica (Iridaceae) rhizome	Ali *et al.* (1983)
Pratensein 7-*O*-(6″-malonylglucoside)	*Bryum capillare* (Bryales) gametophyte	Anhut *et al.* (1984)
5, 7-Dihydroxy-3′, 4′-methylenedioxyisoflavone 7-*O*-glucoside	*Lupinus albus* root	Laman and Oleksyuk (1982)
	L. perennis root	Laman and Oleksyuk (1982)
5, 7-Dihydroxy-3′, 4′-methylenedioxyisoflavone 7-*O*-glucosylglucoside	*Lupinus perennis* root	Laman and Oleksyuk (1982)
Glycitein 7-*O*-glucoside	*Glycine max* seed flour	Eldridge (1982b)
Retusin 7-*O*-glucoside	*Pterocarpus marsupium* heartwood	Mitra and Joshi (1983)
Retusin 7-*O*-neohesperidoside	*Prosopis juliflora* bark	Shukla and Misra (1981)
5, 7-Dihydroxy-6-methoxyisoflavone 7-*O*-rhamnoside[†]	*Pterocarpus marsupium* heartwood	Mitra and Joshi (1983)
Tectorigenin 7-*O*-glucoside (tectoridin)	*Iris milesii* (Iridaceae) rhizome	Agarwal *et al.* (1984a)
Irisolidone 7-*O*-glucoside	*Iris germanica* (Iridaceae) rhizome	Ali *et al.* (1983)
Irisolidone 7-*O*-rhamnoside	*Pterocarpus marsupium* heartwood	Mitra and Joshi (1983)
Iristectorigenin B 7 (or 4′)-*O*-glucoside (iristectorin B)	*Iris milesii* (Iridaceae) rhizome	Agarwal *et al.* (1984a)
Iristectorigenin A 7-*O*-glucoside (iristectorin A)	*Iris spuria* (Iridaceae) rhizome	Shawl *et al.* (1984)
Iristectorigenin A 7-*O*-glucosylglucoside	*Juniperus macropoda* (Cupressaceae) leaf	Sethi *et al.* (1983)
Junipegenin B 7-*O*-glucoside	*Juniperus macropoda* (Cupressaceae) leaf	Sethi *et al.* (1983)
Dalspinin 7-*O*-galactoside	*Dalbergia spinosa* root	Dasan *et al.* (1985)
Irigenin 7-*O*-glucoside (iridin)	*Iris germanica* (Iridaceae) rhizome	Ali *et al.* (1983)
		El-Moghazy *et al.* (1980)
	I. hookeriana rhizome	Shawl *et al.* (1985)
	I. kumaonensis rhizome	Agarwal *et al.* (1984b)
	I. milesii rhizome	Agarwal *et al.* (1984a)
5, 7, 4′-Trihydroxy-6, 3′, 5′-trimethoxyisoflavone 7-*O*-glucoside[†]	*Juniperus macropoda* (Cupressaceae) leaf	Sethi *et al.* (1983)
7-Hydroxy-5, 4′-dimethoxy-8-methylisoflavone 7-*O*-rhamnoside[†]	*Pterocarpus marsupium* heartwood	Mitra and Joshi (1982)
		Mitra and Joshi (1983)
5-Hydroxy-7, 3′, 4′-trimethoxy-8-methylisoflavone 5-*O*-neohesperidoside[†]	*Dolichos biflorus* seed	Mitra *et al.* (1983)

*Plant part was subjected to physiological stress.
[†]Newly reported aglycone.

of erythrinin A (*5.61*) (Suresh *et al.*, 1985). An alternative protective sequence has been employed in the synthesis of the isoflavone (*5.62*) (Fig. 5.3), an intermediate used for the synthesis of the pterocarpan phaseollin (Thomas and Whiting, 1984). In this route, the chromene was protected by radical addition of thiophenol. For deprotection, the thiochroman was oxidized to the corresponding sulphoxide which then gave the chromene in good yield by thermolysis in refluxing toluene.

Selective alkylation and dealkylation reactions are particularly valuable in the synthesis of many isoflavonoid structures, and it is always appropriate to highlight such procedures when they appear in published synthetic sequences. The increased acidity of the 7-hydroxyl relative to hydroxyls in other positions in isoflavones has allowed selective 7-*O*-alkylation or 7-esterification and has been exploited on many occasions. Selective *O*-demethylations at position 2′ using aluminium chloride in acetonitrile have also frequently been employed. This latter reaction appears fairly general, except where additional 3′-methoxyls are present. Under such conditions, 2′, 3′-didemethylation occurs, possibly via stepwise selective demethylations (Al-Ani and Dewick, 1985). With shorter reaction times, the single 2′-demethylation can be obtained (Bhardwaj *et al.*, 1982; Al-Ani and Dewick, 1985). 5-*O*-Benzyl groups may also be removed selectively by heating with hydrochloric acid in acetic acid (Tsukayama *et al.*, 1982, 1985). This reaction can also be applied to isoflavanones (Malik and Grover, 1980), and again the reaction time is important and must be kept very short to achieve the desired selectivity. Catalytic hydrogenolysis is perhaps the cleanest method for removing all *O*-benzyl groups in an isoflavone, but reduction of other functional groups could complicate the deprotection process. A simple, convenient alternative to catalytic hydrogenolysis is catalytic transfer hydrogenation using $Pd(OH)_2$–C and cyclohexene. This debenzylates benzyloxyisoflavones cleanly without the use of gaseous hydrogen in a simple reflux reaction, and can be used with hydrogen-sensitive isoflavonoids, e.g. isoflavanones and pterocarpans (Al-Ani and Dewick, 1984, 1985).

The number of known isoflavonoid glycosides is extremely small compared to the vast range of known flavonoid glycosides. Isoflavone glycosides account for the majority of these, and new reports are listed in Table 5.3. *O*-Glycosides predominate, but several *C*-glycosides are now recognized, including the 8-*C*-glucosides and 6, 8-di-*C*-glucosides of genistein and orobol from *Dalbergia nitidula* bark (Van Heerden *et al.*, 1980). The unusual 5, 7-dihydroxy-6-methoxyisoflavone mentioned earlier was isolated as its 7-*O*-rhamnoside from *Pterocarpus marsupium* heartwood (Mitra and Joshi, 1983), and this source has also yielded the 7-*O*-rhamnoside of 7-hydroxy-5, 4′-dimethoxy-8-*C*-methylisoflavone (Mitra and Joshi, 1982). A further 8-*C*-methyl derivative, 5-hydroxy-7, 3′, 4′-trimethoxy-8-*C*-methylisoflavone has been reported in seeds of *Dolichos biflorus* (Mitra *et al.*, 1983), where it exists as its 5-*O*-neohesperidoside. The isoflavanone ougenin was previously the only known *C*-methylated isoflavonoid. Malonate esters of isoflavone glucosides have been isolated from several sources. The identification of orobol and pratensein 7-*O*-(6″-malonylglucosides) in *Bryum capillare* (Bryales) represents the first report of isoflavonoids in a bryophyte (moss) (Anhut *et al.*, 1984). The 7-*O*-glucosides co-occur, but in much smaller amounts, and aglycones were detected in the dried material, though not in the fresh moss.

Isoflavones, along with isoflavanones, isoflavans and pterocarpans, are frequently reported as stress metabolites from plant tissues challenged with fungi, bacteria or abiotic agents. There is now evidence that isoflavone glycosides are also synthesized as part of the hypersensitive reaction. Suspension cultures of *Vigna angularis* stressed with actinomycin D accumulate daidzin (daidzein 7-*O*-glucoside), daidzein-7, 4′-di-*O*-glucoside and 2′-hydroxydaidzein-7, 4′-di-*O*-glucoside (Kobayashi and Ohta, 1983), and leaves of soybean (*Glycine max*) on treatment with the herbicide acifluorfen synthesize daidzin and ononin (formononetin 7-*O*-glucoside) (Cosio *et al.*, 1985). In the latter studies, the accumulation of isoflavone glycosides was correlated with increased UDP-glucose:isoflavone 7-*O*-glucosyltransferase activity.

5.4 ISOFLAVANONES

Isoflavanones (*5.63*) are much rarer than isoflavones, but new structures continue to be reported (Table 5.4). Most of the new compounds have been isolated via phytoalexin studies, since many isoflavanones have anti-fungal activity, and are also biosynthetic intermediates on the way to other phytoalexins such as pterocarpans and isoflavans. Treatment of *Phaseolus mungo* seedlings with aqueous cupric chloride resulted in the isolation of sixteen isoflavonoids, including ten isoflavanones (Adesanya *et al.*, 1984).

The only new isoflavanone reported that contains novel structural features is bolusanthin (*5.64*), isolated from seeds of *Bolusanthus speciosus* (Asres *et al.*, 1985). This has been identified as a second 3-hydroxyisoflavanone, secondifloran from *Sophora secondiflora* being the only other known natural example.

Several new syntheses of isoflavanones have been described. Since established synthetic routes to this class of isoflavonoid have, in general, been relatively inefficient, these new procedures are welcomed. As with isoflavones, the approach has been to insert a methylene group on to a deoxybenzoin skeleton. Methylene iodide has been employed as a C_1 source but yields are poor, and the reaction leads to a variety of unwanted by-products. The nature of these materials has been established (Malik and Grover, 1980). This reaction can be improved significantly by using a two-phase system with a phase-transfer catalyst

Table 5.4 Isoflavanones

Isoflavanone	*Substituents* 5	6	7	8	2′	3′	4′	5′	6′	*Plant sources*	*Optical activity*	*References*
Dihydroformononetin			OH				OMe			*Zollernia paraensis* wood	±	Ferrari *et al.* (1983)
										Phaseolus coccineus seedling*		Adesanya *et al.* (1985)
2′-Hydroxydihydrodaidzein			OH		OH		OH			*P. mungo* seedling*		Adesanya *et al.* (1984)
Vestitone			OH		OH		OMe			*Medicago rugosa* leaf*	±	Ingham (1982b)
										Trifolium repens leaf*		Woodward (1981a)
Isosativanone			OMe		OH		OMe			*Medicago rugosa* leaf*	±	Ingham (1982b)
Lespedeol C			OH		OMe	OMe	OH			*Lespedeza cyrtobotrya* heartwood	±	Miyase *et al.* (1981)
Onogenin			OH		OMe			OCH_2O		*Ononis* sp. tincture		Kovalev (1983)
										O. arvensis herb		Spilkova and Hubik (1982)
										O. spinosa herb		Spilkova and Hubik (1982)
Onoside			OGlc		OMe			OCH_2O		*Ononis* sp. tincture		Kovalev (1983)
Dalbergioidin	OH		OH		OH		OH			*Desmodium gangeticum* leaf*		Ingham and Dewick (1984)
										Dolichos biflorus leaf*		Keen and Ingham (1980)
										Ougeinia dalbergioides heartwood		Kalidhar and Sharma (1984)
										Phaseolus aureus seedling*		O'Neill *et al.* (1983)
										P. mungo seedling*		Adesanya *et al.* (1984)
Isoferreirin	OH		OH		OMe		OH			*Dolichos biflorus* leaf*		Keen and Ingham (1980)
										Phaseolus coccineus seedling*		Adesanya *et al.* (1985)
										P. mungo seedling*		Adesanya *et al.* (1984)
Ferreirin	OH		OH		OH		OMe			*Diphysa robinioides* leaf*		Ingham and Tahara (1983)
										Ougeinia dalbergioides heartwood		Kalidhar and Sharma (1984)
Homoferreirin	OH		OH		OMe		OMe			*Ougeinia dalbergioides* heartwood		Kalidhar and Sharma (1984)
7, 4′-Dihydroxy-5, 2′-dimethoxy-	OMe		OH		OMe		OH			*Zollernia paraensis* wood	±	Ferrari *et al.* (1984a)
										Phaseolus coccineus seedling*		Adesanya *et al.* (1985)
Cajanol	OH		OMe		OMe		OH			*Cajanus cajan* seed*		Dahiya *et al.* (1984)
Bolusanthin (*5.64*)										*Bolusanthus speciosus* seed		Asres *et al.* (1985)
5-Deoxykievitone (*5.65*)										*Phaseolus aureus* seedling*		O'Neill *et al.* (1983)
										P. mungo seedling*		Adesanya *et al.* (1984)
5-Deoxykievitone hydrate (*5.66*)										*P. mungo* seedling*		Adesanya *et al.* (1984)
Ougenin (*5.67*)										*Ougeinia dalbergioides* heartwood		Kalidhar and Sharma (1984)

Diphysolone (*5.68*)	*Desmodium gangeticum* leaf*		Ingham and Dewick (1984)
	Diphysa robinioides leaf*		Ingham and Tahara (1983)
Diphysolidone (*5.69*)	*Diphysa robinioides* leaf*		Ingham and Tahara (1983)
Kievitone (*5.70*)	*Desmodium gangeticum* leaf*		Ingham and Dewick (1984)
	Diphysa robinioides leaf*		Ingham and Tahara (1983)
	Dolichos biflorus leaf*		Keen and Ingham (1980)
	Phaseolus aureus seedling*		O'Neill *et al.* (1983)
	P. calcaratus seed*, leaf*		Sukumaran and Gnanamanickam (1980)
	P. coccineus seedling*		Adesanya *et al.* (1985)
	P. mungo seedling*		Adesanya *et al.* (1984)
	P. vulgaris cotyledon*		Bailey and Berthier (1981)
			Goosens and Van Laere (1983)
			Whitehead *et al.* (1982)
	P. vulgaris hypocotyl*		Garcia-Arenal *et al.* (1980)
	P. vulgaris various tissues*		Stössel and Magnolata (1983)
4′-*O*-Methylkievitone (*5.71*)	*Phaseolus mungo* seedling*		Adesanya *et al.* (1984)
Kievitone hydrate (*5.72*)	*Phaseolus mungo* seedling*		Adesanya *et al.* (1984)
Cyclokievitone (*5.73*)	*Phaseolus aureus* seedling*		O'Neill *et al.* (1983)
	P. coccineus seedling*		Adesanya *et al.* (1985)
	P. mungo seedling*		Adesanya *et al.* (1984)
Cyclokievitone hydrate (*5.74*)	*Phaseolus mungo* seedling*		Adesanya *et al.* (1984)
Isosophoranone (*5.75*)	*Sophora tomentosa* root		Shirataki *et al.* (1983)
Sophoraisoflavanone B (*5.76*)	*Sophora franchetiana* root	±	Komatsu *et al.* (1981a)

*Plant part was subjected to physiological stress.

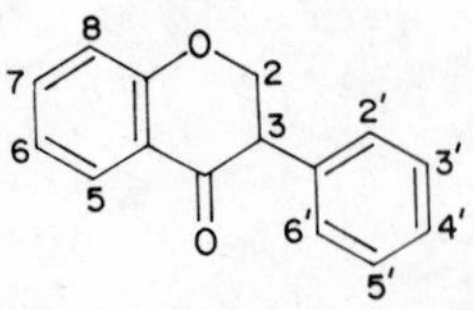

(*5.63*)

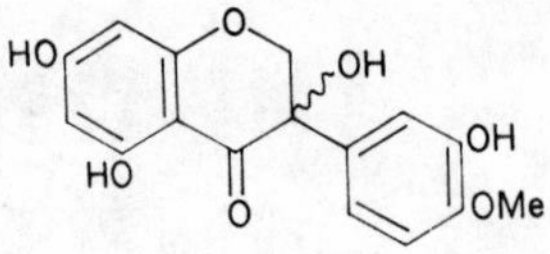

(*5.64*) Bolusanthin

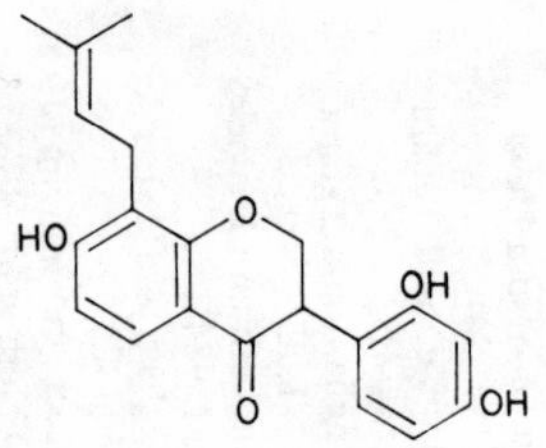

(*5.65*) 5-Deoxykievitone

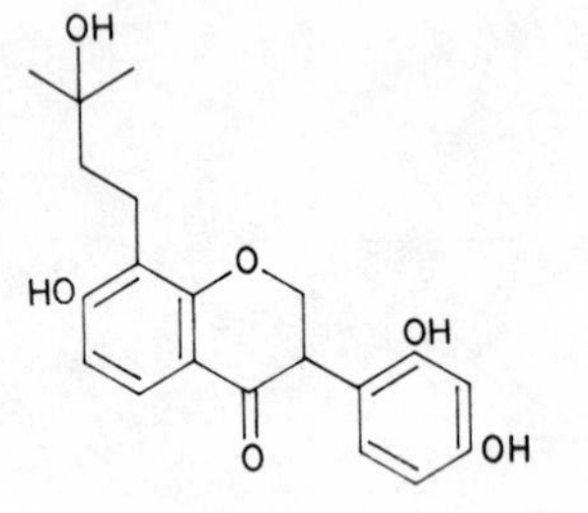

(*5.66*) 5-Deoxykievitone hydrate

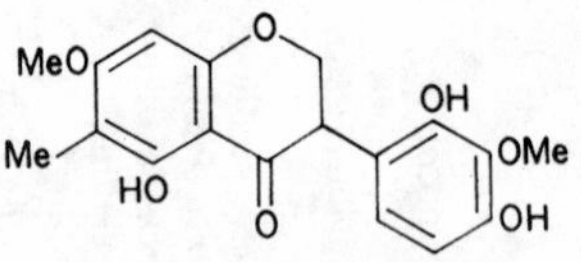

(*5.67*) Ougenin

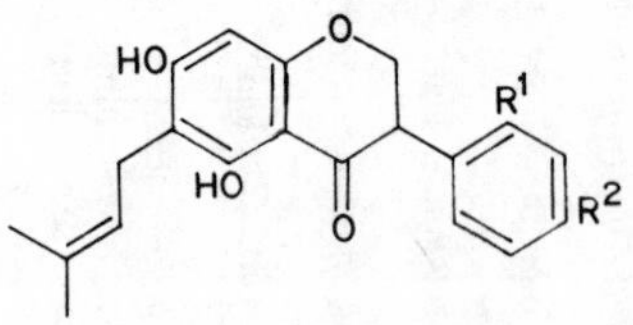

(*5.68*) $R^1 = R^2 = OH$, Diphysolone

(*5.69*) $R^1 = OMe, R^2 = OH$, or $R^1 = OH, R^2 = OMe$, Diphysolidone

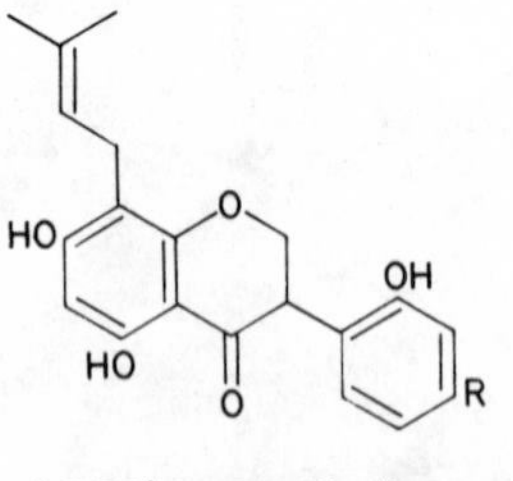

(*5.70*) R = OH, Kievitone

(*5.71*) R = OMe, 4′-*O*-Methylkievitone

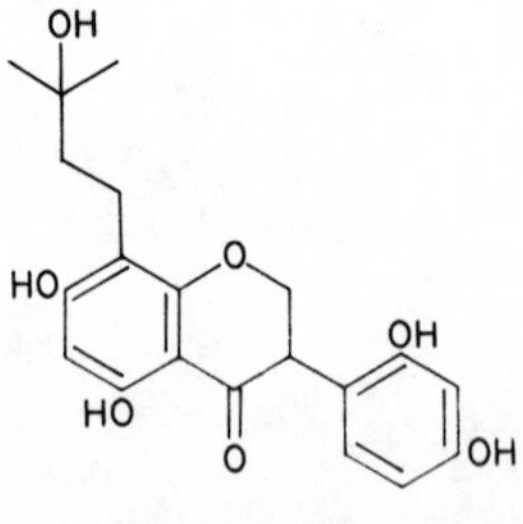

(*5.72*) Kievitone hydrate

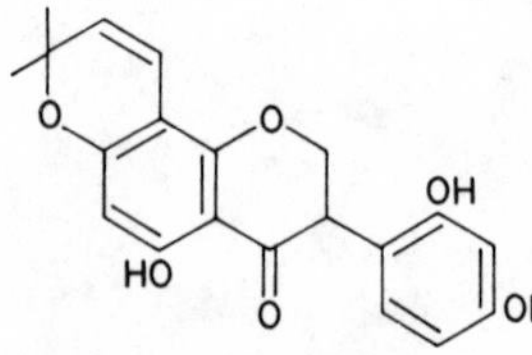

(*5.73*) Cyclokievitone

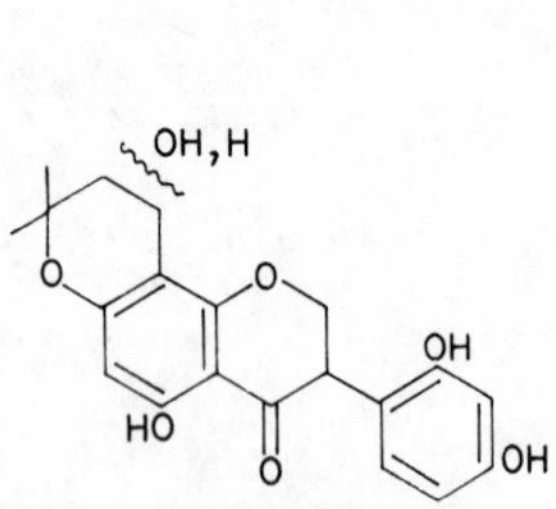

(*5.74*) Cyclokievitone hydrate

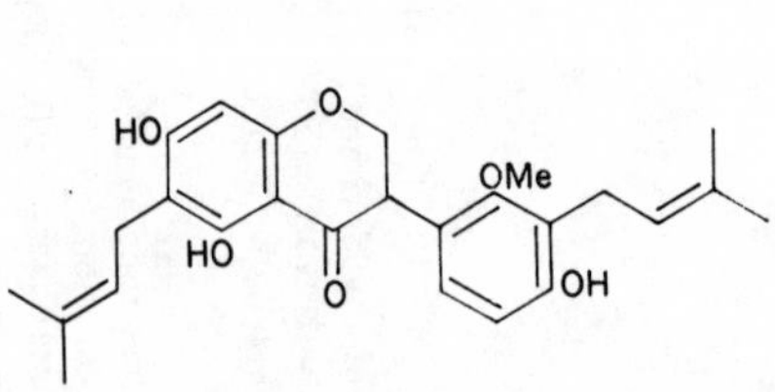

(*5.75*) Isosophoranone

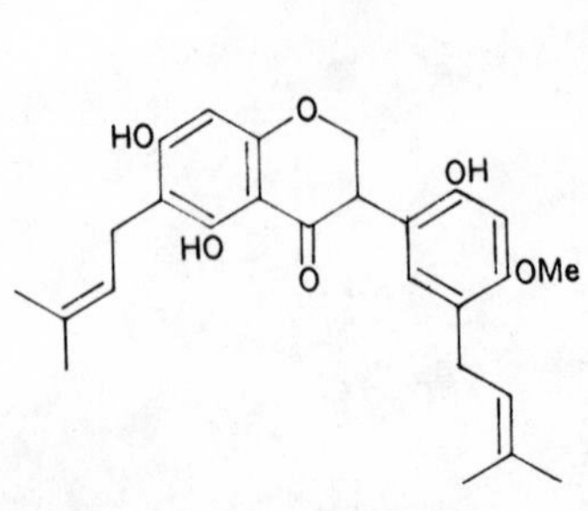

(*5.76*) Sophoraisoflavanone B

(tetra-*n*-butylammonium iodide) instead of the original potassium carbonate – dry acetone conditions. The inclusion of sodium thiosulphate in the reaction mixture traps any iodine formed which would react with the deoxybenzoin to give the unwanted by-products. The overall result is that yields of 60–70% may be obtained (Singh *et al.*, 1982). Hydroxyls must still be protected, however. This comment does not apply to another method using ethoxymethyl chloride as the C_1 source (Jain and Sharma, 1984, 1985), since this reagent also provides *in situ* protection for hydroxyls except the hydrogen-bonded one. The intermediate α-hydroxymethyl derivative (*5.77*) is then cyclized with mild base, and protecting groups removed by treatment with acid (Fig. 5.4). Overall yields of about 50–60% are reported. Formaldehyde, in a two-phase system, similarly gives α-hydroxymethyl deoxybenzoins, which may be cyclized to isoflavanones by treatment with base, giving overall yields of up to 80% (Jain, P.K. *et al.*, 1985). If a phase-transfer catalyst (tetra-*n*-butylammonium hydrogen sulphate) is used,

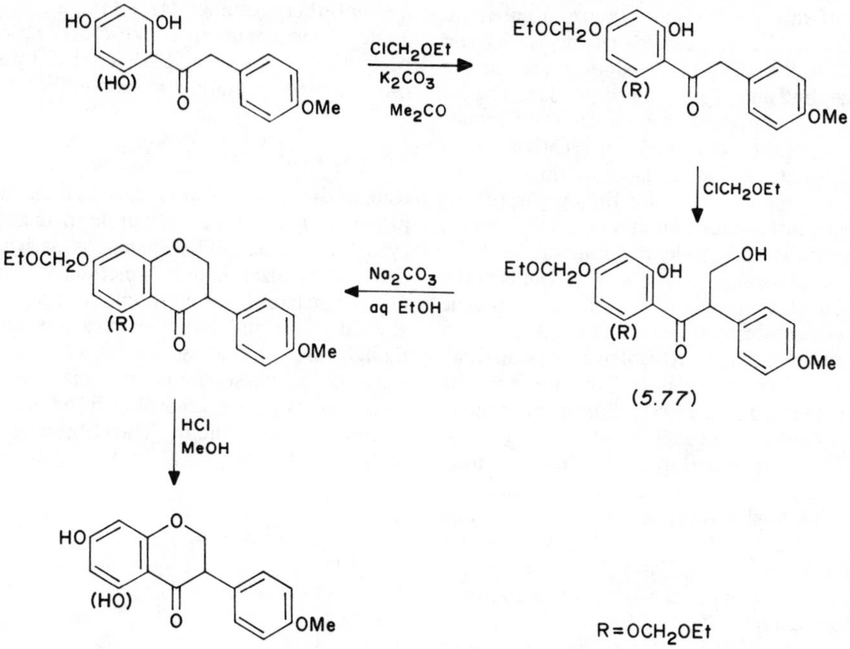

$ClCH_2OEt$, K_2CO_3, Me_2CO; $ClCH_2OEt$; Na_2CO_3, aq EtOH; HCl, MeOH

(*5.77*)

R = OCH_2OEt

Fig. 5.4 Synthesis of dihydroformononetin and dihydrobiochanin A.

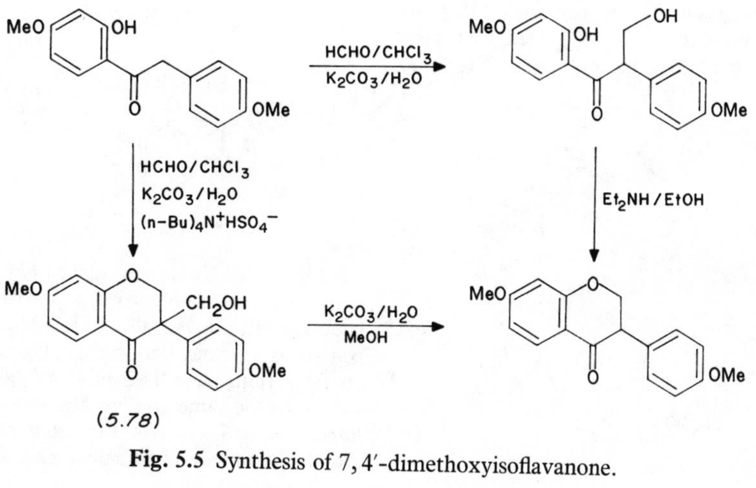

$HCHO/CHCl_3$, K_2CO_3/H_2O; $HCHO/CHCl_3$, K_2CO_3/H_2O, $(n\text{-}Bu)_4N^+HSO_4^-$; $Et_2NH/EtOH$; K_2CO_3/H_2O, MeOH

(*5.78*)

Fig. 5.5 Synthesis of 7, 4′-dimethoxyisoflavanone.

intermediate 3-hydroxymethylisoflavanones (*5.78*) are formed instead, though these also give the required isoflavanone on base treatment (Fig. 5.5). The phase-transfer catalyst must be used for successful synthesis of isoflavanones in the phloroglucinol series, and hydroxyls need protection. Protected deoxybenzoins may also be converted into isoflavanones by the use of paraformaldehyde–diethylamine in an efficient one-pot synthesis (Gandhidasan *et al.*, 1982; Pinkey *et al.*, 1984).

The use of deoxybenzoins for isoflavanone synthesis has the same limitations as indicated for isoflavone synthesis, i.e. production of the required deoxybenzoin substrate may be difficult or inefficient. In many instances, catalytic hydrogenation of isoflavones to isoflavanones may be the method of choice, but since the required product may be reduced further, eventually to the isoflavan, such reductions have to be monitored closely. An alternative, selective reducing agent, di-isobutylaluminium hydride (DIBAH) has now been employed for the preparation of isoflavanones from isoflavones (Antus *et al.*, 1981). This reagent in dry toluene–tetrahydrofuran at $-65°C$ smoothly reduces isoflavones giving 75–90% yields of the isoflavanones. Benzyloxy protecting groups and chromene double bonds are not affected (Vermes *et al.*, 1983).

Isoflavans may be oxidized to isoflavanones in high yield, using DDQ in methanol solution, under a nitrogen atmosphere (Breytenbach *et al.*, 1981). For this reaction to be successful, 2′-hydroxyl groups must be absent, or protected, otherwise cyclization to pterocarpans occurs.

A highly efficient dehydrogenating mixture comprising iodine–dimethyl sulphoxide–sulphuric acid has been used to dehydrogenate a range of flavonoid derivatives, including isoflavanones (Fatma *et al.*, 1984). Isoflavanones are converted into isoflavones via iodination at C-3 followed by dehydrohalogenation.

Trimethylsilylation of five isoflavanones prior to a GLC–MS study led to the formation of three different TMS ethers, except in the case of kievitone where two products were observed. On the basis of their MS fragmentation, these compounds were suggested to be the isoflavanone TMS ether (*5.79*), the enol ether (*5.80*), and

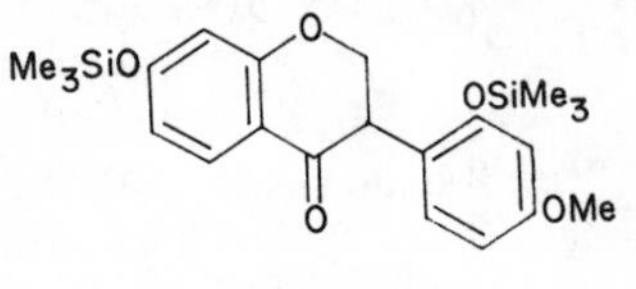

Me3SiO ... OSiMe3 ... OMe

(*5.79*)

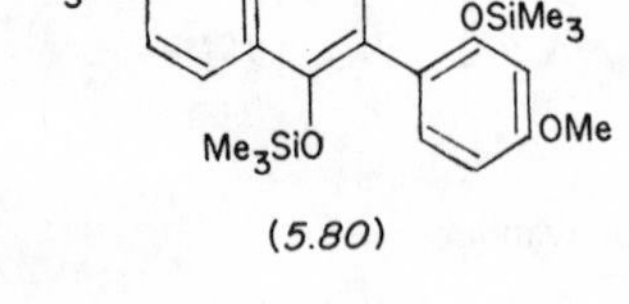

Me3SiO ... OSiMe3 ... Me3SiO ... OMe

(*5.80*)

Me3SiO ... OSiMe3 ... OSiMe3 ... OMe

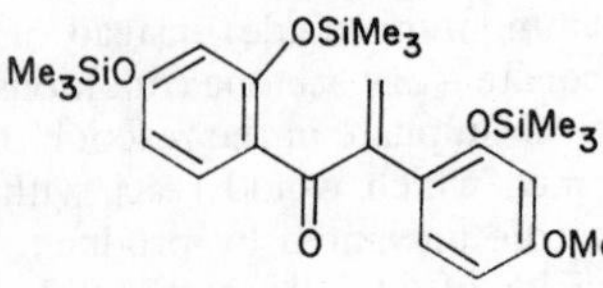

(*5.81*)

the ring-opened compound (*5.81*), using vestitone for illustration (Woodward, 1982). The isoflavanone showed a weak molecular ion, but prominent RDA fragmentation, the enol ether prominent $M-1$ fragments (as with isoflav-3-enes), and compounds like (*5.81*) an intense $M-15$ peak due to loss of a TMS methyl by ring formation involving the carbonyl and its *o*-OTMS neighbour.

5.5 ROTENOIDS

Rotenoids are a class of isoflavonoid characterized by the presence of an extra carbon atom in an additional heterocyclic ring (*5.82*). This system is derived in nature by oxidative cyclization of a 2′-methoxyisoflavone. Systematic nomenclature for the rotenoids has never been generally adopted, and trivial names are used throughout, though the numbering system of (*5.82*) is used. For convenience, these compounds may be subdivided into three major groups according to oxidation levels in the rotenoid ring system, and rotenoids (*5.83*), 12a-hydroxyrotenoids (*5.84*) and dehydrorotenoids (*5.85*)

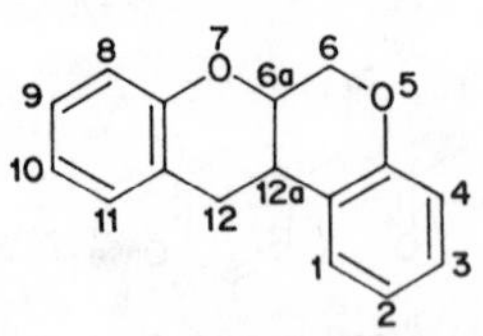

(*5.82*) 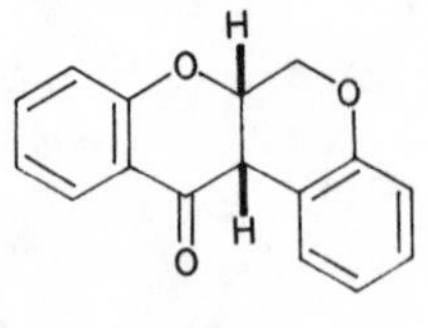 (*5.83*)

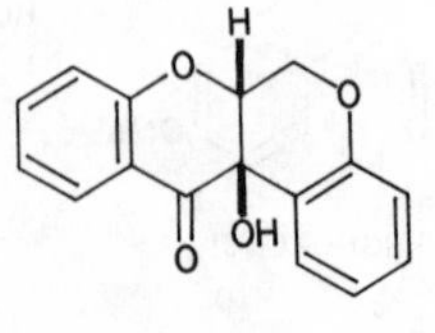

(*5.84*) 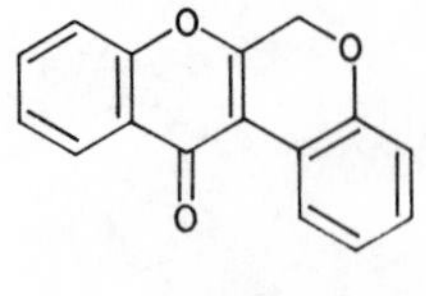 (*5.85*)

form the usual basis for classification. Other variations in oxidation level are regarded as derivatives of one of these subdivisions. Almost all of the known natural rotenoids contain isoprenoid-derived substituents. Where absolute configurations have been reported, all the natural rotenoids have the same configuration as established for rotenone, i.e. 6a*S*, 12a*S* in the main skeleton, and 5′*R* in dihydrofuran side-chain substituents (as appropriate). A

Table 5.5 Rotenoids

Rotenoid	*Plant sources*	*References*
(i) *Rotenoids*		
Rot-2′-enonic acid (*5.86*)	*Millettia pachycarpa* root	Singhal *et al.* (1982a)
Deguelin (*5.87*)	*Derris elliptica* callus	Komada *et al.* (1980)
	Lonchocarpus longifolius root	Braz Filho *et al.* (1980)
	L. salvadorensis seed	Birch *et al.* (1985)
	L. spruceanus root	Menichini *et al.* (1982)
	Piscidia erythrina root bark	Delle Monache *et al.* (1984)
	P. mollis seed	Menichini *et al.* (1982)
	Tephrosia strigosa aerial parts	Kamal and Jain (1980)
	T. sp. seed	Menichini *et al.* (1982)
	T. sp. root	Suarez *et al.* (1980)
Millettone (*5.88*)	*Piscidia erythrina* root bark	Delle Monache *et al.* (1984)
Elliptone (*5.89*)	*Lonchocarpus salvadorensis* seed	Birch *et al.* (1985)
	Tephrosia strigosa aerial parts	Kamal and Jain (1980)
Rotenone (*5.90*)	*Amorpha fruticosa* root bark	Hohmann *et al.* (1982)
	Derris elliptica callus	Komada *et al.* (1980)
	Lonchocarpus longifolius root	Braz Filho *et al.* (1980)
	L. salvadorensis seed	Birch *et al.* (1985)
	L. spruceanus root	Menichini *et al.* (1982)
	Millettia pachycarpa root	Singhal *et al.* (1982a)
	Piscidia erythrina root bark	Delle Monache *et al.* (1984)
	P. mollis seed	Menichini *et al.* (1982)
	Tephrosia strigosa aerial parts	Kamal and Jain (1980)
	T. villosa root	Chandrasekharan *et al.* (1983)
	T. sp. root	Suarez *et al.* (1980)
Isomillettone (*5.91*)	*Piscidia erythrina* root bark	Delle Monache *et al.* (1984)
3-*O*-Demethylamorphigenin (*5.92*)	*Amorpha fruticosa* fruit	Somleva and Ognyanov (1985)
Amorphigenin (*5.93*)	*Amorpha fruticosa* fruit	Somleva and Ognyanov (1985)
	A. fruticosa root bark	Hohmann *et al.* (1982)
	Dalbergia monetaria seed	Abe *et al.* (1985)
Amorphigenin *O*-glucoside (*5.94*)	*Amorpha fruticosa* fruit	Somleva and Ognyanov (1985)
	Dalbergia monetaria seed	Abe *et al.* (1985)
Amorphigenin *O*-vicianoside (amorphin) (*5.95*)	*Amorpha fruticosa* fruit	Somleva and Ognyanov (1985)
	A. fruticosa root bark	Hohmann *et al.* (1982)
Toxicarol (*5.96*)	*Lonchocarpus salvadorensis* seed	Birch *et al.* (1985)
(ii) *12a-Hydroxyrotenoids*		
12a-Hydroxyrot-2′-enonic acid (*5.97*)	*Millettia pachycarpa* root	Singhal *et al.* (1982a)
Tephrosin (12a-hydroxydeguelin) (*5.98*)	*Amorpha fruticosa* root	Somleva and Ognyanov (1985)
	Lonchocarpus longifolius root	Braz Filho *et al.* (1980)
	L. spruceanus root	Menichini *et al.* (1982)
	Piscidia mollis seed	Menichini *et al.* (1982)
	Tephrosia elata root	Lwande *et al.* (1985a)
	T. sp. seed	Menichini *et al.* (1982)
	T. sp. root	Suarez *et al.* (1980)
12a-Hydroxyrotenone (5.99)	*Amorpha fruticosa* root bark	Hohmann *et al.* (1982)
	Lonchocarpus longifolius root	Braz Filho *et al.* (1980)
	L. spruceanus root	Menichini *et al.* (1982)
	Millettia pachycarpa root	Singhal *et al.* (1982a)
	Piscidia mollis seed	Menichini *et al.* (1982)
	Tephrosia sp. root	Suarez *et al.* (1980)
Dalbinol (12a-hydroxyamorphigenin) (*5.100*)	*Amorpha fruticosa* root bark	Hohmann *et al.* (1982)
	Dalbergia monetaria seed	Abe *et al.* (1985)
Dalbinol *O*-glucoside (delbin) (*5.101*)	*Dalbergia monetaria* seed	Abe *et al.* (1985)
	D. nitidula bark	Van Heerden *et al.* (1980)
Dalbinol *O*-vicianoside (12a-hydroxyamorphin) (*5.102*)	*Amorpha fruticosa* fruit	Somleva and Ognyanov (1985)
Volubinol (*5.103*)	*Dalbergia volubilis* branch	Chawla *et al.* (1984)

(*Contd.*)

Table 5.5 (*Contd.*)

Rotenoid	*Plant sources*	*References*
11-Hydroxytephrosin (*5.104*)	*Amorpha fruticosa* fruit	Somleva and Ognyanov (1985)
	Tephrosia viridiflora root, aerial parts	Gomez *et al.* (1985)
Villosinol (12a-hydroxysumatrol) (*5.105*)	*Tephrosia viridiflora* root, aerial parts	Gomez *et al.* (1985)
12-Dihydrodalbinol (*5.106*)	*Dalbergia monetaria* seed	Abe *et al.* (1985)
12-Dihydrodalbinol *O*-glucoside (12-dihydrodalbin) (*5.107*)	*Dalbergia monetaria* seed	Abe *et al.* (1985)
(iii) *Dehydrorotenoids*		
Dehydromillettone (*5.108*)	*Piscidia erythrina* root bark	Delle Monache *et al.* (1984)
Dehydrorotenone (*5.109*)	*Lonchocarpus longifolius* root	Braz Filho *et al.* (1980)
	Tephrosia villosa root	Chandrasekharan *et al.* (1983)
Dehydroamorphigenin (*5.110*)	*Amorpha fruticosa* root bark	Hohmann *et al.* (1982)
6-Hydroxydehydrotoxicarol (*5.111*)	*Amorpha fruticosa* fruit	Somleva and Ognyanov (1985)

A rotenoid Wallichin has been reported in leaves of *Tephrosia wallichi* (Bose and Ganguly, 1981). The structure given (*5.112*) in *Chemical Abstracts* is biosynthetically unrealistic and almost certainly incorrect.

detailed review of the chemistry and biochemistry of rotenoids has been published recently (Crombie, 1984).

5.5.1 Rotenoids

New sources of rotenoids continue to be reported (Table 5.5), but only two new natural products have been described. One of these is rot-2′-enonic acid (*5.86*) which has been isolated from the root of *Millettia pachycarpa* (Singhal *et al.*, 1982a). Rot-2′-enonic acid has been known for many years as a semi-synthetic product prepared from rotenone, and it has been implicated as a biosynthetic precursor of rotenone and amorphigenin (see Dewick, 1982). Indeed, its presence in *Amorpha fruticosa* could be demonstrated by isotope dilution analysis (Crombie *et al.*, 1982). This, however, represents the first conventional isolation of the compound from nature, and supports the suggested role of this compound in the biosynthesis of other rotenoids. 3-*O*-Demethylamorphigenin (*5.92*), isolated from fruits of *Amorpha fruticosa* (Somleva and Ognyanov, 1985), may be regarded as an example of the missing link between 2,3-dimethoxy- and 2,3-methylenedioxy-rotenoids, all previously known natural rotenoids falling into one of these categories. Volubinol

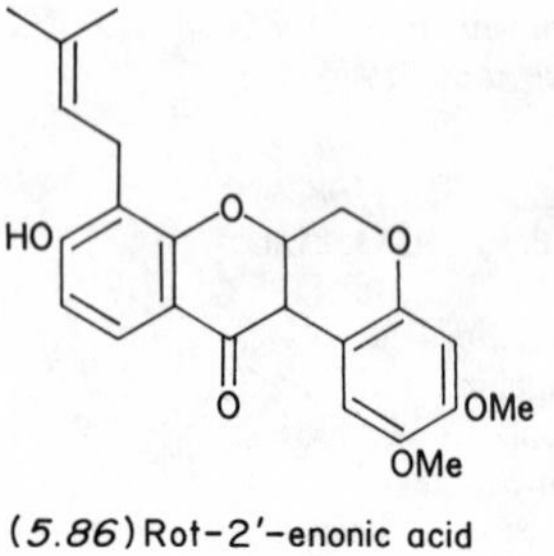

(*5.86*) Rot-2′-enonic acid

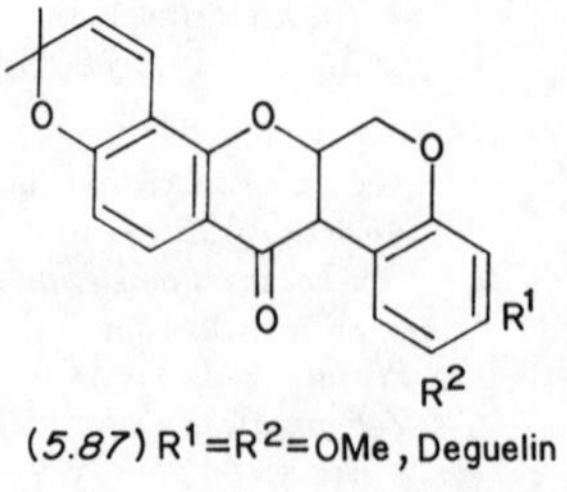

(*5.87*) $R^1=R^2=OMe$, Deguelin
(*5.88*) $R^1R^2=OCH_2O$, Millettone

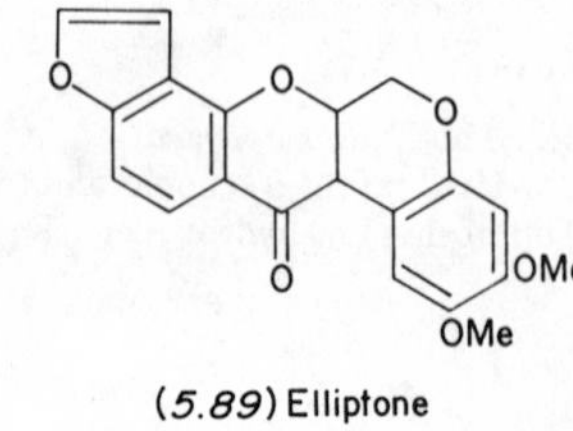

(*5.89*) Elliptone

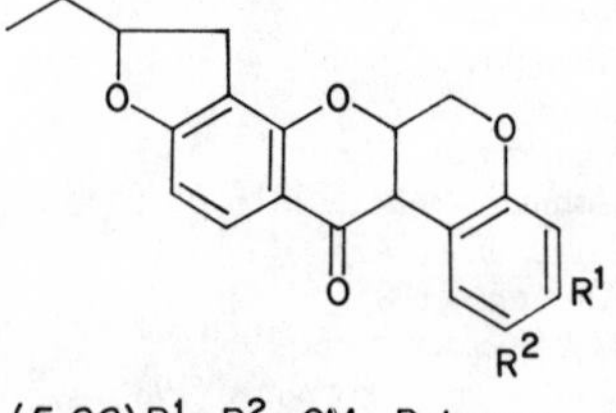

(*5.90*) $R^1=R^2=OMe$, Rotenone
(*5.91*) $R^1R^2=OCH_2O$, Isomillettone

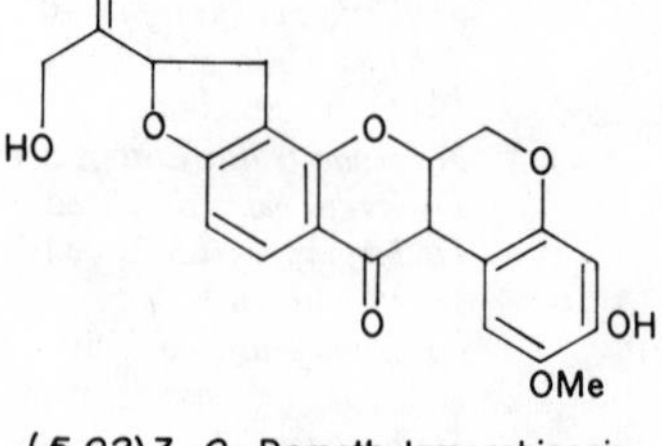

(*5.92*) 3-*O*-Demethylamorphigenin

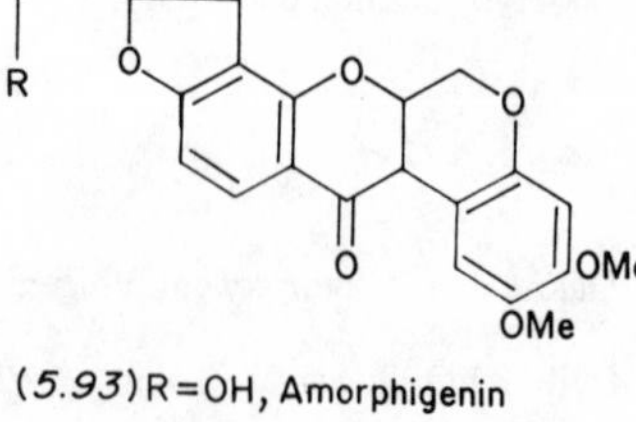

(*5.93*) R=OH, Amorphigenin
(*5.94*) R=OGlc, Amorphigenin-*O*-glucoside
(*5.95*) R=OVicianose, Amorphin

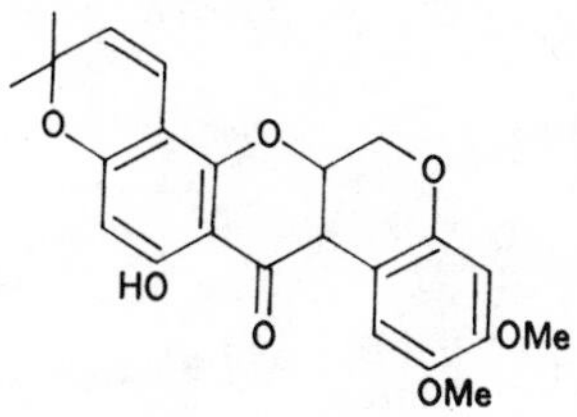

(5.96) Toxicarol

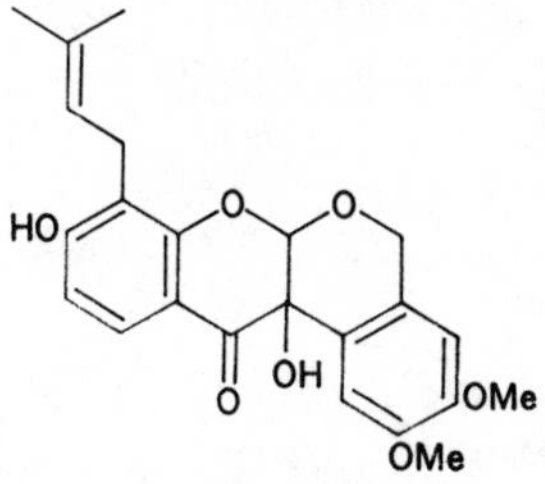

(5.97) 12a-Hydroxyrot-2'-enonic acid

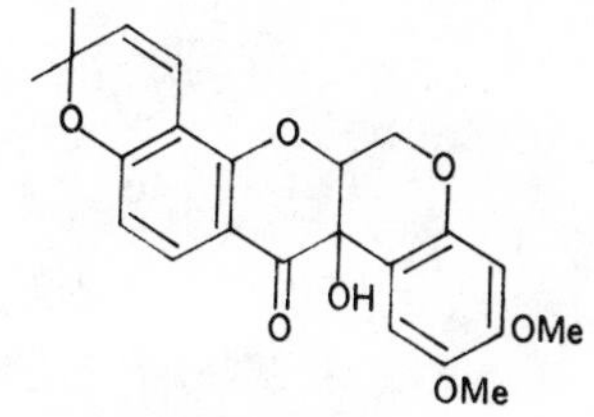

(5.98) Tephrosin

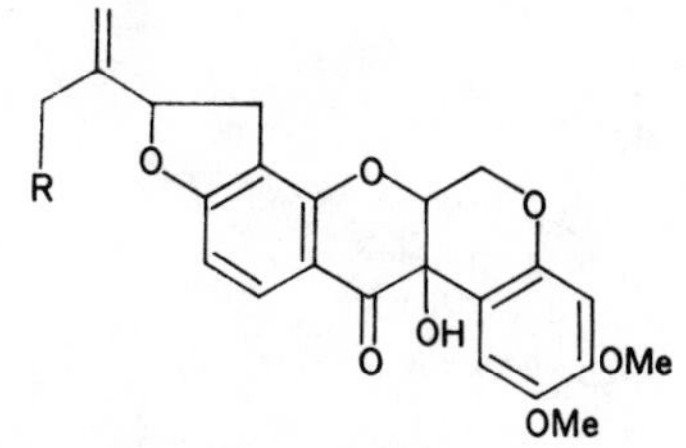

(5.99) R=H, 12a-Hydroxyrotenone
(5.100) R=OH, Dalbinol
(5.101) R=OGlc, Dalbin
(5.102) R=OVicianose, 12a-Hydroxyamorphin

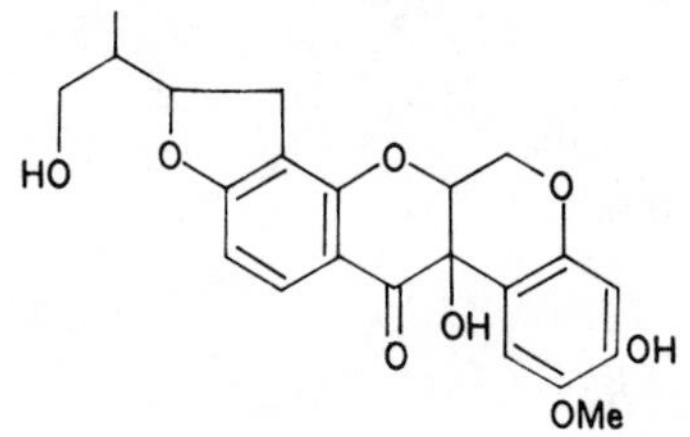

(5.103) Volubinol

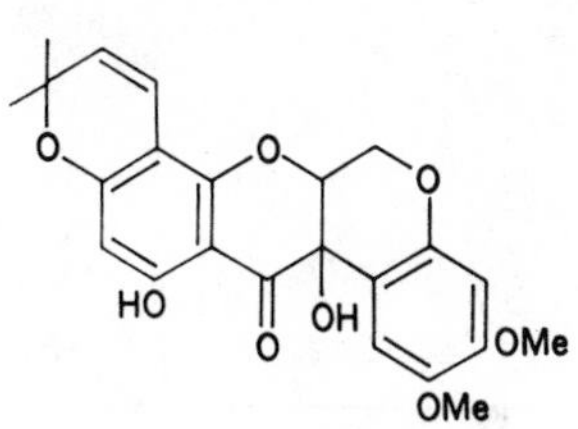

(5.104) 11-Hydroxytephrosin

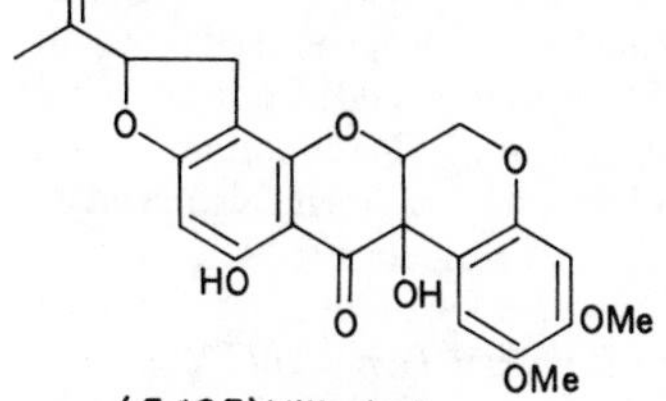

(5.105) Villosinol

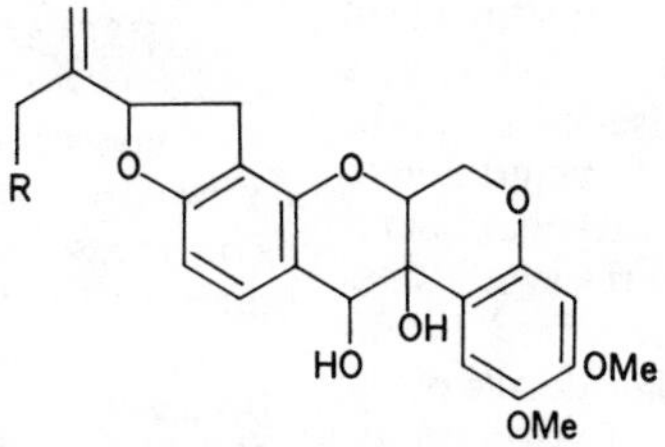

(5.106) R=OH, 12-Dihydrodalbinol
(5.107) R=OGlc, 12-Dihydrodalbin

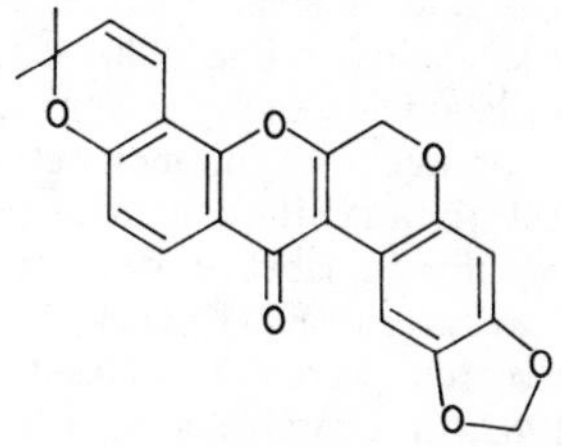

(5.108) Dehydromillettone

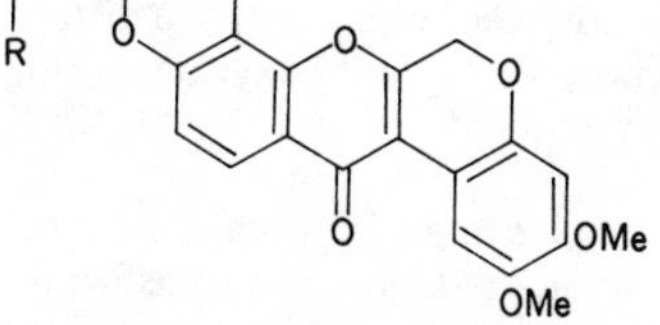

(5.109) R=H, Dehydrorotenone
(5.110) R=OH, Dehydroamorphigenin

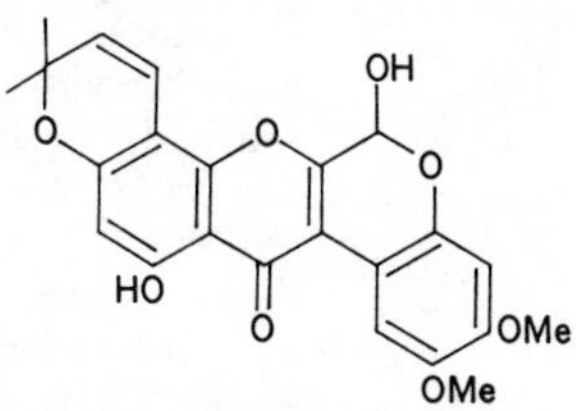

(5.111) 6-Hydroxydehydrotoxicarol

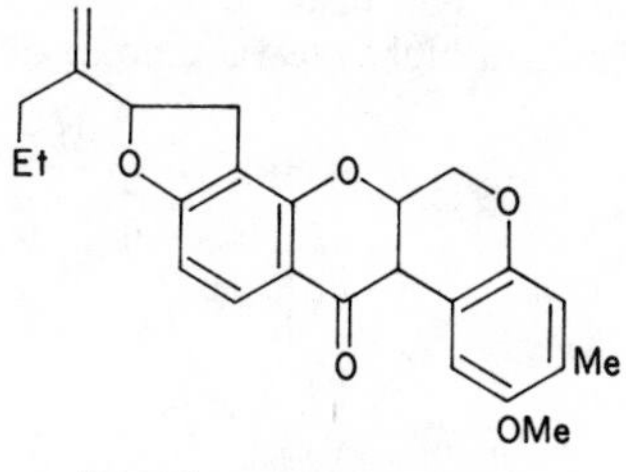

(5.112) Wallichin?

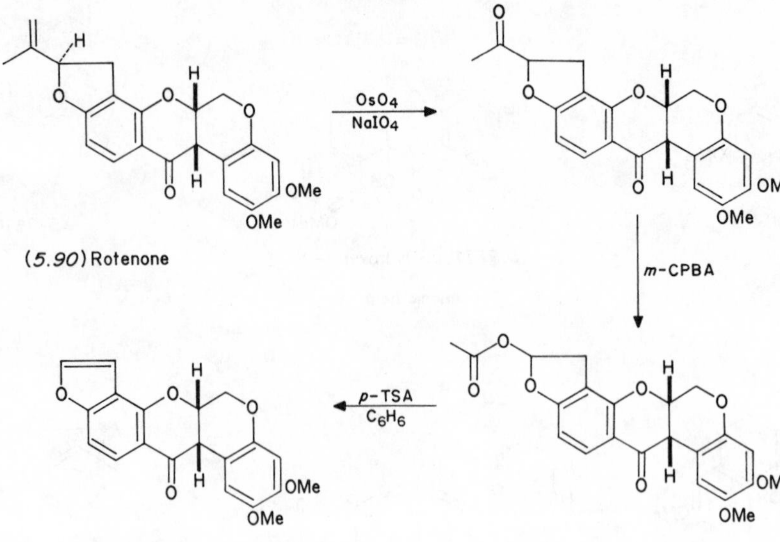

(5.90) Rotenone

OsO_4 / $NaIO_4$; m-CPBA ; p-TSA / C_6H_6

(5.89) Elliptone

Fig. 5.6 Synthesis of elliptone.

(5.103) (see Section 5.5.2) is a second example of a 2-methoxy-3-hydroxy rotenoid derivative.

The chemical conversion of rotenone (*5.90*) into elliptone (*5.89*) has been achieved as shown in Fig. 5.6, utilizing a Baeyer–Villiger oxidation of the ketone formed via osmium tetroxide–sodium periodate treatment of rotenone (Singhal *et al.*, 1982b). Such modification of a side chain obviously has application elsewhere in the isoflavonoids, though isopropenyldihydrofuran groupings seem much less common outside of the rotenoids. Chemical studies in the rotenoid field leading to the synthesis of specifically labelled derivatives were described in the previous review (Dewick, 1982), but fuller details of the research have since been published (Carson *et al.*, 1982a, b; Crombie *et al.*, 1982).

Recent check lists of isoflavonoids (Dewick, 1982; Ingham, 1983) have continued to include myriconol from *Myrica nagi* (Myricaceae) as a rotenoid with the structure (*5.113*). Rather belatedly it should be acknowledged that this unusual structure has been re-evaluated and withdrawn. This plant contains not rotenoids, but a range of bridged biphenyls (Crombie, 1986; Begley *et al.*, 1971).

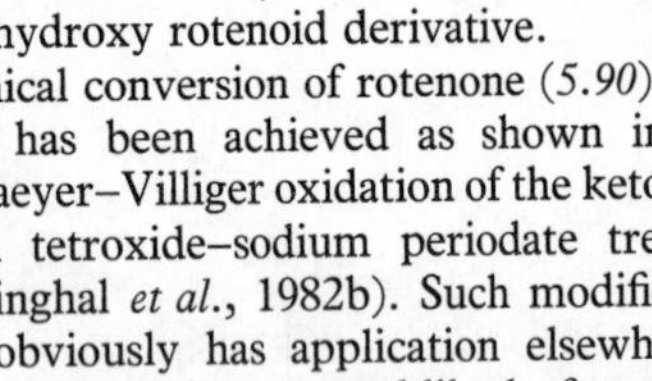

(5.113)

5.5.2 12a-Hydroxyrotenoids

Amongst the 12a-hydroxyrotenoids recently isolated from a variety of sources (Table 5.5), a number of new compounds are reported which contain novel features. Roots of *Millettia pachycarpa* contain 12a-hydroxyrot-2′-enonic acid (*5.97*) as well as rot-2′-enonic acid (*5.86*) (Singhal *et al.*, 1982a). Again, the uncyclized isoprenyl substituent is of biogenetic interest. Volubinol (*5.103*) from branches of *Dalbergia volubilis* (Chawla *et al.*, 1984) contains the rare 2-methoxy-3-hydroxy substitution pattern, also seen in the rotenoid 3-*O*-demethylamorphigenin (*5.92*).

A new variant on the 12a-hydroxyrotenoid skeleton has been isolated from seeds of *Dalbergia monetaria* (Abe *et al.*, 1985). 12-Dihydrodalbinol (*5.106*) is the first natural example of a rotenoid with a reduced carbonyl group. The configuration at C-12 was shown to be *S*, i.e. the two hydroxyl groups are *cis*. Also isolated was the 8′-*O*-glucoside 12-dihydrodalbin (*5.107*). Co-occurring in *D. monetaria* are amorphigenin (*5.93*), dalbinol (12a-hydroxyamorphigenin) (*5.100*) and 7-hydroxy-2′,4′,5′-trimethoxyisoflavone, together with their corresponding glucosides. These compounds seem representative of the main biosynthetic sequence, i.e. 2′-methoxyisoflavone, rotenoid, 12a-hydroxyrotenoid and 12,12a-dihydroxyrotenoid.

Whilst 12a-hydroxyrotenoids may be obtained from rotenoids as artefacts by aerial oxidation, the transformation is better accomplished by means of other oxidizing agents. The use of perbenzoic acid has been recommended (Singhal *et al.*, 1982b). Yields of about 30% of 12a-

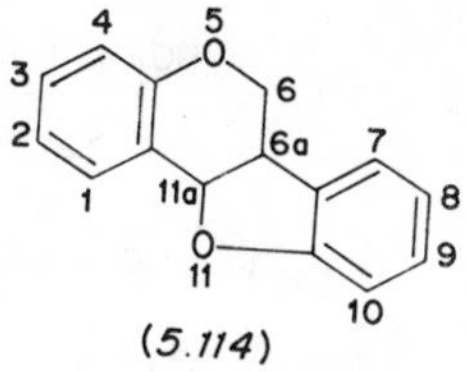

(5.114) 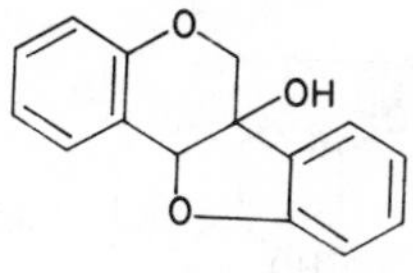 (5.115) 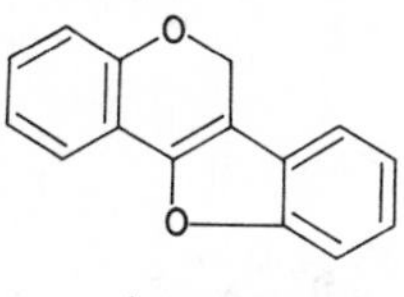 (5.116)

hydroxyrotenone were obtained, with no by-products, the remaining material being unreacted rotenone. Epoxides were not isolated, in contrast to earlier reports. The same 12a-hydroxylation of rotenone has been achieved using an extracellular laccase from *Polyporus anceps* (Sariaslani *et al.*, 1984).

5.5.3 Dehydrorotenoids

Although a number of 6a, 12a-dehydro derivatives of rotenoids are known, and established as natural products rather than artefacts, only four examples have been reported since the last review (Table 5.5). Two new compounds figure, dehydroamorphigenin (*5.110*) from *Amorpha fruticosa* root bark (Hohmann *et al.*, 1982), and 6-hydroxydehydrotoxicarol (*5.111*) from fruits of the same plant (Somleva and Ognyanov, 1985). The latter compound is a further example of a 6-hydroxydehydrorotenoid, and is optically active suggesting it is unlikely to be an artefact.

5.6 PTEROCARPANS

Pterocarpans (*5.114*) contain a tetracyclic ring system derived from the basic isoflavonoid skeleton by an ether linkage between the 4 and 2′ positions. The systematic numbering of (*5.114*) rather than that for simple isoflavonoids is used, however. The majority of natural pterocarpans isolated have arisen from phytoalexin studies, using fungal or abiotically stressed plant tissues, and the number of examples continues to grow, making this the second largest group of isoflavonoids after the isoflavones. For convenience, pterocarpans are subdivided into pterocarpans (*5.114*), 6a-hydroxypterocarpans (*5.115*) and pterocarpenes (*5.116*). New structures, and further sources of known compounds are presented in Table 5.6.

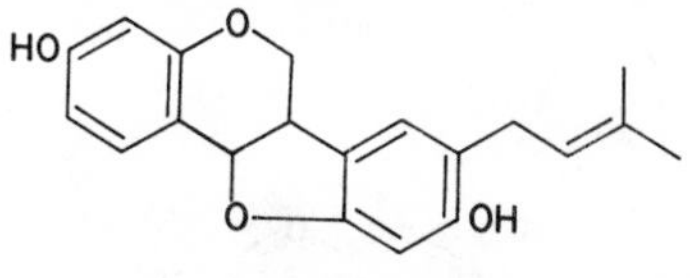

(*5.117*) Sophorapterocarpan A (≡homoedudiol)

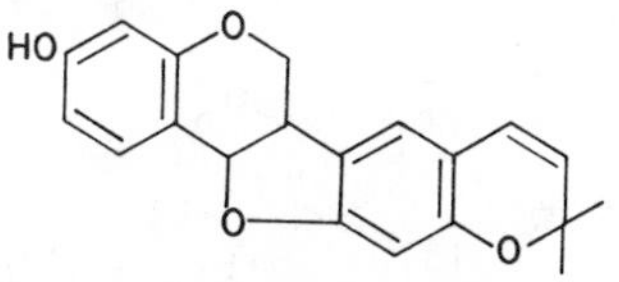

(*5.118*) Isoneorautenol

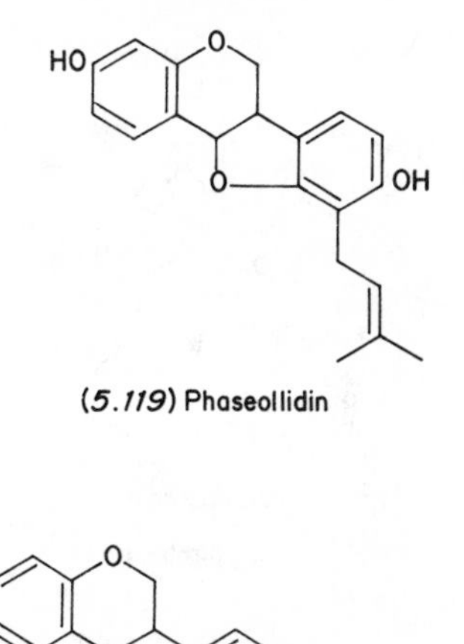

(*5.119*) Phaseollidin

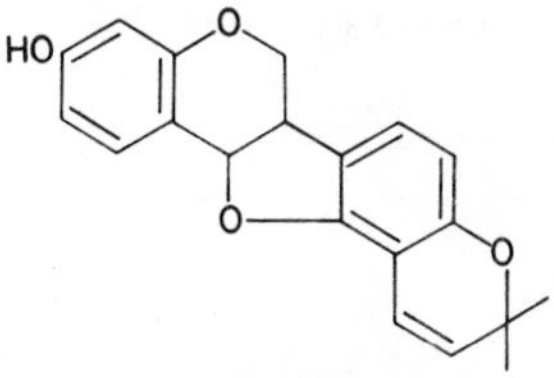

(*5.122*) Phaseollin

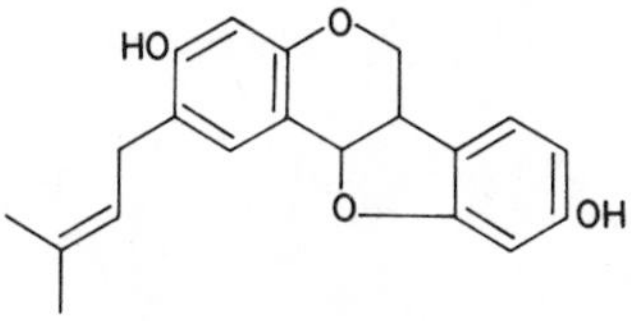

(*5.123*) Calopocarpin

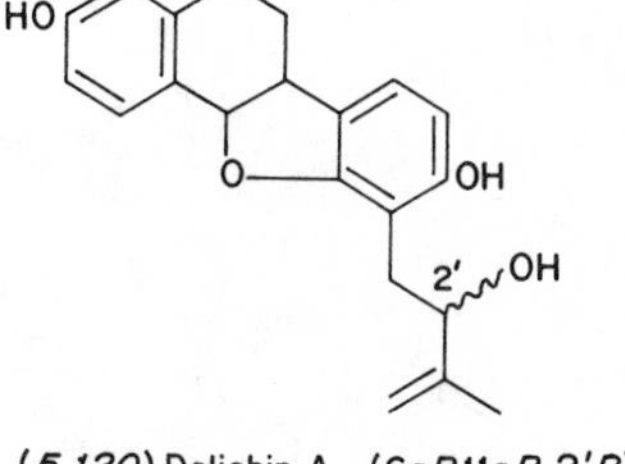

(*5.120*) Dolichin A (6a*R*,11a*R*, 2′*R*)

(*5.121*) Dolichin B (6a*R*,11a*R*, 2′*S*)

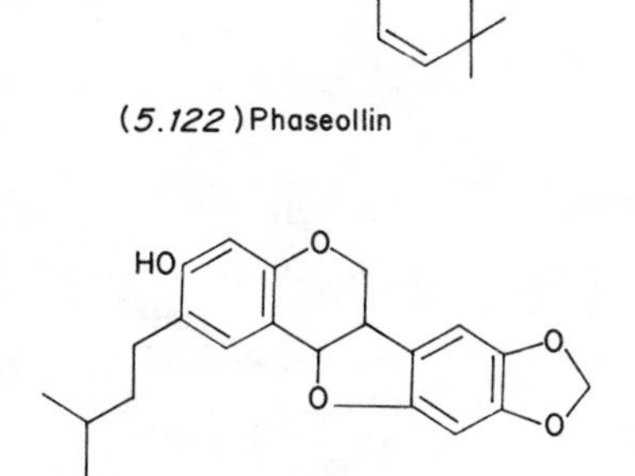

(*5.124*) Cabenegrin A-II

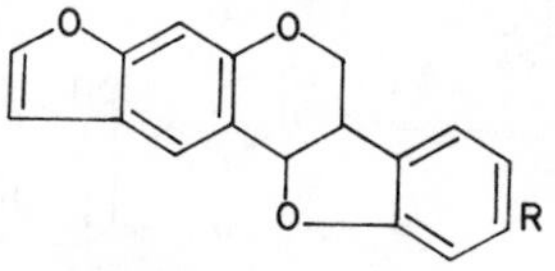

(*5.125*) R=OH, Neodunol

(*5.126*) R=OMe, 9-*O*-Methylneodunol

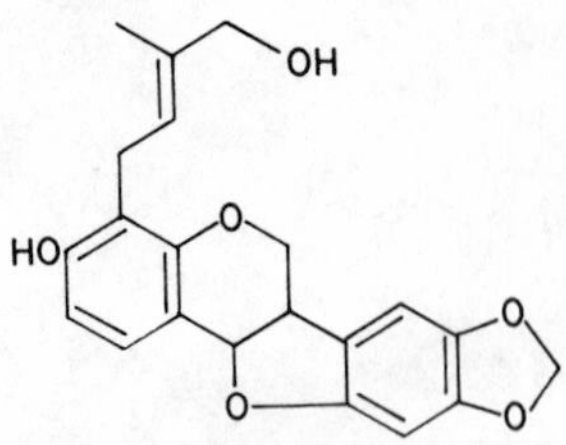

(*5.127*) Cabenegrin A-I

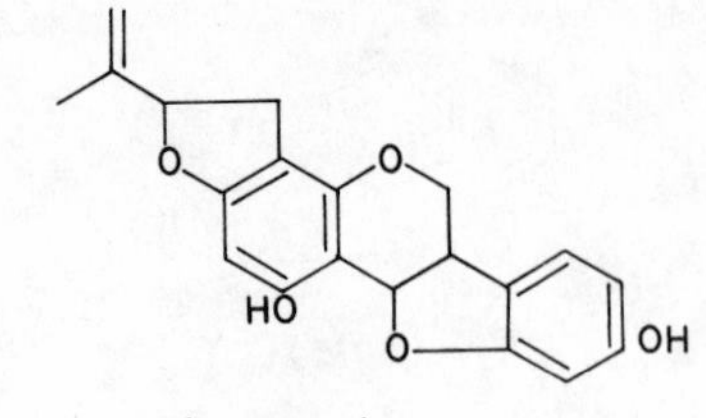

(*5.128*) Apiocarpin

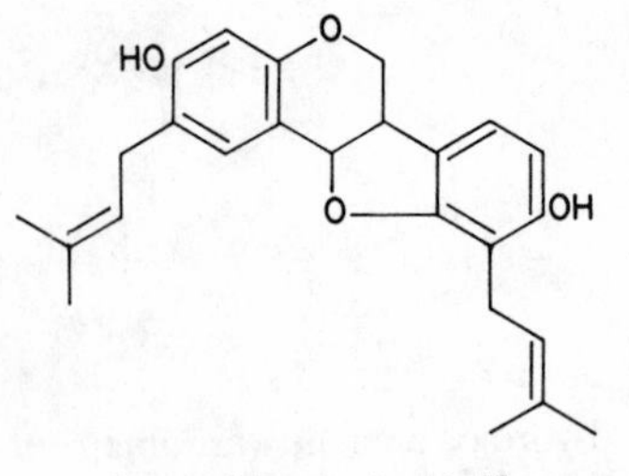

(*5.129*) Erythrabyssin II

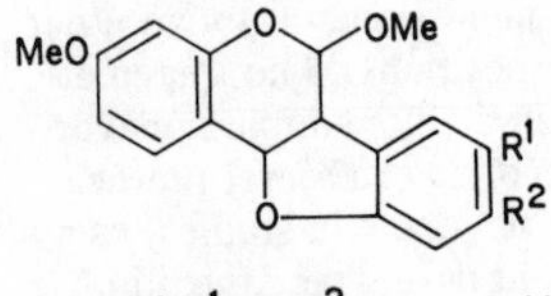

(*5.130*) R^1=H, R^2=OMe, 6-Methoxy-homopterocarpin

(*5.131*) R^1R^2=OCH_2O, 6-Methoxy-pterocarpin

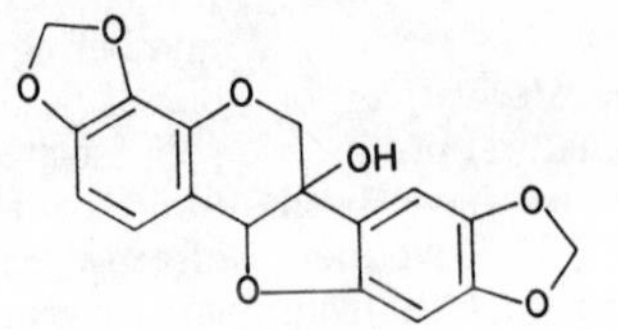

(*5.132*) Acanthocarpan

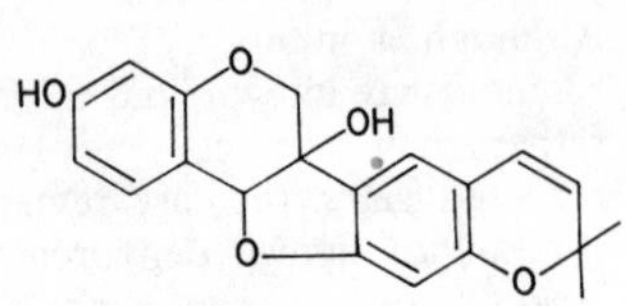

(*5.133*) Tuberosin

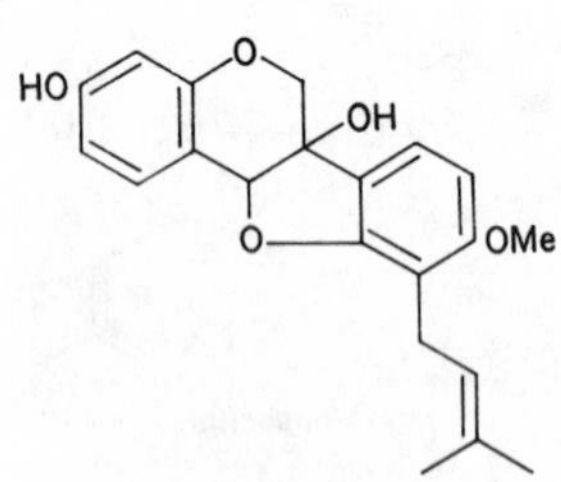

(*5.134*) Cristacarpin (erythrabyssin I)

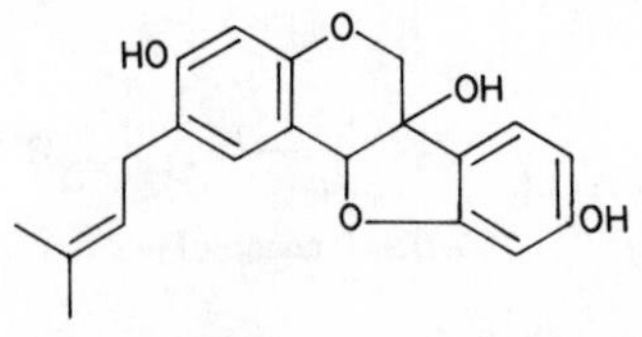

(*5.135*) Glyceocarpin (glyceollidin II)

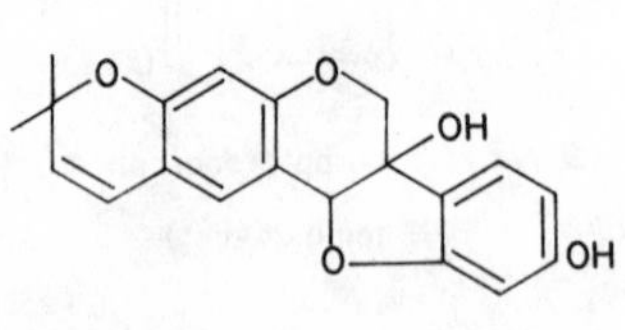

(*5.136*) Glyceollin II

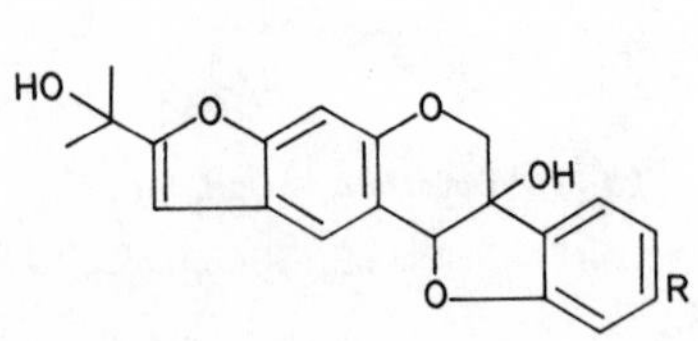

(*5.137*) R=OH, Glyceofuran

(*5.138*) R=OMe, 9-*O*-Methylglyceofuran

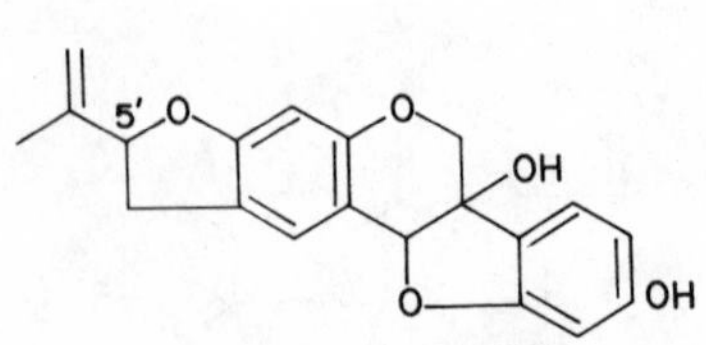

(*5.139*) Canescacarpin (6a*S*,11a*S*,5′*R*)

(*5.140*) Glyceollin III (6a*S*,11a*S*,5′*S*)

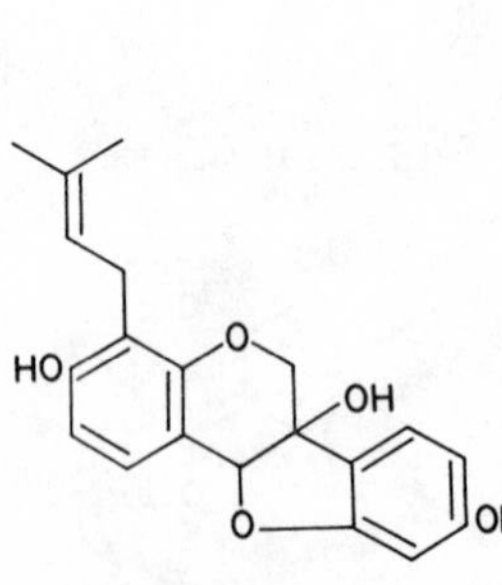

(*5.141*) Glyceollidin I

Table 5.6 Pterocarpans

Pterocarpan	Substituents 1	2	3	4	7	8	9	10	Plant sources	Optical activity/ chirality	References
(i) *Pterocarpans*											
Demethylmedicarpin			OH				OH		*Calopogonium mucunoides* leaf*	–	Ingham and Tahara (1985)
									Trifolium repens leaf*		Woodward (1981a)
Medicarpin			OH				OMe		*Arachis hypogaea* leaf*	+	Strange *et al.* (1985)
									Canavalia bonariensis root		Menichini *et al.* (1982)
									Centrolobium sp.	–	Jurd and Wong (1984)
									Cicer arietinum seed*		Köster *et al.* (1983b)
									Cicer spp. stem		Ingham (1981a)
									Dalbergia odorifera heartwood	±	Goda *et al.* (1985)
									Hedysarum polybotrys root	–	Miyase *et al.* (1984)
									Lathyrus spp. leaf*		Robeson and Harborne (1980)
									Medicago rugosa leaf*	–	Ingham (1982b)
									Ononis spp. leaf*		Ingham (1982a)
									Sophora japonica leaf*	+, –	Van Etten *et al.* (1983)
									Tephrosia purpurea leaf*		Ingham and Markham (1982a)
									T. villosa leaf*		
									Trifolium hybridum root		Fraishtat *et al.* (1981)
									T. pratense root		
									T. repens leaf*		Woodward (1981a)
									T. repens root		Fraishtat *et al.* (1981)
									Trigonella spp. leaf*		Ingham (1981b)
									Vicia faba leaf*		Komives and Casida (1983)
									V. faba root*		Ibrahim *et al.* (1982)
									V. faba cotyledon*		Robeson and Harborne (1980)
									Zollernia paraensis wood	+	Ferrari *et al.* (1983)
Medicarpin *O*-glucoside			OGlc				OMe		*Trifolium hybridum* root		Fraishtat *et al.* (1981)
									T. pratense root		
									T. repens root		
Homopterocarpin			OMe				OMe		*Ononis natrix* aerial parts	–	San Feliciano *et al.* (1983)
									Trifolium hybridum root		Fraishtat *et al.* (1981)
									T. pratense root		
									T. repens root		

(*Contd.*)

Table 5.6 (*Contd.*)

Pterocarpan	*Substituents* 1	2	3	4	7	8	9	10	*Plant sources*	*Optical activity/ chirality*	*References*
Nissicarpin			OH		OH		OMe		*Nissolia fruticosa* leaf*	+	Ingham and Markham (1984)
Fruticarpin			OMe		OH		OMe		*Nissolia fruticosa* leaf*	+	Ingham and Markham (1984)
Kushenin			OH			OMe	OH		*Sophora flavescens* root	–	Wu *et al.* (1985)
Maackiain			OH			OCH_2O			*Cicer arietinum* seed*		Köster *et al.* (1983b)
									Cicer spp. stem*		Ingham (1981a)
									Derris elliptica root	–, +	Obara and Matsubara (1981)
									Euchresta japonica root	–	Shirataki *et al.* (1981, 1982)
									Lathyrus spp. leaf*		Robeson and Harborne (1980)
									Millettia pulchra aerial parts	–	Baruah *et al.* (1984)
									Ononis spp. leaf*		Ingham (1982a)
									Sophora angustifolia root	–	Honda and Tabata (1982)
									S. japonica leaf*	+, –	Van Etten *et al.* (1983)
									Tephrosia bidwilli leaf*	–	Ingham and Markham (1982a)
									T. elata root	–	Lwande *et al.* (1985a)
									T. maxima aerial parts	–	Murthy and Rao (1985a)
									T. purpurea root	–	Rao and Raju (1984)
									T. spp. leaf*		Ingham and Markham (1982a)
									Thermopsis fabacea root	±	Arisawa *et al.* (1980)
									Trifolium hybridum root		Fraishtat *et al.* (1981)
									T. pratense root		
									T. repens root		
									Trigonella spp. leaf*		Ingham (1981b)
Trifolirhizin [(–)-maackiain-*O*-glucoside]			OGlc			OCH_2O			*Euchresta japonica* root		Shirataki *et al.* (1981)
									Ononis sp. tincture		Kovalev (1983)
									Thermopsis fabacea root		Arisawa *et al.* (1980)
									Trifolium hybridum root		Fraishtat *et al.* (1981)
									T. pratense root		
									T. repens root		
									Vexibia alopecuroides root		Yasupova *et al.* (1984)
Pterocarpin			OMe			OCH_2O			*Millettia pulchra* aerial parts	–	Baruah *et al.* (1984)
Methylnissolin			OH				OMe	OMe	*Dalbergia odorifera* heartwood	–	Goda *et al.* (1985)

Methylnissolin O-glucoside			OGlc				OMe	OMe	*Astragalus mongholicus*		Lu *et al.* (1984)
Desmocarpin	OH		OMe				OH		*Desmodium gangeticum* leaf*	–	Ingham and Dewick (1984)
Nissolicarpin		OMe	OH		OH		OMe		*Nissolia fruticosa* leaf*	+	Ingham and Markham (1984)
Melilotocarpan B			OMe	OH			OH		*Melilotus alba* aerial parts	–	Miyase *et al.* (1982)
4-Methoxymedicarpin			OH	OMe			OMe		*Trifolium hybridum* root		Fraishtat *et al.* (1981)
									T. pratense root		
									T. repens root		
Melilotocarpan A			OMe	OH			OMe		*Melilotus alba* aerial parts	–	Miyase *et al.* (1982)
4-Methoxymaackiain			OH	OMe		OCH_2O			*Sophora franchetiana* root	–	Komatsu *et al.* (1981b)
									Tephrosia bidwilli leaf*	–	Ingham and Markham (1982a)
Melilotocarpan D			OMe	OH			OMe	OH	*Dalbergia odorifera* heartwood	–	Goda *et al.* (1985)
									Melilotus alba aerial parts	–	Miyase *et al.* (1982)
Melilotocarpan E			OMe	OH			OH	OMe	*Melilotus alba* aerial parts	–	Miyase *et al.* (1982)
Melilotocarpan C			OMe	OH			OMe	OMe	*Dalbergia odorifera* heartwood	–	Goda *et al.* (1985)
									Melilotus alba aerial parts	–	Miyase *et al.* (1982)
Odoricarpan			OMe	OMe			OMe	OH	*Dalbergia odorifera* heartwood	–	Goda *et al.* (1985)
Trifolian	OH	OMe	OH			OCH_2O			*Trifolium hybridum* root		Fraishtat *et al.* (1981)
									T. pratense root		
									T. repens root		
Sophorapterocarpan A (homoedudiol[†]) (*5.117*)									*Calopogonium mucunoides* leaf*	–	Ingham and Tahara (1985)
Isoneorautenol (*5.118*)									*Sophora franchetiana* root	–	Komatsu *et al.* (1981a)
									Calopogonium mucunoides leaf*	–	Ingham and Tahara (1985)
Phaseollidin (*5.119*)									*Dolichos biflorus* leaf*	–	Keen and Ingham (1980)
									Erythrina abyssinica root	–	Kamat *et al.* (1981)
									Phaseolus aureus seedling*		O'Neill *et al.* (1983)
									P. calcaratus seed*, leaf*		Sukumaran and Gnanamanickam (1980)
									P. coccineus seedling*		Adesanya *et al.* (1985)
									P. vulgaris cotyledon*		Goosens and Van Laere (1983)
									P. vulgaris hypocotyl*		Garcia-Arenal *et al.* (1980)
Dolichin A (*5.120*)									*Dolichos biflorus* leaf*	6a*R*, 11a*R*, 2′*R*	Ingham *et al.* (1981a)
Dolichin B (*5.121*)									*Dolichos biflorus* leaf*	6a*R*, 11a*R*, 2′*S*	Ingham *et al.* (1981a)

(*Contd.*)

Table 5.6 (*Contd.*)

Pterocarpan	*Substituents* 1	2	3	4	7	8	9	10	*Plant sources*	*Optical activity/ chirality*	*References*
Phaseollin (*5.122*)									*Erythrina abyssinica* root	–	Kamat *et al.* (1981)
									Phaseolus coccineus seedling*		Adesanya *et al.* (1985)
									P. vulgaris cotyledon*		Bailey and Berthier (1981)
											Goosens and Van Laere (1983)
											Whitehead *et al.* (1982)
									P. vulgaris leaf*		Komives and Casida (1983)
									P. vulgaris various tissues*		Stössel and Magnolata (1983)
									P. vulgaris hypocotyl*		Garcia-Arenal *et al.* (1980)
Calopocarpin (*5.123*)[‡]									*Alysicarpus* sp. leaf*	–	Ingham and Tahara (1985)
									Calpogonium mucunoides leaf*	–	Ingham and Tahara (1985)
									Pueraria phaseoloides leaf*		Ingham and Tahara (1985)
Cabenegrin A-II (*5.124*)									unidentified root	6a*R*, 11a*R*	Nakagawa *et al.* (1982)
Neodunol (*5.125*)									*Calopogonium mucunoides* leaf*	–	Ingham and Tahara (1985)
9-*O*-Methylneodunol (*5.126*)									*Echinopora lamellosa* (Scleractinidae) marine coral	–	Sanduja *et al.* (1984)
Cabenegrin A-I (*5.127*)									unidentified root	–	Nakagawa *et al.* (1982)
Apiocarpin (*5.128*)									*Apios tuberosa* leaf*	6a*R*, 11a*R*, 5′*S*	Ingham and Mulheirn (1982)
Erythrabyssin II (*5.129*)									*Erythrina abyssinica* root	–	Kamat *et al.* (1981)
6-Methoxyhomopterocarpin (*5.130*)									*Millettia pulchra* aerial parts	6a*S*, 11a*R*, 6*S*	Baruah *et al.* (1984)
6-Methoxypterocarpin (*5.131*)									*Millettia pulchra* aerial parts	6a*S*, 11a*R*, 6*S*	Baruah *et al.* (1984)
6a-Hydroxypterocarpans											
Glycinol			OH				OH		*Glycine max* cotyledon*	–	Weinstein *et al.* (1981)
									G. max seed*		Osswald (1985)
									Phaseolus coccineus seedling*		Adesanya *et al.* (1985)
									P. mungo seedling*		Adesanya *et al.* (1984)

Variabilin		OMe			OMe	*Lathyrus* spp. leaf*		Robeson and Harborne (1980)
						Lens spp. leaf*		Robeson and Harborne (1980)
						Tephrosia purpurea leaf*		Ingham and Markham (1982a)
						T. villosa leaf*		
6a-Hydroxymaackiain		OH		OCH_2O		*Lathyrus sativus* seed*		Robeson and Harborne (1983)
Pisatin		OMe		OCH_2O		*Lathyrus sativus* leaf*		Ingham and Markham (1982b)
								Robeson and Harborne (1983)
						Lathyrus spp. leaf*		Robeson and Harborne (1980)
						Pisum fulvum leaf*		Robeson and Harborne (1980)
						P. sativum pod*	−, +	Banks and Dewick (1982a)
						P. sativum seedling*		Carlson and Dolphin (1981)
						P. sativum leaf*		Komives and Casida (1983)
								Robeson and Harborne (1980)
						Tephrosia bidwilli leaf*	−	Ingham and Markham (1982a)
						T. elata root	+	Lwande *et al.* (1985a)
						T. spp. leaf*		Ingham and Markham (1982a)
Hildecarpin	OMe	OH		OCH_2O		*Tephrosia hildebrandtii* root	−	Lwande *et al.* (1985b)
Lathycarpin	OMe	OMe		OCH_2O		*Lathyrus sativus* leaf*	+	Ingham and Markham (1982b)
Tephrocarpin		OH	OMe	OCH_2O		*Tephrosia bidwilli* leaf*	−	Ingham and Markham (1982a)
Acanthocarpan (*5.132*)[§]		OCH_2O		OCH_2O		*Tephrosia bidwilli* leaf*	−	Ingham and Markham (1982a)
Tuberosin (*5.133*)						*Calopogonium mucunoides* leaf*	−	Ingham and Tahara (1985)
Cristacarpin (erythrabyssin I) (*5.134*)						*Erythrina abyssinica* root	−	Kamat *et al.* (1981)
Glyceocarpin (glyceollidin II) (*5.135*)						*Glycine max* leaf*	−	Ingham *et al.* (1981b)
						G. max cotyledon*	−	Zähringer *et al.* (1981)

(*Contd.*)

Table 5.6 (*Contd.*)

Pterocarpan	*Substituents* 1	2	3	4	7	8	9	10	*Plant sources*	*Optical activity/ chirality*	*References*
Glyceollin II (*5.136*)									*Costus speciosus* (Zingiberaceae) leaf*		Kumar *et al.* (1984)
									Glycine canescens leaf*		Lyne *et al.* (1981)
									G. max leaf*		Cosio *et al.* (1985) Fett (1984) Ingham *et al.* (1981b) Komives and Casida (1983) Osman and Fett (1983)
									G. max seedling*		Komives (1983) Stössel (1982)
									G. max seedling*, pod*		Banks and Dewick (1983b)
									G. max seed*		Osswald (1985)
									G. max various tissues*		Stössel and Magnolata (1983)
Glyceofuran (*5.137*)									*Glycine max* leaf*	–	Cosio *et al.* (1985) Ingham *et al.* (1981b) Komives and Casida (1983)
9-*O*-Methylglyceofuran (*5.138*)									*Glycine max* leaf*	–	Ingham *et al.* (1981b)
Canescacarpin (*5.139*)									*Glycine canescens* leaf*	6a*S*, 11a*S*, 5′*R*	Lyne *et al.* (1981)
Glyceollin III (*5.140*)									*Costus speciosus* (Zingiberaceae) leaf*		Kumar *et al.* (1984)
									Glycine max leaf*		Cosio *et al.* (1985) Fett (1984) Ingham *et al.* (1981b) Komives and Casida (1983) Osman and Fett (1983)
									G. max seedling*		Komives (1983) Stössel (1982)
									G. max seedling*, pod*		Banks and Dewick (1983b)
									G. max seed*		Osswald (1985)
									G. max various tissues*		Stössel and Magnolata (1983)
Glyceollidin I (*5.141*)									*Glycine max* cotyledon*		Zähringer *et al.* (1981)
Glyceollin I (*5.142*)									*Glycine canescens* leaf*		Lyne *et al.* (1981)
									Glycine max leaf*		Cosio *et al.* (1985) Fett (1984) Ingham *et al.* (1981b) Komives and Casida (1983) Osman and Fett (1983)
									G. max seedling*		Komives (1983) Stössel (1982)

			G. max seedling*, pod*		Banks and Dewick (1983b)
			G. max seed*		Osswald (1985)
			G. max various tissues*		Stössel and Magnolata (1983)
Clandestacarpin (*5.143*)			*Glycine clandestina* leaf*	–	Lyne *et al.* (1981)
Hydroxytuberosone (*5.144*)			*Pueraria tuberosa* tuber	6a*R*, 11a*R*	Prasad *et al.* (1984)
(iii) *Pterocarpenes*					
Anhydropisatin	OMe	OCH_2O	*Pisum sativum* seedling*		Carlson and Dolphin (1981)
Anhydrotuberosin (*5.145*)			*Pueraria tuberosa* tuber		Prasad *et al.* (1985)
3-*O*-Methylanhydrotuberosin (*5.146*)			*Pueraria tuberosa* tuber		Prasad *et al.* (1985)
Erycristagallin (*5.147*)			*Erythrina crista-galli* root		Mitscher *et al.* (1984)

*Plant was subjected to physiological stress.
†Revised structure for homoedudiol.
‡Structure previously assigned to homoedudiol.
§Revised structure for acanthocarpin.

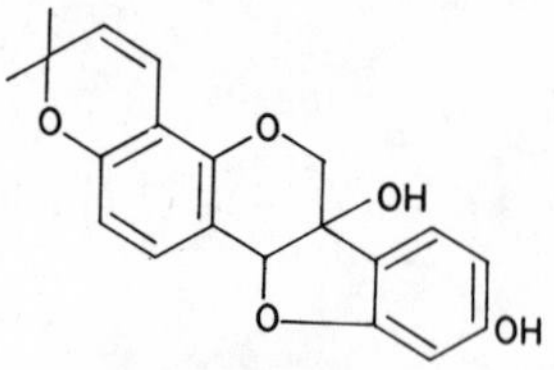

(5.142) Glyceollin I

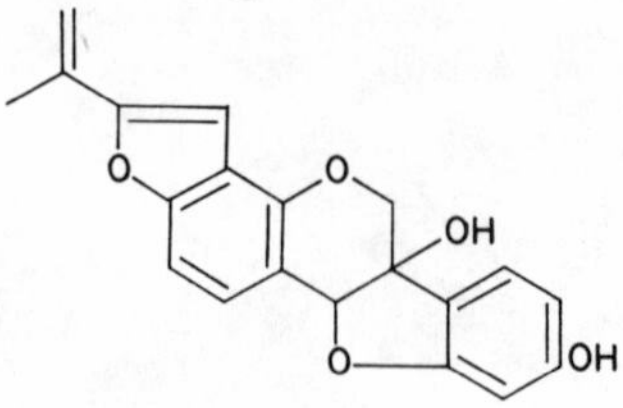

(5.143) Clandestacarpin

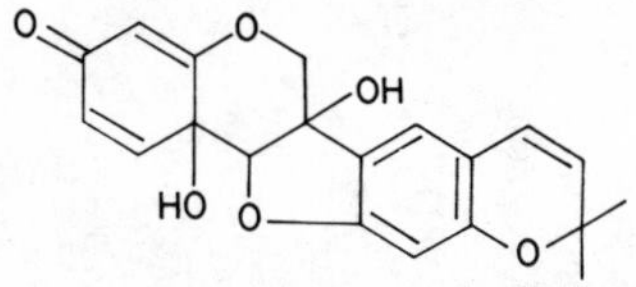

(5.144) Hydroxytuberosone

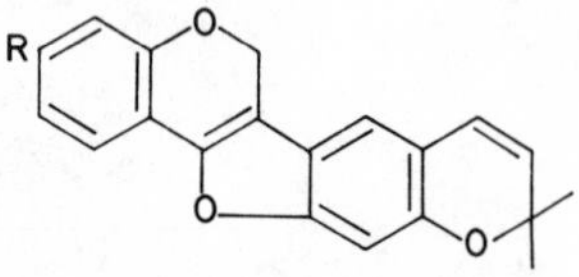

(5.145) R=OH, Anhydrotuberosin

(5.146) R=OMe, 3-*O*-Methyl-anhydrotuberosin

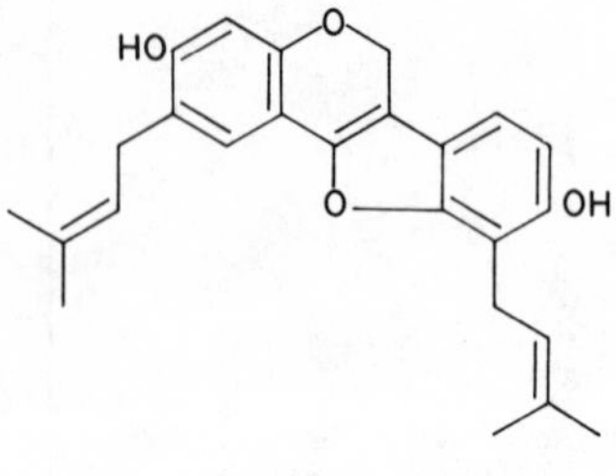

(5.147) Erycristagallin

5.6.1 Pterocarpans

Although pterocarpans contain two chiral centres, only *R,R* and *S,S* configurations are sterically possible. It had been generally accepted that all laevorotatory pterocarpans had the 6a*R*, 11a*R* absolute configuration, and all dextrorotatory ones the 6a*S*, 11a*S* configuration. In the last review (Dewick, 1982), the situation had become somewhat confused because of erroneous reports, so it is now reassuring to be able to confirm this original relationship between optical rotation and absolute configuration. This is possible because of an X-ray crystallographic investigation of (−)-edunol as its 4-bromo-3-*O*-methyl

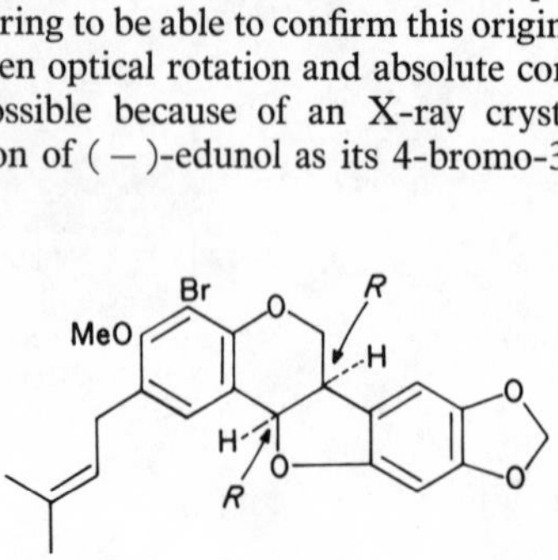

(5.148)

ether (*5.148*) (Breytenbach *et al.*, 1983). Natural pterocarpans are known in laevorotatory, dextrorotatory and racemic forms, and this information, where available, is included in Table 5.6. Most pterocarpans isolated as phytoalexins tend to be laevorotatory, though several that are dextrorotatory are now being reported.

A number of the new structures reported are of particular interest. Cabenegrin A-I (*5.127*) and cabenegrin A-II (*5.124*) were isolated from a South American root 'Cabeca de Negra' used as an oral antidote to snake and spider venoms (Nakagawa *et al.*, 1982). The exact plant extracted is unknown, since some ten different plants may feature under this name. Although the fact that these two pterocarpans do possess potent antidote activity towards snake venom is of especial value, the two structures possess uncommon hydroxyisoprenyl substituents. Syntheses of (±)-cabenegrin A-I from (±)-maackiain (*5.149*) via allylation, Claisen rearrangement and a Wittig reaction (Fig. 5.7), and (±)-cabenegrin A-II from (±)-2-carbomethoxy-3-*O*-benzylmaackiain (*5.150*) (*ex* Li_2PdCl_4-catalysed Heck arylation), again via a Wittig reaction (Fig. 5.8) have been achieved (Ishiguro *et al.*, 1982; Suntry Ltd., 1984). Direct hydroxylation of 4-dimethylallylmaackiain using SeO_2 in pyridine also gave cabenegrin A-I as major product (Baruah, *et al.*, 1984a). A different, and again uncommon, hydroxyisoprenyl substituent is observed in dolichins A and B (*5.120* and *5.121*) isolated as phytoalexins from bacteria-treated leaves of *Dolichos biflorus* (Ingham *et al.*, 1981a). Both compounds possess a 2′-hydroxy-3′-methyl-3′-butenyl substituent, the two compounds differing in the chirality at C-2′, dolichin A being the 6a*R*, 11a*R*, 2′*R* and dolichin B the 6a*R*, 11a*R*, 2′*S* analogue.

Two novel 6-methoxypterocarpans (*5.130*) and (*5.131*) have been isolated from *Millettia pulchra* (Baruah *et al.*, 1984b) and are the first examples of 6-substituted pterocarpans. These compounds have the 6*S*, 6a*S*, 11a*R* configuration and co-occur with (−)-maackiain and (−)-pterocarpin. The isolation of 9-*O*-methylneodunol (*5.126*) from the marine coral *Echinopora lamellosa* (Scleractinidae)

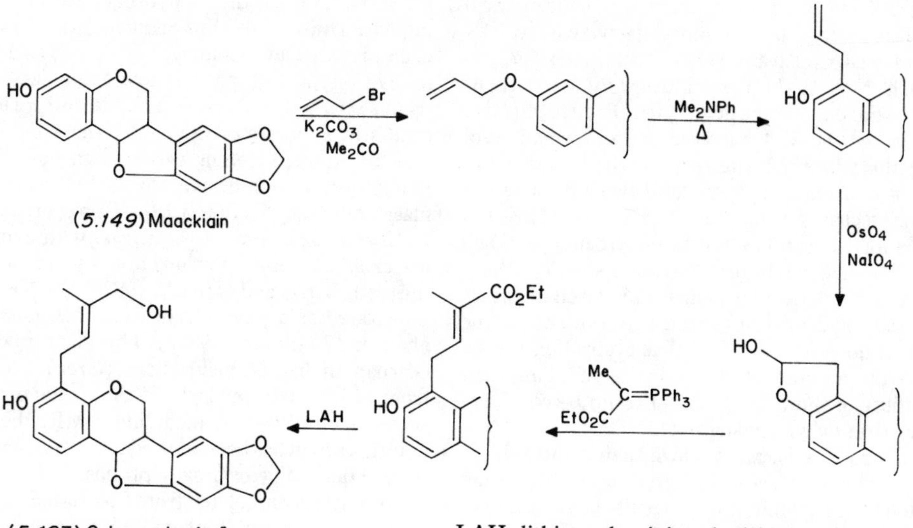

HO, Br, K_2CO_3, Me_2CO, Me_2NPh, Δ, HO, OsO_4, $NaIO_4$, HO, Me, $C{=}PPh_3$, EtO_2C, CO_2Et, HO, LAH, OH, HO

(5.149) Maackiain

(5.127) Cabenegrin A-I

LAH, lithium aluminium hydride.

Fig. 5.7 Synthesis of cabenegrin A-I.

BzO, MeO_2C, LAH, MnO_2, BzO, OHC, $Ph_3P{=}$, OH, BzO, H_2, Pd–C, HO, OH, OH

(5.150)

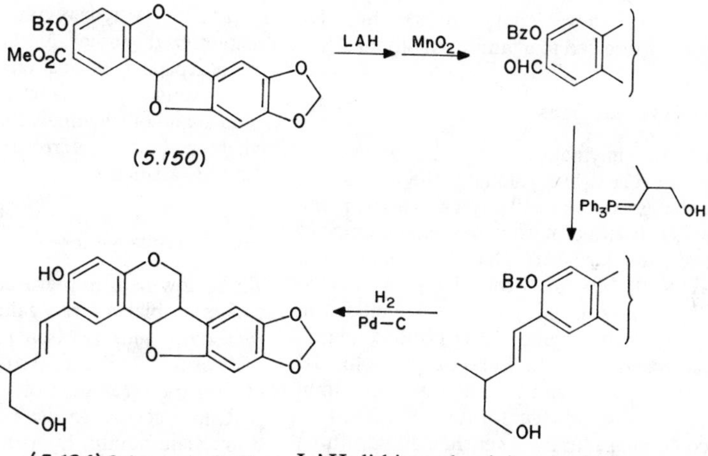

(5.124) Cabenegrin A-II

LAH, lithium aluminium hydride.

Fig. 5.8 Synthesis of cabenegrin A-II.

is remarkable and the first report of an isoflavonoid being found in a marine organism (Sanduja *et al.*, 1984). Its structure is new, although neodunol itself has been found in several plants, and was confirmed by X-ray crystallography.

Structural revision of the pterocarpan homoedudiol originally isolated from *Neorautanenia edulis* has been necessary after detailed study of phytoalexins from fungus-infected leaflets of *Calopogonium mucunoides* (Ingham and Tahara, 1985). This plant produces both 2-dimethylallyl- and 8-dimethylallyl-3,9-dihydroxypterocarpans (*5.123* and *5.117*, respectively). Homoedudiol was originally given the structure (*5.123*), but is now shown to be identical to sophorapterocarpan A (*5.117*), the 8-dimethylallyl derivative. Structure (*5.123*), the 2-dimethylallyl derivative, is now assigned to the second *Calopogonium* phytoalexin, and renamed calopocarpin.

Photolysis of pterocarpans in methanol or acetic acid leads to 4-methoxy- or acetoxy- derivatives of 2′-hydroxy-3, 4-*trans*-isoflavans by ring fission and solvolysis (Breytenbach *et al.*, 1981). Recyclization to pterocarpans has been investigated in some detail and related to effective delocalization of the C-4 carbocation or formation of quinone methide intermediates due to the functionality present in the molecules. The evidence available favours a carbocation mechanism for the cyclization. This has implications for the synthesis of pterocarpans via DDQ oxidation of 2′-hydroxyisoflavans (see Dewick, 1982). Whilst quinone methide intermediates may well be involved where a 7-hydroxyl substituent is present in the isoflavan, it is now demonstrated that cyclization can be achieved in the absence of this substituent. Thus, the synthetic utility of the DDQ reaction could be considerably greater than originally suggested.

The introduction of isoprenyl substituents into a flavonoid or isoflavonoid skeleton by direct alkylation using dimethylallyl halides almost always results in a mixture of products, arising from both *O*- and *C*-alkylations. Typically, *O*-alkylation predominates, unless *meta*-orientated hydroxyls are present, when *C*-alkylation is preferred. In the synthesis of cabenegrin A-I (Baruah *et al.*, 1984a), *C*-isoprenylation of maackiain was made to predominate by using a photochemical reaction with dimethylallyl bromide. Yields were low (about 10%), but *O*-isoprenylation was suppressed to a minimum (about 2%).

5.6.2 6a-Hydroxypterocarpans

For many years, the phytoalexin (+)-pisatin was the only member of this class, but gradually the number of known examples has grown. Several phytoalexins from the genus *Glycine* are 6a-hydroxypterocarpans and a range of different isoprenylated structures has been identified. Canescacarpin (*5.139*), isolated from bacteria-infected leaves of *Glycine canescens* (Lyne *et al.*, 1981), contains an isopropenyldihydrofuran grouping with opposite configuration (*R*) to that observed in the isomeric glyceollin III (*5.140*). Glyceofuran (*5.137*) and 9-*O*-methylglyceofuran (*5.138*) from *Glycine max* represent a further variant on this structure, containing a furan rather than dihydrofuran group (Ingham *et al.*, 1981b), and an uncommon isopropenylfuran is observed in clandestacarpin (5.143) from *Glycine clandestina* (Lyne *et al.*, 1981). The isolation of glyceollin II (*5.136*) and small amounts of glyceollin III (*5.140*) as phytoalexins from fungus-infected leaves of *Costus speciosus* (Zingiberaceae) is of interest, since isoflavonoids do not otherwise occur in this family (Kumar *et al.*, 1984).

As with pterocarpans, the optical activity of 6a-hydroxypterocarpans can be correlated with absolute configuration, laevorotatory compounds having the 6a*S*, 11a*S* configuration. This is, of course, equivalent to the 6a*R*, 11a*R* configuration of laevorotatory pterocarpans, the substitution of the 6a-hydrogen with hydroxyl affecting priorities and thus nomenclature. Most natural examples are laevorotatory, but a range of dextrorotatory 6a-hydroxypterocarpans is now recognized. (+)-Pisatin from *Pisum* spp. is a prime example. Surprisingly, this co-occurs in *P. sativum* with (−)-maackiain, a minor phytoalexin, though feeding experiments have demonstrated that (+)-pisatin is derived by 6a-hydroxylation of (+)-maackiain (see Section 5.18). However, by supplying $CuCl_2$-stressed pods of *P. sativum* with exogenous (−)-maackiain, large proportions of (−)-pisatin are also synthesized (Banks and Dewick, 1982a). (−)-Pisatin can also be isolated as a phytoalexin from *Tephrosia bidwilli* (Ingham and Markham, 1982a). This plant produces acanthocarpan in higher yields than *Caragana acanthophylla*, from which it was first isolated. As a result of more detailed spectroscopic studies, including NMR, the structure of acanthocarpan has been revised to (*5.132*) instead of the *C*-methyl derivative originally proposed. Tuberosin (*5.133*) has recently been isolated from *Calopogonium mucunoides* in laevorotatory form (Ingham and Tahara, 1985), whereas material from *Pueraria tuberosa* was dextrorotatory.

A novel 6a-hydroxypterocarpanone structure has been assigned to hydroxytuberosone (*5.144*) isolated from tubers of *Pueraria tuberosa* (Prasad *et al.*, 1984). The oxidized ring A feature was first described in the phaseollin metabolite la-hydroxyphaseollone and has since been encountered in other microbial transformation products of pterocarpans (see Section 5.17). Hydroxytuberosone is presumably derived from tuberosin (*5.133*) in *P. tuberosa* in an analogous manner, though it cannot be fully certain whether plant or microbial metabolism is responsible for the transformation.

5.6.3 Pterocarpenes

Three new natural pterocarpenes have been isolated, the anti-microbial erycristagallin (*5.147*) from roots of *Erythrina crista-galli* (Mitscher *et al.*, 1984), and anhydrotuberosin (*5.145*) and its methyl ether (*5.146*) from tubers of *Pueraria tuberosa* (Prasad *et al.*, 1985).

Ready access to isoflavanones by the use of di-isobutylaluminium hydride as a selective reducing agent for the 2, 3-double bond of isoflavones has allowed syn-

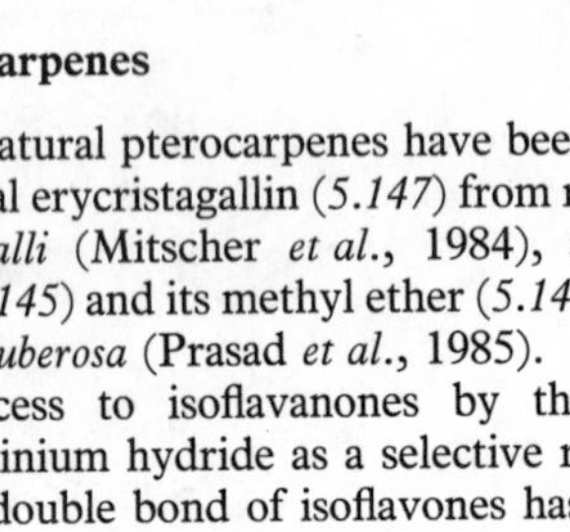

(*5.151*) R=H, ≢Deoxybryaquinone?
(*5.152*) R=OH, ≢Bryaquinone?

Table 5.7 Isoflavans

Isoflavan	Substituents 5	6	7	8	2′	3′	4′	5′	6′	Plant sources	Chirality/ optical activity	References
(i) *Isoflavans*												
7, 3′-Dihydroxy-			OH			OH				Bovine urine – metabolite?	*S*, –	Luk *et al.* (1983)
Equol			OH				OH			Human urine – metabolite?		Axelson *et al.* (1982)
										Bovine urine – metabolite?	*S*, –	Luk *et al.* (1983)
7-Methoxy-4′-hydroxy-			OMe				OH			Bovine urine – metabolite?		Luk *et al.* (1983)
Demethylvestitol			OH		OH		OH			*Lotus angustissimus* leaf*		Ingham and Dewick (1980)
										L. edulis leaf*		
										Phaseolus coccineus seedling*		Adesanya *et al.* (1985)
										P. mungo seedling*		Adesanya *et al.* (1984)
Vestitol			OH		OH		OMe			*Dalbergia odorifera* heartwood	±	Goda *et al.* (1985)
											R	Yahara *et al.* (1985)
										Hedysarum polybotrys root	–	Miyase *et al.* (1984)
										Lotus angustissimus leaf*		Ingham and Dewick (1980)
										L. edulis leaf*		
										Medicago rugosa leaf*	–	Ingham (1982b)
										Trigonella spp. leaf*		Ingham (1981b)
										Zollernia paraensis wood	+	Ferrari *et al.* (1983)
Sativan			OH		OMe		OMe			*Trigonella* spp. leaf*		Ingham (1981b)
Isosativan			OMe		OH		OMe			*Medicago rugosa* leaf*	–	Ingham (1982b)
Laxifloran			OH		OMe	OMe	OH			*Phaseolus vulgaris* root		Biggs *et al.* (1983)
Mucronulatol			OH		OMe	OH	OMe			*Dalbergia odorifera* heartwood	±	Goda *et al.* (1985)
Isomucronulatol			OH		OH	OMe	OMe			*Colutea arborescens* seedling, pod	*R*, –	Al-Ani and Dewick (1985)
Isomucronulatol 7-*O*-glucoside			OGlc		OH	OMe	OMe			*Astragalus mongholicus*		Lu *et al.* (1984)
5′-Methoxyvestitol			OH		OH		OMe	OMe		*Dalbergia odorifera* heartwood	*R*	Yahara *et al.* (1985)
5-Methoxyvestitol	OMe		OH		OH		OMe			*Lotus edulis* leaf*		Ingham and Dewick (1980)

(*Contd.*)

Table 5.7 (*Contd.*)

Isoflavan	*Substituents* 5	6	7	8	2′	3′	4′	5′	6′	*Plant sources*	*Chirality/ optical activity*	*References*
Lotisoflavan	OMe		OMe		OH		OH			*Lotus angustissimus* leaf* *L. edulis* leaf*		Ingham and Dewick (1980)
Duartin			OH	OMe	OMe	OH	OMe			*Dalbergia odorifera* heartwood	+	Goda *et al.* (1985)
Isoduartin			OH	OMe	OH	OMe	OMe			*Dalbergia odorifera* heartwood	±	Goda *et al.* (1985)
6, 8, 2′-Trihydroxy-7, 3′, 4′-trimethoxy-		OH	OMe	OH	OH	OMe	OMe			*Machaerium* sp. heartwood	*S*	Imamura *et al.* (1982)
6, 2′-Dihydroxy-7, 8, 3′, 4′-tetramethoxy-		OH	OMe	OMe	OH	OMe	OMe			*Machaerium* sp. heartwood	*S*	Imamura *et al.* (1982)
Phaseollinisoflavan (*5.154*)										*Phaseolus coccineus* seedling*		Adesanya *et al.* (1985)
										P. vulgaris cotyledon*		Bailey and Berthier (1981)
												Goosens and Van Laere (1983)
										P. vulgaris hypocotyl*		Garcia-Arenal *et al.* (1980)
										P. vulgaris various tissues*		Stössel and Magnolata (1983)
Crotmarine (*5.155*)										*Crotalaria madurensis* leaf, stem	−	Bhakuni and Chaturvedi (1984)
Licoricidin (*5.156*)										*Glycyrrhiza uralensis* root		Chang *et al.* (1983)
(ii) *Isoflavanquinones*												
Claussequinone (*5.157*)										*Dalbergia odorifera* heartwood	*R*	Yahara *et al.* (1985)

*Plant part was subjected to physiological stress.

thesis of several of the bryacarpenes, isolated earlier from the heartwood of *Brya ebenus* (Antus *et al.*, 1982). However, the structures assigned to two related pterocarpene-quinones, bryaquinone (*5.152*) and deoxybryaquinone (*5.51*), isolated from the same source have been shown to be incorrect. Compounds corresponding to these structures were synthesized via silver carbonate oxidation of the quinols, and were different from the natural materials (Antus *et al.*, 1982). Bryaquinone is most probably a pterocarpene-*o*-quinone, but the substitution pattern in the quinone ring has yet to be assigned (Antus *et al.*, 1982; Kolonits *et al.*, 1983).

5.7 ISOFLAVANS

Reported isolations of isoflavans (*5.153*) during the review period are given in Table 5.7. For convenience, isoflavan-quinones are no longer regarded as a separate group of isoflavonoids, but as a subsection of the isoflavans, and are also included in Table 5.7.

The simplest representative of the isoflavans is 7,4′-dihydroxyisoflavan (equol) isolated from mammalian urine, and generally accepted as a metabolite of simple isoflavones like daidzein and formononetin taken in the diet. Equol has also been shown to occur in human urine, and its origin is assigned to isoflavonoid components of soya-based foods (Axelson *et al.*, 1982). In bovine urine, equol and two related compounds, 7-*O*-methylequol and 7,3′-dihydroxyisoflavan have been detected (Luk *et al.*, 1983). The latter compound has an unusual oxygenation pattern, and is logically likely to be derived from a 4′-oxygenated isoflavone such as calycosin ingested via the diet, and then subjected to a deoxygenation process by the animal's microflora.

All plant-derived isoflavans contain a 2′-oxygen substituent, a feature which appears to be a consequence of the close relationship for the biosynthetic pathways to isoflavans and pterocarpans (Dewick, 1982). Also shared with the pterocarpans is the comparative rarity of oxygen substituents at position 5 (position 1 of pterocarpans). Only three examples have so far been reported, the previously known compounds licoricidin (*5.156*) and 5-methoxyvestitol, and the new lotisoflavan (5,7-dimethoxy-2′,4′-dihydroxyisoflavan) from *Lotus angustissimus* and *L. edulis* (Ingham and Dewick, 1980). Lotisoflavan co-occurs with the isomeric 5-methoxyvestitol in *L. edulis*, and with 5-deoxyisoflavans in both plants. The first natural isoflavan glycoside, isomucronulatol 7-*O*-glucoside, has been isolated from *Astragalus mongholicus* (Lu *et al.*, 1984).

No new examples of isoflavanquinones have been isolated, though claussequinone (*5.157*) has recently been reported to occur in the heartwood of *Dalbergia odorifera* (Yahara *et al.*, 1985). One of the five bi-isoflavonoids also isolated from this source contains claussequinone as a monomer unit (see Section 5.15). An X-ray crystallographic examination of (3*R*)-claussequinone confirms the structure assigned, and shows the dihydropyran ring to have a distorted half-chair conformation, with the quinone ring occupying the equatorial position (Gambardella *et al.*, 1983). Syntheses of abruquinones A and B (*5.158* and *5.159* respectively), extractives from *Abrus precatorius*,

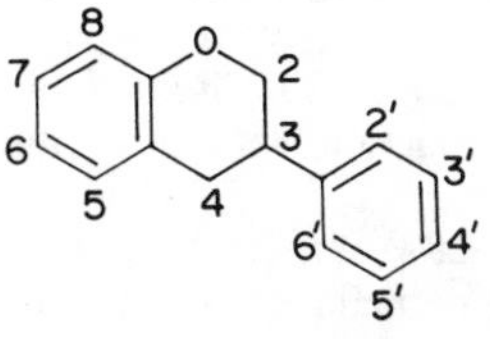

(*5.153*)

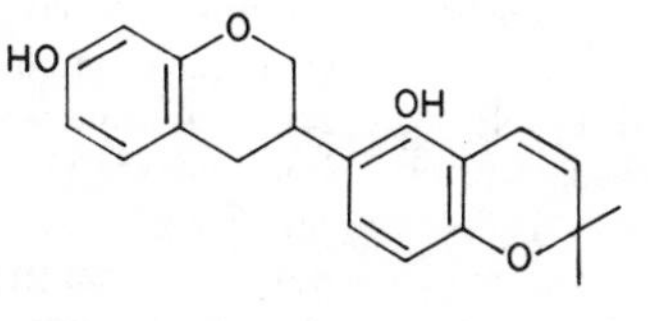

(*5.154*) Phaseollinisoflavan

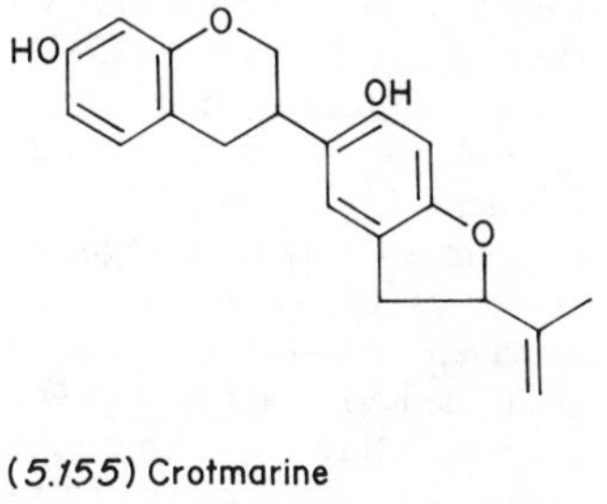

(*5.155*) Crotmarine

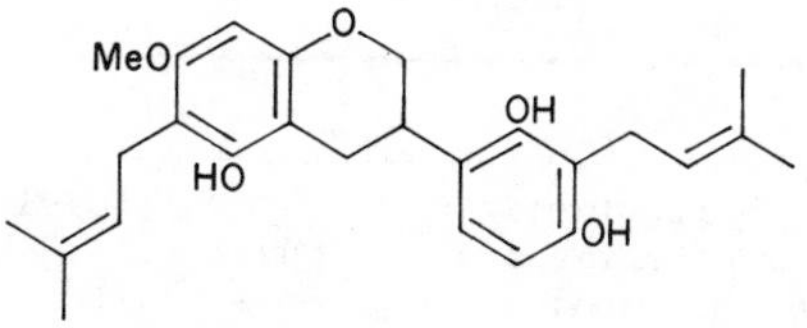

(*5.156*) Licoricidin

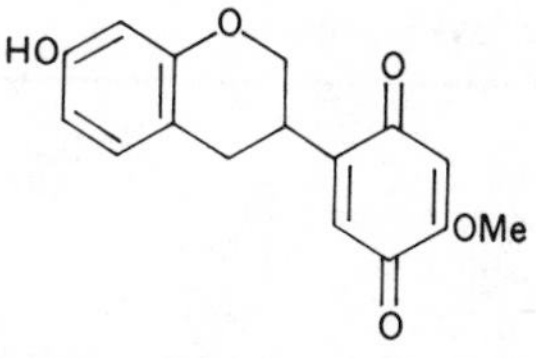

(*5.157*) Claussequinone

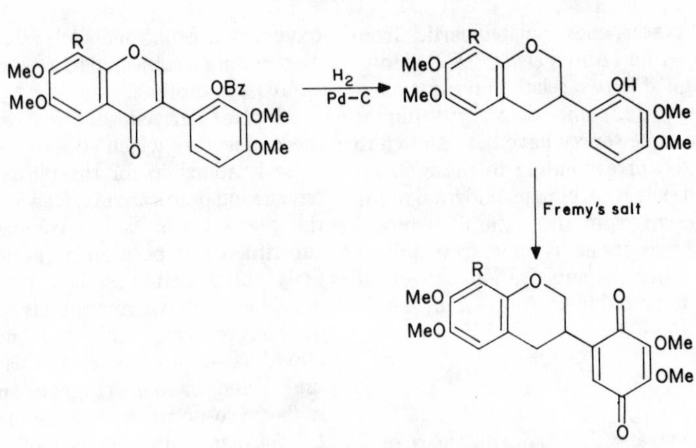

(5.158) R=H, Abruquinone A

(5.159) R=OMe, Abruquinone B

Fig. 5.9 Synthesis of abruquinones A and B.

have been achieved via Fremy's salt oxidation of the corresponding monohydroxyisoflavans (Lupi *et al.*, 1980) (Fig. 5.9).

5.8 ISOFLAV-3-ENES

Isoflav-3-enes (*5.160*) have only been recognized as natural products during the last decade or so, and the reason for this is undoubtedly their high reactivity. As techniques for the isolation and characterization of natural products improve, these labile compounds are being found more frequently. All four isoflav-3-enes reported recently (Table 5.8) are new structures. Haginin C (*5.162*) and haginin D (*5.161*) come from the heartwood of *Lespedeza cyrtobotrya* (Miyase *et al.*, 1981), which earlier yielded two other examples of these compounds, haginins A and B. Odoriflavene (*5.163*) from the heartwood of *Dalbergia odorifera* (Goda *et al.*, 1985) shares the same 7,2′,3′,4′-oxygenation pattern of haginins A and C, and sepiol, but co-occurs with a wide range of variously substituted isoflavans and pterocarpans. An isoflav-3-ene isolated from the heartwood of *Baphia nitida* was assigned one of two structures on spectroscopic evidence (Arnone *et al.*,1981). The correct structure (*5.164*) was confirmed by synthesis from the isoflavanone via lithium aluminium hydride reduction and then acid-catalysed dehydration (Shoukry *et al.*, 1982). The co-occurrence of (*5.164*) with the red pigment santarubin C (*5.165*) allowed the correct substitution pattern in the B-ring of (*5.164*) to be surmised. Santarubin C, together with related pigments, may be envisaged as an isoflavonoid–diarylpropene combination, and could well be derived in the plant from the isoflav-3-ene (*5.164*) (Arnone *et al.*, 1981). It has been demonstrated that small amounts of a santarubin-like derivative (*5.168*) are obtained when isoflavylium salts are reacted with 2,4-diarylpropenes (Afonya *et al.*, 1985a), along with corresponding isoflav-3-ene derivatives (*5.166* and *5.167*) (Fig. 5.10). Per-*O*-methylsantarubin (*5.169*) was obtained from (*5.168*) by catalytic debenzylation followed by oxidation. The isoflav-3-ene structures (*5.166*) and (*5.167*) were confirmed by X-ray crystallography (Afonya *et al.*, 1985b).

A more satisfactory synthetic route from isoflavylium salts to isoflav-3-enes is by reduction with potassium borohydride (Deschamps-Vallet *et al.*, 1983). However, other work shows that the product(s) of reduction depend very much on the nature of the reducing agent used, and the substituents on the isoflavylium skeleton (Liepa, 1981). Thus, 5,7,4′-trihydroxyisoflavylium chloride yielded predominantly the isoflav-3-ene when reduced

Table 5.8 Isoflav-3-enes

Isoflav-3-ene	*Plant sources*	*References*
Haginin D (*5.161*)	*Lespedeza cyrtobotrya* heartwood	Miyase *et al.* (1981)
Haginin C (*5.162*)	*Lespedeza cyrtobotrya* heartwood	Miyase *et al.* (1981)
Odoriflavene (*5.163*)	*Dalbergia odorifera* heartwood	Goda *et al.* (1985)
6, 7, 3′-Trihydroxy-2′, 4′-dimethoxy- (*5.164*)	*Baphia nitida* heartwood	Anone *et al.* (1981)

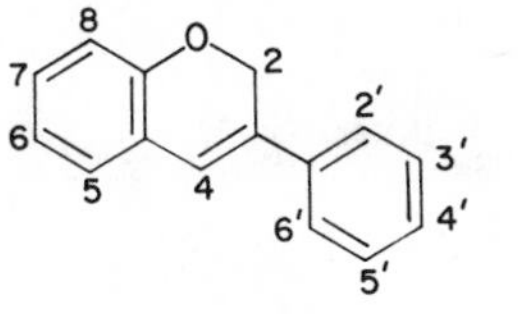

(5.160)

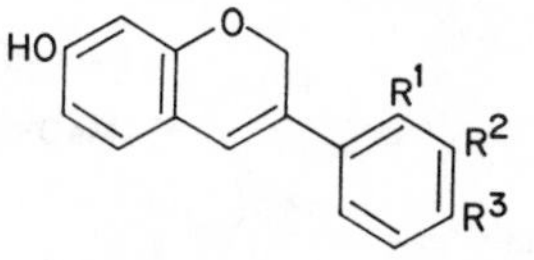

(5.161) $R^1=R^3=OH, R^2=H$, Haginin D
(5.162) $R^1=R^3=OH, R^2=OMe$, Haginin C
(5.163) $R^1=OH, R^2=R^3=OMe$, Odoriflavene

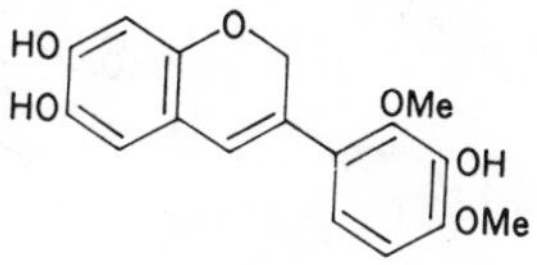

(5.164)

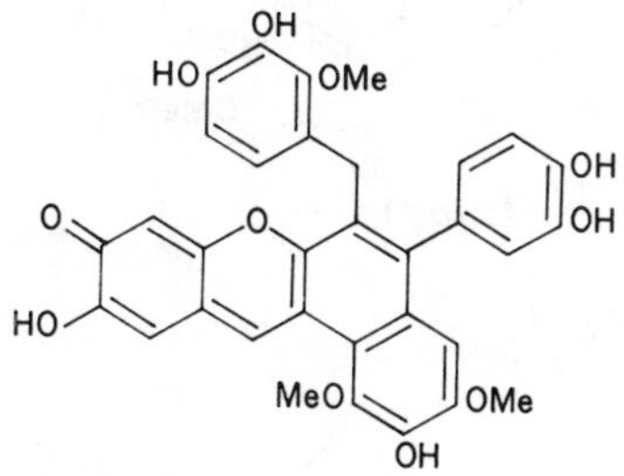

(5.165) Santarubin C

with sodium cyanoborohydride, but the 5,7-dihydroxy-4′-methoxyisoflavylium salt gave a roughly 1:1 mixture of the isoflav-3-ene (*5.173*) and isoflav-2-ene (*5.172*) (Fig. 5.11). Partial catalytic hydrogenation gave the isoflav-3-ene contaminated with the fully hydrogenated isoflavan. The isoflavylium salts (*5.171*) used were obtained by reacting phloroglucinol with arylmalondialdehydes (*5.170*), but this reaction did not proceed with resorcinol. Thus, a synthesis of 7,4′-dihydroxyisoflav-3-ene (*5.175*) was devised by modification of the isoflavone daidzein diacetate (*5.174*) (Fig. 5.12), which illustrates the mild conditions which must be employed to obtain isoflav-3-enes (Liepa, 1981). In particular, the final deacetylation could not be achieved using standard basic or acidic conditions, and imidazole in ethanol was chosen as a mild hydrolysing reagent.

Two types of isoflav-3-ene derivatives are potentially valuable intermediates in the synthesis of other classes of natural isoflavonoids. An enamine-mediated reaction between *N*-styrylmorpholine (*5.176*) and salicylaldehyde gave 2-morpholinoisoflav-3-ene (*5.177*) (Dean and Varma, 1981, 1982). This intermediate may be reduced to an isoflavan (*5.178*), or oxidized to a 3-phenylcoumarin (*5.179*) (Fig. 5.13). Condensation of salicylaldehydes (*5.180*) with arylacetaldehydes (*5.181*), generated *in situ* from an arylglycidate salt (*5.182*), can give, depending on the precise conditions used, either 2,4-dihydroxyisoflavans (*5.183*) or 2-hydroxyisoflav-3-enes (*5.184*) (Liepa, 1984) (Fig. 5.14). The latter products are valuable intermediates, capable of being converted into isoflavylium salts (including 5-deoxy derivatives), isoflavans, isoflav-3-enes, 2-hydroxyisoflavans (and thus into isoflav-2-enes) or 3-arylcoumarins. The sequences of Fig. 5.14 may be used to prepare a useful range of isoflavonoids with hydroxy and methoxy substituents in both rings.

5.9 3-ARYLCOUMARINS

This small group of isoflavonoids (*5.185*) contained only five representatives in 1980. Since then, one compound (*5.186*) has been isolated from a new source, the wood of *Zollernia paraensis*, where it co-occurs with medicarpin and vestitol, pterocarpan and isoflavan analogues (Ferrari *et al.*, 1984a). A sixth naturally occurring example of the 3-arylcoumarins is glycycoumarin (*5.187*), isolated from stem and roots of *Glycyrrhiza uralensis* (Zhu *et al.*, 1984). Glycycoumarin co-occurs with the similarly substituted coumestans glycyrol and isoglycyrol, and is the 7-demethyl derivative of glycyrin (*5.188*), previously isolated from a species of *Glycyrrhiza*.

New approaches to the synthesis of 3-arylcoumarins have been described. A base-catalysed Kostanecki condensation is exploited in the reaction of phenylacetyl chloride with various substituted salicylaldehydes using potassium carbonate in acetone (Rao and Srimannarayana, 1981; Neelakantan *et al.*, 1982), but the sequence has not been applied to synthesize products substituted in the B-ring (Fig. 5.15). Less attractive for general use is a route (Fig. 5.16) in which halogenated α-phenylcinnamic acids (*5.189*) are treated with potassium amide in liquid ammonia (Kessar *et al.*, 1981). Both *cis* and *trans* isomers give the same product, which probably arises via a benzyne intermediate. Good yields of A-ring-substituted 3-phenylcoumarins may be obtained by chromium trioxide oxidation of the appropriate isoflavylium salt (Deschamps-Vallet *et al.*, 1983), and DDQ oxidation of 2-hydroxyisoflav-3-enes (see Section 5.8) offers a high-yielding route to 3-arylcoumarins substituted in both rings (Liepa, 1984).

5.10 3-ARYL-4-HYDROXYCOUMARINS

All newly reported examples of 3-aryl-4-hydroxycoumarins (*5.190*) have been isolated from seeds

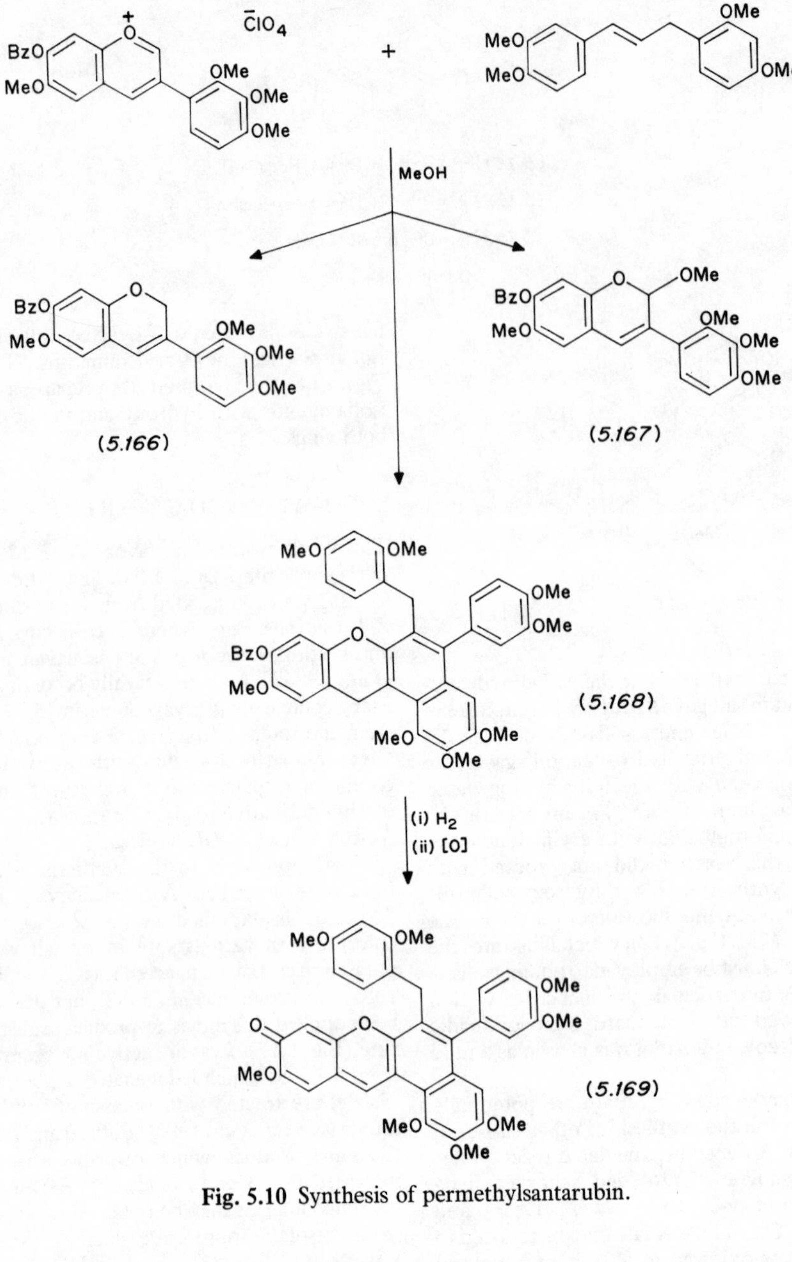

(5.166)

(5.167)

(5.168)

(5.169)

Fig. 5.10 Synthesis of permethylsantarubin.

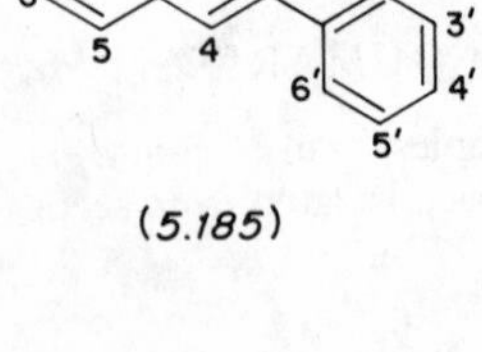

(5.185)

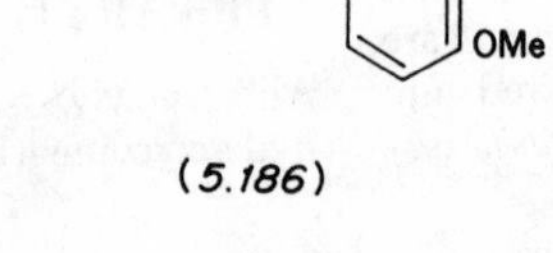

(5.186)

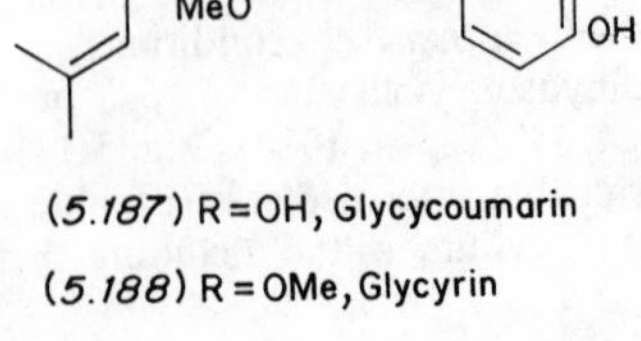

(5.187) R = OH, Glycycoumarin

(5.188) R = OMe, Glycyrin

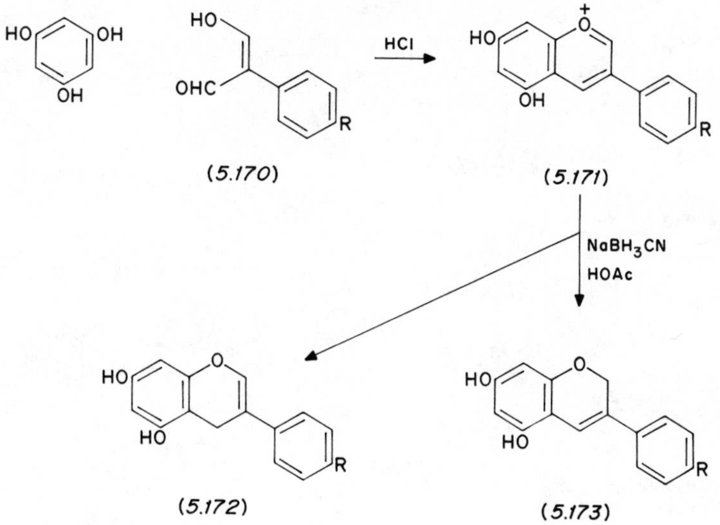

(5.170) (5.171) (5.172) (5.173)

Fig. 5.11 Synthesis of isoflav-2-enes and isoflav-3-enes.

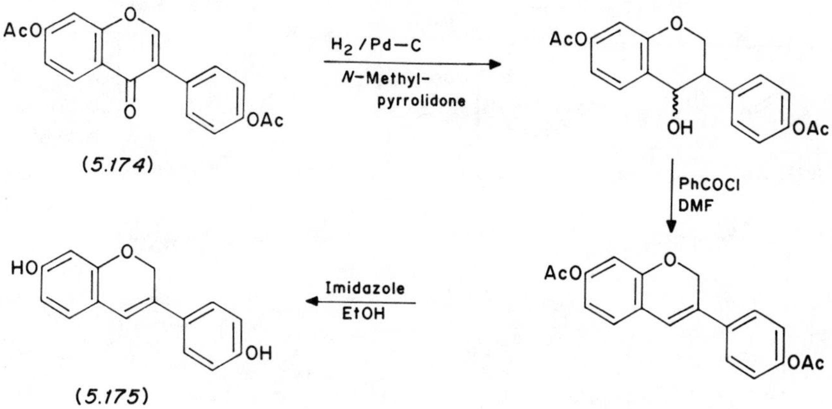

(5.174) (5.175)

Fig. 5.12 Synthesis of 7, 4′-dihydroxyisoflav-3-ene.

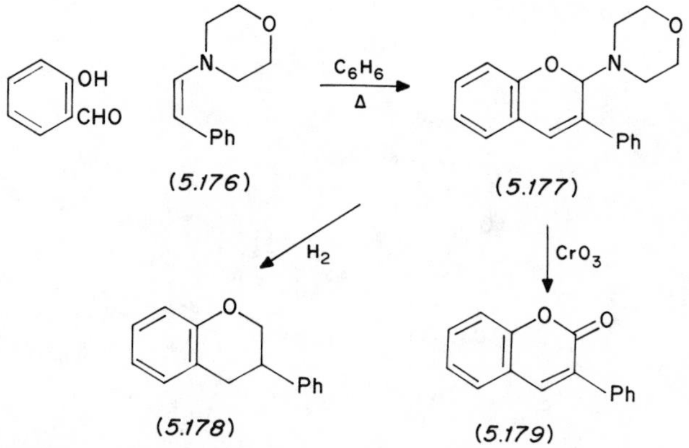

(5.176) (5.177) (5.178) (5.179)

Fig. 5.13 Synthesis of isoflavan and 3-phenylcoumarin.

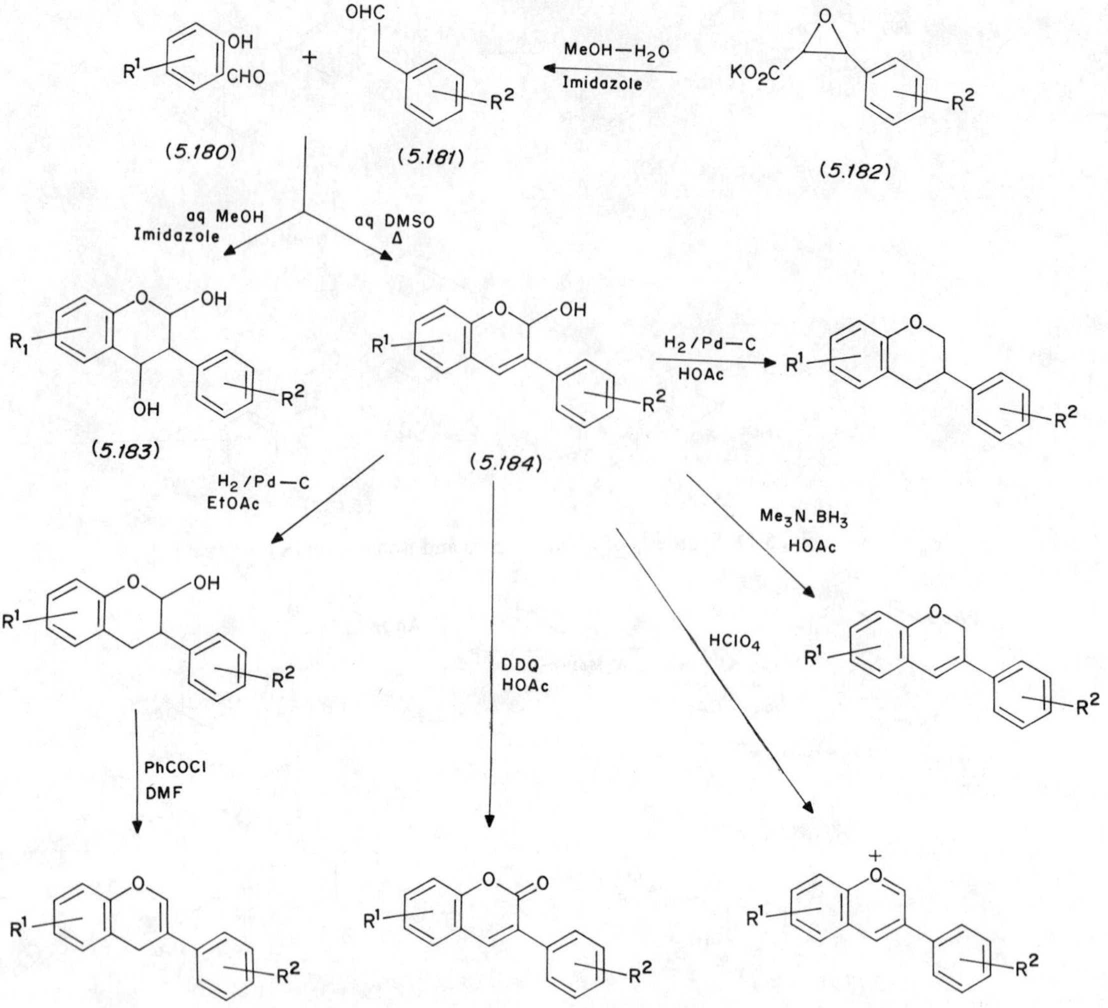

Fig. 5.14 2-Hydroxyisoflav-3-enes as intermediates for the synthesis of other isoflavonoids.

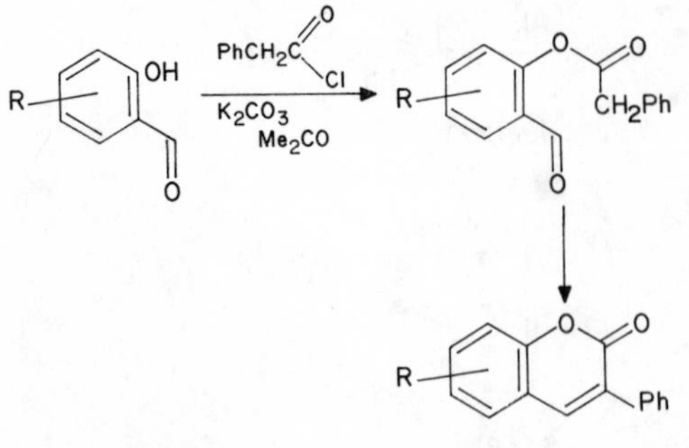

Fig. 5.15 Synthesis of A-ring-substituted arylcoumarins.

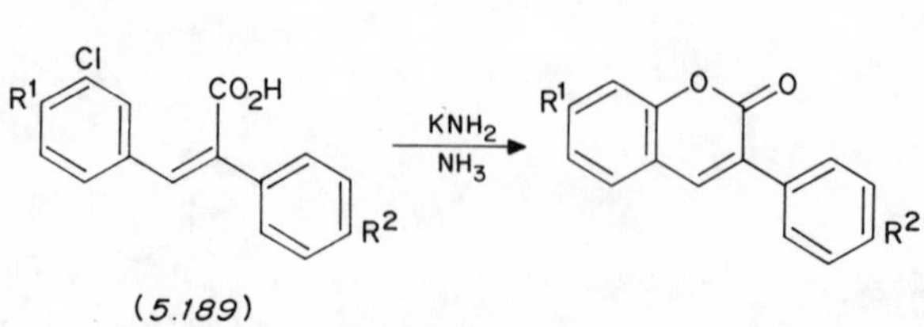

Fig. 5.16 Synthesis of disubstituted 3-arylcoumarins.

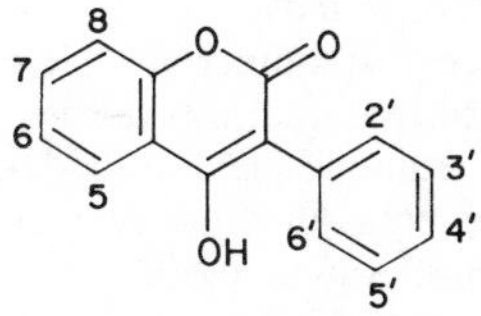

(*5.190*)

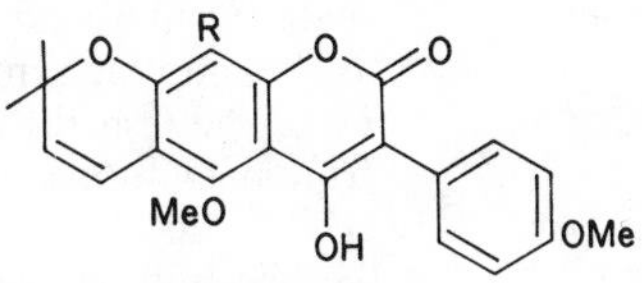

(*5.191*) R=H, Robustic acid

(*5.192*) R=OMe, Thonningine B

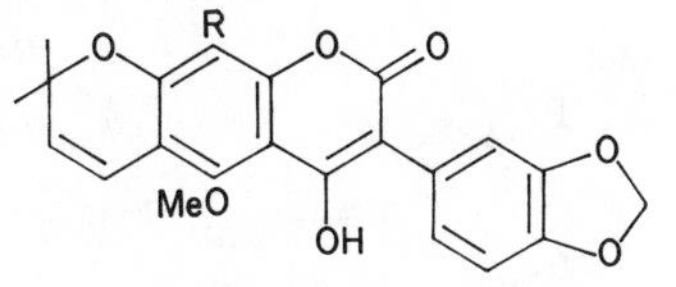

(*5.193*) R=H, Robustin

(*5.194*) R=OMe, Thonningine A

of *Millettia thonningii* by two independent groups (Olivares *et al.*, 1982; Khalid and Waterman, 1983). Robustic acid (*5.191*) is a known compound, but thonningine-A (*5.194*) and thonningine-B (*5.192*) are new structures in this small group. All the known examples have a 5-methoxy substituent.

Robustin (*5.193*), robustic acid and their 4-*O*-methyl ethers have been synthesized by a sequence involving isoprenylation of the basic 3-aryl-4-methoxycoumarin (*5.195*), prepared via the corresponding deoxybenzoin and ethyl chloroformate (Ahluwalia *et al.*, 1981) (Fig. 5.17). The dimethylchromene substituent was introduced via cyclization of the propargyl ether, but this reaction is regioselective, giving the angular isomers (*5.196* and *5.197*). The required linear isomers (*5.194* and *5.198*) were obtained by firstly blocking the 8 position with iodine, which was later removed during the cyclization reaction in dimethylaniline. Partial demethylation of the 4-methoxyl also occurred. This strategy has been applied successfully in the synthesis of several other groups of natural products containing dimethylchromene rings.

5.11 COUMESTANS

Several new coumestans (*5.199*) have been identified, and these are noted in Table 5.9, together with reported isolations of previously known compounds. The structures invite little comment since the substitution patterns are all represented elsewhere in other classes of isoflavonoid. Wedelolactone has been isolated for the first time from a leguminous plant, *Ougeinia dalbergioides* (Kalidhar and Sharma, 1984), where it co-occurs with related isoflavones and isoflavanones. Previously, this coumestan was known only from two plants of the Compositae, *Wedelia calendulacea* and *Eclipta alba*. Re-examination of *W. calendulacea* leaves has confirmed the presence of wedelolactone and norwedelolactone, together with the free acid norwedelic acid (*5.207*) (Govindachari and Premila, 1985). These authors suggest that (*5.207*) is in fact the true natural product, and that the lactones may well be artefacts.

The structure of wairol, isolated from infected foliage of *Medicago sativa* (Biggs and Shaw, 1980) has been confirmed by synthesis, using the conventional route by hydrogen peroxide rearrangement of the appropriate flavylium salt (Shaw *et al.*, 1982). This sequence allowed preparation of the 7-OCD_3-labelled compound and assignment of the mass spectral fragmentation. It was concluded that the presence of a $M-$CHO fragment was diagnostic for 7-methoxycumestans, and a consequence of the proximity of this group to the lactone carbonyl. However, 7-oxygenation is comparatively rare in coumestans, and only three examples are known, repensol and trifoliol (both 7-

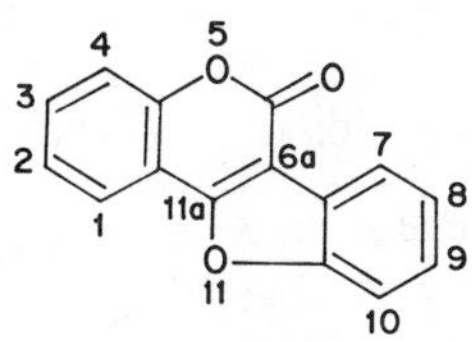

(*5.199*)

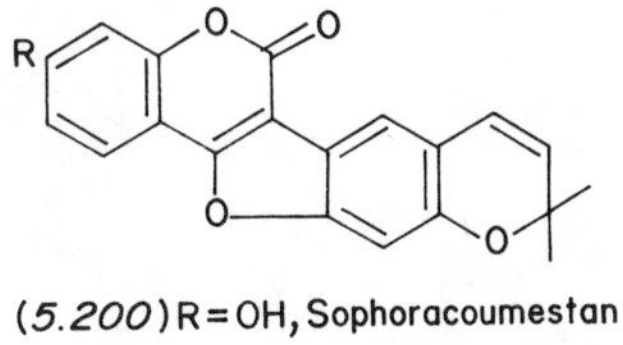

(*5.200*) R=OH, Sophoracoumestan A

(*5.201*) R=OMe, Tuberostan

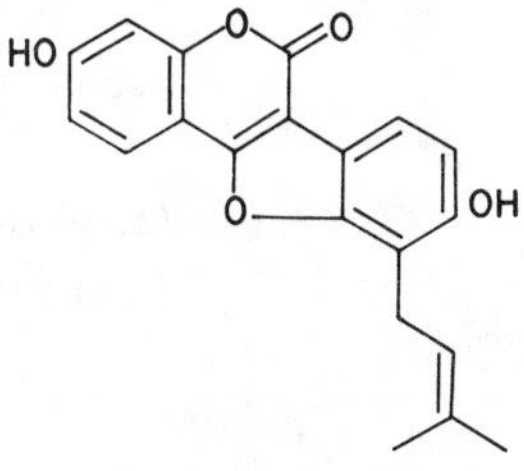

(*5.202*) Isosojagol

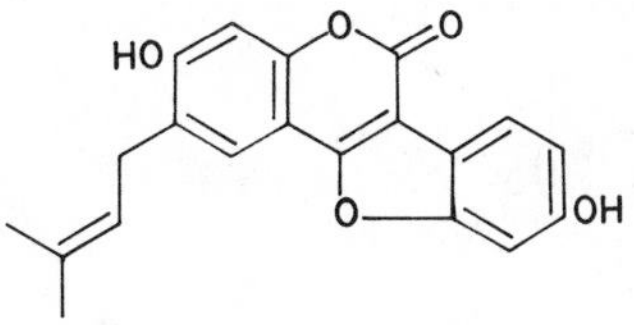

(*5.203*) Psoralidin

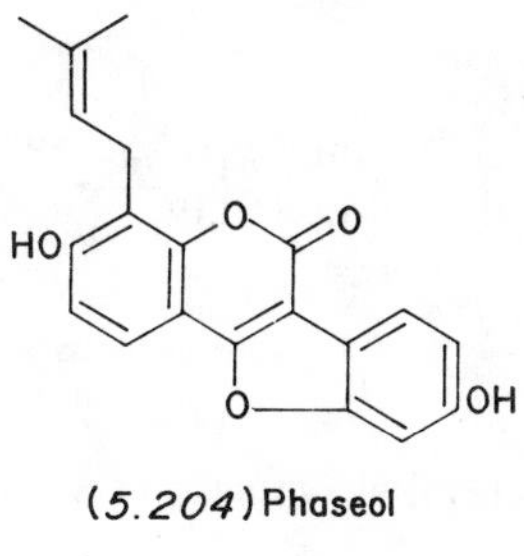

(*5.204*) Phaseol

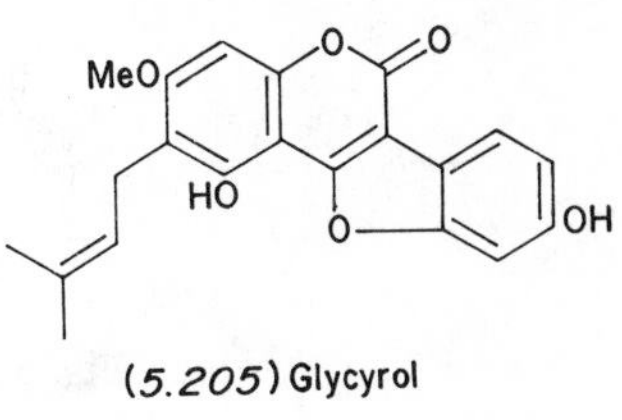

(*5.205*) Glycyrol

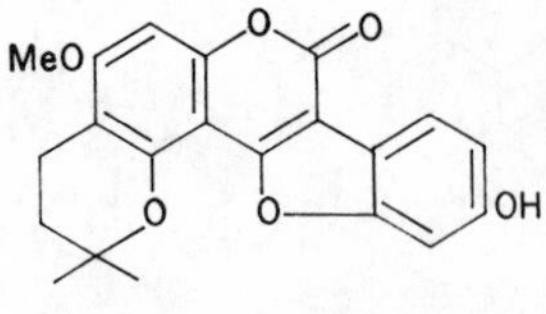

(*5.206*) Isoglycyrol

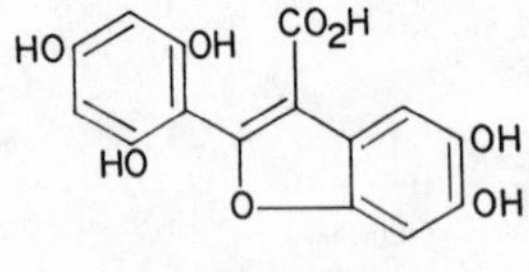

(*5.207*) Norwedelic acid

hydroxylated) and wairol (7-methoxylated).

Variants on synthetic sequences to coumestans have been published. 3-Aryl-4-hydroxycoumarins derived from 4-hydroxycoumarins and *o*-quinone may be oxidatively cyclized to coumestans using potassium ferricyanide (Srihari and Sundaramurthy, 1980). Anodic oxidation of catechol in the presence of a 4-hydroxycoumarin (*5.208*) effectively achieves the same end, and is noteworthy for

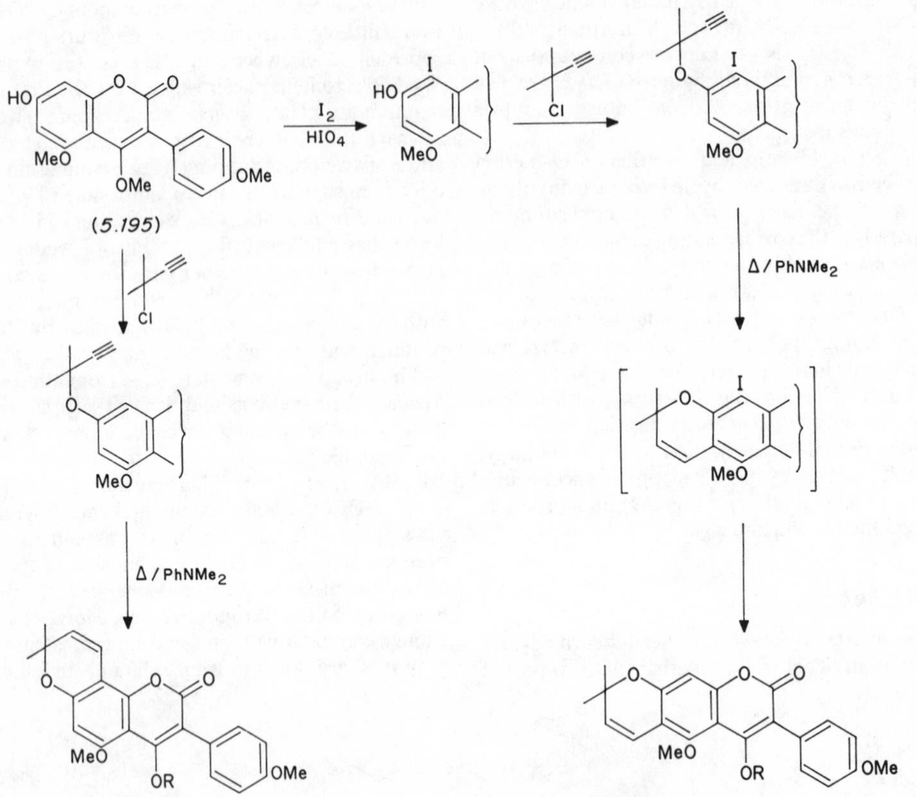

(*5.195*) (*5.196*) R=H (*5.197*) R=Me (*5.191*) R=H, Robustic acid (*5.198*) R= Me

Fig. 5.17 Synthesis of 3-aryl-4-hydroxycoumarins.

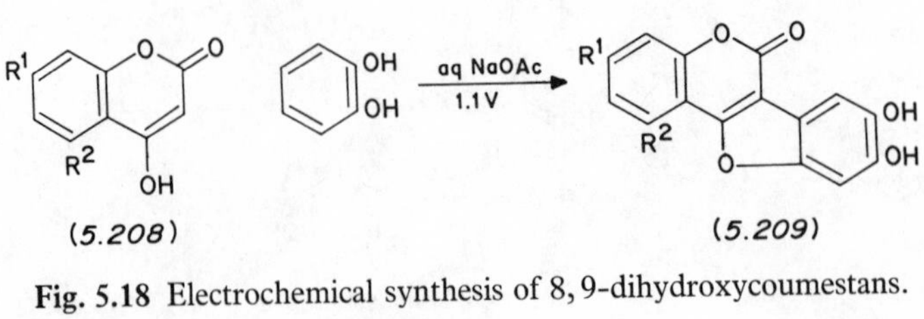

(*5.208*) (*5.209*)

Fig. 5.18 Electrochemical synthesis of 8, 9-dihydroxycoumestans.

Coumestan	Substituents 1	2	3	4	7	8	9	10	Plant sources	References
Coumestrol			OH				OH		*Centrosema pubescens* seed*, leaf*	Sukumaran and Gnanamanickam (1980)
									Dolichos biflorus leaf	Keen and Ingham (1980)
									Glycine max leaf*	Fett (1984)
									G. max root	Le-Van (1984)
										Porter *et al.* (1985)
									G. spp. leaf	Vaughan and Hymowitz (1984)
									Phaseolus calcaratus seed*, leaf*	Sukumaran and Gnanamanickam (1980)
									P. coccineus seedling*	O'Neill *et al.* (1984)
									P. vulgaris cotyledon	Goosens and Van Laere (1983)
										Whitehead *et al.* (1982)
									P. vulgaris leaf*	Beggs *et al.* (1985)
									Pueraria lobata callus	Takeya and Itokawa (1982)
Coumestrin (coumestrol 3-*O*-glucoside)			OGlc				OH		*Trifolium repens* aerial parts	Gil *et al.* (1984)
									Glycine max root	Le-Van (1984)
Wairol			OH		OMe		OMe		*Medicago sativa* leaf*	Biggs and Shaw (1980)
Medicagol			OH			OCH_2O			*Euchresta japonica* root	Shirataki *et al.* (1981, 1982)
Flemichapparin C			OMe			OCH_2O			*Eysenhardtia polystacha* heartwood	Burns *et al.* (1984)
Aureol	OH		OH				OH		*Phaseolus aureus* seedling*	O'Neill (1983)
									P. coccineus seedling*	O'Neill *et al.* (1984)
									P. mungo seedling*	Adesanya *et al.* (1984)
Norwedelolactone (demethyl-wedelolactone)	OH		OH			OH	OH		*Wedelia calendulacea* (Compositae) leaf	Govindachari and Premila (1985)
Wedelolactone	OH		OMe			OH	OH		*Ougeinia dalbergioides* heartwood	Kalidhar and Sharma (1984)
									Wedelia calendulacea (Compositae) leaf	Govindachari and Premila (1985)
Sophoracoumestan B			OH	OMe		OCH_2O			*Sophora franchetiana* root	Komatsu *et al.* (1981b)
Sophoracoumestan A (*5.200*)									*Sophora franchetiana* root	Komatsu *et al.* (1981a)
Tuberostan (*5.201*)									*Pueraria tuberosa* tuber	Prasad *et al.* (1985)
Isosojagol (*5.202*)									*Phaseolus coccineus* seedling*	O'Neill *et al.* (1984)
Psoralidin (*5.203*)									*Dolichos biflorus* leaf	Keen and Ingham (1980)
Phaseol (*5.204*)									*Phaseolus aureus* seedling*	O'Neill (1983)
Glycyrol (*5.205*)									*Glycyrrhiza uralensis* stem, root	Zhu *et al.* (1984)
Isoglycyrol (*5.206*)									*G.* spp.	Hiraga *et al.* (1984)
									Glycyrrhiza uralensis stem, root	Zhu *et al.* (1984)
Norwedelic acid (*5.207*)									*Wedelia calendulacea* (Compositae) leaf	Govindachari and Premila (1985)

*Plant part was subjected to physiological stress.

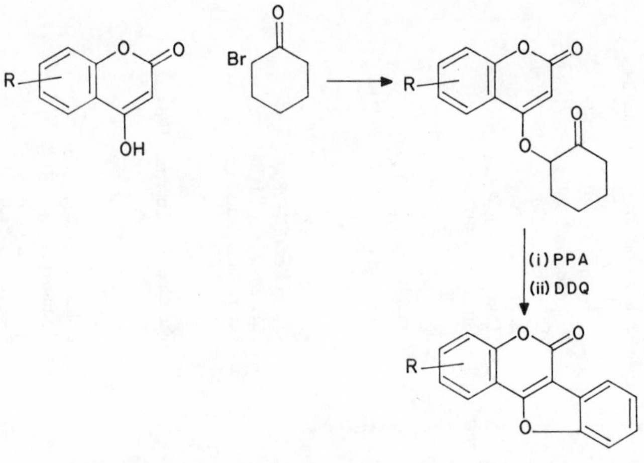

Fig. 5.19 Synthesis of A-ring-substituted coumestans.

the surprisingly high yields (90–95%) of coumestan reported (Tabakovic *et al.*, 1983) (Fig. 5.18). Unfortunately, these sequences result in 8,9-dihydroxycoumestans (*5.209*) and are therefore not applicable to the synthesis of all natural coumestans. A novel coumestan synthesis involves reaction of a 4-hydroxycoumarin with 2-bromocyclohexanone, followed by cyclization with polyphosphoric acid (PPA), then dehydrogenation with DDQ (Singh and Singh, 1985) (Fig. 5.19). It is difficult to envisage this being extended to synthesize naturally occurring examples of coumestans, however.

5.12 COUMARONOCHROMONES

For many years, only a single example of the coumaronochromone (*5.210*) class of isoflavonoid has been recognized. This is lisetin (*5.213*), isolated from *Piscidia erythrina*. It is remarkable therefore to now be able to add six new structures to the group (Table 5.10). Millettin (*5.217*) was isolated from seeds of *Millettia auriculata* (Raju *et al.*, 1981), and five examples were found in roots of white lupin, *Lupinus albus* (Tahara *et al.*, 1985a). Lupinalbin A (*5.211*) is structurally the simplest of these compounds, whereas lupinalbins B–E (*5.215*, *5.216*, *5.212*, *5.214*) all contain isoprenyl substituents. In both *Millettia* and *Lupinus*, these coumaronochromones are known to co-occur with structurally analogous 2′-hydroxyisoflavone derivatives, and it is likely that the 2′-hydroxyisoflavones could be their biosynthetic precursors. Indeed, 2′-hydroxyisoflavones may be cyclized to the corresponding coumaronochromones using a variety of oxidizing agents (Dewick, 1982), and selenium dioxide is a recently reported reagent for this conversion (Chubachi *et al.*, 1983). The use of this reagent allowed the lupinalbins to be correlated with their 2′-hydroxyisoflavone co-constituents (Tahara *et al.*, 1985a).

With further examples available for study, it is now possible to comment on spectroscopic characteristics of this group of isoflavonoids. The compounds behave as modified isoflavones, with prominent UV absorbances at similar wavelengths to the parent isoflavones (Tahara *et al.*, 1985a). The ^{1}H NMR spectra lack the characteristic isoflavone H-2 low-field singlet, and the aromatic proton signals in the B-ring appear at significantly lower field when compared with those of the corresponding 2′-hydroxyisoflavones. Shifts of 0.6–0.7 ppm are typical and

Table 5.10 Coumaronochromones

Coumaronochromone	*Plant sources*	*References*
Lupinalbin A (*5.211*)	*Lupinus albus* root	Tahara *et al.* (1985a)
Lupinalbin D (*5.212*)	*Lupinus albus* root	Tahara *et al.* (1985a)
Lupinalbin E (*5.214*)	*Lupinus albus* root	Tahara *et al.* (1985a)
Lupinalbin B (*5.215*)	*Lupinus albus* root	Tahara *et al.* (1985a)
Lisetin (*5.213*)	*Piscidia erythrina* root bark	Pietta and Zio (1983) Delle Monache *et al.* (1984)
Lupinalbin C (*5.216*)	*Lupinus albus* root	Tahara *et al.* (1985a)
Millettin (*5.217*)	*Millettia auriculata* seed	Raju *et al.* (1981)

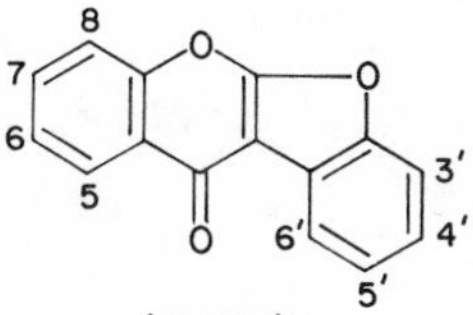

(5.210)

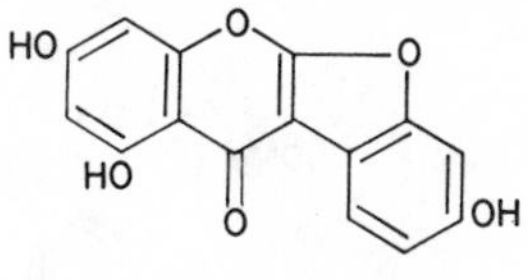

(5.211) Lupinalbin A

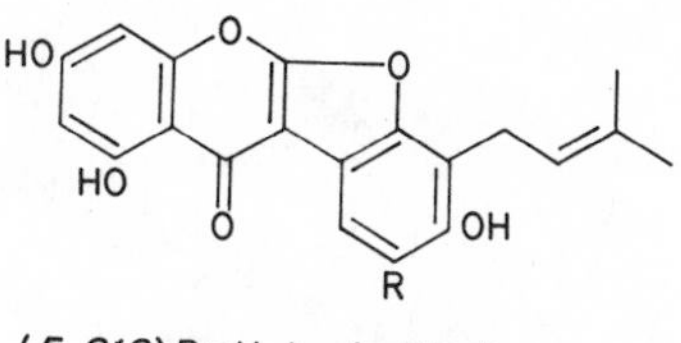

(5.212) R=H, Lupinalbin D

(5.213) R=OMe, Lisetin

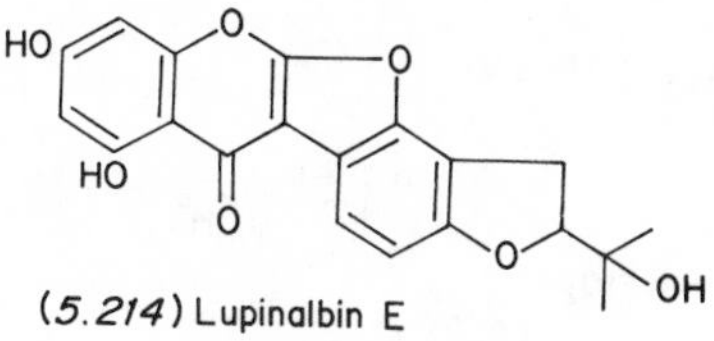

(5.214) Lupinalbin E

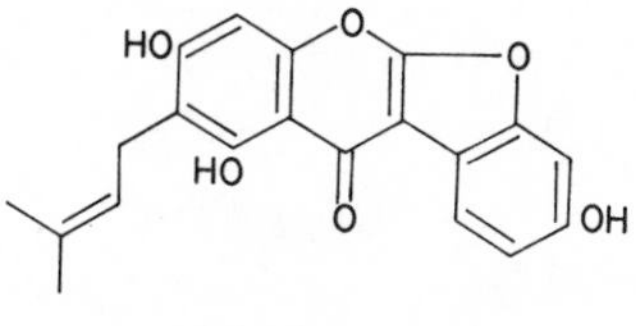

(5.215) Lupinalbin B

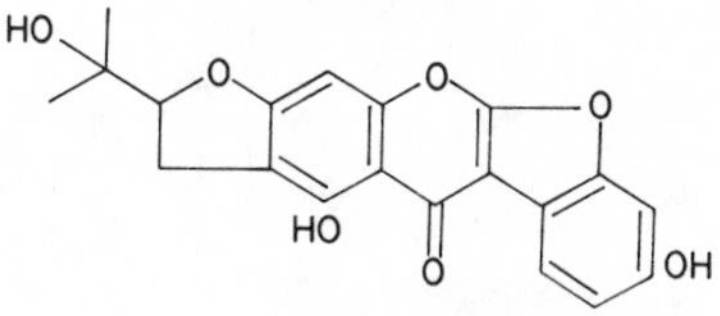

(5.216) Lupinalbin C

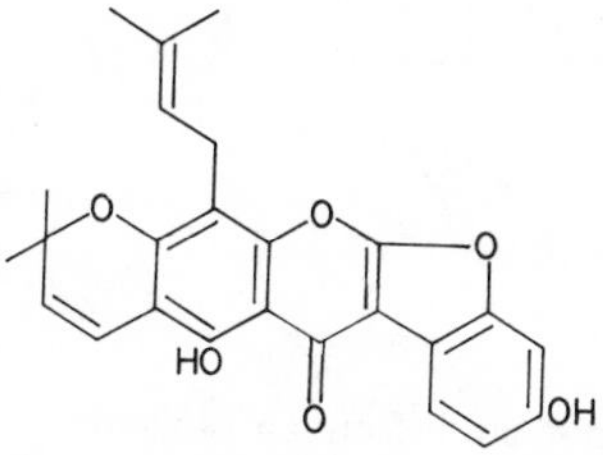

(5.217) Millettin

this effect is presumably due to deshielding by the carbonyl group (Raju *et al.*, 1981; Chubachi *et al.*, 1983; Tahara *et al.*, 1985a). The A-ring protons are only marginally affected.

Almost certainly, the number of natural products in this group of isoflavonoids will continue to grow. Two of the publications quoted here (Chubachi *et al.*, 1983; Tahara *et al.*, 1985a) refer to unpublished work of Chiki and co-workers who have isolated four simple coumaronochromone derivatives from *Beta vulgaris* (Chenopodiaceae), a plant known to synthesize the isoflavone phytoalexin betavulgarin.

5.13 α-METHYLDEOXYBENZOINS

Angolensin (*5.218*) together with some methyl and cadinyl ethers isolated from species of *Pterocarpus* and *Pericopsis* are the representatives of this group, and their co-occurrence with various isoflavonoid derivatives suggests α-methyldeoxybenzoins (*5.219*) could be 'reduced' forms of the isoflavonoid skeleton. Although isoflavones and isoflavanones under certain conditions can be reduced to α-methyldeoxybenzoins, a new observation in which an α-hydroxydihydrochalcone is converted into this skeleton in a photolytic reaction offers an alternative hypothesis for the natural origin of these compounds (Bezuidenhoudt *et al.*, 1981). The α-hydroxydihydrochalcone (*5.220*) was isolated from the heartwood of *Pterocarpus angolensis*, and a synthetic tosylate analogue (*5.221*) of this compound on photolysis in dioxan or dioxan–H_2O gave about 25% yields of the rearranged products (*5.222*) or (*5.224*), respectively, along with other unrearranged flavonoid products. By appropriate reductive sequences (Fig. 5.20), these compounds were both converted into the angolensin analogue (*5.223*).

No new α-methyldeoxybenzoins have been added to the known natural products, but a modified structure marsupol (*5.225*) isolated from heartwood of *Pterocarpus marsupium* has been described as an isoflavonoid glycol (Rao and Mathew, 1982). Whether or not this novel compound is an isoflavonoid, and a variant on the α-methyldeoxybenzoin class remains to be resolved. At first glance, the skeleton is identical, but the fact that both aromatic rings possess a single *para*-hydroxyl group, whereas the acetate-derived ring of isoflavonoids always

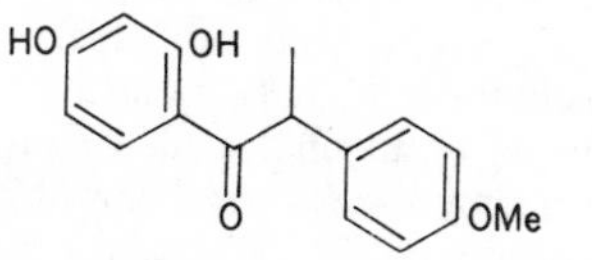

(5.218) Angolensin

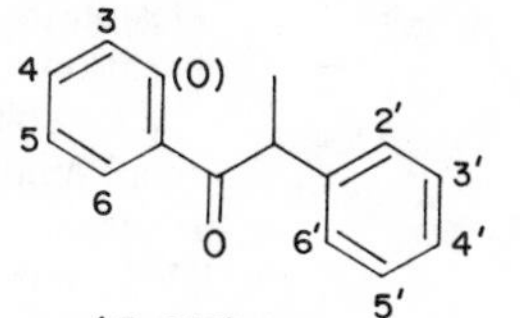

(5.219)

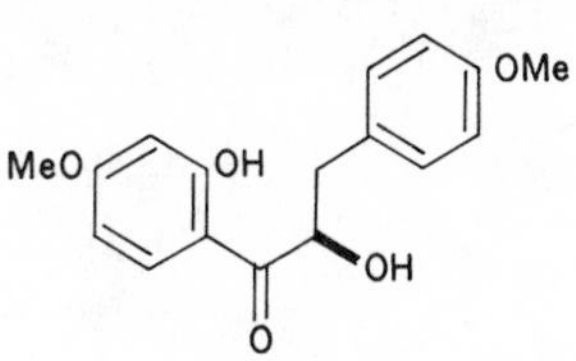

(5.220)

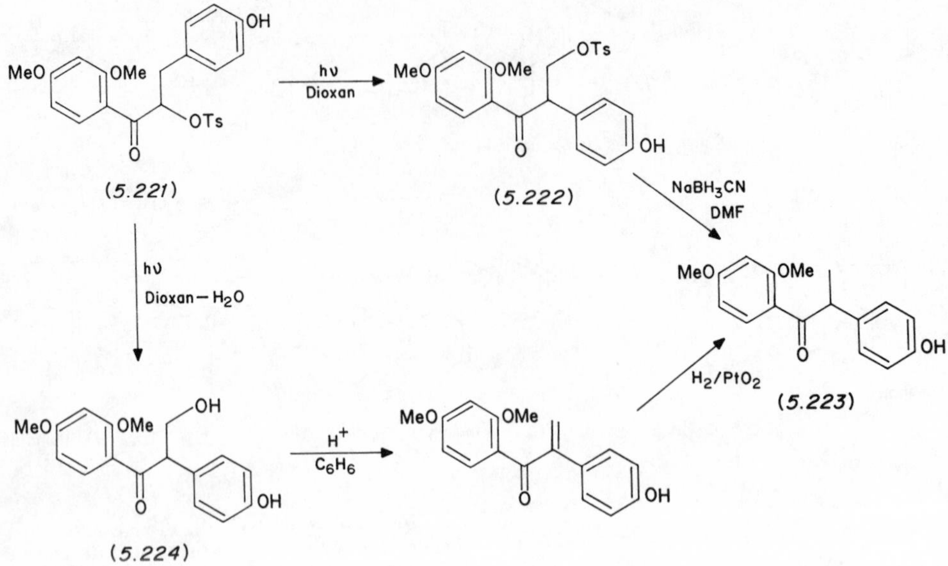

Fig. 5.20 Synthesis of 4′-hydroxy-2, 4-dimethoxy-α-methyldeoxybenzoin.

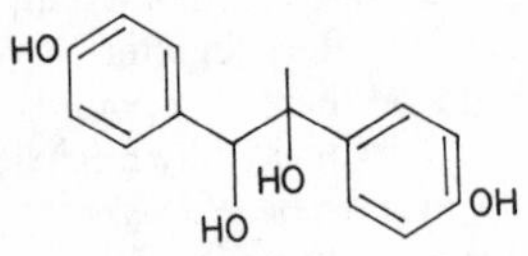

(5.225) Marsupol

contains at least the resorcinol oxygenation pattern, suggests marsupol may belong to some other class of natural product.

5.14 2-ARYLBENZOFURANS

A wide variety of 2-arylbenzofuran (*5.226*) structures is encountered in nature, and several different biosynthetic origins are undoubtedly involved (Dewick, 1982). Some structures are of lignan/neolignan origin and derived from phenylpropane dimers, and others are probably produced by cyclization of stilbenes. Such structures are excluded here, and only those of a third group which appear to be derived by loss of one carbon atom from an isoflavonoid structure are considered. Compounds in this group are derived from leguminous plants, and almost always co-occur with structurally related isoflavonoids. Their possible derivation from the isoflavonoid skeleton is via loss of C-6 from a coumestan, or by a sequence in which the benzofuran moiety is obtained from the acetate-derived ring, rather than the shikimate-derived ring. The only biosynthetic data favour the latter process in the case of vignafuran (Dewick, 1982).

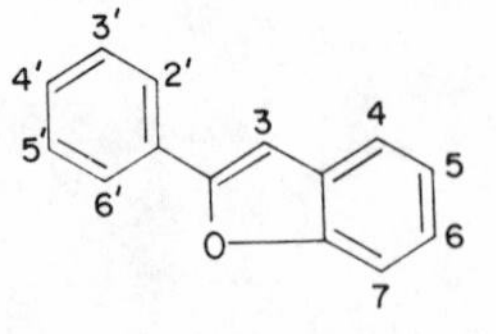

(5.226)

All of the recent reports of such 2-arylbenzofuran structures have yielded new examples (Table 5.11). All compounds appear to be constitutive compounds rather than phytoalexins, although several examples in this group are known to be synthesized as stress metabolites. Ambofuranol (*5.232*) from bulbs of *Neorautanenia amboensis* was the first example of a 2-arylbenzofuran with an oxygen substituent on the heterocyclic ring (Breytenbach and Rall, 1980). A second representative (*5.233*) has been isolated from roots of *Glycyrrhiza uralensis* and confusingly named licobenzofuran and liconeolignan in two different reports (Chang *et al.*, 1981, 1983). The latter name is inappropriate if the compound is isoflavonoid-derived. Despite the 2-aryl substituent containing only *para* oxygenation, an isoflavonoid origin seems quite probable.

The structure of centrolobofuran (*5.228*), isolated in small amounts from heartwood of an unidentified *Centrolobium* species, was confirmed by X-ray crystallography (Jurd and Wong, 1984). Several other 2-arylbenzofurans have been prepared synthetically, thus confirming

Table 5.11 2-Arylbenzofurans

2-Arylbenzofuran	*Plant sources*	*References*
2-(2, 4-Dihydroxyphenyl)-6-hydroxy-benzofuran (*5.227*)	*Lespedeza cyrtobotrya* heartwood	Miyase *et al.* (1981)
Centrolobofuran (*5.228*)	*Centrolobium* sp. heartwood	Jurd and Wong (1984)
Sainfuran (*5.229*)	*Hedysarum polybotrys* root	Miyase *et al.* (1984)
	Hedysarum polybotrys root	Miyase *et al.* (1984)
	Onobrychis viciifolia root	Russell *et al.* (1984)
Methylsainfuran (*5.230*)	*Onobrychis viciifolia* root	Russell *et al.* (1984)
Sophorafuran A (*5.231*)	*Sophora franchetiana* root	Komatsu *et al.* (1981b)
Ambofuranol (*5.232*)	*Neorautanenia amboensis* bulb	Breytenbach and Rall (1980)
Licobenzofuran (liconeolignan) (*5.233*)	*Glycyrrhiza uralensis* root	Chang *et al.* (1983)
	G. sp.	Chang *et al.* (1981)

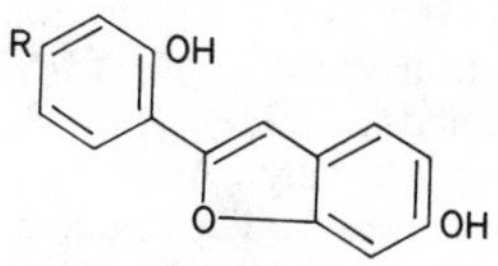

(*5.227*) R=OH

(*5.228*) R=OMe, Centrolobofuran

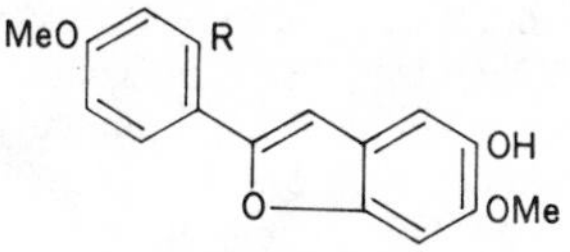

(*5.229*) R=OH, Sainfuran

(*5.230*) R=OMe, Methylsainfuran

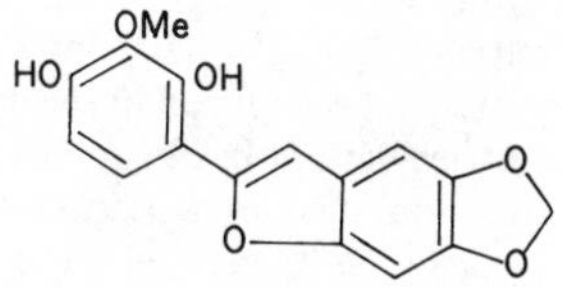

(*5.231*) Sophorafuran A

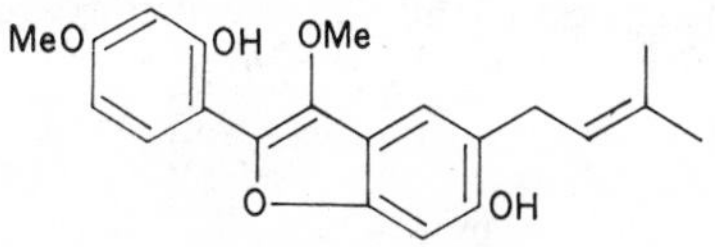

(*5.232*) Ambofuranol

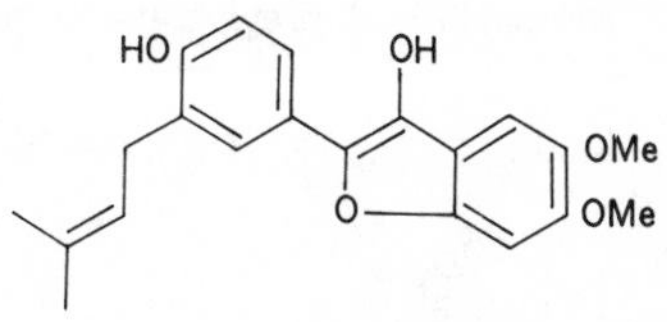

(*5.233*) Licobenzofuran (liconeolignan)

(i) $\overset{+}{P}Ph_3H\overset{-}{Br}$ (ii) Ac_2O

NEt_3/CH_3Ph Δ

$H_2/Pd-C$

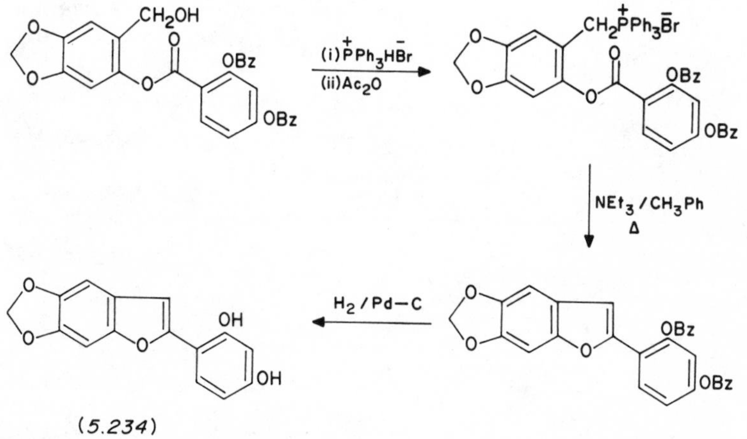

(*5.234*)

Fig. 5.21 Synthesis of 5, 6-methylenedioxy-2-(2′, 4′-dihydroxyphenyl)-benzofuran.

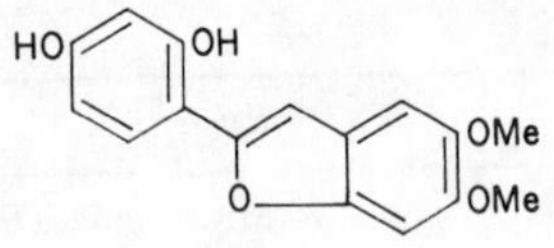

(5.235)

structures proposed from spectroscopic analysis. Synthetic approaches include the frequently utilized reaction of a copper (I) arylacetylide with an *o*-halogenophenol ester (Scannell and Stevenson, 1982; McKittrick *et al.*, 1982), and an intramolecular Wittig reaction of a phenolic ester (McKittrick *et al.*, 1982). The latter route was used to synthesize the dihydroxymethylenedioxy arylbenzofuran (*5.234*) isolated from *Sophora tomentosa* (Fig. 5.21). The similarity of 2-arylbenzofurans to isoflavonoids can be exploited by cleaving out the appropriate carbon atom. This can be achieved in several ways (Dewick, 1982) and a further variant has been described. The acetal derived by thallium nitrate oxidation of a 2-benzyloxychalcone is hydrolysed to lose the acetal carbon, generating a deoxybenzoin which can be cyclized following removal of the benzyl protecting group (Vu *et al.*, 1984). This approach, though not new, has been used to synthesize several 2-arylbenzofurans, and in particular the compound (*5.235*) was shown to be different from the compound isolated from *Myroxylon balsamum* and previously assigned this structure.

5.15 ISOFLAVONOID OLIGOMERS

Flavonoid oligomers including proanthocyanidins and biflavonoids have long been recognized, and have been reviewed as major groups of flavonoid derivatives. Until recently, similar oligomers of isoflavonoids had not been isolated, although there was every reason to believe that these types of compounds should exist in nature. The first bi-isoflavonoid was isolated from the heartwood of *Dalbergia nitidula* in low yield, and characterized as (*5.236*) (Brandt *et al.*, 1982; Bezuidenhoudt *et al.*, 1984). This was found to be a dimeric isoflavan, based on two molecules of (3*S*)-vestitol with a 4 → 5′ linkage and 3,4-*trans* configuration. Examination of the heartwood of *Dalbergia odorifera* resulted in the isolation of five examples of similar isoflavan dimers (Yahara *et al.*, 1985). These compounds all had (3*R*)-vestitol as one component, linked from position 5′ to position 4 of (3*R*)-7,2′,3′-trihydroxy-4′-methoxyisoflavan, (3*R*)-vestitol, (3*R*)-5′-methoxyvestitol, (3*R*)-mucronulatol, or the isoflavanquinone (3*R*)-claussequinone, giving compounds (*5.237*), (*5.238*), (*5.239*), (*5.240*) and (*5.241*) respectively. Again, the compounds had 3,4-*trans* stereochemistry. The structure and stereochemistry of (*5.237*) was confirmed by X-ray crystallography. The vestitol dimer (*5.238*) is, in fact, the enantiomer of (*5.236*) obtained from *D. nitidula*, and this can be related to the co-occurrence of (3*R*)-vestitol in *D. odorifera*, but (3*S*)-vestitol in *D. nitidula*. (3*R*)-Claussequinone and (3*R*)-5′-methoxyvestitol were also isolated from the *D. odorifera* extract.

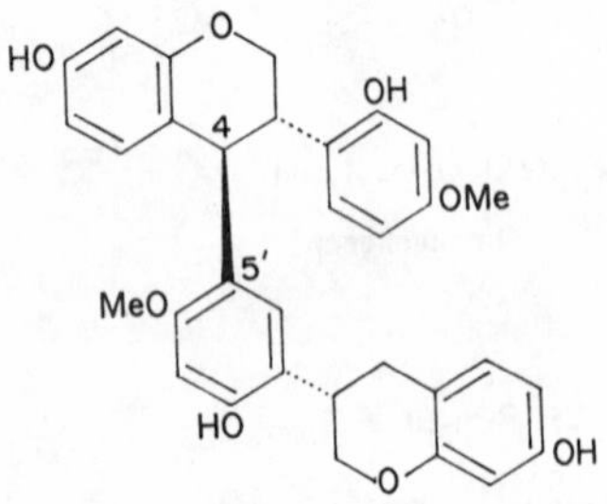

(5.236)

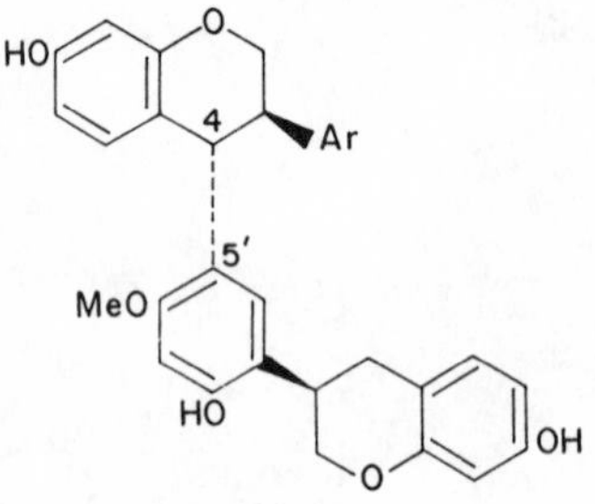

(5.237) Ar=

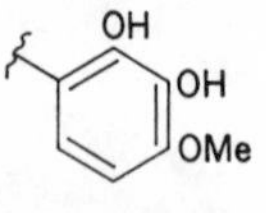

(5.238) Ar=

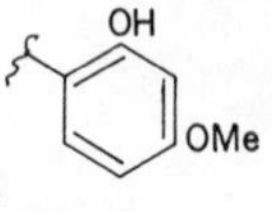

(5.239) Ar=

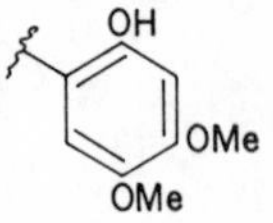

(5.240) Ar=

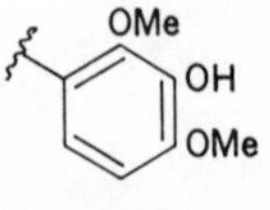

(5.241) Ar=

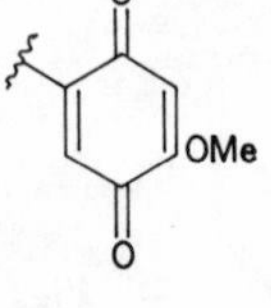

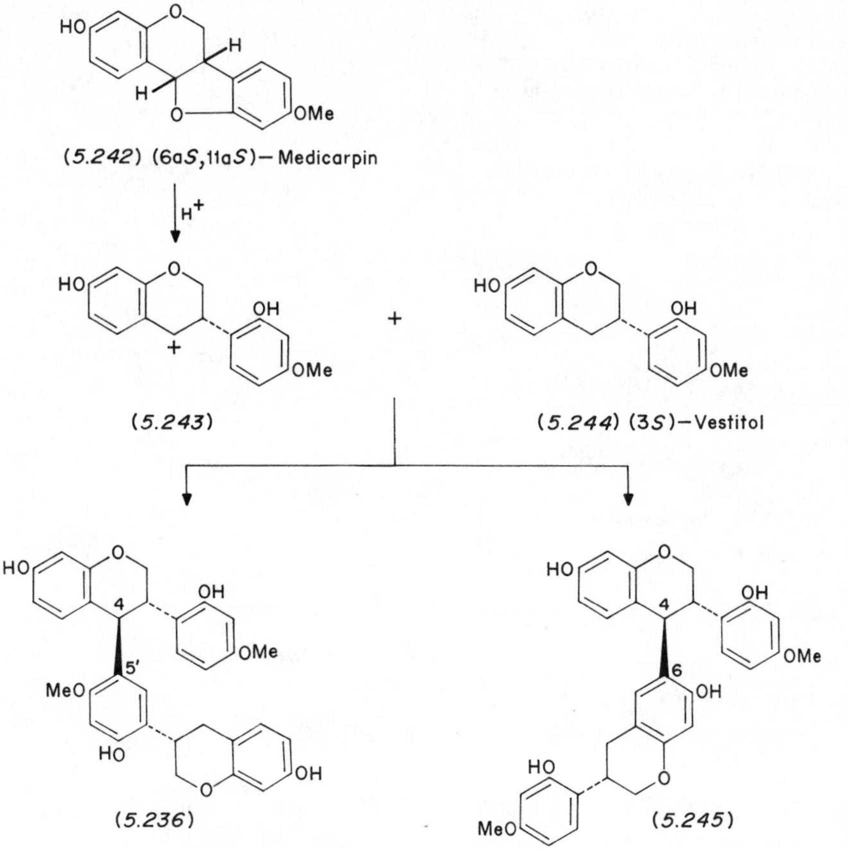

(5.242) (6aS,11aS)–Medicarpin

(5.243)

(5.244) (3S)–Vestitol

(5.236)

(5.245)

Fig. 5.22 Synthesis of isoflavan dimers.

Synthesis of the (3*R*)-vestitol dimer (*5.236*) from *D. nitidula* was achieved by using mild acid conditions in water–ethanol as used for the synthesis of condensed tannins, employing the pterocarpan (6a*S*, 11a*S*)-medicarpin (*5.242*) as source of an isoflavanyl-4-carbocation (*5.243*) and (3*S*)-vestitol (*5.244*) as the nucleophile (Brandt *et al.*, 1982; Benzuidenhoudt *et al.*, 1984). Vestitol served as a bifunctional nucleophile at C-5′ and C-6, yielding two dimers with 4→5′ (*5.236*) and 4→6 (*5.245*) linkages in 20% and 15% yields respectively (Fig. 5.22). Similar results could be obtained by photolysis in ethyl acetate. Further reaction under these conditions resulted in the formation of a single trimeric isoflavan containing 4→5′; 4→6 linkages, with the likely structure (*5.246*). Self-condensation by reacting just the pterocarpan generated dimeric isoflavan–pterocarpan and isoflavan–isoflavan analogues in low yields, so it is possible that other types of oligomer may exist in nature. The intermolecular condensations observed with the isoflavanyl-4-carbocation predominate over intramolecular

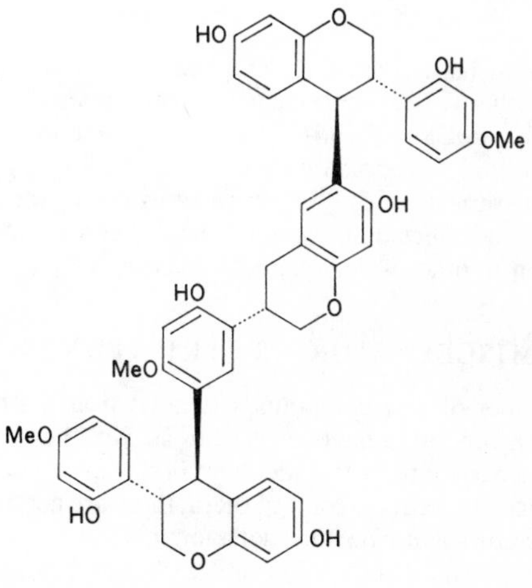

(5.246)

cyclization to regenerate the pterocarpan, presumably because of the ring strain within the pterocarpan system. The 3, 4-*trans* stereochemistry is a consequence of approach of the nucleophile from the least-hindered side and minimization of steric repulsion with the other aryl group.

The santalins (*5.247, 5.248*) and santarubins (*5.249, 5.250, 5.165*) are not strictly isoflavonoid oligomers, but

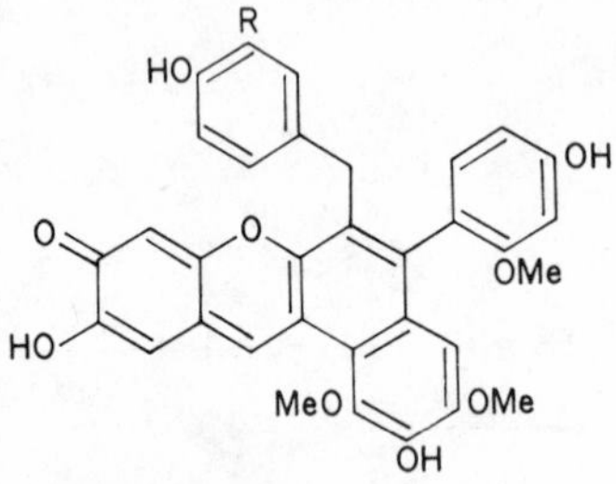

(*5.247*) R=OH, Santalin A

(*5.248*) R=OMe, Santalin B

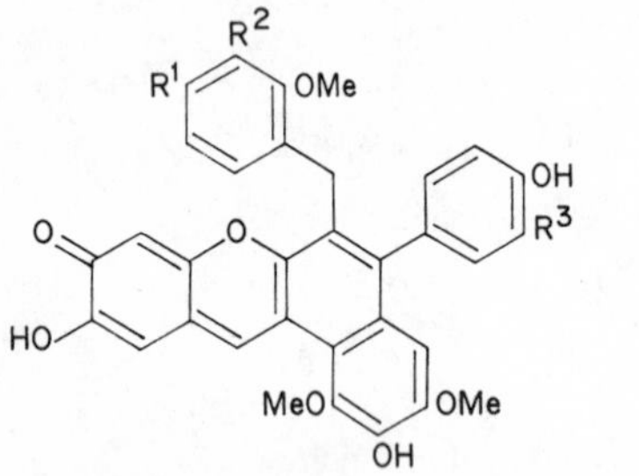

(*5.249*) R^1=OMe, R^2=H, R^3=OH, Santarubin A

(*5.250*) R^1=R^3=OMe, R^2=H, Santarubin B

(*5.165*) R^1=R^2=R^3=OH, Santarubin C

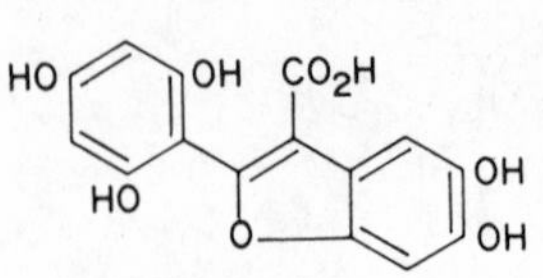

(*5.207*) Norwedelic acid

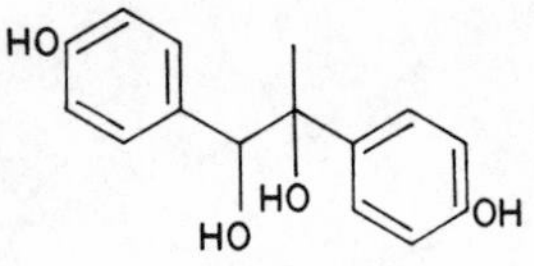

(*5.225*) Marsupol

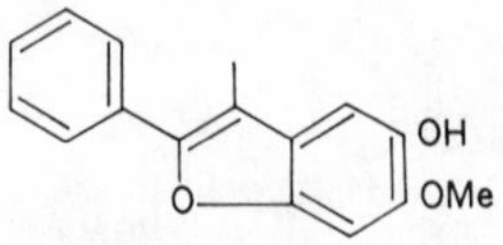

(*5.251*) Parvifuran

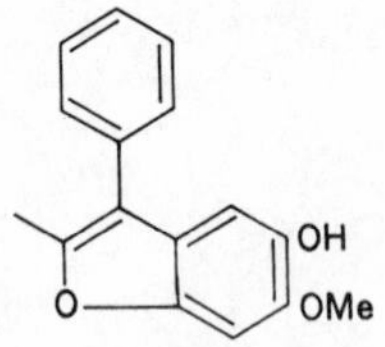

(*5.252*) Isoparvifuran

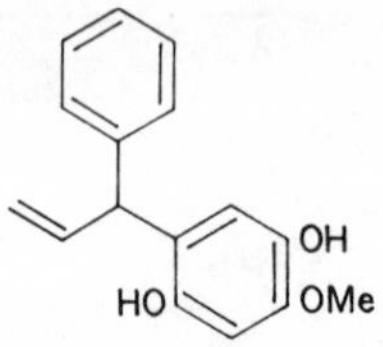

(*5.253*) Obtusaquinol

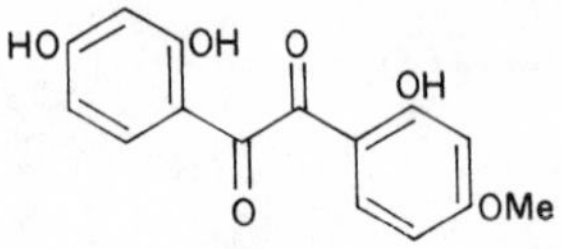

(*5.254*)

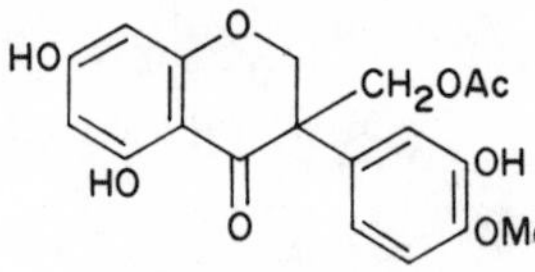

(*5.255*) Comosin

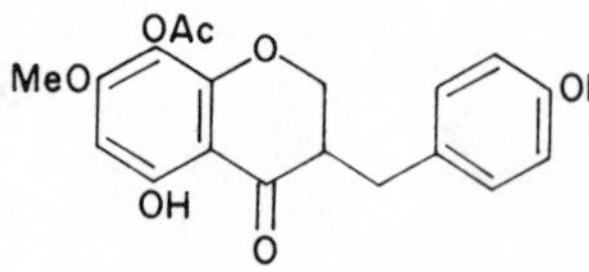

(*5.256*)

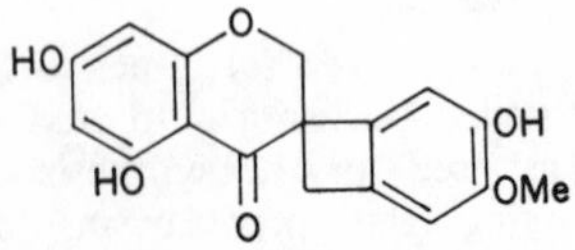

(*5.257*) Muscomosin

appear to be derived from condensation of isoflavylium salts with a 1, 3-diarylpropene (cinnamylphenol) (Afonya *et al.*, 1985a), and can be envisaged as isoflavonoid–flavonoid dimeric structures. These pigmented constituents of the 'insoluble red woods' from *Pterocarpus* and *Baphia* heartwoods have been discussed in relation to other isoflavonoids in Section 5.8.

5.16 MISCELLANEOUS STRUCTURES

A number of new compounds isolated bear a structural relationship to isoflavonoids, but as yet there is little evidence to formally include them in this group of natural products. Indeed, for some of them, there are good reasons to exclude them from the isoflavonoids.

Norwedelic acid (*5.207*) is a 3-carboxyl derivative of a 2-arylbenzofuran, and co-occurs with the coumestans wedelolactone and norwedelolactone in *Wedelia calendulacea* (Compositae) (Govindachari and Premila, 1985). Since it is clearly so closely related to these coumestans, it has been included in that section (Section 5.11). Parvifuran (*5.251*) is superficially similar to norwedelic acid, or α-methyldeoxybenzoins, but it lacks substitution on the 2-aryl substituent, and co-occurs in *Dalbergia parviflora* with isoparvifuran (*5.252*) and neoflavonoids, e.g. obtusaquinol (*5.253*) (Muangnoicharoen and Frahm, 1981). This seems to place the structure along with the neoflavonoids rather than the isoflavonoids.

Marsupol (*5.225*) was isolated from *Pterocarpus marsupium*, a known source of isoflavonoids, and has been classified as an isoflavonoid glycol (Rao and Mathew, 1982). This structure is discussed along with α-methyldeoxybenzoins, but, as pointed out there, it is far from satisfactory to class this as an isoflavonoid without further evidence. The benzil (*5.254*) isolated from the heartwood of *Zollernia paraensis* (Ferrari *et al.*, 1984b) also co-occurs with a wide range of isoflavonoid derivatives, and has the same substitution pattern as a number of them, e.g. vestitol and medicarpin. Clearly, this material could be isoflavonoid in origin, and could be an oxidized isoflavone.

Although comosin (*5.255*) has an isoflavanone skeleton, this material undoubtedly is not an isoflavonoid. It co-occurs in bulbs of *Muscari comosum* (Liliaceae) with 3-benzylchromanone derivatives (so-called homoisoflavonoids) such as (*5.256*) and muscomosin (*5.257*) (Adinolfi *et al.*, 1985). It is highly probable that comosin is biogenetically derived from a benzylchromanone skeleton as (*5.256*), known to arise by cyclization of a methoxy-chalcone, further cyclization to the skeleton of (*5.257*), then ring opening of the cyclobutane. This proposal means the acetoxymethyl group of (*5.255*) ultimately derives from the β-carbon of a chalcone.

5.17 MICROBIAL TRANSFORMATIONS OF ISOFLAVONOIDS

Many micro-organisms possess the capability of metabolizing plant-derived isoflavonoids. In some cases, no identifiable products can be isolated, but in other studies, modified isoflavonoids have been identified as transformation products of the original substrate. Because many of these compounds have yet to be recognized as plant products, it is appropriate to consider them under a separate heading. However, these studies reported have demonstrated that isoflavonoids known as plant products may also be formed as a result of microbial transformation from a given substrate. There is some justification, therefore, for assuming that some of the isoflavonoids listed in the other sections are in fact microbial metabolites of plant isoflavonoids, since few researchers employ aseptically grown plants for their studies. This is particularly likely in the case of phytoalexin studies, where induced isoflavonoids produced as a result of microbial stress are listed as plant products. Under these conditions, the products isolated may arise by a combination of plant and microbial metabolism, and represent part of a detoxification sequence employed by the micro-organism to reduce the antimicrobial properties of the plant isoflavonoid.

The data recorded in Table 5.12 represent the results of deliberate interaction of microbial cultures with isoflavonoid substrates, and demonstrate that a variety of different modification processes may operate. Frequently, the compounds identified represent intermediates in a metabolic sequence, and they may be modified further, ultimately to smaller non-isoflavonoid fragments. However, there are occasionally examples of high-yielding biotransformations which may be employed successfully in synthetic studies (e.g. Banks and Dewick, 1982b, 1983a). Table 5.12 lists the microbial transformations according to the isoflavonoid class of the substrate, but it is appropriate to discuss some of the transformations in terms of the type of modification being effected.

Isoflavonoid glycosides are typically hydrolysed to the aglycones, before further metabolism occurs (McMurchy and Higgins, 1984; Schlieper *et al.*, 1984; Kraft and Barz, 1985). Biochanin A 7-*O*-glucoside is degraded readily by both *Fusarium javanicum* (Schlieper *et al.*, 1984) and

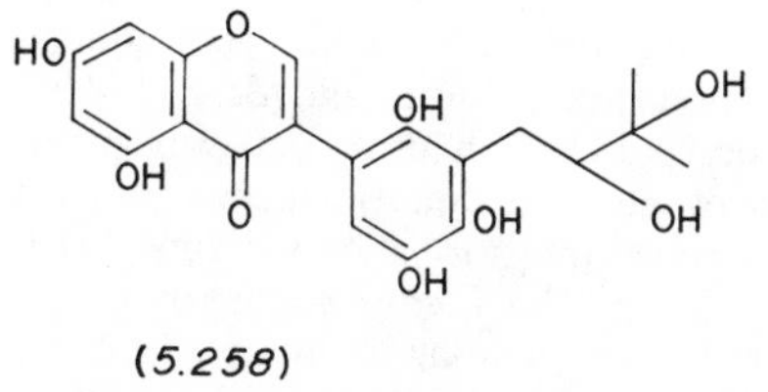

(*5.258*)

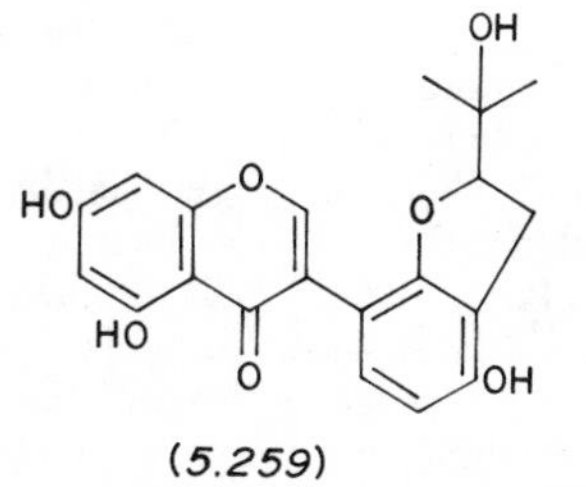

(*5.259*)

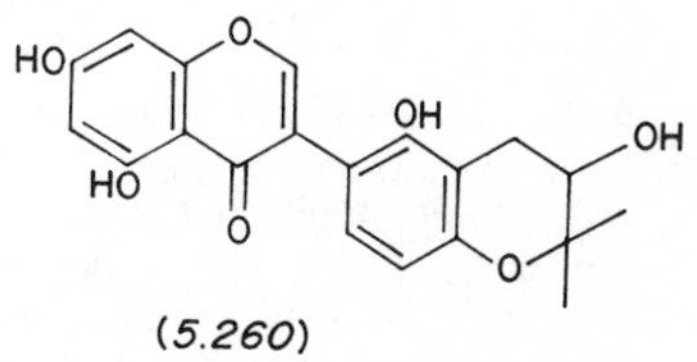

(*5.260*)

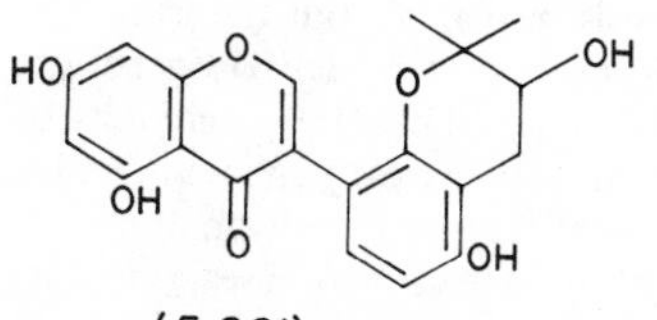

(*5.261*)

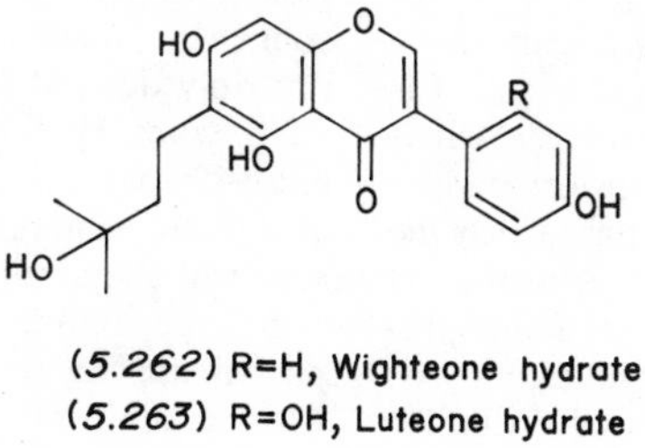

(*5.262*) R=H, Wighteone hydrate
(*5.263*) R=OH, Luteone hydrate

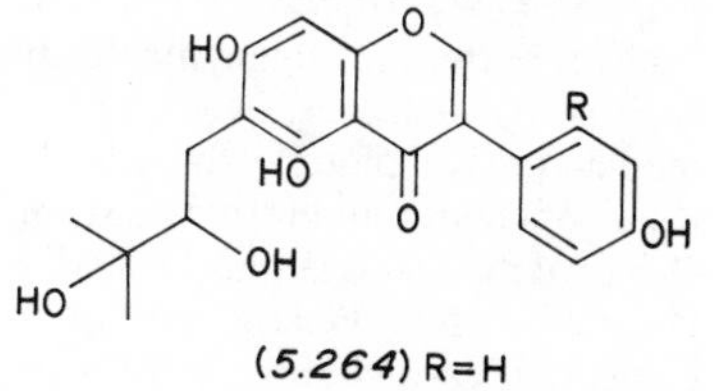

(*5.264*) R=H
(*5.265*) R=OH

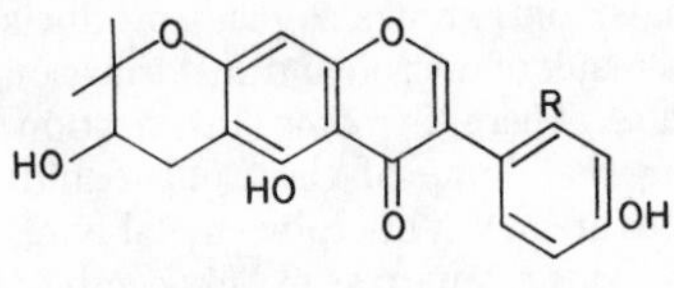

(5.266) R=H

(5.267) R=OH

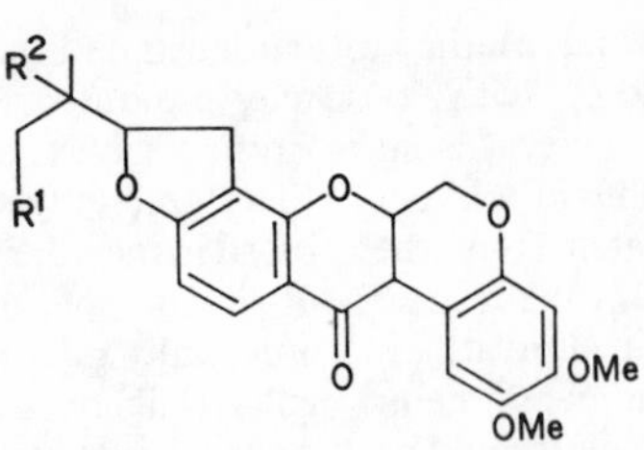

(5.268) $R^1=R^2=OH$, Amorphigenol

(5.269) $R^1=R^2=H$, Dihydrorotenone

(5.270) $R^1=H, R^2=OH$, Dalpanol

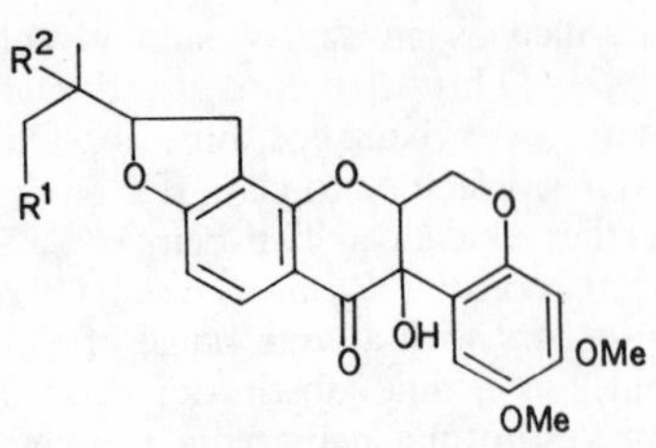

(5.271) $R^1=R^2=H$, 12a-Hydroxydihydrorotenone

(5.272) $R^1=H, R^2=OH$, 12a-Hydroxydalpanol

(5.273) $R^1=OH, R^2=H$, 12a-Hydroxydihydroamorphigenin

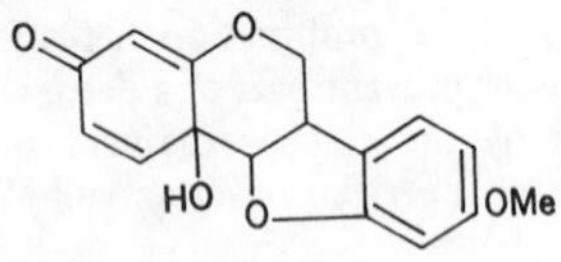

(5.274)

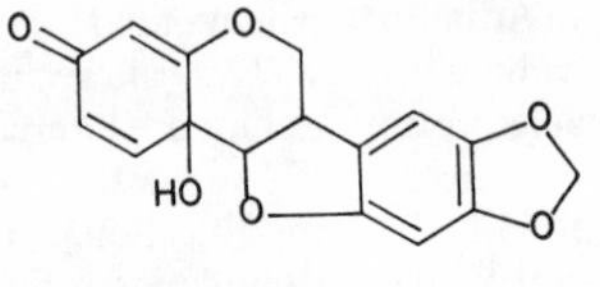

5.275

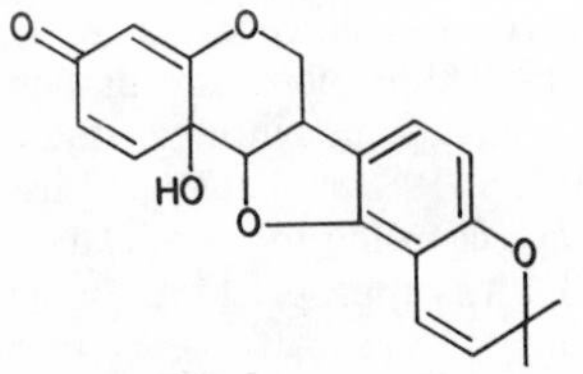

(5.276) 1a-Hydroxyphaseollone

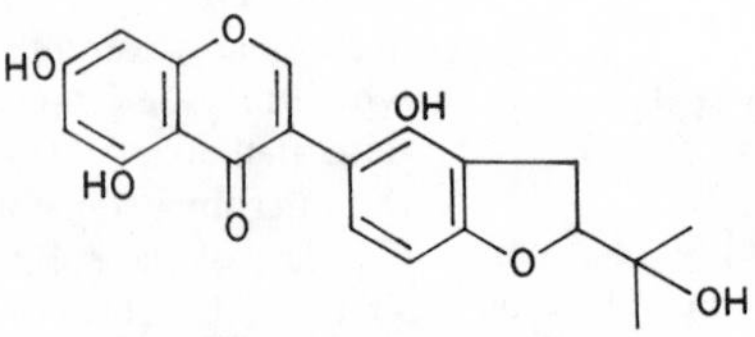

(5.6) Lupinisoflavone D

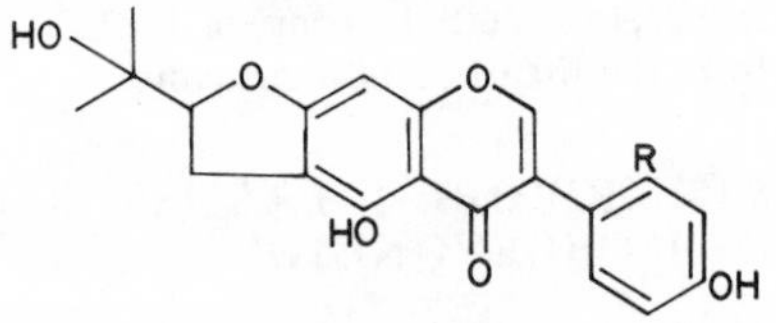

(5.30) R=H, Erythrinin C

(5.31) R=OH, Lupinisoflavone B

Aschochyta rabiei (Kraft and Barz, 1985), but only the former organism shows any significant metabolic activity towards the 6″-*O*-malonate ester of this glucoside, even though *A. rabiei* is a pathogen of chick-pea (*Cicer arietinum*) and the malonate ester is the main phenolic constituent of this plant.

Demethylation of *O*-methyl ethers is commonly encountered since this exposes phenolic functions and ultimately aids breakdown of the isoflavonoid skeleton. Isoflavone 4′-*O*-methyl ethers are demethylated by a number of *Fusarium* species (Willeke and Barz, 1982a; Weltring *et al.*, 1982) and *Aschochyta rabiei* (Kraft and Barz, 1985), though this is dependent on the precise species and fermentation conditions. For example, formononetin may be 4′-*O*-demethylated or 7-*O*-methylated by *Fusarium proliferatum* according to the conditions employed (Weltring *et al.*, 1982). Several organisms have been reported to demethylate pterocarpans and 6a-hydroxypterocarpans. Monomethoxy derivatives such as medicarpin and pisatin may be smoothly transformed into their phenolic counterparts, a process for which no chemical method is applicable because of the sensitivity of the pterocarpan skeleton to typical reagents. Homopterocarpin (3,9-dimethoxypterocarpan), however, may be demethylated at either, or both positions, according to fungus or conditions (Weltring *et al.*, 1981). In contrast, this pterocarpan is not metabolized by *Colletotrichum coccodes*, despite the ability of this organism to demethylate medicarpin (Higgins and Ingham, 1981).

Reduction to isoflavanones appears to be a common process to initiate catabolism of isoflavones by many *Fusarium* fungi (Willeke and Barz, 1982a, b; Willeke *et al.*, 1983; Kraft and Barz, 1985). In some cases, this is preceded by demethylation (Willeke and Barz, 1982a). Reduction of pterocarpans to 2′-hydroxyisoflavans parallels the chemical hydrogenolysis conversion, and has been recognized as a microbial transformation for some years. Degradation of medicarpin by *Fusarium oxysporum* f.sp. *lycopersici* gives the corresponding 2′-hydroxyisoflavan vestitol as one product of metabolism, but this is accompanied by the corresponding isoflav-3-ene and 3-arylcoumarin (Weltring *et al.*, 1983). The arylcoumarin is probably derived from the isoflav-3-ene by allylic oxidation (compare biosynthesis in plants), but the isoflav-3-ene itself may arise from the isoflavan by dehydrogenation, or directly from the pterocarpan. The latter route may be achieved chemically by controlled treatment with acid.

A large number of the microbial transformations reported are the result of oxidative processes, yielding hydroxylated products and compounds derived from them. Simple 4′-methoxy isoflavones such as formononetin and biochanin A may be hydroxylated at the 3′ position by a number of fungi (Weltring *et al.*, 1982; Kraft and Barz, 1985), and the transformations to calycosin or pratensein respectively can be achieved in nearly quantitative yield using *Fusarium oxysporum* f.sp. *lycopersici* (Mackenbrock and Barz, 1983). Metabolism of pterocarpans is markedly dependent on the isolate of fungus used (Denny and Van Etten, 1982), but 6a-hydroxylation is an important and fairly common modification. This reaction is remarkably specific, and *Nectria haematococca*, a sexual stage of *Fusarium solani*, has been demonstrated to 6a-hydroxylate the (−)-(6a*R*, 11a*R*) enantiomers of medicarpin and maackiain, but to give no 6a-hydroxyl derivatives from the (+)-(6a*S*, 11a*S*) enantiomers (Van Etten *et al.*, 1983). This property has been exploited to demonstrate the presence of small amounts of the (−)-isomers of medicarpin and maackiain along with larger amounts of the (+)-isomers as phytoalexins of *Sophora japonica* (Van Etten *et al.*, 1983). Some organisms attack the pterocarpan skeleton at position 1a, rather than 6a. This, of course, means the aromatic character of ring A must be lost, and a 1a-hydroxydienone (1a-hydroxypterocarpanone) structure ensues. This type of metabolite of phaseollin, 1a-hydroxyphaseollone (*5.276*), has been known for some years, but more recently, 1a-hydroxydienone analogues of medicarpin and maackiain produced by *Nectria haematococca* isolates have also been described (Denny and Van Etten, 1982). In the case of phaseollin metabolism by *Fusarium solani* f.sp. *phaseoli*, the metabolic sequence has been shown to proceed via an inducible monooxygenase, in that $^{18}O_2$ but not $H_2{}^{18}O$ is incorporated (Kistler and Van Etten, 1981). This is in keeping with a proposed arene oxide mechanism (Fig. 5.23). With one strain of *Nectria haematococca*, the 1a-hydroxydienone metabolite from medicarpin/maackiain was found to be accompanied by the simple isoflavanone analogue of the pterocarpan (Denny and Van Etten, 1982). This compound is suggested to arise via a similar mono-oxygenase mechanism involving hydroxylation at C-11a and decomposition of the resultant hemiketal. In fact, various strains of *N. haematococca* are capable of metabolizing pterocarpans to 6a-hydroxypterocarpans, isoflavanones, 1a-hydroxydienones or demethylated pterocarpans. All reactions could be mono-oxygenase-mediated, though probably not by the same enzymes (Denny and Van Etten, 1982). The rotenoid skeleton is also susceptible to microbial hydroxylation. An extracellular laccase from *Polyporus anceps* transformed rotenone into 12a-hydroxyrotenone (rotenolone) (Sariaslani *et al.*, 1984), and *Streptomyces griseus* metabolized the semi-synthetic dihydrorotenone (*5.269*) into three major products, all 12a-hydroxyl derivatives (Sariaslani and Rosazza, 1985).

When the isoflavonoid substrate contains isoprenyl or cyclized isoprenyl substituents, this side chain is frequently the site for microbial modification. Simple dimethylallyl groups tend to be oxygenated by hydration, addition of water giving the 3-hydroxy-3-methylbutyl grouping. Thus, kievitone yields kievitone hydrate (*5.72*) with *Fusarium solani* f.sp. *phaseoli* (Zhang and Smith, 1983), and the isoflavones luteone and wighteone give luteone hydrate (*5.263*) and wighteone hydrate (*5.262*) respectively with *Aspergillus flavus*, though not with *Botrytis cinerea* (Tahara *et al.*, 1984b, 1985b). A range of metabolites of luteone, wighteone and licoisoflavone A, obtained by incubation with either *A. flavus* or *B. cinerea* can be envisaged as arising by way of an epoxide of the original dimethylallyl unit, followed by ring opening with water, or alternatively one of the phenolic hydroxyls giving dihydrofuran or dihydropyran derivatives (Tahara, 1984; Tahara *et al.*, 1984b, 1985b,c). The formation of amorphigenin from rotenone by *Cunninghamella blakesleeana* (Sariaslani and

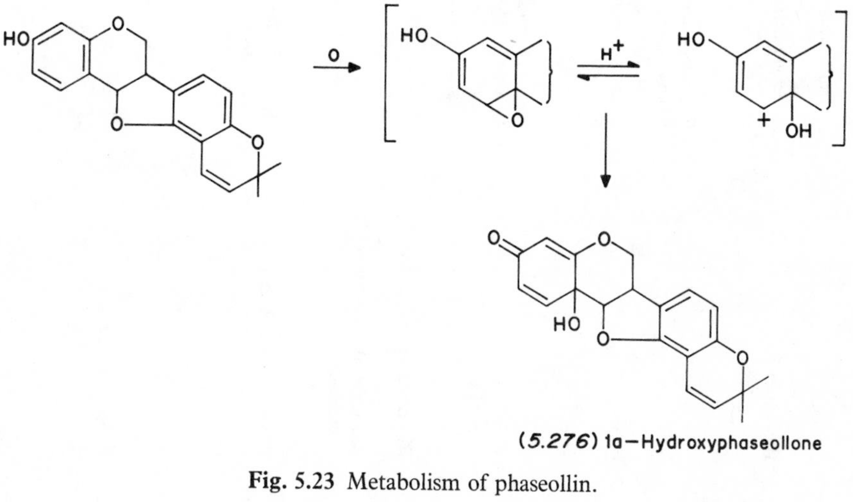

(*5.276*) 1a–Hydroxyphaseollone

Fig. 5.23 Metabolism of phaseollin.

Table 5.12 Microbial transformations of isoflavonoids

Substrate	*Organism*	*Product(s)*	*References*
Isoflavones			
Formononetin	*Fusarium avenaceum*	Calycosin	Weltring *et al.* (1982)
	F. oxysporum f. sp. *lycopersici*	Calycosin	Mackenbrock and Barz (1983)
	F. proliferatum	Daidzein	Weltring *et al.* (1982)
		Dimethyldaidzein	
Biochanin A	*Ascochyta rabiei*	Genistein	Kraft and Barz (1985)
		Prastensein	
		Orobol	
		Dihydrogenistein	
	Fusarium aquaeductum	Genistein	Willeke and Barz (1982a)
		Dihydrobiochanin A	Willeke and Barz (1982b)
	F. javanicum	Genistein	Willeke and Barz (1982a)
	F. moniliforme	Genistein	Willeke and Barz (1982a)
	F. oxysporum f. sp. *apii*	Pratensein	Weltring *et al.* (1982)
	F. oxysporum f. sp. *lini*	Pratensein	Weltring *et al.* (1982)
	F. oxysporum f. sp. *lycopersici*		Mackenbrock and Barz (1973)
	F. oxysporum f. sp. *pisi*	Dihydrobiochanin A	Willeke and Barz (1982a)
	F. solani (Nectria haematococca)	Dihydrobiochanin A	Willeke *et al.* (1983)
	F. solani f. sp. *cucurbitae*	Dihydrobiochanin A	Willeke and Barz (1982a)
	F. solani f. sp. *phaseoli*	Genistein Dihydrogenistein	Willeke and Barz (1982a)
Biochanin A 7-*O*-glucoside	*Ascochyta rabiei*	Biochanin A Pratensein	Kraft and Barz (1985)
Biochanin A 7-*O*-glucoside-6″-*O*-malonate	*Fusarium javanicum*	Biochanin A 7-*O*-glucoside Biochanin A Dihydrobiochanin A	Schlieper *et al.* (1984)
Licoisoflavone A (*5.4*)	*Aspergillus flavus* or *Botrytis cinerea*	2″,3″-Dihydro-2″,3″-dihydroxylicoisoflavone A (*5.258*) Lupinisoflavone D (*5.6*) (*5.259*) (*5.260*) (*5.261*)	Tahara *et al.* (1985c)
Wighteone (*5.16*)	*Aspergillus flavus*	Wighteone hydrate (*5.262*)	Tahara *et al.* (1985b)
	A. flavus or *Botrytis cinerea*	2″,3″-Dihydro-2″,3″-dihydroxywighteone (*5.264*) Erythrinin C (*5.30*) (*5.266*)	Tahara *et al.* (1985b)
Luteone (*5.17*)	*Aspergillus flavus*	Luteone hydrate (*5.263*)	Tahara *et al.* (1984b)
	A. flavus or *Botrytis cinerea*	2″,3″-Dihydro-2″,3″-dihydroxyluteone (*5.265*)	Tahara *et al.* (1984b)
		Lupinisoflavone B (*5.31*) (*5.264*)	Tahara (1984); Tahara *et al.* (1984b)
Isoflavanones			
Kievitone (*5.70*)	*Fusarium solani* f. sp. *phaseoli*	Kievitone hydrate (*5.72*)	Zhang and Smith (1983)

Rotenoids			
Rotenone (*5.90*)	*Cunninghamella blakesleeana*	Amorphigenin (*5.93*)	Sariaslani and Rosazza (1983)
		Amorphigenol (*5.268*)	
	Polyporus anceps – extracellular laccase	12a-Hydroxyrotenone (*5.99*)	Sariaslani *et al.* (1984)
Dihydrorotenone (*5.269*)	*Cunninghamella blakesleeana*	Dalpanol (*5.270*)	Sariaslani and Rosazza (1983)
	Streptomyces griseus	12a-Hydroxydihydro-rotenone (*5.271*)	Sariaslani and Rosazza (1985)
		12a-Hydroxyalpanol (*5.272*)	
		12a-Hydroxydihydroamorphigenin (*5.273*)	
Pterocarpans			
Medicarpin	*Colletotrichum coccodes* (*C. phomoides*)	Demethylmedicarpin	Higgins and Ingham (1981)
	Fusarium oxysporum f. sp. *lycopersici*	Vestitol	Weltring *et al.* (1983)
		7, 2′-Dihydroxy-4′-methoxy-isoflav-3-ene	
		7, 2′-Dihydroxy-4′-methoxy-isoflav-3-en-2-one (*5.186*)	
	F. solani (*Nectria haematococca*)	Vestitone	Denny and Van Etten (1982)
		6a-Hydroxymedicarpin (*5.274*)	
	F. solani f. sp. *phaseoli*	(*5.274*)	Denny and Van Etten (1982)
(−)-Medicarpin	*F. solani* (*Nectria haematococca*)	(−)-6a-Hydroxymedicarpin	Van Etten *et al.* (1983)
Homopterocarpin	*Fusarium anguioides*	Medicarpin	Weltring *et al.* (1981)
		Isomedicarpin	
	F. avenaceum	Medicarpin	Weltring *et al.* (1981)
		Isomedicarpin	
		Vestitol	
	F. graminearum	Medicarpin	Weltring *et al.* (1981)
		Isomedicarpin	
	F. sporotrichioides	Isomedicarpin	Weltring *et al.* (1981)
		Demethylmedicarpin	
	Gibberella saubinetti	Isomedicarpin	Weltring *et al.* (1981)
		Demethylmedicarpin	
Maackiain	*Fusarium solani* (*Nectria haematococca*)	Sophorol	Denny and Van Etten (1982)
		6a-Hydroxymaackiain (*5.275*)	
	F. solani f. sp. *phaseoli*	(*5.275*)	Denny and Van Etten (1982)
(−)-Maackiain	*F. solani* (*Nectria haematococca*)	(−)-6a-Hydroxymaackiain	Van Etten *et al.* (1983)
(−)-Maackiain 3-*O*-glucoside (trifolirhizin)	*Fusarium roseum*	Maackiain	McMurchy and Higgins (1984)
Phaseollin (*5.122*)	*Fusarium solani* f. sp. *phaseoli*	1a-Hydroxyphaseollone (*5.276*)	Kistler and Van Etten (1981)
			Zhang and Smith (1983)
			Denny and Van Etten (1982)
Pisatin	*Botrytis cinerea*	6a-Hydroxymaackiain	Robeson and Harborne (1983)
	Fusarium solani (*Nectria haematococca*)	6a-Hydroxymaackian	Van Etten and Barz (1981)
			Denny and Van Etten (1982)
	Stemphylium botryosum	6a-Hydroxymaackiain	Higgins (1981)
Isoflavans			
Phaseolinisoflavan (*5.154*)	*Fusarium solani* f. sp. *phaseoli*	?Hydrated metabolite	Zhang and Smith (1983)
			Wietor-Orlandi and Smith (1985)

Rosazza, 1983) is formally an allylic hydroxylation, and amorphigenol (*5.268*) produced in the same reaction could then be a hydration product of amorphigenin. Hydroxylation of the isopropyl group of dihydrorotenone can occur at the methyl or methine carbons (Sariaslani and Rosazza, 1983, 1985).

Whilst the metabolic transformation of any particular isoflavonoid is markedly dependent on which organism is employed, it must be stressed that a particular organism can transform isoflavonoid analogues in completely different ways. For example, a *Nectria haematococca* isolate will transform medicarpin (or maackiain) into isoflavanone and la-hydroxydienone derivatives, yet demethylate pisatin (Denny and Van Etten, 1982). This type of phenomenon is well demonstrated by studies with a mixture of phaseollin, kievitone and phaseollinisoflavan being metabolized concurrently by *Fusarium solani* f.sp. *phaseoli* (Zhang and Smith, 1983). The products of metabolism were la-hydroxyphaseollone, kievitone hydrate and an unidentified product from phaseollinisoflavan, 18 mass units heavier and therefore perhaps a hydrated material (Wietor-Orlandi and Smith, 1985).

5.18 BIOSYNTHESIS OF ISOFLAVONOIDS

The isoflavonoids share a common biosynthetic pathway with the flavonoids as far as chalcone–flavanone intermediates, but then a 1,2-aryl migration occurs to produce the rearranged 3-phenylchroman skeleton that differentiates isoflavonoids from other flavonoids. The

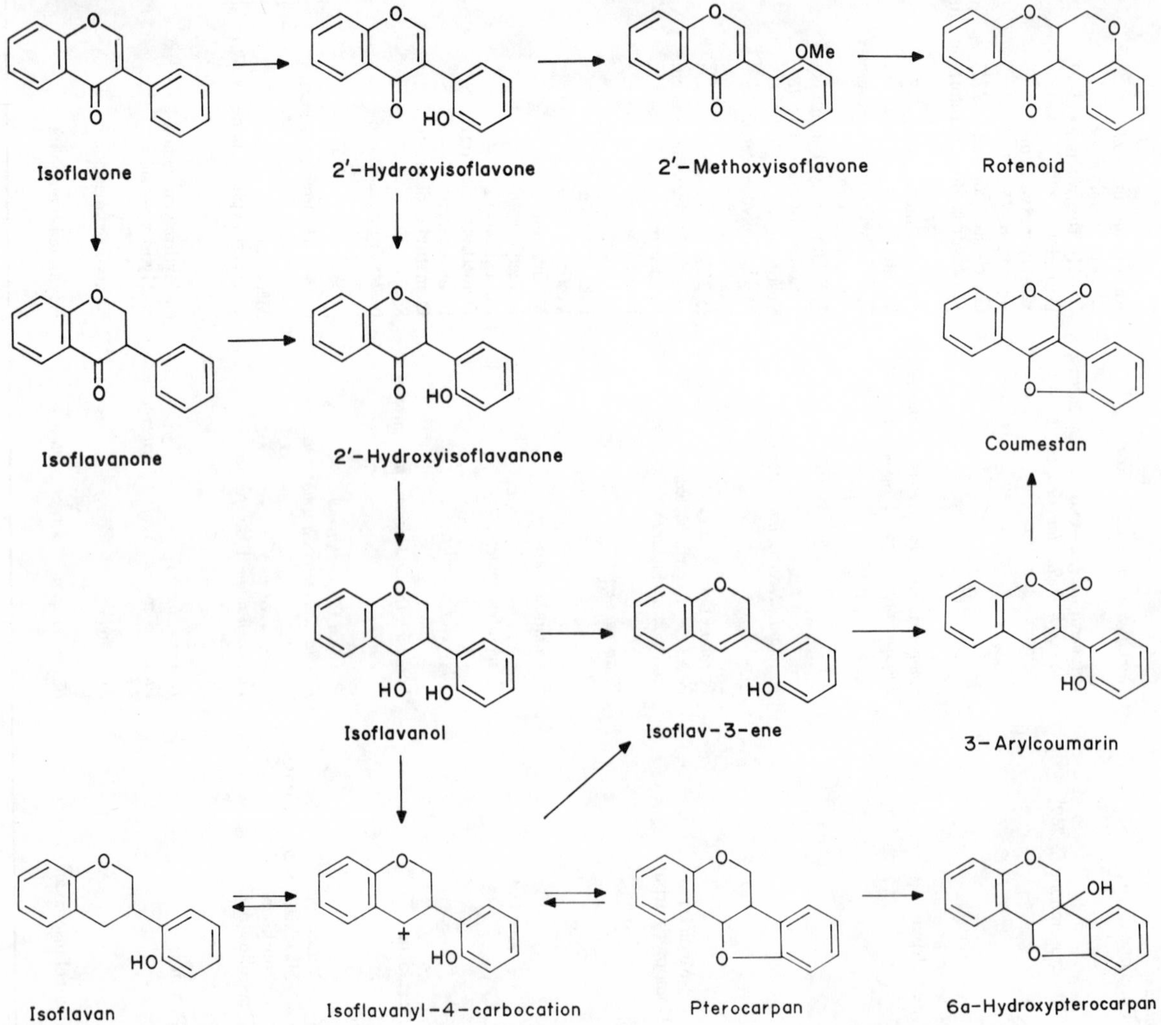

Isoflavone
2′-Hydroxyisoflavone
2′-Methoxyisoflavone
Rotenoid
Isoflavanone
2′-Hydroxyisoflavanone
Coumestan
Isoflavanol
Isoflav-3-ene
3-Arylcoumarin
Isoflavan
Isoflavanyl-4-carbocation
Pterocarpan
6a-Hydroxypterocarpan

Fig. 5.24 General scheme of isoflavonoid biosynthetic relationships.

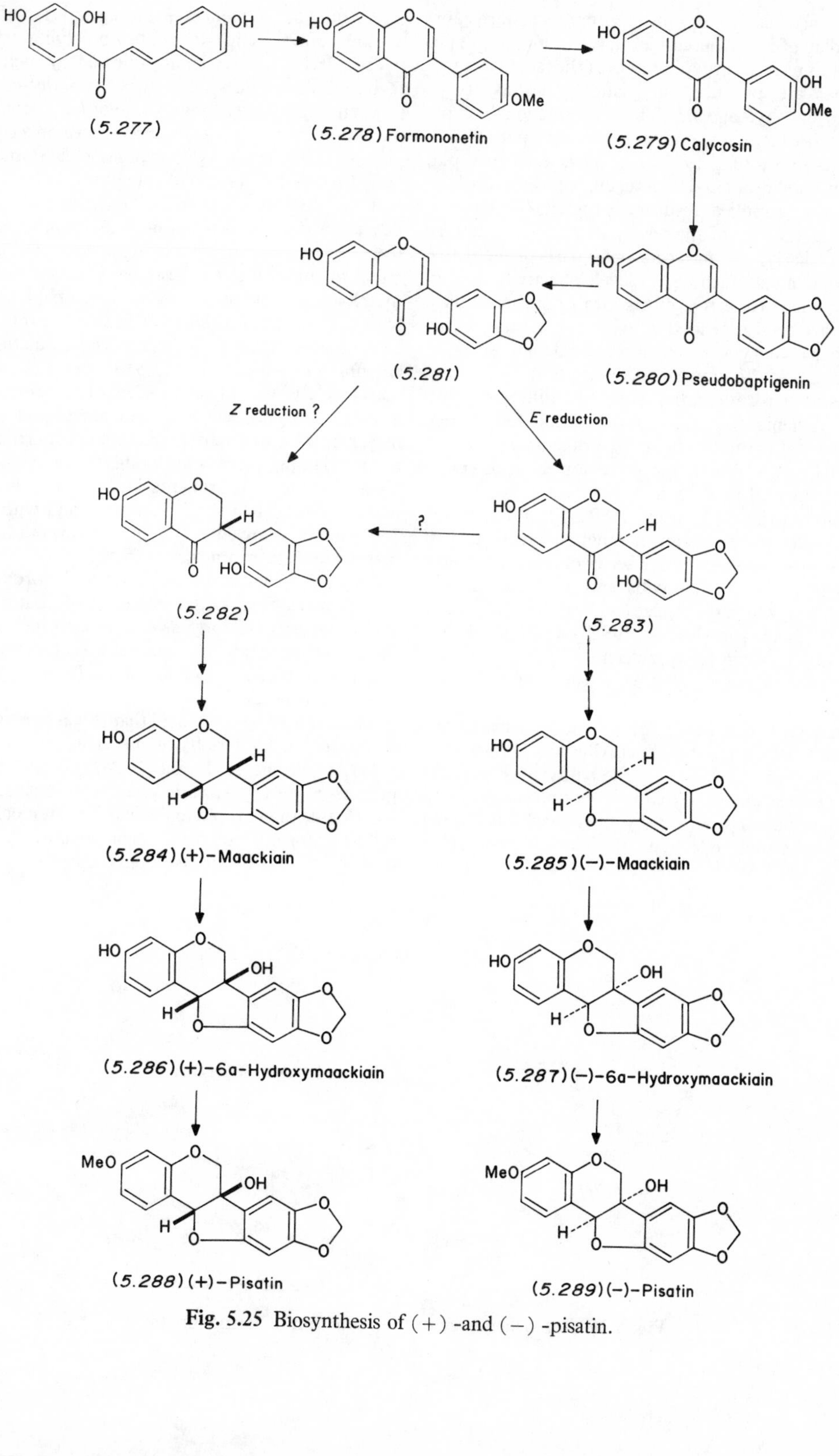

HO OH OH O (5.277) → HO O O OMe (5.278) Formononetin → HO O O OH OMe (5.279) Calycosin → HO O O O O (5.280) Pseudobaptigenin → HO O O HO O O (5.281)

Z reduction ? → HO O H O HO O O (5.282)

E reduction → HO O H O HO O O (5.283) → ? → (5.282)

(5.284) (+)-Maackiain

(5.285) (−)-Maackiain

(5.286) (+)-6a-Hydroxymaackiain

(5.287) (−)-6a-Hydroxymaackiain

(5.288) (+)-Pisatin

(5.289) (−)-Pisatin

Fig. 5.25 Biosynthesis of (+) -and (−) -pisatin.

last decade or so has seen significant progress towards our understanding of isoflavonoid biosynthesis, and the main reason for this must lie in exploitation of the fact that some isoflavonoids are produced in plant tissues as stress metabolites or phytoalexins. Their production can be induced by treatment of plants, or plant parts, with fungi, abiotic inducers or fungal elicitors. This leads to rapid synthesis over a short period of time, during which labelled precursors can be applied, resulting in high incorporations and allowing the use of modern stable isotope methodologies. Similarly, enzymes catalysing parts of the pathway can be found at much higher levels of activity during the stress period. Research efforts into isoflavonoid biosynthesis have been expended in two main areas, how the various isoflavonoid classes are interrelated, and secondly the mechanism of the unusual aryl migration.

As a result of many studies in several different plant systems, a scheme (Fig. 5.24) interrelating ten of the known classes of naturally occurring isoflavonoids can be presented (Dewick, 1982). Evidence for these interrelationships, reported in Volume 1, has been published in fuller detail (Martin and Dewick, 1980). The relationship of 6a-hydroxypterocarpans to the other isoflavonoids has been investigated in an extensive series of feeding experiments using $CuCl_2$-treated pods and seedlings of pea (*Pisum sativum*), which synthesize (+)-pisatin (*5.288*) as a phytoalexin (Banks and Dewick, 1982b). Good incorporations of 2′,4′,4-trihydroxychalcone (*5.277*), the isoflavones formononetin (*5.278*), calycosin (*5.279*), pseudobaptigenin (*5.280*), 7,2′-dihydroxy-4′,5′-methylenedioxyisoflavone (*5.281*), and (±)-maackiain (*5.284* plus *5.285*), but not the methyl ethers of formononetin or maackiain, showed that the pathway (Fig. 5.25) proceeds via maackiain and that methylation is a very late step. In confirmation, (+)-6a-hydroxymaackiain (*5.286*) proved to be a most effective precursor (incorporation 27%). Pterocarp-6a-ene derivatives were poor precursors, ruling out sequences involving addition of water to a double bond, and a route involving direct 6a-hydroxylation of maackiain, followed by methylation, is favoured. Pea plants synthesize both pisatin and maackiain as phytoalexins after fungal or abiotic induction, but it is interesting that these pterocarpans have opposite configurations. The 6a-hydroxylation of maackiain would logically take place with retention of configuration, so it is unlikely that the pea phytoalexin (−)-maackiain (*5.285*) would be a precursor of (+)-pisatin (*5.288*), and incorporations from (±)-maackiain were interpreted as arising by utilization of only the (+)-isomer (*5.284*). When (−)-maackiain was fed to pea pods, substantial incorporations recorded (Banks and Dewick, 1982a) were thus unexpected, and were traced to the induced formation of large amounts of (−)-pisatin (*5.289*) in the presence of (−)-maackiain. Levels of up to 92% of the abnormal (−)-isomer were produced in immature pods fed (−)-maackiain. These observations help to confirm that the 6a-hydroxylation proceeds with retention of configuration at C-6a. Further experiments, using racemic maackiain labelled with ^3H in the (+)-isomer and with ^{14}C in the (−)-isomer, showed preferential incorporation of (+)-maackiain, again implying retention of configuration (Banks and Dewick, 1983a). (+)-6a-Hydroxymaackiain (*5.286*) was a much better precursor of pisatin than (−)-6a-hydroxymaackiain (*5.287*), although incorporation of the (−)-isomer into (−)-pisatin was observed. Isoflavan and isoflav-3-ene analogues of isoflavone (*5.281*) were poorly incorporated.

Thus, it may be concluded that *Pisum sativum* is capable of producing both isoflavanone enantiomers (*5.282* and *5.283*) from the isoflavone (*5.281*). These then act as precursors for (+)-pisatin [via (+)-maackiain] or (−)-maackiain, the two phytoalexins. The stereochemistry of reduction of the isoflavone formononetin (*5.278*) during the biosynthesis of the two pterocarpans (+)-pisatin and

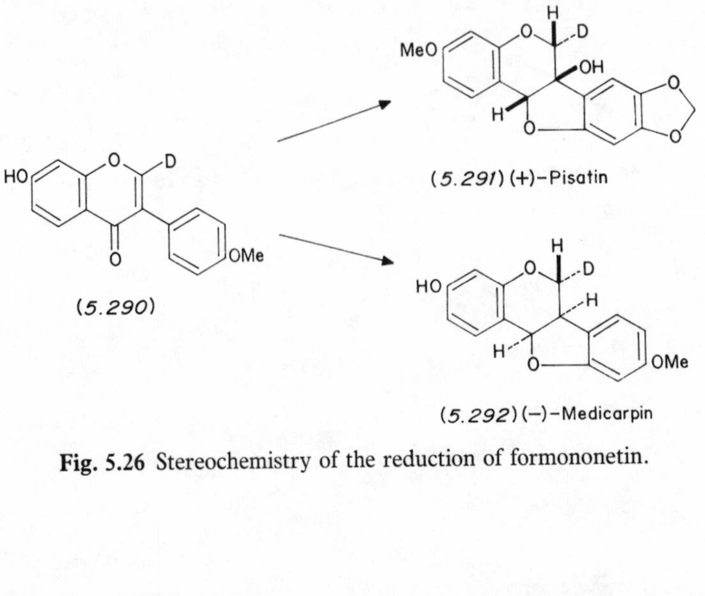

Fig. 5.26 Stereochemistry of the reduction of formononetin.

(−)-medicarpin (*5.292*) has been investigated in $CuCl_2$-treated seedlings of pea and fenugreek (*Trigonella foenum-graecum*) respectively (Banks *et al.*, 1982). Deuterium NMR spectral measurements on the products obtained after feeding [2-^{2}H] formononetin (*5.290*) demonstrated an overall *Z* addition during the biosynthesis of pisatin (*5.291*) (assuming 6a-hydroxylation to occur with retention of configuration), in contrast with an *E* addition for the production of medicarpin (*5.292*) (Fig. 5.26). This means that the stereochemistry of the reduction processes leading to (+)- and (−)-pterocarpans are probably different, although a process of *E* reduction followed by epimerization of the resultant isoflavanone may be mechanistically more acceptable than *Z* reduction of the α, β-unsaturated ketone (Fig. 5.25). Confirmation of the direct hydroxylation mechanism for biosynthesis of 6a-hydroxypterocarpans was obtained by feeding [6, 11a-$^2H_2$] maackiain (*5.293*) to induced pea pods. The ^{2}H NMR spectrum of the pisatin produced showed retention of all ^{2}H labels [i.e. (*5.294*). Fig. 5.27] with relative intensities essentially identical to those of the precursor, thus excluding any pterocarpene intermediates and dehydrogenation/hydration mechanisms (Banks and Dewick, 1983c).

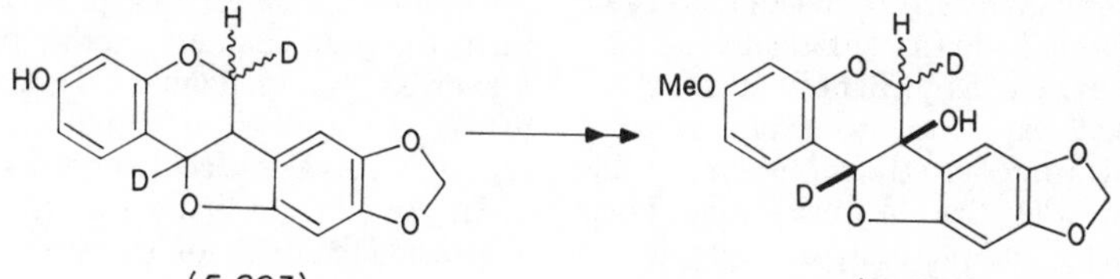

(*5.293*) (*5.294*)

Fig. 5.27 Hydroxylation of maackiain to yield pisatin.

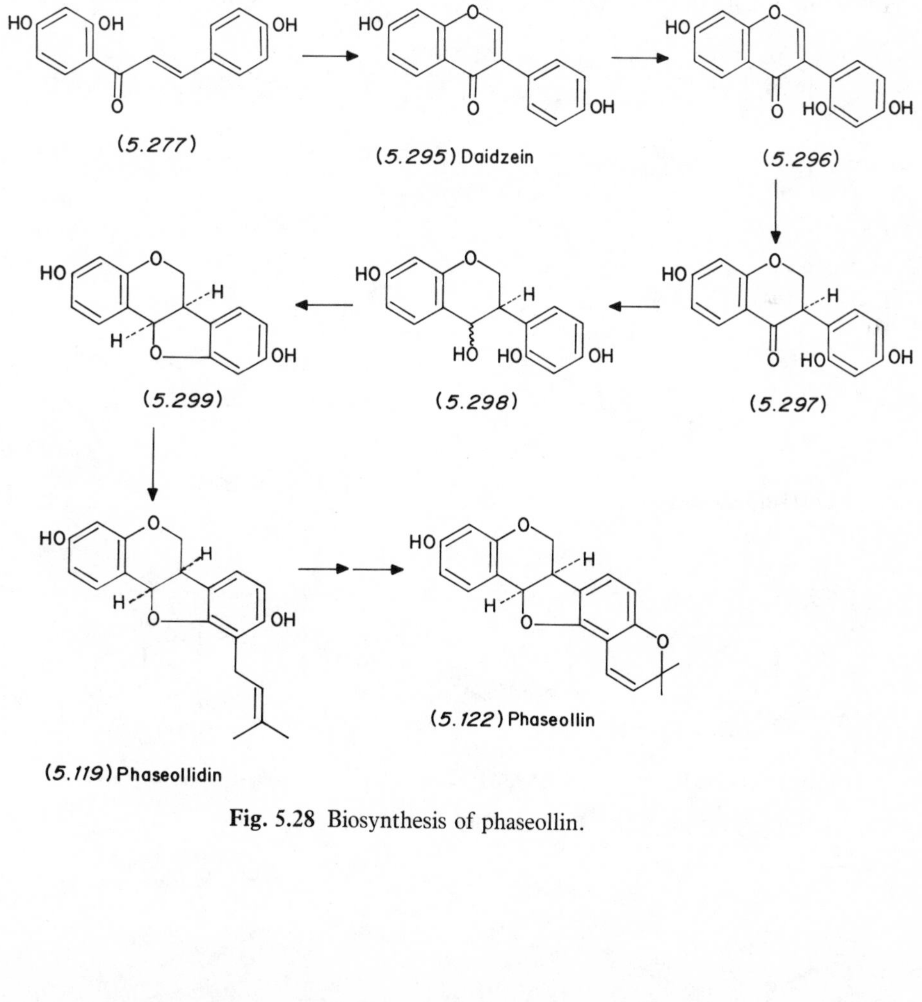

(*5.277*) (*5.295*) Daidzein (*5.296*) (*5.297*) (*5.298*) (*5.299*) (*5.119*) Phaseollidin (*5.122*) Phaseollin

Fig. 5.28 Biosynthesis of phaseollin.

The biosynthesis of phaseollin (*5.122*), a phytoalexin of French bean (*Phaseolus vulgaris*), has been investigated in $CuCl_2$-treated bean seedlings (Dewick and Steele, 1982). Good incorporations of labelled 2′,4′,4-trihydroxychalcone (*5.277*), the isoflavones daidzein (*5.295*) and 7,2′,4′-trihydroxyisoflavone (*5.296*), and the pterocarpans (*5.299*) and phaseollidin (*5.119*) suggested that these compounds represent a logical sequence in the pathway to phaseollin (Fig. 5.28). By analogy with pathways to other pterocarpans, the intermediacy of an isoflavanone (*5.297*) and an isoflavanol (*5.298*) in the reductive cyclization of the isoflavone (*5.296*) is postulated. The results demonstrate that isoprenylation is a late stage in the pathway, and occurs only after the basic pterocarpan skeleton has been produced. Biosynthetic studies in $CuCl_2$-treated seedlings and pods of soybean (*Glycine max*) (Banks and Dewick, 1983b) show that the pathways to the major phytoalexins glyceollins I (*5.142*), II (*5.136*) and III (*5.140*) are, in the early stages, common with that to phaseollin. Thus isoflavones (*5.295*), (*5.296*) and pterocarpan (*5.299*) were all efficient precursors of the glyceollins, as was glycinol (*5.300*). As with phaseollin, this indicates that isoprenylation occurs as a late step, after the basic 6a-hydroxypterocarpan skeleton has been constructed. From earlier studies, the 4- and 2-isoprenylated derivatives of glycinol, i.e. glyceollidins I (*5.141*) and II (*5.135*) respectively, are likely intermediates in the biosynthesis of the glyceollins (Fig. 5.29). The isolation of the dimethylallyltransferase enzyme that catalyses the isoprenylation of glycinol has now been reported in fuller detail (Zähringer *et al.*, 1981). The enzyme fraction isolated from elicitor-induced soybean cotyledons appears to contain two different enzymes, catalysing dimethylallylation at the two aromatic sites, and thus proportions of the various glyceollins produced may be regulated to some extent by these activities. Levels of enzyme activity increase markedly after treatment of cell cultures or hypocotyls of soybean with mycelium of, or with glucan elicitor from, *Phytophthora megasperma* f.sp. *glycinea* (Leube and Grisebach, 1983). In contrast, another potential control point, 3-hydroxy-3-methylglutaryl-CoA reductase, was shown to decrease in activity after induction, suggesting that the supply of mevalonic acid is not a regulatory factor.

An enzyme preparation from elicitor-challenged suspension cultures of soybean cells has been shown to catalyse the 6a-hydroxylation of 3,9-dihydroxypterocarpan to form glycinol (*5.300*) (Hagmann *et al.*, 1984). Essential cofactors were NADPH and molecular oxygen, and the enzyme was specific for the (6a*R*, 11a*R*)-isomer (*5.299*); the (6a*S*, 11a*S*)-isomer from

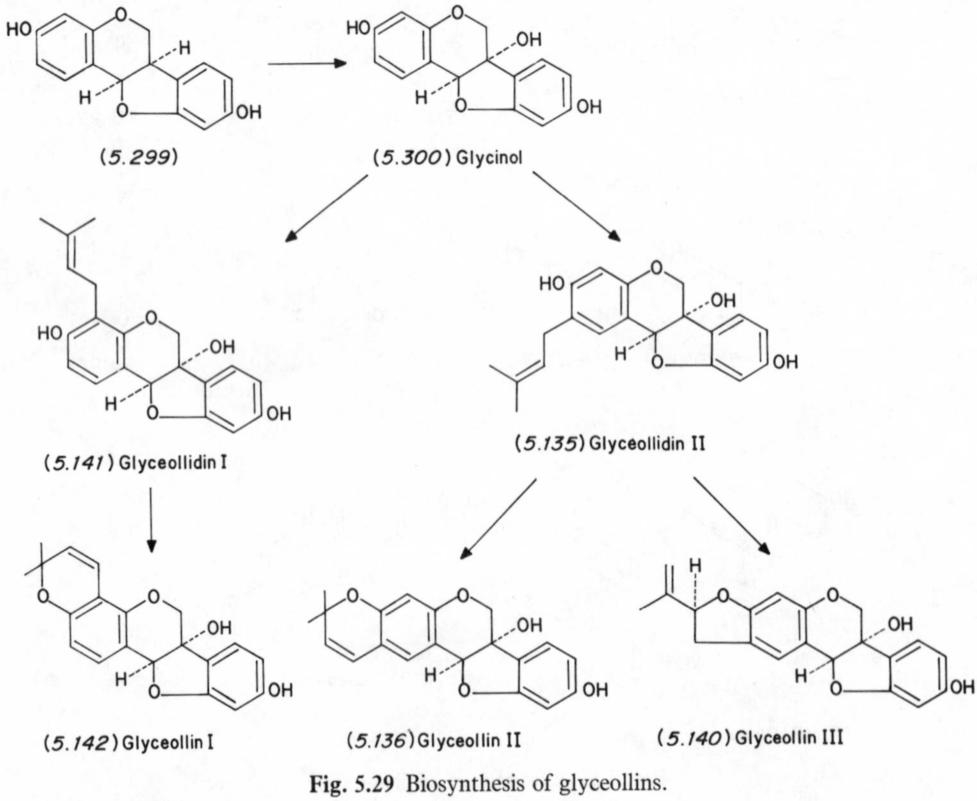

(*5.299*)

(*5.300*) Glycinol

(*5.141*) Glyceollidin I

(*5.135*) Glyceollidin II

(*5.142*) Glyceollin I

(*5.136*) Glyceollin II

(*5.140*) Glyceollin III

Fig. 5.29 Biosynthesis of glyceollins.

racemic substrate did not react. ORD (Optical Rotatory Dispersion) analysis of the product confirmed that the hydroxylation proceeded with retention of configuration, as was also concluded in the pisatin studies. This hydroxylase was very unstable, with a half-life of only 9 minutes at 30°C, and was inhibited by cytochrome *c*, indicating that the enzyme is probably a cytochrome-*p*-450-dependent mono-oxygenase. Enzyme activities were also increased severalfold by treatment with the elicitor.

Results concerning modifications in the isoprene fragment during the biosynthesis of rotenoids in *Amorpha fruticosa* (see Dewick, 1982) have now been published in full (Crombie *et al.*, 1982). Again, in this system, isoprenylation only occurs after the basic rotenoid skeleton has been elaborated.

The biosynthetic pathway to the isoflavan isomucronulatol (*5.303*) has been investigated in seedlings and pods of bladder senna (*Colutea arborescens*), in order to show how the 2′,3′,4′-oxygenation pattern in the B-ring is elaborated (Al-Ani and Dewick, 1985). Feeding experiments demonstrated that the isoflavones formononetin (*5.278*), calycosin (*5.279*), koparin (*5.301*) and 7,2′-dihydroxy-3′,4′-dimethoxyisoflavone (*5.302*) were excellent precursors, especially in pod tissues, whereas 2′-hydroxyformononetin and cladrin (7-hydroxy-3′,4′-dimethoxyisoflavone) were poor substrates. These results suggest that the biosynthetic sequence to isomucronulatol from formononetin (Fig. 5.30) involves 3′-hydroxylation, 2′-hydroxylation, then 3′-*O*-methylation at the isoflavone oxidation level, followed presumably by stereospecific reduction of isoflavone (*5.302*) to the (3*R*)-isoflavan (*5.303*).

A soluble UDP-glucose:isoflavone 7-*O*-glucosyltransferase from roots of chick-pea (*Cicer arietinum*) catalysed the 7-*O*-glucosylation of the 4′-methoxyisoflavones formononetin (*5.278*) and biochanin A (*5.304*) (Köster and Barz, 1981). The corresponding 4′-hydroxyisoflavones daidzein and genistein were poor substrates, and a range of other isoflavones, isoflavanones and flavonoids were not glucosylated. Malonylation of isoflavone 7-*O*-glucosides at position 6 of the glucose, using malonyl-CoA as the acyl donor, has been demonstrated using a malonyltransferase from the same plant source (Köster *et al.*, 1984). Best substrates for the enzyme were formononetin 7-*O*-glucoside and biochanin A 7-*O*-glucoside, giving the products (*5.305*) and (*5.306*), respectively (Fig. 5.31). Isoflavone 4′-*O*-glucosides were not malonylated, and other flavonoid 7-*O*-glucosides proved to be

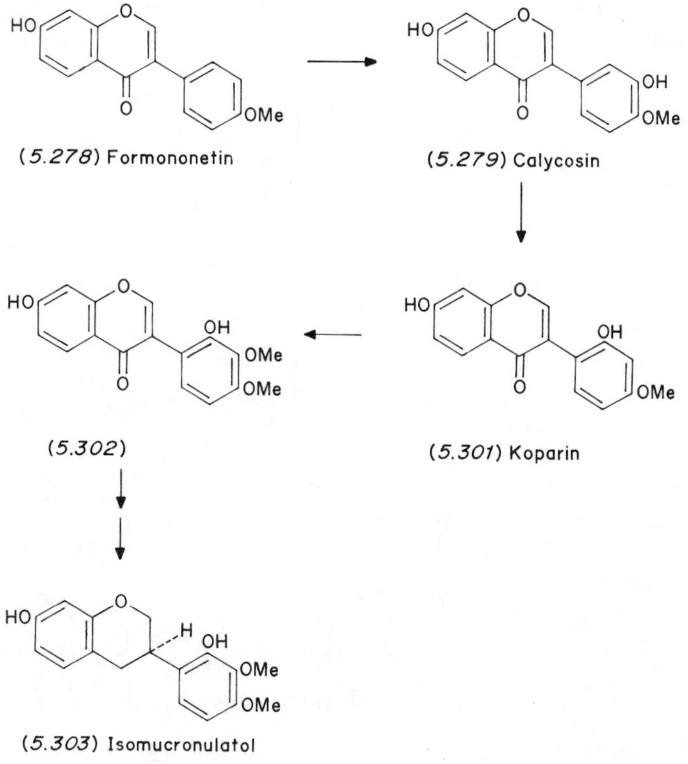

(*5.278*) Formononetin

(*5.279*) Calycosin

(*5.302*)

(*5.301*) Koparin

(*5.303*) Isomucronulatol

Fig. 5.30 Biosynthesis of isomucronulatol.

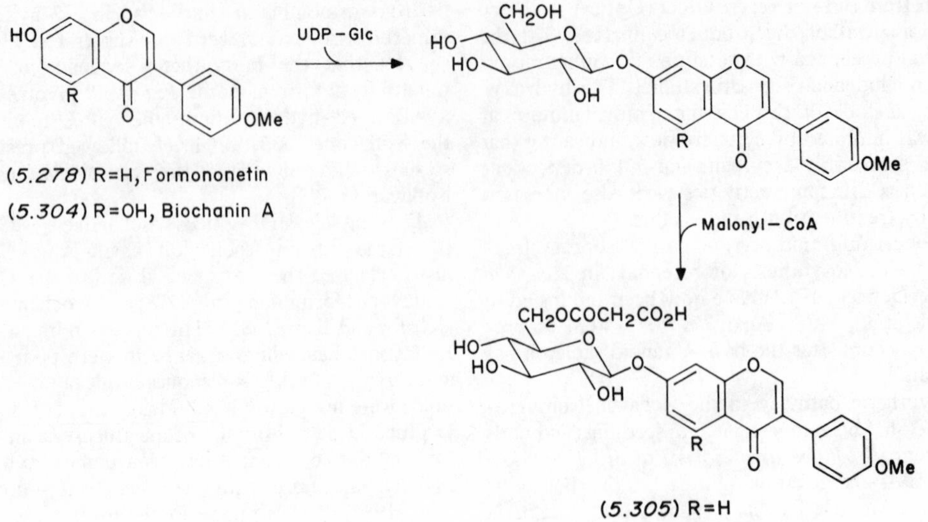

(5.278) R=H, Formononetin

(5.304) R=OH, Biochanin A

(5.305) R=H

(5.306) R=OH

Fig. 5.31 Formation of isoflavone malonylglucosides.

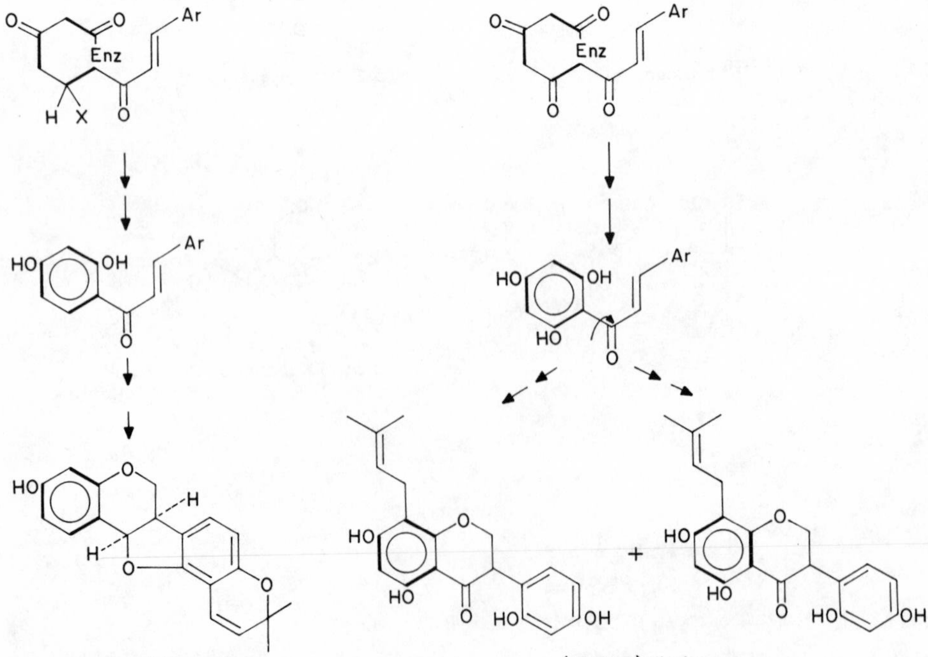

(5.307) Phaseollin

(5.308) Kievitone

Fig. 5.32 Biosynthetic relationships between phaseollin and kievitone.

poorer substrates. Treatment of mature leaves of soybean (*Glycine max*) with the herbicide acifluorfen results in the accumulation of isoflavones, isoflavone glucosides and pterocarpans. The isoflavonoid accumulation is preceded by an induced activity for chalcone synthase, and increased activities for PAL (Phenylalanine ammonia lyase) and UDP-glucose:isoflavone 7-*O*-glucosyltransferase (Cosio *et al.*, 1985). The isoflavone glucosides appear to be synthesized and accumulated specifically in the mesophyll cells.

Labelling patterns in the phytoalexins phaseollin (*5.307*) and kievitone (*5.308*) produced by wounded cotyledons of French bean (*Phaseolus vulgaris*) from sodium [1,2-$^{13}C_2$]acetate were analysed by ^{13}C NMR spectroscopy (Dewick *et al.*, 1982), and have substantiated earlier observations on the formation of the acetate-derived ring of flavonoids/isoflavonoids (see Dewick, 1982). Thus, specific folding of the polyketide chain and reduction of the 'missing' oxygen function prior to ring closure results in the incorporation of three intact acetate units into phaseollin

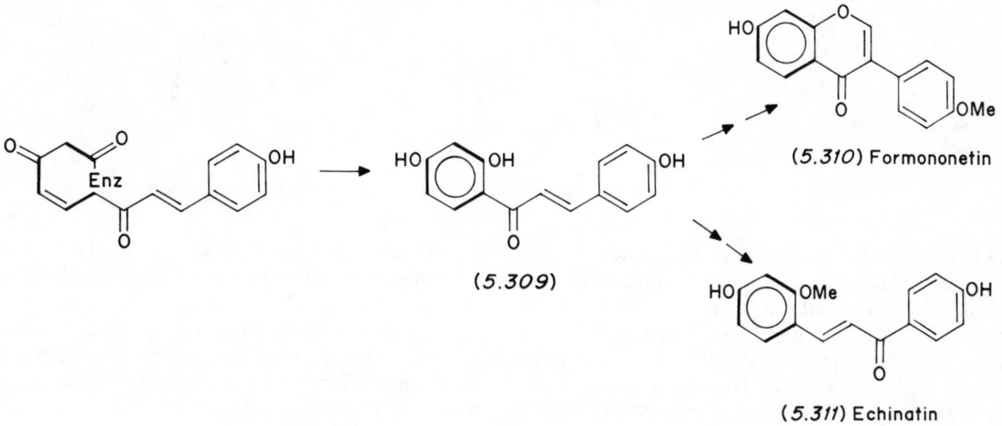

(*5.309*)

(*5.310*) Formononetin

(*5.311*) Echinatin

Fig. 5.33 Biosynthesis of the retrochalcone echinatin.

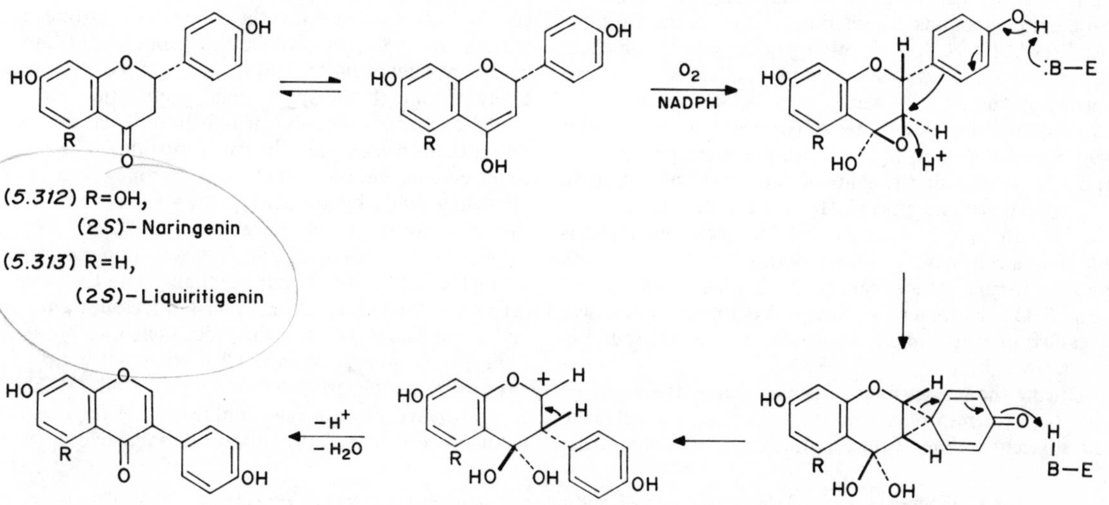

(*5.312*) R=OH, (2*S*)-Naringenin

(*5.313*) R=H, (2*S*)-Liquiritigenin

O_2 NADPH

$-H^+$ $-H_2O$

(*5.314*)

(*5.315*)

(*5.316*) Genistein

(*5.295*) Daidzein

Fig. 5.34 Mechanism of isoflavone synthase.

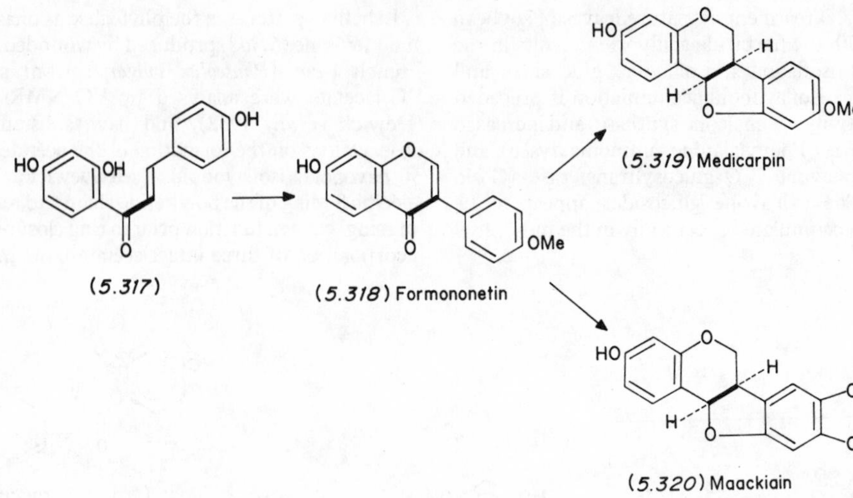

(5.317) (5.318) Formononetin (5.319) Medicarpin (5.320) Maackiain

Fig. 5.35 Intramolecular migration during isoflavonoid biosynthesis.

as shown (Fig. 5.32). In contrast, free rotation of a chalcone intermediate results in randomization of label and two possible labelling patterns in kievitone (*5.308*), as indicated by pairs of satellite signals in its ^{13}C NMR spectrum. The biosynthetic routes to these 5-hydroxy- and 5-deoxy-isoflavonoids thus diverge prior to the formation of chalcones. Despite satisfactory incorporation of label into the aromatic ring, no enrichment was detected in the isoprenoid substituents. In wounded cotyledons, the accumulation of 5-hydroxyisoflavonoids such as kievitone and 2′-hydroxygenistein precedes increases in levels of the 5-deoxy compounds phaseollin and coumestrol (Whitehead *et al.*, 1982). Specific folding of the polyketide chain (as in phaseollin) was also noted in the 5-deoxyisoflavone formononetin (*5.310*) which was isolated as a minor metabolite from cell cultures of *Glycyrrhiza echinata* that had been fed sodium [1, 2-$^{13}C_2$]acetate during a study of the biosynthesis of the unusual retrochalcone echinatin (*5.311*) (Ayabe and Furuya, 1982). The functionality of the C_3 chain in (*5.311*) is reversed during the biosynthesis of this chalcone from 2′, 4′, 4-trihydroxychalcone (*5.309*), and the B-ring, rather than the A-ring, becomes labelled (Fig. 5.33). However, a normal labelling pattern was obtained in formononetin, contrasting with that found in the retrochalcone.

During the biosynthesis of isoflavonoids, the aromatic ring derived from phenylanine/cinnamic acid migrates to the adjacent carbon of the C_3 unit. Over the years, many hypotheses have been proposed for the mechanism of this aryl migration, though none has accommodated all of the available biosynthetic evidence. A significant breakthrough in this field is the long-awaited isolation of the enzyme that catalyses this crucial step. An isoflavone synthase activity was detected in a microsomal preparation from cell suspension cultures of soybean (*Glycine max*) challenged with a glucan elicitor from *Phytophthora megasperma* f.sp. *glycinea* (Hagmann and Grisebach, 1984). The enzyme activity had a half-life of only about 10 minutes at 30°C, required NADPH and molecular oxygen as cofactors, and transformed the flavanone substrates (2*S*)-naringenin (*5.312*) or (2*S*)-liquiritigenin (*5.313*) into the isoflavones genistein (*5.316*) or daidzein (*5.295*), respectively. Although the microsomal preparation also contained chalcone isomerase, it is believed that flavanones rather than chalcones are the true substrates. The cofactor requirements indicate that the enzyme is a monooxygenase, and a hypothetical pathway, via epoxidation of the enol form of the flavanone, has been proposed (Fig. 5.34).* Isoflavone synthase was also detected in untreated cell cultures and seedlings, but the levels of activity increased significantly in both tissues after they had been challenged with the glucan elicitor.

Feeding experiments in $CuCl_2$-treated seedlings of red clover (*Trifolium pratense*) using ^{13}C- and ^{2}H-labelled 2′, 4′, 4-trihydroxychalcone, confirmed that the rearrangement process in the biosynthesis of isoflavonoids was an

*The mechanism proposed in Fig. 5.34 has been criticized, and a more favourable one involving epoxidation of the migrating aryl ring rather than the heterocyclic ring has been suggested (Crombie *et al.*, 1986). However, a spirodienone intermediate analogous to (*5.314*) is included. Further study of the isoflavone synthase preparation has resulted in the isolation of an intermediate, believed to be 5, 7, 2, 4′-tetrahydroxyisoflavanone, in the transformation of (2*S*)-naringenin into genistein (Kochs and Grisebach, 1986). This 2-hydroxyisoflavanone is suggested to be derived from the carbocation (*5.315*), and is dehydrated to yield genistein by an enzyme fraction that requires neither NADPH nor O_2 as cofactors.

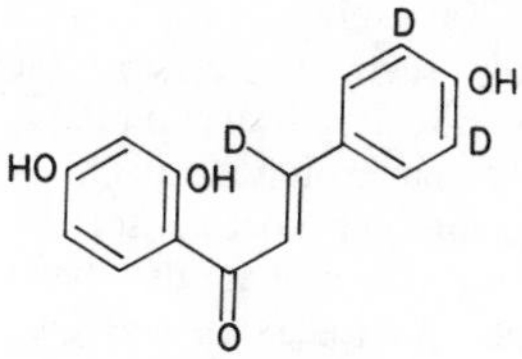

(5.321)

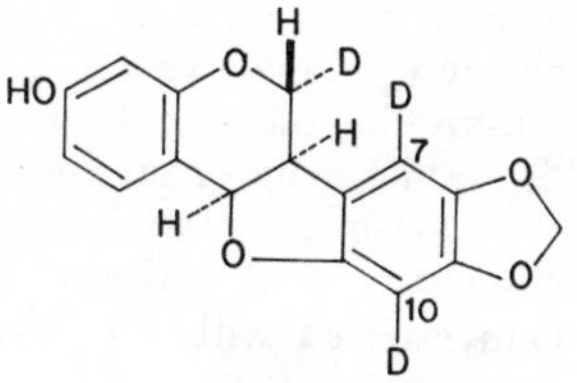

(5.322)

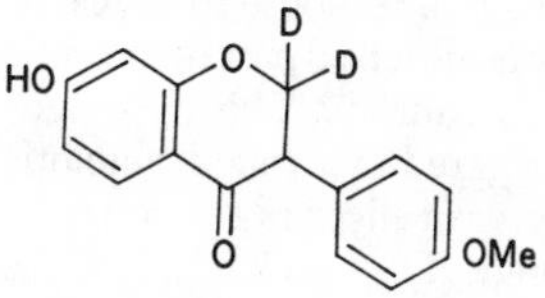

(5.323)

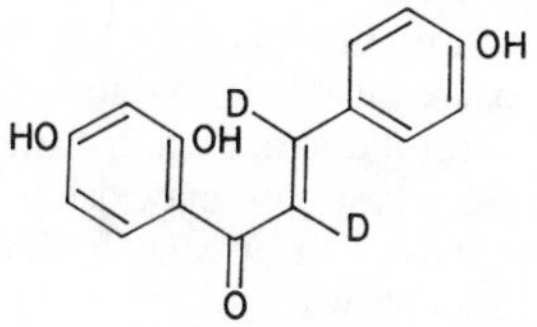

(5.328)

intramolecular migration of an aryl group (Al-Ani and Dewick, 1984). Thus, the [$^{13}C_2$] trihydroxychalcone (*5.317*) was transformed into the labelled isoflavone formononetin (*5.318*) and the corresponding labelled pterocarpan phytoalexins medicarpin (*5.319*) and maackiain (*5.320*) with retention of the intact ^{13}C–^{13}C linkages (Fig. 5.35), demonstrating the aryl migration. The intramolecular nature of the migration was proved by the isolation of $^2H_3$-labelled formononetin and medicarpin after feeding the [$^2H_3$] chalcone (*5.321*) to the plants, as indicated by MS analysis. The sample of maackiain isolated in the same experiment consisted of approximately equal amounts of $^2H_2$- and $^2H_3$-labelled species, the latter molecule (*5.322*) undoubtedly arising via [$^2H_3$] formononetin through an NIH shift during build-up of the more complex substitution pattern in this pterocarpan, and supports the sequence shown in the early part of Fig. 5.25. This was confirmed in a further experiment via the use of ^{2}H NMR spectroscopy, which showed the presence of ^{2}H at both position 7 and position 10. Experiments with the [$^2H_2$] isoflavanone (*5.323*) and subsequent MS analysis showed that this compound was converted into a [$^2H_1$] formononetin and a [$^2H_2$] medicarpin, and gave direct proof of the existence of a metabolic grid of isoflavones and isoflavanones in the sequence to this compound (Fig. 5.36; compare Fig. 5.24). On the basis of incorporation data, the sequence from formononetin (*5.278*) to medicarpin (*5.327*) via the isoflavone (*5.324*) and the isoflavanone (*5.326*) would seem more important than that via the isoflavanones (*5.325*) and (*5.326*). The [$^2H_2$]chalcone (*5.328*) was transformed into formononetin, medicarpin and maackiain with retention of the β-hydrogen, but the α-hydrogen was lost, making route b of Fig. 5.36 more likely than route a. The results are best explained in terms of an oxidative process for the aryl migration, in which an isoflavone is the first isoflavonoid intermediate to be

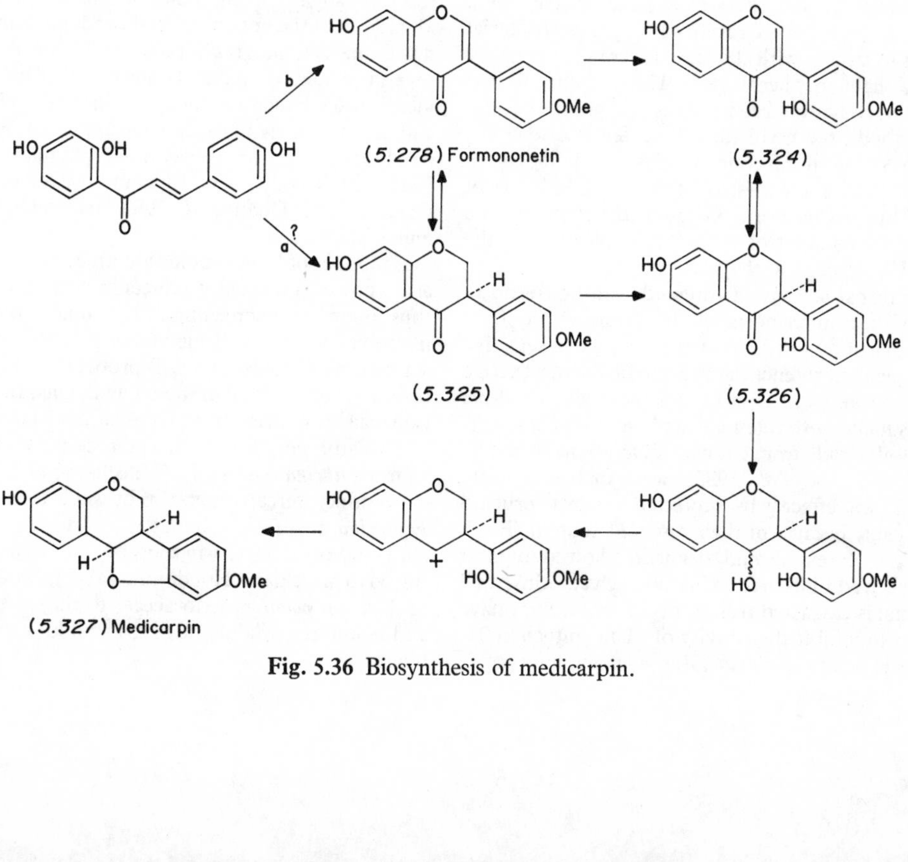

(5.278) Formononetin

(5.324)

(5.325)

(5.326)

(5.327) Medicarpin

Fig. 5.36 Biosynthesis of medicarpin.

formed, and although a chalcone substrate was postulated in these investigations, the data are fully consistent with the enzymic studies in *Glycine max* (Hagmann and Grisebach, 1984).

5.19 BIOLOGICAL PROPERTIES OF ISOFLAVONOIDS

Biological activities reported in the literature for isoflavonoids cover quite a broad range (Ingham, 1983), but at the time of the last review, perhaps only three activities could be regarded as of significant importance to man. These were the oestrogenic activities of simple isoflavones and coumestans, the insecticidal properties of rotenoids and the more recently established anti-fungal and anti-bacterial activities associated with isoflavonoid phytoalexins and related compounds.

The oestrogenic activity of simple isoflavones such as daidzein, formononetin, genistein and biochanin A, and of coumestans, especially coumestrol, is primarily of concern in animal husbandry. These compounds may reach significant levels in forage crops, such as clovers (*Trifolium* spp.) and lucerne (*Medicago sativa*), and techniques for identification and quantitative analysis of these simple isoflavonoids are important in agriculture. Such studies continue (Blanco *et al.*, 1982; Shehata *et al.*, 1982; Gil *et al.*, 1984), and in a review by Smolenski *et al.* (1981), oestrogenic isoflavonoids feature among the important toxic constituents of leguminous plants. It must also be recognized that levels of oestrogenic isoflavones and coumestans may be increased in a plant as part of a phytoalexin response, as a result of fungal or bacterial infection, or treatment with abiotic inducers such as heavy metals, UV light or herbicides. Thus, daidzein and formononetin are synthesized in leaves of soybean (*Glycine max*) treated with the herbicide acifluorfen (Cosio *et al.*, 1985), or on bacterial infection (Osman and Fett, 1983), and UV light induces coumestrol biosynthesis in leaves of the bean (*Phaseolus vulgaris*) (Beggs *et al.*, 1985). In the latter example, coumestrol levels are indicative of the amount of UV damage to the plant system.

The insecticidal activity of rotenoids can be traced to inhibition of the mitochondrial electron transport system, together with inability to detoxify the rotenoid in sensitive systems. In general, rotenone appears to be the most active of the natural rotenoids. Larvicidal activity against *Aedes aegypta* was found with rotenone, and only with amorphigenin 8′-*O*-glucoside from a range of *Dalbergia monetaria* rotenoids tested (Abe *et al.*, 1985). Seeds of *Lonchocarpus salvadorensis* are effectively protected against bruchid beetles probably because of their rotenoid content (Birch *et al.*, 1985). Rotenone and deguelin showed highest mortality when administered to bruchid larvae. Almost all of the rotenoids extracted from roots of *Derris* spp. have been shown to inhibit the activity of plant mitochondria from tubers of potato (*Solanum tuberosum*) and hypocotyls of mung bean (*Phaseolus aureus*) (Ravanel *et al.*, 1984). The presence of the five fused rings seemed necessary for selective inhibition of the potato mitochondrial complex I, deguelin having greater activity than rotenone itself. A rotenone-like inhibition of electron transport in isolated mitochondria of soybean (*Glycine max*) has also been demonstrated with the soybean phytoalexins the glyceollins (Boydston *et al.*, 1983), and this may well be related to structural similarity in the two systems, especially with regard to molecular shape (see below).

Phytoalexins are anti-microbial (especially anti-fungal) compounds produced by plants as a response to fungal or bacterial attack, and may be considered as part of a plant's natural defence against micro-organisms. This hypersensitive response may also be triggered by a range of abiotic inducers, or by carbohydrate glucan elicitors derived from the cell walls of yeasts or fungi. The mechanism of the response is an active research area for plant pathologists, and it is not surprising that isoflavonoid phytoalexins figure prominently in the tabulated data of earlier sections. Many new compounds have been isolated as a consequence of deliberate fungal infection or abiotic stress of leguminous plant tissues. Typically, a range of structurally related isoflavonoids will be produced as stress metabolites, though not all of these will be anti-microbial and thus classified as phytoalexins. Such properties are associated with pterocarpans, isoflavans and some isoflavanones rather than with isoflavones which tend to accumulate as intermediates in the biosynthetic sequence to the anti-microbial metabolites. Some researchers study the full range of compounds produced, whereas others identify only the anti-microbial products using a bioassay technique. A monograph (Bailey and Mansfield, 1982) and several reviews on phytoalexins covering aspects of their identity, accumulation, biosynthesis, metabolism, activity and taxonomic significance have appeared. Reviews with strong emphasis on isoflavonoid phytoalexins include: Rizk and Wood, 1980; Ingham, 1981c; Afzal and Al-Oriquat, 1982; Dixon *et al.*, 1983; Grisebach, 1983; Brooks and Watson, 1985.

A number of isoflavonoids identified as phytoalexins are also known as normal products in other plant species. It thus follows that screening of leguminous plants for anti-microbial activity will inevitably lead to the isolation of active isoflavonoids, and will probably yield other structures not represented in the phytoalexins. In this way, the isoflavan crotmarine from *Crotalaria madurensis* (Bhakuni and Chaturvedi, 1984), the pterocarpene erycristagallin from *Erythrina crista-galli* (Mitscher *et al.*, 1984) and the 6a-hydroxyptercarpan erythrabyssin I (cristacarpin) from *Erythrina abyssinica* (Kamat *et al.*, 1981) were isolated. Anti-microbial activity in roots of *Sophora angustifolia* was traced to maackiain (Honda and Tabata, 1982), and in fruit of *Maclura pomifera* (Moraceae) to the isoflavones osajin and pomiferin (Mahmoud, 1981), all known compounds, however.

It is difficult to generalize on how the anti-microbial activity of isoflavonoids is related to molecular structure. Three-dimensional shape was once held to be important, though several planar isoflavonoids show strong anti-fungal activity, and lipophilicity was also suggested to play an important role (see Dewick, 1982). Studies of potential structure–anti-fungal activity relationships have reached no firm conclusion (Fraile *et al.*, 1982; Stössel, 1985). Fungi differ in sensitivity to the bean (*Phaseolus vulgaris*) phytoalexins and related isoflavonoids, and there is certainly no clear relationship between lipophilicity and anti-fungal activity. Phaseollin and phaseollinisoflavan were more inhibitory than *O*-methyl ethers of phaseollinisoflavan, any methylation of the 7-hydroxyl markedly decreased activity (Stössel, 1985). Of the glyceollins, glyceollin I was twice as effective as glyceollins II and III against one isolate of *Phytophthora megasperma* f.sp. *glycinea*, whereas glyceollin II was most active against zoospore germination (Bhattacharyya and Ward, 1985). *O*-Methylsativan and some dimethoxyisoflavanones have similar levels of activity as the isoprenylated bean phytoalexins (Ravise *et al.*, 1980). Anti-fungal activity of simple isoflavones from *Trifolium pratense* (Popravko *et al.*, 1980), *Glycine max* and *Cicer arietinum* (Krämer *et al.*, 1984) has been recorded, but simple isoflavans appear more active than the corresponding isoflavones or isoflavanones (Krämer *et al.*, 1984). An assay of the anti-fungal activity of a range of pterocarpan and isoflavan phytoalexins against pathogens of five crop plants concludes, however, that the activity levels are too low for isoflavonoid phytoalexins to be used as conventional fungicides (Rathmell and Smith, 1980).

Anti-bacterial activity of a range of isoflavonoid phytoalexins has been shown to be selective towards Gram-positive bacteria, with kievitone being the most toxic (Gnanamanickam and Smith, 1980; Gnanamanickam and Mansfield, 1981). Glyceollin (i.e. the mixture of glyceollins) inhibited growth of most bacteria tested, and was bactericidal to some (Fett and Osman, 1982). The antibacterial activity of glyceollin, glycinol and coumestrol has been shown to be due to general interaction with the bacterial membrane (Weinstein and Albersheim, 1983). The isoflavone daidzein was ineffective.

A more recent biological activity to be associated with isoflavonoids is insect-feeding deterrent activity. This property was first discovered in a number of isoflavonoid phytoalexin structures, and a proposal was made that such compounds may serve two different roles, as phytoalexins and as insect-feeding deterrents (see Dewick, 1982). Further examples have been added to the list of isoflavonoids now known to possess this activity. The two 2-arylbenzofuran derivatives sainfuran and methylsainfuran isolated from roots of sainfoin (*Onobrychis viciifolia*) (Russell *et al.*, 1984), the rotenoid tephrosin from roots of *Tephrosia elata* (Lwande *et al.*, 1985a), and the 6a-hydroxypterocarpan hildecarpin from roots of *Tephrosia hildebrandtii* (Lwande *et al.*, 1985b) are all reported to be insect-feeding deterrents. An extensive range of isoflavonoids has been tested for feeding-deterrent activity in a bioassay against root-feeding larvae of the beetle *Costelytra zealandica* to investigate structure–activity relationships (Lane *et al.*, 1985). The most effective feeding deterrents were phaseollin, phaseollinisoflavan, and rotenone, which reduced feeding significantly at levels of 0.02–0.06 $\mu g\, g^{-1}$. Structural features associated with high activities were a 2′-oxygen function and a cyclic isoprenoid substituent fused to ring B (isoflavone numbering), though these in themselves did not confer activity. It was concluded that activity relates to stereochemistry, and the non-planar shape of phaseollin together with a similar arrangement of polar and lipophilic groups is necessary for activity. If models of phaseollin and rotenone are compared, their three-dimensional shapes are remarkably similar, with the lipophilic isoprenoid substituents in a similar region, the rotenone carbonyl occupying a similar position to phaseollin's dihydrofuran oxygen (the 2′-oxygen) function, and rotenone's methoxyls occurring in the region of phaseollin's polar hydroxyl. Thus, these examples from two different isoflavonoid classes could well interact with the same biological receptor site.

Two biological activities of isoflavonoids reported for the first time are of particular interest. The pterocarpans cabenegrin A-I and cabenegrin A-II, both isoprenylated derivatives of (−)-maackiain, were isolated from the root of an unidentified South American plant 'Cabeca de Negra' (Nakagawa *et al.*, 1982). Some ten plants are collectively included under this name, and are used as an oral antidote to snake and spider venoms. The two pterocarpans isolated were demonstrated to possess quite potent antidote activity towards the venom of the Fer de Lance snake *Bothrops atrox*, and since these compounds have now been synthesized, further application of this unusual property is awaited. The parasitic angiosperm *Agalinis purpurea* (Scrophulariaceae) is attracted to the roots of its host plant as a result of an exogenous chemical signal normally contained in the host root exudate, and this initiates growth of the haustorium and allows the parasite to attach itself to the host. Host-recognition substances have been isolated from the gum exudate of *Astragalus* spp. (gum tragacanth) and termed xenognosins. Xenognosin A was identified as a cinnamylphenol, and xenognosin B as the isoflavone 2′-hydroxyformononetin (Steffens *et al.*, 1982). Formononetin, which co-occurs with the xenognosins, was inactive as a host-recognition substance.

Other biological activities noted with representatives of the isoflavonoids include the following: inhibition of prostaglandin biosynthesis (isoflavans from *Dalbergia odorifera*) (Goda *et al.*, 1985); anti-atherosclerotic activity (genistein 8-*C*-glucoside) (Laman and Oleksyuk, 1982); inhibition of rat heart phosphodiesterase (rotenoids and isoflavonoids) (Petov *et al.*, 1983); inhibition of soybean lipase activity (soybean isoflavones, especially genistein derivatives) (Ohta *et al.*, 1981; Ohta and Mikumo, 1982); inhibition of mammalian mixed-function oxidase systems (pterocarpan and isoflavan phytoalexins) (Gelboin *et al.*, 1981; Friedman *et al.*, 1985); coronary-dilating effect on guinea pig heart (8-*O*-methylretusin) (Chen *et al.*, 1983); diuretic activity (hemerocallone) (Xiu *et al.*, 1982); as antagonists of slow-reacting substance of anaphylaxis and potential use in treatment of asthma (rotenoids) (Ashack *et al.*, 1980); competitive inhibitors for benzodiazepine receptors (isoflavans in bovine urine) (Luk *et al.*, 1983).

REFERENCES

Abe, F., Donnelly, D.M.X., Moretti, C. and Polonksy, J, (1985), *Phytochemistry* **24**, 1071.

Abidi, S.L. (1984), *J. Chromatogr.* **317**, 383.

Adesanya, S.A., O'Neill, M.J. and Roberts, M.F. (1984), *Z. Naturforsch.* **39c**, 888.

Adesanya, S.A., O'Neill, M.J. and Roberts, M.F. (1985), *Phytochemistry* **24**, 2699.

Adinarayana, D., Syamasundar, K.V., Seligmann, O. and Wagner, H. (1982), *Z. Naturforsch.* **37c**, 145.

Adinolfi, M., Barone, G., Belardini, M., Lanzetta, R., Laonigro, G. and Parrilli, M. (1985), *Phytochemistry* **24**, 2423.

Afonya, T.C.A., Epelle, F.B.M., Osman, S.A.A. and Whalley, W.B. (1985a), *J. Chem. Res., Synop.* 305.

Afonya, T.C.A., Epelle, F.B.M., Whalley, W.B., Ferguson, G. and Parvez, M. (1985b), *J. Crystallogr. Spectrosc. Res.* **15**, 289.

Afzal, M. and Al-Oriquat, G. (1982), *Heterocycles* **19**, 1295.

Agarwal, V.K., Thappa, R.K., Agarwal, S.G. and Dhar, K.L. (1984a), *Phytochemistry* **23**, 1342.

Agarwal, V.K., Thappa, R.K., Agarwal, S.G., Mehra, M.S. and Dhar, K.L. (1984b), *Phytochemistry* **23**, 2703.

Ahluwalia, V.K., Prakash, C. and Jolly, R.S. (1981), *J. Chem. Soc., Perkin Trans. I*, 1697.

Al-Ani, H.A.M. and Dewick, P.M. (1984), *J. Chem. Soc., Perkin Trans. I*, 2831.

Al-Ani, H.A.M. and Dewick, P.M. (1985), *Phytochemistry* **24**, 55.

Albuquerque, F.B., Braz Filho, R., Gottlieb, O.R., Magalhaes, M.T., Maia, J.G.S., Oliveira, A.B. de, Oliveira, G.G. de, and Wilberg, V.C. (1981), *Phytochemistry* **20**, 235.

Ali, A.A., El-Emary, N.A., El-Moghazi, M.A., Darwish, F.M. and Frahm, A.W. (1983), *Phytochemistry* **22**, 2061.

Anhut, S., Zinsmeister, H.D., Mues, R., Barz, W., Mackenbrock, K., Köster, J. and Markham, K.R. (1984), *Phytochemistry* **23**, 1073.

Antus, S., Gottsegen, A. and Nogradi, M. (1981), *Synthesis* 574.

Antus, S., Gottsegen, A., Kolonits, P., Nagy, Z., Nogradi, M. and Vermes, B. (1982), *J. Chem. Soc., Perkin Trans. I*, 1389.

Arakawa, Y., Asada, Y., Ishida, H., Chiji, H. and Izawa, M. (1982), *J. Fac. Agric., Hokkaido Univ.* **61**, 1. (Chem. Abstr. 1983, **98**, 197825).

Arisawa, M., Kyozuka, Y., Hayashi, T., Shimizu, M. and Morita, N. (1980), *Chem. Pharm. Bull.* **28**, 3686.

Arnone, A., Camarda, L., Merlini, L., Nasini, G. and Taylor, D.A.H. (1981), *Phytochemistry* **20**, 799.

Ashack, R.J., McCarty, L.P., Malek, R.S., Goodman, F.R. and Peet, N.P. (1980), *J. Med. Chem.* **23**, 1022.

Asres, K., Mascagni, P., O'Neill, M.J. and Phillipson, J.D. (1985), *Z. Naturforsch.* **40c**, 617.

Axelson, M., Kirk, D.N., Farrant, R.D., Cooley, G., Lawson, A.M. and Setchell, K.D.R. (1982), *Biochem. J.* **201**, 353.

Ayabe, S. and Furuya, T. (1982), *J. Chem. Soc., Perkin Trans. I*, 2725.

Babchenko, I.V., Umralina, A.R., Belyaeva, A.V., Moldosanova, T.A., Davydova, M.A., Vasil'Eva, K.V. and Metlitskii, L.V. (1982), *Prikl. Biokhim. Mikrobiol.* **18**, 159. (*Chem. Abstr.* 1982, **96**, 214491).

Bailey, J.A. and Berthier, M. (1981), *Phytochemistry* **20**, 187.

Bailey, J.A. and Mansfield, J.W. (eds) (1982), *Phytoalexins*, Blackie, Glasgow and London.

Banks, S.W. and Dewick, P.M. (1982a), *Phytochemistry* **21**, 1605.

Banks, S.W. and Dewick, P.M. (1982b), *Phytochemistry* **21**, 2235.

Banks, S.W. and Dewick, P.M. (1983a), *Phytochemistry* **22**, 1591.

Banks, S.W. and Dewick, P.M. (1983b), *Phytochemistry* **22**, 2729.

Banks, S.W. and Dewick, P.M. (1983c), *Z. Naturforsch.* **38c**, 185.

Banks, S.W., Steele, M.J., Ward, D. and Dewick, P.M. (1982), *J. Chem. Soc., Chem. Commun.* 157.

Baruah, P., Barua, N.C. and Sharma, R.P. (1984a), *Chem. Ind. (London)* 303.

Baruah, P., Barua, N.C., Sharma, R.P., Baruah, J.N., Kulanthaivel, P. and Herz, W. (1984b), *Phytochemistry* **23**, 443.

Beggs, C.J., Stolzer-Jehle, A. and Wellmann, E. (1985), *Plant Physiol.* **79**, 630.

Begley, M.J., Campbell, R.V.M., Crombie, L., Tuck, B. and Whiting, D.A. (1971), *J. Chem. Soc. (C)* 3634.

Bezuidenhoudt, B.C.B., Brandt, E.V. and Roux, D.G. (1981), *J. Chem. Soc., Perkin Trans. I*, 263.

Bezuidenhoudt, B.C.B., Brandt, E.V. and Roux, D.G. (1984), *J. Chem. Soc., Perkin Trans. I*, 2767.

Bhakuni, D.S. and Chaturvedi, R. (1984), *J. Nat. Prod.* **47**, 585.

Bhardwaj, D.K., Bisht, M.S., Jain, R.K. and Munjal, A. (1982), *Proc. Indian Natl. Sci. Acad., Part A* **48**, 103.

Battacharyya, M.K. and Ward, E.W.B. (1985), *Physiol. Plant Pathol.* **27**, 299.

Biggs, D.R. and Shaw, D.G. (1980), *Phytochemistry* **19**, 2801.

Biggs, D.R., Shaw, G.J., Yates, M.K. and Newman, R.H. (1983), *J. Nat. Prod.* **46**, 742.

Birch, N., Crombie, L. and Crombie, W.M. (1985), *Phytochemistry* **24**, 2881.

Blanco, M.M., Lucena, E.P. and Castro, A.G.G. (1982), *Arch. Zootec.* **31**, 269. (*Chem. Abstr.* 1983, **98**, 195020).

Bohlmann, F., Zdero, C., Robinson, H. and King, R.M. (1981), *Phytochemistry* **20**, 2245.

Bose, P.K. and Ganguly, S.N. (1981), *North Bengal Univ. Rev.* **2**, 79. (*Chem. Abstr.* 1983, **99**, 191644).

Boydston, R., Paxton, J.D. and Koeppe, D.E. (1983), *Plant Physiol.* **72**, 151.

Brandt, E.V., Bezuidenhoudt, B.C.B. and Roux, D.G. (1982), *J. Chem. Soc., Chem. Commun.* 1409.

Braz Filho, R., De Figueiredo, U.S., Gottlieb, O.R. and Mourao, A.P. (1980), *Acta Amazon.* **10**, 843. (*Chem. Abstr.* 1981, **95**, 129337).

Breytenbach, J.C. and Rall, G.J.H. (1980), *Tetrahedron Lett.* **21**, 4535.

Breytenbach, J.C., Van Zyl, J.J., Van der Merwe, P.J., Rall, G.J.H. and Roux, D.G. (1981), *J. Chem. Soc., Perkin Trans. I*, 2684.

Breytenbach, J.C., Leipoldt, J.G., Rall, G.J.H., Roux, D.G. and Levendis, D.C. (1983), *S. Afr. J. Chem.* **36**, 4.

Brooks, C.J.W. and Watson, D.G. (1985), *Nat. Prod. Rep.* **2**, 427.

Brum-Bousquet, M., Lallemand, J.Y., Tillequin, F., Faugeras, G. and Delaveau, P. (1981a), *Plant. Med.* **43**, 367.

Brum-Bousquet, M., Lallemand, Y., Tillequin, F. and Delaveau, P. (1981b), *Tetrahedron Lett.* **22**, 1223.

Burns, D.T., Dalgarno, B.G., Gargan, P.E. and Grimshaw, J. (1984), *Phytochemistry* **23**, 167.

Bushway, R.J. (1984), *J. Chromatogr.* **303**, 263.

Carlson, R.E. and Dolphin, D.H. (1981), *Phytochemistry* **20**, 2281.

Carman, R.M., Russell-Maynard, J.K.L. and Schumann, R.C. (1985), *Aust. J. Chem.* **38**, 485.

Carson, D., Cass, M.W., Crombie, L., Holden, I. and Whiting, D.A. (1982a), *J. Chem. Soc., Perkin Trans. I*, 773.

Carson, D., Crombie, L., Kilbee, G.W., Moffatt, F. and Whiting, D.A. (1982b), *J. Chem. Soc., Perkin Trans. I*, 779.

Casteelle, K.V., Geiger, H. and Van Sumere, C.F. (1982), *J. Chromatogr.* **240**, 81.

Chandrasekharan, I., Amalraj, V.A., Khan, H.A. and Ghanim, A. (1983), *Trans. Indian Soc. Desert Technol. Univ. Cent. Desert Stud.* **8**, 101. (*Chem. Abstr.* 1983, **98**, 176191).

Chang, X., Xu, Q., Zhu, D., Song, G., and Xu, R. (1981), *Zhongcaoyao* **12**, 530. (*Chem. Abstr.* 1982, **97**, 20701).

Chang, X., Xu, Q., Zhu, D., Song, G. and Xu, R. (1983), *Yaoxue Xuebao* **18**, 45. (*Chem Abstr.* 1983, **99**, 67496).

Chawla, H.M., Mittal, R.S. and Rastogi, D.K. (1984), *Indian J. Chem.* 1984, **23B**, 680.

Chen, C.C., Chen, Y.L., Chen, Y.P. and Hsu, H.Y. (1983), *T'ai-wan Yao Hsueh Tsa Chih* **35**, 89. (*Chem. Abstr.* 1983, **99**, 191649).

Chibber, S.S., Sharma, R.P. and Dutt, S.K. (1981), *Curr. Sci.* **50**, 818.

Chubachi, M., Hamada, M. and Kawano, E. (1983), *Agric. Biol. Chem.* **47**, 619.

Cosio, E.G., Weissenböck, G. and McClure, J.W. (1985), *Plant Physiol.* **78**, 14.

Crombie, L. (1984), *Nat. Prod. Rep.* **1**, 3.

Crombie, L. (1986), Private communication.

Crombie, L., Holden, I., Kilbee, G.W. and Whiting, D.A. (1982), *J. Chem. Soc., Perkin Trans. I*, 789.

Crombie, L., Holden, I., Van Bruggen, N., and Whiting, D.A. (1986), *J. Chem. Soc. Chem. Commun.*, 1063.

Dahiya, J.S., Strange, R.N., Bilyard, K.G., Cooksey, C.J. and Garratt, P.J. (1984), *Phytochemistry* **23**, 871.

Daigle, D.J., Ory, R.L. and Branch, W.D. (1985), *Peanut Sci.* **12**, 60. (*Chem. Abstr.* 1986, **104**, 165440).

Da Rocha, A.I. and Zoghbi, M. das G.B. (1982), *Acta Amazon.* **12**, 615. (*Chem. Abstr.* 1983, **99**, 3038).

Dasan, R.G., Nagarajan, N.S., Narayanan, V., Neelakantan, S. and Raman, P.V. (1982), *Indian J. Chem.* **21B**, 385.

Dasan, R.G., Neelakantan, S. and Raman, P.V. (1985), *Indian J. Chem.* **24B**, 564.

Dean, F.M. and Varma, R.S. (1981), *Tetrahedron Lett.* **22**, 2113.

Dean, F.M. and Varma, R.S. (1982), *J. Chem. Soc., Perkin Trans. I*, 1193.

Delle Monache, F., Ferrari, F. and Menichini, F. (1984), *Phytochemistry* **23**, 2945.

Denny, T.P. and Van Etten, H.D. (1982), *Phytochemistry* **21**, 1023.

Deschamps-Vallet, C., Ilotose, J.P. and Meyer-Dayan, M. (1983), *Tetrahedron Lett.* **24**, 3993.

Dewick, P.M. (1982). In *The Flavonoids–Advances in Research*, (eds J.B. Harborne and T.J. Mabry), Chapman and Hall, London and New York, p. 535.

Dewick, P.M. and Steele, M.J. (1982), *Phytochemistry* **21**, 1599.

Dewick, P.M., Steele, M.J., Dixon, R.A. and Whitehead, I.M. (1982), *Z. Naturforsch.* **37c**, 363.

Dixon, R.A., Dey, P.M. and Lamb, C.J. (1983), *Adv. Enzymol. Rel. Areas Mol. Biol.*, 1.

Dziedzic, S.Z. and Dick, J. (1982), *J. Chromatogr.* **234**, 497.

Eldridge, A.C. (1982a), *J. Chromatogr.* **234**, 494.

Eldridge, A.C. (1982b), *J. Agric. Food Chem.* **30**, 353.

El-Moghazy, A.M., Ali, A.A., El-Emary, N.A., and Darwish, F.M. (1980), *Fitoterapia* **51**, 237.

Farmakalidis, E. and Murphy, P.A. (1984), *J. Chromatogr.* **295**, 510.

Fatma, W., Iqbal, J., Manchanda, V., Shaida, W.A. and Rahman, W. (1984), *J. Chem. Res., Synop.* 298.

Ferrari, F., Botta, B. and de Lima, R.A. (1983), *Phytochemistry* **22**, 1663.

Ferrari, F., Botta, B., de Lima, R.A. and Marini-Bettolo, G.B. (1984a), *Phytochemistry* **23**, 708.

Ferrari, F., de Lima, R.A. and Marini-Bettolo, G.B. (1984b), *Phytochemistry* **23**, 2691.

Fett, W.F. (1984), *Physiol. Plant Pathol.* **24**, 303.

Fett, W.F. and Osman, S.F. (1982), *Phytopathology* **72**, 755.

Fomum, Z.T., Ayafor, J.F. and Wandji, J. (1985), *Phytochemistry* **24**, 3075.

Fraile, A., Garcia-Arenal, F., Garcia-Serrano, J.J. and Sagasta, E.M. (1982), *Phytopathol. Z.* **105**, 161.

Fraishtat, P.D., Popravko, S.A. and Vul'fson, N.S. (1980), *Bioorg. Khim.* **6**, 1722. (*Chem. Abstr.* 1981, **94**, 44052).

Fraishtat, P.D., Popravko, S.A. and Vul'fson, N.S. (1981), *Bioorg. Khim.* **7**, 927. (*Chem. Abstr.* 1981, **95**, 165577).

Friedman, F.K., West, D., Dewick, P.M. and Gelboin, H.V. (1985), *Pharmacology* **31**, 289.

Gambardella, M.T.P., Mascarenhas, Y.P. and Santos, R.H.A. (1983), *Acta Crystallogr.* **C39**, 741.

Gandhidasan, R., Neelakantan, S. and Raman, P.V. (1982), *Synthesis* 1110.

Garcia-Arenal, F., Fraile, A. and Sagasta, E.M. (1980), *Ann. Phytopathol.* **12**, 329.

Gelboin, H.V., West, D., Gozukara, E., Natori, S., Nagao, M. and Sugimara, T. (1981), *Nature (London)* **291**, 659.

Gil, L.A., Ramirez, J. and Diaz, J.C. (1984), *Turrialba* **34**, 437. (*Chem. Abstr.* 1985, **103**, 51316).

Gnanamanickam, S.S. and Mansfield, J.W. (1981), *Phytochemistry* **20**, 997.

Gnanamanickam, S.S. and Smith, D.A. (1980), *Phytopathology* **70**, 894.

Goda, Y., Katayama, M., Ichikawa, K., Shibuya, M., Kiuchi, F. and Sankawa, U. (1985), *Chem. Pharm. Bull.* **33**, 5606.

Gomez, F., Calderon, J.S., Quijano, L., Dominguez, M. and Rios, T. (1985), *Phytochemistry* **24**, 1126.

Goosens, J.F. and Van Laere, A.J. (1983), *J. Chromatogr.* **267**, 439.

Govindachari, T.R. and Premila, M.S. (1985), *Phytochemistry* **24**, 3068.

Grisebach, H. (1983). In *Secondary Metabolism and Differentiation in Fungi* (eds J.W. Bennett and A. Ciegler), Mycology Series, Volume 5, Dekker, New York, p. 377.

Gupta, B.B., Bhattacharyya, A., Mitra, S.R. and Adityachaudhury, N. (1983), *Phytochemistry* **22**, 1306.

Hagmann, M. and Grisebach, H. (1984), *FEBS Lett.* **175**, 199.

Hagmann, M.L., Heller, W. and Grisebach, H. (1984), *Eur. J. Biochem.* **142**, 127.

Hahn, M.G., Bonhoff, A. and Grisebach, H. (1985), *Plant Physiol.* **77**, 591.
Hattori, T. and Ohta, Y. (1985), *Plant Cell Physiol.* **26**, 1101.
Higgins, V.J. (1981), *Canad. J. Bot.* **59**, 547.
Higgins, V.J. and Ingham, J.L. (1981), *Phytopathology* **71**, 800.
Hiraga, Y., Endo, H., Takahashi, K. and Shibata, S. (1984), *J. Chromatogr.* **292**, 451.
Ho, L., Xu, Y., Xue, H., Wang, W., Zhou, Z., Chen, P. and Fan, M. (1982), *Shengzhi Yu Biyun* **2**, 23. (*Chem. Abstr.* 1982, **97**, 212692).
Hohmann, J., Rozsa, Z., Reisch, J. and Szendrei, K. (1982), *Herba Hung.* **21**, 179. (*Chem. Abstr.* 1983, **99**, 85141).
Honda, G. and Tabata, M. (1982), *Plant. Med.* **46**, 122.
Hu, X., Xu, Y., Huang, T. and Bai, Y. (1982), *Zhongyao Tongbao* **7**, 29. (*Chem. Abstr.* 1982, **97**, 107032).
Ibrahim, G., Owen, H. and Ingham, J.L. (1982), *Phytopathol. Z.* **105**, 20.
Imamura, H., Ito, M. and Ohashi, H. (1981), *Gifu Daigaku Nogakubu Kenkyu Hokoku*, 77. (*Chem. Abstr.* 1982, **96**, 196524).
Imamura, H., Shinpuka, H., Inoue, H., and Ohashi, H. (1982), *Moduzai Gakkaishi* **28**, 174. (*Chem. Abstr.* 1982, **96**, 177986).
Ingham, J.L. (1981a), *Biochem. Syst. Ecol.* **9**, 125.
Ingham, J.L. (1981b), *Biochem. Syst. Ecol.* **9**, 275.
Ingham, J.L. (1981c). In *Advances in Legume Systematics* (eds R.M. Polhill and P.H. Raven) R. Bot. Gdn. Kew, p. 599.
Ingham, J.L. (1982c), *Biochem. Syst. Ecol.* **10**, 233.
Ingham, J.L. (1982b), *Plant. Med.* **45**, 46.
Ingham, J.L. (1983), *Prog. Chem. Org. Nat. Prod.* **43**, 1.
Ingham, J.L. and Dewick, P.M. (1980), *Phytochemistry* **19**, 2799.
Ingham, J.L. and Dewick, P.M. (1984), *Z. Naturforsch.* **39c**, 531.
Ingham, J.L. and Markham, K.R. (1982a), *Phytochemistry*, **21**, 2969.
Ingham, J.L. and Markham, K.R. (1982b), *Z. Naturforsch.* **37c**, 724.
Ingham, J.L. and Markham, K.R. (1984), *Z. Naturforsch.* **39c**, 13.
Ingham, J.L. and Mulheirn, L.J. (1982), *Phytochemistry* **21**, 1409.
Ingham, J.L. and Tahara, S. (1983), *Z. Naturforsch.* **38c**, 899.
Ingham, J.L. and Tahara, S. (1985), *Z. Naturforsch.* **40c**, 482.
Ingham, J.L., Keen, N.T., Markham, K.R. and Mulheirn, L.J. (1981a), *Phytochemistry* **20**, 807.
Ingham, J.L., Keen, N.T., Mulheirn, L.J. and Lyne, R.L. (1981b), *Phytochemistry* **20**, 795.
Ingham, J.L., Tahara, S. and Harborne, J.B. (1983), *Z. Naturforsch.* **38c**, 194.
Ishiguro, M., Tatsuoka, T. and Nakatsuka, N. (1982), *Tetrahedron Lett.* **23**, 3859.
Jain, A.C. and Sharma, A. (1984), *Indian J. Chem.* **23B**, 451.
Jain, A.C. and Sharma, A. (1985), *J. Chem. Soc., Chem. Commun.* 338.
Jain, A.C., Gupta, S., Gupta, A. and Bambah, P. (1985), *Indian J. Chem.* **24B**, 609.
Jain, P.K., Pinkey, Makrandi, J.K. and Grover, S.K. (1985), *Indian J. Chem.* **24B**, 51.
Jaques, U., Köster, J. and Barz, W. (1985), *Phytochemistry* **24**, 949.
Jha, H.C., Zilliken, F. and Breitmaier, E. (1981), *Angew. Chem. Int. Ed.* **20**, 102.
Jurd, L. and Wong, R.Y. (1984), *Aust. J. Chem.* **37**, 1127.
Kalidhar, S.B. and Sharma, P. (1984), *J. Indian Chem. Soc.* **61**, 561.
Kamal, R. and Jain, S.C. (1980), *Agric. Biol. Chem.* **44**, 2985.
Kamat, V.S., Chuo, F.Y., Kubo, I. and Nakanishi, K. (1981), *Heterocycles* **15**, 1163.
Kazakov, A.L., Kompantsev, V.A. and Leont'eva, T.P. (1980), *Khim. Prir. Soedin* 721. (*Chem. Abstr.* 1981, **94**, 136105).
Keen, N.T. and Ingham, J.L. (1980), *Z. Naturforsch.* **35c**, 923.
Kessar, S.V., Nadir, U.K., Gupta, Y.P., Pahwa, P.S. and Singh, P. (1981), *Indian J. Chem.* **20B**, 1.
Khalid, S.A. and Waterman, P.G. (1983), *Phytochemistry* **22**, 1001.
Kikuchi, M. (1982), *Annu. Rep. Tohoku Coll. Pharm.*, 61. (*Chem. Abstr.* 1983, **99**, 191682).
Kistler, H.C. and Van Etten, H.D. (1981), *Physiol. Plant Pathol.* **19**, 257.
Kitada, Y., Mizobuchi, M., Ueda, Y., and Nakazawa, H. (1985), *J. Chromatogr.* **347**, 438.
Kobayashi, M. and Ohta, Y. (1983), *Phytochemistry* **22**, 1257.
Kobayashi, M., Noguchi, H. and Sankawa, U. (1985), *Chem. Pharm. Bull.* **33**, 3811.
Kochs, G. and Grisebach, H. (1986), *Eur. J. Biochem.* **155**, 311.
Kolar, L. (1981), *Sb. Vys. Sk. Zemed. Praze, Provozne Ekon. Fak. Ceskych. Budejovicich, Rada Biol.* **19**, 17. (*Chem. Abstr.* 1982, **97**, 69287).
Kolonits, P., Major, A. and Nogradi, M. (1983), *Acta Chim. Hung.* **113**, 367. (*Chem. Abstr.* 1984, **100**, 6180).
Komada, T., Yamakawa, T. and Minoda, Y. (1980), *Agric. Biol. Chem.* **44**, 2387.
Komatsu, M., Yokoe, I. and Shirataki, Y. (1981a), *Chem. Pharm. Bull.* **29**, 532.
Komatsu, M., Yokoe, I. and Shirataki, Y. (1981b), *Chem. Pharm. Bull.* **29**, 2069.
Komives, T. (1983), *J. Chromatogr.* **261**, 423.
Komives, T. and Casida, J.E. (1983), *J. Agric. Food Chem.* **31**, 751.
Kononenko, G.P., Popravko, S.A. and Sokolova, S.A. (1983), *S-Kh. Biol.* 133. (*Chem. Abstr.* 1983, **99**, 67625).
Köster, J. and Barz, W. (1981), *Plant. Med.* **42**, 117.
Köster, J., Strack, D. and Barz, W. (1983a), *Plant. Med.* **48**, 131.
Köster, J., Zuzok, A. and Barz, W. (1983b), *J. Chromatogr.* **270**, 392.
Köster, J., Bussmann, R. and Barz, W. (1984), *Arch. Biochem. Biophys.* **234**, 513.
Kovalev, V.N. (1983), *Nauchn. Tr.-Vses. Nauchno-Issled. Inst. Farm.* **20**, 96. (*Chem. Abstr.* 1985, **103**, 68245).
Kovalev, V.N. and Seraya, L.M. (1984), *Khim. Prir. Soedin*, 659.(*Chem Abstr.* 1985, **102**, 42887).
Kraft, B. and Barz, W. (1985), *Appl. Environ. Microbiol.* **50**, 45.
Krämer, R.P., Hindorf, H., Jha, H.C., Kallage, J. and Zilliken, F. (1984), *Phytochemistry* **23**, 2203.
Krishnamurthy, H.G. and Prasad, J.S. (1980), *Phytochemistry* **19**, 2797.
Kumar, S., Shukla, R.S., Singh, K.P., Paxton, J.D. and Husain, A. (1984), *Phytopathology* **74**, 1349.
Laman, N.A. and Oleksyuk, L.P. (1982), *Vestsi Akad. Navuk BSSR, Ser. Biyal. Navuk* 101. (*Chem. Abstr.* 1982, **97**, 3540).
Lane, G.A., Biggs, D.R., Russell, G.B., Sutherland, O.R.W., Williams, E.M., Mac Donald, J.H. and Donnell, D.J. (1985), *J. Chem. Ecol.* **11**, 1713.
Leube, J. and Grisebach, H. (1983), *Z. Naturforsch.* **38c**, 730.
Le-Van, N. (1984), *Phytochemistry* **23**, 1204.
Liepa, A.J. (1981), *Aust. J. Chem.* **34**, 2647.
Liepa, A.J. (1984), *Aust. J. Chem.* **37**, 2545.
Lu, G., Lu, S., Zhang, G., Xu, S., Li, D. and Huang, Q. (1984), *Zhongcaoyao* **15**, 452. (*Chem. Abstr.* 1985, **102**, 50760).
Luk, K., Stern, L., Weigele, M., O'Brien, R.A. and Spirt, N. (1983), *J. Nat. Prod.* **46**, 852.

Luk'Yanchikov, M.S. and Kazakov, A.L. (1982), *Khim. Prir. Soedin* 251. (*Chem. Abstr.* 1982, **97**, 88741).

Luk'Yanchikov, M.S. and Kazakov, A.L. (1983), *Khim. Prir. Soedin* 105. (*Chem. Abstr.* 1983, **98**, 194983).

Lupi, A., Marta, M., Lintas, G. and Marini-Bettolo, G.B. (1980), *Gazz. Chim. Ital.* **110**, 625.

Lwande, W., Greene, C.S. and Bentley, M.D. (1985a), *J. Nat. Prod.* **48**, 1004.

Lwande, W., Hassanali, A., Njoroge, P.W., Bentley, M.D., Delle Monache, F. and Jondiko, J.I. (1985b), *Insect Sci. Appl.* **6**, 537. (*Chem. Abstr.* 1986, **104**, 31710).

Lyne, R.L., Mulheirn, L.J. and Keen, N.T. (1981), *Tetrahedron Lett.* **22**, 2483.

Mackenbrock, K. and Barz, W. (1983), *Z. Naturforsch.* **38c**, 708.

Mahmoud, Z.F. (1981), *Plant. Med.* **42**, 299.

Maksimov, O.B., Krivoshchekova, O.E., Stepanenko, L.V. and Boguslavskaya, L.V. (1985), *Khim. Prir. Soedin*, 775. (*Chem. Abstr.* 1986, **104**, 165370).

Malik, M.L., and Grover, S.K. (1980), *J. Indian Chem. Soc.* **57**, 208.

Marco, J.L., Sanz, J. and Rodriquez, B. (1983), *An. Quim., Ser. C* **79**, 94. (*Chem. Abstr.* 1983, **99**, 191687).

Markham, K.R. and Ingham, J.L. (1980), *Z. Naturforsch.* **35c**, 919.

Martin, M. and Dewick, P.M. (1980), *Phytochemistry* **19**, 2341.

Masuda, Y., Shiraishi, T., Ouchi, S. and Oku, H. (1983), *Nippon Shokubutsu Byori Gakkaiho* **49**, 558. (*Chem. Abstr.* 1985, **103**, 101172).

McCormick, S., Robson, K. and Bohm, B. (1985), *Phytochemistry* **24**, 1614.

McKittrick, B.A., Scannell, R.T. and Stevenson, R. (1982), *J. Chem. Soc., Perkin Trans. I*, 3017.

McMurchy, R.A. and Higgins, V.J. (1984), *Physiol. Plant Pathol.* **25**, 229.

Menichini, F., Delle Monache, F. and Marini Bettolo, G.B. (1982), *Plant. Med.* **45**, 243.

Mitra, J. and Joshi, T. (1982), *Phytochemistry* **21**, 2429.

Mitra, J. and Joshi, T. (1983), *Phytochemistry* **22**, 2326.

Mitra, J., Das, A. and Joshi, J. (1983), *Phytochemistry* **22**, 1063.

Mitscher, L.A., Ward, J.A., Drake, S. and Rao, G.S. (1984), *Heterocycles* **22**, 1673.

Miyase, T., Ueno, A., Noro, T. and Fukushima, S. (1981), *Chem. Pharm. Bull.* **29**, 2205.

Miyase, T., Ohtsubo, A., Ueno, A., Noro, T., Kuroyanagi, M. and Fukushima, S. (1982) *Chem. Pharm. Bull.* **30**, 1986.

Miyase, T., Fukushima, S. and Akiyama, Y. (1984), *Chem. Pharm. Bull.* **32**, 3267.

Moesta, P., Seydel, U., Lindner, B. and Grisebach, H. (1982), *Z. Naturforsch.* **37c**, 748.

Moesta, P., Hahn, M.G. and Grisebach, H. (1983), *Plant Physiol.* **73**, 233.

Muangnoicharoen, N. and Frahm, A.W. (1981), *Phytochemistry* **20**, 291.

Murphy, P.A. (1981), *J. Chromatogr.* **211**, 166.

Murthy, M.S.R. and Rao, E.V. (1985a), *Fitoterapia* **56**, 362.

Murthy, M.S.R. and Rao, E.V. (1985b), *J. Nat. Prod.* **48**, 967.

Nakagawa, M., Nakanishi, K., Darko, L.L. and Vick, J.A. (1982), *Tetrahedron Lett.* **23**, 3855.

Nakanishi, T., Inada, A., Kambayashi, K. and Yoneda, K. (1985), *Phytochemistry* **24**, 339.

Nakov, N. (1983), *Farmatsiya (Sofia)* **33**, 18. (*Chem. Abstr.* 1983, **99**, 155212).

Nakov, N. and Akhtardzhiev, K. (1983), *Pharmazie* **38**, 202.

Neelakantan, S., Raman, P.V. and Tinabaye, A. (1982), *Indian J. Chem.* **21B**, 256.

Negi, R.K.S., Rajagopalan, T.R. and Batra, V. (1985), *Indian J. Chem.* **24B**, 221.

Nicollier, G.F. and Thompson, A.C. (1982), *J. Chromatogr.* **249**, 399.

Obara, Y. and Matsubara, H. (1981), *Meijo Daigaku Nogakubu Gakujutsu Hokoku* **17**, 40. (*Chem. Abstr.* 1981, **95**, 200536).

Ohta, N., and Mikumo, K. (1982), *Kumamoto Joshi Daigaku Gakujutsu Kiyo* **34**, 73. (*Chem. Abstr.* 1982, **97**, 143336).

Ohta, N., Kuwata, G., Akahori, H. and Watanabe, T. (1980), *Nippon Shokuhin Gakkaishi* **27**, 348. (*Chem. Abstr.* 1980, **93**, 219365).

Ohta, N., Mikumo, K., Ikeda, R. and Watanabe, T. (1981), *Kumamoto Joshi Daigaku Gakujutsu Kiyo* **33**, 56. (*Chem. Abstr.* 1981, **95**, 110824).

Olivares, E.M., Lwande, W., Delle Monache, F. and Marini Bettolo, G.B. (1982), *Phytochemistry* **21**, 1763.

O'Neill, M.J. (1983), *Z. Naturforsch.* **38c**, 698.

O'Neill, M.J., Adesanya, S.A. and Roberts, M.F. (1983), *Z. Naturforsch.* **38c**, 693.

O'Neill, M.J., Adesanya, S.A., and Roberts, M.F. (1984), *Phytochemistry* **23**, 2704.

Osman, S.F. and Fett, W.F. (1983), *Phytochemistry* **22**, 1921.

Osswald, W.F. (1985), *J. Chromatogr.* **333**, 225.

Ozimina, I.I. (1981), *Khim. Prir. Soedin*, 242. (*Chem. Abstr.* 1981, **95**, 76862).

Ozimina, I.I. and Bandyukov, A.B. (1983), *Khim. Prir. Soedin*, 712. (*Chem. Abstr.* 1984, **100**, 205909).

Ozimina, I.I. and Bandyukova, V.A. (1985), *Khim. Prir. Soedin*, 507. (*Chem. Abstr.* 1986, **104**, 126523).

Patroni, J.J., Collins, W.J. and Stern, W.R. (1982), *J. Chromatogr.* **247**, 366.

Pereira, M.O. da S., Fantine, E.C. and de Sousa, J.R. (1982), *Phytochemistry* **21**, 488.

Petov, E., Uzunov, P., Kostova, I., Somleva, T., and Ognyanov, I. (1983), *Plant. Med.* **47**, 237.

Pietta, P. and Zio, C. (1983), *J. Chromatogr.* **260**, 497.

Pietta, P., Calatroni, A. and Zio, C. (1983), *J. Chromatogr.* **280**, 172.

Pinkey, Jain, P.K. and Grover, S.K. (1984), *Gazz. Chim. Ital.* **114**, 355.

Pivovarenko, V.G. and Khilya, V.P. (1985), *Dopov. Akad. Nauk Ukr. RSR, Ser. B: Geol., Khim. Biol. Nauki*, **44**. (*Chem. Abstr.* 1986, **104**, 129748).

Pivovarenko, V.G., Khilya, V.P. and Babichev, F.S. (1985) *Dopov. Akad. Nauk Ukr. RSR, Ser. B: Geol. Khim. Biol. Nauki*, 56. (*Chem. Abstr.* 1985, **103**, 141792).

Popravko, S.A., Sokolova, S.A. and Kononeko, G.P. (1980), *Bioorg. Khim.* **6**, 1255. (*Chem. Abstr.* 1981, **94**, 80197).

Porter, P.M., Banwart, W.L. and Hassett, J.J. (1985), *Environ. Exp. Bot.* **25**, 229.

Prasad, A.V.K., Singh, A., Kapil, R.S. and Popli, S.P. (1984), *Indian J. Chem.* **23B**, 1165.

Prasad, A.V.K., Kapil, R.S. and Popli, S.P. (1985), *Indian J. Chem.* **24B**, 236.

Raju, K.V.S., Srimannarayana, G., Ternai, B., Stanley, R. and Markham, K.R. (1981), *Tetrahedron* **37**, 957.

Rao, A.V.S., and Mathew, J. (1982), *Phytochemistry*, **21**, 1837.

Rao, C.P., Vemuri, V.S.S. and Rao, K.V.J. (1982), *Indian J. Chem.* **21B**, 167.

Rao, E.V. and Murthy, M.S.R. (1985), *Phytochemistry* **24**, 875.

Rao, E.V. and Raju, N.R. (1984), *Phytochemistry* **23**, 2339.

Rao, E.V., Murthy, M.S.R. and Ward, R.S. (1984), *Phytochemistry* **23**, 1493.
Rao, G.V. and Rao, P.S. (1985), *Fitoterapia* **56**, 287.
Rao, K.N. and Srimannarayana, G. (1984), *Phytochemistry* **23**, 927.
Rao, P.P. and Srimannarayana, G. (1981), *Synthesis* 887.
Rathmell, W.G. and Smith, D.A. (1980), *Pestic. Sci.* **11**, 568.
Ravanel, P., Tissut, M. and Douce, R. (1984), *Plant Physiol.* **75**, 414.
Ravise, A., Kirkiacharian, B.S., Chopin, J. and Kunesch, G. (1980), *Ann. Phytopathol.* **12**, 335.
Radaelli, C. and Santaniello, E. (1984), *Phytochemistry* **23**, 2976.
Richardson, P.M. (1981), *Biochem. Syst. Ecol.* **9**, 105.
Rizk, A.F. and Wood, G.E. (1980), *C.R.C. Crit. Rev. Food Sci. Nutr.* **13**, 245.
Robeson, D.J. and Harborne, J.B. (1980), *Phytochemistry* **19**, 2359.
Robeson, D.J. and Harborne, J.B. (1983), *Z. Naturforsch.* **38c**, 334.
Russell, G.B., Shaw, G.T., Christmas, P.E., Yates, M.B. and Sutherland, O.R.W. (1984), *Phytochemistry* **23**, 1417.
Sachse, J. (1984), *J. Chromatogr.* **298**, 175.
Sanduja, R., Martin, G.E., Weinheimer, A.J., Alam, M., Hossain, M.B. and Van der Helm, D. (1984), *J. Heterocycl. Chem.* **21**, 845.
San Feliciano, A., Barrero, A.F., Medarde, M., del Corral, J.M.M. and Calle, M.V. (1983), *Phytochemistry* **22**, 2031.
Sariaslani, F.S. and Rosazza, J.P. (1983), *Appl. Environ. Microbiol.* **45**, 616.
Sariaslani, F.S. and Rosazza, J.P. (1985), *Appl. Environ. Microbiol.* **49**, 451.
Sariaslani, F.S., Beale, J.M. and Rosazza, J.P. (1984), *J. Nat. Prod.* **47**, 692.
Scannell, R.T. and Stevenson, R. (1982), *J. Heterocycl. Chem.* **19**, 299.
Schlieper, D., Komossa, D. and Barz, W. (1984), *Z. Naturforsch* **39c**, 882.
Sethi, M.L., Taneja, S.C., Dhar, K.L., and Atal, C.K. (1981), *Phytochemistry* **20**, 341.
Sethi, M.L., Taneja, S.C., Dhar, K.L. and Atal, C.K. (1983), *Phytochemistry* **22**, 289.
Shaw, G.J., Yates, M.K. and Biggs, D.R. (1982), *Phytochemistry* **21**, 249.
Shawl, A.S., Zaman, V.A. and Kalla, A.K. (1984), *Phytochemistry* **23**, 2405.
Shawl, A.S., Dar, B.A. and Vishwapaul (1985), *J. Nat. Prod.* **48**, 849.
Shehata, M.N., Hassan, A. and El-Shazly, K. (1982), *Aust. J. Agric. Res.* **33**, 951.
Shirataki, Y., Komastu, M., Yokoe, I. and Manaka, A. (1981), *Chem. Pharm. Bull.* **29**, 3033.
Shirataki, Y., Manaka, A., Yokoe, I. and Komatsu, M. (1982), *Phytochemistry* **21**, 2959.
Shirataki, Y., Endo, M., Yokoe, I. and Komatsu, M. (1983), *Chem. Pharm. Bull.* **31**, 2859.
Shoukry, M.M., Darwish, N.A. and Morsi, M.A. (1982), *Gazz. Chim. Ital.* **112**, 289.
Shukla, R.V.N. and Misra, K. (1981), *Phytochemistry* **20**, 339.
Singh, H., Jain, P.K., Makrandi, J.K. and Grover, S.K. (1982), *Indian J. Chem.* **21B**, 547.
Singh, R.P. and Singh, D. (1985), *Heterocycles* **23**, 903.
Singhal, A.K., Sharma, R.P., Madhusudanan, K.P., Thyagarajan, G., Herz, W. and Govindan, S.V. (1981), *Phytochemistry* **20**, 803.
Singhal, A.K., Sharma, R.P., Baruah, J.N., Govindan, S.V. and Herz, W. (1982a), *Phytochemistry* **21**, 949.
Singhal, A.K., Sharma, R.P., Baruah, J.N. and Herz, W. (1982b), *Chem. Ind. (London)* 549.
Singhal, A.K., Barua, N.C., Sharma, R.P. and Baruah, J.N. (1983), *Phytochemistry* **22**, 1005.
Smolenski, S.J., Kinghorn, A.D. and Belandrin, M.F. (1981), *Econ. Bot.* **35**, 321.
Somleva, T. and Ognyanov, I. (1985), *Plant. Med.* 219.
Spilkova, J. and Hubik, J. (1982), *Cesk. Farm.* **31**, 24. (*Chem. Abstr.* 1982, **96**, 177960).
Srihari, K. and Sundaramurthy, V. (1980), *Proc. Indian Acad. Sci. [Ser.]. Chem. Sci.* **89**, 405. (*Chem. Abstr.* 1981, **94**, 139555).
Steffens, J.C., Lynn, D.G., Kamat, V.S. and Riopel, J.L. (1982), *Ann. Bot.* **50**, 1.
Stössel, P. (1982), *Phytopathol. Z.* **105**, 109.
Stössel, P. (1985), *Physiol. Plant Pathol.* **26**, 269.
Stössel, P. and Magnolata, D. (1983), *Experientia* **39**, 153.
Strange, R.N., Ingham, J.L., Cole, D.L., Cavill, M.E., Edwards, C., Cooksey, C.J. and Garratt, P.J. (1985), *Z. Naturforsch.* **40c**, 313.
Suarez, L.E.C., Delle Monache, F., Marini Bettolo, G.B. and Menichini, F. (1980), *Farm. Ed. Sci.* **35**, 796. (*Chem. Abstr.* 1980, **93**, 182824).
Sukumaran, K. and Gnanamanickam, S.S. (1980), *Indian J. Microbiol.* **20**, 204.
Suntry Ltd. (Japanese Patent) (1984), *Jpn Kokai Tokkyo Koho JP* 59 13,784 [84 13,784]. (*Chem. Abstr.* 1984, **101**, 72517).
Suresh, R.V., Iyer, C.S.R. and Iyer, P.R. (1985), *Tetrahedron* **41**, 2479.
Tabakovic, I., Grujic, Z. and Bejtovic, Z. (1983), *J. Heterocycl. Chem.* **20**, 635.
Tahara, S. (1984), *Nippon Nogei Kagaku Kaishi* 1984, **58**, 1247. (*Chem. Abstr.* 1985, **102**, 146078).
Tahara, S., Ingham, J.L., Nakahara, S., Mizutani, J. and Harborne, J.B. (1984a), *Phytochemistry* **23**, 1889.
Tahara, S., Nakahara, S., Mizutani, J. and Ingham, J.L. (1984b), *Agric. Biol. Chem.* **48**, 1471.
Tahara, S., Ingham, J.L. and Mizutani, J. (1985a), *Agric. Biol. Chem.* **49**, 1775.
Tahara, S., Nakahara, S., Ingham, J.L. and Mizutani, J. (1985b), *Nippon Nogei Kagaku Kaishi* **59**, 1039. (*Chem. Abstr.* 1986, **104**, 182945).
Tahara, S., Nakahara, S., Mizutani, J. and Ingham, J.L. (1985c), *Agric. Biol. Chem.* **49**, 2605.
Takeya, K. and Itokawa, H. (1982), *Chem. Pharm. Bull.* **30**, 1496.
Taylor, E.C., Conley, R.A., Johnson, D.K., McKillop, A. and Ford, M.E. (1980), *J. Org. Chem.* **45**, 3433.
Thomas, P. and Whiting, D.A. (1984), *Tetrahedron Lett.* **25**, 1099.
Tsukayama, M., Horie, T., Yamashita, Y., Masumura, M. and Nakayama, M. (1980), *Heterocycles* **14**, 1283.
Tsukayama, M., Fujimoto, K., Horie, T., Yamashita, Y., Masamura, M. and Nakayama, M. (1982), *Chem. Lett.* 675.
Tsukayama, M., Iguchi, Y., Horie, T., Masumura, M. and Nakayama, M. (1984), *Heterocycles* **22**, 709.
Tsukayama, M., Fujimoto, K., Horie, T., Masumara, M. and Nakayama, M. (1985), *Bull. Chem. Soc. Jpn* **58**, 136.
Van der Westhuizen, J.H., Ferreira, D. and Roux, D.G. (1980), *J. Chem. Soc., Perkin Trans. I*, 2856.

Van Etten, H.D. and Barz, W. (1981), *Arch. Microbiol.* **129**, 56.
Van Etten, H.D., Matthews, P.S. and Mercer, E.H. (1983), *Phytochemistry* **22**, 2291.
Van Heerden, F.R., Brandt, E.V. and Roux, D.G. (1980), *J. Chem. Soc., Perkin Trans. I*, 2463.
Varma, R.S. (1982), *Chem. Ind. (London)* 56.
Varma, R.S. and Varma, M. (1982a), *Tetrahedron Lett.* **23**, 3007.
Varma, R.S. and Varma, M. (1982b), *Monats. Chem.* **113**, 1469.
Vaughan, D.A. and Hymowitz, T. (1984), *Biochem. Syst. Ecol.* **12**, 189.
Vermes, B., Antus, S., Gottsegen, A. and Nogradi, M. (1983), *Liebigs Ann. Chem.* 2034.
Vilain, C. (1983), *Pharmazie* **38**, 876.
Viscardi, P., Reynaud, J. and Raynaud, J. (1984), *Pharmazie* **39**, 781.
Vivar, R. de, Bratoeff, E.A. and Ontiveros, E. (1982), *Rev. Latinoam. Quim.* **13**, 18. (*Chem. Abstr.* 1982, **97**, 178742).
Vu, B., Mezey-Vandor, G. and Nogradi, M. (1984), *Liebigs Ann. Chem.*, 734.
Waddell, T.G., Thomasson, M.H., Moore, M.W., White, H.W., Swanson-Bean, D., Green, M.E., Van Horn, G.S. and Fales, H.M. (1982), *Phytochemistry* **21**, 1631.
Wang, Z., Ma, Q., Ho, Q., and Go, J. (1983), *Zhongcaoyao* **14**, 97. (*Chem. Abstr.* 1983, **99**, 3072).
Weinstein, L.I. and Albersheim, P. (1983), *Plant Physiol.* **72**, 557.
Weinstein, L.I., Hahn, M.G. and Albersheim, P. (1981), *Plant Physiol.* **68**, 358.
Weltring, K.M., Barz, W. and Dewick, P.M. (1981), *Arch. Microbiol.* **130**, 381.
Weltring, K.M., Mackenbrock, K. and Barz, W. (1982), *Z Naturforsch.* **37c**, 570.
Weltring, K.M., Barz, W. and Dewick, P.M. (1983), *Phytochemistry* **22**, 2883.
Westwood, S.A., Games, D.E. and Sheen, L. (1981), *J. Chromatogr.* **204**, 103.
Whitehead, I.M., Dey, P.M. and Dixon, R.A. (1982), *Planta* **154**, 156.
Wietor-Orlandi, E.A. and Smith, D.A. (1985), *Physiol. Plant Pathol.* **27**, 197.
Willeke, U. and Barz, W. (1982a), *Arch. Microbiol.* **132**, 266.
Willeke, U. and Barz, W. (1982b), *Z. Naturforsch.* **37c**, 861.
Willeke, U., Weltring, K.M., Barz, W. and Van Etten, H.D. (1983), *Phytochemistry* **22**, 1539.
Wolfbeis. O.S., Fürlinger, E., Jha, H.C. and Zilliken, F. (1984), *Z. Naturforsch.* **39b**, 238.
Woodward, M.D. (1981a), *Physiol. Plant Pathol.* **18**, 33.
Woodward, M.D. (1981b), *Phytochemistry* **20**, 532.
Woodward, M.D. (1982), *Phytochemistry* **21**, 1403.
Wu, L.J., Miyase, T., Ueno, A., Kuroyanagi, M., Noro, T. and Fukushima, S. (1985), *Chem. Pharm. Bull.* **33**, 3231.
Xiu, S., Ma, H., Wang, X., Shi, J. and Zhuang, Y. (1982), *Zhongcaoyao* **13**, 1. (*Chem. Abstr.* 1982, **97**, 107026).
Yahara, S., Saijo, R., Nohara, T., Konishi, R., Yamahara, J., Kawasaki, T. and Mirahara, K. (1985), *Chem. Pharm. Bull.* **33**, 5130.
Yusupova, S.S., Batirov, E.K., Abdullaev, S.V. and Malikov, V.M. (1984), *Khim. Prir. Soedin*, 250. (*Chem. Abstr.*, 1984, **101**, 51742).
Zähringer, U., Schaller, E. and Grisebach, H. (1981), *Z. Naturforsch.* **36c**, 234.
Zhang, Y. and Smith, D.A. (1983), *Physiol. Plant Pathol.* **23**, 89.
Zhu, D., Song, G., Jian, F., Chang, X. and Guo, W. (1984), *Huaxue Xuebao* **42**, 1080. (*Chem. Abstr.* 1985, **102**, 75705).

6

Neoflavonoids

DERVILLA M.X. DONNELLY
and M. HELEN SHERIDAN

6.1 INTRODUCTION

The term neoflavonoid refers to a group of C-15 naturally occurring compounds which are related structurally and biogenetically to the flavonoids and to the isoflavonoids. In the higher plants the neoflavonoids have limited taxonomic distribution, being found in the Guttiferae, the Papilionoideae (subfamily Leguminosae) and more recently being identified in the Rubiaceae, the Passifloraceae and the Polypodiaceae. The naturally occurring neoflavonoids have been grouped together in accordance with their structural types and source (Donnelly, 1985). In this chapter we discuss the naturally occurring neoflavonoids isolated since the last review (Donnelly, 1975).

6.2 SPECTROSCOPIC IDENTIFICATION OF NEOFLAVONOIDS

In recent years the use of high-resolution ^{1}H NMR and ^{13}C NMR has become of increasing importance in the structural elucidation of naturally occurring neoflavonoids.

A summary of the spectral data for 5-hydroxy-7-methoxy-4-(3,4-dihydroxyphenyl)-2H-1-benzopyran-2-one, coumarin, 7-methoxycoumarin (Lapper, 1976), 7-methoxy-4-phenylcoumarin and a series of synthetic 4-arylcoumarins (Busteed, 1984) is presented in Table 6.1. A characteristic pattern for the carbons C-3, C-8 and C-10 in the 4-arylcoumarins (*6.1*)–(*6.7*) is obvious, with a doublet *circa* δ 97, a singlet *circa* δ 105 and a doublet at δ 114 due to carbons at the 8-, 10- and 3-positions respectively. From these data it is also clear that 6,7-dioxygenated and 5,6,7-trioxygenated compounds can be distinguished from each other by reference to the signal for C-5. The ^{13}C NMR spectra of the 6,7-substituted coumarins display a signal at δ 117 due to C-5 while in the case of 5-substituted coumarins (*6.4*)–(*6.7*) the signal is in the region δ 150.

^{13}C NMR spectroscopy has also been utilized by Kunesch *et al.* (1983) in the structure determination of calofloride (*6.8*). ^{13}C NMR of this compound was complex due to duplication of the ^{13}C NMR signals as a result of tautomerism. The structural assignment was based on analysis of the ^{13}C NMR spectrum run in pyridine D_6 in which one tautomeric form predominated. The signal values (Fig. 6.1) compared favourably with two synthetic 3,4-dihydro-4-phenylcoumarins (*6.9*) and (*6.10*).

6.3 4-ARYLCOUMARINS

The 4-arylcoumarins (4-aryl-2H-1-benzopyran-2-ones) are of widespread distribution. They are found in the Leguminosae, the Guttiferae, the Rubiaceae and more recently the Passifloraceae. A summary of the structures and sources of new members of this group reported in the literature since the last review is in Table 6.2.

The isolation of a new 5,2′-oxido-4-phenylcoumarin (*6.13*) from *Coutarea hexandra* (Reher and Kraus, 1984) is of interest, as such compounds have not previously been isolated from plant sources.

The Flavonoids. Edited by J.B. Harborne
Published in 1988 by Chapman and Hall Ltd,
11 New Fetter Lane, London EC4P 4EE.

Table 6.1 ^{13}C NMR data for 4-arylcoumarins

Compound	*Coumarin skeleton*															
	C_2	C_3	C_4	C_5	C_6	C_7	C_8	C_9	C_{10}	OCH_3	CH_3	$C{=}O$	C'_1	C'_4	C''_{26}	C''_{35}
	159.9	116.3	142.8	127.4	123.9	131.3	116.4	153.4	118.4	—	—	—	—	—	—	—
CH_3O	160.5	112.1	142.9	128.3	128.3	162.2	100.5	155.4	112.2	55.6	—	—	—	—	—	—
CH_3O	160.8	111.7	155.8	127.8	127.8	162.6	101.1	155.5	112.3	55.7	—	—	134.4	129.3	128.2	—
CH_3O, HO, OH, OH (6.1)	162.6	110.6	157.0	156.8	156.8	159.8	98.2	156.1	102.1	55.5	—	—	130.4	115.5	141.1 118.8	—
CH_3O, HO (6.2)	161.4	110.5	142.4	112.6	112.6	150.1	99.6	155.7	112.4	56.5	—	—	135.6	129.6	128.3*	—

CH_3COO, CH_3COO, O, O (6.3)	160.2	115.4	138.8	121.7	121.7	152.0	112.2	154.9	117.5	—	20.7 20.8	168.0 168.1	134.8	127.5	– 132	—
CH_3O, O, O, O, O (6.4)	163	106.8	150	169	167	150.6	101	152.8	103.9	51	—	—	130.2	123.8	121*	122*
CH_3O, HO, HO, O, O (6.5)	164	116	143.6	153	148	160	95.8	156	105	60	—	—	132.1	134	131.5*	132
CH_3O, CH_3O, CH_3O, O, O (6.6)	160.5	114	139.3	151.6	150.9	156.8	96.2	155.3	101.2	56.2 60.8 60.9	—	—	138.9	127.9	127.1*	127.4*
CH_3O, CH_3O, CH_3COO, O, O (6.7)	159.6	115.4	137.4	153.2	140.2	153.5	98.8	154.9	106.7	56.6	19.0 20.0	167.2 167.6	129.9	128.6	127.7*	128.2*

*Overlapping signals.

Table 6.2 Structures and sources of new neoflavonoids

Compound	*Family, genesis, species*	*References*
4-Arylcoumarins		
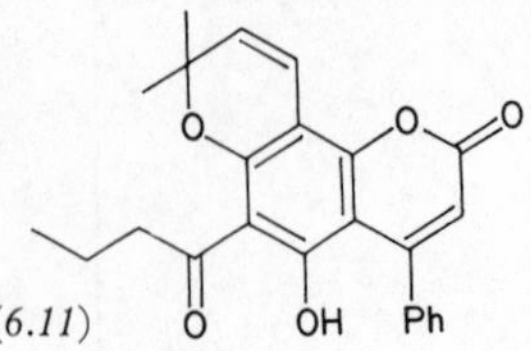 (*6.11*)	Guttiferae *Ochrocarpus siamensis*	Thebtaranonth *et al.* (1981)
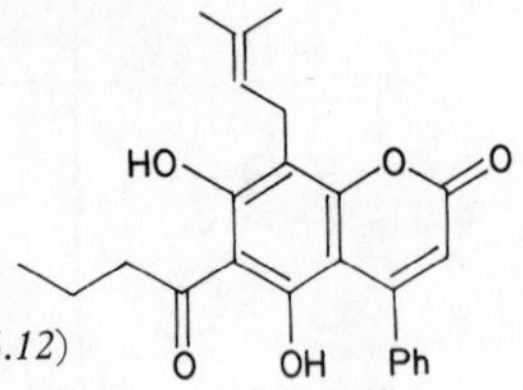 (*6.12*)	*Ochrocarpus siamensis*	
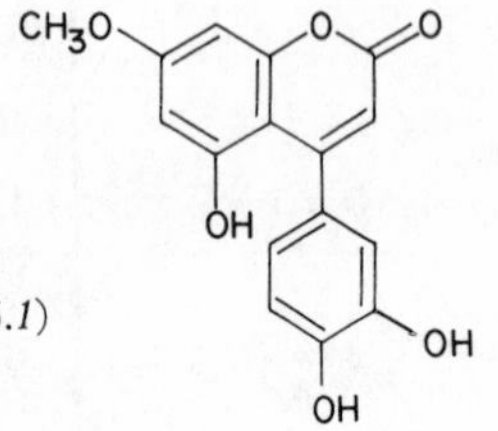 (*6.1*)	Rubiaceae *Coutarea hexandra* Jacq. *Coutarea latiflora*	Reher *et al.* (1983) Iinuma *et al.* (1987)
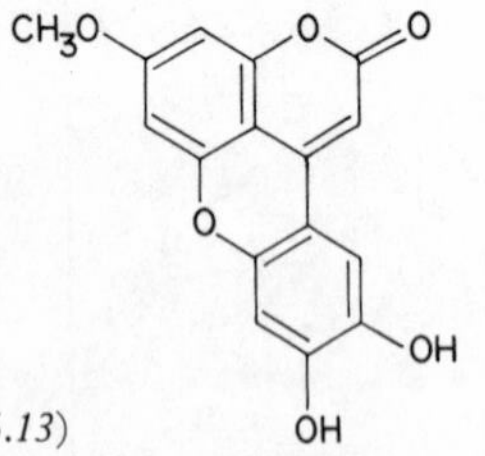 (*6.13*)	*Coutarea latiflora*	Reher and Kravs (1984)
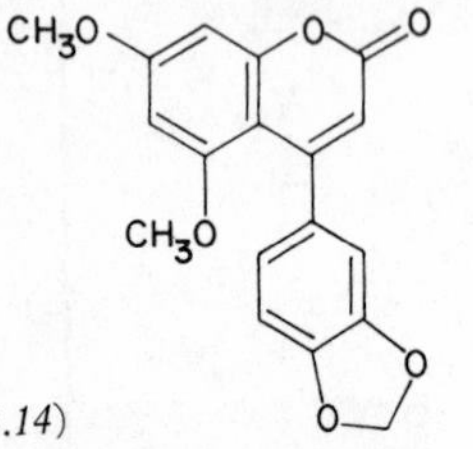 (*6.14*)	*Coutarea hexandra* Jacq.	Monache *et al.* (1984)
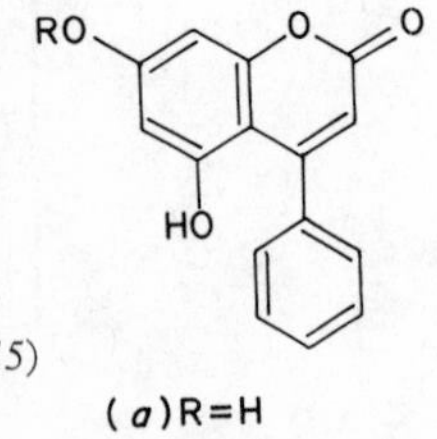 (*6.15*) (*a*) R=H (*b*) R=β-D-Glc	Passifloraceae *Passiflora serratodigitata*	Ulubelen *et al.* (1982)

Table 6.2 (*Contd.*)

Compound	Family, genesis, species	References
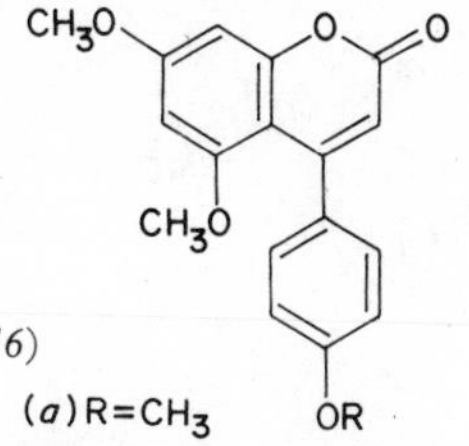 (6.16) (a) R=CH_3 (b) R=H	*C. hexandra* Jacq.	Monache *et al.* (1983)
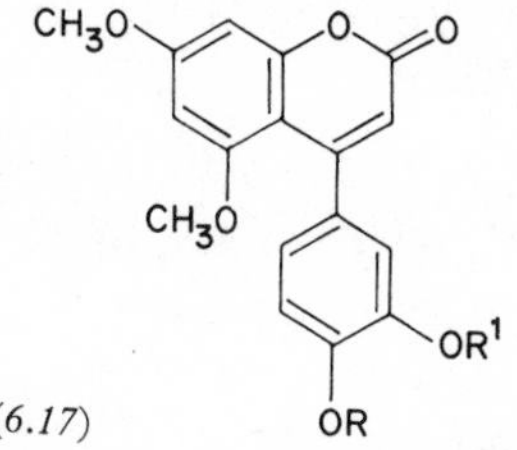 (6.17) (a) R = CH_3, R^1 = H (b) R = H, R^1 = H	*C. hexandra* Jacq.	Monache *et al.* (1983)

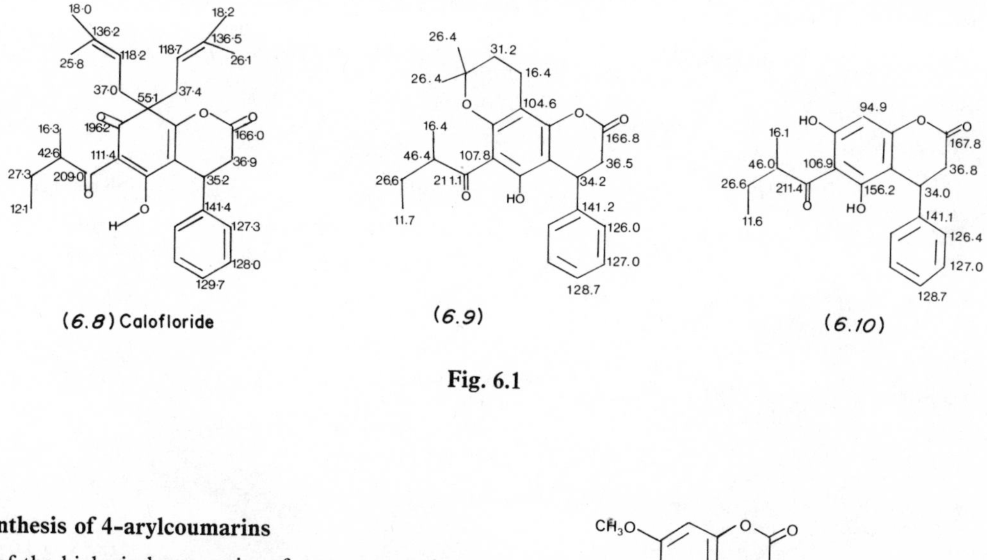

(6.8) Calofloride

(6.9)

(6.10)

Fig. 6.1

6.3.1 Synthesis of 4-arylcoumarins

In view of the biological properties of many coumarins, considerable attention has been focused on their synthesis. Ahluwalia *et al.* (1981) prepared a series of 8-alkyl-5, 7-dimethoxy-4-phenylcoumarins using a Pechmann condensation of the appropriate phenol with ethyl benzoylacetate. Using a similar procedure melannein (*6.18*) was synthesized (Ahluwalia *et al.*, 1982).

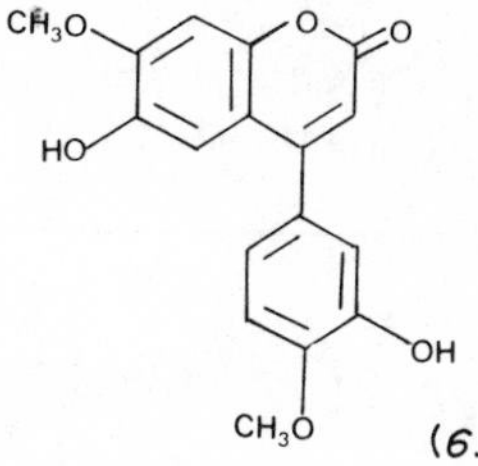

(6.18) Melannein

Table 6.3 Summary of synthetic routes to 4-arylcoumarins

Substrates	*Reagents*	*Product*	*(yield)*	*Reference*
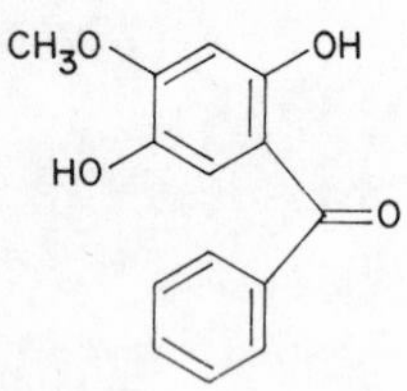	$NaOAc/Ac_2O$	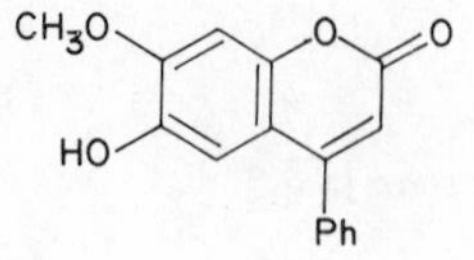	Dalbergin (20–60%)	Donnelly *et al.* (1975)
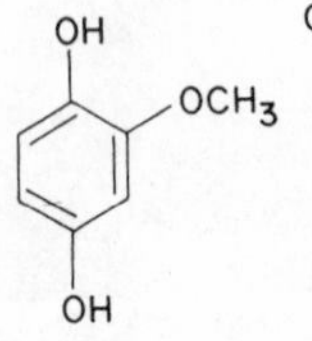 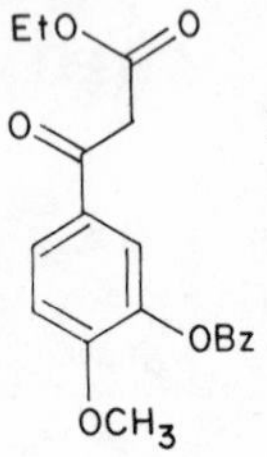	EtOH/HCl or H_2SO_4	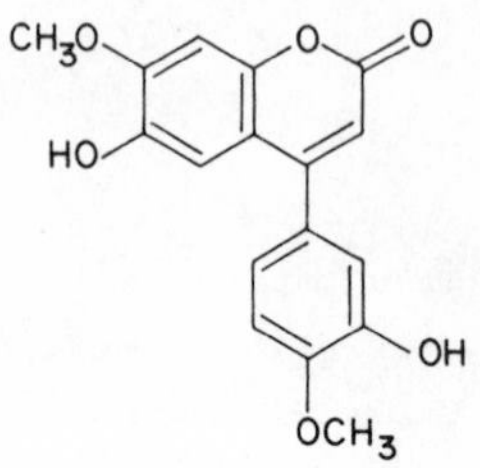	Melannein (20%)	Ahluwalia *et al.* (1982)
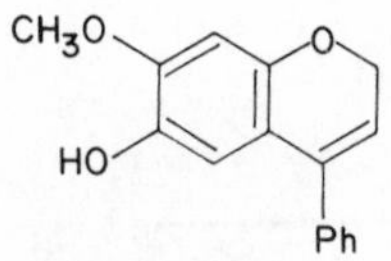	CrO_3/Py or DDQ/C_6H_6	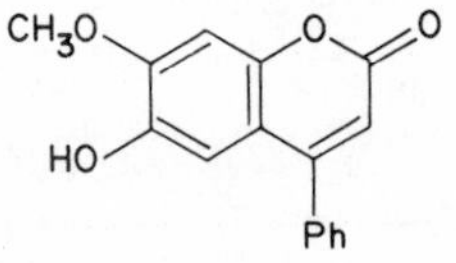	Dalbergin (70%)	Donnelly *et al.* (1975)
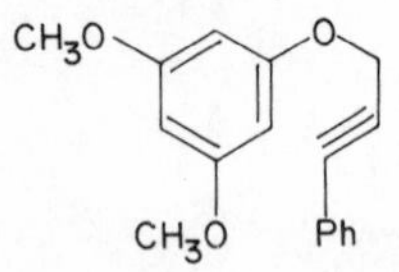	(i) $AgBF_3$ (ii) DDQ/C_6H_6	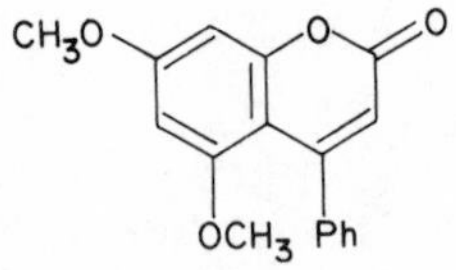	5,7-Dimethoxy-4-phenylcoumarin (90%)	Chaterjee *et al.* (1976)
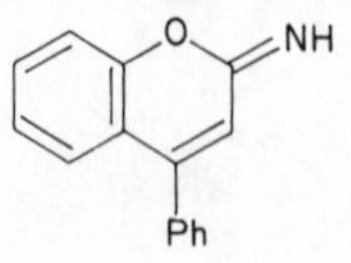	H_3O^+	4-Phenylcoumarin		Shighiro *et al.* (1970)
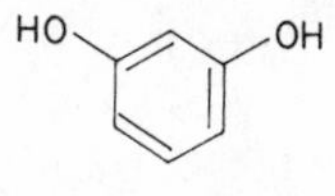 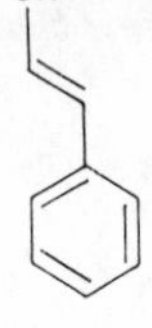	$ZnCl_2$	7-Hydroxy-4-phenylcoumarin		Das Gupta and Paul (1970)
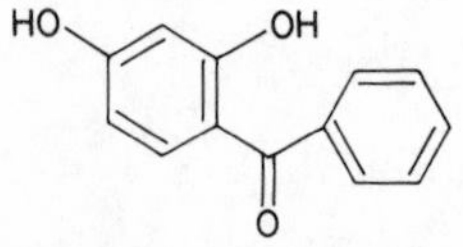	NaH/DMF $(CH_3)_3SiC{=}C{=}O$	7-Hydroxy-4-phenylcoumarin (90%)		Taylor and Cassell (1982)

An interesting route to 4-arylcoumarins via cyclization–elimination using trimethylsilylketene was reported by Taylor and Cassell (1982). Addition of alcohols to trimethylsilylketene affords alkyl α-silylcarboxylates. Since enolate anions derived from such esters can be added to carbonyl compounds with a resultant elimination to give α, β-unsaturated esters, combination of these two reactions in an intramolecular sense would afford a ready synthesis of cyclic unsaturated lactones. This strategy was applied to a one-pot conversion of 2,4-dihydroxybenzophenone to 7-hydroxy-4-phenylcoumarin in 90% yield (Scheme 6.1).

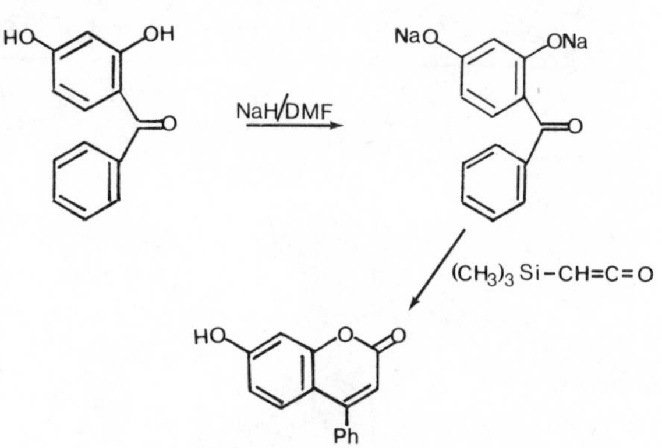

The assigned structures of a series of 4-arylcoumarins (*6.14*)–(*6.16*) isolated from *Coutarea hexandra* (Monache *et al.*, 1983, 1984) were confirmed by synthesis. Perkin cyclization of 2-hydroxy-4, 6, 4′-trimethoxybenzophenone yielded (*6.16a*), as did methylation of 5, 7-dihydroxy-4′-methoxy-4-phenylcoumarin (obtained by Pechmann condensation of ethyl *p*-methoxybenzoylacetate with phloroglucinol). The synthetic products obtained by the two independent routes (the latter in greater yield) and the natural product were coincident in all respects.

Table 6.3 presents a summary of synthetic routes to the 4-arylcoumarins.

Scheme 6.1 DMF, dimethylformamide.

6.4 3,4-DIHYDRO-4-ARYLCOUMARINS

This group of reduced 4-arylcoumarins is represented by several interesting structures, none of which has yet been found in the Leguminosae. Calofloride (*6.8*) isolated from *Calophyllum verticillatum* (Kunesch *et al.*, 1983) is a novel compound which may constitute a missing link in the biogenesis of neoflavonoids (Fig. 6.2).

Another novel group of 3,4-dihydro-4-arylcoumarins are oxygen heterocycles composed of chalcone and neoflavone moieties which have been isolated from *Pityrogramma calomelanos* (Wagner *et al.*, 1979) and *P. trifoliata* (Wollenweber *et al.*, 1980). These and other new members of the 3,4-dihydro-4-arylcoumarins are listed in Table 6.4.

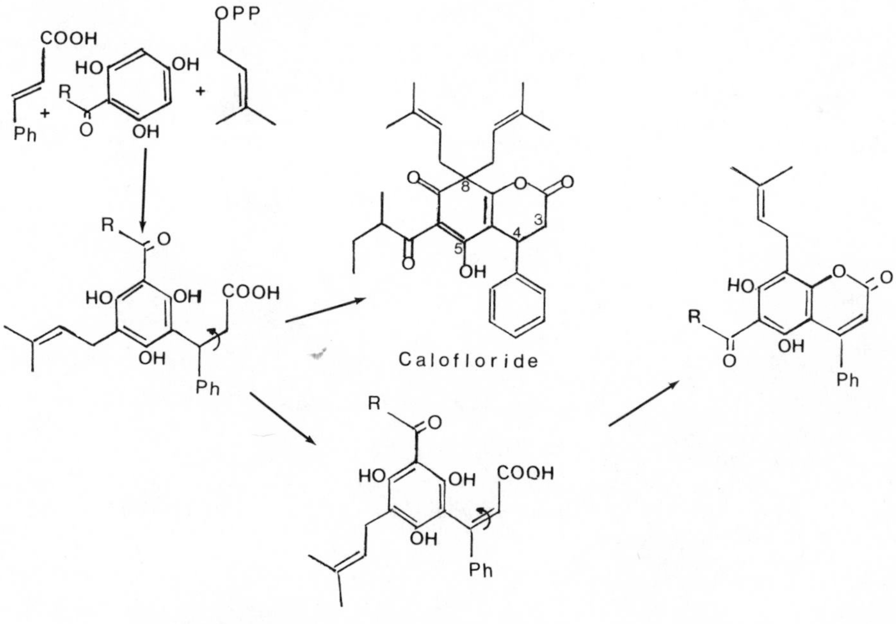

Fig. 6.2 Calofloride and the biogenesis of neoflavonoids.

Table 6.4 Structure and sources of new 3, 4-dihydro-4-phenylcoumarins

Compound	*Family, genesis, species*	*References*
(6.8) 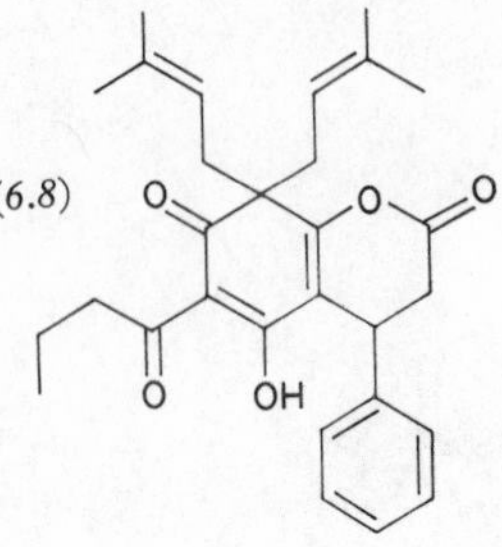	Guttiferae *Calophyllum verticillatum*	Kunesch *et al.* (1983)
(6.19) (a) R=H (b) R=CH_3 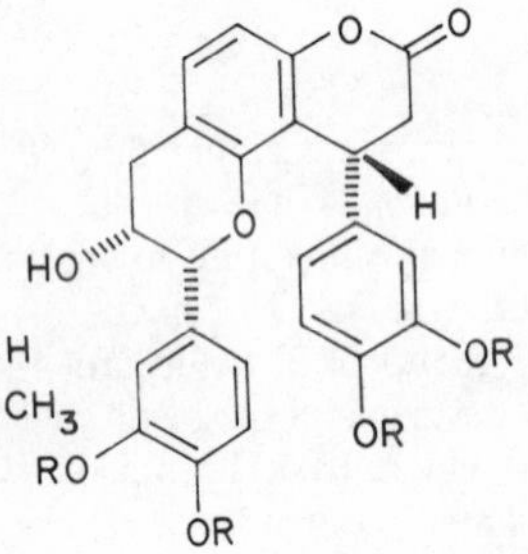	Rubiaceae *Cinchona succirubra*	Nonaka and Nishioka (1982)
(6.20) (a) R=H (b) R=CH_3 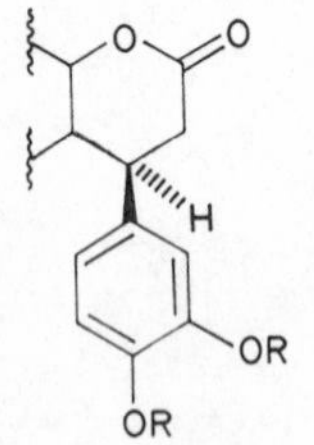	*Cinchona succirubra*	Nonaka and Nishioka (1982)
(6.21) 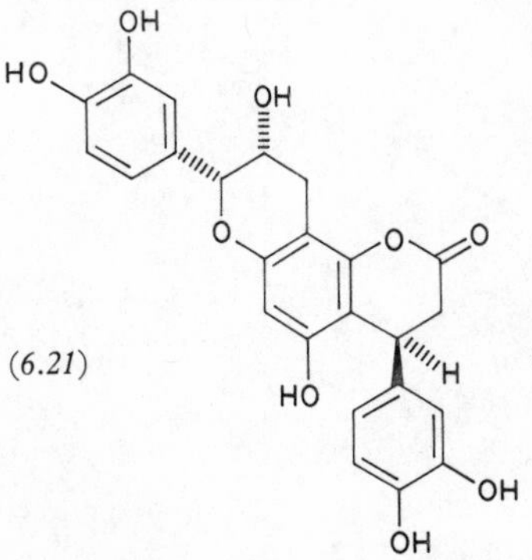	*Cinchona succirubra*	Wollenweber *et al.* (1980)

Table 6.4 (*Contd.*)

Compound	Family, genesis, species	References
(6.22) 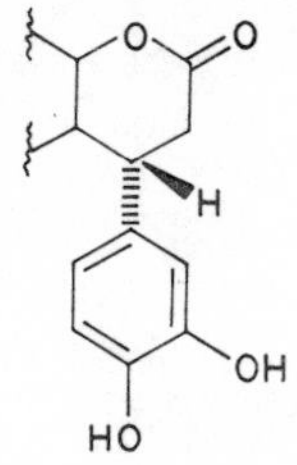	*Cinchona succirubra*	Wollenweber *et al.* (1980)
(6.23) 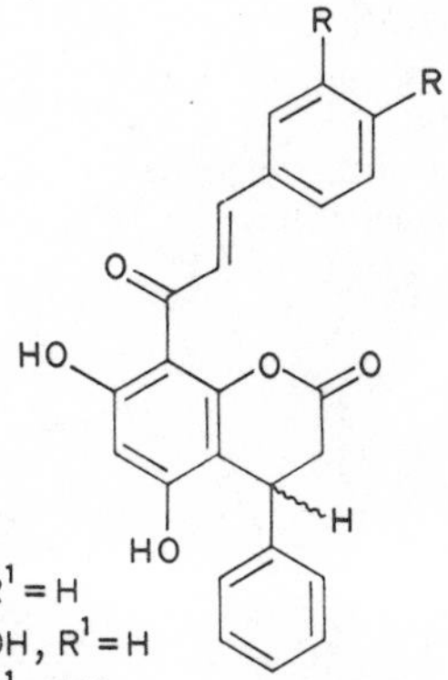 (*a*) R = R[1] = H (*b*) R = OH, R[1] = H (*c*) R = R[1] = OH	*Pityrogramma trifoliata*	Wollenweber *et al.* (1980)
(6.24) 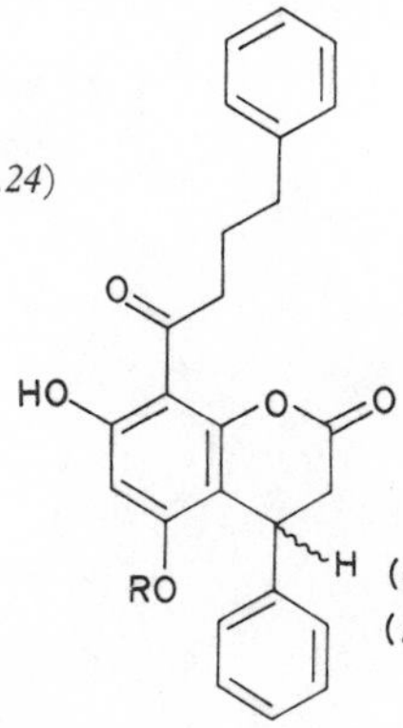 (*a*) R = H (*b*) R = Me	*Pityrogramma calomelanos*	Wagner *et al.* (1979)

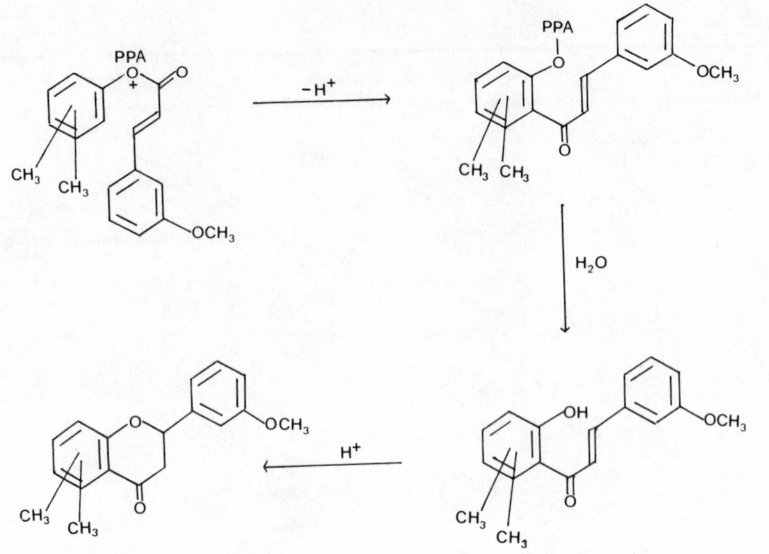

PPA
$-H^+$
OCH$_3$
CH$_3$ CH$_3$
H_2O
H^+
OH

Scheme 6.2 Synthesis of 3,4-dihydro-4-arylcoumarins.

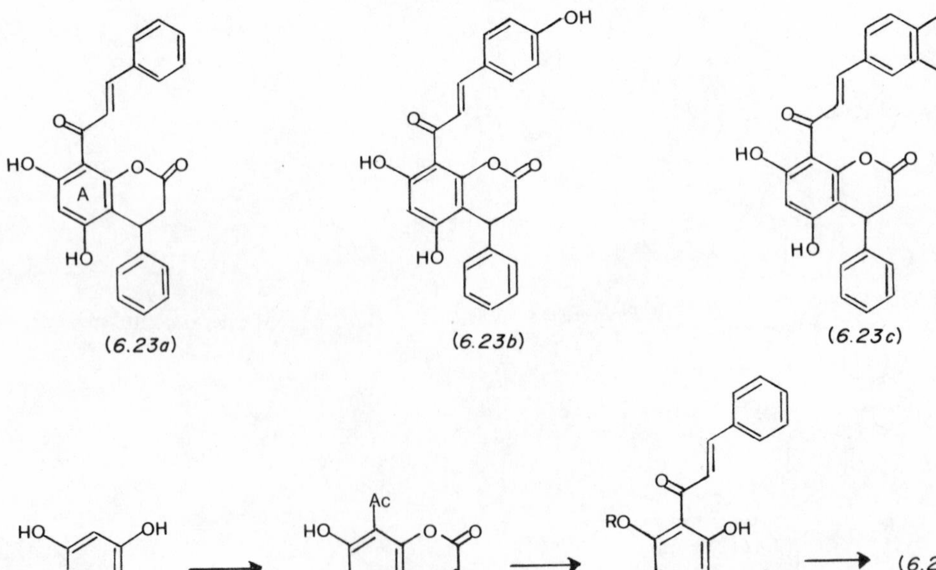

(6.23a) (6.23b) (6.23c)

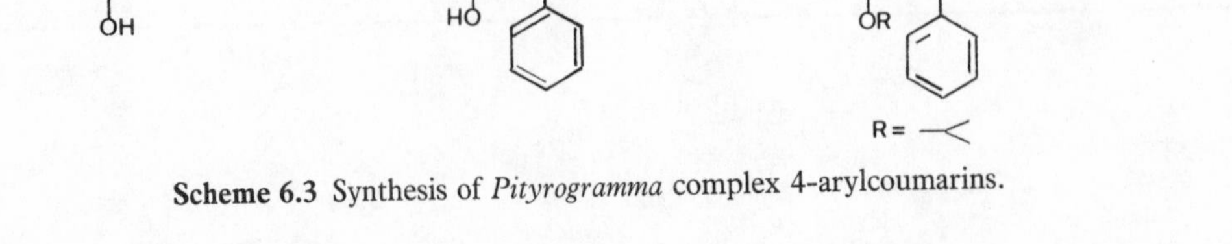

R =

Scheme 6.3 Synthesis of *Pityrogramma* complex 4-arylcoumarins.

6.4.1 Synthesis of 3, 4-dihydro-4-arylcoumarins

Many synthetic approaches to the 3,4-dihydro-4-arylcoumarins have been published. One interesting synthesis involves the reaction of 2- and 4-substituted methoxycinnamic acids with polyphosphoric acid (PPA) and various dimethyl phenols to give 3,4-dihydro-4-phenylcoumarins by direct cycloaddition. The 3-methoxycinnamic acid yielded flavanone formed by cyclization of a 2′-hydroxychalcone which was the intermediate product of a Fries rearrangement (Chenault and Dupin, 1983) (Scheme 6.2).

A different approach (Manimaran and Ramakrishnan, 1979) involved the use of concentrated sulphuric acid to catalyse the reaction of various substituted phenols with cinnamic acid; 3,4-dihydro-4-phenylcoumarins was formed in approximately 20% yield.

Some specific syntheses of 3,4-dihydro-4-arylcoumarins have also been carried out. The structures of the complex 3,4-dihydro-4-arylcoumarins (*6.23a*)–(*6.23c*), isolated from the fern *Pityrogramma trifoliata* (Wollenweber *et al.*, 1980), were assigned on the basis of spectroscopic evidence. The combination of chalcone and neoflavone moieties in these molecules posed an interesting synthetic target molecule. A total synthesis of this series was carried out (Iinuma *et al.*, 1983). The synthetic strategy was based on a Pechmann condensation of acetophenone with benzoylacetic acid to yield a 4-arylcoumarin (neoflavone) with an acetyl substituent in ring A; this neoflavone was then condensed with a benzaldehyde. The Pechmann condensation yielded an equimolar mixture of 6-acetyl- and 8-acetyl-dihydroxy-4-arylcoumarins. The isomers, distinguished by ^{1}H NMR, were separated by column chromatography on silica gel. The 8-acetyl-5,7-dihydroxy-4-arylcoumarin was isopropylated and hydrogenated prior to base-catalysed condensation with the suitably substituted benzaldehydes. Isopropyl groups were removed using BCl_3 to give the required natural products (Scheme 6.3).

The 8-dihydrocinnamoyl-5,7-dihydroxy-4-phenyl-2H-1-3,4-dihydropyran-2-one (*6.24a*) isolated from *Pityrogramma calomelanos* (Wagner *et al.*, 1979) has been synthesized (Donnelly *et al.*, 1986) using a different strategy. The neoflavonoid moiety was constructed via a Perkin condensation and the resulting 5,7-dimethoxy-3,4-dihydro-4-arylcoumarin was acylated with dihydrocinnamoyl chloride in CS_2. A second product was also isolated from the acylation reaction which proved to be dihydrocinnamoyl-7-hydroxy-5-methoxy-4-phenyl-2H-1-3,4-dihydrobenzopyran-2-one (*6.24b*) and was identical with the monomethyl ether of the natural product. Synthesis of the isomeric 6-dihydrocinnamoyl-5,7-dimethoxy-4-aryl-2H-1-3, 4-dihydrobenzopyran-2-one (*6.25*) was carried out. The route employed involved positioning of the cinnamoyl substituent between two methoxy groups and *para* to the hydroxyl function. The

HO, RO, O, O, O, MeO, O, O, O, OMe

(*6.24*) (*a*) R=H
(*b*) R=Me

(*6.25*)

hydroxyl group then acts as the basis for elaboration of the neoflavonoid molecule. The chalcone (*6.26*) was obtained in 96% yield. Two different approaches were employed in building the oxygen heterocycle: (a) cyclization of the phenyl propargyl ether (*6.27*) to the neoflavene (*6.28*) with subsequent allylic oxidation to the 4-arylcoumarin (*6.29*) and (b) synthesis of phenylpropionyloxychalcone (*6.30*) followed by direct cyclization to the 4-arylcoumarin. By route (a) (*6.27*), in the presence of silver tetrafluoroborate in benzene, gave (*6.28*) in 93% yield. Allylic oxidation of the neoflavene by the Sarett method afforded the 6-cinnamoyl-5,7-dimethoxy-4-arylcoumarin in low yield (8.0%). Silver tetrafluoroborate was also used in route (b) to give the desired product which was hydrogenated to give dihydrocoumarin (*6.25*). Partial hydrogenation occurs selectively at the cinnamoyl α,β double bond to give 6-dihydrocinnamoyl-5,7-dimethoxy-4-arylcoumarin which is an intermediate in route (a) (Scheme 6.4).

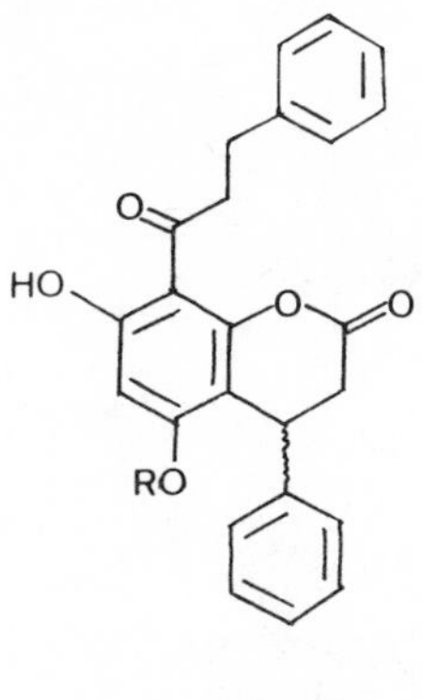

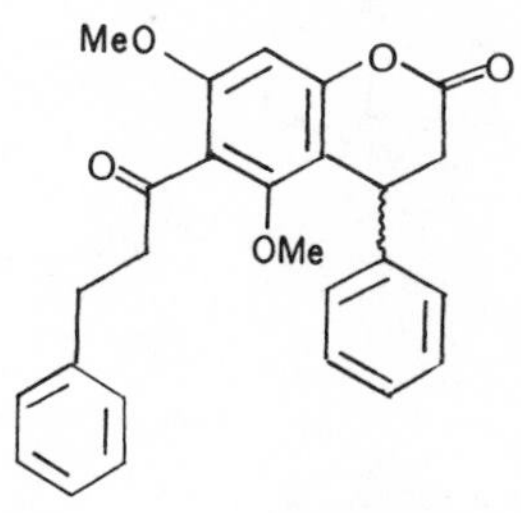

The structure of the neoflavonoid (*6.24a*) isolated from *P. calomelanos* has also been confirmed by an X-ray diffraction study (Donnelly *et al.*, 1986). The natural product is a racemic mixture, in contrast with previous reports (Wollenweber *et al.*, 1980). Compound (*6.24a*) crystallizes from methanol. On standing in air the crystals are destroyed due to loss of solvent from the packing. A wet crystal of approximate size $0.2 \times 0.5 \times 0.1$ mm was sealed in a Lindemann capillary and mounted on a Phillips PW 1100 automatic diffractomer equipped with a graphite monochromator and operating with Cu–K_α radiation (λ – 0.15418 nm). The view of the molecule is given in Fig. 6.3.

6.5 1,1-DIARYLPROPANOIDS

Several examples of open-chain 1,1-diarylpropanoids have been reported.

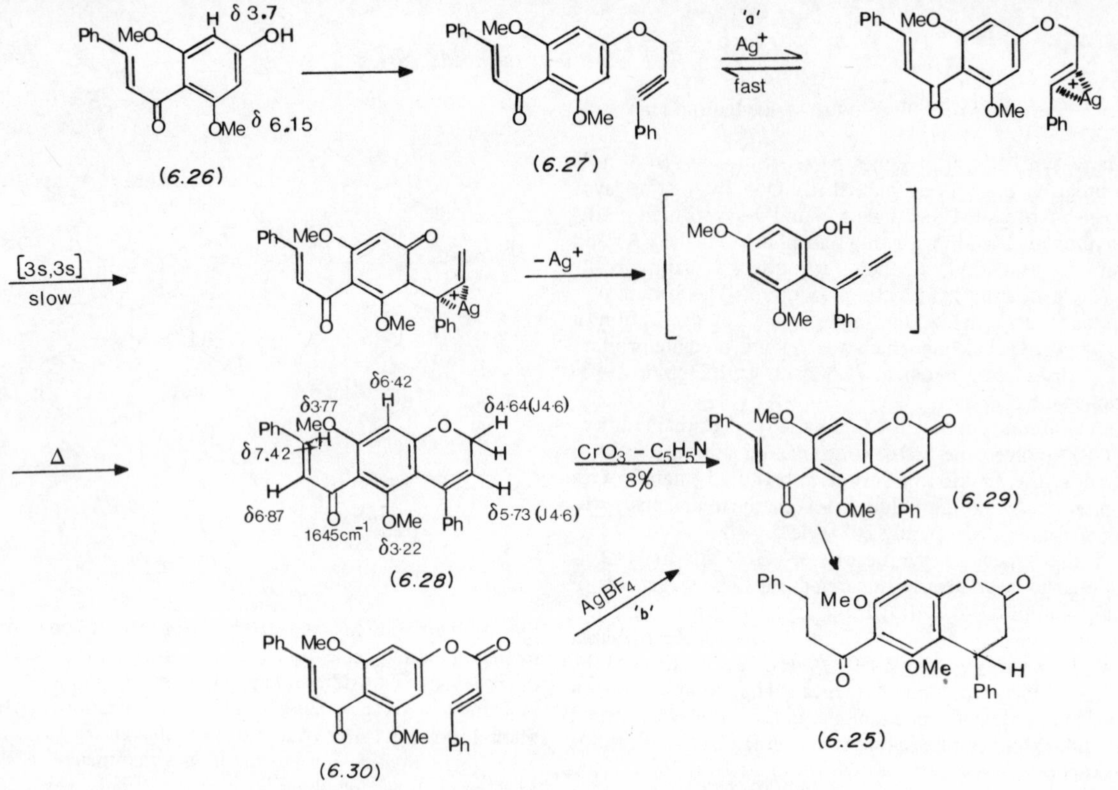

Ph MeO H δ 3.7 OH O OMe δ 6.15

(*6.26*)

Ph MeO O O OMe Ph

(*6.27*)

'a' Ag^+ fast

Ph MeO O O OMe + Ag Ph

[3s,3s] slow

Ph MeO O O OMe + Ag Ph

$-Ag^+$

[MeO OH OMe Ph]

Δ

δ6·42 H δ3·77 Ph MeO δ4·64 (J4·6) H O H δ7.42 H H H δ6·87 O 1645 cm^{-1} OMe δ3·22 Ph δ5·73 (J4·6)

(*6.28*)

$CrO_3 - C_5H_5N$ 8%

Ph MeO O O O OMe Ph (*6.29*)

$AgBF_4$ 'b'

Ph MeO O O O OMe Ph

(*6.30*)

Ph MeO O O O OMe H Ph

(*6.25*)

Scheme 6.4 Synthesis of fern neoflavonoids.

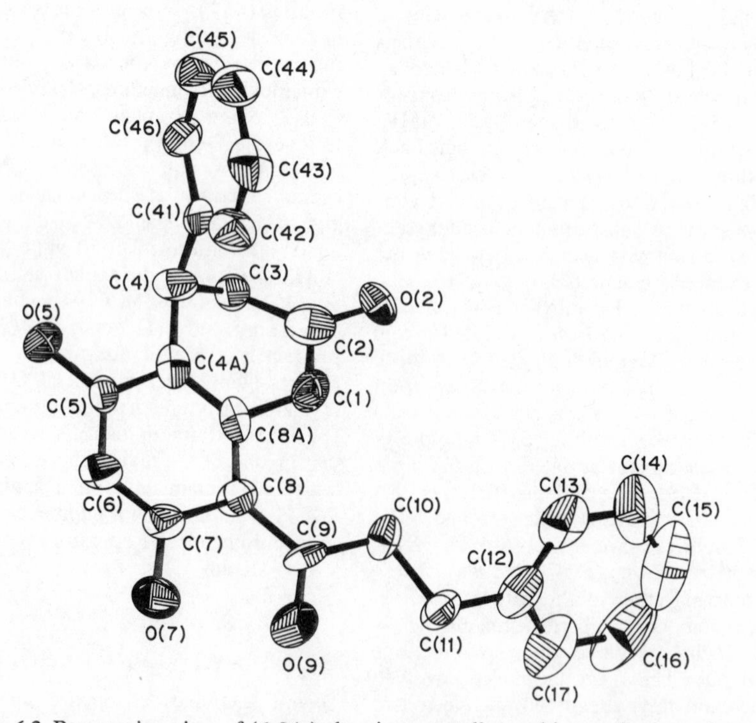

Fig. 6.3 Perspective view of (*6.24a*) showing crystallographic numbering scheme.

6.5.1 Phenylpropanoid-substituted epicatechins – cinchonains

The bark of *Cinchona succirubra* (Rubiaceae), a source of quinine and quinidine, has yielded from the tannin fraction a series of new phenylpropanoid-substituted flavan-3-ols, the cinchonains (*6.19*) and (*6.20*) (Nonaka and Nishioka, 1982). The occurrence of a flavan-3-ol skeleton in the molecule could be easily deduced from the ^{1}H NMR and ^{13}C NMR spectra which resembled those of epicatechin. In addition to the 15 signals for epicatechin the latter spectrum had a methine (δ 35.4 d), a methylene (δ 38.0 t) and signals for a 3, 4-dihydroxy aromatic substituted ring. A signal of δ 168.9 was assigned to the carboxyl function. Methylation gave a hexamethyl ether methyl ester (*6.31*) which was the result of an opening of the lactone ring and subsequent methylation of the carboxylic acid and phenolic hydroxyl group. A significant feature of the electron impact mass spectrum (EI-MS) of this ester (*6.31*) is the appearance of the base peak *m/z* 223 derived from a radical ion of dimethyl dihydrocaffeic acid methyl ester. The location of this unit on the epicatechin structure was deduced from comparisons of ^{13}C NMR spectra A-ring signals with those from the methyl derivatives of substituted catechins, the gambirins. The stereostructures were deduced by analysis of their ^{1}H NMR spectra. In compound (*6.19*) the H-2 and ring-B protons are as in epicatechin but in (*6.20*) the signal H-2 at δ 4.99 is shifted downfield whilst the B-ring protons are shifted upfield. These protons are anisotropically affected by the aromatic ring of the phenylpropanoid moiety (Fig. 6.4). The absolute configuration of the cinchonains was established through synthesis of (*6.19*) and (*6.20*) using epicatechin [a compound of known absolute configurations at the chiral centre $C_2(R)$ and $C_3(R)$] and condensing it with caffeic acid in the presence of *p*-toluenesulphonic acid.

6.5.2 β-(5, 7, 4-Trihydroxy-8-yl)-β-propionic acid

Two polar flavonoids [(*6.32*) and (*6.33*)] with complex chemical structures have been isolated from cultivated

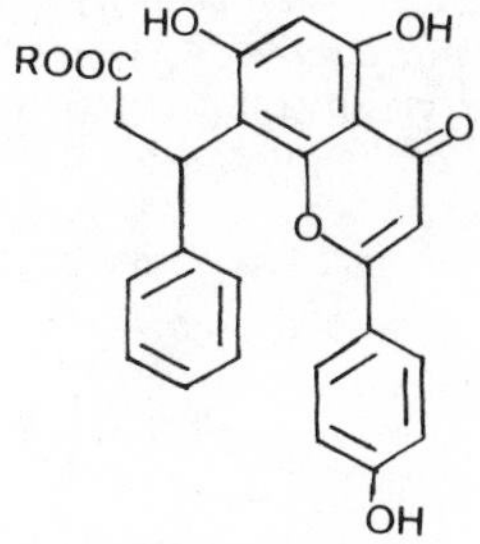

(*6.32*) R = Me

(*6.33*) R = H

plants of *Pityrogramma calomelanos var. aureoflava* (Iinuma *et al.*, 1986). Compound (*6.33*) has also been isolated from some specimens of *P. sulphurea* (Wollenweber *et al.*, 1980).

Neither of these flavonoids possesses a γ-pyrone ring. Some ambiguity exists concerning these compounds as

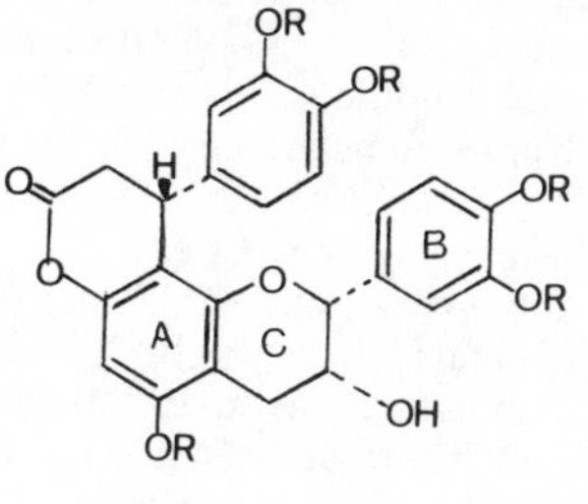

(*6.19*)

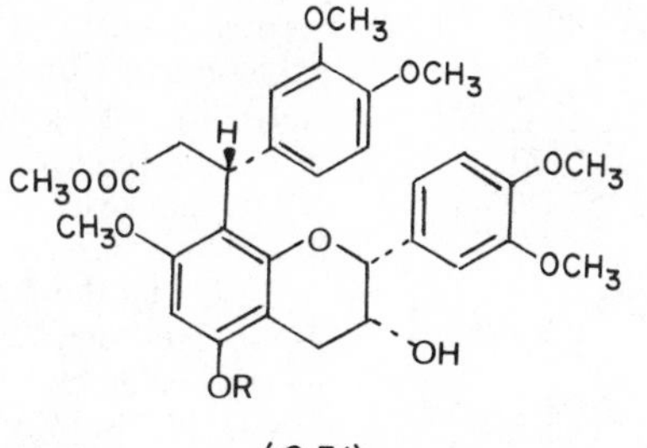

(*6.31*)

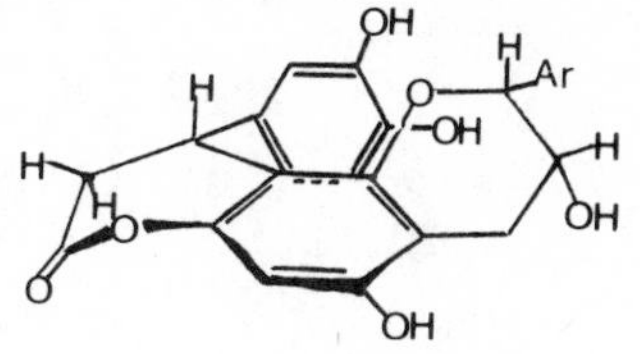

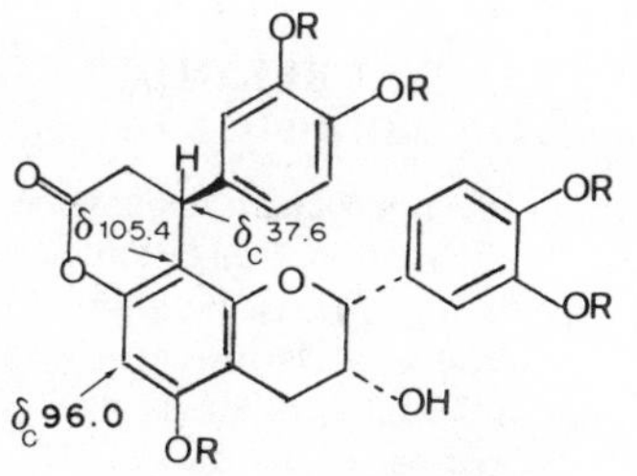

(*6.20*)

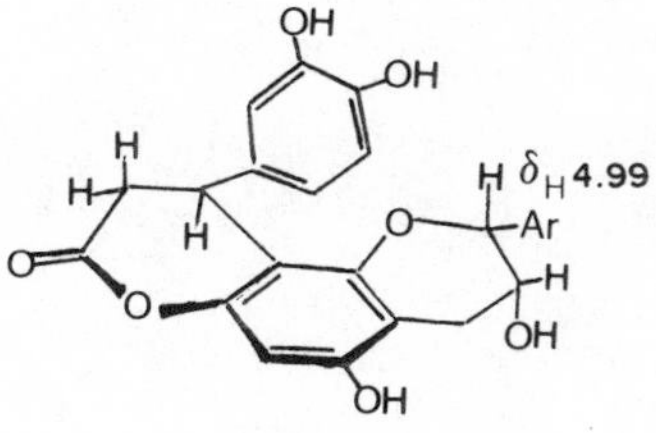

Fig. 6.4 The cinchonains.

they have been reported under different abbreviations, and chemical nomenclature has not been consistent. The structure of these two natural products has been confirmed by unambiguous synthesis.

6.6 4-ARYLFLAVAN-3-OLS

The isolation of the first naturally occurring 4-arylflavan-3-ol (*6.34*) (Kolodziej, 1983, 1984) from the dimeric fraction of the South African succulent *Nelia meyeri* is noteworthy. This 4-arylflavan-3-ol represents an extension of the known series of bineoflavonoids. It also seems reasonable to classify this compound as a neoflavonoid on the basis of its possessing a 2-substituted 4-arylchroman skeleton. The structure of this compound was established by analysis of its heptamethyl ether

(6.34) a) R' = R'' = H

(6.35) b) R' = Me, R'' = Ac

(6.37)

a) R' = R'' = H

b) R' = Me, R'' = Ac

c) R' = R'' = Ac

(6.38)

a) R' = R'' = H

b) R' = Me, R'' = Ac

c) R' = R'' = Ac

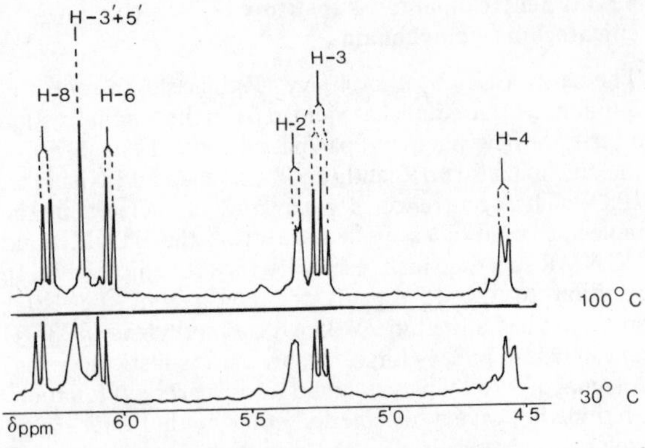

Fig. 6.5 Expanded resonances from ^{1}H NMR spectrum of (*6.35*) recorded at 100° and 36°C.

monoacetate (*6.35*) by mass spectrometry. The molecular ion appeared at m/z 554 [M^+ 1.3%] and was substantiated by loss of acetic acid m/z 494 [$M^+ - 60$, 100%] and the fragment m/z 387 due to elimination of the 4-aryl group. The proton NMR spectrum shows temperature-dependent linewidths due to 'slow' rotation about the C(4)-aryl group bond as evidenced by sharpening of the broadened resonances at elevated temperature (Fig. 6.5). Spin-decoupling experiments established the allocations of resonances. The relative stereochemistry of the molecule was determined by ^{1}H NMR and was confirmed by its positive Cotton effect in the low wavelength of the CD spectrum. The synthesis of this compound was accomplished by condensation of the flavan 3,4-diol (*6.36*), prepared from the tetramethyl ether (−)-epicatechin with phloroglucinol under acid conditions (Scheme 6.5). Co-metabolites of 4-arylflavan-3-ol include the procyanidins (*6.37*) and (*6.38*).

The similarity in the oxygenation pattern of 4-arylflavan-3-ol (*6.34*) and the procyanidins (*6.37*) and (*6.38*) suggests that these compounds are biogenetically related. As an intermediate compound between the 4-arylated neoflavonoids and the procyanidins we can in fact see how (*6.34*) could be derived from either of the proposed biosynthetic pathways leading to these two groups of molecules (Scheme 6.6).

6.7 DALBERGIQUINOLS AND DALBERGIONES

The open-chain neoflavonoids can be subdivided according to their oxidation levels giving the dalbergiquinol (*6.39*), the dalbergione (*6.40*) and the benzophenone groups. It is interesting to compare the compounds of different oxidation levels isolated from one *Dalbergia* species (Fig. 6.6).

The dalbergiones are the most abundant open-chain neoflavonoids and many new sources (Table 6.5) have

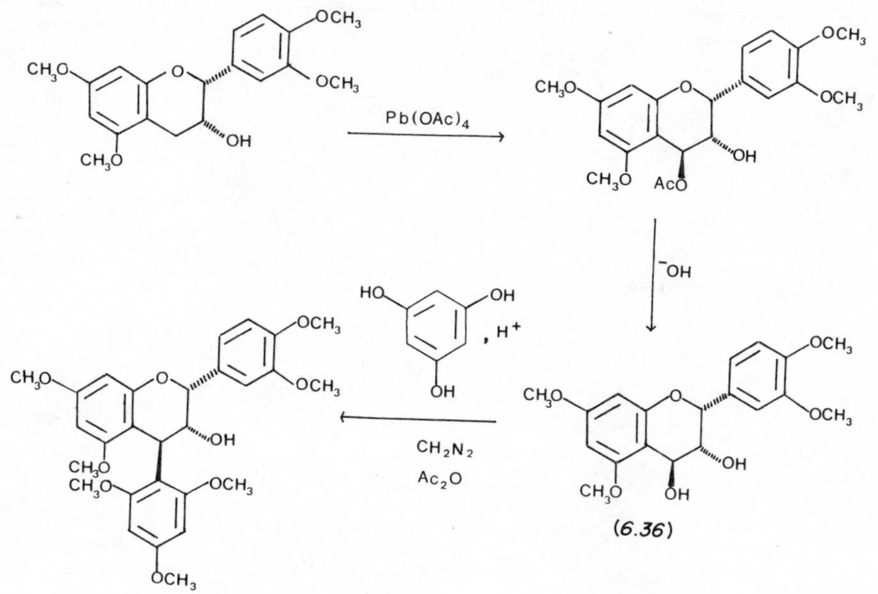

Scheme 6.5 Synthesis of the 4-arylflavan-3-ol (*6.36*).

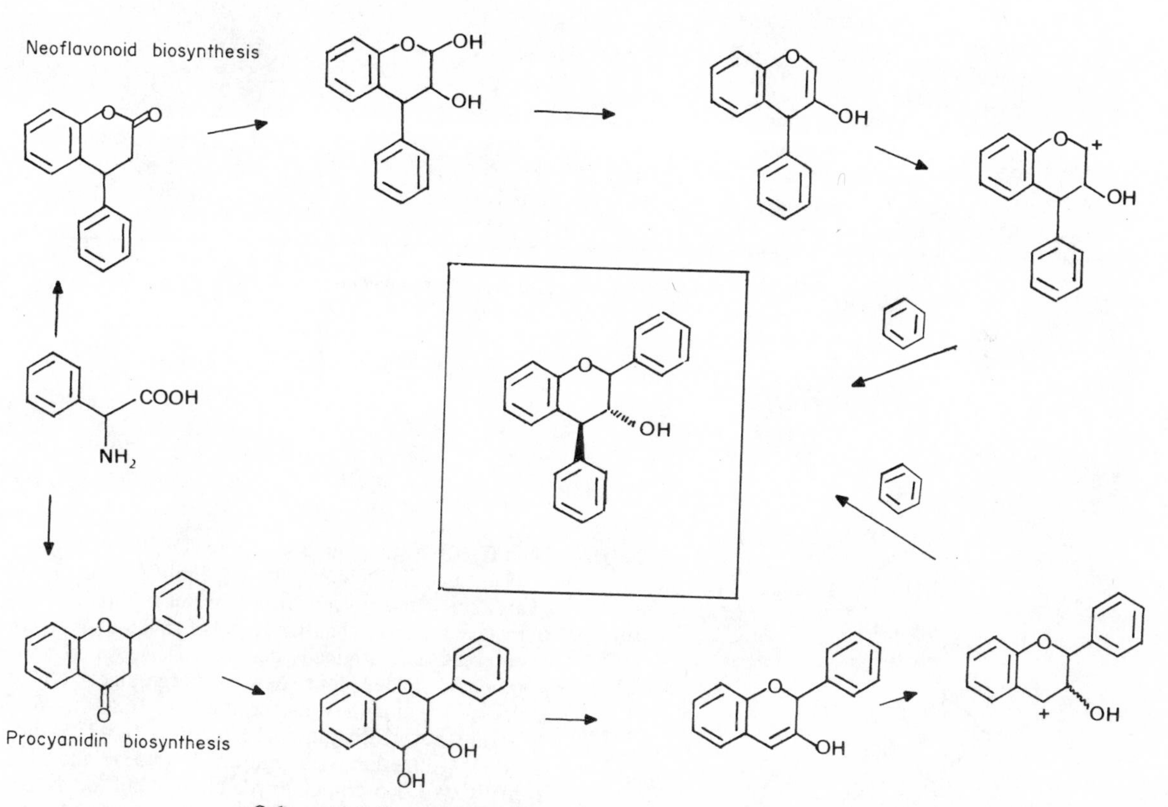

Scheme 6.6 Proposed biosynthetic routes to 4-arylflavan-3-ols.

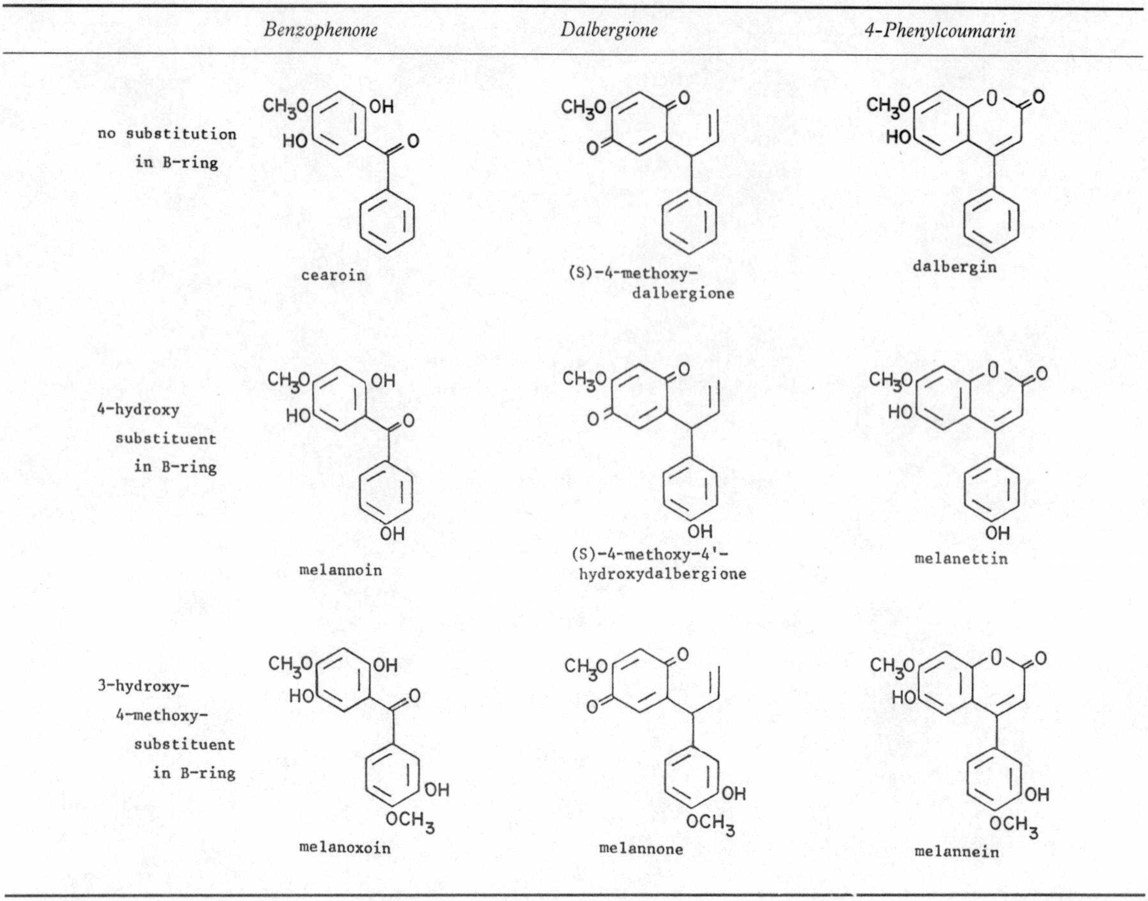

	Benzophenone	*Dalbergione*	*4-Phenylcoumarin*
no substitution in B-ring	cearoin	(S)-4-methoxy-dalbergione	dalbergin
4-hydroxy substituent in B-ring	melannoin	(S)-4-methoxy-4'-hydroxydalbergione	melanettin
3-hydroxy-4-methoxy-substituent in B-ring	melanoxoin	melannone	melannein

Fig. 6.6 Neoflavonoids of *Dalbergia melanoxylon*.

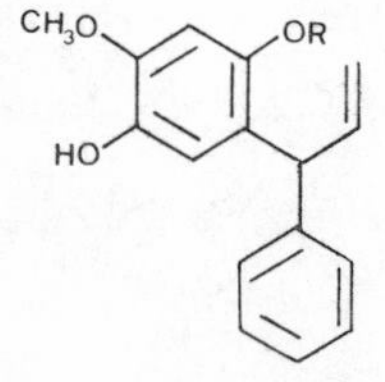

(6.39)

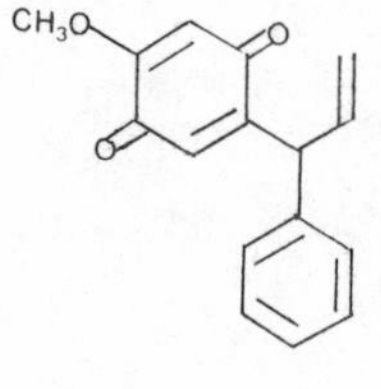

(6.40)

been reported since the last review (Donnelly, 1975).

The dalbergiquinols have limited distribution (Table 6.6) although improved chromatographic techniques have indicated that this group of compounds is more widespread than originally believed. *Dalbergia cochinchinensis* has yielded four new dalbergiquinols: (*R*)-5-hydroxy-2,4-dimethoxydalbergiquinol (*6.51*), (*R*)-2,4,5-trimethoxydalbergiquinol (*6.52*), (*R*)-2′,5′-dihydroxy-2,4-dimethoxydalbergiquinol (*6.53*) and (*R*)-2′-hydroxy-2,4,5-trimethoxydalbergiquinol (*6.54*). The 5-hydroxy-2,4-dimethoxydalbergiquinol was not a new

Table 6.5 Dalbergiones of *Dalbergia* and *Machaerium* species

Compound	*Family, genesis, species*	*References*
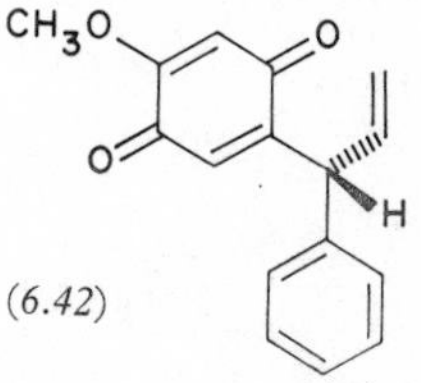 CH_3O, O, O, H; (6.42) (R)-4-Methoxydalbergione	*Dalbergia cochinchinensis* *D. parviflora*	Donnelly *et al.* (1981) Maungroicharoen and Frahm (1982)
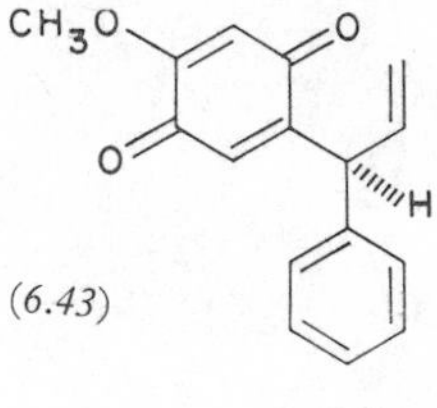 CH_3O, O, O, H; (6.43) (S)-4-Methoxydalbergione	*D. barretoana*	Donnelly *et al.* (1981)
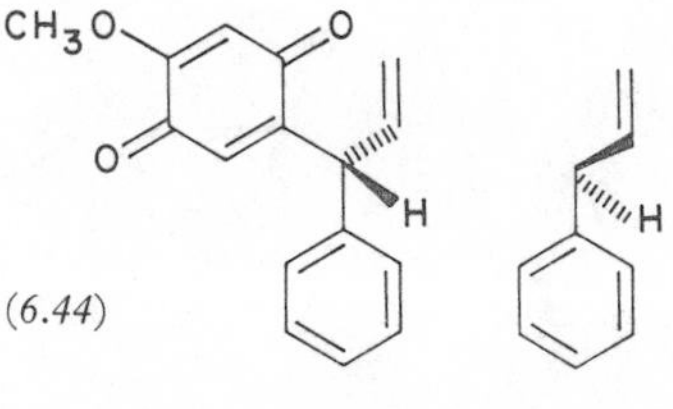 CH_3O, O, O, H, H; (6.44) (R,S)-4-Methoxydalbergione	*D. retusa*	Jurd *et al.* (1972)
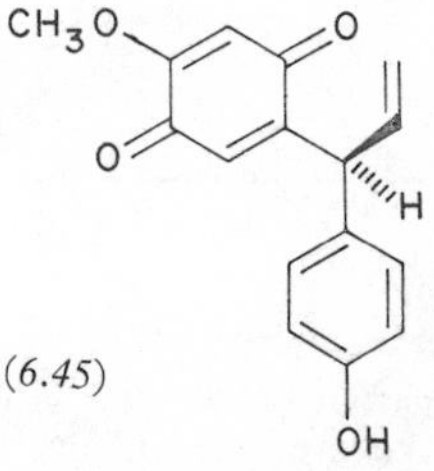 CH_3O, O, O, H, OH; (6.45) (S)-4′-Hydroxy-4-methoxydalbergione	*D. latifolia* *D. miscolobium*	Donnelly *et al.* (1981) Gregson *et al.* (1978)
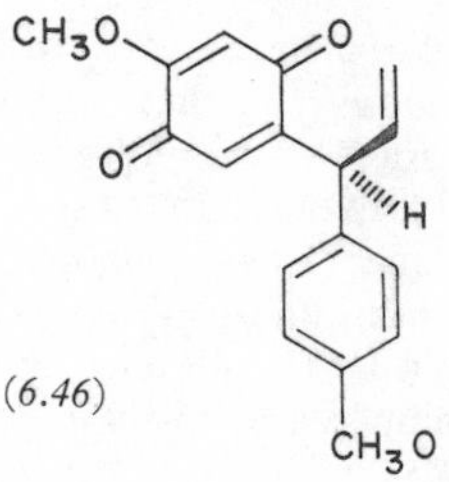 CH_3O, O, O, H, CH_3O; (6.46) (S)-4,4′-Dimethoxydalbergione	*D. nigra*	Eyton *et al.* (1962)

(Contd.)

Table 6.5 (*Contd.*)

Compound	Family, genesis, species	References
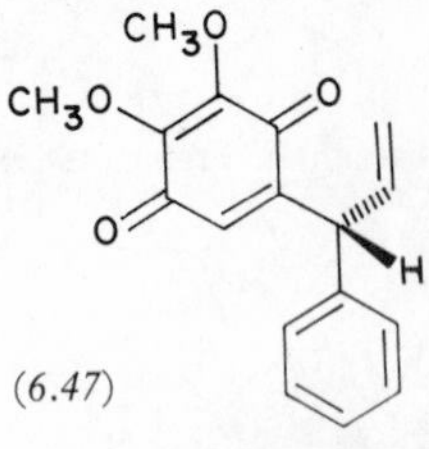 (6.47) (*R*)-3,4-Dimethoxydalbergione	*Machaerium kuhlmanii* *M. nictitans*	Ollis *et al.* (1968)
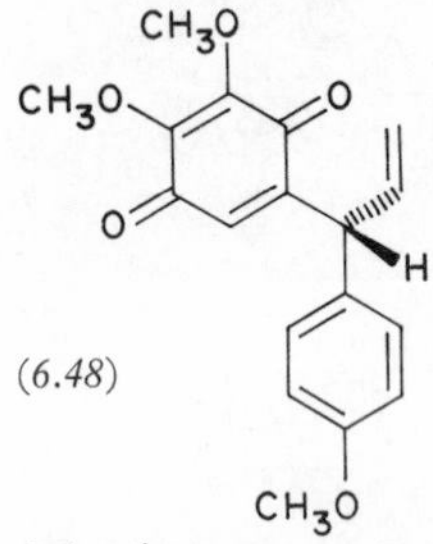 (6.48) (*R*)-4'-Hydroxy-3,4-dimethoxydalbergione	*M. nictitans*	Ollis *et al.* (1968)
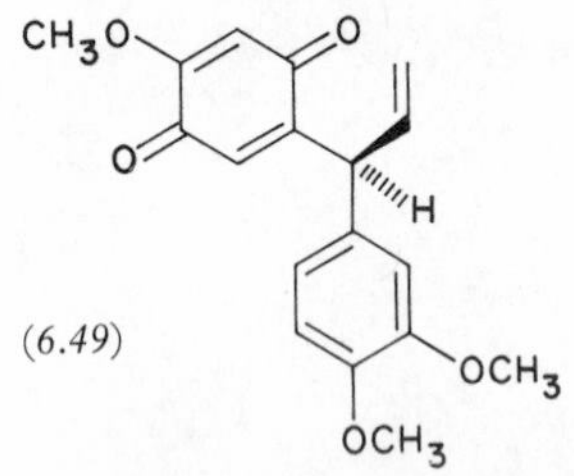 (6.49) (*S*)-3'-Hydroxy-4,4'-dimethoxydalbergione	*D. melanoxylon*	Donnelly (1975)

compound, since it was previously isolated from *D. sissoo* (Kulshrestha *et al.*, 1974) but its absolute configuration was not assigned. The quinol in *D. cochinchinensis* has an *R* configuration (Donnelly *et al.*, 1981) as it afforded (*R*)-4-methoxydalbergione on chromium trioxide oxidation.

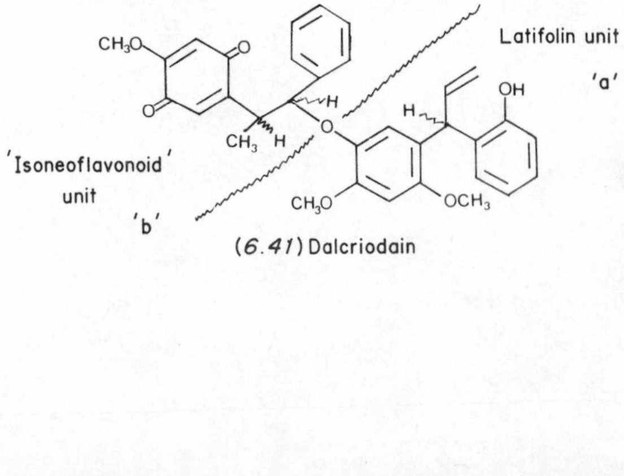

Latifolin unit 'a'

'Isoneoflavonoid' unit 'b'

(*6.41*) Dalcriodain

Binary neoflavonoids have been encountered in *Dalbergia cochinchinensis* and *D. latifolia*. Dalcriodain (*6.41*), a neoflavonoid which arises from linkage of the dalbergiquinol latifolin and an isoneoflavonoid unit, has been identified.

A 4-arylcoumarin moiety is considered to be one unit of cochin B, a bineoflavonoid observed in the GC–MS analysis of an extract of *D. cochinchinensis*, and has a mass spectrum which showed fragmentation of the molecular ion (M^+, m/z 554) to give two neoflavonoid units. The further breakdown of these fragment ions suggested 'cochin B' to be a combination of latifolin (*6.41a*) and dalbergin (6-hydroxy-7-methoxy-4-arylcoumarin). Investigation of this structure was hampered by insufficient material.

Table 6.6 Dalbergiquinols from *Dalbergia* and *Machaerium* species

Compound	*Family, genesis, species*	*References*
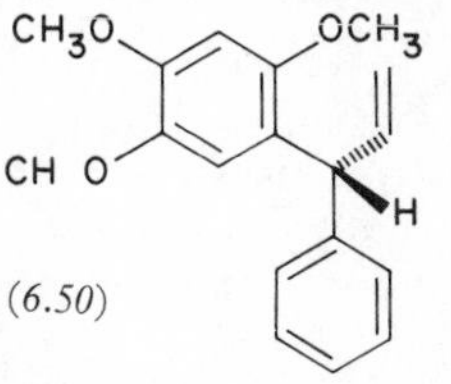 (6.50) (*R*)-5-Hydroxy-2,4-dimethoxydalbergiquinol	*Dalbergia cochinchinensis* *D. sissoo* *D. cultrata*	Donnelly *et al.* (1981) Kulshrestha *et al.* (1974) Donnelly *et al.* (1972)
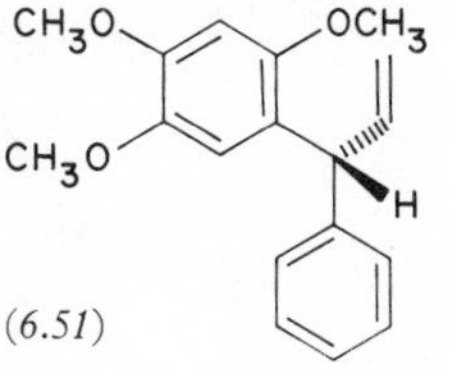 (6.51) (*R*)-2,4,5-Trimethoxydalbergiquinol	*D. cochinchinensis*	Donnelly *et al.* (1981)
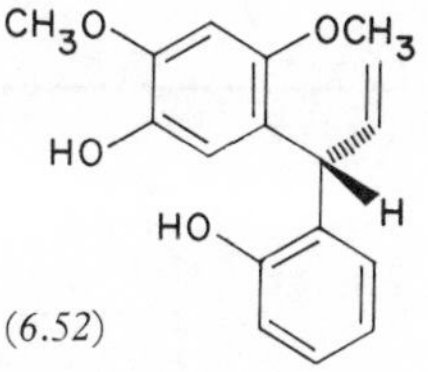 (6.52) (*R*)-2',5-Dihydroxy-2,4-dimethoxydalbergiquinol	*D. cochinchinensis* *D. latifolia*	Donnelly *et al.* (1981)
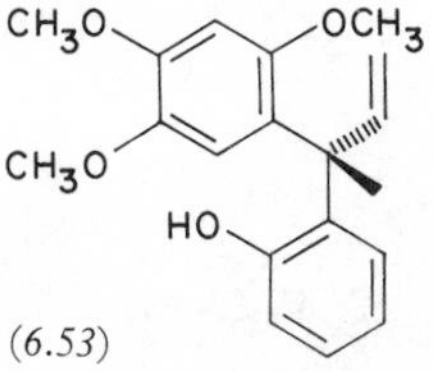 (6.53) (*R*)-2'-Hydroxy-2,4,5-trimethoxydalbergiquinol	*D. cochinchinensis*	Donnelly *et al.* (1981)
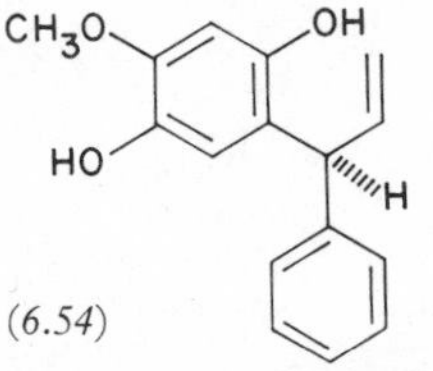 (6.54)	*D. obtusa* *D. retusa*	Jurd *et al.* (1972)
(*R,S*)-2,5-Dihydroxy-4-methoxydalbergiquinol	*Machaerium scleroxylon*	Eyton *et al.* (1965)

(Contd.)

Table 6.6 (*Contd.*)

Compound	Family, genesis, species	References
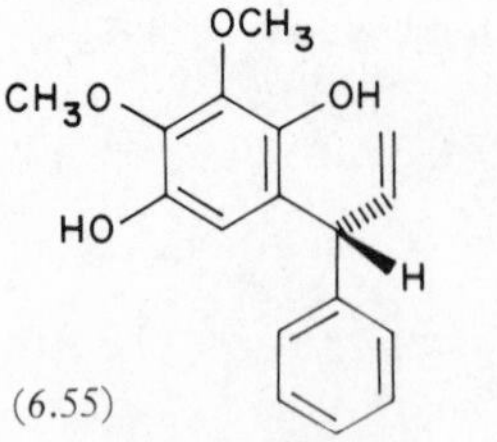 (6.55) (*R*)-2,5-Dihydroxy-3,4-dimethoxydalbergiquinol	*M. kuhlmanii* *M. nictitans*	Ollis *et al.* (1968)
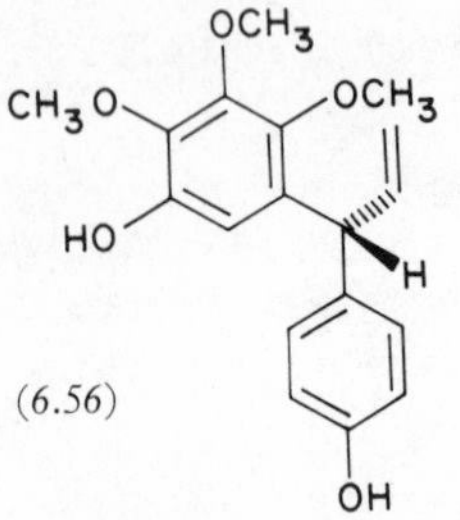 (6.56) (*R*)-4′,5-Dihydroxy-2,3,4′,-trimethoxydalbergiquinol	*M. kuhlmanii* *M. nictitans*	Ollis *et al.* (1968)

(6.58) (6.59) (6.60) (*E*)-(6.61) (6.62) (*S*)-(6.57)

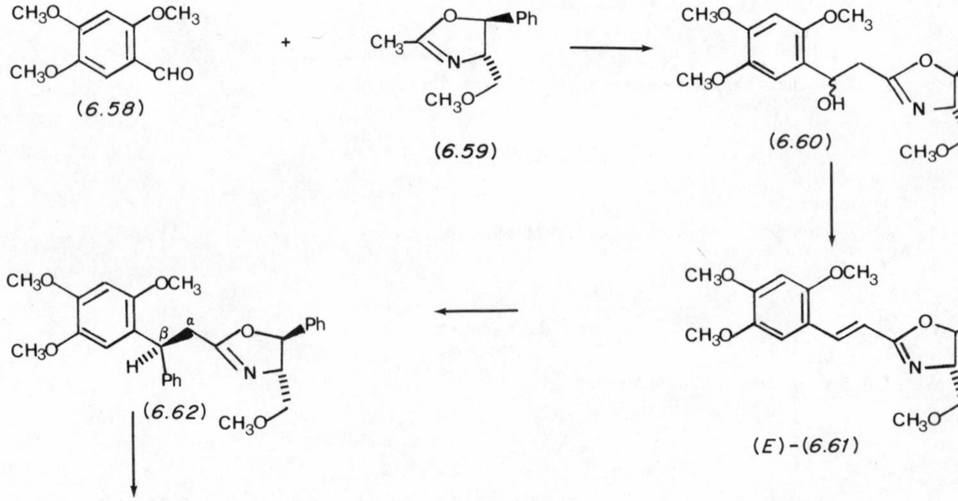

Scheme 6.7 Synthesis of dalbergiquinols.

6.7.1 Synthesis of dalbergiquinols

(*R*)- and (*S*)-2,4,5-Trimethoxydalbergiquinols react with titanium tetrachloride and sodium borohydride to give the corresponding (*R*)- and (*S*)-propan-1-ols, effectively introducing the hydroxyl group in an anti-Markownikov direction. (*S*)-3(2′,4′,5′-Trimethoxybenzene)-3-phenylpropan-1-ol (*6.57*) has been synthesized (O'Sullivan, 1982) using a chiral oxazoline, in greater than 80.2% ee. The synthesis followed the reaction pathway outlined in Scheme 6.7. The condensation of 2,4,5-trimethoxybenzaldehyde (*6.58*) with 2-methyloxazoline (*6.59*) afforded a 1:1 diastereometric mixture of the hydroxyoxazoline (*6.60*). The diastereomeric ratio was determined from the relative intensities of ^{1}H NMR spectra of (*6.60*). Dehydration of the hydroxyoxazoline afforded the (*E*)-vinyloxazoline (*6.61*). The nucleophilic addition of phenyl-lithium to the chiral electrophile afforded the adduct (*6.62*). From the presence of only one methoxyl group signal (δ 3.32) for the methoxymethyl group it would seem that the asymmetric induction occurred with a high degree of stereoselectivity. The diastereomeric adduct was hydrolysed in acidic ethanol to give the ethyl ester which was reduced to give the desired (*S*)-alcohol (*6.58*).

The synthesis of the isomer (*R*)-3-(2′,4′,5′-trimethoxybenzene)-3-phenylpropan-1-ol by application of chiral oxazolines was unsuccessful. The (4*S*, 5*S*)-2-(β-hydroxy-β-phenyl)ethyl-4-methoxymethyl-5-phenyl-oxazoline was formed in 1:1.25 mixture of diastereomers. Again the dehydration proceeded with exclusive formation of the (*E*)-vinyloxazoline but the expected condensation with 1-bromo-2,4,5-trimethoxybenzene did not occur.

6.8 3-ARYLBENZO[*b*]FURANS

Although six-membered heterocyclic neoflavonoids are commonly encountered in *Dalbergia* species the only known true neoflavonoids with a five-membered heterocyclic ring (*6.63*)–(*6.65*) have been isolated from *Dalbergia baroni* (Donnelly *et al.*, 1981) and *D. parviflora* (Muangnoicharoen and Frahm, 1982). The structures (*6.64*) and (*6.65*) were confirmed by independent synthesis via an oxidative rearrangement of dalbergin (Ahluwalia *et al.*, 1985) which unambiguously indicated the presence of a phenyl group at the 3 position and an oxygenated carbon at the 2 position of the benzofuran. This series of 3-phenylbenzofurans demonstrates a possible biogenetic oxidative sequence.

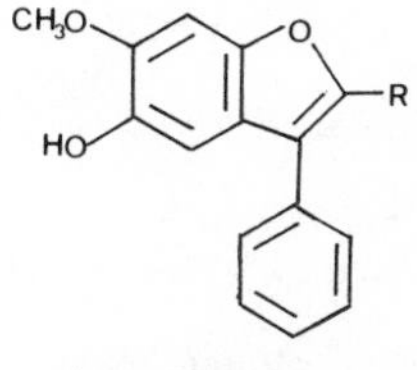

(*6.63*) R=CH_3
(*6.64*) R=CH_2OH
(*6.65*) R=CHO

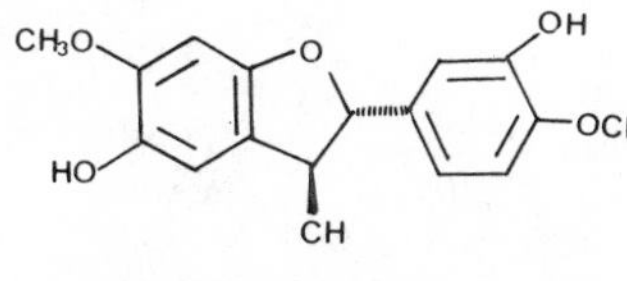

(*6.66*)(2*S*,3*S*)-Melanoxin

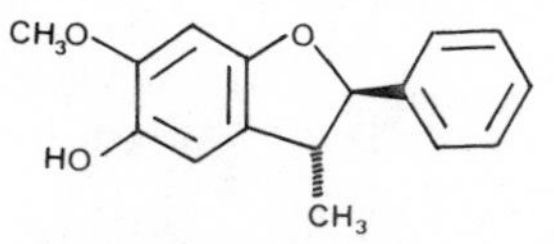

(*6.67*) Obtusafuran

Two examples of 2-aryldihydrobenzo[*b*]furans (2*S*,3*S*)-melanoxin (*6.66*) and obtusafuran (*6.67*) have been reported earlier. These optically active dihydrofurans have been classified with the neoflavonoids (C_6–C_1–C_6) and not as the skeletons suggest in the isoflavonoids (C_6–C_2–C_6). The classification of the neoflavonoids date from the biosynthesis proposals of Ollis and Gottlieb (1968) who suggest that the dihydrobenzo[*b*]furans (*6.66*) and (*6.67*) are formed by alkylation of a phenol cinnamoyl pyrophosphate.

Obtusafuran (*6.67*) has been identified as a unit in the previously described dalcriodain molecule (*6.42*). The opening of the furan ring with subsequent conversion to the quinone form is shown in Scheme 6.8.

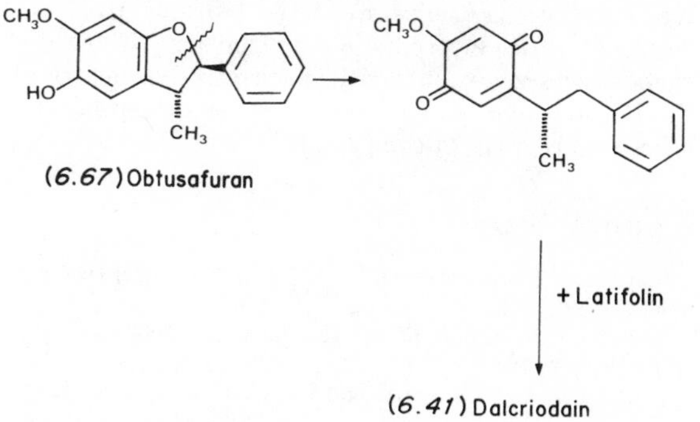

(*6.67*) Obtusafuran

+ Latifolin

(*6.41*) Dalcriodain

Scheme 6.8 Biogenetic relationship between obtusafuran and dalcriodain.

6.9 PHENANTHRA-1, 4-QUINONES

Two examples of phenanthra-1,4-quinones, latinone (*6.68*) and melatinone (*6.69*), have been found as co-metabolites of the neoflavonoids of *D. latifolia* and *D. melonoxylon*. These minor and uncommon quinones have similar oxygenation patterns to the dalbergiquinols and the open-chain neoflavonoids. It has been suggested that the latinones are possibly condensation products of cinnamoyl pyrophosphate with two units of substituted phenol, resulting in the loss of a methoxyl group during the ring formation, and the subsequent decarboxylation. Synthesis of latinone has been reported and the laboratory analogy to

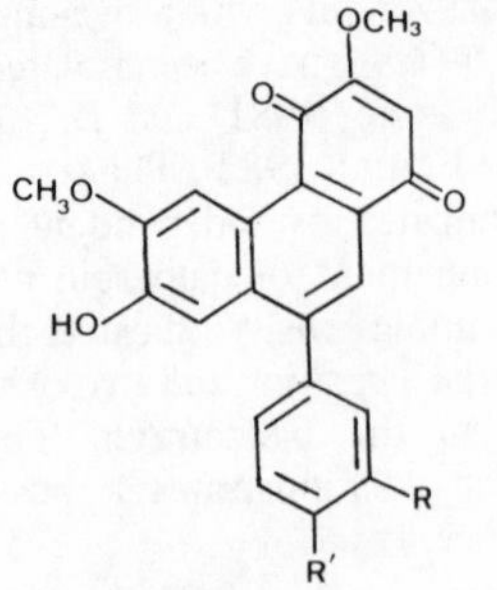

(6.68) R = R[1] = H, Latinone

(6.69) R = OH, R[1] = OCH_3, Melatinone

the loss of a methoxyl group during the phenanthrene skeleton formation has been achieved (Cannon, 1985).

6.10 CONCLUSIONS

Relatively few neoflavonoids have been isolated in the last decade but their occurrence in families other than the Guttiferae and the Leguminosae is being observed. Numerous enzymes involved in flavonoid biosynthesis have been investigated and identified but the pathways formulated by biosynthesis of isoflavonoids (see Chapter 5) and neoflavonoids are derived mainly from the results of *in vivo* incorporation experiments. No biosynthetic studies have been carried out on the neoflavonoids of the Leguminosae and few reports on the biological activities of this group of compounds have appeared.

REFERENCES

Ahluwalia, V.K., Mukherjee, I. and Rani, N. (1981), *Indian J. Chem.* **20B**, 918.
Ahluwalia, V.K., Kapur, K. and Manchanda, S. (1982), *Indian J. Chem.* **21B**, 186.
Ahluwalia, V.K., Mehta, A.C. and Seshadri, T.R. (1985), *Tetrahedron* **4**, 271.
Busteed, L. (1984), M.Sc. thesis, National University of Ireland.
Cannon, P. (1985), Ph.D. thesis, National University of Ireland.
Chenault, J. and Dupin, J.-F.E. (1983), *Heterocycles* **20**, 437.
Chaterjee, A., Ganguly, D. and Sen, R. (1976), *Tetrahedron* **32**, 2407.
Das Gupta, A.K. and Paul, M.S. (1970), *J. Indian Chem. Soc.* **47**, 1017.
Donnelly, D.M.X. (1975). In *The Flavonoids* (eds J.B. Harborne, T.J. Mabry and H. Mabry), Chapman and Hall, London, p. 802.
Donnelly, D.M.X. (1985), *Annu. Proc. Phytochem. Soc. Eur.* **25**, 199.
Donnelly, D.M.X., O'Reilly, J. and Thompson, J. (1972), *Phytochemistry* **11**, 823.
Donnelly, D.M.X., O'Reilly, J. and Whalley, W.B. (1975), *Phytochemistry* **14**, 2287.
Donnelly, D.M.X., O'Criodain, T. and O'Sullivan, M. (1981), *J. Chem. Soc., Chem. Commun.*, 1254.
Donnelly, D.M.X., Fukuda, N., Wollenweber, E., Polonsky, J. and Prangé, T. (1987), *Phytochemistry* **26**, 1143.
Eyton, W.B., Ollis, W.D., Sutherland, I.O., Jackman, L.M., Gottlieb, O.R. and Magalhaes, M.T. (1962), *Proc. Chem. Soc.*, 301.
Eyton, W.B., Ollis, W.D., Fineberg, M., Gottlieb, O.R., Salignac de Souza Guimaraes, I. and Magalhaes, M.T. (1965), *Tetrahedron* **21**, 2697.
Gregson, M., Ollis, W.D., Sutherland, I.O., Gottlieb, O.R. and Magalhes, M.T. (1978a), *Phytochemistry* **17**, 1375.
Gregson, M., Ollis, W.D., Redman, B.T., Sutherland, I.O., Dietrichs, H.H. and Gottlieb, O.R. (1978b), *Phytochemistry* **17**, 1395.
Iinuma, M., Matsuura, S. and Asai, F. (1983), *Heterocycles* **20**, 1923.
Iinuma, M., Hamada, K., Mizuno, M., Asai, F. and Wollenweber, E. (1986), *Z. Naturforsch.* **41c**, 681.
Iinuma, M., Tanaka, T., Hamada, K., Mizuno, M., Asai, F., Reher, G. and Kraus, L. (1987), *Phytochemistry* **26**, 3096.
Jurd, L., Stevens, K. and Manners, G. (1972a), *Tetrahedron Lett.* **21**, 2149.
Jurd, L., Stevens, K. and Manners, G. (1972b), *Phytochemistry* **11**, 3287.
Kolodziej, H. (1983), *Tetrahedron Lett.* **24**, 1825.
Kolodziej, H. (1984), *Phytochemistry* **23**, 1745.
Kulshrestha, S.K., Mukerjee, S.K. and Seshadri, T.R. (1974), *Indian J. Chem.* **12**, 10.
Kunesch, G., Ramiandrasoa, F., Kunesch, B. and Poisson, J. (1983), *Tetrahedron* **39**, 3922.
Lapper, R.D. (1974), *Tetrahedron Lett.*, 4293.
Manimaran, T. and Ramakrishnan, V.T. (1979), *Indian J. Chem.*, 324.
Monache, G.E., Botta, B., Neto, A.S. and de Lima, R.A. (1983), *Phytochemistry* **22**, 1657.
Monache, G.E., Botta, B. and de Lima, R.A. (1984), *Phytochemistry 23*, 1813.
Muangnoicharoen, N. and Frahm, A.N. (1982), *Phytochemistry* **21**, 767.
Nonaka, G. and Nishioka, I. (1982), *Chem. Pharm. Bull.* **30**, 4268.
Ollis, W.D. and Gottlieb, O.R. (1968), *J. Chem. Soc., Chem. Commun.*, 1396.
Ollis, W.D., Redman, B.T., Voberts, R.J. and Sutherland, I.O. (1968), *J. Chem. Soc., Chem. Commun.*, 1392.
O' Sullivan, M. (1982), Ph.D. thesis, National University of Ireland.
Pelter, A., Ward, R.S. and Gray, T.I. (1976), *J. Chem. Soc., Perkin Trans 1*, 2475.
Reher, G. and Kraus, Lj. (1984), *J. Nat. Prod.* **47**, 172.
Reher, G., Kraus, Lj, Sinnwell, V. and Konig, W.A. (1983), *Phytochemistry* **22**, 1524.
Shighiro, A., Sato, K., Asami, T., Amakasu, K., Itutura, T. and Nishio, N. (1970), *Chem. Abstr.* **72**, 7888.
Taylor, R.T. and Cassell, R.A. (1982), *Synthesis*, 672.
Thebtaranonth, C., Imraporn, S. and Padungkul, N. (1981), *Phytochemistry* **20**, 2305.
Ulubelen, A., Kerr, R.R. and Mabry, T.J. (1982), *Phytochemistry* **21**, 1145.
Wagner, H., Seligmann, O., Chari, M.V., Wollenweber, E., Deitz, V.H., Donnelly, D.M.X., Meegan, M. and O'Donnell, B. (1979), *Tetrahedron Lett.*, 4269.
Wollenweber, E., Dietz, V.H., Favre-Bonvin, J. and Gômez, L.D. (1980), *Z. Naturforsch.* **35c**, 36.

7

Flavones and flavonols

E. WOLLENWEBER and M. JAY

7.1 INTRODUCTION

In this chapter, emphasis is given to the natural distribution of flavones and flavonols occurring in the free state. The relevant data are compiled in Tables 7.1–7.7. In all, these tables list over 200 flavones and almost 300 flavonols with simple hydroxyl (or methoxyl) substitution patterns (Tables 7.1 and 7.2), thus stressing the earlier observation that flavonols predominate over flavones. Further, these tables contain more than 160 compounds with extra substituents (including compounds shown in Figs 7.3 and 7.4). Since the data for the previous edition (Wollenweber, 1982a) were compiled, some 40 flavones and 50 flavonols of the usual type have been reported as novel natural products (Tables 7.1 and 7.2), and some 50 novel flavonoids with extra substituents have also been added (Tables 7.3–7.7 and Figs 7.3 and 7.4).

For the sake of completeness, we list in our tables all flavone and flavonol aglycones known as natural products, no matter whether they were originally found in the plant in the free state or as a conjugate. From the literature reports, it is not always clear whether aglycones may not have been liberated during the extraction procedure, intentionally or not. The tables do not contain compounds that are so far only known as products of synthesis, and we do not cite plant sources for aglycones that have obviously been obtained following extraction into acidic medium. Details on the organization of Tables 7.1–7.7 are given in Section 7.1.2.

7.1.1 Occurrence and localization of flavonoid aglycones

In an earlier review on the distribution of flavonoid aglycones in plants (Wollenweber and Dietz, 1981), it was pointed out that, whenever aglycones are encountered as such in the free state, they usually occur externally on the plant surface, or else they are associated with secretory structures. It was also mentioned in the previous edition of this book (Wollenweber, 1982a) that their presence is very often correlated with the production of other lipophilic secondary products, mainly of terpenoid origin, or with 'waxes' in the poorly defined botanical sense. This is particularly true for aglycones which have few hydroxyl groups, be this due to high degree of methylation or to low degree of substitution (e.g. in *Primula* flavones). Their more or less lipophilic nature excludes their accumulation in the aqueous environment of the cell sap, and this leads to external accumulation as a consequence of glandulotropic or epidermal excretion (Hänsel, 1962; Rimpler, 1965). Flavonoid aglycones are, therefore, encountered in oil cells or oil cavities (*Citrus*), in and on glandular trichomes (*Myrica*, *Salvia*, *Primula*), in bud excretions (*Betula*, *Populus*, *Gardenia*), in thin epicuticular layers on leaves (*Escallonia*, *Kalmia*, *Callistemon*), in leaf wax (*Eucalyptus*) and most often in leaf resins (*Baccharis*, *Cistus*, *Dodonaea*, *Larrea* and many others). They have been reported less frequently free in the heartwood of trees.

From these observations, it can be assumed that in the majority of plants reported to contain flavonoid aglycones, these occur externally. When this is known, it is so indicated in Tables 7.1–7.7; unfortunately, in most flavonoid studies, no attempt is made to indicate where the substances are located within the plant. We have checked

The Flavonoids. Edited by J.B. Harborne
Published in 1988 by Chapman and Hall Ltd,
11 New Fetter Lane, London EC4P 4EE.

the relevant plants in a number of cases, at least with herbarium fragments, and have found that our assumption about external flavonoid accumulation has been corroborated almost every time. We therefore suggest that in future phytochemical studies all plant material be checked for external flavonoids prior to extraction of the internal constituents.

It is easy to quickly 'wash' a sample of the plant material under study (either fresh or dry, but not in a withered state) with organic solvent(s) such as acetone and/or diethyl ether, chloroform, hexane, and to test the concentrated solution(s) for the presence of flavonoids. If flavonoids are accumulated externally, it is advantageous to analyse the leaf wash aglycones alone, which can readily be separated from other lipophilic material, e.g. terpenoids, by a passage over Sephadex LH-20, eluting with methanol. Such concentration of the leaf washings provides a means of detecting flavonoid aglycones, even if they are present in low concentrations. This is not the place to discuss the correlations that exist between the external accumulation of flavonoid aglycones and plant habitat. It should be mentioned, though, that the phenomenon of flavonoid excretion so far appears to occur preferentially in plants of arid and semi-arid regions. There are many examples from the deserts of the southern United States and Mexico, but also from the Mediterranean, the mountains in central Europe, South Africa, the Himalayas and Australia (cf. Wollenweber, 1985b, c). Most of the plants concerned show xeromorphic adaptation, such as coarse leaves, reduced leaf surface, and production of 'wax' or leaf resin. These terpenoid/flavonoid mixtures appear to function as a UV screen, for heat reduction, as anti-microbial agents or as insect-feeding deterrents (Proksch and Rodriguez, 1985).

Accumulation of flavonoid aglycones is often correlated with the existence of secretory structures and the production of other lipophilic natural products. Thus exudate flavonoids are observed in distinct families and genera, scattered throughout the plant kingdom. Families for which we at present know with certainty that flavonoid aglycones accumulate externally are: Asteraceae (*Achillea, Artemisia, Baccharis, Brickellia, Chrysothamnus, Ericameria, Eupatorium, Flourensia, Haplopappus, Hazardia, Heterotheca* and many others), Bignoniaceae (*Phyllarthron*), Cistaceae (*Cistus*), Dicrastylidaceae (*Newcastelia*), Ericaceae (*Gaultheria, Kalmia*), Euphorbiaceae (*Bayeria*), Fabaceae (*Ononis*), Gesneriaceae (*Didymocarpus*), Hydrophyllaceae (*Eriodictyon, Wigandia*), Lamiaceae (*Salvia, Sideritis*), Mimosaceae (*Acacia*), Myricaceae (*Comptonia, Myrica*), Myrtaceae (*Eucalyptus*), Penaeaceae (*Saltera*), Rosaceae (*Adenostoma*), Rubiaceae (*Gardenia*), Sapindaceae (*Dodonaea*), Scrophulariaceae (*Mimulus*), Solanaceae (*Lycopersicon, Solanum*) and Zygophyllaceae (*Larrea*). The only monocot so far known to produce exudate flavonoids is *Barbacenia purpurea* (Velloziaceae).

7.1.2 Organization of the tables

In the first column of these tables, the flavonoids are listed according to the increasing number of hydroxyl groups in their basic substitution (1st cipher); the respective methyl ethers are indented (2nd cipher). Following previous practice, the structures are arranged by number and position, in ascending order, of substituents in ring A, followed by those in ring B. We recommend that this now generally adopted convention should always be followed. Where the basic structures themselves have not been found as natural products, they are in parenthesis. Products that do not exist, i.e. have been described erroneously, are cited in square brackets. In order to provide a complete list of all known flavones and flavonols, we also include those compounds that were known previously, though not reported in the period of this review. For the same reason, all compounds listed are numbered throughout the table, including erroneous structures. Aglycones that have so far been found in glycosidic combination only, are marked with a 'G'. Novel compounds, reported for the first time since 1981, are marked with an asterisk.

In the second column we cite trivial names whenever such names have been given. All trivial names are listed together at the end of this volume.

In the third column the plant species follow in alphabetical order. Only one species of a given genus is normally noted for an individual compound from a single publication; if there are more, 'spp.' is used. Families are indicated by abbreviations that should be generally understood. Plant sources of such widespread compounds as apigenin, luteolin or kaempferol, are given only in those cases in which it is clear that they occur externally on the plant surface. In this third column we also refer the reader to revisions as dealt with in Section 7.4; notes on the synthesis of individual compounds are indented. In the next column, the part of plant is indicated whenever this is known.

7.1.3 Comments on trivial names

Care should be taken in using trivial names for flavonoids. Coining new trivial names when these are already available is highly confusing. For instance, there is no need to call scutellarein 6-methyl ether (hispidulin) 6-methoxyapigenin, and other scutellarein derivatives 6-methoxyapigenin-7-methyl ether and 6-methoxyacacetin, as De Luengo and Mabry (1986) have done. The term 3,6-dimethoxyapigenin is quite misleading, since the compound being referred to is the 3,6-dimethyl ether of the flavonol 6-hydroxykaempferol (McCormick and Bohm, 1986). Also the name 8-hydroxychrysoeriol 7-methyl ether is unnecessary when it can be named 8-hydroxyluteolin 7,3′-dimethyl ether or hypolaetin 7,3′-dimethyl ether (Whalen and Mabry, 1979). Again, the use of 6,7-dimethoxyquercetin for quercetagetin 6,7-dimethyl ether (eupatolitin) is just simply incorrect (Yu *et al.*, 1986).

7.2 FLAVONOIDS WITH HYDROXYL AND/OR METHOXYL SUBSTITUTION

From Tables 7.1 and 7.2 it is clear that only a few flavones and flavonols with new types of oxygen substitution have been reported during 1981 to 1985. The 14 novel patterns are: mono-*O*-substitution at C-6, at C-7 and at C-3′ (as the methyl ether), 5,7,2′,3′-tetrahydroxyflavone, 5,7,2′,6′-tetrahydroxyflavone as the 6-methyl ether, four methyl ethers of 5,7,8,2′,6′-pentahydroxyflavone, a trimethyl ether of 5,6,7,8,2′,5′-hexahydroxyflavone, dimethyl ethers of 5,7,8,2′,3′,6′-, of 5,7,8,2′,4′,5′-, and of 5,7,8,2′,5′,6′-hexahydroxyflavones, the hexamethyl ether of 5,6,2′,3′,4′,6′-, hexahydroxyflavone, and three methyl ethers of 5,6,7,8,2′,3′,4′,5′-octahydroxyflavone. In the flavonol series, there are only the 8-methyl ether of 3,7,8-trihydroxyflavone and 3,5,7,2′,6′-pentamethoxyflavone.

Of the new flavonoids 72 are merely methyl ethers (sometimes acylated) of previously known compounds. Methyl ethers of such common substances as apigenin, luteolin, kaempferol and quercetin have been reported with increasing frequency. There are 21 new reports of apigenin methyl ether in plants, 25 for apigenin 4′-methyl ether, 40 for scutellarein 6-methyl ether, 30 for scutellarein 6,3′-dimethyl ether, 24 for kaempferol 3-methyl ether and 22 for quercetin 3-methyl ether, just to give a few examples.

Flavones and flavonols that are *O*-substituted at the 'normal' positions (3,5,7,4′) of the flavonoid molecule in terms of biosynthesis still predominate. A detailed evaluation of the structural diversity of the flavones and flavonols listed in Tables 7.1, 7.2 and 7.3 will be presented in Section 7.5.

7.2.1 Flavone and flavonol esters

Flavonoids in which acylation takes place at the sugar moiety are now well known (see Chapters 1 and 4) as are a series of acylated *C*-glycosides. Acylated aglycones, by contrast, are rather rare. The first natural flavonoids reported of this type were flavanones and dihydroflavonols (Wollenweber, 1985a). The first acylated flavonols reported were the acetate and the butyrate of 7,4′-dimethylherbacetin (*7.4a/b*) found in the farinose frond exudate of the fern *Notholaena affinis* ('NA-1/NA-2'; Jay *et al.*, 1978). Butyryl esters had not been previously described. Similar pairs of flavonol acetate/butyrate (*7.1*)–(*7.7*) have now been found in further *Notholaena* species (Wollenweber, 1982b, 1985a). All these flavonol esters are methylated at position 7 and acylated at position 8. This is true also for compound (*7.2a/b*) isolated recently from *Notholaena sulphurea* which differs from the others by having a hydroxyl group at C-2′ (Arriaga–Giner *et al.*, unpublished). Quercetin 3-*O*-isobutyrate (*7.8*) has been reported from leaf resin of *Traversia baccharoides*, along with the related 3′-*O*-isobutyrate (*7.9*) and 4′-*O*-isobutyrate (*7.10*) (Kulanthaivel and Benn, 1986), while the 4′-*O*-isobutyrate (*7.11*) and the 4′-*O*-isovalerate (*7.12*) of quercetin 3,3′-dimethyl ether occur in *Gutierrezia dracunculoides* (Bohlmann *et al.*, 1981a). The latter quercetin derivatives are the only flavonols so far known to be acylated at the B-ring. Chrysin 7-*O*-benzoate (*7.13*) from the leaf resin of *Baccharis*

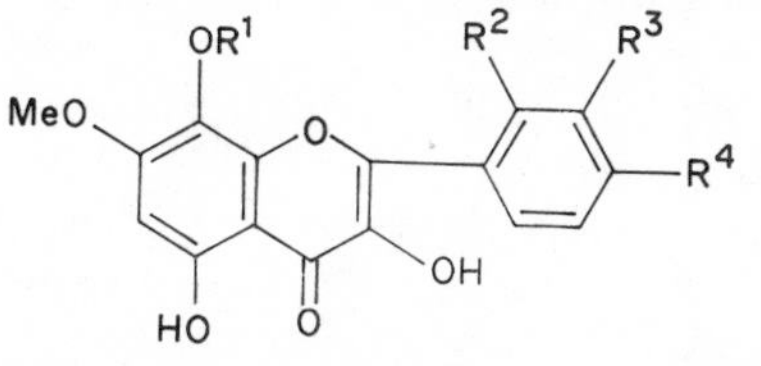

(*7.1*) $R^2=R^3=R^4=H$ (NG-1/2)
(*7.2*) $R^2=OH, R^3=R^4=H$ (NS-II-A/B)
(*7.3*) $R^2=R^3=H, R^4=OH$ (NAS-1/2)
(*7.4*) $R^2=R^3=H, R^4=OMe$ (NA-1/2)
(*7.5*) $R^2=H, R^3=OH, R^4=OMe$ (NA-3/4)

R^1 in (*7.1a*), (*7.2a*), (*7.3a*), (*7.4a*), (*7.5a*) = COPr
R^1 in (*7.1b*), (*7.2b*), (*7.3b*), (*7.4b*), (*7.5b*) = Ac

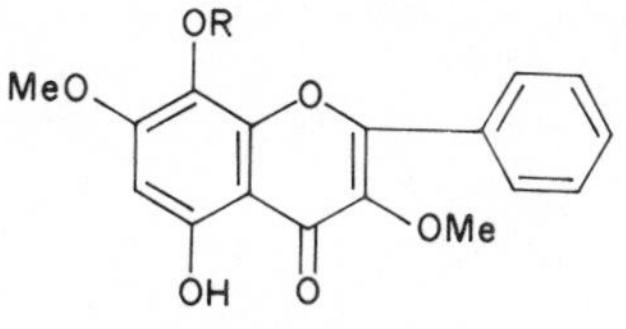

(*7.6*) R = COPr
(*7.7*) R = Ac

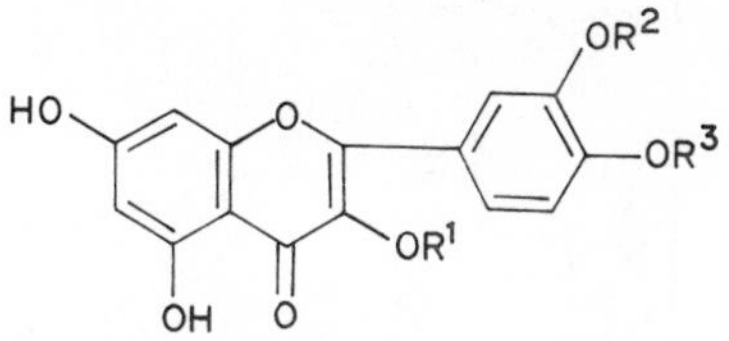

(*7.8*) $R^1=COCHMe_2, R^2=R^3=H$
(*7.9*) $R^1=R^3=H, R^2=COCHMe_2$
(*7.10*) $R^1=R^2=H, R^3=COCHMe_2$
(*7.11*) $R^1=R^2=Me, R^3=COCHMe_2$
(*7.12*) $R^1=R^2=Me, R^3=COCH_2CHMe_2$

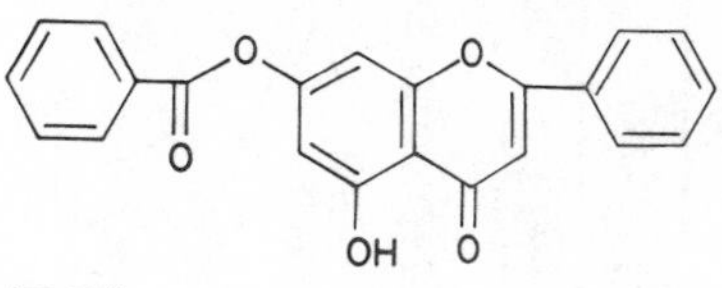

(*7.13*)

Fig. 7.1 Acylated flavonoid aglycones.

Table 7.1 Flavones

No.	Substitution	Trivial name	Plant source[†]		References
1	—	Flavone	*Pimelea decora, P. simplex* (Thymel.)	whole plant	Freeman *et al.* (1981)
	MONO-O-SUBSTITUTED FLAVONES				
2	5-OH	Primuletin			
3	(6-OH)				
3.1	*6-Me		*Pimelea decora* (Thymel.)	whole plant	Freeman *et al.* (1981)
4	(7-OH)				
4.1	*7-Me		*Pimelea simplex* (Thymel.)	whole plant	Freeman et al. (1981)
5	2′-OH				
5.1	*2′-Me		*Pimelea simplex* (Thymel.)	whole plant	Freeman *et al.* (1981)
			Primula kewensis (Prim.)	farinose exudate	Wollenweber and Mann (1986)
6	3′-OH				
6.1	*3′-Me		*Pimelea decora* (Thymel.)	whole plant	Freeman *et al.* (1981)
6A	(4′-OH)				
6A.1	*4′-Me		*Sapindus saponaria* (Sapind.)	seed	Wahab and Selim (1985)
	DI-O-SUBSTITUTED FLAVONES				
7	(5, 6-diOH)				
7.1	6-Me				
7.2	5, 6-diMe				
8	5, 7-diOH	Chrysin	*Acacia* spp. (Fab.)	leaf	Saeedi-Ghomi *et al.* (1984);
			Baccharis bigelovii (as 7-benzoate)	leaf resin	Arriaga–Giner *et al.* (1986)
			Comptonia peregrina (Myric.)	leaf exud.	Wollenweber *et al.* (1985b)
			Eucalyptus spp. (Myrt.)	leaf wax	Wollenweber and Kohorst (1981)
			Flourensia resinosa	leaf exud.	Wollenweber and Mann (1985)
			Hymenoclea salsola	leaf resin	Proksch *et al.* (1983)
			Passiflora serratodigitata (Passif.)	aerial parts	Ulubelen *et al.* (1982)
			Scutellaria ovata (Lam.)	flower	Nicollier *et al.* (1981)
8.1	7-Me	Tectochrysin	*Boesenbergia pandurata* (Zing.)	rhizome	Jaipetch *et al.* (1983)
			Comptonia peregrina (Myric.)	leaf exud.	Wollenweber *et al.* (1985a)
			Flourensia resinosa	leaf resin	Wollenweber and Mann (1985)
8.2	5, 7-diMe		*Boesenbergia pandurata* (Zing.)	rhizome	Jaipetch *et al.* (1983)
			synthesis:		Jain *et al.* (1982)
9	5, 8-diOH	Primetin			
10	5, 2′-diOH				
10.1	*2′-Me		*Primula kewensis* (Prim.)	farinose exudate	Wollenweber and Mann (1986)
11	(6, 3′-diOH)				
11.1	*6. 3′-diMe		*Pimelea decora* (Thymel.)	whole plant	Freemann *et al.* (1981)
12	(6, 4′-diOH)				
12.1	*4′-Me(G)				
13	*7, 4′-diOH		*Pterocarpus marsupium* (Fab.)	heartwood	Maurya *et al.* (1984)
13.1	*7-Me		*Trifolium hybridum* (Fab.)	roots	Fraishtat and Wulfron (1981)
13.2	4′-Me	Pratol			
14	3′, 4′-diOH				

TRI-O-SUBSTITUTED FLAVONES

No.	Substitution	Name	Source	Part	Reference
15	5, 6, 7-triOH	Baicalein	*Austrocylindropuntia subulata* (Cact.)	leaf + thorn	Burret *et al.* (1982)
			Scutellaria spp. (Lam.)	root, whole plant	Liu, Y.-L. *et al.* (1984), Tomimori *et al.* (1984a)
			synthesis:		Iinuma *et al.* (1984g)
15.1	6-Me	Oroxylin A	*Gomphrena martiana* (Amaranth.)	whole plant	Buschi *et al.* (1981)
			Scutellaria rehderiana (Lam.)	root	Liu, Y.-L. *et al.* (1984)
15.2	*7-Me	Negletein	*Centaurea clementei*	aerial parts	Collado *et al.* (1985)
15.3	*5, 6-diMe		*Gomphrena martiana* (Amaranth.)	whole plant	Buschi *et al.* (1981)
15.4	6, 7-diMe				
15.5	5, 6, 7-triMe				
16	5, 7, 8-triOH	Norwogonin	synthesis:		Ryu *et al.* (1985)
			Scutellaria spp. (Lam.)	root	Tomimori *et al.* (1983, 1985)
16.1	8-Me	Wogonin	*Scutellaria* spp. (Lam.)	root	Tomimori *et al.* (1985), Liu, Y.-L. *et al.* (1984), Chou *et al.* (1981)
16.2	*5, 8-diMe		*Scutellaria rivularis* (Lam.)	whole herb	Tomimori *et al.* (1984a)
16.3	7, 8-diMe		*Achyrocline bogotensis*	flower + leaf	Torrenegra *et al.* (1982)
			Andrographis paniculata (Acanth.)	root	Gupta *et al.* (1983)
			Scutellaria spp (Lam.)	root	Tomimori *et al.* (1983, 1985), Chou *et al.* (1981)
16.4	5, 7, 8-triMe		*Achyrocline albicans*	leaf	Mesquita *et al.* (1986)
17	(5, 6, 2′-triOH)				
17.1	5, 6, 2′-triMe		*Sargentia greggii* (Rut.)	root	Meyer *et al.* (1985)
18	(5, 6, 3′-triOH)				
18.1	5, 6, 3′-triMe				
19	*5, 7, 2′-triOH				
19.1	7-Me	Echioidinin	*Scutellaria baicalensis* (Lam.)	root	Tomimori *et al.* (1984c)
20	5, 7, 4′-triOH	Apigenin	Many records, in Asteraceae, Cactaceae, Cistaceae, Fabaceae, Lamiaceae, Liliaceae, Pteridaceae, Scrophulariaceae and Verbenaceae		
20.1	*6-chloro	6-Chloroapigenin	*Equisetum arvense* (Equis.)		Syrchina *et al.* (1980)
20.2	5-Me	Thevetiaflavon			
20.3	7-Me	Genkwanin	Records in Asteraceae, Cistaceae, Eupomataceae, Lamiaceae, Pteridaceae and Saxifragaceae		
20.4	4′-Me	Acacetin	Many records, in Asteraceae, Cistaceae, Clusiaceae, Fabaceae, Liliaceae, Pteridaceae and Scrophulariaceae		

(Contd.)

Table 7.1 (*Contd.*)

No.	*Substitution*	*Trivial name*	*Plant source*[†]		*References*
20.5	5,7-diMe (G)				
20.6	7,4′-diMe		*Baccharis* spp.	aerial parts	Gianello and Giordano (1984), Silva and Mundaca (1971), San Martin *et al.* (1982), Arriaga–Giner *et al.* (1982), González *et al.* (1985), Wollenweber *et al.* (1986)
			Ballota pseudodictamus (Lam.)	aerial parts	Savona *et al.* (1982)
			Boesenbergia pandurata (Zing.)	rhizom	Jaipetch *et al.* (1983)
			Cistus, 4 spp. (Cist.)	leaf resin	Wollenweber and Mann (1984)
			Ericameria, 2 spp.	leaf resin	Clark and Wollenweber (1984)
			Escallonia pulverulenta (Saxif.)	epicut. 1.	Wollenweber (1984a)
			Flourensia resinosa	leaf resin	Wollenweber and Yatskievych (1985)
			Frullania spp. (Hepaticae)	whole plant	Mues *et al.* (1984), Asakawa *et al.* (1980)
			Garcinia kola (Clus.)	seed	Iwu and Igboko (1982)
			Gonystylus bancanus (Thymel.)	heartwood	Ahmad (1983)
			Hieracium intybacium	leaf resin	Wollenweber (1984c)
			Notholaena pallens (Pterid.)	frond exud.	Wollenweber (1984b)
			Rhus undulata (Anacard.)	root	Fourie and Snyckers (1984)
			Salvia palaestina (Lam.)	leaf	Miski *et al.* (1983b)
			Striga asiatica (Scroph.)	whole plant	Nakanishi *et al.* (1985)
			Tasmannia spp. (Wint.)	leaf	Williams and Harvey (1982)
			Thymus piperella (Lam.)	aerial parts	Barberán *et al.* (1985c)
20.7	5,7,4′-triMe		*Boesenbergia pandurata* (Zing.)	rhizome	Jaipetch *et al.* (1983)
			Garcinia kola (Guttif.)	seed	Iwu and Igboko (1982)
21	5,8,2′-triOH				
22	(6,2′,3′-triOH)				
22.1	*6,2′,3′-triMe		*Pimelea decora* (Thymel.)	whole plant	Freeman *et al.* (1981)
23	7,3′,4′-triOH		*Carex longebrachiata* (Cyp.)	leaf	Harborne *et al.* (1985)
			Cyperus spp. (Cyp.)	leaf + infl.	Harborne *et al.* (1982)
			Schoenus spp. (Cyp.)	leaf	Harborne *et al.* (1985)
			Uncinia sp. aff. *compacta* (Cyp.)	leaf	Harborne *et al.* (1985)
23.1	3′-Me	Geraldone			
23.2	7,4′-diMe	Tithonine	*Tithonia*, 3 spp.	vegetative parts	La Duke (1982)
	TETRA-O-SUBSTITUTED FLAVONES				
24	(5,6,7,8-tetraOH)				
24.1	6,7-diMe		*Gnaphalium gaudichaudianum*	aerial parts	Guerreiro *et al.* (1982)
			Helichrysum mimetes	aerial parts	Jakupovic *et al.* (1986)
			Scutellaria baicalensis (Lam.)	root	Tomimori *et al.* (1982a)
			synthesis:		Ryu *et al.* (1985)

No.	Substitution	Trivial name	Source	Plant part	Reference
24.2	[6, 8-diMe]		See Section 7.4.1		
24.3	5, 6, 7-triMe				
24.4.	6, 7, 8-triMe	Alnetin	synthesis:		Ryu *et al.* (1985)
24.5	5, 6, 7, 8-tetraMe				
25	(5, 6, 7, 2′-tetraOH)				
25.1	*6-Me		*Scutellaria baicalensis* (Lam.)	root	Tomimori *et al.* (1983)
			synthesis:		Iinuma *et al.* (1984)
26	5, 6, 7, 4′-tetraOH	Scutellarein	*Asphodeline* spp. (Liliac.)	leaf + fruit	Ulubelen and Tuzlaci (1985)
			Digitalis orientalis (Scroph.)	leaf	Imre *et al.* (1984)
			Pulicaria dysenterica	leaf	Pares *et al.* (1981)
			Scutellaria rivularis (Lam.)	root	Chou *et al.* (1981)
26.1	6-Me	Hispidulin	*Arnica*, 4 spp.	flowerhead	Wolf and Denford (1984), Merfort (1984a, b, 1985).
			Artemisia, 5 spp.	aerial parts	Hurabielle *et al.* (1982) DePasqual–Teresa *et al.* (1986) Liu and Mabry (1981), McCormick and Bohm (1986), Liu and Mabry (1982), Cordano *et al.* (1982), Wollenweber *et al.* (1986)
			Baccharis, 2 spp.	ariel parts	
			Berlandiera texana var. *texana*	aerial parts	Bulinska-Radomska *et al.* (1985b)
			Brickellia baccharoidea	leaf	Timmermann *et al.* (1981)
			Centaurea 4 spp.	aerial parts	Collado *et al.* (1985) Öksüz *et al.* (1984) Ulubelen *et al.* (1982)
			Chenopodium botrys (Chenopod.)	aerial parts	De Pasqual–Teresa *et al.* (1981)
			Citrus sudachii (Rut.)	green peel	Horie and Nakayama (1981)
			Clerodendron phlomidis (Verb.)	flower	Seth *et al.* (1982)
			Decachaeta ovalifolia	leaf + flower	De Luengo and Mabry (1986)
			Digitalis orientalis (Scroph.)	leaf	Imre *et al.* (1984)
			Elytropappus rhinocerotis	leaf resin	P. Proksch *et al.* (1982)
			Eriodictyon angustifolium (Hydrophyll.)	leaf resin	Wollenweber (unpublished)
			Eriocephalus punctulatus	leaf resin	Wollenweber (unpublished)
			Eupatorium 2 spp.	aerial parts	Herz *et al.* (1979, 1981)
			Geigeria burkei	aerial parts	Coleman *et al.* (1984)
			Haplopappus canescens	aerial parts	Öksüz *et al.* (1981)
			Hazardia squarrosa var. *grindelioides*	leaf resin	Clark and Wollenweber (1985)
			Helianthus spp.	leaf + flower	Schilling and Mabry (1981), Schilling (1983)
			Inula viscosa	aerial parts exudate	Grande *et al.* (1985)
			Jasonia tuberosa	aerial parts	González *et al.* (1977)
			Lourteigia ballotrefolia	aerial parts	Triana (1984)
			Nama, 2 spp. (Hydrophyll.)	aerial parts	Bacon *et al.* (1986)
			Salvia plebeja – see Section 7.4.1 (no. 28.1)		

(*Contd.*)

Table 7.1 (*Contd.*)

No.	*Substitution*	*Trivial name*	*Plant source*[†]		*References*
			Santolina chamaecyparissus	aerial parts	Becchi and Carrier (1980)
			Scutellaria ovata (Lam.)	flower	Nicollier *et al.* (1981)
			Stevia 3 spp.		Rajbhandari and Roberts (1984a, 1985a)
			Tithonia, 2 spp.	veg. parts	La Duke (1982)
			synthesis:		Iinuma *et al.* (1984g)
26.2	7-Me	Sorbifolin			
26.3	4′-Me				
26.4	6, 7-diMe	Cirsimaritin	*Achillea depressa*	leaf	Tsankova and Ognyanov (1985)
			Aegialophila pumila	aerial parts	Raffauf *et al.* (1981)
			Arnica 2 spp.		Wolf and Denford (1984)
			Artemisia 3 spp.	aerial parts/ infl.	Bouzid *et al.* (1982a)
					Chandrasekharan *et al.* (1981)
					Chemesova *et al.* (1984)
			Baccharis 4 spp.	aerial parts	Gianello and Giordano (1984) Mesquita *et al.* (1986) Kuroyanagi *et al.* (1985) Wollenweber *et al.* (1986)
			Bromelia pinguin (Brom.)	root + basal stem	Raffauf *et al.* (1981)
			Centaurea urvillei	leaf	Ulubelen and Öksüz (1982)
			Cirsium takaoense – See Section 7.4.1 (no. 28.5)		
			Decachaeta ovalifolia	leaf + flower	De Luengo and Mabry (1986)
			Elytropappus rhinocerotis	leaf resin	P. Proksch *et al.* (1982)
			Heterotheca psammophila	leaf resin	Wollenweber *et al.* (1985c)
			Nama lobbii (Hydrophyll.)	aerial parts	Bacon *et al.* (1986)
			Neurolaena venturano	leaf	Ulubelen *et al.* (1981a)
			Notholaena grayi (Pterid.)	frond exud.	Wollenweber (1984b)
			Orthosiphon spicatus (Lam.)	leaf	Jiyu and Zongshi (1984), Wollenweber and Mann (1985)
			Phacelia ixodes (Hydrophyll.)	leaf	Reynolds and Rodriguez (1981)
			Plectranthus ecklonii (Lam.)	leaf glands	Uchida *et al.* (1980)
			Salvia 2 spp. (Lam.)	leaf	Miski *et al.* (1983b) Ulubelen and Topcu (1984)
			Sideritis 2 spp. (Lam.)	aerial parts	Barberán *et al.* (1984c, 1985b)
			Stevia saturaeifolia	aerial parts	Sosa *et al.* (1984)
			Teucrium gnaphaloides (Lam.)	whole plant, ext.!	Barberán *et al.* (1985e)
			Thymus 2 spp. (Lam.)	aerial parts	Adzet and Martinez–Verges (1980) Ferreres *et al.* (1985b)
26.5	6, 4′-diMe	Pectolinarigenin	*Arnica chamissonis*	flowerhead	Merfort (1984a b, 1985)
			Baccharis vaccinioides	aerial parts, exudate	Wollenweber *et al.* (1986)

No.	Substitution	Trivial name	Source	Organ	Reference
			Clerodendron phlomidis (Verb.)	flower	Seth *et al.* (1982)
			Decachaeta ovalifolia	leaf + flower	De Luengo and Mabry (1986)
			Digitalis orientalis (Scroph.)	leaf	Imre *et al.* (1984)
			Eupatorium 2 spp.	aerial parts	Herz *et al.* (1979, 1981)
			Hazardia squarrosa	leaf resin	Clark and Wollenweber (1985)
			Helenium integrifolium	aerial parts	Ozawa *et al.* (1983)
			Helianthus gracilentus	leaf trichomes	Melek *et al.* (1985)
			Heterotheca psammophila	leaf resin	Wollenweber *et al.* (1985c)
			Nama 2 spp. (Hydrophyll.)	aerial parts	Bacon *et al.* (1986)
			Santolina chamaecyparissus	aerial parts	Becchi and Carrier (1980)
			synthesis:		Horie *et al.* (1985)
26.6	7, 4′-diMe	Ladanein	*Baccharis tucumanensis*	aerial parts	Tonn *et al.* (1982)
			Centaurea clementei	aerial parts	Collado *et al.* (1985)
			Thymus piperella (Lam.)	aerial parts	Barberán *et al.* (1985c)
26.7	*5, 6, 4′-triMe (G)		synthesis:		Horie *et al.* (1985)
26.8	6, 7, 4′-triMe	Salvigenin	*Achillea*, 4 spp.	aerial parts	Valant–Vetschera and Wollenweber (1986), Tsankova and Ognyanov (1985)
			Agastache micrantha (Lam.)	aerial parts	Sanders *et al.* (1980)
			Baccharis articulata	aerial parts	Gianello and Giordano (1984)
			Brickellia baccharoidea	leaf	Timmermann *et al.* (1981)
			Centaurea urvillei	leaf	Ulubelen and Öksüz (1982)
			Chenopodium botrys (Chenopod.)	aerial parts	De Pasqual–Teresa *et al.* (1981)
			Eupatorium semiserratum	aerial parts	Herz *et al.* (1981)
			Heterotheca psammophila	leaf resin	Wollenweber *et al.* (1985c)
			Kickxia spuria (Scroph.)	aerial parts	Tóth *et al.* (1980)
			Ocimum canum (Lam.)	leaf + infl.	Xaasan *et al.* (1980)
			Orthosiphon spicatus (Lam.)	leaf	Jiyu and Zongshi (1984) Wollenweber and Mann (1985)
			Salvia palaestina (Lam.)	leaf	Miski *et al.* (1983b)
			Sideritis congesta (Lam.)	aerial parts	Sezik and Ezer (1984)
26.9	5, 6, 7, 4′-tetraMe		*Teucrium gnaphalodes* (Lam.)	whole plant, (ext.)	Barberán *et al.* (1986e)
			Chenopodium botrys (Lam.)	aerial parts	De Pasqual–Teresa *et al.* (1981)
			Kickxia spuria (Scroph.)	aerial parts	Tóth *et al.* (1980)
			Mandarina satsuma (Rut.)	fruit peel	Ferreres *et al.* (1981)
			Orthosiphon spicatus (Lam.)	leaf	Wollenweber and Mann (1985)
27	(5, 7, 8, 2′-tetraOH)				
27.1	*7-Me		*Scutellaria baicalensis* (Lam.)	root	Takagi *et al.* (1980)
27.2	*8-Me	Scutevulin	*Scutellaria* 3 spp. (Lam.)	root	Tomimori *et al.* (1984b, 1985), Chou *et al.* (1981)
			synthesis:		Iinuma *et al.* (1984g)
27.3	7, 8-diMe	Skullcapflavon I (revised structure)	*Notholaena neglecta* (Pterid.)	farinose exudate	Scheele *et al.* (1987)
			Scutellaria rivularis (Lam.)	whole herb	Tomimori *et al.* (1984a)
			synthesis:		Bhardwaj *et al.* (1981a)
27.4	*8, 2′-diMe		*Scutellaria* 2 spp. (Lam.)	root, whole herb	Tomimori *et al.* (1984a, 1985)
27.5	*5, 8, 2′-triMe				
27.6	7, 8, 2′-triMe		*Scutellaria discolor* (Lam.)	root	Tomimori *et al.* (1986)
			synthesis:		Bhardwaj *et al.* (1981a)

(*Contd.*)

Table 7.1 (*Contd.*)

No.	Substitution	Trivial name	Plant source†		References
28	5, 7, 8, 4′-tetraOH	Isoscutellarein	synthesis:		Horie *et al.* (1983)
			See Section 7.4.1		
28.1	7-Me	[Salvitin]	*Artemisia campestris*	leaf + infl.	Omar and El-Ghazouly (1984)
			Doronicum grandiflorum	aerial parts	Reynaud *et al.* (1983)
28.2	8-Me		*Scutellaria* 2 spp. (Lam.)	root	Tomimori *et al.* (1982a), Chou *et al.* (1981)
			synthesis:		Iinuma *et al.* (1984g)
28.3	4′-Me (G)	Takakin	*Scutellaria baicalensis* (Lam.)	leaf	Takido *et al.* (1976)
			Veronica filiformis (Scroph.)	whole plant	Chari *et al.* (1981)
			synthesis:		Horie *et al.* (1983)
28.4	7, 8-diMe	Bucegin	*Bucegia romanica* (Marchant.; Hepat.)		Markham and Mues (1983)
28.5	8, 4′-diMe (G)	[Cirsitakaogenin]	See Section 7.4.1		Markham (1983)
			synthesis:		Nakayama *et al.* (1983)
28.6	7, 8, 4′-triMe				
28.7	5, 7, 8, 4′-tetraMe				
29	(5, 6, 2′, 6′-tetraOH)				
29.1	6, 2′, 6′-triMe	Zapotinin			
29.2	5, 6, 2′, 6′-tetraMe	Zapotin	*Sargentia greggii* (Rut.)	root	Meyer *et al.* (1985)
30	(5, 6, 3′, 5′-tetraOH)				
30.1	5, 6, 3′, 5′-tetraMe	Cerosillin	*Scutellaria baicalensis* (Lam.)	root	Tomimori *et al.* (1984b)
31	*5, 7, 2′, 3′-tetraOH				
32	5, 7, 2′, 4′-tetraOH	Norartocarpetin			
32.1	7-Me	Artocarpetin			
32.2	5, 7, 2′, 4′-tetraMe	[Argemexitin]	See Section 7.4.1		
33	(5, 7, 2′, 6′-tetraOH)				
33.1	*6′-Me		*Scutellaria baicalensis* (Lam.)	root	Tomimori *et al.* (1982a,1984b)
34	5, 7, 3′, 4′-tetraOH	Luteolin	Many records in Asteraceae, Cistaceae, Fabaceae, Hepaticae, Lamiaceae, Passifloraceae, Scrophulariaceae and Yerbenaceae		
34.1	5-Me		in 27 Austr. Cyperaceae		Harborne *et al.* (1985)
			Cyperus spp. (Cyp.)	leaf + infl.	Harborne *et al.* (1982)
34.2	7-Me		*Baccharis* 2 spp.	aerial parts	Bohlmann *et al.* (1985b)
			Hazardia squarrosa var. *grindelioides*	leaf resin	Clark and Wollenweber (1985)
			Thymus membranaceus ssp. *membranaceus* (Lam.)	aerial parts	Ferreres *et al.* (1985b)
34.3	3′-Me	Chrysoeriol	Many records in Asteraceae, Cyperaceae, Fabaceae, Hydrophyllaceae, Lamiaceae, Passifloraceae, Penaeaceae and Scrophulariaceae		
34.4	4′-Me	Diosmetin	*Agastache aurantiaca* (Lam.)	aerial parts	Exner *et al.* (1981a)
			Arnica 2 spp.	flowerhead	Merfort (1984b), Wolf and Denford (1984)

No.	Substitution	Name	Source	Organ	Reference
			Baccharis bigelovii	aerial parts exudate	Wollenweber *et al.* (1986)
			Haplopappus canescens	aerial parts	Öksüz *et al.* (1981)
			Salvia tomentosa (Lam.)	leaf	Ulubelen *et al.* (1981b)
			Thymbra capitata (Lam.)	aerial parts	Barberán *et al.* (1986b)
34.5	5, 3′-diMe		*Medicago varia* (Fab.)	seed	Gehring and Geiger (1980)
34.6	7, 3′-diMe	Velutin	*Belliolum* spp. (Wint.)	leaf	Williams and Harvey (1982)
			Bubbia spp. (Wint.)	leaf	Williams and Harvey (1982)
			Chromolaena meridensis	aerial parts	Amaro and Méndez (1983)
			Cistus 2 spp. (Cist.)	leaf resin	Wollenweber and Mann (1984)
			Drimys winteri (Wint.)	leaf	Williams and Harvey (1982)
			Escallonia pulverulenta (Saxif.)	epicut. 1.	Wollenweber (1984a)
			Eupomatia laurina (Eupom.)	leaf	Young (1983)
			Exospermum stipitatum (Wint.)	leaf	Williams and Harvey (1982)
			Heterotheca grandiflora	leaf resin	Wollenweber *et al.* (1985c)
			Nama rothrockii	aerial parts	Bacon *et al.* (1986)
			Tasmannia spp. (Wint.)	leaf	Williams and Harvey (1982)
			Trichophorum cespitosum (Cyp.)	stem	Salmenkallio *et al.* (1982)
			Zygogynum spp. (Wint.)	leaf	Williams and Harvey (1982)
34.7	7, 4′-diMe	Pilloin	*Belliolum gracile* (Wint.)	leaf	Williams and Harvey (1982)
			Orthosiphon spicatus (Lam.)	leaf	Wollenweber and Mann (1985)
			Salvia palaestina (Lam.)	leaf	Miski *et al.* (1983b)
			Tasmannia piperita (Wint.)	leaf	Williams and Harvey (1982)
			Zygogynum spp. (Wint.)	leaf	Williams and Harvey (1982)
34.8	3′, 4′-diMe		*Artemisia* 2 spp.	aerial parts	Liu and Mabry (1981, 1982)
			Baccharis 2 spp.	aerial parts exudate	Wollenweber *et al.* (1986) Arriaga–Giner *et al.* (1982)
			Frullania 2 spp. (Hepaticae)	whole plant	Mues *et al.* (1984)
			Striga asiatica (Scroph)	whole plant	Nakanishi *et al.* (1985)
34.9	7, 3, 4′-triMe		*Frullania* 2 spp. (Hepaticae)	whole plant	Mues *et al.* (1984)
			Striga asiatica (Scroph.)	whole plant	Nakanisi *et al.* (1985)
34.10	5, 7, 3′, 4′-tetraMe		*Bauhinia championii* (Fab.)	root	Chen *et al.* (1984)
			Boesenbergia pandurata (Zing.)	rhizome	Jaipetch *et al.* (1983)
			synthesis:		Jain *et al.* (1982)
35	(6, 7, 3′, 4′-tetraOH)				
35.1	6, 4′-diMe	Abrectorin			
35A	*7, 3′, 4′, 5′-tetraOH		*Evolvulus nummularius* (Convolv.)	whole plant	Gupta *et al.* (1984)
	PENTA-O-SUBSTITUTED FLAVONES				
36	(5, 6, 7, 8, 2′-pentaOH)				
36.1	*6, 7-diMe		*Scutellaria baicalensis* (Lam.)	root	Takagi *et al.* (1980)
36.2	*6, 7, 8-TriMe		*Scutellaria* 2 spp. (Lam.)	root	Tomimori *et al.* (1983), Lui *et al.* (1984)
			synthesis:		Ryu *et al.* (1985)
37	(5, 6, 7, 8, 4′-pentaOH)				
37.1	(6, 7-diMe)	(Isothymusin)	(formed on acid treatment of Thymusin)		Ferreres *et al.* (1985a)
37.2	6, 8-diMe	Desmethoxysudachitin	*Ambrosia deltoidea*	leaf resin	Wollenweber (unpublished)
37.3	*7, 8-diMe	Thymusin	*Thymus membranaceus* (Lam.)	aerial parts	Ferreres *et al.* (1985a)
			synthesis:		Horie *et al.* (1980)

(Contd.)

Table 7.1 (*Contd.*)

No.	Substitution	Trivial name	Plant source[†]		References
37.4	6, 7, 8-triMe	Xanthomicrol	*Ambrosia deltoidea*	leaf resin	Wollenweber *et al.* (1987b)
			Baccharis 3 spp.	aerial parts root	Silva *et al.* (1985) Bohlmann *et al.* (1981c) Wollenweber *et al.* (1986) Tonn *et al.* (1982)
			Citrus sudachii (Rut.)	green peel	Horie and Nakayama (1981)
			Mentha piperita (Lam.)	leaf	Jullien *et al.* (1984)
			Sideritis spp. (Lam.)	aerial parts	Barberán *et al.* (1984c, 1985b)
			Thymus spp. (Lam.)	aerial parts	Adzet and Martinez–Verges (1980) Ferreres *et al.* (1985b)
37.5	*6, 7, 4′-triMe	Pedunculin	*Tithonia*, 5 spp.	vegetative parts	La Duke (1982)
37.6	6, 8, 4′-triMe	Nevadensin	*Helianthus*, 5 spp.	leaf, gland. trichomes	Schilling (1983), Herz and Kulanthaivel (1982)
37.7	*7, 8, 4′-triMe		*Stevia berlandieri*	see Section 7.4.1	
			Thymus piperella (Lam.)	aerial parts	Barberán *et al.* (1985c)
37.8	5, 6, 8, 4′-tetraMe				
37.9	6, 7, 8, 4′-tetraMe	Gardenin B	*Citrus reticulata* (Rut.)	fruit peel	Silva and Mundaca (1971)
			Hyptis tomentosa (Lam.)	aerial parts	Kingston *et al.* (1979)
			Mentha piperita (Lam.)	leaf	Jullien *et al.* (1984)
37.10	5, 6, 7, 8, 4′-pentaMe	Tangeretin	*Mandarina satsuma* (Rut.)	fruit peel	Ferreres *et al.* (1981)
38	5, 6, 7, 3′, 4′-pentaOH	6-Hydroxyluteolin			
38.1	6-Me	Nepetin	*Ambrosia* 3 spp.	leaf	Wollenweber *et al.* (1987b)
			Arnica 2 spp.		Merfort (1984b), Wolf and Denford (1984)
			Artemisia 3 spp.	aerial parts	Liu and Mabry (1981, 1982), McCormick and Bohm (1986),
			Asphodeline 2 spp. (Liliac.)	leaf + fruit	Ulubelen and Tuzlaci (1985)
			Berlandiera texana	aerial parts	Bulinska-Radomska *et al.* (1985b)
			Centaurea 4 spp.	leaf/flowerhead	Wollenweber (unpublished), Picher *et al.* (1984), Oksüz *et al.* (1984), Ulubelen and Öksüz (1982)
			Elytropappus rhinocerotis	leaf resin	P. Proksch *et al.* (1982)
			Ericameria 2 spp.	leaf resin	Clark and Wollenweber (1984)
			Geigeria burkei	aerial parts	Coleman *et al.* (1984)
			Haplopappus scrobiculatus	aerial parts	Ates *et al.* (1982)
			Hazardia squarrosa	leaf resin	Clark and Wollenweber (1985)
			Helianthus, 3 spp.	leaf, gland. trichomes	Schilling (1983)
			Lippia nodiflora (Verb.)		Vivar *et al.* (1982)
			Lourteigia ballotrefolia	aerial parts	Triana (1984)
			Phoebanthus teanifolius	leaf	Schilling and Rieseberg (1985)
			Salvia pinnata (Lam.)		Ulubelen and Topcu (1984)
			Santolina chamaecyparissus	aerial parts	Becchi and Carrier (1980)

No.	Substitution	Name	Source	Organ	Reference
38.2	7-Me	Pedalitin	*Frullania* 2 spp. (Hepaticae)	whole plant	Mues *et al.* (1983, 1984)
			Sullivantia spp. (Saxif.)	leaf	Neuman *et al.* (1981)
38.3	3′-Me	Nodifloretin			
38.4	*4′-Me				
38.5	*5, 6-diMe				
38.6	6, 7-diMe	Cirsiliol	*Arnica viscosa*		Wolf and Denford (1984)
			Anisomeles ovata (Lam.)	aerial parts	Rao *et al.* (1983a)
			Aegiophila pumila	aerial parts	El- Masry *et al.* (1981)
			Agastache 2 spp. (Lam.)	aerial parts	Exner *et al.* (1981a), Sanders *et al.* (1980)
			Artemisia 2 spp.	aerial parts	Omar and El-Ghazouly (1984), Chemesova *et al.* (1984)
			Baccharis genistelloides	aerial parts	Kuroyangi *et al.* (1985)
			Centaurea urvillei	leaf	Ulubelen and Öksüz (1982)
			Haplopappus 2 spp.	aerial parts	Ates *et al.* (1982)
			Neurolaena venturana	leaf	Ulubelen *et al.* (1981a)
			Sideritis spp. (Lam.)	aerial parts	Barberán *et al.* (1984c, 1985b)
			Teucrium gnaphalodes (Lam.)	whole plant, (ext.)	Barberán *et al.* (1985e)
38.7	6, 3′-diMe	Jaceosidin	*Arnica* 2 spp.	flowerhead	Merfort (1984a, b)
			Artemisia 3 spp.	aerial parts	Liu and Mabry (1981, 1982), Lao *et al.* (1983)
			Baccharis myrtilloides	aerial parts	Gianello and Giordano (1984)
			Centaurea 4 spp.	aerial parts	Amer *et al.* (1984), Öksüz *et al.* (1984)
			Citrus sudachii (Rut.)	green peel	Horie and Nakayama (1981)
			Digitalis orientalis (Scroph.)	leaf	Imre *et al.* (1984)
			Eriocephalus punctulatus	leaf resin	Wollenweber (unpublished)
			Eupatorium leucolepsis	aerial parts	Herz and Kulanthaivel (1982)
			Helianthus, 3 spp.	leaf, gland. trichomes	Schilling (1983)
			Phoebanthus teanifolius	leaf	Schilling and Rieseberg (1985)
			Tanacetum vulgare	flowerhead	Ognyanov and Todorova (1983)
			Tillandsia utriculata (Brom.)	leaf	Ulubelen and Mabry (1982)
			Volutarella divaricata	aerial parts	Forgacs *et al.* (1981)
38.8	6, 4′-diMe	Desmethoxycentaureidin	*Arnica*, 3 spp.		Wolf and Denford (1984)
			Brickellia baccharidea	leaf	Timmermann *et al.* (1981)
			Haplopappus scrobiculatus	aerial parts	Ates *et al.* (1982)
38.9	7, 3′-diMe		*Nama* 2 spp. (Hydrophyll.)	aerial parts	Bacon *et al.* (1986).
			Thymbra 2 spp. (Lam.)	aerial parts	Barberán *et al.* (1986b), Miski *et al.* (1983a)
			Thymus satureioides (Lam.)	aerial parts	Voirin *et al.* (1985)
38.10	7, 4′-diMe				
38.11	6, 7, 3′-triMe	Cirsilineol	*Aegialophila pumila*	aerial parts	El-Masry *et al.* (1981)
			Anisomeles ovata (Lam.)	aerial parts	Rao *et al.* (1983b)
			Arnica viscosa		Wolf and Denford (1984)
			Artemisia rubripes		Lao *et al.* (1983)
			Heterotheca grandiflora	leaf resin	Wollenweber *et al.* (1985c)
			Salvia tomentosa (Lam.)	leaf	Ulubelen *et al.* (1981b)
			Sideritis spp. (Lam.)	aerial parts	Barberán *et al.* (1984c, 1985b)
			Tanacetum santolinoides	aerial parts	El-Din *et al.* (1985)
			Teucrium gnaphalodes (Lam.)	whole plant (ext.)	Barberán *et al.* (1985e)
			Thymus vulgaris (Lam.)	leaf	Van Den Broucke *et al.* (1982)
			Tillandsia utriculata (Brom.)	leaf	Ulubelen and Mabry (1982)

(*Contd.*)

Table 7.1 (*Contd.*)

No.	Substitution	Trivial name	Plant source†		References
38.12	6,7,4′-triMe	Eupatorin	*Achillea* 3 spp.	aerial parts	Valant–Vetschera and Wollenweber (1986), Sanz *et al.* (1985)
			Baccharis 2 spp.	aerial parts	Kuroyanagi *et al.* (1985), Santos *et al.* (1980)
			Brickellia baccharidea	leaf	Timmermann *et al.* (1981)
			Centaurea 3 spp.	aerial parts	Zapesochnaya *et al.* (1977), Öksüz *et al.* (1984)
			Eupatorium semiserratum	aerial parts	Herz *et al.* (1981)
			Hyptis tomentosa (Lam.)	aerial parts	Kingston *et al.* (1979)
			Orthisiphon spicatus (Lam.)	leaf	Jiyu and Zongshi (1984), Wollenweber and Mann (1985)
			Stevia spp.	aerial parts	Rajbandhari and Roberts (1984a, 1985a), Sosa *et al.* (1984, 1985)
			Teucrium pseudochamaepitys (Lam.)	aerial parts	Savona *et al.* (1979)
			Thymus membranaceus (Lam.)	aerial parts	Ferreres *et al.* (1985b)
38.13	6,3′,4′-triMe	Eupatilin	*Artemisia rubripes*		Lao *et al.* (1983)
			Centaurea alexandrina	aerial parts	Amer *et al.* (1984)
			Eriocephalus punctulatus	leaf resin	Wollenweber (unpublished)
			Salvia palaestina (Lam.)	leaf	Miski *et al.* (1983b)
			synthesis:		Horie *et al.* (1985)
38.14	*7,3′,4′-triMe		*Thymbra* 2 spp. (Lam.)	aerial parts	Barberán *et al.* (1986b), Miski *et al.* (1983a)
			Thymus piperella (Lam.)	aerial parts	Barberán *et al.* (1985c)
38.15	5,6,7,4′-tetraMe				
38.16	6,7,3′,4′-tetraMe		*Achillea*, 7 spp.	aerial parts	Valant–Vetschera and Wollenweber (1986) Khafagy *et al.* (1976) Sanz *et al.* (1985) Wollenweber *et al.* (1987a)
			Aegialophila pumila	aerial parts	El-Masry *et al.* (1981)
			Anisomeles ovata (Lam.)	aerial parts	Rao *et al.* (1983b)
			Artemisia 2 spp.	aerial parts	Esteban *et al.* (1986) Bouzid *et al.* (1982a)
			Centaurea 3 spp.	aerial parts	Öksüz *et al.* (1984) Zapesochnaya *et al.* (1977)
			Chenopodium botrys (Chenopod.)	aerial parts	De Pasqual–Teresa *et al.* (1981)
			Eupatorium microphyllum	flowerhead	Torrenegra *et al.* (1985)
			Hyptis tomentosa (Lam.)	aerial parts	Kingston *et al.* (1979)
			Orthosiphon spicatus (Lam.)	leaf	Wollenweber and Mann (1985)
			Salvia palaestina (Lam.)	leaf	Miski *et al.* (1983b)
			Tanacetum santolinoides	aerial parts	El-Din *et al.* (1985)
			Teucrium pseudochamaepitys (Lam.)	aerial parts	Savona *et al.* (1979)
			Thymus 2 spp. (Lam.)	aerial parts	Ferreres *et al.* (1985b), Voirin *et al.* (1985)
38.17	5,6,7,3′,4′-pentaMe	Sinensetin	*Bauhinia championii* (Fab.)	root	Chen *et al.* (1984)
			Chenopodium botrys (Chenopod.)	aerial parts	De Pasqual–Teresa *et al.* (1981)
			Orthosiphon spicatus (Lam.)	leaf	Jiyu and Zongshi (1984) Wollenweber and Mann (1985)

No.	Substitution	Trivial name	Source	Organ	Reference
39	[5, 6, 8, 2′, 4′-pentaOH]				
39.1	[6, 8-diMe]		See Section 7.4.1		
40	(5, 7, 8, 2′, 3′-pentaOH)	Norwightin			
40.1	7, 8, 2′-triMe	Wightin			
41	[5, 7, 8, 2′, 5′-pentaOH]				
41.1	7, 8-diMe	Rehderianin I (rev. struct.)	*Scutellaria rehderiana* (Lam.), See Section 7.4.1 (no. 39.1) synthesis		Iinuma *et al.* (1985a)
42	(5, 7, 8, 2′, 6′-pentaOH)				
42.1	*7, 8-diMe		*Scutellaria* 2 spp. (Lam.)	root/whole plant	Tomimori *et al.* (1984a, b)
42.2	*8, 6′-diMe		*Scutellaria baicalensis* (Lam.)	root	Tomimori *et al.* (1984c)
42.3	*7, 8, 6′-triMe	Rivularin	*Scutellaria rivularis* (Lam.)	root	Chou (1978)
42.4	*8, 2′, 6′-triMe		*Scutellaria discolor* (Lam.)	root	Tomimori *et al.* (1986)
43	5, 7, 8, 3′, 4′-pentaOH(G)	Hypolaetin			
43.1	8-Me	Onopordin	*Doronicum grandiflorum*	leaf + infl.	Reynaud and Raynaud (1986)
43.2	3′-Me (G)				
43.3	[7, 3′-diMe]		See Section 7.4.1		
43.4	8, 3′-diMe		*Doronicum grandiflorum* aerial parts *Solanum grayi* (Solan.) See section 7.4.1 (no 43.3)	ariel parts	Reynaud *et al.* (1983)
43.5	7, 8, 3′, 4′-tetraMe				
43.6	5, 7, 8, 3′, 4′-pentaMe	Isosinensetin	*Mandarina satsuma* (Rut.)	fruit peel	Ferreres *et al.* (1981)
44	(5, 6, 3′, 4′, 5′-pentaOH)				
44.1	*5, 6, 3′, 4′, 5′-pentaMe	Cerosillin B	*Sargentia greggii* (Rut.)	leaf	Dominguez *et al.* (1976)
45	5, 7, 2′, 4′, 5′-pentaOH (G)	Isoëtin (Hieracin)	synthesis:		Iinuma *et al.* (1984a)
46	5, 7, 3′, 4′, 5′-pentaOH (G)	Tricetin			
46.1	*4′-Me		*Passiflora palmeri* (Passif.)	leaf	Ulubelen *et al.* (1984)
46.2	3′-Me	Selgin	*Artemisia ludoviciana* *Passiflora palmeri* (Passif.) *Poa huecu* (Poac.) *Rhinanthus angustifolius* (*Scroph.*)	leaf leaf whole plant	Liu and Mabry (1982) Ulubelen *et al.* (1984) Rofi and Pomilio (1985) Toth *et al.* (1981)
46.3	3′, 5′-diMe	Tricin	widespread in Cyperaceae *Artemisia* 2 spp. *Phoenix canariensis* (Arec.) *Poa huecu* (Poac.) *Trichophorum cespitosum* (Cyp.) *Wikstroemia indica* (Thymel.)	 aerial parts leaf whole plant stem whole plant	Harborne *et al.* (1985) Liu and Mabry (1981, 1982) Garcia *et al.* (1981) Rofi and Pomilio (1985) Salmenkallio *et al.* (1982) Lee *et al.* (1981)
46.4	*3′, 4′-diMe	Apometzgerin	*Poa huecu* (Poac.)	whole plant	Rofi and Pomilio (1985)
46.5	3′, 4′, 5′-triMe				Theodor *et al.* (1980)
46.6	7, 3′, 4′, 5′-tetraMe	Corymbosin			
46.7	5, 7, 3′, 4′, 5′-pentaMe		*Bauhinia championii* (Fab.)	root	Chen *et al.* (1984)
47	(6, 7, 3′, 4′, 5′-pentaOH)				
47.1	*3′, 4′, 5′-t	Prosogerin E	*Prosopis spicigera* (Fab.)	seed	Bhardwaj *et al.* (1981c)
47.2	6, 3′, 4′, 5′-tetraMe	Prosogerin D	synthesis:		Bhardwaj *et al.* (1981d)
47.3	6, 7, 3′, 4′, 5′-pentaMe	Prosogerin C	synthesis:		Bhardwaj *et al.* (1981c)

(Contd.)

Table 7.1 (*Contd.*)

No.	*Substitution*	*Trivial name*	*Plant source*†		*References*
	HEXA-O-SUBSTITUTED FLAVONES				
48	(5, 6, 7, 8, 2′, 5′-hexaOH)				
48.1	*6, 7, 8-triMe		*Scutellaria baicalensis* (Lam.)	root	Tomimori *et al.* (1984c)
49	(5, 6, 7, 8, 2′, 6′-hexaOH)				
49.1	6, 7, 8, 6′-tetraMe	Skullcapflavon II	synthesis:		Ryu *et al.* (1985)
50	(5, 6, 7, 8, 3′, 4′-hexaOH)				
50.1	*6-Me		*Hemizonia*, 3 spp.	leaf resin	P. Proksch *et al.* (1984)
50.2	*6, 7-diMe (G)	Leucanthoflavone			
50.3	6, 8-diMe		*Gutierrezia* 2 spp.	leaf	Fang *et al.* (1985a, 1986a)
50.4	3′, 4′-diMe				
50.5	6, 7, 8-triMe	Sideritiflavone	*Baccharis* 2 spp.	aerial parts	Silva *et al.* (1985) Wollenweber *et al.* (1984)
			Mentha piperita (Lam.)	leaf	Jullien *et al.* (1984)
			Pteronia incana	leaf resin	Wollenweber (unpublished)
			Sideritis spp. (Lam.)	aerial parts	Barberán *et al.* (1984c, 1985b)
50.6	6, 8, 3′-triMe	Sudachitin [Majoranin]	*Gutierrezia sarothroides* See Section 7.4.1	leaf resin	Hradetzky *et al.* (1986)
			Helianthus strumosus	aerial parts	Herz and Kulanthaivel (1984)
50.7	6, 8, 4′-triMe	Acerosin	*Helianthus strumosus*	aerial parts	Herz and Kulanthaivel (1984)
50.8	*7, 8, 3′-triMe	Thymonin	*Thymus vulgaris* (Lam.) see section 7.4.1	leaf	Van Den Broucke *et al.* (1982)
			synthesis:		Horie *et al.* (1980)
50.9	6, 7, 8, 3′-tetraMe		*Baccharis* 2 spp.	leaf root	Faini *et al.* (1982), Bohlmann *et al.* (1981c)
			Citrus sudachii (Rut.)		Horie and Nakayama (1981)
			Pteronia incana	leaf resin	Wollenweber (unpublished)
			Sideritis spp. (Lam.)	aerial parts	Barberán *et al.* (1984c, 1985b)
			Thymus vulgaris (Lam.)	leaf	Van Den Broucke *et al.* (1982)
50.10	6, 7, 8, 4′-tetraMe	Gardenin D	*Baccharis* 2 spp.	aerial parts	Silva *et al.* (1985), Wollenweber *et al.* (1986)
			Mentha piperita (Lam.)	leaf	Jullien *et al.* (1984)
50.11	6, 8, 3′, 4′-tetraMe	Hymenoxin	*Helianthus*, 6 spp.	aerial parts	Schilling (1983), Herz and Kulanthaivel (1984), Herz *et al.* (1983)
			Phoebanthus teanifolius	leaf	Schilling and Rieseberg (1985)
50.12	*7, 8, 3′, 4′-tetraMe		*Mentha piperita* (Lam.)	leaf	Jullien *et al.* (1984)
			Thymus piperella (Lam.)	aerial parts	Barberán *et al.* (1985c)
50.13	5, 6, 7, 8, 3′-pentaMe		*Eupatorium leucolepis*	aerial parts	Herz and Kulanthaivel (1982)
50.14	6, 7, 8, 3′, 4′-pentaMe	5-Desmethoxynobiletin	*Heteropappus altaicus*	aerial parts	Bohlmann *et al.* (1985a)
			Mentha piperita (Lam.)	leaf	Jullien *et al.* (1984)
			Thymus membranaceus (Lam.)	aerial parts	Ferreres *et al.* (1985b)
50.15	5, 6, 7, 8, 3′, 4′-hexaMe	Nobiletin	*Eupatorium leucolepis*	aerial parts	Herz and Kulanthaivel (1982)
51	[5, 6, 7, 2′, 3′, 5′-hexaOH]				

No.	Compound	Trivial name	Source	Organ	Reference
51.1	[6, 7, 2′, 3′-tetraMe]		See Section 7.4.1		
52	5, 6, 7, 2′, 4′, 5′-hexaOH (G)		*Juniperus thurifera* (Cup.)	leaf	Hassani *et al.* (1985)
52.1	*6, 5′-diMe		*Artemisia ludoviciana*	aerial parts	Liu and Mabry (1982)
			synthesis:		Iinuma *et al.* (1984a)
52.2	*6, 7, 5′-triMe	Arcapillin	*Artemisia capillaris*	flower heads	Namba *et al.* (1983)
				aerial parts	Kiso *et al.* (1982)
			synthesis:		Iinuma *et al.* (1984a)
52.3	[7, 2′, 5′-triMe]	[Isoarcapillin]	See Section 7.4.1		
52.4	6, 2′, 4′, 5′-tetraMe	Tabularin	synthesis:		Iinuma *et al.* (1984a)
53	(5, 6, 7, 3′, 4′, 5′-hexaOH)				
53.1	*6, 5′-diMe		*Artemisia* 2 spp.	aerial parts	Liu and Mabry (1981, 1982)
53.2	6, 3′, 5′-triMe		*Artemisia* 2 spp.	leaf	Liu and Mabry (1982), Bouzid *et al.* (1982b)
53.3	6, 4′, 5′-triMe		*Artemisia ludoviciana*	leaf	Liu and Mabry (1982)
53.4	*6, 7, 3′, 5′-tetraMe		*Artemisia mesatlantica*	aerial parts	Bouzid *et al.* (1982b)
			synthesis:		Iinuma *et al.* (1984d)
53.5	*6, 7, 4′, 5′-tetraMe		*Gardenia* 2 spp. (Rub.)		
			See section 7.4.1 (no.51.1)		
53.6	6, 7, 3′, 4′, 5′-pentaMe		*Gardenia* 2 spp. (Rub.)	bud exudate	Gunatilaka *et al.* (1982)
53.7	*5, 6, 7, 3′, 4′, 5′-hexaMe		*Bauhinia championii* (Fab.)	root	Chen *et al.* (1984)
54	(5, 7, 8, 2′, 3′, 4′-hexaOH)				
54.1	7, 8, 2′, 3′, 4′-pentaMe	Serpyllin			
55	(5, 7, 8, 2′, 3′, 6′-hexaOH)				
55.1	*8, 2′-diMe	Ganhuangenin	*Scutellaria* 2 spp. (Lam.)	root	Tomimori *et al.* (1984c), Liu Y.-L. *et al.* (1984)
		[Viscidulin III]	*Scutellaria viscidula* (Lam.)		
56	(5, 7, 8, 2′, 4′, 5′-hexaOH)		See Section 7.4.1 (sub no. 39.1)		
			synthesis:		Iinuma *et al.* (1985a)
56.1	*8, 5′-diMe		*Gutierrezia microcephala*	aerial parts	Fang *et al.* (1986a)
57	(5, 7, 8, 2′, 5′, 6′-hexaOH)				
57.1	*8, 6′-diMe		*Scutellaria baicalensis* (Lam.)	root	Tomimori *et al.* (1984c)
58	(5, 7, 8, 3′, 4′, 5′-hexaOH)				
58.1	8-Me				
58.2	8, 4′-diMe				
58.3	8, 3′, 4′-triMe				
58.4	7, 8, 4′, 5′-tetraMe				
58.5	8, 3′, 4′, 5′-tetraMe				
58.6	*5, 7, 8, 3′, 4′, 5′-hexaMe		*Murraya paniculata* (Rut.)		De Silva *et al.* (1980)
59	(5, 6, 2′, 3′, 4′, 6′-hexaOH)				
59.1	*5, 6, 2′, 3′, 4′, 6′-hexaMe		*Sargentia greggii* (Rut.)	root	Meyer *et al.* (1985)
	HEPTA-O-SUBSTITUTED FLAVONES				
60	(5, 6, 7, 8, 2′, 4′, 5′-heptaOH)				
60.1	*6, 8, 5′-triMe		*Gutierrezia microcephala*	aerial parts	Fang *et al.* (1986a)
60.2	6, 7, 8, 5′-tetraMe	Agecorynin-D	*Gutierrezia microcephala*	aerial parts	Fang *et al.* 1986a)
			synthesis:		Iinuma *et al.* (1984a)
60.3	5, 6, 7, 8, 2′, 4′, 5′-heptaMe	Agecorynin-C	synthesis:		Iinuma *et al.* (1984a)

(*Contd.*)

Table 7.1 (*Contd.*)

No.	*Substitution*	*Trivial name*	*Plant source*†		*References*
61	(5, 6, 7, 8, 3′, 4′, 5′-heptaOH)				
61.1	6, 7, 8, 4′-tetraMe	Gardenin E			
61.2	6, 8, 3′, 4′-tetraMe	Scaposin			
61.3	8, 3′, 4′, 5′-tetraMe	Trimethoxywogonin			
61.4	6, 7, 8, 4′, 5′-pentaMe	Gardenin C			
61.5	*5, 6, 7, 8, 3′, 5′-hexaMe		*Eupatorium leucolepis*	aerial parts	Herz and Kulanthaivel (1982)
61.6	*5, 6, 7, 8, 4′, 5′-hexaMe		*Eupatorium leucolepis*	aerial parts	Herz and Kulanthaivel (1982)
61.7	6, 7, 8, 3′, 4′, 5′-hexaMe				
61.8	5, 6, 7, 8, 3′, 4′, 5′-heptaMe	Gardenin A			
62	(5, 6, 7, 2′, 3′, 4′, 5′-heptaOH)		synthesis:		Bhardwaj *et al.* (1981b)
62.1	*6, 7, 2′, 4′, 5′-pentaMe	[Brickellin]	See Section 7.4.1		
62.2	*5, 6, 7, 2′, 3′, 4′, 5′-heptaMe	Agehoustin B	*Ageratum houstonianum*	aerial parts	Quijano *et al.* (1982)
			synthesis:		Iinuma *et al.* (1984e)
63	(5, 6, 7, 2′, 3′, 4′, 6′-heptaOH)				
63.1	[6, 7, 2′, 3′-tetraMe]		See Section 7.4.1		
63.2	[6, 7, 2′, 3′, 4′-pentaMe]	[Brickellin]	See Section 7.4.1		
	OCTA-O-SUBSTITUTED FLAVONES				
64	(5, 6, 7, 8, 2′, 3′, 4′, 5′-octaOH)				
64.1	*6, 7, 8, 2′, 4′, 5′-hexaMe	Agehoustin D	*Ageratum houstonianum*	aerial parts	Quijano *et al.* (1985)
64.2	*5, 6, 7, 8, 2′, 4′, 5′-heptaMe	Agehoustin C	*Ageratum houstonianum*	aerial parts	Quijano *et al.* (1985)
64.3	*5, 6, 7, 8, 2′, 3′, 4′, 5′-octaMe	Agehoustin A	*Ageratum houstonianum*	aerial parts	Quijano *et al.* (1982)
			synthesis:		Iinuma *et al.* (1984e)

*Compound newly reported during 1980–1986, or earlier report not included in Wollenweber (1982a).
†Family is Asteraceae, unless otherwise indicated in parentheses.

Table 7.2 Flavonols

No.	*Substitution*	*Trivial name*	*Plant Sources*†		*References*
	DI-O-SUBSTITUTED FLAVONOLS				
1	*3, 7-diOH		*Zuccagnia punctata* (Fab.)	leaf	Pederiva and Giordano (1984)
	TRI-O-SUBSTITUTED FLAVONOLS				
2	3, 5, 7-triOH	Galangin	*Acacia neovernicosa* (Fab.)	exudate	Wollenweber and Seigler (1982)
			Achyrocline satureoides (Lam.)	aerial parts	Ferraro *et al.* (1981)
			Adenostoma sparsifolium (Rosac.)	leaf resin	Proksch *et al.* (1982a)
			Chromolaena chaslaea	aerial parts	Bohlmann *et al.* (1982)
			Comptonia peregrina (Myric.)	leaf exudate	Wollenweber *et al.* (1985a)
			Flourensia resinosa	leaf resin	Wollenweber and Yatskievych (1985)
			Helichrysum platypterum	aerial parts	Jakupovic *et al.* (1986)
2.1	3-Me		*Acacia neovernicosa* (Fab.)	exudate	Wollenweber and Seigler (1982)
			Achyrocline 2 spp.	aerial parts	Norbedo *et al.* (1984) Ferraro *et al.* (1981)
			Andromeda polifolia (Eric.)	epicuticular layer	Wollenweber and Kohorst (1984)
			Flourensia resinosa	leaf resin	Wollenweber and Yatskievych (1985)
			Gaultheria schallon (Eric.)	epicuticular layer	Wollenweber and Kohorst (1984)
			Helichrysum 3 spp.	aerial parts	Çubukçu and Yaksel (1982), Çubukçu and Damadyan (1986), Jakupovic *et al.* (1986)
			Notholaena candida var. *copelandii* (Pterid.)	farinose exudate	Wollenweber (1984b)
2.2	*5-Me		*Pityrogramma triangularis* (Pterid.)	frond exudate	Wollenweber *et al.* (1985d)
2.3	7-Me	Izalpinin	*Pityrogramma triangularis*	frond exudate	Wollenweber (unpublished)
			Comptonia peregrina (Myric.)	leaf exudate	Wollenweber *et al.* (1985a)
			Flourensia resinosa	leaf resin	Wollenweber and Yatskievych (1985)
2.4	3, 7-diMe		*Boesenbergia pandurata* (Zing.)	rhizome	Jaipetch *et al.* (1983)
2.5	*5, 7-diMe		*Pityrogramma triangularis* (Pterid.)	frond exudate	Hitz *et al.* (1982)
2.6	3, 5, 7-triMe		*Boesenbergia pandurata* (Zing.)	rhizome	Jaipetch *et al.* (1983)
			Gomphrena martiana (Amar.)	whole plants	Buschi and Pomilio (1982)
3	(3, 7, 8-triOH)				
3.1	*8-Me		*Zuccagnia punctata* (Fab.)	leaf resin	Pederiva and Giordano (1984)
4	3, 7, 4′,-triOH (G)		*Pterocarpus marsupium* (Fab.)	heartwood	Maurya *et al.* (1984)
	TETRA-O-SUBSTITUTED FLAVONOLS				
5	3, 5, 6, 7-tetraOH	6-Hydroxygalangin	*Cassinia quinquefaria*	leaf resin	Wollenweber (unpublished)
5.1	6-Me	Alnusin	*Baccharis bigelovii*	leaf resin	Arriaga–Giner *et al.* (1986)
			Chromolaena chaslea	aerial parts	Bohlmann *et al.* (1982)
5.2	3, 6-diMe				
5.3	3, 7-diMe				
5.4	*5, 6-diMe		*Adenostoma sparsifolium* (Rosac.)	leaf resin	Proksch *et al.* (1982a)
5.5	5, 7-diMe		*Gomphrena martiana* (Amar.)	whole plant	Buschi and Pomilio (1982)

(*Contd.*)

Table 7.2 (*Contd.*)

No.	Substitution	Trivial name	Plant Sources[†]		References
5.6	3, 6, 7-triMe	Alnustin	*Achyrocline albicans*	leaf	Mesquita *et al.* (1986)
			Gomphrena martiana (Amar.)	whole plant	Buschi and Pomilio (1982)
5.7	3, 5, 6, 7-tetraMe				
6	(3, 5, 7, 8-tetraOH)	(8-Hydroxygalangin)			
6.1	*3-Me		*Achyrocline flaccida*	aerial parts	Norbedo *et al.* (1984)
6.2	7-Me		*Adenostoma sparsifolium* (Rosac.)	leaf resin	Proksch *et al.* (1982a)
6.3	*8-Me		synthesis:		Goudard and Chopin (1976)
6.4	3, 7-diMe	Isognaphalin	*Achyrocline* 2 spp.	aerial parts	Norbedo *et al.* (1984), Hirschmann (1984)
			Pityrogramma triangularis (Pterid.) See Section 7.4.1. (63.1)		
			Notholaena candida (Pterid.) (as esters)	frond exudate	Wollenweber (1985a)
			Notholaena candida var. (Pterid.) (as esters)	frond exudate	Wollenweber (1985a)
6.5	3, 8-diMe	Gnaphalin	*Achyrocline flaccida*	aerial parts	Norbedo *et al.* (1984)
6.6	*7, 8-diMe		*Achyrocline tomentosa*	aerial parts	Ferraro *et al.* (1985)
6.7	*3, 5, 8-triMe		*Achyrocline albicans*	leaf	Mesquita *et al.* (1986)
6.8	3, 7, 8-triMe	Methylgnaphalin	*Achyrocline bogotensis*	leaf + infl.	Torrenegra *et al.* (1982)
6.9	*3, 5, 7, 8-tetraMe		*Achyrocline albicans*	leaf	Mesquita *et al.* (1986)
7	3, 5, 7, 2′-tetraOH	Datiscetin			
7.1	7-Me	Datin			
7.2	2′-Me	Ptaeroxylol			
8	3, 5, 7, 4′-tetraOH	Kaempferol	*Ericameria laricifolia*	leaf resin	Clark and Wollenweber (1984)
			Flindersia australis (Rub.)	stem back	Reisch *et al.* (1984)
			Gutierrezia microcephala	aerial parts	Fang *et al.* (1986a)
			Hazardia squarrosa	leaf resin	Clark and Wollenweber (1984)
			Heterotheca grandiflora	leaf resin	Wollenweber *et al.* (1985c)
			Lupinus spp. (Fab.)		Nichols and Bohm (1983)
			Notholaena standleyi (Pterid.)	frond exudate	Seigler and Wollenweber (1983)
			Pityrogramma triangularis (Pterid.)	frond exudate	Wollenweber *et al.* (1985d)
			Rhamnus pallasii (Rham.)	bark	Sakushima *et al.* (1983)
			Solanum pubescens (Sol.)	leaf	Kumari *et al.* (1985)
8.1	3-Me	Isokaempferide	*Artemisia* 2 spp.	aerial parts	McCormick and Bohm (1986), Liu and Mabry (1982)
			Baccharis, 3 spp.	aerial parts, exudate	Wollenweber *et al.* (1986)
			Centaurea 3 spp.	aerial parts	Collado *et al.* (1985) Öksütz *et al.* (1984)
			Cistus 2 spp. (Cist.)	leaf	Proksch and Gülz (1984) De Pasqual–Teresa *et al.* (1983)
			Cyperus spp. (Cyp.)	leaf + infl.	Harborne *et al.* (1982)
			Ericameria 2 spp.	leaf resin	Clark and Wollenweber (1984)
			Escallonia, 3 spp. (Saxif.)	epicut. layer	Wollenweber (1984a)
			Flourensia resinosa	leaf resin	Wollenweber and Yatskievych (1985)

No.	Substitution	Trivial name	Source	Plant part	Reference
			Frullania densifolia (Hepaticae)	whole plant	Kruijt *et al.* (1986)
			Grindelia, 3 spp.	flowerhead	Pinkas *et al.* (1978)
			Hazardia squarrosa	leaf resin	Clark and Wollenweber (1985)
			Heterotheca 2 spp.	leaf resin	Wollenweber *et al.* (1985c)
			Inula viscosa	aerial parts (exudate)	Grande *et al.* (1985)
			Neurolaena venturana	leaf	Ulubelen *et al.* (1981a)
			Notholaena 2 spp. (Pterid.)	frond exudate	Wollenweber (1984b), Seigler and Wollenweber (1983)
			Opuntia spp. (Cact.)	spines	Burret *et al.* (1982)
			Pityrogramma triangularis (Pterid.)	frond exudate	Wollenweber *et al.* (1985d)
			Pluchea odorata	leaf exudate	Wollenweber *et al.* (1985b)
			Solanum sarrachoides (Sol.)	leaf trichomes	Schilling (1984)
			Wyethia bolanderi	leaf wash	McCormick *et al.* (1985)
8.2	5-Me				
8.3	7-Me	Rhamnocitrin	*Baccharis vaccinoides*	aerial parts, exudate	Wollenweber *et al.* (1986)
			Cistus, 4 spp. (Cist.)	leaf resin	Wollenweber and Mann (1984)
			Ericameria 2 spp.	leaf resin	Clark and Wollenweber (1984)
			Escallonia 2 spp. (Saxif.)	epicut. layer	Wollenweber (1984a)
			Haploppapus scrobiculatus	leaf	Ates *et al.* (1982)
			Hazardia squarrosa	leaf resin	Clark and Wollenweber (1985)
			Heterotheca grandiflora	leaf resin	Wollenweber *et al.* (1985c)
			Inula viscosa	aerial parts exudate	Grande *et al.* (1985)
			Notholaena 2 spp. (Pterid.)	frond exudate	Wollenweber (1984b), Seigler and Wollenweber (1983)
8.4	4′-Me	Kaempferide	*Pityrogramma triangularis* (Pterid.)	frond exudate	Wollenweber *et al.* (1985d)
			Baccharis 3 spp.	aerial parts	Bohlmann *et al.* (1985b), Wollenweber *et al.* (1986)
			Ericameria 2 spp.	leaf resin	Clark and Wollenweber (1984)
			Hazardia squarrosa	leaf resin	Clark and Wollenweber (1984)
			Heterotheca grandiflora	leaf resin	Wollenweber *et al.* (1986c)
			Notholaena standleyi (Pterid.)	frond exudate	Seigler and Wollenweber (1983)
			Pityrogramma triangularis (Pterid.)	frond exudate	Wollenweber *et al.* (1985d)
			Wyethia bolanderi	leaf wash	McCormick *et al.* (1985)
			Linaria dalmatica (Scroph.)	whole plant	Kapoor *et al.* (1985)
8.5	*3, 5-diMe				
8.6	3, 7-diMe	Kumatakenin (Jaranol)	*Baccharis* 2 spp.	aerial parts, exudate	Wollenweber *et al.* (1986)
			Cistus, 8 spp. (Cist.)	leaf resin aerial parts	Wollenweber and Mann (1984), Proksch and Gülz (1984), De Pasqual–Teresa *et al.* (1983)
			Coleus spicatus (Lam.)	aerial parts	Painuly and Tandon (1983a)
			Dodonaea viscosa (Sapind.)	flowerhead	Dreyer (1978)
			Ericameria 2 spp.	leaf resin	Clark and Wollenweber (1984)
			Escallonia 2 spp. (Saxif.)	epicut. layer	Wollenweber (1984a)

(*Contd.*)

Table 7.2 (*Contd.*)

No.	Substitution	Trivial name	Plant Sources[†]		References
			Flourensia resinosa	leaf resin	Wollenweber and Yatskievych (1985)
			Geranium macrorrhizum (Geran.)	leaf	Ognyanov *et al.* (1975)
			Gochnatia foliosa var. *fascicularis*	stem & leaf	Faini *et al.* (1984)
			Grindelia, 5 spp.	flowerhead aerial parts	Pinkas *et al.* (1978), Timmermann *et al.* (1985)
			Hazardia squarrosa	leaf resin	Clark and Wollenweber (1985)
			Heterotheca 2 spp.	leaf resin	Wollenweber *et al.* (1985c)
			Notholaena 2 spp. (Pterid.)	frond exudate	Wollenweber (1984b), Seigler and Wollenweber (1983)
			Pityrogramma triangularis (Pterid.)	frond exudate	Wollenweber *et al.* (1985d)
			Solanum pubescens (Sol.)	leaf	Kumari *et al.* (1985)
			Trichophorum cespitosum (Cyp.)	stem	Salmenkallio *et al.* (1982)
8.7	3,4′-diMe	Ermanin	*Baccharis vaccinioides*	aerial parts, exudate	Wollenweber *et al.* (1986)
			Cistus 8 spp.	leaf resin	Wollenweber and Mann (1984), Proksch & Gülz (1984)
			Ericameria 2 spp.	leaf resin	Clark and Wollenweber (1984)
			Escallonia rubra (Saxif.)	epicut. layer	Wollenweber (1984a)
			Flourensia resinosa	leaf resin	Wollenweber and Yatskievych (1985)
			Geranium macrorrhizum (Geran.)	leaf	Ognyanov *et al.* (1975)
			Hazardia squarrosa	leaf resin	Clark and Wollenweber (1985)
			Heterotheca 2 spp.	leaf resin	Wollenweber *et al.* (1985c)
			Notholaena standleyi (Pterid.)	frond exudate	Seigler and Wollenweber (1983)
			Solanum pubescens (Sol.)	leaf	Kumari *et al.* (1985)
			Wyethia bolanderi	leaf surface	McCormick *et al.* (1985)
8.8	7.4′-diMe		*Baccharis* 2 spp.	aerial parts, exudate	Wollenweber *et al.* (1986)
			Croton pyramidalis (Euphorb.)	leaf & stem	Rodriguez-Hahn *et al.* (1981)
			Escallonia pulverulenta (Saxif.)	epicut.layer	Wollenweber (1984a)
			Hazardia squarrosa	leaf resin	Clark and Wollenweber (1985)
			Notholaena standleyi (Pterid.)	frond exudate	Seigler and Wollenweber (1983)
8.9	3,5,7-triMe		*Artemisia campestris*	leaf + infl.	Omar and El-Ghazouly (1984)
8.10	3,7,4,′-triMe		*Baccharis vaccinioides*	aerial parts, exudate	Wollenweber *et al.* (1986)
			Boesenbergia pandurata (Zing.)	rhizome	Jaipetch *et al.* (1983)
			Cistus, 5 spp. (Cist.)	leaf resin	Wollenweber and Mann (1984)
			Ericameria 2 spp.	leaf resin	Clark and Wollenweber (1984)
			Digitalis windiflora (Scroph.)	leaf	Imre *et al.* (1984)
			Dodonaea viscosa (Sapind.)	flower	Dreyer (1978)
			Hazardia squarrosa	leaf resin	Clark and Wollenweber (1985)
			Heterotheca psammophila	leaf resin	Wollenweber *et al.* (1985c)
			Notholaena bryopoda (Pterid.)	frond exudate	Wollenweber (1984b)
			Pluchea odorata	leaf exudate	Wollenweber *et al.* (1985b)
			Solanum pubescens (Sol.)	leaf	Kumari *et al.* (1985)

9	3, 5, 8, 4′-tetraOH	Pratoletin			
10	(3, 6, 7, 4′-tetraOH)				
10.1	*3, 6-diMe		*Pluchea chingoyo*	aerial parts	Chiang and Silva (1978)
11	3, 7, 8, 4′-tetraOH				
11.1	3-Me				
12	3, 7, 3′, 4′-tetraOH	Fisetin	*Garcinia kola* (Clus.)	seed	Iwu and Igboko (1982)
12.1	3-Me				
12.2	3′-Me	Geraldol			
12.3	4′-Me				
12.4	3, 7, 3′, 4′-tetraMe				
	PENTA-O-SUBSTITUTED FLAVONOLS				
13	(3, 5, 6, 7, 8-pentaOH)				
13.1	6, 8-diMe				
13.2	*3, 6, 7-triMe		*Gnaphalium gaudichaudianum*	aerial parts	Guerreiro *et al.* (1982)
13.3	3, 6, 8-triMe	Araneol	*Gnaphalium elegans*	flowerhead	Torrenegra *et al.* (1980)
13.4	6, 7, 8-triMe		*Achillea nobilis*	leaf + flowerhead	Adekenov *et al.* (1984)
			Achyrocline bogotensis	leaf + infl.	Torrenegra *et al.* (1982)
			Artemisia ludoviciana? See Section 7.4.2		
			Helichrysum graveolens		Hänsel *et al.* (1981)
13.5	*3, 6, 7, 8-tetraMe		confirmed:		
14	3, 5, 6, 7, 4′-pentaOH (G)	6-Hydroxykämpferol	*Helichrysum cephaloideum*	no data/root	Jakupovic *et al.*
14.1	3-Me		*Neurolaena oaxacana*	leaf	Ulubelen *et al.* (1980)
14.2	5-Me (G)	Vogeletin			
14.3	6-Me		*Arnica* 2 spp.	flowerhead	Merfort (1984a, b, 1985)
			Baccharis vaccinioides	aerial parts, exudate	Wollenweber *et al.* (1986)
			Eupatorium areolare	leaf + flowerhead	Yu *et al.* (1986)
			Heterotheca grandiflora	leaf resin	Wollenweber *et al.* (1985c)
			Matricaria chamomilla	flowerhead	Exner *et al.* (1981b)
14.4	4′-Me				
14.5	3, 6-diMe		*Acanthospermum glabratum*	aerial parts	Saleh *et al.* (1980)
			Artemisia ludoviciana	leaf	Liu and Mabry (1982)
			Baccharis 2 spp.	aerial parts, exudate	Wollenweber *et al.* (1986)
			Dodonaea viscosa (Sapind.)	aerial parts	Sachdev and Kulshreshta (1983)
			Gutierrezia 2 spp.	leaf	Fang *et al.* (1985a, 1986a)
			Haplopappus ciliatus	leaf	Bittner and Watson (1982)
			Heterotheca 2 spp.	leaf resin	Wollenweber *et al.* (1985c)
			Parthenium spp.	leaf	Mears (1980)
			Pluchea chingoyo	aerial parts	Chiang and Silva (1978)
			Stevia 2 spp.	leaf	Rajbhandari and Roberts (1984a, 1985a)
			Wyethia bolanderi	leaf surface	McCormick *et al.* (1985)
14.6	3, 7-diMe		*Neurolaena lobota*	leaf	Kerr *et al.* (1981)
			Pulicaria dysenterica	leaf	Pares *et al.* (1981)

(Contd.)

Table 7.2 (*Contd.*)

No.	*Substitution*	*Trivial name*	*Plant Sources*[†]		*References*
14.7	*5, 6-diMe		*Adenostoma sparsifolium* (Rosac.)	leaf resin	M. Proksch *et al.* (1982b)
14.8	6, 7-diMe	Eupalitin	*Artemisia* 2 spp.	aerial parts	McCormick and Bohm (1986), Chandrasekharan *et al.* (1981)
			Baccharis vaccinioides	aerial parts, exudate	Wollenweber *et al.* (1986)
			Eupatorium areolare	leaf & flowerhead	Yu *et al.* (1986)
			Heterotheca 2 spp.	leaf resin	Wollenweber *et al.* (1985c)
			Pluchea odorata		
			See Section 7.4.2 (no. 17.5)		
			Rudbeckia serotina	aerial parts	Waddell and Elkins (1985)
14.9	6, 4′-diMe	Betuletol	*Arnica* 2 spp.	flowerhead	Merfort (1984a, b)
			Baccharis vaccinioides	aerial parts, exudate	Wollenweber *et al.* (1986)
14.10	3, 6, 7-triMe	Penduletin	*Achillea ageratifolia*	aerial parts exudate	Valant-Vetschera and Wollenweber (1986)
			Baccharis, 3 spp.	aerial parts	Wollenweber *et al.* (1986)
			Brickellia 2 spp.	aerial parts	Bulinska–Radomska *et al.* (1985a)
					Goodwin *et al.* (1984)
			Bromelia pinguin (Brom.)	root and stem	Raffauf *et al.* (1981)
			Digitalis thapsii (Scroph.)	leaf	De Pasqual–Teresa *et al.* (1980)
			Dodonea viscosa (Sapind.)	aerial parts	Sachdev and Kulshreshta (1983)
			Eupatorium areolare	leaf and flower	Yu *et al.* (1986)
			Gardenia fosbergii (Rub.)	bud exudate	Gunatilaka *et al.* (1979)
			Heterotheca 2 spp.	leaf resin	Wollenweber *et al.* (1985c)
			Jasonia tuberosa	aerial parts	González *et al.* (1977)
			Vitex agnuscastus (Verb.)	fruit	Wollenweber and Mann (1983)
14.11	3, 6, 4′-triMe	Santin	*Achillea* 2 spp.	aerial parts	Valant–Vetschera and Wollenweber (1986)
			Digitalis orientalis (Scroph.)	leaf	Imre *et al.* (1984)
			Dodonaea 3 spp. (Sapind.)	aerial parts	Sachdev and Kulshreshta (1983), Payne and Jefferies (1973), Dominguez *et al.* (1980)
			Hazardia squarrosa	leaf resin	Clark and Wollenweber (1985)
			Heterotheca grandiflora	leaf resin	Wollenweber *et al.* (1985c)
			Neurolaena macrophylla	leaf	Ulubelen and Mabry (1981)
			Pluchea odorata	leaf exudate	Wollenweber *et al.* (1985b)
			Stevia origanoides	leaf	Rajbhandari and Roberts (1985a)
			Wyethia bolanderi	leaf surface	McCormick *et al.* (1985)
14.12	(3, 7, 4′-triMe)		See Section 7.4.2		
14.13	*5, 6, 7-triMe	Candidol	*Tephrosia candida* (Fab.)	seed	Dutt and Chibber (1983)
14.14	6, 7, 4′-triMe	Mikanin	*Heterotheca psammophila*	leaf exudate	Wollenweber *et al.* (1985c)
14.15	3, 6, 7, 4′-tetraMe		*Achillea* 2 spp.	aerial parts	Valant–Vetschera and Wollenweber (1986)
			Baccharis vaccinioides	aerial parts, exudate	Wollenweber *et al.* (1986)

No.	Substitution	Trivial name	Source	Organ	Reference
			Dodonaea viscosa (Sapind.)	aerial parts	Sachdev and Kulshreshta (1983)
			Heterotheca 2 spp.	leaf resin	Wollenweber *et al.* (1985c)
			Pluchea odorata	leaf exudate	Wollenweber *et al.* (1985b)
			Tillandsia utriculata (Brom.)	leaf	Ulubelen and Mabry (1982)
			Vitex agnuscastus (Verb.)	fruit	Wollenweber and Mann (1983)
15	(3, 5, 6, 8, 4′-pentaOH)				
15.1	[3, 8, 4′-triMe]		See Section 7.4.2		
16	(3, 5, 7, 8, 2′-pentaOH)				
16.1	*3, 7, 8-Me	Dechlorochlorflavonin	*Aspergillus candidus* (Ascomycetes)		Marchelli and Vining (1973)
16.2	*3′-chloro-	Chlorflavonin	*Aspergillus candidus* (Ascomycetes)		Bird and Marshall (1969)
16.3	*3, 7, 8, 2′-tetraMe		*Andrographis paniculata* (Acanth)	root	Gupta *et al.* (1983)
17	3, 5, 7, 8, 4′-pentaOH	Herbacetin	*Eupatorium gracile*	flowerhead	Torrenegra *et al.* (1984)
17.1	*3-Me		*Gutierrezia microcephala*	aerial parts	Roitman and James (1985)
17.2	7-Me	Pollenitin	*Notholaena standleyi* (Pterid.)	frond exudate	Seigler and Wollenweber (1983)
17.3	8-Me	Sexangularetin	*Eupatorium gracile*	flowerhead	Torrenegra *et al.* (1984)
17.4	*4′-Me (G)		See Section 7.4.2		
17.5	3, 7-diMe		*Baccharis sarothroides*	aerial parts, exudate	Wollenweber *et al.* (1986)
17.6	3, 8-diMe		*Geraea canescens*	aerial parts	Proksch *et al.* (1986)
			Gutierrezia microcephala	aerial parts	Roitmann and James (1985), Fang *et al.* (1986a)
			Pityrogramma triangularis (Pterid.)	frond exudate	Wollenweber *et al.* (1985d)
17.7	7, 8-diMe		*Eupatorium gracile*	flowerhead	Torrenegra *et al.* (1984)
17.8	7, 4′-diMe		*Notholaena standleyi* (Pterid.)	frond exudate	Seigler and Wollenweber (1983)
17.9	8, 4′-diMe	Prudomestin	*Pityrogramma triangularis* (Pterid.)	frond exudate	Wollenweber *et al.* (1985d)
17.10	3, 7, 8-triMe		*Larrea tridentata* (Zyg.)	leaf resin	Bernhard and Thiele (1981)
17.11	3, 8, 4′-triMe		*Cistus parviflorus* (Cist.)	leaf resin	Vogt *et al.* (1987)
			Conyza stricta		
			See Section 7.4.2 (no. 15.1) synthesis:		Horie *et al.* (1982)
			Pityrogramma triangularis (Pterid.)	frond exudate	Wollenweber *et al.* (1985d)
17.12	7, 8, 4′-triMe	Tambulin	*Achillea depressa*		
17.13	3, 7, 8, 4′-tetraMe	Flindulatin	See Section 7.4.2		
			Cistus parviflorus (Cist.)	leaf resin	Vogt *et al.* (1987)
18	(3, 6, 7, 8, 4′-pentaOH)				
18.1	3, 6, 7, 8, 4′-pentaMe	Auranetin			
19	3, 5, 7, 2′, 4′-pentaOH	Morin			
20	(3, 5, 7, 2′, 5′-pentaOH)				
20.1	*7, 5′-diMe		*Inula cappa*	aerial parts	Baruah *et al.* (1979)
20.2	[3, 5, 2′-triMe]		See Section 7.4.2		
21	*3, 5, 7, 2′, 6′-pentaOH	Viscidulin I	See Section 7.4.2 (no. 20.1)		
21.1	*3, 5, 7, 2′, 6′-pentaMe		*Scutellaria tenax* (Lam.)		Liu, M.L. *et al.* (1984)
			Scutellaria baicalensis (Lam.)	root	Tomimori *et al.* (1984b)

(*Contd.*)

Table 7.2 (*Contd.*)

No.	*Substitution*	*Trivial name*	*Plant Sources*†		*References*
22	3, 5, 7, 3′, 4′-pentaOH	Quercetin	Records in Asteraceae, Passifloraceae, Rhamnaceae and Solanaceae		
22.1	3-Me		*Achyrocline* 2 spp.	aerial parts	Gutkind *et al.* (1984), Ferraro *et al.* (1981)
			Arnica cordifolia		Wolf and Denford (1984)
			Balsamorhiza macrophylla	leaf wash	McCormick *et al.* (1985)
			Barbacenia purpurea (Velloziac)	aerial parts	Wollenweber (1984a)
			Brickellia glutinosa	aerial parts	Goodwin *et al.* (1984)
			Brickellia monocephala	aerial parts	Norris and Mabry (1985)
			Cistus, 3 spp. (Cist.)	leaf resin	Wollenweber and Mann (1984)
			Cyperus spp. (Cyp.)	leaf + infl.	Harborne *et al.* (1982)
			Ericameria laricifolia	leaf resin	Clark and Wollenweber (1984)
			Grindelia, 3 spp.	flowerhead	Pinkas *et al.* (1978)
			Gutierrezia 2 spp.	leaf	Fang *et al.* (1986a, b)
			Haplopappus 2 spp.	leaf	Bittner and Watson (1982), Ayanoglu *et al.* (1981)
			Hazardia squarrosa	leaf resin	Clark and Wollenweber (1985)
			Hemizonia, 4 spp.	leaf resin	Proksch *et al.* (1984)
			Heterotheca grandiflora	leaf resin	Wollenweber *et al.* (1985c)
			Hymenoclea salsola	leaf resin	Proksch *et al.* (1983)
			Inula viscosa	aerial parts, exudate	Grande *et al.* (1985)
			Lourteigia ballotrefolia	aerial parts	Triana (1984)
			Opuntia spp. (Cact.)	spines	Burret *et al.* (1982)
			Solanum sarrachoides (Sol.)	leaf trichomes	Schilling (1984)
			Wyethia bolanderi	leaf surface	McCormick *et al.* (1985)
22.2	5-Me	Azaleatin			
22.3	7-Me	Rhamnetin	*Artemisia* 2 spp.	aerial parts	Gonzáles *et al.* (1983) Chandrasekharan *et al.* (1981)
			Chromolaena meridensis	aerial parts	Amaro and Mendéz (1983)
			Cistus 2 spp. (Cist.)	leaf resin	Wollenweber and Mann (1984)
			Escallonia pulverulenta (Saxif.)	epicut. layer	Wollenweber (1984a)
			Gnaphalium pellitum	leaf and infl.	Torrenegra *et al.* (1978)
			Hazardia squarrosa	leaf resin	Clark and Wollenweber (1985)
			Hymenoclea salsola	leaf resin	Proksch *et al.* (1983)
			Thymbra spicata (Lam.)	leaf	Miski *et al.* (1983a)
22.4	3′-Me	Isorhamnetin	*Arnica* 2 spp.	flowerhead	Merfort (1984)
			Cistus 5 spp. (Cist.)	leaf resin	Wollenweber and Mann (1984)
			Ericameria laricifolia	leaf resin	Clark and Wollenweber (1984)
			Escallonia pulverulenta (Saxif.)	epicut. layer	Wollenweber (1984a)
			Gochnatia foliosa	stem + leaf	Faini *et al.* (1984)
			Haplopappus 3 spp.	leaf	Bittner and Watson (1982), Ayanoglu *et al.* (1981), Ates *et al.* (1982)

			Hazardia squarrosa	leaf resin	Clark and Wollenweber (1985)
			Hymenoclea salsola	leaf resin	Proksch *et al.* (1983)
			Passiflora palmeri (Passif.)	leaf	Ulubelen *et al.* (1984)
			Rhamnus pallasii (Rham.)	basla	Sakushima *et al.* (1983)
22.5	4′-Me	Tamarixetin	*Achyrocline flaccida*	aerial parts	Norbedo *et al.* (1984)
			Balsamorhiza macrophylla		McCormick *et al.* (1985)
22.6	3, 5-diMe	Caryatin			
22.7	3, 7-diMe		*Brickellia glutinosa*	aerial parts	Goodwin *et al.* (1984)
			Cistus 2 spp. (Cist.)	resin	Wollenweber and Mann (1984)
			Coleus spicatus	aerial parts	Painuly and Tandon (1983a)
			Cyperus spp. (Cyp.)	leaf + infl.	Harborne *et al.* (1982)
			Ericameria laricifolia	leaf resin	Clark and Wollenweber (1984)
			Gochnatia foliosa	stem + leaf	Faini *et al.* (1984)
			Gutierrezia alamanii	aerial parts	Lenherr *et al.* (1986)
			Haplopappus integerrimus	leaf	Ayanoglu *et al.* (1981)
			Hazardia squarrosa	leaf resin	Clark and Wollenweber (1985)
			Hymenoclea salsola	leaf resin	Proksch *et al.* (1983)
22.8	3, 3′-diMe		*Acacia neovernicosa* (Fab.)	exudate	Wollenweber and Seigler (1982)
			Cistus, 4 spp. (Cist.)	leaf resin	Wollenweber and Mann (1984)
			Dittrichia viscosa	aerial parts	Chiappini *et al.* (1982)
			Ericameria laricifolia	leaf resin	Clark and Wollenweber (1984)
			Flourensia resinosa	leaf resin	Wollenweber and Yatskievych (1985)
			Gochnatia foliosa	stem + leaf	Faini *et al.* (1984)
			Grindelia, 5 spp.	aerial parts	Pinkas *et al.* (1978), Timmermann *et al.* (1985, 1986)
			Gutierrezia dracunculoides (in ester form)	aerial parts	Bohlmann *et al.* (1981)
			Haplopappus integerrimus	leaf	Ayanoglu *et al.* (1981)
			Hazardia squarrosa	leaf resin	Clark and Wollenweber (1985)
			Hymenoclea salsola	leaf resin	Proksch *et al.* (1983)
			Inula viscosa	aerial parts, exudate	Grande *et al.* (1985)
			Pluchea 2 spp.	leaf exudate	Wollenweber *et al.* (1985) Vivar *et al.* (1982)
			Solanum pubescens (Sol.)	leaf	Kumari *et al.* (1985)
22.9	3, 4′-diMe		*Balsamorhiza macrophylla*	leaf exudate	McCormick *et al.* (1985)
			Brickellia arguta	aerial parts	Rösler and Goodwin (1984)
22.10	5, 3′-diMe				
22.11	7, 3′-diMe	Rhamnazin	*Cistus* 5 spp. (Cist.)	leaf resin	Wollenweber and Mann (1984)
			Escallonia pulverulenta (Saxif.)	epicut. layer	Wollenweber (1984a)
			Geranium macrorrhizum (Geran.)	leaf	Ognyanov *et al.* (1975)
			Hazardia squarrosa	leaf resin	Clark and Wollenweber (1985)
			Heterotheca grandiflora	leaf resin	Wollenweber *et al.* (1985c)
			Orthosiphon spicatus (Lam.)	leaf	Wollenweber and Mann (1985)
			Passiflora palmeri (Passifl.)	leaf	Ulubelen *et al.* (1984)
22.12	7, 4′-diMe	Ombuin	*Eupatorium areolare*	leaf + flower	Yu *et al.* (1986)
22.13	3′, 4′-diMe	Dillenetin	*Arnica chamissonis*	flowerhead	Merfort (1984b, 1985)
			Baccharis salicifolia	aerial parts, exudate	Wollenweber *et al.* (1986)

(*Contd.*)

Table 7.2 (*Contd.*)

No.	Substitution	Trivial name	Plant Sources[†]		References
22.14	3, 7, 3′-triMe	Pachypodol	*Agastache rugosa* (Lam.)		Ishitsuka *et al.* (1982)
			Begonia glabra (Beg.)	leaf trichomes	Ensemeyer and Langhammer (1982)
			Cistus, 4 spp. (Cist.)	leaf	Wollenweber and Mann (1984)
			Cyperus spp. (Cyp.)	leaf + infl.	Harborne *et al.* (1982)
			Grindelia discoidea	aerial parts	Timmermann *et al.* (1986)
			Hazardia squarrosa	leaf resin	Clark and Wollenweber (1985)
			Heterotheca grandiflora	leaf resin	Wollenweber *et al.* (1985c)
			Jasonia montana	aerial parts	El-Din *et al.* (1985)
			Ledum palustre (Eric.)	leaf	Mikhailova and Rybalko (1980)
			Solanum pubescens (Sol.)	leaf	Kumari *et al.* (1985)
			Trichophorum cespitosum (Cyp.)	stem	Salmenkallio *et al.* (1982)
22.15	3, 7, 4′-triMe	Ayanin	*Distemonanthus benthamianus* (Fab.)	heartwood	Malan and Roux (1979)
			Ericameria laricifolia	leaf resin	Lenherr *et al.* (1986)
			Escallonia 3 spp. (Saxif.)	epicut. layer	Wollenweber (1984a)
			Gutierrezia alamanii	aerial parts	Lenherr *et al.* (1986)
22.16	3, 3′, 4′-triMe		*Achillea santolina*	leaf	Khafagy *et al.* (1976)
			Baccharis salicifolia	aerial parts, exudate	Wollenweber *et al.* (1986)
			Hazardia squarrosa	leaf resin	Clark and Wollenweber (1985)
22.17	7, 3′, 4′-triMe		*Cistus laurifolius* (Cist.)	leaf resin	Wollenweber and Mann (1984)
			Hazardia squarrosa	leaf resin	Clark and Wollenweber (1985)
22.18	*3, 5, 7, 3′-tetraMe		*Astragalus centralpinus* (Fab.)	aerial parts	Paskov and Manchkova (1983)
22.19	3, 5, 7, 4′-tetraMe				
22.20	3, 5, 3′, 4′-tetraMe				
22.21	3, 7, 3′, 4′-tetraMe	Retusin	*Boesenbergia pandurata* (Zing.)	rhizome	Jaipetch *et al.* (1983)
			Cistus, 3 spp. (Cist.)	leaf resin	Wollenweber *et al.* (1984)
			Distemonanthus benthamianus (Fab.)	heartwood	Malan and Roux (1979)
			Hazardia squarrosa	leaf resin	Clark and Wollenweber (1985)
			Parthenium spp.	leaf	Mears (1980)
			Pluchea odorata	leaf exudate	Wollenweber *et al.* (1985b)
			Solanum pubescens (Sol.)	leaf	Kumari *et al.* (1985)
			Sterculia foetida (Ster.)	bark	Anjaneyulu and Murty (1981)
22.22	*5, 7, 3′, 4′-tetraMe				
23	3, 5, 8, 3′, 4′-pentaOH (critical)				
24	3, 6, 7, 3′, 4′-pentaOH	Rhynchosin	*Rhynchosia beddomei* (Fab.)	leaf	Adinarayana *et al.* (1980)
25	3, 7, 8, 3′, 4′-pentaOH	Melanoxetin			
25.1	3-Me	Transilitin			
25.2	8-Me				
25.3	3, 8-diMe				
26	*3, 7, 2′, 3′, 4′-pentaOH (G)				
27	3, 7, 3′, 4′, 5′-pentaOH	Robinetin			

	HEXA-O-SUBSTITUTED FLAVONOLS				
28	(3, 5, 6, 7, 8, 3′-hexaOH)				
28.1	3, 6, 7, 8, 3′-pentaMe	Emmaosunin			
29	(3, 5, 6, 7, 8, 4′-hexaOH)				
29.1.	*3, 6-diMe		*Gutierrezia microcephala*	aerial parts	Fang *et al.* (1986a)
29.2	*3, 6, 8-triMe	Sarothrin	*Baccharis sarothroides*	aerial parts, exudate	Wollenweber *et al.* (1986)
			Gutierrezia 4 spp.	leaf	Fang *et al.* (1985a, 1986a), Roitman and James (1985), Fairchild (1976)
29.3	3, 6, 7, 8-tetraMe	Calycopterin	*Baccharis sarothroides*	leaf, exudate	Wollenweber *et al.* (1986)
			Calycadenia ciliosa	whole plant?	Emerson *et al.* (1986)
			Gardenia fosbergii (Rub.)	bud exudate	Gunatilaka *et al.* (1979)
			Gutierrezia microcephala	aerial parts	Fang *et al.* (1986a)
29.4	3, 6, 8, 4′-tetraMe	Araneosol	*Artemisia ludoviciana*	aerial parts	Liu and Mabry (1982)
29.5	5, 6, 7, 4′-tetraMe	Eriostemin			
29.6	3, 6, 7, 8, 4′-pentaMe	5-Hydroxy auranetin	*Digitalis viridiflora* (Scroph.)	leaf	Imre *et al.* (1984)
			Hyptis tomentosa (Lam.)	aerial parts	Kingston *et al.* (1979)
30	(3, 5, 6, 7, 2′, 4′-hexaOH)				
30.1	3, 6, 7, 2′-tetraMe	Chrysosplin			
31	3, 5, 6, 7, 3′, 4′-hexaOH	Quercetagetin	*Eupatorium gracile*	flowerhead	Torrenegra *et al.* (1984)
31.1	3-Me		*Gutierrezia microcephala*	aerial parts	Roitmann and James (1985)
			Haplopappus rengifoanus	leaf	Ulubelen *et al.* (1981c)
31.2	5-Me				
31.3	6-Me	Patuletin	*Arnica* 3 spp.	flowerhead	Merfort (1984b, 1985) Wolf and Denford (1984)
			Brickellia cylindracea	leaf	Ulubelen *et al.* (1981c)
			Eupatorium gracile	flowerhead	Torrenegra *et al.* (1984)
			Hazardia squarrosa	leaf resin	Clark and Wollenweber (1985)
			Hemizonia, 4 spp.	leaf resin	Proksch *et al.* (1984)
			Tagetes patula	petals	Bhardwaj *et al.* (1980)
31.4	3′-Me (G)				
31.5	*4′-Me (G)				
31.6	3, 6-diMe	Axillarin	*Achillea*, 3 spp.		Wollenweber *et al.* (1987a)
			Artemisia 3 spp.	aerial parts	McCormick and Bohm (1986) Liu and Mabry (1982) Oganesyan *et al.* (1976)
			Balsamorhiza macrophylla	leaf, exudate	McCormick *et al.* (1985)
			Gutierrezia 2 spp.	leaf	Fang *et al.* (1985a, 1986a)
			Hazardia squarrosa	leaf resin	Clark and Wollenweber (1985)
			Hemizonia, 4 spp.	aerial parts	McCormick and Bohm (1986)
			Heterotheca grandiflora	leaf resin	Wollenweber *et al.* (1985c)
			Neurolaena 4 spp.	leaf	Kerr *et al.* (1981) Ulubelen and Mabry (1981), Ulubelen *et al.* (1981a)
			Parthenium spp.	leaf	Mears (1980)
			Pluchea chingoyo	aerial parts	Chiang and Silva (1978)
			Wyethia bolanderi	leaf, exudate	McCormick *et al.* (1985)

(Contd.)

Table 7.2 (*Contd.*)

No.	*Substitution*	*Trivial name*	*Plant Sources*[†]		*References*
31.7	3, 7-diMe	Tomentin	*Haplopappus rengifoanus*	leaf	Ulubelen *et al.* (1981c)
			Neurolaena 4 spp.	leaf	Kerr *et al.* (1981) Ulubelen and Mabry (1981) Ulubelen *et al.* (1981a)
			Pulicaria 2 spp.	leaf	El-Negoumy *et al.* (1982) Pares *et al.* (1981)
31.8	*3, 3′-diMe				
31.9	6, 7-diMe	Eupatolitin	*Artemisia* 2 spp.	aerial parts	Kiso *et al.* (1982), Chandrasekharan *et al.* (1981)
			Brickellia glutinosa	aerial parts	Goodwin *et al.* (1984)
			Chromolaena meridensis	aerial parts	Amaro and Mendez (1983)
			Decachaeta haenkeana	aerial parts	Miski *et al.* (1985)
			Eupatorium areolare	leaf + flower	Yu *et al.* (1986)
			Rudbeckia serotina	aerial parts	Waddel and Elkins (1985)
31.10	6, 3′-diMe	Spinacetin	*Arnica* 3 spp.	flowerhead/ aerial parts	Merfort (1985), Wolf and Denford (1984)
			Artemisia lindleyana	aerial parts	McCormick and Bohm (1986)
			Balsamorhiza macrophylla	leaf, exudate	McCormick *et al.* (1985)
			Decachaeta haenkeana	aerial parts	Miski *et al.* (1985)
			Hazardia squarrosa	leaf resin	Clark and Wollenweber (1985)
			Heterotheca grandiflora	leaf resin	Wollenweber *et al.* (1985c)
31.11	6, 4′-diMe	Laciniatin	*Arnica* 2 spp.	flowerhead	Merfort (1984b)
			Pulicaria arabica	leaf + infl.	El-Negoumy *et al.* (1982)
31.12	*3, 5, 7-triMe				
31.13	3, 6, 7-triMe	Chrysosplenol-D	*Achillea ageratifolia*	aerial parts	Valant-Vetschera and Wollenweber (1986)
			Artemisia lindleyana	aerial parts	McCormick and Bohm (1986)
			Barbacenia purpurea (Velloziac.)	aerial parts	Wollenweber (1984a)
			Brickelia glutinosa	aerial parts	Goodwin *et al.* (1984)
			Heterotheca 2 spp.	leaf resin	Wollenweber *et al.* (1985c)
			Vitex agnuscasti (Verb.)	fruit	Wollenweber and Mann (1983)
31.14	3, 6, 3′-triMe	Jaceidin	*Achillea biebersteinii*	flowerhead	Oskay and Yesilada (1984)
			Artemisia lindleyana	aerial parts	McCormick and Bohm (1986)
			Baccharis gilliessii	aerial parts	Gianello and Giordano (1984)
			Gutierrezia microcephala	aerial parts	Roitman and James (1985), Fang *et al.* (1986a)
			Jasonia montana See section 7.4.2 (no. 31.23)		
			Pluchea odorata	leaf exudate	Wollenweber *et al.* (1985b)
			Stevia cuzcoensis	leaf	Rajbhandari and Roberts (1985b)
			Tanacetum vulgare	flowerhead	Ognyanov and Todorova (1983)
			Wyethia bolanderi	leaf wash	McCormick *et al.* (1985)
31.15	3, 6, 4′-triMe	Centaureidin	*Achillea*, 3 spp.	aerial parts	Valant-Vetschera and Wollenweber (1986)
			Baccharis sarothroides	leaf, twig	Kupchan and Bauerschmidt (1971)
			Balsamorhiza macrophylla	leaf exudate	McCormick *et al.* (1985)
			Brickellia 2 spp.	leaf	Timmermann *et al.* (1981), Timmermann and Mabry (1983)

			Gutierrezia grandis	leaf	Fang *et al.* (1985a)
			Hazardia squarrosa	leaf resin	Clark and Wollenweber (1985)
			Hymenoclea salsola	leaf resin	Proksch *et al.* (1983)
			Pluchea 2 spp.	aerial parts	Chiang and Silva (1978) Wollenweber *et al.* (1985b)
			Stevia 4 spp.	leaf	Rajbhandari and Roberts (1983, 1984b, 1985a, b)
31.16	3, 7, 3′-triMe	Chrysosplenol-C			
31.17	3, 7, 4′-triMe	Oxyayanin-B	*Ambrosia cordifolia*	leaf resin	Wollenweber *et al.* (1987b)
			Distemonanthus benthamianus (Fab.)	heartwood	Malan and Roux (1979)
			Stevia cuzcoensis	leaf resin	Rajbhandari and Roberts (1985b)
31.18	6, 7, 3′-triMe	Veronicafolin	*Rudbeckia serotina*	aerial parts	Waddel and Elkins (1985)
31.19	6, 7, 4′-triMe	Eupatin	*Brickellia baccharidea*	leaf	Timmermann *et al.* (1981)
			Heterotheca grandiflora	leaf resin	Wollenweber *et al.* (1985c)
			Pericome caudata	aerial parts	Wollenweber (unpublished)
31.20	*6, 3′, 4′-triMe		*Arnica chamissonis*	flowerhead	Merfort (1985)
			Decachaeta haenkeana	leaf	Miski *et al.* (1985)
31.21	3, 5, 6, 3′-tetraMe				
31.22	*3, 5, 7, 3′-tetraMe		*Pulicaria arabica*	leaf + infl.	El-Negoumy *et al.* (1982)
31.23	3, 6, 7, 3′-tetraMe	Chrysosplenetin [Polycladin]	*Brickellia arguta*	aerial parts	Rösler and Goodwin (1984)
			Digitalis thapsii (Scroph.) See section 7.4.2	leaf	De Pascual Teresa *et al.* (1980)
			Matricaria chamomilla	flowerhead	Exner *et al.* (1981b)
31.24	3, 6, 7, 4′-tetraMe	Casticin	*Achillea nobilis*	aerial parts	Valant-Vetschera and Wollenweber (1986)
			Brickellia baccharidea	leaf	Timmermann *et al.* (1981)
			Bromelia pinguin (Brom.)	root + basal stem	Raffauf *et al.* (1981)
			Gutierrezia microcephala	aerial parts	Fang *et al.* (1986a)
			Heterotheca psammophila	leaf resin	Wollenweber *et al.* (1985c)
			Jasonia montana	aerial parts	El-Din *et al.* (1985)
			Pluchea odorata	leaf exudate	Wollenweber *et al.* (1983b)
			Tanacetum santolinoides	aerial parts	El-Din *et al.* (1985)
31.25	3, 6, 3′, 4′-tetraMe	Bonanzin	*Artemisia* 3 spp.	aerial parts	Liu and Mabry (1981, 1982) McCormick and Bohm (1986)
			Brickellia 2 spp.	leaf	Timmermann and Mabry (1983) Goodwin *et al.* (1984)
			Stevia cuzcoensis	leaf	Rajbhandari and Roberts (1985b)
31.26	3, 7, 3′, 4′-tetraMe		*Brickellia cylindracea*	leaf	Timmermann and Mabry (1983)
			Distemonanthus benthamianus (Fab.)	heartwood	Malan and Roux (1979)
31.27	5, 6, 7, 4′-tetraMe	Eupatoretin			
31.28	*6, 7, 3′, 4′-tetraMe		*Artemisia lanata*	aerial parts	Esteban *et al.* (1986)
			Pericome caudata	aerial parts	Wollenweber (unpublished)
			Heteromma simplicifolium See section 7.4.2 (no. 34.19)		
31.29	3, 6, 7, 3′, 4′-pentaMe	Artemetin	*Achillea*, 7 spp.	aerial parts	Valant-Vetschera and Wollenweber (1986) Sanz er al. (1985)

(*Contd.*)

Table 7.2 (*Contd.*)

No.	Substitution	Trivial name	Plant Sources[†]		References
			Artemisia 2 spp.	aerial parts	Esteban *et al.* (1986), McCormick and Bohm (1986)
			Brickellia 2 spp.	leaf	Timmermann *et al.* (1981), Timmermann and Mabry (1983)
			Jasonia montana	aerial parts	El-Din *et al.* (1985)
			Pluchea odorata	leaf resin	Arriaga-Giner *et al.* (1983) Wollenweber *et al.* (1985b)
			Stevia procumbens	aerial parts	Sosa *et al.* (1985)
			Tanacetum santolinoides	aerial parts	El-Din *et al.* (1985)
			Verbena officinalis (Verb.)	aerial parts	Makboul (1986)
31.30	3, 5, 6, 7, 3′, 4′-hexaMe				
32	(3, 5, 7, 8, 2′, 4′-hexaOH)				
32.1	3, 7, 8, 2′-tetraMe				
32.2	3, 7, 8, 2′, 4′-pentaMe				
33	(3, 5, 7, 8, 2′, 5′-hexaOH)				
33.1	[7, 2′, 5′-triMe]		*Notholaena aliena* (Pterid.) -as 8-acetate-	frond exudate	Wollenweber (1984b)
34	3, 5, 7, 8, 3′, 4′-hexaOH	Gossypetin	*Gutierrezia microcephala*	aerial parts	Roitman and James (1985)
34.1	*3-Me				
34.2	7-Me (G)	Ranupenin			
34.3	8-Me (G)	Corniculatusin			
34.4	3′-Me (G)				
34.5	3, 7-diMe		*Geraea canescens*	ariel parts	Proksch *et al.* (1986)
34.6	3, 8-diMe		*Gutierrezia* 2 spp.	leaf	Fang *et al.* (1985a, 1986a) Roitman and James (1985)
34.7	*3, 3′-diMe		*Gutierrezia microcephala*	aerial parts	Roitman and James (1985)
34.8	7, 4′-diMe				
34.9	8, 3′-diMe	Limocitrin			
34.10	8, 4′-diMe (G)				
34.11	3, 7, 8-triMe		*Calycadenia ciliosa*	whole plant	Emerson *et al.* (1986)
34.12	3, 7, 3′-triMe		*Notholaena aschenborniana* (Pterid.) -as 8-acetate-	frond exudate	Jay *et al.* (1982)
34.13	3, 8, 3′-triMe		*Gutierrezia* 2 spp.	leaf, ariel parts	Fang *et al.* (1985a)
			Geraea canescens	aerial parts	Proksch *et al.* (1986)
34.14	7, 8, 3′-triMe				
34.15	3, 7, 8, 3′-tetraMe	Ternatin	*Begonia glabra* (Beg.)	aerial parts	Ensemeyer and Langhammer (1982)
			(*Chromolaena morii*) See section 7.4.2		
			Larrea tridentata (Zyg.)	leaf	Bernhard and Thiele (1981)
34.16	3, 7, 8, 4′-tetraMe				
34.17	3, 7, 3′, 4′-tetraMe		*Cistus parviflorus* (Cist.)	leaf exudate	Vogt *et al.* (1986)
34.18	3, 8, 3′, 4′-tetraMe		*Parastrephia quadrangularis*	aerial parts	Loyola *et al.* (1985)

No.	Substitution	Trivial name	Source	Plant part	Reference
34.19	*7, 8, 3′, 4′-tetraMe		*Artemisia lanata*		Esteban *et al.* (1986)
			See section 7.4.2		
34.20	3, 7, 8, 3′, 4′-pentaMe		*Cistus albanicus* (Cist.)	leaf resin	Vogt *et al.* (1987)
			Micromelum zeylanicum (Rut.)	leaf, stem	Bowen *et al.* (1982)
34.21	3, 5, 7, 8, 3′, 4′-hexaMe				
34A	*3, 6, 7, 8, 3′, 4′-hexaOH		*Tagetes erecta*	petal	Bhardwaj *et al.* (1983)
35	(3, 5, 7, 2′, 3′, 4′-hexaOH)				
35.1	(2′, 4′-diMe)	[Viscidulin III]	See section 7.4.1 (no. 41.1)		
35.2	3, 7, 4′-triMe	Apuleidin			
36	3, 5, 7, 2′, 4′, 5′-hexaOH	5′-OH Morin			
36.1	3, 7, 4′-triMe	Oxyayanin-A			
36.2	3, 5, 7, 4′-tetraMe				
36.3	3, 7, 4′, 5′-tetraMe		*Distemonanthus benthamianus* (Fab.)	heartwood	Malan and Roux (1979)
36.4	3, 5, 7, 4′, 5′-pentaMe		*Distemonanthus benthamianus* (Fab.)	heartwood	Malan and Roux (1979)
36.5	3, 7, 2′, 4′, 5′-pentaMe		*Distemonthus benthamianus* (Fab.)	heartwood	Malan and Roux (1979)
36.6	3, 5, 7, 2′, 4′, 5′-hexaMe		*Distemonanthus benthamianus* (Fab.)	heartwood	Malan and Roux (1979)
37	(3, 6, 7, 2′, 4′, 5′-hexaOH)				
37.1	3, 6, 7, 4′-tetraMe	5-*O*-Demethylapulein			
38	3, 5, 7, 3′, 4′, 5′-hexaOH	Myricetin	*Haplopappus canescens*	aerial parts	Öksüz *et al.* (1981)
38.1	3-Me	Annulatin	*Cereus jamacaru, Cleistocerus variispinus, Mammillaria elongata* (Cact.)		Burret *et al.* (1982)
38.2	5-Me (G)				
38.3	7-Me (G)	Europetin			
38.4	3′-Me (G)	Laricitrin			
38.5	4′-Me	Mearnsetin	*Rhamnus pallasii* (Rham.)	bark	Sakushima *et al.* (1983)
38.6	3, 4′-diMe		*Gutierrezia grandis*	leaf	Fang *et al.* (1985a)
38.7	3, 5′-diMe				
38.8	*7, 4′-diMe		*Rhus lancea* (Anacard.)	leaf	Nair *et al.* (1983)
38.9	*3′, 4′-diMe		*Cistus symphytifolius* (Cist.)	leaf resin	Vogt *et al.* (1987)
			Haplopappus integerrimus	leaf	Ayanoglu *et al.* (1981)
38.10	3′, 5′-diMe	Syringetin	*Cistus symphytifolius* (Cist.)	leaf resin	Vogt *et al.* (1987)
38.11	*3, 7, 3′-triMe		*Solanum pubescens* (Sol.)	leaf	Kumari *et al.* (1984)
38.12	3, 7, 4′-triMe		*Cistus* sp. (Cist.)	leaf resin	Vogt *et al.* (1987)
			Gutierrezia alamanii	aerial parts	Lenherr *et al.* (1986)
38.13	*3, 3′, 4′-triMe		*Cistus monspeliensis* (Cist.)	leaf resin	Wollenweber and Mann (1984)
			Haplopappus integerrimus (Ast.)	leaf	Ayanoglu *et al.* (1981)
38.14	7, 3′, 4′-triMe		*Cistus monspeliensis* (Cist.)	leaf resin	Wollenweber and Mann (1984)
38.15	3, 7, 3′, 4′-tetraMe		*Cistus monspeliensis* (Cist.)	leaf resin	Wollenweber and Mann (1984)
			Notholaena candida (Pterid.)	frond exudate	Wollenweber (1984b)
38.16	*3, 7, 3′, 5′-tetraMe		*Ledum palustre* (Eric.)	leaf	Mikhailova and Rybalko (1980)
			Solanum pubescens (Sol.)	leaf	Kumari *et al.* (1984)

(*Contd.*)

Table 7.2 (*Contd.*)

No.	Substitution	Trivial name	Plant Sources†		References
38.17	*3, 3′, 4′, 5′-tetraMe		*Adina cordifolia* (Rub.)	stem	Srivastava *et al.* (1981)
			Cistus sp. (Cist.)	leaf resin	Vogt *et al.* (1987)
			Gutierrezia alamanii	aerial parts	Lenherr *et al.* (1986)
38.18	3, 7, 3′, 4′, 5′-pentaMe	Combretol	*Notholaena candida* var. *candida* (Pterid.)	frond exudate	Wollenweber (1984b)
	HEPTA-O-SUBSTITUTED FLAVONOLS				
39	(3, 5, 6, 7, 8, 2′, 4′-heptaOH)				
39.1	3, 6, 7, 8-tetraMe				
39.2	3, 6, 7, 8, 2′-pentaMe				
39.3	3, 6, 7, 8, 4′-pentaMe				
39.4	3, 6, 7, 8, 2′, 4′-hexaMe				
40	*3, 5, 6, 7, 8, 3′, 4′-heptaOH		*Tagetes erecta*	petal	Bhardwaj *et al.* (1983)
40.1	3, 6-diMe		*Gutierrezia microcephala*	aerial parts	Roitman and James (1985)
40.2	*6, 8-diMe		*Gutierrezia* 2 spp.	leaf	Fang *et al.* (19861b), Roitman and James (1985)
40.3	*3, 6, 8-triMe		*Gutierrezia* 2 spp.	leaf	Fang *et al.* (1985a, 1986a), Roitman and James (1985)
			Hemizonia, 3 spp.	aerial parts	Proksch *et al.* (1984)
40.4	*3, 7, 8-triMe		*Calycadenia ciliosa*	whole plant	Emerson *et al.* (1986)
40.5	*6, 7, 8-triMe		*Gutierrezia microcephala*	aerial parts	Fang *et al.* (1986a)
40.6	6, 8, 3′-triMe	Limocitrol	*Gutierrezia microcephala*	aerial parts	Roitman and James (1985)
40.7	6, 8, 4′-triMe	Isolimocitrol			
40.8	*3, 6, 7, 8-tetraMe		*Calycadenia ciliosa*	whole plant	Emerson *et al.* (1986)
			Gutierrezia 2 spp.	aerial parts	Fang *et al.* (1986a) Bittner *et al.* (1983)
40.9	3, 6, 8, 3′-tetraMe		*Gutierrezia* 2 spp.	aerial parts	Roitman and James (1985), Hradetzky *et al.* (1986)
40.10	*3, 6, 8, 4′-tetraMe		*Gutierrezia* 3 spp.	leaf	Fang *et al.* (1985a, 1986a) Hradetzky *et al.* (1986)
40.11	3, 6, 7, 8, 3′-pentaMe		*Baccharis incarum*	leaf	Faini *et al.* (1982)
			Calycadenia ciliosa	whole plant	Emerson *et al.* (1986)
			Gutierrezia 2 spp.	aerial parts	Roitman and James (1985), Bittner *et al.* (1983)
40.12	*3, 6, 7, 8, 4′-pentaMe		*Gutierrezia* 2 spp.	aerial parts	Fang *et al.* (1986a) Hradetzky *et al.* (1986)
40.13	3, 6, 7, 8, 3′, 4′-hexaMe				
40.14	5, 6, 7, 8, 3′, 4′-hexaMe	Natsudaidain			
40.15	3, 5, 6, 7, 8, 3′, 4′-heptaMe				
41	(3, 5, 6, 7, 2′, 3′, 4′-heptaOH)				
41.1	3, 7, 4′-triMe	Apuleisin			
41.2	3, 5, 6, 7, 2′, 3′, 4′-heptaMe				
43	(3, 5, 6, 7, 2′, 4′, 5′-heptaOH)				
43.1	*3, 6, 4′-triMe		*Gutierrezia grandis*	leaf	Fang *et al.* (1985a)
43.2	*3, 6, 5′-triMe		*Gutierrezia grandis*	leaf	Fang *et al.* (1985a)

43.3	*3, 6, 7, 4′-tetraMe(G)		*Gutierrezia microcephala*	leaf	Fang *et al.* (1985b, 1986a)
43.4	*3, 6, 4′, 5′-tetraMe		*Distemonanthus benthamianus* (Fab.)	wood	Malan and Naidoo (1981)
43.5	*3, 7, 2′, 4′-tetraMe		*Gutierrezia microcephala*	aerial parts	Fang *et al.* (1986a)
43.6	3, 5, 6, 7, 4′-pentaMe	Apulein	Synthesis:	aerial parts	Iinuma *et al.*, (1986a)
43.7	*3, 5, 7, 2′, 4′-pentaMe		*Distemonanthus benthamianus* (Fab.)	wood	Malan and Naidoo (1981)
43.8	3, 5, 7, 4′, 5′-pentaMe				
43.9	*3, 6, 7, 2′, 4′-pentaMe (G)				
43.10	*3, 6, 7, 4′, 5′-pentaMe	Brickellin	*Brickellia glutinosa*	aerial parts	Goodwin *et al.* (1984)
			Brickellia 2 spp.		
			synthesis [See section 7.4.1]		Iinuma *et al.* (1985b)
43.11	3, 7, 2′, 4′, 5′-pentaMe				
43.12	3, 5, 6, 7, 2′, 4′, 5′-heptaMe				
44	(3, 5, 6, 7, 3′, 4′, 5′-heptaOH)	6-OH-myricetin			
44.1	3, 6, diMe				
44.2	3, 4′-diMe		*Gutierrezia* 2 spp.	leaf	Fang *et al.* (1986ab)
44.3	*6, 4′-diMe				
44.4	3, 6, 3′-triMe		*Alluaudia ascendens* (Did.)	stem bark	Rabesa and Voirin (1979)
44.5	3, 6, 4′-triMe		*Gutierrezia grandis*	leaf	Fang *et al.* (1985a)
44.6	*3, 6, 5′-triMe		*Gutierrezia microcephala*	aerial parts	Fang *et al.* (1986a)
44.7	3, 3′, 5′-triMe				
44.8	*6, 3′, 5′-triMe (G)		See Section 7.4.2		
44.9	3, 6, 7, 4′-tetraMe		*Gardenia cramerii* (Rub.)	bud exudate	Gunatilaka *et al.* (1982)
44.10	3, 6, 3′, 4′-tetraMe		*Gutierrezia microcephala*	aerial parts	Fang *et al.* (1986a)
44.11	3, 6, 3′, 5′-tetraMe		*Gutierrezia microcephala*	aerial parts	Fang *et al.* (1986a)
44.12	3, 7, 3′, 4′-tetraMe	Apuleitrin			
44.13	3, 5, 7, 3′, 4′-pentaMe	Apuleirin			
44.14	3, 6, 7, 3′, 4′-pentaMe				
44.15	3, 6, 7, 3′, 5′-pentaMe	Murrayanol			
44.16	[6, 7, 3′, 4′, 5′-pentaMe]		See Section 7.4.2		
44.17	3, 5, 6, 7, 3′, 5′-hexaMe				
44.18	3, 5, 6, 7, 3′, 4′, 5′-heptaMe				
45	(3, 5, 7, 8, 2′, 3′, 4′-heptaOH)				
45.1	*3, 7, 2′, 3′, 4′-pentaMe		*Notholeana aschenborniana* (Pterid.) -as 8-acetate-	farinose frond exudate	Jay *et al.* (1982)
46	(3, 5, 7, 8, 2′, 4′, 5′-heptaOH)				
46.1	*3, 8, 5′-triMe		*Gutierrezia microcephala*	leaf	Fang *et al.* (1985b, 1986a)
46.2	*3, 7, 8, 5′-tetraMe		*Notholaena aschenborniana* (Pterid.)		
			See Section 7.4.1 (Table 7.1, no. 63.1)		
47	3, 5, 7, 8, 3′, 4′, 5′-heptaOH (G)	Hibiscetin			
47.1	3, 7, 4′-triMe				
47.2	*3, 8, 4′-triMe		*Gutierrezia* 2 spp.	leaf	Fang *et al.* (1985a, 1986a)
47.3	3, 7, 8, 4′-tetraMe				
47.4	3, 8, 4′, 5′-tetraMe				
47.5	3, 8, 3′, 4′, 5′-pentaMe	Conyzatin	*Gutierrezia microcephala*	aerial parts	Roitman and James (1985)
47.6	7, 8, 3′, 4′, 5′-pentaMe				
47.7	3, 5, 7, 8, 3′, 4′, 5′-heptaMe		*Murraya paniculata* (Rut.)	leaf	De Silva *et al.* (1980)

(*Contd.*)

Table 7.2 (*Contd.*)

No.	Substitution	Trivial name	Plant Sources[†]		References
	OCTA-O-SUBSTITUTED FLAVONOLS				
48	(3, 5, 6, 7, 8, 2′, 4′, 5′-octaOH)				
48.1	*3, 6, 8, 4′-tetraMe		*Gutierrezia* 2 spp.	leaf	Fang *et al.* (1985a, b, 1986a)
48.2	*3, 6, 8, 5′-tetraMe		*Gutierrezia* 2 spp.		
48.3	*3, 7, 8, 4′-tetraMe		*Gutierrezia microcephala*	aerial parts	Fang *et al.* (1986a)
48.4	*3, 6, 8, 4′, 5′-pentaMe		Gutierrezia microcephala synthesis:	leaf	Fang *et al.* (1985b, 1986a) Iinuma *et al.* (1986b)
48.5	*3, 6, 7, 8, 4′, 5′-hexaMe		*Gutierrezia microcephala*	leaf	Fang *et al.* (1985b, 1986a)
48.6	*3, 5, 6, 7, 8, 2′, 4′, 5′-octaMe	Purpurascenin	*Pogostemon purpurascens* (Lam.)	whole plant	Patwardhan and Gupta (1981)
49	(3, 5, 6, 7, 8, 3′, 4′, 5′-octaOH)				
49.1	*3, 6, 8-triMe		*Gutierrezia* 2 spp.	leaf	Fang *et al.* (1985a, 1986a)
49.2	*3, 6, 8, 4′-tetraMe		*Gutierrezia* 3 spp.	aerial parts	Bohlmann *et al.* (1981a), Fang *et al.* (1985a, 1986a)
49.3	*3, 6, 8, 5′-tetraMe		*Gutierrezia microcephala*	aerial parts	Fang *et al.* (1986a)
49.4	*3, 7, 8, 4′-tetraMe		*Gutierrezia microcephala*	aerial parts	Fang *et al.* (1986a)
49.5	*3, 6, 7, 8, 4′-pentaMe		*Gutierrezia* 3 spp.	aerial parts	Bohlmann *et al.* (1981a), Fang *et al.* (1985a, 1986a)
49.6	*3, 6, 8, 3′, 5′-pentaMe		*Gutierrezia* 2 spp.	leaf	Fang *et al.* (1985a), Roitman and James (1985)
49.7	*3, 6, 8, 4′, 5′-pentaMe		*Gutierrezia* 2 spp.	leaf	Fang *et al.* (1986b), Roitman and James (1985)
49.8	*3, 5, 6, 7, 8, 4′-hexaMe		*Gutierrezia dracunculoides*	aerial parts	Bohlmann *et al.* (1981a)
49.9	3, 6, 7, 8, 4′, 5′-hexaMe	Digicitrin	*Gutierrezia microcephala* *Polygonum orientale* (Polygon.)	aerial parts aerial parts	Fang *et al.* (1986a) Kuroyanagi and Fukushima (1982)
49.10	*3, 6, 8, 3′, 4′, 5′-hexaMe		*Gutierrezia* 2 spp.	leaf	Fang *et al.* (1985a), Roitman and James (1985)
49.11	*3, 5, 6, 7, 8, 4′, 5′-heptaMe		*Polygonum orientale* (Polygon.)	aerial parts	Kuroyanagi and Fukushima (1982)
49.12	*3, 6, 7, 8, 3′, 4′, 5′-heptaMe		*Gutierrezia microcephala*	aerial parts	Fang *et al.* (1986a)
49.13	3, 5, 6, 7, 8, 3′, 4′, 5′-octaMe	Exoticin	*Polygonum orientale* (Polygon.)	aerial parts	Kuroyanagi and Fukushima (1982)

*Compound newly reported during 1981–86, or earlier report not included in Wollenweber (1982a).
[†]Family is Asteraceae, unless otherwise indicated in parentheses.

bigelovii (Arriaga–Giner *et al.*, 1986) is the only representative to date of a compound with aromatic acylation, and so far it is also the only known acylated flavone. Structural formulae of these compounds are shown in Fig. 7.1.

7.2.2. Free flavonoid aglycones in the Hepaticae

The presence of free flavonoid agylcones has been demonstrated for the first time for some liverworts. Apigenin 7,4′-dimethyl ether was detected in *Frullania vethii* by Asakawa *et al.* (1980). The same flavone was found later together with luteolin 3′,4′-dimethyl ether and luteolin 7,3′,4′-trimethyl ether in *Frullania davurica* and in *F. jackii*, while 6-hydroxyluteolin 7-methyl ether (pedalitin) was found in the latter species only (Mues *et al.*, 1984). Luteolin and pedalitin have been found as free aglycones in *F. dilatata* (Mues *et al.*, 1983). Kruijt *et al.* (1986) recently reported kaempferol 3-methyl ether in the free state from *F. densifolia*. According to Mues (pers. comm.) free aglycones are also present in other species of *Frullania*.

These flavones were found in ether and chloroform extracts, hence their occurrence as in the free state can be regarded as certain. This is hardly surprising since these liverworts also produce terpenoids as well as lipophilic aromatic compounds such as 3-methoxy-4′-hydroxybibenzyl; the latter is the major constituent of the lipophilic extract of *F. jackii* (Mues *et al.*, 1984). All these non-polar compounds are presumably localized in the oil bodies that are typical of many liverworts.

The two flavones, apigenin and luteolin, and the isoflavones, orobol and pratensein, have also been reported in the free state from the moss *Bryum capillare* (Anhut *et al.*, 1984). However, in this case they were found in a methanolic extract, so it is possible that the aglycones arose by chemical or enzymic hydrolysis during extraction.

7.3 FLAVONOIDS WITH COMPLEX SUBSTITUTION

7.3.1 *C*-Methylflavonoids

Among the *C*-methylated flavonoids, there are 20 compounds that have been found for the first time as natural products; 41 *C*-methylflavones and *C*-methylflavonols are now known in all. Many of the previously known products have been detected in new plant sources. Until recently, *C*-methylflavones were known only in the Myrtaceae, except that one compound had been found in the genus *Pinus*. The Myrtaceae is still the major source, with reports from the genera *Angophora*, *Eucalyptus*, *Eugenia*, *Lophostemon* and *Syncarpia*; the ericaceous genera *Gaultheria*, *Kalmia* and *Ledum* also contain these compounds. Other sources include one species in the Annonaceae (*Unona lawii*, three formylflavones), one in the Clusiaceae (*Hypericum ericoides*, two flavones) and one in the Fabaceae (*Dalea tuberculata*). Turning to the *C*-methylflavonols, the fern *Pityrogramma triangularis* var. *triangularis* (Pteridaceae) produces ten compounds of this type, nine of which are novel. Five new flavonols have been reported from *Kalmia latifolia* (Ericaceae). Other sources (Table 7.3) include members of the Agavaceae, Didiereaceae, Liliaceae, Meliaceae and Pinaceae. The *C*-methylated flavones and flavonols occur in either leaf wax or lipophilic farinose fern frond exudates. Previously *C*-methylation was thought to always occur at C-6 and/or C-8, but now two flavones have been found which are *C*-methylated at C-3, and one is said to be a 7-*C*-methyl derivative (Table 7.3).

7.3.2 Methylenedioxyflavonoids

Table 7.4, which lists 33 methylenedioxy derivatives, includes five novel flavones and two novel flavonols. The families in which these compounds are found are the Rutaceae, Fabaceae, Asteraceae, Chenopodiaceae and Polygonaceae. Methylenedioxy substitution is still almost exclusively at C-6/C-7 and/or C-3′/C-4′ (except for wharangin with C-7/C-8 substitution).

7.3.3 Flavones and flavonols with isoprene substitution

Compounds with simple C_5- or hydroxy-C_5- and C_{10}-side chains are listed in Table 7.5. Except for one flavone and two flavonols with *O*-prenylation, the isoprenoid side chains are always directly linked through carbon to the aromatic rings. The abbreviation used here to indicate the isoprene side chain does not distinguish between isomeric forms. For this reason in Table 7.5 *cis*- and *trans*-tephrostachin [(*7.16*), (*7.17*)] as well as norartocarpin and mulberrin are cited with one substitution each. The only two flavonols with *C*-prenylation at the B-ring have been reported from *Dodonaea viscosa* from India (occurring probably in the leaf resin). They are 5,7-dihydroxy-3,6,4′-trimethoxy-3′-(3-hydroxymethylbutyl)flavone (Sachdev and Kulshreshtha, 1983) and 5,7,4′-trihydroxy-3,6-dimethoxy-3′-(3-hydroxymethylbutyl)flavone (for unknown reasons designated with the trivial name aliarin; Sachdev and Kulshreshtha, 1982) (Table 7.5). Neither of these two compounds were found in a Mexican population of this plant (Wollenweber *et al.*, 1986), but this is probably because of the considerable intraspecific variability of the *Dodonaea viscosa* species complex. One flavone has the rare *O*-allyl side chain (*7.15*), so far unknown in flavonoids [known e.g. for the neolignan, aurien (Gottlieb *et al.*, 1976), and for the coumarin, lacoumarin (Bhardwaj *et al.*, 1976)].

7.3.4 Pyrano- and furano-flavonoids

Pyranoflavonoids are listed in Table 7.6. The pyran ring is indicated by the abbreviation ODmp for the *O*-linked dimethylallyl unit ($Me_2C—C=CH—$), the first cipher

Table 7.3 *C*-Methylated flavonoids

No.	Substitution	Trivial name	Plant source		References
	C-METHYLFLAVONES				
1	5,7-diOH, 6-Me	6-Methylchrysin Strobochrysin	*Hypericum ericoides* (Clus.)	aerial parts	Cardona and Seoan (1982)
2	*5-OH, 7-OMe, 6-Me		*Dalea tuberculata* (Fab.)	aerial parts	Dominguez *et al.* (1982)
3	*5-OH, 4′-OMe, 7-Me	Saltillin	*Eugenia kurzii* (Myrt.)	leaf	Painuly and Tandon (1983b)
4	*5,7,4′-triOH, 3-Me (G)		*Eucalyptus* spp. (Myrt.)	leaf wax	Wollenweber and Kohorst (1981)
5	*5,4′,-diOH, 7-OMe, 6-Me	8-Desmethyl-sideroxylin	*Gaultheria procumbens* (Eric.)	leaf wax	Wollenweber and Kohorst (1984)
			Hypericum ericoides (Clus.)	aerial parts	Cardona and Seoan (1982)
			Kalmia 3 spp. (Eric.)	leaf wax	Wollenweber and Kohorst (1981, 1984)
			Eucalyptus spp. (Myrt.)	leaf wax	Wollenweber and Kohorst (1981)
6	5,4′-diOH, 7-OMe, 6,8-diMe	Sideroxylin	*Kalmia latifolia* (Eric.)	leaf wax	Wollenweber and Kohorst (1981)
			Eucalyptus spp. (Myrt.)	leaf wax	Wollenweber and Kohorst (1981), Courtney *et al.* (1983)
7	5-OH, 7,4′-diOMe, 6-Me	8-Desmethyl-eucalyptin	*Gaultheria procumbens* (Eric.)	leaf wax	Wollenweber and Kohorst (1984)
			Kalmia 3 spp. (Eric.)	leaf wax	Wollenweber and Kohorst (1981, 1984)
			Ledum palustre (Eric.)	leaf	Zapesochnaya and Pangarova (1980)
			Lophostemon confertus (Myrt.)	leaf wax	Courtney *et al.* (1983)
8	5-OH, 7,4′-diOMe, 6,8-diMe	Eucalyptin	*Angophora hispida* × *bakeri* (Myrt.)	leaf wax	Courtney *et al.* (1983)
			Eucalyptus spp. (Myrt.)	leaf wax	Wollenweber and Kohorst (1981), Courtney *et al.* (1983), Cerecer *et al.* (1974)
			Kalmia latifolia (Eric.)	leaf wax	Wollenweber and Kohorst (1981)
			Syncarpia glomulifera (Myrt.)	leaf wax	Courtney *et al.* (1983)
9	*5,7,3′,4′-tetraOH, 3-Me (G)		*Eugenia kurzii* (Myrt.)	leaf	Painuly and Tandon (1983b)
10	*5,7-diOH-3′,4′-diOMe-6, 8-diMe		*Boerhaavia diffusa* (Nyct.)	root	Gupta and Ahmed (1984)
11	*5,7,3′-triOH-4′,5′-diOMe-6, 8-diMe		*Alluaudiopsis marnieriana* (Didier.)	bark	Rabesa and Voirin (1983)
12	5,7-diOH, 8-Me, 6-CHO	Unonal			
13	5,7-diOH, 6-Me, 8-CHO	Isounonal			
14	5-OH, 7-OMe, 8-Me, 6-CHO	Unonal 7-Methylether			
	C-METHYLFLAVONOLS				
1	*3,5,7-triOH, 8-Me	8-*C*-methyl-galangin	*Pityrogramma triangularis* (Pterid.)	frond exudate	Wollenweber *et al.* (1985d)
2	*3,5,7-triOH, 6,8-diMe		*Pityrogramma triangularis* (Pterid.)	frond exudate	Wollenweber *et al.* (1985d)
3	*5,7-diOH, 3-OMe-, 8-Me		*Pityrogramma triangularis* (Pterid.)	frond exudate	Wollenweber *et al.* (1985d)

6	*3, 5-diOH, 7-OMe, 6, 8-diMe		*Pityrogramma triangularis* (Pterid.)	frond exudate	Wollenweber *et al.* (1985d)
7	3, 5, 7-triOH, 8-OMe, 6-Me	Pityrogrammin	*Pityrogramma triangularis* (Pterid.)	frond exudate	Wollenweber *et al.* (1985d)
8	5, 6, 4′-triOH, 3-OMe, 8-Me	Sylpin	*Pityrogramma triangularis* (Pterid.)	frond exudate	Wollenweber *et al.* (1985d)
9	3, 5, 7, 4′-tetraOH, 6-Me	6-Methyl-kaempferol	*Cedrus deodara* (Pinac.)	leaf	Genderen and Van Schaik (1980)
10	5, 7, 4′-triOH, 3-OMe, 6-Me				
11	5, 7, 4′-triOH, 3-OMe, 6, 8-diMe		*Pityrogramma triangularis* (Pterid.)	frond exudate	Wollenweber *et al.* (1985d)
12	*5, 4′, diOH, 3, 7-diOMe, 6-Me		*Kalmia latifolia* (Eric.)	leaf wax	Wollenweber and Kohorst (1984)
13	*5, 4′-diOH, 3, 7-diOMe, 6, 8-diMe		*Kalmia latifolia* (Eric.)	leaf wax	Wollenweber and Kohorst (1984)
14	*5-OH, 3, 7, 4′-triOMe, 6-Me	8-Desmethyl-kalmiatin	*Kalmia latifolia* (Eric.)	leaf wax	Wollenweber and Kohorst (1984)
15	*5-OH, 3, 7, 4′-triOMe, 6, 8-diMe	Kalmiatin	*Kalmia latifolia* (Eric.)	leaf wax	Wollenweber and Kohorst (1984)
16	*3, 5, 8, 4′-tetraOH, 7-OMe, 6-Me		*Pityrogramma triangularis* (Pterid.)	frond exudate	Wollenweber *et al.* (1985d)
17	3, 5, 7, 3′, 4′-pentaOH, 6-Me	Pinoquercetin			
18	*3, 5, 7, 3′, 4′-pentaOH, 8-Me (G)		*Amoora rohituka* (Meliac.)	stem bark	Jain and Srivastava (1985)
19	3, 5, 7, 3′, 4′-pentaOH, 6, 8-diMe				
20	*5, 7, 3′, 4′-tetraOH, 3-OMe, 6-Me		*Dasylirion acrotrichum* (Agav.)	leaf	Laracine *et al.* (1982)
			Xanthorrhoea hastilis (Lil.)	leaf	Laracine *et al.* (1982)
21	5, 7, 4′-triOH, 3, 3′-diOMe, 6-Me				
22	*5, 7, 4′-triOH-3, 3′-diOMe-6, 8-diMe		*Alluaudia humbertii* (Did.)	aerial parts	Rabesa and Voirin (1985)
23A	(5, 4′-diOH-3, 7, 3′-triOMe-8-Me)		See Section 7.4.3		
23	*5, 4′-diOH-3, 7, 3′-triOMe-6-Me		*Alluaudia humbertii* (Did.)	aerial parts	Rabesa and Voirin (1985)
24	*5, 4′-diOH-3, 7, 3′-triOMe-6, 8-diMe		*Alluaudia humbertii* (Did.)	aerial parts	Rabesa and Voirin (1985)
25	*5, 8, 4′-triOH, 3, 7-diOMe, 6-Me		*Pityrogramma triangularis* (Pterid.)	frond exudate	Wollenweber *et al.* (1985d)
26	3, 5, 7, 3′, 4′, 5′-hexaOH, 6-Me				
27	5, 7, 3′, 4′, 5′-pentaOH, 3-OMe, 6-Me	Alluaudiol			
28	5, 7, 3′, 4′, 5′-pentaOH, 3-OMe, 6, 8-diMe		*Alluaudia humbertii* (Did.)	cortex + spines	Rasamoelisendra *et al.* (1985)
29	3, 5, 7, 3′, 5′-pentaOH, 4′-OMe, 6-Me	Dumosol			
30	3, 5, 7, 3′, 5′-pentaOH, 4′-OMe, 6, 8-diMe				
31	5, 7, 3′, 5′-tetraOH, 3, 4′-diOMe, 6-Me				
32	5, 7, 3′, 5′-tetraOH, 3, 4′-diOMe, 6, 8-diMe				

Table 7.4 Methylenedioxyflavonoids (see also Table 7.7)

No.	Substitution	Trivial name	Plant source		References
	FLAVONES				
1	5-OMe, 7, 8-O_2CH_2		*Millettia hemsleyana* (Fab.)	stem bark	Mahmoud and Waterman (1985)
2	*7-OMe, 3′, 4′-O_2CH_2		*Bauhinia splendens* (Fab.)	wood	Laux *et al.* (1985)
3	*7-OMe, 5, 6-O_2CH_2, 3′, 4′-O_2CH_2				
4	5, 6-diOMe, 3′, 4′-O_2CH_2				
5	7-OH, 6-OMe, 3′, 4′-O_2CH_2	Prosogerin-A	synthesis:		Iinuma *et al.* (1984c)
6	5, 4′-diOH, 6, 7-O_2CH_2	Kanzakiflavon-2	*Bauhinia championii* (Fab.)	root	Chen *et al.* (1984)
7	*5, 7, 5′-triOMe, 3′, 4′-O_2CH_2		synthesis:		Iinuma *et al.* (1984c)
8	5, 8-diOH, 4′-OMe, 6, 7-O_2CH_2	Kanzakiflavon-1	See Section 7.4.3		
9	5, 7-diOH, 6, 8-diOMe, 3′, 4′-O_2CH_2	Lindero-flavone A			
10	5, 6, 7, 8-tetraOMe, 3′, 4′-O_2CH_2	Lindero-flavone B	*Eupatorium leucolepis*	aerial parts	Herz and Kulanthaivel (1982)
11	*5, 6, 7, 5′-tetraOMe, 3′, 4′-O_2CH_2		*Bauhinia championii* (Fab.)	root	Chen *et al.* (1984)
12	5, 6, 7, 8, 5′-pentaOMe, 3′, 4′-O_2CH_2	Eupalestin	*Eupatorium leucolepis*	aerial parts	Herz and Kulanthaivel (1982)
13	*5, 6, 7, 3′, 4′, 5′-hexaOMe, 3′, 4′-O_2CH_2		*Bauhinia championii* (Fab.)	root	Chen *et al.* (1984)
	FLAVONOLS				
1	3, 5-diOMe, 6, 7-O_2CH_2				
2	3, 5-diOMe, 6, 7-O_2CH_2, 3′, 4′-O_2CH_2	Meliternatin			
3	3, 7-diOMe, 3′, 4′-O_2CH_2	Demethoxykanungin			
4	*3, 5, 8-triOMe, 6, 7-O_2CH_2		*Melicope indica* (Rut.)	leaf	Fauvel *et al.* (1981)
5	*3, 5, 8-triOMe, 3′, 4′-O_2CH_2		*Melicope indica* (Rut.)	leaf	Fauvel *et al.* (1981)
6	3, 5, 4′-triOH, 6, 7-O_2CH_2	Gomphrenol			
7	3, 7, 3′-triOMe, 3′, 4′-O_2CH_2	Kanugin	See Section 7.4.3		
8	5-OH, 3, 6, 7-triOMe, 3′, 4′-O_2CH_2	Melisimplin			
9	3, 5, 6, 7-tetraOMe, 3′, 4′-O_2CH_2	Melisimplexin			
10	5-OH, 3, 7, 8-triOMe, 3′, 4′-O_2CH_2				
11	3, 5, 7, 8-tetraOMe, 3′, 4′-O_2CH_2	Meliternin			
12	*3, 5, 8, 3′-tetraOMe, 6, 7, O_2CH_2, 3′, 4′-O_2CH_2		*Polygonum orientale* (Polygon.)	aerial parts	Kuroyanagi and Fukushima (1982)
13	5, 3′, 4′-triOH, 3-OMe, 6, 7-O_2CH_2 (G)		*Spinacia oleracea* (Chenopod.)	leaf	Aritomi and Kawasaki (1984)
14	5, 3′, 4′-triOH, 3-OMe, 7, 8-O_2CH_2	Wharangin	See Section 7.4.3		
15	*5, 4′-diOH, 3, 3′-diOMe, 6, 7-O_2CH_2 (G)		*Spinacia oleracea* (Chenopod.)	leaf	Aritomi and Kawasaki (1984)
16	3, 5, 7-triOH, 6, 8-diOMe, 3′, 4′-O_2CH_2	Melinervin			
17	5-OH, 3, 6, 7, 8-tetraOMe, 3′, 4′-O_2CH_2				
18	3, 5, 6, 7, 8-pentaOMe, 3′, 4′-O_2CH_2	Melibentin			

Table 7.5 Flavones and Flavonols with C_3, C_5, C_5-OH and C_{10} side chains[†] (see also Tables 7.6 and 7.7)

No.	Substitution	Trivial name	Plant source		References
	FLAVONES				
1	7-OMe, 8-C_5-OH	*trans*-Lanceolatin			
2	*5-OH, 7-OMe, 8-C_5	Tephrinone	*Tephrosia villosa* (Fab.)	root	Rao and Srimannarayana (1981)
3	5, 7-diOMe, 8-C_5-OH (see Fig. 7.2)	*cis*-Tephrostachin **trans*-Tephrostachin	*Tephrosia bracteolata* (Fab.) (See Section 7.3.3)	seed	Khalid and Waterman (1981)
4	*5, 7-diOMe, 8-C_5 (see Fig. 7.2)	*trans*-Anhydrotephrostachin	*Tephrosia bracteolata* (Fab.)	seed	Khalid and Waterman (1981)
5	*5-OMe-7, 8-di-O-C_5		*Tephrosia barbigera* (Fab.)	seed	Vilain (1983)
6	*4′-OH, 5-OMe, 7-O-C_5[‡]		*Achyrocline flaccida*	aerial parts	Norbedo *et al.* (1984)
7	*5, 7, 4′-triOH, 3′-C_{10}	Kuwanone S	*Morus lhou* (Morac.)	root bark	Fukai *et al.* (1985)
8	*5, 7-diOH, 4′-OMe, 8, 3′-di-C_5		*Azadirachta indica* (Meliac.)	leaf	Iinuma *et al.* (1983)
9	*6, 7, 4′-triOMe, 5-O-C_3 § (see Fig. 7.2)		*Tinospora malabarica* (Menisp.)	heartwood	Prakash and Zaman (1982)
10	7, 4′, 6′-triOH, 3-C_{10}	Rubraflavone A			
11	7, 4′, 6′-triOH, 3-C_{10}, 8-C_5	Rubraflavone B			
12	5, 7, 2′, 4′-tetraOH, 6-C_5	Artocarpesin			
13	5, 7, 2′, 4′-tetraOH, 6-C_5-OH	Oxidihydroartocarpesin			
14	5, 7, 2′, 4′-tetraOH, 3, 6-diC_5	Norartocarpin Mulberrin	See Section 7.3.3		
15	*5, 7, 2′, 4′-tetraOH, 3, 3′-diC_5	Kuwanone T (see Fig. 7.2)	*Morus lhou* (Morac.)	root bark	Fukai *et al.* (1985)
16	5, 2′, 4′-triOH, 7-OMe, 3-C_5	Integrin			
17	5, 2′, 4′-triOH, 7-OMe, 3, 6-diC_5	Artocarpin			
18	5, 7, 2′, 4′-tetraOH, 3, 8-diC_5	Kuwanone C			
19	5, 7, 3′, 4′-tetraOH, 8-C_5				
20	5, 7, 4′, 6′-tetraOH, 3-C_{10}, 6-C_5	Rubraflavone C			
21	*5, 7, 3′, 4′, 5′-pentaOH, 3-C_5	Asplenetin	*Launaea asplenifolia*		Gupta and Ahmed (1985)
	FLAVONOLS				
1	*3, 5, 7-triOH, 8-C_5	Glepidotin A	*Glycyrrhiza lepidota* (Fab.)	whole plant	Mitscher *et al.* (1983)
2	3, 7, 4′-triOH, 8-C_5-OH				
3	3, 5, 7, 8-tetraOH, 6-C_5				
4	*5-OH, 3, 8-diOMe, 7-O-C_5[‡]		*Achyrocline flaccida*	aerial parts	Norbedo *et al.* (1984)
5	3, 5, 7, 4′-tetraOH, 8-C_5	Noranhydroicaritin			

(*Contd.*)

Table 7.5 (*Contd.*)

No.	Substitution	Trivial name	Plant source		References
6	3, 5, 7, 4′-tetraOH, 8-C_5-OH	Noricaritin	*Glycyrrhiza uralensis* (Fab.)	root	Zhn *et al.* (1984)
6A	*3, 5, 7, 4′-tetraOH, 3′-C_5				
7	3, 5, 4′-triOH, 7-OMe, 8-C_5	Isoanhy-droicaritin			
8	3, 5, 7-triOH, 4′-OMe, 8-C_5-OH	Icaritin			
9	*5, 7, 4′-triOH, 3, 6-diOMe, 3′-C_5-OH	Aliarin	*Dodonaea viscosa* (Sap.)	aerial parts	Sachdev and Kulshreshtha (1982)
10	*5, 7-diOH, 3, 6, 4′-triOMe, 3′-C_5	Viscosol	*Dodonaea viscosa* (Sap.)	aerial parts	Sachdev and Kulshreshtha (1986)
11	*5, 7-diOH, 3, 6, 4′-triOMe, 3′-C_5-OH		*Dodonaea viscosa* (Sap.)	aerial parts	Sachdev and Kulshreshtha (1983)
12	3, 5, 7, 3′, 4′-pentaOH, 6-C_5-OH (G)				
13	*3, 5, 7, 4′-tetraOH-3′-C_5		*Glycyrrhiza uralensis* (Fab.)		Zhu *et al.* (1984)
14	*5, 7, 3′, 4′-tetraOH, 3-OMe, 6, 8-diC_5	Brousso-flavonol B	*Broussonetia papyrifera* (Mor.)	cortex	Matsumoto *et al.* (1985)
15	5, 4′-diOH, 3, 3′-diOMe, 7-O-C_5[‡]				
16	*3, 5, 3′-triOH, 7, 4′-diOMe, 8-C_5	Rhynchospermin	*Rhynchosia cyanosperma* (Fab.)	leaf	Adinarayana *et al.* (1981)
17	*3, 5, 7, 3′, 4′-OH, 8, 2′, 6′-C_5	Brousso-flavonol C	*Broussonetia papyrifera* (Mor.)	root bark	Fukai *et al.* (1986)
18	5, 4′-diOH, 3, 7, 3′-triOMe, 8-C_5				

[†] $C_5 = Me_2CH{-}CH{=}CH_2$ or $Me_2C{=}CH{-}CH_2$; $C_5{-}OH = Me_2C(OH)CH_2{-}CH_2$; $C_{10} = (Me_2C{=}CH{-}CH_2)_2$.

[‡] Note *O*-prenylation.

[§] *O*-Allyl side chain, $-CH_2{-}CH{=}CH_2$.

Table 7.6 Pyranoflavones and pyranoflavonols*

No.	Substitution	Trivial name	Plant source		References
	PYRANOFLAVONES				
1	5-OH, 7, 8-ODmp				
2	5-OMe, 7, 8-ODmp	Isopongaflavon (Candidin)	*Lonchocarpus costaricensis* (Fab.)	seed	Waterman and Mahmoud (1985)
			Tephrosia 2 spp. (Fab.)	seed	Chibber and Dutt (1981), Vilain (1983)
3	5-OH, 6-C_5, 7, 8-ODmp	Fulvinervin B	*Tephrosia fulvinervis* (Fab.)	pods	Venkata Rao *et al.* (1985)
4	5-OH, 6-C_5-OH, 7, 8-ODmp	Fulvinervin C	*Tephrosia fulvinervis* (Fab.)	seed	Venkataratnam *et al.* (1986)
5	*6-OMe, 7, 8-ODmp, 3′, 4′-O_2CH_2 (see Fig. 7.2)	Isopongachromene	*Pongamia glabra* (Fab.)	seed	Pathak *et al.* (1983a)
6	5, 4′-diOH, 7, 6-ODmp				
7	5, 4′-diOH, 7, 6-ODmp, 6′, 3-ODmp	Cyclomulberro-chromene			
8	5, 4′-diOH, 7, 8-ODmp, 6′, 3-ODmp	Cyclomorusin			
9	5, 2′, 4′-triOH, 7, 6-ODmp	Cycloartocarpesin	*Cudrania tricuspidata* (Mor.)	root bark	Fujimoto and Nomura (1985)
10	5, 2′, 4′-triOH, 3-C_5, 7, 6-ODmp	Mulberrochromene			
11	5, 7, 2′-triOH, 3-C_5, 3′, 4′-ODmp	Kuwanone B			
12	5, 7, 4′-triOH, 3-C_5, 2′, 3′-ODmp	Kuwanone A			
13	5, 2′, 4′-triOH, 3-C_5, 7, 8-ODmp	Morusin			
14	5, 2′, 4′-triOH, 3-C_5-OH, 7, 8-ODmp	Oxydihydromorusin			
15	5, 2′, 4′-triOH, 3-C_{10}, 7, 6-ODmp	Rubraflavone D			
16	5, 7, 4′-triOH, 6-C_5, 6′, 3-ODmp	Cyclomulberrin			
17	5, 4′-diOH, 7-OMe, 6-C_5, 6′, 3-ODmp	Cycloartocarpin			
18	5, 3′, 4′-triOH, 6-Me, 7, 8-ODmp	Desmodol			
19	5, 3′, 4′-triOH, 8-C_5, 7, 6-ODmp, 6′, 3-ODmp	Cycloheterophyllin			
20	5, 3′, 4′, 6′-tetraOH, 5′-C_5, 7, 6-ODmp	Artobilichromene			

(*Contd.*)

Table 7.6 (*Contd.*)

No.	Substitution	Trivial name	Plant source		References
	PYRANOFLAVONOLS				
1	3-OMe, 7, 8-ODmp	Karanjachromene			
		Pongaflavone			
2	3, 5-diOH, 8-C_5, 7, 6-ODmp	Sericetin			
3	3, 5-diOMe, 3′, 4′-O_2CH_2, 7, 8-ODmp	Pongachromene			
4	3, 6-diOMe, 7, 8-ODmp				
5	*3, 3′, 4′-triOH, 7, 6-ODmp, 8-C_5	Macaflavon I	*Macaranga indica* (Euphorb.)	leaf	Sultana and Ilyas (1986)
6	*3′, 4′-diOH, 3-OMe, 7, 6-ODmp, 8-C_5	Macaflavon II	*Macaranga indica* (Euphorb.)	leaf	Sultana and Ilays (1986)
7	3, 5, 3′, 4′-tetraOH, 7, 8-ODmp				
8	*5, 3′, 4′-OH, 3-OMe, 8-C_5, 7, 6-ODmp	Broussoflavonol A	*Broussonetia papyrifera* (Mor.)	cortex	Matsumoto *et al.* (1985)
9	3, 5, 7, 4′-OH, 8, 6′-C_5, 3′, 2′-ODmp (see Fig. 7.2)	Broussoflavonol D	*Broussonetia papyrifera* (Mor.)	root bark	Fukai *et al.* (1986)

*Some of the flavonoids listed here could be included in Table 7.5.
ODmp-O-dimethylallyl

Table 7.7 Furanoflavones and flavonols

No.	Substitution	Trivial name	Plant source		References
	FURANOFLAVONES				
1	7, 8-fur	Lanceolatin B			
2	3′, 4′-O_2CH_2, 7, 8-fur	Pongaglabrone			
3	5-OH, 7, 8-fur	Pongaglabol			
4	5-OMe, 7, 8-fur	Pinnatin	*Ochna squarrosa* (Ochnaceae)	stem	Reddy *et al.* (1983)
5	5-OMe, 3′, 4′-O_2CH_2, 7, 8-fur	Gamatin			
6	6-OMe, 7, 8-fur				
7	8-OMe, 7, 6-fur				
8	*2′-OMe, 7, 8-fur		*Pongamia glabra* (Fab.)	seed	Pathak *et al.* (1983b)
9	3′-OH, 7, 8-fur				
10	4′-OH, 7, 8-fur	Isopongaglabol	*Pongamia glabra* (Fab.)	flower	Talapatra *et al.* (1982)
11	*4′-OH, 6-OMe, 7, 8-fur (see Fig. 7.2)	6-Methoxyisopongaglabol	*Pongamia glabra* (Fab.)	flower	Talapatra *et al.* (1982)
12	5, 3′, 6′-triOH, 4′, 5′-OIpf†	Dihydrofurano-artobilichromene b_1 and b_2			
13	5, 3′, 4′-triOH, 6′, 5′-OIpf†	Dihydrofurano-artobilichromene a			
14	5, 4′, 6′-triOH, 3-C_5, 7, 6-OIpf-OH	Mulberranol			
	FURANOFLAVONOLS				
1	3-OMe, 7, 8-fur	Karanjin			
2	3-OMe, 3′, 4′-O_2CH_2, 7, 8-fur	Pongapin			
3	3, 3′-diOMe, 4′, 5′-O_2CH_2, 7, 8-fur	3′-Methoxypongapin			

†OIpf = Isoprenylfuran

always indicating the position of the oxygen. Furano-flavonoids are listed in Table 7.7. The furan ring is abbreviated by fur, the first cipher again indicating the position of the oxygen substitution. Three flavonols bear isoprenylfuran rings, indicated by OIpf. [The abbreviation fur-ODmp, used in Wollenweber (1982a) was incorrect. Also there was a typing error in Fig. 4.1 ascribing 6, 3′-ODmp to the structural formula of cyclomorusin instead of 3, 6′-ODmp substitution.] For other abbreviations used, see also Fig. 7.2.

Compounds in which the pyran or furan ring form a benzochromene or a furanochromene moiety are also called chromenoflavonoids. Hence, what is described here as a 7, 6-ODmp derivative could also be designated as a 6, 7, 2, 2-dimethylchromeno derivative (e.g. in Sultana and Ilyas, 1986).

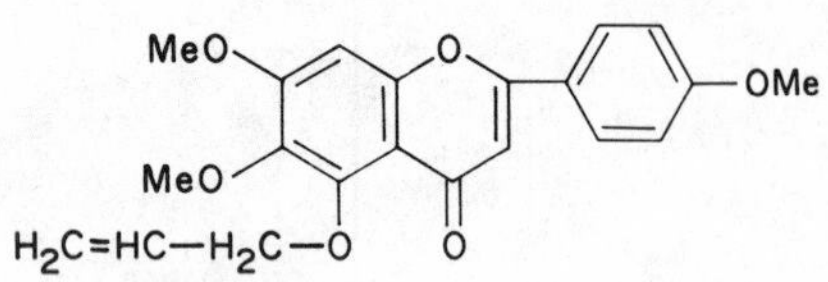

(*7.15*)
6,7,4′-Trihydroxy-5-allyloxyflavone
(5-allyloxysalvigenin)

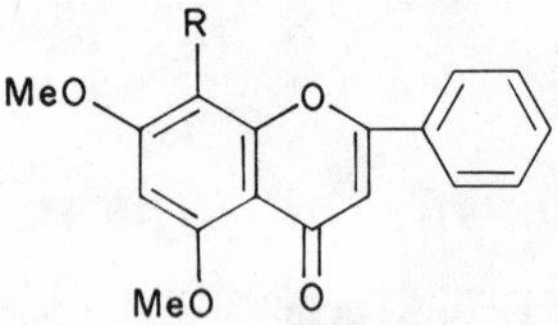

(*7.16*) R=CH$\overset{c}{=}$CHC(OH)Me_2 : *cis*-Tephrostachin
(*7.17*) R=CH$\overset{t}{=}$CHC(OH)Me_2 : *trans*-Tephrostachin
(*7.18*) R=CH$\overset{t}{=}$CHC(Me)=CH_2 : *trans*-Anhydro-tephrostachin

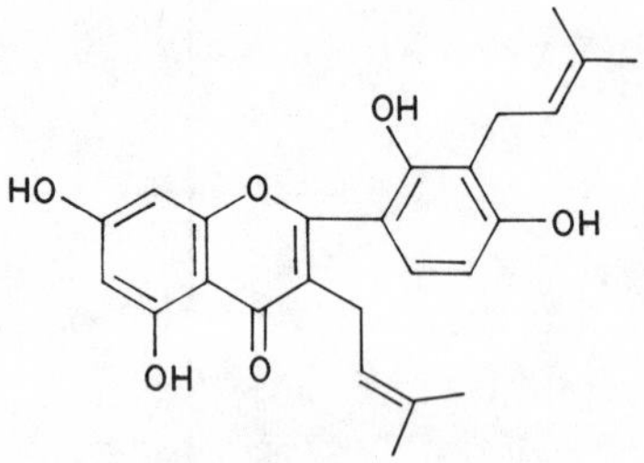

(*7.19*) Kuwanon T
5,7,2′,4′-Tetrahydroxy-3,3′-di-(γ,γ-dimethylallyl)flavone

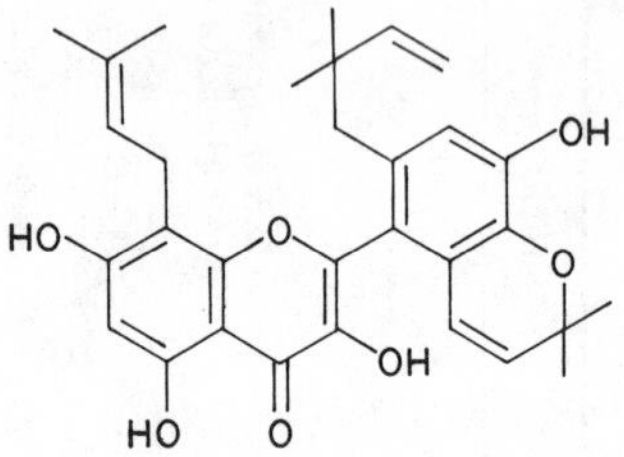

(*7.20*) Broussoflavonol D
3,5,7,4′-tetrahydroxy-8-(3,3-dimethylallyl)-2′-(1,1-dimethylallyl)-5′,6′-ODmp flavone

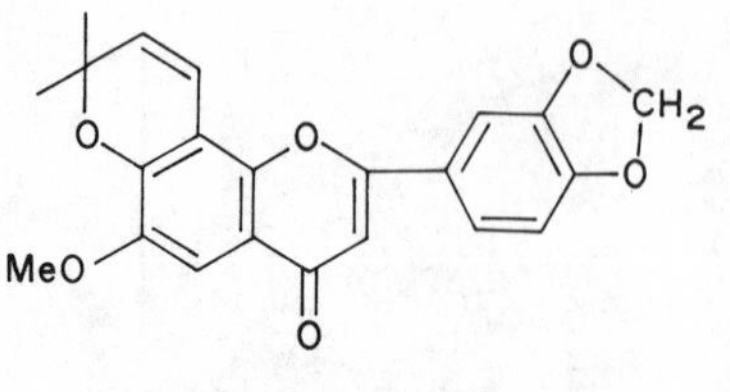

(*7.21*) Isopongachromene
6-Methoxy-6″,6″-dimethyl-3′,4′-methylenedioxychromeno-(7,8,2″,3″)flavone

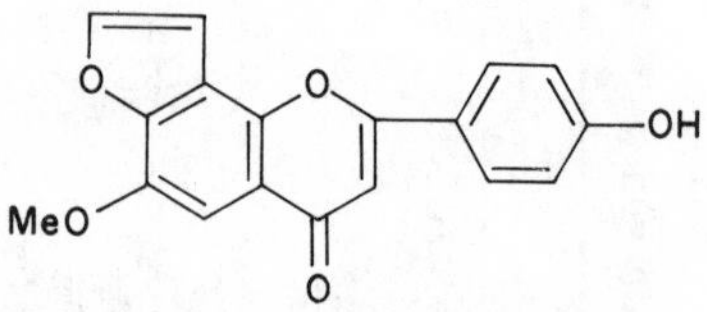

(*7.22*) 6-Methoxyisopongaglabol

Fig. 7.2 Examples of C_3- and C_5-substituted flavones and pyrano- and furano-flavones.

Since Wollenweber's 1982 review, no new flavonoids with oxepine and oxicin rings have been reported. There are still only five compounds of these types: 'compound A' from *Morus*, oxycyclointegrin, chaplashin, isocycloheterophyllin and cyclointegrin from *Artocarpus*.

7.3.5 Chloroflavonoids

Chloroflavonin or 5, 2′-dihydroxy-3, 7, 8-trimethoxy-3′-chloroflavone, was found as early as 1969 in a strain of the fungus *Aspergillus candidus* (Bird and Marshall, 1969). Two chlorine-containing isoflavonoids were described later from a strain of *Streptomyces griseus* (König *et al.*, 1977), namely 6-chlorogenistein and 6, 3′-dichlorogenistein. So far only one chloroflavonoid has been found in higher plants, namely 6-chloroapigenin in *Equisetum arvense* (Syrchina *et al.*, 1980). One might expect that such compounds may eventually be discovered in the Asteraceae, since about half of the chlorine-containing polyacetylenes, thiophenes and sesquiterpene lactones known to date are constituents of this family (Engvild, 1986) and terpenoid-based leaf resins of Asteraceae often contain flavonoid aglycones.

7.3.6 *Tephrosia* flavonoids

In the previous edition of this book, a series of flavonoids from *Tephrosia* (Leguminosae, Papilionoideae) were reported. Since then three new *C*-prenylated flavones (Table 7.5) and two pyranoflavones (Table 7.6) have been described. *Trans*-Tephrostachin and *trans*-anhydrotephrostachin are remarkable for the unusual *trans* configuration in the side chain (Khalid and Waterman, 1981). Three further flavones are shown in Fig. 7.3. Pseudosemiglabrinol (*7.23*) occurs in the 'whole plant' of *T. appolinea* (Ahmad, 1986), tephroglabrin (*7.24*) and tepurindiol (*7.25*) are present in *T. purpurea* root (Pelter *et al.*, 1981).

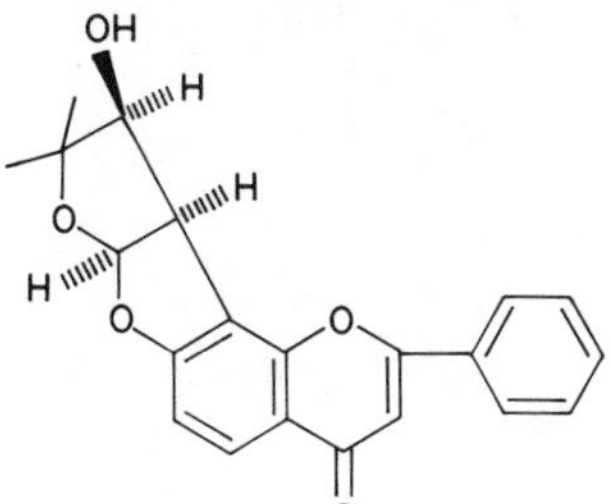

(*7.23*) Pseudosemiglabrinol

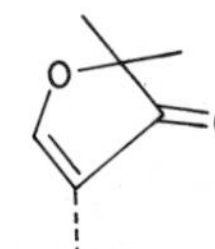

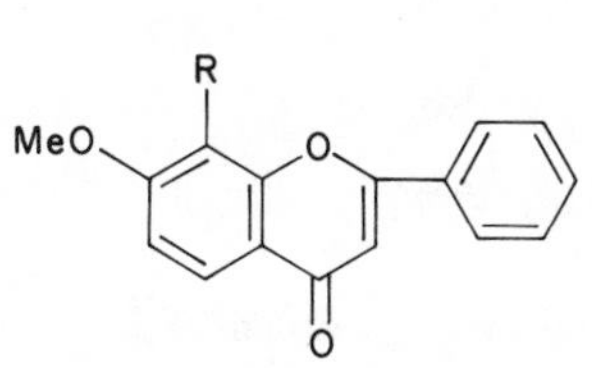

(*7.24*) R=A: Tephroglabrin

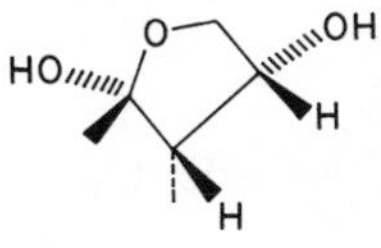

(*7.25*) R=B: Tepurindiol

Fig. 7.3 *Tephrosia* flavonoids.

7.3.7 *Morus* flavonoids

Since the last review (Wollenweber, 1982a), several complex prenylated flavonoids have been reported from two *Morus* species. Their structures are shown in Fig. 7.4. Kuwanon G (*7.26*; Nomura and Fukai, 1980), kuwanon H (*7.27*; Nomura *et al.*, 1980), kuwanon K (*7.31*; Nomura *et al.*, 1983a) and moracein D (*7.29*; Nomura *et al.*, 1981) were isolated from root bark of *M. alba*. Kuwanon M (*7.30*; Nomura *et al.*, 1983b), kuwanon N (*7.32*; Hano *et al.*, 1984) and kuwanon W (*7.28*; Hirakura *et al.*, 1985) come from root bark of *M. lhou*. Two simple prenylated flavones, kuwanones S and T, are also listed in Table 7.5.

7.3.8 *Pityrogramma* flavonoids

A novel type of complex flavonoid from the frond exudate of *Pityrogramma* species (Pteridaceae, Polypodiales) was described in the previous edition (Wollenweber, 1982a). In these compounds, a phenyldihydrocoumarin (or dihydroneoflavonoid) moiety is linked to a flavonoid moiety via a common phloroglucinol ring. Unfortunately, in Fig. 4.3 (Wollenweber, 1982a) the upper structure was ascribed the abbreviation D-1, which refers, in fact, to a dihydrochalcone derivative (now confirmed, by X-ray analysis; Donnelly *et al.*, 1987). 'a:R = H' and 'b:R = OH, refer to D-2/a and D-2/b, which are the flavone and the flavonol derivative designated as B_1 and B_2 by Wagner *et al.*, 1979. The structures of the two flavone derivatives X-1 and X-2 (Favre–Bonvin *et al.*, 1980) in which no additional γ-pyrone ring is present, have now been confirmed by synthesis to be identical with β-(5, 7, 4′-trihydroxy-8-yl)-β-phenylpropionic acid and its methyl ester (Iinuma *et al.*, 1986c). Both compounds were originally isolated from the frond exudate of some cultivated plants of *Pityrogramma calomelanos* var. *aureoflava*. X-2 was also detected in the farina of two specimens of *P. sulphurea* and one specimen of *P. tartarea* (Wollenweber and Dietz, 1980), while compounds D-2/a and D-2/b are part of the typical flavonoid pattern of farinose *Pityrogramma* species. To date no additional compounds of this type have been reported, either from further *Pityrogramma* species or from any other Cheilanthoid fern.

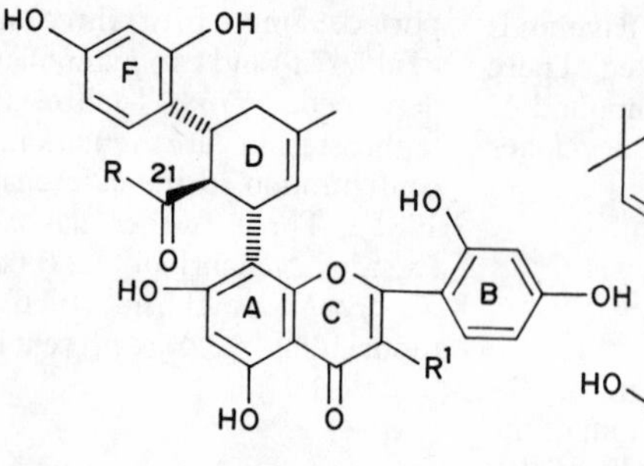

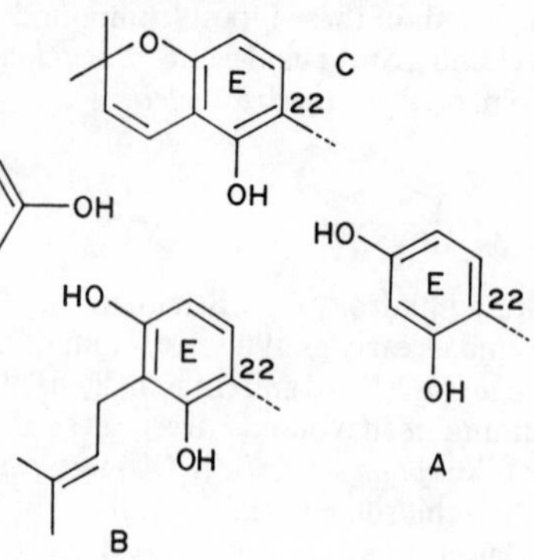

(*7.26*) R = A, $R^1 = C_5$: Kuwanon G
(*7.27*) R = B, $R^1 = C_5$: Kuwanon H
(*7.28*) R = C, $R^1 = C_5$: Kuwanon W
(*7.29*) R = C, $R^1 = C_5$–OH: Moracein D

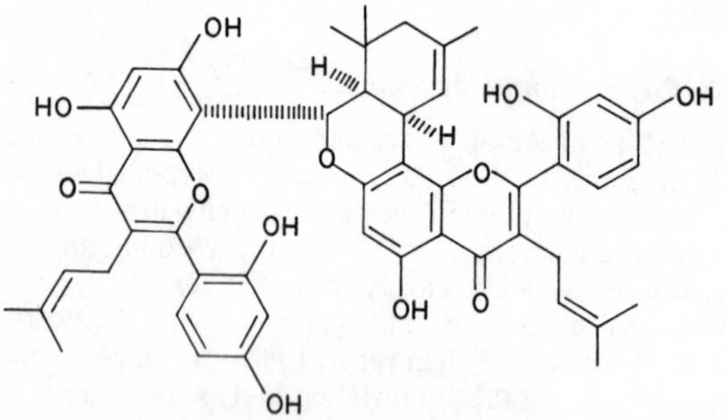

(*7.30*) Kuwanon M

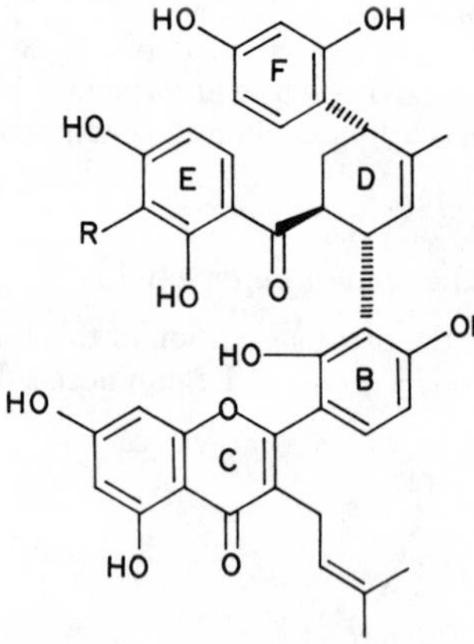

(*7.31*) R = H: Kuwanon K
(*7.32*) R = C_5: Kuwanon N

Fig. 7.4 *Morus* flavonoids.

7.4 REVISIONS AND PROBLEMATICAL STRUCTURES

Except for a methylenedioxyflavonol from spinach (cf. Table 7.6), none of the flavonoids of uncertain structure mentioned by Gottlieb (1975) or Wollenweber (1982a) has, to our knowledge, been confirmed in the meantime. In the following section a few further structures that need either corroboration or revision are mentioned. We also cite all structure revisions that we are aware of, whether or not the erroneous structures were included in the two earlier reviews.

7.4.1 Flavones

5,7-Dihydroxy-6,8-dimethoxyflavone
(Table 7.1, no. 24.2)

This flavone was reported as being constituent of the frond exudate of the fern *Pityrogramma triangularis* var. *triangularis*, ceroptene-type (Dietz, *et al.*, 1981). The structure was later found to be incorrect and was revised to 5,8-dihydroxy-3,7-dimethoxyflavone or isognaphalin (Wollenweber *et al.*, 1985d).

5,8,4′-Trihydroxy-7-methoxyflavone
(Table 7.1, no. 28.1)

'Salvitin' was reported as a new flavone from *Salvia plebeia*, and its structure as 5,8,4′-trihydroxy-7-methoxyflavone was 'confirmed by synthesis' (Gupta *et al.*, 1975). Horie *et al.* (1983) synthesized this same flavone and found that the natural compound was not identical with the synthetic sample. The *Salvia* flavone must be scutellarein 6-methyl ether. This seems reasonable since 6-*O*-methyl ethers are widely present in *Salvia*, whereas 8-*O*-methyl ethers are not (cf. Barberán *et al.*, 1986a). The trivial name salvitin should hence no longer be used.

5,7-Dihydroxy-8,4′-dimethoxyflavone
(Table 7.1, no. 28.5)

Spectral studies by Markham (1983) have shown that this earlier reported structure for a constituent of *Cirsium takaoense*, designated cirsitakaogenin (Lin *et al.*, 1978), should be revised to cirsimaritin or scutellarein 6,7-dimethyl ether. This revision was confirmed by Nakayama *et al.* (1983), who synthesized 5,7-dihydroxy-8,4′-dimethoxyflavone. The latter compound does occur naturally, as the glucuronide in a liverwort, *Bucegia romanica*, and it has been called bucegin (Markham and Mues, 1983).

5,7,2′,6′-Tetrahydroxyflavone
(Table 7.1, no. 33)

Bhardwaj *et al.* (1982) reported this tetrahydroxyflavone from seeds of *Argemone mexicana* and called it argemexitin. Re-examination by Harborne and Williams (1983) revealed that the major seed flavonoid is, in fact, luteolin, and no trace of any novel flavone was detected. From this example, these authors stressed the risk of relying entirely on spectral measurements for identifying new flavonoids.

5,6-Dihydroxy-7,8,4′-trimethoxyflavone
(Table 7.1, no. 37.7)

5,6-Dihydroxy-7,8,4′-trimethoxyflavone was reported as a new flavone from *Stevia berlandieri* without any data being given (Dominguez *et al.*, 1974). A reinvestigation of the leaf resin of this species (Wollenweber, unpublished) revealed only the presence of two methyl derivatives each of kaempferol and of 6-hydroxykaempferol; the identification of this flavone from *S. berlandieri* remains, therefore, uncertain. 5,6-Dihydroxy-7,8,4′-trimethoxyflavone has been found recently, however, in the aerial parts of *Thymus piperella* (Barberán *et al.*, 1985c).

Rehderianin I
(Table 7.1, no. 39.1/no. 41.1)

The structure of 5,2′,4′-trihydroxy-6,8-dimethoxyflavone was ascribed to a flavone 'rehderianin I', isolated from roots of *Scutellaria rehderiana* (Liu, M. *et al.*, 1984a). Iinuma *et al.* (1985a) pointed out that the absence of 7-hydroxylation is very unusual and indeed synthesis showed that this structure was incorrect. The correct structure for rehederianin I is 5,2′,5′-trihydroxy-7,8-dimethoxyflavone. This structural revision is similar to that in the case of skullcapflavone I, which was first thought to be 5,2′-dihydroxy-6,8-dimethoxyflavone but this was later revised, after synthesis, to 5,2′-dihydroxy-7,8-dimethoxyflavone (Takido *et al.*, 1979). Indeed, so far no natural flavonoid is known with 6,8-di-*O*-substitution but which lacks 7-*O*-substitution at the same time.

5-Hydroxy-7,8,2′,3′-tetramethoxyflavone (s.n.)

In Wollenweber (1982a), this flavone was cited erroneously insofar as it is known only as a methylation production of wightin and does not occur naturally itself.

5,8,4′-Trihydroxy-7,3′-dimethoxyflavone
(Table 7.1, no. 43.3)

A flavone isolated from *Solanum grayi* was reported incorrectly as 5,8,4′-trihydroxy-7,3′-dimethoxyflavone (Whalen and Mabry, 1979). A synthetic sample of this flavone was not identical with the natural product and the compound is actually 5,7,4′-trihydroxy-8,3′-dimethoxyflavone (Horie *et al.*, 1983).

5,7,4′-Trihydroxy-6,8,3′-trimethoxyflavone
(Table 7.1, no. 50.6)

A flavone majoranin was isolated from leaves of *Majorana hortensis* (Subramanian *et al.*, 1972) and assigned the structure 5,7,4′-trihydroxy-6,8,3′-trimethoxyflavone. Majoranin was, therefore, listed as a synonym of sudachitin by Wollenweber (1982a). Extensive spectral studies and direct comparisons reveal, however, that the *Majorana* flavone is 5,6,4′-trihydroxy-7,8,3′-trimethoxyflavone and hence is identical to thymonin from *Thymus vulgaris* (Voirin *et al.*, 1984). The trivial name majoranin should, therefore, be deleted.

5,5′-Dihydroxy-6,7,2′,3′-tetramethoxyflavone
(Table 7.1, no. 51.1)

This dihydroxytetramethoxyflavone was reported from bud exudates of two *Gardenia* species (Gunatilaka *et al.*, 1982). By synthesis of this compound as well as of the 6,7,4′,5′-tetramethoxy isomer and comparison of their spectral data with those of the natural product it has become clear that the latter is, in fact, 5,3′-dihydroxy-6,7,4′,5′-tetramethoxyflavone (Iinuma *et al.*, 1984b).

5,6,4′-Trihydroxy-7,2′,5′-trimethoxyflavone
(Table 7.1, no. 52.3)

A new flavonoid, isocarpillin, was found as constituent of *Artemisia capillaris* and the structure was deduced to be 5,6,4′-trihydroxy-7,2′,5′-trimethoxyflavone (Kitagawa *et al.*, 1983). Synthesis of this and two isomeric flavones revealed that none is identical with naturally occurring isocarpillin (Iinuma *et al.*, 1984a). The correct structure of isocarpillin therefore remains unclear.

Brickellin
(Table 7.1, no. 62.1/63.2)

A dihydroxypentamethoxyflavone isolated from *Brickellia chlorolepis* and *B. veronicaefolia* by Roberts *et al.* (1980) was designated brickellin and assigned the structure 5,3′-dihydroxy-6,7,2′,4′,5′-pentamethoxyflavone. As such it was also reported as being a constituent of *B. glutinosa* by Goodwin *et al.* (1984), while Roberts *et al.* (1984) decided its structure should be 5,6′-dihydroxy-6,7,2′,3′,4′-pentamethoxyflavone. However, the natural product did not match a synthetic sample prepared by Iinuma *et al.* (1984f). The ^{13}C NMR assignments were therefore re-examined, further flavonoids were synthesized and brickellin was finally shown to be 5,2′-dihydroxy-3,6,7,4′,5′-pentamethoxyflavone (Iinuma *et al.*, 1985b).

5,4′,6′-Trihydroxy-6,7,2′,3′-tetramethoxyflavone
(Table 7.1, no. 63.1)

A tetra-oxygenated B-ring was thought to be present in flavonoid 'NAS-3', from the frond exudate of the fern *Notholaena aschenborniana*, and the structure 5,4′,6′-trihydroxy-6,7,2′,3′-tetramethoxyflavone was ascribed to it (Jay *et al.*, 1981). On the basis of their experience with brickellin, Roberts *et al.* (1984) questioned this structure, particularly since they found inconsistencies in the interpretation of the UV and MS data. Comparison with a synthetic sample had already shown this structure to be wrong for NAS-3 (Iinuma *et al.*, 1983). Further isomers were synthesized and the natural product was finally shown to be 5,2′,4′-trihydroxy-3,7,8,5′-tetramethoxyflavone (Iinuma *et al.*, 1986d). 2′-Hydroxyflavonol 3-methyl ethers have unusual spectral properties, which make them appear to belong to the flavone rather than the flavonol series.

7.4.2 Flavonols

3,5-Dihydroxy-6,7,8-trimethoxyflavone
(Table 7.2, no. 13.4)

A flavonol from *Artemisia ludoviciana* was ascribed the structure 3,5-dihydroxy-6,7,8-trimethoxyflavone (Dominguez and Cárdenas, 1975). However, its spectral properties differ from those of a flavonol isolated from two *Helichrysum* species, for which the same structure was proposed. ^{13}C NMR spectral studies have subsequently confirmed this structure for the *Helichrysum* flavonol so the exact nature of the *Artemisia* flavonol remains to be determined (Hänsel *et al.*, 1981).

5,6-Dihydroxy-3,7,4′-trimethoxyflavone
(Table 7.2, no. 14.12)

This flavonol was erroneously cited instead of the 3,6,4′-trimethyl ether as a constituent of *Baccharis vaccinoides* leaf resin by Wollenweber *et al.* (1986a).

5,6-Dihydroxy-3,8,4′-trimethoxyflavone
(Table 7.2, no. 15.1)

This structure was ascribed to a flavonol isolated from *Conyza stricta* (Sen *et al.*, 1976). This compound as well as its isomers were synthesized by Horie *et al.* (1982), who demonstrated that the correct structure is 5,7-dihydroxy-3,8,4′-trimethoxyflavone. They thus confirmed an earlier revision of this structure carried out by Tandon and Rastogi (1977).

5,8,4′-Trihydroxy-3,7-dimethoxyflavone
(Table 7.2, no. 17.5)

Herbacetin 3,7-dimethyl ether was reported to occur in *Pluchea odorata* collected in El Salvador (Arriaga–Giner *et al.*, 1983). Our own studies on a Mexican sample of *Pluchea odorata* gave a series of 6-methoxyflavonols, but no 8-*O*-substituted flavonoid. A reinvestigation of the product previously thought to be herbacetin 3,7-dimethyl ether revealed that it was, in fact, 6-hydroxykaempferol 6,7-dimethyl ether (eupalitin) (Wollenweber *et al.*, 1985b).

3,5-Dihydroxy-7,8,4′-trimethoxyflavone
(Table 7.2, no. 17.12) *in Achillea depressa?*

Herbacetin 7,8,4′-trimethyl ether (tambulin) was reported from *Achillea depressa* (Tsankov and Ognyanov, 1985). Our own studies on some 120 samples representing 50 species of *Achillea*, including *A. depressa*, failed to yield any 8-*O*-substituted flavonoid. In fact, 6-hydroxykaempferol 3,6,7-trimethyl ether is the major flavonoid in this species (Wollenweber, unpublished results).

7,5′-Dihydroxy-3,5,2′-trimethoxyflavone
(Table 7.2, no. 20.2)

This substitution pattern was erroneously cited in the earlier review (Wollenweber, 1982a) instead of 3,5,2′-trihydroxy-7,5′-dimethoxyflavone, isolated from *Inula cappa* (Baruah *et al.*, 1979).

5,4′-Dihydroxy-3,6,7,3′-tetramethoxyflavone
(Table 7.2, no. 31.23)

Quercetagetin 3,6,7,3′-tetramethyl ether (chrysosplenetin) was reported as a constituent of aerial parts of *Jasonia montana* (under the trivial name polycladin) (El-Din *et al.*, 1985). The compound turned out later to be identical with jaceidin, i.e. quercetagetin 3,6,3′-trimethyl ether (El-Din, 1986).

5,4′-Dihydroxy-3,7,8,3′-tetramethoxyflavone
(Table 7.2, no. 34.15)

Due to an error in the structural formula, an 8-*C*-methylflavonol was reported from *Chromolaena morii* (Bohlmann *et al.*, 1981b) instead of gossypetin 3,7,8,3′-tetramethyl ether (Bohlmann, 1986). Re-examination of this product now shows that it is quercetagetin 3,6,7,4′-tetramethyl ether (Wollenweber, unpublished results).

Gossypetin 7,8,3′,4′-tetramethyl ether
(Table 7.2, no. 34.19)

In our earlier review, the entry 'gossypetin 3,8,3′-4′-tetramethyl ether from *Heteromma simplicifolium* (Bohlman and Fritz, 1979)' was wrongly transcribed from the literature, since the authors had in fact reported the 7,8,3′,4′-tetramethyl ether (as '8,3′,4′-trimethoxy-izalpinin'). However, Esteban *et al.* (1986) have now found the *Heteromma* constituent to be quercetagetin 6,7,3′,4′-tetramethyl ether. The gossypetin 7,8,3′,4′-tetramethyl ether was isolated separately from *Artemisia lanata*, and the two compounds were shown to be different.

Viscidulin III
(Table 7.2, no. 35.1)

The structure 3,5,7,3′-tetrahydroxy-2′,4′-dimethoxy-flavone was proposed for 'viscidulin III' from *Scutellaria viscidula* (Liu, unpublished, cited in Iinuma *et al.*, 1985a). A flavonol with this structure was synthesized by Iinuma *et al.* (1985a) and found to be different from the natural product. Viscidulin III must, according to these authors, be revised to 5,7,3′,6′-tetra-hydroxy-8,2′-dimethoxyflavone. This compound has previously been reported from *S. baicalensis* (Tomimori *et al.*, 1984c), and from *S. rehderiana* (Liu, Y.-L. *et al.*, 1984), and has been given the trivial name ganhuangenin.

3,5,7,4′-Tetrahydroxy-6,3′,5′-trimethoxyflavone
(Table 7.2, no. 44.8)

In Wollenweber (1982a), this compound was incorrectly listed as the 6,3′,4′-trimethyl ether.

3,5-Dihydroxy-6,7,3′,4′,5′-pentamethoxyflavone
(Table 7.2, no. 44.16)

This flavonol structure was wrongly cited in Wollenweber (1982a) instead of the correct 5-hydroxy-6,7,3′,4′,5′-pentamethoxyflavone for a flavonoid from *Gardenia fosbergii*, due to an error in the original publication (Gunatilaka *et al.*, 1979).

3,5,6,8,3′,4′,5′-Heptamethoxyflavone (s.n.)

This structural formula was proposed for a flavonol from *Murraya exotica*, although a reading of the text in this paper suggests that the authors were dealing with 3,5,7,8,3′,4′,5′-heptamethoxyflavone instead (Chowdhury and Chakraborty, 1971). Indeed, the latter structure has previously been isolated from the same plant (Joshi and Kamat, 1970).

7.4.3 *C*-Methylflavonoids, methylenedioxyflavonoids and pyranoflavonoids

5,4′-Dihydroxy-3,7-dimethoxy-6,8-dimethylflavone and 5,4′-dihydroxy-3,7-dimethoxy-6-methylflavone
(Table 7.3)

These *C*-methylflavonols from *Kalmia latifolia* have been named latifolin and 8-desmethyllatifolin respectively (Wollenweber and Kohorst, 1984). However, latifolin is also the name of an alkaloid and we suggest abandoning these two trivial names.

5,4′-Dihydroxy-3,7,3′-trimethoxy-8-methylflavone
(Table 7.3, flavones no. 22)

This structure was erroneously reported from *Chromolaena morii* (see above), instead of a gossypetin tetramethyl ether, but the compound actually present in *Chromolaena* is quercetagetin 3,6,7,4′-tetramethyl ether (see Section 7.4.2).

Linderoflavone A
(Table 7.4, flavones no. 9)

The structure of this flavone is 5,7-dihydroxy-6,8-dimethoxy-3′,4′-methylenedioxyflavone (the 6-OMe was missing in Wollenweber, 1982a). 3,7,3′-Trimethoxy-3′,4′-methylenedioxyflavone (Table 7.4, flavonols no. 7) or kanugin was earlier listed as the 4′,5′-methylenedioxy derivative but for the sake of consistency it is now cited as above.

5,3′,4′-Trihydroxy-3-methoxy-6,7-methylenedioxyflavone
(Table 7.4 flavonols no. 13)

This structure was proposed for a flavonoid from spinach chloroplasts (Oettmeier and Heupel, 1972), based on incomplete evidence. It has now been corroborated by spectral studies (Aritomi and Kawasaki, 1984). The related 3′-methyl ether also occurs in spinach, both compounds occurring as the 4′-glucuronides.

Praecanson B and phellamuretin

These two compounds were listed in our earlier review (Wollenweber, 1982a) but they are, in fact, chalcone and flavanone respectively (see Chapter 9).

7.5 RELATIVE FREQUENCIES OF CERTAIN SUBSTITUTIONS IN THE FLAVONOID MOLECULE AND NATURAL DISTRIBUTION PATTERNS

Data provided in the lists of flavones and flavonols of the two earlier volumes (Gottlieb, 1975; Wollenweber, 1982a) were analysed to indicate the frequency of particular substitution patterns within the plant kingdom. Here we have not used the common practice of determining the number of species for which the various structures have been reported, but instead we have measured the number of different molecules (structural diversity) present in each taxon and translated this into the nature and frequency of substituents at each carbon of the flavonoid nucleus.

The information obtained from this study comes from a statistical analysis, using the Burt crosstable calculation (a multiple correspondence analysis module in the STAT-ITCF computer software), performed on both chemical and botanical crude data. Botanical data for 32 taxa or taxa clusters were used (Table 7.8) and, for each taxon, two classes or variables have been analysed: A indicates no data in the relevant lists for the compound considered, B shows one or several sources of a given compound. Phytochemical data were arranged according to 10 variables corresponding to the 10 carbon atoms of the flavone nucleus: C-3, C-5, C-6, C-7, C-8, C-2′, C-3′, C-4′, C-5′ and C-6′; each variable was quantified as 1 for a proton substitution, 2 for a hydroxyl group, 3 for a methoxyl group and 4 for *C*-methylation.

Regarding the results presented below one must remember that the available reports on flavonoid occurrence represent only a random sample from the plant kingdom (research tends to concentrate on specific taxa and/or geographical regions). Statistical evaluation is, therefore, limited and trends now observed may change with time.

7.5.1 Structural patterns of the known flavones and flavonols

A general comparison of flavone and flavonol structures is shown in Table 7.9. Conclusions on substitutions at different carbons of the flavonoid skeleton follow. Modifications

Table 7.8 Systematic distribution of known flavone (A) and flavonol (A′) structural diversity expressed for each taxon as number of molecules reported and percent of total chemical diversity

	A-Flavones		*A′-Flavonols*	
	n	*D.I (%)*	*n*	*D.I (%)*
Asteraceae	87	41	153	48
Betulaceae	7	3	30	9
Bromeliaceae	3	1	5	2
Cactaceae	2	1	4	2
Caryophyllales[1]	7	3	4	2
Cistaceae	6	3	24	7
Cyperaceae	6	3	5	2
Didiereaceae	0		25	7
Ericaceae	4	2	7	2
Fabaceae	20	10	29	9
Geraniales[2]	3	2	18	6
Guttiferales[3]	4	2	6	2
Hydrophyllaceae	6	3	0	
Juncaceae	2	1	0	
Lamiaceae*	77	36	23	7
Liliales[4]	4	2	2	1
Myrtales[5]	17	8	1	
Passifloraceae	5	2	5	2
Poaceae	5	2	0	
Polygonaceae	0		6	2
Polycarpicae[6]	7	3	3	1
Primulaceae	3	1	1	
Restionaceae	0		3	1
Rubiaceae	6	3	8	3
Rutaceae	35	17	10	3
Salicaceae	0		11	4
Saxifragaceae	5	3	17	6
Scrophulariaceae	23	11	13	4
Solanaceae†	10	5	15	5
Terebenthales[7]	3	2	14	4
Zingiberaceae	4	2	6	2
Ferns	12	6	52	16

[1]Including Chenopodiaceae, Amaranthaceae and Phytolacaceae.
[2]Including Geraniaceae, Zygophyllaceae and Begoniaceae.
[3]Including Guttiferae, Dilleniaceae and Clusiaceae.
[4]Including Liliaceae, Amaryllidaceae and Velloziaceae.
[5]Including Myrtaceae, Combretaceae and Thymeleaceae.
[6]Including Annonaceae, Lauraceae, Winteraceae and Eupomatiaceae.
[7]Including Vochysiaceae, Anacardiaceae, Hippocastanaceae and Sapindaceae.
*Joined with Verbenaceae.
†Joined with Gesneriaceae, Bignoniaceae and Acanthaceae.
n = number of molecules reported; D.I. = diversity index.

Table 7.9 Frequencies of the classical *O*- and *C*- substitutions at each one of the ten substituted carbon atoms of the flavone/flavonol nucleus

	Frequencies (%)									
	Carbon atoms of the flavonoids nucleus...									
Substitutions	*C-3*	*C-5*	*C-6*	*C-7*	*C-8*	*C-2'*	*C-3'*	*C-4'*	*C-5*	*C-6'*
Flavones										
H	99	10	46	10	57	76	53	30	78	94
OH	0	70	10	43	8	15	22	32	6	2
OMe	0	20	41	47	34	9	25	37	16	4
C-Me	1		3		1					
Flavonols										
H	0	7	48	1	56	85	44	14	70	100
OH	38	79	10	51	11	9	31	45	16	0
OMe	62	14	34	48	28	6	25	41	14	0
C-Me			8		5					

of the basic hydroxyl substitution at C-5 leading to 5-deoxy or 5-methoxy compounds are more frequent among the flavones (30%) than among the flavonols (*ca.* 20%). Methylation of the hydroxyl group at C-7 is as frequent in flavones as in flavonols while 7-deoxy compounds dominate in the flavone series (10%). Compounds without oxygenated substituents in C-6 and C-8, as well as compounds with hydroxy groups at the same positions, are of similar frequency in both groups of flavonoids. However, methoxyl derivatives are slightly more abundant among flavones and *C*-methylation among flavonols. *O*-Substitution at C-2' occurs in 24% of the flavones but only 15% of the flavonols. In both classes hydroxyl groups are more frequent than methoxyl groups.

It is convenient to treat C-3' and C-5' together as they are often responsible for the tri-*O*-substitution of the B-ring. Hydroxyl substitution at these positions is more frequent in flavonols (e.g. 16% for 5'-OH) than in flavones (6% for 5'-OH); this difference can be confirmed by comparing the frequencies of proton substitution at C-3' and C-5' between flavones and flavonols. Unexpectedly, *p*-*O*-substitution at C-4' is found in 86% of the flavonols but in only 69% of the flavones.

7.5.2 Structural diversity and natural distribution

Parts of A and A' of Table 7.8 only give information on the number of different molecules identified at least once in each taxon. This number is transformed in the diversity index (D.I.) or percentage in relation to the total structural diversity: e.g. in Asteraceae 87 flavones have been described, which represents *ca.* 40% of the total known diversity in this class of compounds. This diversity index ranges from 0 to 48% for the flavonols and from 0 to 41% for the flavones. We decided, therefore, to have an absolute limit of 5% significance (for detailed descriptions of some taxa with respect to their structural diversity see Sections 7.5.5 and 7.5.6).

7.5.3 Structural correlations within the substitution patterns of flavonols

Based on our present knowledge of natural flavonol structures Table 7.10 allows precise conclusions with respect to structural correlations between any substituent and any C-atom on the flavonol nucleus. Three examples are given to aid understanding of these tables.

1. Two hypotheses can be considered for the C-3 of flavonols: (a) the transformation of 3-hydroxyl to 3-methoxyl is due to an *O*-methyltransferase and is correlated with other modifications in the flavonol skeleton; and (b) methylation of 3-OH has no effect substitutions at C-5, C-3', C-4' or C-6', but is a result of an increase of *O*-methylation at C-6, C-7, C-8 and C-5', and an increase in *O*-substitution at C-2'.
2. Three structural hypotheses can be proposed for the C-5 of flavonols: (a) basic 5-OH, (b) 5-deoxy compounds and (c) 5-methoxy compounds. The two alternative secondary substitutions (b and c) are correlated with modifications at the other carbon positions as is shown in Table 7.11.
3. Again three substituents are possible at C-8 of flavonols: 8-H, 8-OH and 8-OMe, and are probably related to specific and sequential enzyme activities. The structural correlations between these substituents at C-8 and the substitutions at the other positions are summarized in Table 7.11.

7.5.4 Structural correlations within the substitution patterns of flavones

Table 7.12 utilizes the same format as those presented earlier.

Table 7.10 Substitutional correlations within flavonols reported in Angiosperms and Ferns from 1975 to 1985

C and A rings	r.f. (%)	*Substitutions on the flavonol nucleus*																													
		C-3			*C-5*			*C-6*			*C-7*			*C-8*			*C-2′*			*C-3′*			*C-4′*			*C-5′*			*C-6′*		
		H	*O*	*E*	*H*	*O*	*E*	*H*	*O*	*E*	*H*	*O*	*E*	*H*	*O*	*E*	*H*	*O*	*E*	*H*	*O*	*E*	*H*	*O*	*E*	*H*	*O*	*E*	*H*	*O*	*E*
C-3 OH	38				10	77	12	60	8	23	0	63	35	62	14	18	93	5	2	46	34	20	21	49	30	81	14	5	100	0	0
C-3 OMe	62				4	81	15	41	12	40	1	43	56	52	9	35	82	11	7	43	30	27	11	41	48	63	17	19	99	0	1
C-5 H	7	0	62	38				76	9	15	0	86	14	62	19	19	90	10	0	43	47	10	10	70	20	90	10	0	100	0	0
C-5 OH	79	0	37	63				47	10	34	1	52	47	52	12	30	87	9	4	43	32	25	13	45	42	69	17	14	100	0	0
C-5 OMe	14	0	33	67				42	16	42	0	29	71	76	2	22	80	9	11	49	20	31	22	29	49	64	13	23	98	0	2
C-6 H	48	0	48	52	10	77	12				1	52	47	53	17	27	85	9	6	44	31	25	19	43	37	74	15	11	99	0	1
C-6 OH	10	0	30	70	6	73	21				6	42	52	85	0	12	88	9	3	39	36	24	6	54	40	76	15	9	100	0	0
C-6 OMe	34	0	26	74	3	79	18				0	47	53	54	5	41	83	11	6	43	28	29	10	39	51	63	15	22	100	0	0
C-6 C-Me	8	0	42	58	0	100	0				0	71	29	46	8	4	100	0	0	54	42	4	17	58	25	71	29	0	100	0	0
C-7 H	1	0	33	67	0	100	0	33	67	0				0	33	33	100	0	0	100	0	0	0	67	33	100	0	0	100	0	0
C-7 OH	51	0	48	52	11	81	8	50	9	31				59	10	25	92	7	1	39	39	22	14	56	30	69	18	13	100	0	0
C-7 OMe	48	0	28	72	2	77	21	47	11	37				54	12	32	79	11	10	49	23	28	16	32	52	70	14	16	99	0	1
C-8 H	56	0	42	58	7	73	20	46	16	32	0	54	46				84	10	6	39	32	29	12	46	42	65	20	15	99	0	1
C-8 OH	11	0	50	50	12	85	3	79	0	15	3	44	53				94	0	6	56	29	15	15	65	21	88	9	3	100	0	0
C-8 OMe	28	0	24	76	4	84	11	46	4	49	1	44	54				83	10	7	46	30	24	16	35	49	70	10	20	100	0	0
C-8 C-Me	5	0	47	53	0	100	0	27	7	0	7	67	26				100	0	0	67	33	0	40	40	20	80	20	0	100	0	0
B-ring																															
C-2′ H	86	0	41	59	7	80	13	48	10	33	1	54	45	55	12	27				37	35	28	14	48	38	74	16	10	100	0	0
C-2′ OH	9	0	21	79	7	79	14	46	10	43	0	39	61	68	0	32				89	11	0	11	29	60	28	28	44	100	0	0
C-2′ OMe	5	0	18	83	0	70	30	59	6	36	0	12	88	53	12	35				82	6	12	24	17	59	65	0	35	94	0	6
C-3′ H	44	0	40	60	6	78	16	49	10	33	2	44	54	50	14	30	72	18	10				32	36	32	80	7	13	100	0	0
C-3′ OH	31	0	41	59	10	81	9	47	12	30	0	65	35	57	10	27	96	4	10				0	54	46	61	30	9	100	0	0
C-3′ OMe	25	0	31	69	3	79	18	48	10	40	0	45	55	65	6	28	97	0	3				1	47	51	63	14	23	100	0	0
C-4′ H	14	0	54	46	4	74	22	63	4	29	0	48	52	46	11	30	85	7	9	98	0	2				94	2	4	98	0	2
C-4′ OH	45	0	43	57	11	80	9	48	13	30	1	64	35	57	16	23	92	6	2	36	37	26				78	12	10	100	0	0
C-4′ OMe	41	0	28	72	3	80	17	44	10	41	1	38	61	58	5	34	80	13	7	34	35	1				53	26	21	100	0	0
C-5′ H	70	0	44	56	9	78	13	51	11	30	1	51	48	59	13	28	91	4	5	50	27	23	19	49	32				100	0	0
C-5′ OH	16	0	33	67	4		12	45	10	32	0	59	41	70	6	18	84	16	0	20	59	21	2	31	67				100	0	0
C-5′ OMe	14	0	14	86	0	77	23	39	7	54	0	43	57	57	2	41	60	27	13	41	18	41	4	34	61				100	0	0
C-6′ H	100	0	38	62	6	79	15	48	10	34	1	51	48	56	11	29	86	9	5	44	31	25	14	44	42	70	16	14			
C-6′ OH	0																														
C-6′ OMe	0																														

For each substitution type of the carbon atoms of the A and C rings, the proportions of H-(proton), O-(hydroxy) and E-(methoxy) substitutions on all other positions of the flavonol nucleus were calculated.
r.f., relative frequencies of the various substitutions at each carbon atom.
Underlined numbers indicate a complementary part as *C*-methyl derivatives.

Table 7.11 Effects of modification of the basic pattern at C-5 (a) and C-8 (b) on the substitution type of other carbon atoms on the flavonol nucleus

(a)

Modifications from the basic substitution		
5-H ←	*5-OH*	→ *5-OMe*
	Substitutions on flavonol nucleus	
Decrease in 3-*O*-methylation	C-3	No effect
Decrease in 6-*O*-substitution	C-6	Slight increase in 6-*O*-substitution
Decrease in 7-*O*-methylation	C-7	Increase in 7-*O*-methylation
Decrease in 8-*O*-methylation	C-8	Decrease in 8-*O*-substitution
No effect	C-2′	Slight increase in 2′-*O*-methylation
Decrease in 3′-*O*-methylation	C-3′	No effect
Decrease in 4′-*O*-methylation	C-4′	Decrease in 4′-*O*-substitution
Decrease in 5′-*O*-substitution	C-5′	No effect
No effect	C-6′	No effect

(b)

	Modifications from the basic substitution	
Substitutions on the flavonol nucleus	*8-H* → *8-OH*	→ *8-OMe*
C-3	No effect	Increase in 3-*O*-methylation
C-5	Decrease in 5-*O*-methylation	Slight increase in 5-*O*-methylation
C-6	Decrease in 6-*O*-substitution	Increase in 6-*O*-substitution
C-7	No effect	No effect
C-2′	Decrease in 2′-hydroxylation	Increase in 2′-hydroxylation
C-3′	Decrease in 3′-*O*-substitution	Increase in 3′-*O*-substitution
C-4′	Decrease in 4′-*O*-methylation	Increase in 4′-*O*-methylation
C-5′	Decrease in 5′-*O*-substitution	Increase in 5′-*O*-substitution
C-6′	No effect	No effect

7.5.5 Flavonol diversity and botanical distribution

Taxa showing at least 5% diversity and the substitution frequencies of each carbon atom of the flavonol nucleus are presented in Table 7.13. At C-3, hydroxyl and methoxyl substitutions are of equal frequency in ferns*, Betulaceae, Fabaceae and Saxifragaceae; 3-methoxyl derivatives are more abundant in Asteraceae, Cistaceae and Geraniales and dominate in Didiereaceae, Lamiaceae and Solanaceae. Hydroxylation at C-5 is commonly present among flavonols. Furthermore, 5-deoxyflavonols are restricted to the Asteraceae, Lamiaceae and ferns, and in particular, in Fabaceae.

6-Hydroxyl substitution is poorly represented in most taxa but in contrast 6-*O*-methylation appears to be an important element of flavone diversity. 6-*C*-Methyl derivatives are characteristic for Myrtales and ferns. There are reports of 7-deoxyflavonols in Rutaceae and especially in the Myrtales. Substitution at C-8 is absent or rare except for the flavonols from Asteraceae, Lamiaceae, Rutaceae and Solanaceae.

Lamiaceae, Myrtales and Solanaceae are remarkable for the presence of 2′-*O*-substitution in about 20% of flavonols. Again, Lamiaceae, Myrtales and Solanaceae are important since 40–50% of the flavonols from these three taxa do not have 4′-*O*-substituents. *O*-Substitution at C-5′ is rare in flavonols except for Asteraceae, Fabaceae and Rutaceae in which 20% of the flavonols concerned are 5′-*O*-methyl derivatives. Finally, 6′-*O*-substitution is mainly found in Lamiaceae.

The C-6 position of flavonols is rarely if ever *O*-substituted in Cistaceae, Geraniales and Solanaceae. In contrast, 6-*O*-substitution is found in the flavonoids of Betulaceae, Fabaceae, Lamiaceae and ferns, and is most common in Asteraceae and Didiereaceae. Finally 6-*C*-methyl derivatives are found in Didiereaceae and ferns. *O*-Substitution at C-8 is a rare feature in flavonols, their overall frequency being some 20%, except for Asteraceae and ferns (40%). 8-Hydroxy compounds are predominant

*Here and in the following 'Ferns' refers to flavonoid-excreting cheilanthoid ferns only.

Table 7.12 Substitutional correlations within flavones reported in Angiosperms and ferns from 1975 to 1985

		Substitutions on the flavone nucleus																										
		C-5			*C-6*			*C-7*			*C-8*			*C-2′*			*C-3′*			*C-4′*			*C-5′*			*C-6′*		
A-ring	*r.f. (%)*	*H*	*O*	*E*	*H*	*O*	*E*	*H*	*O*	*E*	*H*	*O*	*E*	*H*	*O*	*E*	*H*	*O*	*E*	*H*	*O*	*E*	*H*	*O*	*E*	*H*	*O*	*E*
C-5 H	8				61	6	33	50	33	17	100	0	0	83	7	10	44	22	34	30	31	39	78	6	16	94	1	5
C-5 OH	72				47	13	35	4	49	45	52	10	36	73	20	7	54	25	21	28	26	36	78	7	15	93	2	5
C-5 OMe	20				35	0	65	13	23	64	57	3	40	80	0	20	52	13	35	30	20	50	75	3	22	98	0	2
C-6 H	46	11	74	15				10	53	35	57	10	33	68	22	10	58	23	19	39	33	28	84	7	9	91	3	6
C-6 OH	10	5	95	0				5	43	52	71	5	24	95	5	0	47	19	33	10	43	47	91	4	5	100	0	0
C-6 OMe	41	7	63	30				11	32	57	53	7	40	78	10	12	44	24	32	25	29	46	67	5	28	97	0	3
C-6 C-Me	3	0	100	0				0	17	83	67	0	0	100	0	0	100	0	0	33	34	33	100	0	0	100	0	0
C-7 H	10	43	33	24	47	5	48				91	9	0	57	14	29	71	5	24	86	9	5	95	5	0	91	0	9
C-7 OH	43	7	83	10	58	10	31				61	9	30	78	18	4	48	30	22	23	45	32	75	10	15	93	2	5
C-7 OMe	47	3	71	26	35	11	48				46	7	44	78	13	9	52	19	29	25	36	49	78	2	20	96	1	3
C-8 H	57	15	66	19	46	12	38	15	45	38				80	13	7	50	24	26	29	36	35	80	4	16	97	1	2
C-8 OH	8	0	95	5	59	6	35	12	47	41				77	23	0	65	23	12	47	35	18	100	0	0	94	6	0
C-8 OMe	34	0	77	23	45	7	48	0	38	62				68	18	14	52	20	28	30	25	45	69	10	21	90	1	9
C-8 C-Me	1	0	100	0	0	0	0	0	0	100				100	0	0	100	0	0	0	50	50	100	0	0	100	0	0
B-ring																												
C-2′ H	76	10	70	20	42	12	42	8	42	48	60	8	30				44	26	30	19	37	44	81	5	14	100	0	0
C-2′ OH	15	3	97	0	69	3	28	9	50	41	47	12	41				91	9	0	72	28	0	72	12	16	72	10	18
C-2′ OMe	9	10	48	42	48	0	52	32	21	47	47	0	53				63	16	21	58	0	42	68	0	32	84	0	16
C-3′ H	53	7	73	20	51	9	34	14	39	46	55	9	33	63	26	11				50	25	25	90	4	6	90	2	8
C-3′ OH	22	9	81	10	47	8	45	2	58	40	62	8	30	87	7	6				8	45	47	68	10	22	98	2	0
C-3′ OMe	25	11	62	27	36	13	51	9	37	54	58	4	38	93	0	7				8	36	56	62	6	32	100	0	0
C-4′ H	30	12	69	19	59	3	34	28	32	40	55	12	33	47	36	17	87	7	6				97	3	0	81	5	14
C-4′ OH	32	6	82	12	47	13	36	3	59	38	62	9	26	87	13	0	41	30	29				77	7	16	100	0	0
C-4′ OMe	38	7	67	26	35	13	49	1	37	60	54	4	40	90	0	10	34	28	38				65	6	29	100	0	0
C-5′ H	78	10	72	18	50	11	35	12	40	46	59	10	29	78	14	8	60	20	20	37	31	32				93	2	5
C-5′ OH	6	0	92	8	58	8	34	8	75	17	42	0	58	67	33	0	33	42	25	17	42	41				92	0	8
C-5′ OMe	16	6	67	27	27	3	70	0	41	59	56	0	44	68	14	18	21	29	50	0	32	68				100	0	0
C-6′ H	94	9	71	20	45	10	41	9	42	47	59	8	32	80	12	8	50	23	27	26	34	40	77	5	18			
C-6′ OH	1	0	100	0	100	0	0	0	67	33	33	34	33	0	100	0	67	33	0	100	0	0	100	0	0			
C-6′ OMe	5	0	89	11	67	0	33	22	44	34	33	0	67	0	67	33	100	0	0	100	0	0	89	11	0			

For each substitution type of the carbon atoms of the A and C rings, the proportions of H-(proton), O-(hydroxy) and E-(methoxy) substitution on all other positions of the flavone nucleus were calculated.
r.f., relative frequencies of the various substitutions at each carbon atom.
Underlined numbers indicate a complementary part as *C*-methyl derivatives.

Table 7.13 Relative frequencies of the proton, hydroxy and methoxy substitutions according to the carbon number on the flavonol nucleus of some representative taxa in angiosperms and ferns

Plant families	*Subst.*	*Relative frequencies*									
		C-3	*C-5*	*C-6*	*C-7*	*C-8*	*C-2′*	*C-3′*	*C-4′*	*C-5′*	*C-6′*
Asteraceae	H	0	1	42	1	57	93	42	14	75	100
	OH	32	93	12	55	7	7	32	48	13	0
	OMe	68	6	45	44	36	0	26	38	12	0
Betulaceae	H	0	0	73	0	100	100	53	3	97	100
	OH	50	100	3	50	0	0	24	47	0	0
	OMe	50	0	24	50	0	0	23	50	23	0
Cistaceae	H	0	0	100	0	83	100	25	0	70	100
	OH	33	100	0	42	0	0	17	46	21	0
	OMe	67	0	0	58	17	0	58	54	9	0
Didiereaceae	H	0	0	20	0	80	100	24	0	48	100
	OH	20	100	4	100	0	0	60	56	48	0
	OMe	80	0	36	0	0	0	16	44	4	0
Fabaceae	H	0	17	72	0	86	69	55	14	73	100
	OH	45	59	17	52	7	14	24	41	3	0
	OMe	55	24	11	48	7	17	21	45	24	0
Geraniales	H	0	0	100	0	72	94	39	0	94	100
	OH	33	100	0	33	17	6	17	83	6	0
	OMe	67	0	0	67	1	0	44	17	0	0
Lamiaceae	H	0	0	60	0	78	91	48	17	91	96
	OH	26	91	0	22	4	0	30	48	4	0
	OMe	74	9	35	78	18	9	22	35	4	4
Saxifragaceae	H	0	0	88	0	100	100	53	24	94	100
	OH	47	100	12	47	0	0	18	59	0	0
	OMe	53	0	0	53	0	0	29	17	6	0
Solanaceae	H	0	0	100	0	80	93	33	7	73	100
	OH	13	100	0	33	7	0	34	53	20	0
	OMe	87	0	0	67	13	7	33	40	7	0
Ferns	H	0	0	77	0	46	82	77	31	92	100
	OH	48	96	0	37	19	6	10	38	2	0
	OMe	52	4	9	63	21	12	13	31	6	0

Underlined numbers indicate a complementary part as *C*-methyl derivatives.

Table 7.14 Relative frequencies of the proton, hydroxy and methoxy substitutions according to the carbon number on the flavone nucleus of some representative taxa in angiosperms and ferns

		Relative frequencies								
Plant families	*Subst.*	*C-5*	*C-6*	*C-7*	*C-8*	*C-2′*	*C-3′*	*C-4′*	*C-5′*	*C-6′*
Asteraceae	H	1	33	0	54	87	43	9	76	100
	OH	80	9	42	6	7	29	46	1	0
	OMe	19	57	58	40	6	18	45	23	0
Fabaceae	H	20	65	0	100	100	40	5	75	100
	OH	55	0	60	0	0	10	30	0	0
	OMe	25	35	<u>35</u>	0	0	50	65	25	0
Lamiaceae	H	0	44	0	53	73	62	40	96	88
	OH	91	16	48	9	23	18	28	3	4
	OMe	9	40	52	38	4	20	31	1	8
Myrtales	H	41	59	35	88	82	71	47	94	100
	OH	53	0	24	0	0	6	29	0	0
	OMe	6	<u>18</u>	41	<u>0</u>	18	23	23	6	0
Rutaceae	H	0	34	9	37	91	37	6	68	97
	OH	63	6	34	3	0	14	29	14	0
	OMe	37	60	57	60	9	48	66	18	3
Scrophulariaceae	H	0	52	0	83	96	65	17	100	96
	OH	87	9	57	4	4	13	39	0	0
	OMe	13	39	43	13	0	22	44	0	4
Solanaceae	H	0	70	0	50	80	80	50	100	100
	OH	100	20	40	10	10	0	40	0	0
	OMe	0	10	60	40	10	20	10	0	0
Ferns	H	0	83	0	92	92	58	17	100	100
	OH	100	0	42	0	8	25	58	0	0
	OMe	0	<u>8</u>	58	8	0	17	25	0	0

Underlined numbers indicate a complementary part as *C*-methyl derivatives.

in Fabaceae and Geraniales, 8-methoxy derivatives are fairly characteristic of Asteraceae, Cistaceae, Lamiaceae and Solanaceae, while *C*-methylation occurs in Didiereaceae and ferns.

Hydroxylation at C-2′ of the flavonol nucleus is found in Asteraceae and Geraniales, methoxylation at the same position in Lamiaceae and Solanaceae, while generally 2′-*O*-substitution is common in Fabaceae and ferns. The *O*-substitution at C-5′ is normally correlated with tri-*O*-substitution of the B-ring, predominantly in Asteraceae, Cistaceae, Didiereaceae, Fabaceae and Solanaceae.

7.5.6 Flavone diversity and botanical distribution

Table 7.14 uses the same parameters for flavones as shown in the previous section for flavonols. 5-Deoxyflavones are found mostly in Fabaceae and Myrtales. At C-5, Solanaceae and ferns exhibit only hydroxyl substitution, i.e. the basic pattern, while all the other taxa show both hydroxyl and methoxyl substitutions (the latter is important in Asteraceae, Fabaceae and Rutaceae).

7.6 CHROMATOGRAPHIC METHODS FOR FLAVONES AND FLAVONOLS

7.6.1 High-performance liquid chromatography

Hostettmann and Hostettmann (1982), in their review of isolation techniques for flavonoids, reviewed the relevant literature on high-performance liquid chromatography (HPLC) up to 1980. Some of the more recent literature is reviewed here.

Daigle and Conkerton (1982) reported t_R values for flavone, tectochrysin, apigenin, acacetin, morin, kaempferol, quercetin, rhamnetin, isorhamnetin and myricetin. Separation was on Bondapak C_{18}, using a methanol–acetic acid–water eluting system with two pumps. Van de Casteele *et al.* (1982) used Lichrosorb RP-18 and a combination of an isocratic and a gradient (aqueous formic acid and methanol) technique. Their table of retention times lists 32 flavones and 20 flavonols. They noted the following correlations between structure and t_R values. Because of strong internal hydrogen bonding between the C-5 hydroxyl and the carbonyl group at C-4, 5-hydroxyflavonoids have higher t_R values than their counterparts lacking a free 5-hydroxy group. The t_R value of flavonol is only slightly higher than that of flavone, due to the much weaker hydrogen bonding between the C-4 carbonyl and the C-3 hydroxyl group. Hence flavones and flavonols with otherwise identical substitution patterns are often 'critical pairs', and are only poorly separated.

Flavonoids which differ in the number of hydroxyl groups at positions other than C-3 and C-5 can easily be separated, since increasing hydroxylation reduces the t_R values considerably (but to a lesser extent if a hydroxyl group is already present *ortho* to the next substitution). This effect is reversed by methylation. Thus flavonoids and their partial methyl ethers can easily be separated (3-*O*-methylflavonols excepted), whereas flavonoids differing in structure by only one methoxyl group are often inseparable. The oxygen of a methoxyl group can be considered as a hydrogen bond acceptor for the mobile phase, while the methyl group contributes to the hydrophobic interaction with the stationary phase. The two effects balance each other so that a permethylated flavone has the same t_R as flavone itself.

The HPLC properties of (poly)methoxylated flavones on Lichrosorb Si 60 with two isocratic solvent systems (heptane–propan-1-ol) was studied by Bianchini and Gaydou (1983). They found a correlation between the capacity factors of flavones and the position of the methoxyl groups on the flavone skeleton. A logarithmic plot of k' versus $\Delta k'$ shows that 15 flavones can be divided into four groups having the same number and position of methoxylation on the A-ring (5, 7, 8; 5, 6, 7; 5, 6, 7, 8; 6, 7, 8). The order of elution of polymethoxylated flavones is related not only to the number of methoxy groups present but, above all, to their position. Barberán *et al.* (1985d) studied the effect of hydroxyl or methoxyl groups located at different positions on the flavone nucleus on the reversed-phase HPLC behaviour of 33 5-hydroxyflavones in which the A-ring was further tri- or tetra-substituted. They used a Perkin–Elmer C_{18} RP column and gradient elution (water–formic acid and acetonitrile). The results confirmed the findings of Van de Casteele *et al.* (1982) and extended them to 6- and 8-*O*-substituted compounds. 6-Hydroxyflavonoids are eluted with a shorter t_R than the 8-hydroxy isomers, due to the internal hydrogen bonding between the hydroxyl groups at C-6 and C-5, which decreases the interaction between the 5-hydroxyl and the 4-keto group. The size of the molecule affects t_R, since the smallest molecules can interact easily with the C_{18} branches of the stationary phase. Thus flavones bearing two methoxyl group in ring B interact less strongly with the stationary phase than their counterparts bearing a single methoxyl group in ring B, although an increase in the number of methoxyl groups should increase t_R. Applications of the HPLC technique to particular analytical problems have been reported, e.g. flavonoid aglycones in propolis (Bankova *et al.*, 1982), flavonoids of *Artemisia* (Tamma *et al.*, 1985) and *Sideritis* (Barberán *et al.*, 1985b).

7.6.2 Thin-layer chromatography (TLC)

The most powerful tool for the TLC of flavonoid aglycones is the use of laboratory-prepared plates based on polyamide DC-11 (Wollenweber, 1982c). Unfortunately, Macherey-Nagel no longer supply this adsorbent, so it is necessary to employ precoated plates, which do not provide quite such good resolution and on which detection with Naturstoff-reagent A (NA) is not so distinctive. It is therefore advisable to check separations on precoated polyamide DC-11, DC-6 or DC-6.6 with separations on silica gel in such solvents as toluene–methyl ethyl ketone (9:1) and toluene–methyl ethyl ketone–acetic acid (18:5:1).

According to Barberán *et al.* (1985a), 6-hydroxyflavones and 8-hydroxyflavones can easily be distinguished by differences in their chromatographic behaviour on cellulose TLC (or PC) with 30% or 60% HOAc; the former have lower R_F values than the latter compounds. These R_F differences, which are due to the internal hydrogen bonding between the hydroxyl at C-5 and that at C-6, provide a means of separating 5, 6-dihydroxy-7, 8-dimethoxyflavones from 5, 8-dihydroxy-6, 7-dimethoxyflavones. Similar observations on PC have been reported by Combier *et al.* (1974) and others. These differences are also observed on polyamide where they may be even more evident due to the fact that the isomeric flavones have different colours. On polyamide, however, while 7-methyl ethers have higher R_F values than 4′-methyl ethers in the 8-hydroxyflavone series, they have lower R_F values than the 4′-methyl ethers if there is hydroxylation at C-6.

One further observation is worth making, namely that flavones and flavonols with a trihydroxy-substituted A-ring, give a special colour with the NA reagent. Compounds such as norwogonin, isoscutellarein, 8-hydroxygalangin, herbacetin and platanetin turn blueish–violet in daylight after spraying with the NA reagent. This reaction is not normally affected by methylation of B-ring hydroxyls or (for flavonols) methylation at C-3 (e.g. herbacetin 4′-methyl ether, gossypetin 3, 3′-dimethyl ether), whereas methylation at C-7 causes atypical brown colours (8-hydroxygalangin 7-methyl ether, herbacetin 7-methyl ether) or sometimes reddish–orange colours (gossypetin 8-methyl ether, gossypetin 3, 8-dimethyl ether). Although gossypetin and its 3-methyl ether diverge from this behaviour (they turn red and reddish–brown/violet, respectively), this reaction may otherwise be considered as characteristic of compounds with 5, 7, 8-trihydroxy substitution.

Barberán *et al.* (1984a) studied the chromatographic

behaviour of 16 5-hydroxyflavones, about half of which were substituted at C-6 and partly also at C-8. When their R_F values on silica in an acidic solvent (benzene–methanol–acetic acid, 45:3:2) were plotted against R_F values in a neutral solvent (chloroform–*n*–hexane–methanol, 40:40:3), the flavones grouped themselves into sections, depending on the number of free hydroxyl groups. From such a chromatographic chart (or, more accurately, from a relative chromatographic chart, using R_F values calculated by giving $R_F = 100$ to e.g. 5,3′,4′-trihydroxy-6,7,8-trimethoxyflavone), the number of free hydroxyl groups in an unknown flavone can be deduced.

When, from the same chromatographic systems, R_M values were plotted against the number of methoxyl groups in each flavone studied, parallel straight lines were obtained for each homologous series, i.e. for groups of flavones with four, five, or six substituents on the flavone nucleus. Such a chart might be used to deduce the number of methoxyl groups in an unknown flavone if the substitution pattern is already known (Barberán *et al.*, 1984a).

When R_M values are plotted against log(number of OH groups/number of OMe groups), a straight-line relationship is observed. Thus it is possible to ascertain the ratio of hydroxyl to methoxyl groups in an unknown flavone and confirm the results obtained with the two preceding methods by calculation of the R_M values.

Flavonoid aglycones can be easily permethylated and the number of possible products compared with partially methylated flavones is thus considerably reduced. Products obtained by permethylation of flavonoids can be compared with authentic permethylated samples to give information about the substitution pattern of the flavone molecule. Various solvents have been used for the TLC of these compounds on silica gel (Barberán *et al.*, 1984a).

Naturstoffreagent A (NA; β-aminoethyl ester of diphenylboric acid) is still unsurpassed as a reagent for the detection and differentiation of flavonoids, due to its sensitivity and the broad spectrum of colours that it produces. Brasseur and Angenot (1986) reported an improvement in its sensitivity for TLC on silica by the addition of 3% PEG (poly-ethylene glycol) 400 to the reagent solution in MeOH.

Several laboratories have good experience now with preparative TLC on a semi-preparative scale via the commercially available Chromatotron for centrifugal TLC (Hostettmann *et al.*, 1986; Nahrstedt, 1986).

One final use of TLC in the identification of 5,6-dihydroxy-7,8-dimethoxy- and 5,8-dihydroxy-6,7-dimethoxy-flavones is worth describing. The identification of these compounds is not easy by means of the classic UV, MS and NMR techniques, and in the past, several of these compounds have been erroneously characterized (see Section 7.4). The well-known Wessley–Moser rearrangement in which 6- or 8-substituted flavones undergo isomerization under acidic conditions has been used recently in the characterization of the glycoside 5,8,3′,4′-tetrahydroxy-6,7-dimethoxyflavone 8-*O*-glucoside (Barberán *et al.*, 1984b) and the aglycone thymusin (5,6,4′-trihydroxy-7,8-dimethoxyflavone) (Ferreres *et al.*, 1985a). The original natural aglycone, or the aglycone obtained on enzymic hydrolysis, should be compared with the reaction products of acidic treatment by TLC or HPLC. TLC on cellulose with 30% or 50% acetic acid as solvent allows the differentiation of the 5,6-dihydroxy and 5,8-dihydroxy isomers since the former show lower R_f values than the latter. Likewise, on HPLC with RP C-18 columns, the 5,6-dihydroxy compounds are eluted with shorter retention times than the isomeric 5,8-dihydroxy compounds (see Sections 7.6.1 and 7.6.2). Comparative studies of the UV spectra and mass spectra of the natural product and the isomer obtained on acidic treatment can also help in the identification of these compounds (cf. Sections 7.7.1 and 7.7.2).

The presence of hydroxyl groups at C-8 in flavones favours the formation of the 6-hydroxy isomer in compounds with a trisubstituted A-ring. In those with a tetrasubstituted A-ring, demethylation is more favoured in acidic conditions than the opening of the pyrone ring to give the isomer, so that 6-hydroxy-8-methoxyflavones exhibit the same activity as 6-methoxy-8-hydroxyflavones. 6,7,8-Trimethoxyflavones, on the other hand, yield only small amounts of the 6-demethyl product even after long acid treatment (Barberán *et al.*, 1985a). Analysis of the demethylation products by the above-mentioned methods can help again in the characterization of the original structure.

7.7 SPECTROSCOPIC METHODS

7.7.1 Ultraviolet absorption spectroscopy

UV spectroscopy is the most useful technique for identifying the flavonoid type, for defining the oxygenation pattern and for determining the positions of phenolic substitution (Mabry *et al.*, 1970; Markham and Mabry, 1975; Markham, 1982; Wollenweber, 1982c). We report here some recent findings on interpretation of spectra which have produced additional information on flavonoid structures.

Sodium methoxide (NaOMe) is used particularly for the detection of free 3- and/or 4′-hydroxy groups. A continual reduction in intensity of Band I indicates not only 3,4′-dihydroxy substitution but also 5,6,7- and 5,7,8-trihydroxy substitution in ring A or 3′,4′,5′-trihydroxy substitution at ring B. The decrease in Band I with time, caused by alkaline decomposition of 3,4′-dihydroxyflavones, has been shown to be significantly influenced by other substituents. For example, methylation of the hydroxyl at C-7 causes a marked decrease in the rate of decomposition (Barberá *et al.*, 1986).

Bacon *et al.* (1976) reported a procedure for the detection of 7-*O*-substitution in 4′-hydroxyflavones and flavonols based on comparing Band I in sodium acetate (NaOMe) spectra. When this band in the NaOAc spectrum is the same as, or appears at longer wavelengths than, Band I in the NaOMe spectrum, the flavonoid contains 7-*O*-substitution. However, for the effect to be observed, all traces of acid must be excluded from the NaOAc reagent and the sample solution. Recent observations (Rösler *et al.*, 1985) indicate that impurities in the reagent from different manufacturers may yield different spectra for the same flavonoid. The authors, therefore, recommend that when a new bottle of NaOAc is used, the pH of a 5% aqueous solution should be between 7.7 and 8.2; the spectrum of quercetin should be recorded in a standard saturated methanolic solution of NaOAc. The absorbances must lie between the following values: 255–259 275.276 sh., 289 sh., 330–335 and 385–394 nm. Where the pH is not in the range of 7.7–8.2, lyophilize or replace the shift reagent.

Apart from the well-known rules for interpretation of the bathochromic or hypsochromic shifts in the presence of $AlCl_3$ with or without HCl, a recent compilation by Viorin (1983) reveals many other uses for these reagents in the structural identification of 5-hydroxy- and 5-hydroxy-3-methoxyflavones. Taking into account the fact that, with neutral or acidic $AlCl_3$, 5-hydroxyflavones give rise to four major peaks Ia and Ib, IIa and IIb, observation of the relative peak heights, bathochromic shifts, peak locations and peak shapes lead to the description of 20 spectral types corresponding to 20 structural patterns (Table 7.15).

There have recently been a number of reports on flavonoids with hydroxyl or methoxyl groups at the 2′- and/or 6′-position of the B-ring (Tomimori *et al.*, 1982b; Matsuura *et al.*, 1983; Kimura *et al.*, 1984; Tanaka *et al.* 1986). It is possible to determine which observations of

Table 7.15 Diagnostic UV spectral properties of some flavone classes in the presence of $AlCl_3$ and $AlCl_3$ + HCl

A Band I giving rise to a single Band Ib (350–381 nm) in the presence of $AlCl_3$–HCl		
Band II in MeOH-	Band I $AlCl_3$–HCl/MeOH	
1 inferior to 279 nm	19–26 nm	Group 1 6-OMeflavones
2 279–281	26–31 nm	Group 2 6-OH, 3-OMeflavones
3 283–286	23–28 nm	Group 3 6-OHflavones
4 292	26 nm	Group 4 6-OH, 8-OMeflavones
A′ Band I giving rise to two peaks Ia and Ib in the presence of $AlCl_3$–HCl		
B Band I with $AlCl_3$ less than 449 nm		
(1) $AlCl_3$–HCl – Band Ia less than 395 nm		Group 5 8-C-Meflavones
		Group 6 6-C-Meflavones
		Group 7 5-OH, 7-ORflavones
(2) $AlCl_3$–HCl – Band Ia more than 395 nm		
(a) and Ia shoulder		Δλ Ia $AlCl_3$–HCl/MeOH less than 68 nm
		Group 8 3,6-diOMeflavones
		Group 9 6-C-Me, 3-OMeflavones
		Δλ Ia $AlCl_3$–HCl/MeOH more than 68 nm
		Group 10 3-OMe, 6,8-diC-Meflavones
		Group 11 3,6,8-triOMeflavones
		Group 12 6,8-diOMeflavones
		Group 13 6,8-diC-Meflavones
(b) and Ia peak		λ max Ia less than 435 nm
		Band II in MeOH less than 271 nm
		Group 14 3-OMeflavones
		Band II in MeOH more than 271 nm
		Group 15 8-OMeflavones
		Group 16 3,8-diOMeflavones
		Group 17 8-OHflavones
		Group 18 8-OH, 7-OMeflavones
		Group 19 3-OMe, 8-C-Meflavones
		λ max Ia more than 435 nm
		Group 20 8-OH, 3-OMeflavones
B′ Band I with $AlCl_3$ more than 449 nm		
Group 16′, 17′ and 20′ homologous derivatives with 3′,4′-diOH		

bathochromic shifts in UV spectra are valuable for structural elucidation of these unusual flavonoids. Normally, a strong base can be used for the detection of 3- and/or 4′-hydroxy groups; a 4′-hydroxy group yields a large bathochromic shift of Band I without a decrease in intensity since the flavone is ionized and, therefore, forms a stable planar structure. A similar reaction would be expected for 2′-hydroxyflavones. From Table 7.16 it is evident that the UV bands of 5,7,4′-trihydroxyflavone (apigenin) and 5,7,2′-trihydroxyflavone are similar. However, the intensity of Band I in 2′-hydroxyflavone is weaker than that of the 4′-hydroxyflavone in MeOH. In both cases, the addition of NaOMe causes a large bathochromic shift of Band I (+55–64 nm) and an increase in intensity, but the amount is smaller for the 2′-hydroxyflavone. Flavones possessing a 2′-hydroxy group and lacking 6′-*O*-substitution (Table 7.17) produce large shifts of Band I in the presence of NaOMe and the flavonoid structure will decompose in the case of compounds having a hydroquinone moiety (e.g. no. 11, 12, 13). In MeOH, 2′-hydroxyflavones possessing 6′-*O*-substitution show a weak band I leading to a UV spectrum similar in shape to that of flavanones. In the presence of NaOMe (Table 7.17) these flavones produce small bathochromic shifts of 10–30 nm.

Zapesochnaya and Sokol'skaya (1984) have studied the spectral properties of seven *C*-methylflavones: noreucalyptin (I), eucalyptin (II), sideroxylin (III), silpin (IV) and their demethylation products: 6-*C*-methylapigenin (V from I), 6,8-di-*C*-methylapigenin (VI from II and III), demethylsilpin (VII from IV). The spectral features reported in MeOH show that a *C*-methyl group at C-6 has the same influence on the UV spectrum as a 6-*O*-substitution. With $AlCl_3$/HCl, the authors observed a small bathochromic shift of about 20 nm as compared with MeOH. These latter observations do not agree with the spectral features from Voirin's (1983) compilation which give a bathochromic shift in the range 60–80 nm for 6-*C*- and 6,8-di-*C*-methylflavones.

Table 7.17 Bathochromic shifts of Band I in the UV spectra of 2′-oxygenated flavones in the presence of NaOMe reagent

Compound No.	*Position of OH/OMe*	*Shift NaOMe/MeOH* (nm)
1	4′/	61
2	2′/	60
3	2′/3′	80
4	2′/4′	65
5	2′, 5, 7/8	73
6	2′, 5, 7/6	60
7	2′/5, 6, 7, 8	72
8	2′, 5/6, 7, 8	67
9	2′/5, 7/4′, 5′, 6	64
10	2′, 5/3′, 5′, 6, 7	84
11	2′, 5′, 5/7, 8	dec*
12	2′, 5′, 5/6, 7	dec*
13	2′, 5′, 5/6, 7, 8	dec*
14	2′, 5, 7/6′	27
15	3′, 6′, 5, 7/2′, 8	16
16	2′, 6′, 5, 7/3′, 8	28

*Decomposed.

7.7.2 Mass spectroscopy

Electron impact mass spectrometry (EI–MS) serves as a valuable aid in determining the structures of flavones and flavonols, especially when only very small quantities are available. Most aglycones are sufficiently volatile at probe temperatures of 100–300°C to allow successful mass spectroscopy without derivatization. The general rules for interpretation of the molecular formula, type and number of substitutions in the three rings of the flavonoid skeleton, and in some cases of exact location of these substitutions, have been treated *in extenso* in Markham and Mabry (1975) and in Markham (1982). Some further information has been obtained in the last five years.

Relative ion abundance and abundance ratios have been measured for 18 flavonoids by Madhusudanan *et al.* (1985), and for 10 polymethoxyflavones by Rizzi and Boeing (1984); some of these results can be compared with those of Goudard *et al.* (1978, 1979). Rizzi and Boeing have used M^+ $[M-\mathrm{Me}]^+$ and ions due to intact A- and B-ring fragments for the identification and/or quantification of these flavones in foods and in biological fluids. A mechanism for the formation of $[M-1]^+$ ions in flavones has been discussed by Guidugli *et al.* (1984).

It is generally agreed that flavonols or 3-methoxyflavones with either 2′-hydroxy or 2′-methoxy groups

Table 7.16 Comparison of UV absorption properties of apigenin with 2′-oxygenated flavones

	λ_{max} nm (log ε)	
Flavones	*MeOH*	*MeOH + NaOMe*
5, 7, 4′-TriOH flavone (apigenin)	268(4.11) 333(4.15)	275(4.19) 388(4.49)
5, 7, 2′-TriOH flavone	267(4.35) 335(3.83)	273(4.35) 399(3.94)
5, 7, 2′, 6′-TetraOH flavone	259(4.23) 308(4.00) 330 sh	266(4.29) 335(4.00)

undergo unique fragmentations leading to the formation of an internal five-membered ring by loss of hydroxyl $[M^+ - 17]$ or methoxyl $[M^+ - 31]$ radicals. Recently it has been shown that these ions are characterstic but are not specific for 2′-hydroxy- or 2′-methoxyflavones. In Table 7.18, compound 1, which has no 2′-hydroxyl group, shows an intense $[M^+ - 31]$ ion. However, only 2′-oxygenated flavonols (compounds 8 and 16) give base peaks at $[M^+ - 17]$ and $[M^+ - 31]$ respectively.

Field desorption (FD) mass spectrometry, the first ionization technique used for the study of thermolabile compounds without derivatization, has been applied to the study of flavonoids by Schulten and Games (1974), Zapesochnaya *et al.* (1984) and Domon and Hostettmann (1985). Another soft ionization method which has been developed is desorption chemical ionization (DCI) mass spectrometry which uses a probe of electrically heated tungsten wire which is introduced into the chemical ionization source (Arpino and Devant, 1979; Hostettmann *et al.*, 1981). Conventionally the solid sample is deposited on the tungsten wire, but this practice gives poor results for ions of high mass. When the sample is solubilized in glycerol before deposition, the relative quantity of ions of higher mass is considerably increased (Arpino and Guiochon, 1982; Domon and Hostettmann, 1985). Usually the reactant gas is ammonia, but recently there are some reports of the use of amines for the reaction, in particular for flavonoid aglycones (Bankova *et al.*, 1986). Sakushima *et al.* (1984) reported on the MS of some flavonoids in the negative-ion mode DCI using methane at low pressures as reagent.

Table 7.18 Relative intensities of ions $[M^+ - 17]$ and $[M^+ - 31]$ in the mass spectra of 2′-oxygenated flavones

		r.i. (%)	
Compound No.	*Flavone*	$[M^+ - 17]$	$[M^+ - 31]$
1	4′-OH	10	
2	2′-OH	17	
3	2′-OH, 3′-OMe	12	
4	2′, 5, 7-triOH, 8-OMe	–	
5	3′, 6′, 5, 7-tetraOH, 2′, 8-diOMe	6	
6	2′, 5-diOH, 3′, 7, 8-triOMe	–	
7	2′, 6′-diOH	12	
8	2′, 3-diOH	100	
9	3′, 5′-diOMe		11
10	2′-OMe		4
11	2′, 3′-diOMe		4
12	2′, 5′-diOMe		5
13	2′, 3′, 4′-diOMe		3
14	2′, 3′, 5′-triOMe		–
15	2′, 6′-diOMe		5
16	2′, 3-diOMe		100

r.i. (%) relative intensities.

A third new technique is fast atomic bombardment (FAB) mass spectroscopy for use with neutral atoms. Here, a sample is solubilized in a polar matrix (e.g. glycerol) and deposited on a copper target, which is bombarded with energized atoms inducing desorption and ionization. The ability of FAB-MS to produce useful structural information for flavonoids has been tested (Sakushima *et al.*, 1984; De Koster *et al.*, 1985; Domon and Hostettmann, 1985; Crow *et al.*, 1986). De Koster *et al.* (1985) showed that in the MS of flavonols the protonated aglycones provide the base peak, and various aglycones reveal structure-specific fragmentation reactions which are more distinct in their metastable ions and collisional activation spectra, thus allowing discrimination between isomeric forms. Though sometimes used for flavonoid aglycones, these new techniques are chiefly of interest for more complex flavonoids (see Chapter 8).

7.7.3 Nuclear magnetic resonance spectroscopy

^{1}H and ^{13}C nuclear magnetic resonance spectrometry are now well-established methods of flavonoid structure analysis. The most general technique is ^{1}H NMR spectroscopy, but with recent technical advances and the greater availability of ^{13}C NMR spectroscopy, the past five years have seen a great increase in the use of ^{13}C NMR methods. This powerful technique, giving information on the carbon skeleton of the molecule, may be regarded as complementary to ^{1}H NMR spectroscopy since it relates more closely to the structural environment of the flavonoid nucleus. In the case of aglycones, the use of ^{13}C NMR spectroscopy may not be needed for fine structural identifications but it is a valuable method for examining naturally occurring complex flavonoid glycosides (Gonzaléz and Pomilio, 1982).

Basic information for interpretation of ^{1}H NMR spectra is available in Mabry *et al.* (1970), Markham and Mabry (1975), and Markham (1982), and in Markham and Chari (1982) and Markham (1982) for ^{13}C NMR spectroscopy. Two additional features may be mentioned: the effects of the ionization of hydroxyl groups on the carbon chemical shifts (Mendez *et al.*, 1982) and magnesium-induced shifts in the ^{1}H NMR spectra.

ACKNOWLEDGEMENTS

E.W. wishes to thank Drs F.A. Tomas-Barberan (Murcia, Spain), M. Iinuma (Gifu, Japan), T. Nomura (Chiba, Japan), P.G. Waterman (Glasgow, UK) for reviewing individual sections or for special information. Thanks are due to G. Yatskievych (Bloomington, Ind.) for checking the scientific plant names and in particular to Mrs K. Mann (Darmstadt, FRG) for her inestimable help with preparation of Tables 7.1–7.7. M.J. thanks Dr J.-F. Gonnet (Lyon, France) for statistical analysis of the flavonoid data.

REFERENCES

Adekenov, S.M., Mucahmetzhanov, M.N., Kagarlitskii, A.D., and Turmuchambetov, A.Z. (1984), *Khim. Prir. Soedin.* 5, 603.

Adinarayana, D., Gunasekar, D., Seligman, O. and Wagner, H. (1980), *Phytochemistry* **19**, 483.

Adinarayana, D., Ramachandraiah, P., Seligmann, O. and Wagner, H. (1981), *Phytochemistry* **20**, 2058.

Adzet, T. and Martinez-Verges, F. (1980), *Plant. Med. Phytothér.* **14**, 8.

Adzet, T. and Martinez, F. (1981), *Biochem. Syst. Ecol.* **9**, 293.

Ahmad, S. (1983), *Plant. Med.* **48**, 62.

Ahmad, S. (1986), *Phytochemistry* **25**, 955.

Amaro, J.M. and Méndez, A.M. (1983), *Rev. Latinoam. Quim.* **14**, 86.

Amer, M.M., Salama, O.M. and Omar, A.A. (1984), *Acta Pharm. Jugosl.* **34**, 257.

Anhut, S., Zinsmeister, H.D., Mues, R., Barz, W., Mackenbrock, K., Köster, J. and Markham, K.R. (1984), *Phytochemistry* **23**, 1073.

Anjaneyulu, A.S.R. and Murty, V.S. (1981), *Indian J. Chem.* **20B**, 87.

Aritomi, J. and Kawasaki, T. (1984), *Phytochemistry* **23**, 2043.

Arpino, P.J. and Devant, G. (1979), *Analysis* **7**, 348.

Arpino, P.J. and Guiochon, G. (1982), *J. Chromatogr.* **251**, 153.

Arriaga-Giner, F.J., Borges del Castillo, J., Manresa-Ferrero, M.T., Pena de Recinos, S. and Luis, F.R. (1982), *Rev. Latinoam. Quim.* **13**, 47.

Arriaga-Giner, F.J., Borges del Castillo, J., Manresa-Ferraro, M.T., Vázquez-Bueno, P., Rodriguez-Luis, F. and Valdes-Iraheta, S. (1983), *Phytochemistry* **22**, 1767.

Arriaga-Giner, F.J., Wollenweber, E. and Hradetzky, D. (1986), *Z. Naturforsch.* **41c**, 946.

Asakawa, Y, Tokunaga, N., Takemoto, T., Hattori, S., Mizutani, M. and Suire, C. (1980), *J. Hattori Bot. Lab.* **47**, 153.

Ates, N., Ulubelen, A., Clark, W.D., Brown, G.K., Mabry, T.J., Dellamonica, G. and Chopin, J. (1982), *J. Nat. Prod.* **45**, 189.

Avanoglu, E., Ulubelen, A., Clark, W.D., Brown, G.K., Kerr, R.R. and Mabry, T.J. (1981), *Phytochemistry* **20**, 1715.

Bacon, J.D., Mabry, T.J. and Mears, J.A. (1976), *Rev. Latinoamer. Quim.* 7, 83.

Bacon, J.D., Fang, N. and Mabry, T.J. (1986), *Plant. Syst. Evol.* **151**, 223.

Bankova, V.S., Popov, S.S. and Marekov, N.L. (1982), *J. Chromatogr.* **242**, 135.

Bankova, V.S., Mollova, N.N. and Popov, S.S. (1986), *Org. Mass Spectr.* **21**, 109.

Barberá, O., Sanz, J.F. and Marco, J.A. (1986), *J. Nat. Prod.* **49**, 702.

Barberán, F.A.T. (1986), *Fitoterapıa* **57**, 67.

Barberán, F.A.T., Tomás, F. and Ferreres, F. (1984a), *J. Chromatogr.* **315**, 101.

Barberán, F.A.T., Tomás, F. and Ferreres, F. (1984b), *Phytochemistry* **23**, 2112.

Barberán, F.A.T., Tomás, F. and Gil, M.I. (1984c), *Farm. Tijdschr. Belg.* **61**, 397.

Barberán, F.A.T., Ferreres, F. and Tomás, F. (1985a), *Tetrahedron* **41**, 5733.

Barberán, F.A.T., Nunez, J.M. and Tomás, F. (1985b), *Phytochemistry* **24**, 1285.

Barberán, F.A.T., Hernández, L., Ferreres, F. and Tomás, F. (1985c), *Plant. Med.*, 452.

Barberán, F.A.T. Tomás, F., Hernandez, L. and Ferreres, F. (1985b), *J. Chromatogr.* **347**, 443.

Barberán, F.A.T., Gil, M.I., Tomás, F., Ferreres, F. and Arques, A. (1985e), *J. Nat. Prod.* **48**, 859.

Barberán, F.A.T., Hernandez, L. and Tomás, F. (1986a), *Phytochemistry* **25**, 561.

Barberán, F.A.T., Ferreres, F., Tomás, F. and Guirado, A. (1986b), *Phytochemistry* **25**, 923.

Baruah, N.C., Sharma, R.P., Thyagarajan, G., Herz, W. and Govindan, S.V. (1979), *Phytochemistry* **18**, 2003.

Becchi, M. and Carrier, M. (1980), *Plant. Med.* **38**, 267.

Bernhard, H.O. and Thiele, K. (1981), *Plant. Med.* **41**, 100.

Bhardwaj, D.K., Murari, R., Seshadri, T.R. and Singh, R. (1976), *Phytochemistry* **15**, 1789.

Bhardwaj, D.K., Bisht, M.S., Uain, S.C., Mahta, C.K. and Sharma, G.C. (1980), *Phytochemistry* **19**, 713.

Bhardwaj, D.K., Gupta, A.K., Chaud, R. and Jain, K. (1981a), *Curr. Sci.* **50**, 750.

Bhardwaj, D.K., Gupta, A.K., Jain, R.K. and Rawi, A. (1981b), *Curr. Sci.* **50**, 491.

Bhardwaj, D.K., Gupta, A.K., Jain, R.K. and Sharma, G.C. (1981c), *J. Nat. Prod.* **44**, 656.

Bhardwaj, D.K., Gupta, A.K., Jain, R.K. and Munjal, A. (1981d), *Indian J. Chem.* **20B**, 446.

Bhardwaj, D.K., Bisht, M.S., Jain, R.K. and Munjal, A. (1982), *Phytochemistry* **21**, 2154.

Bhardwaj, D.K., Jain, R.K. and Kohli, R.M. (1983), *Natl. Sci. Acad.* **49A**, 408.

Bianchini, J.P. and Gaydou, E.M. (1983), *J. Chromatogr.* **259**, 150.

Bird, A.E. and Marshall, A.C. (1969), *J. Chem. Soc. (C)*, 2418.

Bittner, M. and Watson, W.H. (1982), *Rev. Latinoam. Quim.* **13**, 27.

Bittner, M., Silva, M., Vargas, J. and Bohlmann, F. (1983), *Phytochemistry* **22**, 1523.

Bohlmann, F. (1986), *Private communication.*

Bohlmann, F. and Fritz, U. (1979), *Phytochemistry* **18**, 1080.

Bohlmann, F., Grenz, M., Dhar, A.K. and Goodman, M. (1981a), *Phytochemistry* **20**, 105.

Bohlmann, F., Gupta, R.K., King, R.M. and Robinson, H. (1981b), *Phytochemistry* **20**, 1417.

Bohlmann, F., Kramp, W., Grenz, M., Robinson, H. and King, R.M. (1981c), *Phytochemistry* **20**, 1907.

Bohlmann, F., Singh, D., Jakupovic, J., King, R.M. and Robinson, H. (1982), *Phytochemistry* **21**, 371.

Bohlmann, F., Zdero, C. and Huneck, S. (1985a), *Phytochemistry* **24**, 1027.

Bohlmann, F., Banerjee, S., Jakupovic, J., *et al.* (1985b), *Phytochemistry* **24**, 511.

Bouzid, N., Moulis, C. and Fouraste, I. (1982a), *Plant. Med.* **44**, 157.

Bouzid, N., Fouraste, I., Voirin, B., Favre-Bonvin, J. and Lebreton, Ph. (1982b), *Phytochemistry* **21**, 803.

Bowen, J.H. and Perera, K.P.W.C. (1982), *Phytochemistry* **21**, 433.

Brasseur, T. and Angenot, T. (1986), *J. Chromatogr.* **351**, 351.

Bulinska-Radomska, Z., Norris, J.A. and Mabry, T.J. (1985a), *J. Nat. Prod.* **48**, 144.

Bulinska-Radomska, Z., Norris, J.A., Hosage, D.A. and Mabry, T.J. (1985b), *J. Nat. Prod.* **48**, 667.

Burret, F., Lebreton, Ph. and Voirin, B. (1982), *J. Nat. Prod.* **45**, 687.

Buschi, C.A. and Pomilio, A.B. (1982), *J. Nat. Prod.* **45**, 557.

Buschi, C.A., Pomilio, A.B. and Gros, E.G. (1981), *Phytochemistry* **20**, 1178.

Cardona, M.L. and Seoan, E. (1982), *Phytochemistry* **21**, 2759.

Cerecer, M.J., Santos, E. and Crabbé, P. (1974), *Rev. Soc. Quim. Mex.* **18**, 269.

Chandrasekharan, I., Khan, H.A. and Ghanim, A. (1981), *Plant. Med.* **43**, 310.

Chari, M.M., Grayer-Barkmeijer, R.J., Harborne, J.B. and Österdahl, B.-G. (1981), *Phytochemistry* **20**, 1977.

Chemesova, I.I., Belenovskaya, L.N. and Markova, L.P. (1984), *Khim. Prir. Soedin*, 789.

Chen, C.-C., Chen, Y.-P., Hsu, H.-Y. and Chen, Y.-L. (1984), *Chem. Pharm. Bull.* **32**, 166.

Chiang, M.T. and Silva, M. (1978), *Rev. Latinoam. Quim.* **9**, 102.

Chiappini, I., Fardella, G., Menghini, A. and Rossi, C. (1982), *Plant. Med.* **44**, 159.

Chibber, S.S. and Dutt, S.K. (1981), *Phytochemistry* **20**, 1460.

Chou, C.-J. (1978), *J. Taiwan Pharm. Assoc.* **30**, 36.

Chou, C., Liu, K. and Yang, T. (1981), *Annu. Rep. Natl. Res. Inst. Chin. Med.*, 91.

Chowdhury, B.K. and Chakraborty, D.B., (1971), *J. Indian Chem. Soc.* **48**, 80.

Clark, W.D. and Wollenweber, E. (1984), *Z. Naturforsch.* **39c**, 1184.

Clark, W.D. and Wollenweber, E. (1985), *Phytochemistry* **24**, 1122.

Coleman, P.C., Potgieter, D.J.J., Van Aswegen, C.H. and Vermeulen, N.M.J. (1984), *Phytochemistry* **23**, 1202.

Collado, I.G., Macias, F.A., Massanet, G.M. and Luis, F.R. (1985), *J. Nat. Prod.* **48**, 819.

Combier, H., Jay, M., Voirin, B. and Lebreton, P. (1974), *C.R. Groupe Polyphenols* **5**.

Cordano, G.F., dela Pena, A.M. and Medina, J.J. (1982), *J. Nat. Prod.* **45**, 653.

Courtney, J.L., Lassak, E.V. and Speirs, G.B. (1983), *Phytochemistry* **22**, 947.

Crow, F.W., Tomer, K.B., Looker, J.H. and Gross, M.L. (1986), *Anal. Biochem.* **155**, 286.

Çubukçu, B. and Damadyan, B. (1986), *Fitoterapia* **57**, 124.

Çubukçu. B. and Yuksel, V. (1982), *J. Nat. Prod.* **45**, 137.

Daigle, D.L. and Conkerton, E.J. (1982), *J. Chromatogr.* **240**, 202.

De Koster, C.G., Heerma, W., Dijkstra, G. and Niemann, G.J. (1985), *Biomed. Mass Spectrom.* **12**, 596.

De Luengo, D.H. and Mabry, T.J. (1986), *J. Nat. Prod.* **49**, 183.

De Pascual-Teresa, J., Diaz, F., Sanchez, F.J. *et al.* (1980) *Plant. Med.* **38**, 271.

De Pascual-Teresa, J., Gonzalez, M.S., Vicente, S. and Bellido, I.S. (1981), *Plant. Med.* **41**, 389.

De Pascual-Teresa, J., Urones, J.G., Marcos, I.S., Núñez, L. and Basabe, P. (1983), *Phytochemistry* **22**, 2805.

De Pascual-Teresa, J., Gonzalez, M.S., Muriel, M.R., Arcocha, A.D. and Bellido, I.S. (1986), *J. Nat. Prod.* **49**, 177.

De Silva, L.B., De Silva, U.L., Mahendran, M. and Jennings, R. (1980), *J. Natl. Sci. Counc. Sri Lanka* **8**, 123.

Dietz, V.H., Wollenweber, E., Favre-Bonvin, J. and Smith, D.M. (1981), *Phytochemistry* **20**, 1181.

Dominguez, X.A. and Cárdenas, G.E. (1975), *Phytochemistry* **14**, 2511.

Dominguez, X.A., Gonzáles, A., Zanudio, M.A. and Garza, A. (1974), *Phytochemistry* **13**, 2001.

Dominguez, X.A., Villegas, D., Rodriguez, V.M. and Zamora, G. (1976), *Rev. Latinoam. Quim.* 7, 45.

Dominguez, X.A. and Franco, R., Cano, C.G. and Chávez, C.N. (1980), *Rev. Latinoam. Quim.* **11**, 150.

Dominguez, X.A., Franco, R., Marroquin, J., Merijanian, A., and Gonzales, Q.J.A. (1982), *Rev. Latinoam. Quim.* **13**, 39.

Domon, B. and Hostettmann, K. (1985), *Phytochemistry* **24**, 575.

Donnelly, D.M.X., Fukuda, N., Wollenweber, E., Polonsky, J., and Prangé, T. (1987), *Phytochemistry* **26**, 1143.

Dreyer, D.L. (1978), *Rev. Latinoam. Quim.* **9**, 97.

Dutt, S.K. and Chibber, S.S. (1983), *Phytochemistry* **22**, 325.

El-Din, A.S. (1986), Private communication.

El-Din, A.S., El-Sebakhy, N. and El. Ghazouly, M. (1985), *Acta Pharmacol. Jugosl.* **35**, 283.

El-Masry, S., Omar, A.A., Abonshoer, M.I.A. and Saleh, M.R.I., (1981), *Plant. Med.* **42**, 199.

El-Negoumy, S.I., Mansour, R.M.A. and Saleh, N.A.M. (1982), *Phytochemistry* **21**, 953.

Emerson, J.K., Carr, R.L., McCormick, S. and Bohm, B.A. (1986), *Biochem. Syst. Ecol.* **14**, 29.

Engvild, K.C. (1986), *Phytochemistry* **25**, 781.

Ensemeyer, M. and Langhammer, L. (1982), *Plant. Med.* **46**, 254.

Esteban, M.D., Gonzáles Collado, I., Macias, F.A., Massanet, G.M. and Rodriguez Luis, F. (1986), *Phytochemistry* **25**, 1502

Exner, J., Ulubelen, A. and Mabry, T.J. (1981a), *Rev. Latinoam. Quim.* **12**, 37.

Exner, J., Reichling, R., Cole, T.C.H. and Becker, H. (1981b), *Plant. Med.* **41**, 198.

Faini, F.A., Castillo, M. and Torres, M.R. (1982), *J. Nat. Prod.* **45**, 501.

Faini, F., Torres, R. and Castillo, M. (1984), *J. Nat. Prod.* **47**, 552.

Fairchild, E.H. (1976), PhD Thesis, Ohio State University, Part III.

Fang, N., Leidig, M. and Mabry, T.J. (1985a), *Phytochemistry* **24**, 2693.

Fang, N., Leidig, M., Mabry, T.J. and Iinuma, M. (1985b), *Phytochemistry* **24**, 3029.

Fang, N., Leidig, M. and Mabry, T.J. (1986a), *Phytochemistry* **25**, 927.

Fang, N., Mabry, T.J. and Le-Van, N. (1986b), *Phytochemistry* **25**, 235.

Fauvel, M.Th., Gleye, J., Moulist, C., Blasco, F. and Stanislas, E. (1981), *Phytochemistry* **20**, 2059.

Favre-Bonvin, J., Jay, M., Wollenweber, E. and Dietz, V.H. (1980), *Phytochemistry* **19**, 2043.

Ferraro, G.E., Norbedo, C. and Coussio, J.D. (1981), *Phytochemistry* **20**, 2053.

Ferraro, G.E., Martino, V.S., Villar, S.I. and Coussio, J.D. (1985), *J. Nat. Prod.* **48**, 817.

Ferreres, F., Tomás, F. and Guirado, A. (1981), *Rev. Agroquim. Tecnol. Aliment.* **20**, 285.

Ferreres, F., Barberán, F.A.T. and Tomás, F. (1985a), *Phytochemistry* **24**, 1869.

Ferreres, F., Tomás, F., Barberán, F.A.T. and Hernández, L. (1985b), *Plant.. Méd. Phytothér.* **19**, 89.

Forgacs, P., Desconcloid, J.F. and Dubec, J. (1981), *Plant. Med.* **42**, 284.

Fourie, T.G. and Snyckers, F.O. (1984), *J. Nat. Prod.* **47**, 1057.

Fraishtat, P.D. and Wulfson, N.S. (1981), *Khim. Prir. Soedin.*, 663.

Freeman, P.W., Murphy, S.T., Nemorin, J.E. and Taylor, W.C. (1981), *Aust. J. Chem.* **34**, 1779.

Fujimoto, T. and Nomura, T. (1985), *Plant. Med.*, 190.

Fukai, T., Hano, Y., Hirakura, K., Nomura, T. and Uzawa, J. (1985), *Chem. Pharm. Bull.* **33**, 4288.

Fukai, T., Ikuta, J. and Nomura, T. (1986), *Chem. Pharm. Bull.* **34**, 1987.

Garcia, B., Marco, J.A., Seoane, E. and Tordajada, A. (1981), *J. Nat. Prod.* **44**, 111.

Gehring, E. and Geiger, H. (1980), *Z. Naturforsch.* **35c**, 380.

Genderen, H.H. Van and Van Schaik, J. (1980), *Z. Naturforsch.* **35c**, 342.
Gianello, J.C. and Giordano, O.S. (1984), *Rev. Latinoam. Quim.* **15**, 84.
Gonzaléz, M.D. and Pomilio, A.B. (1982), *An. Asoc. Quim. Argent.* **70**, 145.
González, A.G., Bermejo, J., Dominguez, B., Massanet, G.M., Amaro, J.M. and De la Rosa, A.D. (1977), *An. Quim.* **73**, 460.
González, A.G., Bermejo, J., Estévez, F. and Velázquez, R. (1983), *Phytochemistry* **22**, 1515.
González, A., Galindo, A., Mansilla, H., Gutierrez, A., Palenzuela, J.A., Afonso, M.M., Trigos, A.R., Kesternich, V.H. and De Armas, Y.J.C. (1985), *Comm. at XI Reunion Bienal de Quimica Organica*, Valladolid.
Goodwin, R.S., Rösler, K.H.A., Mabry, T.J. and Varma, S.D. (1984), *J. Nat. Prod.* **47**, 711.
Gottlieb, O.R. (1975). In *The Flavonoids* (eds J.B. Harborne, T.J. Mabry, and H. Mabry), Chapman and Hall, London.
Gottlieb, O.R., Maia, J.G.S. and Mourao, J.C. (1976), *Phytochemistry* **15**, 1289.
Goudard, M. and Chopin, J. (1976), *C.R. Acad. Sci.* **282**, 683.
Goudard, M., Favre-Bonvin, J., Lebreton, P. and Chopin, J. (1978), *Phytochemistry* **17**, 145.
Goudard, M., Favre-Bonvin, J., Strelisky, J., Nogradi, M. and Chopin, J. (1979), *Phytochemistry* **18**, 186.
Grande, M., Piera, F., Cuenca, A., Torres, P. and Bellido, I.S. (1985), *Plant. Med.* **43**, 414.
Guerreiro, E. Kavka, J. and Giordano, O.S. (1982), *Phytochemistry* **21**, 2601.
Guidugli, F.H., Pestchanker, M.J., Kavka, J. and Joseph-Nathan, P. (1984), *Org. Mass Spectrom.* **19**, 502.
Gunatilaka, A.A.L., Sirimanne, S.R., Sotheeswaran, S. and Nakanishi, T. (1979), *J. Chem. Res. (S), 216.*
Gunatilaka, A.A.L., Sirimanne, S.R., Sotheeswaran, S. and Sriyani, H.T.B. (1982), *Phytochemistry* **21**, 805.
Gupta, D.R. and Ahmed, B. (1984), *Indian J. Chem.* **23B**, 682.
Gupta, D.R. and Ahmed, B. (1985), *Phytochemistry* **24**, 873.
Gupta, D.R., Ahmed, B. and Dhimani, R.P. (1984), *Shoyakugaku Zasshi* **38**, 341.
Gupta, H.C., Ayengar, K.N.N. and Rangaswami, S. (1975), *Indian J. Chem.* **13**, 215.
Gupta, K.K., Tanejoa, S.C., Dhar, K.L. and Atal, C.K. (1983), *Phytochemistry* **22**, 314.
Gutkind, G., Norbedo, C., Mollerach, M., Ferraro, G., Coussio, J.D. and De Torres, R. (1984), *J. Ethnopharm.* **10**, 319.
Hänsel, R. (1962), *Plant. Med.* **10**, 361.
Hänsel, R., Khaliffi, F. and Pelter, A. (1981), *Z. Naturforsch.* **36B**, 1171.
Hano, Y., Hirakura, H., Nomura, T., Terada, S. and Fukushima, K. (1984), *Plant. Med.*, 127.
Harborne, J.B. and Williams, C.A. (1983), *Phytochemistry* **22**, 1520.
Harborne, J.B., Williams, C.A. and Wilson, K.L. (1982), *Phytochemistry* **21**, 2491.
Harborne, J.B., Williams, C.A. and Wilson, K.L. (1985), *Phytochemistry* **24**, 751.
Hassani, M.I., Favre-Bonvin, J. and Lebreton, P. (1985), *C.R. Acad. Sci. Paris. Ser. III.* **300**, 1.
Herz, W. and Kulanthaivel, P. (1982), *Phytochemistry* **21**, 2363.
Herz, W. and Kulanthaivel, P. (1984), *Phytochemistry* **23**, 1453.
Herz, W., De Groote, R., Murari, R. and Kumar, N. (1979), *J. Org. Chem.* **44**, 2784.
Herz, W., Govindan, S.V. and Kumar, N. (1981), *Phytochemistry* **20**, 1343.
Herz, W., Kulanthaivel, P. and Watanabe, K. (1983), *Phytochemistry* **22**, 2021.
Hirakura, K., Fukai, T., Hano, Y. and Nomura, T. (1985), *Phytochemistry* **24**, 159.
Hirschmann, G.S. (1984), *Rev. Latinoam. Quim.* **15**, 134.
Hitz, C., Mann, K. and Wollenweber, E. (1982), *Z. Naturforsch.* **37c**, 337.
Horie, T. and Nakayama, M. (1981), *Phytochemistry* **20**, 337.
Horie, T., Kourai, H. and Nakayama, M. (1980), *Chem. Soc. Jpn (9)*, 1397.
Horie, T., Kourai, H., Osaka, H. and Nakayama, M. (1982), *Bull. Chem. Soc. Jpn* **55**, 2933.
Horie, T., Kourai, H. and Fujita, N. (1983), *Bull. Chem. Soc. Jpn* **56**, 3773.
Horie, T., Kourai, H., Tsukayama, M., Masumura, M. and Nakayama, M. (1985), *Yakugaku Zasshi* **105**, 232.
Hostettmann, K. and Hostettman, M. (1982). In *The Flavonoids – Advances in Research* (eds J.B. Harborne and Mabry, T.J.), Chapman and Hall, London and New York.
Hostettman, K., Doumas, J. and Hardy, M. (1981), *Helv. Chim. Acta* **64**, 297.
Hostettmann, K., Hostettman, M. and Marston, A. (1986), *Preparative Chromatography Techniques*, Springer, Berlin.
Hradetzky, D., Wollenweber, E. and Roitman, J.N. (1986), *Z. Naturforsch.* **42c**, 73.
Hurabielle, M., Eberle, J. and Paris, M. (1982), *Plant. Med.* **46**, 124.
Iinuma, M., Tanaka, T. and Matsuura, S. (1983), *Heterocycles* **20**, 2425.
Iinuma, M., Iwashima, K. and Matsuura, S. (1984a), *Chem. Pharm. Bull.* **32**, 4935.
Iinuma, M., Matsuura, S. and Tanaka, T. (1984b), *Chem. Pharm. Bull.* **32**, 1472.
Iinuma, M., Tanaka, T. and Matsuura, S. (1984c), *Chem. Pharm. Bull.* **32**, 1006.
Iinuma, M., Tanaka, T. and Matsuura, S. (1984d), *Chem. Pharm. Bull.* **32**, 2296.
Iinuma, M., Tanaka, T. and Matsuura, S. (1984e), *Chem. Pharm. Bull.* **32**, 3354.
Iinuma, M., Iwashima, M., Tanaka, T. and Matsuura, S. (1984f), *Chem. Pharm. Bull.* **32**, 4217.
Iinuma, M., Tanaka, T., Iwashima, K. and Matsuura, S. (1984g), *Yakugaku Zasshi* **104**, 691.
Iinuma, M., Tanaka, T., Mizuno, M. and Min, Z.-D. (1985a), *Chem. Pharm. Bull.* **33**, 3982.
Iinuma, M., Roberts, M.F., Matlin, S.A., Stacey, V.E., Timmermann, B.N., Mabry, T.J. and Brown, R. (1985b), *Phytochemistry* **24**, 1367.
Iinuma, M., Matoba, Y., Tanaka, T. and Mizuno, M. (1986a), *Chem. Pharm. Bull.* **34**, 1656.
Iinuma, M., Tanaka, T., Mizuno, M. and Mabry, T.J. (1986b), *Chem. Pharm. Bull.* **34**, 2228.
Iinuma, M., Hamada, K., Mizuno, M., Asai, F. and Wollenweber, E. (1986c), *Z. Naturforsch.* **41c**, 681.
Iinuma, M., Fang, N., Tanaka, T., Mabry, T.J., Wollenweber, E., Favre-Bonvin, J., Voirin, B. and Jay, M. (1986d), *Phytochemistry* **25**, 1257.
Imre, S., Islimyeli, S., Oztzunc, A. and Buyuktimkin, B. (1984), *Plant. Med.*, 360.
Ishitsuka, H., Ohsawa, C., Ohiwa, T., Umeda, I. and Suhara, Y. (1982), *Antimicrob. Ag. Chemother.* **22**, 611.
Iwu, M. and Igboko, O. (1982), *J. Nat. Prod.* **45**, 650.

Jain, S.A. and Srivastava, S.K. (1985), *J. Nat. Prod.* **48**, 299.
Jain, P.K., Makrandi, J.K. and Grover, S.K. (1982), *Synthesis* **3**, 221.
Jaipetch, T., Reutrakui, V., Tuntiwachwuttikul, P. and Santisuk, T. (1983), *Phytochemistry* **22**, 625.
Jakupovic, J., Kuhnke, J., Schuster, A., Metwally, M.A. and Bohlman, F. (1986), *Phytochemistry* **25**, 1133.
Jay, M., Wollenweber, E. and Favre-Bonvin, J. (1979), *Phytochemistry* **18**, 153.
Jay, M., Favre-Bonvin, J., Viricel, M.-R. and Wollenweber, E. (1981), *Phytochemistry* **20**, 2307.
Jay, M., Viricel, M.-R., Favre-Bonvin, J., Voirin, B. and Wollenweber, E. (1982), *Z. Naturforsch.* **37c**, 721.
Jiyu, Z. and Zongshi, W. (1984), *Acta Bot. Yunnanica* **6**, 344.
Johnson, N.D. (1983), *Biochem. Syst. Ecol.* **11**, 221.
Joshi, B.S. and Kamat, V.N. (1970), *Phytochemistry* **9**, 889.
Jullien, F., Voirin, B., Bernillon, J. and Favre-Bonvin, J. (1984), *Phytochemistry* **23**, 2972.
Kapoor, R., Rishi, A.K. and Atal, C.K. (1985), *Fitoterapia* **56**, 296.
Kerr, K.M., Mabry, T.J. and Yoser, S. (1981), *Phytochemistry* **20**, 791.
Khafagy, S.M., Sabri, N.N., Soliman, F.S.G., Abou-Donia, A.H. and Mosandi, A. (1976), *Die Pharmazie* **31**, 894.
Khalid, S.A. and Waterman, P.G. (1981), *Phytochemistry* **20**, 1719.
Kimura, Y., Okuda, H., Taira, Z., Shouji, N., Takemoto, T. and Arichi, S. (1984), *Plant. Med.* **46**, 290.
Kingston, D.G.I., Rao, M.M. and Zucker, W.V. (1979), *J. Nat. Prod.* **42**, 496.
Kiso, Y., Sasaki, K., Oshima and Hikino, H. (1982), *Heterocycles* **19**, 1615.
Kitagawa, I., Fukuda, Y., Yoshihara, M. and Yoshikawa, M. (1983), *Chem. Pharm. Bull.* **31**, 352.
König, W.A., Krauss, C. and Zähner, H. (1977), *Helv. Chim. Acta* **60**, 2071.
Kraft, R., Otto, A., Makower, A. and Etzold, G. (1981), *Analyt. Biochem.* **113**, 193.
Kruijt, R.Ch., Niemann, G.J., De Koster, C.G. and Heerma, W. (1986), *Crypt. Bryol. Lichenol.* **7**, 165.
Kullanthaivel, P. and Benn, M.H. (1986), *Canad. J. Chem.* **64**, 514.
Kumari, G.N.K., Rao, L.J.M. and Rao, N.S.P. (1984), *Phytochemistry* **23**, 2701.
Kumari, G.N.K., Rao, L.J.M. and Rao, N.S.P. (1985), *J. Nat. Prod.* **48**, 149.
Kupchan, S.M. and Bauerschmidt, E. (1971), *Phytochemistry* **10**, 664.
Kuroyanagi, M. and Fukushima, S. (1982), *Chem. Pharm. Bull.* **30**, 1163.
Kuroyanagi, M., Fujita, K., Kazaoka, M., Matsumoto, S., Ueno, A. Fukushima, S. and Katsuoka, M. (1985), *Chem. Pharm. Bull.* **33**, 5075.
La Duke, J.C. (1982), *Am. J. Bot.* **69**, 784.
Lao, A., Fujimoto, Y. and Tatsuno, T. (1983), *Yakugaku Zasshi* **103**, 696.
Laracine, C., Favre-Bonvin, J. and Lebreton, P. (1982), *Z. Naturforsch.* **37c**, 335.
Laux, D.O., Stefani, G.M. and Gottlieb, O.R. (1985), *Phytochemistry* **24**, 1081.
Lee, K.-H., Tagahara, K., Suzuki, H., Wu, R.-Y., Haruna, M., Hall, I.H., Huang, H.-C., Ito, K., Iida, T. and Lai, J.-S. (1981), *J. Nat. Prod.* **44**, 530.
Lenherr, A., Fang, N. and Mabry, T.J. (1986), *J. Nat. Prod.* **49**, 185.
Lin, C.-N., Arisawa, M., Shimizu, M. and Morita, N. (1978), *Chem. Pharm.* **26**, 2036.
Liu, M., Li, M., Wang, F. and Liang, X. (1984a), *Bull. Chin. Mat. Med.* **9**, 76.
Liu, M., Li, M. and Wang, F. (1984b), *Yaoxue Xuebao* **19**, 545.
Liu, Y.-L., Mabry, T.J. (1981), *Phytochemistry* **20**, 1389.
Liu, Y.-L. and Mabry, T.J. (1982), *Phytochemistry* **21**, 209.
Liu Y.-L., Song, W., Ji, Q. and Bai, Y. (1984), *Acta Pharm. Sin.* **19**, 830.
Loyola, L.A., Naranjo, S.J. and Morales, B.G. (1985), *Phytochemistry* **24**, 1871.
Mabry, T.J., Markham, K.R. and Thomas, M.B. (1970), *The Systematic Identification of Flavonoids*, Springer-Verlag, Berlin.
Madhusudanan, K.P., Sachdev, K., Harrison, D.A. and Kulshreshta, D.K. (1985), *J. Nat. Prod.* **48**, 319.
Mahmoud, E.N. and Waterman, P.G. (1985), *Phytochemistry* **24**, 369.
Makboul, A.M. (1986), *Fitoterapia* **57**, 50.
Malan, E. and Naidoo, S. (1981), *S. Afr. J. Chem.* **34**, 91.
Malan, E. and Roux, D.G. (1979), *J. Chem. Soc. Perkin Trans I*, 2696.
Marchelli, E. and Vining, L.C. (1973), *Canad. J. Biochem.* **51**, 1624.
Markham, K.R. and Mabry, T.J. (1975). In *The Flavonoids – Advances in Research* (eds J.B. Harborne, T.J. Mabry and H. Mabry), Chapman and Hall, London and New York.
Markham, K.R. (1982), *Techniques of Flavonoid Identification*, Academic Press, London, 113.
Markham, K.R. (1983), *Phytochemistry* **22**, 316.
Markham, K.R. and Chari, V.M. (1982). In *The Flavonoids – Advances in Research* (eds J.B. Harborne and T.J. Mabry), Chapman and Hall, London and New York.
Markham, K.R. and Mues, R. (1983), *Phytochemistry* **22**, 143.
Matsumoto, J., Fujimoto, T., Takino, C., Saitoh, M., Hano, Y., Fukai, T. and Nomura, T. (1985), *Chem. Pharm. Bull.* **33**, 3250.
Matsuura, S., Tanaka, T., Iinuma, M., Tanaka, T. and Himura, N. (1983), *Yakugaku Zasshi* **103**, 997.
Maurya, R., Ray, A.B., Duah, F.K., Slatkin, D.J. and Schiff Jr., P.L. (1984), *J. Nat. Prod.* **47**, 179.
McCormick, S. and Bohm, B. (1986), *J. Nat. Prod.* **49**, 167.
McCormick, S., Robson, K. and Bohm, B. (1985), *Phytochemistry* **24**, 2133.
Mears, J.A. (1980), *J. Nat. Prod.* **43**, 708.
Melek, F.R., Ahmed, A.A. and Mabry, T.J. (1985), *Rev. Latinoam. Quim.* **16**, 27.
Mendez, B., Martinez, R.A. and Hurtado, J. (1982), *Acta Cient. Venez.* **33**, 198.
Merfort, I. (1984a), *Plant. Med.*, 107.
Merfort, I. (1984b), *Farm. Tijdschr. Belg.* **61**, 244.
Merfort, I. (1985), *Plant. Med.*, 136.
Mesquita, A.A.L., De B. Corrêa, D., De Padua, A.P., Guedes, M.L.O. and Gottlieb, O.R. (1986), *Phytochemistry* **25**, 1255.
Meyer, B.N., Wall, M.E., Wani, M.C. and Taylor, H.L. (1985), *J. Nat. Prod.* **48**, 952.
Mikhailova, N.S. and Rybalko, K.S. (1980), *Khim. Prir. Soedin.*, 175.
Miski, M., Ulubelen, A. and Mabry, T.J. (1983a), *Phytochemistry* **22**, 2093
Miski, M., Ulubelen, A., Johansson, C. and Mabry, T.J. (1983b),

J. Nat. Prod. **46**, 874.
Miski, M., Gage, D.A. and Mabry, T.J. (1985), *Phytochemistry* **24**, 3078.
Mitscher, L.A., Rao, G.S.R., Khanna, I., Veysoglu, T. and Drake, S. (1983), *Phytochemistry* **22**, 573.
Mues, R. (1987), Private communication.
Mues, R., Strassner, A. and Zinsmeister, H.D. (1983), *Cryptogamie Bryol. Lichenol.* **4**, 111.
Mues, R., Hattori, S., Asakawa, Y. and Grolle, R. (1984), *J. Hattori Bot. Lab.* **56**, 227.
Nahrstedt, A. (1986), Private communication.
Nair, R.A.G., Kotiyal, J.P. and Bhardwaj, D.K. (1983), *Phytochemistry* **22**, 318.
Nakanishi, T., Ogaki, J., Inada, A., Murata, H., Nishi, M., Iinuma, M. and Yoneda, K. (1985), *J. Nat. Prod.* **48**, 491.
Nakayama, M., Horie, T., Lin, C.-N., Arisawa, M., Shimizu, M. and Morita, N. (1983), *Yakugaku Zasshi* **103**, 675.
Namba, T., Hattori, M., Takehana, Y., Tsunezuka, M., Tomimori, T. Kizu, H. and Miyaichi, Y. (1983), *Phytochemistry* **22**, 1057.
Neuman, P., Mabry, T.J. and Kerr, K.M. (1981), *J. Nat. Prod.* **44**, 50.
Nicholls, K.W. and Bohm, B.A. (1983), *Canad. J. Bot.* **61**, 708.
Nicollier, G.F., Thompson, A.C. and Salin, M.L. (1981), *J. Agric. Food Chem.* **29**, 1179.
Nomura, T. and Fukai, T. (1980), *Chem. Pharm. Bull.* **28**, 2548.
Nomura, T., Fukai, T. and Narita, T. (1980), *Heterocycles* **14**, 1943.
Nomura, T., Fukai, T., Sato, E. and Fukushima, K. (1981), *Heterocycles* **16**, 983.
Nomura, T., Fukai, T., Hano, Y., Nemoto, K., Terada, S. and Kuramochi, T. (1983a). *Plant. Med.* **47**, 151.
Nomura, T., Fukai, T., Hano, Y. and Ikuta, H. (1983b), *Heterocycles* **20**, 585.
Norbedo, C., Ferraro, G. and Coussio, J.D. (1984), *Phytochemistry* **23**, 2698.
Norris, J.A. and Mabry, T.J. (1985), *J. Nat. Prod.* **48**, 668.
Numata, A., Nabae, M. and Uemura, E. (1984), *Chem. Pharm. Bull.* **32**, 1174.
Oettmeier, W. and Heupel, A. (1972), *Z. Naturforsch.* **27b**, 177.
Oganesyan, E.T., Smirnova, L.P., Dzhumyrko, S.F. and Kechantova, N.A. (1976), *Khim. Prir. Soedin.* **12**, 599.
Ognyanov, I. and Todorova, M. (1983), *Plant. Med.* **48**, 181.
Ognyanov, I.V., Ivantcheva, S.V. and Zapesotchnaya, G.G. (1975), *C.R. Acad. Bulg. Sci.* **28**, 1621.
Öksüz, S., Ulubelen, A., Clark, W.D., Brown, G.K. and Mabry, T.J. (1981), *Rev. Latinoam. Quim.* **12**, 12.
Öksüz, S., Ayyildiz, H. and Johansson, C. (1984), *J. Nat. Prod.* **47**, 902.
Omar, A.A. and El-Ghazouly, M.G. (1984), Poster, presented at *Biochemistry of Plant Phenolics*, Ghent, Aug. 1984.
Oskay, E. and Yesilada, A. (1984), *J. Nat. Prod.* **47**, 742.
Ozawa, A.T., Rivera, P.A. and De Vivar, A.R. (1983), *Rev. Latinoam. Quim.* **14**, 40.
Painuly, P. and Tandon, J.S. (1983a), *J. Nat. Prod.* **46**, 285.
Painuly, P. and Tandon, J.S. (1983b), *Phytochemistry* **22**, 243.
Palacios, P., Mitsakos, A. Bodden, J. and Wollenweber, E. (1986), *Phytochemistry*, **25**, 2367.
Pares, J.O., Öksüz, S., Ulubelen, A. and Mabry, T.J. (1981), *Phytochemistry* **20**, 2057.
Paskov, D. and Marichkova, L. (1983), *Probl. Farm.* **11**, 36.
Pathak, V.P., Saini, T.R. and Khanna, R.N. (1983a), *Phytochemistry* **22**, 308.
Pathak, V.P., Saini, T.R. and Khanna, R.N. (1983b), *Plant. Med.* **49**, 61.
Pathwardhan, S.A. and Gupta, A.S. (1981). *Phytochemistry* **20**, 1458.
Payne, T.C. and Jefferies, D.R. (1973), *Tetrahedron*, **29**, 2575.
Pederiva, R. and Giordano, O.S. (1984), *Phytochemistry* **23**, 1340.
Pelter, A., Ward, R.S., Rao, E.V. and Raju, N.R. (1981), *J. Chem. Soc. Perkin Trans. I*, 2491.
Picher, M.T., Seoane, E. and Tortajada, A. (1984), *Phytochemistry* **23**, 1995.
Pinkas, M., Didry, N., Torqu, M., Bezanger, L. and Cazin, J., (1978), *Ann. Pharm. France* **36**, 97.
Prakash, S. and Zaman, A. (1982), *Phytochemistry* **21**, 2992.
Proksch, M., Weissenboeck, G. and Rodriguez, E. (1982a), *Phytochemistry* **21**, 2893.
Proksch, M., Proksch, P., Weissenboeck, G. and Rodriguez, E. (1982b), *Phytochemistry* **21**, 1835.
Proksch, P. and Gülz, P.-G. (1984), *Phytochemistry* **23**, 470.
Proksch, P. and Rodriguez, E. (1985), *Biologie in unserer Zeit* **15**, 75.
Proksch, P., Proksch, M., Rundel, P.W. and Rodriguez, E., (1982), *Biochem. Syst. Ecol.* **10**, 49.
Proksch, P., Wollenweber, E. and Rodriguez, E. (1983), *Z. Naturforsch.* **38c**, 668.
Proksch, P., Budzikiewicz, H., Tanowitz, B.D. and Smith, D.M., (1984), *Phytochemistry* **23**, 679.
Proksch, P., Mitsakos, A., Bodden, J. and Wollenweber, E. (1986), *Phytochemistry* **25**, 2367.
Quijano, L., Calderón, J.S., Goméz, G.F. and Rios, T. (1982), *Phytochemistry* **21**, 2965.
Quijano, L., Calderón, J.S., Goméz, G.F., Escobar, E. and Ríos, T. (1985), *Phytochemistry* **24**, 1085.
Rabesa, Z. and Voirin, B. (1979), *Phytochemistry* **18**, 360.
Rabesa, Z. and Voirin, B. (1983), *Phytochemistry* **22**, 2092.
Rabesa, Z. and Voirin, B. (1985), *C.R. Seances Acad. Sci. Ser III* **301**, 351.
Raffauf, R.F., Menachery, M.D., Le Quesne, P.W., Arnold, E.V., and Clardy, J. (1981), *J. Org. Chem.* **46**, 1094.
Rajbhandari, A. and Roberts, M.F. (1983), *J. Nat. Prod.* **46**, 194.
Rajbhandari, A. and Roberts, M.F. (1984a), Poster, presented at *Biochemistry of Plant Phenolics*, Ghent, Aug. 1984.
Rajbhandari, A. and Roberts, M.F. (1984b), *J. Nat. Prod.* **47**, 559.
Rajbhandari, A. and Roberts, M.F. (1985a), *J. Nat. Prod.* **48**, 502.
Rajbhandari, A. and Roberts, M.F. (1985b), *J. Nat. Prod.* **48**, 858.
Rao, L.J.M., Kumari, G.N.K. and Rao, N.S.P. (1983a), *Phytochemistry* **22**, 1522.
Rao, L.J.M., Kumari, G.N.K. and Rao, N.S.P. (1983b), *J. Nat. Prod.* **46**, 595.
Rao, P.P. and Srimannarayana, G. (1981), *Curr. Sci. India* **50**, 319.
Rasamoelisendra, R., Voirin, B., Favre-Bonvin, J., Andrianstiferana, M. and Rabesa, Z. (1985), *Die Pharmazie* **40**, 59.
Reddy, K.C., Kumar, K.A. and Srimannarayana, G. (1983), *Phytochemistry* **22**, 800.
Reisch, J., Hussain, R.A. and Mester, I. (1984), *Phytochemistry* **23**, 2114.
Reynaud, J. and Raynaud, J. (1986), *Biochem. Syst. Ecol.*, **14**, 191.
Reynaud, J., Raynaud, J. and Voirin, B. (1983), *Die Pharmazie* **38**, 628.
Reynolds, G.W. and Rodriguez, E. (1981), *Plant. Med.* **43**, 187.
Rimpler, H. (1965), *Plant. Med.* **13**, 412.
Rizzi, G.P. and Boeing, S.S. (1984), *J. Agric. Food Chem.* **32**, 551.

Roberts, M.F., Timmermann, B.N. and Mabry, T.J. (1980), *Phytochemistry* **19**, 127.

Roberts, M.F., Timmermann, B.N., Mabry, T.J., Brown, R. and Matlin, S.A. (1984), *Phytochemistry* **23**, 163.

Rodríguez-Hahn, L., Valencia, A., Saucedo, R. and Díaz, E., (1981), *Rev. Latinoam. Quim.* **12**, 16.

Rofi, R.D. and Pomilio, A.B. (1985), *Phytochemistry* **24**, 2131.

Roitman, J.N. and James, L.F. (1985), *Phytochemistry* **24**, 835.

Rösler, K.-H.A. and Goodwin, R.S. (1984), *J. Nat. Prod.* **47**, 316.

Rösler, K.-H.A., Wong, D.P.C. and Mabry, T.J. (1985), *J. Nat. Prod.* **48**, 837.

Ryu, S.H., Yoo, B.T., Ahn, B.Z. and Pack, M.Y. (1985), *Arch. Pharm.* **318**, 659.

Sachdev, K. and Kulshreshta, D.K. (1982), *Indian J. Chem.* **21B**, 789.

Sachdev, K. and Kulshreshta, D.K. (1983), *Phytochemistry* **22**, 1253.

Sachdev, K. and Kulshrestha, D.K. (1986), *Phytochemistry* **25**, 1967.

Saeedi-Ghomi, M.H., Hurtado, L.M., Vega, P. and Maldonado, R. (1984), *Rev. Latinoam. Quim.* **14**, 148.

Sakushima, A., Coskun, M., Hisada, S. and Nishibe, S. (1983), *Phytochemistry* **22**, 1677.

Sakushima, A., West, H. and Brandenberger, H. (1984), *Iyo Masu Kenkyukai Koenshu* **9**, 217.

Saleh, A.A., Cordell, G.A. and Farnsworth, N.R. (1980), *J. Chem. Soc. Perkin Trans. I*, 1090.

Salmenkallio, M., Mc.Cormick, S., Mabry, T.J., Dellamonica, G. and Chopin, J. (1982), *Phytochemistry* **21**, 2990.

Sanders, R., Ulubelen, A. and Mabry, T.J. (1980), *Rev. Latinoam. Quim.* **11**, 139.

San Martin, A., Rovirosa, J. and Castillo, M. (1982), *Bol. Soc. Chil. Quim.* **27**, 252.

Santos, F.D. Dos, Sarti, S.J., Vichnewski, W., Bulhoes, M.S., and De Freitas, Leitao Fho, H. (1980), *Rev. Fac. Farm. Odontal. Rebereio Preto* **17**, 43.

Sanz, J., Martinezcastro, I. and Pinar, M. (1985), *J. Nat. Prod.* **48**, 993.

Savona, G., Paternostro, M., Piozzi, F. and Rodriguez, B. (1979), *An. Quim.* **75c**, 433.

Savona, G., Bruno, M., Piozzi, F. and Barbagalloc, C. (1982), *Phytochemistry* **21**, 2132.

Scheele, C., Wollenweber, E. and Arriaga-Giner, F.J. (1987), *J. Nat. Prod.* **50**, 181.

Schilling, E.E. (1983), *Biochem. Syst. Ecol.* **11**, 341.

Schilling, E.E. (1984), *Biochem. Syst. Ecol.* **12**, 403.

Schilling, E.E. and Mabry, T.J. (1981), *Biochem. Syst. Ecol.* **9**, 161.

Schilling, E. and Rieseberg, L.H. (1985), *Biochem. Syst. Ecol.* **13**, 403.

Schulten, H.R. and Games, D.E. (1974), *Biomed. Mass Spectrom.* **1**, 120.

Seigler, D.S. and Wollenweber, E. (1983), *Am. J. Bot.* **70**, 790.

Sen, A.K., Mahato, S.B. and Dutta, N.L. (1976), *Indian J. Chem.* **14B**, 849.

Seth, K.K., Pandey, V.B. and Dasgupta, B. (1982), *Die Pharmazie* **37**, 74.

Sezik, E. and Ezer, N. (1984), *Acta Pharm. Turcica* **26**, 4.

Silva, G.A.B., Henriques, A. and Alice, C.B. (1985), *J. Nat. Prod.*, **48**, 861.

Silva, M. and Mundaca, J.M. (1971), *Phytochemistry* **10**, 1942.

Sinha, N.K., Seth, K.K., Pandey, V.B., Dasgupta, B. and Shah, A.H. (1981), *Planta Med.* **42**, 296.

Sosa, V.E., Oberti, J.C., Prasad, J.S. and Herz, W. (1984), *Phytochemistry* **23**, 1515.

Sosa, V.E., Gil, R., Oberti, J.C., Kulanthaivel, P. and Herz, W. (1985), *J. Nat. Prod.* **48**, 340.

Srivastava, S.K., Gupta, R.K. and Srivastava, S.D. (1981), *Indian J. Chem.* **20**, 833.

Subramanian, S.S., Nair, A.G., Rodriguez, E. and Mabry, T.J. (1972), *Curr. Sci.* **41**, 202.

Sultana, S. and Ilyas, M. (1986), *Phytochemistry* **25**, 953.

Syrchina, A.I., Zapesochnaya, G.G., Tyukavkina, N.A. and Voronkov, M.G. (1980), *Khim. Prir. Soedin.* 499.

Takagi, S., Yamaki, M. and Inoue, K. (1980), *Yakugaku Zasshi* **100**, 1220.

Takido, M., Aimi, M., Yamanouchi, S., Yasukawa, K. Torii, H. and Takahashi, S. (1976), *Yakugaku Zasshi* **96**, 381.

Takido, M., Yasukawa, K., Matsuura, S. and Iinuma, M. (1979), *Yakugaku Zasshi* **99**, 443.

Talapatra, S.K., Mallik, A.K. and Talapatra, B. (1982), *Phytochemistry* **21**, 761.

Tamma, R.V., Miller, G.C. and Everett, R. (1985), *J. Chromatogr.* **322**, 236.

Tanaka, T., Iinuma, M. and Mizuno, M. (1986), *Chem. Pharm. Bull.* **34**, 1667.

Tandon, S. and Rastogi, R.P. (1977), *Phytochemistry* **16**, 1455.

Theodor, R., Zinsmeistel, H.D., Mues, R. and Markham, K.R. (1980), *Phytochemistry* **19**, 1695.

Timmermann, B.N. and Mabry, T.J. (1983), *Biochem. Syst. Ecol.* **11**, 37.

Timmerman, B.N., Graham, S.A. and Mabry, T.J. (1981), *Phytochemistry* **20**, 1762.

Timmermann, B.N., Hoffmann, J.J., Jolad, S.D. and Schram, K.H. (1985), *Phytochemistry* **24**, 1031.

Timmermann, B.N., Hoffmann, J.J., Jolad, S.D., Bates, R.B. and Siahaan, T.J. (1986), *Phytochemistry* **25**, 723.

Tomimori, T., Miyaichi, Y. and Kizu, H. (1982a), *Yakugaku Zasshi* **102**, 388.

Tomimori, T., Miyaichi, Y., Imoto, Y., Kizu, H. and Tanabe, T. (1982b), *Yakugaku Zasshi* **102**, 524.

Tomimori, T., Miyaichi, Y., Imoto, Y., Kizu, H. and Tanabe, Y. (1983), *Yakugaku Zasshi* **103**, 607.

Tomimori, T., Miyaichi, Y., Imoto, Y. and Kizu, H. (1984a), *Shoyakugaku Zasshi* **38**, 249.

Tomimori, T., Miyaichi, Y., Imoto, Y., Kizu, H. and Suzuki, C. (1984b), *Yakugaku Zasshi* **104**, 529.

Tomimori, T., Miyaichi, Y., Imoto, Y., Kizu, H. and Tanabe, Y. (1984c), *Yakugaku Zasshi* **104**, 524.

Tomimori, T., Miyaichi, Y., Imoto, Y., Kizu, H. and Namba, T. (1985), *Chem. Pharm. Bull.* **33**, 4457.

Tomimori, T., Miyaichi, Y., Imoto, Y., Kizu, H. and Namba, T. (1986), *Chem. Pharm. Bull.* **34**, 406.

Tonn, C.E., Rossomando, C. and Giordano, O.S. (1982), *Phytochemistry* **21**, 2599.

Torrenegra, G., R.D., Escarria, R.S. and Dominguez, X.A. (1978), *Rev. Latinoam. Quim.* **9**, 101.

Torrenegra, G., R.D., Escarria, R.S., Raffelsberger, B. and Achenbach, H. (1980), *Phytochemistry* **19**, 2795.

Torrenegra, G., R.D., Escarria, R.S. and Tenorio, E. (1982), *Rev. Latinoam. Quim.* **13**, 75.

Torrenegra, G., R.D., Pedrezo, J.A. and Escarria, S. (1984), *Rev. Latinoam. Quim.* **15**, 129.

Torrenegra, G., R.D., Bautista, A.R. and Pedrozo, J.A. (1985), *Rev. Latinoam. Quim.* **16**, 64.

Tóth, L., Kokovay, K., Bujitás, Gy. and Pápay, V. (1980),

Pharmazie **35**, 334.
Tóth, L., Bulyaki, M. and Bujitas Gy. (1981), *Pharmazie* **36**, 577.
Triana, J. (1984), *Phytochemistry* **23**, 2072.
Tsankov, E. and Ognyanov, J. (1985), *Plant. Med.*, 180.
Uchida, M., Rüedi, P. and Eugster, C.H. (1980), *Helv. Chim. Acta* **63**, 225.
Ulubelen, A. and Mabry, T.J. (1981), *J. Nat. Prod.* **44**, 457.
Ulubelen, A. and Mabry, T.J. (1982), *Rev. Latinoam. Quim.* **13**, 35.
Ulubelen, A. and Ösküz, S. (1982), *J. Nat. Prod.* **45**, 373.
Ulubelen, A. and Topcu, G. (1984), *J. Nat. Prod.* **47**, 1068.
Ulubelen, A. and Tuzlaci, E. (1985), *Phytochemistry* **24**, 2923.
Ulubelen, A., Kerr, K.M. and Mabry, T.J. (1980), *Phytochemistry* **19**, 1761.
Ulubelen, A., Kerr, K.M. and Mabry, T.J. (1981a), *Plant. Med.* **43**, 95.
Ulubelen, A., Miski, M. and Mabry, T.J. (1981b), *J. Nat. Prod.* **44**, 586.
Ulubelen, A., Clark, W.D. and Brown, G.K. (1981c), *J. Nat. Prod.* **44**, 294.
Ulubelen, A., Kerr, R.R. and Mabry, T.J. (1982), *Phytochemistry* **21**, 1145.
Ulubelen, A., Mabry, T.J., Dellamonica, G. and Chopin, J. (1984), *J. Nat. Prod.* **47**, 384.
Valant-Vetschera, K.M. and Wollenweber, E. (1986). In *Flavonoids and Bioflavonoids* (eds L. Farkas, M. Gábor and F. Kállay), Akademiai Kiado, Budapest, pp. 213–220.
Van de Casteele, K., Geiger, H. and Van Sumere, C.F. (1982), *J. Chromatogr.* **240**, 81.
Van Den Broucke, C., Domisse, R., Esmans, E. and Lemli, J. (1982), *Phytochemistry* **21**, 2581.
Venkata Rao, E., Venkataratnam, G. and Vilain, C. (1985), *Phytochemistry* **24**, 2427.
Venkataratnam, G., Venkata Rao, E. and Vilain, C. (1986), *Phytochemistry* **25**, 1507.
Vilain, C. (1983), *Pharmazie* **38**, 876.
Vivar, A.De., Reyes, B., Delgado, G. and Schlemper, E.O. (1982), *Chem. Lett.*, 957.
Vogt, T., Proksch, P., Gülz, P.-G. and Wollenweber, E. (1987), *Phytochemistry* **26**, 1027.
Voirin, B. (1983), *Phytochemistry* **22**, 2107.
Voirin, B., Favre-Bonvin, J., Indra, V. and Nair, A.G.R. (1984), *Phytochemistry* **23**, 2973.
Voirin, B., Viricel, M.R., Favre-Bonvin, J., Van den Broucke, C.O. and Lemli, J. (1985), *Planta Med.* 523.
Waddell, T.G. and Elkins, S.K. (1985), *J. Nat. Prod.* **48**, 163.
Wagner, H., Seligmann, O. and Chari, M.V. (1979), *Tetrahedron* **44**, 4269.
Wahab, S.M.A. and Selim, M.A. (1985), *Fitoterapia* **56**, 167.
Waterman, P.G. and Mahmoud, E.N. (1985), *Phytochemistry* **24**, 571.
Watson, W.A.F. (1981), *Mut. Res. Letts.* **103**, 145.
Whalen, M.D. and Mabry, T.J. (1979), *Phytochemistry* **18**, 263.
Williams, C.A. and Harvey, W.J. (1982), *Phytochemistry* **21**, 329.
Wolf, S.J. and Denford, K.E. (1984), *Biochem. Syst. Ecol.* **12**, 183.
Wollenweber, E. (1974), *Biochem. Physiol. Pflanzen* **166**, 419.
Wollenweber, E. (1982a). In *The Flavonoids–Advances in Research* (eds.J.B. Harborne and T.J. Mabry), Chapman and Hall, London and New York.
Wollenweber, E. (1982b). In *The Plant Cuticle* (*Linn. Soc. Symp.* no. 10) (eds D.F. Cutler, K.L. Alvin and C.E. Price), Academic Press, London.
Wollenweber, E. (1982c), *Supplement Chromatographie* 1982, 50. GIT-Verlag, Darmstadt.
Wollenweber, E. (1984a). In *Biology and Chemistry of Plant Trichomes* (eds E. Rodriguez, P.L. Healey and I. Mehta), Plenum Press, New York and London.
Wollenweber, E. (1984b), *Rev. Latinoan. Quim.* **15**, 3.
Wollenweber, E. (1984c), *Z. Naturforsch.* **39c**, 833.
Wollenweber, E. (1985a), *Phytochemistry* **24**, 1493.
Wollenweber, E. (1985b), *Plant. Syst. Evol.* **150**, 83.
Wollenweber, E. (1985c), *3rd Int. Conf. Chem. Biotechn. Biol. Act. Nat. Prod.* (Bulg. Acad. Sci.), Sofia, 1985.
Wollenweber, E. and Dietz, V.H. (1980), *Biochem. Syst. Ecol.* **8**, 21.
Wollenweber, E. and Kohorst, G. (1981), *Z. Naturforsch.* **36c**, 913.
Wollenweber, E. and Kohorst, G. (1984), *Z. Naturforsch.* **39c**, 710.
Wollenweber, E. and Mann, K. (1983), *Plant. Med.* **48**, 126.
Wollenweber, E. and Mann, K. (1984), *Z. Naturforsch.* **39c**, 303.
Wollenweber, E. and Mann, K. (1985), *Plant. Med.*, 459.
Wollenweber, E. and Mann, K. (1986), *Biochem. Physiol. Pflanzen*, **181**, 665.
Wollenweber, E. and Seigler, D.S. (1982), *Phytochemistry* **21**, 1063.
Wollenweber, E. and Yatskievych, G. (1985), *Rev. Latinoam. Quim.* **16**, 45.
Wollenweber, E., Kohorst, G., Mann, K. and Bell, J.M. (1985a), *J. Plant. Physiol.* **117**, 423.
Wollenweber, E., Mann, K., Arriaga, F.J. and Yatskievych, G. (1985b), *Z. Naturforsch.* **40c**, 321.
Wollenweber, E., Schober, I., Clark, W.D. and Yatskievych, G. (1985c), *Phytochemistry* **24**, 2129.
Wollenweber, E., Dietz, V.H., Schilling, G., Favre-Bonvin, J. and Smith, D.M. (1985d), *Phytochemistry* **24**, 965.
Wollenweber, E., Schober, I., Dostal, P., Hradetzky, D., Arriaga-Giner, F.J. and Yatskievych, G. (1986), *Z. Naturforsch.* **41c**, 87.
Wollenweber, E., Valant-Vetschera, K.M. and Ivancheva, S., Kuzmanov, B. (1987a), *Phytochemistry* **26**, 181.
Wollenweber, E., Hradetzky, D., Mann, K., Roitman, J.N., Yatskievych, G., Proksch, M. and Proksch, P. (1987b), *J. Plant Physiol.* **131**, 37.
Wollenweber, E., Asakawa, Y., Schillo, D., Lehmann, U. and Weigel, H. (1987c), *Z. Naturforsch.* **42c**, 1030.
Xaasan, C.C., Ciilmi, C.X. and Faarax, M.X. (1980), *Phytochemistry* **19**, 2229.
Young, D.A. (1983), *Biochem. Syst. Ecol.* **11**, 209.
Yu, S.G., Gage, D.A., Fang, N. and Mabry, T.J. (1986), *J. Nat. Prod.* **49**, 181.
Zapesochnaya, G.G. and Pangarova, T.T. (1980), *Compt. Rend. Acad. Bulg. Sci.* **33**, 393.
Zapesochnaya, G.G. and Sokolskaya, T.A. (1984), *Khim. Prir. Soedin.*, 306.
Zapesochnaya, G.G., Evstratova, R.I. and Mukhametzhanov, M.N. (1977), *Khim. Prir. Soedin.*, 706.
Zapesochnaya, G.G., Stepanov, A.N. and Perov, A.A. (1984), *Khim. Prir. Soedin.*, 573.
Zhu, D., Song, G., Jian, F., Chang, X. and Guo, W. (1984), *Huaxue Xuebao* **42**, 1080.

8

Flavone and flavonol glycosides

JEFFREY B. HARBORNE and CHRISTINE A. WILLIAMS

8.1 INTRODUCTION

By 1975, some 360 flavone and flavonol glycosides were known to occur in the plant kingdom (Harborne and Williams, 1975). In the following five years, this number doubled so that some 720 structures were listed in our first supplement (Harborne and Williams, 1982). The pace of discovery during 1981–1985 has not slowed down and we now list at the end of this book nearly a thousand substances. This is a conservative estimate in the sense that a number of partly characterized glycosides, which probably differ in mode of linkage or in the type of sugar configuration from those reported here, have been described but are not included in our lists. It is to be regretted that some investigators still do not characterize their glycosides with the necessary rigour to establish the structures unambiguously. As before, the term 'glycoside' is used in its widest sense and we include here flavones and flavonols which are acylated and sulphated as well as glycosylated.

An impressive array of novel conjugates has been described during 1981 to 1985 and a few of these are illustrated in Fig. 8.1. The discovery of luteolin linked to three glucuronic acid residues is notable. This compound, the 7-diglucuronide-4′-monoglucuronide occurs in primary leaves of rye *Secale cereale*, where it is specifically located in the mesophyll cells (Schulz *et al.*, 1985). The apigenin analogue was reported earlier from *Conocephalum conicum* (Markham *et al.*, 1976) but it only occurs in European races of this liverwort. The most usual position of sugar substitution in flavone glycosides is the 7-hydroxyl, so that reports of luteolin 3′-xyloside, luteolin 3′-rhamnoside and tricetin 3′, 5′-diglucoside (see Fig. 8.1) are all new. These substances have been uncovered in the foliage of gymnosperms and especially in members of the Podocarpaceae (Markham *et al.*, 1985). Further derivatives of this type can be confidently expected in the future from related sources.

The regular application of ^{13}C NMR spectroscopy to flavonoid glycosides has revealed an increasing number of acylated glycosides. New acylating groups reported during 1981–1985 (see Section 8.4.4) include 2-hydroxypropionic, 3-hydroxy-3-methylglutaric and quinic acids. The latter acid occurs widely in the free state in plants but is most frequently encountered linked to caffeic acid, as the ubiquitous chlorogenic acid. The discovery of quinic acid linked via glucose to the 7-hydroxyl of herbacetin in leaves of *Ephedra alata* (Ephedraceae) (Nawwar *et al.*, 1984b) appears to be the first report of quinic acid occurring in association with flavonoids.

Relatively few novel oligosaccharides have been fully characterized as occurring attached to flavones and flavonols and compounds with more than three sugar residues are still quite rare. However, the range of intersugar linkages among disaccharides has been extended. Disaccharides based on glucose and rhamnose are most commonly rutinose [rhamnosyl($\alpha1 \rightarrow 6$)glucose], with neohesperidose [rhamnosyl($\alpha1 \rightarrow 2$)glucose] and rungiose [rhamnosyl($\alpha1 \rightarrow 3$)glucose] being relatively rare. The disaccharide glucosyl($\beta1 \rightarrow 4$)rhamnose has been reported once linked to kaempferol in *Rosa multiflora* (cf. Harborne and Williams, 1982) and we can now add glucosyl ($\beta1 \rightarrow 3$)rhamnose, which occurs attached to kaempferol in *Cinnamomum sieboldii* leaf (Nakano *et al.*, 1983a). Only two isomers in the series remain to be found; rhamnosyl($\alpha1 \rightarrow 4$)glucose and glucosyl($\beta1 \rightarrow 2$)rhamnose.

The Flavonoids. Edited by J.B. Harborne
Published in 1988 by Chapman and Hall Ltd,
11 New Fetter Lane, London EC4P 4EE.

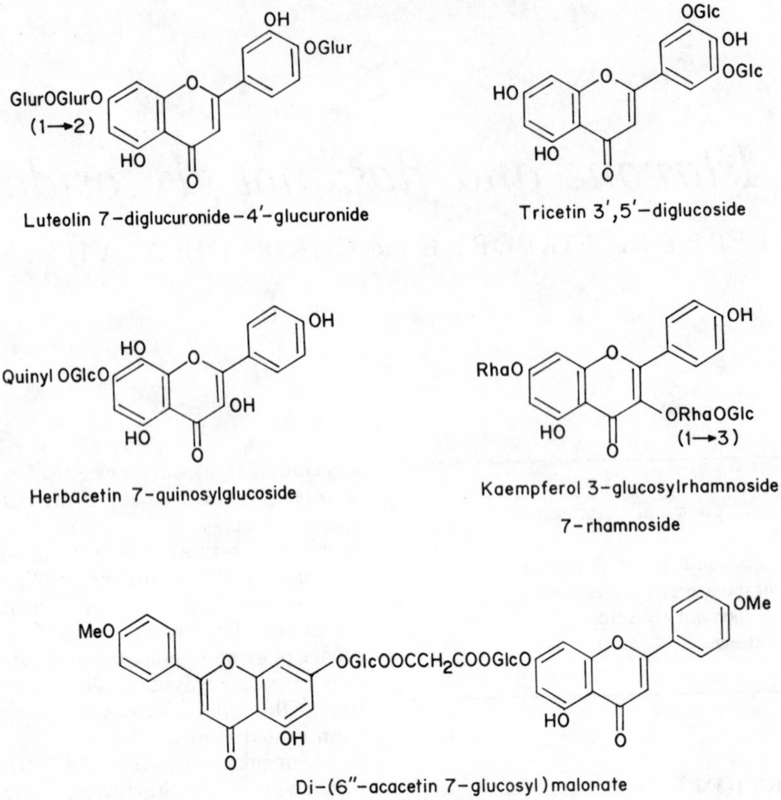

Luteolin 7-diglucuronide-4′-glucuronide

Tricetin 3′,5′-diglucoside

Herbacetin 7-quinosylglucoside

Kaempferol 3-glucosylrhamnoside-7-rhamnoside

Di-(6″-acacetin 7-glucosyl)malonate

Fig. 8.1 Representative structures of new flavone and flavonol glycosides.

One specific novelty must finally be mentioned here; the occurrence of flavone glycosides disubstituting an organic dicarboxylic acid. The first representation of this type, namely di-(6″-acacetin-7-glucosyl)malonate (see Fig. 8.1) has been reported from *Agastache rugosa* (Labiatae) (Itokawa *et al.*, 1981c). Since malonated glycosides are by no means uncommon in nature, further examples of disubstituted malonates will undoubtedly reveal themselves during future studies.

8.2 SEPARATION, PURIFICATION AND QUANTIFICATION

A major development since 1980 has been the more widespread application of high-performance liquid chromatography (HPLC) techniques to separation and quantification of flavone and flavonol glycosides. HPLC now takes its place alongside the older methods of PC, TLC and CC for the purification of these substances (Daigle and Conkerton, 1983: Harborne, 1983, 1985). Reverse phase separations are most commonly achieved on a C_8 or C_{18} Partisil column which is eluted with a methanol–acetic acid–water gradient with monitoring by UV absorption at 280 and 330 nm. HPLC has the advan tage of being complementary to other chromatographi techniques and it may well resolve mixtures of glycoside which run together in other systems. For example, it ha proved useful for separating the mixtures of partl

Table 8.1 HPLC retention and R_F values of some myı icetin, larycitrin and syringetin monoglycosides

Flavonol glycoside	t_R(min)*	R_F(× 100) *in BAW*†
Myricetin 3-galactoside	9.38	37
Myricetin 3-rhamnoside	11.06	60
Larycitrin 3-galactoside	11.64	43
Syringetin 3-galactoside	13.65	51
Larycitrin 3-rhamnoside	13.92	65
Syringetin 3-arabinoside	15.56	64
Syringetin 3-rhamnoside	16.32	83

*On a C_8 column (250 × 5 mm) and gradient elution with wate and methanol–acetic acid–water (90:5:5).

†Measured on Whatman no. 1 paper, BAW = butan-1-ol–aceti acid–water (4:1:5, top).

methylated flavonol 3-galactosides, 3-rhamnosides and 3-arabinosides present in stems of *Chondropetalum* species (Harborne et al., 1985). Some typical retention times and R_F values are indicated in Table 8.1.

Bernardi *et al.* (1984) have similarly found HPLC on a Bondapak RP-18 column eluted with a *t*-butanol–water gradient to be helpful in resolving mixtures of 8-methoxylated flavonol galactosides and arabinosides present in *Dryas octopetala*. HPLC combined with an ion-pairing reagent such as 0.01 M tetrabutylammonium phosphate is an excellent technique for distinguishing flavonol glycosides from sulphate conjugates. Without ion-pairing, sulphates have a shorter retention time than simple glycosides such as rutin, whereas with ion-pairing they are retained on the column and are eluted well after rutin (Table 8.2) (Harborne and Boardley, 1984). When mixtures of flavonol 3-rutinosides and 3-sulphates occur together as in *Oenanthe crocata*, they are well resolved and can be quantified using this reagent. Ion-pairing with HPLC has also been proposed for the separate determination of 6-hydroxyflavones and their glucuronides in *Scutellaria* root (Sagara *et al.*, 1985).

The amounts of flavonoid glycosides in crude plant extracts can be determined accurately (after pre-column clean up) by HPLC and this is now almost a standard procedure (e.g. Harborne *et al.*, 1985). Additionally, the substances can be separated, quantified *and identified* in one and the same operation if the HPLC apparatus is linked up to a UV diode array detector. Hostettmann *et al.* (1984) have successfully modified the usual UV spectral reagents (e.g. aluminium chloride, etc.) so that they can be applied to obtain spectral shifts on the different glycosides as they separate off the HPLC column. Siewek and Galensa (1984) have proposed that the degree of glycosylation in an unknown glycoside should be determined in such a set up, by benzoylating the compound before applying it to the column and then by measuring the ratio of absorbances at 231 and 301 nm of the benzoates. However, from the data

Table 8.2 HPLC retention times of flavonol sulphates with and without ion-pairing

	t_r(min)*	
Flavonol derivative	*Without ion-pairing*	*With ion-pairing*
Quercetin 3-sulphate	6.0	18.2
Kaempferol 3-sulphate	7.2	19.3
Isorhamnetin 3-sulphate	8.4	19.7
Quercetin 3′-sulphate	9.2	22.5
Rutin	14.2	14.2
Quercetin 3-rhamnoside	16.2	16.2

*On a C_{18} column eluted with a gradient of methanol–acetic acid–water. Ion-pairing in the presence of 0.01 M tetrabutylammonium phosphate.

available for kaempferol glycosides, the method does not seem to be particularly reliable unless the position of substitution remains the same.

Preparative scale separations of flavonoid glycosides are now readily achievable on HPLC columns, using similar procedures to analytical studies. For example, the flavonol glycosides of *Securidaca diversifolia* were well separated on the 25–35 mg scale on a Lobar Lichroprep RP-8 column (27 × 2.5 cm) eluted with methanol–water–formic acid (25:73:2) (Hamburger *et al.*, 1985). It is common practice to pass a flavonoid extract over Sephadex LH-20 either in initial fractionation or as a final stage of purification. An alternative procedure proposed by Rösler and Goodwin (1984) is to use an Amberlite XAD-2 resin column instead. For a review of other chromatographic procedures used in the flavonoid field, and especially for DCCC, see Hostettmann (1985).

8.3 IDENTIFICATION

The most important advances in the mass spectrometry of flavonoid glycosides have been the introduction of desorption chemical ionization (DCI) and fast atom bombardment (FAB) techniques (Domon and Hostettmann, 1985). These two techniques supplement field desorption (FD) mass spectrometry, which has been in use previously (see e.g. Geiger and Schwinger, 1980). In earlier studies of flavonoid glycosides, it was necessary to derivatize (e.g. by permethylation) in order to obtain a molecular ion and even then the molecular ion was often weak and difficult to detect because of its high molecular weight (see Harborne and Williams, 1982). With these newer techniques it is not only possible to obtain strong molecular ions without derivatization, but also fragmentation patterns indicative of the location of the glycosidic and acyl substituents (see Table 8.3). For example, DCI mass spectrometry (which uses a tungsten wire and ammonia as a reactant gas) of rutin and of quercetin 3-rhamnoside-7-glucoside clearly differentiates the two isomers, since only the latter compound exhibits ions indicating separate loss of glucose and rhamnose. It is thus possible by DCI-MS to distinguish between glucose as a terminal or as inner sugar residue. Measurements can be made either in a positive or negative ion mode, the latter being more successful with flavonol glycosides.

For FAB-MS, glycerol can be used to solubilize the sample and addition of trace amounts of inorganic salt (e.g. NaCl) may optimize production of a cationic molecular ion. Contamination of a sample with too much salt, however, can be fatal for a successful analysis. FAB-MS is particularly useful for determining the molecular ions of glucuronides (Domon and Hostettmann, 1985) and sulphates.

These newer techniques have proved to be useful for studying acylated glycosides, since they often (but not infallibly) allow the detection of acyl residues, which might

Table 8.3 DCI- and FAB-mass spectrometry of flavonol glycosides

Flavonol glycoside	*MS technique*	*Fragmentation*
Kaempferol 3-rhamnosylgalactoside-7-rhamnoside (robinin)	DCI negative mode or FAB negative mode	Molecular ion,* loss of Rha and of Rha + Gal
Quercetin 3-rhamnosylglucoside (rutin)	DCI negative mode	Molecular ion, loss of Rha
Quercetin 3-rhamnoside-7-glucoside	DCI negative mode	Molecular ion, loss of Rha and of Glc
Quercetin 3-glucuronide	FAB	Molecular ion, loss of Glur

*Relative intensities of molecular ion of robinin by DCI (glycerol matrix): PI 0%, NI 13%, FAB:PI 11%, NI 100%, FD ($+Na^+$): 100% (PI = positive ion, NI = negative ion).
Data from Domon and Hostettmann (1985)

otherwise be unsuspected, particularly those that are aliphatic in nature. A detailed study has been made of the FD-MS of glycoside acetates (Zapesochnaya *et al.*, 1984a) and of glycosides substituted with aromatic acids (Zapesochnaya *et al.*, 1984b). In the mass spectra of monoacetylated monoglycosides, the molecular ion appears as the parent ion and fragmentations corresponding to the aglycone and the acetylated anhydrosugar are also present. Di- and tri-acetates give peaks corresponding to the expected di- and tri-acetyl anhydrosugars. In acetylated flavone biosides, the fragmentation ions are not so apparent, but in favourable cases, FD-MS will indicate the sugar to which the acyl group is linked. Thus in the spectrum of 4‴-acetylpectolinarin, a fragment corresponding to acetylrhamnose is obtained, confirming the location of the acetyl group on the terminal rhamnose of the disaccharide moiety.

In the case of aromatic acylated derivatives, FD-MS is not quite so diagnostic as with acetates, and there is the added danger of transacylation occurring in the mass spectrometer. However, molecular ions are usually obtainable and deacylation is observable with the deacylated glycoside giving a reasonably intense ion.

FD-MS also works for flavones (and flavonols) acylated with malonic acid, but it does not indicate the position of attachment of the organic acid to the sugar residue. Other procedures (e.g. ^{13}C NMR spectroscopy) have to be employed. In determining the structure of 6-hydroxyluteolin 7-(6″-malonylglucoside), Stein *et al.* (1985) used FD-MS and also HPLC comparison with a synthetic malonate, produced from the corresponding glucoside and malonyl-coenzyme A in the presence of a malonyltransferase specific for the 6-hydroxyl of glucose.

8.4 SUGARS OF FLAVONE AND FLAVONOL GLYCOSIDES

8.4.1 Monosaccharides

The number of monosaccharides known to occur in combination with flavones and flavonols is nine (Table 8.4), with one sugar, mannose, on the 1982 list now of uncertain status (see below). Some comments on the rarer sugars would seem to be in order. Apiose is still uncommon. Although a number of new disaccharides containing apiose have been recently characterized (see Section 8.4.2), there is only one report of this sugar in monoglycosidic combination with 6-hydroxyluteolin (see Harborne and Williams, 1982).

Table 8.4 Monosaccharides of flavone and flavonol glycosides

Pentoses	*Hexoses*	*Uronic acids*
D-Apiose	D-Allose	D-Galacturonic acid†
L-Arabinose*	D-Galactose	D-Glucuronic acid†
L-Rhamnose	D-Glucose	
D-Xylose		

*Known to occur in both pyranose and furanose forms; all other sugars (except apiose) are normally in the pyranose form.
†Also reported to occur as the methyl and ethyl esters.

Only one new glycoside containing galacturonic acid has been described (Mabry *et al.*, 1984), but a variety of novel allosides have been identified. These include three apigenin 7,4′-glycosides from *Thalictrum thunbergii*; the 7,4′-bisalloside, the 7-(6″-acetylalloside)-4′-alloside and the 7-(4″,6″-diacetylalloside)-4′-alloside (Shimizu *et al.*, 1984). In addition, allose has been found in disaccharide combination with glucose as isoscutellarein 7-allosyl-(1→2)glucoside in *Sideritis leucantha* (Barberan *et al.*, 1985). Also isoscutellarein and 8-hydroxyluteolin 4′-methyl ether 7-(6″-acetylallosyl(1→2)glucoside) were identified, together with the known isoscutellarein 4′-methyl ether derivative, in *Stachys recta* (Lenherr *et al.*, 1984a).

Thus, the number of allose-containing glycosides has been significantly increased in the last five years. Work in progress in several laboratories suggests furthermore that flavones with allosyl(1→2)glucose attachments are fairly frequent in the family Labiatae. They have now been found in *Teucrium* as well as *Sideritis* and *Stachys*, and in the latter genus, they appear to be regular constituents. This contrasts with repeated reports in the Russian literature between 1976 and 1982 of mannosylglucose derivatives in *Stachys* species from Caucasia (cf. Lenherr *et al.*, 1984a). Since mannose could be confused with allose in hydrolysates of these flavone glycosides, it now appears probable that the nine compounds reported by the Russian workers as mannose-containing are probably based on allose and glucose. Further work will be needed to confirm this.

Other reports of mannose-containing flavone glycosides also seem doubtful. Thus, the description of an apigenin 7-galactosylmannoside from the seed of *Daucus carota* (Gupta and Niranjan, 1982) cannot be regarded as complete, since the identification of the sugars was based on PC comparison in a single solvent system (BAW), which is known not to separate glucose from galactose. Additionally, it is an unlikely finding since earlier workers (see Harborne, 1971) have reported apigenin 7-rutinoside and 7-diglucoside from this source and the Indian workers made no mention of these compounds as being present.

The need for care in the identification of the monosaccharides obtained on acid hydrolysis of flavonoid glycosides should always be borne in mind. Allose and mannose can be readily distinguished from each other and from other hexoses by paper chromatography, but only if at least four different solvent systems are used (Harborne, 1984). Clearly if an uncommon sugar is thought to be present, it is essential to use other methods of comparison as well – TLC, GLC and HPLC procedures for monosaccharide separations are all available (Harborne, 1984) and usefully complement an identification based on PC. The configuration of the sugar substituents in flavonoid glycosides can also be confirmed by ^{13}C NMR spectroscopy, since the different hexoses exhibit distinctive shift values (Markham and Chari, 1982).

Recent reports of flavonoid methyl and/or ethyl glucuronides (Ferreres and Tomas, 1980; Sinha *et al.*, 1981; Rao and Rao, 1982; Hiermann, 1982; Nawwar *et al.*, 1984a) have to be regarded with some suspicion since esters of uronic acids are easily produced as artifacts from the parent acids when plant material is extracted in methanol or ethanol. The report of kaempferol 3-glucuronide ethyl ester in flowers of *Tamarix indica* (Nawwar *et al.*, 1984a) may be correct, since the only solvents apparently used in extraction were acetone and methanol. However, any system of purification involving prolonged exposures of glucuronides to organic solvents could give rise to ester formation and it is advisable to extract under very mild conditions when glucuronides are suspected to be present.

Finally in this section, it is worth noting that a number of hitherto unreported but simple monoglycosides have now been found in nature. These include, for example, the 5-glucuronide, 7-xyloside, 3′-xyloside and 3′-rhamnoside of luteolin and the 5-galactoside and 4′-arabinoside of apigenin (see Table 8.9). Three new flavonol glycosides, which fill in gaps, are kaempferol 5-rhamnoside from *Callitris glauca*, myricetin 3′-xyloside from *Ledum palustre* and syringetin 3-arabinoside from *Chondropetalum* (see Table 8.10).

8.4.2 Disaccharides

In the last five years 12 new disaccharides have been fully characterized as components of flavone or flavonol glycosides. These are listed together with the 21 previously recorded disaccharides in Table 8.5. The number of new reports reflects a considerable advance in the determination of linkages in oligosaccharides using techniques such as ^{13}C NMR spectroscopy and FAB-MS. There are still however recent publications describing what may be new disaccharides, which have not been fully characterized either because of lack of plant material or availability of the necessary technology.

Among the new disaccharides, the characterization of the first diglucuronide, luteolin 7-glucuronosyl(1 → 2)-glucuronide from *Elodea canadensis* (Mues, 1983) is of especial note. This sugar has also been found attached to the 7-position of luteolin 4′-glucuronide in a triglycoside from rye seedlings, *Secale cereale*, by Schulz *et al.* (1985). Similarly, three disaccharides with apiose as one of the sugars have been fully characterized for the first time: apiosyl(1 → 2)arabinose, apiosyl(1 → 2)xylose and apiosyl(1 → 2)galactose. All three, together with apiosyl-(1 → 2)glucose, were found attached at the 3-position of quercetin in the leaf of *Securidaca diversifolia*, Polygalaceae (Hamburger *et al.*, 1985).

A number of new isomers of previously recorded disaccharides have also been found, i.e. (glucosyl-(1 → 2)galactose, galactosyl(1 → 6)glucose and glucosyl-(1 → 3)rhamnose. Reports of disaccharides with mannose, e.g. mannosyl(1 → 2)glucose, are excluded here because of doubts about their identification.

Finally, an unusual bioside in which a methylpentose is attached to the aglycone and a hexose is the terminal sugar has been discovered. Thus glucosyl(1 → 3)rhamnose occurs linked to the 7-position of herbacetin in rhizomes of *Rhodiola rosea*, Crassulaceae (Zapesochnaya and Kurkin, 1983) and also at the 3-position of kaempferol 7-rhamnoside in the leaf of *Cinnamomum sieboldii*, Lauraceae (Nakano *et al.*, 1983a). The linkage and order of sugars in a reported genkwanin 4′-glucosyl(1 → 3)xyloside by Bhandari *et al.* (1981) was not satisfactorily proved and hence this disaccharide is not tabulated here.

8.4.3 Trisaccharides

No new trisaccharides have been fully characterized in the period 1981–85. However, three linear trisaccharides, primflasine, 2′-rhamnosyl-laminaribiose and 6′-maltosyl-

Table 8.5 Disaccharides of flavone and flavonol glycosides

Structure	Trivial name
Pentose–pentose	
O-α-L-Rhamnosyl(1→2) arabinose*	—
O-α-L-Rhamnosyl(1→4) xylose	—
O-β-D-Xylosyl(1→2) rhamnose*†	—
O-β-D-Apiofuranosyl(1→2) xylose*	—
Pentose–hexose	
O-α-L-Arabinosyl(1→6) glucose	Vicianose
O-α-L-Arabinosyl(1→6) galactose	—
O-β-D-Xylosyl(1→2) glucose	Sambubiose
O-β-D-Xylosyl(1→2) galactose	Lathyrose
O-β-D-Apiofuranosyl(1→2) glucose*	—
O-β-D-Apiofuranosyl(1→2) galactose*	—
O-α-L-Rhamnosyl(1→2) glucose	Neohesperidose
O-α-L-Rhamnosyl(1→3) glucose	Rungiose
O-α-L-Rhamnosyl(1→6) glucose	Rutinose
O-α-L-Rhamnosyl(1→2) galactose	—
O-α-L-Rhamnosyl(1→6) galactose	Robinobiose
Hexose–pentose	
O-β-D-Glucosyl(1→3) rhamnose*‡	—
O-β-D-Glucosyl(1→4) rhamnose	—
O-β-D-Galactosyl(1→4) rhamnose	—
Hexose–hexose	
O-β-D-Glucosyl(1→2) glucose	Sophorose
O-β-D-Glucosyl(1→3) glucose	Laminaribiose
O-β-D-Glucosyl(1→6) glucose	Gentiobiose
O-β-D-Glucosyl(1→2) galactose*	—
O-β-D-Glucosyl(1→6) galactose	—
O-β-D-Galactosyl(1→4) glucose	Lactose
O-β-D-Galactosyl(1→6) glucose*§	—
O-β-D-Galactosyl(1→4) galactose	—
O-β-D-Galactosyl(1→6) galactose*	—
O-β-D-Allosyl(1→2) glucose	—
Hexose–uronic acid	
O-α-L-Rhamnosyl(1→2) galacturonic acid*	—
Uronic acid–uronic acid	
O-β-D-Glucuronosyl(1→2) glucuronic acid*	—

*Newly discovered in diglycosidic combination, in the period 1981–1985.
†Present as 4′-rhamnoside.
‡Also present as the 7-rhamnoside.
§Present as 7-dirhamnoside.

glucose were inadvertently omitted from the 1982 list and these are now included in the new table of 11 linear and 9 branched trisaccharides (Table 8.6). A recent report of gentiotriose in the leaf flavonoids of *Tribulus terrestris* by Saleh *et al.* (1982b) requires further substantiation; the authors detected small amounts of what they suggest were isorhamnetin 3-gentiotrioside and 3-gentiotrioside-7-glucoside but their evidence was based only on comparative R_F and R_G values. Another partially characterized, probably new, trisaccharide, quercetin 3-glucosyl-(1→4)galactosylrhamnose was reported from the fern *Cheilanthes fragrans* (Imperato, 1985a). A number of other noval trisaccharides await further investigation, e.g. the sugar of a chrysoeriol 7-triglucuronide from *Medicago sativa* (Saleh *et al.*, 1982a).

8.4.4 Acylated derivatives

In the period 1981–85 some 66 new acyl derivatives have been reported. The identification and determination of the positioning of the acyl groups is largely dependent on ^{13}C NMR spectroscopy but FAB-MS is proving a useful additional technique in this field.

Nine aliphatic and eight aromatic acids have now been characterized as acyl substituents in flavones and flavonols (Table 8.7). There are three new aliphatic acids: 2-hydroxypropionic, 3-hydroxy-3-methylglutaric and quinic. Thus, kaempferol 3-(2-hydroxypropionyl-glucoside)-4′-glucoside was determined in leaves of *Sisymbrium galliesii* (Cruciferae) by Aguinagalde and del Pero Martinez (1982), sudachitin (5,7,4′-trihydroxy-6,8,3′-trimethoxyflavone)-7-(3-hydroxy-3-methylglutarate)-4′-glucoside was identified in the peel of *Citrus sudachi* (Rutaceae) (Kumamoto *et al.*, 1985) and herbacetin 7-(6″-quinylglucoside) was characterized in

Trivial name
Primflasine
2′-Rhamnosylrutinose
2′-Rhamnosyl-laminaribiose
Sophorotriose
2′-Glucosylgentiobiose
6′-Maltosylglucose
Sorborose
Rhamninose
Sugar of alaternin
Sugar of faralatroside★
—
2^{G}-Apiosylrutinose
2^{G}-Rhamnosylrutinose
2^{G}-Glucosylrutinose
3^{G}-Glucosylneohesperidose
2^{G}-Rhamnosylgentiobiose
2^{G}-Glucosylgentiobiose
4^{G}-Rhamnosylneohesperidose
2^{Gal}-Rhamnosylrobinobiose
4^{Gal}-Rhamnosylrobinobiose

Table 8.7 Acylating acids found in flavone and flavonol derivatives

Organic acid	*Example of occurrence*	*Reference*
Aliphatic acids		
Acetic	Apigenin 7-(2″-acetylglucoside)	Redaelli *et al.* (1980)
Malonic	Quercetin 3-(malonylglucoside)★	Woeldecke and Herrmann (1974)
†2-Hydroxypropionic	Kaempferol 3-(2-hydroxypropionylglucoside)-4′-glucoside†	Aguinagalde and del Pero Martinez (1982)
Succinic	Kaempferol 7-(6″-succinylglucoside)	Hiraoka and Maeda (1979)
Butyric	Herbacetin 8-butyrate	Wollenweber *et al.* (1978)
2-Methylbutyric	Acacetin 7-[2‴-(2 methylbutyryl)rutinoside]	Chari *et al.* (1977)
Tiglic	Quercetin 7-(6″-tiglyglucoside)	Ogawa and Ogihara (1975)
†3-Hydroxy-3-methyl-glutaric	Sudachitin 7-(3-hydroxy-3-methylglutarate)-4′-glucoside	Kumamoto *et al.* (1985)
†Quinic	Herbacetin 7-(6″-quinylglucoside)†	Nawwar *et al.* (1984b)
Aromatic acids		
Benzoic	Kaempferol 3-(benzoylglucoside)★	Schonsiegel *et al.* (1969)
p-Hydroxybenzoic	Kaempferol 3-(*p*-hydroxybenzoylglucoside)★	Schonsiegel *et al.* (1969)
Gallic	Quercetin 3-(6″-gallylglucoside)	Collins *et al.* (1975)
p-Coumaric	Apigenin 7-(4″-*p*-coumarylglucoside)	Karl *et al.* (1976)
Caffeic	Apigenin 7-glucoside-4′-caffeate	Gella *et al.* (1967)
Ferulic	Luteolin 7-glucuronide-3′-ferulylglucoside★	Markham *et al.* (1978)
Isoferulic	Quercetin 3-(isoferulylglucuronide)★	El Ansari *et al.* (1976)
Sinapic	Kaempferol 3-(sinapylsophoroside)-7-glucoside★	Stengel and Geiger (1976)

★In this example, the position of attachment of acyl group to sugar has not been established.
†Newly discovered in the period 1981–1985.

Ephedra alata (Ephedraceae) by Nawwar *et al.* (1984b).

The most common acylating acids are still acetic, gallic, *p*-coumaric and ferulic acids. Whilst monoacylation remains the norm a number of new diacylated derivatives have been characterized including several new isomers. For example, the 2″,6″-,3″,6″- and 4″,6″-di-*p*-coumarates of apigenin 7-glucoside (Rao, L. J.M. *et al.*, 1982, 1983) from *Anisomeles ovata* (Labiatae) and the 2″,3″- and 3″,4″-diacetates of apigenin 7-glucoside from *Matricaria chamomilla* (Compositae) (Redaelli *et al.*, 1982).

However, the most interesting find is that of the first triacylated glycoside: the 8-(2″,3″,4″-triacetylxyloside) of 8-hydroxykaempferol, together with the 8-(2″,3″-diacetylxyloside) and 8-(3″-acetylarabinoside) in roots of *Rhodiola algida* (Pangarova and Zapesochnaya, 1975).

Malonic acid, which has recently been shown to be a common acylating acid in anthocyanin pigments (see Chapter 1), has also been reported as the 7-(6″-malonylglucoside) of luteolin, diosmetin and 6-hydroxyluteolin in the moss *Bryum capillare* (Stein *et al.*, 1985) and at the 4′-hydroxyl of apigenin 7-glucuronide in flowers of *Centaurea cyanus* (Compositae) (Tamura *et al.*, 1983). The most unusual new malonate is the flavonoid from *Agastache rugosa* which has two moieties of acacetin 7-glucoside substituting the malonic acid residue (Itokawa *et al.*, 1981c) (see Fig. 8.1). Malonated flavones and flavonols may well have been overlooked during the classical period of flavonoid chemistry because of their instability and many such compounds may well remain to be detected in plants in future investigations.

8.4.5 Sulphate conjugates

By 1980 54 sulphate conjugates had been characterized. However, during the last five years only a further twelve structures have been described, bringing the total number of known conjugates to 65 (Table 8.8). The numbers do not quite tally as the previously described kaempferol 3-glucoside sulphate has been replaced by three compounds

Table 8.8 Sulphate conjugates of flavones and flavonols

Type of conjugate	*Known aglycones*
Flavones	
7-Sulphate	Apigenin, luteolin, chrysoeriol, diosmetin
8-Sulphate	Hypolaetin
3′-Sulphate	Luteolin, diosmetin, tricetin
4′-Sulphate	Luteolin
7, 3′-Disulphate	Luteolin, diosmetin, tricetin
7-Sulphatoglucoside	Apigenin, luteolin, chrysoeriol, tricin
7-Sulphatoglucuronide	Tricin*, luteolin*
7-Disulphatoglucuronide	Tricin*
7-Sulphatorutinoside	Luteolin
8-Glucoside-3′-sulphate	Hypolaetin, hypolaetin 4′-methyl ether*
7-Sulphate-3′-glucoside	Luteolin
7-Sulphate-3′-rutinoside	Luteolin
Flavonols	
3-Sulphate	Rhamnocitrin, kaempferol 7, 4′-dimethyl ether, kaempferol*, quercetin, isorhamnetin, tamarixetin, rhamnetin, rhamnazin, gossypetin, patuletin, eupatoletin, eupatin, veronicafolin
7-Sulphate	Kaempferol, isorhamnetin, patuletin, quercetagetin 3-methyl ether
3, 7-Disulphate	Kaempferol, isorhamnetin
3, 4′-Disulphate	Quercetin
3, 5, 4′-Trisulphate	Rhamnetin
3, 7, 3′-Trisulphate	Quercetin*
3, 7, 4′-Trisulphate	Quercetin, kaempferol*, isorhamnetin*
3, 7, 3′, 4′-Tetrasulphate	Quercetin
3-(3″-Sulphatoglucoside)	Quercetin*, kaempferol*
3-(6″-Sulphatoglucoside)	Kaempferol*
3-Sulphatorhamnoside	Kaempferol, quercetin, myricetin
3-Sulphatorutinoside	Quercetin, isorhamnetin, kaempferol*
3-(6″-Sulphatogentiobioside)	Kaempferol
3-Glucoside-7-sulphate	Patuletin
3-Glucuronide-7-sulphate	Kaempferol, quercetin, isorhamnetin
8-Glucuronide-3-sulphate	Gossypetin*
3′-Glucuronide-3, 5, 4′-trisulphate	Rhamnetin

*Newly discovered sulphate conjugate in the period 1981–1985.

in which the linkage of sulphate to glucose is known, i.e. kaempferol 3-(3″-sulphatoglucoside) and 3-(6″-sulphatoglucoside) from the fern *Cystopteris fragilis* (Imperato, 1983a) and quercetin 3-(3″-sulphatoglucoside) from another fern, *Asplenium septentrionale* (Imperato, 1983b). Other new flavonol derivatives have sulphate attached directly to the hydroxyl groups of the aglycone. These include quercetin 3,7,3′-trisulphate from *Flaveria bidentis* (Compositae) (Cabrera *et al.*, 1985) and kaempferol and isorhamnetin 3,7,4′-trisulphates and kaempferol 3-sulphate from *Acrotema uniflorum* (Dilleniaceae) (Gurni *et al.*, 1981). Quercetin 3,7,4′-trisulphate and 12 other flavonol 3-sulphates were known previously (Harborne and Williams, 1982).

Three unusual flavone glucuronide sulphates: tricin 7-monosulphato- and disulphato-glucuronides from *Cyperus polystachyos* (Harborne *et al.*, 1982) and luteolin 7-sulphatoglucuronide from *Fuchsia excorticata* and *F. procumbens* (Williams *et al.*, 1983) are described for the first time but the linkage between the sulphate and glucuronic acid moieties was not established.

8.5 NEW REPORTS OF FLAVONE GLYCOSIDES

Some 90 new flavone glycosides have been discovered in the period 1981–1985. These are listed in Table 8.9 with plant source and reference. A complete check list of all the known flavone glycosides is given at the end of the book. The number of compounds has increased by about 25% since 1980 to give a total of some 330 flavone glycosides. These include a further 18 apigenin, 19 luteolin and three tricin glycosides bringing their totals to 58, 72 and 17, respectively. A number of other aglycones have been found in glycosidic combination for the first time, e.g. 6-hydroxy-4′-methoxyflavone 6-arabinoside from *Cassia spectabilis* (Leguminosae) (Singh and Singh, 1985a), 7,3′,4′,5′-tetrahydroxyflavone 7-rhamnoside and 7-glucoside from *Evolvulus nummularius*, Cucurbitaceae (Gupta *et al.*, 1984) and 8-hydroxyluteolin 4′-methyl ether as the 7-(6″-acetylallosyl(1→2)glucoside) from *Stachys recta* leaves, Labiatae (Lenherr *et al.*, 1984).

Reports of wogonin 7-glucoside (Bekirov *et al.*, 1974) and luteolin 7-glucosylrhamnoside and 7-rhamnosyldiglucoside (Pacheco, 1957) were accidently left out of our earlier tabulations (Harborne and Williams, 1975, 1982). A previous entry for luteolin 7-*p*-coumarylglucoside has been replaced by one in which the linkage is shown to be (1→2) (Gupta and Saxena, 1984). The structure for the new glycoside, scutellarein 7-rhamnosylarabinoside (Srinivasan and Subramanian, 1981) has recently been corrected to scutellarein 6-xyloside-7-rhamnoside by Nair *et al.* (1986).

Some of the most interesting new glycosides are glucuronic acid derivatives: apigenin, luteolin and chrysoeriol 7-glucuronosyl(1→2)glucuronides from *Elodea canadensis* (Mues, 1983), luteolin 7-glucuronosyl(1→2)glucuronide-4′-glucuronide from the grass, *Secale cereale* (Schulz *et al.*, 1985). 7-Triglucuronides of chrysoeriol and tricin have been provisionally reported from *Medicago sativa* (Saleh *et al.*, 1982a), but they have yet to be fully characterized.

8.6 NEW REPORTS OF FLAVONOL GLYCOSIDES

Some 193 new flavonol glycosides have been reported in the period 1981–85 and these are shown in Table 8.10 with plant source and reference. A check list of the 650 known flavonol glycosides is given at the end of the book. Since 1980 the rate of discovery of new glycosides has fallen but a number of the original entries, where the interglycosidic linkage was not known, have been replaced by fully characterized glycosides. For example, kaempferol and quercetin 3-xylosyl(1→2) galactosides from *Armoracia rusticana* (Cruciferae) (Larsen *et al.*, 1982) have replaced a kaempferol 3-xylosylgalactoside from *Lysichiton camtschatcense* (Araceae) (Williams *et al.*, 1981) and quercetin 3-xylosylgalactoside from flowers of a *Hibiscus* sp. (Malvaceae) (Makhsudova *et al.*, 1967) respectively. Until the latter glycosides are completely characterized there is of course no way of knowing whether they are identical or different from their replacements but it would be too confusing to give both in the check list. Similarly, quercetin 3-(6″-malonylglucoside) (Buschi and Pomilio, 1982) and kaempferol 3-apiosyl(1→2)glucoside (Hamburger *et al.*, 1985) have replaced previous incomplete entries.

A few di- and tri-glycosides, where the positions of sugars or linkage is unknown, have been included in the list but a number of reports of apparently new glycosides, where no attempt was made to determine the order or linkage of sugars, mostly in taxonomic papers, have been omitted (e.g. Doyle, 1983; Parker *et al.*, 1984; Bohm and Bhat, 1985). Similarly, reports of kaempferol 3-xylofuranosylgalactopyranoside and glucofuranosyl-(1→6)galactopyranoside (Tsiklauri, 1979), of quercetin 3-galactopyranosyl(1→6)glucofuranoside and 3-galactopyranosyl(1→6)xylofuranoside (Tsiklauri *et al.*, 1979) and of quercetin 3-glucoside-7-rhamnofuranoside (Alania, 1983) are not included because insufficient evidence was presented to prove that the sugars were in the furanose form (cf. Harborne and Williams, 1982).

Quercetagetin 3,6,4′-trimethyl ether 7-glucoside (centaurein) from *Centaurea juncea*, Compositae (Farkas *et al.*, 1964) was inadvertently omitted from both previous check lists (Harborne and Williams, 1975, 1982) but is now included in Table 8.10.

Among the new flavonol glycosides are a large number of acylated, some sulphated derivatives and five rare diglycosides based on apiose, which have all been discussed already. However, the most novel substitution pattern is that of laricytrin 3,7,5′-triglucoside, which was found together with the corresponding 3,5′-diglucoside in

Table 8.9 New flavone glycosides

Glycoside	*Source*	*References*
5,7-Dihydroxyflavone (chrysin)		
5-Xyloside	*Ixora arborea* stem; Rubiaceae	Chauhan *et al.* (1984)
7-Gentiobioside	*Spartium junceum* flowers; Leguminosae	De Rosa and De Stefano (1983)
6-Hydroxy-4′-methoxyflavone		
6-Arabinoside	*Cassia spectabilis* seeds; Leguminosae	Singh and Singh (1985a)
5,6,7-Trihydroxyflavone (baicalein)		
5,6-Dimethyl ether		
7-Glucoside	*Scutellaria ovata* roots, stems and leaves; Labiatae	Nicollier *et al.* (1981)
5,7-Dihydroxy-8-methoxyflavone (wogonin)		
5-Glucoside	*Scutellaria baicalensis*; Labiatae	Takagi *et al.* (1980)
7-Glucoside	*Scutellaria orientalis* Labiatae	Bekirov *et al.* (1974)
7,3′,4′-Trihydroxyflavone		
7-Galactoside	*Callistemon* spp. leaf; Myrtaceae	Hashim *et al.* (1980)
Apigenin		
5-Galactoside	*Ixora arborea* stem; Rubiaceae	Chauhan *et al.* (1982a)
4′-Arabinoside	*Bommeria hispida* frond; Pteridaceae	Haufler and Giannasi (1982)
7-Methylglucuronide	*Adenocalymma alliaceum* flowers; Bignoniaceae	Rao and Rao (1982)
7-(6″-Ethylglucuronide)	*Centaurea aspera*; Compositae	Ferreres and Tomas (1980)
7-Glucuronosyl(1 → 2)glucuronide	*Elodea canadensis* leaf; Hydrocharitaceae	Mues (1983)
7,4′-Bisalloside	*Thalictrum thunbergii* aerial parts; Ranunculaceae	Shimizu *et al.* (1984)
7-(6″-*p*-Coumarylglucoside)	*Pogostemon cablin*; Labiatae	Itokawa *et al.* (1981a)
7-Caffeylglucoside	*Perilla frutescens* leaf; Labiatae	Ishikura (1981)
7-(3″,6″-Di-*p*-coumarylglucoside)	*Anisomeles ovata*; aerial parts; Labiatae	Rao *et al.* (1982)
7-(2″,6″-Di-*p*-coumarylglucoside) 7-(4″,6″-Di-*p*-coumarylglucoside)	*Anisomeles ovata* aerial parts; Labiatae	Rao, L.J.M. *et al.* (1983)
7-(2″,3″-Diacetylglucoside) 7-(3″,4″-Diacetylglucoside)	*Matricaria chamomilla* flowers; Compositae	Redaelli *et al.* (1982)
7-(6″-Acetylalloside)-4′-alloside 7-(4″,6″-Diacetylalloside)-4′-alloside	*Thalictrum thunbergii* aerial parts; Ranunculaceae	Shimizu *et al.* (1984)
4′-(6″-Malonylglucoside)-7-glucuronide	*Centaurea cyanus* flowers; Compositae	Tamura *et al.* (1983)
7-Sulphatogalactoside	*Tetracera stuhlmanniana*; Dilleniaceae	Gurni *et al.* (1981)
Apigenin 4′-methyl ether (acacetin)		
7-Galactoside	*Chrysanthemum indicum* flowers; Compositae	Chatterjee *et al.* (1981)
7-(6″-Methylglucuronide)	*Clerodendrum infortunatum* flowers; Verbenaceae	Sinha *et al.* (1981)

Table. 8.9 (*Contd.*)

Glycoside	*Source*	*References*
7-Arabinosylrhamnoside	*Premna latifolia* leaf; Verbenaceae	Rao and Raju (1981)
7-(6″-Acetylglucoside)	*Agastache rugosa*; Labiatae	Zakharova *et al.* (1979)
7-(2″-Acetylglucoside) Di-(6″-acacetin-7-glucosyl)malonate	*Agastache rugosa*; Labiatae	Itokawa *et al.* (1981c)
6-Hydroxyapigenin (scutellarein)		
6-Glucoside	*Haplopappus rengifoanus* leaf; Compositae	Ulubelen *et al.* (1981a)
7-Neohesperidoside	*Barleria prionitis* flowers; Acanthaceae	Nair and Gunasegaran (1982)
6-Xyloside-7-rhamnoside	*Triumfetta rhomboidea* leaf; Tiliaceae	Nair *et al.* (1986)
7-(6″-Ferulylglucuronide)	*Holmskioldia sanguinea* flowers; Verbenaceae	Nair and Mohandoss (1982)
Scutellarein 4′-methyl ether		
7-Glucoside	*Veronica* spp.; Scrophulariaceae	Frolova and Dzhumyrko (1984)
8-Hydroxyapigenin (isoscutellarein)		
7-Allosyl(1 → 2)glucoside	*Sideritis leucantha*; Labiatae	Barberan *et al.* (1985)
7-[6‴-Acetylallosyl(1 → 2)glucoside]	*Stachys recta* leaf; Labiatae	Lenherr *et al.* (1984a)
8-Hydroxyapigenin 8, 4′-dimethyl ether		
7-Glucuronide	*Bucegia romanica* whole plant; Marchantiaceae	Markham and Mues (1983)
5, 7, 4′-Trihydroxy-6, 8-dimethoxyflavone		
7-Glucoside	*Sideritis leucantha* whole plant; Labiatae	Tomas and Ferreres (1980)
7, 3′, 4′, 5′-Tetrahydroxyflavone		
7-Rhamnoside 7-Glucoside	*Evolvulus nummularius* whole plant; Cucurbitaceae	Gupta *et al.* (1984)
Luteolin		
5-Glucuronide	*Torilis arvensis* leaf/stem; Umbelliferae	Saleh *et al.* (1983a)
7-Xyloside	*Thymus membranaceus* aerial parts; Labiatae	Tomas *et al.* (1985)
7-Rhamnoside	*Phyllocladus* spp. leaf; Podocarpaceae	Markham *et al.* (1985)
7-Methylglucuronide	*Digitalis lanata* leaf; Scrophulariaceae	Hiermann (1982)
3′-Xyloside	*Podocarpus nivalis* foliage; Podocarpaceae	Markham *et al.* (1984)
3′-Rhamnoside	*Phyllocladus* spp. leaf; Podocarpaceae	Markham *et al.* (1985)
7-Sambubioside	*Thymus membranaceus* aerial parts; Labiatae	Tomas *et al.* (1985)
7-Glucosylrhamnoside	*Wisteria sinensis* leaf; Leguminosae	Pacheco (1957)
7-Glucuronosyl(1 → 2)glucuronide	*Elodea canadensis* leaf; Hydrocharitaceae	Mues (1983)
7-Glucoside-3′-xyloside	*Podocarpus nivalis* foliage; Podocarpaceae	Markham *et al.* (1984)
7, 4′-Diglucoside	*Launaea* spp. leaf and stem; Compositae	Mansour *et al.* (1983)

(*Contd.*)

Table 8.9 (*Contd.*)

Glycoside	*Source*	*References*
7-Rhamnosyldiglucoside	*Wisteria sinensis* leaf; Leguminosae	Pacheco (1957)
7-Gentiobioside-4′-glucoside	*Launaea* spp. leaf and stem; Compositae	Mansour *et al.* (1983)
7-Glucuronosyl(1 → 2)glucuronide-4′-glucuronide	*Secale cereale* primary leaf mesophyll; Gramineae	Schulz *et al.* (1985)
7-(2″-*p*-Coumarylglucoside)	*Barleria prionitis*; Acanthaceae	Gupta and Saxena (1984)
7-Caffeylglucoside	*Perilla frutescens* leaf; Labiatae	Ishikura (1981)
7-(6″-Malonylglucoside)	*Bryum capillare*; Bryaceae	Stein *et al.* (1985)
7-(3″-Acetylapiosyl)(1 → 2)xyloside	*Campanula patula*; Companulaceae	Belenovskaya *et al.* (1980)
7-Sulphatoglucuronide	*Fuchsia excorticata* & *F. procumbens* leaf; Onagraceae	Williams *et al.* (1983)
Luteolin 7-methyl ether		
5-Glucoside	*Daphne sericea* aerial parts; Thymelaeaceae	Ulubelen *et al.* (1982b)
Luteolin 3′-methyl ether (chrysoeriol)		
7-Rhamnosylgalactoside	*Vepris heterophylla* leaf; Rutaceae	Gomes *et al.* (1983)
7-Glucuronosyl(1 → 2)glucuronide	*Elodea canadensis* leaf; Hydrocharitaceae	Mues (1983)
Luteolin 4′-methyl ether (diosmetin)		
7-(6″-Malonylglucoside)	*Bryum capillare*; Bryaceae	Stein *et al.* (1985)
Luteolin 7, 3′-dimethyl ether		
5-Rhamnoside	*Cassia nodosa* seed; Leguminosae	Tiwari and Sinha (1982)
5-Glucoside	*Daphne sericea* aerial parts; Thymelaeaceae	Ulubelen *et al.* (1982b)
6-Hydroxyluteolin		
5-Glucoside	*Salvia tomentosa* leaf; Labiatae	Ulubelen *et al.* (1981b)
7-Gentiobioside	*Lomatogonum carinthiacum* aerial parts; Gentianaceae	Schaufelberger and Hostettmann (1984)
7-(6″-Malonylglucoside)	*Bryum capillare*; Bryaceae	Stein *et al.* (1985)
6-Hydroxyluteolin 7-methyl ether		
7-Glucuronide } 7-Methylglucuronide }	*Digitalis lanata* leaf; Scrophulariaceae	Hiermann (1982)
6-Hydroxyluteolin 4′-methyl ether		
7-Allosyl(1 → 2)glucoside	*Sideritis hirsuta* Labiatae	Martin–Lomas *et al.* (1983)
8-Hydroxyluteolin (hypolaetin)		
8-Glucoside	*Sideritis mugronensis*; Labiatae	Villar *et al.* (1985)
7-Allosyl(1 → 2)glucoside	*Sideritis leucantha*; whole plant; Labiatae	Barberán *et al.* (1984)
8-Hydroxyluteolin 3′-methyl ether		
7-Allosyl(1 → 2)glucoside	*Sideritis leucantha*; Labiatae	Barberán and Tomas (1985)

Table 8.9 *(Contd.)*

Glycoside	*Source*	*References*
8-Hydroxyluteolin 4′-methyl ether		
7-[6‴-Acetylallosyl(1 → 2)glucoside]	*Stachys recta* leaf; Labiatae	Lenherr *et al.* (1984a)
8-Glucoside 8-Glucoside-3′-sulphate	*Althaea officinalis* leaf; Malvaceae	Gudej (1985)
5, 8, 3′, 4′-Tetrahydroxy-6, 7-dimethoxyflavone		
8-Glucoside	*Sideritis leucantha* whole plant; Labiatae	Barberán *et al.* (1984)
5, 7, 4′-Trihydroxy-6, 8, 3′-trimethoxyflavone (sudachitin)		
4′-Glucoside	*Citrus sudachi* green peel; Rutaceae	Horie *et al.* (1982)
7-(3-Hydroxy-3-methyl-glutarate)-4′-glucoside	*Citrus sudachi* peel; Rutaceae	Kumamoto *et al.* (1985)
Tricetin		
7-Diglucoside	*Lathyrus pratensis* leaf; Leguminosae	Reynaud *et al.* (1981)
3′, 5′-Diglucoside	*Dacrycarpus dacrydioides* foliage; Podocarpaceae	Markham and Whitehouse (1984)
Tricetin 3′, 5′-dimethyl ether (tricin)		
7-Rhamnosyl(1 → 2)galacturonide	*Saccharum* spp. mill syrup; Gramineae	Mabry *et al.* (1984)
7-Sulphatoglucuronide 7-Disulphatoglucuronide	*Cyperus polystachyos* leaf Cyperaceae	Harborne *et al.* (1982)
3-(3-Methylbutyl)tricetin		
5-Neohesperiodoside	*Launaea asplenifolia* Compositae	Gupta and Ahmed (1985)
3-C-Methylapigenin		
5-Rhamnoside	*Euglenia kurzii* aerial parts; Myrtaceae	Painuly and Tandon (1983)
5,7,4′-Trihydroxy-3′-C-methylflavone		
4′-Rhamnoside	*Cassia javanica* immature leaves; Leguminosae	Chakrabarty *et al.* (1984)

flowers of *Medicago arborea* (Leguminosae) (Torek *et al.*, 1983).

Rutin, quercetin 3-rhamnosyl(1 → 6)glucoside, is probably the most frequently occurring flavonol glycoside; its isomer quercetin 3-rhamnosyl(1 → 2)glucoside or neohesperidoside and other neohesperidosides are quite rare in nature. It is therefore interesting that three new neohesperidosides of rather unusual aglycones have been reported since 1980. These include the 7-neohesperidosides of gossypetin 3′-methyl ether and 8, 3′-dimethyl ether (Yuldashev *et al.*, 1985a, b) and the 3-neohesperidoside of 3, 7, 2′, 3′, 4′-pentahydroxyflavone (Ferguson and Lien, 1982). Rutin has, incidentally, recently been isolated from a fossil plant leaf of *Liriodendron* dating back to the Miocene era (Niklas *et al.*, 1985).

8.7 DISTRIBUTION PATTERNS

In this chapter, the emphasis has been on new plant glycosides and the many continuing reports in the literature of new sources of known glycosides have not been summarized here. However, the natural distribution of both flavone and flavonol glycosides will be discussed *inter alia* later in this book in Chapters 12–15. Their various occurrences in lower plants, gymnosperms and angiosperms are considered from a systematic viewpoint. Here,

Table 8.10 New flavonol glycosides

Glycoside	*Source*	*Reference*
3, 7, 3′, 4′-Tetrahydroxyflavone (fisetin)		
4′-Glucoside	*Phyllanthus nirurii*; Euphorbiaceae	Gupta and Ahmed (1984)
Kaempferol		
3-(6″-Ethylglucuronide)	*Tamarix nilotica* flowers; Tamaricaceae	Nawwar *et al.* (1984a)
5-Rhamnoside	*Callitris glauca* leaf; Cupressaceae	Ansari *et al.* (1981)
3-Xylosyl(1 → 2)galactoside	*Armoracia rusticana* leaf; Cruciferae	Larsen *et al.* (1982)
3-Apiosyl(1 → 2)glucoside	*Securidaca diversifolia* leaf; Polygalaceae	Hamburger *et al.* (1985)
3-Glucosyl(1 → 2)galactoside	*Lilium candidum* perianth; Liliaceae	Nagy *et al.* (1984)
3, 5-Digalactoside	*Indigofera hirsuta* leaf; Leguminosae	Rao *et al.* (1984)
3-α-L-Arabinofuranoside-7-α-L-rhamnopyranoside	*Cinnamomum sieboldii* leaf; Lauraceae	Nakano *et al.* (1983a)
3-Xyloside-7-glucoside	*Zinnia elegans*; Compositae	Dembinska-Migas *et al.* (1983)
3-Rhamnoside-7-glucoside	*Betula* spp. leaf; Betulaceae	Pawlowska (1980)
3-Galactoside-7-rhamnoside	*Vicia faba* epidermis; Leguminosae	Vierstra *et al.* (1982)
3-Glucoside-7-galactoside	*Asplenium bulbiferum* aerial parts; Aspleniaceae	Imperato (1985b)
3-Glucuronide-7-glucoside	*Euphorbia sanctae-catharinae*; Euphorbiaceae	Saleh (1985)
3, 4′-Dixyloside	*Pancratium maritimum* leaf; Amaryllidaceae	Ali *et al.* (1981)
7, 4′-Dirhamnoside	*Crotalaria verrucosa* flower; Leguminosae	Rao and Rao (1985)
4′-Rhamnosyl(1 → 2)[rhamnosyl (1 → 6)galactoside]	*Rhamnus nakaharai* fruits; Rhamnaceae	Lin *et al.* (1982)
3-β-D-Apiofuranosyl(1 → 2)-α-L-arabinofuranosyl-7-α-L-rhamnopyranoside 3-Glucosyl(1 → 3)-α-L-rhamnoside-7-α-L-rhamnoside	*Cinnamomum sieboldii* leaf; Lauraceae	Nakano *et al.* (1983a)
3-Neohesperidoside-7-glucoside	*Paris verticillata*; Trilliaceae	Nakano *et al.* (1981)
3-Gentiobioside-7-glucoside	*Tribulus pentandrus* leaf; Tribulaceae	Saleh *et al.* (1982b)
3-Rutinoside-4′-glucoside	*Geranium yemens* and *G. rotundifolium* whole plant; Geraniaceae	Saleh *et al.* (1983b)
3-Glucoside-7, 4′-dirhamnoside	*Coronilla emerus* flowers; Leguminosae	Harborne and Boardley (1983a)
3-Galactosyl(1 → 6)glucoside-7-dirhamnoside	*Melilotus alba* Leguminosae	Nicollier and Thompson (1982)
3-Sophorotrioside-7-glucoside	*Asplenium septentrionale*; Aspleniaceae	Imperato (1984a)
3-(6″-Malonylglucoside) 3-(6″-Malonylgalactoside)	*Ceterach officinarum* fronds; Filicales	Imperato (1981)
7-Galloylglucoside	*Acacia farnesiana*; Leguminosae	El-Negoumy and El-Ansari (1981)

Table 8.10 (*Contd.*)

Glycoside	*Source*	*Reference*
3-(2″, 4″-Di-*p*-coumarylglucoside)	*Quercus ilex*; Fagaceae	Romussi *et al.* (1983)
3-(2″, 6″-Di-*p*-coumarylglucoside)	*Quercus ilex*; Fagaceae	Romussi *et al.* (1984)
3-(2^G-*p*-Coumarylrutinoside)	*Castanea sativa*; Fagaceae	Romussi *et al.* (1981)
3-[2″-Acetyl-α-L-arabinopyranosyl(1→6)-β-D-galactopyranoside]	*Trillium tschonoskii* aerial parts; Liliaceae	Nakano *et al.* (1983b)
3-α-L-(3″-Acetylarabinofuranoside)-7-α-L-rhamnopyranoside	*Woodsia polystichoides* frond; Aspidiaceae	Murakami *et al.* (1984)
3-Glucoside-7-(*p*-coumarylglucoside) } 3-(Caffeylglucoside)-7-glucoside }	*Aconitum noveboracence* leaf; Ranunculaceae	Young and Sterner (1981)
3-Ferulylglucoside-7-glucoside	*Crambe fruticosa* leaf; Cruciferae }	Aguinagalde and del Pero Martinez (1982)
3-*p*-Coumarylglucoside-4′-glucoside	*Crambe tartaria* leaf; Cruciferae }	
3-(2-Hydroxypropionylglucoside)-4′-glucoside	*Sisymbrium galliesii* leaf; Cruciferae }	
3-*p*-Coumarylrutinoside-7-glucoside	*Aconitum columbianum* leaf; Ranunculaceae	Young and Sterner (1981)
3-*p*-Coumarylglucoside-7, 4′-diglucoside	*Crambe cordifolia*; leaf; Cruciferae	Aguinagalde and del Pero Martinez (1982)
3-(*p*-Coumarylferulyldiglucoside)-7-rhamnoside	*Crambe scaberrima* leaf; Cruciferae	Aguinagalde and del Pero Martinez (1982)
3-Gentiobioside-7-(caffeyl-arabinosylrhamnoside)	*Aconitum noveboracence* leaf; Ranunculaceae	Young and Sterner (1981)
3-β-(3″-Sulphatoglucoside)	*Cystopteris fragilis* fronds; Athyriaceae	Imperato (1983a)
3-β-(6″-Sulphatoglucoside)		
3-Sulphatorutinoside	*Adiantum capillusveneris*; Adiantaceae	Imperato (1982a)
3-Sulphate } 3, 7, 4′-Trisulphate }	*Acrotema uniflorum*; Dilleniaceae	Gurni *et al.* (1981)
Kaempferol 3-methyl ether		
7-Rhamnoside	*Carduus getulus*; Compositae	Abdel–Salam *et al.* (1982)
7-Glucoside	*Haplopappus foliosus* whole plant; Compositae	Ulubelen *et al.* (1982a)
Kaempferol 7-methyl ether		
3-Rhamnoside	*Loranthus europaeus*; Loranthaceae	Harvala *et al.* (1984)
3-Galactoside	*Anthyllis onobrychioides* aerial parts; Leguminosae	Marco *et al.* (1985)
5-Glucoside	*Podocarpus nivalis* foliage; Podocarpaceae	Markham *et al.* (1984)
Kaempferol 4′-methyl ether		
3-Galactoside	*Prosopis juliflora* bark; Leguminosae	Shukla and Misra (1981)
3-Glucoside-7-rhamnoside	*Asplenium bulbiferum* aerial parts; Aspleniaceae	Imperato (1984b)
3-Rhamnoside-7-glucoside } 3, 7-Diglucoside }	*Asplenium bulbiferum*; Aspleniaceae	Imperato (1984c)
Kaempferol 7, 4′-dimethyl ether		
3-(6″-(*E*)-*p*-Coumarylglucoside)	*Phlomis spectabilis*; Labiatae	Kumar *et al.* (1985)
6-Hydroxykaempferol 3-methyl ether		
7-Sulphate	*Neurolaena lobata* leaf; Compositae	Kerr *et al.* (1981)

(*Contd.*)

Table 8.10 (*Contd.*)

Glycoside	*Source*	*Reference*
6-Hydroxykaempferol 6-methyl ether (eupafolin)		
3-Robinobioside	*Brickellia arguta* whole plant; Compositae	Rosler *et al.* (1984)
6-Hydroxykaempferol 6, 7-dimethyl ether (eupalitin)		
3-Galactosylrhamnoside	*Ipomopsis polycladon* whole plant; Polemoniaceae	Kiredjian and Smith (1985)
3-Glucosylgalactoside	*Brickellia monocephala* aerial parts; Compositae	Norris and Mabry (1985)
6-Hydroxykaempferol 6, 7, 4′-trimethyl ether		
3-Glucoside	*Sesuvium portulacastrum* whole plant; Aizoaceae	Khajuria *et al.* (1982)
8-Hydroxykaempferol (herbacetin)		
7-Rhamnoside	*Rhodiola rosea* rhizomes; Crassulaceae	Zapesochnaya and Kurkin (1983)
8-α-L-Arabinopyranoside	*Rhodiola algida* roots; Crassulaceae	Pangarova and Zapesochnaya (1975)
7-Glucosyl(1 → 3)rhamnoside	*Rhodiola rosea* rhizomes; Crassulaceae	Zapesochnaya and Kurkin (1983)
3-Glucoside-8-xyloside 7-Rhamnoside-8-glucoside	*Rhodiola rosea* aerial parts; Crassulaceae	Kurkin *et al.* (1984)
8, 4′-Dixyloside	*Rhodiola krylovii*; Crassulaceae	Krasnov and Demidenki (1984)
3-Sophoroside-8-glucoside	*Equisetum hyemale*; Equisetaceae	Geiger *et al.* (1982)
7-(6″-Quinylglucoside)	*Ephedra alata* whole plant; Ephedraceae	Nawwar *et al.* (1984b)
8-(3″-Acetyl-α-L-arabinopyranoside) 8-(2″, 3″-Diacetylxyloside) 8-(2″, 3″, 4″-Triacetylxyloside)	*Rhodiola algida* roots; Crassulaceae	Pangarova and Zapesochnaya (1975)
Herbacetin 8-methyl ether (sexangularetin)		
3-Galactoside	*Dryas octopeta*; Rosaceae	Servettaz *et al.* (1984)
3-Glucoside-7-rhamnoside	*Gossypium hirsutum* immature flower buds; Malvaceae	Elliger (1984)
3-Glucoside-7-rutinoside	*Ephedra alata* whole plant; Ephedraceae	Nawwar *et al.* (1984b)
Quercetin		
3-Alloside	*Wagneriopteris formosa* frond; Filicinae	Murakami *et al.* (1984)
3-(6″-Methylglucuronide) 3-(6″-Ethylglucuronide)	*Tamarix nilotica*; Tamaricaceae	Nawwar *et al.* (1984a)
7-Xyloside	*Adansonia digitata* stem; Bombacaceae	Chauhan *et al.* (1982b)
3-Arabinosylxyloside	*Quercus rubra* leaf; Fagaceae	McDougal and Parks (1984)
3-Rhamnosyl(1 → 2)arabinoside	*Brassica nigra* seeds; Cruciferae	Geiger *et al.* (1983)
3-Xylosyl(1 → 2)galactoside	*Armoracia rusticana* leaf; Cruciferae	Larsen *et al.* (1982)
3-Apiofuranosyl(1 → 2)arabinoside 3-Apiofuranosyl(1 → 2)xyloside 3-Apiofuranosyl(1 → 2)glucoside 3-Apiofuranosyl(1 → 2)galactoside	*Securidaca diversifolia* leaf; Polygalaceae	Hamburger *et al.* (1985)
3-Galactosyl(1 → 4)rhamnoside	*Anogeissus latifolia* root; Combretaceae	Chatuveda and Saxena (1985)

Table 8.10 (*Contd.*)

Glycoside	*Source*	*Reference*
3-Sambubioside	*Hibiscus mutabilis* var. *versicolor* flowers; Malvaceae	Ishikura (1982)
3-Glucosyl(1 → 2)galactoside	*Corylus avellana* pollen; Betulaceae	Strack *et al.* (1984)
3-Glucosyl(1 → 6)galactoside	*Astrantia major*; Umbelliferae	Hiller *et al.* (1984)
7-Glucosylrhamnoside	*Capparis spinosa*; Capparaceae	Artmieva *et al.* (1981)
3-Glucuronide-7-glucoside	*Euphorbia sanctaecatharinae*; Euphorbiaceae	Saleh (1985)
3-Glucosyl(1 → 4)galactosylrhamnoside	*Cheilanthes fragrans* frond; Sinopteridaceae	Imperato (1985a)
3-Rutinoside-7-rhamnoside	*Lepidium syraschicum* epigeal parts; Cruciferae	Zaitsev and Fursa (1984)
3-Xylosyl(1 → 2)rhamnoside-4′-rhamnoside	*Ziziphus spina-christi* leaf; Rhamnaceae	Nawwar *et al.* (1984c)
3-Rutinoside-4′-glucoside	*Erodium pulverulentum* *E. malacoides* whole plant; Geraniaceae	Saleh *et al.* (1983b)
3-(6″-Caffeylgalactoside)	*Hydrocotyle sibthorpioides* whole plant; Umbelliferae	Shigematsu *et al.* (1982)
3-(2″-Galloylrhamnoside)	*Polygonum filiforme*; Polygonaceae	Isobe *et al.* (1981)
3-(6″-*p*-Hydroxybenzoylgalactoside)	*Ledum palustre*; Ericaceae	Zapesochnaya and Pangarova (1980)
3-(6″-Malonylglucoside)	*Salicornia europaea*; Chenopodiaceae	Geslin and Verbist (1985)
3-(6″-Malonylgalactoside)	*Adiantum capillusveneris* frond; Adiantaceae	Imperato (1982b)
3-(3″-Acetylgalactoside) 3-(6″-Acetylgalactoside)	*Ledum palustre*; Ericaceae	Zapesochnaya and Pangarova (1980)
3-Ferulylglucoside-7-glucoside	*Crambe fruticosa* leaf; Cruciferae	Aguinagalde and del Pero Martinez (1982)
3-Ferulylglucoside-4′-glucoside	*Crambe tataria* leaf; Cruciferae	
3-Malonylglucoside-4′-glucoside	*Sisymbrium gilliesii* leaf; Cruciferae	
3-Ferulylglucoside-7, 4′-diglucoside	*Crambe cordifolia* leaf; Cruciferae	
3-(3‴-Benzoylsophoroside)-7-rhamnoside 3-(x″-Benzoyl-x″-xylosylglucoside) 3-(x″ or x‴-Benzoyl-x″-glucosylglucoside)	*Delphinium carolinianum* leaf; Ranunculaceae	Warnock *et al.* (1983)
3-(3″-Sulphatoglucoside)	*Asplenium septentrionale* frond; Aspleniaceae	Imperato (1983b)
3, 7, 3′-Trisulphate	*Flaveria bidentis*; Compositae	Cabrera *et al.* (1985)
Quercetin 3-methyl ether		
7-Rhamnoside 3′-Xyloside 4′-Glucoside	*Dacrycarpus dacrydioides* foliage; Podocarpaceae	Markham and Whitehouse (1984)
	Neochilenia spp. tepals; Cactaceae	Iwashina *et al.* (1984)
7-Rhamnosyl-3′-xyloside	*Dacrycarpus dacrydioides* foliage; Podocarpaceae	Markham and Whitehouse (1984)
Quercetin 7-methyl ether (rhamnetin)		
5-Glucoside	*Podocarpus nivalis* foliage; Podocarpaceae	Markham *et al.* (1984)
3-Neohesperidoside	*Cassia occidentalis* pods; Leguminosae	Singh and Singh (1985b)

(*Contd.*)

Table 8.10 (*Contd.*)

Glycoside	*Source*	*Reference*
3-Galactosyl(1 → 4)galactoside 3-Galactosyl(1 → 6)galactoside	*Cassia laevigata* flowers; Leguminosae	Singh (1982)
3-Galactoside-3′-rhamnoside	*Coprosma* sp. leaf; Rubiaceae	Wilson (1984)
3-Rhamnosyl(1 → 3)-4″-acetyl-rhamnosyl(1 → 6)galactoside	*Rhamnus saxatilis* ssp. *saxatilis*; Rhamnaceae	Riess-Maurer and Wagner (1982)
Quercetin 3′-methyl ether (isorhamnetin)		
7-Rhamnoside	*Sedum caucasicum*; Crassulaceae	Zaitsev *et al.* (1983)
7-Glucoside	*Coprosma* spp. leaf; Rubiaceae	Wilson (1984)
3-Neohesperidoside	*Parietaria officinalis* leaf and flower; Urticaceae	Budzianowski *et al.* (1985)
3-Robinobioside	*Gomphrena martiana*; Amaranthaceae	Buschi and Pomilio (1982)
3-Gentiobioside-7-glucoside	*Tribulus terrestris* leaf; Tribulaceae	Saleh *et al.* (1982b)
3-Rutinoside-4′-rhamnoside	*Cucurbita pepo* flowers; Cucurbitaceae	Itokawa *et al.* (1981b)
3-*p*-Coumarylglucoside	*Tribulus pentandrus* and *T. terrestris* leaf; Tribulaceae	Saleh *et al.* (1982b)
3, 7, 4′-Trisulphate	*Acrotema uniflorum*; Dilleniaceae	Gurni *et al.* (1981)
Quercetin 4′-methyl ether (tamarixetin)		
3-Rhamnoside	*Flemingia stricta* leaf; Leguminosae	Rao, C.P. *et al.* (1983)
3-Digalactoside	*Thevetia neriifolia* flowers; Apocynaceae	Gunasegan and Nair (1981)
Quercetin 3, 3′-dimethyl ether		
7-Glucoside	*Desmanthodium perfoliatum* whole plant; Compositae	Bohm and Stuessy (1981)
4′-Glucoside	*Typha latifolia* leaf; Typhaceae	Woo *et al.* (1983)
Quercetin 7, 3′-dimethyl ether (rhamnazin)		
4′-Glucoside	*Hibiscus furcatus*; Malvaceae	Nair *et al.* (1981)
3-Rutinoside	*Cucurbita pepo* flowers; Cucurbitaceae	Itokawa *et al.* (1981b)
Quercetin 7, 4′-dimethyl ether (ombuin)		
3-Rutinoside-5-glucoside	*Erythroxylon argentinum* aerial parts; Erythroxylaceae	Inigo and Pomilio (1985)
Quercetin 7, 3′, 4′-trimethyl ether		
3-Arabinoside	*Cassia spectabilis* seeds; Leguminosae	Singh and Singh (1985a)
Quercetin 5, 7, 3′, 4′-tetramethyl ether		
3-Arabinoside	*Psophocarpus tetragonolobus* seeds; Leguminosae	Lachman *et al.* (1982)
Quercetagetin 6-methyl ether (patuletin)		
5-Glucoside	*Linanthus* spp. leaf; Polemoniaceae	Smith *et al.* (1982)
3-Robinobioside	*Brickellia arguta* whole plant; Compositae	Rosler *et al.* (1984)

Table 8.10 (*Contd.*)

Glycoside	*Source*	*Reference*
3-Galactosylrhamnoside	*Ipomopsis polycladon* whole plant; Polemoniaceae	Kiredjian and Smith (1985)
3-Digalactoside	*Brickellia cylindracea* leaf; Compositae	Timmermann and Mabry (1983)
3-Digalactosylrhamnoside	*Ipomopsis polycladon* whole plant; Polemoniaceae	Kiredjian and Smith (1985)
Quercetagetin 4′-methyl ether		
3-Arabinoside	*Stevia nepetifolia* leaf; Compositae	Rajbhandari and Roberts (1984)
Quercetagetin 3, 6-dimethyl ether		
4′-Glucuronide	*Spinacia oleracea*; Chenopodiaceae	Wagner *et al.* (1977)
Quercetagetin 6, 7-dimethyl ether (eupatolitin)		
3-Glucosylgalactoside	*Brickellia monocephala* aerial parts; Compositae	Norris and Mabry (1985)
Quercetagetin 3, 3′-dimethyl ether		
7-Glucoside	*Clibadium terebinthinaceum*; Compositae	Bohm and Stuessy (1985)
Quercetagetin 3, 6, 3′-trimethyl ether (jaceidin)		
7-Neohesperidoside	*Cassia occidentalis* pods; Leguminosae	Singh and Singh (1985b)
4′-Sulphate	*Brickellia glutinosa* aerial parts; Compositae	Goodwin *et al.* (1984)
Quercetagetin 3, 6, 4′-trimethyl ether		
7-Glucoside (centaurein)	*Centaurea juncea*; Compositae	Farkas *et al.* (1964)
Quercetagetin 6, 3′, 4′-trimethyl ether		
3-Sulphate	*Decachaeta haenkeana* aerial parts; Compositae	Miski *et al.* (1985)
8-Hydroxyquercetin (gossypetin)		
7-Rhamnoside	*Rhodiola rosea* aerial parts; Crassulaceae	Kurkin *et al.* (1984)
8-Rhamnoside	*Gossypium arboreum* flowers; Malvaceae	Waage and Hedin (1984)
8-Glucuronide	*Hibiscus vitifolius* petal; Malvaceae	Nair and Subramanian (1974)
7-Rhamnoside-8-glucoside	*Rhodiola rosea* aerial parts; Crassulaceae	Kurkin *et al.* (1984)
3-Sophoroside-8-glucoside	*Equisetum hyemale*; Equisetaceae	Geiger *et al.* (1982)
8-Glucuronide-3-sulphate	*Malva sylvestris* leaf; Malvaceae	Nawwar and Buddrus (1981)
Gossypetin 7-methyl ether		
3-Rutinoside	*Ruta graveolens* flowers; Rutaceae	Harborne and Boardley (1983b)
Gossypetin 8-methyl ether		
3-α-L-Arabinofuranoside	*Dryas octopetala*; Rosaceae	Servettaz *et al.* (1984)
3-Glucoside	*Humata pectinata* whole plant; Davalliaceae	Wu and Furukawa (1983)
	Lotus corniculatus stem/leaf; Leguminosae	Reynaud *et al.* (1982)

(*Contd.*)

Table 8.10 (*Contd.*)

Glycoside	Source	Reference
Gossypetin 3′-methyl ether		
7-Glucoside	*Haplophyllum* spp. aerial parts; Rutaceae	Batirov *et al.* (1981)
7-Neohesperidoside (haploside-F)	*Haplophyllum perforatum* aerial parts; Rutaceae	Yuldashev *et al.* (1985a)
7-(6″-Acetylrhamnosyl(1 → 2)glucoside)	*Haplophyllum* spp. aerial parts; Rutaceae	Batirov *et al.* (1981)
Gossypetin 8, 3′-dimethyl ether (limocitrin)		
3-Rhamnoside	*Citrus unshiu* peel; Rutaceae	Matsubara *et al.* (1985)
3-Glucoside	*Haplophyllum pedicellatum*; Rutaceae	Ulubelen *et al.* (1984)
7-Neohesperidoside } 7-(6″-Acetylglucoside) }	*Haplophyllum perforatum*; Rutaceae	Yuldashev *et al.* (1985b)
Myricetin		
3′-Xyloside	*Ledum palustre*; Ericaceae	Zapesochnaya and Pangarova (1980)
3-Dirhamnoside	*Azara microphylla*; Flacourtaceae	Sagareishvili *et al.* (1983)
3-Rhamnoside-7-glucoside	*Quercus rubra* leaf; Fagaceae	McDougal and Parks (1984)
3-Glucosylrutinoside	*Lithophragma* spp.; Saxifragaceae	Nicholls and Bohm (1984)
3-(2″-Galloylrhamnoside) } 3-(4″-Galloylrhamnoside) }	*Desmanthus illinoenis* Leguminosae	Nicollier and Thompson (1983)
Myricetin 3-methyl ether		
7-Rhamnoside	*Dacrycarpus dacrydioides* foliage; Podocarpaceae	Markham and Whitehouse (1984)
3′-Xyloside	*Ledum palustre*; Ericaceae	Zapesochnaya and Pangarova (1980)
3-Rhamnoside-3′-xyloside	*Dacrycarpus dacrydioides* foliage; Podocarpaceae	Markham and Whitehouse (1984)
Myricetin 3′-methyl ether (larycitrin)		
7-Glucoside	*Limonium sinuatum*; Plumbaginaceae	Ross (1984)
3, 5′-Diglucoside } 3, 7, 5′-Triglucoside }	*Medicago arborea* flowers; Leguminosae	Torek *et al.* (1983)
Myricetin 3, 4′-dimethyl ether		
3′-Xyloside } 7-Rhamnoside-3′-xyloside }	*Dacrycarpus dacrydioides* foliage; Podocarpaceae	Markham and Whitehouse (1984)
Myricetin 7, 4′-dimethyl ether		
3-Galactoside	*Rhus lancea* leaf; Anacardiaceae	Nair *et al.* (1983)
Myricetin 3′, 5′-dimethyl ether (syringetin)		
3-Arabinoside	*Chondropetalum* spp. stem; Restionaceae	Harborne *et al.* (1985)
3, 5, 7, 2′-Tetrahydroxyflavone (datiscetin)		
3-Glucoside	*Datisca cannabina*; Datiscaceae	Zapesochnaya *et al.* (1982)
3, 7, 2′, 3′, 4′-Pentahydroxyflavone		
3-Neohesperidoside	*Jacaranda acutifolia* bark; Bignoniaceae	Ferguson and Lien (1982)

Table 8.10 (*Contd.*)

Glycoside	*Source*	*Reference*
5, 6, 5′-Trihydroxy-3, 7, 4′-trimethoxyflavone		
5′-Glucoside	*Chrysosplenium americanum*; Saxifragaceae	Collins *et al.* (1981)
5, 5′-Dihydroxy-3, 6, 7, 4′-tetramethoxyflavone		
5′-Glucoside	*Chrysosplenium americanum*; Saxifragaceae	Collins *et al.* (1981)
5, 8, 4′-Trihydroxy-3, 7, 3′-trimethoxyflavone		
8-Acetate	*Notholaena ashenborniana* frond exudate; Filicales	Jay *et al.* (1982)
5, 8-Dihydroxy-3, 7, 2′, 3′, 4′-pentamethoxyflavone		
8-Acetate	*Notholaena ashenborniana* frond exudate; Filicales	Jay *et al.* (1982)
5, 2′, 5′-Trihydroxy-3, 6, 7, 4′-tetramethoxyflavone		
5′-Glucoside	*Chrysosplenium americanum*; Saxifragaceae	Collins *et al.* (1981)
5, 5′-Dihydroxy-3, 6, 7, 2′, 4′-pentamethoxyflavone		
5′-Glucoside	*Chrysoplenium americanum*; Saxifragaceae	Collins *et al.* (1981)

it is appropriate to consider three more general aspects of their natural distribution: their concentrations in plant tissues, their subcellular localization and their differential distribution within the plant.

With the wide availability today of HPLC equipment, it is possible to determine for the first time with a high degree of accuracy the actual concentrations of plant glycosides in living tissues. Unfortunately, the opportunity has not been taken to any extent to make such measurements, except for a few medicinal or ornamental plants. It is to be hoped that more efforts will be made to determine (a) the total concentration of flavonoid glycosides and (b) the percentage of the different glycosides relative to that total concentration. Two recent HPLC studies indicate the potentialities of such measurements in chemotaxonomic investigation. In *Chondropetalum*, qualitative analyses of the flavonol glycosides present (see Table 8.1) in a group of seven species indicated that four were very similar in glycosidic type. However, quantitative measurements via HPLC indicated clear differences between all species. In particular, the two otherwise similar species *C. nudum* and *C. rectum* were well separated by large amounts of larycitrin 3-galactoside in the former (47% as against 23%) and by large amounts of syringetin 3-galactoside in the latter (31% as against 12%) (Harborne *et al.*, 1985). A comparable study of flavonoid glycosides and iridoids in a group of ten labiate species of the *Stachys erecta* complex indicated that almost all taxa could be distinguished by the quantitative data produced by automatic HPLC sampling of the plant extracts (Lenherr *et al.*, 1985b).

Until recently, the main interest in the subcellular localization of flavone and flavonol glycosides lay in the possibility that these vacuolar constituents might also occur in chloroplasts. A variety of experiments in the 1970s indicated that it was possible to isolate rutin and other glycosides from chloroplast preparations (for review, see McClure, 1975). However, the flavonoids detected could have come from the vacuole as a result of irreversible adsorption on the outer chloroplast membrane during the organelle preparation. Charriere-Ladreik and Tissut (1981) were able to prepare spinach chloroplasts essentially free of flavonoids, in spite of earlier reports of flavonoids being present in them. It now appears, therefore, that leaf flavonoids are not present in chloroplasts; the considerable water solubility of most of these glycosides has always argued for their presence exclusively in the vacuoles.

Nevertheless, it is conceivable that within the growing leaf, there will be distinctive gradients within the cells in the vacuolar flavonoid content. A variety of investigations, mainly of legume or cereal plants, now indicate that there are considerable differences in the localization of these substances. Indeed, in three legume species investigated (Table 8.11), flavonol glycosides have been found to be restricted almost entirely to the vacuoles of epidermal cells. In *Vicia faba*, for example, the guard cells of the upper epidermis contain no less than 59% of the total leaf flavonoids, while the lower epidermal cells contain only 7% (Weissenbock *et al.*, 1984). In the case of grasses (McClure, 1986), both epidermal and mesophyll cells contain flavonoids, but the compounds present vary from

Table 8.11 Differential compartmentalization of flavonoid glycosides in plant leaves

Plant	*Epidermal constituents*	*Mesophyll constituents*
GRAMINEAE		
Hordeum vulgare	Isovitexin 7-glucoside	Iso-orientin 7-glucoside
Avena sativa	Glycosylflavones	Glycosylflavones and kaempferol glycosides
Secale cereale	Glycosylflavones	Luteolin glucuronides and cyanidin glycosides
LEGUMINOSAE		
Pisum sativum	Kaempferol and quercetin *p*-coumaryltriglucosides and anthocyanins	None
Glycine max	Flavonol glycosides	None
Vicia faba	Kaempferol 3,7-diglycosides*	None

*Also present in guard cells of the upper epidermis.

species to species. The greatest difference in cell content is in the rye plant, since glycosylflavones are restricted to the epidermal cells, whereas the luteolin glucuronides are only present in the mesophyll (Schulz *et al.*, 1985). The physiological significance of these different patterns is under active investigation (McClure, 1986).

In petals, too, flavonoid glycosides may be restricted to the vacuoles of epidermal cells. Indeed, a survey of the petals of 201 species from 60 families showed that these glycosides were specifically located in the epidermis in the great majority of cases (Kay *et al.*, 1981). This is related to their UV-absorbing properties in white and yellow petals and to their role as co-pigments in cyanic petals. In flowers with UV-patterning, the glycosides may be further restricted in their distribution to the inner portions of the petals. For example, in *Potentilla* UV-patterning is determined by specific location in epidermal cells of either quercetin glycosides (the 3-glucoside, the 3-glucuronide and 3′-glucoside) or of the chalcone glucoside, isosalipurposide (Harborne and Nash, 1984). In *Coronilla valentina*, both yellow visible coloration and UV patterning are provided by the 3-rutinosides of two gossypetin methyl ethers (Harborne, 1981) while in *Coronilla emerus*, colourless kaempferol and quercetin glycosides provide the UV patterning (Harborne and Boardley, 1983a).

Finally, there is the interesting question of organ specificity of flavonoid glycosides. In Tables 8.9 and 8.10, the plant sources of new glycosides are included, together with the part of plant examined. It is not always clear from the literature which parts have been examined. Even when this is stated, authors rarely indicate whether the same compound occurs or is lacking in other parts of the same plant. This is unfortunate, since the differential distribution of these glycosides within the various organs is always of interest, particularly with regard to their possible function. Organ specificity is obviously related to tissue distribution of the enzymes of biosynthesis and this is an area of present investigation (see Chapter 11). Transport within the plant is always possible, although flavone and flavonol glycosides are usually considered to be formed *in situ*, close to the vacuolar membrane.

At present, it is difficult to make much sense of distribution patterns within the plant, where these have been determined. There are so many variations from plant to plant. In *Crocus*, for example, kaempferol 3-sophoroside occurs equally in petal and in leaf, whereas the 3-rutinoside-7-glucoside of the same flavonol is restricted to the petal (Harborne and Williams, 1984). Differences in the flavonol glycosides of leaf and fruit are also often pronounced (see e.g. Harborne *et al.*, 1986) but these differences are difficult to rationalize in terms of function. Determination of the raison d'être of flavone and flavonol glycosides requires detailed information regarding their localization within the plant. Until we have adequate information on this topic, it will be that much more difficult to establish their value to the plant.

REFERENCES

Abdel-Salam, N.A., Mahmoud, Z.F., Abdel-Hamid, R. and Khafagy, S.M. (1982), *Egypt J. Pharm. Sci.* **23**, 199.
Aguinagalde, I, and del Pero Martinez, M.A. (1982), *Phytochemistry* **21**, 2875.
Alania, M.D. (1983), *Khim. Prir. Soedin* **19**, 645.
Ali, A.A., El-Moghazy, A.M., Ross, S.A. and Shanawary, M.A. (1981), *Fitoterapia* **52**, 209.
Ansari, F.R., Ansari, W.H., Rahman, W., Okigawa, M. and Kawano, N. (1981), *Indian J. Chem.* **20B**, 724.
Artmieva, M.V., Karryer, M.O., Meshcheryatov, A.A. and Gordienko, V.P. (1981), *Isv. Akad. Nauk. Turkin. Khim. Geol. Nauk* (3), 123.
Barberán, F.A.T. and Tomás, F. (1985), *Rev. Latinoam Quim.* **16**, 47.
Barberán, F.A.T., Tomás, F. and Ferreres, F. (1984), *Phytochemistry* **23**, 2112.
Barberán, F.A.T., Tomás, F. and Ferreres, F. (1985), *J. Nat. Prod.* **48**, 28.

Batirov, E.K., Malikov, V.M. and Mirzamatov, R.T. (1981), *Khim. Prir. Soedin.* **17**, 836.

Bekirov, E.P., Nasudari, A.A., Popova, T.P. and Litvinenko, V.J. (1974), *Khim. Prir. Soedin.* **10**, 663.

Belenovskaya, L.M., Markova, L.P. and Kapranova, G.I. (1980), *Khim. Prir. Soedin.* **16**, 835.

Bernardi, M., Uberti, E. and Vidari, G. (1984), *J. Chromatogr.* **248**, 269.

Bhandari, P., Tandon, S. and Rastogi, R.P. (1981), *Plant. Med.* **41**, 407.

Bohm, B.A. and Bhat, U.G. (1985), *Biochem. Syst. Ecol.* **13**, 437.

Bohm, B.A. and Stuessy, T.F. (1981), *Phytochemistry* **20**, 1573.

Bohm, B.A. and Stuessy, T.F. (1985), *Phytochemistry* **24**, 2134.

Budzianowski, J., Skrzypczak, I. and Walkowiak, D. (1985), *J. Nat. Prod.* **48**, 336.

Buschi, C.A. and Pomilio, A.B. (1982), *J. Nat. Prod.* **45**, 557.

Cabrera, J.L., Juliani, H.R. and Gross, E.G. (1985), *Phytochemistry* **24**, 1394.

Chakrabarty, K., Chawla, H.M. and Rastogi, D.K. (1984), *Indian J. Chem.* **23B**, 543.

Chari, V.M., Jordan, M., Wagner, H. and Thies, P.W. (1977), *Phytochemistry* **16**, 1110.

Charriere-Ladreix, Y. and Tissut, M. (1981), *Planta* **151**, 309.

Chatterjee, A., Sarker, S. and Saha, S.K. (1981), *Phytochemistry* **20**, 1760.

Chatuveda, S.K. and Saxena, V.K. (1985), *J. Indian Chem. Soc.* **62**, 76.

Chauhan, J.S., Kumar, S. and Chaturvedi, R. (1982a), *Curr. Sci.* **51**, 1069.

Chauhan, J.S., Chaturvedi, R. and Kumar, S. (1982b), *Indian J. Chem.* **21B**, 254.

Chauhan, J.S., Kumar, S. and Chaturvedi, R. (1984), *Phytochemistry* **23**, 2404.

Collins, F.W., Bohm, B.A. and Wilkins, C.K. (1975), *Phytochemistry* **14**, 1099.

Collins, F.W., De Luca, V., Ibrahim, R.K., Voirin, B. and Jay, M. (1981), *Z. Naturforsch.* **36c**, 730.

Daigle, D.J. and Conkerton, E.J. (1983), *J. Liq. Chromatogr.* **6**, 105.

Dembinska-Migas, W., Gill, S., Handke, E. and Worobiec, S. (1983), *Herb Pol.* **29**, 197.

De Rosa, S. and De Stefano, S. (1983), *Phytochemistry* **22**, 2323.

Domon, B. and Hostettmann, K. (1985), *Phytochemistry* **24**, 575.

Doyle, J.J. (1983), *Am. J. Bot.* **70**, 1085.

El Ansari, M.A., Nawwar, M.A.M., El Dein, A., El Sherbeiny, A. and El Sissi, H.T. (1976), *Phytochemistry* **15**, 231.

Elliger, C.A. (1984), *Phytochemistry* **23**, 1199.

El-Negoumy, S.L. and El Ansari, M.A. (1981), *Egypt. J. Chem.* **24**, 471.

Farkas, L., Hörhammer, L., Wagner, H., Rosler, H. and Gurniak, R. (1964), *Chem. Ber.* **97**, 1666.

Ferguson, N.M. and Lien, E.J. (1982), *J. Nat. Prod.* **45**, 523.

Ferreres, F. and Tomás, F. (1980), *An. Quim. Ser. C* **76**, 92.

Frolova, V.I. and Dzhumyrko, S.F. (1984), *Khim. Prir. Soedin.* (5), 655.

Geiger, H. and Schwinger, G. (1980), *Phytochemistry* **19**, 897.

Geiger, H., Reichert, S. and Markham, K.R. (1982), *Z. Naturforsch.* **37b**, 504.

Geiger, H., Maier, H. and Markham, K.R. (1983), *Z. Naturforsch.* **38c**, 490.

Gella, E.V., Makarova, G.V. and Borisyuk, Y.G. (1967), *Farmatsert. Zh. (Kiev)* **22**, 80.

Geslin, M. and Verbist, J.F. (1985), *J. Nat. Prod.* **48**, 111.

Gomes, E., Dellamonica, G., Gleye, J., Moulis, Cl., Chopin, J. and Stanisclas, E. (1983), *Phytochemistry* **22**, 2628.

Goodwin, R.S., Rosler, R.H.A., Mabry, T.J. and Varma, S.D. (1984), *J. Nat. Prod.* **47**, 711.

Gudej, J. (1985), *Acta Pol. Pharm.* **42**, 215.

Gunasegan, R. and Nair, A.G.R. (1981), *Indian J. Chem.* **20B**, 832.

Gupta, D.R. and Ahmed, B. (1984), *Shoyakugaku Zasshi* **38**, 213.

Gupta, D.R. and Ahmed, B. (1985), *Phytochemistry* **24**, 873.

Gupta, D.R., Ahmed, B. and Dhiman, R.P. (1984), *Shoyakugaku Zasshi* **38**, 341.

Gupta, H.M. and Saxena, V.K. (1984), *Natl. Acad. Sci. Lett. (India)* **7**, 187.

Gupta, K.R. and Niranjari, G.S. (1982), *Plant. Med.* **46**, 240.

Gurni, A.A., König, W.A. and Kubitzki, K. (1981), *Phytochemistry* **20**, 1057.

Hamburger, M., Gupta, M. and Hostettmann, K. (1985), *Phytochemistry* **24**, 2689.

Harborne, J.B. (1971). In *The Biology and Chemistry of the Umbelliferae* (ed. V.H. Heywood), Academic Press, London, pp. 293–314.

Harborne, J.B. (1981), *Phytochemistry* **20**, 1117.

Harborne, J.B. (1983). In *Chromatography, Part B. Applications, Journal of Chromatography Library Series*, Vol. 22B (ed. E. Heftmann), Elsevier, Amsterdam pp. 407–434.

Harborne, J.B. (1984). In *Phytochemical Methods*, 2nd edn., Chapman and Hall, London, pp. 222–242.

Harborne, J.B. (1985). In *Advances in Medicinal Plant Research* (eds Vlietinck, A.J. and Dommisse, R.A.), Wissenschaftliche Verlagsgesellschaft, Stuttgart, pp. 135–151.

Harborne, J.B. and Boardley, M. (1983a), *Phytochemistry* **22**, 622.

Harborne, J.B. and Boardley, M. (1983b), *Z. Naturforsch.* **38c**, 148.

Harborne, J.B. and Boardley, M. (1984), *J. Chromatogr.* **299**, 377.

Harborne, J.B. and Nash, R.J. (1984), *Biochem. Syst. Ecol.* **12**, 315.

Harborne, J.B. and Williams, C.A. (1975). In *The Flavonoids* (eds J.B. Harborne, T.J. Mabry and H. Mabry), Chapman and Hall, London, pp. 376–441.

Harborne, J.B. and Williams, C.A. (1982). In *The Flavonoids – Advances in Researach* (eds J.B. Harborne and T.J. Mabry), Chapman and Hall, London, pp. 261–311.

Harborne, J.B. and Williams, C.A. (1984), *Z. Naturforsch.*, **39c**, 18.

Harborne, J.B., Williams, C.A. and Wilson, K.L. (1982), *Phytochemistry* **21**, 2491.

Harborne, J.B., Boardley, M. and Linder, P. (1985), *Phytochemistry* **24**, 273.

Harborne, J.B., Heywood, V.H. and Chen, X.Y. (1986), *Biochem. Syst. Ecol.* **14**, 81.

Harvala, E., Exner, J. and Becker, H. (1984), *J. Nat. Prod.* **47**, 1054.

Hashim, F.H., El-Shamy, A.M. and Shehata, A.H. (1980), *Bull. Fac. Pharm. (Cairo Univ.)* **19**, 131.

Haufler, C.H. and Giannasi, D.E. (1982), *Biochem. Syst. Ecol.* **10**, 107.

Hiermann, A. (1982), *Plant. Med.* **45**, 59.

Hiller, K., Jähnert, W. and Habisch, D. (1984), *Die Pharmazie* **39**, 51.

Hiraoka, A. and Maeda, M. (1979), *Chem. Pharm. Bull.* **27**, 3130.

Horie, T., Tsukayama, M. and Nakayama, M. (1982), *Bull. Chem. Soc. Jpn* **55**, 2928.
Hostettmann, K. (1985), *Ann. Proc. Phytochem. Soc. Eur.* **25**, 1.
Hostettmann, K., Domon, B., Schaufelberger, D. and Hostettmann, M. (1984), *J. Chromatogr.* **283**, 137.
Imperato, F. (1981), *Chem. Ind. (London)* 696.
Imperato, F. (1982a), *Chem. Ind. (London)*, 957.
Imperato, F. (1982b), *Chem. Ind. (London)*, 604.
Imperato, F. (1983a), *Chem. Ind. (London)* 204.
Imperato, F. (1983b), *Chem. Ind. (London)* 390.
Imperato, F. (1984a), *Am. Fern J.* **74**, 14.
Imperato, F. (1984b), *Chem. Ind. (London)* 667.
Imperato, F. (1984c), *Chem. Ind. (London)* 186.
Imperato, F. (1985a), *Chem. Ind. (London)* 709.
Imperato, F. (1985b), *Phytochemistry* **24**, 2136.
Inigo, R.P.A. and Pomilio, A.B. (1985), *Phytochemistry* **24**, 347.
Ishikura, N. (1981), *Agric. Biol. Chem.* **45**, 1855.
Iskikura, N. (1982), *Agric. Biol. Chem.* **46**, 1705.
Isobe, T., Kanazawa, K., Fujimura, M. and Noda, Y. (1981), *Bull. Chem. Soc. Jpn* **54**, 3249.
Itokawa, H., Suto, K. and Takeya, K. (1981a), *Chem. Pharm. Bull.* **29**, 254.
Itokawa, H., Oshida, Y., Ikuta, A., Inatomi, H. and Ikegami, S. (1981b), *Phytochemistry* **20**, 2421.
Itokawa, H., Suto, K. and Takeya, K. (1981c), *Chem. Pharm. Bull.* **29**, 1777.
Iwashina, T., Ootani, S. and Hayashi, K. (1984), *Bot. Mag. Tokyo* **97**, 23.
Jay, M., Viricel, M.R., Favre-Bonvin, J., Voirin, B. and Wollenweber, E. (1982), *Z. Naturforsch.* **37c**, 721.
Karl, C., Müller, G. and Pedersen, P.A. (1976), *Phytochemistry* **15**, 1084.
Kay, Q.O.N., Daoud, H.S. and Stirton, C.H. (1981), *Bot. J. Linn. Soc.* **83**, 57.
Kerr, K.M., Mabry, T.J. and Yoser, S. (1981), *Phytochemistry* **20**, 791.
Khajuria, R.K., Suri, K.A., Suri, O.P. and Atal, C.K. (1982), *Phytochemistry* **21**, 1179.
Kiredjian, M. and Smith, D.M. (1985), *Biochem. Syst. Ecol.* **13**, 141.
Krasnov, E.A. and Demidenko, L.A. (1984), *Khim. Prir. Soedin.* **20**, 43.
Kumamoto, H., Matsubara, Y., Lizuka, Y., Okamoto, K. and Yokoi, K. (1985), *Agric. Biol. Chem.* **49**, 2797.
Kumar, R., Bhan, S., Kalla, A.K. and Dhar, K.L. (1985), *Phytochemistry* **24**, 1124.
Kurkin, V.A., Zapesochnaya, G.G. and Schchavlinskii, A.N. (1984), *Khim. Prir. Soedin.* **20**, 657.
Lachman, J., Rehakova, V., Hubacek, J. and Pivec, V. (1982), *Sci. Agric. Bohemoslov* **14**, 265.
Larsen, L.M., Nielsen, J.K. and Sorensen, H. (1982), *Phytochemistry* **21**, 1029.
Lenherr, A., Lahloub, M.F. and Sticher, O. (1984a), *Phytochemistry* **23**, 2343.
Lenherr, A., Meier, B. and Sticher, O. (1984b), *Plant. Med.* **50**, 403.
Lin, C.N., Arisawa, M., Shimizu, M. and Monta, N. (1982), *Phytochemistry* **21**, 1466.
Mabry, T.J., Liu, Y.L., Pearce, J., Dellamonica, G., Chopin, J., Markham, K.R., Paton, N.H. and Smith, P. (1984), *J. Nat. Prod.* **47**, 127.
Makhsudova, B., Pakudina, E.O. and Sadykov, A.S. (1967), *Khim. Prir. Soedin.* **3**, 11.
Mansour, R.M.A., Ahmed, A.A. and Saleh, N.A.M. (1983), *Phytochemistry* **22**, 2630.
Marco, J.A., Barberá, O., Sanz, J.F. and Sanchez-Parareda (1985), *Phytochemistry* **24**, 2471.
Markham, K.R. and Chari, V.M. (1982). In *The Flavonoids – Advances in Research* (eds J.B. Harborne and T.J. Mabry), Chapman and Hall, London, pp. 19–134.
Markham, K.R. and Mues, R. (1983), *Phytochemistry* **22**, 143.
Markham, K.R. and Whitehouse, L.A. (1984), *Phytochemistry* **23**, 1931.
Markham, K.R., Porter, L.J., Mues, R., Zinsmeister, H.D. and Brehm, B.G. (1976) *Phytochemistry* **15**, 147.
Markham, K.R., Zinsmeister, H.D. and Mues, R. (1978), *Phytochemistry* **17**, 1601.
Markham, K.R., Webby, R.F. and Vilain, C. (1984), *Phytochemistry* **23**, 2049.
Markham, K.R., Vilain, C. and Molloy, P.J. (1985), *Phytochemistry* **24**, 2607.
Martin-Lomas, M., Rabanal, R.M., Rodriguez, B. Valverde, S. (1983), *An. Quim Ser. C.* **79**, 230.
Matsubara, Y., Kumamoto, H., Iizuka, Y., Murakami, T., Okamoto, K., Niyake, H. and Yokoi, K. (1985), *Agric. Biol. Chem.* **49**, 909.
McClure, J.W. (1975). In *The Flavonoids* (eds J.B. Harborne T.J. Mabry and H. Mabry), Chapman and Hall, London, pp. 970–1055.
McClure, J.W. (1986). In *Plant Flavonoids in Biology and Medicine* (eds V. Cody, E. Middleton and J.B. Harborne), Alan R. Liss, New York, pp. 77–86.
McDougal, K.M. and Parks, C.R. (1984), *Am. J. Bot.* **71**, 301.
Miski, M., Gage, D.A. and Mabry, T.J. (1985), *Phytochemistry* **24**, 3078.
Mues, R. (1983), *Biochem. Syst. Ecol.* **11**, 261.
Murakami, T., Satake, T., Hirasawa, C., Ikeno, Y., Saiki, Y. and Chen, C.M. (1984), *Yak. Zasshi* **104**, 142.
Nagy, E., Seres, I., Verzar-Petri, G. and Neszmelyi, A. (1984), **39**, 1813.
Nair, A.G.R. and Gunasegaran, R. (1982), *Indian J. Chem.* **21B**, 1135.
Nair, A.G.R. and Mohandoss, S. (1985), *Indian J. Chem., Sect. B.* **24**, 323.
Nair, A.G.R. and Subramanian, S.S. (1974), *Indian J. Chem.* **12**, 890.
Nair, A.G.R., Kotiyal, J.P. and Bhardwaj, D.K. (1983), *Phytochemistry* **22**, 318.
Nair, A.G.R., Seetharaman, T.R., Voirin, B. and Favre-Bonvin, J. (1986), *Phytochemistry* **25**, 768.
Nair, G.A., Joshua, C.P. and Nair, A.G.R. (1981), *Indian J. Chem.* **20B**, 939.
Nakano, K., Murakami, K., Nohara, T., Tomimatsu, T. and Kawasaki, T. (1981), *Chem. Pharm. Bull.* **29**, 1445.
Nakano, K., Takatani, M., Tomimatsu, T. and Nohara, T. (1983a), *Phytochemistry* **22**, 2831.
Nakano, K., Maruhashi, A., Nohara, T., Tomimatsu, T., Imamura, N. and Kawasaki, T. (1983b), *Phytochemistry*, **22**, 1249.
Nawwar, M.A. and Buddrus, J. (1981), *Phytochemistry* **20**, 2446.
Nawwar, M.A.M., Souleman, A.M.A., Buddrus, J. and Linscheid, M. (1984a), *Phytochemistry* **23**, 2347.
Nawwar, M.A.M., El-Sissi, H.I. and Barakat, H.B. (1984b), *Phytochemistry* **23**, 2937.
Nawwar, M.A.M., Ishak, M.S., Michael, H.N. and Buddrus, J. (1984c), *Phytochemistry* **23**, 2110.

Nicholls, K.W. and Bohm, B.A. (1984), *Canad. J. Bot.* **62**, 1636.
Nicollier, G.F. and Thompson, A.C. (1982), *J. Agric. Food Chem.* **30**, 760.
Nicollier, G. and Thompson, A.C. (1983), *J. Nat. Prod.* **46**, 112.
Nicollier, G.F., Thompson, A.C. and Salin, M.L. (1981), *J. Agric. Food Chem.* **29**, 1179.
Niklas, K.J., Giannasi, D.E. and Baghai, N.L. (1985), *Biochem. Syst. Ecol.* **13**, 1.
Norris, J.A. and Mabry, T.J. (1985), *J. Nat. Prod.* **48**, 668.
Ogawa, M. and Ogihara, Y. (1976), *Yakugaku Zasshi* **95**, 655.
Pacheco, H. (1957), *Bull. Soc. Chim. Biol.* **39**, 971.
Painuly, P. and Tandon, J.S. (1983), *Phytochemistry* **22**, 243.
Pangarova, T.T. and Zapesochnaya, G.G. (1975), *Khim Prir. Soedin.* **11**, 712.
Parker, W.H., Maze, J., Bennett, F.E., Cleveland, T.A. and McLachlan, D.G. (1984), *Taxon* **33**, 1.
Pawlowska, L. (1980), *Acta Soc. Bot. Pol.* **49**, 297.
Rajbhandari, A. and Roberts, M.F. (1984), *J. Nat. Prod.* **47**, 559.
Rao, C.B. and Raju, G.V.S. (1981), *Curr. Sci.* **50**, 180.
Rao, C.P., Hanumaiah, T., Vemuri, V.S.S. and Rao, K.V.J. (1983), *Phytochemistry* **22**, 621.
Rao, E.V. and Rao, M.A. (1982), *Curr. Sci.* **51**, 1040.
Rao, G.V. and Rao, P.S. (1985), *Fitoterapia* **LVI**, 175.
Rao, J.U.M., Hanumaiah, T., Rao, B.K. and Rao, K.V.J. (1984), *Indian J. Chem.* **23B**, 91.
Rao, L.J.M., Kumari, G.N.K. and Rao, N.S.P. (1982), *Heterocycles* **19**, 1655.
Rao, L.J.M., Kumari, G.N.K. and Rao, N.S.P. (1983), *Phytochemistry* **22**, 1058.
Redaelli, C., Formentini, L. and Santaniello, E. (1980), *Phytochemistry* **19**, 985.
Redaelli, C., Formentini, L. and Santaniello, E. (1982), *Phytochemistry* **21**, 1828.
Reiss-Maurer, I. and Wagner, H. (1982), *Tetrahedron* **38**, 1269.
Reynaud, J., Ismaili, A. and Jay, M. (1981), *Phytochemistry* **20**, 2052.
Reynaud, J., Jay, M. and Raynaud, J. (1982), *Phytochemistry* **21**, 2604.
Romussi, G., Most, L. and Bignardi, G. (1981), *Liebigs Ann. Chem.* (5), 761.
Romussi, G., Cafaggi, S. and Ciarallo, G. (1983), *Liebigs Ann. Chem.* (2), 334.
Romussi, G., Parodi, B. and Sancassan, F. (1984), *Liebigs Ann. Chem.* (11), 1867.
Rösler, K.H.A. and Goodwin, R.S. (1984), *J. Nat. Prod.* **47**, 188.
Rösler, K.H.A., Goodwin, R.S., Mabry, T.J., Varma, S.D. and Norris, J. (1984), *J. Nat. Prod.* **47**, 316.
Ross, S.A. (1984), *J. Nat. Prod.* **47**, 862.
Sagara, K., Ito, Y. Oshima, T., Misaki, T. and Murayama, H. (1985), *J. Chromatog.* **328**, 289.
Sagareishvili, T.G., Alaniar, M.D. and Kemertelidze, P. (1983), *Khim. Prir. Soedin.* **19**, 288.
Saleh, N.A.M. (1985), *Phytochemistry* **24**, 371.
Saleh, N.A.M., Boulos, L., El-Negoumy, S.I. and Abdella, M.I. (1982a), *Biochem. Syst. Ecol.* **10**, 33.
Saleh, N.A.M., Ahmed, A.A. and Abdella, M.F. (1982b), *Phytochemistry* **21**, 1995.
Saleh, N.A.M., El-Negoumy, S.I., El-Hahidi, M.N. and Hosni, H.A. (1983a), *Phytochemistry* **22**, 1417.
Saleh, N.A.M., El-Karemy, A.R., Mansour, R.M.A. and Fayed, A.A.A. (1983b), *Phytochemistry* **22**, 2501.
Schaufelberger, D. and Hostettmann, K. (1984), *Phytochemistry* **23**, 787.
Schonsiegel, I., Egger, K. and Keil, M. (1969), *Z. Naturforsch.* **24B**, 1213.
Schulz, M., Strack, D., Weissenbock, G., Markham, K.R., Dellamonica, G. and Chopin, J. (1985), *Phytochemistry* **24**, 343.
Scrinivasan, K.K. and Subramanian, S.S. (1981), *Fitoterapia* **52**, 285.
Servettaz, O., Colombo, M.L., Debernardi, M., Uberti, E., Viders, G. and Vitafinzi, P. (1984), *J. Nat. Prod.* **47**, 809.
Shigematsu, N., Kouno, I. and Kawano, N. (1982), *Phytochemistry* **21**, 2156.
Shimizu, E., Tomimatsu, T. and Nohara, T. (1984), *Chem. Pharm. Bull.* **32**, 5023.
Shukla, R.V. and Misra, K. (1981), *Phytochemistry* **20**, 339.
Siewek, F. and Galensa, R. (1984), *J. Chromatogr.* **294**, 385.
Singh, J. (1982), *Phytochemistry* **21**, 1832.
Singh, M. and Singh, J. (1985a), *Z. Naturforsch.* **40B**, 550.
Singh, M. and Singh, J. (1985b), *Plant. Med.* **46**, 525.
Sinha, M.K., Seth, K.K., Pandey, V.B., Dasgupta, B. and Shah, A.H. (1981), *Plant. Med.* **42**, 296.
Smith, D.M., Glennie, C.W. and Harborne, J.B. (1982), *Biochem. Syst. Ecol.* **10**, 37.
Stein, W., Anhul, S., Zinsmeister, H.D., Mues, R., Barz, W. and Köster, J. (1985), *Z. Naturforsch.* **40**, 469.
Stengel, B. and Geiger, H. (1976), *Z. Naturforsch.* **31C**, 622.
Strack, D., Meurer, B., Wray, V., Grotjahn, L., Austenfeld, F.A. and Wiermann, R. (1984), *Phytochemistry* **23**, 2970.
Takagi, S., Yamaki, M. and Inoue, K. (1980), *J. Pharm. Soc. Jpn* **100**, 1220.
Tamura, H., Kondo, T., Kato, Y. and Goto, T. (1983), *Tetrahedron Lett.* **24**, 5749.
Timmermann, B.N. and Mabry, T.J. (1983), *Biochem. Syst. Ecol.* **11**, 37.
Tiwari, R.T. and Sinha, K.S. (1982), *Indian Chem. Soc.* **59**, 526.
Tomas, F. and Ferreres, F. (1980), *Phytochemistry* **19**, 2039.
Tomas, F., Hernández, L., Barberán, F.A.T. and Ferreres, F. (1985), *Z. Naturforsch. C* **40**, 583.
Torek, M., Pinkas, M., Jay, M. and Favre-Bonvin, J. (1983), *Die Pharmazie* **38**, 783.
Tsiklauri, G.G. (1979), *Khim. Prir. Soedin.* **15**, 98.
Tsiklauri, G.G., Shalashvili, A.G. and Litvinenko, V.I. (1979), *Izv. Akad. Nauk, Gruz. SSR, Ser. Khim.* **5**, 118.
Ulubelen, A., Clark, W.D., Brown, G.K. and Mabry, T.J. (1981a), *J. Nat. Prod.* **44**, 294.
Ulubelen, A., Miski, M. and Mabry, T.J. (1981b), *J. Nat. Prod.* **44**, 586.
Ulubelen, A., Ayanoglu, A., Clark, W.D., Brown, G.K. and Mabry, T.J. (1982a), *J. Nat. Prod.* **45**, 363.
Ulubelen, A., Bucker, R. and Mabry, T.J. (1982b), *Phytochemistry* **21**, 801.
Ulubelen, A., Öksüz, S., Halfron, B., Aynehchi, Y., Mabry, T.J. and Matlin, S.A. (1984), *Phytochemistry* **23**, 2941.
Vierstra, R.D., John, T.R. and Poff, K.L. (1982), *Plant Physiol.* **69**, 522.
Villar, A., Gasco, M.A., Alcaraz, M.J., Manez, S. and Cortes, D. (1985), *Plant. Med.* **45**, 70.
Waage, S.K. and Hedin, P.A. (1984), *Phytochemistry* **23**, 2509.
Wagner, H., Maurer, J., Farkas, L. and Strelisky, J. (1977), *Tetrahedron* **33**, 1405.
Warnock, M.J., Liu, Y.L. and Mabry, T.J. (1983), *Phytochemistry* **22**, 1834.
Weissenbock, G., Schnobl, H., Sachs, G., Elbert, C. and Heller, F.O. (1984), *Physiol. Plant.* **62**, 356.
Williams, C.A., Harborne, J.B. and Mayo, S.J. (1981), *Phytochemistry* **20**, 217.

Williams, C.A., Fronczyk, J.H. and Harborne, J.B. (1983), *Phytochemistry* **22**, 1953.
Wilson, R.D. (1984), *N.Z.J. Bot.* **22**, 195.
Woeldecke, M. and Herrmann, K. (1974), *Z. Naturforsch.* **29C**, 355.
Wollenweber, E., Favre-Bonvin, J. and Jay, M. (1978), *Bull. Liaison Groupe Polyphenols* **8**, 341.
Woo, W.S., Choi, J.S. and Kang, S.S. (1983), *Phytochemistry* **22**, 2881.
WU, T-S. and Furukawa, H. (1983), *Phytochemistry* **22**, 1061.
Young, D.A. and Sterner, R.W. (1981), *Phytochemistry* **20**, 2055.
Yuldashev, M.P., Batirov, E.K. and Malikov, V.M. (1985a), *Khim. Prir. Soedin.* **21**, 269.
Yuldashev, M.P., Batirov, E.K. and Malikov, V.M. (1985b), *Khim. Prir. Soedin.* **21**, 192.
Zaitsev, V.G. and Fursa, N.S. (1984), *Khim. Prir. Soedin* (1984) **20**, 661.
Zaitsev, V.G., Fursa, N.S. and Belyaeva, L.E. (1983), *Khim. Prir. Soedin* **19**, 527.
Zakharova, O.I., Zakahrov, G.V. and Glyzin, V.I. (1979), *Khim. Prir. Soedin.* **15**, 642.
Zapesochnaya, G.G. and Kurkin, V.A. (1983), *Khim. Prir. Soedin.* **19**, 23.
Zapesochnaya, G. and Pangarova, T. (1980), *Dokl. Bolg. Akad. Nauk.* **33**, 933.
Zapesochnaya, G.G., Tjukavkina, N.A. and Shervashidze, I.V. (1982), *Khim. Prir. Soedin.* **18**, 176.
Zapesochnaya, G.G., Stepanov, A.N. and Perov, A.A. (1984a), *Khim. Prir. Soedin.* **20**, 573.
Zapesochnaya, G.G., Stepanov, A.N. and Perov, A.A. (1984b), *Khim. Prir. Soedin.* **20**, 582.

9

The minor flavonoids

BRUCE A. BOHM

9.1 GENERAL INTRODUCTION

It is the purpose of this chapter to discuss the five remaining flavonoid structural types, here collectively known as the minor flavonoids, chalcones, aurones, dihydrochalcones, flavanones and dihydroflavonols. In order to orientate the reader it should be mentioned that chalcones, aurones and dihydrochalcones were discussed in Chapter 9 of *The Flavonoids* (1975) while flavanones and dihydroflavonols were treated in Chapter 11. In *The Flavonoids–Advances in Research* (1982) the five structural types were brought together in Chapter 6 under the present name. As far as is practicable I have presented the material here in the same order and format as used in the earlier two volumes. This makes it possible for the reader to compare current material easily and directly with material in the corresponding sections in the other two books. For the most part each structural type is presented in terms of increasing structural complexity: the primary classification is based upon B-ring oxygenation with the presentation of compounds in order of degree of A-ring oxygenation, *O*-alkylation and *C*-alkylation. Some exceptions to this order will be found especially in cases where a group of related compounds occur in one organism or in a group of related organisms. Only newly discovered compounds are presented in the text except in instances where comparisons with known compounds is considered meaningful. Reports of all compounds, known and new alike, and all sources of compounds are included in the appropriate tables. New family records are included in the text. A few references overlooked in the preparation of the earlier reviews are also included. In order to facilitate cross reference to the earlier two reviews I will simply refer to references or structures in the 1975 volume by means of a roman I along with a page number; reference to the second book will use roman II and a page number.

9.2 CHALCONES

9.2.1 Introduction

Chalcones, and dihydrochalcones, are C_6–C_3–C_6 compounds that lack a central heterocyclic ring. Positions on these compounds, unfortunately, are identified using a numbering system unique to these groups. Chalcones were apparently recognized as being structurally related to acetophenones whose ring carbons were identified by primed numbers. Hence, chalcone (and dihydrochalcone) A-ring carbons are also identified with primed numbers; the B-ring carbons are identified with unprimed numbers. This system is illustrated below.

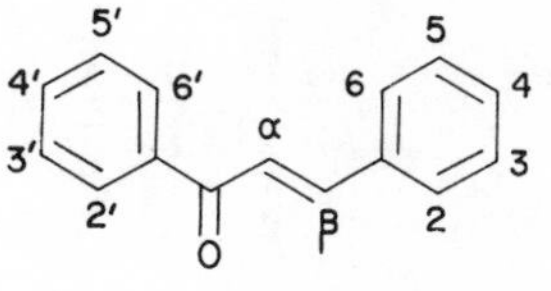

9.2.2 Structures of naturally occurring chalcones

The structures of all naturally occurring chalcones are summarized in Table 9.1.

The Flavonoids. Edited by J.B. Harborne
Published in 1988 by Chapman and Hall Ltd,
11 New Fetter Lane, London EC4P 4EE.

Table 9.1 Structures of naturally occurring chalcones

Substituents			No. in text	Trivial name	Source*		References
OH	OCH_3	Other					
4′		4′-*O*-Glucose	(9.2)		*Shorea robusta*	Dip	Jain *et al.* (1982)
2′, 4′			(9.3)		*Acacia neovernicosa*	Leg	Wollenweber and Seigler (1982)
4′	2′		(9.4)		*Acacia neovernicosa*		Wollenweber and Seigler (1982)
2′, 4′	3′		(9.5)		*Acacia neovernicosa*		Wollenweber and Seigler (1982)
2′		4′-*O*-Prenyl	(9.7)		*Helichrysum rugulosum*	Com	Bohlmann and Misra (1984)
4′	2′	4′-*O*-Prenyl	(9.6)		*Helichrysum rugulosum*		Bohlmann and Misra (1984)
2′, 4′		3′-*C*-Prenyl	(9.8)	Flemistrictin-A	*Helichrysum rugulosum*		Bohlmann and Misra (1984)
					Flemingia stricta	Leg	Subrahmanyam *et al.* (1982)
	2′	Furano(2″, 3″, 4′, 3′); 4-*C*-methyl	(9.17)	Purpuritenin	*Tephrosia purpurea*	Leg	Sinha *et al.* (1982)
	2′	Furano(2″, 3″, 4′, 3′); β-OCH_3	(9.18)	*O*-Methylpongamol	*Tephrosia purpurea*	Leg	Pelter *et al.* (1981)
	2′	6″, 6″,-Dimethylpyrano-(2″, 3″, 4′, 3′); β-OH	(9.19)		*Tephrosia purpurea*	Leg	Rao and Raju (1984)
2′, 4′, 6′			(9.20)		*Chromolaena chasleae*	Com	Bohlmann *et al.* (1982)
					Hymenoclea salsola	Com	Proksch *et al.* (1983)
2′, 4′	6′		(9.22)		*Alpinia katsumadai*	Zin	Kuroyanagi *et al.* (1983)
					A. speciosa	Zin	Itokawa *et al.* (1981)
					Boesenbergia pandurata	Zin	Jaipetch *et al.* (1982)
					Comptonia peregrina	Myr	Wollenweber *et al.* (1985b)
					Myrica pensylvanica	Myr	Wollenweber *et al.* (1985b)
2′, 6′	4′		(9.24)		*Boesenbergia pandurata*	Zin	Jaipetch *et al.* (1982)
					Cajanus cajan	Leg	Cooksey *et al.* (1982)
					Onychium siliculosum	Fil	Wollenweber (1982)
					Piper aduncum	Pip	Burke and Nair (1986)
					P. hispidum	Pip	Burke and Nair (1986)
2′	4′, 6′		(9.23)	Flavokawin-B	*Alpinia speciosa*	Zin	Itokawa *et al.* (1981)
					Didymocarpus corchorifolia	Ges	Wollenweber *et al.* (1981)
					Myrica pensylvanica	Myr	Wollenweber *et al.* (1985b)
					Pityrogramma triangularis var. *pallida*	Fil	Wollenweber *et al.* (1981)
2′, 4′	6′	5′-*C*-Methyl	(9.27)		*Comptonia peregrina*	Myr	Wollenweber *et al.* (1985b)
					Myrica pensylvanica	Myr	Wollenweber *et al.* (1985b)
2′, 4′	6′	3′-*C*-Methyl	(9.25)		*Comptonia peregrina*	Myr	Wollenweber *et al.* (1985b)
					Sterculia urens	Ste	Anjaneyulu and Raju (1984)
2′	4′, 6′	3′-*C*-Methyl		Aurentiacin	*Pityrogramma triangularis* var. *pallida*	Fil	Wollenweber *et al.* (1981)

					Comptonia peregrina	Myr	Wollenweber *et al.* (1985b)
2′, 6′	4′	3′, 5′-Di-*C*-methyl	(*9.26*)	Myrigalon-B	*Myrica pensylvanica*	Myr	Wollenweber *et al.* (1985b)
					Comptonia peregrina	Myr	Wollenweber *et al.* (1985b)
2′, 3′	4′, 6′		(*9.29*)		*Piper hispidum*	Pip	Vieira *et al.* (1980)
2′	3′, 4′, 6′		(*9.30*)		*P. hispidum*	Pip	Vieira *et al.* (1980)
3′	2′, 4′, 5′, 6′		(*9.31A*)		*Didymocarpus pedicellata*	Ges	Rathore *et al.* (1981a)
2′	4′, 5′, 6′		(*9.31*)		*Didymocarpus pedicellata*	Ges	Rathore *et al.* (1981b)
2′, 4′, 6′		3′-*C*-Prenyl			*Helichrysum athrixiifolia*	Com	Bohlmann and Ates(Gören) (1984)
					Pleiotaxis cf. *rugosa*	Com	Bohlmann and Zdero (1982)
2′, 4′, 6′		4′-*O*-Prenyl	(*9.32*)		*Helichrysum athrixiifolia*	Com	Bohlmann and Ates(Gören) (1984)
					Pleiotaxis cf. *rugosa*	Com	Bohlmann and Zdero (1982)
2′, 6′	4′	3′-*O*-Prenyl			*Pleiotaxis* cf. *rugosa*	Com	Bohlmann and Zdero (1982)
2′	4′, 6′	3′-*C*-Prenyl		Ovalichalcone	*Tephrosia candida*	Leg	Roy *et al.* (1986)
2′	6′	6″, 6″-Dimethylpyrano-(2″, 3″, 4′, 3′)	(*9.33*)	Pongachalcone-I	*T. candida*	Leg	Chibber *et al.* (1981)
2′	6′	6″, 6″-Dimethylpyrano-(2″, 3″, 4′, 5′)	(*9.34*)	Oaxacacin	*T. woodii*	Leg	Dominguez *et al.* (1983)
2′	6′	6″, 6″-Dimethylpyrano-(2″, 3″, 4′, 3′); β-OH	(*9.35*)		*Lonchocarpus costaricensis*	Leg	Waterman and Mahmoud (1985)
2′		6″, 6″-Dimethylpyrano-(2″, 3″, 4′, 3′)-6‴, 6‴-dimethylpyrano-(2‴, 3‴, 6′, 5′)	(*9.36*)	Flemiculosin	*Flemingia fruticulosa*	Leg	Khattri *et al.* (1984)
2′, 4		4-*O*-Glucose	(*9.41*)		*Adhatoda vasica*	Aca	Bhartiya and Gupta (1982)
2′, 4, 4′			(*9.42*)	Isoliquiritigenin	*Helianthus longifolius*	Com	Schilling (1983)
					Tithonia brachypappa	Com	LaDuke (1982)
					Zollernia paraensis	Leg	Ferrari *et al.* (1983)
		2′-*O*-(Rhamnose-4-glucose)			*Viburnum cortinifolium*	Cap	Srivastava *et al.* (1981b)
4, 4′	2′		(*9.43*)		*Pisum sativum*	Leg	Carlson and Dolphin (1982)
2′, 6′	4′				*Caesalpinia pulcherrima*	Leg	McPherson *et al.* (1983)
4, 4′	2′	5′-*C*-Formyl	(*9.44*)		*Psoralea corylifolia*	Leg	Gupta *et al.* (1980)
2′, 4, 4′		3, 5-Di-*C*-prenyl	(*9.48*)	Abyssinone-V	*Erythrina abyssinica*	Leg	Kamat *et al.* (1981)
2′	4	6″, 6″-Dimethylpyrano-(2″, 3″, 4′, 3′)	(*9.45*)		*Millettia pachycarpa*	Leg	Singhal *et al.* (1983)
4, 4′		6″, 6″-Dimethyl-4″, 5″-dihydropyrano(2″, 3″, 2′, 3′)	(*9.46*)		*Crotalaria madurensis*	Leg	Bhakuni and Chaturved (1984)
2, 2′, 4′		3′-*C*-Lavandulyl	(*9.49*)		*Ammothamnus lehmanni*	Leg	Sattikulov *et al.* (1983)
2′, 4, 4′, 6′			(*9.54*)	Chalcononaringenin	*Lycopersicon esculentum*	Sol	Hunt and Baker (1980)
		2′-*O*-Glucose	(*9.55*)	Isosalipurposide	*Acacia dealbata*	Leg	Imperato (1980)
					Helichrysum armenium	Com	Çubukçu and Yüksel (1982)
					Potentilla species	Ros	Harborne and Nash (1984)
		2′-*O*-Xylose			*Acacia dealbata*	Leg	Imperato (1980)
		4′-*O*-Glucose			*Sorghum* sp.	Gra	Gujer *et al.* (1986)
		2′-*O*-(Xylose-4-rhamnose)			*Acacia dealbata*	Leg	Imperato (1982b)

(*Contd.*)

Table 9.1 (*Contd.*)

Substituents							
OH	*OCH_3*	*Other*	*No. in text*	*Trivial name*	*Source**		*References*
2′, 4, 6′	4′			Neosakuranetin	*Pityrogramma calomelanos*	Fil	Hitz *et al.* (1982)
2′, 4, 4′	6′		(*9.56*)		*Achryocline flaccida*	Com	Norbedo *et al.* (1982)
2′, 4′, 6′	4		(*9.57*)		*Wyethia glabra*	Com	McCormick *et al.* (1985)
2′, 4, 4′, 6′		4′-*O*-Prenyl	(*9.58*)		*Helichrysum athrixiifolium*	Com	Bohlmann and Ates (Gören) (1984)
3, 4		4-*O*-Galactose-arabinose	(*9.59*)		*Bauhinia purpurea*	Leg	Bhartiya and Gupta (1981)
2′, 3, 4			(*9.60*)		*Semecarpus vitiensis*	Ana	Pramono *et al.* (1981)
2′, 4	2		(*9.76*)	Echinatin	*Glycyrrhiza echinata*	Leg	Ayabe *et al.* (1980a)
2′, 4, 4′	2	5′-*C*-Lavandulyl	(*9.78*)	Kushenol-D	*Sophora flavescens*	Leg	Wu *et al.* (1985a)
2		3′, 4′-Methylenedioxy-6″, 6″-dimethylpyrano (2″, 3″, 4, 3)	(*9.77*)		*Pongamia glabra*	Leg	Saini *et al.* (1983)
3, 4, 2′, 4′		4′-*O*-Glucose		Coreopsin	*Coreopsis* sect. *Coreopsis*	Com	Crawford and Smith (1985)
					Helianthus heterophyllus	Com	Schilling (1983)
					Viguiera series *Viguiera*	Com	Rieseberg and Schilling (1985)
4, 2′, 4′	3			3-Methylbutein	*Tithonia fruticosa*	Com	LaDuke (1982)
3, 4, 2′, 3′, 4′		4′-*O*-Glucose		Marein	*Coreopsis* sect. *Coreopsis*	Com	Crawford and Smith (1985)
					Thelesperma megapotanicum	Com	Ateya *et al.* (1982)
					Viguiera dentata	Com	Rieseberg and Schilling (1985)
		4′-*O*-Glucose-arabinose	(*9.62*)		*Abies pindrow*	Cnf	Tiwari and Minocha (1980)
3, 4, 2′, 4′ 3′		4′-*O*-Glucose		Lanceolin	*Coreopsis* sect. *Coreopsis*	Com	Crawford and Smith (1985)
2′, 3, 3′, 4′4		4′-*O*-Glucose	(*9.63*)		*Bidens torta*	Com	McCormick *et al.* (1984)
2′, 3, 3′, 4′4		4′-*O*-Glucose-X″-acetate	(*9.64*)		*B. torta*	Com	McCormick *et al.* (1984)
2′, 3′, 4′	3, 4	4′-*O*-Glucose	(*9.65*)		*B. torta*	Com	McCormick *et al.* (1984)
2′, 4′	3, 3′, 4	4′-*O*-Glucose	(*9.66*)		*B. torta*	Com	McCormick *et al.* (1984)
2′	3, 3′, 4, 4′		(*9.67*)		*B. torta*	Com	McCormick *et al.* (1984)
2′, 4, 4′, 6′3		2′-*O*-Glucose	(*9.68*)		*Acacia dealbata*	Leg	Imperato (1982a)
2′, 4′	2, 3, 6′		(*9.75*)		*Scutellaria discolor*	Lab	Tomimori *et al.* (1985)
2′, 5′	4′, 6′	3, 4-Methylenedioxy	(*9.69*)		*Ageratum strictum*	Com	Quijano *et al.* (1982)
2′	2, 4, 5	6″, 6″-Dimethylpyrano-(2″,3″,4′,3′)	(*9.80*)		*Pongamia glabra*	Leg	Pathak *et al.* (1983)

*Key: Aca = Acanthaceae, Ana = Anacardiaceae, Cap = Caprifoliaceae, Cnf = Coniferae, Com = Compositae, Dip = Dipterocarpaceae, Fil = Filicinae, Ges = Gesneriaceae, Gra = Gramineae, Lab = Labiatae, Leg = Leguminosae, Myr = Myricaceae, Pip = Piperaceae, Ros = Rosaceae, Sol = Solanaceae, Ste = Sterculiaceae, Zin = —Zingiberaceae.

(a) *Chalcones lacking B-ring oxygenation*

As I noted in 1975 (I, 444) chalcone itself has not been found as a natural product. At that time I commented that the natural occurrence of 4′-methoxychalcone (*9.1*) in *Citrus limon* (I, 444) should be confirmed. I am unaware of any further work on that system but a recent report of 4′-hydroxychalcone, as the glucoside (*9.2*), in heartwood of *Shorea robusta* (Dipterocarpaceae) (Jain *et al.*, 1982) indicates that such simple oxygenation patterns do exist.

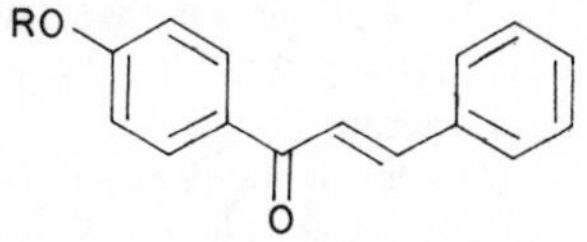

(*9.1*) R=CH_3
(*9.2*) R=Glucose

The picture rapidly becomes more complex when chalcones based on the resorcinol A-ring pattern are considered. The parent compound (*9.3*) along with its 2′-methyl ether (*9.4*) and its 3′-methoxy derivative (*9.5*) were reported to occur in *Acacia neovernicosa* by Wollenweber and Seigler (1982). In the five populations of this plant that these authors studied only (*9.4*) was present in all. Four chalcones, one of them a new natural product, based upon this parent system, have been described from *Helichrysum rugulosum* by Bohlmann and Misra (1984). The new compound is the 2′-*O*-methyl-4′-*O*-prenyl* derivative (*9.6*). Known derivatives are the variously methylated and/or prenylated compounds (*9.7*), (*9.8*) and (*9.9*).

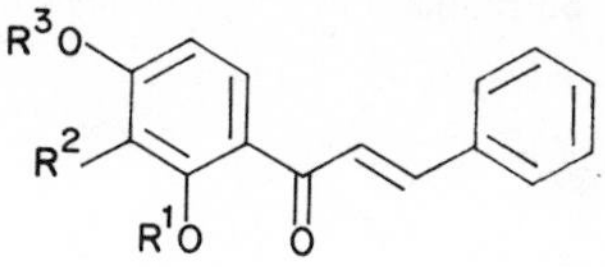

(*9.3*) $R^1 = R^2 = R^3 = H$
(*9.4*) $R^1 = CH_3; R^2 = R^3 = H$
(*9.5*) $R^1 = R^3 = H; R^2 = OCH_3$
(*9.6*) $R^1 = CH_3; R^2 = H; R^3 =$ prenyl
(*9.7*) $R^1 = R^2 = H; R^3 =$ prenyl
(*9.8*) $R^1 = R^3 = H; R^2 =$ prenyl
(*9.9*) $R^1 = H; R^2 =$ prenyl ; $R^3 = CH_3$

Members of the legume genus *Flemingia* exhibit a rich and varied flavonoid chemistry as reference to the earlier reviews will attest. New compounds are continually being reported; Subrahmanyam and co-workers (1982) described five chalcones belonging to this subgroup from *F. stricta*, of which two are new. The known natural products include flemistrictin-A (*9.8*) which clearly serves as the starting material for formation of the others. The other two known compounds are flemistrictin-B (*9.10*) and flemistrictin-C (*9.11*). In both of these compounds ring closure has occurred involving the 4′-hydroxyl, in one case yielding a furan derivative (*9.10*), and in the other (*9.11*) a pyran derivative. The new compounds are isomeric to this pair involving cyclization with the 2′-hydroxyl to yield compounds (*9.12*) and (*9.13*), 'flemistrictin-E' and 'flemistrictin-F' respectively. Further examples of the biosynthetic versatility of *Flemingia* were revealed in a study of *F. wallichii* (Sivarambabu *et al.*, 1985).

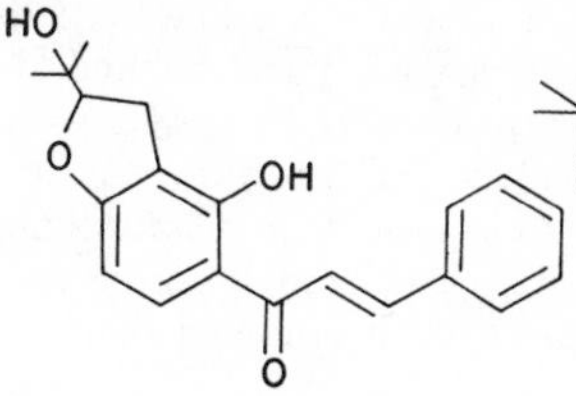

(*9.10*) 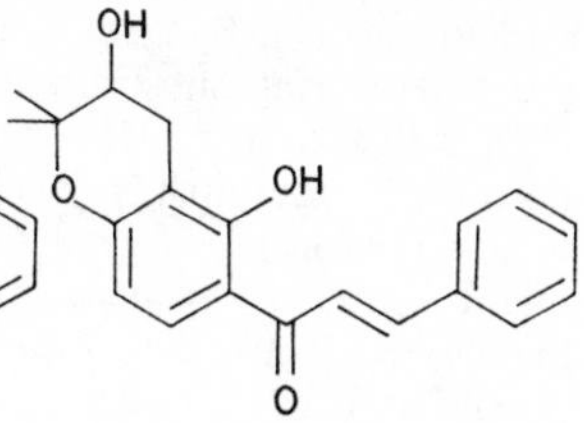 (*9.11*)

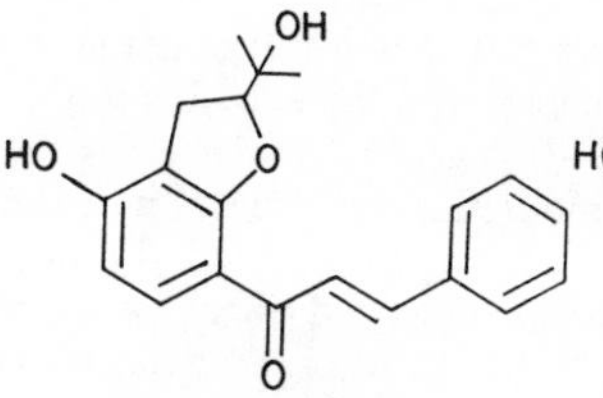

(*9.12*) 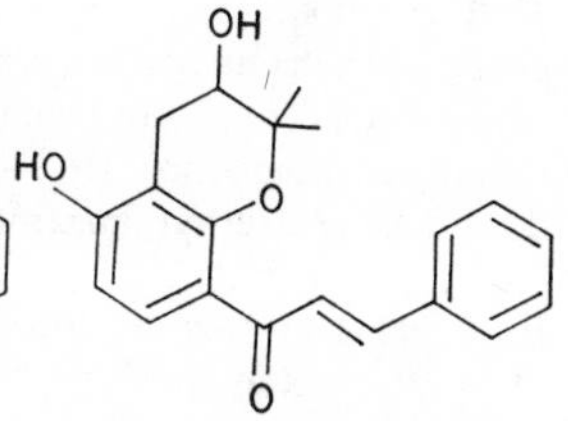 (*9.13*)

'Flemiwallichin-F' exhibits the fundamental 2′,4′-dioxygenation pattern in addition to which we see extra hydroxylation at C-5′ and alkylation at C-3′. The 10-carbon alkyl function, derived almost certainly from a geranyl moiety, has undergone cyclization, hydroxylation and repositioning of the distal hydroxyl group. The result is compound (*9.14*). Two other compounds from this plant were described in the same paper: compounds (*9.15*) and (*9.16*) have more complex B-ring substitution but are obviously related to (*9.14*).

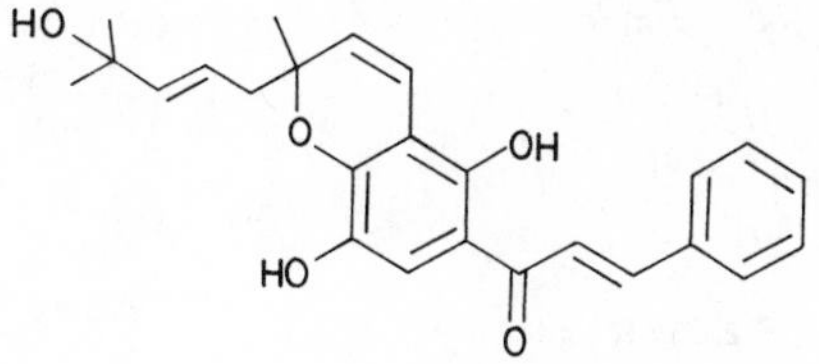

(*9.14*)

*Prenyl is used throughout to represent 3, 3-dimethylprop-2-enyl (= γ, γ-dimethylallyl).

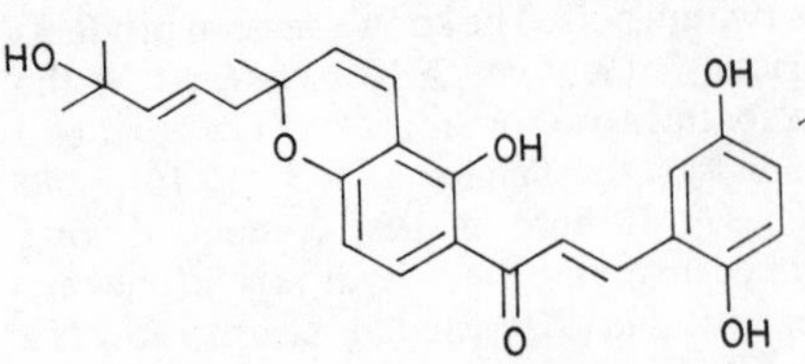

(9.15) 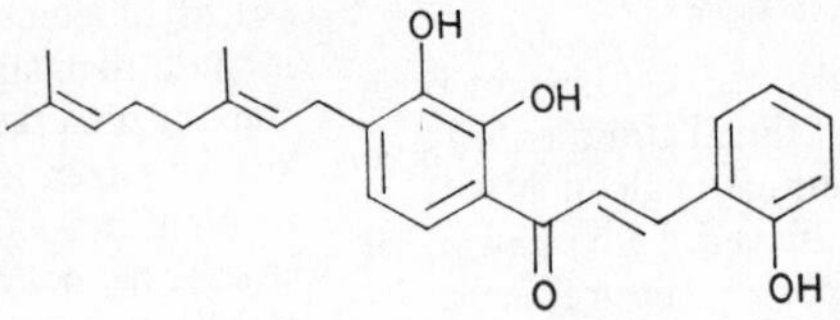 (9.16) 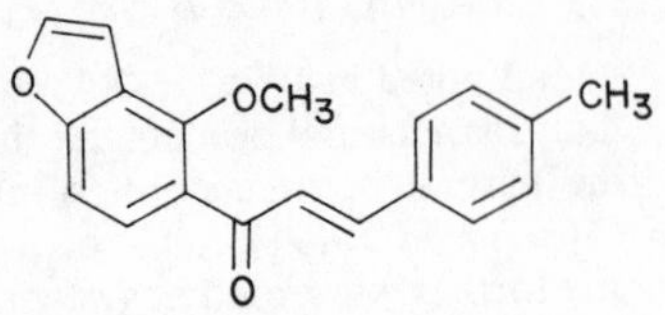 (9.17)

A most interesting observation was made by Sinha and co-workers (1982) in their study of the flavonoid constituents of *Tephrosia purpurea* seeds. A new chalcone, named 'purpuritenin', was isolated and shown to have the structure (*9.17*). Pongamol, a known α,β-diketone with the same A-ring substitutions and an unsubstituted B-ring, was also found (II, 347). Sinha *et al.* (1982) were prompted, presumably because of the unusual *C*-methylation in the B-ring, to suggest that purpuritenin might represent a secondary fungal metabolite produced in the seed as a result of infection. It is hoped that this possibility is being pursued.

'*O*-Methylpongamol' (*9.18*) has been described in *Tephrosia purpurea* by Pelter and his associates (1981). This compound is one of a comparatively uncommon group of chalcone derivatives that have β-oxygenation. β-Hydroxychalcones can be considered the enol form of β-diketones. Pongamol (II, 347) is generally written as the diketone. *O*-Methylpongamol (*9.18*) is 'locked' into the enol form by virtue of its existence as the methyl ether. Enol forms do exist in nature as evidenced by the isolation of (*9.19*) by Rao and Raju (1984), also from *T. purpurea*.

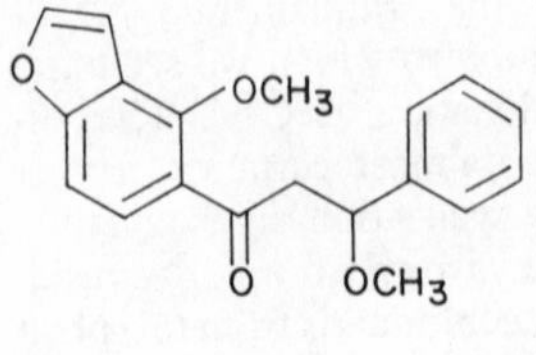

(9.18) 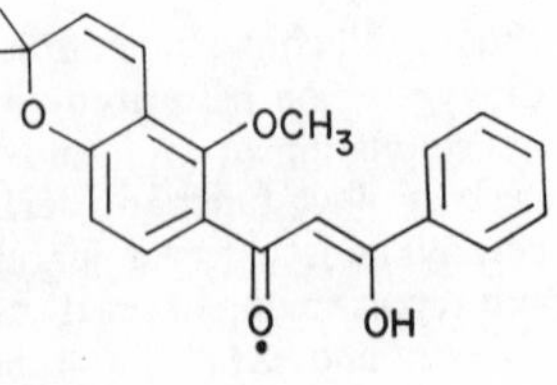 (9.19)

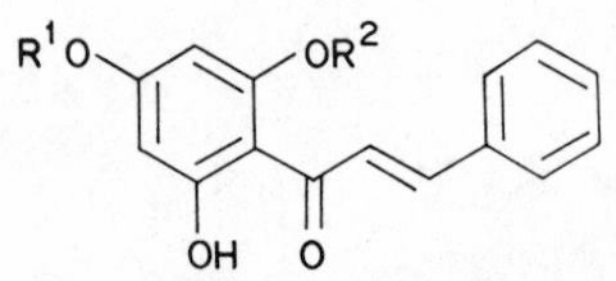

(9.20) $R^1 = R^2 = H$
(9.21) $R^1 = CH_3$; $R^2 = H$
(9.22) $R^1 = H$; $R^2 = CH_3$
(9.23) $R^1 = R^2 = CH_3$

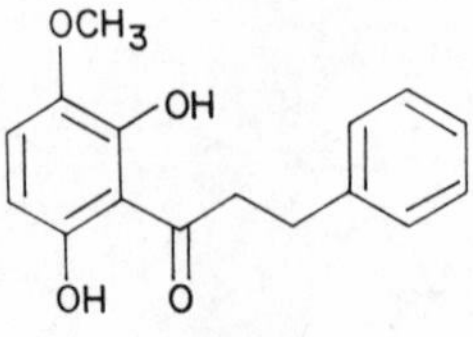

(9.24)

The next group of chalcones is based upon the phloroglucinol-type A-ring. The parent compound (*9.20*), 2′,6′-dihydroxy-4′-methoxychalcone (*9.21*), 2′,4′-dihydroxy-6′-methoxychalcone (*9.22*) and 2′-hydroxy-4′,6′-dimethoxychalcone (also called flavokawin-B) (*9.23*) are all known compounds. Recent records of these compounds are summarized in Table 9.1. The only new family record is the finding of (*9.22*) and (*9.23*) in the Myricaceae (Wollenweber *et al.*, 1985b). Wollenweber (1982) reported (*9.21*) and the previously unknown 2′,6′-dihydroxy-3′-methoxychalcone (*9.24*) from the farina of the fern *Onychium siliculosum*.

Another level of complexity is reached in *Comptonia peregrina* and *Myrica pensylvanica* (both Myricaceae) where Wollenweber and co-workers (1985b) found, in addition to the simpler compounds mentioned above, a series of chalcones that combine *O*- and *C*-methylation. Compounds (*9.25*) and (*9.26*) (= 'myrigalon-B') were isolated from *Comptonia peregrina* while (*9.27*) and (*9.28*) (= 'myrigalon-D') were isolated from both species. Compound (*9.25*) has also been reported to occur in *Sterculia urens* (Sterculiaceae) by Anjaneyulu and Raju (1984).

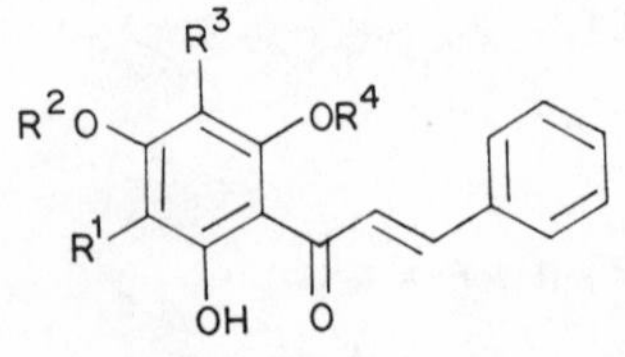

(9.25) $R^1 = R^4 = CH_3$; $R^2 = R^3 = H$
(9.26) $R^1 = R^2 = R^3 = CH_3$; $R^4 = H$
(9.27) $R^1 = R^2 = H$; $R^3 = R^4 = CH_3$
(9.28) $R^1 = R^3 = R^4 = CH_3$; $R^2 = H$

A combination of extra hydroxylation and *O*-methylation provide compounds (*9.29*) and (*9.30*) in *Piper hispidum* var. *obliquum* (Vieira *et al.*, 1980) and in *Didymocarpus pedicellata* (Rathore *et al.*, 1981a, b). Compound (*9.29*) appears to be a new natural product, and (*9.30*) is new to the genus *Piper* but was previously identified from a member of *Popowia* (II, 316). The genus *Didymocarpus* appears to be characterized by highly substituted and *O*-methylated A-ring flavonoids (II, 316–317). Compound

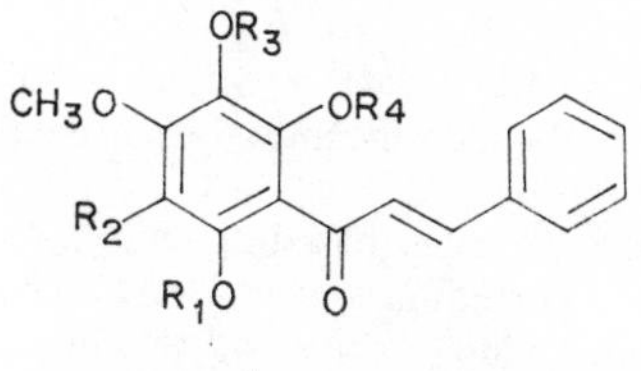

(9.29) $R^1 = CH_3$, $R^2 = R^3 = R^4 = H$

(9.30) $R^1 = R^3 = CH_3$; $R^2 = R^4 = H$

(9.31) $R^1 = R^2 = H$; $R^3 = R^4 = CH_3$

(9.31A) $R^1 = R^3 = R^4 = CH_3$; $R^2 = OH$

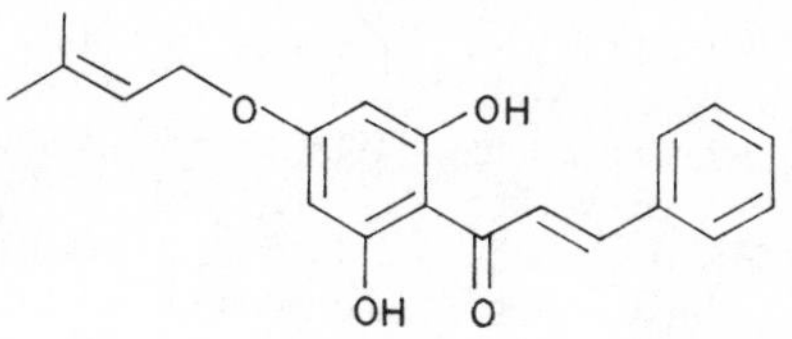

(9.32)

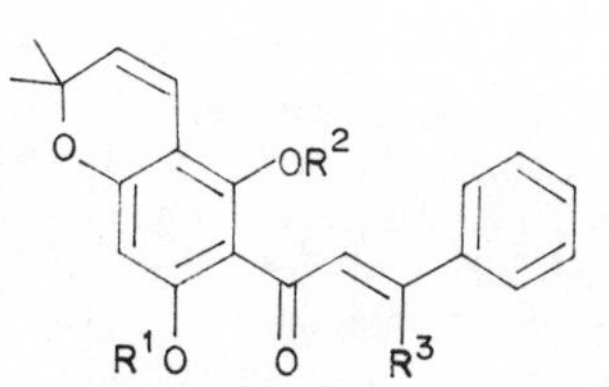

(9.33) $R^1 = CH_3$; $R^2 = R^3 = H$

(9.34) $R^1 = R^3 = H$; $R^2 = CH_3$

(9.35) $R^1 = CH_3$; $R^2 = H$; $R^3 = OH$

(9.31) is new to *Didymocarpus*, but is known from *Helichrysum* (I, 316). Compound *(9.31A)*, also from *D. pedicellata*, appears to be a new natural product (Rathore *et al.*, 1981a).

The following small group of compounds is characterized by *C*-alkylation on a phloroglucinol A-ring. Newly reported occurrences of known compounds involving prenylation and *O*-methylation of the parent compound are to be found in Table 9.1. The only new addition to the list within this group is the 4′-prenyl ether *(9.32)* reported from the roots of *Pleiotaxis* cf. *rugosa* (Compositae; Mutisieae) (Bohlmann and Zdero, 1982) and more recently from *Helichrysum athrixiifolium* (Bohlmann and Ates (Gören), 1984) where it occurs with the corresponding 4-hydroxychalcone.

Pongachalcone-I *(9.33)* is a known compound, originally reported from *Pongamia glabra* (I, 323). More recently, Chibber and colleagues (1981) reported it from another legume, *Tephrosia candida*. A compound isomeric to pongachalcone-I has now been identified from *T. woodii* by Dominguez and co-workers (1983). The new compound is called 'oaxacacin' and is shown as *(9.34)*. A related chalcone having a β-hydroxyl function *(9.35)* was isolated from seeds of *Lonchocarpus costaricensis* (Waterman and Mahmoud, 1985). A new prenylated chalcone from *Flemingia fruticulosa*, called 'flemiculosin,' was shown by Khattri and co-workers (1984) to be *(9.36)*. Careful attention was given to the exact nature of the ring closures in this system since alternative structures are possible.

The Lauraceae have provided phytochemists with several highly unusual and challenging structures, noteworthy amongst which is rubranine *(9.37)* (I, 324). Rubranine has now been identified from a completely unrelated source: *Boesenbergia* (*Kaempferia*) *pandurata* of the Zingiberaceae (Tuntiwachwuttikul *et al.*, 1984). *Boesenbergia* has yielded other compounds of comparable complexity as well as two simple *O*-methyl ethers 2′,6′-dihydroxy-4′-methoxychalcone *(9.21)* and 2′,4′-dihydroxy-6′-methoxychalcone *(9.22)*. Jaipetch and co-workers (1982) described 'boesenbergin-A' *(9.38)* along with the two simple chalcones. Boesenbergin-A is related to chalcone *(9.21)* through geranylation at C-3′ followed by cyclization with the 2′-hydroxyl group. Alkylation of

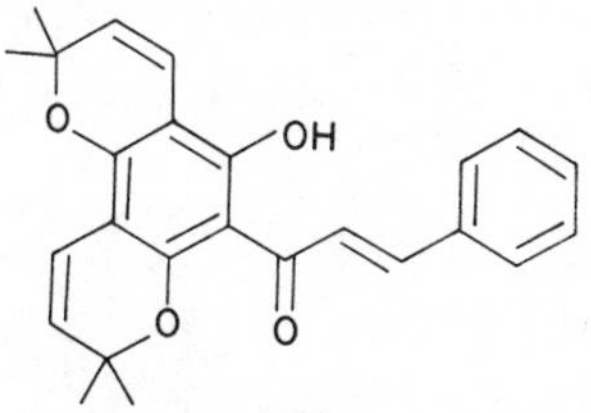

(9.36)

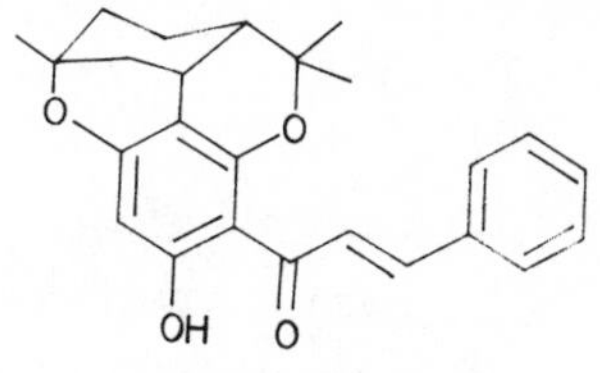

(9.37)

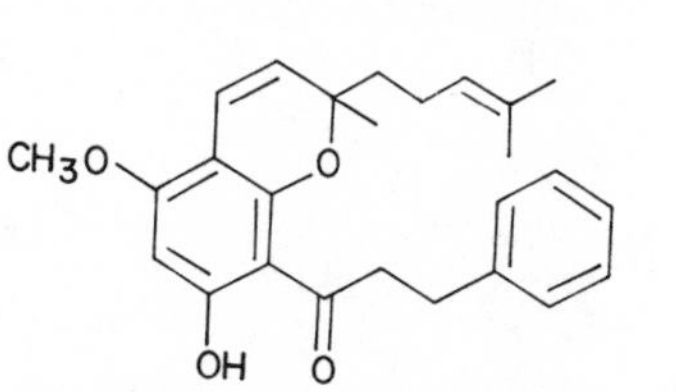

(9.38)

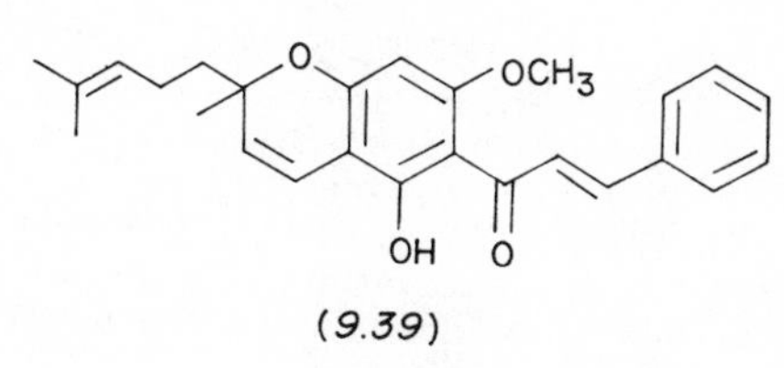

(9.39)

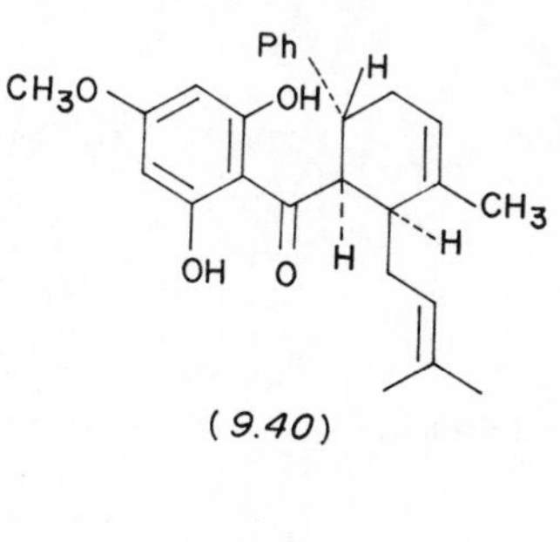

(9.40)

simple chalcone (*9.22*) with the geranyl function at C-3′ followed by cyclization with the 4′-hydroxyl group would yield the isomeric 'boesenbergin-B′ (*9.39*) whose presence in *B. pandurata* was reported by Mahidal and co-workers (1984). Mahidal and co-workers (1984) also described a new compound, 'panduratin-A' (*9.40*), which is not a chalcone but is thought to have arisen by a Diels-Alder-type condensation between a simple chalcone, compound (*9.21*) has the correct substitution pattern, and a 10-carbon unsaturated compound. Several other presumptive Diels-Alder-type products will be discussed in more detail later in this chapter.

(b) *Chalcones having one B-ring oxygenation*

One of the simplest chalcones to be identified as a natural product is 2,4′-dihydroxychalcone 4′-*O*-*β*-D-glucoside (*9.41*) which Bhartiya and Gupta (1982) reported from *Adhatoda vasica* of the Acanthaceae. Chalcones based upon isoliquiritigenin are well known. The parent compound itself (*9.42*) is known principally as a constituent of legumes with one or two reports from the Compositae. Its presence in *Zollernia paraensis* (Ferrari *et al.*, 1983) represents yet another generic source from amongst the legumes. A newly described natural derivative of isoliquiritigenin is 2′-methoxy-4,4′-dihydroxychalcone (*9.43*) which Carlson and Dolphin (1982) found, along with two cinnamyl phenols, in *Pisum sativum* maintained under stress conditions. The existence of neobavachalcone, 2′,4-dihydroxy-4′-methoxy-5′-formylchalcone, in *Psoralea corylifolia* was commented on earlier (II, 328). The isomeric compound 'isoneobavachalcone' has also been reported from this plant and shown to be 2′-methoxy-4,4′-hydroxy-5′-formylchalcone (*9.44*) (Gupta *et al.*, 1980). Synthesis confirmed the identification (Sharma and Gupta, 1984).

Several new *C*-alkylated chalcones based upon 2′,4,4′-oxygenation have been reported in recent years. The simplest of these is the compound (*9.45*) in which a linear cyclization of the alkyl group has occurred; this compound was isolated from *Millettia pachycarpa* by Singhal and co-workers (1983). The heterocyclic ring in (*9.46*) has been further modified by reduction. This compound was isolated from *Crotalaria madurensis* (Bhakuni and Chaturved, 1984). Such saturated systems are comparatively uncommon amongst chalcones; the products of hydration, such as seen in bavachromenol (II, 327), are more common. Gupta and co-workers (1982) isolated the derived furanochalcone (*9.47*) from seeds of *Psoralea corylifolia*. Chalcone (*9.47*) exhibits both a saturated heterocyclic ring and an alcoholic hydroxyl group that may have arisen through hydration of a terminal double bond in the side chain. A series of mono- and di-prenylflavanones and the related 3,5-di-*C*-prenychalcone (*9.48*) were isolated from *Erythrina abyssinica* by Kamat and co-workers (1981) who called the chalcone 'abyssinone-V.' One of the flavanones, 'abyssinone-IV,' has a substitution pattern corresponding to chalcone (*9.48*). The most unusual member of this small group is compound (*9.49*) reported as being a constituent of the legume *Ammothamnus lehmanni* (Sattikulov *et al.*, 1983). Two structural features distinguish it: 2-hydroxylation and the C-3′-alkyl substituent. 2-Hydroxychalcones are a rare group; two were included in the first review (I, 464), none in the second. The alkyl substituent at C-3′ is

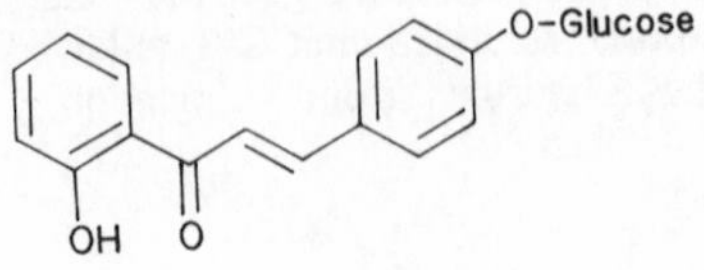

(*9.41*)

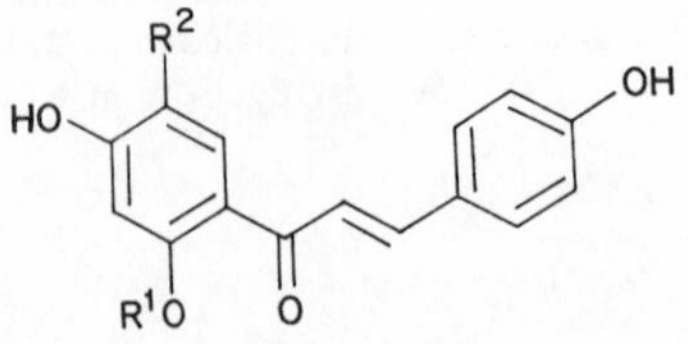

(*9.42*) $R^1 = R^2 = H$
(*9.43*) $R^1 = CH_3$; $R^2 = H$
(*9.44*) $R^1 = CH_3$; $R^2 = CHO$

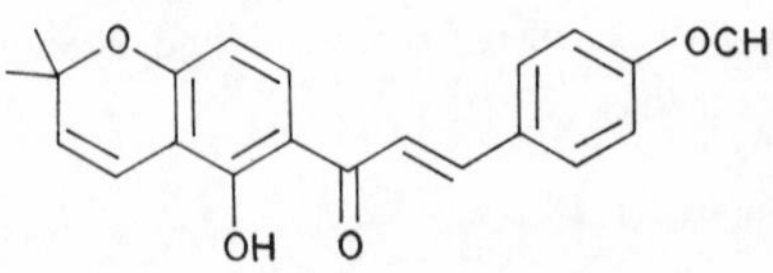

(*9.45*)

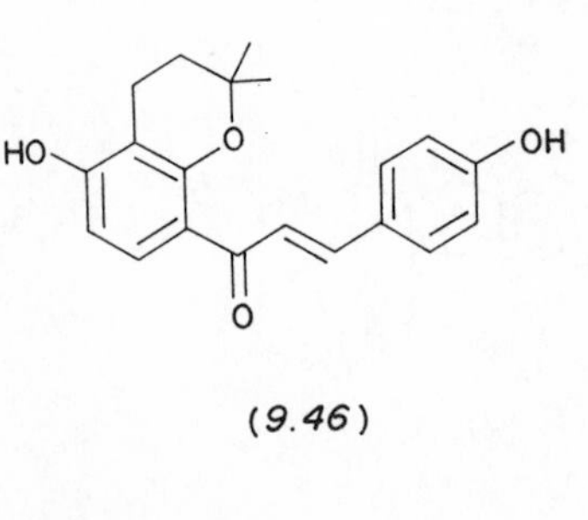

(*9.46*)

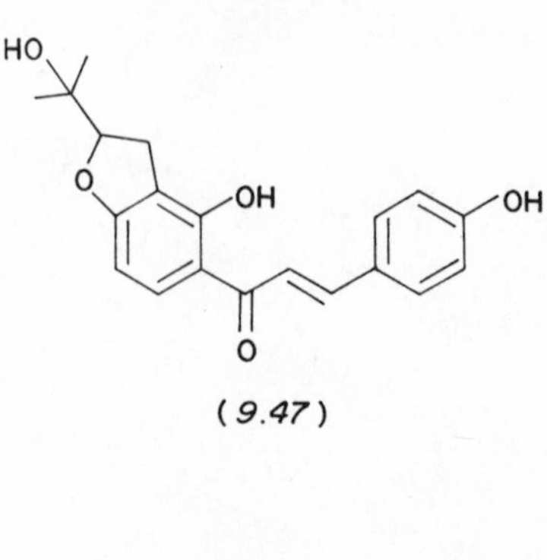

(*9.47*)

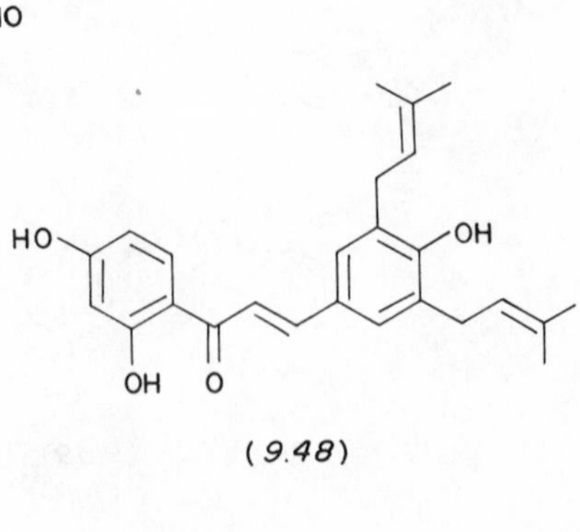

(*9.48*)

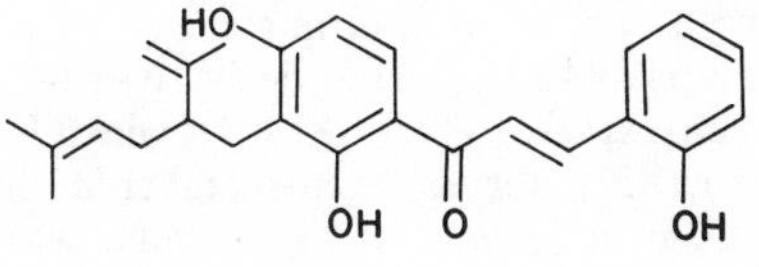

(9.49)

also of rare occurrence. Other appearances of the lavandulyl group (lavandulol is 2,6-dimethyl-3-hydroxymethylhepta-1,5-diene) in members of the minor flavonoids can be seen in compounds (*9.78, 9.222, 9.223, 9.316, 9.320, 9.322*).

2-Hydroxylation has been reported in several compounds isolated from members of the genus *Uvaria*, noteworthy examples of which can be seen in those compounds having 2-hydroxybenzyl substituents, e.g. (*9.141*). Two new additions to this group were described by Hufford and co-workers (1981) as a result of their study of *U. afzelii*. One of these is an ordinary flavanone, 5,7,2′-trihydroxy-6,8-di-*C*-methylflavanone (*9.219*), while the other is a quite extraordinary chalcone, 2-hydroxy-7,8-dehydrograndiflorone (*9.50*). The chalcone is not stable, however, and converts rapidly to emorydone (*9.51*). The formation of (*9.51*) could be rationalized by a mechanism of the sort suggested by Dreyer and co-workers (1975) for the formation of dalrubrone and methoxydalrubrone, constituents of *Dalea* species that bear strong structural similarities to emorydone (*9.51*). Another, related, compound, vafzelin (*9.52*), has been isolated from *Uvaria afzelii* (Hufford *et al.*, 1980). It was suggested that vafzelin can be derived by '*C*-alkylation of syncarpic acid (*9.53*) by an *O*-hydroxycinnamoyl moiety followed by conjugate

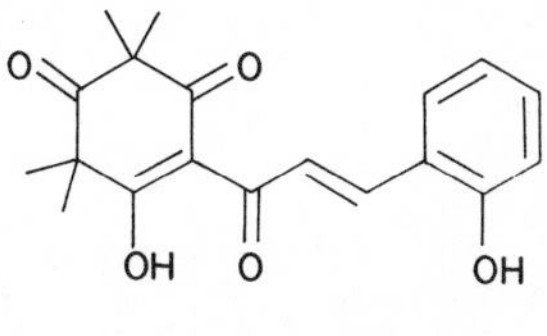

(9.50) 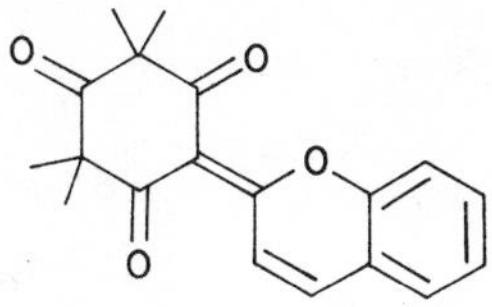 (9.51)

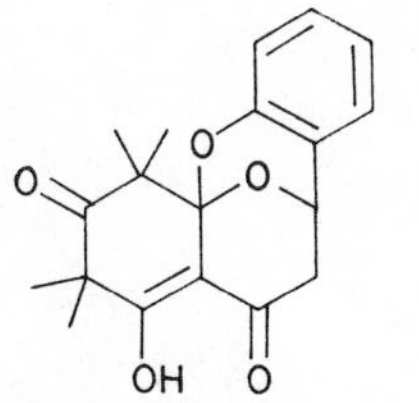

(9.52) 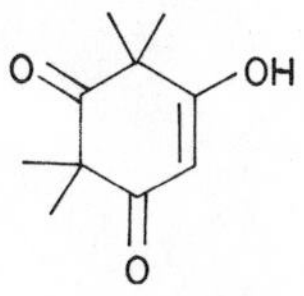 (9.53)

additions.' These authors go on to suggest that, since these compounds were all optically inactive, they may have arisen by non-enzymic reactions. A more fundamental question may be asked: do any of these flavonoid-like compounds arise by the normal biosynthetic pathway or do some come about by cinnamylation of some A-ring precursor? Biosynthetic studies would be most welcome.

The parent compound of this next section is the familiar 2′,4,4′,6′-tetrahydroxychalcone (*9.54*), known as 'chalcononaringenin.' The compound is rare as the free phenol, however. It was recorded from pollen of two unrelated species in the second review (II, 328) and has been reported as being a component of tomato fruit cuticle (Hunt and Baker, 1980). Previously known glycosidic derivatives are summarized in Table 9.1. A new family record, the Rosaceae, come from the work of Harborne and Nash (1984) who found isosalipurposide (*9.55*) in 11 species of *Potentilla*.

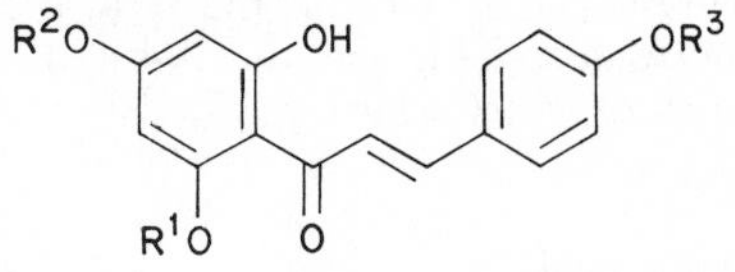

(9.54) $R^1 = R^2 = R^3 = H$

(9.55) R^1 = glucose; $R^2 = R^3 = H$

(9.56) $R^1 = CH_3$; $R^2 = R^3 = H$

(9.57) $R^1 = R^2 = H$; $R^3 = CH_3$

(9.58) $R^1 = R^3 = H$; R^2 = prenyl

Recently reported simple ethers of chalcononaringenin are the 6′-*O*-methyl derivative (*9.56*) from *Achryocline flaccida* (Norbedo *et al.*, 1982), the 4-*O*-methyl derivative (*9.57*) from *Wyethia glabra* (McCormick *et al.*, 1985), and the 4′-*O*-prenyl derivative (*9.58*) from *Helichrysum athrixiifolium* (Bohlmann and Ates (Gören), 1984). Several related compounds from *H. athrixiifolium* were included in the last mentioned paper.

(c) *Chalcones having two B-ring oxygenations*

The simplest member of this oxygenation subgroup has no A-ring substitution: 3,4-dihydroxychalcone 4-*O*-β-arabinopyranosyl-*O*-β-D-galactopyranoside (*9.59*) was isolated from *Bauhinia purpurea* by Bhartiya and Gupta (1981). A compound with the structure (*9.60*), 2′,3,4-trihydroxychalcone, occurs in *Semecarpus vitiensis* (Anacardiaceae) (Pramono *et al.*, 1981).

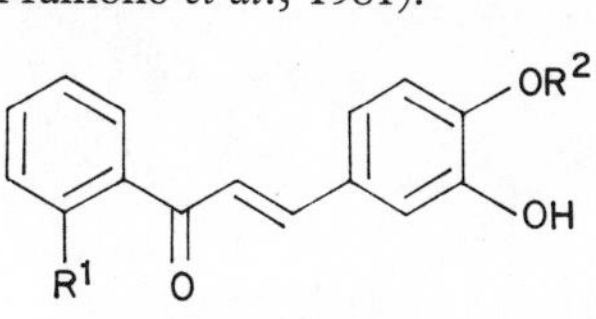

(9.59) $R^1 = H$; R^2 = galactose–arabinose

(9.60) $R^1 = OH$; $R^2 = H$

The most common chalcones with catecholic B-rings are butein, 2′,3,4,4′-tetrahydroxychalcone, and okanin, 2′,3,3′,4,4′-pentahydroxychalcone (*9.61*). Recent reports of these compounds and their known simple ethers and glycosides are summarized in Table 9.1. Noteworthy amongst the reports of okanin glycosides is the finding of okanin 4′-(L-arabinofuranosyl-α-1→4-β-D-glucopyranoside) (*9.62*) in *Abies pindrow* (Tiwari and Minocha, 1980), which represents the first report of an okanin derivative, and one of only a very few reports of chalcones in general from gymnosperms. A rich array of flavonoids was encountered in our study of the genus *Bidens* in Hawaii (Bohm, unpublished results). One of the most complex members within this group is *B. torta* from which five okanin derivatives were obtained: 4-*O*-methyl-4′-*O*-glucoside (*9.63*) and its monoacetate (*9.64*), the 3,4-di-*O*-methyl-4′-*O*-glucoside (*9.65*), the 3,4,3′-tri-*O*-methyl-4′-*O*-glucoside (*9.66*), and the 3,4,3′,4′-tetra-*O*-methyl ether (*9.67*) (McCormick *et al.*, 1984).

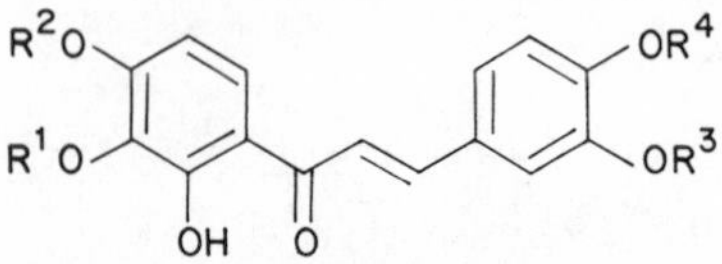

(*9.61*) $R^1 = R^2 = R^3 = R^4 = H$

(*9.62*) $R^1 = R^3 = R^4 = H$; R^2 = glucose–arabinose

(*9.63*) $R^1 = R^3 = H$; R^2 = glucose; $R^4 = CH_3$

(*9.64*) $R^1 = R^3 = H$; R^2 = glucose–X″–acetate; $R^4 = CH_3$

(*9.65*) $R^1 = H$; R^2 = glucose; $R^3 = R^4 = CH_3$

(*9.66*) $R^1 = R^3 = R^4 = CH_3$; R^2 = glucose

(*9.67*) $R^1 = R^2 = R^3 = R^4 = CH_3$

Comparatively few natural chalcones have been reported that are based upon the 2′,3,4,4′,6′-pentahydroxychalcone. Three compounds with this substitution pattern have been reported, however. Imperato (1982b) reported 2′,4,4′,6′-tetrahydroxy-3-methoxychalcone 2′-*O*-glucoside (*9.68*) from *Acacia dealbata* where it occurs with the known aurone glucoside cernuoside (4,6,3′,4′-tetrahydroxyaurone-4-*O*-glucoside). Extra A-ring hydroxylation and *O*-alkylation yields (*9.69*) described by Quijano and co-workers (1982) from *Ageratum strictum*. *C*-Prenylation of flavonoids is common with cyclization to form a variety of derivatives frequently seen. Most of these derivatives have six- or five-membered rings. In two species of which I am aware cyclization occurs to form a seven-membered heterocyclic system as seen in (*9.70*). This compound was recently described as a component of the non-polar leaf-surface flavonoids of *Wyethia angustifolia* (McCormick *et al.*, 1986). This ring system was also found in a dihydrochalcone and a bibenzyl from the liverwort genus *Radula* (II, 344). Also see structure (*9.312*) on p. 378.

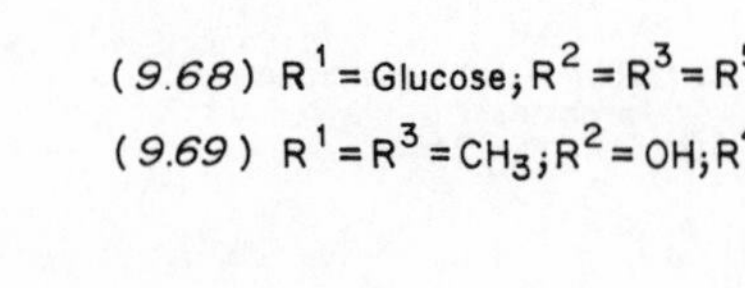

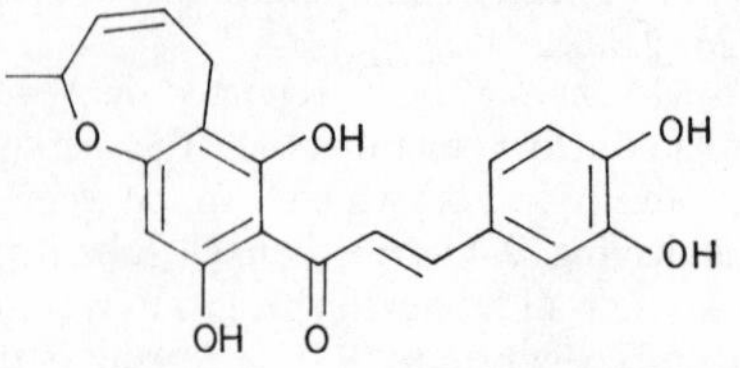

(*9.68*) R^1 = Glucose; $R^2 = R^3 = R^5 = H$; $R^4 = CH_3$

(*9.69*) $R^1 = R^3 = CH_3$; $R^2 = OH$; $R^4 + R^5 = –CH_2–$

(*9.70*)

Pyrano derivatives, e.g. (*9.45*), would be formed by reaction between C-3″ of a 3-methylbuten-2-yl side chain and a phenolic hydroxyl as shown in the conversion of model compound (*9.71*) to (*9.72*). (A five-membered heterocycle, e.g. (*9.47*), would be formed by the reaction at C-2″ of the side chain.) Formation of the seven-membered ring system seen in the *Radula* and *Wyethia* compounds can be rationalized by considering a reaction between a phenolic group and the terminal carbon of the double bond of a 3-methylbuten-3-yl side chain as shown in the conversion of model compound (*9.73*) to (*9.74*).

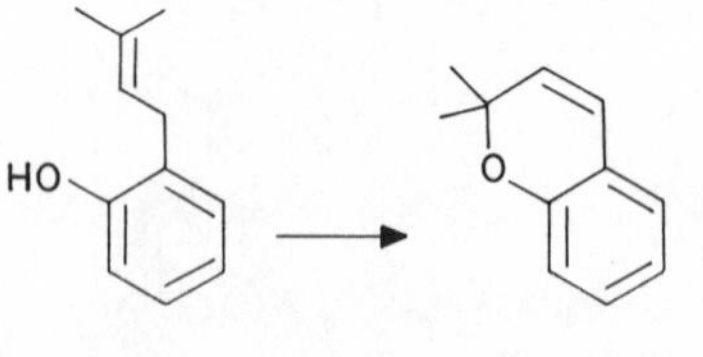

(*9.71*) (*9.72*)

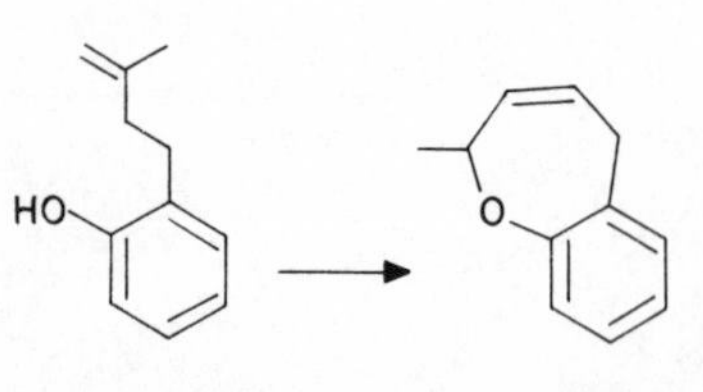

(*9.73*) (*9.74*)

Members of the genus *Scutellaria* are known to contain several flavonoids with unusual B-ring hydroxylation patterns. Several flavanones are known that are characterized by 2′-hydroxylation or 2′,6′-dihydroxylation

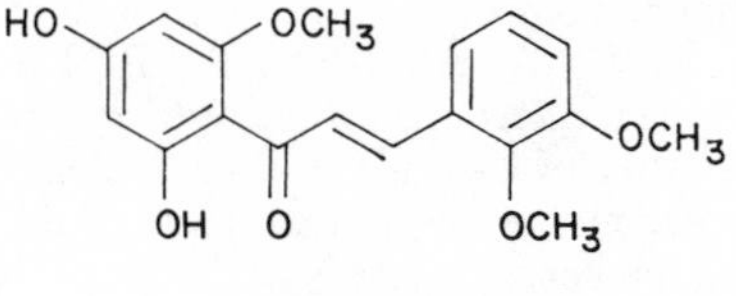

(*9.75*)

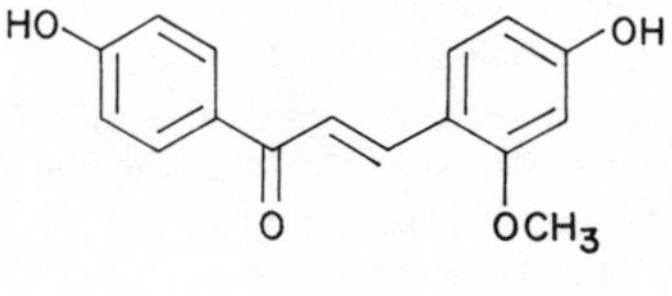

(*9.76*)

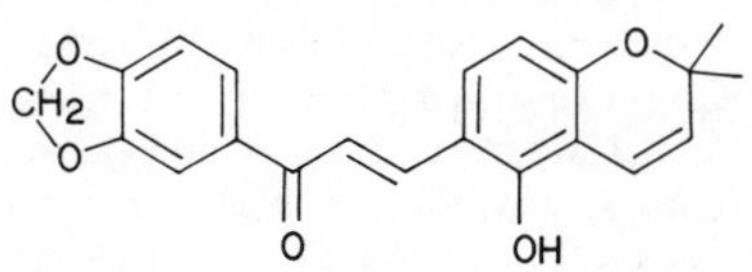

(*9.77*)

(*9.183–9.188*). An unusual chalcone with 2,3-dioxygenation in the B-ring was recently described by Tomimori and co-workers (1985) as a constituent of *S. discolor*. The compound is 2′,4′-dihydroxy-2,3,6′-trimethoxychalcone (*9.75*).

The term 'retrochalcone' was coined to recognize the existence of compounds, such as echinatin (*9.76*) from *Glycyrrhiza echinata*, whose oxygenation patterns appear to be reversed relative to common flavonoids. Biosynthetic studies described in the earlier review (II, 330) showed that the A-ring, as written in (*9.76*), was derived from a phenylpropanoid precursor and that the B-ring was the acetate-derived ring. Further more detailed comments on that system have been published by Ayabe and co-workers (1980a, b). Although no new retrochalcones have been described *per se* it seems reasonable to suggest that other such compounds do indeed exist. A case in point involves a compound that Saini and co-workers (1983) isolated from *Pongamia glabra*. Compound (*9.77*) has an A-ring with 3′,4′-dioxygenation and 2,4-dioxygenation in ring-B. The A-ring substitution as written in (*9.77*) is that normally seen in B-rings (quercetin-type pattern) and the 2,4-dioxygenation could easily be rationalized as a normal, A-ring resorcinol pattern. The intermediacy of dibenzoylmethane derivatives in the formation of retrochalcones has been considered (Ayabe *et al.*, 1980a, b). It is of interest, then, that *Pongamia glabra* also makes such compounds; 2,4,4′-dibenzoylmethane (II, 348) is known from *P. glabra*. Biosynthetic studies of the *Pongamia* compounds would be most welcome.

A chalcone bearing the unusual lavandulyl side chain has been identified from *Sophora flavescens* by Wu and co-workers (1985). These authors showed that 'kushenol-D' has structure (*9.78*). Two related flavanones, 'kushenol-A' (*9.222*) with 2′-oxygenation, and 'kushenol-B' (*9.318*) with 2′,4′-dioxygenation occur with the chalcone in this species.

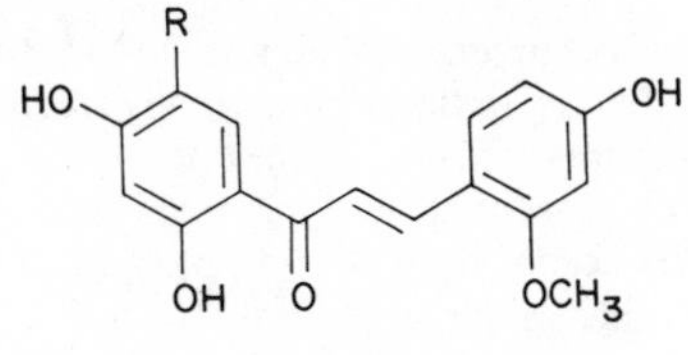

(*9.78*) R = lavandulyl

During the past several years a number of remarkable flavonoids have been isolated from *Morus* (Moraceae), the mulberry, and identified. In addition to relatively simple compounds showing five- or ten-carbon alkyl side chains there have also surfaced compounds that are essentially flavonoid dimers although not in the sense of the word used to describe biflavonoids. The section below on flavanones includes several of these unusual structures. Chalcones belonging to this class are also known. From callus cultures of *Morus alba* Ueda and co-workers (1982) obtained a compound they called 'kuwanon-J' (*9.79*). Compounds of

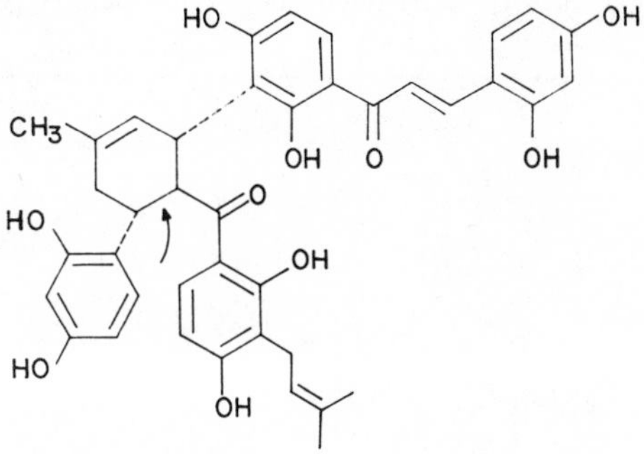

(*9.79*)

this class are considered to have arisen by a Diels-Alder-type condensation, in this case between two molecules of 2,2′,4,4′-tetrahydroxy-3′-*C*-prenylchalcone. Classical

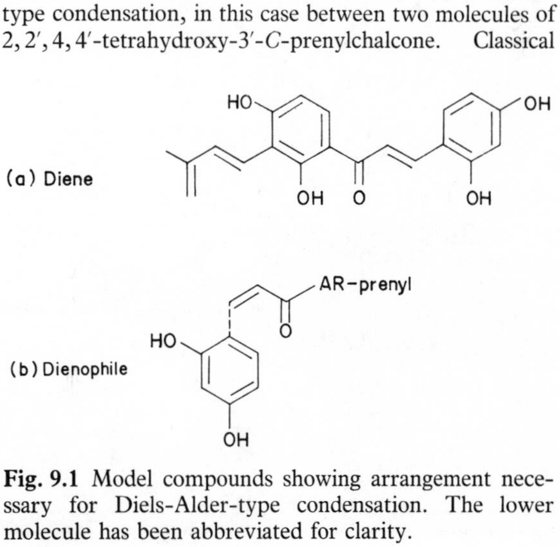

(a) Diene

(b) Dienophile

Fig. 9.1 Model compounds showing arrangement necessary for Diels-Alder-type condensation. The lower molecule has been abbreviated for clarity.

Diels-Alder reactions occur between a diene and a dienophile which is frequently attached to an electron-withdrawing group. Such an arrangement is seen in Fig. 9.1 where reaction would occur between the side chain diene in molecule (a) and the α, β-double bond of the chalcone (b) which serves as the dienophile. The electron-withdrawing group in this case is the carbonyl function. The *trans* stereochemistry about the marked bond of kuwanon-J is concordant with the *trans* stereochemistry of the chalcone α, β-double bond.

In a recent very detailed account of the flavonoids of *M. alba* cell cultures Ikuta and co-workers (1986) discuss the assignment of structures to four such compounds, including kuwanon-J. In this set of compounds two monomeric chalcones are involved, A and B, structures of which are shown in Table 9.2. Chalcone-A has 2,4-dihydroxylation; chalcone-B has 4-hydroxylation. There are four possible combinations of these monomers all of which are seen: A + A, B + A, A + B and B + B. The products of these condensations are summarized in the table. The stereochemistry of C-3″ is consistently alpha, that is, the chalcone moiety lies beneath the plane of the paper with respect to the benzoyl function.

Table 9.2 Structures of kuwanons J, Q, R and V and the combination of precursor chalcones from which they could be formed

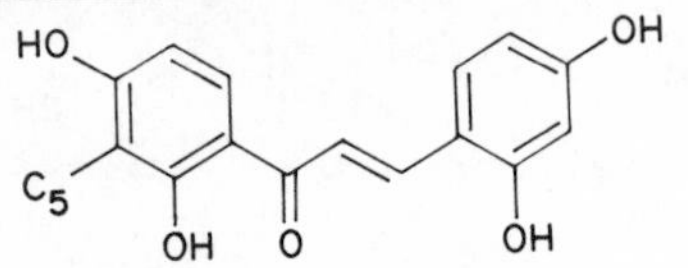

Chalcone-A

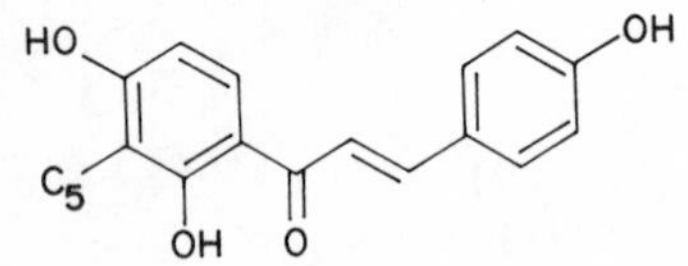

Chalcone-B

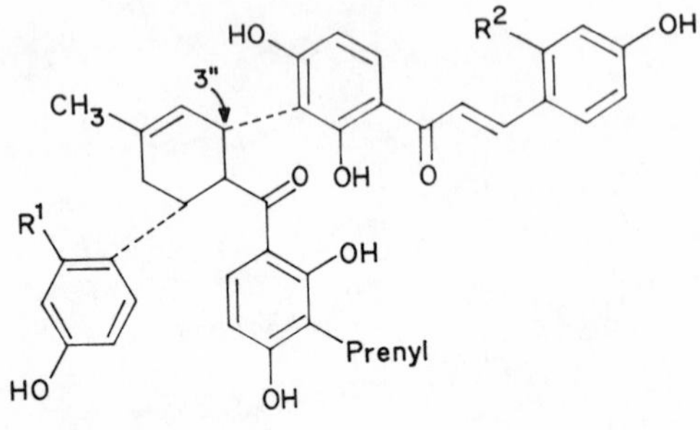

Kuwanon	R_1	R_2	*Combination*
J	OH	OH	A + A
Q	H	OH	B + A
R	OH	H	A + B
V	H	H	B + B

(d) *Chalcones having three B-ring oxygenations*

Only one compound with a trioxygenated B-ring has been reported since the last review. In a study of *Pongamia glabra* Pathak and co-workers (1983) reported chalcone (*9.80*) along with pongachalcone-I (*9.33*) a well-known component of the genus.

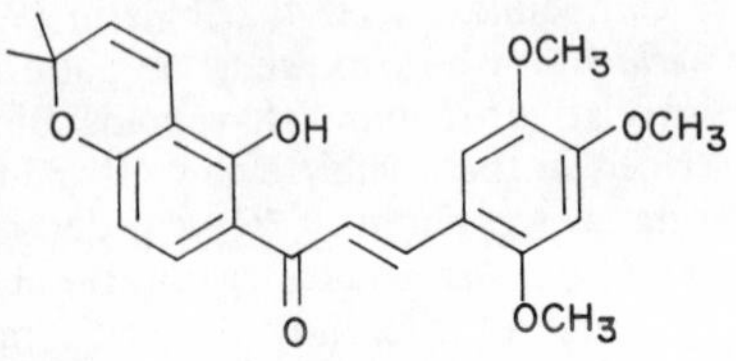

(9.80)

9.3 AURONES

9.3.1 Introduction

Aurones are based upon 2-benzylidenecoumaranone, shown below. Positions in aurones are identified using primed numbers for the B-ring and unprimed for the A-ring; aurones are in this regard 'normal' flavonoids. However, position-4 in aurones is biosynthetically equivalent to position-5 in normal flavonoids. Included for convenience in this section are comments on 'auronols', compounds that are very closely related to aurones but have an hydroxylated heterocyclic ring.

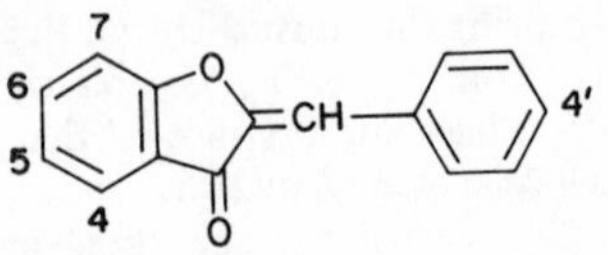

9.3.2 Structures of naturally occurring aurones

Structures and occurrences of all aurones are summarized in Table 9.3. Owing to the comparatively small number of reports of aurones during the last five years or so the material to be presented here has not been divided into separate parts according to B-ring oxygenation as is the case for the other minor flavonoids. No new aurone aglycones have been reported since the last review (II, 334-339). New occurrences of derivatives of the well-known aurones sulfuretin (*9.81*), maritimetin (*9.82*), leptosidin

Table 9.3 Structures of naturally occurring aurones (including an auronol)

Substituents: *OH*	*OCH$_3$*	*Other*	*No. in text*	*Trivial name*	*Source**		*References*
6, 3′, 4′		6-*O*-Glucose	(*9.81*)	Sulfurein	*Coreopsis* sect. *Coreopsis*	Com	Crawford and Smith (1985)
					Coreopsis sect. *Euleptosyne*	Com	Crawford and Smith (1984)
					Coreopsis sect. *Pugiopappus*	Com	Crawford and Smith (1984)
					Helianthus heterophyllus	Com	Schilling (1983)
					Tithonia koelzii	Com	LaDuke (1982)
					Viguiera series *Viguiera*	Com	Rieseberg and Schilling (1985)
					Zinnia linearis	Com	Harborne *et al.* (1983)
6, 7, 3′, 4′		6-*O*-Glucose	(*9.82*)	Maritimein	*Coreopsis* sect. *Coreopsis*	Com	Crawford and Smith (1985)
					Viguiera dentata	Com	Rieseberg and Schilling (1985)
					Zinnea linearis	Com	Harborne *et al.* (1983)
6, 3′, 4′	7	6-*O*-Glucose	(*9.83*)	Leptosin	*Coreopsis* sect. *Coreopsis*	Com	Crawford and Smith (1985)
					Vaccinium oxycoccos	Eri	Jankowski and Jocelyn-Paré (1983)
4, 6, 3′, 4′		6-*O*-Rhamnose-glucose	(*9.84*)		*Cyperus scarious*	Cyp	Bhatt *et al.* (1981b)
		6-Arabinose-xylose			*C. scarious*	Cyp	Bhatt *et al.* (1984)
				Aureusidin	*Cyperus* species	Cyp	Harborne *et al.* (1982)
		4-*O*-Glucose		Cernuoside	*Acacia dealbata*	Leg	Imperato (1982a)
		6-*O*-Glucose			*Mussaenda hirsutissima*	Rub	Harborne *et al.* (1983a)
		4, 6-Di-*O*-glucose		Aureusin	*Mussaenda hirsutissima*	Rub	Harborne *et al.* (1983)
					M. hirsutissima	Rub	Harborne *et al.* (1983)
2, 6, 4′	4		(*9.85*)	Carpusin	*Pterocarpus marsupium*	Leg	Maurya *et al.* (1982); Mathew and Subba Rao (1983)

*Key: Com = Compositae, Cyp = Cyperaceae, Eri = Ericaceae, Leg = Leguminosae, Rub = Rubiaceae.

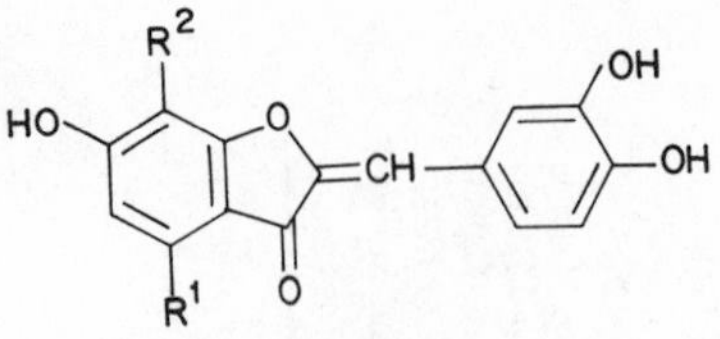

(*9.81*) $R^1 = R^2 = H$

(*9.82*) $R^1 = H$; $R^2 = OH$

(*9.83*) $R^1 = H$; $R^2 = OCH_3$

(*9.84*) $R^1 = OH$; $R^2 = H$

(*9.83*) and aureusidin (*9.84*) are given in Table 9.3. Some new glycosidic forms and first records of interest are worth mentioning. Harborne and co-workers (1983) reported the co-existence of the novel aureusidin 4, 6-di-*O*-glucoside with aureusin (the 6-*O*-glucoside) and cernuoside (the 4-*O*-glucoside) in the orange flowers of *Mussaenda hirsutissima*. This represents the first report of an aurone from the Rubiaceae. Although other aglycone types have been reported from members of the Leguminosae the finding of cernuoside, the 4-*O*-glucoside of (*9.84*), in *Acacia dealbata* by Imperato (1982a) represents a first report for the family. This is of interest because of the low proportion of phloroglucinol-based A-ring flavonoids obtained from legumes. Leptosin, the 6-*O*-glucoside of (*9.83*), formerly known only from *Coreopsis*, has now been reported to occur in trace amounts in cranberries (*Vaccinium oxycoccos*) by Jankowski and Paré (1983). This represents the first report of an aurone from the Ericaceae. The presence of two unusual diglycosides of leptosidin in leaves of *Cyperus scarious* was reported by Bhatt and co-workers (1981b, 1984). The first of these was 6-*O*-β-D-glucopyranosyl-(1→4)-*O*-α-L-rhamnopyranoside, the second 6-*O*-β-D-xylopyranosyl-(1→4)-*O*-β-D-arabinopyranoside. For information about the broad distribution of aurones in the Cyperaceae the reader is referred to the large surveys of the genus published by Harborne and co-workers (1982, 1985).

Localization of flavonoid derivatives in floral tissue has attracted attention in recent years because of the involvement of floral pattern with pollination vector attraction. As part of an ongoing study of *Viguiera*, Rieseberg and Schilling (1985) have studied the floral pigment patterns in *V. dentata* where the marein–maritimein chalcone–aurone pair was found, and in *Viguiera* series *Viguiera* all of whose nine species contain the coreopsin–sulfurein pair. Degot and co-workers (1983) found several different flavonoid types including aurones in *Linaria macroura* but only aurones in the flowers. Floral pigments of *Mussaenda hirsutissima* were shown to be glucosides of aureusidin by Harborne and co-workers (1983) who reported in the same paper that flowers of *Zinnia linearis* accumulate sulfurein and the marein–maritimein pair.

'Auronols' are 2-hydroxy-2-benzylcoumaranones. This is a small set of aurone derivatives in which the benzylidene unsaturation has undergone hydration (speaking chemically, the biosynthetic origin of these compounds is unknown). Five different aglycones were described in our first review (I, 474). No members of this group, new or otherwise, appeared in the 1982 review. Since that time only one addition has been made to the list. The new compound is 2, 6-dihydroxy-2-(4-hydroxybenzyl)-4-methoxycoumaranone (*9.85*) and was reported from

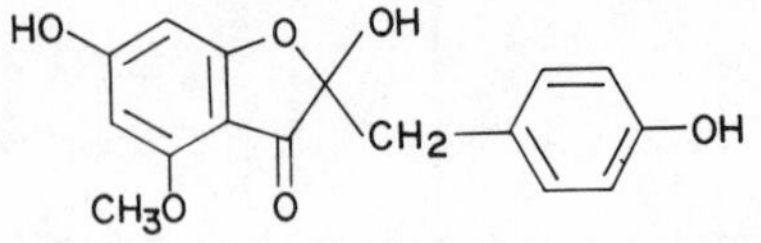

(*9.85*)

Pterocarpus marsupium (Leguminosae) by two sets of workers (Maurya *et al.*, 1982; Mathew and Subba Rao, 1983). Mathew and Subba Rao chose to call the new compound 'carpusin.' It could have been named 4-methylmaeopsin as it is a derivative of maeopsin, 2, 4, 6-trihydroxy-2-(4-hydroxybenzyl)coumaranone.

9.4 DIHYDROCHALCONES

9.4.1 Introduction

Identification of positions in the dihydrochalcones follows the pattern described for chalcones. This section also treats dihydrochalcones that have substituents on the α- and β-carbons of the bridge.

In addition to the chemical studies of dihydrochalcones described below there have been several studies focusing on other aspects of these compounds. The utilization of certain dihydrochalcone derivatives and related compounds as sweetening agents has been reviewed recently by Horowitz (1986). One aspect of this research has been an attempt to learn something about the structure of the sweet-sensing receptor site. Wong and Horowitz (1986) studied the crystal structure of neohesperidin dihydrochalcone sweetener (2′, 3, 4′, 6-tetrahydroxy-4-methoxydihydrochalcone 4′-*O*-neohesperidoside). Such information would give clues to the shape and/or size of the binding site. Likewise, Cody (1984) studied the crystal structure of phloridzin dihydrate in order to gain some insight into details of its inhibitory interaction with iodothyronine deiodinase. Interactions between dihydrochalcones and chalcones with isolated mitochondria showed that these compounds can act as decouplers in electron transport (Ravanel *et al.*, 1982). Readers interested in details of the physiological activities of dihydrochalcones, and other flavonoid types, should refer to the recent review volume edited by Cody *et al.* (1986).

9.4.2 Structures of naturally occurring dihydrochalcones

Recently reported dihydrochalcones are summarized in Table 9.4.

(a) *Dihydrochalcones lacking B-ring oxygenation*

The phloroglucinol-based A-ring member of this group (*9.86*) has been isolated from *Lindera umbellata* along with the 4′-methyl ether (*9.87*). The 2′-methyl ether, uvangolatin (*9.88*), was isolated from *Uvaria angolensis* by Hufford and Oguntimein (1982) and was synthesized by Bhardwaj and co-workers (1982). 2′,4′-Dimethyldihydrochalcone (*9.89*) has been isolated from two sources: *Alpinia speciosa* (Itokawa *et al.*, 1981) and *Uvaria angolensis* (Hufford and Oguntimein, 1982).

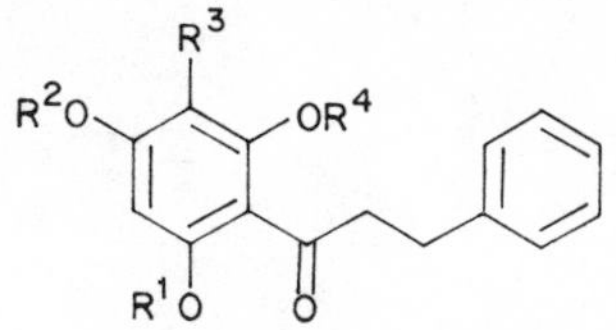

(*9.86*) $R^1 = R^2 = R^3 = R^4 = H$

(*9.87*) $R^1 = R^3 = R^4 = H$; $R^2 = CH_3$

(*9.88*) $R^1 = CH_3$; $R^2 = R^3 = R^4 = H$

(*9.89*) $R^1 = R^2 = CH_3$; $R^3 = R^4 = H$

(*9.90*) $R^1 = R^2 = R^4 = H$; R^3 = prenyl

(*9.91*) $R^1 = R^3 = H$; $R^2 = CH_3$; R^4 = prenyl

C-Prenylation of the base molecule would lead to the formation of compound (*9.90*) which Bohlmann and co-workers (1984a) have isolated from *Helichrysum argyrolepis*. *Radula* is a genus of liverworts known to produce *C*-prenylated flavonoids and bibenzyls (II, 344). Asakawa and co-workers (1982) reported a simple dihydrochalcone from *R. perrottetii* and showed it to be (*9.91*). Both of these compounds are newly reported as natural products. *C*-Prenylation is frequently seen in flavonoids isolated from members of the genus *Flemingia*. Yet another new compound was described by Subrahamanyam and co-workers (1982) from *F. stricta*; 'flemistrictin-D' was shown to have structure (*9.92*). It occurs in *F. stricta* along with a series of chalcones and flavanones most of which vary principally in the manner of ring closure of the prenyl group

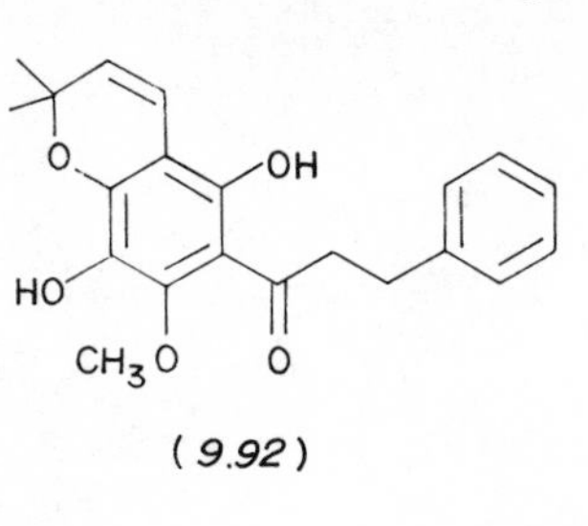

(*9.92*)

Unusual *C*-alkylation products have been identified as constituents of *Lindera umbellata* var. *lancea* by Tanaka and co-workers (1984a). 'Linderatin', whose structure is shown as (*9.93*), is characterized by a monoterpene substituent at position C-3′. A flavanone with the same substitution pattern, including the monoterpene side chain, has also been reported from this species (Tanaka *et al.*, 1985); 'linderatone' is compound (*9.160*).

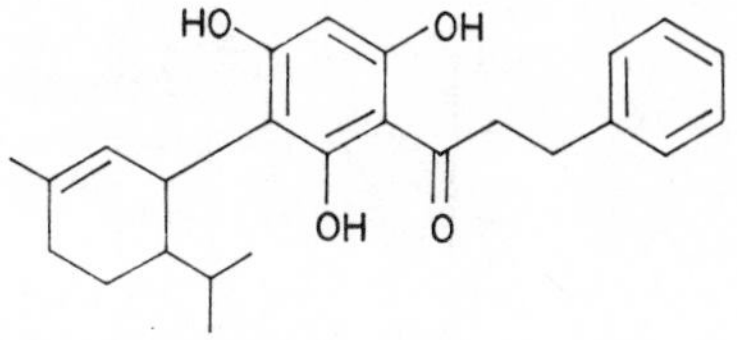

(*9.93*)

Ten-carbon alkyl side chains are known from members of several genera including *Helichrysum* (Compositae), *Angelica* and *Bonannia* (Umbelliferae), *Ammothamnus*, *Flemingia*, *Lespedeza* and *Sophora* (Leguminosae), *Morus* (Moraceae), *Boesenbergia* (Zingiberaceae) and *Aniba* (Lauraceae). Most of the derivatives seen in these taxa are linear, the geranyl group being the most common. A few have undergone cyclization to yield 6″-methyl-6″-isohexenylpyran derivatives. Prior to the description of linderatin (*9.93*), however, only one ten-carbon substituted flavonoid (at least amongst the minor flavonoids) was known to exist in the cyclized monoterpene form: rubranine (*9.94*) is known from *Aniba rosaeodora* (II, 324) of the Lauraceae. It is of interest that the newest monoterpene-flavonoids, lineratin and linderatone, also come from a member of the Lauraceae. A biosynthetic relationship exists between linderatin and rubranine. Both contain a *C*-linked monoterpene ring; in the case of rubranine two further bonds exist joining the side chain to the chalcone A-ring via the 2′- and 4′-oxygens. Biosynthetic studies would be welcome.

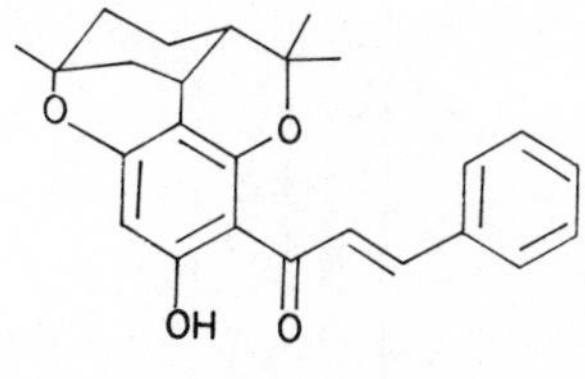

(*9.94*)

The next group of compound is distinguished by the presence of the highly unusual *o*-hydroxybenzyl function. Compound (*9.95*) is a known constituent of the genus *Uvaria* (II, 345). The related compounds (*9.96*) and (*9.97*) have been isolated for the first time from *U. angolensis* by

Table 9.4 Structures of naturally occurring dihydrochalcones

Substituents			No. in text	Trivial name	Source*		References
OH	OCH_3	Other					
2′, 4′, 6′			(9.86)		*Lindera umbellata*	Lau	Tanaka *et al.* (1984b)
2′, 6′	4′		(9.87)		*L. umbellata*	Lau	Tanaka *et al.* (1984b)
4′, 6′	2′		(9.88)	Uvangolatin	*Uvaria angolensis*	Ano	Hufford and Oguntimein (1980)
6′	2′, 4′		(9.89)	Dihydroflavokawin-B	*Alpinia speciosa*	Zin	Itokawa *et al.* (1981)
					Uvaria angolensis	Ano	Hufford and Oguntimein (1980)
2′, 4′, 6′		3′-*C*-Prenyl	(9.90)		*Helichrysum argyrolepis*	Com	Bohlmann *et al.* (1984c)
2′, 6′	4′	5′-*C*-Prenyl	(9.91)		*Radula perrottetii*	Lvr	Asakawa *et al.* (1982)
2′, 4′	6′	3′-*C*-(2-Hydroxybenzyl); 5′-*C*-methyl	(9.98)	Anguvetin	*Uvaria angolensis*	Ano	Hufford and Oguntimein (1982)
2′, 4′	6′	3′-*C*-(2-Hydroxybenzyl)	(9.95)		*Uvaria angolensis*	Ano	Muhammad and Waterman (1985)
2′, 4′	6′	3′, 5′-Di-*C*-(2-hydroxybenzyl)	(9.96)		*Uvaria angolensis*	Ano	Muhammad and Waterman (1985)
2′, 6′		Di-(2-hydroxybenzyl) but one is cyclized	(9.97)		*Uvaria angolensis*	Ano	Muhammad and Waterman (1985)
2′, 5′	6′	6″, 6″-Dimethylpyrano(2″, 3″, 4′, 3′)	(9.92)	Flemistrictin-D	*Flemingia stricta*	Leg	Subrahmanyam *et al.* (1982)
2′, 4′, 6′		3′-*C*-(3-Methyl-6-isopropyl-cyclohex-2-enyl)	(9.93)	Linderatin	*Lindera umbellata*	Lau	Tanaka *et al.* (1984a)
2′, 4, 4′		4′-*O*-Glucose	(9.101)	Confusoside	*Symplocos confusa*	Sym	Tanaka *et al.* (1982)
2′, 4, 4′		α-Hydroxy	(9.114)		*Zollernia paraenis*	Leg	Ferrari *et al.* (1983)
2′, 4, 4′		3′-*C*-Glucose; α-hydroxy	(9.115)	Coatline-A	*Eysenhardtia polystachya*	Leg	Beltrami *et al.* (1982)

2′, 3, 4, 4′		3′-*C*-Glucose; α-hydroxy	(*9.116*)	Coatline-B	*E. polystachya*	Leg	Beltrami *et al.* (1982)
2′	4, 4′	α-Hydroxy	(*9.112*)	Odoratol	*Lathyrus odoratus*	Leg	Fuchs *et al.* (1984)
					Pterocarpus angolensis	Leg	Bezuidenhoudt *et al.* (1981)
		See text for structure	(*9.99*)	Grenoblone	*Platanus acerifolia*	Pla	Kaouadji (1986)
		See text for structure	(*9.100*)	Hydroxygrenoblone	*P. acerifolia*	Pla	Kaouadji *et al.* (1986a)
	2′, 4, 4′	α-Hydroxy	(*9.113*)	Methylodoratol	*Lathyrus odoratus*	Leg	Fuchs *et al.* (1984)
2′, 4, 4′, 6′		α-Hydroxy	(*9.111*)	Nubigenol	*Podocarpus nubigena*	Cnf	Bhakuni *et al.* (1973)
2′, 4, 4′, 6′		2′-*O*-Glucose	(*9.104*)		*Symplocos lancifolia*	Sym	Tanaka *et al.* (1980)
					S. spicata	Sym	Tanaka *et al.* (1980)
					Lithophragma affine	Sax	Nicholls and Bohm (1984)
		4′-*O*-Glucose	(*9.103*)		*Symplocos microcalyx*	Sym	Tanaka *et al.* (1980)
2′, 4, 6′	4′		(*9.105*)	Lyonogenin	*Lyonia formosa*	Eri	Shukla *et al.* (1973)
				Asebogenin	*Pityrogramma calomelanos*	Fil	Hitz *et al.* (1982)
		2′-*O*-Glucose		Lyonotin	*Lyonia formosa*	Eri	Shukla *et al.* (1973)
2′, 4′	4, 6′		(*9.106*)		*Iryanthera grandis*	Myt	Vieira *et al.* (1983)
2′, 6′	4, 3′	4′, 5′-Methylenedioxy	(*9.107*)		*Pityrogramma ebenea*	Fil	Miraglia *et al.* (1985)
2′, 4, 6′	4′	3′, 5′-Di-*C*-methyl	(*9.108*)		*P. triangularis*	Fil	Wollenweber *et al.* (1985a)
2′, 3, 4, 4′, 6′		2′-*O*-Glucose			*Astilbe* X *arendsii*	Sax	Bohm and Bhat (1985)
		2′-*O*-Galactose			*Astilbe* X *arendsii*	Sax	Bohm and Bhat (1985)
2′	5′, 6′	Furano(2″, 3″, 4′, 3′); 3, 4-methylenedioxy	(*9.110*)		*Derris araripensis*	Leg	do Nascimento and Mors (1981)
	2′, 4′, β	3, 4-Methylenedioxy	(*9.117*)	Dihydromilletenone methyl ether	*Millettia hemsleyana*	Leg	Mahmoud and Waterman (1985b)
	2, 4, β	3′, 4′-Methylenedioxy	(*9.118*)	Dihydroisomilletenone methyl ether	*M. hemsleyana*	Leg	Mahmoud and Waterman (1985b)

*Key: Ano = Annonaceae, Com = Compositae, Cnf = Coniferae, Eri = Ericaceae, Fil = Filicinae, Lau = Lauraceae, Leg = Leguminosae, Lvr = liverwort, Myt = Myristicaceae, Sax = Saxifragaceae, Sym = Symplocaceae, Zin = Zingiberaceae.

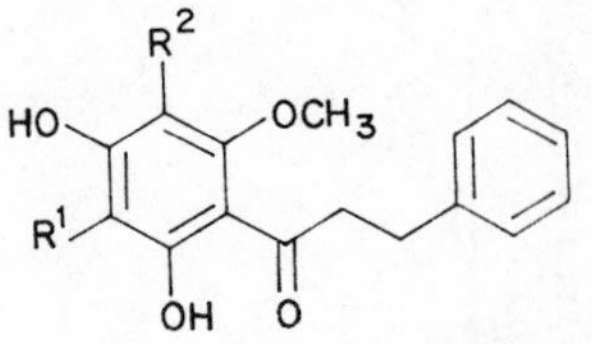

(9.95) R^1 = 2-hydroxybenzyl; R^2 = H
(9.96) R^1 = R^2 = 2-hydroxybenzyl
(9.98) R^1 = 2-hydroxybenzyl; R^2 = CH_3

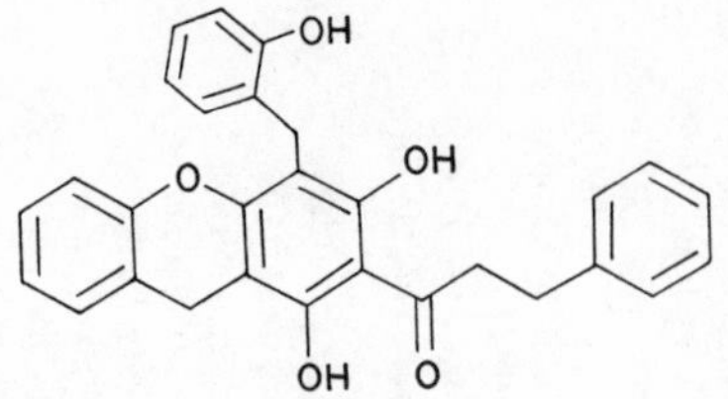

(9.97)

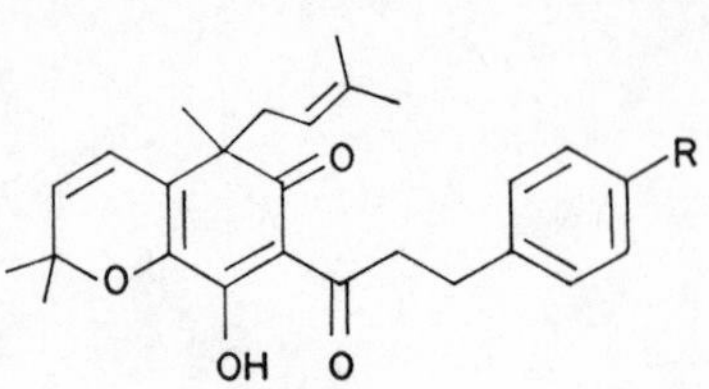

(9.99) R = H
(9.100) R = OH

Muhammad and Waterman (1985). Compound (*9.97*) results from cyclization of the 2-hydroxybenzyl function and the C-4′ hydroxyl group. Hufford and Oguntimein (1982) reported (*9.98*) from *U. angolensis*. Other 2-hydroxylated flavonoids and related compounds, e.g. (*9.50*) and (*9.51*) have been discussed above.

In three recent papers Kaouadji and co-workers (Kaouadji, 1986; Kaouadji *et al.*, 1986a, b) described several compounds that they obtained from buds of *Platanus acerifolia* (Platanaceae). Three of these are simple prenylated flavanones (*9.146*), (*9.147*) and (*9.153*). The other two compounds, (*9.99*) and (*9.100*), share a most unusual A-ring. 'Grenoblone' and '4-hydroxygrenoblone' exhibit the rare 2′, 3′, 6′-trihydroxylation pattern as well as double alkylation at C-5′. The latter feature prevents aromatization, of course.

(b) *Dihydrochalcones having one B-ring oxygenation*

The simplest member of this group, 'confusoside,' is structure (*9.101*) which was isolated from *Symplocos confusa* (Symplocaceae) by Tanaka and co-workers (1982). It is a pity that their trivial name has been assigned to such a simple compound. There are numerous examples in this chapter more deserving of such a descriptive title! Compound (*9.101*) is isomeric to the known davidoside (II, 345) which is the 2′-*O*-glucoside.

The dimeric form of 2′, 4, 4′-trihydroxydihydrochalcone, where the linkage connects the two α-carbons, has been found by Drewes and Hudson (1983) as a natural constituent of *Brackenridgea zanguebarica* (Ochnaceae). So far as I am aware, this is the first reported natural biflavonoid consisting of two dihydrochalcone moieties.

Earlier studies by Tanaka and co-workers (1980) had established the presence of other dihydrochalcones in *Symplocos: S. microcalyx* had 0.6% phloretin 4′-*O*-β-D-glucoside (*9.103*); *S. lancifolia* and *S. spicata* had 0.6% and 0.7%, respectively, of phloretin 2′-*O*-β-D-glucoside (*9.104*). Compound (*9.104*) was reported by Nicholls and Bohm (1984) to be a variable component of the flavonoid profile of *Lithophragma affine* (Saxifragaceae).

The appearance of a paper by Hitz *et al.* (1982) in which they reported the presence of asebogenin (*9.105*) in the leaf exudate of *Pityrogramma calomelanos*, purportedly as the first report of the aglycone as a natural product, prompts me to review the history of this compound. It also allows me to correct an oversight in my 1982 review. Eykman (1883) first reported the isolation of a crystalline compound, which he called asebotin, from *Pieris japonica* (*Andromeda japonica*). (Asebogenin would, therefore, be the logical name for the aglycone). Several years later Bourquelot and Fichtenholz (1912) described a crystalline glycoside from leaves of *Kalmia latifolia* and showed that it was the same as the compound isolated by Eykman. They (Bourquelot and Fictenholz, 1912) noted the much higher

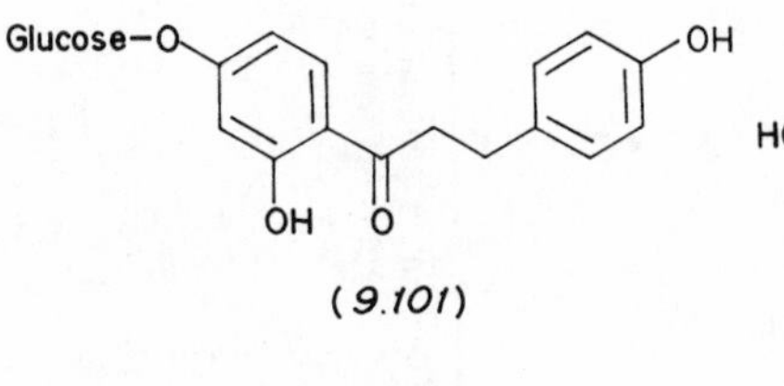

(9.101)

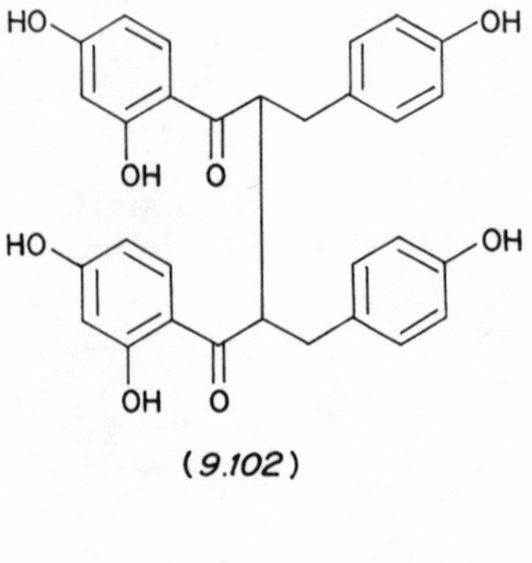

(9.102)

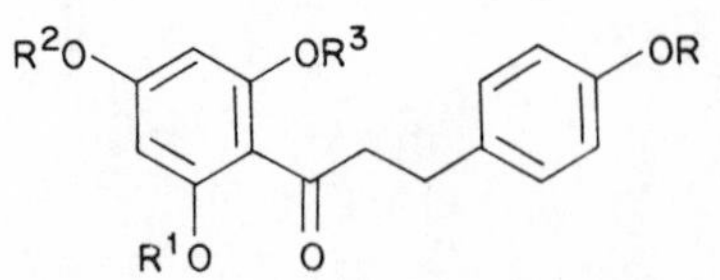

(9.103) R^1 = R^3 = R^4 = H; R^2 = glucose
(9.104) R^1 = glucose; R^2 = R^3 = R^4 = H
(9.105) R^1 = R^3 = R^4 = H; R^2 = CH_3
(9.106) R^1 = R^4 = CH_3; R^2 = R^3 = H

concentration of asebotin in *Kalmia*, 26.6 g kg^{-1} of leaves compared to 4.15 g kg^{-1} in *Andromeda*.

Murakami and Takeuchi (1936) methylated asebotin and showed that the methylated aglycone was identical to phloretin trimethyl ether. This served to establish that both compounds are 2′-*O*-glucosides (glucose had been established as the sugar). In the same year Tamura (1936) fused asebotin with KOH and recovered phloroglucinol monomethyl ether and phloretic acid (3-(4-hydroxyphenyl) propionic acid). Thus, asebotin has a methyl group at either position 2′ or position 4′. Using the Hoesch synthesis Tamura (1936) obtained the 2′-methyl ether whose properties differed from asebogenin, thus confirming that asebogenin must have a methyl ether at position-4′. Zemplén and Mester (1942) confirmed these methylation results by showing that total methylation of asebotin and acid hydrolysis gave the same product as total methylation of phloridzin followed by acid hydrolysis. They (Zemplen and Mester, 1942) then described an unequivocal synthesis of the natural product.

The next appearance of asebotin in the literature was the report by Williams (1964) that it occurs in *Kalmia angustifolia*. A report appeared in 1973, which I overlooked in my earlier review, that described the presence of both the glucoside and the aglycone in the leaves of *Lyonia formosa* (Shukla *et al.*, 1973). This plant, also a member of the Ericaceae, occurs in the northeastern Himalayas and is known to be toxic to mules. Shukla *et al.* (1973) apparently unaware of the earlier work, chose to call their compounds 'lyonotin' and 'lyonogenin' respectively. Two years later Mabry and co-workers (1975) reported asebotin and its aglycone from three species of *Rhododendron* and claimed that that was the first report of the aglycone as a natural product. The same claim has now been registered by Hitz and co-workers (1982) as a result of their finding of asebogenin in the leaf exudate of a single specimen of *Pityrogramma calomelanos*. The record from *Pityrogramma* certainly is the first report of the compound outside of the Ericaceae. However, the credit for first recognizing asebogenin as a natural product must go to Shukla and co-workers (1973). Since asebogenin would be the logical name for the aglycone of asebotin (the name originally given by Eykman), the names lyonotin and lyonogenin should not be used.

2′,4′-Dihydroxy-4,6′-dimethoxydihydrochalcone (*9.106*), first found as a component of *Iryanthera laevis* (II, 346), has been synthesized by Bhardwaj and co-workers (1982) and more recently found in another species of *Iryanthera*, *I. grandis*, by Vieira and co-workers (1983).

Members of the fern genus *Pityrogramma* continue to yield interesting and new natural flavonoids. Two new compounds, both of which exhibit totally substituted A-rings, are shown as structures (*9.107*) and (*9.108*). Miraglia and co-workers (1985) found (*9.107*) in the farinose exudate of *P. ebenea*, while compound (*9.108*) was reported by Wollenweber and his collaborators (1985a) as yet another constituent of the farinose chemistry of *P. triangularis*.

(c) *Dihydrochalcones having two B-ring oxygenations*

Only two members of this group appear to have been reported since the last review. One of these is the very simple 2′,3,4,4′,6′-pentahydroxydihydrochalcone

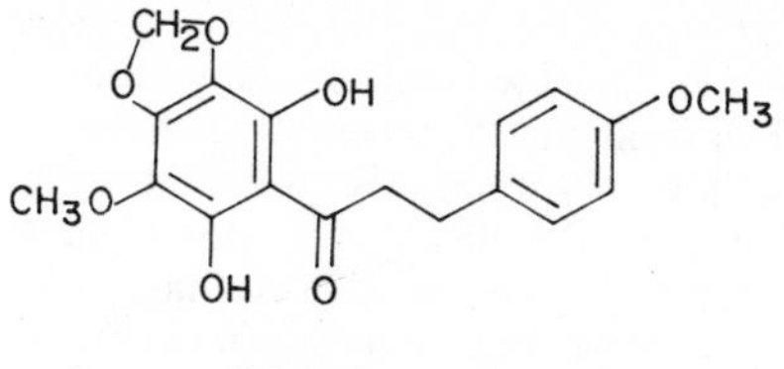

(*9.107*)

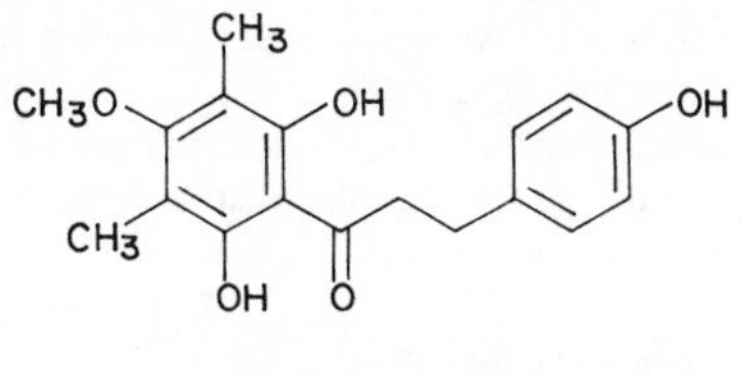

(*9.108*)

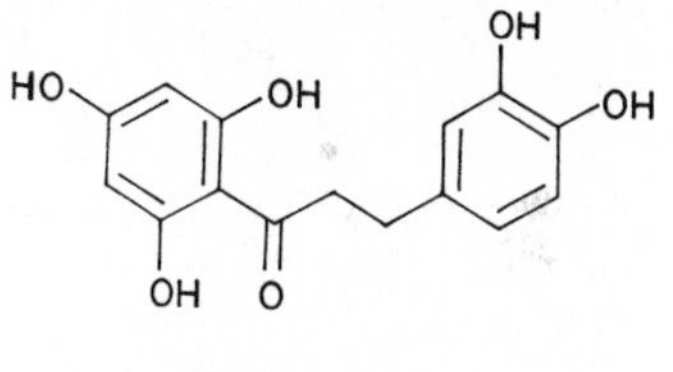

(*9.109*)

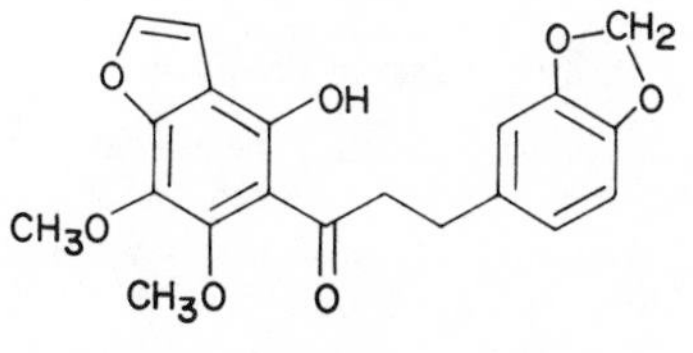

(*9.110*)

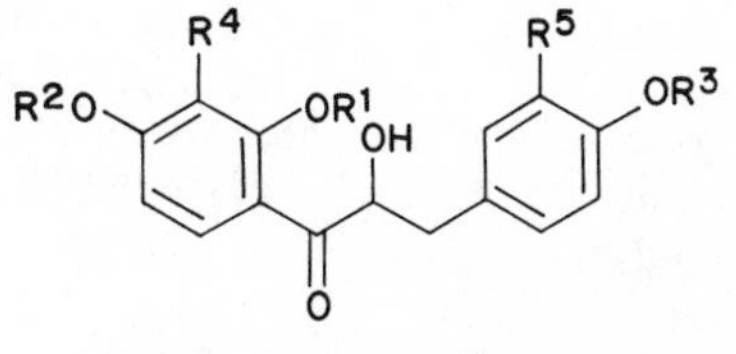

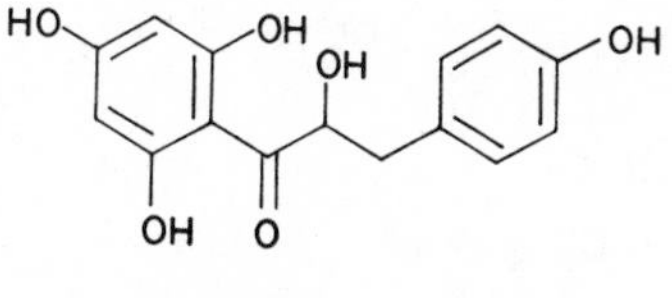

(*9.111*)

(*9.112*) $R^1 = R^4 = R^5 = H$; $R^2 = R^3 = CH_3$

(*9.113*) $R^1 = R^2 = R^3 = CH_3$; $R^4 = R^5 = H$

(*9.114*) $R^1 = R^2 = R^3 = R^4 = R^5 = H$

(*9.115*) $R^1 = R^2 = R^3 = R^5 = H$; R^4 = glucose

(*9.116*) $R^1 = R^2 = R^3 = H$; R^4 = glucose; $R^5 = OH$

(*9.109*) which we (Bohm and Bhat, 1985) found as a mixture of 2′-*O*-glucoside and 2′-*O*-galactoside in *Astilbe* × *arendsii* (Saxifragaceae). The second compound, structure (*9.110*), was found in *Derris araripensis* by do Nascimento and Mors (1981) along with a flavone, four 3-*O*-methylflavonols, a flavanone, a 3-*O*-methyldihydroflavonol and a flavan, several of which have the furano and methylenedioxy functions seen in (*9.110*).

(d) *Dihydrochalcones having substituents on bridge carbons*

Dihydrochalcones bearing an oxygen substituent adjacent to the carbonyl function on the bridge are rare in nature. The first such compound to be discovered in nature is α, 2′, 4, 4′, 6′-pentahydroxydihydrochalcone (*9.111*), 'nubigenol,' which Bhakuni and co-workers (1973) isolated from aerial parts of the conifer *Podocarpus nubigena*. This remained the sole representative of this type until Bezuidenhoudt and co-workers (1981) identified α, 2′-dihydroxy-4,4′-dimethoxydihydrochalcone (*9.112*) from the heartwood of the legume *Pterocarpus angolensis*. Compound (*9.112*), 'odoratol,' along with 'methylodoratol' (*9.113*), were observed in *Lathyrus odoratus* that had been stressed by either infection by *Phytophthora megasperma* var. *sojae* or by treatment with mercuric acetate (Fuchs *et al.*, 1984). The simplest member of this group, compound (*9.114*), was found, along with isoliquiritigenin, liquiritigenin and 7-hydroxy-4′-methoxyisoflavone, in *Zollernia paraensis* (Ferrari *et al.*, 1983). Two *C*-glucosyl α-hydroxydihydrochalcones were identified as constituents of the legume *Eysenhardtia polystachya* (Beltrami *et al.*, 1982): 'coatline-A' is compound (*9.115*), 'coatline-B' is the 3, 4-dihydroxy analogue (*9.116*). It should be noted that the compounds from legumes, structures (*9.112–9.116*) have a resorcinol-based A-ring which is quite common within the family. The only non-legume source of these compounds, *Podocarpus*, afforded a compound with a phloroglucinol-based A-ring (*9.111*).

Several β-oxygenated dihydrochalcones were discussed in the earlier review (II, 347–349). Only one additional report of compounds with this substitution pattern has appeared since then. Mahmoud and Waterman (1985b) described compounds (*9.117*) and (*9.118*) in stem bark of *Millettia hemsleyana* where they occur with the β-hydroxychalcone milletenone (*9.119*). Compounds (*9.117*) and (*9.118*) can be thought of as representing the two keto–enol tautomers of milletenone (*9.119*) that have been trapped into their respective forms by reduction of the α, β-unsaturation and/or *O*-methylation (Mahmoud and Waterman, 1985b). The base peak in the EI-mass spectrum of milletenone arises from the 2′, 4′-dimethoxy A-ring form which suggests that structure (*9.119*) is the predominant form. On this basis they suggest that structure (*9.117*) corresponds to the milletenone structure and should be called 'dihydromilletenone.' Structure (*9.118*) would then be called 'dihydroisomilletenone.' Other examples of 'retroflavonoids' are discussed on page 339.

9.5 FLAVANONES

9.5.1 Introduction

The past five years or so have seen a large number of flavanones reported as natural products many for the first time. Flavanone itself has not yet been reported to occur naturally but several very simple derivatives thereof, including 7-hydroxyflavanone, are known. Structural diversity ranges from the monohydroxy derivative to exceedingly complex Diels-Alder-type compounds found in mulberry root. This is without a doubt the richest array of compounds within the 'minor flavonoids' category. There may come a time when flavanones will have earned the right to their own chapter in this series!

The stereochemistry of naturally occurring flavanones was discussed in the more recent review (II, 349). C-2 is a centre of asymmetry; the phenyl substituent at that position can be either in the (2*S*) configuration with that group indicated as being below the plane of the page or (2*R*) with the group indicated as being above the plane of the page. The former configuration is considered to be the natural one. Many, if not most, reports on flavanones do not comment on the stereochemistry of the compounds. This is usually because of the very small amounts of material available; in some cases identifications are assumed on the basis of chromatographic comparisons.

9.5.2 Structures of naturally occurring flavanones

All reported occurrences of flavanones since the last review are summarized in Table 9.5.

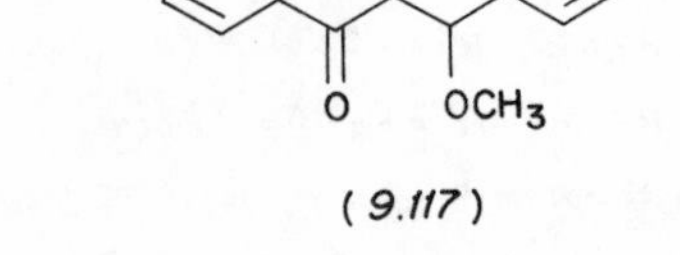

(*9.117*)

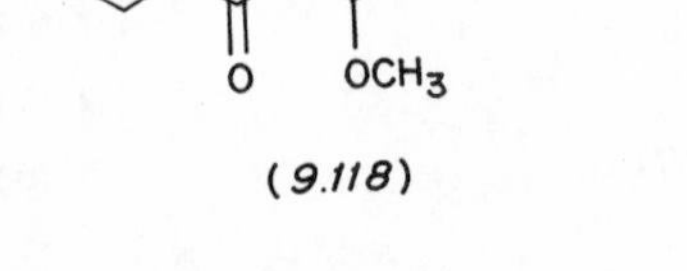

(*9.118*)

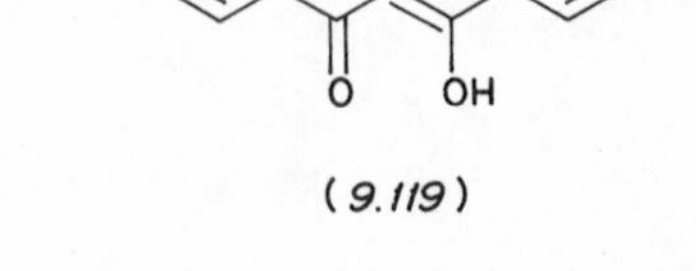

(*9.119*)

Table 9.5 Structures of naturally occurring flavanones

Substituents							
OH	OCH$_3$	Other	No. in text	Trivial name	Source*		References
7			(9.120)		*Acacia neovernicosa*	Leg	Wollenweber and Seigler (1982)
7		8-*C*-Prenyl	(9.121)		*Tephrosia falciformis*	Leg	Khan *et al.* (1986)
	7	8-*C*-(3-Hydroxy-3-methyl-*trans*-buten-1-yl)	(9.122)	Falciformin	*T. falciformis*	Leg	Khan *et al.* (1986)
		6″, 6″-Dimethylpyrano (2″, 3″, 7, 8)		Isolongocarpin	*T. purpurea*	Leg	Pelter *et al.* (1981)
		6″, 6″-Dimethylpyrano(2″, 3″, 7, 6); 8-*C*-prenyl	(9.123)		*Lannea acida*	Ana	Sultana and Ilyas (1986)
5, 7			(9.124)	Pinocembrin	*Acacia neovernicosa*	Leg	Wollenweber and Seigler (1982)
					Alpinia katsumadai	Zin	Kuroyanagi *et al.* (1983)
					Artemisia campestris	Com	Hurabielle *et al.* (1982)
					Baccharis concinna	Com	Bohlmann *et al.* (1981b)
					B. oxydonta	Com	Bohlmann *et al.* (1981b)
					Chromolaena chasleae	Com	Bohlmann *et al.* (1982)
					Comptonia peregrina	Myr	Wollenweber *et al.* (1985b)
					Dodonaea viscosa	Sap	Sachdev and Kulshreshta (1983)
					Glycyrrhiza lepidota	Leg	Mitscher *et al.* (1983)
					Myrica pensylvanica	Myr	Wollenweber *et al.* (1985b)
		7-*O*-Neohesperidose-2-*O*‴-acetate			*Nierembergia hippomanica*	Sol	González *et al.* (1981)
		3‴ and 6‴-*O*-acetates			*N. hippomanica*	Sol	González and Pomilio (1982)
		4‴-*O*-acetate			*N. hippomanica*	Sol	Ripperger (1981)
7	5		(9.125)	Alpinetin	*Alpinia katsumadai*	Zin	Kuroyanagi *et al.* (1983)
5	7		(9.126)	Pinostrobin	*A. speciosa*	Zin	Itokawa *et al.* (1981)
					Boesenbergia pandurata	Zin	Jaipetch *et al.* (1983)
					Myrica pensylvanica	Myr	Wollenweber *et al.* (1985b)
					Onychium siliculosum	Fil	Wu *et al.* (1981)
					Piper aduncum	Pip	Burke and Nair (1986)
					P. hispidum	Pip	Burke and Nair (1986)
	5, 7		(9.127)		*Boesenbergia pandurata*	Zin	Jaipetch *et al.* (1983)
5, 6, 7			(9.128)	Dihydrobaicalein	*Scutellaria epilobifolia*	Lab	Watkin (1960)
					S. baicalensis	Lab	Tomimori *et al.* (1983)
					S. galericulata	Lab	Popova *et al.* (1976)
		7-*O*-Glucuronic acid			*S. galericulata*	Lab	Popova *et al.* (1976)
5	6, 7		(9.135)		*Onychium siliculosum*	Fil	Wu *et al.* (1981)
6	5, 7		(9.132)		*Piper hispidum* var. *obliquum*	Pip	Vieira *et al.* (1980)

(*Contd.*)

Table 9.5 (*Contd.*)

Substituents			No. in text	Trivial name	Source*		References
OH	OCH_3	*Other*					
5, 7, 8			(9.129)	Dihydronorwogonin	*Scutellaria galericulata*	Lab	Popova *et al.* (1976)
		7-*O*-Glucuronic acid			*S. galericulata*	Lab	Popova *et al.* (1976)
5, 8	7	8-Acetate	(9.130)		*Notholaena neglecta*	Fil	Wollenweber and Yakskievych (1982)
5	7, 8		(9.131)		*Andrographis paniculata*	Aca	Gupta *et al.* (1983)
					Onychium siliculosum	Fil	Wollenweber (1982)
8	5, 7		(9.133)		*Piper hispidum* var. *obliquum*	Pip	Vieira *et al.* (1980)
	5, 7, 8		(9.134)		*P. hispidum* var. *obliquum*	Pip	Vieira *et al.* (1980)
5	6, 7, 8		(9.137)		*Didymocarpus pedicellata*	Ges	Rathore *et al.* (1981a, b)
	5, 6, 7, 8		(9.136)		*D. pedicellata*	Ges	Rathore *et al.* (1981a, b)
5		7-*O*-(3-Methyl-2, 3-epoxybutyl)	(9.151)		*Achyrocline flaccida*	Com	Norbedo *et al.* (1984)
5, 7		6-*C*-Methyl		Strobopinin	*Comptonia peregrina*	Myr	Wollenweber *et al.* (1985b)
7	5	6-*C*-Methyl	(9.139)		*C. peregrina*	Myr	Wollenweber *et al.* (1985b)
5, 7		8-*C*-Methyl		Cryptostrobin	*C. peregrina*	Myr	Wollenweber *et al.* (1985b)
					Myrica pensylvanica	Myr	Wollenweber *et al.* (1985b)
5	7	8-*C*-Methyl	(9.138)		*M. pensylvanica*	Myr	Wollenweber *et al.* (1985b)
					Pityrogramma triangularis	Fil	Wollenweber *et al.* (1985a)
5, 7		6, 8-Di-*C*-methyl		Desmethoxy-matteucinol	*Myrica pensylvanica*	Myr	Wollenweber *et al.* (1985b)
					Uvaria afzelii	Ano	Hufford *et al.* (1981)
5	7	6, 8-Di-*C*-methyl			*Myrica pensylvanica*	Myr	Wollenweber *et al.* (1985b)
7	5	6, 8-Di-*C*-methyl	(9.140)		*Uvaria angolensis*	Ano	Hufford and Oguntimein (1982)
5, 7		8-*C*-(2-Hydroxybenzyl)		Chamanetin	*Uvaria ferruginea*	Ano	Kolpinid *et al.* (1985)
7	5	8-*C*-(2-Hydroxybenzyl)	(9.141)	5-Methylchamanetin	*U. angolensis*	Ano	Hufford and Oguntimein (1982)
5, 7		8-*C*-Prenyl		Glabranin	*Glycyrrhiza lepidota*	Leg	Mitscher *et al.* (1983)
					Piscidia erythrina	Leg	Labbiento *et al.* (1986)
5, 7		7-*O*-Prenyl; 8-*C*-prenyl	(9.148)		*Helichrysum rugulosum*	Com	Bohlmann and Misra (1984)
7	5	8-*C*-Prenyl	(9.154)		*H. rugulosum*	Com	Bohlmann and Misra (1984)
7	5	7-*O*-Prenyl; 8-*C*-prenyl	(9.149)		*Lonchocarpus costaricensis*	Leg	Waterman and Mahmoud (1985)
5, 7		7-*O*-Prenyl; 8-*C*-(3-hydroxy-3-methyl-*trans*-buten-1-yl)	(9.150)		*L. costaricensis*	Leg	Waterman and Mahmoud (1985)
5	7	8-*C*-(3-Hydroxy-3-methyl)butyl	(9.152)	Tephrowatsin-C	*Tephrosia watsoniana*	Leg	Gomez *et al.* (1985)

	5, 7	8-*C*-Prenyl	(*9.156*)	Candidone	*Lonchocarpus costaricensis*	Leg	Waterman and Mahmoud (1985)
					Tephrosia candida	Leg	Roy *et al.* (1986)
					T. elata	Leg	Lwande (1985)
5, 7		6-*C*-Prenyl			*Helichrysum thapsus*	Com	Bohlmann and Zdero (1983)
					H. rugulosum	Com	Bohlmann and Misra (1984)
5, 7		6-*C*-Prenyl; 8-*C*-methyl	(*9.153*)		*Platanus acerifolia*	Pla	Kaouadji *et al.* (1986b)
5, 7		6, 8-Di-*C*-Prenyl	(*9.154*)		*Helichrysum rugulosum*	Com	Bohlmann and Misra (1984)
5, 7		7-*O*-Prenyl	(*9.146*)		*Helichrysum athrixiifolium*	Com	Bohlmann and Ates (Gören) (1984)
5, 7		7-*O*-Prenyl; 8-*C*-methyl	(*9.147*)		*H. rugulosum*	Com	Bohlmann and Misra (1984)
					Platanus acerifolia	Pla	Kaouadji *et al.* (1986b)
					P. acerifolia	Pla	Kaouadji *et al.* (1986b)
	5	6″, 6″-Dimethylpyrano(2″, 3″, 7, 8)	(*9.157*)	Obovatin methyl ether	*Tephrosia bracteolata*	Leg	Khalid and Waterman (1981)
				Mixtecacin	*T. woodiii*	Leg	Dominguez *et al.* (1983)
5		6-*C*-Prenyl; 6″, 6″-dimethylpyrano(2″, 3″, 7, 8)	(*9.158*)		*T. fulvinervis*	Leg	Rao *et al.* (1985)
7, 4′			(*9.161*)	Liquiritigenin	*Centrolobium* species	Leg	Jurd and Wong (1984)
					Umtiza listerana	Leg	Burger *et al.* (1983)
					Zollernia paraensis	Leg	Ferrari *et al.* (1983)
		4′-*O*-Glucose			*Glycyrrhiza uralensis*	Leg	Nakanishi *et al.* (1985)
		4′-*O*-Glucose-apiose			*G. uralensis*	Leg	Nakanishi *et al.* (1985)
7, 4′		8-*C*-Prenyl		Isobavachin	*Sophora tomentosa*	Leg	Komatsu *et al.* (1978)
7, 4		8, 3′, 5′-Tri-*C*-prenyl	(*9.166*)		*Millettia pulchra*	Leg	Baruah *et al.* (1984)
4′	7		(*9.162*)	Methyl-liquiritigenin	*Lychnophora phylicaefolia*	Com	Bohlmann *et al.* (1980)
	7, 4′	6-*C*-Methyl	(*9.165*)		*Adina cordifolia*	Rub	Srivastava and Gupta (1983)
7, 3′, 4′			(*9.164*)	Butin	*Umtiza listerana*	Leg	Burger *et al.* (1983)
3′	7, 4′		(*9.163*)		*Lychnophora phylicaefolia*	Com	Bohlmann *et al.* (1980)
7, 4′		3′-*C*-Prenyl	(*9.167*)	Abyssinone-II	*Erythrina abyssinica*	Leg	Kamat *et al.* (1981)
7, 4′		3′, 5′-Di-*C*-prenyl	(*9.168*)	Abyssinone-IV	*E. abyssinica*	Leg	Kamat *et al.* (1981)
7		6″, 6″-Dimethylpyrano-(2″, 3″, 4′, 3′)	(*9.169*)	Abyssinone-I	*E. abyssinica*	Leg	Kamat *et al.* (1981)
7		5′-*C*-Prenyl; 6″, 6″-dimethylpyrano(2″, 3″, 4′, 3′)	(*9.170*)	Abyssinone-III	*E. abyssinica*	Leg	Kamat *et al.* (1981)
5, 7, 4′			(*9.172*)	Naringenin	*Acinos alpinus*	Lab	Venturella *et al.* (1980)
					Artemisia campestris	Com	Hurabielle *et al.* (1982)
					A. dracunculus	Com	Balza *et al.* (1985)
					A. glutinosa	Com	Gonzalez *et al.* (1983); Teresa *et al.* (1986)
					Baccharis salzmannii	Com	Bohlmann *et al.* (1981c)
					B. varians	Com	Bohlmann *et al.* (1981c)
					Calamintha nepeta	Lab	Bellino *et al.* (1980)
					Camellia sinensis	The	Chkhikvishvili *et al.* (1984)
					Couepia paraensis	Chr	Sanduja *et al.* (1983)
					Flemingia stricta	Leg	Rao *et al.* (1982)

(*Contd.*)

Table 9.5 (*Contd.*)

Substituents							
OH	*OCH$_3$*	*Other*	*No. in text*	*Trivial name*	*Source**		*References*
					Helichrysum armenicum	Com	Çubukçu and Yüksel (1982)
					H. bracteatum	Com	Forkmann (1983)
					Lycopersicon esculentum	Sol	Hunt and Baker (1980)
					Mammillaria elongata	Cac	Burret *et al.* (1982)
					Micromeria species	Lab	Bellino *et al.* (1980)
					Monopteryx inpae	Leg	Albuquerque *et al.* (1981)
					Podocarpus hallii	Cnf	Markham *et al.* (1985)
					Rhodocactus grandifolius	Cac	Burret *et al.* (1982)
					Salix caprea	Sal	Malterud *et al.* (1985)
					Semecarpus prainii	Ana	Ahmad *et al.* (1981)
					Silybum marianum	Com	Fiebig and Wagner (1984)
					Tanacetum sibiricum	Com	Stepanova *et al.* (1981)
					Teucrium gnaphalodes	Lab	Barberán *et al.* (1985a); Baberán (1986)
					Thymus piperella	Lab	Barberán *et al.* (1985b)
					T. membranaceus	Lab	Ferreres *et al.* (1985)
					T. vulgaris	Lab	Stoess (1972)
					Verbena hybrida	Ver	Stotz *et al.* (1984)
					Wyethia angustifolia	Com	McCormick *et al.* (1986)
		5-*O*-Glucose			*Helichrysum armenium*	Com	Çubukçu and Yuksel (1982)
		5-*O*-Rhamnose			*Prunus cerasoides*	Ros	Shrivastava *et al.* (1982)
		5-*O*-Neohesperidose			*Adina cordifolia*	Rub	Srivastava and Gupta (1983)
		7-*O*-Glucose		Prunin	*Abies lasiocarpa*	Cnf	Parker and Maze (1984)
					Erythroxylum ulei	Ery	Bohm *et al.* (1981)
					Lycopersicon esculentum	Sol	Hunt and Baker (1980)
					Philadelphus lewisii	Hyd	Bohm and Chalmers (1986)
					Pinus massoniana	Cnf	Shen and Theander (1985)
					Podocarpus hallii, *P. nivalis*	Cnf	Markham *et al.* (1985)
					Potentilla species	Ros	Harborne and Nash (1984)
					Robinsonia gracilis	Com	Pacheco *et al.* (1985)
					Salix caprea	Sal	Malterud *et al.* (1985)
		7-*O*-(3″-*p*-Coumarylglucose)			*Mabea caudata*	Eup	Barros *et al.* (1982)
		7-*O*-(3″, 6″-Di-*p*-coumarylglucose)			*M. caudata*	Eup	Barros *et al.* (1982)
		7-*O*-Neohesperidose		Naringin	*Ceterach officinarum*	Fil	Imperato (1983)
					Citrus paradisi	Rut	Shalashvili and Mkhavanadze (1985)
					Origanum vulgare	Lab	Antonescu *et al.* (1982)
		7-*O*-Glucose-arabinose			*Ceterach officinarum*	Fil	Imperato (1983)
		7-*O*-Glucose-galactose			*Dillenia pentagyna*	Dil	Srivastava (1981)
		6-*C*-Glucose	(*9.173*)		*Acacia retinoide*	Leg	Lorente *et al.* (1982)

7, 4′	5		(*9.175*)		*Achryocline flaccida*	Com	Norbedo *et al.* (1982)
5, 4′	7		(*9.174*)	Sakuranetin	*Artemisia campestris*	Com	Hurabielle *et al.* (1982); Teresa *et al.* (1986)
					Baccharis intermixta	Com	Bohlmann *et al.* (1981c)
					B. leptocephala	Com	Bohlmann *et al.* (1981b)
					B. serrulata	Com	Bohlmann *et al.* (1981c)
					Betula pubescens	Com	Pokhilo *et al.* (1983)
					Brickellia vernicosa	Com	Ahmed *et al.* (1986)
					Eupatorium odoratum	Com	Metwally and Ekejiuba (1981)
					Hieracium intybaceum	Com	Wollenweber (1984)
					Iris milesii	Iri	Agarwal *et al.* (1984)
					Wyethia glabra	Com	McCormick *et al.* (1985)
5, 7	4′		(*9.176*)	Isosakuranetin	*Artemisia campestris*	Com	Teresa *et al.* (1986)
					Baccharis leptocephala	Com	Bohlmann *et al.* (1981b)
					Tanacetum sibiricum	Com	Stepanova *et al.* (1981)
					Wyethia angustifolia	Com	McCormick *et al.* (1986)
					W. helenioides	Com	McCormick *et al.* (1986)
		7-*O*-Xylose			*Prunus cerasoides*	Ros	Shrivastava (1982)
		7-*O*-Rutinose			*Acinos alpinus*	Lab	Venturella *et al.* (1980)
		7-*O*-Neohesperidose			*Calamintha nepeta*	Lab	Bellino *et al.* (1980)
5	7, 4′		(*9.177*)		*Hieracium intybaceum*	Com	Wollenweber (1984)
					Mamillaria elongata	Cac	Burret *et al.* (1982)
					Rhodocactus grandifolius	Cac	Burret *et al.* (1982)
5, 6, 7, 4′			(*9.179*)	Isocarthamidin	*Scutellaria baicalensis*	Lab	Takido *et al.* (1976)
5, 7, 8, 4′			(*9.178*)	Carthamidin	*S. baicalensis*	Lab	Takido *et al.* (1976)
5, 7, 4′	6		(*9.182*)		*S. baicalensis*	Lab	Takagi *et al.* (1980); Kimura *et al.* (1982)
	5, 6, 7, 4′		(*9.190*)		*Tanacetum sibiricum*	Com	Stepanova *et al.* (1981)
					Polygonum nepalense	Pol	Rathore *et al.* (1986)
	5, 4′	6, 7-Methylenedioxy	(*9.191*)		*P. nepalense*	Pol	Rathore *et al.* (1986)
5, 7	8, 2′		(*9.185*)		*Scutellaria discolor*	Lab	Tomimori *et al.* (1985)
7	5, 8, 2′		(*9.186*)		*S. discolor*	Lab	Tomimori *et al.* (1985)
5, 2′		6″, 6″-Dimethylpyrano-(2″, 3″, 7, 6); 8-*C*-prenyl	(*9.220*)		*Lonchocarpus minimiflorus*	Leg	Mahmoud and Waterman (1985a)
5, 4′		6″, 6″-Dimethylpyrano-(2″, 3″, 7, 6); 8-*C*-prenyl	(*9.221*)		*L. minimiflorus*	Leg	Mahmoud and Waterman (1985a)
5, 4′	7	6-*C*-Methyl			*L. minimiflorus*	Leg	Mahmoud and Waterman (1985a)
5, 7	8, 4′	6-*C*-Methyl	(*9.192*)		*Blechnum regnellianum*	Fil	Miraglia *et al.* (1985)
5	7, 4′	6, 8-Di-*C*-methyl; 5-*O*-galactose	(*9.193*)		*Prosopis juliflora*	Leg	Malhotra and Misra (1983)
5, 7, 4′		6-*C*-Prenyl	(*9.198*)	Sophora-flavanone-B	*Sophora tomentosa*	Leg	Komatsu *et al.* (1978)
					Wyethia angustifolia	Com	McCormick *et al.* (1986)
					W. glabra	Com	McCormick *et al.* (1985)

(*Contd.*)

Table 9.5 (*Contd.*)

Substituents			No. in text	Trivial name	Source*		References
OH	*OCH*$_3$	*Other*					
5, 7, 4′		8-*C*-Prenyl	(*9.197*)		*Wyethia angustifolia*	Com	McCormick *et al.* (1986)
					W. athrixiifolium	Com	Bohlmann and Ates(Gören) (1984)
					W. glabra	Com	McCormick *et al.* (1985)
					W. helenioides	Com	Bohlmann *et al.* (1981); McCormick *et al.* (1986)
5, 7, 4′		7-*O*-Prenyl	(*9.196*)		*W. athrixiifolium*	Com	Bohlmann and Ates(Gören) (1984)
5, 7, 4′		6, 3′-Di-*C*-prenyl	(*9.203*)	Euchresta-flavanone-A	*Euchresta japonica*	Leg	Shirataki *et al.* (1981)
5, 4′		6″, 6″-Dimethylpyrano-(2″, 3″, 7, 8); 6-*C*-prenyl	(*9.201*)	Erythrisenegalone	*Erythrina senegalensis*	Leg	Fomum *et al.* (1985)
				Cajaflavanone	*Cajanus cajan*	Leg	Bhanumati *et al.* (1978)
5, 4′		6″, 6″-Dimethylpyrano-(2″, 3″, 7, 6); 8-*C*-prenyl	(*9.202*)	Lupinifolin	*Tephrosia lupinifolia*	Leg	Smalberger *et al.* (1974)
5, 7	4′	8, 3′-Di-*C*-prenyl	(*9.205*)	Nimbaflavone	*Azadirachta indica*	Mel	Garg and Bhakuni (1984)
5, 7, 4′		8, 3′, 5′-Tri-*C*-prenyl	(*9.206*)	Hydroxysophoranone	*Millettia pulchra*	Leg	Baruah *et al.* (1984)
5, 7, 4′		6, 8, 3′-Tri-*C*-prenyl	(*9.208*)	Amorilin	*Amorpha fruticosa*	Leg	Rózsa *et al.* (1982b)
5, 7, 2′		8-*C*-Lavandulyl	(*9.222*)	Kushenol-A	*Sophora flavescens*	Leg	Wu *et al.* (1985a)
5, 6, 7, 4′		8-*C*-Prenyl; 5-*O*-rutinose	(*9.199*)	Nirurin	*Phyllanthus niruri*	Eup	Gupta and Ahmed (1984)
5, 7, 4′		6-*C*-Prenyl; 3′-*C*-(3-methyl-2, 3-epoxybutyl)	(*9.204*)	Flemiflavanone-D	*Flemingia stricta*	Leg	Mitscher *et al.* (1985)
5, 7, 4′		6-*C*-Geranyl	(*9.214*)	Bonannione-A	*Bonannia graeca*	Umb	Bruno *et al.* (1985)
				Mimulone	*Diplacus aurantiacus*	Scr	Lincoln and Walla (1986)
5, 7, 4′		8-*C*-Geranyl	(*9.215*)	Sophora-flavanone-A	*Sophora tomentosa*	Leg	Shirataki *et al.* (1983) Bruno *et al.* (1985)
5, 7, 4′		3′-*C*-Geranyl	(*9.216*)	Kuwanon-S	*Morus lhou*	Mor	Fukai *et al.* (1985)
	6, 3′4′		(*9.228*)		*Macaranga peltata*	Eup	Anjaneyulu and Reddy (1981)
7, 8, 3′, 4′		7-*O*-Glucose	(*9.226*)	Isoökanin-glucoside	*Abies pindrow*	Cnf	Tiwari and Minocha (1980)
		7-*O*-Rhamnose	(*9.227*)	Isoökanin-rhamnoside	*Alstonia scholaris*	Apo	Chauhan *et al.* (1985)
5, 7, 3′, 4′			(*9.229*)	Eriodictyol	*Argemone mexicana*	Pap	Harborne and Williams (1983)
					Baccharis genistelloides	Com	Kuroyanagi *et al.* (1985)
					B. reticularia	Com	Bohlmann *et al.* (1981c)
					B. varians	Com	Bohlmann *et al.* (1981c)
					Brickellia vernicosa	Com	Ahmed *et al.* (1986)
					Garcinia conrauana	Gut	Hussain and Waterman (1982)
					Hazardia squarrosa	Com	Clark and Wollenweber (1985)

					Helichrysum bracteatum	Com	Forkmann (1983)
					Hemizonia increscens	Com	Proksch *et al.* (1984)
					H. lobbii, H. pentactis	Com	Proksch *et al.* (1984)
					Hymenoclea salsola	Com	Proksch *et al.* (1983)
					Mentha species	Lab	Burzanska-Hermann (1978)
					Rhamnus pallasii	Rhm	Sakushima *et al.* (1983)
					Silybum marianum	Com	Fiebig and Wagner (1984)
					Sorghum sp.	Gra	Gujer *et al.* (1986)
					Thymus piperella	Lab	Barberán *et al.* (1985b)
					Verbena hybrida	Ver	Stotz *et al.* (1984)
					Wyethia angustifolia	Com	McCormick *et al.* (1986)
					W. glabra	Com	McCormick *et al.* (1985)
		5-*O*-Glucose			*Sorghum* sp.	Gra	Gujer *et al.* (1986)
		7-*O*-Glucose			*Lasianthus japonica*	Rub	Ishikura *et al.* (1979)
					Mentha aquatica	Lab	Baslas (1983)
					Philadelphus lewisii	Hyd	Bohm and Chalmers (1986)
		3′-*O*-Glucose			*Robinsonia gracilis*	Com	Pacheco *et al.* (1985)
					Pinus massoniana	Cnf	Shen and Theander (1985)
		7-*O*-Rutinose			*Mentha piperita*	Lab	Hoffmann and Lunder (1984)
7, 3′, 4′	5	7-*O*-Galactose-xylose			*Myoporum tenuifolium*	Myo	Tomas *et al.* (1985)
		7-*O*-Arabinose-xylose			*Diplazium esculentum*	Fil	Srivastava *et al.* (1981a)
5, 3′, 4′	7		(*9.230*)		*Sapium sebiferum*	Eup	Bhatt *et al.* (1981a)
					Artemisia campestris	Com	Teresa *et al.* (1986)
					A. glutinosa	Com	González *et al.* (1983)
					Hazardia squarrosa	Com	Clark and Wollenweber (1985)
					Hemizonia species	Com	Proksch *et al.* (1984)
					Heterotheca grandiflora	Com	Wollenweber *et al.* (1985c)
5, 7, 4′	3′			Homoeriodictyol	*Wyethia glabra*	Com	McCormick *et al.* (1985)
					Artemisia campestris	Com	Teresa *et al.* (1986)
5, 7, 3′	4′				*Tanacetum sibiricum*	Com	Stepanova *et al.* (1981)
					Brickellia vernicosa	Com	Ahmed *et al.* (1986)
					Citrus tankan	Rut	Cheng *et al.* (1985)
		7-*O*-Glucose			*Mentha* species	Lab	Burzanska-Hermann (1978)
		7-*O*-Rutinose			*Mentha aquatica*	Lab	Baslas (1983)
					Citrus unshiu	Rut	Shalashvili and Tsiklauri (1984)
					Hyssopus species	Lab	Zotov *et al.* (1976)
		7-*O*-Neohesperidose			*Mentha aquatica*	Lab	Baslas (1983)
					Citrus unshiu	Rut	Shalashvili and Tsiklauri (1984)
7, 3′	5, 4′	7-*O*-Glucose	(*9.232*)		*Cleome viscosa*	Cpr	Srivastava (1982)
5, 4′	7, 3′				*Brickellia vernicosa*	Com	Ahmed *et al.* (1986)
					Hazardia squarrosa	Com	Clark and Wollenweber (1985)
5, 7, 4′	8, 3′		(*9.234*)		*Achyrocline tomentosa*	Com	Ferraro *et al.* (1985)
5, 3′	6, 7, 4′		(*9.235*)		*Vitex negundo*	Ver	Achari *et al.* (1984)

(*Contd.*)

Table 9.5 (*Contd.*)

Substituents			No. in text	Trivial name	Source*		References
OH	*OCH₃*	*Other*					
5, 3′	7, 8, 4′		(9.236)		*V. negundo*	Ver	Achari *et al.* (1984)
6, 4′	5, 7, 3′		(9.237)		*Ageratum strictum*	Com	Quijano *et al.* (1982)
6	5, 7, 3′, 4′		(9.238)		*A. strictum*	Com	Quijano *et al.* (1982)
6	5, 7	3′, 4′-Methylenedioxy	(9.239)		*A. strictum*	Com	Quijano *et al.* (1982)
5	6	Furano(2″, 3″, 7, 8); 3′, 4′-methylenedioxy	(9.233)		*Derris araripensis*	Leg	do Nascimento and Mors (1981)
7, 3′, 4′	5	6-*C*-Methyl; 7-*O*-glucose	(9.246)		*Prosopis juliflora*	Leg	Malhotra and Misra (1983)
3′	5, 7, 4′	8-*C*-Methyl	(9.245)		*Adina cordifolia*	Rub	Srivastava and Gupta (1983)
5, 4′	3′	6, 8-Di-*C*-methyl	(9.247)		*Qualea labouriauana*	Voc	Corrêa *et al.* (1981)
5, 7, 3′, 4′		6-*C*-Prenyl	(9.249)		*Wyethia angustifolia*	Com	McCormick *et al.* (1986)
					W. arizonica	Com	Bohlmann *et al.* (1984b)
					W. glabra	Com	McCormick *et al.* (1985)
					W. helenioides	Com	Bohlmann *et al.* (1981d); McCormick *et al.* (1986)
5, 7, 3′, 4′		8-*C*-Prenyl	(9.248)		*Wyethia angustifolia*	Com	McCormick *et al.* (1986)
					W. glabra	Com	McCormick *et al.* (1985)
					W. helenioides	Com	Bohlmann *et al.* (1981d); McCormick *et al.* (1986)
5, 7, 3′, 4′		5′-*C*-Prenyl	(9.250)	Sigmoiden-B	*Erythrina sigmoidea*	Leg	Fomum *et al.* (1986)
5, 7 3′, 4′		2′, 5′-Di-*C*-prenyl	(9.251)	Sigmoiden-A	*E. sigmoidea*	Leg	Fomum *et al.* (1983)
5, 7, 3′		6″, 6″-Dimethylpyrano-(2″, 3″, 4′, 5′)	(9.252)	Sigmoiden-C	*E. sigmoidea*	Leg	Fomum *et al.* (1986)
5, 3′, 4′	7	6, 8-Di-*C*-prenyl	(9.209)	Amoradicin	*Amorpha fruticosa*	Leg	Rózsa *et al.* (1984)
5, 4′	7, 3′	6, 8-Di-*C*-prenyl	(9.210)	Amoradinin	*A. fruticosa*	Leg	Rózsa *et al.* (1984)
5, 7, 3′, 4′		6, 8, 5′-Tri-*C*-prenyl	(9.211)	Amorisin	*A. fruticosa*	Leg	Rózsa *et al.* (1982b)
5, 7, 4′	3′	6, 8, 5′-Tri-*C*-prenyl	(9.212)	Amoritin	*A. fruticosa*	Leg	Rózsa *et al.* (1982b)
5, 7, 3′		6, 8-Di-*C*-prenyl; 6″, 6″-dimethylpyrano(2″, 3″, 4′, 5′)	(9.213)	Amorinin	*A. fruticosa*	Leg	Rózsa *et al.* (1982a)
5, 7, 3′, 4′		6-*C*-Geranyl	(9.217)	Diplacone	*Diplacus aurantiacus*	Scr	Lincoln (1980)
5, 7, 3′	4′	6-*C*-Geranyl	(9.218)	4′-Methyldiplacone	*D. aurantiacus*	Scr	Lincoln and Walla (1986)
5, 7, 2′, 4′		5′-*C*-Geranyl	(9.260)	Kuwanon-E	*Morus alba*	Mor	Nomura and Fukai (1981a, b)
5, 7	2′	6″, 6″-Dimethylpyrano-(2″, 3″, 4′, 3′)	(9.253)	Sophoranol	*Sophora tomentosa*	Leg	Komatsu *et al.* (1978)
5, 7, 2′		6″, 6″-Dimethylpyrano-(2″. 3″. 4′. 3′)	(9.259)	Sangganon-F	'*Morus* root barks'	Mor	Nomura *et al.* (1983)

5, 7, 4′		6″, 6″-Dimethylpyrano-(2″, 3″, 2′, 3′)	(9.258)	Sangganon-H	'Morus root barks'	Mor	Hano and Nomura (1983)
5, 7, 2′, 4′		6, 5′-Di-C-prenyl	(9.255)	Euchresta-flavanone-B	*Euchresta japonica*	Leg	Shirataki *et al.* (1982)
					Maclura pomifera	Mor	Delle Monache *et al.* (1984)
5, 7, 2′		8-C-Prenyl; 6″, 6″-dimethylpyrano(2″, 3″, 4′, 5′)	(9.257)	Euchresta-flavanone-C	*Euchresta japonica*	Leg	Shirataki *et al.* (1982)
					Cudrania tricuspida	Mor	Fujimoto and Nomura (1984)
					Maclura pomifera	Mor	Delle Monache *et al.* (1984)
5, 7, 2′		6-C-Prenyl; 6″, 6″-dimethylpyrano(2″, 3″, 4′, 5′)	(9.256)	Cudraflavanone-A	*Cudrania tricuspida*	Mor	Fujimoto and Nomura (1984)
5, 2′	4′	6-C-Prenyl; 6″, 6″-dimethylpyrano(2″, 3″, 7, 8)	(9.254)	Fleminone	*Flemingia macrophylla*	Leg	Rao and Srimannarayana (1983)
5, 7, 2′, 4′		6-C-Lavandulyl	(9.224)	Vexibinol	*Vexibia alopecuroides*	Leg	Batirov *et al.* (1985)
5, 7, 4′	2′	6-C-Lavandulyl	(9.225)	Vexibidin	*V. alopecuroides*	Leg	Batirov *et al.* (1985)
5, 7, 2′, 4′		6-C-Prenyl; 8-C-lavandulyl	(9.223)	Kushenol-B	*Sophora flavescens*	Leg	Wu *et al.* (1985a)
5, 7, 2′, 5′					*Tanacetum sibiricum*	Com	Stepanova *et al.* (1981)
5, 7, 2′, 5′	6				*T. sibiricum*	Com	Stepanova *et al.* (1981)
5, 7, 2′, 6′			(9.183)		'Scutellaria Radix'	Lab	Kubo *et al.* (1981)
					Scutellaria baicalensis	Lab	Ikram *et al.* (1979); Kimura *et al.* (1982)
7, 2′, 6′	5		(9.184)		*S. baicalensis*	Lab	Tomimori *et al.* (1984)
5, 2′	6, 7, 6′		(9.188)		*S. discolor*	Lab	Tomimori *et al.* (1985)
5, 2′	7, 8, 6′		(9.187)		*S. discolor*	Lab	Tomimori *et al.* (1985)
5, 7, 3′, 4′, 5′			(9.273)		*Helichrysum bracteatum*	Com	Forkmann (1983)
					Verbena hybrida	Ver	Stotz *et al.* (1984)
		3′-O-Glucose	(9.274)	Plantagoside	*Plantago major* var. *japonica*	Pln	Endo *et al.* (1981)
5, 3′, 4′	7, 5′		(9.275)		*Hazardia squarrosa*	Com	Clark and Wollenweber (1985)
5, 7, 3′	4′, 5′	6, 8-Di-C-methyl	(9.276)		*Alluaudiopsis marnieriana*	Did	Rabesa and Voirin (1983)
	5, 6, 7, 3′	4′, 5′-Methylenedioxy	(9.277)		*Ageratum corymbosum*	Com	Quijano *et al.* (1980)
	5, 6, 7, 8, 3′	4′, 5′-Methylenedioxy	(9.278)		*A. corymbosum*	Com	Quijano *et al.* (1980)
	5, 6, 7, 2′, 3′, 4′, 5′		(9.279)		*Polygonum nepalense*	Pol	Rathore *et al.* (1986)

*Key: Aca = Acanthaceae, Ana = Annacardiaceae, Ano = Anonaceae, Apo = Apocynaceae, Cac = Cactaceae, Chr = Chrysobalanaceae, Cnf = Coniferae, Com = Compositae, Cpr = Capparaceae, Did = Didieraceae, Dil = Dilleniaceae, Ery = Erythroxylaceae, Eup = Euphorbiaceae, Fil = Filicinae, Ges = Gesneriaceae, Gra = Gramineae, Gut = Guttiferae, Hyd = Hydrangeaceae, Iri = Iridaceae, Lab = Labiatae, Leg = Leguminosae, Mel = Meliaceae, Mor = Moraceae, Myo = Myoporaceae, Myr = Myrtaceae, Pip = Piperaceae, Pln = Plantaginaceae, Pla = Platanaceae, Pol = Polygonaceae, Rhm = Rhamnaceae, Ros = Rosaceae, Rub = Rubiaceae, Rut = Rutaceae, Sal = Salicaceae, Sap = Sapindaceae, Scr = Scrophulariaceae, Sol = Solanaceae, The = Theaceae, Umb = Umbelliferae, Ver = Verbenaceae, Voc = Vochysiaceae, Zin = Zingiberaceae.

(a) *Flavanones lacking B-ring oxygenation*

The simplest known flavanone is (*9.120*), 7-hydroxyflavanone. It was originally discovered in members of the Leguminosae (I, 568) but was later found to occur in a member of the Compositae as well (II, 350). Wollenweber and Seigler (1982) found it as a variable constituent of the flavonoid profile of *Acacia neovernicosa*. *C*-Prenylated derivatives of this compound are fairly well known (II, 350). The parent *C*-prenyl derivative (*9.121*), known from the genus *Tephrosia*, has now been isolated from another member, namely *T. falciformis*, where it occurs with the new natural product 'falciformin' (*9.122*) (Khan *et al.*, 1986).

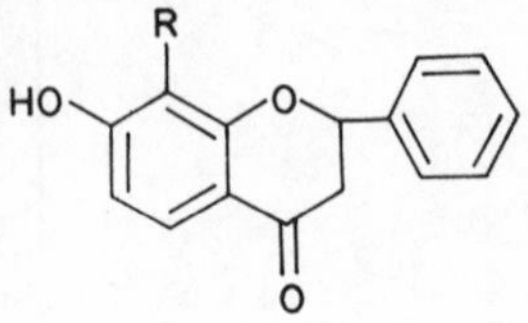

(*9.120*) R = H

(*9.121*) R = prenyl

(*9.22*) R = CH=CHC(OH)Me_2

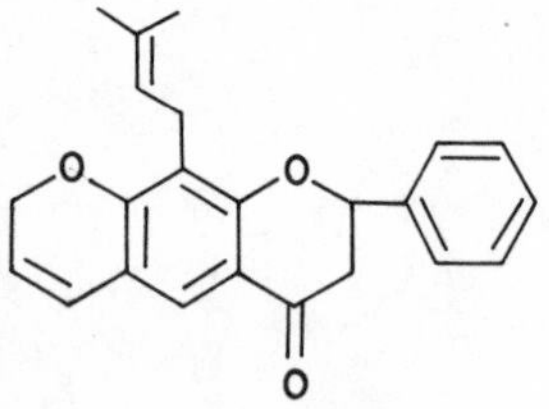

(*9.123*)

The parent compound with two prenyl groups has been isolated from *Lannea acida*, a member of the Anacardiaceae, by Sultana and Ilyas (1986). Prenylation has occurred at positions-6 and -8 followed by cyclization to yield the linear pyranoflavanone (*9.123*). This appears to be the first report of a prenylated flavanone in a member of the Anacardiaceae.

By far the largest number of compounds in this subgroup is based upon pinocembrin, 5,7-dihydroxyflavanone (*9.124*). The free phenol is known to occur naturally, of course, as are several glycosidic derivatives and the three possible methyl ethers. Only new family records will be mentioned unless some structural feature warrants special note; all occurrences can be found in Table 9.5. The discovery of pinocembrin in *Dodonaea viscosa*, Sapindaceae, by Sachdev and Kulshreshta (1983) and in *Comptonia peregrina* and *Myrica pensylvanica*, both Myricaceae, by Wollenweber and co-workers (1985b) represent first reports from these families. The finding of pinocembrin in seeds of *Alpinia katsumadai* (Kuroyanagi *et*

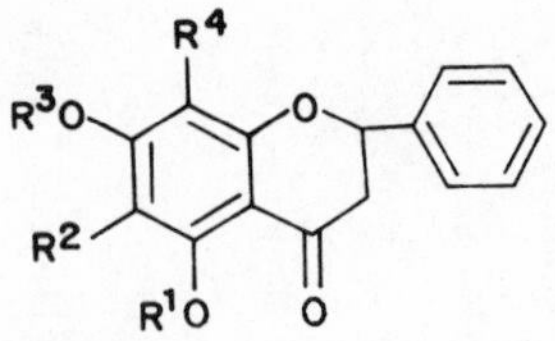

(*9.124*) $R^1 = R^2 = R^3 = R^4 = H$

(*9.125*) $R^1 = CH_3$; $R^2 = R^3 = R^4 = H$

(*9.126*) $R^1 = R^2 = R^4 = H$; $R^3 = CH_3$

(*9.127*) $R^1 = R^3 = CH_3$; $R^2 = R^4 = H$

(*9.128*) $R^1 = R^3 = R^4 = H$; $R^2 = OH$

(*9.129*) $R^1 = R^2 = R^3 = H$; $R^4 = OH$

(*9.130*) $R^1 = R^2 = H$; $R^3 = CH_3$; $R^4 = O{-}\overset{O}{\overset{\|}{C}}{-}CH_3$

(*9.131*) $R^1 = R^2 = H$; $R^3 = R^4 = CH_3$

(*9.132*) $R^1 = R^3 = CH_3$; $R^2 = OH$; $R^4 = H$

(*9.133*) $R^1 = R^3 = CH_3$; $R^2 = H$; $R^4 = OH$

(*9.134*) $R^1 = R^3 = CH_3$; $R^2 = H$; $R^4 = OCH_3$

(*9.135*) $R^1 = R^4 = H$; $R^2 = OCH_3$; $R^3 = CH_3$

(*9.136*) $R^1 = R^3 = CH_3$; $R^2 = R^4 = OCH_3$

(*9.137*) $R^1 = H$; $R^2 = R^4 = OCH_3$; $R^3 = CH_3$

al., 1983) represents a new record in Zingiberaceae, although 5-*O*-methylpinocembrin (*9.125*), commonly known as alpinetin, has long been known from the family. The 7-*O*-neohesperidoside of pinocembrin was described originally by Pomilio and Gros (1979) as a component of *Nierembergia hippomanica* (Solanaceae). Three recent reports described further studies of this taxon. In 1981 pinocembrin 7-*O*-β-(2‴-*O*-acetyl)neohesperidoside was reported by González and co-workers (1981) while the 4‴-*O*-acetyl derivative was found by Ripperger (1981). The following year saw the description of the remaining two members of this group, namely the 3‴-*O*-acetyl and 6‴-*O*-acetyl derivatives, by González and Pomilio (1982).

Pinostrobin, 7-*O*-methylpinocembrin, (*9.126*), also a well-known compound, has now been reported for the first time in a fern (*Onychium siliculosum*) (Wu *et al.*, 1981); in the Piperaceae (*Piper* spp.) (Bruke and Nair, 1986); in the Myricaceae (*Myrica pensylvanica*) (Wollenweber *et al.*, 1985b); and in the Zingiberaceae (*Boesenbergia pandurata*) (Jaipetch *et al.*, 1983). These last workers also found 5,7-di-*O*-methylpinocembrin (*9.127*) in *B. pandurata*.

Extra hydroxylation of the A-ring, often coupled with *O*-methylation, provides another level of structural variation. The parent compounds, 5,6,7-trihydroxyflavanone (*9.128*), known as 'dihydrobaicalein,' and 5,7,8-trihydroxyflavanone (*9.129*), known as 'dihydronorwogonin,' were first reported by Popova and co-workers (1976) as constituents of *Scutellaria galericulata* (Labiatae).

The former was observed to occur both as the free phenol and as the 7-β-D-glucuronide, the latter only as the glucuronide.

The 6-methyl ether of (*9.128*) and the 8-methyl ether of (*9.129*) were discussed earlier (I, 570). The first report of 5, 8-dihydroxy-7-methoxyflavanone was as a component of the farina of the fern *Notholaena neglecta*, where it occurs as the 8-acetyl derivative (*9.130*) (Wollenweber and Yatskievych, 1982). Occurring with it in the farina is the corresponding flavonol. Newly reported as natural products are the 7, 8-dimethyl ether (*9.131*) from *Andrographis paniculata* (Gupta *et al.*, 1983) and *Onychium siliculosum* (Wollenweber, 1982), and the two 5, 7-dimethyl ethers (*9.132*) and (*9.133*) from *Piper hispidum* var. *obliquum* (Vieira *et al.*, 1980). The totally *O*-methylated compound (*9.134*) was also reported by these workers (Vieira *et al.*, 1980) from this *Piper* variety but it is not a new compound (II, 362). 5-Hydroxy-6, 7-dimethoxyflavanone (*9.135*) is also a known compound but its report from *Onychium siliculosum* (Wu *et al.*, 1981) represents its first record from a fern.

5, 6, 7, 8-Tetrahydroxyflavanone has not been reported as a natural product but several of its methyl ethers do occur naturally. The tetramethyl ether (*9.136*), initially described from *Popowia* (II, 362), has also been found in *Didymocarpus pedicellata* (Rathore *et al.*, 1981a) where it occurs with the newly reported trimethyl ether (*9.137*). Fully substituted A-ring chalcones from *D. pedicellatus* were discussed earlier (I, 452).

C-Methylation of pinocembrin yields the very well known strobopinin (6-*C*-methyl), cryptostrobin (8-*C*-methyl), and desmethoxymatteucinol (6, 8-di-*C*-methyl). New records of these compounds and their *O*-methyl derivatives are summarized in Table 9.5. Members of this structural group newly reported as natural products are 7-*O*-methyl-8-*C*-methylpinocembrin (= 7-*O*-methyl-cryptostrobin) (*9.138*) found in the farina of *Pityrogramma triangularis* (Wollenweber *et al.*, 1985a) and as a minor component in the leaf resin of *Myrica pensylvania* (Wollenweber *et al.*, 1986b). The latter workers reported 5-*O*-methyl-6-*C*-methylpinocembrin (= 5-*O*-methylstrobopinin) (*9.139*), as a major component of the resinous exudate of *Comptonia peregrina* leaves. Several other *O*- and/or *C*-methylated compounds of known

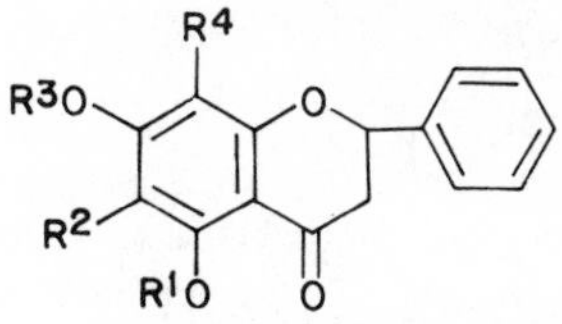

R4
R3O
R2
R1O O

(*9.138*) $R^1 = R^2 = H; R^3 = R^4 = CH_3$

(*9.139*) $R^1 = R^2 = CH_3; R^3 = R^4 = H$

(*9.140*) $R^1 = R^2 = R^4 = CH_3; R^3 = H$

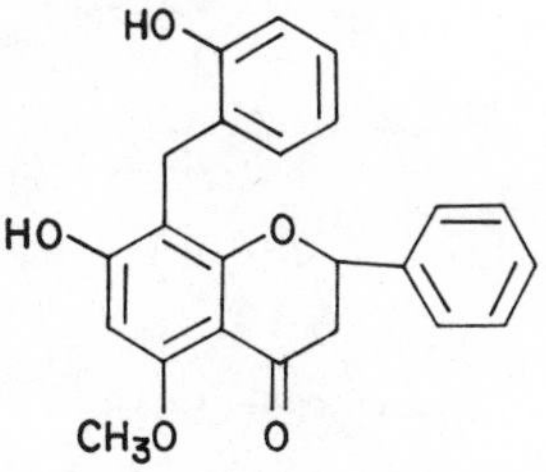

HO
HO
O
CH_3O O

(*9.141*)

structure were also reported in that study of *Comptonia* and *Myrica*. One other new relative of these compounds was reported by Hufford and Oguntimein (1982) as part of their continuing study of biologically active flavonoids of *Uvaria*. In the paper of interest to us here they report the occurrence of 5-*O*-methyl-6, 8-di-*C*-methylpinocembrin (*9.140*) as well as 5-*O*-methyl-8-*C*-(2-hydroxybenzyl) pinocembrin (*9.141*) which they called '(±)-chamanetin 5-methyl ether,' from *U. angolensis*. The flavanones were accompanied by dihydrochalcones that also bear 2-hydroxybenzyl groups (*9.96*), (*9.97*) and (*9.98*).

Byrne and her colleagues (1982) pointed out that, while no argument exists about strobopinin (*9.142*) and cryptostrobin (*9.143*) being mono-*C*-methylated A-ring flavanones, unequivocal evidence as to which is the C-6 and which the C-8 isomer was lacking. By means of two independent methods these workers have now established the structures of these compounds beyond doubt. Using long-range coupling between the C-5 hydroxyl proton and the C-6 carbon atom they were able to show that strobopinin has the *C*-methyl group at C-6 as shown correctly in (*9.142*). In the second study the authors

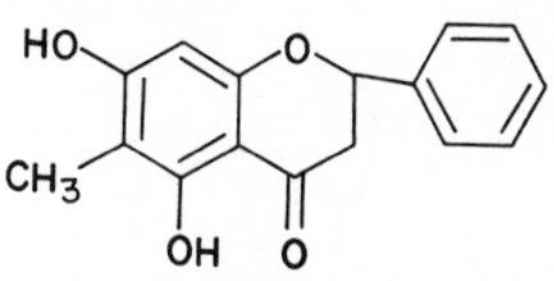

HO O
CH_3
OH O

(*9.142*)

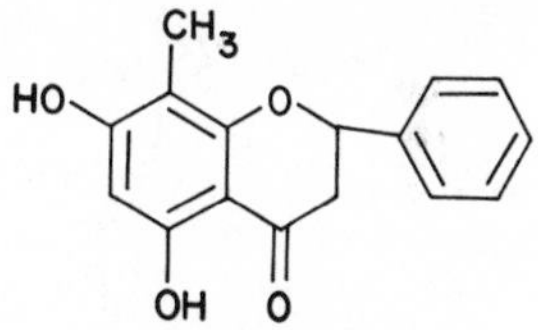

CH_3
HO O
OH O

(*9.143*)

subjected optically active cryptostrobin to bromination and subjected the product to X-ray diffraction analysis. The results clearly established bromine at position 6. Since bromination of (−)-cryptostrobin yielded an optically active product it is certain that ring opening and isomerization (the Wessely–Moser rearrangement) had not occurred. The correct structures are then those shown. These results also allowed the structure of (±)-lawinal, originally described as 6-formyl-8-*C*-methylpinocembrin (*9.144*) by Joshi and Gawad (1974), to be corrected. The correct formulation of lawinal requires exchanging the formyl and methyl groups. The proper structure is shown

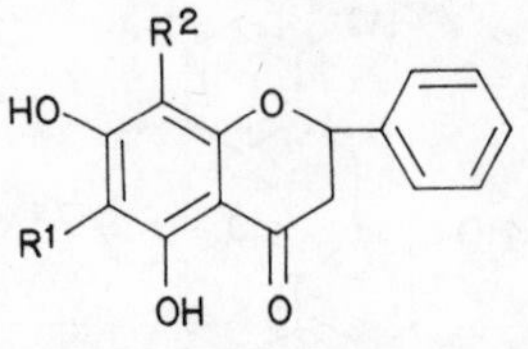

(*9.144*) $R^1 = CHO$; $R^2 = CH_3$

(*9.145*) $R^1 = CH_3$; $R^2 = CHO$

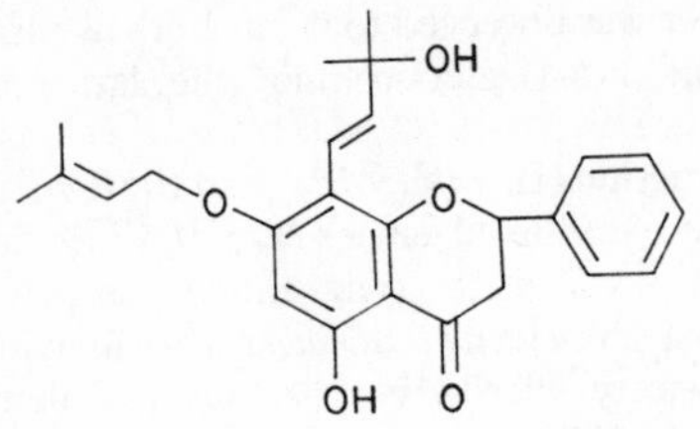

as (*9.145*). Reference to lawinal, with the incorrect structure, can be found in the earlier review (II, 363).

Two *C*-prenylated pinocembrin derivatives were described in the first review; about a half-dozen new structures were added in the second. Only one pyrano derivative was seen in each. Since the time of the second review several new compounds based upon prenylation of the parent compound have been described. Prenylated pinocembrin derivatives were first obtained from members of two plant families, Leguminosae and Compositae. It is not surprising, then, that these two families continue to yield new and interesting members of this group of flavonoids.

Amongst the newly reported natural products are several prenyl ethers. The simplest of these, 7-*O*-prenylpinocembrin (*9.146*), has been isolated from *Helichrysum athrixiifolium* (Bohlmann and Ates (Gören), 1984), *H. rugulosum* (Bohlmann and Misra, 1984), and from the buds of *Platanus acerifolia* (Kaouadji *et al.*, 1986b). *Platanus acerifolia* also yielded 7-*O*-prenyl-8-*C*-methylpinocembrin (*9.147*), and the doubly *C*-alkylated compound 6-*C*-prenyl-8-*C*-methylpinocembrin (*9.153*), both new compounds.

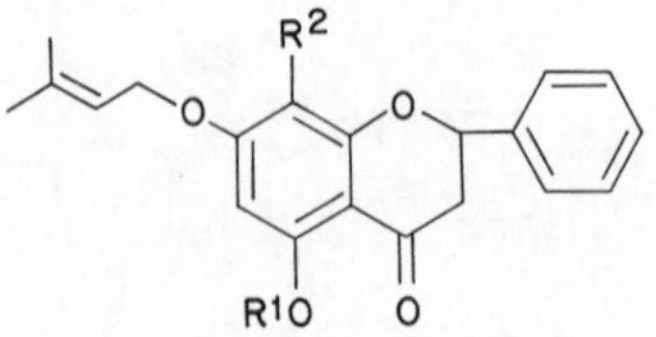

(*9.146*) $R^1 = R^2 = H$

(*9.147*) $R^1 = H$; $R^2 = CH_3$

(*9.148*) $R^1 = H$; R^2 = prenyl

(*9.149*) $R^1 = CH_3$; R^2 = prenyl

In *Helichrysum rugulosum* compound (*9.146*) was accompanied by 7-*O*-prenyl-8-*C*-prenylpinocembrin (*9.148*) and other newly discovered prenylated flavanones, (*9.154*) and (*9.155*) (Bohlmann and Misra, 1984). Greater levels of complexity are seen in the prenyl ethers isolated from seeds of *Lonchocarpus costaricensis* by Waterman and Mahmoud (1985): compound (*9.149*), which is the 5-methyl ether of (*9.148*), and compound (*9.150*), where the

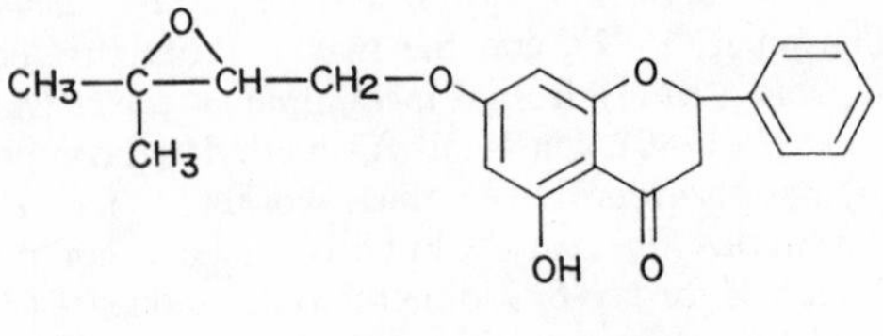

(*9.150*)

(*9.151*)

C-prenyl function has been modified through hydroxylation and double bond shift to yield the 3-hydroxy-3-methyl-*trans*-buten-1-yl group. This modified side chain was also seen in 'falciformin,' which was shown above as (*9.122*). Compound (*9.151*), where the prenyl group has been epoxidized to yield the 3-methyl-2,3-epoxybutyl function, was isolated from *Achyrocline flaccida* (Compositae) by Norbedo and co-workers (1984). Hydration of the prenyl function also occurs to yield compounds of the sort seen in (*9.152*); this compound was isolated from *Tephrosia watsonia* by Gomez and co-workers (1985) who called it 'tephrowatsin-C.' Various other *C*-alkylated flavanones involving methyl and/or prenyl groups, some of which also exhibit *O*-methylation, expand our list of structural variants even further. The occurrence of 6-*C*-prenyl-8-*C*-methyl derivative (*9.153*) in *Platanus* was mentioned above. The 8-*C*-prenyl (*9.154*) and 6,8-di-*C*-prenyl (*9.155*) compounds occur in the group of flavonoids obtained from *Helichrysum rugulosum* (Bohlmann and Misra, 1984). We can also add the newly reported

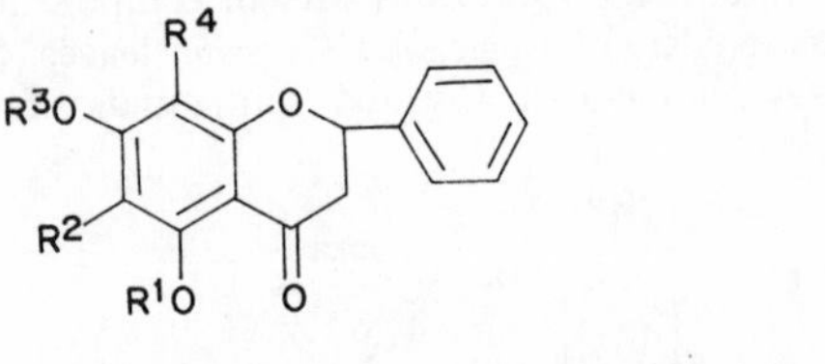

(*9.152*) $R^1 = R^2 = H$; $R^3 = CH_3$; $R^4 = CH_2CH_2C(OH)Me_2$

(*9.153*) $R^1 = R^3 = H$; R^2 = prenyl; $R^4 = CH_3$

(*9.154*) $R^1 = CH_3$; $R^2 = R^3 = H$; R^4 = prenyl

(*9.155*) $R^1 = R^3 = H$; $R^2 = R^4$ = prenyl

(*9.156*) $R^1 = R^3 = CH_3$; $R^2 = H$; R^4 = prenyl

compound 'candidone' (*9.156*) to the list of derived flavonoids found in the genus *Tephrosia*; Lwande (1985) reported it, along with the known obovatin methyl ether (*9.157*) from *T. elata* while Roy and co-workers (1986) reported candidone from *T. candida* (from which the common name derives). A newly discovered pyranoflavanone is compound (*9.158*), called 'fulvinervin,' which was isolated from *T. fulvinervis* (Rao *et al.*, 1985).

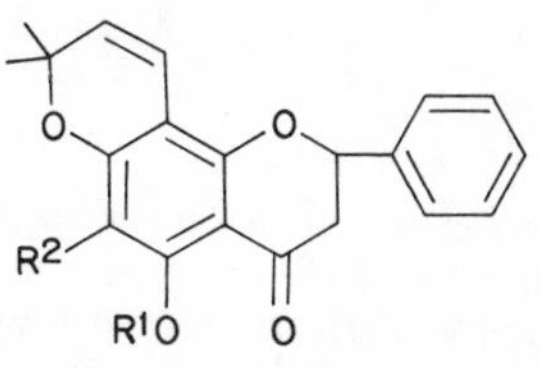

(*9.157*) $R^1 = CH_3$; $R^2 = H$

(*9.158*) $R^1 = H$; $R^2 =$ prenyl

C-Geranylflavanones are comparatively common, especially in members of the Leguminosae. Several structural modifications of the ten-carbon function are known as well. Cyclizations to yield pyranoflavanones are also comparatively common. Carbocyclic terpenic derivatives, on the other hand, are rare amongst the minor flavonoids. The only one reported until recently was rubranine (*9.159*) which was originally isolated from *Aniba rosaeodora* (Lauraceae) (II, 324), and more recently from *Boesenbergia pandurata* (Zingiberaceae) by Tuntiwachwuttikul and co-workers (1984). A recently reported study of *Lindera umbellata* by Tanaka and co-workers (1985) described the new flavanone 'linderatone' whose structure is shown as (*9.160*). *Lindera* is also a member of the Lauraceae.

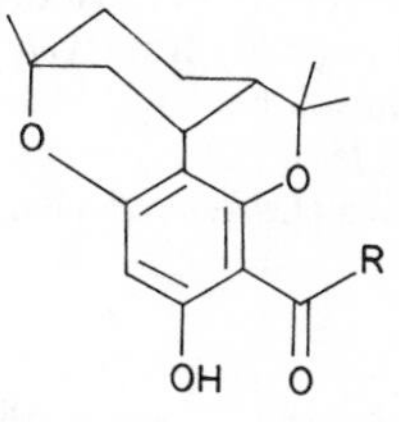

(*9.159*)

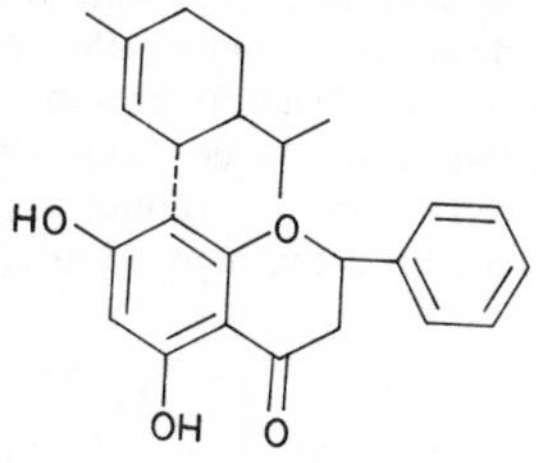

(*9.160*)

(b) *Flavanones having one B-ring oxygenation*

We can initiate discussion of this group with a consideration of liquiritigenin (*9.161*) and its derivatives. The parent flavanone is well known as the free phenol, simple glycosides, or as more complex alkylated derivatives. In the earlier reviews most of the occurrences of these compounds involved members of the Leguminosae with one report from each of the Amaranthaceae and Compositae. Members of the Leguminosae again dominate; eight genera of legumes are represented. There is a new generic record from the Compositae and one from the Rubiaceae. New records for liquiritigenin and its 4'-glucoside are summarized in Table 9.5. A new glycosidic form of liquiritigenin is worth comment, however; Nakanishi and co-workers (1985) found that isoliquiritigenin 4'-*O*-glucoside was accompanied in *Glycyrrhiza uralensis* by the 4'-*O*-β-D-apiofuranosyl-(1→2)-β-D-glucopyranoside. This report adds another member to the list of plant species that exhibit glycosides involving the branched chain sugar apiose.

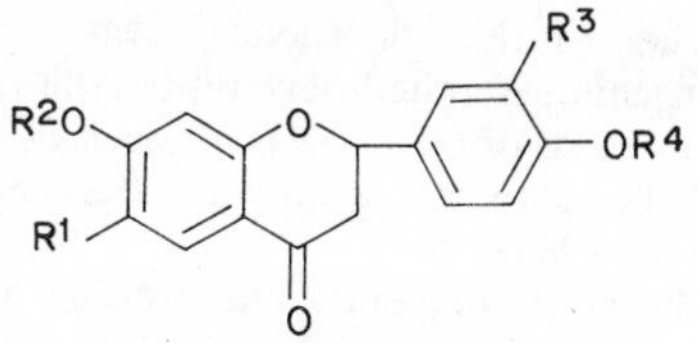

(*9.161*) $R^1 = R^2 = R^3 = R^4 = H$

(*9.162*) $R^1 = R^3 = R^4 = H$; $R^2 = CH_3$

(*9.163*) $R^1 = R^3 = H$; $R^2 = R^4 = CH_3$

(*9.164*) $R^1 = R^2 = R^4 = H$; $R^3 = OH$

(*9.165*) $R^1 = R^2 = R^4 = CH_3$; $R^3 = H$

7-*O*-Methyl-liquiritigenin (*9.162*) and 7,4'-dimethoxy-3'-hydroxyflavanone (*9.163*), both newly reported natural products, were isolated from *Lychnophora phylicaefolia* (Compositae) by Bohlmann and co-workers (1980). 7,3',4'-Trihydroxyflavanone (*9.164*), 'butin,' along with liquiritigenin, was reported as a constituent of *Umtiza listerana* (Burger *et al.*, 1983). The first report of liquiritigenin from the Rubiaceae was the finding of 7,4'-di-*O*-methyl-6-*C*-methylflavanone (*9.165*) in the heartwood of *Adina cordifolia* by Srivastava and Gupta (1983).

No group of flavonoids seems immune from prenylation in the legumes. Several examples are found in this subgroup. 6-Prenyl-, 8-prenyl, 6,8-diprenyl and 8,3',5'-triprenyl derivatives were recorded in the earlier reviews. The triprenyl compound (*9.166*), originally found in *Sophora* (I, 572-573), from which its common name 'sophoranone' comes, has recently been described from another legume, *Millettia pulcha* (Baruah *et al.*, 1984). Five new B-ring-prenylated flavanones, four of which are liquiritigenin derivatives, were reported from *Erythrina abyssinica* by Kamat and co-workers (1981). These compounds, 'abyssinones-I, -II, III and IV,' are shown as the set of structures (*9.167–9.170*). Abyssinones I and III were optically active, abyssinones II and IV were not, which suggests that the latter two compounds may have been formed from the corresponding chalcones during the isolation procedures. The chalcone corresponding to abyssinone-IV (*9.168*) was also isolated from this

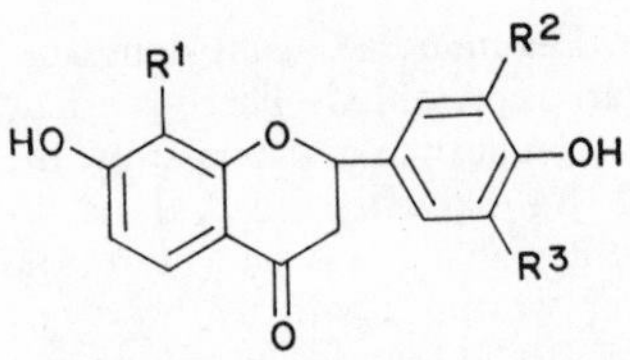

(9.166) $R^1 = R^2 = R^3$ = prenyl

(9.167) $R^1 = R^3 = H$; R^2 = prenyl

(9.168) $R^1 = H$; $R^2 = R^3$ = prenyl

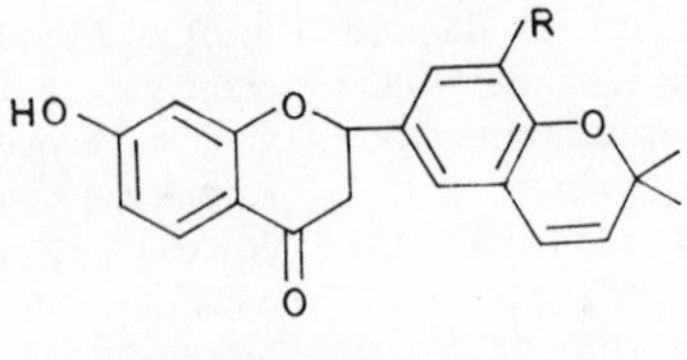

(9.169) R=H

(9.170) R=prenyl

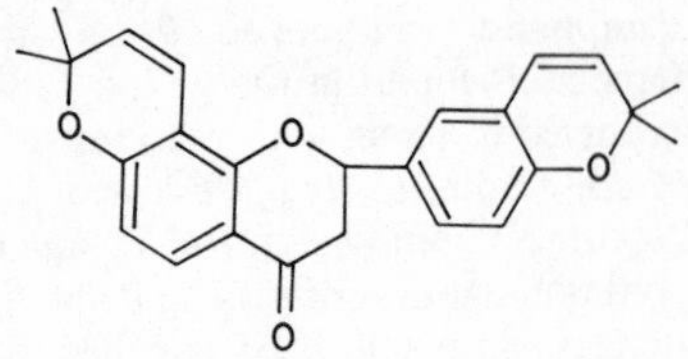

(9.171)

plant. The fifth flavanone was 3′,5′-di-*C*-prenylnaringenin. Another new diprenylflavanone was isolated from seeds of *Calopogonium mucunoides* (Leguminosae) and shown to be the dipyrano derivative (*9.171*) (Pereira *et al.*, 1982).

By far the most frequently encountered flavanone is naringenin (*9.172*). It is common as the free phenol, occurs in a wide variety of glycosylated forms, has been found in all possible *O*-methylated forms, and is the base molecule for numerous *C*-alkylations. Combinations of these structural features are also common. Approximately 40 citations involving the natural occurrence of the free phenol and its various glycosides alone can be found in Table 9.5. Worth noting is the fact that naringenin or one of its glycosides has been reported from members of the following families for the first time, references for which can be found in Table 9.5: Cactaceae (*Rhodocactus* and *Mammillaria*), Chrysobalanaceae (*Couepia*), Dilleniaceae (*Dillenia*), Erythroxylaceae (*Erythroxylum*), Euphorbiaceae (*Mabea*), Hydrangeaceae (*Philadelphus*), Rubiaceae (*Adina*), and Verbenaceae (*Verbena*).

In contrast to the wide occurrence of *O*-glycosides of naringenin, *C*-glycosyl derivatives are quite rare. The most recently reported example is 6-*C*-glucosylnaringenin (*9.173*) which Lorente and co-workers (*1982*) isolated from flowers of *Acacia retinoide*. The only other *C*-glycosylflavanones of which I am aware are 6-*C*- and 8-*C*-glucosylnaringenin in *Eucalyptus hemiphloia* (I, 576), 8-*C*-galactosyl-liquiritigenin in *Aerva persica* (II, 366), and 6-*C*-glucosylnaringenin in *Tulipa gesneriana* (II, 367).

Simple *O*-methyl derivatives of naringenin are well-known natural products. Recent reports of the occurrence of 7-*O*-methylnaringenin (sakuranetin) (*9.174*), 5-*O*-methylnaringenin (*9.175*), 4′-*O*-methylnaringenin (isosakuranetin) (*9.176*) and 7,4′-di-*O*-methylnaringenin (*9.177*) are summarized in Table 9.5. These reports include two new family records: sakuranetin in the Iridaceae (*Iris*) (Agarwal *et al.*, 1984), and 7,4′-di-*O*-methylnaringenin in the Cactaceae (*Rhodocactus* and *Mammillaria*) (Burret *et al.*, 1982).

Compounds resulting from extra hydroxylation of naringenin at position 6 and position 8 have been isolated from several sources; many of these compounds are newly reported as natural products. In a paper overlooked in my earlier review Takido and co-workers (1976) described isolation of 8-hydroxynaringenin (*9.178*), which they called 'carthamidin,' and 6-hydroxynaringenin (*9.179*), which they called 'isocarthamidin,' from *Scutellaria baicalensis* (Labiatae). Apparently this was the first report of these compounds in nature. Early structural problems dealing with pigments from *Carthamnus tinctorius* were discussed by Dean (1963, p. 337), who pointed out that the correct structure for the colourless pigment from *Carthamnus*, called isocarthamin, must be 5,6,7,4′-tetrahydroxyflavanone 5-*O*-glucoside (*9.180*). The coloured pigment from this plant was the related chal-

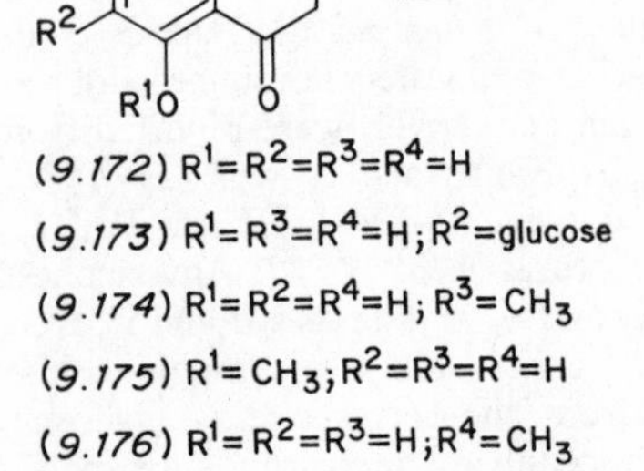

(9.172) $R^1 = R^2 = R^3 = R^4 = H$

(9.173) $R^1 = R^3 = R^4 = H$; R^2 = glucose

(9.174) $R^1 = R^2 = R^4 = H$; $R^3 = CH_3$

(9.175) $R^1 = CH_3$; $R^2 = R^3 = R^4 = H$

(9.176) $R^1 = R^2 = R^3 = H$; $R^4 = CH_3$

(9.177) $R^1 = R^2 = H$; $R^3 = R^4 = CH_3$

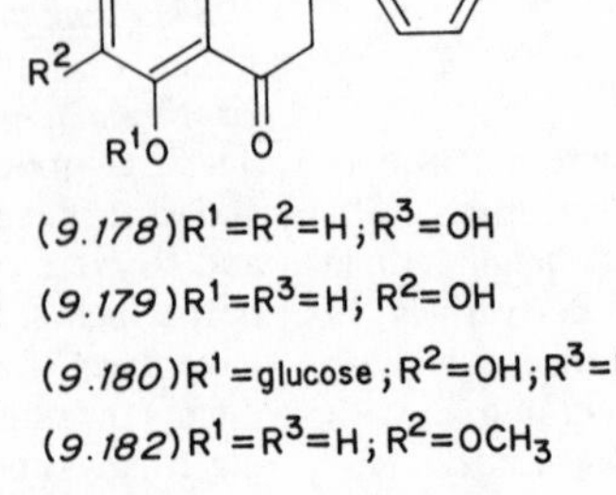

(9.178) $R^1 = R^2 = H$; $R^3 = OH$

(9.179) $R^1 = R^3 = H$; $R^2 = OH$

(9.180) R^1 = glucose; $R^2 = OH$; $R^3 = H$

(9.182) $R^1 = R^3 = H$; $R^2 = OCH_3$

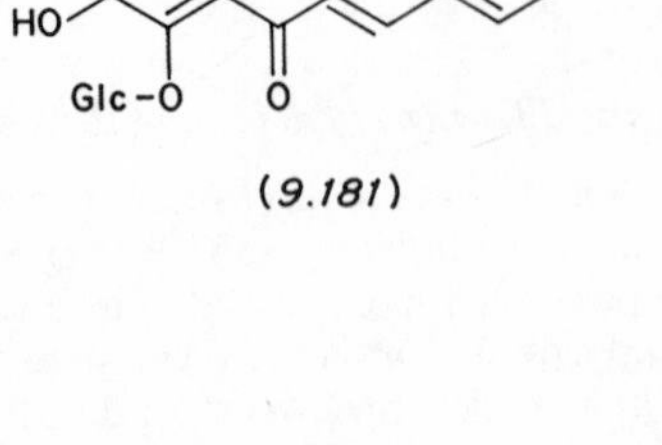

(9.181)

cone 4,2′,3′,4,′,6′-pentahydroxychalcone 2′-O-glucoside (*9.181*). From the trivial names used by the Japanese workers it is obvious that they were aware of the *Carthamnus* work (I have not seen their original paper). Isocarthamidin (*9.179*) would be the expected aglycone from isocarthamin (9.180). Hydrolysis of the chalcone glucoside 'carthamin' (*9.181*) would yield an aglycone that would be correctly called 'carthamidin.' Since the chalcone aglycone would have unsubstituted hydroxyl groups at positions 2′ and 6′, cyclizations would be expected from which two flavanones would result: if cyclization involved the C-2′ hydroxyl group then 'carthamidin' would result, while 'isocarthamidin' would be the product if cyclization involved the C-6′ hydroxyl group.

Two groups of workers (Takagi *et al.*, 1980; Kimura *et al.*, 1982) have subsequently reported isolation of 5,7,4′-trihydroxy-6-methoxyflavanone (*9.182*) from *Scutellaria baicalensis* (the compound could be called 6-methylisocarthamidin). The simple 5,6,7-trihydroxy-flavanone 'dihydrobaicalein,' was reported more recently in the same species by Tomimori and co-workers (1983), as well as many years ago by Watkin (1960) from *Scutellaria epilobifolia*.

It seems reasonable to continue with flavanones from *Scutellaria* even if some of them have two B-ring oxygenations. Unusual B-ring patterns are common in this genus. The first of these to be reported was the 2′,6′-hydroxylated flavanone (*9.183*) isolated by Ikram and co-workers (1979) from *S. baicalensis*. A confirming report of this compound from the same source was published three years later (Kimura *et al.*, 1982). Kubo and co-workers (1981) reported compound (*9.183*) from extracts of the drug 'Scutellaria Radix.' The 5-O-methyl derivative (*9.184*) has also been reported from *S. baicalensis* (Tomimori *et al.*, 1984). In the most recent study Tomimori and co-workers (1985) reported four compounds from *S. discolor* that exhibit combinations of the unusual A-ring and B-ring patterns. The simplest of the four are the 2′-methoxyflavanones (*9.185*) and (*9.186*) both of which were obtained in optically active form. Compounds (*9.187*) and (*9.188*) combine 2′,6′-dioxygenation with the two possible A-ring methoxy forms. Both of these compounds, however, were obtained in optically inactive form which suggests that they may have arisen from cyclization of chalcone derivatives during isolation. Both could arise by cyclization of the chalcone (*9.189*) which could either be a natural product (not yet detected) or the intermediate in the Wesseley–Moser rearrangement of either (or both) flavanones (*9.187*) and (*9.188*).

(*9.189*)

(*9.183*) R=H
(*9.184*) R=CH_3

(*9.185*) R=H
(*9.186*) R=CH_3

(*9.187*) R^1=H; R^2=OCH_3
(*9.188*) R^1=OCH_3; R^2=H

In a study of flavanones from Polygonaceae, Rathore and co-workers (1986) described three compounds that exhibit extra hydroxylation on the A-ring and complete O-alkylation of the resulting compound. They described 5,6,7,4′-tetramethoxyflavanone (*9.190*), 5,4′-dimethoxy-6,7-methylenedioxyflavanone (*9.191*), and the heptamethoxyflavanone (*9.279*) from *Polygonum nepalense*. Extra hydroxylation at position 8 coupled with O-methylation is seen in 5,7-dihydroxy-6-C-methyl-8,4′-dimethoxyflavanone (*9.192*) reported by Miraglia and co-workers (1985) from the fern *Blechnum regnellianum*. The last member of this small group of compounds is new but is closely related to the well-known flavanones 'farrerol' (6,8-di-C-methylnaringenin) and 'matteucinol' (the corresponding 4′-methyl ether). The new member was isolated from roots of *Prosopis juliflora* and shown to be the 5-O-galactoside of 6,8-di-C-methylnaringenin 7,4-dimethyl ether (*9.193*) (Malhotra and Misra, 1983).

(*9.190*) R^1=R^2=R^3=CH_3; R^4=H
(*9.191*) R^1=CH_3; R^2 and R^3=$-CH_2-$; R^4=H
(*9.192*) R^1=R^3=H; R^2=CH_3; R^4=OCH_3

The next two compounds might well be disqualified from consideration in this chapter but their relationship to flavanones prompted me to include them. Tanaka and co-

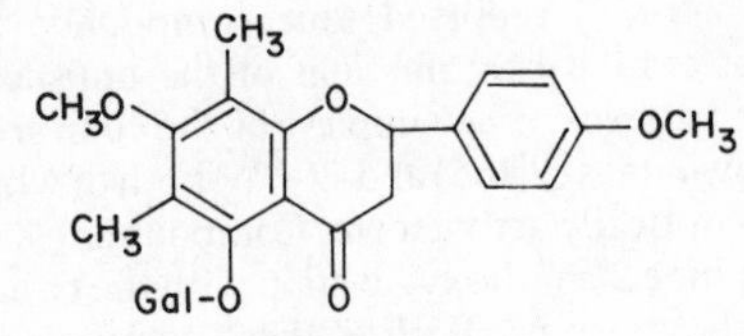

(9.193)

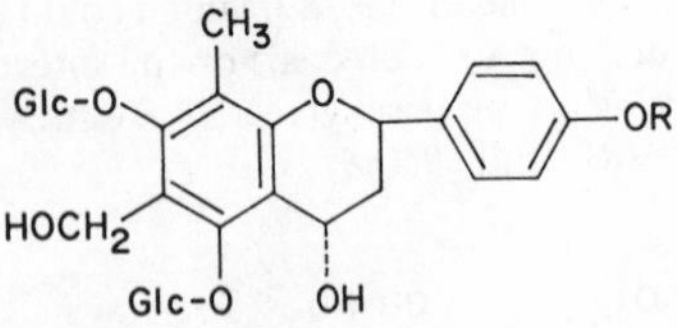

(9.194) R = CH_3

(9.195) R = H

workers (1985) isolated both compounds, 'triphyllin-A' (*9.194*) and 'triphyllin-B' (*9.195*), from *Pronephrium triphyllum*. The 4-hydroxyl group is the most unusual feature of these compounds, of course. It would be of considerable interest to learn whether there is a close biochemical relationship between these compounds and flavanones. The hydroxymethyl group is also rare but is only one reduction step removed from the formyl function which is known in several flavonoids. The 5, 7-diglycosylation pattern is unusual but offers no particular challenge to the imagination concerning its origin.

Two prenylated derivatives of naringenin were reported in the initial review; only two new ones were listed in the second. The picture has changed dramatically since then. The first report of naringenin 7-prenyl ether (*9.196*) appeared only recently (Bohlmann and Ates (Gören), 1984); it occurs in *Helichrysum athrixiifolium* along with the well known 8-*C*-prenyl derivative. Several new reports of 8-prenylnaringenin and its 6-isomer (*9.198*), known as 'sophoraflavanone-B', are summarized in Table 9.5. The first report of this compound was that of Komatsu and co-workers (1978) who isolated it from *Sophora tomentosa*. (That work was overlooked in preparation of the earlier review.) Sherif and co-workers (1982) have described a synthesis of 8-prenylnaringenin. A rare combination of prenylation and *O*-glycosylation within a single flavonoid was reported from *Phyllanthus nirurii* by Gupta and Ahmed (1984). The compound is 6-hydroxy-8-*C*-prenylnaringenin 5-*O*-rutinoside, 'nirurin,' the aglycone of which, called 'nirurinetin,' is shown as (*9.199*). This is the first record of a prenylated flavanone from a member of the Euphorbiaceae.

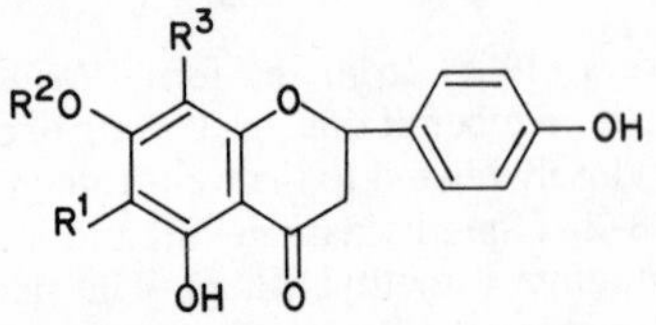

(9.196) $R^1=R^3=H$; R^2=prenyl

(9.197) $R^1=R^2=H$; R^3=prenyl

(9.198) R^1=prenyl; $R^2=R^3=H$

(9.199) $R^1=OH$; $R^2=H$; R^3=prenyl

Multi *C*-prenylation is a commonly met feature of flavanones (and other flavonoid types as well) in the Leguminosae. All but one of the examples that follow come from that family. The B-ring diprenylflavanone (*9.200*) occurs in *Erythrina abyssinica* (Kamat *et al.*, 1981) along with the set of related flavanones (*9.167–9.170*) and the chalcone (*9.48*). Compound (*9.200*) is somewhat unusual in that its A-ring has the phloroglucinol hydroxylation pattern while others exhibit the resorcinol pattern. *Erythrina senegalensis* exhibits double A-ring prenylation in 'erythrisenegalone' (*9.201*) (Fomun *et al.*, 1985). A flavanone having the same structure, (*9.201*), was described earlier by Bhanumati and

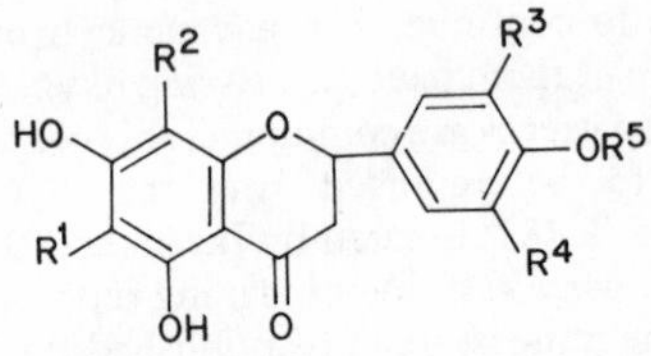

(9.200) $R^1=R^2=R^5=H$; $R^3=R^4$=prenyl

(9.203) $R^1=R^3$=prenyl; $R^2=R^4=R^5=H$

(9.204) R^1=prenyl; $R^2=R^4=R^5=H$; $R^3=CH_2CH$—$C(O)CH_3$ (CH_3)

(9.205) $R^1=R^4=H$; $R^2=R^3$=prenyl; $R^3=CH_3$

(9.206) $R^1=R^5=H$; $R^2=R^3=R^4$=prenyl

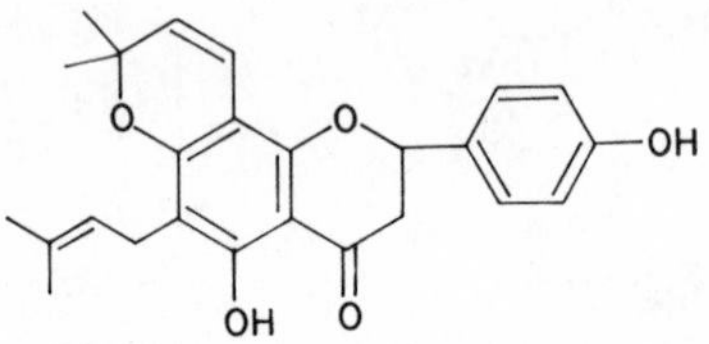

(9.201)

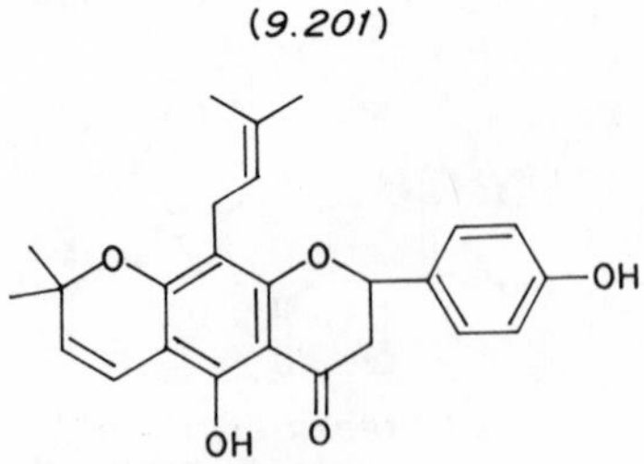

(9.202)

co-workers (1978) from the legume *Cajanus cajan*. Those workers called their product 'cajaflavanone.'

Alternative ring closure would yield the linear pyranoflavanone (*9.202*) which was reported some years ago by Smalberger and co-workers (1974) as a constituent of *Tephrosia lupinifolia*. 'Lupinifolin,' as the compound was called, occurs with the corresponding dihydroflavonol (*9.298*).

The simple 6, 3′-di-*C*-prenylflavanone (*9.203*) has been isolated from *Euchresta japonica* (Shirataki *et al.*, 1981) where it occurs with the 2′, 4′-dihydroxyflavanones (*9.255*) and (*9.256*) to be treated below. Recently, Mitscher and co-workers (1985) showed that the correct structure for 'flemiflavanone-D', from *Flemingia stricta*, is (*9.204*) instead of the structure bearing a modified geranyl substituent as originally proposed (Rao *et al.*, 1982). The only non-legume source of a prenylated flavanone in this series is *Azadirachta indica* of the Meliaceae from which Garg and Bhakuni (1984) obtained 6, 3′-di-*C*-prenyl-4′-*O*-methylnaringenin (*9.205*). The compound was given the unfortunate name 'nimbaflavone' by those authors presumably before it was recognized to be a flavanone.

Some of the most highly *C*-alkylated flavanones known have been found, not surprisingly, in members of the Leguminosae. Accompanying the known flavanone sophoranone in *Millettia pulchra* is the hitherto unreported triprenylated '5-hydroxysophoranone' (*9.206*) (Baruah *et al.*, 1984). A series of flavanones having two or three prenyl groups has been isolated from *Amorpha fruticosa*. Despite five of the seven having 3′, 4′-dioxygenation it is convenient to present them all at this time. All seven share the 6, 8-diprenyl substitution. The simplest compound identified was 'amoridin' which is 6, 8-diprenyl-7-*O*-methylnaringenin (*9.207*) (Rózsa *et al.*, 1984). 'Amorilin' (*9.208*) is 6, 8, 3′-triprenylnaringenin (Rózsa *et al.*, 1982b). Biosynthetically related to 'amoridin' (*9.207*) are

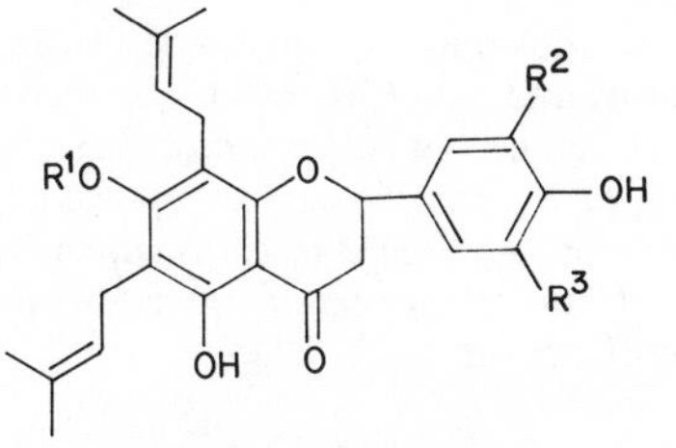

(*9.207*) $R^1=CH_3$; $R^2=R^3=H$

(*9.208*) $R^1=R^3=H$; R^2=prenyl

(*9.209*) $R^1=CH_3$; $R^2=H$; $R^3=OH$

(*9.210*) $R^1=CH_3$; $R^2=H$; $R^3=OCH_3$

(*9.211*) $R^1=H$; R^2=prenyl; $R^3=OH$

(*9.212*) $R^1=H$; R^2=prenyl; $R^3=OCH_3$

'amoradicin' (*9.209*) and 'amoradinin' (*9.210*) (Rózsa *et al.*, 1984). The same stepwise elaboration of the B-ring, possibly starting from (*9.208*), can be seen in 'amorisin' (*9.211*) and 'amoritin' (*9.212*) (Rózsa *et al.*, 1982b). 'Amorinin' (*9.213*), the last of this group, results from cyclization of the B-ring prenyl function with the 4′-hydroxyl group. Kinetic studies of formation of these compounds, if such were technically possible, would add immeasurably to our understanding of how and in what order these compounds are formed.

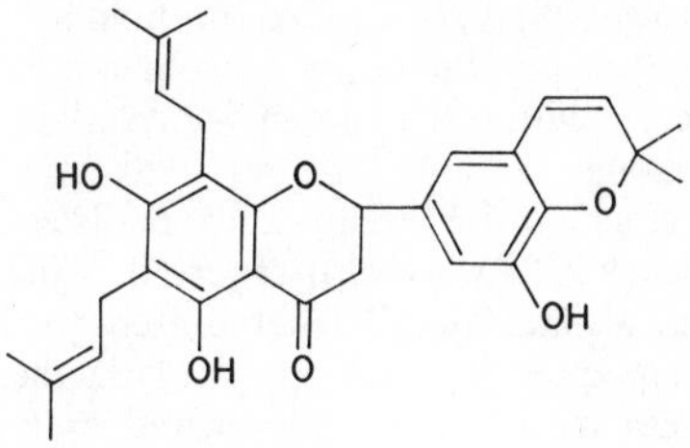

(*9.213*)

We can finish off naringenin-based flavanones by considering a small group of compounds distinguished by the presence of *C*-linked geranyl groups. The first of these was isolated from *Sophora tomentosa* by Shiritaki and co-workers (1983) and shown to be (2*S*)-6-geranylnaringenin (*9.214*). Two years later Bruno and his co-workers (1985) isolated geranylnaringenin from *Bonannia graeca* (Umbelliferae) and established its structure as (2*S*-6-geranylnaringenin but noted that its physical characteristics differed from those published by the Shiritaki group. There is a note in Bruno *et al.* (1985) that the compound from *Sophora* is the 8-geranyl isomer (*9.215*). The trivial name given to the *Sophora* compound originally was 'sophoraflavanone-A' which can serve just as well for the corrected structure. The 6-geranyl compound from *Bonannia*, which represents a rare report of a flavanone from the Umbelliferae, was called 'bonannione-A' (Bruno *et al.*, 1985). Bonannione-A occurs with two closely related *C*-geranyl dihydroflavonols (*9.299*) and (*9.300*). 'Kuwanon-S', 3′-geranylnaringenin (*9.216*), was identified as a constituent of *Morus lhou* (Moraceae) by Fukai and

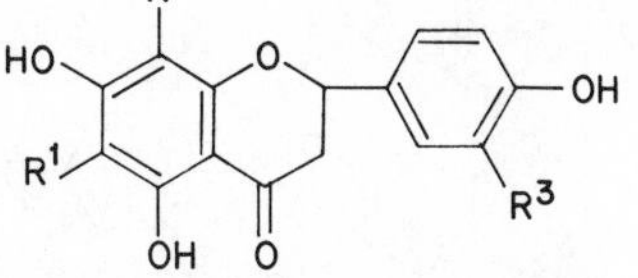

(*9.214*) R^1=geranyl; $R^2=R^3=H$

(*9.215*) $R^1=R^3=H$; R^2=geranyl

(*9.216*) $R^1=R^2=H$; R^3=geranyl

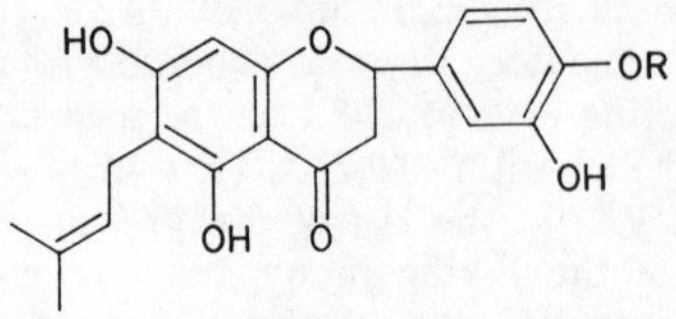

(*9.217*) R = H

(*9.218*) R = CH_3

co-workers (1985). The final report is the finding of 6-geranylnaringenin (*9.214*), called 'mimulone,' in the leaf resin of *Diplacus aurantiacus* by Lincoln and Walla (1986) where it occurs with related geranyl derivatives of eriodictyol, 'diplacone' (*9.217*) (Lincoln, 1980) and 'diplacone 4′-methyl ether' (*9.218*). Dihydroflavonols equivalent to (*9.217*) and (*9.218*) were also reported (Lincoln, 1980; Lincoln and Walla, 1986). These represent the first report of such compounds from the Scrophulariaceae.

Flavanones with B-rings oxygenated only at the 2′-position are quite rare in nature. Two such compounds, (*9.185*) and (*9.186*), from *Scutellaria discolor* were described above. Other additions to the list of these come from members of the Leguminosae and Annonaceae. In the latter case *Uvaria afzelii* was shown (Hufford *et al.*, 1981) to accumulate 5,7,2′-trihydroxy-6,8-di-*C*-methylflavanone (*9.219*) along with the unusual chalcone

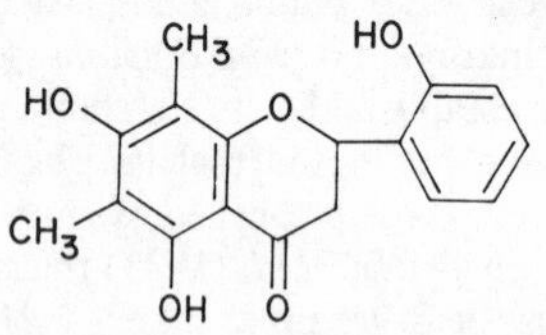

(*9.219*)

derivatives (*9.50–9.52*). Mahmoud and Waterman (1985) isolated a prenylated flavanone from seeds of *Lonchocarpus minimiflorus* with the structure shown in (*9.220*). The 4′-isomer (*9.221*) was also present. A pair of unusual flavanones was isolated from *Sophora flavescens* and described by Mu and co-workers (1985a). 'Kushenol-A

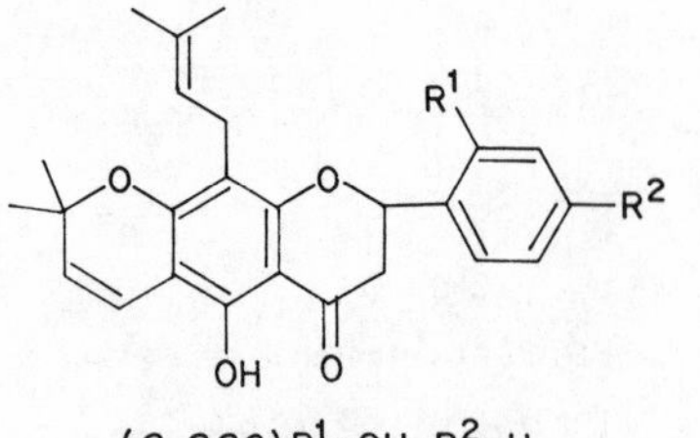

(*9.220*) R^1 = OH; R^2 = H

(*9.221*) R^1 = H; R^2 = OH

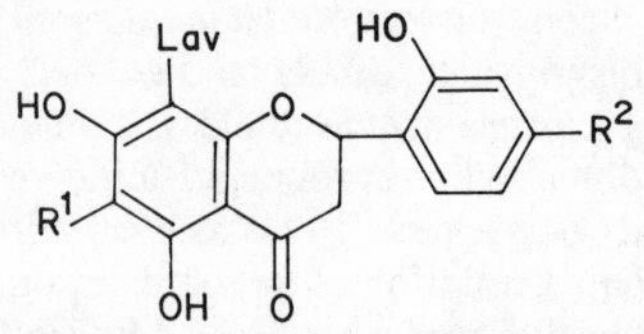

(*9.222*) R^1 = R^2 = H; Lav = lavandulyl

(*9.223*) R^1 = prenyl; R^2 = OH; Lav = lavandulyl

(*9.222*) is the 2′-hydroxy member of the pair; 'kushenol-B' has 2′,4′-dihydroxylation (*9.223*). Both flavanones, as well as a chalcone called 'kushenol-D' (*9.78*), have the 2-isopropyl-5-methylhex-4-enyl (= lavandulyl) side chain. Compounds with similar substitutions, including the lavandulyl group, have been reported from *Vexibia* (*Sophora*) *alopecuriodes* by Batirov and co-workers (1985) who described 'vexibinol' (*9.224*) and 'vexibidin' (*9.225*), and from *Ammothamnus lehmanni* by Sattikulov and co-workers (1983) who identified the new chalcone (*9.49*). The co-occurrence of two such unusual substituents, the lavandulyl group and 2′-oxygenation, is intriguing. Further information on flavonoids of related taxa would be of considerable interest from the chemosystematic point of view.

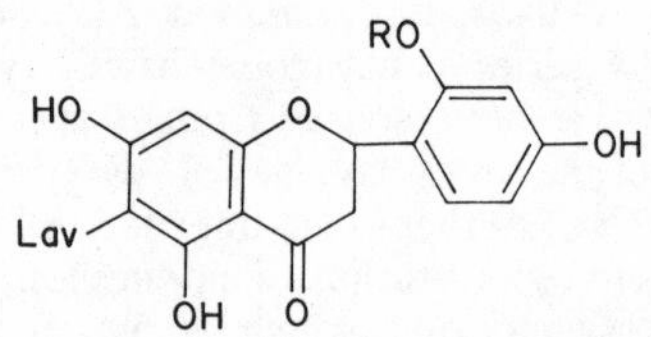

(*9.224*) R = H; Lav = lavandulyl

(*9.225*) R = CH_3; Lav = lavandulyl

(c) *Flavanones having two B-ring oxygenations*

The report of isoökanin 7-*O*-glucoside (*9.226*) in *Abies pindrow* (Tiwari and Minocha, 1980) represents the first record of isoökanin in a gymnosperm. The 7-*O*-rhamnoside (*9.227*), previously unreported, has been isolated from roots of *Alstonia scholaris* which represents a first report of this compound from a member of the Apocynaceae (Chauhan *et al.*, 1985).

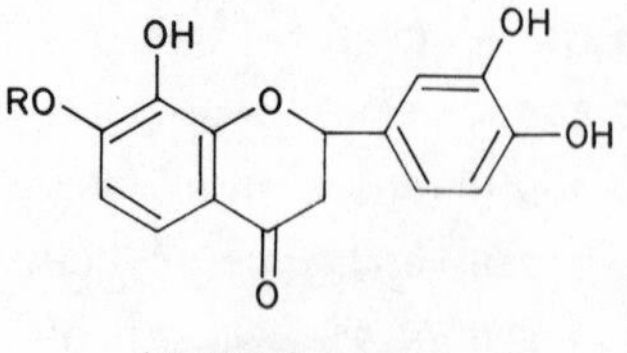

(*9.226*) R = Glucose

(*9.227*) R = Rhamnose

An unusual substitution pattern was reported (Anjaneyulu and Reddy, 1981) for a flavanone (*9.228*) from the bark of *Macaranga peltata* of the Euphorbiaceae. Whereas 5-deoxyflavonoids are quite common 7-deoxyflavonoids are not. To gain this particular A-ring pattern requires loss of oxygen from positions 5 and 7 and hydroxylation at position 6. The structural assignment seems to be correct, however, since these authors compared the natural product with an authentic sample of 6,3′,4′-trimethoxyflavanone (synthetic) and found their physical characteristics to be in agreement.

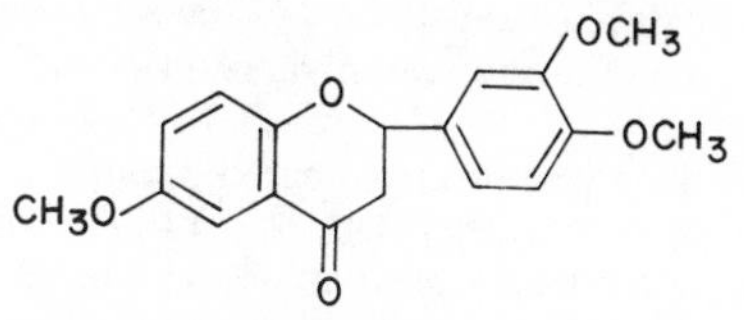

(9.228)

Eriodictyol, 5,7,3′,4′-tetrahydroxyflavanone (*9.229*), enjoys a fairly wide distribution in the plant kingdom either as the free phenol or in a variety of glycosidic forms. See Table 9.5 for all recent reports of glycoside occurrences. New family records amongst these are: Hydrangeaceae (*Philadelphus*) (Bohm and Chalmers, 1986); Gramineae (*Sorghum*) (Gujer *et al.*, 1986); Myoporaceae (*Myoporum*) (Tomas *et al.*, 1985); Papaveraceae (*Argemone*) (Harborne and Williams, 1983); Rhamnaceae (*Rhamnus*) (Sakushima *et al.*, 1983); and Verbenaceae (*Verbena*) (Stotz *et al.*, 1984). Additional reports of the known 3′-*O*-methyl (homoeriodictyol), 4′-*O*-methyl-, and 7,3′-dimethyleriodictyols can be found in Table 9.5. New as a natural product is 7-*O*-methyleriodictyol (*9.230*) which has been reported from genera of the Compositae: *Artemisia*, *Hazardia*, *Hemizonia*, *Heterotheca* and *Wyethia* (see Table 9.5 for references). In most of those cases the compound was found as a component of non-polar glandular exudates. 5-*O*-Methyleriodictyol (*9.231*) was reported to occur in the seeds of *Sapium sebiferum* (Euphorbiaceae) as the 7-*O*-β-

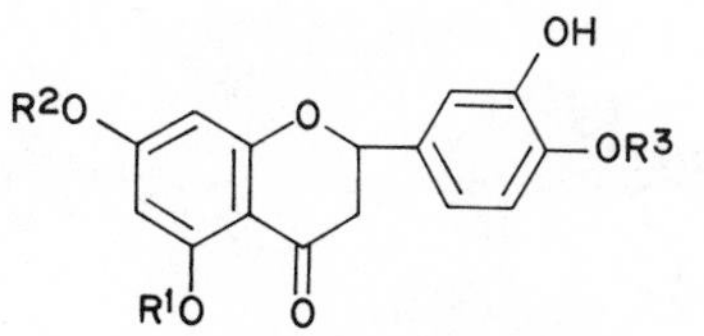

(9.229) $R^1 = R^2 = R^3 = H$

(9.230) $R^1 = R^3 = H; R^2 = CH_3$

(9.231) $R^1 = CH_3; R^2 = R^3 = H$

(9.232) $R^1 = R^3 = CH_3; R^2 = H$

D-xylosylarabinoside (Bhatt *et al.*, 1981a), and in extracts of whole plant of the fern *Diplazium esculentum* (Srivastava *et al.*, 1981a) where it occurs as the 7-*O*-β-xylosylgalactoside. A new dimethyl ether, 5,4′-di-*O*-methyleriodictyol (*9.232*), occurs as the 7-β-D-glucoside in *Cleome viscosa* (Capparaceae) (Srivastava, 1982).

Seven eriodictyol derivatives with extra hydroxylation of the A-ring have been reported recently; all of them are new natural products. The first of these to be reported, compound (*9.233*), exhibits a 6-methoxy group as well as a furano ring system. This compound was found as a component of *Derris araripensis* (do Nascimento and Mors, 1981) where it occurs with the corresponding dihydroflavonol (*9.314*). Compound (*9.234*) occurs in *Achyrocline tomentosa* (Compositae) (Ferraro *et al.*, 1985), the isomeric pair (*9.235*) and (*9.236*) occur in *Vitex negundo* (Verbenaceae) (Achari *et al.*, 1984), and the trio (*9.237*), (*9.238*) and (*9.239*), which show increasing *O*-alkylation in the B-ring, were found to occur in *Ageratum strictum* (Quijano *et al.*, 1982) where they are accompanied by the chalcone (*9.69*) corresponding to (*9.239*).

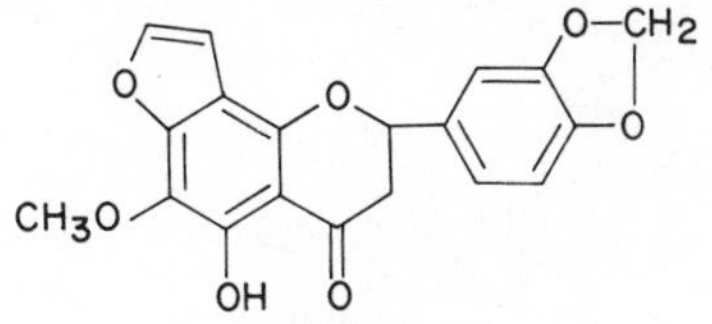

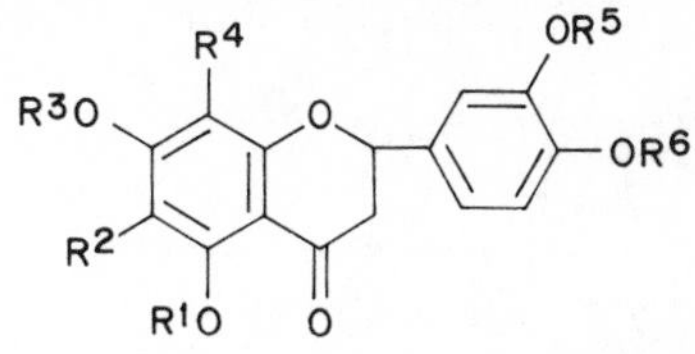

(9.233)

(9.234) $R^1 = R^2 = R^3 = R^6 = H; R^4 = OCH_3; R^5 = CH_3$

(9.235) $R^1 = R^4 = R^5 = H; R^2 = OCH_3; R^3 = R^6 = CH_3$

(9.236) $R^1 = R^2 = R^5 = H; R^3 = R^6 = CH_3; R^4 = OCH_3$

(9.237) $R^1 = R^3 = R^5 = CH_3; R^2 = OH; R^4 = R^6 = H$

(9.238) $R^1 = R^3 = R^5 = R^6 = CH_3; R^2 = OH; R^4 = H$

(9.239) $R^1 = R^3 = CH_3; R^2 = OH; R^4 = H; R^5$ and $R^6 = -CH_2-$

The anti-hepatotoxic flavonololignans from *Silybum* were discussed in both of the earlier reviews (I, 593; II, 379–380). Recent studies from the same laboratories have resulted in the identification of five additional biologically active compounds. These new compounds are 3-

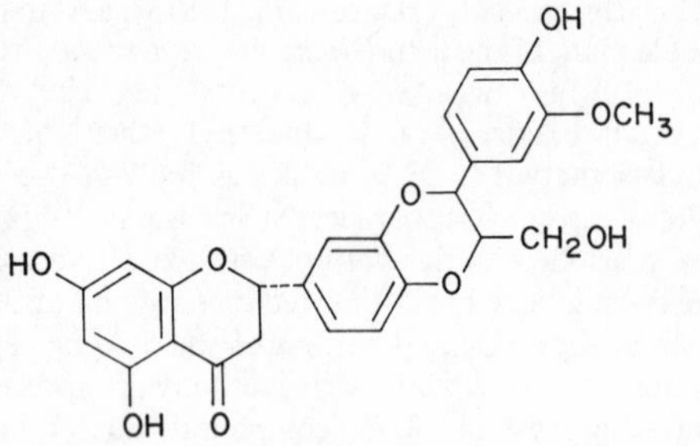

(9.240)

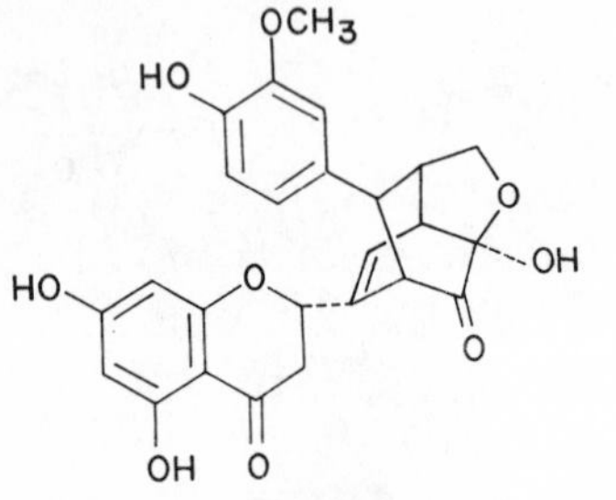

(9.241)

deoxyflavonololignans, or, flavanones. Two of these were described by Szilági and co-workers (1981) as (−)-silandrin (*9.240*) and (+)-silymonin (*9.241*). Compound (*9.241*) is most closely related to silydanin (I, 593) both of which have undergone dearomatization of the B-ring. More recently, Fiebig and Wagner (1984) identified three more members of this group from a white-flowered variant of *Silybum*. 'Silyhermin' (*9.242*) exhibits a reduced furan ring fused to ring-B which resulted from condensation of the coniferyl alcohol moiety with C-5′ of eriodictyol rather than with the other phenolic hydroxyl group. The stereochemistry of the attachment of the 3-methoxy-4-hydroxyphenyl and hydroxymethyl groups was not established. 'Neosilyhermin-A' (*9.243*) and 'neosilyhermin-B' (*9.244*) are similar to (*9.242*) but are based on condensation at C-2′ of the eriodictyol B-ring. They differ with regard to the stereochemistry at C-2: neosilyhermin-A has the (2*R*) configuration; neosilyhermin-B has the (2*S*) configuration.

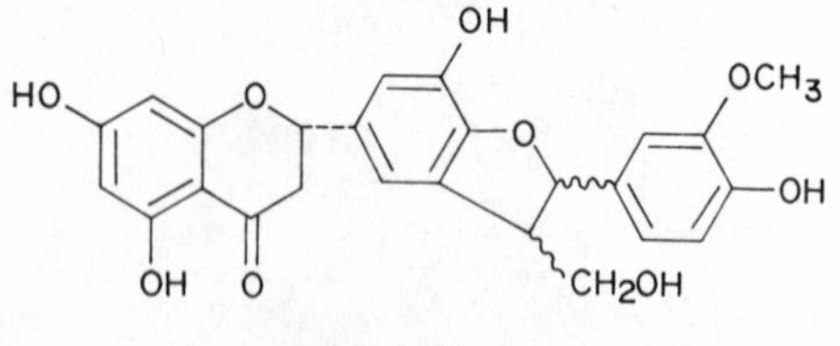

Three new natural products based upon *C*-methylated eriodictyol have been reported. 5,7,4′-Tri-*O*-methyl-8-*C*-methyleriodictyol (*9.245*) occurs in *Adina cordifolia* (Rubiaceae) (Srivastava and Gupta, 1983). Malhotra and Misra (1983) reported the glucoside (*9.246*) from roots of *Prosopis juliflora*. The last of these, and the most derived, is 'cyrtominetin 3′-methyl ether' (*9.247*) which Corrêa and co-workers (1981) isolated from two species of *Qualea* (Vochysiaceae).

A single *C*-prenylated eriodictyol derivative was available for inclusion in the earlier reviews. The situation has changed considerably in recent years. 8-Prenyleriodictyol (*9.248*) and 6-prenyleriodictyol (*9.249*) have been isolated from *Wyethia helenioides* (Bohlmann *et al.*, 1981d; McCormick *et al.*, 1986), *W. glabra* (McCormick *et al.*, 1985), and *W. angustifolia* (McCormick *et al.*, 1986). The 6-isomer only was reported from *W.* arizonica (Bohlmann *et al.*, 1984b). Prenylation at C-5′ occurs in *Erythrina sigmoidea* to yield 'sigmoiden-B' (*9.250*) (Fomum *et al.*, 1983) while prenylation at positions 2′ and 5′ affords 'sigmoiden-A' (*9.251*). 'Sigmoiden-C' (*9.252*), which would be expected to result from cyclization of (*9.250*), was also reported from *E. sigmoidea* by these workers (Fomum *et al.*, 1986).

A particularly rich array of flavanones having 2′,4′-dioxygenation has been reported in the last few years. A few of these were treated above in association with 2′-

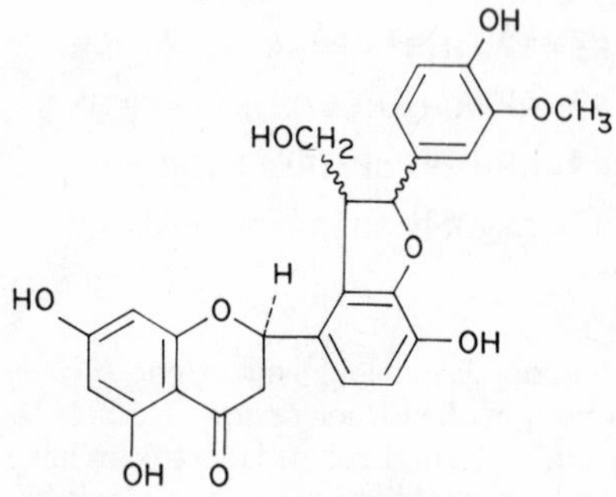

(9.242)

(9.243)

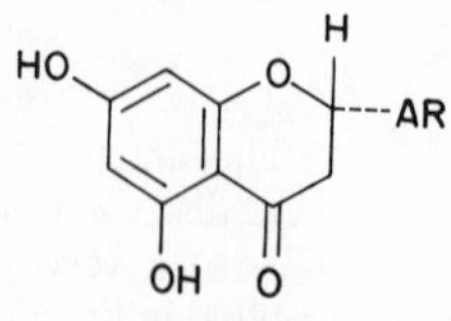

(9.244)

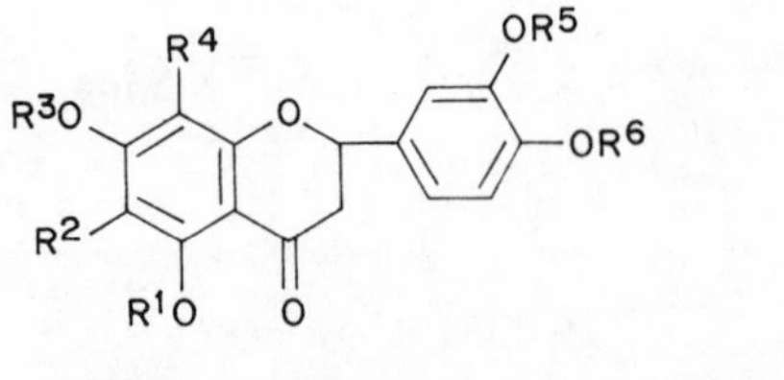

(9.245) $R^1 = R^3 = R^4 = R^6 = CH_3$; $R^2 = R^5 = H$

(9.246) $R^1 = R^2 = CH_3$; R^3 = Glucose; $R^4 = R^5 = R^6 = H$

(9.247) $R^1 = R^3 = R^6 = H$; $R^2 = R^4 = R^5 = CH_3$

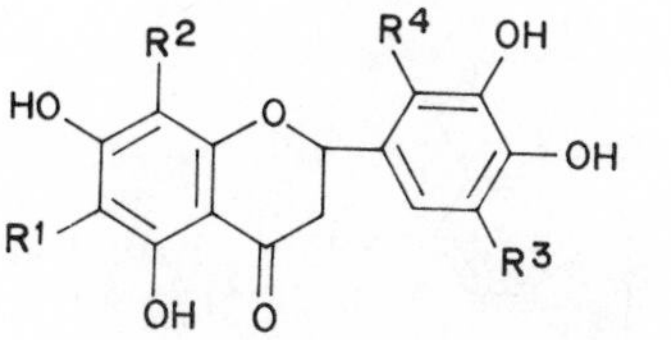

(9.248) $R^1 = R^3 = R^4 = H$; R^2 = prenyl
(9.249) R^1 = prenyl; $R^2 = R^3 = R^4 = H$
(9.250) $R^1 = R^2 = R^4 = H$; R^3 = prenyl
(9.251) $R^1 = R^2 = H$; $R^3 = R^4$ = prenyl

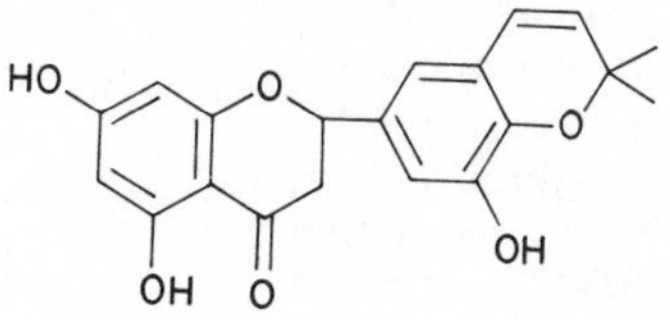

(9.252)

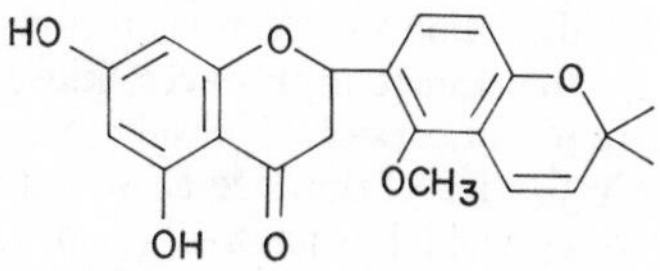

(9.253)

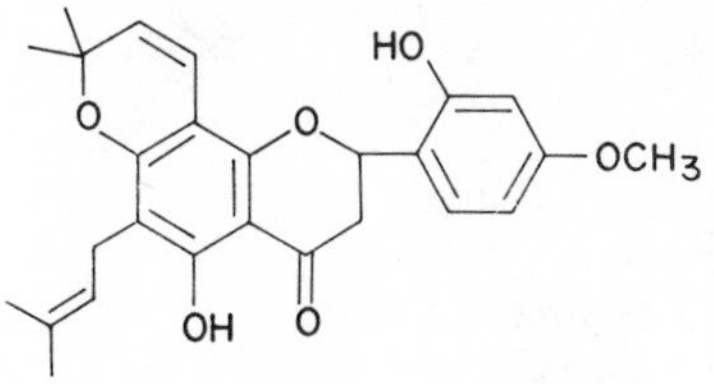

(9.254)

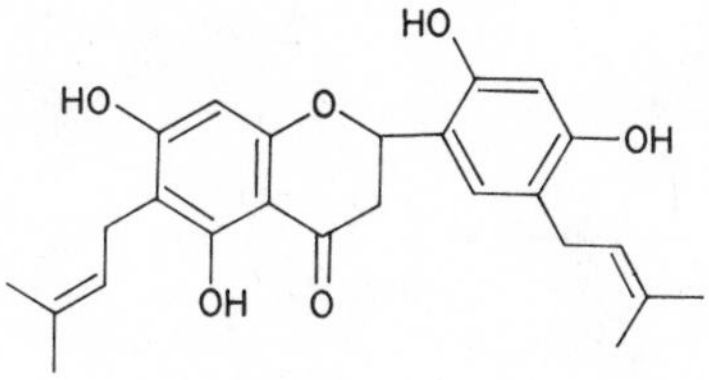

(9.255)

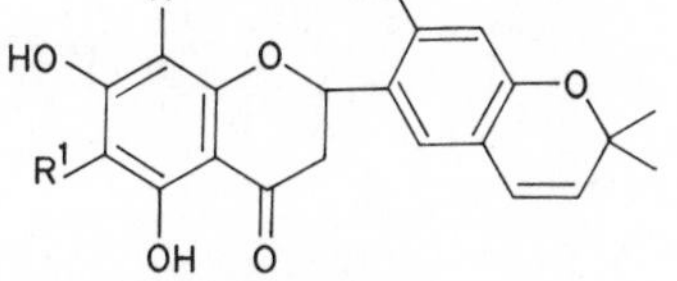

(9.256) R^1 = prenyl; R^2 = H
(9.257) R^1 = H; R^2 = prenyl

oxygenated compounds. A reference that I overlooked in preparation of the earlier review is Komatsu and co-workers' (1978) description of the structure of 'sophoranol' (*9.253*) from aerial parts of *Sophora tomentosa*. *Flemingia macrophylla* yielded a compound, 'fleminone,' that Rao and Srimannaryana (1983) proved to be the diprenylated flavanone (*9.254*).

Three C-prenylated flavanones were isolated from *Euchresta japonica* by Shirataki and co-workers (1982). One of these is a derivative of naringenin (*9.203*) and was mentioned above. The other two compounds were reported to have the structures shown as (*9.255*) for 'euchrestaflavanone-B' and (*9.256*) for 'euchrestaflavanone-C.'

The same two compounds, (*9.255*) and (*9.256*), were also reported from *Maclura pomifera* (Moraceae) by Delle Monache and co-workers (1984). In the same year Fujimoto and Nomura (1984) reported the results of their study of *Cudrania tricuspidata*, also a member of the Moraceae. These authors described two compounds, the first of which, 'cudraflavanone-A', is identical to 'euchrestaflavanone-C' as described by Shirataki and co-workers (1982). The second compound was shown to be (*9.257*), 5,7,2′-trihydroxy-8-*C*-prenyl-6″,6″-dimethylpyrano (2″,3″,4′,5′) flavanone. Its properties were identical to those of the original 'euchrestaflavanone-C' which was assigned a structure having the prenyl group at position 6. Thus, 'euchrestaflavanone-C' must be written as shown in (*9.257*) and not as shown in (*9.256*) (Fujimoto and Nomura, 1984).

Over the past several years a considerable volume of work has appeared dealing with the flavonoids of preparations of *Morus*, whether defined specifically as *M. alba* or as the Chinese crude drug 'sāng-bái-pi' which is identified as 'Morus root barks.' Much of the earlier work was reviewed by Nomura and Fukai (1981a). The flavanones encountered are all based upon 2′,4′-dioxygenated B-rings. The simplest members are based on cyclization of a 3′-*C*-prenyl group to either the 2′- or 4′-hydroxyl group. Cyclization with the 2′-hydroxyl group would yield 'sangganon-H' (*9.258*) (Hano and Nomura, 1983). Cyclization at position-4′ would lead to 'sangganon-F' (*9.259*) (Nomura *et al.*, 1983a, b).

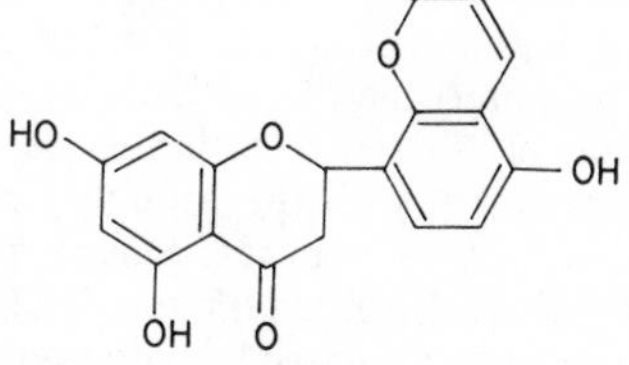

(9.258)

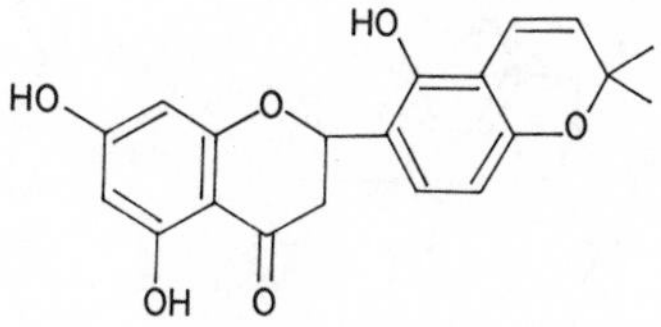

(9.259)

Alkylation of the parent compound at C-5′ with the geranyl function produces 'kuwanon-E' (*9.260*) isolated from root bark of *M. alba* by Nomura and Fukai (1981b). Cyclization of this side chain with the 4′-hydroxyl group would yield 'kuwanon-F' (*9.261*) which was isolated from the same source (Nomura and Fukai, 1981b). Nomura and Fukai, (1981b) have also reported a third compound, 'kuwanon-D', to which they tentatively assigned structure (*9.262*) on the basis of extensive spectral information.

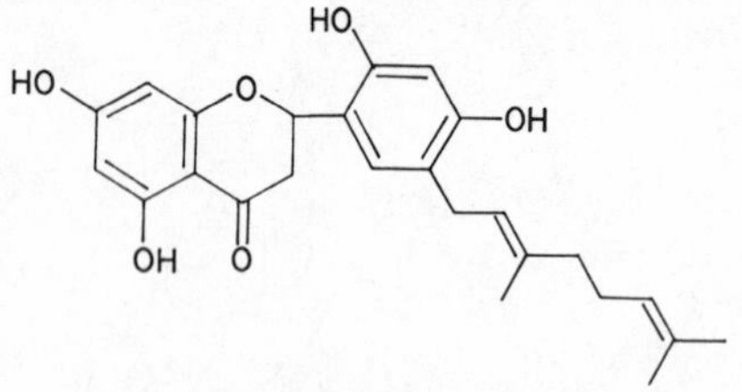

(9.260)

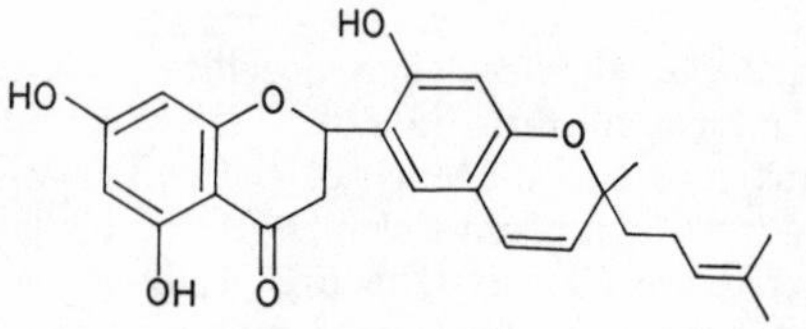

(9.261)

Formation of such an unusual flavonoid can be rationalized comparatively easily by rewriting kuwanon-F (B-ring only) as shown in (*9.263*). In this configuration interaction between the two double bonds would yield the observed cyclobutyl ring. Formation of cyclobutyl ring systems by carbonium ion-based mechanisms can be found in a discussion of sesquiterpene biosynthesis by Cane (1981).

If alkylation of the parent flavanone (5,7,2′,4′-tetrahydroxyflavanone) with the geranyl group were to occur at position 3′, rather than at C-5′ as seen above, then cyclization with either hydroxyl group would be possible. Compounds presumably derived by both cyclizations have been isolated from the crude drug. Cyclization with the 2′-

(9.262)

(9.263)

(9.264)

(9.265)

hydroxyl group would give 'sangganon-I' (*9.264*) (Hano and Nomura, 1983); cyclization with the 4′-hydroxyl would yield 'sangganon-N' (*9.265*) (Hano *et al.*, 1984).

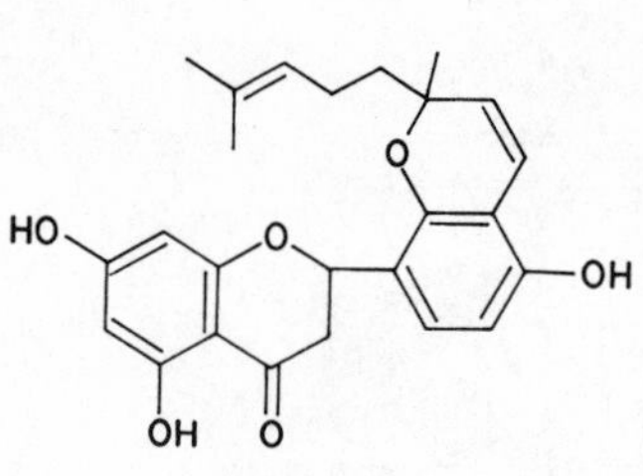

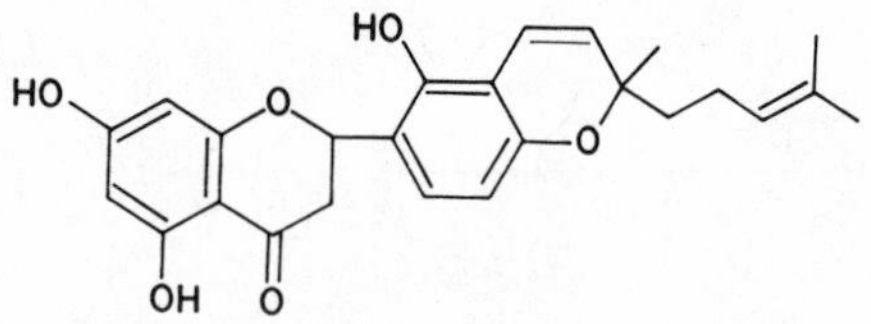

A highly unusual type of flavanone derivative involving *C*-alkylation at C-3 was first described in 1980 (Nomura *et al.*, 1980, 1983). In addition to 3-prenylation 'sangganon-A' (*9.266*) exhibits other unusual structural features: hydroxylation at C-2 and ring closure between C-3 and the 2′-hydroxyl group. The A-ring has also been alkylated. Since the first report of these compounds several others of this general sort have been described that differ only in the position and degree of A-ring alkylation. 'Sangganon-M' (*9.267*) is an isomer of compound (*9.266*) and differs only

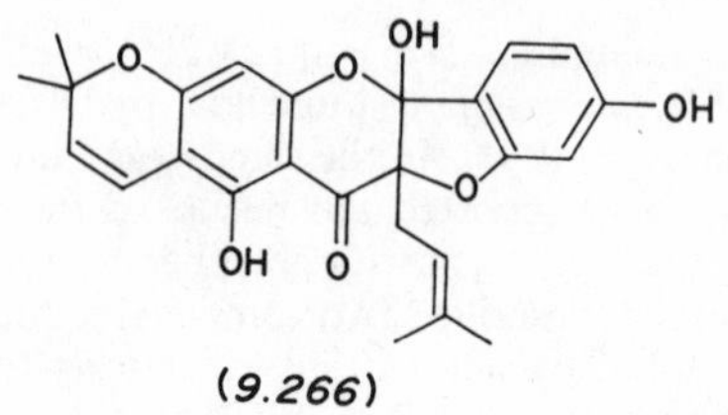

(9.266)

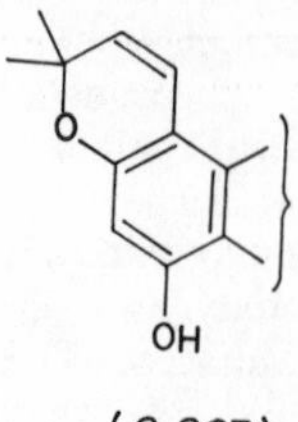

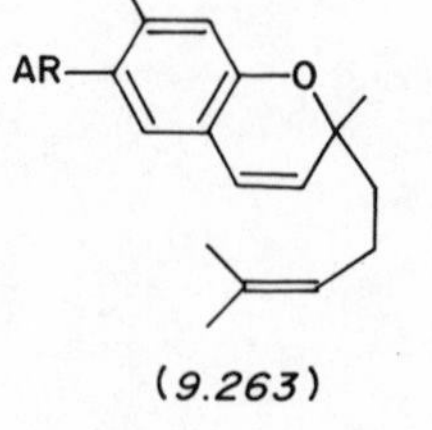

(9.267)

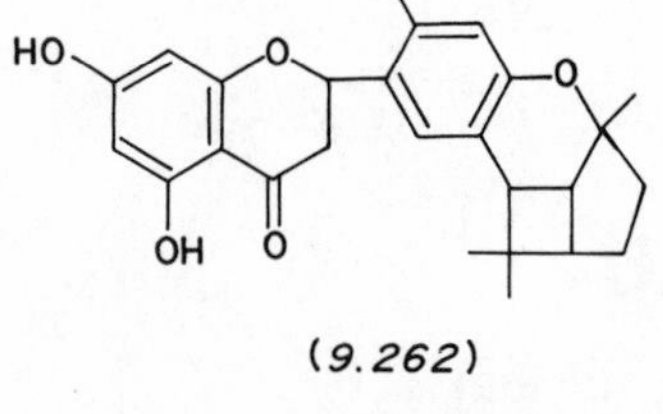

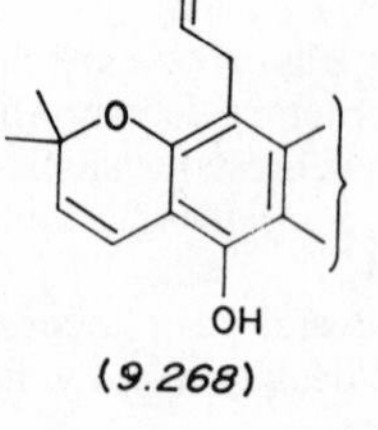

(9.268)

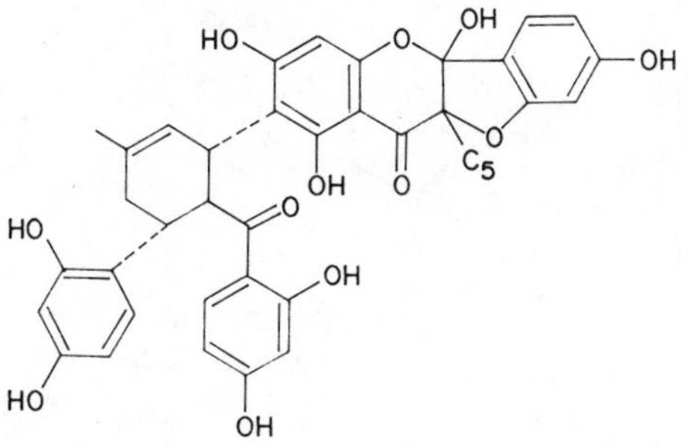

(9.269)

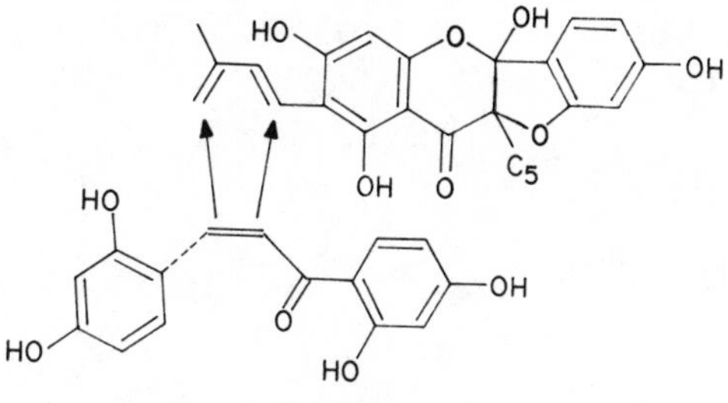

(9.270)

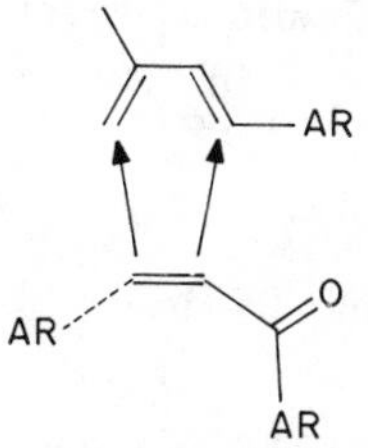

(9.271)

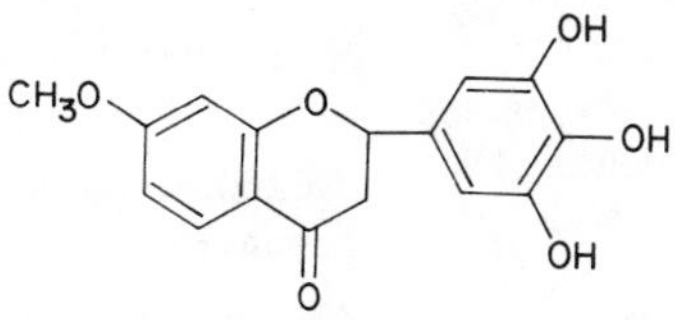

(9.272)

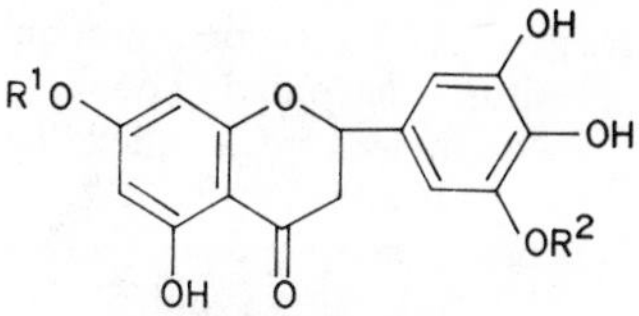

(9.273) $R^1=R^2=H$

(9.274) $R^1=H$; $R^2=$glucose

(9.275) $R^1=R^2=CH_3$

in that it is based on cyclization of an 8-prenyl group. Compound (*9.268*), 'sangganon-L', exhibits prenylation at both positions 6 and 8 with cyclization having occurred with the 6-prenyl group. Both of these compounds have been identified in extracts of the crude drug (Hano *et al.*, 1984).

A highly substituted and exceptionally complex flavanone was isolated by Nomura and co-workers (1982), again from the crude Morus root bark preparation, that showed hypotensive activity. 'Sangganon-D' (*9.269*) consists, essentially, of two subunits, the parent flavanone system seen in compounds (*9.266*)–(*9.268*) coupled, via its 6-prenyl group, with a C_6-C_3-C_6 system that is thought to have arisen from a chalcone. The structures written in (*9.270*) show the two presumed parent molecules written in such a way as to bring the chalcone α, β-double bond and flavanone 6-prenyl group into focus. The suggestion has been made by Fukai *et al.* (1983) that (*9.269*) and related compounds arise through a Diels-Alder-type reaction which is generalized in (*9.271*).

(d) *Flavanones having three B-ring oxygenations*

The first compound in this subgroup presents a problem. Chemesova and co-workers (1984) reported structure (*9.272*) for a flavanone from *Artemisia xanthochroa*. There is nothing intrinsically troublesome with this as a natural flavonoid; the problem arises in trying to reconcile the structure with the reported characteristics, namely, the alkali fusion product and the melting point. Alkali degradation of flavanones usually yields the B-ring as a benzoic acid. Hence, the reported formation of protocatechuic acid (3, 4-dihydroxybenzoic acid) would suggest dihydroxylation of the B-ring, not trihydroxylation as reported. The authors state that *O*-methylation of their unknown afforded a compound whose melting point (157°C) agreed with the value reported by Wollenweber and co-workers (1980) for 5-hydroxy-7, 3′, 4′, 5′-tetramethoxyflavanone. Reference to the Wollenweber paper shows that they did indeed describe that compound but did *not* report its melting point! They did, however, report a melting point of 156°C for 5-hydroxy-7, 3′, 4′-trimethoxyflavanone. Could the Russian's compound have been 5, 3′, 4′-trihydroxy-7-methoxyflavanone? Such a structure would be concordant with their reported observations. Unfortunately, UV, IR and PMR data, although apparently gathered, were not reported in their paper. Since I did not have access to the Russian paper in the original language I have no way of knowing whether the error was made at source or whether it arose in the translation.

The rest of the compounds in this subgroup are based upon the 5, 7, 3′, 4′, 5′-pentahydroxylation pattern which has been further derivatized to greater or lesser extent in several of them. The parent compound, (*9.273*), was described as a constituent of the bracts of *Helichrysum bracteatum* (Forkmann, 1983) where it occurred along with naringenin and eriodictyol. It was also found to occur in flowers of *Verbena hybrida* (Stotz *et al.*, 1984), again along with the simpler flavanones, but also with the dihydroflavonols aromadendrin, taxifolin and ampelopsin.

A new flavanone glycoside 'plantagoside' (*9.274*) was shown to occur in seeds of *Plantago major* var. *japonica* by Endo and co-workers (1981), and the structure shown as (*9.275*) was assigned to one of the compounds present in

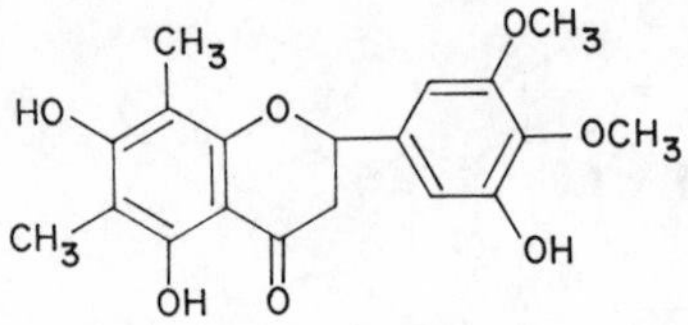

(9.276) 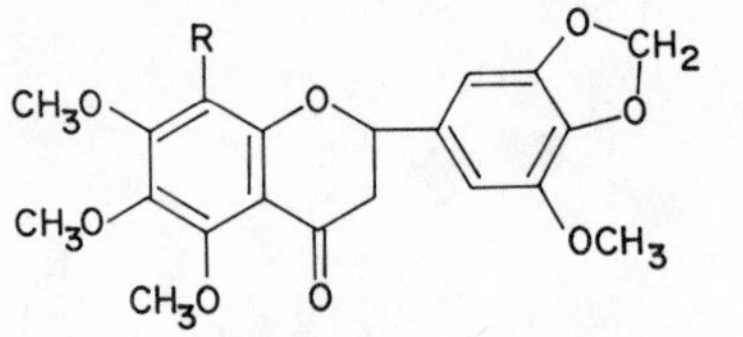 (9.277) R=H (9.278) R=OCH_3 (9.279)

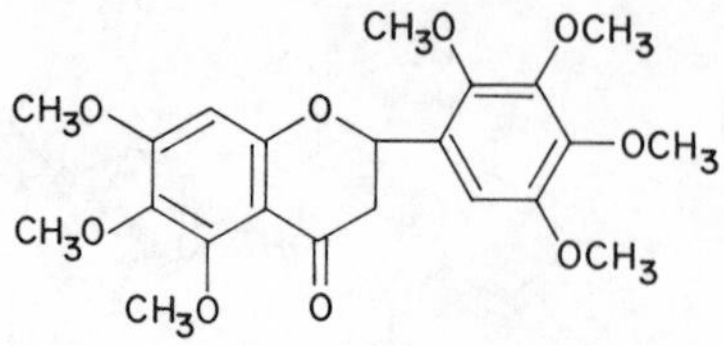

the leaf resin of *Hazardia squarrosa* var. *grindelioides* (Clark and Wollenweber, 1985). The last compounds to be treated in this section are amongst the most highly substituted flavonoids known. 5, 7, 3′-Trihydroxy-6, 8-di-*C*-methyl-4′, 5′-dimethoxyflavanone (*9.276*) was found by Rabesa and Voirin (1983) as a component of *Alluaudiopsis marnieriana* (Didiereaceae). 'Agecorynin-A' (*9.277*) and 'agecorynin-B' (*9.278*) were identified by Quijano and co-workers (1980) as constituents of *Ageratum corymbosum*.

(e) *Flavanones having four B-ring oxygenations*

Tetrasubstituted B-ring flavonoids, excluding *C*-alkylations, are scarce in the plant kingdom. A very recent addition to this list is compound (*9.279*), 5, 6, 7, 2′, 3′, 4′, 5′-heptamethoxyflavanone, which Rathore and co-workers (1986) isolated from whole-plant extracts of *Polygonum nepalense*. It occurs with two other flavanones characterized by total *O*-alkylation, 5, 4′-dimethoxy-6, 7-methylenedioxyflavanone (*9.191*) and 5, 6, 7, 4′-tetramethoxyflavanone (*9.190*).

9.6 DIHYDROFLAVONOLS

9.6.1 Introduction

The stereochemistry of dihydroflavonols was discussed in some detail in our first review (I, 594–599) so it is not necessary to go into any further detail here. The reader should simply remember that dihydroflavonols have two centres of asymmetry, C-2 and C-3, and that four possible structures exist with regard to the phenyl and hydroxyl substituents. Three of those possible types are represented in the compounds presented below. The most common form by far is (2*R*:3*R*); 25 of the 30 compounds for which stereochemical data were available exhibit this arrangement of groups. Of the remaining five compounds two exhibited (2*S*:3*S*) stereochemistry and three (2*R*:3*S*) stereochemistry. The three types are shown below in simplified form. In a large proportion of the papers that describe dihydroflavonols there is no information given on the stereochemistry of the compound(s) in question. In the text that follows stereochemical information will be given when it is available.

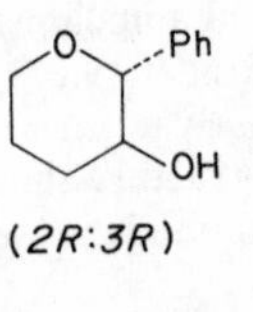

(2*R*:3*R*) 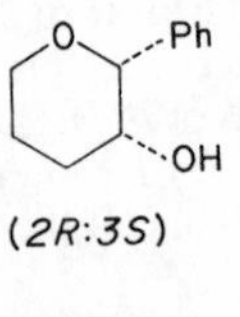 (2*R*:3*S*) 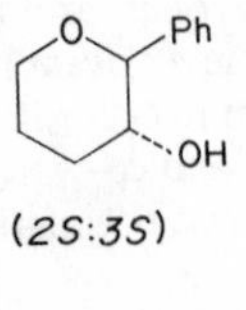 (2*S*:3*S*)

9.6.2 Structures of naturally occurring dihydroflavonols

All reported occurrences of dihydroflavonols have been summarized in Table 9.6.

(a) *Dihydroflavonols lacking B-ring oxygenation*

The simplest member of this group to appear in the literature since the last review is pinobanksin, 3, 5, 7-trihydroxyflavanone (*9.280*). It was found as the free phenol in species of *Eremophila* (Myoporaceae) (Jeffries *et al.*, 1982) and in *Baccharis oxydonta* where it occurs with the corresponding flavanone (Bohlmann *et al.*, 1981b). Pinobanskin and its 6-methoxyderivative (*9.281*), both with (2*R*:3*R*) stereochemistry, were reported as components of *Chromolaena chasleae* (Bohlmann et *al.*, 1982). Kuroyanagi and co-workers (1982) found pinobanksin and the new dihydroflavonol (*9.283*), (2*R*:3*R*)-5-methoxy-6, 7-methylenedioxydihydroflavonol, in *Polygonum nodosum*. Chauhan and co-workers (1984) found pinobanksin as the 5-*O*-β-D-galactopyranosyl(1 → 4)-β-D-glucopyranoside in *Adansonia digitata* (Bombacaceae). Wollenweber and his colleagues (1985b) reported pinobanksin 3-acetate in leaf washes of *Comptonia peregrina* and *Myrica pensylvanica* (both Myricaceae).

5-*O*-Methylpinobanksin (*9.282*) was isolated from

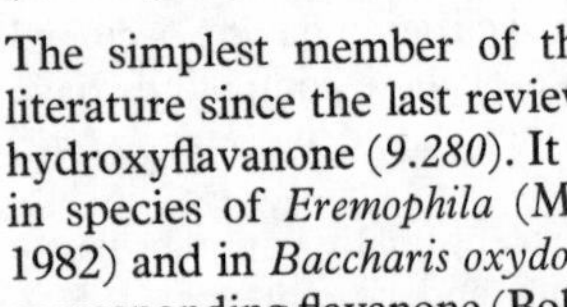

(9.280) $R^1=R^2=H$
(9.281) $R^1=H$; $R^2=OCH_3$
(9.282) $R^1=CH_3$; $R^2=H$
(9.283)

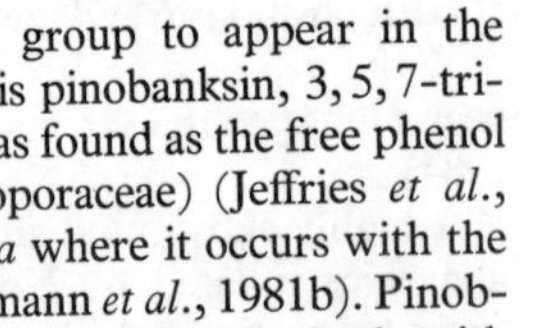

Table 9.6 Naturally occuring dihydroflavonols

Substituents							
OH	*OCH_3*	*Other*	*No. in text*	*Trivial name*	*Source**		*References*
5,7			(9.280)	Pinobanksin	*Baccharis oxydonta*	Com	Bohlmann *et al.* (1981b)
					Chromolaena chasleae	Com	Bohlmann *et al.* (1982a)
					Eremophila alternifolia	Myo	Jeffries *et al.* (1982)
					E. ramosissima	Myo	Jeffries *et al.* (1982)
					Polygonum nodosum	Pol	Kuroyanagi *et al.* (1982)
		3-Acetate			*Comptonia peregrina*	Myr	Wollenweber *et al.* (1985b)
					Myrica pensylvanica	Myr	Wollenweber *et al.* (1985b)
		5-Glucose-galactose			*Adansonia digitata*	Bom	Chauhan *et al.* (1984)
7	5		(9.282)		propolis		Bankova *et al.* (1983)
5,7		6-*C*-Prenyl	(9.287)		*Helichrysum thapsus*	Com	Bohlmann and Zdero (1983)
5,7		6-*C*-Geranyl	(9.288)		*H. thapsus*	Com	Bohlmann and Zdero (1983)
5,7		8-*C*-Prenyl	(9.286)	Glepidotin-B	*Glycyrrhiza lepidota*	Leg	Mitscher *et al.* (1983)
5,7	6		(9.281)		*Chromolaena chasleae*	Com	Bohlmann *et al.* (1982a)
	5	6,7-Methylenedioxy	(9.283)		*Polygonum nodosum*	Pol	Kuroyanagi *et al.* (1982)
7	4′	7-*O*-Xylose-glucose	(9.289)	Kushenol-J	*Sophora flavescens*	Leg	Wu *et al.* (1985c)
7,4′		8,3′,5′-Tri-*C*-prenyl	(9.290)		*Millettia pulchra*	Leg	Baruah *et al.* (1984)
5,7,4′			(9.291)	Aromadendrin	*Abies lasiocarpa*	Cnf	Parker and Maze (1984)
					Acacia catechu	Leg	Deshpande and Patil (1981); Burret *et al.* (1982)
					Camellia sinensis	The	Chkhikvishvili *et al.* (1984)
					Coleus blumei	Lab	Baslas and Kumar (1981)
					Eucalyptus citriodora	Myr	Dayal (1982)
					Podocarpus species	Cnf	Markham *et al.* (1985)
					Rhamnus pallasii	Rhm	Sakushima *et al.* (1983)
					Salix caprea	Sal	Malterud *et al.* (1985)
		3-*O*-Galactose			*Abies lasiocarpa*	Cnf	Parker and Maze (1984)
		3-*O*-Glucose			*Podocarpus* species	Cnf	Markham *et al.* (1985)
		3-*O*-Rhamnoside			*Vitis vinifera*	Vit	Trousdale and Singleton (1983)
		7-*O*-Glucose			*Podocarpus* species	Cnf	Markham *et al.* (1985)
		3-Acetate			*Aframomum pruinosum*	Zin	Ayafor and Connolly (1981)
					Baccharis varians	Com	Bohlmann *et al.* (1981c)
5,7	4′		(9.294)		*Salix caprea*	Sal	Malterud *et al.* (1985)
5,4′	7		(9.292)		*Artemisia campestris*	Com	Hurabielle *et al.* (1982)
					A. dracunculus	Com	Balza *et al.* (1985)
					A. scoparia	Com	Chandrasekharan *et al.* (1981)

(*Contd.*)

Table 9.6 (*Contd.*)

Substituents: OH	OCH$_3$	Other	No. in text	Trivial name	Source*		References
					Dittrichia viscosa	Com	Chiappini *et al.* (1982)
					Eucalyptus citriodora	Myr	Dayal (1982)
					Podocarpus nivalis	Cnf	Markham *et al.* (1984)
		3-Acetate			*Aframomum pruinosum*	Zin	Ayafor and Connolly (1981)
5	7,4′		(*9.293*)		*A. pruinosum*	Zin	Ayafor and Connolly (1981)
		3-*O*-Rhamnose		Aurapin	*Acinos alpinus*	Lab	Venturella *et al.* (1980)
5,4′		7-*O*-Prenyl			*Pterocaulon virgatum*	Com	Bohlmann *et al.* (1981a)
5,7,4′		8-Glucosyloxy	(*9.295*)		*Calluna vulgaris*	Eri	Jalal *et al.* (1982)
5,7,4′		6-*C*-Methyl; 7-*O*-glucose	(*9.297*)		*Pinus massoniana*	Cnf	Shen and Theander (1985)
7,4′	5	6-*C*-Methyl	(*9.296*)		*Qualea labouriauana*	Voc	Corrêa *et al.* (1981)
5,7,4′		6-*C*-Geranyl	(*9.299*)	Bonanniol-A	*Bonannia graeca*	Umb	Bruno *et al.* (1985)
7,4′	5	6-*C*-Geranyl	(*9.300*)	Bonanniol-B	*B. graeca*	Umb	Bruno *et al.* (1985)
5,4′		6″,6″-Dimethylpyrano-(2″,3″,7,6); 8-*C*-prenyl	(*9.298*)		*Tephrosia lupinifolia*	Leg	Smalberger *et al.* (1974)
7,3′,4′			(*9.301*)	Fustin	*Umtiza listerana*	Leg	Burger *et al.* (1983)
5,7,3′,4′			(*9.303*)	Taxifolin	*Acacia catechu*	Leg	Deshpande and Patil (1981)
					Acinos alpinus	Lab	Venturella *et al.* (1980); Burret *et al.* (1982)
					Cactaceae, various genera		
					Erythroxylum ulei	Ery	Bohm *et al.* (1981)
					Illicium munipurense	Ill	Williams and Harvey (1982)
					Polygonum nodosum	Pol	Kuroyanagi *et al.* (1982)
					Salix caprea	Sal	Malterud *et al.* (1985)
					Winteraceae, various genera		Williams and Harvey (1982)
		7-*O*-Glucose			*Podocarpus nivalis*	Cnf	Markham *et al.* (1984)
					Podocarpus species	Cnf	Markham *et al.* (1985)
					Robinsonia gracilis	Com	Pacheco *et al.* (1985)
					Sorghum spp.	Gra	Gujer *et al.* (1986)
		5-*O*-Glucose			*Podocarpus nivalis*	Cnf	Markham *et al.* (1984)
7,8,3′,4′			(*9.302*)	8-Hydroxyfustin	*Acacia melanoxylon*	Leg	Foo (1986)
		5-*O*-Galactose			*Dillenia pentagyna*	Dil	Srivastava (1981)
		3-*O*-Rhamnose			*Vitis vinifera*	Vit	Trousdale and Singleton (1983)
		3-*O*-Glucose			*Pinus massoniana*	Cnf	Shen and Theander (1985)
					Podocarpus species	Cnf	Markham *et al.* (1985)
					Taxillus kaempferi		Sakurai *et al.* (1982)
		3-*O*-Glucose-6″-*O*-phenyl acetate			*Pinus massoniana*	Cnf	Shen and Theander (1985)
		3-*O*-Glucose-6″-*O*-gallate			*Taxillus kaempferi*		Sakurai and Okumura (1983)

		3-*O*-Rhamnose-glucose			*Diospyros peregrina*	Ebe	Chauhan *et al.* (1982)
		3-Acetate			*Baccharis varians*	Com	Bohlmann *et al.* (1981c)
5,7,4′	3′			Dihydro-isorhamnetin	*Artemisia dracunculus*	Com	Balza *et al.* (1985)
5,3′,4′	7		(*9.306*)	Padmatin	*A. glutinosa*	Com	González *et al.* (1983)
		3-Acetate		3-Acetyl padmatin	*Inula viscosa*	Com	Grande *et al.* (1985)
5,7,3′	4′		(*9.304*)		*Blumea balsamifera*	Com	Ruangrungsi *et al.* (1981)
		3-*O*-Glucose-rhamnose			*Anacheilium* sp.	Orc	Ferreira *et al.* (1986)
5,3′	7,4′		(*9.305*)		*Blumea balsamifera*	Com	Ruangrungsi *et al.* (1981)
5,4′	7,3′				*Artemisia campestris*	Com	Teresa *et al.* (1984, 1986)
					A. glutinosa	Com	González *et al.* (1983)
					Grindelia discoidea	Com	Timmermann *et al.* (1986)
	3,5,6	Furano(2″,3″,7,8); 3′,4′-methylenedioxy	(*9.314*)		*Derris araripensis*	Leg	do Nascimento and Mors (1981)
5,7′,3′,4′		7-*O*-Prenyl	(*9.309*)		*Pterocaulon virgatum*	Com	Bohlmann *et al.* (1981a)
5,7,4′	3′	8-*C*-Prenyl	(*9.313*)		*Wyethia angustifolia*	Com	McCormick *et al.* (1986)
5,7,3′,4′		6-*C*-Geranyl	(*9.310*)	Diplacol	*Diplacus aurantiacus*	Scr	Lincoln (1980)
5,7,3′	4′	6-*C*-Geranyl	(*9.311*)	Diplacol methyl ether	*D. aurantiacus*	Scr	Lincoln and Wall (1986)
		See text for structure	(*9.312*)		*Wyethia angustifolia*	Com	McCormick *et al.* (1986)
5,7,2′,4′			(*9.315*)	Dihydromorin	*Scutellaria baicalensis*	Lab	Takagi *et al.* (1981)
5,7,2′,4′		6,8-Di-*C*-prenyl	(*9.318*)	Kushenol-L	*Sophora flavescens*	Leg	Wu *et al.* (1985c)
7,2′,4′	5	8-*C*-Lavandulyl (2*R*:3*R*)	(*9.316*)	Kushenol-I	*S. flavescens*	Leg	Wu *et al.* (1985b)
7,2′,4′	5	8-*C*-Lavandulyl (2*R*:3*S*)	(*9.321*)	Kushenol-N	*S. flavescens*	Leg	Wu *et al.* (1986)
5,7,2′,4′		6-*C*-Prenyl; 8-*C*-lavandulyl	(*9.319*)	Kushenol-M	*S. flavescens*	Leg	Wu *et al.* (1985c)
7,2′,4′	5	8-*C*-(2-Isopropenyl-5-methyl-5-hydroxyhexyl)	(*9.317*)	Kushenol-H	*S. flavescens*	Leg	Wu *et al.* (1985b)
5,7,3′,4′,5′				Ampelopsin	*Mahonia siamensis*	Ber	Ruangrungsi *et al.* (1984)
					Rhamnus pallasii	Rhm	Sakushima *et al.* (1983)
		3′-*O*-Glucose			*Semecarpus vitiensis*	Ana	Pramono *et al.* (1981)
		3-*O*-Rhamnose			*Catha edulis*	Cel	Gellért *et al.* (1981); Al-Meshal *et al.* (1986)
5,7,3′,5′	4′		(*9.323*)		*Rhamnus pallasii*	Rhm	Sakushima *et al.* (1983)
5,7,3′,4′,5′		6,8-Di-*C*-methyl	(*9.324*)		*Alluaudia humbertii*	Did	Voirin *et al.* (1986)
5,7,2′,4′,5′		3-*O*-Gallate; 2′-$OSO_3^-K^+$	(*9.325*)		*Myrica rubra*	Myr	Nonaka *et al.* (1983)

*Key: Ana = Anacardiaceae, Ber = Berberidaceae, Bom = Bombacaceae, Cel = Celastraceae, Com = Compositae, Cnf = Coniferae, Did = Didiereaceae, Dil = Dilleniaceae, Ebe = Ebenaceae, Eri = Ericaceae, Ery = Erythoxylaceae, Gra = Gramineae, Ill = Illiciaceae, Lab = Labiatae, Leg = Leguminosae, Myo = Myoporaceae, Myr = Myrtaceae, Orc = Orchidaceae, Pol = Polygonaceae, Rhm = Rhamnaceae, Sal = Salicaceae, Scr = Scrophulariaceae, The = Theaceae, Umb = Umbelliferae, Vit = Vitaceae, Voc = Vochysiaceae, Zin = Zingiberaceae.

propolis, a waxy substance found in bees' nests, by Bankova and co-workers (1983) along with the 2-hydroxyflavanone shown as structure (*9.284*). This unusual compound was found as a natural component in the exudate of *Populus nigra* buds (Wollenweber and Egger, 1971). The relationship between 2-hydroxyflavanones and dibenzoylmethanes was discussed in the more recent review (II, 347–348).

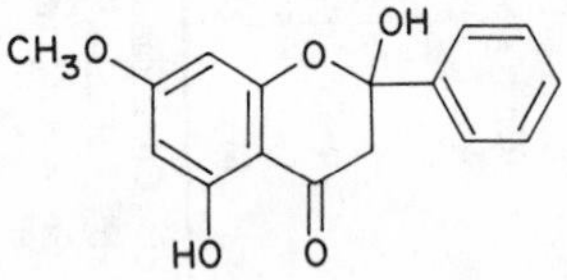

(*9.284*)

C-Alkylated pinocembrin derivatives involving either prenyl or geranyl functions are known. Unusual stereochemistry has been described for two of these compounds. 6-Prenylpinobanksin (*9.285*) was found as one of the components of *Helichrysum thapsus* (Bohlmann and Zdero, 1983); (*9.286*), the 8-prenyl isomer, was found in *Glycyrrhiza lepidota* by Mitscher and co-workers (1983). In both cases the normal stereochemistry was reported. Two other compounds were identified from *H. thapsus*, however, that exhibited the less common (2*S*:3*S*) configuration (Bohlmann and Zdero, 1983). Compound (*9.287*) is epimeric to (*9.285*) at C-3. Compound (*9.288*) is 6-*C*-geranylpinobanksin.

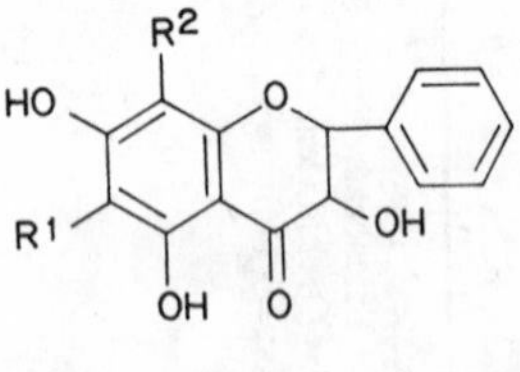

(*9.285*) R^1 = prenyl; R^2 = H
(*9.286*) R^1 = H; R^2 = prenyl

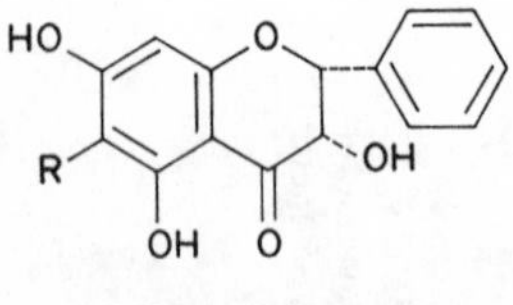

(*9.287*) R = prenyl
(*9.288*) R = geranyl

(b) *Dihydroflavonols having one B-ring oxygenation*

The simplest member of this subgroup is 'kushenol-J' which Wu and co-workers (1985c) isolated from *Sophora flavescens*. Kushenol-J, which has the structure shown as (*9.289*), occurs along with a series of quite complex 2′,4′-dihydroxydihydroflavonols which will be commented upon below. Another 5-deoxydihydroflavonol recently reported is the previously unknown triprenylated structure (*9.290*) obtained by Baruah and co-workers (1984) from *Millettia pulchra*. This new compound has (2*R*:3*R*) stereochemistry.

The most commonly encountered member of this subgroup is dihydrokaempferol (*9.291*), which is also known as aromadendrin. Occurrences of this compound as the free phenol, as glycosides, and, in two cases, as the 3-acetate are listed in Table 9.6. 7-*O*-Methylaromadendrin (*9.292*) has been found as the free phenol in some *Artemisia* species (Balza and Towers, 1984; Chandrasekharan *et al.*, 1981; Hurabielle *et al.*, 1982), in *Eucalyptus citriodora* (Dayal, 1982), and in *Dittrichia* (*Inula*) *viscosa* (Chiappini *et al.*, 1982). 7-*O*-Methylaromadendrin 5-*O*-β-D-glucopyranoside was identified from *Podocarpus nivalis* (Markham *et al.*, 1984), and as the 3-acetate in *Aframomum pruinosum* (Ayafor and Connolly, 1981). The latter compound is a new natural product and occurs with aromadendrin 3-acetate and the known 7,4′-di-*O*-methylaromadendrin (*9.293*). These represent the first reports of dihydroflavonols from *Afromomum* although not the first from a member of the Zingiberaceae. The 3-*O*-rhamnoside of (*9.293*), (2*R*:3*R*)-5-hydroxy-7,4′-dimethoxydihydroflavonol 3-*O*-rhamnoside, was named 'aurapin' by Venturella and co-workers (1980) who isolated it from *Acinos alpinus* (Labiatae). 4′-*O*-Methylaromadendrin (*9.294*) was reported as a constituent of *Salix caprea* (Malterud *et al.*, 1985) which represents a first record from the genus.

Only a single example of an aromadendrin derivative with extra A-ring hydroxylation has been reported recently. Jalal and co-workers (1982) described compound (*9.295*), 8-glucosyloxyaromadendrin, from *Calluna vulgaris* (Ericaceae). This appears to be the first report of this compound as a natural product.

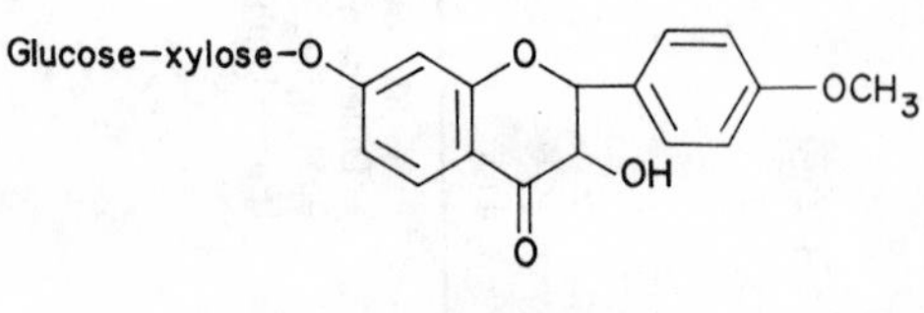

(*9.289*)

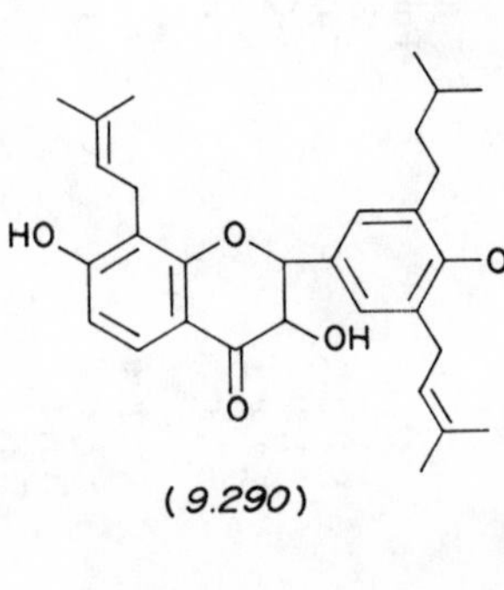

(*9.290*)

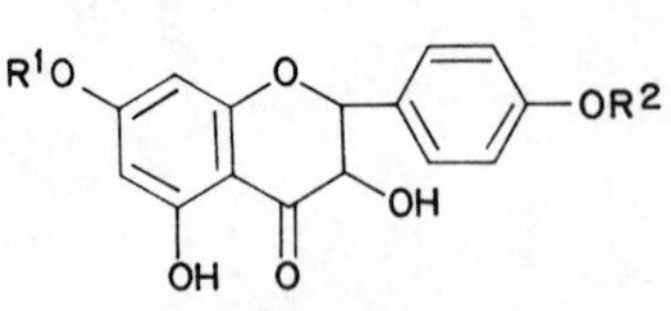

(*9.291*) $R^1 = R^2 = H$
(*9.292*) $R^1 = CH_3$; $R^2 = H$
(*9.293*) $R^1 = R^2 = CH_3$
(*9.294*) $R^1 = H$; $R^2 = CH_3$

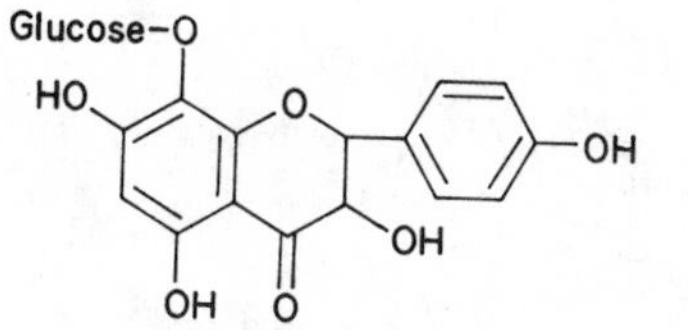

(9.295)

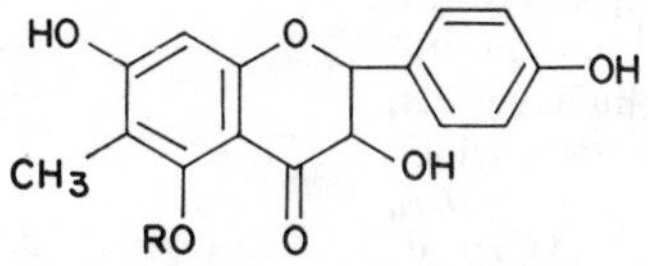

(9.296) R = CH_3

(9.297) R = H

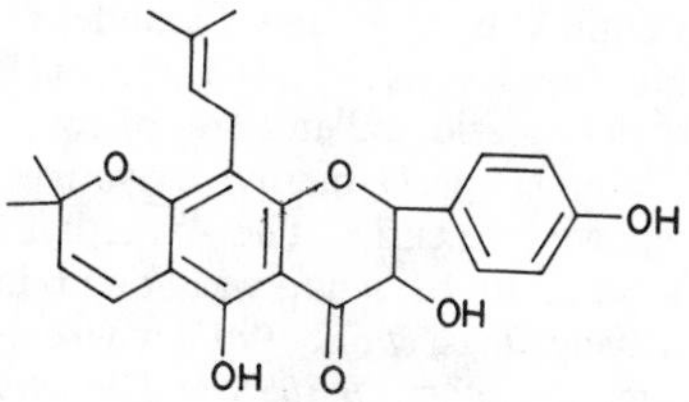

(9.298)

C-Methylated dihydroflavonols are comparatively rare in nature; the 1975 review listed only two, one of which was a tentative structure while the 1982 review also had two only one of which was new. Two new members of this small group have been reported recently. The first of these was described by Corrêa and co-workers (1981) as a constituent of *Qualea labouriauna* of the Vochysiaceae. (2*S*:3*S*)-7, 4′-Dihydroxy-5-methoxy-6-*C*-methyl-dihydroflavonol (*9.296*) is the first dihydroflavonol to be identified from the family. It also exhibits the rare (2*S*:3*S*) stereochemistry. A recent report by Shen and Theander (1985) described the related compound (*9.297*) as a constituent of needles of *Pinus massoniana* where it occurs as the 7-*O*-glucoside. No mention was made of the stereochemistry of the compound by those authors.

C-Prenylated dihydroflavonols are a somewhat better known group of natural products. Compound (*9.298*), which exhibits two *C*-prenyl groups one of which has been further modified through cyclization with the 7-hydroxyl function, was first reported by Smalberger and co-workers (1974) in a paper overlooked in preparation of the earlier reviews. Those workers found the compound in *Tephrosia lupinifolia* and showed that it has (2*R*:3*R*) stereochemistry. The same compound was reported in the 1982 review (II, 377). Two new natural products were recently described from studies of the umbel *Bonannia graeca* (Bruno *et al.*, 1985). Both of these compounds have a *C*-geranyl function at position 6: 'bonanniol-A' is (*9.299*); 'bonanniol-B' (*9.300*) is the corresponding 5-methyl ether. Both of these compounds have (2*R*:3*R*) stereochemistry. Occurring with them is the flavanone equivalent of bonanniol-A (see (*9.214*)).

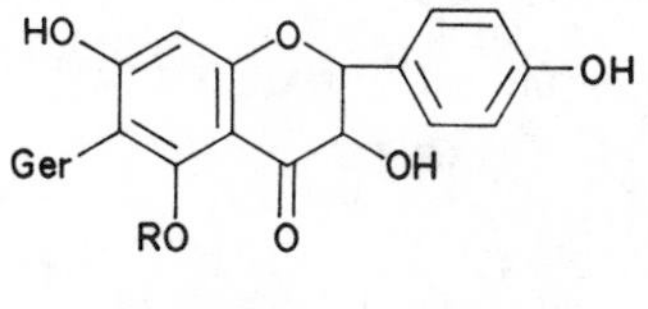

(9.299) R = H; Ger = geranyl

(9.300) R = CH_3; Ger = geranyl

(c) *Dihydroflavonols having two B-ring oxygenations*

The simplest member of this subgroup to appear in the recent literature is (+)-fustin (*9.301*) which was described as a constituent of *Umtiza listerana* (Leguminosae) by Burger and co-workers (1983). This compound is a known constituent of members of both the Leguminosae and the Anacardiaceae (II, 375). 8-Hydroxyfustin (*9.302*) was reported as a constituent of *Acacia melanoxylon* recently by Foo (1986). This compound has been reported from several species of *Acacia* (II, 375).

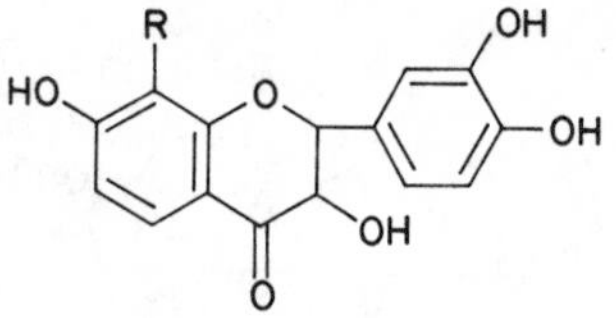

(9.301) R = H

(9.302) R = OH

The major compound in this subgroup is 5,7,3′,4′-tetrahydroxydihydroflavonol (*9.303*) commonly known as dihydroquercetin or taxifolin. All recently recorded sources are summarized in Table 9.6. These listings include several first family records: Cactaceae (Burret *et al.*, 1982); Compositae (Bohlmann *et al.*, 1981c); Ebenaceae (Chauhan *et al.*, 1982); Erythroxylaceae (Bohm *et al.*, 1981); Illiciaceae (Williams and Harvey, 1982); Poaceae (Gujer *et al.*, 1986); Polygonaceae (Kuroyanagi *et al.*, 1982); Salicaceae (Malterud *et al.*, 1985); Vitaceae (Trousdale and Singleton, 1983); and Winteraceae (Williams and Harvey, 1982). Taxifolin occurs as the free phenol in many plants but is known as well in several glycosidic forms and occurs as the acetate derivatives in a few plants as well. Most of these are well-known compounds and have been disposed of by way of Table 9.7. A few are novel and deserve at least brief recognition. The first record of a dihydroflavonol 'linear' diglycoside in nature would appear to be the report by Chauhan and co-workers (1982) of taxifolin 3-*O*-β-D-glucopyranosyl-(1→4)-α-L-rhamnopyranoside in *Diospyros peregrina*. The only other naturally occurring diglycoside of which I am aware is taxifolin 3,5-di-*O*-rhamnoside in *Cordia obliqua* (II, 376). Two acylated glucosides have also been described recently. (2*R*:3*S*)-Taxifolin 3-*O*-(6″-*O*-galloyl)glucose was reported from the mistletoe *Taxillus kaempferi* (Sakurai and Okumura, 1983). The 6″-*O*-phenylacetyl derivative of taxifolin 3-glucoside was identified by Shen and Theander (1985) from needles of *Pinus massoniana*. Phenylacetic acid is a comparatively rare acylating acid; gallates are, of course, very common.

Four simple *O*-methyl derivatives of taxifolin have been reported recently. The 4′-methyl ether (*9.304*), which appears to be new, was found in *Blumea balsamifera* (Ruangrungsi *et al.*, 1981), where it occurs with the 7,4′-dimethyl ether (*9.305*), and in *Anacheilium* sp. (Orchidaceae) where it occurs as the 3-rhamnosylglucoside (Ferreira *et al.*, 1986). The other two methylated derivatives are 7-methyltaxifolin (= padmatin) (*9.306*) which occurs as the 3-acetate in *Inula viscosa* (Grande *et al.*, 1985), and as the free phenol in *Artemisia glutinosa* (González *et al.*, 1983). (2*R*:3*R*)-7,3′-Dimethyltaxifolin (*9.307*) was isolated from *Grindelia discoidea* (Timmermann *et al.*, 1986) and from several species of *Artemisia* (Balza and Towers, 1984; Teresa *et al.*, 1984, 1986; González *et al.*, 1983).

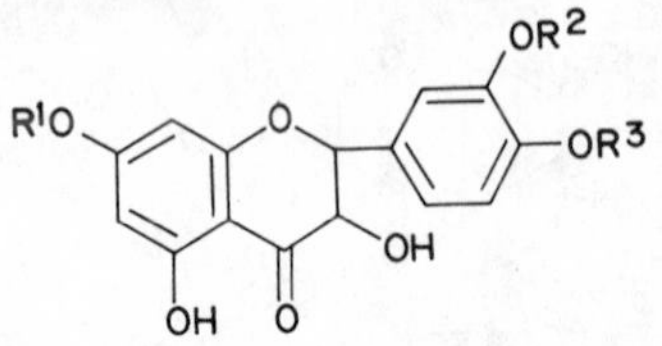

(*9.303*) $R^1 = R^2 = R^3 = H$

(*9.304*) $R^1 = R^2 = H; R^3 = CH_3$

(*9.305*) $R^1 = R^3 = CH_3; R^2 = H$

(*9.306*) $R^1 = CH_3; R^2 = R^3 = H$

(*9.307*) $R^1 = R^2 = CH_3; R^3 = H$

Bohlmann and co-workers (1981a) isolated and identified two taxifolin derivatives from *Pterocaulon virgatum* (Compositae) that are characterized by prenyl ether linkages at position 7. The compounds are shown as (*9.308*) and (*9.309*). *C*-Geranyl derivatives of taxifolin have been reported as constituents of the glandular exudate of *Diplacus aurantiacus* (Scrophulariaceae) (Lincoln, 1980; Lincoln and Walla, 1986). These compounds, 'diplacol' (*9.310*) and 'diplacol 4′-methyl ether' (*9.311*), are accompanied in the plant by the corresponding flavanones. This appears to be the first report of these compounds in nature

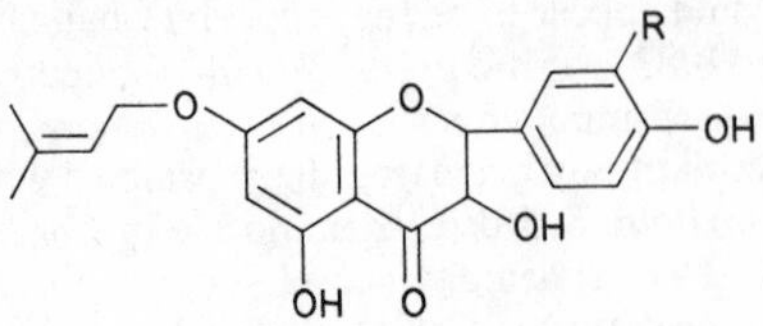

(*9.308*) R = H

(*9.309*) R = OH

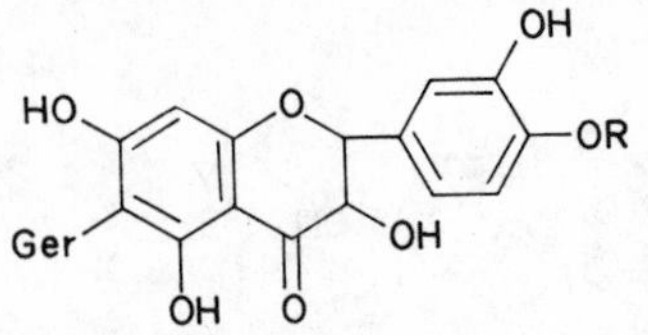

(*9.310*) R = H ; Ger = geranyl

(*9.311*) $R = CH_3$; Ger = geranyl

and so far as I am aware, the first report of a dihydroflavonol in a member of the Scrophulariaceae.

An unusual *C*-prenylated taxifolin derivative was recently isolated from the dichloromethane leaf wash of *Wyethia angustifolia* and shown to possess structure (*9.312*) (McCormick *et al.*, 1986). The corresponding

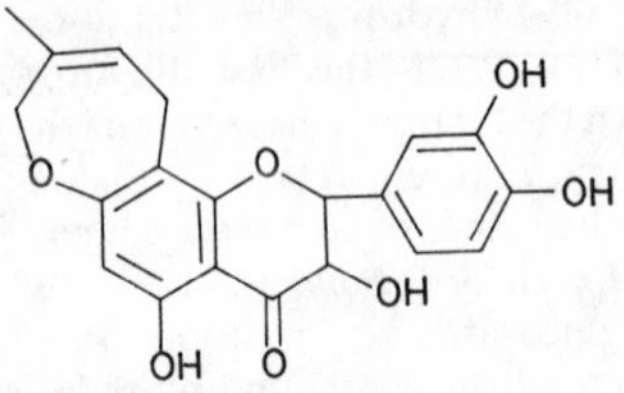

(*9.312*)

flavanone was also obtained. These compounds exhibit a seven-membered ring system seen otherwise only in the liverwort genus *Radula* which has a dihydrochalcone and a bibenzyl with this feature (II, 344). 8-*C*-Prenyldihydroisorhamnetin (*9.313*) was also identified from the *W. angustifolia* leaf wash fraction (McCormick *et al.*, 1986).

A group of highly *O*-alkylated flavonoids has been isolated from *Derris araripensis* by do Nascimento and

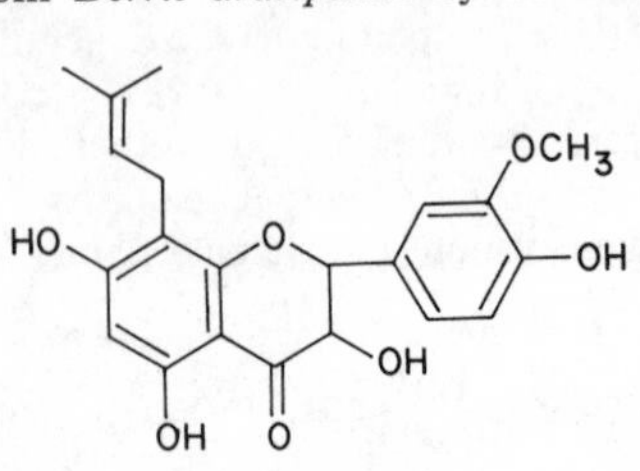

(*9.313*)

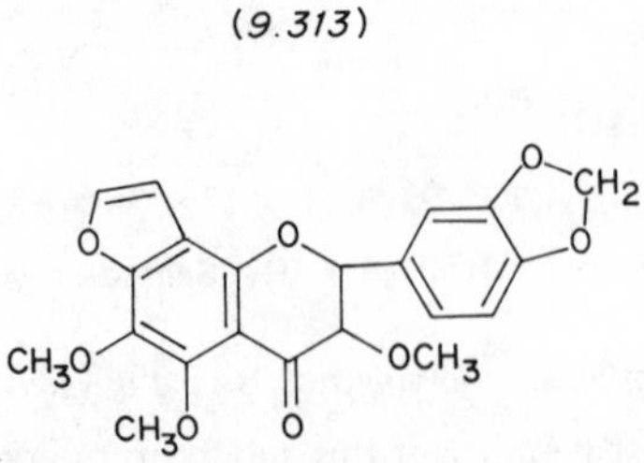

(*9.314*)

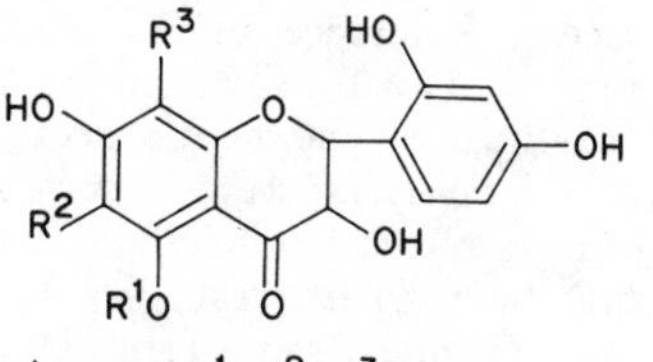

(9.315) $R^1=R^2=R^3=H$

(9.316) $R^1=CH_3$; $R^2=H$; R^3=lavandulyl

(9.317) $R^1=CH_3$; $R^2=H$; R^3=2-isopropenyl-5-methyl-5-hydroxyhexyl

(9.318) $R^1=H$; $R^2=R^3$=prenyl

(9.319) $R^1=H$; R^2=prenyl; R^3=lavandulyl

(9.320) $R^1=CH_3$; $R^2=H$; R^3=2-isopropenyl-5-methyl-5-hydroxyhexyl

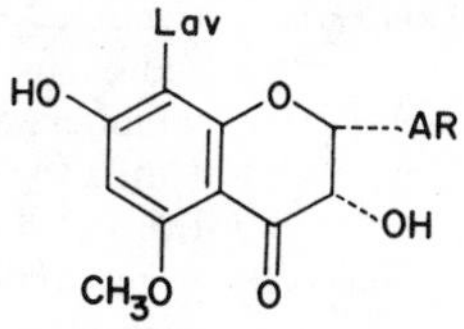

(9.321)

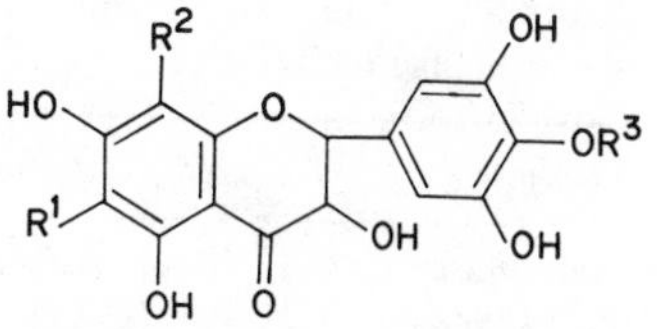

(9.322) $R^1=R^2=H$; $R^3=CH_3$

(9.323) $R^1=R^2=CH_3$; $R^3=H$

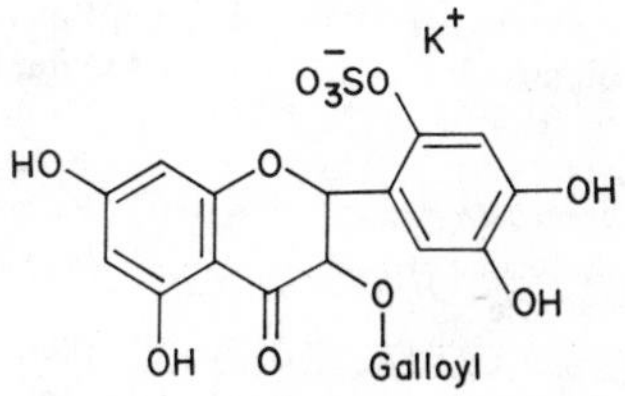

(9.324)

Mors (1981). Relevant to our present discussion is compound (*9.314*) whose complexity is matched by a dihydrochalcone (*9.110*) and a flavanone (*9.233*).

Dihydromorin (*9.315*), isomeric to taxifolin in having the 2′,4′-B-ring substitution pattern, has long been known as a component of the Moraceae where it occurs in several genera (I, 586; II, 376). It has now been recorded as a component of *Scutellaria baicalensis*, a member of the Labiatae, by Takagi and co-workers (1981). The dihydromorin base structure is also seen in several flavonoids isolated from members of the legume genus *Sophora* (e.g. *S. angustifolia*; I, 581). Very recent studies of *S. flavescens* have shown the existence of an extremely complex array of compounds several of which are dihydroflavonols. All members of this group exhibit *C*-alkylation involving either the prenyl group, the lavandulyl group, a modified lavandulyl group or some combination of these. Further variation is provided by occasional *O*-methylation. 'Kushenol-I' (*9.316*) and 'kushenol-H' (*9.317*) were described in the second of a series of papers by Wu and his co-workers (1985b). 'Kushenol-L' (*9.318*), 'kushenol-K' (*9.320*) and 'kushenol-M' (*9.319*) were described in the third paper (Wu *et al.*, 1985c). Two flavanones, (*9.222*) with only 2′-hydroxylation and (*9.223*) which has the same substitution pattern as 'kushenol-M', were described above. A seventh compound from this plant has also been described: 'kushenol-N' (*9.321*) (Wu *et al.*, 1986) is epimeric to 'kushenol-I' at carbon-3. In other words, 'kushenol-N' has (2*R*:3*S*) stereochemistry compared to the (2*R*:3*R*) stereochemistry of the other dihydroflavonols in this group.

(d) *Dihydroflavonols having three B-ring oxygenations*

The most commonly encountered member of this group is dihydromyricetin which is also known as ampelopsin. Its newly reported occurrences are summarized in Table 9.6. The only noteworthy one of these is the report of the 3′-*O*-glucoside in *Semecarpus vitiensis* (Anacardiaceae) (Pramono *et al.*, 1981). An unusual methylated derivative, 4′-*O*-methylampelopsin (*9.322*), was reported from *Rhamnus pallasii*, where it occurs with aromadendrin and taxifolin (Sakushima *et al.*, 1983). 6,8-Di-*C*-methylampelopsin (*9.323*) was found recently by Voirin and co-workers (1986) as a constituent of *Alluaudia humbertii* (Didiereaceae). The most unusual member of this subgroup is compound (*9.324*) which Nonaka and co-workers (1983) reported from *Myrica rubra* bark. The 2′,4′,5′-trihydroxylation pattern is rare and the occurrence of dihydroflavonol bisulphates are unknown to this point as far as I am aware. The gallyl function at position 3 is more common and is reminiscent of tannins. The compound was shown to have (2*R*:3*R*) stereochemistry.

9.7 COMMENTS ON OCCURRENCES OF THE MINOR FLAVONOIDS

It would likely lead to incorrect conclusions to attempt to establish taxonomic correlations based only upon the flavonoid types described in this chapter. Some observations on general distribution are in order, however, and a few tentative suggestions would seem justified if only to stimulate others with access to larger data sets to undertake such comparisons as might be possible. First, let us look at the family occurrences of the five flavonoid types. These have been summarized in Tables 9.7 to 9.11. Data for these tables were taken from the appropriate chapters of the 1975 and 1982 reviews and from the current information. The 1975 chapter included the information presented by Shimokoriyama (1962) in his chapter in *The Chemistry of Flavonoid Compounds* edited by Geissman. It is possible to see at a glance which families have yielded the particular compound type in question during the period covered by each review. Thus, Table 9.7 tells us that during the past five years chalcones have been reported for the first time from the Caprifoliaceae, Dipterocarpaceae,

Table 9.7 Family occurrences of chalcones (Coniferae and Filicinae included)

Family	*Source**	*Family*	*Source*
Acanthaceae	I II III	Liliaceae	I
Anacardiaceae	I II III	Loranthaceae	II
Anonaceae	II	Malvaceae	I
Betulaceae	II	Myricaceae	II III
Canabinnaceae	I	Myricaceae	II
Caprifoliaceae	III	Oleaceae	I
Caryophyllaceae	I	Onagraceae	II
Combretaceae	II	Piperaceae	I III
Compositae	I II III	Plumbaginaceae	
Coniferae	I	Ranunculaceae	I
Cyperaceae	I	Rosaceae	I II
Dipterocarpaceae	III	Rutaceae	I II
Euphorbiaceae	I	Salicaceae	I II
Fagaceae	II	Solanaceae	II III
Filicinae	I II III	Sterculiaceae	III
Gesneriaceae	I II III	Umbelliferae	II
Gramineae	III	Xanthorrhoeaceae	I
Lauraceae	I II	Zingiberaceae	I III
Leguminosae	I II III		

*I = 1975 review; II = 1982 review; III = current review.

Table 9.8 Family occurrences of aurones (including auronols)

Family	*Source**
Anacardiaceae	I II
Compositae	I II III
Cyperaceae	I III
Ericaceae	III
Gesneriaceae	I
Leguminosae	I II III
Oxalidaceae	I
Plumbaginaceae	I
Rubiaceae	III
Scrophulariaceae	I II

*I = 1975 review; II = 1982 review; III = current review.

Table 9.9 Family occurrences of dihydrochalcones (Hepaticae, Coniferae and Filicinae included)

Family	*Source**	*Family*	*Source**
Annonaceae	II III	Leguminosae	I II III
Balanophoraceae	II	Liliaceae	I
Caprifoliaceae	I	Myricaceae	II
Compositae	II	Myristicaceae	II III
Coniferae	III	Platanaceae	III
Ericaceae	I II III	Rosaceae	I II
Fagaceae	I	Salicaceae	I
Filicinae	I II III	Saxifragaceae	III
Hepaticae	II III	Symplocaceae	III
Lauraceae	III	Zingiberaceae	III

*I = 1975 review; II = 1982 review; III = current review.

Gramineae and Sterculiaceae. Reference to Table 9.8 shows that aurones were reported for the first time from members of two families, the Ericaceae and the Rubiaceae. Similarly, dihydrochalcone derivatives were reported from *Podocarpus* and members of the following angiosperm families for the first time: Platanaceae, Saxifragaceae, Symplocaceae and Zingiberaceae (Table 9.9). Table 9.10 shows the following new family records for flavanones: Apocynaceae, Cactaceae, Chrysobalanaceae, Didiereaceae, Erythroxylaceae, Hydrangeaceae, Iridaceae, Myoporaceae, Myricaceae, Plantaginaceae, Platanaceae, Sapindaceae and Verbenaceae. And, finally, Table 9.11 records that dihydroflavonols have been reported for the first time from the Berberidaceae, Bombacaceae, Celastraceae, Didiereaceae, Erythroxylaceae, Illiciaceae, Myricaceae, Orchidaceae, Polygonaceae, Sali-

Table 9.10 Family occurrences of flavanones (Coniferae included)

Family	*Source**	*Family*	*Source**
Acanthaceae	I	Juglandaceae	I
Aizoaceae	II	Labiatae	I II III
Amaranthaceae	II	Lauraceae	I II
Anacardiaceae	I II III	Leguminosae	I II III
Annonaceae	II III	Liliaceae	II
Apocynaceae	III	Malvaceae	II
Araceae	I	Meliaceae	II III
Balanophoraceae	I	Moraceae	I II III
Betulaceae	I II	Myoporaceae	III
Bignoniaceae	I	Myricaceae	III
Boraginaceae	II	Myristicaceae	II
Cactaceae	III	Myrtaceae	I II
Capparaceae	II III	Piperaceae	I III
Chrysobalanaceae	III	Plantaginaceae	III
Clusiaceae	II III	Platanaceae	III
Cochlospermaceae	II	Rosaceae	I II III
Combretaceae	II	Rubiaceae	II III
Compositae	I II III	Rutaceae	I II III
Coniferae	I II	Salicaceae	I II III
Coriariaceae	II	Sapindaceae	III
Crassulaceae	I	Scrophulariaceae	I
Cruciferae	I	Solanaceae	II III
Cyperaceae	I	Theaceae	II III
Didiereaceae	III	Umbelliferae	I III
Ericaceae	I	Verbenaceae	III
Erythroxylaceae	III	Zingiberaceae	I III
Eucryphiaceae	I		
Euphorbiaceae	I II III		
Fagaceae	I II		
Filicinae	I II III		
Gentianaceae	II		
Gesneriaceae	I II III		
Hydrangeaceae	III		
Hydrophyllaceae	I		
Iridaceae	III		

*I = 1975 review; II = 1982 review; III = current review.

Table 9.11 Family occurrences of dihydroflavonols (Coniferae included)

Family	*Source**			*Family*	*Source**		
Anacardiaceae	I	II	III	Moraceae	I	II	
Berberidaceae			III	Myoporaceae	I		III
Betulaceae	I			Myricaceae			III
Bombacaceae			III	Myrtaceae	I		III
Boraginaceae		II		Orchidaceae			III
Capparaceae		II		Platanaceae	I		
Celastraceae			III	Polygonaceae			III
Cercidiphyllaceae	I			Primulaceae	I		
Compositae		II	III	Rhamnaceae		II	III
Coniferae	I	II	III	Rosaceae	I	II	
Cruciferae	I	II		Rubiaceae	I		
Didiereaceae			III	Rutaceae	I	II	
Dilleniaceae		II	III	Salicaceae			III
Ebenaceae		II	III	Santalaceae	I		
Ericaceae	I	II	III	Sapotaceae	I		
Erythroxylaceae			III	Saxifragaceae	I		
Eucryphiaceae		II		Scrophulariaceae			III
Fagaceae	I			Solanaceae	I		
Graminae			III	Theaceae			III
Hamamelidaceae	I			Tiliaceae	I		
Illiciaceae			III	Ulmaceae	I		
Juglandaceae	I			Umbelliferae			III
Labiatae		II	III	Vitaceae	I		III
Lauraceae	I	II		Vochysiaceae			III
Leguminosae	I	II	III	Winteraceae			III
Liliaceae	I			Zingiberaceae	I		
Meliaceae	I	II					

*I = 1975 review; II = 1982 review; III = current review.

caceae, Scrophulariaceae, Theaceae, Umbelliferae, Vochysiaceae and Winteraceae.

For the most part, these reported occurrences are of limited taxonomic value. Although some of the studies had taxonomic goals in mind the scope of the surveys or the availability of comparatively valuable information is limited. More detailed surveys of some, perhaps most, of the groups could be of potential value, especially in instances where unusual compounds are involved. For example, the very interesting chemistry already noted for members of the Annonaceae (i.e. 2-hydroxybenzylated flavonoids) is based on a comparatively small sampling of the family. I recognize that the primary thrust of studies of *Uvaria* and *Unona* is pharmacognostic, but others with principally comparative goals in mind could perform the service of surveying more widely in the family using the excellent information already available on the flavonoids of the family. The same sort of problem exists with other families, e.g. Lauraceae, Moraceae, Piperaceae and Zingiberaceae. Each of these produce fascinating flavonoid derivatives but the sampling in each has been very limited.

A particularly conspicuous source of unusual compounds amongst the small group of families just mentioned is the Lauraceae. Compounds, a few at least, from this family are characterized by the presence of cyclic monoterpene derivatives that are seen as *C*-linked substituents. In addition to desiring a better understanding of the distribution of these compounds in the family, it would be of biosynthetic interest to determine whether the terpenic function is attached to the flavonoid nucleus preformed or whether the cyclization event occurs after alkylation. Further, a study of the biosynthesis of rubranine (*9.37*) in *Aniba* (Lauraceae) and *Boesenbergia* (Zingiberaceae) would be welcome. Do these two unrelated taxa do the job differently?

The obvious value of wide sampling can be seen in the many studies of flavonoids of the Leguminosae and Compositae that have been published. One of the striking features of the pigments from these families is the high frequency of alkylation, in particular involving prenyl and geranyl groups or structural modifications thereof. The value of such comparisons can be seen by examining the occurrence of alkylated compounds in the three subfamilies of the Leguminosae. Using data taken exclusively from the minor flavonoid treatments of the three reviews I found that 65 genera were represented, seven mimosoid, 11 caesalpinioid and 47 papilionate. (Despite the size of the sampling of the family note the lack of balance between subfamilies.) No prenylated or geranylated derivatives were found in members of the mimosoid genera and only one caesalpinoid genus had a member that could perform the requisite biochemical step(s). No less than 19 genera of the Papilionatae had members that produce these *C*-alkyl derivatives. In other words, of the 20 genera in which these compounds were found 19 belong to the same subfamily. Despite the many objections to such a generalization that could be raised it is nonetheless clear that such a distribution pattern deserves a closer look! Table 9.12 summarizes the occurrence of the various *C*-alkyl derivatives found in the Leguminosae to which reference has been made.

Prenylated derivatives have been found in seven genera of the Compositae. Sampling within some of these genera is quite small while sampling in others, e.g. *Helichrysum* and *Wyethia*, is a bit more representative. *C*-Prenyl compounds (of minor flavonoid types) are known from *Flourensia, Helichrysum, Marshallia, Pleiotaxis* and *Wyethia*. *O*-Prenyl compounds occur in *Achyrocline, Helichrysum* and *Pterocaulon*. It would be of interest to know if this limitation of *O*-prenylation to members of the Inuleae is real or just an accident of limited sampling.

An interesting distinction between prenyl derivatives of the Leguminosae and Compositae can be made. Cyclization of both types of derivatives to yield variously substituted pyran and furan-type ring systems is common amongst the legumes but comparatively uncommon in the Compositae. Broader surveys and attention to all flavonoid types, of course, are in order.

C-Prenylation and *C*-geranylation are not restricted to the two families mentioned above. Such compounds are

Table 9.12 A summary of *C*-alkylated minor flavonoids in genera of legumes

C-*Prenyl*		C-*Ger/Lav**		C-*Methyl*	
Amorpha	(P)†	*Ammothamnus*	(P)	*Dalea*	(P)
Cajanus	(P)	*Flemingia*	(P)	*Flemingia*	(P)
Crotalaria	(P)	*Lespedeza*	(P)	*Lonchocarpus*	(P)
Derris	(P)	*Sophora*	(P)	*Prosopis*	(M)
Erythrina	(P)	*Vexibia*	(P)		
Euchresta	(P)				
Flemingia	(P)				
Glycyrrhiza	(P)				
Lespedeza	(P)				
Lonchocarpus	(P)				
Millettia	(P)				
Peltogyne	(C)				
Piscidia	(P)				
Pongamia	(P)				
Psoralea‡	(P)				
Rhynchosia	(P)				
Sophora	(P)				
Tephrosia	(P)				

*Ger = geranyl; Lav = lavandulyl.
†P = Papilionatae, M = Mimosoideae, C = Caesalpinioideae.
‡*Psoralea* also exhibits *C*-formylflavonoids.

known in members of *Radula* (a liverwort), *Platanus* (Platanaceae), *Lannea* (Anacardiaceae), *Phyllanthus* (Euphorbiaceae), *Angelica* (Umbelliferae) and *Phellodendron* (Rutaceae). *C*-Geranylation is now known to occur in members of *Bonannia* (Umbelliferae), *Diplacus* (Scrophulariaceae), and *Boesenbergia* (Zingiberaceae). Daring to sound repetitive, further surveys in these and related taxa are clearly indicated.

C-Methylation is widespread in the plant kingdom showing no particular pattern. It occurs in the following families of angiosperms: Annonaceae, Didiereaceae, Ericaceae, Gesneriaceae, Myricaceae, Myrtaceae, Platanaceae, Rubiaceae and Vochysiaceae. *C*-Methylflavonoids are also well known from several conifers and from a number of genera of pteridophytes. The significance of these observations awaits comparative biochemical studies of the enzymes responsible for the alkylation steps. Little, in fact, is known about how, or even when, most flavonoid *C*-alkylation is done.

9.8 BIOLOGICAL ACTIVITY OF MINOR FLAVONOIDS

A thoroughgoing review of the biological activity of flavonoids was published recently under the editorship of Cody, Middleton and Harborne (1986). Although the primary thrust of the book is pharmacological in nature some information on natural biological activity was included. Natural functions of flavonoids were discussed briefly by Harborne (pp. 21–22) and by Swain (pp. 5–6) who presented his views in terms of factors influencing flavonoid evolution. Somewhat larger discussions of flavonoids in relation to insect resistance by plants by Hedin and Waage (pp. 87–100), and the specific action of isoflavonoids as phytoalexins by Smith and Banks (pp. 113–124) were also presented. The above discussions involved representatives of several flavonoid structural classes including members of the minor flavonoid group. A few recent papers dealing specifically with biological activity of various minor flavonoids are summarized below. Some of these involved 'biological activity' in the broadest sense while others treat situations where the activity of the compounds could be considered more or less ecologically relevant.

An interesting biological activity of natural dihydrochalcones was described by Dreyer and Jones (1981). These authors found that dihydrochalcones related to phloretin (*9.104*), neohesperidin dihydrochalcone and naringin dihydrochalcone were effective feeding deterrents against the aphids *Schizaphis graminum* and *Myzus persicae*. Moreover, these compounds are biologically active at concentrations that approximate naturally occurring levels in the plant. Crude extracts of *Amorpha fruticosa* also show feeding deterrence along with insecticidal, anti-parasitic, anti-microbial and hypotensive activities (Rózsa *et al.*, 1982b). Flavonoids isolated from this plant are illustrated as (*9.207*) to (*9.213*).

Kamat and co-workers (1981) commented on the antimicrobial activity of the 'abyssinones', e.g. (*9.167*), which they isolated from *Erythrina abyssinica*. These compounds were active against *Staphylococcus aureus*, *Bacillus subtillus*, and to a lesser extent, *Micrococcus lysodeikticus*. Significant activity was also recorded against the yeasts *Saccharomyces cerevisae* and *Candida utilis*, and the fungi *Sclerotina libertiana, Mucor mucedo* and *Rhizopus chinensis*. Extracts of *Glycyrrhiza lepidota*, which yielded the flavanone 'glabranin' (8-*C*-prenylpinocembrin), the flavonol 'glepidotin-A' and the dihydroflavonol 'glepidotin-B' (*9.286*), were active against *Staphylococcus aureus, Mycobacterium smegmatis, Candida albicans* and *Klebsiella pneumoniae* but not against *E. coli* or *Salmonella gallinarum* (Mitscher *et al.*, 1983). In a later work Mitscher and co-workers (1985) reported that 'flemiflavanone-D' (*9.204*), from *Flemingia stricta*, has significant activity against *S. aureus* and *M. smegmatis*. An anti-bacterial substance, that was found to be 5, 7, 2′, 6′-tetrahydroxyflavanone (*9.183*), was isolated from the drug 'Scutellariae Radix' by Kubo and associates (1981). Leading references to similar studies of anti-microbial and cytotoxic flavonoids in *Uvaria* can be found in Hufford and Oguntimein (1982) and in Hufford *et al.*, (1981).

Malterud and co-workers (1985) showed that the wood rot fungi *Coniophora puteana, Sporotrichum pulverulentum* and *Trichoderma viride* were inhibited by some of the compounds dealt with in this chapter. Naringenin (*9.172*) was active against all three organisms, aromadendrin (*9.291*) and taxifolin 4′-methyl ether (*9.304*) were active against *C. puteana* and prunin (naringenin 7-*O*-glucoside) was active against *S. pulverulentum*. Taxifolin (*9.303*) was inactive against all three test organisms but had been shown many years ago (Kennedy, 1956) to be active against *Fomes annosus* and *Lentinus lepideus*. Ten bacteria and non-wood-rot fungi were only slightly inhibited by the flavonoids so the effect is not a general microbiological one (Malterud *et al.*, 1985).

Treatment of pods of *Lathyrus odoratus* with either a solution of mercuric acetate or with *Phytophthora megasperma* var. *sojae* induced the formation of two dihydrochalcone derivatives: 2′, α-dihydroxy-4, 4′-dimethoxydihydrochalcone, 'odoratol' (*9.112*) and α-hydroxy-2′, 4, 4′-trimethoxydihydrochalcone, 'methylodoratol' (*9.113*). This was the first report of α-hydroxydihydrochalcones as stress metabolites in a member of the Vicieae (Fuchs *et al.*, 1984).

Carlson and Dolphin (1982) showed that copper-(II)-stressed peas (*Pisum sativum*) produced three unusual compounds. The simple but previously unknown 4,4′-dihydroxy-2′-methoxychalcone was illustrated above as structure (*9.43*). The other two compounds are the 1, 3-diphenylpropene derivatives 'obtustyrene' (*9.325*) and 'xenognosin' (*9.326*). The structural similarities of the three compounds is obvious; it would be of interest to learn how they are related biosynthetically.

'Purpuritenin' (*9.17*) was reported to be present in seeds of *Tephrosia purpurea* by Sinha and co-workers (1982). The interest in this compound lies in the highly unusual placement of the *C*-methyl function. *C*-Alkylation of flavonoids is very common, especially in members of the

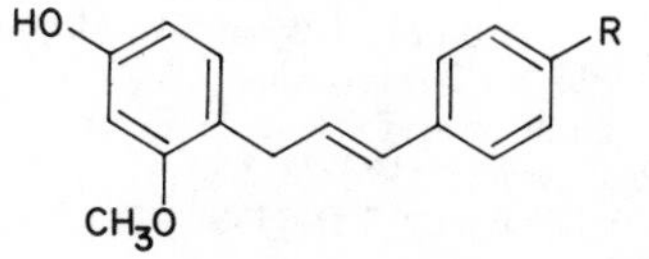

HO, R, CH_3O

(*9.325*) R=H

(*9.326*) R=OH

Leguminosae. 4′-*C*-Methylation is very rare. These authors suggested that the compound may have resulted from secondary modification of a plant product by fungi on or in the seeds. This interesting possibility needs to be pursued.

And finally we can note that Ravanel and co-workers (1982) have studied the uncoupling activities of chalcones and dihydrochalcones on isolated mitochondria from potato tubers and mung bean hypocotyls. Of the 23 compounds tested, several of which are either naturally occurring or closely related to natural compounds, 12 totally uncoupled oxidative phosphorylation in potato mitochondria while 16 did so in the mung bean preparations.

ACKNOWLEDGEMENTS

Preparation of this manuscript and work from our laboratory cited herein were supported by grants from the Natural Sciences and Engineering Research Council of Canada. Their support over the years has been very much appreciated. Dr Elijah Tannen deserves no less than mention in these acknowledgements.

REFERENCES

Achari, B., Chowdhury, U.S., Dutta, P.K. and Pakrashi, S.C. (1984), *Phytochemistry* **23**, 703.

Agarwal, V.K., Thappa, R.K., Agarwal, S.G. and Dhar, K.L. (1984), *Phytochemistry* **23**, 1342.

Ahmad, I., Ishratullah, K., Ilyas, M., Rahman, W., Seligmann, O. and Wagner, H. (1981) *Phytochemistry* **20**, 1169.

Ahmed, A.A., Norris, J.A. and Mabry, T.J. (1986), *Phytochemistry* **25**, 1501.

Albuquerque, F.B., Braz, F.R., Gottlieb, O., Magalhães, M.T., Maia, J.G.S., de Oliveira, A.B., de Oliveira, G.G. and Wilberg, V.C. (1981), *Phytochemistry* **20**, 235.

Al-Meshal, I.A., Hifnawy, M.S. and Nasir, M. (1986), *J. Nat. Prod.* **49**, 172.

Anjaneyulu, A.S.R. and Raju, S.N. (1984), *Indian J. Chem.* **23B**, 1010.

Anjaneyulu, A.S.R. and Reddy, D.S. (1981), *Indian J. Chem.* **20B**, 251.

Antonescu, V., Sommer, L., Predescu, I. and Barza, P. (1982), *Farmacia* **30**, 201.

Asakawa, Y., Takikawa, K., Toyota, M. and Takemotto, T. (1982), *Phytochemistry* **21**, 2481.

Ateya, A.-M.A., Okarter, T.U., Knapp, J.E., Schiff, Jr., P.L. and Slatkin, D.J. (1982), *Plant. Med.* **45**, 247.

Ayabe, S.-I., Kobayashi, M., Hikichi, M., Matsumoto, K. and Furuya, T. (1980a), *Phytochemistry* **19**, 2179.
Ayabe, S.-I., Yoshikawa, T., Kobayashi, M. and Furuya, T. (1980b), *Phytochemistry* **19**, 2331.
Ayafor, J.F. and Connolly, J.D. (1981), *J. Chem. Soc., Perkin Trans. I*, 2563.
Balza, F. and Towers, G.H.N. (1984), *Phytochemistry* **23**, 2333.
Balza, F., Jamieson, L. and Towers, G.H.N. (1985) *J. Nat. Prod.* **48**, 339.
Bankova, V.S., Popov, S.S. and Marekov, N.L. (1983), *J. Nat. Prod.* **46**, 471.
Barberán, F.A.T. (1986), *Fitoterapia* **57**, 67.
Barberán, F.A.T., Gil, M.I., Tomas, F., Ferreres, F. and Argues, A. (1985a), *J. Nat. Prod.* **48**, 859.
Barberán, F.A.T., Hernández, L., Ferreres, F. and Tomás, F. (1985b), *Plant. Med.* **51**, 452.
Barros, D.A.D., DeAlvarenga, M.A., Gottlieb, O.R. and Gottlieb, H.E. (1982), *Phytochemistry* **21**, 2107.
Baruah, P., Barua, N.C., Sharma, R.P., Baruah, J.N., Kulanthaivel, P. and Herz, W. (1984), *Phytochemistry* **23**, 443.
Baslas, R.K. (1983), *Herba Hung.* **22**, 85.
Baslas, R.K. and Kumar, P. (1981), *Herba Hung.* **20**, 213.
Batirov, E.Kh., Yusupova, S.S., Abdullaev, Sh.V., Vdouin, A.D., Malikov, V.M. and Yagudaev, M.R. (1985), *Khim. Prir. Soed.* **21**, 32.
Bellino, A., Venturella, P. and Marleno, C. (1980), *Fitoterapia* **51**, 163.
Beltrami, E., de Bernardi, M., Fronza, G., Mellerio, G., Vidari, F. and Vita-Finzi, P. (1982), *Phytochemistry* **21**, 2931.
Bezuidenhoudt, B.C.B., Brandt, E.V. and Roux, D.G. (1981), *J. Chem. Soc., Perkin Trans. I*, 263.
Bhakuni, D.S. and Chaturved, R. (1984), *J. Nat. Prod.* **47**, 585.
Bhakuni, D., Bittner, M., Silva, M. and Sammes, P.G. (1973), *Phytochemistry* **12**, 2777.
Bhanumati, S., Chhabra, S.C., Gupta, R.S. and Krishnamoorthy (1978), *Phytochemistry* **17**, 2045.
Bhardwaj, D.K., Jain, R.K., Munjal, A. and Prasher, M. (1982), *Indian J. Chem.* **21B**, 476.
Bhartiya, H.P. and Gupta, P.C. (1981), *Phytochemistry* **20**, 2051.
Bhartiya, H.P. and Gupta, P.C. (1982), *Phytochemistry* **21**, 247.
Bhatt, S.K., Dixit, V. and Singh, K.V. (1981a), *Phytochemistry* **20**, 2442.
Bhatt, S.K., Saxena, V.K. and Singh, K.U. (1981b), *Phytochemistry* **20**, 2605.
Bhatt, S.K., Sthapak, J.K. and Singh, K.U. (1984), *Fitoterapia* **55**, 370.
Bohlmann, F. and Ates (Gören), N. (1984), *Phytochemistry*, **23**, 1338.
Bohlmann, F. and Misra, L.N. (1984), *Plant. Med.* **50**, 271.
Bohlmann, F. and Zdero, C. (1982), *Phytochemistry* **21**, 1434.
Bohlmann, F. and Zdero, C. (1983), *Phytochemistry* **22**, 2877.
Bohlmann, F., Zdero, C., Robinson, H. and King, R.M. (1980), *Phytochemistry* **19**, 2381.
Bohlmann, F., Abraham, W.-R., King, R.M. and Robinson, H. (1981a), *Phytochemistry* **20**, 825.
Bohlmann, F., Kramp, W., Grenz, M., Robinson, H. and King, R.M. (1981b), *Phytochemistry* **20**, 1907.
Bohlmann, F., Zdero, C., Grenz, M., Dhar, A.K., Robinson, H. and King, R.M. (1981c), *Phytochemistry* **20**, 281.
Bohlmann, F. Zdero, C., Robinson, H. and King, R.M. (1981d), *Phytochemistry* **20**, 2245.
Bohlmann, F., Singh, P., Jakupovic, J., King, R.M. and Robinson, H. (1982), *Phytochemistry* **21**, 371.
Bohlmann, F., Misra, L.N. and Jakupovic, J. (1984a), *Plant. Med.* **50**, 174.
Bohlmann, F., Zdero, C., King, R.M. and Robinson, H. (1984b), *Plant. Med.* **50**, 195.
Bohm, B.A. and Bhat, U.G. (1985), *Biochem. Syst. Ecol.* **13**, 437.
Bohm, B.A. and Chalmers, G. (1986), *Biochem. Syst. Ecol.* **14**, 79.
Bohm, B.A., Phillips, D.W. and Ganders, F.R. (1981), *J. Nat. Prod.* **44**, 676.
Bourquelot, E. and Fichtenholz, A. (1912), *Compt. Rend. Acad. Sci.* **154**, 526.
Bruno, M., Savona, G., Lamartina, L. and Lentini, F. (1985), *Heterocycles* **23**, 1147.
Burger, A.P.N., Brandt, E.V., and Roux, D.G. (1983), *Phytochemistry* **22**, 2813.
Burke, B. and Nair, M. (1986), *Phytochemistry* **25**, 1427.
Burret, F., LeBreton, Ph. and Voirin, B. (1982), *J. Nat. Prod.* **45**, 687.
Burzanska-Hermann, Z. (1978), *Acta Pol. Pharm.* **35**, 673.
Byrne, L.T., Cannon, J.R., Gawad, D.H., Joshi, B.S., Skelton, B.W., Toia, R.F. and White, A.H. (1982), *Aust. J. Chem.* **35**, 1851.
Cane, D.E. (1981). In *Biosynthesis of Isoprenoid Compounds* (eds J.W. Porter and S.L. Spurgeon), Wiley-Interscience, New York, Vol. 1, pp. 284–374.
Carlson, R.E. and Dolphin, D.H. (1982), *Phytochemistry* **21**, 1733.
Chandrasekharan, I., Khan, H.A. and Ghanin, A. (1981), *Plant. Med.* **43**, 310.
Chauhan, J.S., Saraswat, M. and Kumar, G. (1982), *Indian J. Chem.* **21B**, 169.
Chauhan, J.S., Kumar, S. and Chaturvedi, R. (1984), *Plant. Med.*, **50**, 113.
Chauhan, J.S., Chaturvedi, R. and Kumar, S. (1985), *Indian J. Chem.* **24B**, 219.
Chemesova, I.I., Belenovskaya, L.M. and Markova, L.P. (1984), *Khim. Prir. Soed.* **20**, 748.
Cheng, Y.-S., Lee, C.-S., Chan, C.T. and Chang, C.-F. (1985), *J. Chin. Chem. Soc.* **32**, 85.
Chiappini, I., Fardella, G., Meghini, A. and Rossi, C. (1982), *Plant. Med.* **44**, 159.
Chibber, S.S., Dutt, S.K., Sharma, R.P. and Sharma, A. (1981), *Indian J. Chem.* **20B**, 626.
Chkhikvishvili, I.D., Kurkin, V.A. and Zaprometov, M.N. (1984) *Khim. Prir. Soed.* **20**, 629.
Clark, W.D. and Wollenweber, E. (1985), *Phytochemistry* **24**, 1122.
Cody, V. (1984), *Amer. Cryst. Assoc. Absts.* **12**, 54.
Cody, V., Middleton, V. and Harborne, J.B. (eds) (1986), *Plant Flavonoids in Biology and Medicine*, Alan R. Liss, New York.
Cooksey, C.J., Dahiya, J.S., Garratt, P.J. and Strange, R.N. (1982), *Phytochemistry* **21**, 2935.
Corrêa, D. de B., Guerra, L.F.B., Gottlieb, O.R. and Maia, J.G.S. (1981), *Phytochemistry* **20**, 305.
Crawford, D.J. and Smith, E.B. (1984), *Madroño* **31**, 1.
Crawford, D.J. and Smith, E.B. (1985), *Biochem. Syst. Ecol.* **13**, 115.
Çubukçu, B. and Yüksel, V. (1982), *J. Nat. Prod.* **45**, 137.
Dayal, R. (1982), *J. Indian Chem. Soc.*, **59**, 1008.
Dean, F.M. (1963), *Naturally Occurring Oxygen Ring Compounds*, Butterworths, London.
Degot, A.V., Fursa, N.S., Zaitsev, V.G., Chaika, E.A., Popova, T.P., Litvinenko, V.I. and Kornievskii, Yu. I. (1983), *Khim. Prir. Soed.* **19**, 370.

Delle Monache, F., Ferrari, F. and Pomponi, M. (1984), *Phytochemistry* **23**, 1489.
Deshpande, V.H. and Patil, A.D. (1981), *Indian J. Chem.* **20B**, 628.
Dominguez, X.A., Tellez, O. and Ramirez, E. (1983), *Phytochemistry* **22**, 2047.
do Nascimento, M.C. and Mors, W.B. (1981), *Phytochemistry* **20**, 147.
Drewes, S.E. and Hudson, N.A. (1983), *Phytochemistry* **22**, 2823.
Dreyer, D.L. and Jones, K.C. (1981), *Phytochemistry* **20**, 2489.
Dreyer, D.L., Munderloh, K.P. and Theissen, W.E. (1975), *Tetrahedron* **31**, 287.
Endo, T., Taguchi, H. and Yosioka, I. (1981), *Chem. Pharm. Bull.* **29**, 1000.
Eykman, J.F. (1883), *Rec. Trav. Chim.* **2**, 99.
Ferrari, F., Botta, B. and De Lima, A.R. (1983), *Phytochemistry* **22**, 1663.
Ferraro, G.E., Martino, V.S., Villar, S.I. and Coussio, J.D. (1985), *J. Nat. Prod.* **48**, 817.
Ferreira, V.F., Parente, J.P., Pinto, M.C.F.R., Pinto, A.U., Porter, B., Silva, M.M. and Montinho, J.L. (1986), *Biochem. Syst. Ecol.* **14**, 199.
Ferreres, F., Tomas, F., Barberan, F.A.T. and Hernandez, L. (1985), *Plant. Med. Phytother.* **19**, 89.
Fiebig, M. and Wagner, H. (1984), *Plant. Med.* **51**, 310.
Fomum, Z.T., Ayafor, J.F. and Mbafor, J.T. (1983), *Tetrahedron Lett.* **24**, 4127.
Fomum, Z.T., Ayafor, J.F. and Wandji, J. (1985), *Phytochemistry* **24**, 3075.
Fomum, Z.T., Ayafor, J.F., Mbafor, J.T. and Mbi, C.M. (1986), *J. Chem. Soc., Perkin Trans. I*, 33.
Foo, L.Y. (1986), *J. Chem. Soc. Chem. Commun.*, 675.
Forkmann, G. (1983), *Z. Naturforsch.* **38c**, 891.
Fuchs, A., De Vries, F.W., Landheer, C.A. and van Veldhuizen, A. (1984), *Phytochemistry* **23**, 2199.
Fujimoto, T. and Nomura, T. (1984), *Heterocycles* **22**, 997.
Fukai, T., Hano, Y., Fujimoto, T. and Nomura, T. (1983), *Heterocycles* **20**, 611.
Fukai, T., Hano, Y., Hirakura, K., Nomura, T. and Uzawa, J. (1985), *Chem. Pharm. Bull.* **33**, 4288.
Garg, H.S. and Bhakuni, D.S. (1984), *Phytochemistry* **23**, 2115.
Gellért, M., Szendrei, K. and Reisch, J. (1981), *Phytochemistry* **20**, 1759.
Gomez, F., Quijano, L., Calderón, J.S., Rodriquez, C. and Rios, T. (1985), *Phytochemistry* **24**, 1057.
González, M.D. and Pomilio, A.B. (1982), *Phytochemistry* **21**, 757.
González, M.D., Pomilio, A.B. and Gros, E.G. (1981), *Phytochemistry* **20**, 1174.
González, A.G., Bermejo, J., Estevez, F. and Velazquez, R. (1983), *Phytochemistry* **22**, 1515.
Grande, M., Piera, F., Cuenca, A., Torres, P. and Bellido, I.S. (1985), *Plant. Med.* **52**, 414.
Gujer, R., Magnolato, D. and Self, R. (1986), *Phytochemistry* **25**, 1431.
Gupta, B.K., Gupta, G.K., Dhar, K.L. and Atal, C.K. (1980), *Phytochemistry* **19**, 2034.
Gupta, D.R. and Ahmed, B. (1984), *J. Nat. Prod.* **47**, 958.
Gupta, G.K., Suri, J.L., Gupta, B.K. and Dhar, K.L. (1982), *Phytochemistry* **21**, 2149.
Gupta, K.K., Taneja, S.C., Dhar, K.L. and Atal, C.K. (1983), *Phytochemistry* **22**, 314.
Hano, Y. and Nomura, T. (1983), *Heterocycles* **20**, 1071.
Hano, Y. and Nomura, T. (1985), *Heterocycles* **23**, 2499.
Hano, Y., Itoh, M., Koyama, N. and Nomura, T. (1984), *Heterocycles* **22**, 1791.
Hano, Y., Itoh, M., Fukai, Y., Nomura, T. and Urano, S. (1985), *Heterocycles* **23**, 1691.
Harborne, J.B. and Nash, R.J. (1984), *Biochem. Syst. Ecol.* **12**, 315.
Harborne, J.B. and Williams, C.A. (1983), *Phytochemistry* **22**, 1520.
Harborne, J.B., Williams, C.A. and Wilson, K.L. (1982), *Phytochemistry* **21**, 2491.
Harborne, J.B., Girija, A.R., Devi, H.M. and Lakshmi, N.K.M. (1983), *Phytochemistry* **22**, 2741.
Harborne, J.B., Williams, C.A. and Wilson, K.L. (1985), *Phytochemistry* **24**, 751.
Hitz, C., Mann, K. and Wollenweber, E. (1982), *Z. Naturforsch.* **37c**, 337.
Hoffmann, B.G. and Lunder, L.T. (1984), *Plant. Med.* **50**, 361.
Horowitz, R.M. (1986). In *Plant Flavonoids in Biology and Medicine.* (eds V. Cody, E. Middleton and J.B. Harborne), Alan R. Liss, New York, pp. 163–175.
Hufford, C.D. and Oguntimein, B.O. (1982), *J. Nat. Prod.* **45**, 337.
Hufford, C.D., Oguntimein, B.O., Van Engen, D., Muthard, D. and Clardy, J. (1980), *J. Am Chem. Soc.* **102**, 7365.
Hufford, C.D., Oguntimein, B.O. and Baker, J.K. (1981), *J. Org. Chem.* **46**, 3073.
Hunt, G.M. and Baker, E.A. (1980), *Phytochemistry* **19**, 1415.
Hurabielle, M., Eberle, J. and Paris, M. (1982), *Plant. Med.* **46**, 124.
Hussain, R.A. and Waterman, P.G. (1982), *Phytochemistry* **21**, 1393.
Ikram, M., Jehangir, S., Razaq, S. and Kawano, M. (1979), *Plant. Med.* **37**, 189.
Ikuta, J., Fukai, T., Nomura, T. and Ueda, S. (1986), *Chem. Pharm. Bull.* **34**, 2471.
Imperato, F. (1980), *Chem. Ind. (London)* **19**, 786.
Imperato, F. (1982a), *Experientia* **38**, 67.
Imperato, F. (1982b), *Phytochemistry* **21**, 480.
Imperato, F. (1983), *Phytochemistry* **22**, 312.
Ishikura, N., Sugahara, K. and Kurosawa, K. (1979), *Z. Naturforsch.* **35c**, 628.
Itokawa, H., Morita, M. and Mihasi, S. (1981), *Phytochemistry* **20**, 2503.
Jain, A.C. and Tyagi, O.D. (1986), *Indian J. Chem.* **25B**, 754.
Jain, A., Bhartiya, H.P. and Vishwakarma, A.N. (1982), *Phytochemistry* **21**, 957.
Jaipetch, T., Kanghae, S., Pancharoen, O., Patrick, V.A., Reutrakul, V., Tuntiwachwuttikul, P. and White, A.H. (1982), *Aust. J. Chem.* **35**, 351.
Jaipetch, T., Reutrakul, V., Tuntiwachwuttikul, P. and Santisuk, T. (1983), *Phytochemistry* **22**, 625.
Jalal, M.A.F., Read, D.J. and Haslam, E. (1982), *Phytochemistry* **21**, 1397.
Jankowski, K. and Jocelyn-Pairé, J.R. (1983), *J. Nat. Prod.* **46**, 190.
Jeffries, P.R., Knox, J.R. and Middleton, E.J. (1982), *Aust. J. Chem.* **18**, 532.
Joshi, B.S. and Gawad, D.H. (1974), *Indian J. Chem.* **12B**, 1033.
Jurd, L. and Wong, R.Y. (1984), *Aust. J. Chem.* **37**, 1127.
Kamat, V.S., Chuo, F.Y., Kubo, I. and Nakanishi, K. (1981), *Heterocycles* **15**, 1163.
Kaouadji, M. (1986), *J. Nat. Prod.* **49**, 500.

Kaouadji, M., Morand, J.-M. and Gilly, C. (1986a), *J. Nat. Prod.* **49**, 508.

Kaouadji, M., Ravanel, P. and Mariotte, A.M. (1986b), *J. Nat. Prod.* **49**, 153.

Kennedy, R.W. (1956), *For. Prod. J.* **6**, 80.

Khalid, S.A. and Waterman, P.G. (1981), *Phytochemistry* **20**, 1719.

Khan, H.A., Chandrasekharan, I. and Ghanim, A. (1986), *Phytochemistry* **25**, 767.

Khattri, P.S., Sahai, M., Dasgupta, B. and Ray, A.B. (1984), *Heterocycles* **22**, 249.

Kimura, Y., Okuda, H., Tani, T. and Arichi, S. (1982), *Chem. Pharm Bull.* **30**, 1792.

Kodpinid, M., Thebtaranonth, C. and Thebtaranonth, Y. (1985), *Phytochemistry* **24**, 3071.

Komatsu, M., Yokoe, I. and Shiritaki, Y. (1978), *Chem. Pharm. Bull.* **26**, 3863.

Krishnamurti, M. and Parthasarathi, J. (1981), *Indian J. Chem.* **20B**, 247.

Kubo, M., Kimura, Y., Odnai, T., Tani, T. and Namba, K. (1981), *Plant. Med.* **43**, 194.

Kuroyanagi, M., Yamamoto, Y., Fukushima, S., Veno, A., Noro, T. and Miyau, T. (1982), *Chem. Pharm. Bull.* **30**, 1602.

Kuroyanagi, M., Noro, T., Fukushima, S., Aiyama, R., Ikuta, A., Itokawa, H. and Morita, M. (1983), *Chem. Pharm. Bull.* **31**, 1544.

Kuroyanagi, M., Fujita, K., Kazaoka, M., Matsumoto, S., Veno, A., Fukushima, S. and Katsuoka, M. (1985), *Chem. Pharm. Bull.* **33**, 5075.

Labbiento, L., Menichini, F. and Delle Monache, F. (1986), *Phytochemistry* **25**, 1505.

LaDuke, J.C. (1982), *Am. J. Bot.* **69**, 784.

Lincoln, D.E. (1980), *Biochem. Syst. Ecol.* **8**, 397.

Lincoln, D.E. and Walla, M.D. (1986), *Biochem. Syst. Ecol.* **14**, 195.

Lorente, F.T., Ferreres, F. and Barberan, F.A.T. (1982), *Phytochemistry* **21**, 1461.

Lwande, W. (1985), *J. Nat. Prod.* **48**, 1004.

Mabry, T.J., Sakakibara, M. and King, B. (1975), *Phytochemistry* **14**, 1448.

Mahidal, C., Tuntiwachwuttikul, P., Reutrakul, V. and Taylor, W.C. (1984), *Aust. J. Chem.* **37**, 1739.

Mahmoud, E.N. and Waterman, P.G. (1985a), *J. Nat. Prod.* **48**, 648.

Mahmoud, E.N. and Waterman, P.G. (1985b), *Phytochemistry* **24**, 369.

Malhotra, S. and Misra, K. (1983), *Plant. Med.* **47**, 46.

Malterud, K.E., Bremnes, T.E., Faegri, A., Moe, T. and Dugstad, E.K.S. (1985), *J. Nat. Prod.* **48**, 559.

Makham, K.R., Webby, R.F. and Vilain, C. (1984), *Phytochemistry* **23**, 2049.

Markham, K.R., Webby, R.F., Whitehouse, L.A., Molloy, B.P.T., Vilain, C. and Mues, R. (1985), *N. Z. J. Bot.* **23**, 1.

Mathew, J. and Subba Rao, A.V. (1983), *Phytochemistry* **22**, 794.

Maurya, R., Ray, A.B., Duah, F.K., Slatkin, D.J. and Schiff, Jr., P.L. (1982), *Heterocycles* **19**, 2103.

McCormick, S.P., Bohm, B.A. and Ganders, F.R. (1984), *Phytochemistry* **23**, 2400.

McCormick, S., Robson, K. and Bohm, B.A. (1985), *Phytochemistry* **24**, 1614.

McCormick, S., Robson, K. and Bohm, B.A. (1986), *Phytochemistry* **25**, 1723.

McPherson, D.D., Cordell, G.A., Soejarto, D.D., Pezzuto, J.M. and Fong, H.H.S. (1983), *Phytochemistry* **22**, 2835.

Metwally, A.M. and Ekejiuba, E.C. (1981), *Plant. Med.* **42**, 403.

Miraglia, do C.M., de Padua, A.P., Mesquita, A.A.L. and Gottlieb, O.R. (1985), *Phytochemistry* **24**, 1120.

Mitscher, L.A., Rao, G.S.R., Khanna, I., Veysoglu, T. and Drake, S. (1983), *Phytochemistry* **22**, 573.

Mitscher, L.A., Gollapudi, S.R., Khanna, I.K., Drake, S.D., Hanumaiah, T., Ramaswamy, T. and Rao, K.V.J. (1985), *Phytochemistry* **24**, 2885.

Muhammad, I. and Waterman, P.G. (1985), *J. Nat. Prod.* **48**, 571.

Murakami, S. and Takeuchi, S. (1936), *J. Pharm. Soc. Jpn.* **56**, 649.

Nakanishi, T., Inada, A., Kamabayashi, K. and Yoneda, K. (1985), *Phytochemistry* **24**, 339.

Nicholls, K.W. and Bohm, B.A. (1984), *Canad. J. Bot.* **62**, 1036.

Nomura, T. and Fukai, T. (1981a), *Heterocycles* **15**, 1531.

Nomura, T. and Fukai, T. (1981b), *Plant. Med.* **42**, 79.

Nomura, T., Fukai, T., Hano, Y., Sugaya, Y. and Hosoyo, T. (1980), *Heterocycles* **14**, 1785.

Nomura, T., Fukai, T., Hano, Y. and Uzawa, J. (1982), *Heterocycles* **17**, 381.

Nomura, T., Fukai, T. and Hano, Y. (1983a), *Plant. Med.* **47**, 30.

Nomura, T., Fukai, T., Hano, Y., Nemoto, K., Tereda, S. and Kuramochi, T. (1983b), *Plant. Med.* **47**, 151.

Nonaka, G.-I., Muta, M. and Nishioka, I., (1983), *Phytochemistry* **22**, 237.

Norbedo, C., Ferraro, G. and Coussio, J.D. (1982), *J. Nat. Prod.* **45**, 635.

Norbedo, C., Ferraro, G. and Coussio, J.D. (1984), *Phytochemistry* **23**, 2698.

Pacheco, P., Crawford, D.J., Stuessy, T.F. and Silva, M. (1985), *Am. J. Bot.* **72**, 989.

Parker, W.H. and Maze, J. (1984), *Am. J. Bot.* **71**, 1051.

Pathak, V.P., Saini, T.R. and Khanna, R.N. (1983), *Phytochemistry* **22**, 1303.

Pelter, A., Ward, R.S., Rao, E.V. and Ranga Raju, N. (1981), *J. Chem. Soc., Perkin Trans. I*, 2491.

da S. Pereira, M.O., Fantine, E.C. and de Sousa, J.R. (1982), *Phytochemistry* **21**, 488.

Pokhilo, N.D., Denisenko, V.A., Makharikov, V.V. and Uvarova, N.I. (1983), *Khim. Prir. Soed.* **19**, 374.

Pomilio, A.B. and Gros, E.G. (1979), *Phytochemistry* **18**, 1410.

Popova, T.P., Pakaln, D.A., Chernykh, N.A., Zoz, I.K. and Litvinenko, V.I. (1976), *Rast. Resur.* **12**, 232.

Pramono, S., Gleye, J., Dehray, M. and Stanislas, E. (1981), *Plant. Med. Phytother.* **15**, 224.

Proksch, P., Wollenweber, E. and Rodriguez, E. (1983), *Z. Naturforsch.* **38c**, 668.

Proksch, P., Budzikiewicz, H., Tanowitz, B.D. and Smith, D.M. (1984), *Phytochemistry* **23**, 679.

Quijano, L., Calderón, J.S., Gómez, G. F., Soria, I.E. and Rios, T. (1980), *Phytochemistry* **19**, 2439.

Quijano, L., Calderón, J.S., Gómez, G.F., and Rios, T. (1982), *Phytochemistry* **21**, 2575.

Rabesa, Z. and Voirin, B. (1983), *Phytochemistry* **22**, 2092.

Rao, C.P., Vemuri, V.S.S. and Rao, V.K.J. (1982), *Indian J. Chem.* **21B**, 167.

Rao, E.V. and Ranga Raju, N. (1984), *Phytochemistry* **23**, 2339.

Rao, E.V., Venkataratnam, G. and Vilain, C. (1985), *Phytochemistry* **24**, 2427.

Rao, K.N. and Srimannarayana, G. (1983), *Phytochemistry* **22**, 2287.

Rathore, J.S., Garg, S.K. and Gupta, S.R. (1981a), *Phytochemistry* **20**, 1755.

Rathore, J.S., Garg, S.K., Nagar, A., Sharma, N.D. and Gupta, S.R. (1981b), *Plant. Med.* **43**, 86.

Rathore, A., Sharma, S.C. and Tondon, J.S. (1986), *Phytochemistry* **25**, 2223.

Ravanel, P., Tissut, M. and Douce, R. (1982), *Phytochemistry* **21**, 2845.

Rieseberg, L.H. and Schilling, E.E. (1985), *Am. J. Bot.* **72**, 999.

Ripperger, H. (1981), *Phytochemistry* **20**, 1757.

Roy, M., Mitra, S.R., Bhattacharyya, A. and Adityachaudhury, N. (1986), *Phytochemistry* **25**, 961.

Rózsa, Zs., Hohmann, J., Reisch, J., Mester, I. and Szendrei, K. (1982a), *Phytochemistry* **21**, 1827.

Rózsa, Zs., Hohmann, J., Szendrei, K., Reisch, J. and Mester, I. (1982b), *Heterocycles* **19**, 1793.

Rózsa, Zs., Hohmann, J., Szendrei, K., Mester, I. and Reisch, J. (1984), *Phytochemistry* **23**, 1818.

Ruangrungsi, N., Tappayuthipijarn, P. and Tantivatana, P. (1981), *J. Nat. Prod.* **44**, 541.

Ruangrungsi, N., De-Eknamkul, W. and Lange, G.L. (1984), *Plant. Med.* **50**, 432.

Sachdev, K. and Kulshreshta, D.K. (1983), *Phytochemistry* **22**, 1253.

Saini, T.R., Pathak, U.P. and Khanna, R.N. (1983), *J. Nat. Prod.* **46**, 936.

Sakurai, A. and Okumura, Y. (1983), *Bull. Chem. Soc. Jpn.* **56**, 542.

Sakurai, A., Okada, K. and Okumura, Y. (1982), *Bull. Chem. Soc. Jpn.* **55**, 3051.

Sakushima, A., Coskun, M., Hisada, S. and Nishibe, S. (1983), *Phytochemistry* **22**, 1677.

Sanduja, R., Alam, M. and Euler, K.L. (1983), *J. Nat. Prod.* **46**, 149.

Sattikulov, A., Abdullaev, Sh.V., Batirov, E.Kh., Kurbatov, Yu.V., Malikov, V.M., Vdovin, A.D. and Yagudaev, M.R. (1983), *Khim. Prir. Soed.* **19**, 417.

Schilling, E.E. (1983), *Biochem. Syst. Ecol.* **11**, 341.

Shalashvili, A.G. and Mkhavanadze, V.V. (1985), *Khim. Prir. Soed.* **21**, 383.

Shalashvili, A.G. and Tsiklauri, G.Ch. (1984), *Khim. Prir. Soed.* **20**, 622.

Sharma, V.K. and Gupta, S.R. (1984), *Indian J. Chem.* **23B**, 780.

Shen, Z. and Theander, O. (1985), *Phytochemistry* **24**, 155.

Shimokoriyama, M. (1962), In *The Chemistry of Flavonoid Compounds*, (ed. T.A. Geissman), Pergamon Press, Oxford, pp. 286–316.

Shiritaki, Y., Komatsu, M., Yokoe, I. and Manaka, A. (1981), *Chem. Pharm. Bull.* **29**, 3033.

Shiritaki, Y., Manaka, A. Yokoe, I. and Komatsu, M. (1982), *Phytochemistry* **21**, 2959.

Shiritaki, Y., Endo, M., Yokoe, I. and Komatsu, M. (1983), *Chem. Pharm. Bull.* **31**, 2859.

Shrivastava, S.P. (1982), *Phytochemistry* **21**, 1464.

Shrivastava, S.P., Srivastava, S.K. and Nigam, S.S. (1982), *Indian J. Chem.* **21B**, 604.

Shukla, Y.N., Tandon, J.S. and Dhar, M.M. (1973), *Indian J. Chem.* **11**, 720.

Singhal, A.K., Barua, N.C., Sharma, R.P. and Baruah, J.N. (1983), *Phytochemistry* **22**, 1005.

Sinha, B., Natu, A.A. and Nanavati, D.D. (1982), *Phytochemistry* **21**, 1468.

Sivarambabu, S., Vemuri, V.S.S., Rao, C.P. and Rao, K.V.J. (1985), *Indian J. Chem.* **24B**, 217.

Smalberger, T.M., Vleggaar, R. and Weber, J.C. (1974), *Tetrahedron* **30**, 3927.

Srivastava, S.D. (1981), *Phytochemistry* **20**, 2445.

Srivastava, S.D. (1982), *Indian J. Chem.* **21B**, 165.

Srivastava, S.K. and Gupta, H.O. (1983), *Plant. Med.* **48**, 58.

Srivastava, S.K., Srivastava, S.D., Saksena, V.K. and Nigam, S.S. (1981a), *Phytochemistry* **20**, 862.

Srivastava, S.K., Srivastava, S.D. and Tiwari, K.P. (1981b), *Indian J. Chem.* **20B**, 347.

Stepanova, T.A., Sheichenko, V.I., Smirnova, L.P. and Glyzin, V.I. (1981), *Khim. Prir. Soed.* **17**, 520.

Stoess, G. (1972), Inaugural Doctoral Dissertation, Münster, Germany. Cited by Van Den Broucke, C.O., Dommisse, R.A., Esmans, E.L. and Lemli, J.A. (1982), *Phytochemistry* **21**, 2581.

Stotz, G., Spribille, R. and Forkmann, G. (1984), *J. Plant Physiol.* **116**, 173.

Subrahmanyam, K., Rao, J.M., Vemuri, V.S.S., Sivarambabu, S., Roy, C.P., Rao, K.V.J. and Merlini, L. (1982), *Indian J. Chem.* **21B**, 895.

Sultana, S. and Ilyas, M. (1986), *Phytochemistry* **25**, 963.

Szilagyi, I., Tetenyi, P., Antus, S., Seligmann, O., Chari, V.M., Seitz, M. and Wagner, H. (1981), *Plant. Med.* **43**, 121.

Takagi, S., Yamaki, M. and Inoue, K. (1980), *Yakugaku Zasshi* **100**, 1220.

Takagi, S., Yamaki, M. and Inoue, K. (1981), *Yakugaku Zasshi* **101**, 899.

Takido, M., Aimi, M., Yamonouchi, S., Yasukawa, K., Torri, H. and Takahashi, S. (1976), *Yakugaku Zasshi* **96**, 381.

Tamura, K. (1936), *J. Chem. Soc. Jpn.* **57**, 1141.

Tanaka, H., Ichino, K. and Ito, K. (1984a), *Chem. Pharm. Bull.* **32**, 3747.

Tanaka, H., Ichino, K. and Ito, K. (1984b), *Phytochemistry* **23**, 1198.

Tanaka, H., Ichino, K. and Ito, K. (1985), *Chem. Pharm. Bull.* **33**, 2602.

Tanaka, T., Yamasaki, K., Kohda, H., Tanaka, O. and Mahato, S.B. (1980), *Plant. Med. Suppl.*, 81.

Tanaka, T., Kawamura, K., Kohda, H., Yamasaki, K. and Tanaka, O. (1982), *Chem. Pharm. Bull.* **30**, 2421.

Teresa, J. De P., Gonzalez, M.S., Muriel, M.R. and Bellido, I.S. (1984), *Phytochemistry* **23**, 1819.

Teresa, J. De P., Gonzalez, M.S., Muriel, M.R., Arcocha, A.D. and Bellido, I.S. (1986), *J. Nat. Prod.* **49**, 177.

Timmermann, B.N., Hoffmann, J.J., Jolad, S.D., Bates, R.B. and Siahaan, T.J. (1986), *Phytochemistry* **25**, 723.

Tiwari, K.P. and Minocha, P.K. (1980), *Phytochemistry* **19**, 2501.

Tomas, F., Ferreres, F. and Barberan, F.A.T. (1985), *J. Nat. Prod.* **48**, 506.

Tomimori, T., Miyaichi, Y., Imoto, Y., Kizu, H. and Tanabe, Y. (1983), *J. Pharm. Soc. Jpn* **103**, 607.

Tomimori, T., Miyaichi, Y., Imoto, Y., Kizu, H. and Suzuki, C. (1984), *J. Pharm. Soc. Jpn* **104**, 529.

Tomimori, T., Miyaichi, Y., Imoto, Y., Kizu, H. and Namba, T. (1985), *Chem. Pharm. Bull.* **33**, 4457.

Trousdale, E.K. and Singleton, V.L. (1983), *Phytochemistry* **22**, 619.

Tuntiwachwuttikul, P., Pancharoen, O., Reutrakul, V. and Byrne, L.T. (1984), *Aust. J. Chem.* **37**, 449.

Ueda, S., Nomura, T., Fukai, T. and Matsumoto, J. (1982), *Chem. Pharm. Bull.* **30**, 3042.

Venturella, P., Bellino, A., Marino, M.L. and Sorrentino, M. (1980), *Heterocycles* **14**, 1979.

Vieira, P.C., DeAlvarenga, M.A., Gottlieb, O.R. and Gottlieb, H.E. (1980), *Plant. Med.* **39**, 153.

Vieira, P.C., Gottlieb, O.R. and Gottlieb, H.E. (1983), *Phytochemistry* **22**, 2281.

Voirin, B., Rasamoelisendra, R., Favre-Bonvin, J., Adriantsiferana, M. and Rabesa, Z. (1986), *Phytochemistry* **25**, 560.

Waterman, P.G. and Mahmoud, E.N. (1985), *Phytochemistry* **24**, 571.

Watkin, J.E. (1960), *Proccedings of Plant Phenolics Group of North America*, Fort Collins, Colorado, 39.

Williams, A.H. (1964), *Nature (London)* **202**, 824.

Williams, C.A. and Harvey, W.J. (1982), *Phytochemistry* **21**, 329.

Wollenweber, E. (1982), *Phytochemistry* **21**, 1462.

Wollenweber, E. (1984),, *Z. Naturforsch.* **39c**, 833.

Wollenweber, E. and Egger, K. (1971), *Phytochemistry* **10**, 225.

Wollenweber, E. and Seigler, D.S. (1982), *Phytochemistry* **21**, 1063.

Wollenweber, E. and Yatskievych, G. (1982), *J. Nat. Prod.* **45**, 216.

Wollenweber, E., Rehre, C. and Dietz, V.H. (1981), *Phytochemistry* **20**, 1167.

Wollenweber, E. Dietz, V.H., Schillo, D. and Schilling, G. (1980), *Z. Naturforsch.* **35c**, 685.

Wollenweber, E., Dietz, V.H., Schilling, G., Favre-Bonvin, J. and Smith, D. (1985a), *Phytochemistry* **24**, 965.

Wollenweber, E., Kohorst, G., Mann, K. and Bell, J.M. (1985b), *J. Plant Physiol.* **117**, 423.

Wollenweber, E., Schober, I., Clark, W.D., and Yatskievych, G. (1985c), *Phytochemistry* **24**, 2129.

Wong, R.Y. and Horowitz, R.M. (1986), *J. Chem. Soc., Perkin Trans. I*, 843.

Wu, L.J., Miyase, T., Veno, A., Kuroyanagi, M., Noro, T. and Fukushima, S. (1985a), *Chem. Pharm. Bull.* **33**, 3231.

Wu, L.J., Miyase, T., Veno, A., Kuroyanagi, M., Noro, T. and Fukushima, S. (1985b), *J. Pharm. Soc. Jpn* **105**, 736.

Wu, L.J., Miyase, T., Veno, A., Kuroyanagi, M., Noro, T., Fukushima, S. and Sasaki, S. (1985c), *Yakugaku Zasshi* **105**, 1034.

Wu, L.J., Miyase, T., Veno, A., Kuroyanagi, M., Noro, T., Fukushima, S. and Sasaki, S. (1986), *Yakugaku Zasshi* **106**, 22.

Wu, T.-S., K, C.-S., Ho, S.-T, Yang, M.-S. and Lee, K.K. (1981), *Phytochemistry* **20**, 527.

Zemplen, G. and Mester, L. (1942), *Ber.* **75B**, 1298.

Zotov, E.P., Litvinenko, V.I. and Khazanovich, R.L. (1976), Deposited Doc., VINITI, 252 (9 pages); *Chem. Abstr.* 88:71449f (1978).

10

Miscellaneous flavonoids

HANS GEIGER

10.1 INTRODUCTION

This chapter is intended to cover compounds which do not fit into the preceding chapters of this book. Since the authors of the present volume have taken a rather broad view of their topics, not too much is left for this chapter. I am very grateful to the authors of other chapters, especially to Dr Lawrence Porter, for furnishing me with references. Even so, the following lists will not be comprehensive, because for most of these compounds there are no generally accepted generic names. This makes a systematic search impossible, and I must apologize for any omissions.

10.2 DIARYLPROPANES

From the oxygenation patterns of the naturally occurring diarylpropanes, it may be inferred that they are biogenetically related to flavonoids. Long before their discovery in some members of the Myristicaceae, they had been synthesized by reduction of chalcones and hydrogenolysis of catechins. The early work on diarylpropanes has been summarized by Gottlieb (1977). The detection of these compounds in species of Lauraceae, Leguminosae, Moraceae and Myristicaceae has only been possible by isolation and determination of structure. Because of the difficulty of differentiating them from the other simple phenolics on chromatograms, no phytochemical screening has so far been performed. In Table 10.1, the known diarylpropanes, are listed.

10.3 CINNAMYLPHENOLS

This is a small group of flavonoid-related phenolics which possess various biological activities. For a summary of the earlier work see Gottlieb (1977). The cinnamylphenols are listed in Table 10.2. They have all been isolated from members of the Leguminosae.

10.4 HOMOISOFLAVONOIDS

The term homoisoflavonoid is a misnomer, since the compounds discussed under this heading are biogenetically unrelated to isoflavonoids. I have, however, retained this widely used name, because the alternative name – benzylchromanoid – advocated by Dewick (1975) is equally misleading, if one looks at some of the recently discovered structures.

It is only five years since these compounds were comprehensively reviewed by Heller and Tamm (1981). Meanwhile, the number of known homoisoflavonoids has more than doubled. Nothing basically new concerning the isolation, structural, elucidation, and synthesis has been reported since this review. All known homoisoflavonoids are listed in Table 10.3. The names in this list are those derived from the primary literature.

With the exception of four structures which occur in Leguminosae, all other homoisoflavonoids have been isolated from Liliaceae. Since no surveys have been carried out, it is not yet clear whether or not these compounds have a wider distribution.

The biosynthesis of the homoisoflavonoids has been studied by Dewick (1975): the carbon-2 stems from the *S*-methyl of methionine, and the rest of the skeleton from cinnamic acid and three acetate units. The compounds are most likely formed via a 2′-methoxychalcone intermediate.

The Flavonoids. Edited by J.B. Harborne
Published in 1988 by Chapman and Hall Ltd,
11 New Fetter Lane, London EC4P 4EE.

Table 10.1 List of naturally occurring diarylpropanes

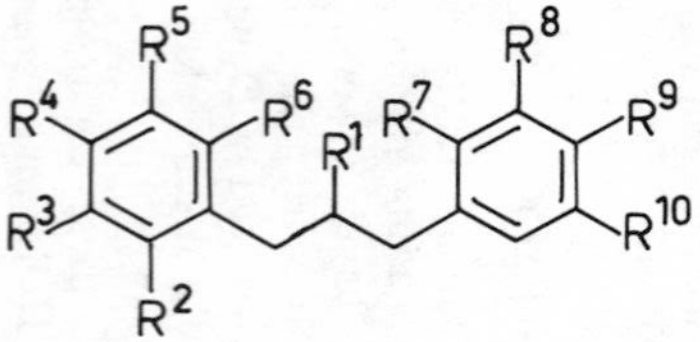

	Trivial name	R^1	R^2	R^3	R^4	R^5	R^6	R^7	R^8	R^9	R^{10}	References not in Gottlieb (1977)
1A.1	Propterol	OH	H	H	OH	H	H	H	H	OH	H	Rao *et al.* (1984)
1A.2	Broussonin C	H	OH	H	OH	H	H	H	X	OH	H	Takasugi *et al.* (1980b)
1A.3	Broussonin A	H	OH	H	OMe	H	H	H	H	OH	H	Takasugi *et al.* (1980a)
1A.4	Broussonin B	H	OMe	H	OH	H	H	H	H	OH	H	Takasugi *et al.* (1980a)
1A.5	Broussonin D	OH	OH	H	OMe	H	H	H	H	OH	H	Takasugi *et al.* (1984)
1A.6	Kazinol F	H	OH	H	OH	H	H	X	X	OH	OH	Ikuta *et al.* (1986)
1A.7	Kazinol C	H	OH	H	OH	Y	H	X	X	OH	OH	Ikuta *et al.* (1986)
1A.8	Kazinol J	H	OMe	H	OH	H	H	X	X	OH	OH	Kato *et al.* (1986)
1A.9	Kazinol G	H	OH	H	OH	Y	H	X	X	OH	OX	Ikuta *et al.* (1986)
1A.10		H	OMe	H	OH	H	H	H	H	OMe	OH	
1A.11	Broussonin F	H	OMe	H	OH	H	H	H	H	OH	OMe	Braz Filho *et al.* (1980a); Takasugi *et al.* (1984)
1A.12	Broussonin E	H	OH	H	OMe	H	H	H	H	OMe	OH	Takasugi *et al.* (1984)
1A.13		H	OH	H	OMe	H	H	H	H	OH	OMe	
1A.14		H	OH	H	OH	H	H	H	H	—OCH_2O—		
1A.15	Virolane	H	OH	H	OMe	H	H	H	H	—OCH_2O—		Kijjoa *et al.* (1981)
1A.16		H	H	H	OMe	H	H	OH	H	—OCH_2O—		
1A.17	Virolanol B	OH	OH	H	OH	H	H	H	H	OH	OMe	Kijjoa *et al.* (1981)
1A.18	Virolanol C	OH	OH	H	OMe	H	H	H	H	OH	OMe	Kijjoa *et al.* (1981)
1A.19	Virolanol	OH	OH	H	OMe	H	H	H	H	—OCH_2O—		Kijjoa *et al.* (1981); Fernandez *et al.* (1980)
1A.20		H	OMe	H	OMe	H	OH	H	H	OH	OMe	Diaz and De Diaz (1986)
1A.21		H	OH	H	OH	H	OMe	H	H	—OCH_2O—		
1A.22		H	OH	Me	OH	Me	H	OH	H	—OCH_2O—		
1A.23		H	OMe	H	OH	Me	H	OH	H	—OCH_2O—		Braz Filho *et al.* (1980b)
1A.24		H	OH	Me	OH	H	H	OMe	H	—OCH_2O—		Almeida *et al.* (1979)
1A.25		H	OMe	H	OMe	H	OH	H	H	—OCH_2O—		Diaz and DeDiaz (1986)
1A.26		OH	OMe	H	OH	H	OH	H	H	OH	OMe	Morimoto *et al.* (1985)

X = CH=CH—CMe_2

Y = $C(Me)_2CH{=}CH_2$

1B.1 Kazinol D Ikuta *et al.* (1986)

1B.2	Kazinol M	$R^1 = R^2 = H$	Kato *et al.* (1986)
1B.3	Kazinol N	$R^1 = CH_3$, $R^2 = C(Me)_2CH{=}CH_2$	Kato *et al.* (1986)
1B.4	Kazinol K	$R^1 = H$, $R^2 = C(Me)_2CH{=}CH_2$	Ikuta *et al.* (1986)

1B.5 Kazinol L Kato *et al.* (1986)

1C.1 Virolaflorin Kijjoa *et al.* (1981)

IC.2 Spiroelliptin Braz Filho *et al.* (1980a)

Table 10.2 List of naturally occurring cinnamylphenols

Trivial name		R^1	R^2	R^3	R^4	R^5	R^6	CH=CH	References not in Gottlieb (1977)
2A.1	Obtusastyrene	H	H	OH	H	H	H	E	Gregon *et al.* (1978a)
2A.2	Obtustyrene	OMe	H	OH	H	H	H	E	Gregson *et al.* (1978a); Carlson and Dolphin (1982)
2A.3	Mucronustyrene	OMe	OMe	OH	H	H	H	E	Kurosawa *et al.* (1978)
2A.4		OMe	H	OH	OH	H	H	E	
2A.5	Violastyrene	OMe	H	OH	OMe	H	H	E	Gregson *et al.* (1978b)
2A.6	Isoviolastyrene	OMe	H	OMe	OH	H	H	E	Gregson *et al.* (1978b)
2A.7	Xenognosin	OMe	H	OH	H	H	OH	E	Lynn *et al.* (1981); Carlson and Dolphin (1982)
2A.8	Mucronulastyrene	OMe	OMe	OH	H	OH	H	Z	Kurosawa *et al.* (1978)
2A.9	Villostyrene	OMe	OMe	OH	H	OMe	H	Z	Kurosawa *et al.* (1978)
2A.10	Petrostyrene	OMe	OMe	OMe	OH	OH	H	E	Ollis *et al.* (1978b)
2A.11	Kuhlmannistyrene	OMe	OMe	OMe	OH	OH	H	Z	Ollis *et al.* (1978a)

2B.1 Obtusaquinone Gregson *et al.* (1978a)

Table 10.3 List of known homoisoflavonoids

3A.1: R = H: 7-Hydroxy-3-(4-hydroxybenzyl)chromone
3A.2: R = OMe: 7-Hydroxy-8-methoxy-(4-hydroxybenzyl)-chromone
} Camarda *et al.* (1983)

		R^1	R^2	R^3	R^4	R^5	R^6	*References not in Heller and Tamm (1981)*
3B.1	7-Hydroxy-3-(4-hydroxybenzyl)-chroman-4-one	H	H	OH	H	H	OH	Camarda *et al.* (1983)
3B.2	4′-Demethyl-3, 9-dihydroeucomin	OH	H	OH	H	H	OH	Camarda *et al.* (1983); Adinolfi *et al.* (1985a)
3B.3	5-*O*-Methyl-4′-demethyl-3, 9-dihydroeucomin	OMe	H	OH	H	H	OH	
3B.4	3, 9-Dihydroeucomin	OH	H	OH	H	H	OMe	
3B.5	7-*O*-Methyl-3, 9-dihydroeucomin	OH	H	OMe	H	H	OMe	
3B.6	3′-Hydroxy-3, 9-dihydroeucomin	OH	H	OH	H	OH	OMe	Adinolfi *et al.* (1985a)
3B.7	5, 7-Dihydroxy-6-methyl-3-(4-hydroxybenzyl)chroman-4-one	OH	Me	OH	H	H	OH	Camarda *et al.* (1983)
3B.8	Ophiopogonanone B	OH	Me	OH	H	H	OMe	Kaneda *et al.* (1983)
3B.9	Ophiopogonanone A	OH	Me	OH	H	—OCH_2O—		Tada *et al.* (1980b)
3B.10	Methylophiopogonanone B	OH	Me	OH	Me	H	OMe	
3B.11	Methylophiopogonanone A	OH	Me	OH	Me	—OCH_2O—		Tada *et al.* (1980a)
3B.12	3, 9-Dihydroeucomnalin	OH	OMe	OH	H	H	OH	
3B.13	8-*O*-Demethyl-7-*O*-methyl-3, 9-dihydropunctatin	OH	H	OMe	OH	H	OH	Adinolfi *et al.* (1984)
3B.14	8-*O*-Demethyl-8-*O*-acetyl-7-O-methyl-3, 9-dihydropunctatin	OH	H	OMe	OAc	H	OH	
3B.15	3, 9-Dihydropunctatin	OH	H	OH	OMe	H	OH	Adinolfi *et al.* (1985b)
3B.16	7-*O*-Methyl-3, 9-dihydropunctatin	OH	H	OMe	OMe	H	OH	Adinolfi *et al.* (1984)
3B.17	4′-*O*-Methyl-3, 9-dihydropunctatin	OH	H	OH	OMe	H	OMe	
3B.18	Muscomin	OH	OMe	OMe	OH	H	OH	Adinolfi *et al.* (1985a)
3B.19	6-Formylisoophiopogonanone B	OH	CHO	OH	Me	H	OMe	
3B.20	7-*O*-Methyl-6-formyliso-ophiopogonanone B	OH	CHO	OMe	Me	H	OMe	
3B.21	6-Formylisoophiopogonanone A	OH	CHO	OH	Me	—OCH_2O—		Kaneda *et al.* (1983)
3B.22	7-*O*-Methyl-6-formyliso-ophiopogonanone A	OH	CHO	OMe	Me	—OCH_2O—		

(*Contd.*)

Table 10.3 (*Contd.*)

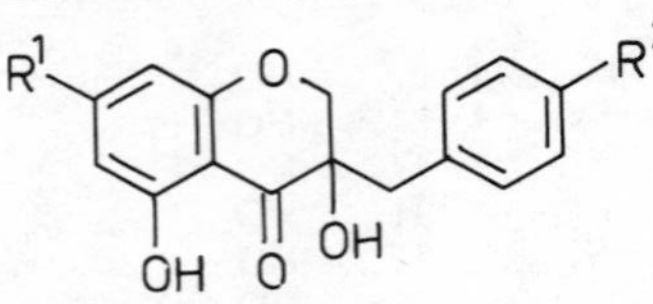

	Compound	R^1	R^2		Reference
3C.1	Demethyleucomol	OH	OH		Heller and Tamm (1981)
3C.2	Eucomol	OH	OMe		
3C.3	7-*O*-Methyleucomol	OMe	OMe		

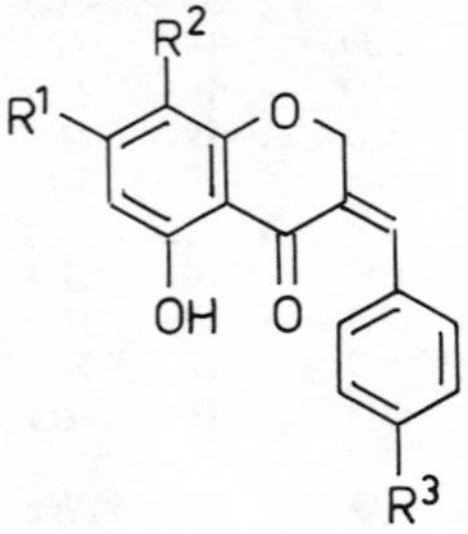

	Compound	R^1	R^2	R^3	Reference
3D.1	(*Z*)-Eucomin	OH	H	OMe	Heller and Tamm (1981)
3D.2	(*Z*)-4′-*O*-Methylpunctatin	OH	OMe	OMe	

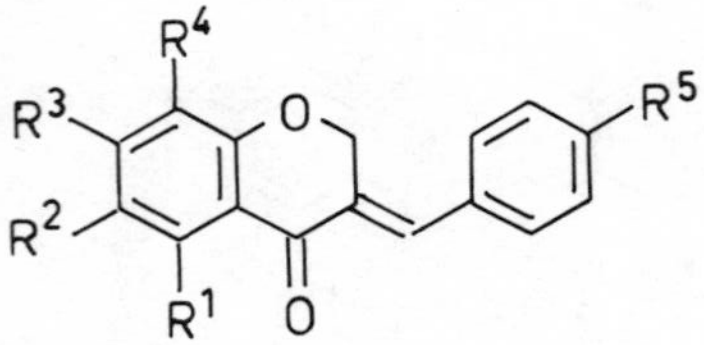

	Compound	R^1	R^2	R^3	R^4	R^5	*References not in Heller and Tamm (1981)*
3E.1	Bonducellin	H	H	OMe	H	OH	Purushothaman *et al.* (1982); McPherson *et al.* (1983)
3E.2	8-Methoxybonducellin	H	H	OMe	OMe	OH	McPherson *et al.* (1983)
3E.3	(*E*)-4′-Demethyleucomin	OH	H	OH	H	OH	
3E.4	(*E*)-Eucomin	OH	H	OH	H	OMe	
3E.5	(*E*)-7-*O*-Methyleucomin	OH	H	OMe	H	OMe	
3E.6	(*E*)-Eucomnalin	OH	OMe	OH	H	OH	
3E.7	(*E*)-Punctatin	OH	H	OH	OMe	OH	
3E.8	(*E*)-4′-*O*-Methylpunctatin	OH	H	OH	OMe	OMe	

	R^1	R^2	R^3	R^4	*References not in Heller and Tamm (1981)*
3F.1 Muscomosin	H	OH	OMe	OH	Adinolfi *et al.* (1985b)
3F.2 Scillascillin	H	OH	—OCH_2O—		
3F.3 2-Hydroxy-7-*O*-methyl-scillascillin	OH	OMe	—OCH_2O—		

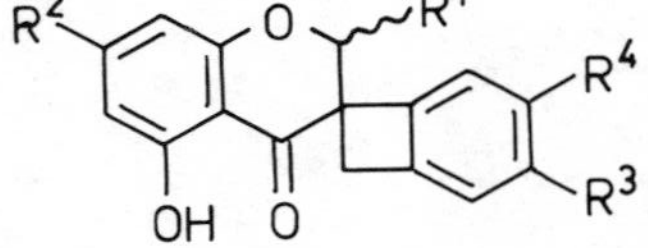

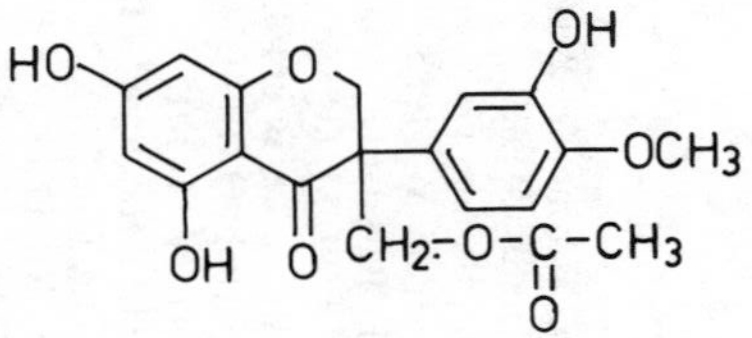

3G.1 Comosin (Adinolfi *et al.*, 1985b)

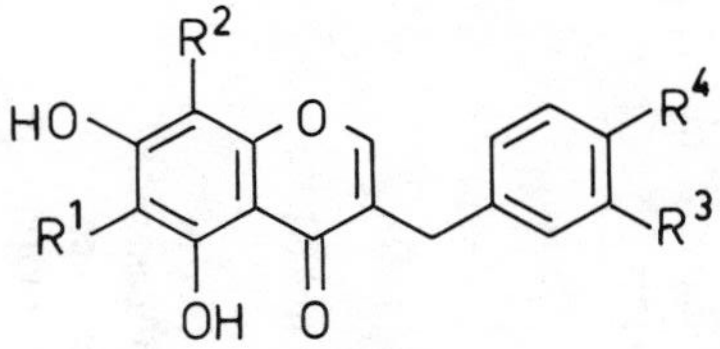

	R^1	R^2	R^3	R^4	
3H.1 Ophiopogonone B	Me	H	H	OMe	Tada *et al.* (1985b)
3H.2 Ophiopogonone A	Me	H	—OCH_2O—		Tada *et al.* (1985b)
3H.3 Desmethy-lisoophiopogonone B	H	Me	H	OH	Tada *et al.* (1985b)
3H.4 Isoophiopogonone A	H	Me	—OCH_2O—		Tada *et al.* (1985b)
3H.5 Methylophiopogonone B	Me	Me	H	OMe	Tada *et al.* (1985a)
3H.6 Methylophiopogonone A	Me	Me	—OCH_2O—		Tada *et al.* (1985a)

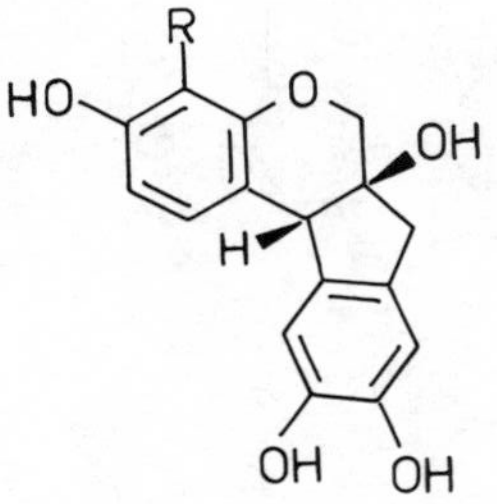

3I.1 R = H: Brazilin
3I.2 R = OH: Haematoxylin

Heller and Tamm (1981)

10.5 SPHAGNORUBINS

Some *Sphagnum* (peat moss) species turn red when subjected to low temperatures (Rudolph and Jöhnk, 1982). This red colour is due to sphagnorubins which are, unlike the anthocyanins, not present in the cell sap, but are bound to the cell wall. They are linked so tenaciously to the membranes that they are only extracted with great difficulty. The structures of three sphagnorubins A–C have so far been elucidated.

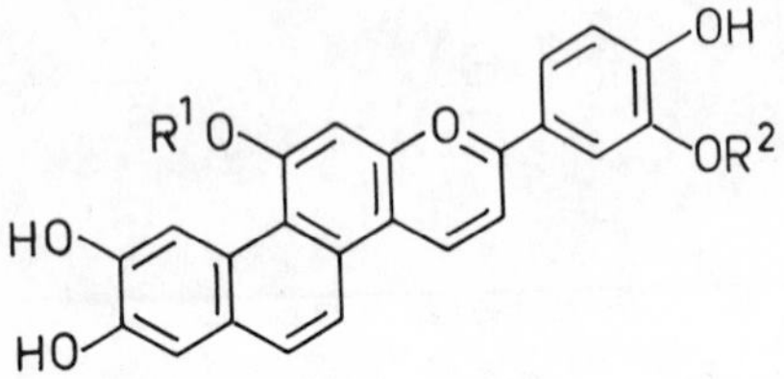

Sphagnorubin A:	$R^1{=}R^2{=}H$	Vowinkel(1975)
Sphagnorubin B:	$R^1{=}Me, R^2{=}H$	Mentlein and Vowinkel(1984)
Sphagnorubin C:	$R^1{=}R^2{=}Me$	Mentlein and Vowinkel(1984)

10.6 REARRANGED AND DEGRADED FLAVONOIDS

Diaroylmethanes and chalcones related to the 3→8 linked biflavonoids (see Chapter 4) may undergo ring closure to benzofurans or dihydrobenzofurans respectively, instead of forming flavanones or flavones. Only two such cases (compounds 3 and 4 in Drewes *et al.*, 1984) have so far been reported; however similar compounds may be more widespread.

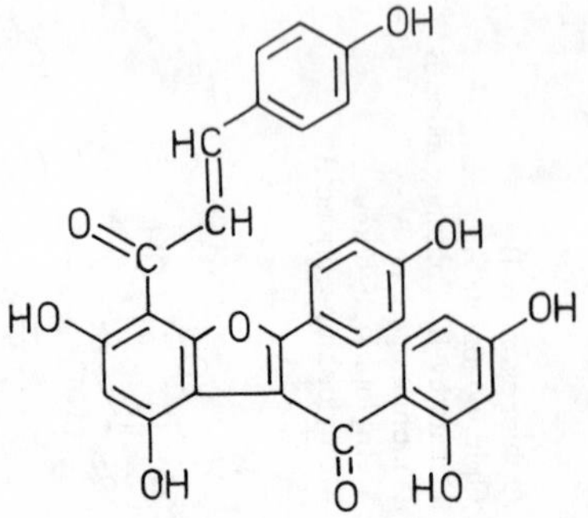

'Compound 3'

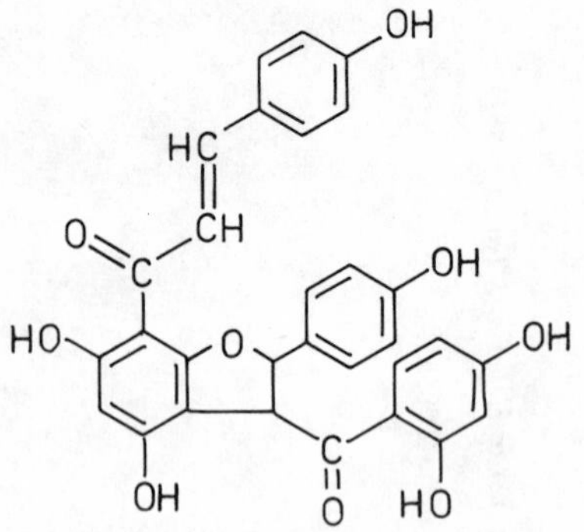

'Compound 4'

A 3→8-linked flavanone–chromone has been isolated along with fukugetin (compound 2B.3 in Chapter 4) from the leaves of *Garcinia dulcis*(Roxb.) Kurz. Since the second B-ring in fukugetin, which is missing in the flavanone–chromone, is a catechol ring, it is likely that the chromone moiety is formed by oxidative degradation of that ring (Ansari *et al.*, 1976).

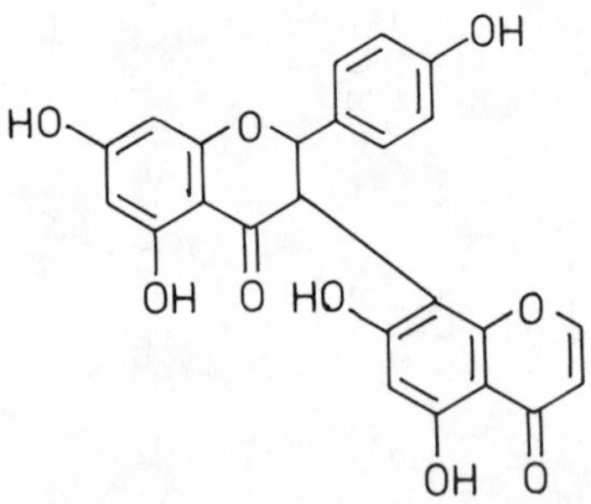

Flavanone–chromone GD IV

REFERENCES

Adinolfi, M., Barone, G., Belardini, M., Lanzetta, R., Laonigro, G. and Parrilli, M. (1984), *Phytochemistry* **23**, 2091.

Adinolfi, M., Barone, G., Lanzetta, R., Laonigro, G., Mangoni, L. and Parrilli, M. (1985a), *Phytochemistry* **24**, 624.

Adinolfi, M., Barone, G., Belardini, M., Lanzetta, R., Laonigro, G. and Parrilli, M. (1985b), *Phytochemistry* **24**, 2423.

Almeida, M.E.L. de, Braz Filho, R., Bülow, M.V. von, Corrêa, J.J.L., Gottlieb, O.R., Maia, J.G.S. and Silva, M.S. da (1979), *Phytochemistry* **18**, 1015.

Ansari, W.H., Rahman, W., Barraclough, D., Maynard, R. and Scheinmann, F. (1976), *J. Chem. Soc., Perkin I*, 1458.

Braz Filho, R., Diaz D., P.P. and Gottlieb, O.R. (1980a), *Phytochemistry* **19**, 455.

Braz Filho, R., Da Silva, M.S. and Gottlieb, O.R. (1980b), *Phytochemistry* **19**, 1195.
Camarda, L., Merlini, L. and Nasini, G. (1983), *Heterocycles* **20**, 39.
Carlson, R.E. and Dolphin, D.H. (1982), *Phytochemistry* **21**, 1733.
Dewick, P.M. (1975), *Phytochemistry* **14**, 983.
Diaz D., P.P. and De Diaz, A.M.P. (1986), *Phytochemistry* **25**, 2395.
Drewes, S.E., Hudson, N.A., Bates, R.B. and Linz, G.S. (1984), *Tetrahedron Lett.* **25**, 105.
Fernandez, J.B., Nilce de S. Ribeiro, M., Gottlieb, O.R. and Gottlieb, H.E. (1980), *Phytochemistry* **19**, 1523.
Gottlieb, O.R. (1977) *Isr. J. Chem.* **16**, 45.
Gregson, M., Ollis, W.D., Redman, B.T., Sutherland, I.O., Dietrichs, H.H. and Gottlieb, O.R. (1978a), *Phytochemistry* **17**, 1395.
Gregson, M., Ollis, W.D., Sutherland, I.O., Gottlieb, O.R. and Magalhães, M.T. (1978b), *Phytochemistry* **17**, 1375.
Heller, W. and Tamm, Ch. (1981) *Fortschr. Chem. Org. Naturstoffe* **40**, 105.
Ikuta, J., Hano, Y., Nomura, T., Kawakami, Y. and Sato, T. (1986), *Chem. Pharm. Bull.* **34**, 1968.
Kaneda, N., Nakanishi, H., Kuraishi, T. and Katori, T. (1983), *J. Pharm. Soc. Jpn* **103**, 1133.
Kato, S., Fukai, T., Ikuta, J. and Nomura, T. (1986), *Chem. Pharm. Bull.* **34**, 2448.
Kijjoa, A., Giesbrecht, A.M., Gottlieb, O.R. and Gottlieb, H.E. (1981), *Phytochemistry* **20**, 1385.
Kurosawa, K., Ollis, W.D., Sutherland, I.O., Gottlieb, O.R. and De Oliveira, A.B. (1978), *Phytochemistry* **17**, 1389.
Lynn, D.G., Steffens, J.C., Kamut, V.S., Graden, D.W., Shabanowitz, J. and Riopel, J.L. (1981), *J. Am. Chem. Soc.* **103**, 1868.
Mentlein, R. and Vowinkel, E. (1984), *Liebigs Ann. Chem.*, 1024.
Morimoto, S., Nonaka, G.-I., Nishioka, I., Ezaki, N. and Takizawa, N. (1985), *Chem. Pharm. Bull.* **33**, 2281.
Ollis, W.D., Redman, B.T., Roberts, R.J., Sutherland, I.O., Gottlieb, O.R. and Magalhães, M.T. (1978a), *Phytochemistry* **17**, 1383.
Ollis, W.D., Redman, B.T., Sutherland, I.O. and Gottlieb, O.R. (1978b), *Phytochemistry* **17**, 1379.
McPherson, D.D., Cordell, G.A., Soejarto, D.D., Pezzuto, J.M. and Fong, H.H.S. (1983), *Phytochemistry* **22**, 2835.
Purushothaman, K.K., Kalyani, K. and Subramaniam, K. (1982), *Indian J. Chem.* **21B**, 383.
Rao, A.V.S., Mathew, J. and Sankaram, A.V.B. (1984), *Phytochemistry* **23**, 897.
Rudolph, H. and Jöhnk, A. (1982), *J. Hattori Bot. Lab.* **53**, 195.
Tada, A., Kasai, R., Saitoh, T. and Shoji, J. (1980a), *Chem. Pharm. Bull.* **28**, 1477.
Tada, A., Kasai, R., Saitoh, T. and Shoji, J. (1980b), *Chem. Pharm. Bull.* **28**, 2044.
Takasugi, M., Anetai, M., Masamune, T., Shirata, A. and Takahashi, K. (1980a), *Chem. Lett.*, 339.
Takasugi, M., Kumagai, Y., Nagao, Sh., Masamune, T., Shirata, A. and Takahashi, K. (1980b), *Chem. Lett.*, 1459.
Takasugi, M., Niino, N., Nagao, Sh., Anetai, M., Masamune, T., Shirata, A. and Takahashi, K. (1984), *Chem. Lett.*, 689.
Vowinkel, E. (1975), *Chem. Ber.* **108**, 1166.

11

Biosynthesis

WERNER HELLER and GERT FORKMANN

11.1 INTRODUCTION

In the past few years, considerable progress has been made in elucidating the biosynthesis of flavonoids. In particular, our knowledge of the enzymology has developed rapidly. With the exception of the anthocyanins, where a few reactions still remain unknown, the essential steps of the biosynthetic pathway of the main flavonoid classes are now clear.

There are two main reasons for this rapid development. The improvement in chalcone synthase preparation and handling on the one hand and the realization that flowers are rich sources of enzymes for further reactions in flavonoid biosynthesis on the other, have provided the means for producing ^{14}C-labelled substrates, particularly (*S*)-naringenin and (2*R*, 3*R*)-dihydrokaempferol, in enantiomeric pure form and with high specific activity. Furthermore, studies in cell cultures are being supplemented or replaced by studies in flowers from intact plants. In this context, the extensive genetic information already available about flavonoid biosynthesis in flowering plants has proved to be of great advantage. The detection of enzymes in cell-free extracts has also been improved considerably and high-performance liquid chromatography has been more and more applied to the separation, identification and quantification of substrates and products.

In the first part of this chapter we will give a general overview of the origins of flavonoid precursors and the interlocking of the individual reactions leading to the various flavonoid classes. Since all flavonoids derive their carbon skeletons from two basic compounds, malonyl-CoA and the CoA ester of a hydroxycinnamic acid, the biosynthesis of these compounds is discussed in the second part of this chapter. In the third and major section, we will describe the flavonoid enzymes in detail and discuss some aspects of their regulation.

11.2 GENERAL ASPECTS

The origins of the direct flavonoid precursors 4-coumaroyl-CoA and malonyl-CoA and the biosynthetic interrelations of the various flavonoid classes are demonstrated in Fig. 11.1, with particular reference to substances with a single hydroxyl group in the B-ring. The enzymes involved in their biosynthetic pathway are summarized in Table 11.1.

Both flavonoid precursors are derived from carbohydrates. Malonyl-CoA is synthesized from the glycolysis intermediate acetyl-CoA and carbon dioxide, the reaction being catalysed by acetyl-CoA carboxylase. The supply of 4-coumaroyl-CoA is more complex. It involves the shikimate/arogenate pathway, the main route to the aromatic amino acids phenylalanine and tyrosine in higher plants (Jensen, 1985). Subsequent transformation of phenylalanine to *trans*-cinnamate is catalysed by phenylalanine ammonia-lyase which provides the link between primary metabolism and the phenylpropanoid pathway. Aromatic hydroxylation of cinnamate by cinnamate 4-hydroxylase leads to 4-coumarate which is further transformed to 4-coumaroyl-CoA by action of 4-coumarate: CoA ligase.

The central step in flavonoid biosynthesis is the condensation of three molecules of malonyl-CoA with a suitable hydroxycinnamic acid CoA ester, ordinarily 4-coumaroyl-CoA, to the C_{15} chalcone intermediate

The Flavonoids. Edited by J.B. Harborne
Published in 1988 by Chapman and Hall Ltd,
11 New Fetter Lane, London EC4P 4EE.

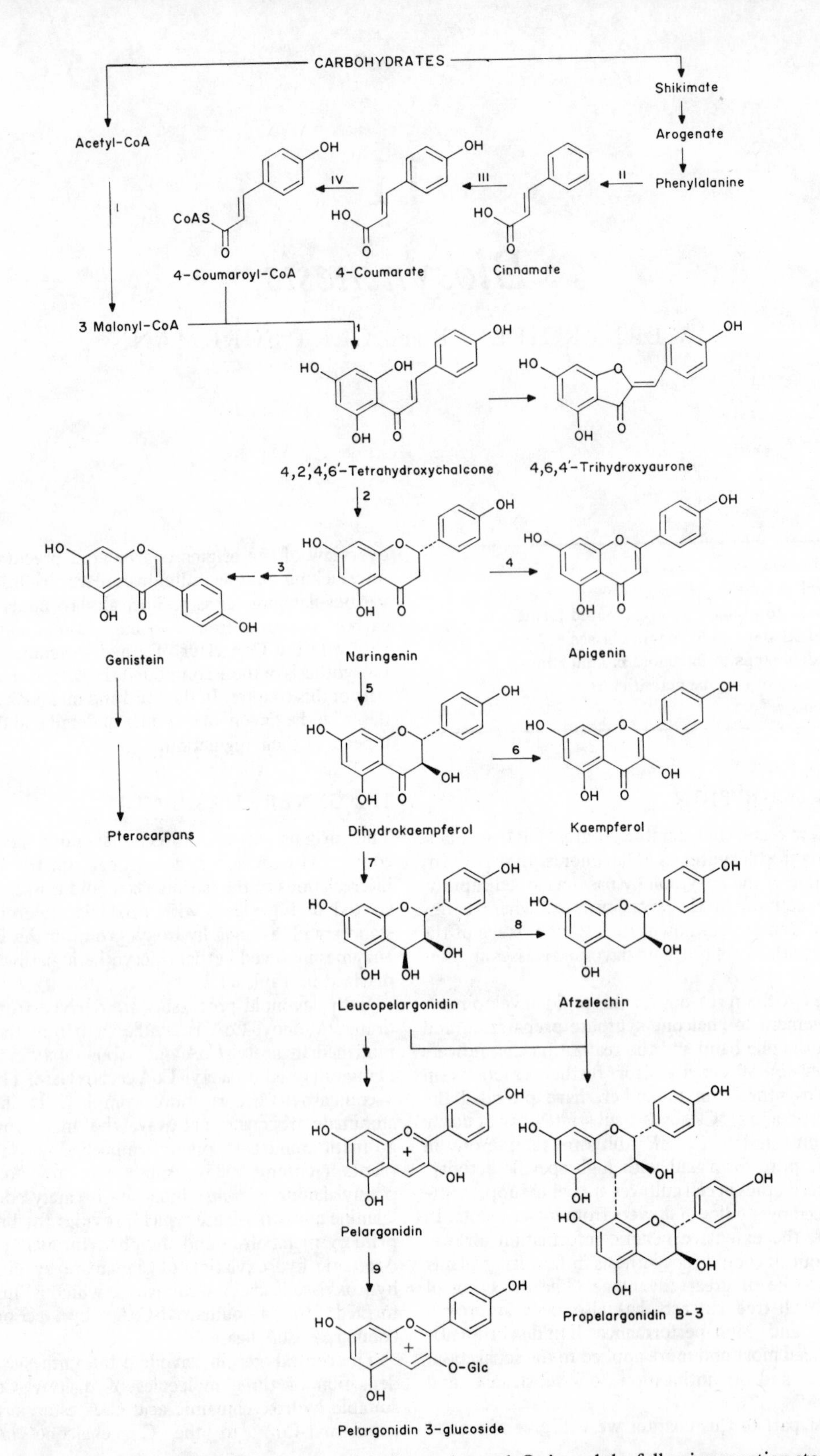

Fig. 11.1 Scheme illustrating the pathways to phenylalanine and acetyl-CoA, and the following reaction steps leading to the various flavonoid classes. The enzymes marked with numbers are listed in Table 11.1. Enzymes of the preflavonoid pathways are marked with roman numerals, enzymes of the flavonoid pathway with arabic numerals.

Table 11.1 List of enzymes leading to various flavonoid classes

	Enzymes	*EC number*
Non-flavonoid precursors		
I	Acetyl-CoA carboxylase	6.4.1.2
II	Phenylalanine ammonia-lyase	4.3.1.5
III	Cinnamate 4-hydroxylase	1.14.13.11
IV	4-Coumarate:CoA ligase	6.2.1.12
Flavonoid classes		
1	Chalcone synthase	2.3.1.74
2	Chalcone isomerase	5.5.1.6
3	2-Hydroxyisoflavanone synthase	
4	Flavone synthase	
5	(2*S*)-Flavanone 3-hydroxylase	1.14.11.9
6	Flavonol synthase	
7	Dihydroflavonol 4-reductase	
8	Flavan-3, 4-*cis*-diol 4-reductase	
9	Anthocyanidin/flavonol 3-*O*-glucosyltransferase	2.4.1.91

(4, 2′, 4′, 6′-tetrahydroxychalcone, see Fig. 11.1). The reaction is catalysed by chalcone synthase. Flavonoids, aurones and other diphenylpropanoids are derived from the chalcone intermediate. Transformation by stereospecific action of chalcone isomerase provides the first flavonoid, a (2*S*)-flavanone (naringenin).

Oxidative rearrangement of the flavanone, involving a 2, 3-aryl shift, yields an isoflavone (genistein). The reaction is catalysed by 'isoflavone synthase'. Introduction of a double bond between C-2 and C-3 of flavanone leads to the abundant class of flavones (apigenin). Two different enzymes are known to catalyse this reaction: a dioxygenase and a mixed-function mono-oxygenase. Dihydroflavonols (dihydrokaempferol) are formed by direct hydroxylation of flavanones in the 3 position. This reaction is catalysed by the dioxygenase, flavanone 3-hydroxylase.

Dihydroflavonols are biosynthetic intermediates in the formation of flavonols, catechins, proanthocyanidins and anthocyanidins. The large class of flavonols (e.g. kaempferol) is formed by introduction of a double bond between C-2 and C-3 of dihydroflavonols. Flavonol synthase, the enzyme catalysing this reaction, is also a dioxygenase. Reduction of the carbonyl group of dihydroflavonols in the 4 position gives rise to flavan 2, 3-*trans*-3, 4-*cis*-diols (leucopelargonidin). These compounds, also named leucoanthocyanidins, are the immediate precursors for the synthesis of catechins and proanthocyanidins. Catechins (afzelechin) are synthesized from leucoanthocyanidins by action of flavan 3, 4-*cis*-diol reductase. Proanthocyanidins (propelargonidin B-3) are formed by a condensation of catechins and leucoanthocyanidins.

The reaction steps from leucoanthocyanidins to anthocyanidins (pelargonidin) are still unknown. An obligatory reaction in the sequence is a glycosylation, usually a glucosylation, in the 3 position of the anthocyanidin or of a suitable intermediate. This reaction leads to the first stable anthocyanin (e.g. pelargonidin 3-glucoside).

Modification by hydroxylation of the A- and, in particular, the B-ring, methylation of hydroxyl groups as well as glycosylation and acylation reactions result in the immense diversity of flavonoids found in nature. Numerous enzymes catalysing these modifications have been described. Some enzymes can act on both intermediates (flavanone or dihydroflavonol) and end products (flavone, isoflavone, flavonol or anthocyanidin 3-glycoside), others exclusively on the end products.

As a result of extensive genetic studies in a wide variety of plants, mutants are available for each step in flavonoid biosynthesis from chalcone formation up to complex modifications of the anthocyanin molecule. In recent years, this immense amount of genetic information has increasingly been used in the elucidation of flavonoid biosynthesis. Flowers of genetically defined plants have proved to be very valuable for supplementation experiments with potential precursors, as well as for correlating single genes with particular enzymes. Such correlations clearly establish that an enzyme capable of catalysing a particular step *in vitro* has the same function *in vivo*.

11.3 PATHWAYS TO PRECURSORS OF FLAVONOID FORMATION

Malonyl-CoA as one of the two substrates for chalcone synthase is formed from acetyl-CoA and CO_2. Acetyl-CoA carboxylase, which catalyses the reaction in the presence of ATP and Mg^{2+}, has been studied extensively in relation to fatty acid biosynthesis (Lowenstein, 1981). The data on the plant carboxylase have also been summarized (Nielsen, 1981). The enzyme involved in flavonoid biosynthesis has only been studied in detail in UV-irradiated parsley cell suspension cultures. From this source it has been purified to apparent homogeneity, and enzyme kinetic studies have been performed. Molecular weights of about 420 000 for the native enzyme and 220 000 for the subunit have been reported (see Ebel and Hahlbrock, 1982; Egin-Bühler and Ebel, 1983). Recently, acetyl-CoA carboxylase from developing soyabean seeds has been purified and characterized (Charles and Cherry, 1986). A subunit molecular weight of 240 000 was determined.

The second substrate, most frequently 4-coumaroyl-CoA, is supplied via the hydroxycinnamic acid pathway. A great number of phenolic metabolites in plants are derived from hydroxycinnamic acids (Fig. 11.2). In many plants, the individual pathways leading to specific classes of phenolics can be stimulated by exogenous factors. Thus, UV irradiation induces flavonoid glycoside synthesis in parsley cell cultures (see Ebel and Hahlbrock, 1982), whereas furanocoumarin phytoalexins (which are derived from hydroxycinnamic acids by a still unknown reaction sequence) are formed on treatment of parsley cells with

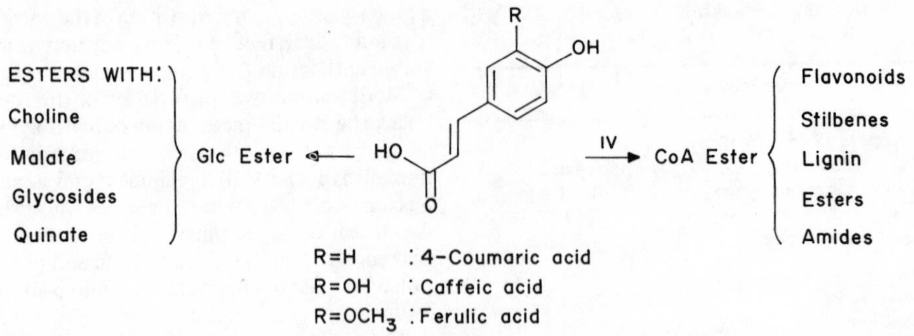

ESTERS WITH: Choline, Malate, Glycosides, Quinate } Glc Ester ← HO(C=O)–CH=CH–(3-R-4-OH-phenyl) —IV→ CoA Ester { Flavonoids, Stilbenes, Lignin, Esters, Amides

R=H : 4-Coumaric acid
R=OH : Caffeic acid
R=OCH_3 : Ferulic acid

Fig. 11.2 Scheme illustrating the central position of activated hydroxycinnamic acids, i.e. 1-*O*-glucose and CoA esters, in the biosynthesis of various phenylpropanoid matabolites.

fungal cell wall preparations (so-called elicitors) (Tietjen *et al.*, 1983; Kombrink and Hahlbrock, 1986). In soybean and many other legumes, the same elicitors give rise to the formation of isoflavonoid phytoalexins (see Grisebach *et al.*, 1985; see also Chapter 5), and in *Arachis* of stilbenoids (Fritzemeir and Kindl, 1981).

Evidence is increasing that isoenzymes may be involved in the separate pathways of phenylpropanoid metabolism. Different enzyme forms have been frequently reported for hydroxycinnamate: CoA ligase (see Ebel and Hahlbrock, 1982) and, very recently, for phenylalanine ammonia-lyase from *Phaseolus vulgaris* cell cultures (Bolwell *et al.*, 1985). Moreover, during molecular genetic studies with this plant system, several different genomic clones have been detected for the latter enzyme (Lamb, 1986).

Phenylalanine ammonia-lyase, which channels the shikimate-derived L-phenylalanine into the various phenylpropanoid pathways (Table 11.1, Figs. 11.1 and 11.2), is one of the best studied enzymes in phenolic metabolism. The reaction to *trans*-cinnamate proceeds via elimination of ammonia from L-phenylalanine (see Ebel and Hahlbrock, 1982). The properties, as well as the various possible functions, of this enzyme in plants have been reviewed by Hanson and Havir (1981) and Jones (1984). Phenylalanine ammonia-lyase has been purified to apparent homogeneity from various sources (Hanson and Havir, 1981). More recently, a homogeneous enzyme was prepared from tulip anthers in a single step by immunoaffinity chromatography (Kehrel and Wiermann, 1985). A molecular weight of 320 000 was determined for the native enzyme, and 83 000 for the subunit. These values are consistent with others reported (Hanson and Havir, 1981). Similar data have been reported for the enzyme from buckwheat (Belunis and Hrazdina, 1986).

An *in vivo* test for phenylalanine ammonia-lyase (which is based on the release of label from the *pro*-3*S*-position of L-[2,3-^{3}H]phenylalanine) was earlier reported (Amrhein *et al.*, 1976). It was later found, however, that a specific release of label is indeed observed with buckwheat hypocotyl segments but not with leaf discs of either buckwheat or sunflower (Holländer and Amrhein, 1981). This technique is not, therefore, of general applicability.

Several studies of the kinetic properties of phenylalanine ammonia-lyase showed apparent co-operativity between enzyme and substrate (see Hanson and Havir, 1981), but the enzyme from mustard cotyledons (Gupta and Acton, 1979) as well as those of microbial origin (Hodgins, 1971) exhibited normal Michaelis–Menten characteristics with a single K_m value. Similar results have recently been obtained with four partially purified forms of the *Phaseolus* enzyme, separated by chromatofocusing (Bolwell *et al.*, 1985). In contrast to the mixture of all four forms, which exhibits apparent negative rate co-operativity with respect to the substrate, the individual activities display normal Michaelis–Menten kinetics. The presence of multiple phenylalanine ammonia-lyase forms with identical molecular weight but with different pI is also reflected in two-dimensional gel electrophoresis of immunoprecipitated ^{35}S-labelled enzyme *in vivo* and *in vitro* from *Phaseolus* (Bolwell *et al.*, 1985; Bell *et al.*, 1986), parsley (Kuhn *et al.*, 1984) and soybean (Grab *et al.*, 1985).

The second enzyme of hydroxycinnamic acid metabolism, cinnamate 4-hydroxylase, catalyses the 4-hydroxylation of *trans*-cinnamate to *trans*-4-coumarate using molecular oxygen and NADPH (Table 11.1, Fig. 11.1). The reaction mechanism involves a hydrogen shift from position 4 to 3, known as an NIH shift. The enzyme system is located at the endoplasmic membrane and consists essentially of a cytochrome *P*-450 and NADPH–cytochrome *P*-450 (cytochrome *c*) reductase. The involvement of cytochrome *P*-450 has been frequently substantiated by CO inhibition, which is reversible by irradiation with light at 450 nm. Cytochrome *c* inhibits the reaction by competing for the electrons supplied by the reductase. Although NADH alone cannot mediate the reaction, simultaneous incubation with NADPH leads to enhanced activities. This synergistic effect indicates an association of the enzyme with cytochrome b_5 and NADH–cytochrome *b* reductase (see West, 1980; Peterson and Prough, 1986).

Cytochrome *P*-450 inhibitors, such as ketoconazole,

tetcyclacis, SKF 525-A, metyrapon and ancymidol (Schwinn, 1984; Coulson *et al.*, 1984), also inhibit cinnamate 4-hydroxylase. Different inhibition characteristics of this hydroxylase from various sources have led to the postulation of the existence of different enzyme forms (Fritsch *et al.*, 1986). A specific inactivation of cinnamate 4-hydroxylase was reported with 1-aminobenzotriazole which is thought to be a suicide inhibitor of this enzyme (Reichhart *et al.*, 1982).

Cinnamate 4-hydroxylase has frequently been used as a reference enzyme in studies with other cytochrome *P*-450-dependent mono-oxygenases (Hagmann *et al.*, 1983; Maule and Ride, 1983; Grand, 1984; Heller and Kühnl, 1985; Kochs and Grisebach, 1986). Recently, cinnamate 4-hydroxylase activity was successfully reconstituted from purified NADPH–cytochrome *P*-450 (cytochrome *c*) reductase and partially purified cytochrome *P*-450, both from Jerusalem artichoke, in the presence of dilauroyl phosphatidylcholine as lipid phase (Benveniste *et al.*, 1986).

All lines of evidence suggest that 4-coumarate is the main or even exclusive physiological precursor for flavonoids in many plants. In a few cases, however, caffeate may be utilized instead (see Section 11.4.1). Formation of caffeic acid derivatives from 4-coumarate substrates has been observed with three different enzyme systems: phenolases, flavin-dependent mono-oxygenases and a cytochrome *P*-450-dependent mono-oxygenase. Phenolases, ubiquitous in plants and quite unspecific towards their phenolic substrate, are, however, usually involved in catabolic rather than biosynthetic pathways. Such enzymes usually utilize ascorbate as reductant (Butt and Lamb, 1981).

An enzyme catalysing the 3-hydroxylation of 4-coumaroyl-CoA has been described from *Silene dioica* flowers (Kamsteeg *et al.*, 1981). It requires NADPH and FAD for activity. A similar activity has been reported from particulate preparations of potato tubers (Boniwell and Butt, 1986) which mediates the 3-hydroxylation of 4-coumarate in the presence of NADH/NADPH and FAD/FMN. However, the inhibition by KCN and diethyldithiocarbamate and a rather low substrate specificity suggest that this enzyme is a phenolase, even though it does not use ascorbate as a reducing cofactor.

Recently, microsomal preparations from elicitor-induced parsley cell suspension cultures were found to

Table 11.2 Hydroxycinnamate:CoA ligases from various sources

		Substrate specificity: Relative rate of conversion with 4-coumarate = 100 (app. K_m value × 10^6 mol l^{-1})					
Source of enzyme	*10^{-3} × Molecular weight*	*4-Coumarate*	*Caffeate*	*Ferulate*	*5-Hydroxy-ferulate*	*Sinapate*	*Reference*
*Avena** Leaf							Knogge *et al.* (1981a)
Erythrina crista-galli[§] Xylem	40	100[†]	34	71	12	34	Kutsuki *et al.* (1982)
Picea abies Cambial sap	63	100(11)[‡]	91	106(10)	19	0	Lüderitz *et al.* (1982)
Populus × euramericana Stem							Grand *et al.* (1983)
Isoenzyme I		100(4.2)	0	42(2)	50(1.7)	25(2.2)	
Isoenzyme II		100(1)	0	158(1.1)	0	0	
Isoenzyme III		100(1.3)	171(1)	143(1.5)	0	0	
Salix babylonica Stem	57	100(31)	43(4.7)	61(46)	?[‖]	?[‖]	Feutry and Letouzé (1984)
Triticum Leaf		100	82	77	34	0	Maule and Ride (1983)

*Age- and tissue-dependent substrate specificity.
[†]Other acids tested: isoferulic, 71; 3,4-dimethoxycinnamic, 60; 4-methoxycinnamic, 49.
[‡]Other acid tested: cinnamic, 15.
[§]Substrate specificities for crude enzyme preparations of another 13 woody species reported (4 gymnosperms, 3 monocots, 8 dicots).
[‖]Data not reported.

catalyse the formation of *trans*-5-*O*-caffeoylshikimate from *trans*-5-*O*-(4-coumaroyl)shikimate (Heller and Kühnl, 1985). The reaction bears all the signs of a cytochrome *P*-450-dependent mono-oxygenase, such as inhibition by carbon monoxide with reversion by light, inhibition by cytochrome *c*, and tetcyclacis (Fritsch *et al.*, 1986), and a pronounced synergistic effect between NADPH and NADH.

Activation of hydroxycinnamic acids can be achieved in two ways, either by a glucosyl transfer or by formation of a CoA ester (Fig. 11.2). The glucosyl esters are the intermediates for a small group of hydroxycinnamoyl derivatives with choline (Gräwe and Strack, 1986), 1-*O*-(hydroxycinnamoyl)glucose (Dahlbender and Strack, 1986) and malate (Strack, 1982). The CoA esters are the central substrates for a large variety of enzymes leading to flavonoids (see Ebel and Hahlbrock, 1982), stilbenes (see Kindl, 1985), lignin precursors and several ester and amide derivatives (see Gross, 1985).

The enzyme responsible for CoA ester synthesis is hydroxycinnamate:CoA ligase (Table 11.1, Figs. 11.1 and 11.2), which is often called 4-coumarate:CoA ligase, referring to its preferred substrate. The reaction depends strictly on ATP as a co-substrate and this cannot be substituted for by any other nucleoside triphosphate, and it requires Mg^{2+} as cofactor. The reaction proceeds via an acyl-AMP intermediate which then reacts with CoA-SH to form the thioester (see Ebel and Hahlbrock, 1982). HPLC has been used to identify the products, *trans*-4-coumaroyl-CoA and AMP (Knogge *et al.*, 1981b). For most enzyme preparations, a molecular weight of approximately 55 000 has been observed, but higher or lower values have also been reported (see Ebel and Hahlbrock, 1982). Reinvestigation of the parsley enzyme revealed different molecular weights depending on whether they were determined by gel chromatography on Sephadex G-200 (54 000) or by SDS–polyacrylamide gel electrophoresis (60 000) (Ragg *et al.*, 1981). Similar relations were obtained for the ligase from *Picea abies*, 57 000 by gel chromatography on Sephadex G-100 and 63 000 by gel electrophoresis (Lüderitz *et al.*, 1982).

Isoenzymes for the CoA ligase reaction have been isolated from various plants by the combination of ion-exchange chromatography and disc-gel electrophoresis (see Ebel and Hahlbrock, 1982). More recently, three distinct CoA ligases were separated by chromatofocusing from crude extracts of *Populus* x *euramericana* stems. The three activities showed different substrate specificities and tissue distribution (Grand *et al.*, 1983) and it was assumed that these isoenzymes control the monomeric composition of lignins rather than being involved in different biosynthetic pathways. The relationship of the various CoA ligase isoenzymes to specific pathways of the phenylpropanoid metabolism is still an open question. The data for various crude, partially purified and homogeneous ligase preparations are summarized in Table 11.2.

11.4 INDIVIDUAL STEPS TO FLAVONOID CLASSES

11.4.1 Chalcone synthase

The formation of chalcone is catalysed by the well-characterized enzyme, chalcone synthase (see Ebel and Hahlbrock, 1982). Since chalcone is the central C_{15} intermediate for all flavonoids, chalcone synthase can be regarded as the key enzyme of flavonoid biosynthesis. Moreover, the interesting mode of regulation of chalcone synthase makes this reaction and the genes coding for this enzyme in different plants an attractive system for studying the regulation of gene expression in plants.

The enzyme has no cofactor requirements. Substrates for chalcone formation are malonyl-CoA and the CoA ester of a hydroxycinnamic acid. Short-chain fatty acid CoA esters can also prime the reaction (Schüz *et al.*, 1983).

For all chalcone synthases tested so far, 4-coumaroyl-CoA is the best substrate. The immediate product formed is naringenin chalcone. For the enzyme from *Matthiola incana*, *Antirrhinum majus* (Spribille and Forkmann, 1981, 1982a) and *Phaseolus vulgaris* (Whitehead and Dixon, 1983), 4-coumaroyl-CoA is the exclusive substrate. Chalcone synthases from other plants studied also accept caffeoyl-CoA or even feruloyl-CoA (see Fig. 11.2). In these cases, eriodictyol chalcone and homoeriodictyol chalcone, respectively, are the immediate products (see Ebel and Hahlbrock, 1982). Recent examples where caffeoyl-CoA can act as substrate for chalcone synthase were reported from petals of *Dianthus caryophyllus* (Spribille and Forkmann, 1982b), *Verbena hybrida* (Stotz *et al.*, 1984), *Callistephus chinensis*, bracts of *Helichrysum bracteatum* (Spribille and Forkmann, unpublished), flowers and cell cultures from *Daucus carota* (Hinderer *et al.*, 1983; Hinderer and Seitz, 1985) and soybean cell cultures (Welle, 1986), and where feruloyl-CoA can act as substrate from petals of *Cosmos sulphureus* (Sütfeld and Wiermann, 1981), tulip anthers (Sütfeld *et al.*, 1978; Sütfeld and Wiermann, 1981), spinach leaves (Beerhues and Wiermann, 1985) and buckwheat hypocotyls (Hrazdina *et al.*, 1986).

There is much evidence that 4-coumaroyl-CoA is the main physiological substrate for chalcone formation. Up to now, evidence for an *in vivo* use of caffeoyl-CoA in addition to 4-coumaroyl-CoA was only provided for chalcone synthases from *Verbena* (Stotz *et al.*, 1984) and tulip (Sütfeld *et al.*, 1978; Sütfeld and Wiermann, 1981). Flowers of scarlet *Verbena hybrida* varieties, albeit exhibiting no measurable flavonoid 3′-hydroxylase activity, contain cyanidin as well as pelargonidin derivatives (see Fig. 11.5 and Section 11.5.1) (Stotz *et al.*, 1984). In *Dianthus*, caffeoyl-CoA can be ruled out as the physiological substrate. Although caffeoyl-CoA serves as substrate for *in vitro* chalcone formation, 4-coumaroyl-CoA is exclusively used in competition experiments. This agrees with chemogenetic studies on *Dianthus* and with the occurrence of a flavonoid-specific 3′-hydroxylase (Spribille and Forkmann, 1982b). As will be discussed in Section 11.5.1, genetic evidence and the presence of specific flavonoid B-ring hydroxylases and methyltransferases indicate that B-ring substitutions occur at the C_{15} level.

In chalcone synthase assays containing 2-mercaptoethanol, side products of the reaction are usually observed (see Ebel and Hahlbrock, 1982; Schüz *et al.*, 1983). The use of oxygen-free buffers and ascorbate essentially prevents formation of these products (Britsch and Grisebach, 1985). Moreover, with 4-coumaroyl-CoA as substrate, these conditions lead to a considerable increase in the formation of naringenin, which is one of the most important substrates for enzymic studies in flavonoid biosynthesis.

Although two and more genes for chalcone synthase have been found in some plants (see below), two different chalcone synthase activities have, as yet, only been demonstrated for spinach (Beerhues and Wierman, 1985). The activities could be separated by DEAE ion-exchange chromatography and chromatofocusing. The proteins have the same molecular weight (subunit 45 000), exhibit only slight differences in substrate specificity, but differ in their pI values (4.65 and 4.30). Chalcone synthase has also been purified from carrot (Ozeki *et al.*, 1985), *Phaseolus vulgaris* (Whitehead and Dixon, 1983), buckwheat (Hrazdina *et al.*, 1986) and tulip (Kehrel and Wiermann, 1985). Molecular weights for the native enzymes range between 77 000 and 85 000.

Chalcone synthases from several plants are more or less strongly inhibited by the flavanone formed in the reaction (see Ebel and Hahlbrock, 1982; Hinderer and Seitz, 1985). Interestingly, the enzyme from *Phaseolus vulgaris* is strongly inhibited by the isoflavanone kievitone. A possible role for this effect may be in protecting the cell from phytotoxic concentrations of this product (Whitehead and Dixon, 1983). 3′-Dephospho-CoA derivatives, which are formed by action of a specific 3′-nucleotidase, are also inhibitory *in vitro* (Hinderer and Seitz, 1986). In carrot cell cultures such a nucleotidase activity is induced on addition of gibberellic acid to the culture medium which leads to suppression of anthocyanin formation (Hinderer *et al.*, 1984).

Chalcone synthase activity is genetically controlled in *Matthiola* (gene *f*), *Antirrhinum* (gene *niv*) and *Zea mays* (gene *c2*). Genotypes with recessive alleles of the genes completely lack chalcone synthase activity (Spribille and Forkmann, 1981, 1982a; Dooner, 1983). In maize, three further genes have been identified that influence chalcone synthase activity (Dooner, 1983). Control by transposable elements has also been observed for the *nivea* locus of *Antirrhinum* (Bonas *et al.*, 1984; Upadhyaya *et al.*, 1985; Sommer *et al.*, 1985). Three transposable elements (*Tam1*, *Tam2* and *Tam3*), residing at this locus, have been identified. The unstable alleles *niv-53* (*Tam1*) and *niv-98* (*Tam3*), both leading to spots and flakes with anthocyanin on a colourless background, and the stable white flowering mutant with the allele *niv-44* (*Tam2*) result from these insertions.

A conditional mutant was characterized in *Petunia* (cv. 'Red Star'). Here, with increasing light intensity, the flowers show white sectors, where the gene for chalcone synthase is not expressed. Neither chalcone synthase protein nor translationally active chalcone synthase mRNA was found to be present (Mol *et al.*, 1983). Using a parsley chalcone synthase antiserum a dramatic increase in chalcone synthase protein in petals of a wild-type line of *Matthiola* could be demonstrated in the course of flower development. In the recessive mutant (*ff*), a reduced and nearly constant level of inactive enzyme protein was observed (Rall and Hemleben, 1984).

Chalcone synthase cDNA and genomic clones have recently been described (Reimold *et al.*, 1983; Bonas *et al.*, 1984; Reif *et al.*, 1985; Wienand *et al.*, 1986). Molecular genetic methods have revealed the existence of two genes for chalcone synthase in *Petunia* (Reif *et al.*, 1985) and five genes in *Phaseolus* (Lamb, 1986).

The structure of the *Antirrhinum* chalcone synthase gene (*niv*) has recently been elucidated (Sommer and Seadler, 1986). The gene codes for a protein subunit with a molecular weight of 42 665. In the case of parsley, sequence analysis of the cDNA gives an enzyme subunit with a molecular weight of 43 682 (Reimold *et al.*, 1983).

In cell cultures of *Picea excelsa*, capable of forming both stilbenes and flavonoids, stilbene synthase and chalcone synthase can be differentiated. Although both enzymes use malonyl-CoA and 4-coumaroyl-CoA as substrates and exhibit similar molecular properties, they are clearly two different enzymes (Rolfs and Kindl, 1984).

In spite of the extensive progress in assaying chalcone synthase, *in vitro* formation of 6′-deoxychalcones or of the respective 5-deoxyflavanones has not yet been demonstrated. In all cases tested so far, only the formation of phloroglucinol-type chalcones or flavanones has been observed.

11.4.2 Chalcone isomerase

With only a few exceptions, the various flavonoid classes (e.g. flavones, flavonols, anthocyanins and isoflavones) are derived from a flavanone intermediate (Fig. 11.1). Flavanones are formed from chalcones by isomerization. In the case of a 2′, 4′, 6′-trihydroxylation pattern (phloroglucinol type), this reaction proceeds spontaneously under physiological conditions, leading to a racemic product. This reaction is considerably repressed by elevated protein concentrations (Mol *et al.*, 1985), which might explain the considerable stability of chalcones in tulip anthers and in pollen of various plants, provided chalcone isomerase is absent (Chmiel *et al.*, 1983).

Results to date suggest that only flavanones with an (*S*)-

configuration act as substrates for the enzymes of the flavonoid pathway. Thus, highly efficient chalcone isomerases exist which catalyse the stereospecific formation of (2*S*)-flavanones from the respective chalcone substrates (Heller *et al.*, 1979). In contrast to the chemical *anti*-addition, the enzymic ring-closure reaction is an overall *syn*-addition to the *E*-double bond (see Ebel and Hahlbrock, 1982). On the basis of the pH profile of the reaction, a mechanism with participation of an amino group (Dixon *et al.*, 1982) or a histidyl residue (Boland and Wong, 1979) has been proposed.

Two types of chalcone isomerase are known, which are classified with regard to their substrate specificity. In plants containing 5-deoxy and 5-hydroxyflavonoids, e.g. *Phaseolus vulgaris* (Dixon *et al.*, 1982) and *Cosmos sulphureus* (Chmiel *et al.*, 1983), the enzyme is able to isomerize both 2′,4′-dihydroxy- (resorcinol-type) and 2′,4′,6′-trihydroxy- (phloroglucinol-type) chalcones. In all cases, the enzyme exhibits a higher specificity (V_{max}/K_m) towards the phloroglucinol-based substrate. In plants devoid of 5-deoxyflavonoids, e.g. tulip (Chmiel *et al.*, 1983), *Callistephus* (Kuhn *et al.*, 1978) and *Dianthus* (Forkmann and Dangelmayr, 1980), only phloroglucinol-type chalcones are substrates for the enzyme. Two isoenzymes are present in *Petunia* (van Weely *et al.*, 1983). They are, however, selectively expressed in flower and anther tissue. The proteins differ distinctly in molecular weight (62 500, and 44 000 respectively) and isoelectric point (5.3 and 4.5, respectively). Molecular weight and pI value reported for both the tulip and the *Cosmos* enzymes are 31 000 and 5.0 respectively (Chmiel *et al.*, 1983).

The *Phaseolus* isomerase has been purified to homogeneity by a final application of a chromatofocusing step (Robbins and Dixon, 1984). The homogeneous protein has a subunit molecular weight of 27 000 and an apparent pI of 5.0. No isoenzymes have been observed. Both 5-hydroxy- and 5-deoxy-flavonoids and isoflavonoids act as competitive inhibitors, the most potent being kievitone (K_i 9.2 μM) and coumestrol (K_i 2.5 μM) (Dixon *et al.*, 1982).

Evidence exists in several plants that chalcone isomerase activity is under genetic control (see Ebel and Hahlbrock, 1982). For instance, mutants of *Dianthus caryophyllus* with recessive alleles of the gene *I* lack isomerase activity and accumulate chalcone (Forkmann and Dangelmayr, 1980). Spontaneous isomerization of some chalcone to flavanone *in vivo* may lead to the formation of moderate amounts of anthocyanins leading to an orange colour in the flowers.

11.4.3 Isoflavone formation

The central step in isoflavone formation is the 2,3-migration of the aryl side chain of a flavanone/chalcone intermediate. An enzymic activity catalysing this transformation was recently found in microsomal preparations from elicitor-challenged soybean cell suspension cultures and etiolated seedlings. It transforms (2*S*)-naringenin into genistein and (2*S*)-liquiritigenin into daidzein (Hagmann and Grisebach, 1984) (see Fig. 11.3).

It was found that two enzymic steps are involved in the transformation of naringenin to genistein. The first step comprises oxidation and rearrangement of naringenin to 2-hydroxy-2,3-dihydrogenistein. This step is strictly

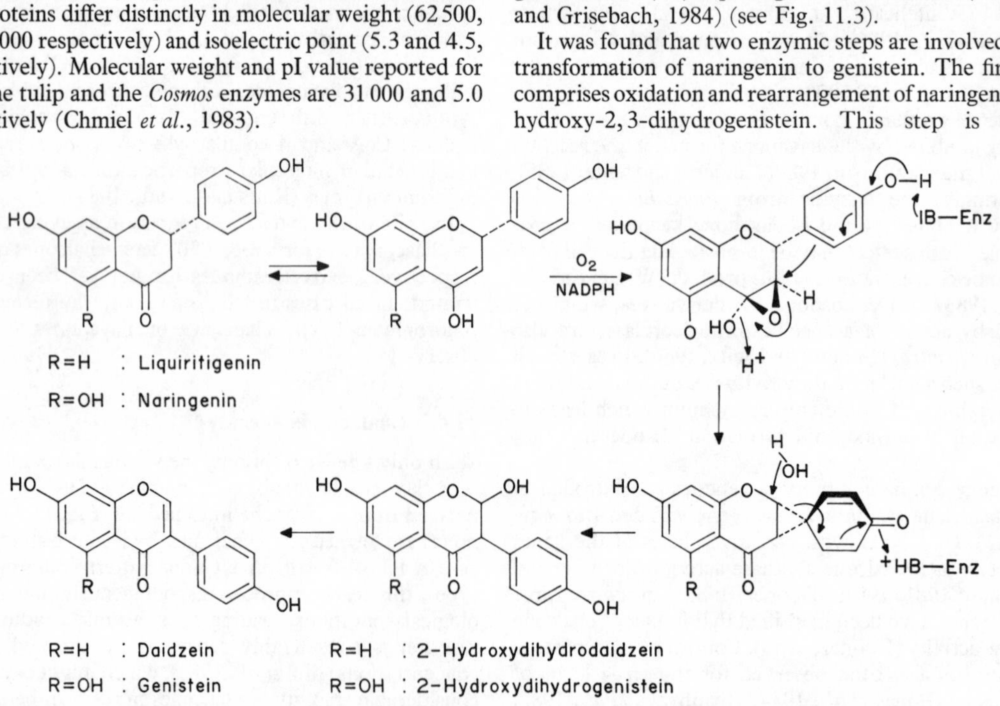

Fig. 11.3 Proposed reaction mechanism of isoflavone formation from a flavanone (suggested by L. Crombie, Nottingham). The reaction proceeds via a stable 2-hydroxyisoflavanone intermediate. Two enzymes act in sequence, a particulate cytochrome *P*-450-dependent mono-oxygenase and a soluble dehydratase.

dependent on NADPH and molecular oxygen. Inhibition by cytochrome *c*, by carbon monoxide and its reversal by light as well as inhibition with typical cytochrome *P*-450 inhibitors indicate the participation of a cytochrome *P*-450-dependent enzyme system (Kochs and Grisebach, 1986; Fritsch *et al.*, 1986). The structure of this intermediate is based on ultraviolet (λ_{max} 288 nm), mass spectrometric (chemical ionization: m/z 288 [M^-]; electron impact: m/z 289 [$M^+ + 1$]) and chemical data. Under acidic conditions, the compound is non-enzymically transformed to genistein. On a Percoll gradient, the mono-oxygenase-containing membrane fraction coincides with marker enzymes for the endoplasmic reticulum and with the position of cytochrome *P*-450. The second enzyme which catalyses the elimination of water from the 2-hydroxyisoflavanone is in the soluble fraction and has not yet been characterized (Kochs and Grisebach, 1986).

For the following reasons, (2*S*)-flavanones and not the chalcones are very probably the actual substrates: (a) a stereospecific incorporation of (2*S*)-naringenin into biochanin A (5, 7-dihydroxy-4′-methoxyisoflavone) has been described earlier (Patschke *et al.*, 1966); (b) only the (2*S*)- but not the (2*R*)-enantiomer acts as a substrate *in vitro* (Kochs and Grisebach, 1986); (c) the equilibrium of 4, 2′, 4′, 6′-tetrahydroxychalcone is at least 1000:1 in favour of the flavanone (Boland and Wong, 1975).

The hypothetical pathway from (2*S*)-flavanone to isoflavone as shown in Fig. 11.3 is consistent with the participation of NADPH and molecular oxygen. The sequence of events may be initiated by an epoxidation step analogous to that described for cinnamate 4-hydroxylase (see Butt and Lamb, 1981). Protonation and subsequent cleavage of the epoxide would render a positive charge to the B-ring. Keto–enol tautomerism, as indicated by proton exchange at C-3 in flavanones (Grisebach and Zilg, 1968), allows homoallylic interaction between C-3 and C-1′, and rearrangement of the structure then takes place. Addition of hydroxyl ion to C-2 leads to the 2-hydroxyisoflavanone intermediate which is transformed to the isoflavone by elimination of a molecule of water.

An additional product observed in the complete assay was identified as 2′-hydroxygenistein which indicates the presence of a 2′-hydroxylase in soybean (Kochs and Grisebach, 1986). 2′-Hydroxyisoflavones are intermediates in the biosynthesis of pterocarpans (see Chapter 5). The observation that these enzyme activities are increased severalfold in elicitor-treated soybean cell cultures and seedlings suggests that they play a role in the regulation of pterocarpan phytoalexins in this plant (see Grisebach *et al.*, 1985).

11.4.4 Flavone formation

In vitro conversion of flavanones to flavones was first observed in parsley plants (Sutter *et al.*, 1975). The reaction has been studied in more detail in parsley cell suspension cultures (Britsch *et al.*, 1981) and in *Antirrhinum* flowers (Stotz and Forkmann, 1981).

The parsley enzyme requires 2-oxoglutarate, Fe^{2+} and possibly ascorbate as cofactors. This cofactor requirement would classify it as belonging to the 2-oxoglutarate-dependent dioxygenases (Abbot and Udenfriend, 1974). Ascorbate stimulates this and other 2-oxoglutarate-dependent dioxygenases involved in the flavonoid pathway and, in addition, exhibits a remarkable stabilizing effect on the enzyme activity (Britsch and Grisebach, 1986).

In *Antirrhinum* (Stotz and Forkmann, 1981) and in flavone-producing flowers of a range of other plants including *Verbena* (Stotz *et al.*, 1984), *Dahlia*, *Streptocarpus* and *Zinnia* (Forkmann and Stotz, 1984), desaturation of flavanone to flavone is catalysed by a microsomal enzyme requiring NADPH as cofactor. Although mutants recessive with respect to flavone formation are not, as yet, known, evidence for the enzyme being involved in flavone formation is provided by the fact that flowers of plants that naturally lack flavones are also devoid of this NADPH-dependent microsomal enzyme activity. A similar enzyme activity was recently observed in soybean cell cultures treated with 0.4 M glucose (Kochs and Grisebach, 1987).

Both the parsley and the flower enzyme catalyse the reaction from (2*S*)-naringenin to apigenin and from (2*S*)-eriodictyol to luteolin. 5, 7, 3′, 4′, 5′-Pentahydroxyflavanone, tested with microsomal preparations from *Verbena* flowers, was found to be a poor substrate. It is of interest that the same transformation is mediated by two different enzyme systems. This fact could be of significance in considering the evolution of flavonoid biosynthesis in plants.

The mechanism of double-bond formation is still unclear. It has been suggested that a 2-hydroxyflavanone is formed in the first step, and that water is then eliminated via a dehydratase (Britsch *et al.*, 1981; Stotz and Forkmann, 1981). However, no such 2-hydroxy intermediate has, as yet, been observed even with a nearly homogeneous enzyme protein (Britsch, 1986). On the other hand, 2-hydroxyflavanones certainly exist as plant metabolites and indeed are the substrates in *C*-glycosylflavone formation (Kerscher and Franz, 1987; see Section 11.5.5).

11.4.5 Flavanone 3-hydroxylase

Clear evidence for *in vitro* hydroxylation of naringenin and eriodictyol in the 3 position to the respective dihydroflavonols has been obtained with *Matthiola* flower extracts (Forkmann *et al.*, 1980). The enzyme occurs in the soluble fraction and requires 2-oxoglutarate, Fe^{2+} and, for full activity, ascorbate as cofactors. Accordingly, like the parsley flavone synthase, it is a 2-oxoglutarate-dependent dioxygenase (Abbott and Udenfriend, 1974).

Flavanone 3-hydroxylase has also been detected in

enzyme preparations from UV-irradiated parsley cell suspension cultures (Britsch *et al.*, 1981) and in flower extracts from defined genotypes of *Antirrhinum* (Forkmann and Stotz, 1981), and *Petunia* (Froemel *et al.*, 1985) as well as from flowers of *Dahlia, Streptocarpus, Zinnia* and *Verbena* (Forkmann and Stotz, 1984).

Although flavanone 3-hydroxylase is rather unstable under normal conditions, the enzyme was recently purified to apparent homogeneity from young flowers of a red *Petunia* cultivar (Britsch and Grisebach, 1986). The molecular weight determined for the native enzyme is about 74 000. The stoichiometric formation of dihydroflavonol and carbon dioxide, the stimulation of the reaction by catalase and the strong inhibition by pyridine 2,4- and 2,5-dicarboxylate confirm the classification of flavanone 3-hydroxylase as a 2-oxoglutarate-dependent dioxygenase. (2*S*)-Naringenin, but not the (2*R*)-enantiomer, is a substrate for the enzyme. The product formed was unequivocally identified as (2*R*,3*R*)-dihydrokaempferol. (2*S*)-Eriodictyol is converted to (2*R*,3*R*)-dihydroquercetin. 5,7,3′,4′,5′-Pentahydroxyflavanone does not serve as a substrate in *Petunia*. Enzyme extracts from *Verbena*, however, catalyse the 3-hydroxylation of pentahydroxyflavanone to dihydromyricetin at a considerable rate (Forkmann and Stotz, 1984).

With the exception of *Matthiola*, where mutants involving 3-hydroxylation are not yet known, genetic control of flavanone 3-hydroxylase activity has been demonstrated for white flowering genotypes of all plants mentioned above. Enzyme activity was greatly reduced in flowers of genetically defined *Petunia* lines with the multiple allele *an3-1* at the *An3* locus which is involved in 3-hydroxylation (Froemel *et al.*, 1985). Moreover, a two-element system causing instability at the *An3* locus has been reported (Gerats *et al.*, 1985), but the unstable mutants and their revertants have not as yet been characterized enzymically. Very recently, a similar instability in *Verbena* and *Dianthus* was found to be correlated with flavanone 3-hydroxylase activity (Koch and Forkmann, unpublished).

11.4.6 Flavonol formation

Enzymic conversion of dihydroflavonols to flavonols was first observed with enzyme preparations from parsley cell suspension cultures (Britsch *et al.*, 1981). As in flavone formation in parsley, synthesis of flavonols was found to be catalysed by a soluble 2-oxoglutarate-dependent dioxygenase.

Flavonol synthesis most probably proceeds via a 2-hydroxy intermediate with subsequent dehydration, similar to the scheme proposed for the conversion of flavanones to flavones. Such intermediates (2-hydroxydihydrokaempferol and 2-hydroxydihydroquercetin) have been shown spontaneously to eliminate water, giving rise to the respective flavonols (Hauteville *et al.*, 1979).

Flavonol synthase has also been demonstrated in flower extracts from *Matthiola* (Spribille and Forkmann, 1984) and *Petunia* (Forkmann *et al.*, 1986). As in parsley, flavonol formation in these flowers is catalysed by a soluble 2-oxoglutarate-dependent dioxygenase. The enzyme activity in flowers at various developmental stages is clearly related to their flavonol content. Dihydrokaempferol and dihydroquercetin are substrates for the *in vitro* formation of kaempferol and quercetin, respectively. Dihydromyricetin was tested with *Petunia* flower extracts and found to be a poor substrate. This agrees with the fact that myricetin only occurs in these flowers in minor amounts (Gerats *et al.*, 1982).

In *Petunia* mutants with recessive alleles (*f1f1*) of the gene *F1*, flavonol formation is greatly reduced when compared to the wild-type plant. This situation is reflected in the high flavonol synthase activity in the presence of the wild-type allele *F1*, and by low activity in genotypes with recessive alleles (Forkmann *et al.*, 1986).

11.4.7 Dihydroflavonol 4-reductase

The first evidence for a NADPH-dependent dihydroflavonol 4-reductase involved in proanthocyanidin biosynthesis was obtained by Stafford and Lester (1982) with enzyme preparations from *Pseudotsuga menziesii* cell suspension cultures. In the presence of NADPH, but not NADH or ascorbate, (2*R*,3*R*)-dihydroquercetin is transformed to (2*R*,3*S*,4*S*)-leucocyanidin, the respective flavan-3,4-*cis*-diol (Stafford and Lester, 1984).

With enzyme preparations from *Ginkgo biloba*, as well as *Pseudotsuga* cell cultures, *cis*-leucocyanidin and *cis*-leucodelphinidin are formed from the respective dihydroflavonols (Stafford and Lester, 1985). The products have been identified on the basis of their chemical behaviour: (i) formation of anthocyanidin on treatment with a mineral acid; (ii) formation of a vanillin adduct; (iii) formation of a proanthocyanidin dimer on addition of catechin; and (iv) chromatographic comparison with leucoanthocyanidins prepared by borohydride reduction of (2*R*,3*R*)-dihydroflavonols. Under controlled conditions, the chemical reduction yields exclusively 3,4-*trans*-diols which can be isomerized to the 3,4-*cis*-form with mild acid. This epimerization reaction has been directly observed with *trans*-leucocyanidin by proton NMR spectroscopy (Stafford *et al.*, 1985). These results have been confirmed by detailed studies of the reductase from barley grains (Kristiansen, 1984, 1986).

Supplementation experiments with white flowering *Matthiola* mutants indicated that flavan-3,4-diols are also intermediates in anthocyanin biosynthesis (see Section 11.4.9). The experiment also suggested that the gene *e* regulates the conversion of dihydroflavonols to flavan-3,4-diols (Heller *et al.*, 1985a). Biochemical studies revealed an enzyme activity in *Matthiola* flower extracts which catalyses, in analogy to *Pseudotsuga* and barley, the stereospecific reduction of (2*R*, 3*R*)-dihydrokaempferol to 3,4-*cis*-leucopelargonidin. Dihydroquercetin and dihydromyricetin were also found to be converted to the respective 3,4-*cis*-diols (for substitution patterns see Fig. 11.5). In agreement with earlier chemogenetic work and with the feeding experiments, the enzyme activity is controlled by the gene *e*. These findings prove the involvement of a dihydroflavonol 4-reductase in anthocyanin biosynthesis (Heller *et al.*, 1985b). Similar results were obtained with *Callistephus chinensis* (gene *F*) and *Dianthus caryophyllus* (gene *A*) (Forkmann, unpublished). The barley *ant18* mutant which also accumulates dihydroquercetin instead of the usual proanthocyanidin derivatives also seems to be blocked in this step (Kristiansen, 1984; von Wettstein *et al.*, 1985).

Using transposable elements as gene tags, the *a1* locus of *Zea mays* (O'Reilly *et al.*, 1985) and the *pallida* locus of *Antirrhinum* (Coen *et al.*, 1986) have been cloned. Moreover, evidence was recently provided that the *in vitro* expressed maize *a1* cDNA catalyses the NADPH-dependent reduction of (+)-dihydroquercetin to the respective 3,4-*cis*-diol (Rohde *et al.*, 1986). Clear homology of both the *a1* and *pallida* locus at the DNA and amino acid level (Schwarz-Sommer *et al.*, 1987; Rohde, 1986) and feeding experiments (Stickland and Harrison, 1974) indicate that the *pallida* locus also concerns dihydroflavonol 4-reductase.

Interestingly, reductases from *Matthiola* and barley are not specific for NADPH as cofactor, exhibiting as they do activities of up to 90% with NADH (Heller *et al.*, 1985b; Kristiansen, 1986). The *Matthiola* and *Pseudotsuga* enzymes show pH optima of 6 and 7.4 respectively. The observation of two optima (pH 6.0 and 7.0) for the barley enzyme, led to the suggestion that tissue-specific isoenzymes exist, one specific for anthocyanin (pH 6.0, *palea-lemma* tissue), the other for proanthocyanidin (pH 7.0, *testa* layer) formation (Kristiansen, 1986).

Very recently, enzymic reduction of (2*S*)-naringenin to apiferol (4,5,7,4′-tetrahydroxyflavan) was demonstrated with crude extracts from *Sinningia* (syn. *Rechsteineria*) *cardinalis* flowers, which contain exclusively 3-deoxyanthocyanins (Stich and Forkmann, unpublished).

11.4.8 Catechin and proanthocyanidin formation

The formation *in vitro* of catechin was first demonstrated with a crude extract from *Pseudotsuga* cell suspension cultures using 2,3-*trans*-3,4-*cis*-leucocyanidin as substrate and NADPH (Stafford and Lester, 1984). A similar reductase activity was later found in *Ginkgo* cell culture extracts (Stafford and Lester, 1985). Both enzymes transform *cis*-leucocyanidin and *cis*-leucodelphinidin to the respective flavan-2,3-*trans*-3-ols, catechin and gallocatechin. When crude extracts which contain both dihydroflavonol and leucoanthocyanidin 4-reductases are used, these products can be directly obtained from dihydroquercetin and dihydromyricetin, respectively (Stafford and Lester, 1985).

A similar two-step reaction has been observed in maturing barley grain extracts, which transform [^{14}C]dihydroquercetin to [^{14}C]catechin (Kristiansen, 1986). When NADH is used instead of NADPH in this incubation, only [^{14}C]-*cis*-leucocyanidin but no [^{14}C]catechin is formed, indicating that the second reaction step is strictly NADPH-dependent. Preliminary experiments also showed that chemically prepared [4-^{3}H]-*cis*-leucocyanidin is transformed to labelled catechin (Kristiansen, 1986).

The postulated condensing enzyme in the synthesis of oligomeric procyanidins, linking e.g. 2,3-*trans*-3,4-*cis*-leucocyanidin to catechin to form procyanidin dimer B3, and adding this dimer to another leucocyanidin monomer to form the procyanidin trimer C2, has not yet been detected (Kristiansen, 1986). This condensation reaction spontaneously occurs by acid- and base-catalysed mechanisms with retention of configuration at both positions 2 and 3 (see Section 11.4.7). This would suggest a carbocation or quinonemethide intermediate as opposed to the flav-3-en-3-ol intermediate postulated earlier (Stafford, 1983; see Chapter 2).

The role of catechin in proanthocyanidin biosynthesis is clearly demonstrated by genetic evidence. The barley mutant *ant19* is essentially free of proanthocyanidins as well as catechin, but anthocyanin biosynthesis is not impaired in the plant (Kristiansen, 1984).

11.4.9 Formation of anthocyanidin 3-*O*-glucoside

In contrast to flavone and flavonol aglycones, anthocyanidins are, if at all, rarely found in nature. The first stable products of anthocyanin biosynthesis are the anthocyanidin 3-*O*-glycosides with glucose as the most common substituent. The enzymes catalysing 3-*O*-glycosylation should therefore not be regarded as modifying enzymes but rather as enzymes more directly involved in anthocyanin formation.

Good evidence that leucoanthocyanidins are precursors of anthocyanin biosynthesis has come from supplementation experiments on the genetically defined white flowering lines 17, 18 and 19 of *Matthiola* (Heller *et al.*, 1985a). The lines 17 (recessive *ee*) and 19 (recessive *gg*) have different genetic blocks between dihydroflavonol and anthocyanin, while line 18 (recessive *ff*) is blocked in the chalcone synthase gene. Administration of dihydroflavonols to white flowers of these lines leads to anthocyanin formation in flowers of line 18, whereas flowers of lines 17 and 19 remain acyanic (Forkmann, 1977a). Supplementation with leucoanthocyanidins, however, initiates anthocyanin formation in flowers of both line 18 and line 17. Flowers of line 19 again remain acyanic (Heller *et al.*, 1985a). Similar results have been obtained with acyanic mutants of *Dianthus* and *Callistephus* (Forkmann, unpublished). Incorporation of [4-^{3}H]leucopelargonidin into pelargonidin derivatives using flowers of *Matthiola* line 18 has further confirmed the role of leucoanthocyanidins in anthocyanin synthesis (Heller *et al.*, 1985a).

Since the 3,4-*cis*-diol is formed by dihydroflavonol 4-reductase from all plants studied so far (see Section 11.4.7), it can be assumed that this compound is the actual precursor for anthocyanin formation. The 3,4-*cis*-diol was also found to be the substrate for catechin synthesis (see Section 11.4.8).

Except for UDP-glucose: flavonoid 3-*O*-glucosyltransferase, which has been extensively studied (see below), enzymes catalysing the further conversion of leucoanthocyanidins to anthocyanins are not yet known. Essentially four reaction steps seem to be involved: a hydroxylation in the 2 position, two dehydrations and the glycosylation of the hydroxyl group in the 3 position. There are several possible sequences (Fig. 11.4). Glycosylation, however, seems to occur at a late stage since neither dihydroflavonols nor leucoanthocyanidins were found to be glucosylated in the 3 position in respective assays with enzyme extracts from *Matthiola* flowers (Heller and Forkmann, unpublished).

Although mutants exist where the biosynthetic pathway between leucoanthocyanidins and anthocyanidin 3-*O*-glycosides is blocked, no accumulation of leucoanthocyanidins or of any of the postulated aglycone intermediates has yet been observed. It can therefore be assumed that these intermediates are as unstable as are leucoanthocyanidins. The first detectable intermediates are pseudobases (see Fig. 11.4) of often highly modified anthocyanins. This indicates that pseudobases may serve as substrates for 3-*O*-glycosylation as well as for modifying enzymes. Interestingly, in *Pisum* flowers, a gene *am* is known which controls the conversion of anthocyanin pseudobases to anthocyanins (Crowden, 1982).

In vitro formation of anthocyanidin 3-*O*-glucosides was first demonstrated with pollen extracts from maize (Larson and Coe, 1968; Larson and Lonergan, 1973). The enzyme catalyses the transfer of glucose from UDP-glucose to the hydroxyl group at the 3 position of anthocyanidins and flavonols. In the meantime, a complex genetic control by the structural gene for this reaction, *Bz*, including the involvement of controlling elements, and by at least seven different regulatory genes of maize has been reported (Larson and Coe, 1977; Dooner and Nelson, 1977, 1979a, 1979b; Dooner 1981, 1983; Klein and Nelson, 1983; Gerats *et al.*, 1983a, 1984a).

Apart from maize, enzyme activity for 3-*O*-glucosylation has been demonstrated in crude preparations

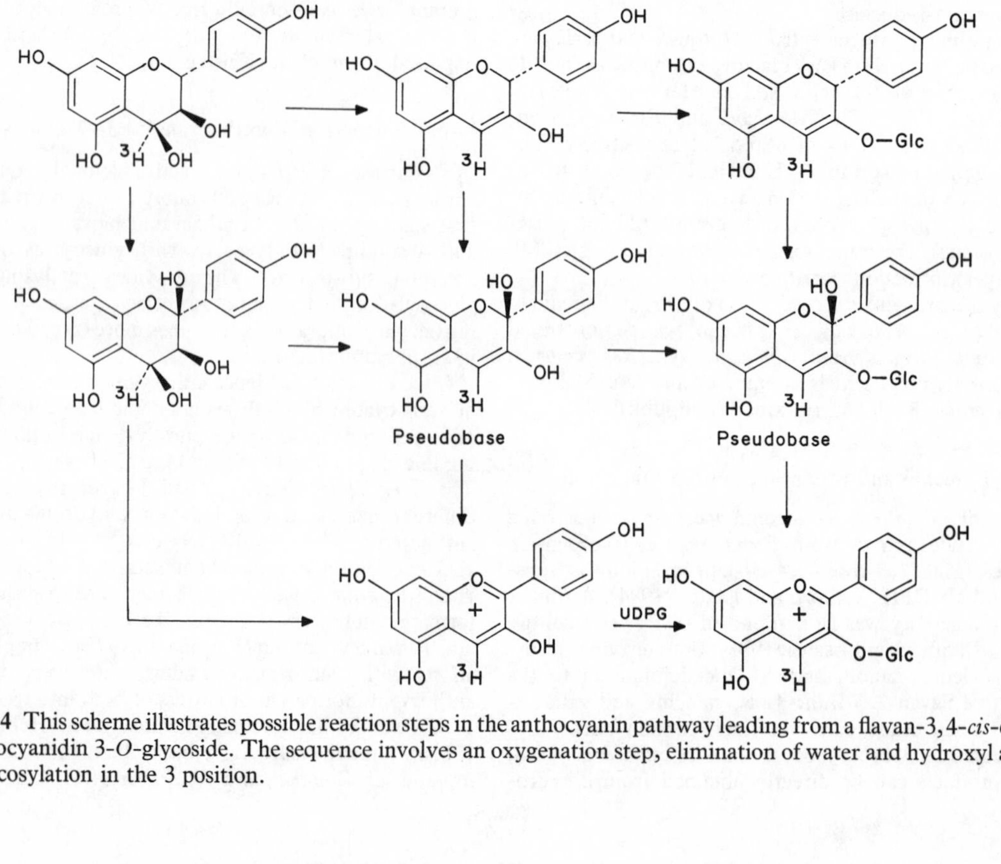

HO OH OH HO ^{3}H OH
HO OH OH HO ^{3}H
HO OH O–Glc HO ^{3}H
HO HO OH OH HO ^{3}H OH
HO HO OH OH HO ^{3}H
Pseudobase
HO HO OH O–Glc HO ^{3}H
Pseudobase
HO OH + OH HO ^{3}H
UDPG
HO OH + O–Glc HO ^{3}H

Fig. 11.4 This scheme illustrates possible reaction steps in the anthocyanin pathway leading from a flavan-3,4-*cis*-diol to an anthocyanidin 3-*O*-glycoside. The sequence involves an oxygenation step, elimination of water and hydroxyl anion, and glycosylation in the 3 position.

from various sources (see Ebel and Hahlbrock, 1982). More recent studies were performed on flowers of genetically defined lines of *Petunia* (Gerats *et al.*, 1983b, 1984b, 1985; Jonsson *et al.*, 1984c), *Matthiola* (Seyffert, 1982; Teusch *et al.*, 1986a) and *Callistephus* (Teusch and Forkmann, unpublished). These enzymes require UDP-glucose as sugar donor, have a distinct positional specificity for the 3-hydroxyl group and use anthocyanidins and flavonols as substrates.

In several plants, recessive alleles of genes connected with the biosynthetic pathway between dihydroflavonols and anthocyanins cause a clearly reduced 3-*O*-glucosyltransferase activity. In *Petunia* and *Matthiola* this has been studied in more detail. In *Petunia* the genes *An1* and *An2* influence 3-*O*-glucosyltransferase activity, while the genes *An6* and *An9* seem to exert no effect. Recessive mutants of *An1* or *An2* show an activity of at most 20% of the wild-type value and in the double mutant (*an1an1*, *an2an2*) only about 5% activity remains (Gerats *et al.*, 1983b). Investigations of a multiple series of both gene *An1* and *An2* showed a positive correlation between anthocyanin concentration and 3-*O*-glucosyltransferase activity. Both SAM: anthocyanin 3′, 5′-methyltransferase (SAM, S-adenosyl methionine) and 5-*O*-glucosyltransferase activities are also influenced by these genes (Gerats *et al.*, 1984b; Jonsson *et al.*, 1984d). A further gene, *An4*, which controls a late step of the anthocyanin pathway in *Petunia* anthers might be the structural gene for the 3-*O*-glucosyltransferase (Gerats *et al.*, 1985). This was concluded from thermal inactivation experiments.

In *Matthiola*, the 3-*O*-glucosyltransferase is influenced by the genes *e*, *g* and *z* which are also concerned with later steps of the anthocyanin pathway. Single recessive mutants of these genes show enzyme activities between 10% and 40% of that of the wild-type value (Teusch *et al.*, 1986a). But genetic and biochemical studies have shown that none of these genes can be regarded as the structural gene for 3-*O*-glucosyltransferase. Gene *e* correlates with dihydroflavonol 4-reductase activity and gene *g* most probably exerts a regulatory function (Heller *et al.*, 1985a, 1985b; Teusch *et al.*, 1986a).

It is noteworthy that, in *Petunia* and *Matthiola*, the mutants blocked in biosynthetic steps before dihydroflavonol formation possess wild-type activity of 3-*O*-glucosyltransferase (Gerats *et al.*, 1983b, Teusch *et al.*, 1986a). A possible induction of gene expression by anthocyanidins or other late intermediates in their biosynthesis can therefore be excluded.

The influence of several genes localized late in anthocyanin biosynthesis on 3-*O*-glucosyltransferase activity suggest that either these genes exert a regulatory effect or that the enzymes for the last steps in this pathway are closely associated in a functional complex, in which a defective portion might effect all other enzyme activities.

11.5 INDIVIDUAL STEPS TO FLAVONOID MODIFICATIONS

Hydroxylation, in particular of the B-ring, and methylation of hydroxyl functions lead to the various aglycone structures within each flavonoid class. Further frequent modification of the flavonoid skeleton is due to glycosylation of hydroxyl groups and to acylation of these sugar moieties. Other modifications, such as *C*-glycosylation and sulphation, are restricted to a few flavonoid classes. Some important structural modifications and the enzymes involved are depicted in Fig. 11.5.

11.5.1 B-ring hydroxylation

The B-ring substitution pattern may be determined in two ways: (1) incorporation of already substituted hydroxycinnamic acid derivatives during synthesis of the C_{15} skeleton or (2) substitution at the C_{15}-stage by specific hydroxylases and methyltransferases (see Hahlbrock and Grisebach, 1975). In the first case, 4-coumarate would be the precursor for 4′-hydroxy-, caffeate for 3′, 4′-dihydroxy- and ferulate for 4′-hydroxy-3′-methoxyflavonoids and so on. In the second case, 4-coumarate would be the precursor for all the differently substituted flavonoids. Most of the accumulated evidence favours the second route.

Enzymic hydroxylation of naringenin and dihydrokaempferol to eriodictyol and dihydroquercetin, respectively (see Fig. 11.5), was first demonstrated in microsomal preparations from *Haplopappus* cell cultures (Fritsch and Grisebach, 1975). The reaction requires NADPH as cofactor and molecular oxygen. Flavonoid 3′-hydroxylases have also been found in microsomal fractions prepared from flowers of defined genotypes of *Matthiola* (Forkmann *et al.*, 1980), *Antirrhinum* (Forkmann and Stotz, 1981), *Dianthus* (Spribille and Forkmann, 1982b), *Verbena* (Stotz *et al.*, 1984) and *Petunia* (Stotz *et al.*, 1985) as well as from maize seedlings (Larson and Bussard, 1986). The same activity has been demonstrated in a microsomal fraction from irradiated parsley cell suspension cultures (Britsch *et al.*, 1981).

In parsley, the flavonoid 3′-hydroxylase was studied in more detail (Hagmann *et al.*, 1983). The enzyme catalyses the hydroxylation of naringenin to eriodictyol, dihydrokaempferol to dihydroquercetin, kaempferol to quercetin and apigenin to luteolin. As was found for *Matthiola* (Forkmann *et al.*, 1980), the parsley enzyme did not catalyse the hydroxylation of 4-coumaric acid to caffeic acid. The enzyme reaction requires molecular oxygen and NADPH, and is inhibited by carbon monoxide as well as by cytochrome *c* and $NADP^+$. Similar data have been reported for the enzyme from maize seedlings (Larson and Bussard, 1986). Flavonoid 3′-hydroxylase is therefore, like cinnamate 4-hydroxylase, a cytochrome *P*-450-dependent mono-oxygenase (see Section 11.3).

The demonstration of flavonoid 3′-hydroxylase in flowers of defined genotypes of several plants has meant

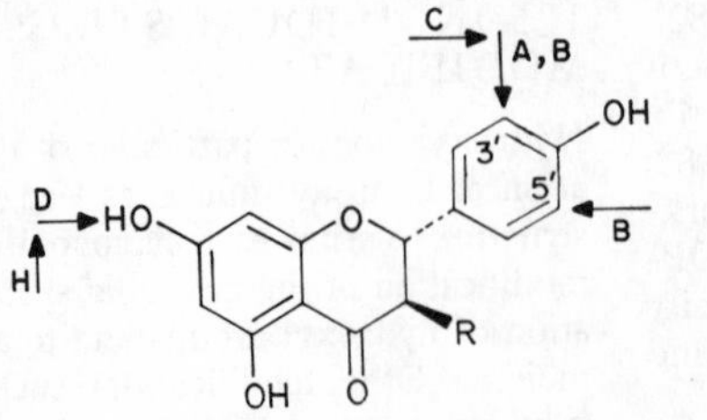

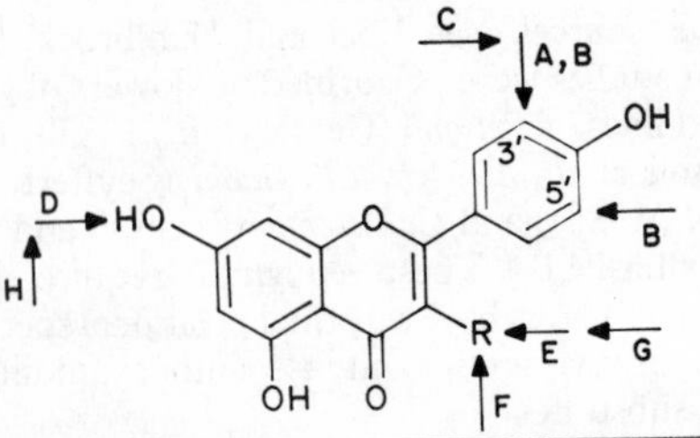

3′	5′	FLAVANONE (R=H)	DIHYDROFLAVONOL (R=OH)	FLAVONE (R=H)	FLAVONOL (R=OH)
H	H	Naringenin	Dihydrokaempferol	Apigenin	Kaempferol
OH	H	Eriodictyol	Dihydroquercetin	Luteolin	Quercetin
OH	OH	—	Dihydromyricetin	Tricetin	Myricetin

R_1	R_2	ANTHOCYANIDIN (3-GLUCOSIDE)
H	H	Pelargonidin
OH	H	Cyanidin
OH	OH	Delphinidin
OCH_3	H	Peonidin
OCH_3	OH	Petunidin
OCH_3	OCH_3	Malvidin

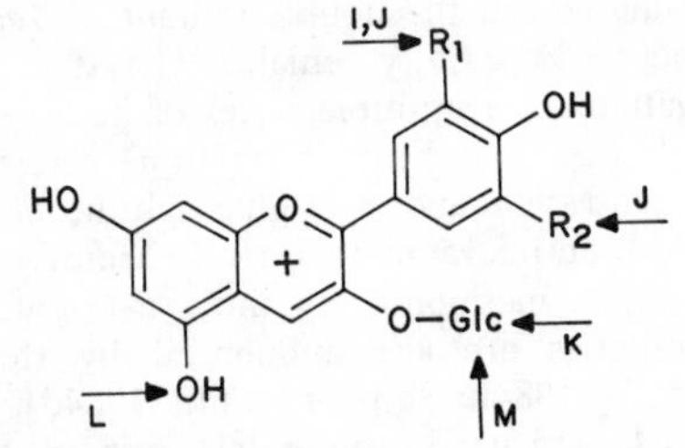

Fig. 11.5 Scheme illustrating the positions of frequent modification reactions observed with selected flavonoid classes. The enzymes are marked with capital letters: (A) flavonoid 3′-hydroxylase, (B) flavonoid 3′,5′-hydroxylase, (C) flavonoid 3′-*O*-methyltransferase, (D) flavonoid 7-*O*-glycosyltransferase, (E) flavonol 3-*O*-glycosyltransferase, (F) flavonol 3-*O*-methyltransferase, (G) flavonol 3-*O*-glucoside 6″-*O*-malonyltransferase, (H) flavonoid 7-*O*-glucoside 6″-*O*-malonyltransferase, (I) anthocyanin 3′-*O*-methyltransferase, (J) anthocyanin 3′,5′-*O*-methyltransferase, (K) anthocyanidin 3-*O*-glucoside glycosyltransferase, (L) anthocyanidin 3-*O*-glycoside 5-*O*-glucosyltransferase, (M) anthocyanidin 3-*O*-glycoside acyltransferase.

that the genetic control of this enzyme activity could be studied. In the respective plants, genes are known which regulate the presence of pelargonidin and cyanidin derivatives. All plant species investigated showed high flavonoid 3′-hydroxylase activity in enzyme preparations from genotypes with dominant alleles (cyanidin-type; Fig. 11.5) but no activity, when the enzyme extracts were prepared from genotypes with recessive alleles (pelargonidin-type; Fig. 11.5). This shows clearly that the 3′-hydroxyl group is introduced into the flavonoid nucleus. Yet more support for this interpretation is provided by the fact that the genes involved in 3′-hydroxylation exert no influence on the substrate specificity of chalcone synthase (Spribille and Forkmann, 1981, 1982a, 1982b). A genetically controlled change in the substrate specificity of this enzyme (i.e. use of caffeoyl-CoA instead of 4-coumaroyl-CoA) would be expected, if the B-ring hydroxylation pattern was determined during synthesis of the flavonoid skeleton.

There are, however, a few examples where hydroxycinnamic acids other than 4-coumaric acid are incorporated *in vivo*. Tulip pollen transiently accumulates naringenin, eriodictyol and homoeriodictyol chalcones. In agreement with this, the chalcone synthase from this source uses 4-coumaroyl-CoA, caffeoyl-CoA and feruloyl-CoA (for structures see Figs. 11.2 and 11.5) as substrates for the condensation leading to chalcone (Sütfeld and Wiermann, 1981). A somewhat different situation was reported for *Verbena*. Red flowering strains contain 4′- and 3′,4′-hydroxylated flavonoids in varying amounts. The latter substitution pattern can be attained in two different ways. A small percentage is formed by incorporation of caffeic acid as well as of 4-coumaric acid into the flavonoid skeleton. This basic level can be considerably enhanced by a later hydroxylation of 4′-hydroxylated flavonoids in the 3′ position by action of flavonoid 3′-hydroxylase (Stotz *et al.*, 1984 see also Section 11.4.1). *Silene* is a special case: genetic and biochemical studies indicate that the hydroxylation in ring B is determined at the 4-coumaroyl-CoA stage. Gene *P* was found to control the activity of an enzyme which catalyses the hydroxylation of 4-coumaroyl-CoA to caffeoyl-CoA. The latter compound is presumably used as precursor for the synthesis of cyanidin glycoside and its final acylation with caffeic acid (Kamsteeg *et al.*, 1981).

Enzyme activity for hydroxylation of flavonoids in both the 3′ and 5′ positions was first demonstrated with flower extracts from delphinidin-producing *Verbena* strains (Stotz and Forkmann, 1982). As for flavonoid 3′-hydroxylase, the enzyme activity for 3′,5′-hydroxylation is localized in the microsomal fraction and the reaction requires NADPH and molecular oxygen. The microsomal preparation catalyses hydroxylation of naringenin and dihydrokaempferol to 5,7,3′,4′,5′-pentahydroxyflavanone and dihydromyricetin, respectively. The 3′,4′-hydroxylated flavonoids eriodictyol and dihydroquercetin can also be hydroxylated in the 5′ position yielding the compounds mentioned above (Fig. 11.5). In *Verbena*, earlier chemogenetic studies had shown that the hydroxylation in ring B of anthocyanins in both the 3′ and 5′ position is controlled by one gene only (Beale, 1940; Beale *et al.*, 1940). This fact and the above biochemical data indicate that only one enzyme is involved in 3′,5′-hydroxylation of flavonoids.

Flavonoid 3′,5′-hydroxylase has also been identified in microsomal preparations from *Lathyrus*, *Callistephus* and *Petunia* flowers (Stotz *et al.*, 1983). In these plants, and in *Verbena*, enzyme activity for 3′,5′-hydroxylation was found to be strictly correlated with the presence of 3′,4′,5′-hydroxylated flavonoids in the flowers. Moreover, the enzyme activity is clearly controlled by the gene *R* in *Callistephus* (Forkmann, 1977b) and by the gene *Hf1* in *Petunia* (Tabak *et al.*, 1981), which are responsible for the formation of delphinidin derivatives in the flowers of the respective plants. Thus, at least in the species studied so far, substitution of the B-ring in the 3′ and 5′ positions is also achieved by specific hydroxylation of C_{15} intermediates.

11.5.2 Methylation reactions

Many enzymes catalysing a methyl transfer from *S*-adenosylmethionine to the various hydroxyl groups of flavonoid substrates have been described (Ebel and Hahlbrock, 1982). Some of them require Mg^{2+} as an obligatory cofactor. *S*-Adenosylhomocysteine formed in the reaction is an inhibitor of these enzymes. The question as to whether aglycones or glycosides are the actual substrates has not been unequivocally answered for most *O*-methyltransferases. There is good evidence that the final B-ring substitution pattern (see Fig. 11.5), i.e. hydroxylation and methoxylation in positions 2′,3′ and 5′ is commonly determined at the flavonoid level (see Sections 11.4.1 and 11.5.1).

Extensive studies of anthocyanin methylation, such as cyanidin to peonidin and delphinidin to petunidin and malvidin (Fig. 11.5), have been carried out with *Petunia hybrida* flowers (Jonsson *et al.*, 1982). Methylation of ring B hydroxyl groups is a late event in the biosynthetic pathway. Anthocyanidin 3-*O*-(4-coumaroyl)-rutinosido-5-*O*-glucosides have been found to be the only efficient substrates (Jonsson *et al.*, 1982). On the basis of genetic studies, four genes related to anthocyanin methylation have been postulated: *Mt1* and *Mf1*, *Mt2* and *Mf2* (Jonsson *et al.*, 1983). Biochemical studies reveal at least four methyltransferases with pI values ranging between 4.8 and 5.4, each one being controlled by one of these genes (Table 11.3). In agreement with the anthocyanin metabolites observed *in vivo*, all four enzymes catalyse both 3′- and 5′-*O*-methylation, but double methylation is more pronounced with the Mf enzymes (Jonsson *et al.*, 1984b). Antiserum raised against a partially purified *Mf2* enzyme precipitated three of the four transferases, indicating a common origin for these proteins (Jonsson *et al.*, 1984b).

A group of *O*-methyltransferases responsible for the formation of polymethoxylated flavonol glucosides in *Chrysosplenium americanum* has been studied in detail (Collins *et al.*, 1981). Four methylating activities with high positional and substrate specificity were identified in crude enzyme extracts using quercetin and partially methylated derivatives of quercetin and quercetagetin (6-hydroxyquercetin) as substrates (Table 11.3). The data indicate the sequential methylation of the hydroxyl groups 3, 7 and 4′, hydroxylation at C-6 and methylation. The four enzymes have been partially purified by a four-step procedure including chromatography on hydroxyapatite and chromatofocusing (DeLuca and Ibrahim, 1985a). Due to pH-sensitivity, the 7-*O*-methylating activity requires a salt gradient elution from the polybuffer column at a pH above its pI (Khouri *et al.*, 1986). Another methyltransferase, which co-eluted with this enzyme, catalyses the formation of both 2′-methoxy-3,6,7,4′-tetra-*O*-methylquercetagetin 5′-glucoside and 2′-glucosyloxy-3,7,4′-tri-*O*-methylquercetin from the respective 2′- and 5′-hydroxyl compounds. These are the last steps in this pathway (Khouri *et al.*, 1986). The 6-*O*-methyltransferase differs somewhat from the other transferases on account of its absolute Mg^{2+} requirement and a severe substrate inhibition at concentrations close to its Michaelis constant. It is noteworthy that some activity for the 8 position of 8-hydroxykaempferol was also observed with this enzyme (DeLuca and Ibrahim, 1985a).

A 8-hydroxyflavonol 8-*O*-methyltransferase was reported from *Lotus corniculatus* flowers (Jay *et al.*, 1983, 1985). The biochemical properties are very similar to those mentioned for the 6-position-specific enzyme from *Chrysosplenium* (Table 11.3). *Lotus* flowers also contain 3- and 3′-*O*-methyltransferases which have been shown to methylate the flavone luteolin and the flavonols quercetin, myricetin and isorhamnetin in the relevant positions (Jay *et al.*, 1983).

Table 11.3 Methyltransferases from various sources

Source of enzyme	*Compound converted*	*Position or sequence of methylation*	*pH optimum; pI*	*10^{-3} × Molecular weight*	*Effects of Mg^{2+} ions and EDTA*	*Remarks**	*Reference*
Apple[†] Cell cultures	Quercetin, 3-*O*-methylquercetin	3–7–3′	7.3, 8.8	47	Mg^{2+} required for full activity, inhibition by EDTA	Inhibition by 4CMBS	Macheix and Ibrahim (1984)
Avena sativa Leaf	Vitexin 2″-*O*-rhamnoside, apigenin	7	7.5	52	None		Knogge and Weissenböck (1984)
Chrysosplenium americanum Shoot	Quercetin, isorhamnetin	3	8–8.5; 4.8	57/65	None	Inhibition by 4CMB, NEM, SAH	DeLuca and Ibrahim (1982, 1985a, 1985b)
	3-*O*-Methylquercetin	7	8–9; 5.7	57/65	None	?[‡]	DeLuca and Ibrahim (1982, 1985a)
	3, 7-Di-*O*-methylquercetin, 3, 7-di-*O*-methylquercetagetin	4′	8–8.5; 5.4	57/65	Mg^{2+} required for full activity	4CMB, NEM, SAH	Deluca and Ibrahim (1982, 1985a, 1985b)
	3, 7-Di-*O*-methylquercetagetin, 3, 7, 3′-tri-*O*-methylquercetagetin, quercetagetin, (8-hydroxykaempferol)	6, (8)	8–9; 5.7	57/65	Mg^{2+} required for activity, inhibition by EDTA	Inhibition by 4CBM, NEM, SAH, substrate	DeLuca and Ibrahim (1982, 1985a, 1985b)
	2′-Hydroxy-3, 6, 7, 4′-tetra-*O*-methylquercetagetin 5′-glucoside, (2′-hydroxy-3, 7, 4′-tri-*O*-methylquercetin 2′-glucoside)	2′, (5′)	7; 5	57	None	Inhibition by PMA, NEM, Zn^{2+} and other divalent cations	Khouri *et al.* (1986)
Glycyrrhiza glabra Cell cultures	Licodione, 2″-hydroxylicodione, isoliquiritigenin	2′	8.0		None	Inhibition by Cu^{2+} and Zn^{2+}	Ayabe *et al.* (1980)

Lotus corniculatus Flower	8-Hydroxykaempferol, 8-hydroxyquercetin, (6-hydroxyapigenin)	8, (6)	7.5–8.4; 5.5	55	Mg^{2+} required for full activity, inhibition by EDTA	Requirement for SH group protectors	Jay *et al.* (1983, 1985)
	Quercetin, isorhamnetin	3	7.7; 5.1	?‡	?‡	?‡	Jay *et al.* (1983)
	8-Hydroxyquercetin, quercetin, myricetin, luteolin	3′	7.7; 5.1	?‡	Mg^{2+} required for full activity, inhibition by EDTA	?‡	Jay *et al.* (1983)
Petunia hybrida Flower	Anthocyanidin 3-*O*-(4-coumaroyl)-rutinosido-5-*O*-glucoside	3′ and 5′		45–50	Mg^{2+} required for full activity, no inhibition by EDTA	Inhibition by SAH	Jonsson *et al.* (1984a)
Mt1§			7.7				
Mt2			7.7				
Mf1			7.7; 4.8				
Mf2			7.7; 5.7				
Pisum sativum Shoot	(+)-6a-Hydroxymaackiain	3	(7.5)	?‡	?‡	?‡	Sweigard *et al.* (1986)
Robinia pseudacacia† Shoot	Apigenin, naringenin	4′	9	?‡	None	?‡	Kuroki and Poulton (1981)
	Quercetin	?‡			Mg^{2+} required for full activity, inhibition by EDTA	?‡	Kuroki and Poulton (1981)
Silene alba† Leaf	Iso-orientin 2″-*O*-rhamnoside, iso-orientin	3′	8–8.5	?‡	Mg^{2+} required for full activity, some inhibition by EDTA	Activation by Co^{2+} and Mn^{2+}	van Brederode and Kamps-Heinsbroke (1981a) van Brederode *et al.* (1982)
Spinach	Quercetin	3–3′–4′,	(8.2)	?‡	?‡	?‡	Charrière-Ladreix *et al.* (1981)
Chloroplast†		3–7–(4′)–3′					
Leaf	Quercetin 3, 7-di-*O*-methylquercetagetin	3–7–4′ 4′, 6–4′		?‡	?‡	?‡	Thresh and Ibrahim (1985)

*4CBM = 4-chloromercuribenzoate; 4CMBS = 4-chloromercuribenzene sulphonate; NEM = *N*-ethylmaleimide; PMA = phenylmercuriacetate; SAH = *S*-adenosylhomocysteine.
†Caffeate 3′-*O*-methyltransferase activity differentiated.
‡Data not reported.
§Designation of enzyme form.

Table 11.4 Glycosyltransferases from various sources

Source of enzyme	*Substrates*[†]	*Glycosyl donor*	*pH optimum; pI*	*10^{-3} × Molecular weight*	*Remarks*[‡]	*Reference*
Silene pratensis	*Flavones*					
Petals, green parts						
(1) 2″-*O*-Xylosyltransferase	Vitexin	UDP-xylose	7.8	–	Zn^{2+}, Hg^{2+} (i); Mg^{2+}, Mn^{2+} (s)	van Brederode and Kamps-Heinsbroek (1981b)
(2) 7-*O*-Galactosyltransferases (two different enzymes)	Isovitexin Isovitexin 2″-glycosides	UDP-galactose	7.0	–	Zn^{2+}, Hg^{2+} (i)	van Brederode and Steyns (1983) Steyns and van Brederode (1986a)
(3) 7-*O*-Glucosyltransferase	Isovitexin	UDP-glucose	7.0–7.5	45	Zn^{2+}, Hg^{2+} (i); Mg^{2+}, Mn^{2+} (s)	Steyns and van Brederode (1986b)
(4) 2″-*O*-Rhamnoside glucosyltransferase	Isovitexin 2″-*O*-rhamnoside	UDP-glucose	7.0–7.2	45	Zn^{2+}, Hg^{2+} (i); Mg^{2+}, Mn^{2+} (s)	Steyns and van Brederode (1986b)
Cicer arietinum	*Isoflavones*					
Roots 7-*O*-Glucosyltransferase	Biochanin A, formononetin	UDP-glucose	8.5–9.0; 5.4	50		Köster and Barz (1981)
Chrysosplenium americanum	*Flavonols*					
Shoots Flavonol ring B-*O*-glucosyltransferase (position 2′ and 5′)	Partially methylated flavonols	UDP-glucose	7.5–8.0; 5.1	42	Cu^{2+}, Fe^{2+}, Zn^{2+}, UDP and both reaction products (i)	Bajaj *et al.* (1983) Khouri and Ibrahim (1984)
Anethum graveolens						
Cell cultures 3-*O*-Glucuronosyltransferase	Quercetin	UDP-glucuronic acid	–	–	–	Möhle *et al.* (1985)
Pisum sativum						
Flowers (1) 3-*O*-Glucosyltransferase	Kaempferol, quercetin, myricetin	UDP-glucose	8.0–9.0	–	UDP(i)	Jourdan and Mansell (1982)
(2) 3-*O*-Glucoside glucosyltransferase	3-*O*-Glucosides of kaempferol, quercetin and myricetin	UDP-glucose	8.0–9.0	–	UDP(i)	
(3) 3-*O*-Diglucoside glucosyltransferase	3-*O*-Diglucosides of kaempferol, quercetin and myricetin	UDP-glucose	8.0–9.0	–	UDP(i)	

Tulipa Anthers						
(1) 3-*O*-Glucosyltransferase	Kaempferol, quercetin, isorhamnetin,	UDP-Glucose	8.5–9.0	40	Zn^{2+}, Cu^{2+} Mn^{2+} (i)	Kleinehollenhorst *et al.* (1982)
(2) 3-*O*-Glucoside rhamnosyltransferase	Flavonol 3-*O*-glucoside and diglycoside	UDP-rhamnose	8.5–9.0	40	Zn^{2+}, Cu^{2+}(i)	
(3) 3-*O*-Glucoside xylosyltransferase	Flavonol 3-*O*-glucoside, galactoside and diglycoside	UDP-xylose	8.5–9.0	30	NH_4^+, Ca^{2+} (s)	
Matthiola incana Flowers	*Anthocyanins*					
(1) 3-*O*-Glucosyltransferase	Pelargonidin, cyanidin, delphinidin, kaempferol, quercetin	UDP-glucose	8.5 (pelargonidin) 9.5 (quercetin)	–	Fe^{2+}, Cu^{2+}(i)	Teusch *et al.* (1986a)
(2) 3-*O*-Glucoside xylosyltransferase	Anthocyanidin 3-*O*-glucosides and 3-*O*-(4‴-coumaroyl) glucoside	UDP-xylose	6.5	–	Fe^{2+}, Cu^{2+}, Zn^{2+}(i), Ca^{2+}(s)	Teusch (1986)
(3) 3-*O*-Glycoside 5-*O*-glucosyltransferase	Anthocyanidin 3-*O*-(4‴-coumaroyl) glucoside and 3-*O*-(4‴-coumaroyl) xylosylglucoside	UDP-glucose	7.5	–	Fe^{2+}, Cu^{2+}, Zn^{2+}, Co^{2+}(i)	Teusch *et al.* (1986b)
Petunia hybrida Flowers						
3-*O*-Glycoside 5-*O*-glucosyltransferase	Delphinidin and petunidin 3-*O*-(4-coumaroyl)-rutinoside	UDP-glucose	8.3	–	–	Jonsson *et al.* (1984d)

[†]The position of glycosylation is evident from the name of the individual enzyme.
[‡](i) inhibition; (s) stimulation.

An enzyme which catalyses the 4′-*O*-methylation of apigenin to acacetin has been identified in extracts from *Robinia pseudacacia* (Kuroki and Poulton, 1981). *O*-Methyltransferases specific for *C*-glycosylflavones have been described from *Silene pratensis* (Syn. *S. alba*) flowers and *Avena sativa* leaves. The *Silene* enzyme converts iso-orientin and iso-orientin 2″-*O*-rhamnoside to isoscoparin and its 2″-*O*-rhamnoside derivative respectively. A much lower affinity for iso-orientin (K_m 0.32 × 10^{-3} M) suggests that the 2″-*O*-rhamnoside derivative (K_m 7 × 10^{-6} M) is the real substrate of the enzyme. The *Avena* enzyme catalyses the *O*-methylation of vitexin 2″-*O*-rhamnoside in the 7 position (Knogge and Weissenböck, 1984). Neither the flavonoids naringenin, vitexin and vitexin 2″-arabinoside nor caffeate were substrates. Interestingly, apigenin was converted by the homogeneous enzyme to 7-*O*-methylapigenin, although at a low rate. *S*-Adenosylhomocysteine–Sepharose chromatography was a highly efficient step in the purification of this enzyme.

Enzymatic *O*-methylation of licodione (1,3-dioxo-2′,4′,4″-trihydroxy-1,3-diphenylpropane), an intermediate in the biosynthesis of echinatin (4,4′-dihydroxy-2-methoxychalcone), was observed in extracts from *Glycyrrhiza echinata* (Ayabe *et al.*, 1980). Isoliquiritigenin, a precursor of licodione *in vivo*, is also methylated by the enzyme, but only at one-tenth of the rate for licodione which is assumed to be the actual substrate *in vivo*.

3-*O*-Methylation of the pterocarpan (+)-6a-hydroxymaackiain is the last step in the biosynthesis of the stress metabolite (+)-pisatin in *Pisum sativum* (see Chapter 5). The transferase activity shows a strict substrate specificity for (+)-6a-hydroxymaackiain and does not accept either the (−)-enantiomer or the maackiain enantiomers (Sweigard *et al.*, 1986). The isoflavone *O*-methyltransferase from *Cicer arietinum*, which had been described as transforming daidzein and genistein to formononetin and biochanin A respectively (Wengenmayer *et al.*, 1974), was reinvestigated. It was found that the enzyme methylates isoflavones in the 7 rather than in 4′ position (Hagmann and Grisebach, 1984; see Grisebach, 1985).

Extensive enzyme kinetic studies have been performed with the *O*-methyltransferases from *Petunia* (Jonsson *et al.*, 1984a), *Avena* (Knogge and Weissenböck, 1984), *Chrysosplenium* (DeLuca and Ibrahim, 1985b) and *Lotus* (Jay *et al.*, 1985). The collected data suggest a *mono-iso* Theorell–Chance mechanism for the *Avena* and *Lotus* enzymes, whereas an ordered mechanism has been proposed for the *Chrysosplenium* transferases.

11.5.3 Glycosylation

The vast number of flavonoid glycosides found in nature suggests the occurrence of a great range of glycosyltransferases with varying substrate specificities. Indeed there is a multitude of well-known glycosyltransferases (see Hahlbrock and Grisebach, 1975; Hösel, 1981; Ebel and Hahlbrock, 1982) and many more have been characterized in recent years (Table 11.4).

Novel flavonol *O*-glycosyltransferases were demonstrated in enzyme preparations from tulip anthers (Kleinehollenhorst *et al.*, 1982), *Pisum* flowers (Jourdan and Mansell, 1982), *Chrysosplenium* shoots (Bajaj *et al.*, 1983; Khouri and Ibrahim, 1984) and *Anethum* cell cultures (Möhle *et al.*, 1985). These enzymes exhibit a pronounced specificity with regard to the substrate, the position and the sugar transferred. The isoflavone 7-*O*-glucosyltransferase isolated from *Cicer* shows a similar high specificity (Köster and Barz, 1981). By means of combined affinity chromatography on UDP-glucuronic agarose and Brown-dye ligand columns, two enzymes could be separated from *Chrysosplenium* extracts, which glycosylate the B-ring of highly methoxylated flavonols in the 2′ and 5′ positions (Ibrahim, 1986).

A particularly interesting situation has been found during extensive genetic and biochemical studies of the glycosylation of isovitexin (6-*C*-glucosylapigenin) in *Silene pratensis* and *Silene dioica* (Steyns *et al.*, 1984, and references in Table 11.4). Eleven functional alleles, spread over six loci, have now been identified. They code for different glycosyltransferases which catalyse glucosylation, galactosylation and xylosylation of the 7-hydroxyl group or glucosylation, rhamnosylation, xylosylation and arabinosylation of the 2″-hydroxyl group of the carbon-bound glucose of isovitexin. The enzymes act in the petals and/or different developmental stages of the green parts. Serological studies show that the various isovitexin *O*-glycosyltransferases are evolutionarily related (van Brederode *et al.*, 1986; van Brederode, 1986).

Enzymes catalysing the formation of anthocyanidin 3-*O*-glucosides are considered in Section 11.4.9. As already found for most anthocyanidin 3-*O*-glucosyltransferases described earlier (see Ebel and Hahlbrock, 1982), the respective enzymes from flowers of *Petunia, Matthiola* and *Callistephus* also efficiently glucosylate flavonols. Extensive studies on different genotypes of *Petunia hybrida* (Jonsson *et al.*, 1984c) and *Matthiola incana* (Teusch *et al.*, 1986a) have shown that the 3-*O*-glucosyltransferases for flavonols and anthocyanidins are identical. Interestingly, however, the flavonol 3-*O*-glucosyltransferase from tulip anthers does not glucosylate anthocyanidins but is highly specific for flavonols (Kleinehollenhorst *et al.*, 1982).

Glycosylation reactions leading to anthocyanidin 3-biosides and 3,5-di-*O*-glucosides were first found with flower extracts from genetically defined lines of *Silene dioica*. The enzyme activity for rhamnosylation of anthocyanidin 3-*O*-glucosides is controlled by the gene *N* (Kamsteeg *et al.*, 1980a), while the activity for 5-*O*-glucosylation of anthocyanidin 3-*O*-glycosides is governed by the gene *M* (Kamsteeg *et al.*, 1978, 1980b). An enzyme for 5-*O*-glucosylation was also demonstrated in flower extracts from defined genotypes of *Petunia* (Jonsson *et al.*,

1984d). This enzyme glucosylates the 3-*O*-(4-coumaroyl)-rutinoside of delphinidin and petunidin. The 3-*O*-glucoside and 3-*O*-rutinoside of delphinidin, and cyanidin 3-*O*-rutinoside did not serve as substrates. Surprisingly, the enzyme activity is not controlled by the gene *Gf* which was thought to be responsible for 5-*O*-glucosylation.

Biochemical studies on genetically defined *Matthiola* lines revealed the presence of two *O*-glycosylating enzyme activities in the flower extracts. One enzyme catalyses the xylose transfer from UDP-xylose to both anthocyanidin 3-*O*-glucosides and 3-*O*-(4-coumaroyl)glucosides, the latter being the better substrates. Interestingly, kaempferol and quercetin 3-glucosides were also found to be xylosylated (Teusch, 1986). The other enzyme catalyses 5-*O*-glucosylation of anthocyanins (Teusch *et al.*, 1986b). The best substrate for this reaction is the pelargonidin or cyanidin 3-*O*-xylosylglucoside (sambubioside) acylated with 4-coumarate followed by the non-acylated 3-biosides and the 3-*O*-(4-coumaroyl)glucosides.

An enzyme catalysing 5-*O*-glucosylation of anthocyanins was also found in flower extracts of *Callistephus chinensis*. The 3-*O*-glucosides of pelargonidin, cyanidin and delphinidin, but not 3-*O*-(malonyl)glucosides, were accepted as substrates.

Very recently, evidence for the *in vitro* formation of the *C*-glucosides vitexin and isovitexin was obtained with buckwheat enzyme preparations. 2-Hydroxyflavanones, but neither flavones nor flavanones, are the substrates for this glucosyltransferase activity (Kerscher and Franz, 1987).

11.5.4 Acylation

Acylated flavonoids, including anthocyanins, are widespread in nature (see Chapters 1, 5 and 8). Two classes of acyl moieties have been found: (1) aromatic, mostly hydroxycinnamic acids, and (2) aliphatic acids, e.g. malonic acid.

Two *O*-malonyltransferases have been isolated and extensively purified from UV-irradiated parsley cell cultures, which accumulate malonated flavone and flavonol glycosides (Matern *et al.*, 1981, 1983a, b). Both enzymes appear to be specific for flavonoid glycosides. They catalyse malonyl transfer from malonyl-CoA to the primary hydroxyl group of the glucosyl moiety. The enzymes show distinct substrate specificity in accordance with the pattern of malonated flavonoids found in parsley cell cultures (see Ebel and Hahlbrock, 1982). Antibodies raised against the purified 3-*O*-glucoside malonyltransferase did not inhibit the activity of the 7-*O*-glucoside-specific enzyme. The fact that the malonated flavonoid glycosides of parsley, which are synthesized in the cytoplasm, are located exclusively within the vacuoles and the observation of pH-dependent conformational changes in malonated apigenin 7-*O*-glucoside led to the assumption that malonylation is of importance for the transport of these compounds into the vacuole (Matern *et al.*, 1983b, 1986).

A malonyltransferase specific for the 6″ position of isoflavone 7-*O*-glucoside has been isolated from seedlings of *Cicer arietinum* (Köster *et al.*, 1984). The enzyme shows a pH optimum of 8.0, a pI of 5.3 and a molecular weight of 112 000. A pronounced specificity was found for the endogenous 7-*O*-glucosides of formononetin and biochanin A. The 7-*O*-glucosides of other isoflavonoids and flavonoids and a chalcone 4′-*O*-glucoside were much poorer substrates.

In recent years, the occurrence of anthocyanins acylated with aliphatic acids has often been reported (Harborne and Boardley, 1985; Takeda *et al.*, 1986). Very recently a corresponding enzyme activity was detected in protein extracts from *Callistephus* flowers (Teusch and Forkmann, 1987). This catalyses the acylation of pelargonidin, cyanidin and delphinidin 3-*O*-glucoside with aliphatic acids. Cyanidin 3, 5-bis-*O*-glucoside but not the 3-*O*-xylosylglucoside was also a substrate. Malonyl-CoA is the most efficient acyl donor but can be replaced to a reasonable extent by the CoA esters of methylmalonate, succinate and glutarate. As might be expected from the presence of a malonyltransferase, *Callistephus* flowers are known to contain aliphatic acylated anthocyanins (Forkmann, 1977b). One of them was recently identified as pelargonidin 3-*O*-(6″-*O*-malonyl)glucoside (Takeda *et al.*, 1986).

Although aromatic acylation of anthocyanins has long been known, a corresponding acyltransferase was first demonstrated in 1980 in *Silene* (Kamsteeg *et al.*, 1980c). This enzyme catalyses the transfer of the 4-coumaroyl or caffeoyl moiety of the respective CoA esters to the rhamnosyl 4-position of anthocyanidin 3-*O*-rutinosides or 3-*O*-rutinosido-5-*O*-glucosides. In confirmation of chemogenetic work, this transferase activity is controlled by the gene *Ac*.

A similar acyltransferase was recently found in soluble enzyme preparations from flowers of genetically defined lines of *Matthiola* (Teusch *et al.*, 1986c). The enzyme catalyses the acylation of anthocyanidin 3-*O*-glucosides and 3-*O*-xylosylglucosides with 4-coumaric and caffeic acid respectively. 5-*O*-Glucosylated anthocyanins, however, were not acylated. In *Matthiola* flowers the gene *u* governs the presence of anthocyanidin 3-glucosides and 3-xylosylglucosides acylated with 4-coumaric or caffeic acid. In agreement with this fact, acyltransferase activity was only found in enzyme extracts from flowers of lines with the dominant allele u^+.

11.5.5 Pterocarpan 6a-hydroxylase

The stereoselective introduction of a hydroxyl group in the 6a position of the pterocarpan skeleton has been postulated from results of experiments *in vivo* with pea, using specifically deuterated isoflavone and pterocarpan intermediates (see Chapter 5).

Elicitor-challenged soybean cell cultures and seedlings, which accumulate pterocarpan phytoalexins, have been used for the study of this reaction *in vitro* (Hagmann *et al.*, 1984). Microsomal enzyme fractions of this plant system catalyse a NADPH- and oxygen-dependent hydroxylation of 3, 9-dihydroxypterocarpan to 3, 6a, 9-trihydroxypterocarpan (glycinol) which are both intermediates of glyceollin biosynthesis (Fig. 11.6). Cofactor requirement and inhibition pattern, including inhibition by ketoconazole (Fritsch *et al.*, 1986), are typical for a cytochrome *P*-450-dependent enzyme system.

In the hydroxylation reaction no more than 50% of the chemically prepared 3,9-dihydroxy[6, 11a-^{3}H]pterocarpan was converted to the 3, 6a, 9-trihydroxypterocarpan, and dihydroxypterocarpan reisolated from such reaction mixtures was not transformed upon repeated incubation (Hagmann *et al.*, 1984). Obviously, only one enantiomeric form of dihydroxypterocarpan is substrate for this hydroxylase. Optical rotatory dispersion spectroscopy was used to identify the stereochemistry of the reaction, which involves retention of configuration. Like the naturally occurring trihydroxypterocarpan, the enzymically formed product exhibits a maximal positive Cotton effect at 287 nm. This proves the (6a*S*, 11a*S*)-configuration for the enzymic product (Ingham and Markham, 1980). In contrast, the unreactive dihydroxypterocarpan showed a negative Cotton effect at 285 nm and a maximal positive rotation at 240 nm which indicates (6a*S*, 11a*S*)-configuration* for this structure.

11.5.6 Prenyltransferases

Prenylated isoflavonoids frequently occur as stress-induced metabolites in members of the Leguminosae and often possess fungicidal and bactericidal activity (see Chapter 5). The pterocarpan phytoalexins glyceollin I–IV are examples of such compounds and are formed in soybean challenged with elicitor preparations, mycelia or zoospores from the pathogen *Phytophthora megasperma* f. sp. *glycinea* and by many other stress-inducing compounds (see Section 11.6 and Chapter 5). Minor quantities of 2, 6a, 9-trihydroxypterocarpan (glycinol) and the glyceollidins I and II (glyceocarpin) have also been identified (Fig. 11.6) (Grisebach *et al.*, 1985).

The flavonoid-specific prenyltransferase activity was first observed in particulate enzyme preparations from elicitor-treated soybean cotyledons (Zähringer *et al.*, 1979) and cell suspension cultures (Zähringer *et al.*, 1981).

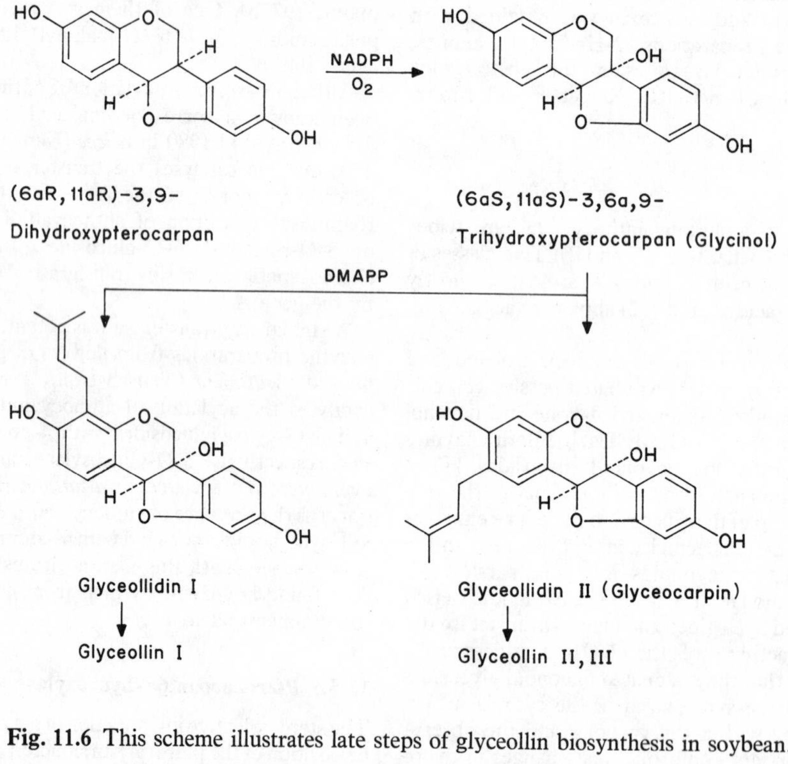

Fig. 11.6 This scheme illustrates late steps of glyceollin biosynthesis in soybean.

*Note that an oxygen function in the 6a position reverses the specification of the chirality.

Two separate enzymes have been described, which catalyse a dimethylallyl transfer from dimethylallylpyrophosphate to the 3, 6a, 9-trihydroxypterocarpan 2 or 4 position in a ratio of 3 : 1. Both enzymes need Mn^{2+} for full activity.

Another prenyltransferase, prepared from wounded *Lupinus albus* hypocotyls, catalyses the introduction of a dimethylallyl substituent into position 6 of the isoflavones genistein and 2′-hydroxygenistein yielding wighteone and luteone respectively (Schröder *et al.*, 1979).

11.5.7 Interaction of modifying enzymes reactions

Because of the generally high substrate specificity of modifying enzymes, the substitution of flavonoids by hydroxylation, methylation, glycosylation and acylation proceeds in a defined sequential order.

In *Pisum* flowers, flavonol 3-*O*-triglucosides are accumulated. Their formation apparently occurs by means of three sequential single-step transfers of glucose, each catalysed by a specific enzyme (Jourdan and Mansell, 1982). A similar sequential order of glycosylation reactions is found in tulip anthers. Glucosylation of flavonols in the 3 position by a specific enzyme is a prerequisite for the action of a rhamnosyltransferase and/or a xylosyltransferase (Kleinehollenhorst *et al.*, 1982).

With regard to enzymes which methylate hydroxyl groups, it is commonplace that hydroxylation precedes methylation. But previous methylation of a specific position, glycosylation and even acylation may also be necessary before methylation of a specific hydroxyl group can occur. Thus, in *Chrysosplenium*, formation of penta-*O*-methylquercetagetin glucoside (see Section 11.5.2) proceeds via the following sequence from quercetin: sequential methylation in the positions 3,7 and 4′, hydroxylation and subsequent methylation at C-6, then hydroxylation at C-6′ (which then becomes C-2′), glucosylation in the 5′ position and finally methylation of the remaining 2′-hydroxyl group (DeLuca and Ibrahim, 1985a; Khouri *et al.*, 1986).

In *Petunia*, the substrate for methylation of the hydroxyl groups in the 3′ and 5′ positions is the anthocyanidin 3-*O*-(4-coumaroyl)rutinosido-5-*O*-glucoside which itself is probably provided by the following sequential reactions: anthocyanidin 3-*O*-glucoside → 3-*O*-rutinoside → acylated 3-*O*-rutinoside → acylated 3-*O*-runinosido-5-*O*-glucoside (Jonsson *et al.*, 1982, 1984d). A similar sequence of events can be derived from chemogenetic and biochemical studies for the formation of anthocyanidin 3-*O*-(hydroxycinnamoyl)xylosylglucosido-5-*O*-glucoside in *Matthiola* flowers. Although minor pathways may exist, the main biosynthetic route is as follows: anthocyanidin 3-*O*-glucoside → acylated 3-*O*-glucoside → acylated 3-*O*-xylosylglucoside → acylated 3-*O*-xylosylglucosido-5-*O*-glucoside. This sequence is based on both the substrate specificities *in vitro* of all enzymes involved and the amounts of anthocyanin intermediates present in mutant flowers lacking activity of one or more of these enzymes (Seyffert, 1982; Teusch *et al.*, 1986a, b, c; Teusch, 1986).

11.6 REGULATION OF ENZYME ACTIVITIES

Many flavonoids can be considered to be constitutive in plants, their appearance being controlled by endogenous factors during normal development. This endogenous control is reflected in the appearance of the relevant enzymes. During bud and flower development in *Matthiola* and *Petunia*, for example, early flavonol and subsequent anthocyanin formation and modification correlates with appearance of enzyme activities of the respective pathways (Dangelmayr *et al.*, 1983; Forkmann *et al.*, 1986; Froemel *et al.*, 1985; Spribille and Forkmann, 1984; Teusch, 1986; Teusch *et al.*, 1986b, c).

In many cases, however, flavonoid synthesis is induced by external factors. The best studied cases of induced flavonoid synthesis are photocontrol and induction by stress. Photocontrol of flavonoid synthesis can involve up to three photoreceptors, the red/far-red absorbing phytochrome system, a blue UV-A light receptor (cryptochrome) and a UV-B photoreceptor. In some cases, activation of a single receptor is sufficient but in many cases complex co-actions between the three photoreceptors may occur. Photocontrol of flavonoid synthesis has been reviewed by Beggs *et al.* (1986). Unfortunately, in almost all cases the effect of light on total flavonoid or anthocyanin amount was studied; thus it is not known to what extent activation of the different photoreceptors might lead to formation of different flavonoid patterns. In one case, *Sinapis alba* cotyledons, it has been shown that, in response to light, quercetin glycosides accumulate preferentially in the upper epidermis and anthocyanins in the lower epidermis. Formation of anthocyanin is favoured by red light pulses whereas quercetin glycoside synthesis occurs only in response to continuous far-red light (Wellmann, 1974). It is probable, however, that both responses are mediated via phytochrome.

The most detailed study of the chain of events involved in light-induced flavonoid synthesis is that by Hahlbrock and co-workers for photocontrol of flavonoid synthesis in *Petroselinum* cell suspension cultures. These workers correlated flavonoid appearance with a co-ordinated transient increase in the transcription rates of the genes for both phenylalanine ammonia-lyase and chalcone synthase. These increases in transcription rates correlated with appearance of mRNAs for the enzymes, the enzymes themselves and the flavonoid products (see Chappell and Hahlbrock, 1984; Kuhn *et al.*, 1984). These experiments were carried out using white light, so it is not possible to separate the possible functions of the different photoreceptors. In *Petroselinum*, UV-B is known to be obligatory for flavonoid synthesis (Wellmann, 1971, 1975) but both phytochrome and the blue light receptor can modulate the response to UV-B strongly (Duell-Pfaff and Wellmann, 1982). Recent measurements of chalcone synthase mRNA accumulation in the same system appear to show that activation of the blue light receptor and the phytochrome system and changes in the fluence rate of UV-B do not modify the mRNA accumulation rate itself but can modify both the lag phase in mRNA appearance and the duration of the transient increase in mRNA (Bruns *et al.*, 1986). Unfortunately, there is no information available as to how any of these photoreceptors might activate gene transcription. There are many other examples of light-induced increases in the activity of phenylalanine ammonia-lyase, chalcone synthase and chalcone isomerase and some of light-induced increases in activities of other enzymes of flavonoid biosynthesis (see Beggs *et al.*, 1986).

In leguminous plants, isoflavonoids are often formed in response to stress (although these compounds are also constitutive in this family). Many stress factors (see West, 1981; Loschke *et al.*, 1983; Creasy, 1985), such as challenge with microbial organisms and cell wall preparations, certain proteins, nucleic acids, heavy metal ions, elevated medium pH (Hattori and Ohta, 1985), ozone, sulphur dioxide, certain herbicides (Cosio *et al.*, 1985; Rubin *et al.*, 1986) and many other agents, may induce this type of response. Short-wavelength UV-B and UV-C radiation has been shown to induce coumestrol in *Phaseolus vulgaris* (Beggs *et al.*, 1985), pisatin in *Pisum* (Hadwiger and Schwochau, 1971) and glyceollin I in *Glycine* (Bridge and Klarman, 1973). These responses can be reversed by subsequent or simultaneous irradiation with blue but not red light thus suggesting participation of a photoreactivation process. This suggests that DNA itself is the target of this radiation and that (probably unspecific) DNA damage somehow activates the genes necessary for isoflavonoid synthesis. Very specific responses to unspecific DNA damage have been well characterized in bacteria (Walker, 1984). These responses to UV-B are not the same as UV-induced flavonoid synthesis (e.g. in *Petroselinum*). Such responses cannot be reversed by blue light and show an action spectrum extending to longer wavelengths.

The most extensively studied system of stress-induced isoflavone synthesis is the response to microbial infection and so-called biotic elicitors (fungal cell wall preparations). These isoflavonoids probably have phytoalexin functions. As with light-induced flavonoid synthesis it has been possible to correlate product formation with transient increases in *de novo* synthesized mRNAs for biosynthetic enzymes and in the enzymes themselves (see Loschke *et al.*, 1983; Schmelzer *et al.*, 1984; Grab *et al.*, 1985; Dixon 1986; Ebel, 1986). A plasma membrane receptor for cell wall fragments from *Phytophthora megasperma* f. sp. *glycinea* has been described from *Glycine* (Yoshikawa *et al.*, 1983; Schmidt and Ebel, 1986). Again little is known concerning the transduction chain from elicitor or infection to gene activation. In the context of stress responses, it should be noted that anthocyanins are often synthesized in response to stress in many plant families (e.g. in *Zea*; Heim *et al.*, 1983). In several families, phytoalexins which are not flavonoids but are also derived from the general phenylpropanoid pathway are induced by the same agents as induce isoflavonoids in the Leguminosae. In some cases, increases in phenylalanine ammonia-lyase, cinnamate 4-hydroxylase and 4-hydroxycinnamate:CoA ligase activities have also been reported. Examples of such responses are stilbene formation in *Arachis* (Fritzemeier and Kindl, 1981) and furanocoumarin formation in parsley (Hahlbrock *et al.*, 1981; Tietjen *et al.*, 1983; Kombrink and Hahlbrock, 1986; Wendorff and Matern, 1986).

REFERENCES

Abbot, M.T. and Udenfriend, S. (1974). In *Molecular Mechanism of Oxygen Activation* (ed. O. Hayaishi), Academic Press, New York, pp. 187–214.

Amrhein, N., Gödeke K.H. and Gerhart, J. (1976), *Planta* **131**, 33.

Ayabe, S.-I., Yoshikawa, T., Kobayashi, M. and Furuya, T. (1980), *Phytochemistry* **19**, 2331.

Bajaj, K.L., DeLuca, V., Khouri, H. and Ibrahim, R.K. (1983), *Plant Physiol.* **72**, 891.

Beale, G.H. (1940), *J. Genet.* **40**, 337.

Beale, G.H., Price, J.R. and Scott-Moncrieff, R. (1940), *J. Genet.* **41**, 65.

Beerhues, L. and Wiermann, R. (1985), *Z. Naturforsch.* **40c**, 160.

Beggs, C.J., Stolzer-Jehle, A. and Wellmann, E. (1985), *Plant Physiol.* **79**, 630.

Beggs, C.J., Wellmann, E. and Grisebach, H. (1986). In *Photomorphogenesis in Plants* (eds R.E. Kendrick and G.H.M. Kronenberg), Martinus Nijhoff/Dr. W. Junk Publ., Dordrecht, pp. 467–499.

Bell, J.N., Ryder, T.B., Wingate, V.P.M., Bailey, J.A. and Lamb, C.J. (1986), *Mol. Cell. Biol.* **6**, 1615.

Belunis, C.J. and Harazdina, G. (1986), *Plant Physiol.* **80**, 40S.

Benveniste, I., Gabriac, B. and Durst, F. (1986), *Biochem. J.* **235**, 365.

Boland, M.J. and Wong, E. (1975), *Eur. J. Biochem.* **50**, 383.

Boland, M.J. and Wong, E. (1979), *Bioorg. Chem.* **8**, 1.

Bolwell, G.P., Bell, J.N., Cramer, C.L., Schuch, W., Lamb, C.J. and Dixon, R.A. (1985), *Eur. J. Biochem.* **149**, 411.

Bonas, U., Sommer, H., Harrison, B.J. and Saedler, H. (1984), *Mol. Gen. Genet.* **194**, 138.

Boniwell, J.M. and Butt, V.S. (1986), *Z. Naturforsch.* **41c**, 56.

Bridge, M.A. and Klarman, W.L. (1973), *Phytopathology* **63**, 606.

Britsch, L. (1986), Private communication.

Britsch, L. and Grisebach, H. (1985), *Phytochemistry* **24**, 1975.

Britsch, L. and Grisebach, H. (1986), *Eur. J. Biochem.* **156**, 569.

Britsch, L., Heller, W. and Grisebach, H. (1981), *Z. Naturforsch.* **36c**, 742.

Bruns, B., Hahlbrock, K. and Schäfer, E. (1986), *Planta* **169**, 393.

Butt, V.S. and Lamb, C.J. (1981). In *The Biochemistry of Plants* (eds P.K. Stumpf and E.E. Conn), Academic Press, New York, Vol. 7, pp. 627–665.

Chappell, J. and Hahlbrock, K. (1984), *Nature (London)* **311**, 76.

Charles, D.J. and Cherry, J.H. (1986), *Phytochemistry* **25**, 1067.
Charriere-Ladreix, Y., Douce, R. and Joyard, J. (1981), *FEBS Lett.* **133**, 55.
Chmiel, E., Sütfeld, R. and Wiermann, R. (1983), *Biochem. Physiol. Pflanzen* **178**, 139.
Coen, E.S., Carpenter, R. and Martin, C. (1986), *Cell* **47**, 285.
Collins, F.W., DeLuca, V., Ibrahim, R.K., Voirin, B. and Jay, M. (1981), *Z. Naturforsch.* **36c**, 730.
Cosio, E.G., Weissenböck, G. and McClure, J.W. (1985), *Plant Physiol.* **78**, 14.
Coulson, C.J., King, D.J. and Wiseman, A. (1984), *Trends Biochem. Sci.* **9**, 446.
Creasy, L.L (1985), *Recent Adv. Phytochem.* **19**, 47.
Crowden, R.K. (1982), *Phytochemistry* **21**, 2989.
Dahlbender, B. and Strack, D. (1986), *Phytochemistry* **25**, 1043.
Dangelmayr, B., Stotz, G., Spribille, R. and Forkmann, G. (1983), *Z. Naturforsch.* **38c**, 551.
DeLuca, V. and Ibrahim, R.K. (1982), *Phytochemistry* **21**, 1537.
DeLuca, V. and Ibrahim, R.K. (1985a), *Arch. Biochem. Biophys.* **238**, 596.
DeLuca, V. and Ibrahim, R.K. (1985b), *Arch. Biochem. Biophys.* **238**, 606.
Dixon, R.A. (1986), *Biol. Rev.* **61**, 239.
Dixon, R.A., Dey, P.M. and Whitehead, I.M. (1982), *Biochim. Biophys. Acta* **715**, 25.
Dooner, H.K. (1981), *Cold Spring Harbor Symp. Quant. Biol.* **45**, 457.
Dooner, H.K. (1983), *Mol. Gen. Genet.* **189**, 136.
Dooner, H.K. and Nelson, O.E. (1977), *Proc. Natl. Acad. Sci. USA* **74**, 5623.
Dooner, H.K. and Nelson, O.E. (1979a) , *Proc. Natl. Acad. Sci. USA* **76**, 2369.
Dooner, H.K. and Nelson, O.E. (1979b), *Genetics* **91**, 309.
Duell-Pfaff, N. and Wellmann, E. (1982), *Planta* **136**, 213.
Ebel, J. (1986), *Annu. Rev. Phytopathol.* **24**, 235.
Ebel, J. and Hahlbrock, K. (1982). In *The Flanonoids – Advances in Research* (eds J.B. Harborne and T.J. Mabry), Chapman and Hall, London, pp. 641–679.
Egin-Bühler, B. and Ebel, J. (1983), *Eur. J. Biochem.* **133**, 335.
Feutry, A. and Letouzé, R. (1984), *Phytochemistry* **23**, 1557.
Forkmann, G. (1977a), *Planta* **137**, 159.
Forkmann, G. (1977b), *Phytochemistry* **16**, 299.
Forkmann, G. and Dangelmayr, B. (1980), *Biochem. Genet.* **18**, 519.
Forkmann, G. and Stotz, G. (1981), *Z. Naturforsch.* **36c**, 411.
Forkmann, G. and Stotz, G. (1984), *Planta* **161**, 261.
Forkmann, G., Heller, W. and Grisebach, H. (1980), *Z. Naturforsch.* **35c**, 691.
Forkmann, G., de Vlaming, P., Spribille, R., Wiering, H. and Schram, A.W. (1986), *Z. Naturforsch.* **41c**, 179.
Fritsch, H. and Grisebach, H. (1975), *Phytochemistry* **14**, 2437.
Fritsch, H., Kochs, G., Hagmann, M., Heller, W., Jung, J. and Grisebach, H. (1986), *6th Int. Congr. Pestic. Chem.*, IUPAC, August 10–15th, Ottawa, Abstract 3B-16.
Fritzemeier, K.H. and Kindl, H. (1981), *Planta* **151**, 48.
Froemel, S., de Vlaming, P., Stotz, G., Wiering, H., Forkmann, G. and Schram, A.W. (1985), *Theor. Appl. Genet.* **70**, 561.
Gerats, A.G.M., de Vlaming, P., Doodeman, M., Al, B. and Schram, A.W. (1982), *Planta* **155**, 364.
Gerats, A.G.M., Groot, S.P.C., Peterson, P.A. and Schram, A.W. (1983a), *Mol. Gen. Genet.* **190**, 1.
Gerats, A.G.M., Wallroth, M., Donker-Koopman, W., Groot, S.P.C. and Schram, A.W. (1983b), *Theor. Appl. Genet.* **65**, 349.
Gerats, A.G.M., Bussard, J., Coe, Jr., E.H. and Larson, R. (1984a), *Biochem. Genet.* **22**, 1161.
Gerats, A.G.M., Farcy, E., Wallroth, M., Groot, S.P.C. and Schram, A.W. (1984b), *Genetics* **106**, 501.
Gerats, A.G.M., Vrijlandt, E., Wallroth, M. and Schram, A.W. (1985), *Biochem. Genet.* **23**, 591.
Grab, D., Loyal, R. and Ebel, J. (1985), *Arch. Biochem. Biophys.* **243**, 523.
Grand, C. (1984), *FEBS Lett.* **169**, 7.
Grand, C., Boudet, A. and Boudet A.M. (1983), *Planta* **158**, 225.
Gräwe, W. and Strack, D. (1986), *Z. Naturforsch.* **41c**, 28.
Grisebach, H. (1985). In *The Biochemistry of Plant Phenolics* (eds van Sumere C.F. and P.J. Lea), *Annu. Proc. Phytochem. Soc. Eur.*, Clarendon Press, Oxford, Vol. 25, pp. 183–198.
Grisebach, H. and Zilg, H. (1968), *Z. Naturforsch.* **23b**, 499.
Grisebach, H., Börner, H., Hagmann, M., Hahn, M.G., Leube, J. and Moesta, P. (1985). In *Cellular and Molecular Biology of Plant Stress* (eds J.L. Key and T. Kosuge) Alan R. Liss, New York, pp. 275–290.
Gross, G.G. (1985). In *Biosynthesis and Biodegradation of Wood Components* (ed. T. Higuchi), Academic Press, Orlando, pp. 229–271.
Gupta, S. and Acton, J. (1979), *Biochim. Biophys. Acta* **570**, 187.
Hadwiger, L.A. and Schwochau, M.E. (1971), *Plant Physiol.* **47**, 588.
Hagmann, M. and Grisebach, H. (1984), *FEBS Lett.* **175**, 199.
Hagmann, M., Heller, W. and Grisebach, H. (1983), *Eur. J. Biochem.* **134**, 547.
Hagmann, M., Heller, W. and Grisebach, H. (1984), *Eur. J. Biochem.* **142**, 127.
Hahlbrock, K. and Grisebach, H. (1975). In *The Flavonoids* (eds J.B. Harborne, T.J. Mabry and H. Mabry), Chapman and Hall, London, pp. 866–915.
Hahlbrock, K., Lamb, C.J., Purwin, C., Ebel, J., Fautz, E. and Schäfer, E. (1981), *Plant Physiol.* **67**, 768.
Hanson, K.R. and Havir, E.A. (1981). In *The Biochemistry of Plants* (eds P.K. Stumpf and E.E. Conn), Academic Press, New York, Vol. 7, pp. 577–625.
Harborne, J.B. and Boardley, M. (1985), *Z. Naturforsch.* **40c**, 305.
Hattori, T. and Ohta, Y. (1985), *Plant Cell Physiol.* **26**, 1101.
Hauteville, M., Chadenson, M. and Chopin, J. (1979), *Bull. Soc. Chim. Fr.*, II-125.
Heim, D., Nicholson, R.L., Pascholati, S.F., Hagerman, A.E. and Billett, W. (1983), *Phytopathology* **73**, 424.
Heller, W. and Kühnl, T. (1985), *Arch. Biochem. Biophys.* **241**, 453.
Heller, W., Egin-Bühler, B., Gardiner, S.E., Knobloch, K.-H., Matern, U., Ebel, J. and Hahlbrock, K. (1979), *Plant Physiol.* **64**, 371.
Heller, W., Britsch, L., Forkmann, G. and Grisebach, H. (1985a), *Planta* **163**, 191.
Heller, W., Forkmann, G., Britsch, L. and Grisebach, H. (1985b), *Planta* **165**, 284.
Hinderer, W. and Seitz, H.U. (1985), *Arch. Biochem. Biophys.* **240**, 265.
Hinderer, W. and Seitz, H.U. (1986), *Arch. Biochem. Biophys.* **246**, 217.
Hinderer, W., Noé, W. and Seitz, H.U. (1983), *Phytochemistry* **22**, 2417.
Hinderer, W., Petersen, M. and Seitz, H.U. (1984), *Planta* **160**, 544.

Hodgins, D.S. (1971), *J. Biol. Chem.* **246**, 2977.
Holländer, H. and Amrhein, N. (1981), *Planta* **152**, 374.
Hösel, W. (1981). In *The Biochemistry of Plants* (eds P.K. Stumpf and E.E. Conn), Academic Press, New York, Vol. 7, pp. 725–753.
Hrazdina, G., Lifson, E. and Weeden, N.F. (1986), *Arch. Biochem. Biophys.* **247**, 414.
Ibrahim, R.K. (1986), Private communication.
Ingham, J.L. and Markham, K.R. (1980), *Phytochemistry* **19**, 1203.
Jay, M., De Luca, V. and Ibrahim, R.K. (1983), *Z. Naturforsch.* **38c**, 413.
Jay, M., De Luca, V. and Ibrahim, R.K. (1985), *Eur. J. Biochem.* **153**, 321.
Jensen, R.A. (1985), *Physiol. Plant.* **66**, 164.
Jones, D.H. (1984), *Phytochemistry* **23**, 1349.
Jonsson, L.M.V., Aarsman, M.E.G., Schram, A.W. and Bennink, G.J.H. (1982), *Phytochemistry* **21**, 2457.
Jonsson, L.M.V., De Vlaming, P., Wiering, H., Aarsman, M.E.G. and Schram, A.W. (1983), *Theor. Appl. Genet.* **66**, 349.
Jonsson, L.M.V., Aarsman, M.E.G., Poulton, J.E. and Schram, A.W. (1984a), *Planta* **160**, 174.
Jonsson, L.M.V., Aarsman, M.E.G., De Vlaming, P. and Schram, A.W. (1984b), *Theor. Appl. Genet.* **68**, 459.
Jonsson, L.M.V., Aarsman, M.E.G., Bastiaannet, J., Donker-Koopman, W.E., Gerats, A.G.M. and Schram, A.W. (1984c), *Z. Naturforsch.* **39c**, 559.
Jonsson, L.M.V., Aarsman, M.E.G., van Diepen, J., de Vlaming, P., Smit, N. and Schram, A.W. (1984d), *Planta* **160**, 341.
Jourdan, P.S. and Mansell, R.L. (1982), *Arch. Biochem. Biophys.* **213**, 434.
Kamsteeg, J., van Brederode, J. and van Nigtevecht, G. (1978), *Biochem. Genet.* **16**, 1045.
Kamsteeg, J., van Brederode, J. and van Nigtevecht, G. (1980a), *Z. Naturforsch.* **35c**, 249.
Kamsteeg, J., van Brederode, J. and van Nigtevecht, G. (1980b), *Z. Pflanzenphysiol.* **96**, 87.
Kamsteeg, J., van Brederode, J., Hommels, C.H. and van Nigtevecht, G. (1980c), *Biochem. Physiol. Pflanzen* **175**, 403.
Kamsteeg, J., van Brederode, J., Verschuren, P.M. and van Nigtevecht, G. (1981), *Z. Pflanzenphysiol.* **102**, 435.
Kehrel, B. and Wiermann, R. (1985), *Planta* **163**, 183.
Kerscher, F. and Franz, G. (1987), *Z. Naturforsch.* **42c**, 519.
Khouri, H. and Ibrahim, R.K. (1984), *Eur. J. Biochem.* **142**, 559.
Khouri, H.E., Ishikura, N. and Ibrahim, R.K. (1986), *Phytochemistry* **25**, 2475.
Kindl, H. (1985). In *Biosynthesis and Biodegradation of Wood Components* (ed. T. Higuchi), Academic Press, Orlando, pp. 349–277.
Klein, A.S. and Nelson, O.E., Jr. (1983), *Phytochemistry* **22**, 2634.
Kleinehollenhorst, G., Behrens, H., Pegels, G., Srunk. N. and Wiermann, R. (1982), *Z. Naturforsch.* **37c**, 587.
Knogge, W. and Weissenböck, G. (1984), *Eur. J. Biochem.* **140**, 113.
Knogge, W., Beulen, C. and Weissenböck, G. (1981a), *Z. Naturforsch.* **36c**, 389.
Knogge, W., Weissenböck, G. and Strack, D. (1981b), *Z. Naturforsch.* **36c**, 197.
Kochs, G. and Grisebach, H. (1986), *Eur. J. Biochem.*, **155**, 311.
Kochs, G. and Grisebach, H. (1987), *Z. Naturforsch.* **42c**, 343.
Kombrink, E. and Hahlbrock, K. (1986), *Plant Physiol.* **81**, 216.
Köster, J. and Barz, W. (1981), *Arch. Biochem. Biophys.* **212**, 98.
Köster, J., Bussmann, R. and Barz, W. (1984), *Arch. Biochem. Biophys.* **234**, 513.
Kristiansen, K.N. (1984), *Carlsberg Res. Commun.* **49**, 503.
Kristiansen, K.N. (1986), *Carlsberg Res. Commun.* **51**, 51.
Kuhn, B., Forkmann, G. and Seyffert, W. (1978), *Planta* **138**, 199.
Kuhn, D.N., Chapell, J., Boudet, A. and Hahlbrock, K. (1984), *Proc. Natl. Acad. Sci. USA* **81**, 1102.
Kuroki, G. and Poulton, J.E. (1981), *Z. Naturforsch.* **36c**, 916.
Kutsuki, H., Shimada, M. and Higuchi, T. (1982), *Phytochemistry* **21**, 267.
Lamb, C.J. (1986), Private communication.
Larson, R.L. and Bussard, J.B. (1986), *Plant Physiol.* **80**, 483.
Larson, R.L. and Coe, E.H., Jr. (1968), *Proc. XII Int. Congr. Genet.* **1**, 131.
Larson, R.L. and Coe, E.H., Jr. (1977), *Biochem. Genet.* **15**, 153.
Larson, R.L. and Lonergan, C.M. (1973), *Cereals Res. Commun. (Hungary)* **1**, 13.
Loschke, D.C., Hadwiger, L.A. and Wagoner, W. (1983), *Physiol. Plant Pathol.* **23**, 163.
Lowenstein, J.M. (ed.) (1981), *Methods Enzymol.* **71**, 5.
Lüderitz, T., Schatz, G. and Grisebach, H. (1982), *Eur. J. Biochem.* **123**, 583.
Macheix, J.-J. and Ibrahim, R.K. (1984), *Biochem. Physiol. Pflanzen* **179**, 659.
Matern, U., Potts, J.R.M. and Hahlbrock, K. (1981), *Arch. Biochem. Biophys.* **208**, 233.
Matern, U., Feser, C. and Hammer, D. (1983a), *Arch. Biochem. Biophys.* **226**, 206.
Matern, U., Heller, W. and Himmelspach, K. (1983b), *Eur. J. Biochem.* **133**, 439.
Matern, U., Reichenbach, C. and Heller, W. (1986), *Planta* **167**, 183.
Maule, A.J. and Ride, J.P. (1983), *Phytochemistry* **22**, 1113.
Möhle, B., Heller, W. and Wellmann, E. (1985), *Phytochemistry* **24**, 465.
Mol, J.N.M., Schram, A.W., de Vlaming, P., Gerats, A.G.M., Kreuzaler, F., Hahlbrock, K., Reif, H.J. and Veltkamp, E. (1983), *Mol. Gen. Genet.* **192**, 424.
Mol, J.N.M., Robbins, M.P., Dixon, R.A. and Veltkamp, E. (1985), *Phytochemistry* **24**, 2267.
Nielsen, N.C. (1981), *Methods Enzymol.* **71**, 44.
O'Relly, C., Shepherd, N.S., Pereira, A., Schwarz-Sommer, Zs., Bertram, I., Robertson, D.S., Peterson, P.A. and Saedler, H. (1985), *EMBO J.* **4**, 877.
Ozeki, Y., Sakano, K., Komamine, A., Tanaka, Y., Noguchi, H., Sankawa, U. and Suzuki, T. (1985), *J. Biochem. Tokyo* **98**, 9.
Patschke, L., Barz, W. and Grisebach, H. (1966), *Z. Naturforsch.* **21b**, 45.
Peterson, J.A. and Prough, R.A. (1986). In *Cytochrome P-450. Structure, Mechanism, and Biochemistry* (ed. P.R. Ortiz de Montellano), Plenum Press, New York, pp. 89–117.
Ragg, H., Kuhn, D.N. and Hahlbrock, K. (1981), *J. Biol. Chem.* **256**, 10061.
Rall, S. and Hemleben, V. (1984), *Plant Mol. Biol.* **3**, 137.
Reichhart, D., Simon, A., Durst, F., Mathews, J.M. and Ortiz de Montellano, P.R. (1982), *Arch. Biochem. Biophys.* **216**, 522.

Reif, H.J., Niesbach, U., Deumling, B. and Saedler, H. (1985), *Mol. Gen. Genet.* **199**, 208.
Reimold U., Kröger, M., Kreuzaler, F. and Hahlbrock, K. (1983), *EMBO J.* **2**, 1801.
Robbins, M.P. and Dixon, R.A. (1984), *Eur. J. Biochem.* **145**, 195.
Rohde, W. (1986), Private communication.
Rohde, W., Barzen, E., Marocco, A., Saedler, H. and Salamini, F. (1986), *Barley Genet.*, Vol. 5 (in press).
Rolfs, C.-H. and Kindl, H. (1984), *Plant Physiol.* **75**, 489.
Rubin, B., Penner, D. and Saettler, A.W. (1983), *Environ. Toxicol. Chem.* **2**, 295.
Schmelzer, E., Börner, H., Grisebach, H., Ebel, J. and Hahlbrock, K. (1984), *FEBS Lett.* **172**, 59.
Schmidt, W.E. and Ebel, J. (1986), *Biol. Chem. Hoppe-Seyler* **367**, 202S.
Schröder, G., Zähringer, U., Heller, W., Ebel, J. and Grisebach, H. (1979), *Arch. Biochem. Biophys.* **194**, 635.
Schüz, R., Heller, W. and Hahlbrock, K. (1983), *J. Biol. Chem.* **258**, 6730.
Schwinn, F.J. (1984), *Pestic. Sci.* **15**, 40.
Seyffert, W. (1982), *Biol. Zbl.* **101**, 465.
Sommer, H. and Saedler, H. (1986), *Mol. Gen. Genet.* **202**, 429.
Sommer, H., Carpenter, R., Harrison, B.J. and Saedler, H. (1985), *Mol. Gen. Genet.* **199**, 225.
Spribille, R. and Forkmann, G. (1981), *Z. Naturforsch.* **36c**, 619.
Spribille, R. and Forkmann, G. (1982a), *Phytochemistry* **21**, 2231.
Spribille, R. and Forkmann, G. (1982b), *Planta* **155**, 176.
Spribille, R. and Forkmann, G. (1984), *Z. Naturforsch.* **39c**, 714.
Stafford, H.A. (1983), *Phytochemistry* **22**, 2643.
Stafford, H.A. and Lester, H.H. (1982), *Plant Physiol.* **70**, 695.
Stafford, H.A. and Lester, H.H. (1984), *Plant Physiol.* **76**, 184.
Stafford, H.A. and Lester, H.H. (1985), *Plant Physiol.* **78**, 791.
Stafford, H.A., Lester, H.H. and Porter, L.J. (1985), *Phytochemistry* **24**, 333.
Steyns, J.M. and van Brederode, J. (1986a), *Z. Naturforsch.* **41c**, 9.
Steyns, J.M. and van Brederode, J. (1986b), *Biochem. Genet.* **24**, 349.
Steyns, J.M., Mastenbroek, O., van Nigtevecht, G. and van Brederode, J. (1984), *Z. Naturforsch.* **39c**, 568.
Stickland, R.G. and Harrison, B.J. (1974), *Heredity* **33**, 108.
Stotz, G. and Forkmann, G. (1981), *Z. Naturforsch.* **36c**, 737.
Stotz, G. and Forkmann, G. (1982), *Z. Naturforsch.* **37c**, 19.
Stotz, G., Spribille, R. and Forkmann, G. (1983), *Jahrestagung der Gesellschaft für Genetik*, 2.-3. Juni, Bayreuth, Poster Abstract.
Stotz, G., Spribille, R. and Forkmann, G. (1984), *J. Plant Physiol.* **116**, 173.
Stotz, G., de Vlaming, P., Wiering, H., Schram, A.W. and Forkmann, G. (1985), *Theor. Appl. Genet.* **70**, 300.
Strack, D. (1982), *Planta* **155**, 31.
Sutter, A., Poulton, J.E. and Grisebach, H. (1975), *Arch. Biochem. Biophys.* **170**, 547.
Sütfeld, R. and Wiermann, R. (1981), *Z. Naturforsch.* **36c**, 30.
Sütfeld, R., Kehrel, B. and Wiermann, R. (1978), *Z. Naturforsch.* **33c**, 841.
Sweigard, J.A., Matthews, D.E. and VanEtten, H.D. (1986), *Plant Physiol.* **80**, 277.
Tabak, A.J.H., Schram, A.W. and Bennink G.J.H. (1981), *Planta* **153**, 462.
Takeda, K., Harborne, J.B. and Self, R. (1986), *Phytochemistry* **25**, 1337.
Teusch, M. (1986), *Planta* **169**, 559.
Teusch, M. and Forkmann, G. (1987), *Phytochemistry* **26**, 2181.
Teusch, M., Forkmann, G. and Seyffert, W. (1986a), *Z. Naturforsch.* **41c**, 699.
Teusch, M., Forkmann, G. and Seyffert, W. (1986b), *Planta* **168**, 586.
Teusch, M., Forkmann, G. and Seyffert, W. (1987), *Phytochemistry* **26**, 991.
Thresh, K. and Ibrahim, R.K. (1985), *Z. Naturforsch.* **40c**, 331.
Tietjen, K.G., Hunkler, D. and Matern U. (1983), *Eur. J. Biochem.* **131**, 401.
Upadhyaya, K.C., Sommer, H., Krebbers, E. and Saedler, H. (1985), *Mol. Gen. Genet.* **199**, 201.
van Brederode, J. (1986), Private communication.
van Brederode, J. and Kamps-Heinsbroek, R. (1981a), *Z. Naturforsch.* **36c**, 486.
van Brederode, J. and Kamps-Heinsbroek, R. (1981b), *Z. Naturforsch.* **36c**, 484.
van Brederode, J. and Steyns, J.M. (1983), *Z. Naturforsch.* **38c**, 549.
van Brederode, J. Kamps-Heinsbroek, R. and Mastenbroek, O. (1982), *Z. Pflanzenphysiol.* **106**, 43.
van Brederode, J., Kamps-Heinsbroek, R. and Steyns, J. (1987), *Experientia* **43**, 202.
van Weely, S., Bleumer, A., Spruyt, R. and Schram, A.W. (1983), *Planta* **159**, 226.
von Wettstein, D., Nilan, R.A., Ahrenst-Larsen, B., Erdal, K., Ingversen, J., Jende-Strid, B., Kristiansen, K.N., Larsen, J., Outtrup, H. and Ullrich, S.E. (1985), *MBBA Tech. Q.* **22**, 41.
Walker, G.C. (1984), *Microbiol. Rev.* **48**, 60.
Welle, R., (1986), Private communication.
Wellmann, E. (1971), *Planta* **101**, 283.
Wellmann, E. (1974), *Ber. Dtsch. Bot. Ges.* **87**, 275.
Wellmann, E. (1975), *FEBS Lett.* **51**, 105.
Wendorff, H. and Matern, U. (1986), *Eur. J. Biochem.* **161**, 391.
Wengenmayer, H., Ebel, J. and Grisebach, H. (1974), *Eur. J. Biochem.* **50**, 135.
West, C.A. (1980). In *The Biochemistry of Plants* (eds P.K. Stumpf and E.E. Conn), Academic Press, New York, Vol. 2, pp. 317–364.
West, C.A. (1981), *Naturwissenschaften* **68**, 447.
Whitehead, J.M. and Dixon, R.A. (1983), *Biochim. Biophys. Acta* **747**, 298.
Wienand, U., Weydemann, U., Niesbach-Klösgen, U., Peterson, P.A. and Saedler, H. (1986), *Mol. Gen. Genet.* **203**, 202.
Yoshikawa, M., Keen, N.T. and Wang, M.-C. (1983), *Plant Physiol.* **73**, 497.
Zähringer, U., Ebel, J., Mulheirn, L.J. Lyne, R.L. and Grisebach, H. (1979), *FEBS Lett.* **101**, 90.
Zähringer, U., Schaller, E. and Grisebach, H. (1981), *Z. Naturforsch.* **36c**, 234.

12

Distribution of flavonoids in the lower plants and its evolutionary significance

KENNETH R. MARKHAM

12.1 INTRODUCTION

The present chapter is the first compilation in which the flavonoid chemistry of all lower plants has been brought together in a comprehensive manner. In a chapter of this size it is not possible to be exhaustive in a historical sense, and reference to early work is only made when such work is considered to be of continuing significance. Nevertheless this summary attempts to establish a reasonably complete database representing work published prior to early 1986 and upon which future updates can be based. The emphasis throughout is on published work in which flavonoid structures have been defined. Thus some extensive surveys in which (for example) only presence or absence data appear are not fully represented. Discussion is focused primarily on the distribution of flavonoids and the possible phylogenetic significance (if evident) of this distribution. Limited space precludes all but passing comment on the myriad of taxonomic implications addressed in the literature.

Data for all major plant groups are presented in four large tables which have been structured to reveal the variety of flavonoids in a particular taxon rather than where flavonoid types are found. This approach has been chosen in order to highlight biochemical similarities and/or differences between plant groups. In each table, the major ordinal or family groups are arranged alphabetically as also are the genera within each group. This should not only aid the non-botanist to find listings, but has also overcome the problem faced by the author of having to select from the bewildering array of different phylogenetically arranged groupings proposed in the botanical literature. The sheer volume of data in these tables has necessitated the use of symbols for the potentially lengthy descriptions of flavonoid structures. Structure information in Tables 1–4 is presented in three columns, the first symbolizing the flavonoid type and oxygenation pattern, the second the glycosylation type (if any) and the third the locations and types of substituents. Symbols employed are defined as follows:

Flavonoid type

F, flavone; HF, dihydroflavone (flavanone); FL, flavonol; HFL, dihydroflavonol (flavanonol); I, isoflavone; A, aurone; C, chalcone; HC, dihydrochalcone; B, biflavonyl (symbols following, define the two flavonoids involved and the linkage, e.g. amentoflavone = BF574 (3′8″) F574); AN, anthocyanidin; P, proanthocyanidin; FN, flavan.

Oxygenation pattern

Oxygenated sites are listed in numerical order, within the sequence, C-ring sites first, A-ring sites second and B-ring sites third. Thus quercetin is represented by FL35734. C-ring substitution in chalcones is designated by α or β.

Glycosylation type

O and C represent *O*- and *C*-glycosylation respectively.

The Flavonoids. Edited by J.B. Harborne
Published in 1988 by Chapman and Hall Ltd,
11 New Fetter Lane, London EC4P 4EE.

Substituent type

For di- and tri-saccharides, sugars are listed in sequence from the flavonoid out.

glu,	glucose	gluu,	glucuronic acid
gal,	galactose	galu,	galacturonic acid
ara,	arabinose	xyl,	xylose
rha,	rhamnose	all,	allose
rut,	rutinose	soph,	sophorose
gent,	gentiobiose	pent,	pentose
hex,	hexose	gly,	undefined sugar
Me,	methyl	neoh,	neohesperidose
+,	links co-occurring components		
/,	separates alternative substituents (e.g. 7Me3glu/3gal = 7Me3glu + 7Me3gal)		
X2,	two compounds		
(1/2),	one species out of two studied		

12.2 FLAVONOID DISTRIBUTION IN THE ALGAE

Early related claims that flavonols and anthocyanins had been found in *Clamydomonas* (Kuhn and Low, 1949; Moewus, 1950; Birch *et al.*, 1953) are now thought to be spurious (Ryan, 1955; Harborne, 1967). Other pre-1967 reports also appear to have alternative explanations (Harborne, 1967). Since 1967 two further reports, describing the occurrence of *C*-glycosylflavones in the Charophyceae, have appeared (Markham and Porter, 1969; Swain, 1975). Swain's comments refer to unpublished work on *Chara* by Mabry which remains unpublished and has proved to be unreproducible (Mabry, 1982). The other report describes the partial characterization of several apigenin and luteolin di-*C*-glycosides from *Nitella hookeri*. These were found at low levels, but the identifications were well supported with physical evidence and the voucher specimen has recently been confirmed as *N. hookeri*. The uniqueness of this finding prompted the present author to seek confirmation of this earlier work and to extend the investigation to include a range of fresh- and salt-water algae. Representatives of the following genera were included: *Chara, Nitella, Codium, Enteromorpha, Ulva, Caulerpa, Chaetomorpha, Cladophora, Rhizoclonium, Trentephohia, Acetabularia, Ectocarpus, Griffithsia, Fritschiella* and *Hydrodictyon*. All were surveyed by two-dimensional paper chromatography and proved negative for flavonoids at the levels investigated. The *Nitella* sample was collected from a deep-water shady location, whereas that used in the 1969 study was harvested from the surface of a densely packed tub of *Nitella* in full sunlight. This original source was unfortunately no longer available, but it is possible that *Nitella* growing under such conditions could have responded to light (and other?) stress by producing protective levels of flavonoids. However, this is only speculation and until confirmation of the original finding is achieved, the ability of *Nitella*, or any algae, to biosynthesize flavonoids can only be accepted with reservation.

12.3 FLAVONOID DISTRIBUTION IN THE BRYOPHYTES

Since the late 1960s it has been increasingly evident that flavonoids are of widespread occurrence in the liverworts and mosses but are apparently absent from the Anthocerotales. Although a number of reviews/surveys of the distribution of flavonoids in these groups have appeared in recent years, they vary widely in style and comprehensiveness, e.g. for bryophytes Suire (1975), Markham and Porter (1978b), Suire and Asakawa (1979), Zinsmeister and Mues (1980), Huneck (1983) and for liverworts only Spencer (1979), Asakawa (1982) and Mues (1986). That much work still remains to be done is evident from the statistics quoted by Mues (1986) in his compilation of presence or absence data. Only about 6% of the 7000 presumed valid liverwort species have so far been investigated. The level for mosses is not available but of the 14 000–25 000 known species (Bold, 1973), it is probable that less than 2% have been studied. Broad-based chromatographic surveys (e.g. McClure and Miller, 1967; Mues, 1986) reveal that of the species so far studied, about 41% of liverworts and 48% of mosses have been found to accumulate flavonoids.

The survey data presented here in Tables 12.1 and 12.2 are ordered largely in terms of the classification systems of Grolle (1983) for the liverworts (Hepaticae) and Vitt (1983) for the mosses (Musci). Presented in alphabetical order the classification system used is as follows:

Class: Anthocerotales
Class: Hepaticae (Table 12.1)
  Orders: Calobryales (Table 12.1.1)
    Jungermanniales (Table 12.1.2)
    Marchantiales – including Sphaerocarpales (Table 12.1.3)
    Metzgeriales (Table 12.1.4)
    Monocleales (Table 12.1.5)
    Takakiales (Table 12.1.6)
Class: Musci (Table 12.2)
  Order: Andreales (Table 12.2.1)
    Bryales – including 15 suborders (Table 12.2.2)
    Polytrichales (Table 12.2.3)
    Sphagnales (Table 12.2.4)
    Tetraphidales (Table 12.2.5)

12.3.1 Class Anthocerotales

Numerous investigators have failed to find flavonoids in any members of this order, the genera *Anthoceros*, *Megaceros* and *Phaeoceros* having been examined (Brehm and Comp, 1967; Markham and Porter, 1978b; Mues, 1986;

Markham, unpublished). Mendez and Sanz-Cabanilles (1979) have reported the isolation of methyl caffeate and methyl *p*-coumarate from *Anthoceros/Phaeoceros* and Mues (1986) has recently confirmed the occurrence of cinnamic acids (but not their esters) in these and other genera.

12.3.2 Class Hepaticae (Table 12.1)

A wide range of flavonoids which includes flavone *C*- and *O*-glycosides, flavonols, dihydroflavones, dihydrochalcones and even aurones has been shown to occur in liverworts. As yet, however, no authenticated identifications of isoflavones, chalcones, biflavonyls or proanthocyanidins have been recorded. Flavonoids do not appear to be evenly distributed throughout the liverworts. Mues (1986) has reported that while 70% of the species in the Marchantiales/Sphaerocarpales accumulate detectable levels of flavonoids, only 27% of the Metzgeriales and 38% of the Jungermanniales appear to do so.

(a) *Order Calobryales*
(Table 12.1.1)

Only one species of this order, *Haplomitrium gibbsiae*, has been studied in detail and *H. mnioides* is thought to contain flavonoids from two-dimensional paper chromatographic analysis. *H. gibbsiae* is rich in flavonoids but accumulates only flavone *O*-glucosides and their acylated derivatives (Markham, 1977). These data if typical indicate no close affinity with any other order, including the Takakiales with which the Calobryales are often linked (Grolle, 1972; Bold, 1973; Schuster, 1983). Other chemical studies confirm this distinction (Asakawa *et al.*, 1979).

(b) *Order Jungermanniales*
(Table 12.1.2)

The occurrence of flavonoids in families of this order is scattered (Mues, 1986). Two families have been shown to accumulate them consistently, the Pleuroziaceae and the Radulaceae, and they are probably also widespread in the Lejeuneaceae (Kruijt *et al.*, 1986). In contrast, some families may not produce flavonoids at all (Mues, 1986). Biosynthetically 'complex' flavone *C*-glycosides, i.e. containing a range of sugars and/or additional *O*-glycosylation, oxygenation, methylation etc., are the predominant flavonoids found in this order. However, flavone *O*-glycosides and dihydrochalcones also appear occasionally and flavonols have been reported once.

A number of the complex flavone *C*-glycosides are of particular interest. For example, several di-*C*-glycosides of tricetin and its methyl ethers have been isolated and characterized for the first time, including the 6,8-di-*C*-glucoside and 6-*C*-arabinosyl-8-*C*-glucoside of tricetin from *Plagiochila* and *Radula* and the 6,8-di-*C*-glucoside of tricin from *Frullania*. The isolation of substantial quantities of the rare neoschaftoside from *Radula complanata* permitted for the first time its full structural elucidation as 6-*C*-β-D-glucopyranosyl-8-*C*-β-L-arabinopyranosyl-apigenin (Besson *et al.*, 1984). 'Neo' isomers in the luteolin and tricetin series have also been detected.

(c) *Order Marchantiales – including Sphaerocarpales*
(Table 12.1.3)

Flavone uronic acid derivatives, and particularly *O*-glucuronides, occur throughout this order, the only well-documented exception being *Corsinia* in which flavonols are the sole flavonoids. Flavonols are rare in liverworts generally, and their presence in *Corsinia* sets this genus (possibly together with *Exomortheca* and *Targionia*) apart from the bulk of the Marchantialean genera, including *Carrpos* with which it has in the past been closely aligned (Schuster, 1966). Also appearing sporadically in species of this order are biosynthetically 'simple' flavone *C*-glycosides, i.e. di-*C*-glucosides of apigenin and luteolin.

The Sphaerocarpalean liverworts (genera *Riella*, *Sphaerocarpos* and *Carrpos*) are indistinguishable from those of the Marchantiales in terms of flavonoid content, except in so far as they appear to consistently lack flavone *C*-glycosides. While the biosynthetic diversity displayed is more limited, the predominant presence of apigenin and luteolin *O*-glucuronides and the occurrence of the aurone aureusidin 6-*O*-glucuronide align this taxon very closely with the Marchantiales, e.g. as in the Marchantiidae (Schuster, 1983). The minute desert-living liverwort *Carrpos*, whose taxonomic placement has been disputed, fits comfortably into the same order on the basis of its flavonoid chemistry (Markham, 1980).

(d) *Order Metzgeriales*
(Table 12.1.4)

Relatively little work has been reported on species of this order and the bulk of the chemistry that has appeared relates to the genera *Hymenophyton* and *Metzgeria/Apometzgeria*. As with the Jungermanniales, the characteristic flavonoids appear to be 'complex' flavone *C*-glycosides and again some novel examples have been identified. These include apigenin 6,8-di-*C*-arabinosides from *Hymenophyton*, apigenin 6-*C*-rhamnoside-7-*O*-glucoside (isofurcatain 7-glucoside) from *Metzgeria*, tricetin 3′,4′-dimethyl ether (apometzgerin)-6,8-di-*C*-arabinoside from *Apometzgeria* and tricin 6,8-di-*C*-glycosides from *Metzgeria*. The single example of the occurrence of flavonols proved to be of taxonomic value in that it permitted for the first time a clear distinction of *Hymenophyton leptopodum* from *H. flabellatum* and so led to reinstatement of the former as a separate species (Campbell *et al.*, 1975).

Table 12.1 Flavonoids in the Hepaticae (liverworts)

Species	*Flavonoids found*			*References*
12.1.1 ORDER CALOBRYALES				
Haplomitrium gibbsiae	F574	O	7glu + 74′glu + monoacyl derivs	Markham (1977)
Haplomitrium gibbsiae	F5734	O	7glu + acyl deriv.	
12.1.2 ORDER JUNGERMANNIALES				
20 Families	HFL? HF?		Dihydroflavones/ flavonols characterize one major group	Mues (1982b)
20 Families	F		Flavones characterize one major group	
20 Families			− ve? One major group (cinnamics dominant)	
224 Species			+ ve (131)	Mues (1986)
69 Species			+ ve(19), − ve(50)	Mues and Zinsmeister (1977)
Blepharostoma minus	F5734	C	68glu	Mues (1986)
Blepharostoma trichophyllum	F574	C	68glu + 6ara8glu	Mues (1982a)
Blepharostoma trichophyllum	F5734	C	68glu + 6ara8glu + 6glu8ara	
Chiloscyphus pallescens	F574	CO	6glu7glu?	Dombray (1926)
Frullania davurica	F574		74′Me	Mues *et al.* (1984)
Frullania davurica	F5734		73′4′Me + 3′4′Me	
Frullania davurica	F5674	O	7diglu + 7gluxyl/ gluxylglu? + 6xyl7glu?	
Frullania davurica	F56734	O	7soph + 6xyl7glu? + 6glu7gluxyl?	
Frullania davurica	F56734	O	3′Me7soph?7gluxyl?/ 6xyl7glu?	
Frullania davurica	F56734	O	3′Me7gluxyl6glu?	
Frullania davurica	F57345	C	3′5′Me68glu	
Frullania dilatata	F56734		7Me	
Frullania dilatata	F5734	O	7glu + 7glu(1–6)glu + 74′glu?	Mues *et al.* (1983)
Frullania dilatata	F56734	O	7glu + 74′glu?	Mues *et al.* (1983)
Frullania jackii	F574		74′Me	Mues *et al.* (1984)
Frullania jackii	F5734		73′4′Me + 3′4′Me	
Frullania jackii	F56734	O	7Me + 7soph + 6xyl7glu? + 6diglu?	
Frullania jackii	F56734	O	7glurha? + 7glurha6glu?	
Frullania jackii	F56734	O	3′Me7soph?	
Frullania jackii	F57345	C	3′5′Me68glu	
Frullania jackii	F5674	O	6glu7glurha?	
Frullania jackii	F56	O	7soph prob = gent (m.p. data)	Klimek (1987)
Frullania tamarisci	F5734	O	7glu(acyl) + 74′glu(acyl)	Mues *et al.* (1983)
Frullania tamarisci	F574	O	gly	Mues and Zinsmeister (1977)
Frullania vethii	F574		74′Me	Asakawa *et al.* (1980)
Frullanoides densifolia	FL3574		3Me	Kruijt *et al.* (1986)
Herbertus			− ve	Mues (1982a)
Lejeunea cavifolia	F	C	glycs	Mues and Zinsmeister (1977)
Lophocolea bidentata			+ ve	Mues and Zinsmeister (1973)
Mastigophora			− ve	Mues (1982a)
Plagiochila asplenioides	F57345	C	68glu + 68glu8hex + 6pent8hex + 8pent6hex	Mues and Zinsmeister (1976)
Plagiochila asplenioides	F57345	C	3′Me6hex8pent	
Plagiochila asplenioides	F5734	C	6xyl8glu + 6glu8xyl + 68glu	Mues and Zinsmeister (1975)
Plagiochila asplenioides	F574	C	68hex	Mues and Zinsmeister (1976)
Porella baueri	F574	CO	6glu7glu	Nilsson (1969a)
Porella bolanderii	F574	CO	6glu7glu	
Porella capensis	F574	CO	6glu7glu	
Porella cordaeana	F574	CO	6glu7glu	Nilsson (1969a)
Porella platyphylla	F574	CO	6glu7glu	

Table 12.1 *(Contd.)*

Species	*Flavonoids found*			*References*
Porella platyphylla	F574	C	6glu8ara + 68glu + 6glu	Mues (1982a)
Porella platyphylla	F574	C	68glu	Nilsson (1973a)
Porella platyphylla	F574	CO	8glu7glu (spot test)	Molisch (1911)
Porella platyphylla	F574	CO	8glu7glu	Tjukavkina *et al.* (1970)
Porella platyphylloidea	F574	CO	6glu7glu	Nilsson (1969)
Porella spp.	F	C	detected	Mues (1982a)
Ptilidium			− ve	Mues (1982a)
Radula buccinifera	F57345	C	6ara8glu + 6glu8ara + 'neo' + 68glu	Mues (1984)
Radula caringtonii	F57345	C	68glu + 6glu8ara + 6glu	Mues (1984)
Radula complanata	F574	C	6glu8arab(neo), first str. detn.	Besson *et al.* (1984)
Radula complanata	F57345	C	6glu8ara	Markham and Mues (1984)
Radula complanata	F574	C	68glu + 6glu8ara + 'neo'	Mues (1984)
Radula complanata	F5734	C	6glu8ara + neocarlinoside	Mues (1984)
Radula complanata	F57345	C	6ara8glu + 68glu − Orha	Mues (1984)
Radula grandis	F57345	C	68glu − Ogly + 6glu8ara(neo)	Mues (1984)
Radula grandis	F5734	C	6glu8ara(neo)	Mues (1984)
Radula lindbergiana (mid-European)	F57345	C	6glu8ara + 'iso' + 'neo' + 6glu + 68glu	Mues (1984)
Radula lindbergiana (mid-European)	F5734	C	6glu8ara(neo)	Mues (1984)
Radula lindbergiana (mid-European)	F574	C	68glu + 6glu8ara + neo-schaftoside	Mues (1984)
Radula lindbergiana (mediterranean)	F57345	C	68glu + 6glu8ara + 'iso' + 'neo' + 6glu	Mues (1984)
Radula nudicaulis	F57345	C	68glu − Ogly + 6glu8ara + 'neo'	Mues (1984)
Radula plicata	F57345	C	68glu + 6glu8ara + 'neo'	Mues (1984)
Radula spp.	HC246	C	4Me3prenyl + prenyl closed to 7OH	Asakawa *et al.* (1982)
Radula tasmanica	F57345	C	68glu + 6glu8ara	Mues (1984)
Radula uvifera	F57345	C	6ara8glu + 6glu8ara + 'neo' + 68glu	Mues (1984)
Radula uvifera	F574	C	68glu	Mues (1984)
Radula variabilis	HC246		4′O-prenyl($OCH_2CH(Me)=CHCH_2$) to C3′	Asakawa *et al.* (1978)
Radula wichurae	F57345	C	68glu + 6glu + other	Mues (1984)
Saponaria officinalis	F574	CO	8glu7glu (spot test)	Barger (1902)
Scapania undulata	F5734	C	68glu	Mues (1986)
Trichocolea hatcheri			+ ve, low levels, unidentified	Mues (1982c)
Trichocolea mollissima			+ ve, low levels, unidentified	Mues (1982c)
Trichocolea rigida			+ ve, low levels, unidentified	Mues (1982c)
Trichocolea spp.	F	C	detected in some	Mues (1982a)
Trichocolea tomentella	F574	C	68glu	Mues (1982c)
Trichocolea tomentella	F5734	C	68glu + 3′Me68glu	Mues (1982c)
12.1.3 ORDER MARCHANTIALES (INCLUDING SPHAEROCARPALES)				
28 Species			all + ve (PC analysis)	Schier (1974)
Asterella australis	F574	O	7galu(tentative) + 7galurha	Markham and Porter (1978b)
Asterella australis	F5784	O	8galu + 84′galu (tentative)	Markham and Porter (1978b)
Asterella tenera	F574	O	7galu + 7galurha (tentative)	Markham and Porter (1978b)
Asterella tenera	F5734	O	3′Me7galu (tentative)	Markham and Porter (1978b)

(Contd.)

Table 12.1 (*Contd.*)

Species	*Flavonoids found*			*References*
Bucegia romanica (Czech)	F5784	O	8gluu + 4′Me8gluu + 84′Me7gluu	
Bucegia romanica (Poland)	F5784	O	8gluu + 8gluugly + 4′Me8gluu	
Bucegia romanica (Poland)	F5784	O	84′Me7gluu	Markham and Mues (1983)
Bucegia romanica (Rumania)	F5784	O	8gluu + 8gluu7gly + 8gluugly(many)	
Carrpos sphaerocarpos	F574	O	7gluu	Markham (1980)
Carrpos sphaerocarpos	F5734	O	7gluu	
Carrpos sphaerocarpos	A4634	O	6gluu	
Conocephalum conicum	F574	O	7digluu + 7gluu4′rha	
Conocephalum conicum	F574	C	6glu8ara + 6glu8rha + 6rha8glu	
Conocephalum conicum	F5734	O	7gluu3′rha + 7gluu (acyl)3′rha	Porter (1981)
Conocephalum conicum	F5734	O	3′Me74′gluu + 3′Me7digluu4′gluu	
Conocephalum conicum	F5734	C	3′Me68glu	
Conocephalum conicum	F5734		luteolin	Brehm and Comp (1967)
Conocephalum conicum (Ger)	F5734	O	74′gluu	
Conocephalum conicum (Ger)	F5734	O	74′gluu + 7digluu-4′gluu	
Conocephalum conicum (US + Ger)	F574	OC	7gluu + 7gluu4′rha + 68glu	Markham *et al.* (1976b)
Conocephalum conicum (US + Ger)	F5734	OC	7gluu4′rha + 3′Me7gluu4′rha + 68glu	
Conocephalum conicum (USA)	F5734	O	7gluu3′4′dirha + 73′gluu	
Conocephalum supradecompositum	A4634	O	6gluu	Markham and Porter (1978a)
Conocephalum supradecompositum	F574	OC	7gluu + 7digluu + 68glu	Porter (1981)
Conocephalum supradecompositum	F5734	O	7gluu + 74′gluu + 5gluu	
Corsinia coriandrina	FL3574	O	3galugly + 3trigly(not uronic A)	
Corsinia coriandrina	FL35734	O	3galu + 3galugly + 7Me3galugly + 3′Me3galugly	Markham (1980)
Corsinia coriandrina	FL3574		Kampferol + (4′Me?)	Reznik and Wiermann (1966)
Dumortiera hirsuta	F5734	O	5gluu	Campbell *et al.* (1979)
Exomortheca pustulosa	FL35734?	O	3glurha?	Schier (1974)
Lunularia cruciata	F5734	O	3′gluu + 3′4′gluu	Markham and Porter (1974a)
Mannia fragrans	F574	O	4′Me7galurha (tentative)	Markham and Porter (1978b)
Mannia fragrans	FL35734?	O	3glurha?	Schier (1974)
Marchantia alpestris	F574	O	7gluu	
Marchantia alpestris	F5734	O	7 + 3′ + 73′ + 74′ + (tent) 3′4′gluu	Markham and Porter (1978b)
Marchantia aquatica	F5734	O	7 + 3′ + 73′ + 74′ + (tent) 3′4′ + 73′4′gluu	
Marchantia aquatica	F574	O	7 + 74′gluu	
Marchantia berteroana	F574	O	7gluu + 7galu	
Marchantia berteroana	F5734	O	7 + 3′ + 73′ + 74′ + 3′4′ gluu/galu	Markham and Porter (1975b)
Marchantia berteroana	F5784	O	8gluu	

Table 12.1 (*Contd.*)

Species	*Flavonoids found*			*References*
Marchantia berteroana	F57834	O	8gluu + 84′gluu	Markham and Porter (1975b)
Marchantia berteroana	A4634	O	6gluu	Markham and Porter (1978a)
Marchantia berteroana	F574	O	4′Me7gluu/7galu	Markham *et al.* (1978a)
Marchantia cataractarum	F574	O	74′gluu	
Marchantia cataractarum	F574	C	68glu	
Marchantia domingensis	F574	O	7glu + 7gluu	Campbell *et al.* (1979)
Marchantia domingensis	F5734	OC	3′Me7gluu + 68glu	
Marchantia foliacea (a)	F574	OC	7gluu + 7gluurha + 68glu	
Marchantia foliacea (a)	F5734	O	3′Me7gluu/7gluurha	
Marchantia foliacea (a)	F57345	O	3′5′Me7gluu/ 7gluurha	Markham and Porter (1973)
Marchantia foliacea (b)	F574	OC	7glu + 7gluu + 7gluurha + 68glu	
Marchantia foliacea (b)	F5734	OC	7glu + 3′Me7gluu/7gluurha + 68glu	Campbell *et al.* (1979)
Marchantia geminata	F5734	O	7gluu	
Marchantia geminata	F574	OC	7gluu + 68glu	
Marchantia macropora (a)	F574	O	7gluu	
Marchantia macropora (a)	F5734	O	7 + 3′ + 73′gluu + 3′Me7gluu	Markham and Porter (1975c)
Marchantia macropora (b)	F574	OC	7gluu + 68glu	
Marchantia macropora (b)	F5734	OC	7 + 73′gluu + 3′Me7gluu + 68glu	
Marchantia palmatoides	F5734	OC	7 + 73′ + 74′gluu + 68glu	Campbell *et al.* (1979)
Marchantia palmatoides	F574	OC	7 + 74′gluu + 7digluu + 68glu	
Marchantia polymorpha	F5734	O	7 + 73′ + 74′ + 3′4′ + 73′4′gluu	Markham and Porter (1974b)
Marchantia polymorpha	F574	O	7gluu + 74′gluu	Markham and Porter (1974b)
Marchantia polymorpha	A4634	O	6gluu	Markham and Porter (1978a)
Marchantia polymorpha	F574	O	apigenin + glycs	
Marchantia polymorpha	F5734	O	luteolin + glycs	Brehm and Comp (1967)
Monoselenium tenerum	F574	O	7galurha4′rha + 7gluurha4′rha	
Monosolenium tenerum	F574	O	4′Me7gluurha/7galurha + 7gluurha/7galurha	
Monosolenium tenerum	F5734	O	7gluurha4′rha + 7galurha4′rha	Campbell *et al.* (1979)
Neohodgsonia mirabilis	F5734	O	7 + 3′ + 73′gluu	
Neohodgsonia mirabilis	F574	O	7gluu	
Oxymitra paleacea	F574	O	7 + 74′gluu	Markham and Porter (1978b)
Plagiochasma rupestre			apig/kamp/quer glycs? (pc indent only)	Schier (1974)
Preissia quadrata (a)	F574	O	7gluu + 4′rha + 7gluu4′rha (+ acyl)	
Preissia quadrata (a)	F5734	O	7 + 4′ + 74′gluu(+ acyl)	
Preissia quadrata (a)	F5734	O	3′Me7gluu(+ acyl)- 3′Me7gluu4′rha	
Preissia quadrata (b)	F5734	O	7 + 4′ + 74′gluu + 3′Me7gluu (+ 4′rha)	Campbell *et al.* (1979)
Preissia quadrata (b)	F574	O	7gluu + 7digluu + 7gluu 4′rha	
Reboulia hemispherica	F574	OC	7galurha + 4′Me8glu	Markham and Mabry (1972)
Reboulia hemispherica	F574	O	7 + 4′ + 74′galu(tentative)	
Reboulia hemispherica	F5734	O	73′4′galu(tentative)	Markham and Porter (1978b)
Reboulia hemispherica	F5784	O	7galu(tentative)	
Riccia bicaranata	F574	O	7glu	
Riccia bicarinata	HF5734?	O	4′Me7glurha?(PC only)	Schier (1974)

(*Contd.*)

Table 12.1 (*Contd.*)

Species	*Flavonoids found*			*References*
Riccia bifurca	F5734	O	7 + 3′ + 73′ + 73′4′gluu	Markham and Porter (1978b)
Riccia bifurca	F574	O	7gluu	
Riccia crystallina	HF574	O	7glu + aglyc	Markham and Porter (1975a)
Riccia crystallina	F574	O	7glu + 7gluu + acyl derivs	
Riccia crystallina	F5734	O	7gluu	
Riccia crystallina	HF5734	O	7glu	
Riccia duplex (Aust)	F5734	O	7gluu + 3′Me7gluu	Markham *et al.* (1978b)
Riccia duplex (Aust)	F574	O	7gluu	
Riccia fluitans	F5734	CO	7gluu + 7gluu4′rha + 68glu	Vanderkerkhove (1978a)
Riccia fluitans	F574	O	7gluu	
Riccia fluitans (Ger)	F574	O	7gluu + 4′gluu + 74′gluu	Markham *et al.* (1978b)
Riccia fluitans (Ger)	F5734	OC	7 + 4′ + 74′gluu + 68glu	
Riccia fluitans (Ger)	F5734	O	7gluu3′glu + 74′gluu3′glu/3′ferulyl glucoside	
Riccia fluitans (Ger)	F5734	O	7gluu3′(2-OH propionyl-glu) + 7gluu + 3′4′(2-OH propionylglu)?	
Riccia fluitans (NZ)	F5734	O	7gluu3′ferulylglu only	
Riccia gangetica	F5734	O	3′rha(tentative)	Markham and Porter (1978b)
Ricciocarpus natans	F5734	O	7gluu + 73′gluu	Markham and Porter (1975a)
Ricciocarpus natans	F574	O	7gluu	
Riella americana	F5734	O	7gluu + 3′Me7gluu + 3′gluu	Markham *et al.* (1976a)
Riella affinis	F5734	O	7gluu + 3′Me7gluu + 3′gluu	Markham *et al.* (1976a)
Riella americana	F574	O	7gluu	
Riella affinis	F574	O	7gluu	
Reilla helicophylla	F574		apigenin	Viell (1980)
Reilla helicophylla	F5734		luteolin	
Sphaerocarpos texanus	F5734	O	7gluu + 74′gluu	Markham *et al.* (1976a)
Targionia spp.	FL?	O	3glyc7diglyc + 3diglyc (pc only)?	Schier (1974)
Wiesnerella denudata	F574	OC	7rha + 4′Me7rha + 68pent	Campbell *et al.* (1979)
Wiesnerella denudata	F5734	O	5gluu	
12.1.4 ORDER METZGERIALES				
Aneura			– ve (or undetected)	Mues (1982b)
Apometzgeria pubescens	F574	C	68ara + 6ara8glu/8 (2-ferulylglu)	Theodor *et al.* (1981)
Apometzgeria pubescens	F57345	C	68glu + 6ara8glu	
Apometzgeria pubescens	F57345	C	3′4′Me68ara/6hex8pent	
Apometzgeria pubescens	F57345	C	3′5′Me68ara + 68glu + 6ara8glu	
Blasia pusilla	F	OC	flavone O- and C-glycs	Mues (1982b)
Fossombronia brasiliensis			+ ve for flavonoids	
Hymenophyton flabellatum	F574	C	68ara	Markham *et al.* (1976c)
Hymenophyton flabellatum	F574	C	68ara + 6glu8ara + 6ara8glu(+ neo + neoiso)	
Hymenophyton leptopodum	F574	C	68ara + 6glu8ara + 6ara8glu(+ neo + neoiso)	
Hymenophyton flabellatum	F574	O	7gluu + 4′Me7gluu	
Hymenophyton leptopodum	F574	O	7gluu + 4′Me7gluu	
Hymenophyton flabellatum	HC2464	C	3′5′gly	
Hymenophyton leptopodum	HC2464	C	3′5′gly	
Hymenophyton leptopodum	F574	C	68ara	
Hymenophyton leptopodum	FL3574	O	3glu7rha + 37rha + 3glurha7rha	

Table 12.1 *(Contd.)*

Species	*Flavonoids found*			*References*
Metzgeria chilensis	F57345	C	3′5′Me68glu	Mues *et al.* (1984)
Metzgeria conjugata	F574	C	68glu + 6ara8glu/ 8(2-ferulylglu)	Theodor *et al.* (1981)
Metzgeria conjugata	F57345	C	3′5′Me68glu/6glu8ara/ 6ara8glu	
Metzgeria furcata	F574	C	68glu + 6glu8ara + 8glu6ara + 6xyl8hex	Theodor *et al.* (1983)
Metzgeria furcata	F5734	C	6ara8hex + 6hex	
Metzgeria furcata	F57345	C	68glu	
Metzgeria furcata ulvulva	F574	OC	6rha7glu	Markham *et al.* (1982)
Metzgeria furcata ulvulva	F57345	C	68glu + 6glu8ara + 6ara8glu	Theodor *et al.* (1983)
Metzgeria furcata ulvulva	F5734	C	68glu	
Metzgeria furcata ulvulva	F574	C	68glu + 6glu8ara + 6ara8glu/8(2-ferulylglu	
Metzgeria hamata	F574	C	68glu	Markham and Porter (1978b)
Metzgeria (Leptoneura?)	F57345	C	3′5′Me6xyl8hex + tricetin C-glys?	Theodor *et al.* (1981)
Metzgeria simplex	F57345	C	3′5′Me6ara8glu/ 6glu8ara/68glu	Theodor (1981)
Metzgeria simplex	F574	C	6ara8glu(ferulyl)	
Moerckia			− ve (or undetected)	Mues (1982b)
Pallavicinia			+ ve?	Campbell *et al.* (1975)
Pellia			− ve (or undetected)	Mues (1982b)
Pellia epiphylla			phenolic pattern only	Szweykowski and Krzakowa (1977)
Pellia borealis			phenolic pattern only	
Riccardia			− ve (or undetected)	Mues (1982b)
Symphyogyna			+ ve?	Campbell *et al.* (1975)
12.1.5 ORDER MONOCLEALES				
Monoclea forsteri	F57834	O	8Me74′galu-oligosacch.	Markham (1972)
12.1.6 ORDER TAKAKIALES				
Takakia ceratophylla	F574	C	68ara	Markham and Porter (1979)
Takakia ceratophylla	F5784	O	8gluu + 4′Me8gluu	
Takakia ceratophylla	F5734	C	68arab	
Takakia lepidozioides	F574	CO	68ara + 68ara-Ogly	
Takakia lepidozioides	F574	C	6glu8ara + 68glu	
Takakia lepidozioides	F5784	O	8gluu + 4′Me8gluu /8gluuxyl	
Takakia lepidozioides	F574	O	7glu + 7gluu	
Takakia lepidozioides	F57345	C	6glu8ara + 6ara8glu + 6 + 8mono?	
Takakia lepidozioides	F5734	CO	68ara(+ Oglyc) + 6glu8ara + 68glu	
Takakia lepidozioides	FL3574	O	3glu7xyl	
Takakia lepidozioides	FL35734	O	3glu	

(e) *Order Monocleales*
(Table 12.1.5)

This monotypic order exhibits some affinity with the Marchantiales in that *Monoclea forsteri* has been shown to accumulate flavone glycuronides (Markham, 1972). The predominant compound has been described as a flavone-polysaccharide in which 8-methoxyluteolin is glycosylated at the 7- and 4′-positions with galacturonic acid. The structure is not fully defined but clearly contains a larger number of sugar residues than is normally encountered. Whether *Monoclea* should be in its own order has been discussed at length (Campbell, 1984, and references therein), but proposed alignments with the Marchantiales

(Grolle, 1972; Schuster, 1983), e.g. with the Sphaerocarpales in the Marchantiidae, find some support in these data.

(f) *Order Takakiales*
(Table 12.1.6)

This order consists of only two species, both of which have been studied (Markham and Porter, 1979). They are rich in flavonoids and contain a remarkably wide range of flavonoid types including flavone *C*-glycosides, *O*-glucosides and *O*-glucuronides, and flavonol *O*-glycosides. It is thought that *T. ceratophylla* is distinguished from *T. lepidozioides* in lacking flavonols, but the analysed sample of the former was small and from a herbarium specimen.

12.3.3 Class Musci (Table 12.2)

A remarkably wide range of flavonoid types has been found amongst the very small proportion of extant mosses investigated to date. The range surpasses that so far encountered in the Hepaticae and includes flavone *O*-and *C*-glycosides, biflavones, aurones, isoflavones and 3-deoxyanthocyanins. As yet no convincing evidence has been presented to support the presence of flavonols, dihydroflavonols or 3-hydroxyanthocyanins, and proanthocyanidins do not seem to be biosynthesized (Bendz *et al.*, 1966). The biosynthetic pathway which leads to dihydroflavonols via 3-oxygenation, and thence to the other 3-oxygenated flavonoids, thus appears to be essentially inactive in mosses.

The bulk of published flavonoid work relates to the one major order, Bryales, and for this reason flavonoid typing of the other orders is not possible. Flavonoids have not yet been detected in the Andreales, and in the Sphagnales the red cell-wall-bound pigment, sphagnorubin (12.1) and its

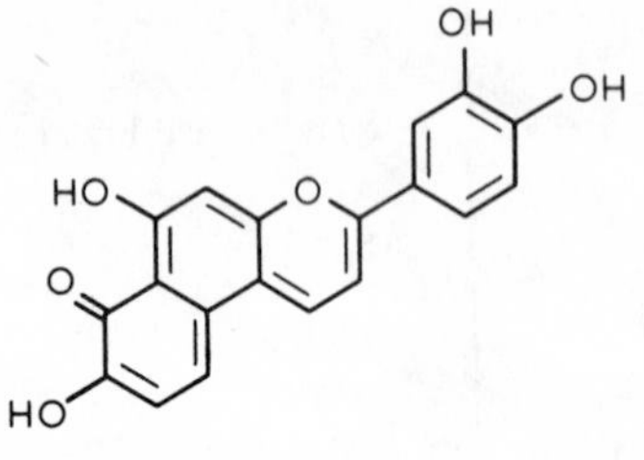

(*12.1*) Sphagnorubin

methyl ethers are the only 'flavonoids' (Vowinkel, 1975) so far identified. The Tetraphidales are reported to contain dihydroflavonols but the meagre published data need to be substantiated in view of the significance of such an occurrence. Flavonoids have been detected in the Polytrichales (McClure and Miller, 1967) by paper chromatography, but this too is in need of confirmation.

(a) *Order Bryales*

Flavonoid data for all 15 suborders of the Bryales are scanty and only in the Bryineae have a number of substantial investigations been carried out. Characterization of most suborders in terms of flavonoid content is thus near impossible. A characteristic of the suborder Dicranineae, however, may well be the accumulation of biflavones. The first biflavone isolated from a non-vascular plant, 5′8′-biluteolin, was found as the major flavonoid ($10\,mg\,kg^{-1}$) in *Dicranum scoparium* (Lindberg *et al.*, 1974). This was accompanied by smaller amounts of luteolin 7-rhamnoglucoside and the novel luteolin-7-*O*-(2″,4″-dirhamnosyl)glucoside. Interestingly, these three flavonoids have been isolated recently also from *Hylocomium* (suborder Hypnineae). The biflavones in *Dicranoloma*, however, are not accompanied by flavone *O*-glycosides. Two of these contain the 2′–6‴ interflavonoid linkage (Markham, in press).

The only flavonoid identified so far from the suborder Funariineae is the aurone, bracteatin. This was the first aurone ever isolated from a bryophyte (Weitz and Ikan, 1977).

Flavone *O*- and *C*-glycosides seem to predominate in the suborder Bryineae, although 3-deoxyanthocyanins and quite remarkably, isoflavones, have also been isolated from *Bryum*. The finding of isoflavones was the first, and to date the only, recorded occurrence of isoflavonoids in plants other than gymnosperms or angiosperms. The report of flavonols in *Rhizomnium* is unconvincing.

12.4 FLAVONOID DISTRIBUTION IN THE FERN ALLIES

The classification system used in this section is essentially that outlined by Bold (1973) in which the various groupings of fern allies are assigned ordinal status. In Table 12.3 these groupings are arranged as follows:

Orders: Equisetales (Table 12.3.1)
Isoetales (Table 12.3.2)
Lycopodiales (Table 12.3.3)
Psilophytales (Table 12.3.4)
Selaginellales (Table 12.3.5)

(a) *Order Equisetales*
(Table 12.3.1)

The predominant flavonoids accumulated by *Equisetum*, the sole genus in this order, are flavonols, representatives of which have been found in all investigated species. The biosynthetically related proanthocyanidins are also widespread, while flavones have a markedly more restricted distribution. Two particularly unusual flavonoids have been isolated from *Equisetum arvense*, protogenkwanin 4′-*O*-glucoside (*12.2*) and 6-chloroapigenin.

Table 12.2 Flavonoids in the Musci (Mosses)

Species	*Flavonoids found*			*References*
12.2.1 ORDER ANDREALES				
Andreaea rupestris			– ve	McClure and Miller (1967)
12.2.2 ORDER BRYALES				
Suborder Bryineae				
Bryum argenteum	F574	O	7glu + 7glu(6″ malonate)	Markham (in press)
Bryum argenteum	F5734	O	7glu + 7glu(6″ malonate)	
Bryum argenteum	F5784	O	8glu	
Bryum argenteum	F57834	O	8glu	
Bryum capillare	I5734	O	7glu + 4′Me7glu + 6″malonates of both (gametophytes)	Anhut *et al.* (1984)
Bryum capillare	F574	O	7glu + 6″malonate (sporophyte only)	Stein *et al.* (1985)
Bryum capillare	F5734	O	7glu + 7glu6″malonate + 4′Me equivalents	
Bryum capillare	F56734	O	7glu + 7glu6″malonate	
Bryum capillare	BI5734F5734	O	5′-8″ and 5′-6″ links	Geiger *et al.* (1987)
Bryum cryophilum	AN5734	O	5glu + 5diglu	Bendz *et al.* (1962)
Bryum cryophilum	AN5734	O	5glu + 5diglu	Bendz and Martensson (1961)
Bryum rutilans	AN5734	O	5glu + 5diglu (probable)	Bendz and Martensson (1963)
Bryum weigelii	AN5734	O	5glu + 5diglu (probable)	
Bryum weigelii	F5674	O	7glu?	Nilsson (1969b)
Mnium affine (?)	F5734	C	3′Me68glu?	Melchert and Alston (1965)
Mnium affine (?)	F574	CO	8glu + 68glu + Oglycs of each?	
Mnium affine (?)	F5734	CO	3′Me8glu + 3′Me68glu + Ogly of each?	
Mnium affine (?)	F5734	CO	8glu + 68glu + Oglycs(all?)	Alston (1968)
Mnium affine (?)	F5734	C	8glu + 68glu + 3′Me8glu + Oglycs (all?)	Melchert and Alston (1965)
Mnium arizonicum	FL35734	O	3diglyc? (no evidence)	
Mnium cuspidatum	F574	CO	8glu7glu?	
Mnium cuspidatum	FL35734	O	3digly?	
Mnium cuspidatum	F574	CO	8glu7glu by spot test	Kozlowski (1921)
Mnium undulatum	F5734	C	3′Me68glu	Österdahl (1979)
Mnium undulatum	F574	C	8glu7glu + 68glu + 6glu8ara + iso/neoschaftoside	
Mnium undulatum	F5734	C	3′Me6ara8hex	
Plagiomnium affine	F5734	OC	6glu3′soph/3′neohesperi-doside	Freitag *et al.* (1986)
Plagiomnium cuspidatum	F574	OC	6glu7glu	Koponen and Nilsson (1977)
Plagiomnium drumondii			+ ve for flavonoids	
Plagiomnium ellipticum			+ ve for flavonoids	
Plagiomnium japonicum			+ ve for flavonoids	
Plagiomnium tezukae			+ ve for flavonoids	
Plagiomnium undulatum	F574	OC	6glu7glu	
Plagiomnium undulatum	F574	C	6glu8ara + iso/neoiso-schaftoside + 68glu	Österdahl (1979)
Plagiomnium undulatum	F5734	C	6ara8hex	
Plagiomnium acutum			+ ve for flavonoids	Koponen and Nilsson (1977)
Plagiomnium insigne			+ ve for flavonoids	
Plagiomnium elatum			+ ve for flavonoids	
Plagiomnium ciliare			+ ve for flavonoids	
Plagiomnium medium			+ ve for flavonoids	
Pohlia wahlenbergii			caffeic acid diglu ester only	Martensson and Nilsson (1974)

(*Contd.*)

Table 12.2 (*Contd.*)

Species	*Flavonoids found*			*References*
Rhizomnium magnifolium	F5734	O	7gluu + 73′gluu + 3′Me7gluu	Mues *et al.* (1986)
Rhizomnium magnifolium	F57345	O	3′Me75′gluu + 3′4′Me7gluu + 73/gluu	
Rhizomnium pseudopunctatum	F5734	O	7gluu + 73′gluu + 3′Me7gluu	
Rhizomnium pseudopunctatum	F57345	O	73′gluu + 3′Me75′gluu + 3′4′ Me7gluu	
Rhizomnium hattorii	FL?		yellow spot in uv	Koponen and Nilsson (1977)
Rhizomnium striatulum	FL?		yellow spot in uv	
Rhizomnium (9spp.)			– ve for flavonoids	
Suborder Buxbaumiinae				
Diphysium foliosum			– ve (probably)	McClure and Miller (1967)
Suborder Dicranineae				
Ceratodon purpureus	F5734		luteolin (sporophytes)	Vanderkerkhove (1978b)
Dicranoloma robustum	BF5734–F5734		2′6″, 5′8″ and 5′6″ links	Markham (in press)
Dicranoloma robustum	BHF5734–F5734		2′6″ and 65″ links	
Dicranum scoparium	F574	O	7glu2″, 4″dirha	Nilsson *et al.* (1973)
Dicranum scoparium	BF5734(5′8″)F5734			Lindberg *et al.* (1974)
Dicranum scoparium	F5734	O	4′Me7glu2″4″dirha	Österdahl (1978a)
Dicranum scoparium	BF5734(5′8″) F5734			Österdahl (1983)
Dicranum scoparium	F5734	O	7glurha	Nilsson (1973b)
Suborder Funariineae				
Funaria hygrometrica	A46345			Weitz and Ikan (1977)
Suborder Hypnineae (= Hypnobryales)				
Drepanocladus pseudomentosus			red pigments (sphagnorubin + 3 deoxyanthoc?)	Bendz *et al.* (1962)
Hylocomium splendens	F574	O	7glurha alone	Vanderkerkhove (1977b)
Hylocomium splendens	BF5734(5′6″) F5734		(= 5′3‴-dihydroxyrobusta-flavone)	Becker *et al.* (1986)
Hylocomium splendens	BF5734(5′8″) F5734			
Hylocomium splendens	BHF5734–HF5734			
Hylocomium splendens	F574	O	4′Me7glurha + 7glu2″4″rha	
Hylocomium splendens	F5734	O	4′Me7glurha + 7glurha	
Pleurozium schreberi	F574	O	apigenin + 7glurha + 4′ di(or 74′gly)	Vanderkerkhove (1980)
Suborder Leucodontineae (= Isobryales)				
Hedwigia ciliata	F5734	C	68glu	Österdahl (1978b)
Hedwigia ciliata	F5734	O	74′di(glu(1–2)rha)	Österdahl (1976)
Hedwigia ciliata	F5734	O	7glu(1–2)rha-4′glu(1–2)glu	Österdahl and Lindberg (1977)
Hedwigia ciliata	F574	C	68glu	Österdahl (1978b)
Hedwigia ciliata	BF5734(5′8″) F5734		unconfirmed	Österdahl *et al.* (1976)
Suborder Splachnineae				
Splachnum rubrum	AN5734		derivs	Nilsson and Bendz (1973)
Splachnum vasculosum	AN5734		derivs	
12.2.3 ORDER POLYTRICHALES				
Dawsonia sp.			lignin present	Siegel (1969)
Dendroligotrichum sp.			lignin present	
Polytrichum			+ ve	McClure and Miller (1967)
Atricum			+ ve	
12.2.4 ORDER SPHAGNALES				
Sphagnum fimbriatum			– ve	McClure and Miller (1967)
Sphagnum girgensohnii			unidentified flavonoid glycs?	Efimenko and Dzenis (1962)
Sphagnum magellanicum			Unidentified flavonoid glycs	

Table 12.2 (*Contd.*)

Species	*Flavonoids found*			*References*
Sphagnum magellanicum	AN734		sphagnorubin	Vowinkel (1975)
Sphagnum nemoreum	AN734		sphagnorubin + 7/73′Me	Rudolph and Johnk (1982)
Sphagnum plumulosum	AN734		sphagnorubin + 7/73′Me	Rudolph and Johnk (1982)
Sphagnum rubellum	AN734		sphagnorubin + 7/73′Me	Rudolph and Johnk (1982)
Sphagnum warnstorfii	AN734		sphagnorubin + 7/73′Me	Rudolph and Johnk (1982)
12.2.5 ORDER TETRAPHIDALES				
Tetraphis (= *Georgia*) *pellucida*			+ ve	McClure and Miller (1967)
Tetraphis pellucida	HFL?	O	5-hydroxy?	Geiger (1986)
Tetraphis pellucida	HFL?	O	5-deoxy?	Vanderkerkhove (1977a)

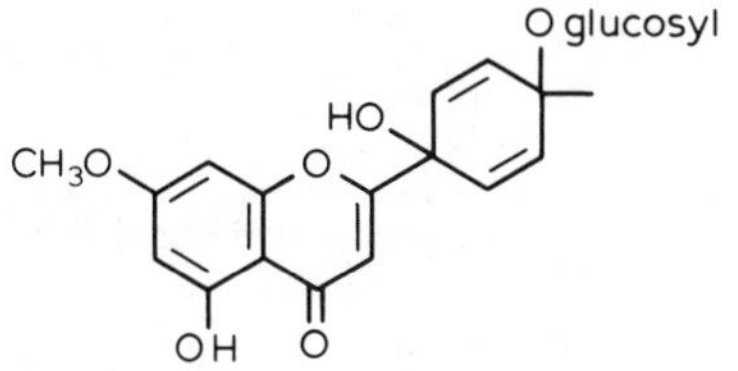

O glucosyl
HO
CH_3O
O
OH O

(*12.2*) Protogenkwanin 4′-*O*-glucoside

(b) *Order Isoetales*
(Table 12.3.2)

All flavonoid data available for the order Isoetales derive from the work of Voirin and co-workers. Flavones appear to be the only flavonoid type accumulated and no evidence for *C*-glycosylation has been found. Biosynthetic modification of the basic flavones is restricted to additional oxygenation of the 2′, 5′, 6 or 8 positions and to selective methylation and glycosylation. To date little work has been reported on the flavone glycosides which abound (Markham, unpublished) in both the *Isoetes* and the *Stylites*.

(c) *Order Lycopodiales*
(Table 12.3.3)

Flavones appear to be the only flavonoids accumulated by Lycopodialean species, biflavones, proanthocyanidins and flavonols having been shown to be absent (Voirin and Jay, 1978a). Flavone *C*-glycosides are rare but have been encountered once in *Lycopodium* and in *Phylloglossum* (Markham *et al.*, 1983). Several attempts have been made to use flavonoid characters to help resolve long-standing taxonomic problems in this order (Voirin and Jay, 1978b; Markham *et al.*, 1983).

(d) *Order Psilotales*
(Table 12.3.4)

The predominant flavonoids in both genera of the Psilotales are biflavones of the amentoflavone type. An early tentative identification of hinokiflavone in *Psilotum triquetum* is claimed to have been confirmed (Voirin, 1986), but this compound was not found in *P. nudum* (Wallace and Markham, 1978b). Proanthocyanidins and flavonols are absent. Although the major flavonoid component (by far) of both *Psilotum* and *Tmesipteris* is the aglycone amentoflavone, small amounts of apigenin and amentoflavone glycosides have also been isolated. The amentoflavone *O*-glycosides found are the only known amentoflavone glycosides, and the major representative of this group is the 7, 4′, 4‴-tri-*O*-glucoside (Markham, 1984). *Psilotum* and *Tmesipteris* also accumulate high levels (2–5%) of psilotin, a unique biologically active phenolic glycoside biosynthetically related to the flavonoids (Arnason *et al.*, 1986).

(e) *Order Selaginellales*
(Table 12.3.5)

Limited data reported by a number of workers indicate that the only flavonoid types accumulated by *Selaginella* (the sole genus) are biflavones. Good evidence has been presented to support the widespread occurrence of both amentoflavone and hinokiflavone and their methyl ethers.

12.5 FLAVONOID DISTRIBUTION IN THE FERNS

A vast number of classification systems have been proposed for the approximately 900 known species of ferns (see for example Pichi Sermolli, 1973, 1977, and Tryon and Tryon, 1982), and successive authors have recognized anything from 12 to more than 50 families. The classification system adopted for the compilation in Table 12.4 is

Table 12.3 Flavonoids in the fern allies

Species	*Flavonoids found*			*References*
12.3.1 ORDER EQUISETALES				
Survey	P		present in 43%	
Survey	FL		present in 100%	Voirin and Lebreton (1971)
Survey	F		found in 14%	
Equisetum (10 spp.)	FL3574	O	3glu(6/10) + 7glu(10/10) + 7diglu(2/10)	
Equisetum (10 spp.)	FL3574	O	7glu(6/10) + 3rut(3/10) + 3soph(7/10)	
Equisetum (10 spp.)	FL3574	O	3soph(4/10) + 3rut(2/10)	Saleh *et al.* (1972)
Equisetum (10 spp.)	FL3574	O	3soph(2/10) + 3glu(2/10) + 7glu(1/10)	
Equisetum (10 spp.)	FL35734	O	3diglu7glu(3/10) + 37diglu(1/10)	
Equisetum arvense	F574	O	5glu	Syrchina *et al.* (1974)
Equisetum arvense	F574		6-chloroapigenin	Syrchina *et al.* (1980a)
Equisetum arvense	F574	C	6glu + apigenin	Syrchina *et al.* (1973)
Equisetum arvense	F574	CO	6glu + 5glu	
Equisetum arvense	F5734	O	5glu	Syrchina *et al.* (1980b)
Equisetum arvense	FL35734	O	3glu	
Equisetum arvense	FL3574	O	3soph	
Equisetum arvense	FL3574	O	3diglu	
Equisetum arvense	FL35734	O	3glu	Nakamura and Hukuti (1940)
Equisetum arvense	F5734	O	5glu	
Equisetum arvense	HFL35734			
Equisetum arvense	HFL3574			Syrchina *et al.* (1975)
Equisetum arvense	HF574			
Equisetum arvense	HF5714	O	7Me4′glu (protogenkwanin 4′glu)	Hauteville *et al.* (1980)
Equisetum arvense	FL357834	O	7glu	Hauteville *et al.* (1981)
Equisetum arvense	F574	O	4′glu	Geiger (1986)
Equisetum arvense	F574		7Me	Syrchina *et al.* (1978a)
Equisetum arvense	HFL3574			
Equisetum arvense	HFL35734			
Equisetum arvense	HF574			Syrchina *et al.* (1978b)
Equisetum arvense	F574			
Equisetum arvense	F5734			
Equisetum fluviatile	FL357834	O	7glu	
Equisetum fluviatile	FL35784	O	7glu	Saleh *et al.* (1972)
Equisetum fluviatile	F574	O	4′glu	
Equisetum hyemale	FL3574	O	7glu3soph	Markham *et al.* (1978c)
Equisetum hyemale	FL35784	O	3soph8glu	Geiger *et al.* (1982)
Equisetum hyemale	FL357834	O	3soph8glu	
Equisetum palustre	FL3574	O	7glu + 7glu3rut	Markham *et al.* (1978c) Beckman and Geiger (1963)
Equisetum ramosissimum	FL3574	O	3soph7glu(no luteolin5glu or kaempferol37glu)	Saleh and Abdalla (1980)
Equisetum spp. (10)	FL3574	O	3glu/rut/soph + 7glu + 37glu + 3soph7glu	Saleh (1975)
Equisetum spp. (10)	FL35734	O	3soph	
Equisetum spp. (10)	F574	O	4′glu	
Equisetum sylvaticum	FL35784	O	7glu	Saleh *et al.* (1972)
Equisetum sylvaticum	FL3574	O	3glurha + 7rha + 3rut7rha	Aly *et al.* (1975) Markham *et al.* (1978c)
Equisetum sylvaticum	FL35734	O	3glu7rha + 7rha + 3rut7rha	Aly *et al.* (1975)
Equisetum telmateia	FL357834	O	7glu	Kutney and Hall (1971)
Equisetum telmateia	FL3574	O	7glu + 7rha + 3glu + 7rha3glu + 7rha3glu(6OAc)	Hall (1967) Markham *et al.* (1978c) Geiger *et al.* (1978)
Equisetum telmateia	FL3574	O	7glu/6″OAc + 3rut + 3rut7glu	Geiger *et al.* (1978)
Equisetum variegatum	FL35784	O	7glu	Saleh *et al.* (1972)

Table 12.3 (*Contd.*)

Species	*Flavonoids found*			*References*
12.3.2 ORDER ISOETALES				
Survey (4 spp.)	F		Flavones only	Voirin and Jay (1978a)
Survey	P		absent in all	Voirin and Lebreton (1971)
Isoetes delilei	F57245	O	isoetin + 5′glu?	Voirin and Jay (1978a); Voirin *et al.* (1975)
Isoetes delilei	F574			Voirin (1972)
Isoetes delilei	F5734			
Isoetes delilei	F5734		3′Me	Voirin and Jay (1978a)
Isoetes durieui	F57245	O	isoetin + 5′glu?	Voirin *et al.* (1975)
Isoetes durieui	F574			Voirin (1972)
Isoetes durieui	F5734			
Isoetes durieui	F56734?		(56834?)	Voirin and Jay (1978a)
Isoetes durieui	F57345		3′5′Me	
Isoetes lacustris	F574			
Isoetes lacustris	F5734		luteolin + 3′Me	
Isoetes lacustris	F57345		5′Me (selgin)	
Isoetes velata	F574			
Isoetes velata	F5734		luteolin + 3′Me	
Isoetes velata	F57345		5′Me (selgin)	
12.3.3 ORDER LYCOPODIALES				
Survey	P		absent from all	Voirin and Lebreton (1971)
Surveys (9 spp.)	F		only	Voirin and Jay (1978a)
Survey	B		absent from all	Voirin and Jay (1978b)
Diphasium alpinum	F574			Voirin and Jay (1978b)
Diphasium alpinum	F5734		3′Me	
Diphasium alpinum	F57345?		3′5′Me	
Diphasium tristachium	F57345?		3′5′Me	
Diphasium tristachium	F5734		3′Me	
Diphasium tristachium	F574?			
Huperzia selago	F574			
Huperzia selago	F5734		luteolin + 3′Me	
Huperzia selago	F57345		3′Me + 3′5′Me	
Lepidotis carolinianum	F5734		luteolin + 3′Me	
Lepidotis carolinianum	F574			
Lepidotis carolinianum	F57345		3′5′Me	
Lepidotis cernua	F574			
Lepidotis cernua	F5734		luteolin + 3′Me	
Lepidotis cernua (= *selago*?)	F5734			Voirin (1972)
Lepidotis cernua (= *selago*?)	F574			
Lepidotis convoluta	F574			Voirin and Jay (1978b)
Lepidotis convoluta	F5734		luteolin + (3′Me?)	
Lycopodium (4/12 spp.)	F57345	O	3′5′Me7glu + acyl derivs	Markham *et al.* (1983)
Lycopodium (9 spp.)	P		absent	Cambie *et al.* (1961)
Lycopodium (10/12 spp.)	F5734	O	3′Me7glu + acyl derivs	Markham *et al.* (1983)
Lycopodium (11/12 spp.)	F574	O	7glu + acyl derivs	
Lycopodium (19/24 spp.)	F	O	7glu + acyl derivs	
Lycopodium (8/12 spp.)	F5734	O	7glu + acyl derivs	
Lycopodium (9/24 spp.)	F	O	4′glu + acyl derivs	
Lycopodium annotinum	F5734		luteolin + 3′Me	Voirin and Jay (1978b)
Lycopodium annotinum	F57345		3′5′Me	
Lycopodium cernuum	F574	C	6glu8ara + 6ara8glu	Markham and Moore (1980)
Lycopodium clavatum	F574	O	4′glu(2″4″di-*p*-coumarate)	Ansari *et al.* (1979)
Lycopodium clavatum	F57345		3′5′Me	Voirin and Jay (1978b)
Lycopodium clavatum	F5734		3′Me	

(*Contd.*)

Table 12.3 (*Contd.*)

Species	*Flavonoids found*			*References*
Lycopodium scariosum	F5734	O	3′Me5glu/5diglu/4′glu/54′glu/74′glu	Markham and Moore (1980)
Lycopodium scariosum	F57345	O	3′5′Me5glu/5diglu/4′glu/7glu/57glu	Markham and Moore (1980)
Lycopodium obscurum-complex	F5734		3′Me	Fusiak (1982)
Phylloglossum drummondii	F5734	CO	68glu + 3′Me7glu	Markham *et al.* (1983)
Phylloglossum drummondii	F57345	O	7glu + (3′Me7glu?)	Markham *et al.* (1983)
12.3.4 ORDER PSILOPHYTALES				
Survey (2 spp.)	BF		only biflavones	Voirin and Jay (1978a)
Survey	P		absent from all	Voirin and Lebreton (1971)
Survey (4 spp.)	P		absent from all	Cooper-Driver (1977)
Psilotum complanatum	BF574(3′8″)-F574		amentoflavone	Cooper-Driver (1977)
Psilotum nudum	BF574(3′8″)-F574	O	amentoflavone + glucosides	Wallace and Markham (1978b)
Psilotum nudum	F574	CO	7glurha + 7glurha4′glu + 68ara?	Wallace and Markham (1978b)
Psilotum nudum	P		absent	Cambie *et al.* (1961)
Psilotum nudum	BF574(3′8″)-F574	O	74′4‴glu	Markham (1984)
Psilotum nudum	BF574(3′8″)-F574	O	4′4‴glu + 74′4‴glu	Markham (1984)
Psilotum nudum	F574	C	68glu	Markham (1984)
Psilotum triquetrum	BF574(3′8″)-F574			Voirin and Lebreton (1966)
Psilotum triquetrum	BF574(4′6″)-F5674(?)		hinokiflavone	Voirin and Lebreton (1966)
Psilotum triquetrum	BF574(4′6″)-F5674		hinokiflavone	Voirin (1986)
Psilotum triquetrum	BHF–F(?)			Voirin (1986)
Tmesipteris elongata	F574	O	7glurha	Wallace and Markham (1978b)
Tmesipteris elongata	BF574(3′8″)-F574		amentoflavone + glucosides	Wallace and Markham (1978b)
Tmesipteris tannensis	BF574(3′8″)-F574			Voirin and Jay (1977)
Tmesipteris tannensis	F574	CO	7glurha + 7glurha + 4′glu + 68glu + 74′glu	Wallace and Markham (1978b)
Tmesipteris tannensis	BF574(3′8″)-F574		amentoflavone + glucosides	Wallace and Markham (1978b)
Tmesipteris tannensis	P		absent	Cambie *et al.* (1961).
12.3.5 ORDER SELAGINELLALES				
Survey (20 spp.)	BF		only biflavonoids	Voirin and Jay (1978a)
Survey	P		absent from all	Voirin and Lebreton (1971)
Selaginella	BF574(3′8″)-F574		amentoflavone	Voirin (1972)
Selaginella (19 spp.)	BF574(4′6″)-F5674		hinokiflavone + 7″Me	Voirin (1972)
Selaginella (spp. 16)	BF574(3′8″)-F574		amentoflavone in all	Voirin (1972)
Selaginella nipponica	BF574(3′8″)-F574		amentoflavone (+ Me ethers?)	Okigawa *et al.* (1971)
Selaginella pachystachys	BF574(3′8″)-F574		amentoflavone (+ Me ethers?)	Okigawa *et al.* (1971)
Salaginella tamariscina	BF574(3′8″)-F574		amentoflavone	Okigawa *et al.* (1971)
Selaginella tamariscina	BF574(4′6″)-F5674		hinokiflavone + 7″Me	Okigawa *et al.* (1971)

essentially that of Crabbe *et al.* (1975), but updated by consultation with Dr A.C. Jermy of the British Museum and Dr D.R. Given of Botany Division, DSIR, New Zealand. The families in Table 12.4 are arranged as follows: Adiantaceae (Table 12.4.1); Aspidaceae (Table 12.4.2); Aspleniaceae (Table 12.4.3); Athyriaceae (Table 12.4.4); Azollaceae (Table 12.4.5); Blechnaceae (Table 12.4.6); Cyatheaceae – including Dicksoniaceae (Table 12.4.7); Davalliaceae – including Oleandraceae (Table 12.4.8); Dennstaedtiaceae – including Lindsaeaceae (Table 12.4.9); Dicksoniaceae – see Cyatheaceae; Dryopteridaceae – see Aspidaceae and Athyriaceae; Gleicheniaceae (Table 12.4.10); Grammitidaceae (Table 12.4.11); Hymenophyllaceae (Table 12.4.12); Lindsaeaceae – see Dennstaedtiaceae; Lomariopsidaceae (Table 12.4.13); Loxsomaceae (Table 12.4.14); Marattiaceae (Table 12.4.15); Marsileaceae (Table 12.4.16); Oleandraceae – see Davalliaceae; Ophioglossaceae (Table 12.4.17); Osmundaceae (Table 12.4.18); Polypodiaceae (Table 12.4.19); Pteridaceae (Table 12.4.20); Salviniaceae – with Azollaceae; Schizaeaceae (Table 12.4.21); Sinopteridaceae (Table 12.4.22); Stromatopteridaceae (Table 12.4.23); Thelypteridaceae (Table 12.4.24).

Previous reviews of the flavonoids (and other components) in ferns vary widely in approach and level of completeness. Notable amongst these are Berti and Bottari (1968), Voirin (1970), Swain and Cooper–Driver (1973), Cooper–Driver (1980), Soeder (1985) and Hegnauer (1986).

Most flavonoid types have been isolated from ferns, the notable exceptions to date being aurones, isoflavones and surprisingly, in view of the widespread occurrence of flavonols, dihydroflavonols. Biflavonyls have only been isolated recently (from *Osmunda* and *Alsophila*). Anthocyanin representatives are almost entirely of the 3-deoxy type, only one 3-hydroxylated example having been recorded and confirmed (from *Davallia*). In contrast, the proanthocyanidins appear to be all of the 3-hydroxylated type (Harborne, 1966; Voirin, 1970; Cooper–Driver, 1977). 3-Deoxyproanthocyanidins are in fact exceedingly rare in nature, only two examples having been found and never in ferns (see Chapter 2).

A summary of the vast amount of flavonoid data published on ferns is presented in Table 12.4. Families for which sufficient data are available are discussed below and salient features highlighted.

(a) *Family Adiantaceae*
(Table 12.4.1) – see also *F. Sinopteridaceae*

Proanthocyanidins, 3-deoxyanthocyanins, flavonols, flavones, dihydroflavones, chalcones and dihydrochalcones have all been found in the one genus (*Adiantum*) so far investigated. The last three types however were found amongst the 'external' flavonoids (see Sinopteridaceae). A wide survey of *Adiantum* species (Cooper–Driver and Swain, 1977) indicates possible subdivision of this genus into five chemical groups. Two of these are rich in sulphate esters, presumably of the type first isolated by Imperato (1982a, c). Overall the predominant internal flavonoids appear to be flavonol 3-glycosides (and 3-deoxyanthocyanins?), sporadically sulphated or acylated.

(b) *Family Aspidaceae*
(Table 12.4.2)

Hiraoka (1978) and others have found proanthocyanidins, 3-deoxyanthocyanins, flavones, dihydroflavones and flavonols in this family, all as glycosides involving a range of sugars. Apigenin and luteolin 8-*C*-glucosides are common, as also are kaempferol and quercetin 3-*O*-glycosides. Dihydroflavones are found in many genera, usually as 7-*O*-glycosides, and the aglycones of some of these are reminiscent of 'external' flavonoids found in the Sinopteridaceae. Two unusual flavonoids have been found, one a 5-acetoxyflavan from *Dryopteris* (Karl *et al.*, 1981) in which the methyl group of the acetoxyl is further bonded to C-4 on the C-ring, and the other, a 1′-hydroxylated flavanone, protofarrerol (*12.3*) from *Leptorumohra* (Noro *et al.*, 1969). The flavonoid distribution in this family has been

CH_3 HO HO OH CH_3 OH O

(*12.3*) Protofarrerol

found useful in defining generic limits (Hiraoka, 1978), e.g. *Arachniodes* (flavones only) and *Leptorumohra* (flavonols only). Overall an alignment of this family with the Athyriaceae is indicated.

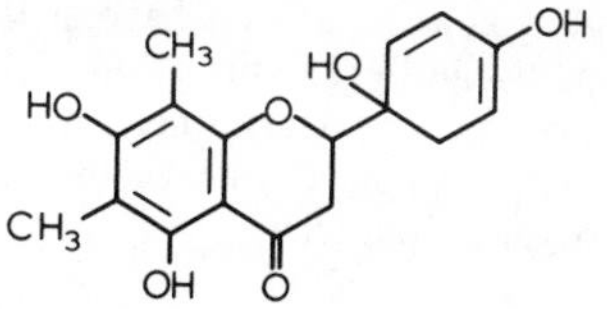

(c) *Family Aspleniaceae*
(Table 12.4.3)

Proanthocyanidins and flavone *C*- and *O*-glycosides have not yet been found in this family. The predominant flavonoids are kaempferol and quercetin *O*-glycosides with some methylation, sulphation, acylation and additional glycosylation being evident.

(d) *Family Athyriaceae*
(Table 12.4.4)

The range and type of the major flavonoids accumulated by the Athyriaceae are very similar to those of the Aspidaceae, e.g. apigenin and luteolin 8-*C*-glucosides, kaempferol and quercetin 3- and 7-*O*-glycosides. Flavone *O*-glycosides have not been found but dihydroflavone *O*-glycosides have, especially in *Matteuccia* where they are 6- and 8-*C*-methylated. This flavonoid composition is consistent with the close alignment of the Athyriaceae and Aspidaceae suggested by Tryon and Tryon (1982) who include both families within the Dryopteridaceae.

(e) *Family Cyatheaceae*
(Table 12.4.7)

Flavonol *O*-glycosides, flavone *C*-glycosides and proanthocyanidins appear to be widespread in this family while flavone *O*-glycosides and dihydroflavones are conspicuously absent. The isolation and full identification of two biflavonoids, 2,3-dihydro-3‴-hydroxy-6,6″-biapigenin and 2,3-dihydro-6,6″-biluteolin, from *Alsophila* (Wada *et al.*, 1985) is of particular significance, since it appears to resolve the recent controversy over the existence of biflavonoids in ferns (Wallace and Hu, 1985) that resulted from the earlier report of biflavonoids in *Osmunda japonica* (Okuyama *et al.*, 1979).

(f) *Family Davalliaceae – includes Oleandraceae*
(Table 12.4.8)

From the few data available it is evident that flavonol glycosides and proanthocyanidins have at least restricted distribution in this family. Confirmation of the occurrence of monardein in *Davallia* (Harborne, 1966) has provided the only known example of the accumulation of a 3-hydroxyanthocyanin in a fern.

(g) *Family Dennstaedtiaceae*
(Table 12.4.9)

Flavonol glycosides appear to occur widely in this family, proanthocyanidins have restricted distribution, and the only examples of *O*-glycosylated flavones and dihydroflavones possess the unusual 5,7-dioxygenation pattern. Preliminary work on *Sphenomeris* appears to align it more with the Cyatheaceae than with this family.

(h) *Family Gleicheniaceae*
(Table 12.4.10)

The flavonoids of his family are remarkably uniform. Proanthocyanidins are accumulated, and flavonols 3-*O*-glycosylated with only glucose and/or rhamnose are the only flavonoid glycosides detected to date.

(i) *Family Hymenophyllaceae*
(Table 12.4.12)

Kaempferol and quercetin 3-*O*-glycosides are common in this putatively primitive leptosporangiate family, and flavones when they occur are predominantly apigenin and luteolin *C*-glycosides. However, one example of the rare group of tricetin *C*-glycosides, tricetin-8-*C*-glucoside, has been identified in *Trichomanes* as also has apigenin-7,4′-*O*-glucoside. Unusually, *T. petersii* accumulates flavone *C*- and *O*-glycosides as its only flavonoids. *C*-glycosylxanthones occur sporadically in the Hymenophyllaceae (Wallace *et al.*, 1982), *Hymenophyllum dilatatum* accumulating a unique *C*-allosylated example (Markham and Woolhouse, 1983) and *H. recurvum* being characterized by a range of benzoylated xanthones and a complete absence of flavonoids (Markham and Wallace, 1980). The variety and distribution of flavonoid and xanthone types shows potential to aid in the resolution of long standing taxonomic problems in this family (see e.g. Wallace, 1986).

(j) *Family Loxsomaceae*
(Table 12.4.14)

Both genera of this small disjunct family produce flavonol *O*-glycosides as their only flavonoids. *Loxsoma* is distinguished from *Loxsomopsis* by its ability to glycosylate at the 7-hydroxyl group with arabinose, a sugar not found in the *Loxsomopsis* glycosides.

(k) *Family Marattiaceae*
(Table 12.4.15)

Four of the seven genera in this small tropical to subtropical family have so far been investigated. An impressive range of flavone *C*-glycosides, including violanthin, vicenin-1, schaftoside and their 'iso' forms, vicenin-2 and apigenin-6,8-di-*C*-arabinoside, has been found in *Angiopteris* which lacks flavonols (Wallace *et al.*, 1981). Most of these compounds are unique to this genus in the pteridophytes. In contrast, the other genera investigated do accumulate flavonol *O*-glycosides. The structures of these have not been defined but are said to be atypical (Wallace *et al.*, 1984). *Danaea*, the new world genus, accumulates both flavonols and flavone *C*-glycosides.

(l) *Family Marsileaceae*
(Table 12.4.16)

Wallace *et al.* (1984) examined all three genera in this family for both flavonoids and xanthones and concluded that marsileaceous species show an affinity with 'primitive' leptosporangiate ferns, e.g. Hymenophyllaceae, rather than with the eusporangiate ferns. Supporting this contention is the predominant accumulation of flavonol 3-*O*-glycosides and the sporadic occurrences of xanthone and flavone *C*-glycosides (in the major genus *Marsilea*). The finding of myricetin in *Marsilea* is notable, as this flavonol is rarely encountered in ferns. *Regnellidium* is reported to accumulate only rutin, and does not appear to be a complexion or reduction intermediate between *Pilularia* and *Marsilea* as has previously been suggested.

(m) *Family Ophioglossaceae*
(Table 12.4.17)

Together with the Marattiaceae this family constitutes the eusporangiate ferns, but the little published data give no indication of any affinities.

(n) *Family Osmundaceae*
(Table 12.4.18)

Representatives of all three genera have been examined and the only flavonoid glycosides so far identified are flavonol 3-*O*-glycosides (including the first flavanoid *O*-alloside) and one 3-deoxyanthocyanin. Flavones are reported to occur in *Leptopteris* but the evidence is equivocal. The first biflavonoids ever found in ferns, a series of amentoflavone methyl ethers, were recently isolated from *Osmunda japonica* (Okuyama *et al.*, 1979) and their structures supported with substantial physical evidence. However, some doubt was cast on this finding when Wallace and Hu (1985) subsequently reported that they were unable to detect biflavones in this species or in some 200 others also investigated. Assuming that the plants were correctly identified in both cases, one can only assume that seasonal or chemotypic variation must account for these differences. In any event there is now no doubt that biflavonoids are accumulated by some ferns, since their presence in an unrelated species, *Alsophila spinulosa* (Cyatheaceae), has now been firmly established (Wada *et al.*, 1985).

(o) *Families Schizaeaceae and Stromatopteridaceae*
(Tables 12.4.21, 12.4.23)

These two putatively primitive leptosporangiate families both accumulate only kaempferol and quercetin glycosides. Rhamnose and glucose appear to be the only glycosylating sugars. Proanthocyanidins, including prodelphinidins, are common.

(p) *Family Sinopteridaceae*
(Table 12.4.22)

Most of the flavonoids reported from this large family are 'external' flavonoids, i.e. lipophilic flavonoid aglycones which occur on the outside of the fronds. These 'farinose exudates' are largely confined to the gymnogrammoid ferns and especially *Notholaena*, *Pityrogramma* and *Cheilanthes*, but also appear occasionally in *Pterozonium*, *Bommeria*, *Hemionitis*, *Pellaea*, *Negripteris*, *Sinopteris*, *Onychium* and *Adiantum*. This feature is consistent with a close taxonomic alignment between the Sinopteridaceae and the Adiantaceae, much as is encompassed in the family Pteridaceae of Tryon and Tryon (1982). A summary of the taxonomic significance of these flavonoids has appeared (Wollenweber, 1978) and has been updated to some extent in more recent publications, e.g. Wollenweber *et al.* (1985), Giannasi (1980), Smith (1980) and Wollenweber (1984). Studies of external flavonoids have also demonstrated the existence of chemical races, e.g. in *Pityrogramma triangularis* (Smith, 1980; Wollenweber *et al.*, 1985) and *Notholaena* (Wollenweber *et al.*, 1982).

The range of flavonoid aglycones found is extensive and includes flavones, dihydroflavones, flavonols (but no dihydroflavonols), chalcones and dihydrochalcones, all often *C*- or *O*-methylated and sometimes acylated. Internal flavonoids in contrast are largely unmethylated kaempferol and quercetin *O*-glycosides or apigenin and luteolin *C*-glycosides; proanthocyanidins have not yet been reported. These marked differences have prompted the suggestion that internal and external flavonoids may originate from different biochemical pools (Wallace, 1986). Of particular interest amongst the external flavonoids are the *Pityrogramma triangularis* constituents, ceroptene (*12.4*) and its co-occurring 'isoceroptene'. Isoceroptene, originally thought to be a monomer (Markham *et al.*, 1985), is now known to be a dimer. '(α-β', α'-β)-diceroptene', of type (*12.5*) (Vilain, Hubert and Markham, 1987). Also of interest is the biosynthetically un-

CH_3O CH_3 CH_3 OH O O

(*12.4*) Ceroptene

CH_3 CH_3 CH_3O O OH O O OH O OCH_3 CH_3 CH_3

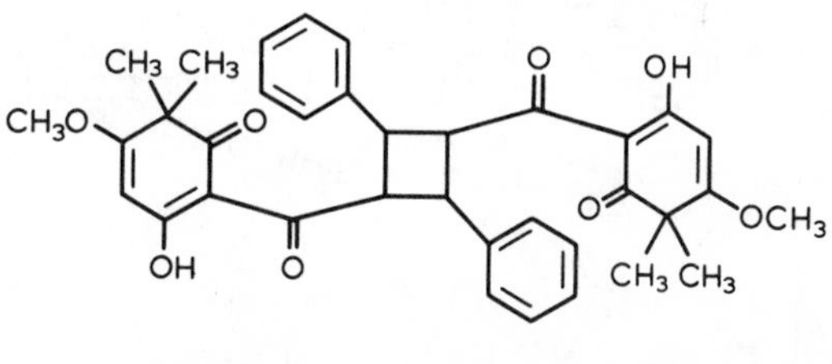

(*12.5*) Diceroptene (='Isoceroptene')

usual 5,2',4'-trihydroxy-3,7,8,5'-tetramethoxyflavone from *Notholaena aschenborniana* the structure of which was recently corrected (Iinuma *et al.*, 1986) from the originally proposed 5,4',6'-trihydroxy-6,7,2',3'-tetramethoxyflavone (Jay *et al.*, 1981). Structurally intriguing β-phenylpropionyl esters, e.g. (*12.6*), which can be considered dihydroneoflavonoids if viewed from a different perspective, e.g. (*12.7*), have been isolated from

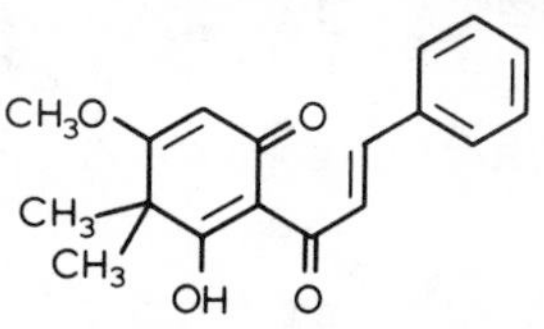

Table 12.4 Flavonoids in the ferns

Species	Flavonoids found			References
12.4.1 FAMILY ADIANTACEAE				
Adiantum (2 spp.)	P		present in 1/2	Cambie *et al.* (1961)
Adiantum (53 spp.)	P35734		present in 17/53	Cooper–Driver and Swain (1977)
Adiantum (53 spp.)	P357345		present in 14/53	
Adiantum (53 spp.)	FL3574	O	3glu(14/53) + rut(9/53) + 3gluu(10/53) + 7glu(1/53)	
Adiantum (53 spp.)	FL35734	O	3glu(14/53) + 3rut(10/53) + 3gluu(10/53)	
Adiantum (53 spp)	FL35734	O	acyl(caffeic)glycs widespread	
Adiantum (53 spp.)	FL3574	O	acyl(caffeic)glycs widespread	
Adiantum (53 spp.)	F574	C	68gly?	
Adiantum aethiopicum	HF574	O	7glu + 7glurha	Hasegawa and Akabori (1968)
Adiantum aethiopicum	FL3574	O	3glu	
Adiantum aethiopicum	FL35734	O	3glu	
Adiantum capillusveneris	FL3574		3sulphate	Imperato (1982a)
Adiantum capillusveneris	FL35734	O	3glu(6″malonyl)	Imperato (1982b)
Adiantum capillusveneris	FL3574	O	3sulphate	
Adiantum capillusveneris	FL3574	O	37glu	
Adiantum capillusveneris	FL3574	O	3glu + 3glurha + 3gluu	Akabori and Hasegawa (1969)
Adiantum capillusveneris	FL35734	O	3glu + 3glurha + 3gluu	
Adiantum capillusveneris	FL3574	O	3glu + 3rut + 3gal? + 3rut-sulphate	Imperato (1982c)
Adiantum caudatum	FL3574	O	3gly?(based on UV spec only)	Saunders and McClure (1976)
Adiantum caudatum	FL35734	O	3gly?(based on UV spec only)	Saunders and McClure (1976)
Adiantum cuneatum	FL35734	O	3gluu	Akabori and Hasegawa (1969)
Adiantum cuneatum	FL3574	O	3glu + 3gluu	
Andiantum cunninghamii	F5734		luteolin	Voirin (1970)
Adiantum monochlamys	HF574	O	7glu	Hasegawa and Akabori (1968)
Adiantum monochlamys	FL35734	O	3gal + 3glu	
Adiantum monochlamys	FL3574	O	3gal + 3glu	
Adiantum pendatum cv. *Klondike*	AN5734	O	5glu	Harborne (1966)
Adiantum pendatum cv. *Klondike*	AN574	O	5glu	
Adiantum poiretii var. *sulphureum*	C2464		4′Me	Wollenweber (1976a)
Adiantum poiretii var. *sulphureum*	HC2464		4′Me	
Adiantum poiretii var. *sulphureum*	FL357		357OH + 7Me	
Adiantum sulphureum	C246		2′4′6′OH + 2′6′OH4′OMe	Wollenweber (1979)
Adiantum sulphureum	HC246		2′6′OH4′OMe	
Adiantum sulphureum	FL357		357OH + 35OH7OMe	
Adiantum veitchianum	AN574	O	5glu + ″adiantum-1″	Harborne (1966)
Adiantum veitchianum	AN5734	O	5glu + ″adiantum-1″	
12.4.2 FAMILY ASPIDACEAE (= SUBFAMILY OF DRYOPTERIDACEAE?)				
Arachnoides (4 spp.)	F5734	C O	8glu(4/4) + 7glu(1/4)	Hiraoka (1978)
Arachnoides (4 spp.)	F574	C O	8glu(4/4) + 7glu(1/4)	
Ctenitis velutina	P		present	
Cyrtomium (3 spp.)	FL3574	O	3glu + 3ara	
Cyrtomium (3 spp.)	FL35734	O	3glu	
Cyrtomium (3 spp.)	F574	C	8glu	
Cyrtomium (3 spp.)	F5734	C	8glu	
Cyrtomium (3 spp.)	HF	O	7glu C–Me(2, unident)	

Cyrtomium (6 spp.)	HF5734	O	68Me(glu)	Kishimoto (1956b)
Cyrtomium (6 spp.)	HF574	O	68Me(glu)	
Cyrtomium (3 spp.)	FL3574	O	7glu6″succinyl ester	Hiraoka and Maeda (1979)
Cyrtomium clavicola	HF574		68Me	Arthur and Kishimoto (1956)
Cyrtomium clavicola	HF5734		68Me	
Cyrtomium falcatum	FL3574	O		Kishimoto (1956a)
Cyrtomium falcatum	HF5734	O	68Me	
Cyrtomium falcatum	HF574	O	68Me	
Cyrtomium falcatum	FL35734	O	glu	Hattori (1962)
Cyrtomium fortunei	HF574	O	68Me	Kishimoto (1956a)
Cyrtomium fortunei	HF5734	O	68Me	
Cyrtomium fortunei	FL3574	O		
Cyrtomium fortunei	FL35734	O		
Dryopteris (4/18 spp.)	FL3574	O	7glu6″succinyl ester	Hiraoka and Maeda (1979)
Dryopteris erythrosora	AN35734	O	gly(prob = 3deoxy)	Harborne (1966)
Dryopteris erythrosora	AN574	O	gly(dryopteris-1)	
Dryopteris erythrosora	AN5734	O	gly(dryopteris-2)	
Dryopteris erythrosora	AN57345?	O	gly(dryopteris-3)	Saunders and McClure (1976)
Dryopteris erythrosora	F574	O	7glu?(based on UV spec only)	Hayashi and Abe (1955)
Dryopteris filix-mas	FN35734		5OAc(Me linked to C-4)	Saunders and McClure (1976) Karl *et al.* (1981)
Dryopteris intermedia	FL3574	O	3gly (gametophyte + sporophyte)	Petersen and Fairbrothers (1980)
Dryopteris intermedia	FL35734	O	3gly (gametophyte + sporophyte)	
Dryopteris marginalis	FL35734	O	3gly (gametophyte + sporophyte)	
Dryopteris marginalis	FL3574	O	3gly (gametophyte + sporophyte)	
Dryopteris oligophlebia	FL3574	O	3glurha	Kobayashi and Hayashi (1952)
Dryopteris setigera	F574		apigenin	Voirin (1970)
Dryopteris (erthrovariae)	F574	C	8glu(10/10 spp) + 8glu 7glu(4/10 spp.)	Hiraoka (1978)
Dryopteris (erthrovariae)	F5734	C	8glu(10/10 spp.)	
Dryopteris (erthrovariae)	FL3574	O	3glu(10/10 spp.) + 7ara (3/10 spp.)	
Dryopteris (erthrovariae)	FL3574	O	7xyl(5/10 spp.) + 3glurha (4/10 spp.)	
Dryopteris (erthrovariae)	FL3574	O	3gluxyl(4/10) + 7glurha(5/10)	
Dryopteris (erthrovariae)	FL3574	O	3gluglurha(4/10)	
Dryopteris (erthrovariae)	FL35734	O	3glu(10/10 spp.) + 3glurha (3/10 spp.)	
Dryopteris (erthrovariae)	HF5734	O	7glu(10/10 spp.) + 7glurha (10/10 spp.)	
Dryopteris (sec. dryopteris)	FL3574	O	3glu(7/8 spp.) + 7rha(4/8 spp.)	
Dryopteris (sec. dryopteris)	FL3574	O	3glurha(6/8 spp.) + 3galrha 7rha(6/8 spp.)	
Dryopteris (sec. dryopteris)	FL3574	O	3ararha7ara(6/8 spp.)	
Dryopteris (sec. dryopteris)	FL35734	O	3glu(6/8 spp.)	
Dryopteris (sec. dryopteris)	HF5734	O	7glu(3/8 spp.)	
Dryopteris (sec. dryopteris)	F5734	O	7glu(3/8 spp.)	
Dryopteris (sec. dryopteris)	F5734	C	8glu(3/8 spp.)	
Dryopteris (sec. dryopteris)	F574	C	8glu(3/8 spp.)	
Leptorumohra miqueliana	FL3574	O	3glu + 3glurha + 3galrha	
Leptorumohra miqueliana	FL35734	O	3glu	

(*Contd.*)

Table 12.4 (*Contd.*)

Species	*Flavonoids found*			*References*
Leptorumohra miqueliana	HF5714		571′4′OH68Mel′2′dihydro	Noro *et al.* (1969)
Polystichum (1/4 spp.)	FL3574	O	7glu6″succinyl ester	Hiraoka and Maeda (1979)
Polystichum (3 spp.)	P		present	
Polystichum (5 spp.)	HF57	O	7rha(1/5)	
Polystichum (5 spp.)	HF5734	O	7glu(2/5)	
Polystichum (5 spp.)	F5734	C	8glu(5/5)	Hiraoka (1978)
Polystichum (5 spp.)	F574	C	8glu(5/5)	
Polystichum (5 spp.)	FL3574	O	3glu(3/5) + 7ara(1/5)	
Polystichum (5 spp.)	FL35734	O	3glu(3/5) + 3glurha(1/5)	
Polystichum (5 spp.)	FL3574	O	3gluglu/glurha/gluglurha(1/5)	
Polystichum aculeatum	FL35734		3Me	Voirin (1970)
Polystichum setiferum	FL35734		3Me	
12.4.3 FAMILY ASPLENIACEAE				
Asplenium (hybrids)	FL		flavonols additive	
Asplenium (10 spp.)	FL3574		(1/10)	Harborne *et al.* (1973)
Asplenium (10 spp.)	FL35734		(2/10)	
Asplenium (10 spp.)	FL357345		(1/10)	
Asplenium (7 spp.)	P		absent (6/7)	Cambie *et al.* (1961)
Asplenium (15 spp.)	P		present (13/15)	Voirin (1970)
Asplenium bulbiferum	FL3574	O	37glu + 3rha7glu + 3glu 7rha + 3glu7gal	Imperato (1985a)
Asplenium bulbiferum	FL3574	O	4′Me37diglu + 4′Me3rha7glu	Imperato (1984a)
Asplenium bulbiferum	FL3574	O	4′Me3glu7rha	
Asplenium diplazisorum	FL3574		4′Me	Voirin (1967)
Asplenium montanum	FL3574	O	glycs(unidentified)	
Asplenium platyneuron	FL3574	O	37glu/acyl + 34′Me7glu	Harborne *et al.* (1973)
Asplenium rhizophyllum	FL3574	O	7/37/74′glu(all caffeoylated)	
Asplenium rhizophyllum	FL3574	O	3soph7glu(caffeoylated)	
Asplenium septentrionale	FL35734	O	3glu + 3glu3″sulphate	Imperato (1983c)
Asplenium septentrionale	FL3574	O	3glu-sulphate	
Asplenium septentrionale	FL3574	O	3sophorotrioside7glu	Imperato (1984c)
Asplenium trichomanes	FL3574	O	3rha7rha + 3rha7ara + 3ara7rha	Imperato (1979)
Asplenium viride	FL3574		kaempferol	Voirin (1970)
Asplenium viride	FL35734		quercetin + 3Me	Voirin and Jay (1974)
Ceterach officinarum (= *Asplenium?*)	HF574	O	7glu(1–6)ara(pyr)	Imperato (1983a)
Ceterach officinarum	FL35734	O	3glu + 3gent	Imperato (1981)
Ceterach officinarum	FL3574	O	3glu/gal-6″malonyl	Imperato (1981)
Cystopteris fragilis	FL3574	O	34′glu + 3glu3″sulphate + 3glu6″sulphate	Imperato (1983b)

12.4.4 FAMILY ATHYRIACEAE (=SUBFAMILY OF DRYOPTERIDACEAE?)

Athyrium (5 spp.)	FL3574	O	3glu	Hiraoka (1978)
Athyrium (5 spp.)	FL35734	O	3glu	
Cystopteris fragilis (complex)			flavonoids found (no strs)	Cooper–Driver and Haufler (1983)
Diplazium (4 spp.)	FL35734	O	3glu	Hiraoka (1978)
Diplazium (4 spp.)	FL3574	O	3glu	
Diplazium esculentum	HF5734	O	5Me7galxyl	Srivastava *et al.* (1981)
Lunathyrium (= *Deparia*) 7spp.	FL3574	O	3glu	
Lunathyrium (= *Deparia*) 7spp.	FL35734	O	3glu	Hiraoka (1978)
Lunathyrium (= *Deparia*) 7spp.	F574	C	8glu	
Lunathyrium (= *Deparia*) 7spp.	F5734	C	8glu	
Matteuccia (2 spp.)	FL3574	O	7glu-6″succinyl ester	Hiraoka and Maeda (1979)
Matteuccia (2 spp.)	F5734	C	8glu	
Matteuccia (2 spp.)	F574	C	8glu	
Matteuccia (2 spp.)	FL3574	O	3glu + 7ara	Hiraoka (1978)
Matteuccia (2 spp.)	FL35734	O	3glu + 3gal	
Matteuccia (2 spp.)	HF574	O	68Me4′Me7glu (matteucinol 7glu)	
Matteuccia (2 spp.)	HF	O	7glu C-Me unidentified	
Matteuccia orientalis	HF57		57OH68Me(+ 574′OH68Me)	Fujise (1930)
Matteuccia orientalis	HF572		572′OH68Me	Mohri *et al.* (1982)
Matteuccia orientalis	HF5725		572′OH5′OMe58Me	
Onoclea sensibilis	F5734	C	8glu	
Onoclea sensibilis	F574	C	8glu	
Onoclea sensibilis	FL35734	O	3glu + 7gal + 3rha	
Onoclea sensibilis	FL3574	O	3glu + 7ara	Hiraoka (1978)
Woodsia (2 spp.)	FL3574	O	3glu/glurha/galrha(1/2)	
Woodsia (2 spp.)	FL3574	O	3gluxyl + 3rhaxyl	
Woodsia (2 spp.)	FL35734	O	3glu + 3rha + 3gal(1/2)	
Woodsia polystichoides	FL3574	O	3arafur(3″OAc)7rha + 3ara7rha	Murakami *et al.* (1984)
Woodsia polystichoides	FL35734	O	3glu	

12.4.5 FAMILY AZOLLACEAE AND F. SALVINIACEAE

Azolla filiculoides	P		present	Voirin (1970)
Azolla mexicana	AN5734	O	5glu	Holst (1977)
Azolla rubra	P		absent	Cambie *et al.* (1961)
Salvinia auriculata	P		present	Voirin (1970)

12.4.6 FAMILY BLECHNACEAE

Blechnum (4 spp.)	FL3574		Kaempferol(2/4)	Voirin (1970)
Blechnum (9 spp.)	P		present (7/9)	Cambie *et al.* (1961)
Blechnum (4 spp.)	FL35734		quercetin (2/4)	Voirin (1970)
Blechnum brasiliense	FL3574		Kaempferol (gametophyte + sporophyte)	Laurent (1966)
Blechnum brasiliense var. *corcovadense*	AN574	O	5glu	Harborne (1966)
Blechnum brasiliense var. *corcovadense*	AN5734	O	5glu	

(*Contd.*)

Table 12.4 (*Contd.*)

Species	*Flavonoids found*			*References*
Blechnum procerum	AN574	O	5glu/diglu/glurha + 7glu/diglu	Crowden and Jarman (1974)
Blechnum procerum	AN574	O	aglyc + 5glu7glu	
Blechnum procerum	AN5734	O	aglyc + 5glu7glu	
Blechnum procerum	AN5734	O	5glu/diglu/glurha + 7glu/diglu	
Blechnum regnellianum	HF574		68Me4′Me (= (−)matteucinol)	Miraglia *et al.* (1985)
Doodia media	P		present	Cambie *et al.* (1961)
12.4.7 FAMILY CYATHEACEAE (INCLUDES DICKSONICEAE)				
Alsophila bryophila	F574	C	8glu	Conant and Cooper-Driver (1980)
Alsophila bryophila	FL3574	O	3glu	
Alsophila dryopteroides	FL574	C	6glu	
Alsophila dryopteroides	FL3574	O	3glu	
Alsophila spinulosa	BHF574-(66‴) F5734			Wada *et al.* (1985)
Alsophila spinulosa	BHF5734-(66‴) F5734			
Culcita dubia	FL3574	O	3glys	Markham (unpublished)
Culcita dubia	FL35734	O	3glys	
Cyathea (5/5 spp.)	FL3574	O	7glu6″succinyl ester	Hiraoka and Maeda (1979)
Cyathea (6 spp.)	P		present (2/6)	Cambie *et al.* (1961)
Cyathea (5/5 spp.)	F574	C	8glu	Hiraoka and Hasegawa (1975)
Cyathea (5/5 spp.)	F5734	C	8glu	
Cyathea contaminans	FL3574	O	7glu6″succinyl ester	Hiraoka and Maeda (1979)
Cyathea contaminans	FL3574	O	kaempferol + 3glu + 3rha + 3soph + 7glurha	
Cyathea divergens	F574	C	6glu	Soeder and Babb (1972)
Cyathea fauriei	FL3574	O	3glu/rha/ara/soph + 7glurha	Hiraoka and Hasegawa (1975)
Cyathea fauriei	F574	C	8glu	Ueno *et al.* (1963)
Cyathea hancockii	FL3574	O	3glu/rha/ara/gal/glurha	Hiraoka and Hasegawa (1975)
Cyathea kermadecensis	FL3574	O	3gly	Markham and Given (1979)
Cyathea leichhardtiana	FL3574	O	3glu/rha/ara/soph + 7glurha	Hiraoka and Hasegawa (1975)
Cyathea mertensiana	FL3574	O	3glu/rha/ara/soph + 7glurha	
Cyathea milnei	FL3574	O	3gly	Markham and Given (1979)
Cyathea onusta	F574	C	8glu	Soeder and Babb (1972)
Cyathea podophylla	FL3574	O	3glu/rha/ara/gal/glurha	Hiraoka and Hasegawa (1975)
Cyathea spinulosa	F574	C	8glu + 6glu	Seshadri and Rangaswami (1974)
Cyathea spinulosa	F5734	C	8glu	
Cyathea tueckheimii	F574	C	8glu	Soeder and Babb (1972)
Dicksonia (3 spp.)	P		present (2/3)	Cambie *et al.* (1961)
Dicksonia gigantea	FL35734	O	3glu	Satake *et al.* (1984)
Nephelia portoricensis	F574	C	6glu	Conant and Cooper–Driver (1980)
Nephelia portoricensis	FL3574	O	3glu	

12.4.8 FAMILY DAVALLIACEAE (INCLUDES OLEANDRACEAE)

Davallia divaricata	AN3574	O	5glu3(p-coumaroyl)glu(= monardein)	Harborne (1966)
Davallia divaricata	FN35734	O	(−)epicatechin-3all + 3all2″/3″cinnamoyl	Murakami *et al.* (1985)
Humata pectinata	FL35784	O	8Me3glu	Wu and Furukawa (1983)
Humata pectinata	FL357834	O	8Me3glu	
Rumohra hispida	P		absent	Cambie *et al.* (1961)
Rumohra adiantiformis	F5734		luteolin	Voirin (1970)
Nephrolepis auriculata	FL3574	O	3glu	Murakami *et al.* (1985)
Nephrolepis cordifolia	P		present	Cambie *et al.* (1961)

12.4.9 FAMILY DENNSTAEDTIACEAE (INCLUDES LINDSAEACEAE)

Dennstaedtia wilfordii	HF57	O	7glu/7glurha (neohesperidoside?)	
Dennstaedtia wilfordii	FL35734	O	3glu	Akabori and Hasegawa (1970)
Dennstaedtia wilfordii	FL3574	O	3glu	
Histiopteris incisa	P		absent	
Hypolepis distans	P		absent	
Leptolepia novazelandiae	P		absent	Cambie *et al.* (1961)
Lindsaea viridis	P		absent	
Lonchitis tisserantii	F57	O	6Me + 6Me7gly	Voirin and Lebreton (1967)
Paesia anfractuosa	FL35734	O	3rut	Satake *et al.* (1984)
Paesia anfractuosa	FL3574	O	3rut	
Paesia scaberula	P		present	Cambie *et al.* (1961)
Pteridium aquilinum	FL3574	O	5glu	Wij (1978)
Pteridium aquilinum	FL35734	O	glu	Hattori (1962)
Pteridium aquilinum	FL3574	O	3glu(*p*-coumaroyl)	Want *et al.* (1973)
Pteridium aquilinum	FL35734	O	7glu	Nakabayashi (1955b)
Pteridium aquilinum	FL3574	O	3glu + 3glu(*p*-coumaroyl)	Wang *et al.* (1973)
Pteridium aquilinum	FL35734	O	3glu + 3rut	Nakabayashi (1955a)
Pteridium aquilinum	FL3574	O	3glurha	Cooper–Driver (1976)
Pteridium aquilinum	P35734		present	Voirin (1970)
Pteridium aquilinum	P357345		present	
Sphenomeris chusana	F574	C	8glu	Ueno *et al.* (1963)

12.4.10 FAMILY GLEICHENIACEAE

Dicranopteris linearis	FL3574	O	3glu + 3glurha	
Dicranopteris linearis	FL35734	O	3glu + 3glurha	
Dicranopteris pectinata	FL35734	O	3glu	
Dicranopteris pectinata	FL3574	O	3glu + 3glurha	Wallace *et al.* (1983)
Gleichenia bifida	FL3574	O	3glurha	
Gleichenia bifida	FL35734	O	3glurha	
Gleichenia circinata	P		present	Cambie *et al.* (1961)
Gleichenia costaricensis	FL35734	O	3glu + 3glurha	Wallace *et al.* (1983)
Gleichenia costaricensis	FL3574	O	3glu + 3glurha	
Gleichenia cunninghamii	FL35734	O	3glu + 3rha + 3glurha	Wallace and Markham (1978a)
Gleichenia cunninghamii	P		present	Cambie *et al.* (1961)
Gleichenia intermedia	FL3574	O	3glurha	Wallace *et al.* (1983)
Gleichenia linearis	P35734		procyanidin	Cooper–Driver (1977)
Gleichenia linearis	P357345		prodelphinidin	

(*Contd.*)

Table 12.4 (*Contd.*)

Species	*Flavonoids found*			*References*
Gleichenia pallescens	FL35734	O	3glurha	
Gleichenia pallescens	FL3574	O	3glurha	
Gleichenia underwoodiana	FL3574	O	3glu + 3glurha	
Gleichenia underwoodiana	FL35734	O	3glu + 3glurha	
Hichriopteris bancroftii	FL35734	O	3glu + 3glurha + 3diglu + 3triglu	Wallace *et al.* (1983)
Hichriopteris bancroftii	FL3574	O	3glu + 3glurha	
Sticheris cunninghamii (= *Gleichenia*?)	FL3574	O	3rha	
Sticheris retroflexum	FL3574	O	3glurha	
Sticheris retroflexum	FL35734	O	3glurha	
12.4.11 FAMILY GRAMMITIDACEAE				
Grammitis heterophylla	P		present	Cambie *et al.* (1961)
12.4.12 FAMILY HYMENOPHYLLACEAE				
Cardiomanes reniforme	FL35734	O	3glu	Wallace and Markham (1978a)
Cardiomanes reniforme	FL3574	O	3glu	
Hymenophyllum (8 spp.)	P		present	Cambie *et al.* (1961) Harada and Saiki (1955)
Hymenophyllum barbatum	F574		apigenin	Wallace and Markham (1978a)
Hymenophyllum demissum	FL35734	O	3glurha	
Hymenophyllum demissum	FL3574	O	3glurha (– ve flavones)	
Hymenophyllum demissum	FL35734	O	3glurha (– ve flavones)	Markham and Wallace (1980)
Hymenophyllum dilatatum	F5734	C	trace	
Hymenophyllum dilatatum	F574	C	trace	
Hymenophyllum dilatatum	FL35734	O	3gal + 3ara	
Hymenophyllum polyanthose	FL3574		kaempferol	Harada and Saiki (1955)
Hymenophyllum recurvum			xanthones only, no flavonoids	Markham and Wallace (1980)
Hymenophyllum spp.	FL3574		kaempferol	Voirin (1970)
Hymenophyllum spp.	FL35734		quercetin	
Hymenophyllum tunbridgense	FL35734	O	3glu + 3glurha (– ve flavones)	
Hymenophyllum tunbridgense	FL3574	O	3glu + 3glurha (– ve flavones)	
Trichomanes javanicum	FL3574		kaempferol	
Trichomanes petersii	F574	CO	6glu + 8glu + 74′glu (– ve flavonols)	Markham and Wallace (1980)
Trichomanes petersii	F5734	CO	6glu + 8glu (– ve flavonols)	
Trichomanes reniforme	FL3574	O	3glu(– ve flavones)	
Trichomanes reniforme	FL35734	O	3glu(– ve flavones)	
Trichomanes reniforme	P		present	Cambie *et al.* (1961) Voirin (1970)
Trichomanes thysanostoma	FL3574		kaempferol	Harada and Saiki (1955)
Trichomanes venosum	F574	C	6glu + 8glu/2″?ara(fur)	
Trichomanes venosum	F5734	C	6glu + 8glu2″?ara(fur)	Markham and Wallace (1980)
Trichomanes venosum	F57345	C	8glu(= affinetin)	
12.4.13 FAMILY LOMARIOPSIDACEAE				
Bolbitis subcordata	FN34574		54′OMe68Me	Tanaka *et al.* (1980)

12.4.14 FAMILY LOXSOMACEAE				
Loxsoma cunninghamii	FL35734	O	3glu + 3rut + 3glu7ara	Markham and Given (1979)
Loxsoma cunninghamii	FL3574	O	3glu + 3rut + 3glu7ara	
Loxsomopsis costaricensis	FL3574	O	3rut + 3glu	
12.4.15 FAMILY MARATTIACEAE				
Angiopteris (3 spp.)	F5734	C	not further ident	Wallace *et al.* (1981)
Angiopteris evecta	FL35734		not verified, Wallace and Story (1978)	Voirin (1970)
Angiopteris evecta	P		trace	
Angiopteris evecta	F574	C	6glu8rha(violanthin) + 6rha8glu	Wallace *et al.* (1979)
Angiopteris evecta	F574	CO	glycs(no flavonols)	Wallace and Story (1978)
Angiopteris evecta(2 spp.)	F574	C	6glu8rha + 6rha8glu + (6xyl8glu + 6glu8xyl + 6glu8ara?) + 68glu	Wallace *et al.* (1981)
Angiopteris hypoleuca	F574	CO	6glu8rha/8xyl/8ara + 68glu + 7glu	
Angiopteris hypoleuca	F574	C	68ara	
Angiopteris lygodiifolia	F574	C	68glu + 6glu8ara + 6ara8glu + 68ara(no violanthin)	
Angiopteris palmaformis	F574	C	similar to other spp.	
Angiopteris spp.	B		not found	
Angiopteris spp.	FL		not found	
Christensenia	FL3574	O	gly	Cooper-Driver (1980)
Danaea	F574	C	gly	
Danaea	FL	O	gly	
Marattia	P		absent (2/2)	Voirin (1970)
Marattia excavata	FL35734			
Marattia excavata	FL3574			
Marattia salicina	FL3574			
Marattia salicina	FL3574	O	glurha (two)	Wallace and Story (1978)
Marattia salicina	P		present	Cambie *et al.* (1961)
12.4.6 FAMILY MARSILEACEAE				
Marsilea aegyptiaca	FL35734	O	3gal + 3glu	Wallace *et al.* (1984)
Marsilea difusa	FL35734	O	3gal + 3diglu	
Marsilea drummondii	FL357345	O	3glu + 3gal	Eades *et al.* (1984)
Marsilea drummondii	FL35734	O	3glycs	
Marsilea elata	FL3574	O	3glu	Wallace *et al.* (1984)
Marsilea elata	FL35734	O	3gal + 3ara	
Marsilea minuta	FL35734	O	3gal + 3glu	
Marsilea minuta	FL3574	O	3glu	
Marsilea mucronata	FL3574	O	3rha	
Marsilea mucronata	FL35734	O	3glu/gal/ara/rha + 3glurha	
Marsilea mucronata	F574	C	68glu	
Marsilea mucronata	F5734	C	68glu	
Marsilea mutica	FL35734	O	3gal + 3glu	
Marsilea vestita	FL35734	O	3glu + 3rha + 2diglu	
Marsilea vestita	F574	C	68glu	
Pilularia americana	FL35734	O	3glu + 3ara + 3rha + 3diglu + (3lyx?)	
Pilularia americana	FL3574	O	3glu	
Regnellidium diphyllum	FL35734	O	3glurha	

(*Contd.*)

Table 12.4 (*Contd.*)

Species	*Flavonoids found*			*References*
12.4.17 FAMILY OPHIOGLOSSACEAE				
Botrychium lunaria	FL35734		quercetin predominant	Voirin (1970)
Ophioglossum (3 spp.)	P		absent (3/3)	
Ophioglossum lusitanicum	FL35734		3Me	Voirin and Jay (1974)
Ophioglossum reticulatum	FL35734		3Me	Voirin (1970)
Ophioglossum vulgatum	FL35734	O	3Me7diglu4′glu	Markham *et al.* (1969)
Ophioglossum vulgatum	FL35734		37Me?	Voirin (1970)
12.4.18 FAMILY OSMUNDACEAE				
Leptopteris superba	F?	O	glyc present (unident)	Sobel and Whalen (1983)
Leptopteris superba	FL		absent	
Osmunda asiatica	FL3574	O	3all	Okuyama *et al.* (1978)
Osmunda asiatica	FL3574	O	kaempferol + 3glu	
Osmunda cinnamomea	FL3574	O	3glu + 3glurha	Sobel and Whalen (1983)
Osmunda cinnamomea	FL35734	O	3glu + 3glurha	
Osmunda cinnamomea	F		absent	
Osmunda japonica	FL3574	O	3glu	Okuyama *et al.* (1979)
Osmunda japonica	BF574(3′8″)F574		4′4‴Me + 4′7″4‴Me	
Osmunda japonica	BF574(3′8″)F574		74′4‴Me + 77″4′4‴Me	
Osmunda japonica	BF		absent	Wallace and Hu (1985)
Osmunda (3 spp.)	BF		absent	
Osmunda regalis	AN5734	O	gly("pteris-1")	Harborne (1966)
Todea barbara	FL3574	O	3glu	Sobel and Whalen (1983)
Todea barbara	FL35734	O	3glu + 3glurha	
Todea barbara	F?	O	glyc present(unident)	
Todea barbara	P		present	Cambie *et al.* (1961)
Todea hymenophylloides	P		present	
12.4.19 FAMILY POLYPODIACEAE				
Leptochilis zeylanicus	FL357345		myricetin	Voirin (1970)
Phlebodium aureum	FL35734	O	3glu/glurha/ara	Gomez and Wallace (1986)
Phlebodium aureum	FL3574	O	3glurha + 3ara	
Phlebodium decumanum	FL3574	O	3glu/ara/diara/gal/(ribose?)	
Phlebodium decumanum	FL35734	O	3glu + 3glurha + 3ara	
Phymatodes (2spp.)	P		absent	Cambie *et al.* (1961)
Polypodium vulgare	FN35734		catechin-7apioside/7ara(fur)	Karl *et al.* (1982)
Polypodium vulgare	FL3574	O	3glu + 7rha + 3ara7rha + 3ararha	
Polypodium vulgare	F574	C	6ara8glu + 6glu8ara	
Polypodium vulgare	FL35734	O	quercetin + 3rut + 3trioside	Grzbek (1976)
Polypodium vulgare	FL3574	O	3glurha	

Taxon				Reference
12.4.20 FAMILY PTERIDACEAE				
Pteris cretica	F574		apigenin	Voirin (1970)
Pteris cretica	F5734		luteolin	
Pteris grandifolia	FL35734	O	3rha3″OAc/4″OAc	Tanaka *et al.* (1978)
Pteris longipinnula (= *Dryopteris?*)	AN574	O	5glu	Harborne (1966)
Pteris longipinnula	AN5734	O	5glu + 'pteris-1'	
Pteris quadriaurita	AN5734	O	5glu + 'pteris-1 + 2'	
Pteris tremula	P		absent	Cambie *et al.* (1961)
Pteris vittata	AN5734	O	5glu	Harborne (1966)
12.4.21 FAMILY SCHIZAEACEAE				
Anemia (3 spp.)	P35734		procyanidin	Cooper-Driver (1977)
Anemia (3 spp.)	P357345		prodelphinidin	
Anemia (3 spp.)	FL3574		kaempferol	
Anemia (3 spp.)	FL35734		quercetin	
Lygodium articulatum	FL3574	O	3gly7gly	Markham and Given (1979)
Lygodium articulatum	FL3574	O	3glu	Markham (unpublished)
Lygodium articulatum	P		absent	Cambie *et al.* (1961)
Lygodium japonicum	FL3574		kaempferol	Cooper-Driver (1977)
Lygodium japonicum	FL35734		quercetin	
Lygodium japonicum	P35734		procyanidin	
Lygodium japonicum	P357345		prodelphinidin	
Lygodium palmatum	FL3574	O	3glurha7glu	Wallace (1986)
Mohria caffrorum	FL3574		kaempferol	Cooper-Driver (1977)
Mohria caffrorum	FL35734		quercetin	
Mohria caffrorum	P35734		procyanidin	
Mohria caffrorum	P357345		prodelphinidin	
Schizaea bifida	FL35734	O	3glu + 3glurha	Wallace and Markham (1978a)
Schizaea bifida	FL3574	O	3glu + 3glurha	
Schizaea dichotoma	P		present	Cambie *et al.* (1961)
Schizaea pectinata	P35734		procyanidin	Cooper-Driver (1977)
Schizaea pectinata	P357345		prodelphinidin	
12.4.22 FAMILY SINOPTERIDACEAE				
Bommeria (4 spp.)	FL35734	O	3rut + 3gly(x5)	Gianassi (1980)
Bommeria (4 spp.)	FL3574	O	3rut + 3glu	
Bommeria (4 spp.)	F5734	C	68gly	
Bommeria (4 spp.)	F574	CO	68gly + 4′gly	
Bommeria (all 4 spp. in genus)	FL35734	O	3glu + 3glurha(4/4) + 3gly(3/4)	Haufler and Gianassi (1982)
Bommeria (all 4 spp. in genus)	FL3574	O	3glu + 3glurha(4/4) + 3gly(1/4)	
Bommeria (all 4 spp. in genus)	F574	O	apigenin + 4′ara(2/4)	
Bommeria (all 4 spp. in genus)	F5734	O	luteolin + 3′glu(1/4)	
Bommeria (all 4 spp. in genus)	F574	C	6glu(4/4) + 8glu(2/4)	
Bommeria (all 4 spp. in genus)	F5734	C	8glu(2/4) + 68glu(1/4)	
Cheilanthes argentea	HF5674		67Me + 674′Me	Wollenweber *et al.* (1980)
Cheilanthes argentea	HF5784		78Me + 784′Me	
Cheilanthes argentea	HF56784		784′Me + 678Me + 6784′Me	

(*Contd.*)

Table 12.4 (*Contd.*)

Species	Flavonoids found			References
Cheilanthes farinosa	FL3574		74′Me + 374′Me	Erdtman *et al.* (1966)
Cheilanthes farinosa	F574		74′Me	
Cheilanthes farinosa	FL3574		37Me	Rangaswamy and Iver (1969)
Cheilanthes farinosa	FL3574		374′Me	
Cheilanthes fragrans	FL35734	O	3rhagal(1–4)glu	Imperato (1985b)
Cheilanthes kaulfussii	FL357		357OH + 3Me + 37Me	Wollenweber (1979)
Cheilanthes kaulfussii	FL3574		37Me + 74′Me	
Cheilanthes kuhnii var. *brandtii*	FL3574		3/37/34′Me + (major)74′/374′Me	Serizawa and Wollenweber (1977)
Cheilanthes kuhnii var. *brandtii*	F574		apigenin + 4′(+7?)Me	
Cheilanthes longissima	FL3574		37Me + 374′Me	Sunder *et al.* (1974)
Cheilanthes longissima	F574		7Me	
Cheilanthes viscida	F574		apigenin + 7/4′/74′Me	Wollenweber (1979)
Cheilanthes (and others)			exudate flavonoids (review)	
Gymnopteris = *Hemionitis*? (5/5 spp.)	FL35734	O	3rut + 3gly(x3) + 34′gly (x5) + 37gly	Gianassi (1980)
Gymnopteris (5/5 spp.)	F5734	C	68gly	
Gymnopteris (5/5 spp.)	F574	C	68gly	
Gymnopteris (1/5 spp.)	FL35734?	O	34′diglys	Gianassi and Mickel (1979)
Gymnopteris (1/5 spp.)	F5734	C	68glu	
Gymnopteris (1/5 spp.)	F574	C	68glu	
Gymnopteris (4/5 spp.)	FL35734	O	3glu + 3rut + 3gly	
Gymnopteris (4/5 spp.)	FL3574	O	3glu + 3rut	
Hemionitis = *Gymnopteris*? (5/5 spp.)	FL35734	O	3rut + 3gly(x3) + 34′gly (x5) + 37gly	Gianassi (1980)
Hemionitis (5/5 spp.)	F574	C	68gly	
Hemionitis (5/5 spp.)	F5734	C	68gly	
Hemionitis (2/5 spp.)	FL3574	O	3glu + 3rut	Gianassi and Mickel (1979)
Hemionitis (2/5 spp.)	FL35734	O	3glu + 3rut + 3gly	
Hemionitis (3/5 spp.)	FL35734?	O	34′diglys	
Hemionitis (3/5 spp.)	FL35734	O	37diglys(x2)	
Hemionitis (3/5 spp.)	F574	C	68glu	
Hemionitis (3/5 spp.)	F5734	C	68glu	
Notholaena (and others)			Exudate flavonoids-review	Wollenweber (1978)
Notholaena spp.	FL357834		74′/373′Me(butyrate + acetate)	Wollenweber (1984)
Notholaena spp.	FL3578		7Me(butyrate + acetate)	
Notholaena spp.	FL35784		7/74′Me(butyrate + acetate)	
Notholaena spp.	FL3578234		372′3′4′Me(butyrate + acetate)	Jay *et al.* (1982)
Notholaena affinis	FL3567824		3678/36782′/36784′/36782′4′Me	Jay *et al.* (1979b)
Notholaena affinis	FL35784		74′Me(butyrate + acetate)	Wollenweber *et al.* (1978b)
Notholaena affinis	FL357824		3782′Me + 3782′4′Me	Jay *et al.* (1979a)
Notholaena aliena	FL3578		7Me8acetate/butyrate	Wollenweber (1984)
Notholaena aliena	FL357825		72′5′Me/8acetate	
Notholaena aliena	FL357823		72′3′Me	
Notholaena aliena	FL35784		7Me8acetate/8butyrate	
Notholaena aschenborniana	FL3578245		3785′Me (corrected from 5672′3′4′6′ oxidin. pattern)	Iinuma *et al.* (1986)

Notholaena aschenborniana	FL3578234	372′3′4′Me8acetate	Jay *et al.* (1982)
Notholaena aschenborniana	FL35734	373′Me8acetate	
Notholaena aschenborniana	FL35784	7Me8butyrate/8acetate	Wollenweber *et al.* (1978a)
Notholaena aschenborniana	FL3578	7Me8butyrate/8acetate	
Notholaena aschenborniana	F5672346	672′3′Me[incorrect, see Iinuma *et al.* (1986)]	Jay *et al.* (1981)
Notholaena bryopoda	FL3574	3/7/37/74′/374′Me	Wollenweber (1984)
Notholaena bryopoda	FL35734	37/374′Me	
Notholaena bryopoda	F574	7Me(sporadic occ, weak)	
Notholaena bryopoda	FL35734	37/374′Me (trace only)	
Notholaena bryopoda	F574	7Me (trace)	
Notholaena californica	FL3578	7Me8acetate/butyrate	
Notholaena californica	F574	7Me + 74′Me	Wollenweber *et al.* (1982)
Notholaena californica	F5734	7/3′/4′/73′/74′Me	
Notholaena californica	FL35734	3/37/374′Me	
Notholaena californica	FL3574	3/7/37/374′Me	
Notholaena californica	FL35784	8acetate + 8butyrate	
Notholaena californica	FL3578	7Me8acetate/butyrate (yellow form)	Wollenweber (1984)
Notholaena californica	F574	7/74′Me(white form)	
Notholaena californica	FL3574	7/3/37/374′Me (white form)	
Notholaena californica	FL35734	3/37/374′Me(white form)	
Notholaena californica	F5734	7/3′/4′/73′/74′Me (white form)	
Notholaena candida var. *copelandii*	FL357	galangin + 3Me(major)	
Notholaena candida var. *copelandii* and *candida*	FL3574	3Me	
Notholaena candida var. *candida*	FL3578	37Me(acetate + butyrate)	Wollenweber (1985)
Notholaena candida var. *candida*	FL357345	373′4′5′Me + 373′4′Me (major)	
Notholaena candida var. *candida*	FL357	3Me(major) + galangin	Wollenweber (1984)
Notholaena fendleri	HF5734	74′Me + 73′4′Me	Wollenweber (1985)
Notholaena fendleri	HF574	7Me + 4′Me	Wollenweber *et al.* (1980)
Notholaena fendleri	HF5734	74′Me + 73′4′Me + 4′Me	
Notholaena fendleri	HF574	4′Me + 7Me	
Notholaena fendleri	HF5734	74′Me + 73′Me + 7Me + 4′Me	Wollenweber (1981)
Notholaena fendleri	F5734	74′Me + 73′Me(vetulin) traces of each	
Notholaena fendleri	FL35734	37Me + 373′Me(traces of each)	
Notholaena fendleri	F574	4′Me(traces)	
Notholaena galapagensis	FL3578	7Me8acetate + butyrate	
Notholaena grayi	F574	apigenin + 7Me + 4′Me(all major)	Wollenweber *et al.* (1978a)
Notholaena grayi	FL3578	7Me8acetate(or butyrate) minor	Wollenweber (1984)
Notholaena lemmonii	HC246	2′6′OH4′OMe(typical)	
Notholaena lemmonii	HF5734	7/73′/74′/73′4′Me (typical)	
Notholaena lemmonii	HF57345	73′/73′4′/73′5′/73′4′5′Me (typical)	
Notholaena lemmonii	HC	absent(= chemotype?)	
Notholaena lemmonii var. *lemmonii*	HF57345	73′4′5′Me	Wollenweber *et al.* (1980)
Notholaena limitania var. *mexicana*	HF5734	74′Me + 73′4′Me	
Notholaena neglecta	HF578	8-OAc	Wollenweber and Yatskievych (1982)
Notholaena neglecta	FL3578	7Me8acetate/butyrate	
Notholaena neglecta	FL35734	73′Me(rhamnazin, trace only)	
Notholaena neglecta	FL3574	7OH/7OMe(traces only) + 8Ac + 8butyrate	

(*Contd.*)

Table 12.4 (*Contd.*)

Species	*Flavonoids found*			*References*
Notholaena pallens	F574		apigenin + 7/4′/74′Me	
Notholaena pallens	FL35734		33′Me(v. weak, 1/10 samples)	Wollenweber (1984)
Notholaena pallens	FL3574		3/37/34′/74′Me	
Notholaena pallens	F574		apigenin + 7/4′/74′Me (all major)	
Notholaena standleyi	FL3574		kaempferol (major, L. yellow race) + 7/4′/37/34′Me	Seigler and Wollenweber (1983)
Notholaena standleyi	FL35784		7Me + 74′Me(yell-grn race)	
Notholaena standleyi	FL3574		3Me + 4′Me(major, yell-grn race) + 37/34′/74′Me	Wollenweber (1976a)
Notholaena standleyi	FL3574		7Me(major in golden race)	
Notholaena standleyi	FL35784		7/74′Me	
Notholaena sulphurea	C246		2′6′OH4′OMe(major in yell. var.)	
Notholaena sulphurea	C246		2′6′OH4′OMe(in white var.)	Wollenweber (1984)
Notholaena sulphurea	C2464		2′6′OH44′OMe(major in white var.)	
Notholaena sulphurea	HC246		2′6′OH4′OMe(white form)	
Notholaena sulphurea	HC2464		2′6′OH44′OMe(white form)	
Onychium auratum (= *siliculosum?*)	C246		2′6′OH4′OMe	
Onychium auratum	HF57		7Me(pinostrobin)	Ramakrishnan *et al.* (1974)
Onychium auratum	C2456		2′6′OH4′5′OMe (pashanone)	
Onychium auratum	C246		2′6′OH4′OMe	
Onychium japonicum	FL3574	O	37rha	Kobayashi (1944)
Onychium siliculosum	C246		2′6′OH4′OMe	Wollenweber (1982)
Onychium siliculosum	C2456		2′6′OH4′5′OMe (pashanone)	
Pityrogramma austroamericana	C2464		2′6′OH44′Me	Wollenweber (1978)
Pityrogramma austroamericana	HC2464		2′6′OH44′Me	
Pityrogramma calomelanos var. *aureoflava*	F574		8-β(β-phenylpropionic A) + Me ester	Favre-Bonvin *et al.* (1980)
Pityrogramma calomelanos	C2464		2′6′OH4′OMe (neosakuranetin)	Hitz *et al.* (1982)
Pityrogramma calomelanos	HC2464		2′6′4OH4′OMe (asebogenin)	Mabry (1973)
Pityrogramma calomelanos	HC2464		44′Me in the white form	
Pityrogramma calomelanos	HC246		4′(β-phenylpropionyl-ester)-β-C3′ linked	
Pityrogramma calomelanos	F57		7(β-phenylpropionyl-ester)-β-C6 linked	Wagner *et al.* (1979)
Pityrogramma calomelanos	FL357		7(β-phenylpropionyl-ester)-β-C6 linked	
Pityrogramma candida var. *copelandii*	FL3574		3Me	Arriaga-Giner and Wollenweber (1986)
Pityrogramma candida var. *copelandii*	FL357		3Me	
Pityrogramma chrysoconia	FL357		57OH + 7Me	Wollenweber (1972)
Pityrogramma chrysoconia	FL357		7Me(izalpin)	Star (1980)

Pityrogramma chrysoconia	HC246		4′Me	Wollenweber (1979)
Pityrogramma chrysoconia	C246		4′Me	
Pityrogramma chrysoconia	FL357		galangin + 7Me	Wollenweber (1972)
Pityrogramma chrysophylla	C246		4′Me	Nilsson (1961)
Pityrogramma chrysophylla	C2464		2′6′OH44′Me	
Pityrogramma ebenea	HC234564		43′Me4′5′OCH_2O	Miraglia *et al.* (1985)
Pityrogramma lehmanii	HC2464		2′6′OH44′Me	Wollenweber (1976b)
Pityrogramma tartarea	HC2464		2′6′OH4′Me	Mabry (1983)
Pityrogramma tartarea	FL3574		7Me	
Pityrogramma tartarea	F574		7Me	
Pityrogramma triangularis car. *pallida*	HF57		6Me + 68Me + 8Me(in dec, order of yield)	Wollenweber *et al.* (1981b)
Pityrogramma triangularis var. *pallida*	HF57		57OH + 57Me8Me + 5Me8Me + 5Me	Markham *et al.* (1987)
Pityrogramma triangularis var. *pallida*	C246		6′OMe5′Me/3′Me + 6′OMe + 4′6′OMe	Wollenweber *et al.* (1981b)
Pityrogramma triangularis	F5678		68Me	Dietz *et al.* (1981)
Pityrogramma triangularis	HC246		4′OMe3′5′Me	Wollenweber *et al.* (1985)
Pityrogramma triangularis	C246		3′diMe4′OMe6′C = O (ceroptene)	
Pityrogramma triangularis	FL3578		37Me(not F57OH68OMe as prev publ)	
Pityrogramma triangularis	FL3578		37Me6Me + 7Me6Me	
Pityrogramma triangularis	FL357		8Me + 3Me8Me + 7Me8Me	
Pityrogramma triangularis	FL35784		38Me + 84′Me + 384′Me	
Pityrogramma triangularis	FL357		68Me + 3Me68Me + 7Me68Me	
Pityrogramma triangularis	C246		4′OMe3′Me (triangularin) + 2′4′OMe	
Pityrogramma triangularis	HF57		6Me + 8Me(ex. chalcone cyclization)	
Pityrogramma triangularis	HC2464		2′6′OH4′OMe3′5′Me	
Pityrogramma triangularis	FL3578		8Me6Me(pityrogrammin) + 7Me6Me	Star *et al.* (1975a)
Pityrogramma triangularis	HF57		‘isoceroptene’ structure since revised to $\alpha\beta'$, $\alpha'\beta$-diceroptene (Vilain, Hubert and Markham, 1987)	Markham *et al.* (1985)
Pityrogramma triangularis	FL357		7Me + 57Me	Hitz *et al.* (1982)
Pityrogramma triangularis	FL35734	O	3glu + 3glurha(2n + 4n, ceroptene chemotype)	Star *et al.* (1975b)
Pityrogramma triangularis	FL3574	O	7Me3glurha(2n + 4n, c. chemotype)	
Pityrogramma triangularis	FL3574	O	7Me3glu(4n, c. chemotype)	
Pityrogramma triangularis	FL35734	O	3glu + 3glurha(4n, kaempferol methyl ether chemotype)	
Pityrogramma triangularis	FL35734	O	3di/trigly(2n + 4n, kme chemotype)	
Pityrogramma triangularis	FL35734	O	3di/trigly(2n + 4n, c. chemotype)	
Pityrogramma triangularis	HC2464		2′6′4OH4′OMe3′Me (in var. viscosa)	Wollenweber *et* al. (1979)
Pityrogramma triangularis	C246		4′6′OMe + 4′6′OMe3′Me	Wollenweber *et al.* (1981a)
Pityrogramma triangularis	FL3574		3/5/4′/34′/74′/374′Me	Smith (1980)
Pityrogramma triangularis	FL357		3/37Me	
Pityrogramma trifoliata	C246		4′(β-phenyl propionyl ester)-β-3′-linked	Dietz *et al.*, (1980)

(*Contd.*)

Table 12.4 (*Contd.*)

Species	*Flavonoids found*			*References*
Pityrogramma trifoliata	C2464		4′(β-phenyl propionyl ester)-β-3′-linked	Dietz *et al.* (1980)
Pityrogramma trifoliata	C24634?		as above but tentative	Dietz *et al.* (1980)
Pityrogramma williamsii	FL357		galangin	} Star (1980)
Pityrogramma (16 spp.)	C246		2′4′OH4′Me(4/16)	
Pityrogramma (16 spp.)	C2464		2′6′OH44′Me(5/16)	
Pityrogramma (16 spp.)	HC2464		2′6′OH44′Me(8/16)	
Pityrogramma (16 spp.)	HC246		2′6′OH4′Me(4/16)	
Pterozonium brevifrons	C2464		4′Me	Wollenweber (1978)
Pterozonium brevifrons	C2464		2′6′OH44′Me	Wollenweber (1979)
Trachypteris	FL35734	O	3rut	} Gianassi (1980)
Trachypteris	FL3574	O	3rut	
12.4.23 FAMILY STROMATOPTERIDACEAE				
Stromatopteris moniliformis	FL3574	O	3glu + 3glurha	} Wallace and Markham (1978a)
Stromatopteris moniliformis	FL35734	O	3glu + 3glurha	
Stromatopteris moniliformis	P35734		procyanidin	} Cooper–Driver (1977)
Stromatopteris moniliformis	P357345		prodelphinidin	
12.4.24 FAMILY THELYPTERIDACEAE				
Glaphyropteridopsis erubescens	FL3574	O	3rha	} Tanaka *et al.* (1984)
Glaphyropteridopsis erubescens	FL35734	O	3rha	
Glaphyropteridopsis erubescens	FN4574	O	684′Me57glu + 6844′Me57glu	
Glaphyropteridopsis erubescens	FN4574	O	684′Me5glu(also linked 2″-4)	
Phegopteris polypodioides	F574	O	4′glu	} Ueno (1962)
Phegopteris polypodioides	FL3574	O	3gluara	
Pronephrium triphyllum	FN4574	O	4′Me6(CH_2OH)8Me 57diglu	} Tanaka *et al.* (1985)
Pronephrium triphyllum	HF574	O	6(CH_2OMe)8Me4′Me7glu	
Thelypteris (2 spp.)	P		present (1/2)	Cambie *et al.* (1961)
Wagneriopteris nipponica	FL3574	O	3glu + 3all	} Murakami *et al.* (1985)
Wagneriopteris nipponica	FL35734	O	3glu + 3all	

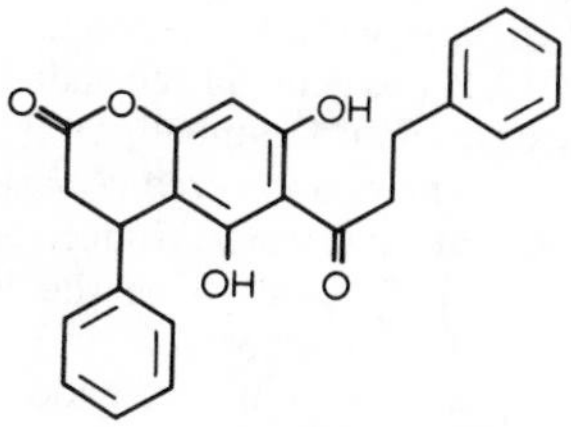

(12.6)

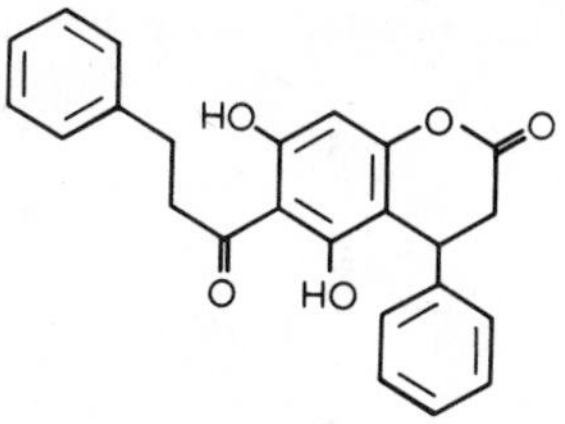

(12.7)

Pityrogramma calomelanos (Wagner *et al.*, 1979) and from *P. trifoliata* (Dietz *et al.*, 1980).

(q) *Family Thelypteridaceae*
(Table 12.4.24)

Flavonol and dihydroflavone glycosides and flavans have all been recorded in the Thelypteridaceae. The family has not been studied extensively, but biosynthetic features such as *C*- and *O*-methylation and unusual *C*- and *O*-linked side chains are apparent and hint at a level of biosynthetic diversity not found in many fern families.

12.6 PHYLOGENETIC CONSIDERATIONS

12.6.1 General

Previous attempts to interpret flavonoid distribution data in terms of their phylogenetic significance (e.g. Crawford, 1978; Giannasi, 1978; Campbell *et al.*, 1979; Swain and Cooper-Driver, 1981; Richardson, 1983) have met with varied success. It is clear, however, that there is no generally accepted method for such interpretations. Accordingly, this section of the chapter is necessarily speculative. Complex factors such as whether the taxonomic groupings used are phylogenetically relevant, whether a major taxon is monophyletic or whether particular biosynthetic characteristics are 'primitive' or 'advanced' for example may have a major bearing on the interpretation. Further, attempting to elucidate phylogeny by consideration of secondary compound distribution in isolation from other data, presumably just as relevant, could well be misleading. Thus in the present discussion cognizance is taken where possible of commonly accepted botanical concepts, and reference is made to published botanical comment to which the flavonoid data are relevant.

The definition of 'primitive' and 'advanced' becomes of prime importance in assessing flavonoid distribution (or any other set of characters) in phylogenetic terms. In this respect it is important to stress that since all plant groups studied are *extant* taxa, no one can truly be considered a primitive version of another. 'Primitive' sets of flavonoid characters are defined here as those least modified from the ancestral state, and because different taxa may have different ancestors, it follows that advanced characters in one taxon are not necessarily exhibited by advanced members of another taxon.

It is likely that flavonoid accumulation was developed in early green plants as a protective screen against the intense levels of potentially destructive ultraviolet radiation that existed on earth prior to the build-up of oxygen and ozone in the atmosphere. At the time plants began to colonize the land it is estimated that the oxygen level was only about 10% of today's level. Flavonoids, with very strong absorption in the 265 and 340 nm regions, would thus have provided exceptionally effective protection for susceptible and vital UV-absorbing biomolecules such as tyrosine and thiamine ($\lambda_{max} \sim 260$ nm) and NAD/NADP (λ_{max}340 nm). It is thus conceivable that successful colonization of the land was initially dependent upon, amongst other things, the ability to biosynthesize highly effective screening compounds such as flavonoids. If so it is perhaps not surprising that virtually all land-based green plants produce flavonoids, albeit now often for reasons other than UV protection. The lack of flavonoids in extant plants within flavonoid-producing taxa is likely therefore to be a derived condition.

With the possible exception of the charophyte, *Nitella*, flavonoids have not been found in algae. They are, however, produced by a high proportion of the bryophytes, virtually all ferns, and all fern allies, gymnosperms and angiosperms. Each of these groups has been subjected to a variety of different selective pressures in the course of evolution and each has responded in its own way with respect to flavonoid 'evolution'. In the bryophytes, ferns and fern allies there seems to have been a general tendency at about the ordinal (or familial) level, to restrict flavonoid biosynthesis to the production of one predominant flavonoid type. However, unless the plant group has been isolated for a long period of time, the loss of ability to produce other flavonoid types is rarely absolute. Trace (and sporadic?) occurrences of the other flavonoids may thus be indicative of past evolutionary relationships. Biochemical simplifications such as these may well be a general phenomenon in the evolution of plants (Smith, 1976; Pirie, 1985). Further responses to selective pressures within each plant group would then be constrained by this inherited biosynthetic simplification. Thus evolutionary advancement is frequently evidenced by further modification of the basic 'speciality' flavonoid type, e.g. by methylation, acylation, additional oxygenation and/or glycosylation etc. Less commonly, it is evidenced by development of a new (or latent) flavonoid type, e.g. flavones in the ferns; aurones and isoflavones in the bryophytes. These general concepts result from the more detailed considerations of particular plant groups presented below.

12.6.2 Bryophytes – Hepaticae

Assuming that the class Hepaticae is a monophyletic group, the ordinal relationships as indicated by comparative flavonoid chemistry can be represented in cladogram form as in Fig. 12.1. It is generally considered on botanical grounds (Schuster, 1966; Campbell, 1971; Hattori *et al.*, 1974) that the recently discovered *Takakia* is the most primitive of extant liverworts. Indeed its discoverer, Hattori, considered *Takakia* to be the liverwort most closely related to the ancestors of modern bryophytes. Intriguingly, *Takakia* accumulates a wide range of flavonoid types, in fact all major types that are found in the other orders. On this basis it is suggested here that liverwort evolution may have progressed by a process of reduction from *Takakia*-like ancestral stock rich in flavonoid types,

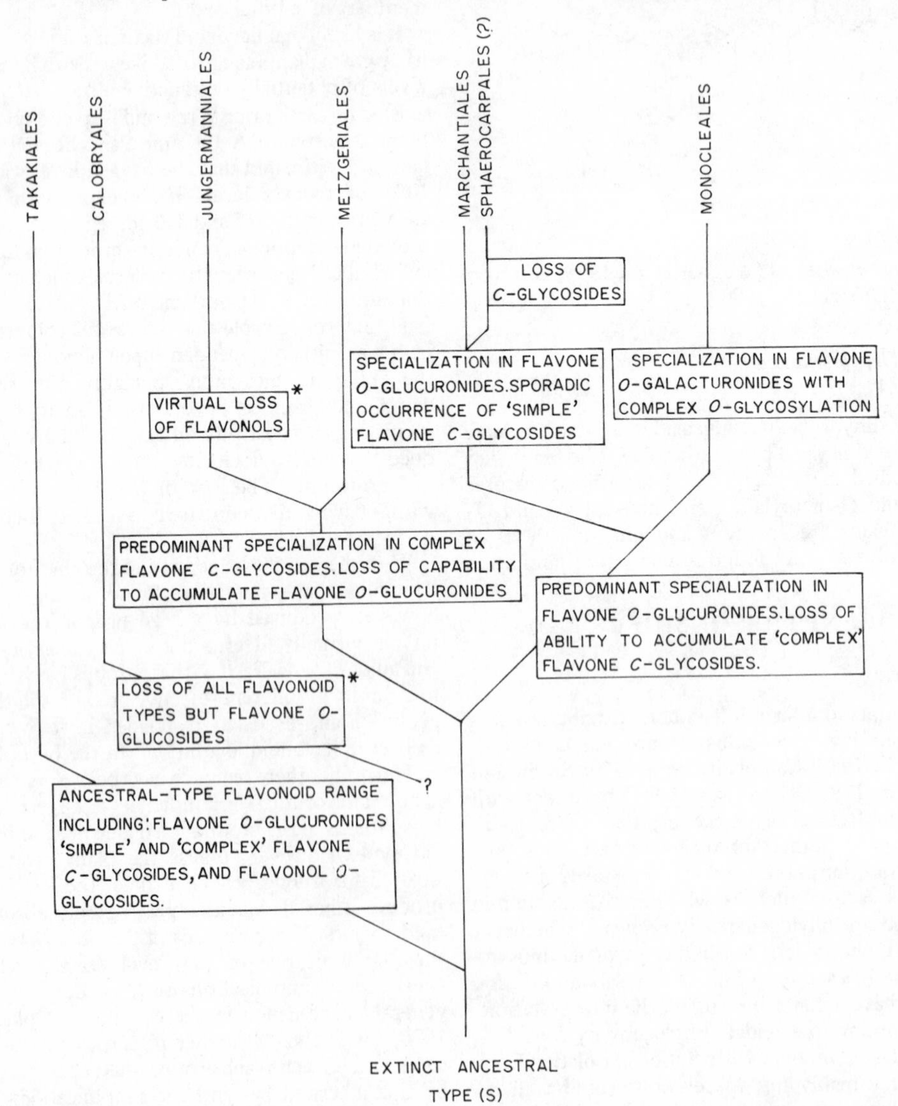

Fig. 12.1 Proposed relationships between liverwort orders based on available flavonoid distribution data. *Yet to be confirmed – data inadequate.

to more advanced taxa in which the range of flavonoid types is reduced (see Fig. 12.1). Thus the Takakiales have retained most of the biosynthetic capabilities of the ancestral stock, while the Metzgeriales/Jungermanniales have largely lost the capability to accumulate flavone *O*-glycosides and have specialized in flavone *C*-glycoside elaboration and accumulation. In contrast, the Marchantiales have largely lost the ability to elaborate flavone *C*-glycosides and have specialized in the production and accumulation of flavone *O*-glucuronides. The flavonoid chemistry of the Monocleales is consistent with an alignment near the Marchantiales as proposed by some botanists (e.g. Grolle, 1972; Schuster, 1983), but there is evidence of a biosynthetic gulf which may indicate separation from the mainstream Marchantialean line at an early stage. This could also be true for the Sphaerocarpales if they are considered a separate order. The level at which the Calobryales branch point occurs is uncertain due to the limited data available.

Evolution within each order generally seems to be evidenced by the degree of biosynthetic versatility shown within the 'chosen' flavonoid type. Thus in the Marchantiales, families like the Marchantiaceae and Conocephalaceae which contain flavones with a variety of advanced (Harborne, 1967) biosynthetic features such as complex and unusual *O*-glycosylation, 8-oxygenation, *O*-methylation, acylation and aurone production, are considered to represent the more advanced. The least advanced are likely to be families with limited biosynthetic capabilities like the Sphaerocarpaceae and possibly families like the Corsiniaceae which accumulate what appear to be relic flavonoid types (in this case, flavonols). Substantial chemical data relating to the Metzgeriales and Jungermanniales are available only for a few families. Amongst these, the Metzgeriaceae and possibly the Frullaniaceae appear to be in the advanced category. The evolutionary status of taxa which apparently lack flavonoids is unclear, but the condition would seem to be derived.

12.6.3 Bryophytes – Musci

The very limited number of flavonoid data available for most moss orders or suborders preclude all but the most general comment at this stage. Only one of the orders, the Andreales, appears to lack flavonoids completely (McClure and Miller, 1967) and it is likely that this order separated from the others at an early stage in their evolution. The same may also be said of the Sphagnales if the flavonoids in this order prove only to be of the sphagnorubin type. The other orders appear to accumulate 'normal' flavonoids and are perhaps more closely related to one another than to the Andreales and Sphagnales, a relationship recently suggested by Vitt (1983). The biosynthetic complexity evident in the Bryales and in particular the suborder Bryineae (= Eubryales or 'true mosses'), e.g. additional 6, 8- and 5′-oxygenation, malonyl acylation, isoflavone production etc., is suggestive of more active recent evolution within this group.

12.6.4 Bryophytes – Anthocerotales

The Anthocerotales, lacking the ability to biosynthesize flavonoids at all, appear to occupy an isolated position relative to the Hepaticae and Musci, both of which biosynthesize a related range of flavonoids. There is thus some support in this distribution for the hypothesis made by a number of authors (e.g. Schuster, 1977; Crandall-Stotler, 1980) that the Anthocerotales arose from stock at best distantly related to other primitive bryophytes. The evolutionary affinity between the mosses and liverworts proposed by Schuster also finds some general support (in the broadest sense) in these data.

12.6.5 Fern allies

The dominant biosynthetic capabilities displayed by each of the five orders of fern allies are summarized in Table 12.5. The sharp demarcations evident between most of these groups and the relative 'biosynthetic purity' within them suggests that they may have been isolated in a reproductive sense for a long period of time. The 'biosynthetic purity' displayed by the Psilotales in accumulating essentially only one flavonoid aglycone, amentoflavone, is particularly striking. All other orders display some level of biosynthetic versatility within their 'chosen' flavonoid group. The trace quantities of apigenin *O*- and *C*-glycosides (and hinokiflavone?) found in the Psilotales, may well represent relic biosynthetic capabilities, and if so, hint at a distant relationship with ancestral stock capable of biosynthesizing such compounds. It follows therefore that such stock might also have given rise independently to lines currently represented by the Selaginellales (accumulating amentoflavone and hinokiflavone), Lycopodiales (flavone *O*- and *C*-glycosides) and the Isoetales (elaborated flavone *O*-glycosides). These three groups are in fact frequently aligned with one another in phylogenetic classifications (Pichi-Sermolli, 1977; Stewart, 1983), and indeed have at times been jointly separated with the Psilotales at the ordinal level from the Equisetales (see Pichi-Sermolli, 1973, and references therein).

Development of the biosynthetic pathway leading to the 3-oxygenated flavonoids, proanthocyanidins and flavonols, is seen only in the Equisetales. This feature sets the Equisetales apart from the other orders and may indicate an alignment with the ferns, the bulk of which use this pathway extensively. It is possible that the pathway developed in a specialized line of ancestral stock which ultimately gave rise to both the Equisetales and the ferns. Such a relationship has previously been proposed on other grounds (Stewart, 1983).

Table 12.5 Dominant flavonoids accumulated by fern allies

Order	*Flavonols + Proanthocyanidins*	*Flavones*		*Biflavones*	
		4'/3'4'/3'4'5'-OH	*Additional 6, 8 or 2'-OH*	*Amentoflavone*	*Hinokiflavone*
Equisetales	+				
Isoetales		+	+		
Lycopodiales		+			
Psilophytales				+	
Selaginellales				+	+

+, dominant flavonoid type present.

12.6.6 Ferns

Comment on the phylogenetic significance of the distribution of flavonoids in ferns is made particularly difficult by two factors; (i) the lack of any general agreement amongst pteridologists on the number and composition of fern families, and (ii) the inadequate flavonoid data available for most putative families.

It is probable that all fern families accumulate flavonols, and because flavonols are so widespread, they probably represent a character inherited from ancestral stock. Further elaboration of the basic kaempferol and quercetin 3-glycosides, e.g. by glycosylation with sugars other than rhamnose and glucose, glycosylation at other sites, methylation, additional oxygenation or substitution, acylation and sulphation, is likely therefore to be indicative of phylogenetic advancement. The significance of the flavones is more difficult to rationalize as their occurrence is scattered and they occasionally appear as the sole flavonoid type, e.g. in the genus *Angiopteris* and the species *Trichomanes petersii*. The accumulation of flavones (or dihydroflavones) and their *O*-glycosides has been observed in the Adiantaceae, Sinopteridaceae, Aspidaceae, Aspleniaceae, Athyriaceae, Dennstaedtiaceae, Hymenophyllaceae and possibly the Marattiaceae and the Osmundaceae. Flavones could be a relic character which is in the process of disappearing or alternatively could represent a developing character of more recent origin. The latter seems the more likely however since in general the families in which flavones appear are not those that exhibit primitive flavonol characters. Many in fact are families which also show advanced biosynthetic capabilities in the production of their external flavonoids, e.g. *C*-methylation, additional oxygenation, loss of B-ring oxygenation, dihydrochalcone formation etc. On this basis, families such as are defined in this classification as Adiantaceae, Sinopteridaceae, Aspidaceae, Aspleniaceae and the Athyriaceae, in which flavones and dihydroflavones constitute a significant portion of the total internal or external flavonoids, are likely to be amongst the more advanced, i.e. furthest removed in a biosynthetic sense from the ancestral type or types. Flavone *C*-glycosides have a similar distribution except that they are also found in the Cyatheaceae and Marsiliaceae, but to date not in the Dennstaedtiaceae or Osmundaceae. Where they do occur they tend to be more widespread within the family. The biosynthetic lineage of flavone *C*-glycosides is quite distinct from that of other flavones and the two are not necessarily related in an evolutionary sense. In the ferns, their status as evolutionary markers is still unclear.

On the basis of the above reasoning, the fern families most closely related to the primitive stock would be those that lack flavones and accumulate only biosynthetically basic flavonol *O*-glycosides, such as Stromatopteridaceae, Gleicheniaceae and Schizaeaceae (and Loxsomaceae?). It is with this group of ferns that Bierhorst (1971, 1973) has suggested the Psilotaceae should be aligned. However, as has been stated by several authors (Cooper-Driver, 1977; Wallace and Markham, 1978b), the predominant accumulation of biflavones and the complete absence of flavonols in the Psilotaceae is not consistent with this hypothesis. The recent finding of biflavones in ferns is unlikely to be relevant since they have not been found in the primitive ferns.

REFERENCES

Akabori, Y. and Hasegawa, M. (1969), *Bot. Mag. Tokyo* **82**, 294.
Akabori, Y. and Hasegawa, M. (1970), *Bot. Mag. Tokyo* **83**, 263.
Alston, R.E. (1968). In *Recent Advances in Phytochemistry* (eds. T.J. Mabry, R.E. Alston and V.C. Runeckles), Vol 1, Appleton-Century-Crofts, New York, p. 305.
Aly, H.-F., Geiger, H., Schucker, U., Waldrum, H., Vander-Velde, G. and Mabry, T.J. (1975), *Phytochemistry* **14**, 1613.
Anhut, S., Zinsmeister, H.D., Mues, R., Barz, W., Machenrock, K., Koster, J. and Markham, K.R. (1984), *Phytochemistry* **23**, 1073.
Ansari, F.R., Ansari, W.H., Rahman, W., Seeligman, O., Chari, V.M. Wagner, H. and Osterdahl, B.G. (1979), *Plant. Med.* **36**, 196.
Arnason, J.T., Philogene, B.J.R., Donskov, N., Muir, A. and Towers, G.H.N. (1986), *Biochem. Syst. Ecol.* **14**, 287.
Arriaga-Giner, F.J. and Wollenweber, E. (1986), *Phytochemistry* **25**, 735.

Arthur, H.R. and Kishimoto, Y. (1956), *Chem. Ind.*, 738.
Asakawa, Y. (1982), *Prog. Ort. Nat. Prod.* **42**, 1.
Asakawa, Y., Toyota, M. and Takemoto, T. . (1978), *Phytochemistry* **17**, 2005.
Asakawa, Y., Hattori, S., Mizutani, M., Tokunaga, N. and Takemoto, T. (1979), *J. Hattori Bot. Lab.* **46**, 77.
Asakawa, Y., Tokunaga, N., Takemoto, T., Hattori, S., Mizutani, M. and Suire, C. (1980), *J. Hattori Bot. Lab.* **47**, 153.
Asakawa, Y., Takikawa, K., Toyota, M. and Takemoto, T. (1982), *Phytochemistry* **21**, 2481.
Barger, C. (1902), *Chem. Ber.* **35**, 1296.
Becker, R., Mues, R., Zinsmeister, H.D., Herzog, F. and Geiger, H. (1986), *Z. Naturforsch.* **41c**, 507.
Beckman, S. and Geiger, H. (1963), *Phytochemistry* **2**, 281.
Bendz, G. and Martensson, O. (1961), *Acta Chem. Scand.* **15**, 1185.
Bendz, G. and Martensson, O. (1963), *Acta Chem. Scand.* **17**, 266.
Bendz, G., Martensson, O. and Tenerius, L. (1962), *Acta Chem. Scand.* **16**, 1183.
Bendz, G., Martensson, O. and Nilsson, E. (1966), *Acta Chem. Scand.* **20**, 277.
Berti, G. and Bottari, F. (1968). In *Progress in Phytochemistry* (eds L. Reinhold and Y. Liwschitz), Interscience, London, Vol. 1, p. 589.
Besson, E., Chopin, J., Markham, K.R., Mues, R., Wong, H. and Bouillant, M.-L. (1984), *Phytochemistry* **23**, 159.
Bierhorst, D.W. (1971) *Morphology of Vascular Plants*, MacMillan, New York.
Bierhorst, D.W. (1973), *Bot, J. Linn. Soc.* **67**, 45.
Birch, A.J. Donovan, F.W. and Moewus, F. (1953), *Nature (London)* **172**, 902.
Bold, H.C. (1973), *Morphology of Plants*, Harper and Rowe, New York.
Brehm, B.G. and Comp, P.C. (1967), *Amr. J. Bot.* **54**, 660.
Cambie, R.C., Cain, B.F. and La Roche, S. (1961), *N. Z. J. Sci.* **4**, 707.
Campbell, E.O. (1971), *N.Z.J. Bot.* **9**, 678.
Campbell, E.O. (1984), *J. Hattori Bot. Lab.* **55**, 315.
Campbell, E.O., Markham, K.R. and Porter, L.J. (1975), *N.Z.J. Bot.* **13**, 593.
Campbell, E.O., Markham, K.R., Moore, N.A., Porter, L.J. and Wallace, J.W. (1979), *J. Hattori Bot. Lab.* **45**, 185.
Conant, D.S. and Cooper-Driver, G. (1980), *Amr. J. Bot.* **67**, 1269.
Cooper-Driver, G.A. (1976), *Bot. J. Linn. Soc.* **73**, 35.
Cooper-Driver, G.A. (1977), *Science* **198**, 1260.
Cooper-Driver, G.A. (1980), *Bull. Torrey Bot. Club.* **107**, 116.
Cooper-Driver, G.A. and Haufler, C. (1983), *Fern Gazz.* **12**, 283.
Cooper-Driver, G.A. and Swain, T. (1977), *Bot. J. Linn. Soc.* **74**, 1.
Crabbe, C.A., Jermy, A.C. and Mickel, J.T. (1975), *Fern Gaz.* **11**, 141.
Crandall-Stotler, B. (1980), *Bioscience* **30**, 580.
Crawford, D.J. (1978), *Bot. Rev.* **44**, 431.
Crowden, R.K. and Jarman, J. (1974), *Phytochemistry* **13**, 1947.
Dietz, V.H., Wollenweber, E., Favre-Bonvin, J. and Gomez, P.L.D. (1980) *Z. Naturforsch.* **35c**, 36.
Dietz, V.H., Wollenweber, E., Favre-Bonvin, J. and Smith, D.M. (1981), *Phytochemistry* **20**, 1181.
Dombray, P. (1926), Dr Sc. Thesis, National University of Paris, Ser. A., 1045: pp. 1–211.
Eades, D., Wallace, J.W. and Bhardwaja, T.N. (1984), *Amr. J. Bot.* **71**, 132.
Efimenko, O.M. and Dzenis, A. Ya. (1962), *Chem. Abstr.* **57**, 3786.
Erdtman, H., Novotny, L. and Romanuk, M. (1966), *Tetrahedron Suppl.* **8**, 71.
Favre-Bonvin, J., Jay, M.J., Wollenweber, E. and Dietz, V.H. (1980), *Phytochemistry* **19**, 2043.
Freitag, P. Mues, R., Brill-Fess, C., Stoll, M., Zinsmeister, H.D. and Markham, K.R. (1986), *Z. Naturforsch.* **25**, 699.
Fujise, S. (1930), *Chem. Abstr.* **24**, 3238.
Fusiak, F. (1982), *Am. Fern. J.* **72**, 96.
Geiger, H. (1986), Private communication.
Geiger, H., Lang, U., Britsch, E., Mabry, T.J., Suhr-Schucker, U., Vander-Velde, G. and Waldrum, H. (1978), *Phytochemistry* **17**, 336.
Geiger, H., Reichert, S. and Markham, K.R. (1982), *Z. Naturforsch.* **37b**, 504.
Geiger, H., Stein, W., Mues, R. and Zinsmeister, H.Ó. (1987) *Z. Naturforsch* **42c**, 863.
Giannasi, D.E. (1978), *Bot. Rev.* **44**, 399.
Gianassi, D.E. (1980), *Bull. Torrey Bot. Club* **107**, 128.
Gianassi, D.E. and Mickel, J.T. (1979), *Brittonia* **31**, 405.
Gomez, L.D. and Wallace, J.W. (1986), *Syst. Biochem. Ecol.* **14**, 407.
Grolle, R. (1972), *J. Bryol.* **7**, 201.
Grolle, R. (1983), *Acta Bot. Fenn.* **121**, 1.
Grzbek, J. (1976), *Acta Biol. Cracov. Ser. Bot.* **19**, 69.
Hall, J.E. (1967), PhD Thesis, Chemistry Department; University of British Columbia.
Harada, T. and Saiki, Y. (1955) *Pharm. Bull. (Tokyo)* **3**, 469.
Harborne, J.B. (1966), *Phytochemistry* **5**, 589.
Harborne, J.B. (1967), *Comparative Biochemistry of the Flavonoids*, Academic Press, London, p. 313.
Harborne, J.B., Williams, C.A. and Smith, D.M. (1973), *Biochem. Syst.* **1**, 51.
Hasegawa, M. and Akabori, Y. (1968), *Bot. Mag. Tokyo* **81**, 469.
Hattori S. (1962). In *The Chemistry of Flavonoid Compounds* (ed. T.A. Geissman), Pergamon Press, Oxford, p. 317.
Hattori, S., Iwatsuki, Z., Mizutani, M. and Inoue, S. (1974), *J. Hattori Bot. Lab.* **38**, 115.
Haufler, C.H. and Giannasi, D.E. (1982), *Biochem. Syst. Ecol.* **10**, 107.
Hauteville, M., Chopin, J., Geiger, H. and Schuler, L. (1980), *Tetrahedron. Lett.* **21**, 1227.
Hauteville, M., Chopin, J., Geiger, H. and Schuler, L. (1981), *Tetrahedron* **37**, 377.
Hayashi, K. and Abe, Y. (1955), *Bot. Mag. Tokyo* **68**, 299.
Hegnauer, R. (1986), *Chemotaxonomie der Pflanzen*, band 7, Berhauser Verlag, Basel, p. 416.
Hiraoka, A. (1978), *Biochem. Syst. Ecol.* **6**, 171.
Hiraoka, A. and Maeda, M. (1969), *Chem. Pharm. Bull.* **27**, 3130.
Hiraoka, A. and Hasegawa, M. (1975), *Bot. Mag. Tokyo* **88**, 127.
Hitz, C., Mann, K. and Wollenweber, E. (1982), *Z. Naturforsch.* **37c**, 337.
Holst, R.W. (1977), *Am. Fern. J.* **67**, 99.
Huneck, S. (1983). In *New Manual of Bryology* (ed. R.M. Schuster), Hattori Botanical Laboratory, Japan, Vol. 1, p. 1.
Iinuma, M., Fang, N., Tanaka, T., Mizuno, M., Mabry, T.J., Wollenweber, E., Favre-Bonvin, J., Voirin, B. and Jay, M. (1986), *Phytochemistry* **25**, 1257.
Imperato, F. (1979), *Experientia* **35**, 1134.

Imperato, F. (1981), *Chem. Ind.*, 695.
Imperato, F. (1982a), *Chem. Ind.*, 604.
Imperato, F. (1982b), *Phytochemistry* **21**, 2158.
Imperato, F. (1982c), *Chem. Ind.*, 957.
Imperato, F. (1983a), *Phytochemistry* **22**, 312.
Imperato, F. (1983b), *Chem Ind.*, 204.
Imperato, F. (1983c), *Chem. Ind.*, 390.
Imperato, F. (1984a), *Chem. Ind.*, 186.
Imperato, F. (1984b), *Chem. Ind.*, 667.
Imperato, F. (1984c), *Am. Fern. J.* **74**, 14.
Imperato, F. (1985a), *Phytochemistry* **24**, 2136.
Imperato, F. (1985b), *Chem. Ind.*, 799.
Jay, M., Favre-Bonvin, J. and Wollenweber, E. (1979a), *Canad. J. Chem.* **57**, 1901.
Jay, M., Wolleweber, E. and Favre-Bonvin, J. (1979b), *Phytochemistry* **18**, 153.
Jay, M., Favre-Bonvin, J., Voirin, B., Viricel, M.-L. and Wollenweber, E. (1981), *Phytochemistry* **20**, 2307.
Jay, M., Viricel, M.-R., Favre-Bonvin, J. and Voirin, B. (1982), *Z. Naturforsch.* **37c**, 721.
Karl, C., Pedersen, P.A. and Muller, G. (1981), *Z. Naturforsch.* **36c**, 607.
Karl, C., Muller, G. and Pedersen, P.A. (1982), *Z. Naturforsch.* **37c**, 148.
Kishimoto, Y. (1956a), *J. Pharm. Soc. Jpn.* **76**, 246.
Kishimoto, Y. (1956b), *J. Pharm. Soc. Jpn.* **76**, 250.
Klimek, B. (1988) *Phytochemistry* **27**, 255.
Kobayashi, K. (1944), *J. Pharm. Soc. Jpn.* **64**, 35.
Kobayashi, K. and Hayashi, K. (1952), *J. Pharm. Soc. Jpn.* **72**, 3.
Koponen, T. and Nilsson, E. (1977), *Bryophyt. Bibl.* **13**, 411.
Kozlowski, A. (1921), *C.R. Acad. Sci. Paris* **173**, 429.
Kruijt, R. Ch., Niemann, G.J., de Koster, C.G. and Heerma, W. (1986), *Cryptogamie, Bryol. Lichenol.* **7**, 165.
Kuhn, R. and Low, I. (1949), *Chem. Ber.* **82**, 481.
Kutney, J.P. and Hall, J.E. (1971), *Phytochemistry* **10**, 3287.
Laurent, S. (1966), Ph.D. Thesis, University of Paris.
Lindberg, G., Osterdahl, B-G. and Nilsson, E. (1974), *Chem. Script.* **5**, 140.
Mabry, T.J. (1973), *Pure Appl. Chem.* **34**, 377.
Mabry, T.J. (1982), Private communication.
Markham, K.R. (1972), *Phytochemistry* **11**, 2047.
Markham, K.R. (1977), *Phytochemistry* **16**, 617.
Markham, K.R. (1980), *Biochem. Syst. Ecol.* **8**, 11.
Markham, K.R. (1984), *Phytochemistry* **23**, 2053.
Markham, K.R. and Given, D.R. (1979), *Biochemical Syst. Ecol.* **7**, 91.
Markham, K.R. and Mabry, T.J. (1972), *Phytochemistry* **11**, 2875.
Markham, K.R. and Moore, N.A. (1980), *Biochem. Syst. Ecol* **8**, 17.
Markham, K.R. and Mues, R. (1983), *Phytochemistry* **22**, 143.
Markham, K.R. and Mues, R. (1984), *Z. Naturforsch.* **39c**, 309.
Markham, K.R. and Porter, L.J. (1969), *Phytochemistry* **8**, 1777.
Markham, K.R. and Porter, L.J. (1973), *Phytochemistry* **12**, 2007.
Markham, K.R. and Porter, L.J. (1974a), *Phytochemistry* **13**, 1553.
Markham, K.R. and Porter, L.J. (1974b), *Phytochemistry* **13**, 1937.
Markham, K.R. and Porter, L.J. (1975a), *Phytochemistry* **14**, 199.
Markham, K.R. and Porter, L.J. (1975b), *Phytochemistry* **14**, 1093.
Markham, K.R. and Porter, L.J. (1975c), *Phytochemistry* **14**, 1641.
Markham, K.R. and Porter, L.J. (1978a), *Phytochemistry* **17**, 159.
Markham, K.R. and Porter, L.J. (1978b), *Progr. Phytochem.* **5**, 181.
Markham, K.R. and Porter, L.J. (1979), *Phytochemistry* **18**, 611.
Markham, K.R. and Wallace, J.W. (1980), *Phytochemistry* **19**, 415.
Markham, K.R. and Woolhouse, A.D. (1983), *Phytochemistry* **22**, 2827.
Markham, K.R., Mabry, T.J. and Voirin, B. (1969), *Phytochemistry* **8**, 469.
Markham, K.R., Porter, L.J. and Miller, N.G. (1976a), *Phytochemistry* **15**, 151.
Markham, K.R., Porter, L.J. Mues, R., Zinsmeister, H.D. and Brehm, B.G. (1976b), *Phytochemistry* **15**, 147.
Markham, K.R., Porter, L.J., Campbell, E.O., Chopin, J. and Bouillant, M-L. (1976c), *Phytochemistry* **15**, 1517.
Markham, K.R., Moore, N.A. and Porter, L.J. (1978a), *Phytochemistry* **17**, 911.
Markham, K.R., Zinsmeister, H.D. and Mues, R. (1978b), *Phytochemistry* **17**, 1601.
Markham, K.R., Ternai, B., Stanley, R., Geiger, H. and Mabry, T.J. (1978c), *Phytochemistry* **34**, 1389.
Markham, K.R., Theodor, R., Mues, R. and Zinsmeister, H.D. (1982), *Z. Naturforsch.* **37c**, 562.
Markham, K.R., Moore, N.A. and Given, D.R. (1983), *N.Z.J. Bot.* **21**, 113.
Markham, K.R., Vilain, C., Wollenweber, E., Dietz, V.H. and Schilling, G. (1985), *Z. Naturforsch.* **40c**, 317.
Markham, K.R., Wollenweber, E. and Schilling, G. (1987) *J. Plant Physiol.* **131**, 45.
Martensson, O. and Nilsson, E. (1974), *Lindbergia* **2**, 145.
McClure, J.W. and Miller, H.A. (1967), *Nova Hedwigia* **14**, 111.
Melchert, T.E. and Alston, R.E. (1965), *Science* **150**, 1170.
Mendez, J. and Sanz-Cabanilles, F. (1979), *Phytochemistry* **18**, 1409.
Miraglia, M.D.C.M., De Padua, A.P., Mesquita, A.A.L. and Gottlieb, O.R. (1985), *Phytochemistry* **24**, 1120.
Moewus, F. (1950), *Angew. Chem.* **62**, 469.
Mohri, K., Takemoto, T. and Kondo, Y. (1982) *Yakugaku Zasshi* **102**, 310.
Molisch, C. (1911), *Ber. Dtsch. Bot. Ges.* **29**, 487.
Mues, R. (1982a), *J. Hattori Bot. Lab.* **53**, 271.
Mues, R. (1982b), *Ber. Dtsch. Bot. Ges.* **95**, 115.
Mues, R. (1982c), *J. Hattori Bot. Lab.* **51**, 61.
Mues, R. (1984). In *Proceedings of the Third Meeting of the Bryologists from Central and East Europe*, Praha, 14–18 June 1982, p. 37.
Mues, R. (1985). In *Proceedings of the Fourth Meeting of the Bryologists from Central and East Europe*, Eger, 12–13 August, 1985; *Abstracta Botanica* **9**, Suppl 2, 171.
Mues, R. and Zinsmeister, H.D. (1973), *Osterr. Bot. Z.* **121**, 151.
Mues, R. and Zinsmeister, H.D. (1975), *Phytochemistry* **14**, 577.
Mues, R. and Zinsmeister, H.D. (1976), *Phytochemistry* **15**, 1757.
Mues, R. and Zinsmeister, H.D. (1977), *Bryophyt. Bibl.* **13**, 399.
Mues, R., Strassner, A. and Zinsmeister, H.D. (1983), *Cryptogamie. Bryol. Lichenol.* **4**, 111.
Mues, R., Hattori, S., Asakawa, Y. and Grolle, R. (1984), *J. Hattori Bot. Lab.* **56**, 227.
Mues, R., Leidinger, G., Lauck, V., Zinsmeister, H.D., Koponen, T. and Markham, K.R. (1986), *Z. Naturforsch.* **41c**, 971.
Murakami, T., Satake, T. Hirasawa, C., Ikeno, Y., Saiki, Y. and Chen, C-M. (1984), *Yakugaku Zasshi* **104**, 142.
Murakami, T., Tanaka, N., Kuraishi, T., Saiki, Y. and Chen,

C-M. (1985), *Yakugaku Zasshi* **105**, 649
Nakabayashi, T. (1955a), *Bull. Agric. Chem. Soc. Jpn* **19**, 104.
Nakabayashi, T. (1955b), *Vitamins Kyoto* **8**, 410.
Nakamura, H. and Hukuti, G. (1940), *J. Pharm. Soc. Jpn* **60**, 449.
Nilsson, E. (1961), *Acta Chem. Scand.* **15**, 211.
Nilsson, E. (1969a), *Acta Chem. Scand.* **23**, 2910.
Nilsson, E. (1969b), *Ark. Kemi* **31**, 475.
Nilsson, E. (1973a), *Phytochemistry* **12**, 722.
Nilsson, E. (1973b), Abstracts of Uppsala Dissertations from the Faculty of Science, No. 239, Uppsala.
Nilsson, E. and Bendz, G. (1973), In *Nobel 25 Symposium (1973) Chemistry in Botanical Classification* (eds G. Bendz and J. Santesson), Academic Press, New York, p. 117.
Nilsson, E., Lindberg, G. and Österdahl, B-G. (1973), *Chem. Script.* **4**, 66.
Noro, T., Fukushima, S., Saiki, Y., Uemo, A. and Akahori, Y. (1969), *Yakugaku Zasshi* **89**, 851.
Nung, V.N., Bezanger-Beauquesne, L. and Torck, M. (1971), *Plant. Med. Phytother.* **5**, 177.
Okigawa, M., Hwa, C.W. and Kawano, N. (1971), *Phytochemistry* **10**, 3286.
Okuyama, T., Hosoyama, K., Hiraga, Y., Kurono, G. and Takemoto, T. (1978), *Chem. Pharm. Bull.* **26**, 3071.
Okuyama, T., Ohta, Y. and Shibata, S. (1979), *Shoyakugaku Zasshi* **33**, 185.
Österdahl, B-G. (1976), *Acta Chem. Scand.* **B30**, 867.
Österdahl, B-G. (1978a), *Acta Chem. Scand.* **B32**, 714.
Österdahl, B-G. (1978b), *Acta Chem. Scand.* **B32**, 93.
Österdahl, B-G. (1979), *Acta Chem. Scand.* **B33**, 400.
Österdahl, B-G. (1983), *Acta Chem. Scand.* **B37**, 69.
Österdahl, B-G. and Lindberg, G. (1977), *Acta Chem. Scand.* **B31**, 293.
Österdahl, B-G., Bendz, G. and Fredga, A. (1976), *10th International Symposium on the Chemistry of Natural Products (IUPAC)*, Dunedin, New Zealand, Paper No. D14.
Petersen, R.L. and Fairbrothers, D.E. (1980), *Am. Fern J.* **70**, 93.
Pichi Sermolli, R.E.G. (1973). In *The Phylogeny and Classification of the Ferns* (eds A.C. Jermy, J.A. Crabbe and B.A. Thomas), Academic Press, London, p. 11.
Pichi Sermolli, R.E.G. (1977), *Webbia* **31**, 313.
Pirie, N.W. (1985), *Origins of Life* **15**, 207.
Porter, L.J. (1981), *Taxon* **30**, 739.
Ramakrishnan, G., Banjeri, A. and Chadha, MS. (1974), *Phytochemistry* **13**, 2317.
Rangaswamy, S. and Iver, R.T. (1969), *Indian J. Chem.* **7**, 526.
Reznik, H. and Wiermann, R. (1966), *Naturwissenschaft* **9**, 19.
Richardson, P.M. (1983), In *Advances in Cladistics* (eds N.J. Platnick and V.A. Funk), Columbia University Press, New York, Vol 2, p. 115–124.
Rudolph, H. and Johnk, A. (1982), *J. Hattori Bot. Lab.* **53**, 195.
Ryan, F.J. (1955), *Science* **122**, 470.
Saleh, N. (1975), *Phytochemistry* **14**, 286.
Saleh, N.A.M. and Abdalla, M.F. (1980), *Phytochemistry* **19**, 987.
Saleh, N.A.M., Majak, W. and Towers, G.H.N. (1972), *Phytochemistry* **11**, 1095.
Satake, T., Murakami, T., Saiki, Y., Chen, C-M. and Gomez P.L.P. (1984), *Chem. Pharm. Bull.* **32**, 4620.
Saunders, J.A. and McClure, J.W. (1976), *Phytochemistry* **15**, 809.
Schier, W. (1974), *Nova Hedwigia* **25**, 549.
Schuster, R.M. (1966), *The Hepaticae and Anthocerotae of North America*, Columbia University Press, New York.
Schuster, R.M. (1977). In *Beiträge zur Biologie der niederen Pflanzen* (eds W. Frey, H. Hurka and F. Overwinkler), Gustav Fischer, Stuttgart, p. 107.
Schuster, R.M. (ed.) (1983). In *New Manual of Bryology* Hattori Botanical Laboratory, Japan, Vol. 2, p. 892.
Seigler, D.S. and Wollenweber, E. (1983), *Am. J. Bot.* **70**, 790.
Serizawa, S. and Wollenweber, E. (1977) *Am. Fern J.* **67**, 107.
Seshadri, T.R. and Rangaswamy, S. (1974), *Indian J. Chem.* **12**, 783.
Siegel, S.M. (1969), *Am. J. Bot.* **56**, 175.
Smith, D.M. (1980), *Bull. Torrey Bot. Club.* **107**, 134.
Smith, D.M. and Harborne, J.B. (1971), *Phytochemistry* **10**, 2117.
Smith, P.M. (1976), *The Chemotaxonomy of Plants*, Edward Arnold, London, p. 62.
Sobel, G.L. and Whalen, M.D. (1983), *Fern Gaz.* **12**, 295.
Soeder, R.W. (1985), *Bot. Rev.* **51**, 442.
Soeder, R.W. (1986). In *Morphology of the Pteridophyta* (eds S.C. Verma and B.M. Johri) (in press).
Soeder, R.W. and Babb, M.S. (1972), *Phytochemistry* **11**, 3079.
Spencer, K.C. (1979), *Phytochem. Bull.* **12**, 1.
Srivastava, S.K., Srivastava, S.D., Saksena, V.K. and Nigam, S.S. (1981), *Phytochemistry* **20**, 462.
Star, A.E. (1980), *Bull. Torrey Bot. Club.* **107**, 146.
Star, A.E., Rosler, H., Mabry, T.J. and Smith, D.M. (1975a), *Phytochemistry* **14**, 2275.
Star, A.E., Seigler, D.S., Mabry, T.J. and Smith, D.M. (1975b), *Biochem. Syst. Ecol.* **2**, 109.
Stein, W., Anhut, S., Zinsmeister, H.D., Mues, R., Barz, W. and Koster, J. (1985), *Z. Naturforsch.* **40c**, 469.
Stewart, W.N. (1983), *Paleobotany and Evolution of Plants*, Cambridge University, New York.
Suire, C. (1975), *Rev. Bryol. Lichenol.* **41**, 105.
Suire, C. and Asakawa, Y. (1979). In *Bryophyte Systematics* (eds G.C.S. Clarke and J.G. Ducket), Academic Press London, p. 447.
Sunder, R.K., Ayengar, N.N. and Rangaswamy, S. (1974), *Phytochemistry* **13**, 1610.
Swain, T. (1975). In *The Flavonoids* (eds J.B. Harborne, T.J. Mabry and H. Mabry), Chapman and Hall, London, p. 1096.
Swain, T. and Cooper-Driver, G. (1973). In *The Phylogeny and Classification of the Ferns* (eds A.C. Jermy, J.A. Crabbe and B.A. Thomas), Academic Press, London, p. 111.
Swain, T. and Cooper-Driver, G. (1981). In *Paleobotany, Paleoecology and Evolution* (ed. K.J. Niklas), Vol. 1, p. 103–122.
Syrchina, A.I., Voronkov, M.G. and Tyukavkina, N.A. (1973), *Chem. Nat. Comp.* **9**, 640.
Syrchina, A.I., Voronkov, M.G. and Tyukavkina, N.A. (1974), *Chem. Nat. Compds.* **10**, 683.
Syrchina, A.I., Voronkov, M.G. and Tynkavkina, N.A. (1975), *Chem. Nat. Compds.* **11**, 439.
Syrchina, A.J., Voronkov, M.G. and Tyukavkina, N.A. (1978a), *Chem. Nat. Compds.* **14**, 691.
Syrchina, A.I., Voronkov, M.G. and Tyukavkina, N.A. (1978b), *Chem. Nat. Compds.* **14**, 685.
Syrchina, A.I., Zapesochnaya, G.G. Tyukavkina, N.A. and Voronkov, M.G. (1980a), *Chem. Nat. Compds.* **16**, 356.
Syrchina, A.I., Gorokhova, V.G., Tyukavkina, N.A., Babkin, V.A. and Voronkov, M.G. (1980b), *Chem. Nat. Compds.* **16**, 245.
Szweykowski, J. and Krzakowa, M. (1977), *Bull. Acad. Pol. Sci.* **15**, 251.
Tanaka, N., Murakami, T., Saiki, Y., Chen, C-M. and Gomez, L.D. (1978), *Chem. Pharm. Bull.* **26**, 3580.
Tanaka, N., Komazawa, Y., Obara, K., Murakami, T., Saiki, Y. and Chen, C-M. (1980), *Chem. Pharm. Bull.* **28**, 1884.

Tanaka, N., Sada, T., Murakami, T., Saiki, Y. and Chen, C-M. (1984), *Chem. Pharm. Bull.* **32**, 490.
Tanaka, N., Murakami, T., Wada, H., Gutierrez, A.B., Saiki, Y. and Chen, C-M. (1985), *Chem. Pharm. Bull.* **33**, 5231.
Theodor, R. (1981), Ph.D., Dissertation, University of Saarbrucken.
Theodor, R., Markham, K.R., Mues, R. and Zinsmeister, H.D. (1981), *Phytochemistry* **20**, 1457.
Theodor, R., Mues, R., Zinsmeister, H.D. and Markham, K.R. (1983), *Z. Naturforsch.* **38c**, 165.
Tjukavkina, N.A., Benesova, V. and Herout, V. (1970), *Coll. Czech. Chem. Commun.* **35**, 1306.
Tryon, R.M. and Tryon, A.F. (1982), *Ferns and Allied Plants*, Springer-Verlag, New York.
Ueno, A. (1962), *Yakugaku Zasshi* **82**, 1479.
Ueno, A., Oguri, N., Saiki, Y. and Harada, T. (1963), *Yakugaku Zasshi* **83**, 420.
Vanderkerkhove, O. (1977a), *Z. Pflanzenphysiol.* **82**, 455.
Vanderkerkhove, O. (1977b), *Z. Pflanzenphysiol.* **85**, 135.
Vanderkerkhove, O. (1978a), *Z. Pflanzenphysiol.* **86**, 217.
Vanderkerkhove, O. (1978b), *Z. Pflanzenphysiol.* **86**, 279.
Vanderkerkhove, O. (1980), *Z. Pflanzenphysiol.* **100**, 369.
Viell, B. (1980), *Z. Pflanzenphysiol.* **98**, 419.
Vitt, D.H. (1983). In *New Manual of Bryology* (ed. R.M. Schuster), Hattori Botanical Laboratory, Japan, Vol. 2, p. 696.
Voirin, B. (1967), *C.R. Acad. Sci. Ser. D* **264**, 665.
Voirin, B. (1970), Ph.D. Thesis, University of Lyon.
Voirin, B. (1972), *Phytochemistry* **11**, 257.
Voirin, B. (1986), Private communication.
Voirin, B. and Jay, M. (1974), *Phytochemistry* **13**, 275.
Voirin, B. and Jay, M. (1977), *Phytochemistry* **16**, 2043.
Voirin, B. and Jay, M. (1978a) *Biochem. Syst. Ecol.* **6**, 99.
Voirin, B. and Jay, M. (1978b), *Biochem. Syst. Ecol.* **6**, 95.
Voirin, B. and Lebreton, P. (1966), *C.R. Acad. Sci. Ser. D.* **262**, 707.
Voirin, B. and Lebreton, P. (1967), *Bull. Soc. Chim. Biol.* **49**, 1402.
Voirin, B. and Lebreton, P. (1971), *Boissera* **19**, 259.
Voirin, B., Jay, M. and Hauteville, M. (1975), *Phytochemistry* **14**, 257.
Vowinkel, E. (1975), *Chem. Ber.* **108**, 1166.
Wada, H., Satake, T., Murakami, T., Kojima, T., Saiki, Y. and Chen, C-M. (1985), *Chem. Pharm. Bull.* **33**, 4182.
Wagner, H., Seligmann, O., Chari, M.V., Wollenweber, E., Dietz, V.H., Donnelly, D.M.X., Meegan, M.J. and O'Donnell, B. (1979), *Tetrahedron Lett.*, 4269.
Wallace, J.W. (1986). In *Morphology of the Pteridophyta* (eds S.C. Verma and B.M. Johri) (in press).
Wallace, J.W. and Hu, Z. (1985), *Am. J. Bot.* 72, 928.
Wallace, J.W. and Markham, K.R. (1978a), *Am. J. Bot.* **65**, 965.
Wallace, J.W. and Markham, K.R. (1978b), *Phytochemistry* **17**, 1313.
Wallace, J.W. and Story, D.T. (1978), *J. Nat. Prod.* **41**, 651.
Wallace, J.W., Story, D.T., Besson, E. and Chopin, J. (1979), *Phytochemistry* **18**, 1077.
Wallace, J.W., Yopp, D.L., Besson, E. and Chopin, J. (1981), *Phytochemistry* **20**, 2701.
Wallace, J.W., Markham, K.R., Gianassi, D.E., Mickel, J.T., Yopp, D.L., Gomez, L.D., Pittillo, J.D. and Soeder, R. (1982), *Am. J. Bot.* **69**, 356.
Wallace, J.W., Pozner, R.S. and Gomez, L.D. (1983), *Am. J. Bot.* **70**, 207.
Wallace, J.W., Chapman, M., Sullivan, J.E. and Bhardwaja, T.N. (1984), *Am. J. Bot.* **71**, 660.
Wang, C-Y., Mahir Pumakcu, A. and Bryan, G.T. (1973), *Phytochemistry* **12**, 2298.
Weitz, S. and Ikan, R. (1977), *Phytochemistry* **16**, 1108.
Wij, M. (1978), *Indian J. Chem.* **14B**, 644.
Wollenweber, E. (1972), *Phytochemistry* **11**, 425.
Wollenweber, E. (1976a), *Phytochemistry* **15**, 2013.
Wollenweber, E. (1976b), *Z. Pflanzenphysiol.* **78**, 344.
Wollenweber, E. (1978), *Am. Fern. J.* **68**, 13.
Wollenweber, E. (1979), *Flora* **168**, 138.
Wollenweber, E. (1981), *Z. Naturforsch.* **36c**, 604.
Wollenweber, E. (1982), *Phytochemistry* **21**, 1462.
Wollenweber, E. (1984), *Rev. Latinoamer. Quim.* **15**, 3.
Wollenweber, E. (1985), *Phytochemistry* **24**, 1493.
Wollenweber, E. and Yatskievych, G. (1982), *J. Nat. Prod.* **45**, 216.
Wollenweber, E., Favre-Bonvin, J. and Jay, M. (1978a), *Z. Naturforsch.* **33c**, 831.
Wollenweber, E., Favre-Bonvin, J. and Lebreton, P. (1978b), *Phytochemistry* **17**, 1684.
Wollenweber, E., Dietz, V.H., Smith, D.M. and Seigler, D.S. (1979), *Z. Naturforsch.* **34c**, 876.
Wollenweber, E., Dietz, V.H., Schillo, D. and Schilling, G. (1980), *Z. Naturforsch.* **35c**, 685.
Wollenweber, E., Rehse, C. and Dietz, V.H. (1981a), *Phytochemistry* **20**, 1167.
Wollenweber, E., Walter, J. and Schilling, G. (1981b), *Z. Pflanzenphysiol.* **104**, 161.
Wollenweber, E., Smith, D.M. and Reeves, T. (1982). In *Flavonoids and Bioflavonoids, 1981* (eds L. Farkas, F. Kallay, M. Gabor and H. Wagner), Elsevier, Amsterdam, p. 221.
Wollenweber, E., Dietz, V.H., Schilling, G., Favre-Bonvin, J. and Smith, D.M. (1985), *Phytochemistry* **24**, 965.
Wu, T-S. and Furukawa, H. (1983), *Phytochemistry* **22**, 1061.
Zinsmeister, H.D. and Mues, R. (1980), *Rev. Latinoam. Quim.* **11**, 23.

13

Distribution and evolution of the flavonoids in gymnosperms

GERARD J. NIEMANN

13.1 GENERAL INTRODUCTION

The gymnosperms form a relatively small group of about 800 species of mainly evergreen trees or shrubs. They have been fairly well investigated for phenolic constituents and often appear to accumulate a rich variety of flavonoids. Most attention has been paid to the leaves or leafy twigs, mainly with the object of determining the distribution from species to species. Flavonoid biochemistry has proven its worth in solving various taxonomic problems. The now classic example of heartwood flavonoid patterns being correlated with the morphological subdivision of the genus *Pinus* into Haploxylon and Diploxylon pines (Erdtman, 1963) has been followed by many others (Section 13.3). Exploration of flavonoid patterns at the populational level has also received recent attention, revealing for example interesting correlations with plant geography (Section 13.4).

This chapter provides a general picture of the distribution of these pigments in the gymnosperms; it summarizes some of the taxonomic highlights but also demonstrates the variability that can occur at the species level. The distribution of flavonoids in the gymnosperms has been earlier reviewed by Hegnauer (1962) and Harborne (1967). More recent partial reviews, considering e.g. one family or one type of flavonoid, and the various surveys that have been carried out are listed in Table 13.1.

13.2 DISTRIBUTION PATTERNS WITHIN THE PLANT AND WITHIN THE CELL

Within the plant considerable differences in flavonoid pattern may be found in the various organs and even in different organelles within the cell (Weissenbock, 1973; Hrazdina *et al.*, 1982; Niemann *et al.*, 1983; Weissenbock *et al.*, 1984). Evaluation of distribution patterns, therefore, also largely depends on aspects of internal location, which will be considered in this section.

13.2.1 Wood and bark

Many conifers yield timber of great importance and this means that most attention has been given to the analysis of heartwoods of economically important species (Hegnauer, 1962; Harborne, 1967). It appears that heartwoods are a rich source of flavonoid aglycones, with *Pinus*, with its many dihydroflavonols and *C*-methylated compounds, as one of the best examples (Wollenweber and Dietz, 1981). Also, infraspecific variation in flavonoid patterns of the heartwoods is low, which may be advantageous for intraspecific comparison of flavonoid patterns. In several aspects heartwood may differ from sapwood and bark; heartwoods lack the monomeric proanthocyanidins and catechins (Hergert, 1960) and differ in both the oxidation pattern of the flavonoid skeleton (Hergert, 1962) and in the glycosidic pattern (Sasaya *et al.*, 1980) from sapwood and bark. Apart from some incidental explorations on *Cedrus* wood (Agrawal and Rastogi, 1984) and bark (Ragunathan *et al.*, 1974) and on *Larix* wood (Lapteva *et al.*, 1974; Sasaya *et al.*, 1980), these plant parts have hardly been reinvestigated during the last decade and half.

The Flavonoids. Edited by J.B. Harborne
Published in 1988 by Chapman and Hall Ltd,
11 New Fetter Lane, London EC4P 4EE.

Table 13.1 Reviews and general surveys of the gymnosperm flavonoids

Taxa	*Type*	*Character**	*References*
Gymnosperms	General	R	Hegnauer (1962) (see also Hegnauer, 1986)
	General	R	Harborne (1967)
	Flavonoids wood/bark	R	Hergert (1960, 1962)
	Biflavonoids	R	Geiger and Quinn (1975, 1982)
	C-Glycosylflavonoids	R	Chopin and Bouillant (1975); Chopin *et al.* (1982)
	C-Glycosylflavonoids	S/R	Niemann and Miller (1975)
	Proanthocyanidins	S/R	Lebreton *et al.* (1980)
Cycadales	General	S/R	Lebreton (1980)
	Biflavonoids	S/R	Dossaji *et al.* (1975)
Gnetopsida	*C*-Glycosylflavonoids	S/R	Wallace (1979); Wallace *et al.* (1982)
	General	R/S	Thivend *et al.* (1979)
Coniferales	Biflavonoids	S/R	Cambie and James (1967)
	Anthocyanins	R	Timberlake and Bridle (1975)
	C-Glycosylflavonoids	S	Lebreton *et al.* (1978a)
	Flavonol/flavanonol aglycones	S	Takahashi *et al.* (1960)
Podocarpaceae	Anthocyanins	S	Lowry (1972)
Araucariaceae	Biflavonoids	R	Ilyas *et al.* (1977b)
Cupressaceae (+ Taxodiaceae)	General	R/S	Lebreton (1982)
Cupressoideae	Biflavonoids	S/R	Gadek and Quinn (1985)
Callitroideae	Biflavonoids	S/R	Gadek and Quinn (1983)
Taxodiaceae (+ Cupressaceae)	General	R/S	Lebreton (1982)
Pinaceae	General	R	Norin (1972)
	Needles	R	Niemann (1979)
	Flavonol aglycones	S	Lebreton and Sartre (1983)
	Heartwood	R	Erdtman *et al.* (1966)
	Conelet anthocyanins	S/R	Santamour (1966)
Cedrus	General		Agrawal and Rastogi (1984)

*R, review; S, survey.

13.2.2 The leaf

Leaves generally have a much larger array of flavonoids, with many different glycosides, than wood and bark (Hegnauer, 1962; Niemann, 1979). Also the flavonoid composition may be much more variable, because of the possible rapid turnover in these organs. Thus, Dittrich (1970) found that the flavonol 3-glycosides in young needles of *Picea* were not present in the older needles. Quantitative seasonal variations of the same compounds were observed in *Larix* needles (Niemann, 1976) and both seasonal variation and a conspicuous difference in phenolic constituents between current and 1-year-old needles were found in *Pinus taeda* (Chen, 1979). For biflavones, on the other hand, Gadek and Quinn (1982) found an identical pattern in juvenile and adult leaves of *Callitris macleagana* and there was also a very close similarity in juvenile and adult leaves in *Dacrydium kirkii* (Quinn and Gadek, 1981). Autumnal leaves of *Ginkgo biloba* show a fourfold increase in quantity of the four main biflavones in comparison with the leaves in Spring, but the relative composition remains practically the same (Briançon–Scheid *et al.*, 1983).

These differences in behaviour between biflavones and flavonol glycosides may be due to the fact that they have different locations within the cell. In general, there are three ways that toxic secondary metabolites may be removed from the protoplast (Schnepf, 1976): (1) transport through the plasmalemma into the cell wall; (2) accumulation in an internal non-cytoplasmic compartment (mainly the vacuole); and (3) transport to the extracytoplasmic space by exocytosis. The lipophilic biflavone aglycones are thus located at a different site in (or on) the cell from the water-soluble glycosides. From fluorescence studies, Gadek *et al.* (1984) showed that the biflavones in a number of gymnosperm species are primarily located in the cuticle, sometimes extending some distance in and along the anticlinal walls of the epidermis. In two species of the Pinaceae, which do not accumulate biflavonoids, no fluorescence was observed in the cuticle. In *Pinus radiata* leaves a yellow fluorescence was not only visible in the protoplast (vacuole) of the epidermis, but also, and even more intensely, in the epidermal cell wall and the primary walls of the hypodermal cells. In *Cedrus deodara* the fluorescence was confined to the epidermal walls and the primary anticlinal walls of the hypodermis.

There is good evidence that certain flavonoids may be synthesized and stored entirely within the epidermis, whereas others may be preferentially confined to the mesophyll (Hrazdina *et al.*, 1982; Strack *et al.*, 1982). In many cases the epidermis is the major location for flavonoid accumulation (Tronchet, 1968; Tissut and Ravanel, 1980; Niemann *et al.*, 1983). Cell-wall-located flavonoid aglycones (Charriere–Ladreix, 1976) and glycosides (Brisson *et al.*, 1986) and vacuolar *O*-glycosides and *C*-glycosylflavones (van Genderen *et al.*, 1983) may be responsible for the fluorescence noted in the Pinaceae (see above). The biflavonoids, on the other hand, seem to constitute a part of the complex waxes associated with the cuticle. Normally, the composition and morphology of the wax is comparatively constant (Juniper and Jeffree, 1983), and this explains why there is so little variation in the biflavonoid fraction.

Within the cell of the leaf, flavonoids could conceivably be located in the plastids (Charriere–Ladreix, 1976), and more especially in the chloroplasts (Weissenbock, 1973; Saunders and McClure, 1976). However, such reports have to be viewed with some scepticism since the chloroplasts are easily, and to a high degree, contaminated with vacuolar flavonoids during the isolation procedure (Charriere–Ladreix and Tissut, 1981). Nevertheless, selective enrichment of certain flavonoids in the chloroplast in comparison with the rest of the leaf, and the detection of some flavonoid-synthesizing enzymes in the purified chloroplast fraction, point to a possible role of the chloroplasts in flavonoid biosynthesis (Weissenbock *et al.*, 1976). From gymnosperm chloroplasts Saunders and McClure (1976) isolated naringenin (*Pinus nigra*) and kaempferol and quercetin 3-glycoside(s) (*Ginkgo biloba*); in both these plants, however, the chromatographic pattern of the chloroplast fraction was essentially identical to that of whole leaf extracts, so there is some doubt about the validity of these findings. For a recent review of the subject of compartmentation of phenolic biosynthesis, see Hrazdina and Wagner (1985).

13.2.3 Pollen, seed testa, cones and roots

Roots have been comparatively little investigated. No flavonoids could be detected in roots of three *Ephedra* species by Taraskina *et al.* (1973). Roots of *Pseudotsuga mensiezii* contain, in addition to the main leaf flavanonol dihydroquercetin and its 3′-glycosides (Barton, 1967), fairly large quantities of the 6-*C*-methylated flavanol, poriol, and its 7-glucoside (Hills and Ishikura, 1969; Barton, 1972); none of these compounds occurs in other parts of the tree. The major wood and leaf constituents, such as quercetin and its 3′-glucoside, could not be detected in either rootwood or rootbark of 34 samples (Barton, 1967).

Pollen may be a very special source of flavonoids. Ohmoto and Yoshida (1983) isolated nine flavonoids from pollen grains of *Cryptomeria japonica* which were a completely different type to those found in the leaves. The pollen biflavones were based on amentoflavone whereas the leaf biflavones were based on hinokiflavone. In the case of the flavonols, a different glycosylation pattern was observed in pollen and leaf. However, the results of the pollen analyses were compared with (comparatively old) leaf analyses by other authors so that the differences reported here really need checking. A complex mixture of flavonols and flavones has also been identified in *Taxus baccata* pollen (Harborne, 1967). Some flavonol glycosides were isolated from pollen of two *Pinus* species (Katsumata *et al.*, 1974) and of *Cedrus deodara* (Agrawal and Rastogi, 1984).

Little similarity was also found between the biflavonoid patterns of the seed testa and of the leaves in eight species of the Cycadales (compare Gadek, 1982 and Dossaji *et al.*, 1975). Surprisingly, two species of *Macrozamia* contained a large amount of cupressuflavone in the testa, a compound never reported from the leaves of these species or any other species of the Cycadales, being mainly restricted to the Cupressaceae and Araucariaceae.

The production of anthocyanins in gymnosperms is largely confined to the reproductive structures (Santamour, 1967). Glycosides of cyanidin and, in several cases, of delphinidin have been reported from conelets of 35 species of six genera of the Pinaceae (Santamour, 1966), of several species of *Chamaecyparis* (Santamour, 1967) and, accompanied by pelargonidin glycosides, of five species of the Podocarpaceae (Crowden and Grubb, 1971).

These examples given above certainly underline the necessity of using homologous tissues for chemotaxonomic work.

13.3 NATURAL DISTRIBUTION

13.3.1 Ginkgoales, Gnetales and Cycadales

Table 13.2 summarizes the major distribution patterns of the flavonoids found in species of the orders Ginkgoales, Gnetales and Cycadales. The absence of biflavones, otherwise characteristic gymnosperm constituents, from the three families of the Gnetales–Ephedraceae, Gnetaceae and Welwitschiaceae – has been confirmed. Instead, species of these families appear to accumulate *C*-glycosylflavones (Niemann and Miller, 1975; Wallace, 1979; Thiven *et al.*, 1979; Wallace *et al.*, 1982; Ouabonzi *et al.*, 1983), but not to the exclusion of other flavonoids, as suggested by Wallace (1979). Chumbalov and Chekneneva (1976) identified 8-methylherbacetin glucoside from *Ephedra equisetina*, whereas Nawwar *et al.* (1984) found a

Table 13.2 Natural distribution of flavonoids in Ginkgoales, Gnetales and Cycadales

Order, family genus, species	*Biflavones*	*Proanthocyanidins*	*C-Glycoflavones*	*Flavonols, flavones*	*Leading references*
Ginkgoales					
Ginkgoaceae					
Ginkgo biloba	+ +	+	−	+ +	Dossaji *et al.* (1975); Lebreton *et al.* (1980); Niemann and Miller (1975); Lebreton (1980); Geiger (1979)
Gnetales					
Ephedraceae					
Ephedra (up to 6 spp.)	−	+ +	+ +	+	Sawada (1958); Thivend *et al.* (1979); Niemann and Miller (1975); Wallace (1979); Wallace *et al.* (1982); Chumbalov and Chekneneva (1976); Nawwar *et al.* (1984)
Gnetaceae					
Gnetum (up to 3 spp.)	−	− /tr	+ +	− / +	Wallace (1979); Lebreton *et al.* (1980); Wallace *et al.* (1982); Ouabonzi *et al.* (1983)
Welwitschiaceae					
Welwitschia mirabilis	−	−	+ +	tr/ −	Wallace (1979); Lebreton *et al.* (1980); Thivend *et al.* (1979); Wallace *et al.* (1982)
Cycadales					
Cycadaceae					
Cycas (3 to 9 spp.)	+ +	±	+		Dossaji *et al.* (1975); Lebreton *et al.* (1980); Niemann and Miller (1975); Lebreton (1980)
Stangeriaceae					
Stangeria eriopus (Kunze) Nash = *S. paradoxa* Th. Moore	−	tr	(+)		Dossaji *et al.* (1975); Lebreton *et al.* (1980); Lebreton (1980)
Zamiaceae					
Lepidozamia (2 spp.)	+ +				Dossaji *et al.* (1975); Lebreton *et al.* (1980); Niemann and Miller (1975); Lebreton (1980); Carson and Wallace (1972)
Dioon (up to 6 spp.)	+ +	±	− /tr		
Encephalartos (up to 28 spp.)	+ +	+	−		
Zamia (up to 14 spp.)	+ +		−		
Macrozamia (up to 14 spp.)	+ +	+	−		
Bowenia (2 spp.)	+ +				
Microcycas (1 spp.)	+ +				
Ceratozamia (up to 5 spp.)	+ +	±	−		

+ +, major component(s); + and ±, present in increasingly smaller amounts; tr, trace; −, absent; − /tr, noted as absent and trace amount by different authors; (+), not present in all analyses by the same author(s).

number of glycosides of the same aglycone and kaempferol and quercetin 3-rhamnoside in *Ephedra alata*, in addition to *C*-glycosylflavones. In *Gnetum africanum* leaves apigenin 7-neohesperidoside is present as the major flavonoid, in addition to a number of 8-*C*-glycosylflavones (Ouabonzi *et al.*, 1983). The major occurrence of *C*-glycosylflavones forms a chemical base characterizing the Gnetales. The Ephedraceae, in addition, accumulate large amounts of proanthocyanidins (Thivend *et al.*, 1979).

To some extent, those families or species which lack the ability to synthesize biflavones seem to accumulate *C*-glycosylflavones and possibly flavones and/or flavonols instead (see also Section 13.5). In this connection the flavonoid pattern of the sole member of the Cycadales lacking biflavones, *Stangeria eriopus*, deserves reinvestigation. Apart from one report, by Lebreton (1980), indicating the incidental occurrence of *C*-glycosylflavones, this species remained uninvestigated for flavonoids other than biflavones. The biflavonoids in the Cycadales (and in *Ginkgo biloba*) have systematically been surveyed by

Dossaji *et al.* (1975) and are discussed by Geiger and Quinn (1982; see also Chapter 4). The leaf biflavones isolated are all of the amentoflavone type, but in their pattern there are significant differences between genera. Remarkably, the biosynthetic potential of the seed testa as far as the biflavones are concerned, is quite different from that of the leaves (see also Section 13.2.3). Gadek (1982) found that three species of the genus *Macrozamia* are characterized by the occurrence of amento- and cupressu-flavone-based compounds in the testa. Two species of *Cycas* contained only amentoflavone-based compounds whereas single species of the genera *Encephalartos*, *Lepidozamia* and *Zamia* only had very little biflavonoid or none at all in the seed testa.

As far as they have been investigated, species of the different genera of the Cycadales appear to accumulate proanthocyanidins to some extent (Lebreton *et al.*, 1980), and are distinctly different in the accumulation of *C*-glycosylflavones, which are present in *Cycas* and *Stangeria*, but not in the Zamiaceae (Table 13.2). Only in *Dioon spinulosum* have orientin, and probably vitexin, been detected as minor leaf constituents (Carson and Wallace, 1972).

Ginkgo biloba is alone in accumulating, in addition to its biflavones, such compounds as larycitrin (3′-*O*-methylmyricetin; Geiger, 1979) and kaempferol coumaroylglucosylrhamnoside (Nasr *et al.*, 1986), which are otherwise only known in the gymnosperms from the Pinaceae (Niemann, 1979; Section 13.3.2.f).

13.3.2 Coniferopsida

(a) *Cephalotaxaceae*

Leaves of *Cephalotaxus* species have mainly been investigated for their biflavonoids; it seems that this monogeneric family is characterized by the presence of only amentoflavone and its methylated derivatives (Ishratulla *et al.*, 1981). Aqil *et al.* (1981) identified a 6-*C*-methylbiflavone from *Cephalotaxus harringtonia*, which is the first example of a naturally occurring *C*-methylated biflavonoid. The few species investigated contain an average amount of proanthocyanidin in their leaves (two species; Lebreton *et al.*, 1980), and are devoid of major *C*-glycosylflavones (two species; Niemann and Miller, 1975; Lebreton *et al.*, 1978a). From *C. drupacea* Sieb. et Zucc. (*C. harringtonia* var. *drupaceae* (Sieb. et Zucc.) Koidzumi) leaves, Takahashi *et al.*, (1960) isolated taxifolin and found no major amount of flavonol glycosides. The genus has not been further investigated for this type of compound.

(b) *Podocarpaceae*

Dallimore and Jackson (1974) recognize seven genera in this family. Two of these genera, *Podocarpus* and *Dacrydium*, appear to be rather artificial and thus for the classical genus *Podocarpus* de Laubenfels (1969) proposed that it should be subdivided into five genera. Also for the genus *Dacrydium* the generic boundaries have been redefined and three new genera recognized (Quinn, 1982; Geiger and Quinn, 1982). For the new genus *Lepidothamnus* Phil. this was partly based on the fact that the three species of this genus contain cupressuflavone derivatives as their major biflavonoid constituents instead of amentoflavone derivatives, which are present in most species of the classic genus *Dacrydium* (Quinn and Gadek, 1981; Geiger and Quinn, 1982). In their survey of the biflavones of *Dacrydium sensu lato*, Quinn and Gadek (1981) found marked discontinuities in the biflavonyl patterns, which provided the basis for further taxonomic revision within the group.

Recently, strong support was also obtained for de Laubenfels' recognition of three genera within *Podocarpus sensu lato* from the distribution of flavonoid glycosides in representative New Zealand species (Markham *et al.*, 1985b). Whereas all species investigated contain biflavonoids as major constituents, the three *Prumnopitys* species practically lack *C*-glycosylflavones, but have major amounts of flavonol 3-*O*-diglycosides. Both *Dacrycarpus* (3 spp.) and *Podocarpus sensu stricto* (5 spp.) have major amounts of *C*-glycosylflavones, but for the *Dacrycarpus* species there is an additional accumulation of 3-methoxyflavonol glycosides and they are also rich in flavonoids with trihydroxylation in the B-ring (Markham and Whitehouse, 1984; Markham *et al.*, 1985b).

From *Podocarpus nivalis* foliage several 5-glucosides of 7-*O*-methylated flavonols and dihydroflavonols were isolated (Markham *et al.*, 1984). Previously, Rizvi and Rahman (1974) isolated the related 7, 4′-dimethyldihydrokaempferol and its 5-glucoside from leaves of *Podocarpus nerriifolius* D. Don. 5-Glycosylation, especially in the flavonol series, is otherwise very rare (Harborne and Williams, 1982).

The genus *Phyllocladus* is remarkably homogeneous in its flavonoid chemistry and quite distinct from other genera studied in the Podocarpaceae. In addition to a number of tentatively identified biflavones (Cambie and James, 1967) there appears to be a predominance in the accumulation of flavone *O*-glycosides (in particular luteolin derivatives) accompanied by lower levels of flavonol glycosides (Markham *et al.*, 1985a).

Older reports on the occurrence of isoflavones in the heartwood of *Podocarpus spicatus* R.Br. (Harborne, 1967) have not been followed up by any later discoveries of isoflavones in the Podocarpaceae. The only other isoflavones reported in gymnosperms were isolated from *Juniperus macropoda* (Sehti *et al.*, 1981). Anthocyanins have been found in seeds, seed coat, receptacle or young leaves of many Podocarpaceae (Crowden and Grubb, 1971; Lowry, 1972). Like practically all other species of the Coniferales, the three species of *Podocarpus* investigated accumulate proanthocyanidins (Silva *et al.*, 1972; Lebreton, 1980).

(c) *Araucariaceae*

Species of the two genera, *Agathis* and *Araucaria*, have mainly been investigated for their biflavonoids (Ilyas *et al.*, 1977b; Geiger and Quinn, 1982). The biflavone pattern in the genus *Araucaria* is the most complex yet known. The few species of *Araucaria* investigated are rich in proanthocyanidins (Hida, 1958; Hegnauer, 1962; Lebreton *et al.*, 1980) but do not accumulate *C*-glycosylflavones in any quantity (Niemann and Miller, 1975; Lebreton *et al.*, 1978a).

(d) *Cupressaceae*

The two subfamilies of the Cupressaceae, Callitroideae and Cupressoideae, have recently been systematically investigated for their biflavonoid patterns (Gadek and Quinn, 1983, 1985). The major biflavonoids are based on amentoflavone, cupressuflavone and hinokiflavone. The uneven distribution among the different genera generally follows generic divisions, although some genera, e.g. *Chamaecyparis*, appear chemically to be rather heterogeneous (Gadek and Quinn, 1985). Affinities between genera, suggested by biflavone distribution patterns, do not correlate with currently recognized tribal groupings and a taxonomic revision of the family based on a broad range of characters including those derived from the chemistry is under way (see Chapter 4).

Like many other accumulators of biflavones, the species of Cupressaceae investigated have no *C*-glycosylflavones (Niemann and Miller, 1975; Lebreton *et al.*, 1978a), but all contain proanthocyanidins with some interesting generic differences in quantities (Lebreton *et al.*, 1980). A classification of the Cupressaceae mainly based on flavonoid content (cupressuflavone/prodelphinidin/myricetin) was proposed by Lebreton (1982). The recognition of three subfamilies partly agrees with the groupings proposed by Gadek and Quinn (1983, 1985) based on biflavone patterns alone. A biochemically close relationship between *Cupressus* and *Juniperus*, a special place for *Chamaecyparis nootkatensis* separate from other species of this genus, and a relation between *Thujopsis* and the other *Chamaecyparis* species emerge from both studies.

A screening programme for flavonols by Takahashi *et al.* (1960) suggests that the Cupressaceae may be comparatively rich in flavonol glycosides, but only a few species have been investigated in any detail. In addition to the common flavones and flavonols (apigenin, kaempferol and quercetin), Lamer and Bodalski (1968) found myricetin and tricetin in *Thuja*. 6- or 8-Hydroxylated compounds also occur characteristically. Hassani *et al.* (1985) isolated the new flavone 6-hydroxyisoetin (5, 6, 7, 2′, 4′, 5′-hexahydroxyflavone) in addition to 6-hydroxyapigenin, 6-hydroxyluteolin, apigenin, kaempferol and quercetin from leaves of *Juniperus thurifera* L. Ilyas *et al.* (1977a) found quercetagetin (6-hydroxyquercetin) in addition to kaempferol and quercetin in *J. macropoda* Boiss., and 6-hydroxyapigenin and 6-hydroxyluteolin (and probably the 6-methylated derivative of the latter) were isolated from *J. communis* L. (Lamer–Zarawska, 1977). From berries of *J. macropoda*, Siddiqui and Sen (1971) isolated 8-hydroxyluteolin, a compound also found in *Callitris glauca* leaves (Ansari *et al.*, 1981). The last species also contains the rare kaempferol 5-rhamnoside (see Section 13.3.2.b).

From *J. macropoda* leaves Sehti *et al.* (1980, 1981) isolated several isoflavones. Among the conifers the Podocarpaceae are the only other family known to contain isoflavones (Section 13.3.2.b).

(e) *Taxodiaceae*

From a phytochemical point of view, the Taxodiaceae are rather heterogeneous (Lebreton, 1982). Six of the ten genera have both amentoflavone and hinokiflavone derivatives in the leaves, *Cryptomeria* and *Glyptostrobus* only contain the latter, whereas *Sciadopitys* lacks hinokiflavone derivatives (Geiger and Quinn, 1982). *Metasequoia* and *Taiwania* have additional rare biflavones, namely dihydroamentoflavone (Beckman *et al.*, 1971) and taiwaniaflavone derivatives (Kamil *et al.*, 1981). Pollen of *Cryptomeria japonica* D. Don contains both amentoflavone and dihydroamentoflavone (Ohmoto and Yoshida, 1983) and therefore may fall within the general pattern and even be closer related to *Metasequoia* than expected from its presently known leaf constituents.

Sequoia is the only genus which contains large amounts of *C*-glycosylflavones (Niemann and Miller, 1975; Lebreton *et al.*, 1978a). All genera investigated contain proanthocyanidins, with a heterogeneous distribution of prodelphinidin and procyanidin (Lebreton *et al.*, 1980). Apart from *Sciadopitys*, the Taxodiaceae mainly contain glycosides of the common flavonols (Takahashi *et al.*, 1960). Infraspecific variability (see Section 13.4) probably accounts for conflicting results in *Sequoia* (Lebreton, 1982) and *Cunninghamia* (Hsiao, 1984) and may be in *Metasequoia* (Takahashi *et al.*, 1960; Beckman and Geiger, 1968; Lebreton, 1982). Leaves of species investigated in more detail like *Metasequoia glyptostroboides* (Beckman and Geiger, 1968) and *Taxodium distichum* (Geiger and De Groot-Pfleiderer, 1979) were found to have a mixture of flavone and flavonol glycosides, which is rather complex in *T. distichum*. Both species contain tricetin or its rare 3′-methyl ether. *Sequoia* proanthocyanidins have recently been identified by Stafford and Lester (1986).

(f) *Pinaceae*

The absence of biflavones (Sawada, 1958) forms one of the special characteristics of the Pinaceae. However, the recent isolation of a new biflavone, abiesin (5, 3″, 7″-trihydroxy-7, 4′, 4‴-trimethoxy-(3′, 6″)-biflavone) by Chatterjee *et al.* (1984) from leaves of *Abies webbiana* Lindl. and the report of three known biflavonoids from

Picea smithiana (Wall.) Boiss. by Azam *et al.* (1985) may well be the start of a reconsideration of this concept. Otherwise the Pinaceae are 'pronounced accumulators of flavonoids' (Hegnauer, 1962). Heartwood of *Pinus* species is characterized by a complex mixture of related flavones, flavanones and flavanols of which the flavanones can be 6- and/or 8-methylated (Erdtman *et al.*, 1966). The chemical differences found between Diploxylon and Haploxylon pines substantiate the division of the genus into these two subgenera. 6-Methylated dihydroflavonols such as cedeodarin (6-methyldihydroquercetin) and cedrin (6-methyldihydromyricetin) and its glucoside have been isolated from *Cedrus deodara* wood (Agrawal and Rastogi, 1984). The same species also contains dihydroquercetin and dihydromyricetin, compounds which have also been isolated from the heartwood of *Larix leptolepis* (Sasaya *et al.*, 1980).

Leaves of Pinaceae species are rich in proanthocyanidins (Hida, 1958; Krugman, 1959; Lebreton *et al.*, 1980) and flavonol glycosides (Takahashi *et al.*, 1960; Niemann, 1979; Lebreton and Sartre, 1983). Some genera (e.g. *Abies*, *Larix*, *Tsuga*) have *C*-glycosylflavones (Niemann and Miller, 1975; Lebreton *et al.*, 1978a; Parker *et al.*, 1979) others contain *C*-methylated components (e.g. *Cedrus*, *Pinus*; van Genderen and van Schaik, 1980; Higuchi and Donnelly, 1978; Niemann, 1978; Shen and Theander, 1985). 6-*C*-Methylation has also been found in *Pseudotsuga mensiezii* roots (Barton, 1972; Section 13.2.3). From leaves of three genera (*Larix*, *Picea*, *Pinus*) *O*-acylated flavonoids have been isolated (Niemann, 1975, 1979; Zapesochnaja *et al.*, 1978a, b, c, 1980; Higuchi and Donnelly, 1978).

In spite of the relatively few extensive investigations and large infraspecific variations (see Section 13.4), flavonoid chemistry has indicated (Niemann and van Genderen, 1980; Lebreton and Sartre, 1983) the existence of certain relationships (e.g. an affinity between *Larix*, *Abies* and *Cedrus*) among the different genera of the Pinaceae.

(g) *Taxaceae*

The Taxaceae has been relatively little investigated. Biflavones have been reported from three genera and they are all of the amentoflavone series (cp Geiger and Quinn, 1982). One species of *Taxus* contains proanthocyanidins, whereas one species of *Torreya* lacks them (Lebreton *et al.*, 1980). All species investigated were found to be devoid of *C*-glycosylflavones (Niemann and Miller, 1975; Lebreton *et al.*, 1978a), while flavonols and/or dihydroflavonols were rarely detected (Takahashi *et al.*, 1960). Some more modern investigations of this family would be welcome.

13.4 SEASONAL VARIATION, GEOGRAPHICAL DISTRIBUTION AND HYBRIDIZATION

The variability within a species or among closely related species can easily obscure the more general implications of chemical relationships. One aspect, the variation with season and age, has already been discussed in Section 13.2.2. Obviously, both juvenile and adult leaf samples should always be checked during more detailed investigations.

The existence of chemical races, the influence of habitat and the possibility of intra- and infra-specific hybridization may all affect the flavonoid pattern of a given species. Seedlings of *Larix leptolepis* grown under identical environmental circumstances show large quantitative variation in flavonoid composition of their needles (Niemann and Baas, 1978). Different clones of the same species, grown at the same location, similarly vary greatly in the phenolic composition of both their defoliated shoots (Nomura and Kishida, 1978) and their needles (Niemann, 1980), whereas only minor differences occur between trees of the same clone. Sauer *et al.* (1973) found very significant differences in morphology and phenolic composition in a number of clones of *Picea abies*. The influence of the site of growth on the composition of a number of unidentified phenolics (probably derivatives of aromatic hydroxy acids) was very low. Chen (1979) considered the variation in phenolic compounds in needles of progenies of *Pinus taeda* to be the result of genetic differences rather than of environmental factors. On the other hand, Forrest (1975) found, in addition to large variations within a provenance of *Picea sitchensis*, highly significant differences between provenances and sites in concentrations and total weights of polyphenol fractions. Whether these differences can be attributed to differences in environmental conditions or in genotype is not clear.

Polymorphism in flavonoid composition was also found in *Pinus sylvestris* (Laracine, 1984; Laracine and Lebreton, 1985). On the basis of the relative composition of proanthocyanidins, Lebreton *et al.* (1978b) and Lebreton and Thivend (1981) distinguished distinct geographical taxa within *Pinus laricio*, *Cupressus sempervirens* and *Juniperus phoenicea*. Parker and Maze (1984) found a segregation of Eastern and Western groups of *Abies lasiocarpa* populations in both needle morphology and flavonoid composition. Both groups were furthermore arranged in north/south arrays. Characteristic *A. balsameae* flavonoids were prevalent in the Eastern *A. lasiocarpa* populations, suggesting introgression with this species.

Interspecific hybridization is not uncommon among the Coniferae (Dallimore and Jackson, 1974) and it may lead to an additive flavonoid composition, involving components of both parents. Dominant-recessive inheritance, or induced polyploidy, however, may alter simple flavonoid complementation in the composition of a hybrid (cp Crawford and Giannasi, 1982). Considerable variation occurs in F_1 phenolics in *Pinus rigida* × *P. taeda* (Feret *et al.*, 1980). In some cases new substances are produced not found in either parent; in others, substances produced in the parents are not found in progeny. Also, when crossed with different individuals, a particular parent does not pass on the same number or type of markers in the progeny. By contrast, in the hybrid *Dacrydium laxifolium intermedium* there is an almost perfect complement of the phenolic glycosides of the parental species (Quinn and Rattenbury, 1972). Webby *et al.* (1986) analysed several hybrids of New Zealand species of *Podocarpus* and found the flavonoid chemistry, in conjunction with morphological characters and geographical considerations, of considerable value in the recognition of these natural hybrids. If the two parental species are similar in their flavonoids, then such hybrids, e.g. *P. halii* × *totara*, cannot be identified through their phenolic profiles.

From the above, it might appear that infraspecific chemical variation is the rule rather than the exception. Other species, however, are more constant in their chemical characters. Hsiao (1984) studied the essential oils and flavonoids of 121 specimens of *Cunninghamia lanceolata* (Lamb.) Hook. and of *C. konishii* Hayata grown in the same place from seed collected from various locations. The results showed that *Cunninghamia* is a monotypic genus in which *C. konishii* the Luanta fir is really only a variety of the China fir (*C. lanceolata*). Data on the volatile oils of gymnosperms also clearly show that the degree of infraspecific and populational variation varies considerably from species to species (Von Rudloff, 1975).

13.5 EVOLUTIONARY TRENDS

Assignment of a flavonoid structure as primitive or advanced in a phylogenetic context is relative, since it depends on its distribution in a given phylogenetic scheme and furthermore, it appears that certain structures, considered primitive, reappear in taxa which are thought to be much more advanced (Crawford and Giannasi, 1982).

One of the special characteristics of the gymnosperms, Gnetales and Pinaceae excepted, is the general occurrence of biflavonoids. The universal occurrence of these compounds throughout the gymnosperms lends support to the opinion that the ancestral gymnosperm must have been capable of forming biflavones and that this ability was lost in the evolution of the Pinaceae and the Gnetales (Geiger and Quinn, 1975). The amentoflavone (and possibly also the hinokiflavone) group may be considered as a primitive character (Geiger and Quinn, 1975, 1982) whereas the more complex mixtures found in, for example, the Araucariaceae points to a more recent development. In addition to the agathisflavone group, this family also accumulates robustaflavone, a compound which might be formed by rearrangement from amentoflavone. The recent isolation of another 3′,6″ biflavone, abiesin, from *Abies webbiana* (Chatterjee *et al.*, 1984), however, favours the idea that the biosynthesis of robustaflavone is independent of that of amentoflavone.

Of the other flavonoids encountered in the gymnosperms the *C*-glycosylflavones, found equally in ferns and bryophytes, are of a very ancient stock (Swain, 1975). 3- and 8-Hydroxylation, introduction of B-ring trihydroxylation and *C*-methylation occurred later on, whereas 6-hydroxylation and formation of isoflavones appear to be more recent developments. As for biflavones, the absence of *C*-glycosylflavones may point to loss of the ability to form these compounds. On the other hand a separate, more or less simultaneous development of biflavones versus glycosylflavones could have occurred as well. It is tempting to speculate on evolutionary aspects of development of flavonoid biochemistry and accumulation based on their functional significance. Accumulation in cuticle and/or epidermis (see Section 13.2.2), essential as a UV screen (Lee and Lowry, 1980) and as a general defence against microbial or insect attack (Fraenkel, 1959; Nomura and Kishida, 1978, 1979; Ravise and Chopin, 1978, 1981; Elliger *et al.*, 1980), could be derived from cuticular biflavones *or* from cell-wall-located flavonoid aglycones and glycosides *or* from vacuolar *C*-glycosylflavones (and flavonol glycosides). Indeed, often a certain preference for one or other of these alternatives is found (Section 13.3.1, 13.3.2.f). If such developments really took place, it has to be considered as part of a general evolutionary trend which includes changes in morphological features and in other secondary metabolites.

Meyen (1984) arranged the Ginkgoales and Ephedrales in the Ginkgoopsida which he considers as a monophyletic line, co-ordinate with the Cycadopsida and Coniferopsida. This concept is still highly disputed (Beck, 1985; Miller, 1985) and is certainly not supported by the flavonoid chemical data of present species (Table 13.2). From morphological characteristics there is insufficient evidence to establish true phylogenetic links between the conifer families, and it seems conceivable that several families developed independently from the ancient Voltziaceae (Meyen, 1984). Flavonoid data partly support this view. It seems probable that for instance the occurrence of *C*-methylation and of isoflavones in certain genera or species is related to the evolutionary development of those particular taxa rather than indicating a phylogenetic relationship with other taxa. For these compounds, however, as well as for the flavone and flavonol glycosides, our present data base is very limited and these ideas may well change as more results accumulate.

13.6 CONCLUSIONS

Gymnosperms have been comparatively well investigated for biflavones, proanthocyanidins and, to a lesser extent, *C*-glycosylflavones, but for flavone and flavonol glycosides our knowledge is remarkably fragmentary. Only some genera (e.g. *Podocarpus*, *Prumnopitys*, *Phyllocladus*, *Larix*, *Abies*, *Pinus*) have been investigated in any detail. The recent isolations of isoflavones from *Juniperus macropoda* and of a new biflavone from *Abies webbiana* indicate the possibility of uncovering a yet wider range of special flavonoids.

From the taxonomic angle, the flavonoids have proven their value. In combination with morphological characteristics, flavonoid patterns in many cases have led to a re-evaluation of existing tribal groupings. In addition to the flavonoids, the lower terpenoids are very important gymnosperm characters. There is remarkable little integration of the comparative data, however, when one considers taxonomic position or geographical distribution. The flavonoid data should certainly be combined with terpenoid data, whenever possible, in future systematic studies of these plants.

REFERENCES

Agrawal, P.K. and Rastogi, R.P. (1984), *Biochem. Syst. Ecol.* **12**, 133.

Ansari, F.R., Ansari, W.H., Rahman, W., Okigawa, M. and Kawano, N. (1981), *Indian J. Chem.* **20B**, 724.

Aqil, M., Rahman, W., Hasaka, N., Okigawa, M. and Kawano, N. (1981), *J. Chem. Soc. Perkin Trans I*, 1389.

Azam, A. Qasim, M.A. and Khan, M.S.Y. (1985), *J. Indian Chem. Soc.* **52**, 788.

Barton, G.M. (1967), *Canad. J. Bot.* **45**, 1545.

Barton, G.M. (1972), *Phytochemistry* **11**, 426.

Beck, C.B. (1985), *Bot. Rev.* **51**, 273.

Beckmann, S. and Geiger, H. (1968), *Phytochemistry* **7**, 1667.

Beckmann, S., Geiger, H. and de Groot Pfleiderer, W. (1971), *Phytochemistry* **10**, 2465.

Briancon-Scheid, F., Lobstein-Guth, A. and Anton, R. (1983), *Plant. Med.* **49**, 204.

Brisson, L., Vacka, W.E.K. and Ibrahim, R.K. (1986), *Plant. Sci.* **44**, 175.

Cambie, R.C. and James, M.A. (1967), *N.Z. J. Sci.* **10**, 918.

Carson, J.L. and Wallace, J.W. (1972), *Phytochemistry* **11**, 842.

Charriere-Ladreix, Y. (1976), *Planta* **129**, 167.

Charriere-Ladreix, Y. and Tissut, M. (1981), *Planta* **151**, 309.

Chatterjee, A., Kotoky, J., Das, K.K., Banerji, J. and Chakraborty, T. (1984), *Phytochemistry* **23**, 704.

Chen, C.C. (1979), *Diss. Abstr. Intern.* **B40**, 1003.

Chopin, J. and Bouillant, M.L. (1975). In *The Flavonoids* (eds J.B. Harborne, T.J. Mabry and H. Mabry), Chapman and Hall, London, pp. 632–691.

Chopin, J., Bouillant, M.L. and Besson, E. (1982). In *The Flavonoids – Advances in Research* (eds J.B. Harborne and T.J. Mabry), Chapman and Hall, London and New York, pp. 449–503.

Chumbalov, T.K. and Chekneneva, L.N. (1976), *Khim. Prir. Soedin.* **12**, 543.

Crawford, D.J. and Giannasi, D.E. (1982), *Bio. Sci.* **32**, 114.

Crowden, R.K. and Grubb, M.J. (1971), *Phytochemistry* **10**, 2821.

Dallimore, W. and Jackson, A.B. (1966/74), *A Handbook of Coniferae and Ginkgoaceae* (reviewed S.G. Harrison), Edward Arnold, London.

de Laubenfels, D.J. (1969), *J. Arnold Arbor.* **50**, 274.

Dittrich, P. (1970), Untersuchungen über den Umsatz sekundärer Pflanzenstoffe in den Nadeln von *Picea abies* L. Dissertation Ludwig Maximilians Universität München.

Dossaji, S.F., Mabry, T.J. and Bell, E.A. (1975), *Biochem. Syst. Ecol.* **2**, 171.

Elliger, C.A., Chan, B.G., Waiss, Jr., A.C., Lundin, R.E. and Haddon, W.F. (1980), *Phytochemistry* **19**, 293.

Erdtman, H. (1963). In *Chemical Plant Taxonomy* (ed. T. Swain) Academic Press, London, pp. 89–126.

Erdtman, H., Kimland, B. and Norin, T. (1966), *Bot. Mag. Tokyo* **79**, 499.

Feret, P.P., Mayes, R.A., Twokoski, T.J. and Kreh, R.E. (1980), *Silvae Gen.* **29**, 155.

Forrest, G.I. (1975), *Canad. J. For. Res.* **5**, 46.

Fraenkel, G.S. (1959), *Science* **129**, 1466.

Gadek, P.A. (1982), *Phytochemistry* **21**, 889.

Gadek, P.A. and Quinn, C.J. (1982), *Phytochemistry* **21**, 248.

Gadek, P.A. and Quinn, C.J. (1983), *Phytochemistry* **22**, 969.

Gadek, P.A. and Quinn, C.J. (1985), *Phytochemistry* **24**, 267.

Gadek, P.A., Quinn, C.J. and Ashford, A.E. (1984), *Aust. J. Bot.* **32**, 15.

Geiger, H. (1979), *Z. Naturforsch.* **34c**, 878.

Geiger, H. and de Groot-Pfleiderer, W. (1979), *Phytochemistry* **18**, 1709.

Geiger, H. and Quinn, C.J. (1975). In *The Flavonoids* (eds J.B. Harborne, T.J. Mabry and H. Mabry), Chapman and Hall, London, pp. 692–742.

Geiger, H. and Quinn, C.J. (1982). In *The Flavonoids – Advances in Research* (eds J.B. Harborne and T.J. Mabry), Chapman and Hall, London and New York, pp. 505–534.

Harborne, J.B. (1967), *Comparative Biochemistry of the Flavonoids*, Academic Press, London and New York.

Harborne, J.B. and Williams, C.A. (1982). In *The Flavonoids – Advances in Research* (eds J.B. Harborne and T.J. Mabry), Chapman and Hall, London and New York., pp. 261–311.

Hassani, M.I., Favre-Bonvin, J. and Lebreton, P. (1985), *C.R. Acad. Sci. Paris, T.* **300**, III Ser. 1.

Hegnauer, R. (1962), *Chemotaxonomie der Pflanzen*, Vol. 1, Birkhäuser Verlag, Basel.

Hegnauer, R. (1986), *Chemotaxonomie der Pflanzen*, Vol. 7, Birkhäuser Verlag, Basel.

Hergert, H.L. (1960), *For. Prod. J.* **10**, 610.

Hergert, H.L. (1962). In *the Chemistry of Flavonoid Compounds* (ed. T.A. Geissman), Pergamon Press, Oxford, pp. 553–592.

Hida, M. (1958), *Bot. Mag. Tokyo* **71**, 425.

Higuchi, R. and Donnelly, D.M.X. (1978), *Phytochemistry* **17**, 787.

Hillis, W.E. and Ishikura, N. (1969), *Aust. J. Chem.* **22**, 483.

Hrazdina, G. and Wagner, G.J. (1985), *Annu. Proc. Phytochem. Soc. Eur.* **25**, 119.

Hrazdina, G., Marx, G.A. and Hoch, H.C. (1982), *Plant Physiol.* **70**, 745.

Hsiao, J-Y. (1984), *Proc. Natl. Res. Counc. ROC* **8**, 104.

Ilyas, M., Ilyas, N. and Wagner, H. (1977a), *Phytochemistry* **16**, 1456.

Ilyas, M., Seligmann, O. and Wagner, H. (1977b), *Z. Naturforsch.* **32c**, 206.

Ishratulla, K., Rahman, W., Okigawa, M. and Kawano, N. (1981), *Indian J. Chem.* **20B**, 935.

Juniper, B.E. and Jeffree, C.E. (1983), *Plant Surfaces*, Edward Arnold, London.

Kamil, M., Ilyas, M., Rahman, W., Hasaka, N., Okigawa, M. and Kawano, N. (1981), *J. Chem. Soc. Perkin Trans I*, 533.

Katsumata, T., Nakamura, S. and Togasawa, Y. (1974), *J. Fac. Agric. Iwate. Univ.* **12**, 21.

Krugman, S.L. (1959), *For. Sci.* **5**, 169.

Lamer, E. and Bodalski, T. (1968), *Diss. Pharm. Pharmacol.* **20**, 623.

Lamer-Zarawska, E. (1977), *Roczn. Chem.* **51**, 2131.

Lapteva, K.I., Lutkii, V.I. and Tjukavkina, N.A. (1974), *Khim. Prir. Soedin.* **10**, 97.
Laracine, C. (1984), Etude de la variabilité flavonique infraspécifique chez deux Conifères: le Pin sylvestre et le Genévrier commun, Thèse de l'Université de Lyon I.
Laracine, C. and Lebreton, P. (1985), 7 Reunion Groupe Gen. Biol. Pop., Montpellier, 2–6 Sept.
Lebreton, P. (1980), *Rev. Gen. Bot.* **87**, 133.
Lebreton, P. (1982), *Candollea* **37**, 243.
Lebreton, P. and Sartre, J. (1983), *Canad. J. For. Res.* **13**, 145.
Lebreton, P. and Thiven, S. (1981), *Naturalia Monspeliensa*, Ser. Bot. **47**, 1.
Lebreton, P., Boutard, B. and Thivend, S. (1978a), *C.R. Acad. Sci. Paris.* **287D.**, 1255.
Lebreton, P., Boutard, B. and Sartre, J. (1978b), *Bull. Inst. Sc. Rabat* 155.
Lebreton, P., Thiven, S. and Boutard, B. (1980), *Plant. Med. Phytother.* **14**, 105.
Lee, D.W. and Lowry, J.B. (1980), *Biotropica* **12**, 75.
Lowry, J.B. (1972), *Phytochemistry* **11**, 725.
Markham, K.R. and Whitehouse, L.A. (1984), *Phytochemistry* **23**, 1931.
Markham, K.R., Webby, R.F. and Vilain, C. (1984), *Phytochemistry* **23**, 2049.
Markham, K.R., Vilain, C. and Molloy, B.P.J. (1985a), *Phytochemistry* **24**, 2607.
Markham, K.R., Webby, R.F., Whitehouse, L.A., Molloy, B.P.J., Vilain, C. and Mues, R. (1985b), *N.Z. J. Bot.* **23**, 1.
Meyen, S.V. (1984), *Bot. Rev.* **50**, 1.
Miller Jr., C.N. (1985), *Bot. Rev.* **51**, 295.
Nasr, C., Haag-Berrurier, M. and Lobstein-Guth, A. (1986), *Phytochemistry* **25**, 770.
Nawwar, M.A.M., El-Sissi, H.I. and Barakat, H.H. (1984), *Phytochemistry* **23**, 2937.
Niemann, G.J. (1975), *Phytochemistry* **14**, 1437.
Niemann, G.J. (1976), *Acta Bot. Neerl.* **25**, 349.
Niemann, G.J. (1978), *Z. Naturforsch.* **33c**, 777.
Niemann, G.J. (1979), *Acta Bot. Neerl.* **28**, 73.
Niemann, G.J. (1980), *Canad. J. Bot.* **58**, 2313.
Niemann, G.J. and Baas, W.J. (1978), *Acta Bot. Neerl.* **27**, 229.
Niemann, G.J. and Miller, H.J. (1975), *Biochem. Syst. Ecol.* **2**, 169.
Niemann, G.J. and van H.H. Genderen, (1980), *Biochem. Syst. Ecol.* **8**, 237.
Niemann, G.J., Koerselman-Kooy, J., Steijns, J.M.J.M. and van Brederode, J. (1983), *Z. Pflanzenphysiol.* **109**, 105.
Nomura, K. and Kishida, A. (1978), *J. Jpn For. Soc.* **60**, 273.
Nomura, K. and Kishida, A. (1979), *J. Jpn For. Soc.* **61**, 1.
Norin, T. (1972), *Phytochemistry* **11**, 1231.
Ohmoto, T. and Yoshida, O. (1983), *Chem. Pharm. Bull.* **31**, 919.
Ouabonzi, A., Bouillant, M.L. and Chopin, J. (1983), *Phytochemistry* **22**, 2632.
Parker, W.H. and Maze, J. (1984), *Am. J. Bot.* **71**, 1051.
Parker, W.H., Maze, J. and MacLachlan, D.G. (1979), *Phytochemistry* **18**, 508.
Quinn, C.J. (1982), *Aust. J. Bot.* **30**, 311.
Quinn, C.J. and Gadek, P. (1981), *Phytochemistry* **20**, 677.
Quinn, C.J. and Rattenbury, J.A. (1972), *N. Z. J. Bot.* **10**, 427.
Ragunathan, K., Rangaswami, S. and Seshadri, T.R. (1974), *Indian J. Chem.* **12**, 1126.
Ravisé, A. and Chopin, J. (1978), *C.R. Acad. Sciences. Ser. D* **286**, 1885.
Ravisé, A. and Chopin, J. (1981), *Phytopathology Z.* **100**, 257.
Rizvi, S.H.M. and Rahman, W. (1974), *Phytochemistry* **13**, 2879.
Santamour Jr., F.S. (1966), *For. Sci.* **12**, 429.
Santamour Jr., F.S. (1967), *Morris Arbor. Bull.* **18**, 41.
Sasaya, T., Takehara, T., Miki, K. and Sakakibara, A, (1980), *Res. Bull. Coll. Exp. For., Coll. Agric., Hokkaido Univ.* **37**, 837.
Sauer, A., Kleinschimt, J. and Lunderstädt, J. (1975), *Silvae Gen.* **22**, 173.
Saunders, J.A. and McClure, J.W. (1976), *Phytochemistry* **15**, 809.
Sawada, T. (1958), *J. Pharm. Soc. Jpn* **78**, 1023.
Schnepf, E. (1976). In *Secondary Metabolism and Coevolution* (eds M. Luckner, L. Mothes and L. Nover), Deutsch. Akad. Naturforsch., Leopoldina, Halle, pp. 23–44.
Sehti, M.L., Taneja, S.C., Dhar, K.L. and Atal, C.K. (1981), *Phytochemistry* **20**, 341.
Sehti, M.L., Taneja, S.C., Agarwal, S.G. and Dhar, K.L. (1980), *Phytochemistry* **19**, 1831.
Shen, Z. and Theander, O. (1985), *Phytochemistry* **24**, 155.
Siddiqui, S.A. and Sen, A.B. (1971), *Phytochemistry* **10**, 434.
Silva, M., Hoeneisen, M. and Sammes, P.G. (1972), *Phytochemistry* **11**, 433.
Stafford, H.A. and Lester, H.H. (1986) *Amer. J. Bot.* **93**, 1555.
Strack, D., Meurer, B. and Weissenböck, G. (1982), *Z. Pflanzenphysiol.* **108**, 131.
Swain, T. (1975). In *The Flavonoids* (eds. J.B. Harborne, T.J. Mabry and H. Mabry), Chapman and Hall, London, pp. 1096–1129.
Takahashi, M., Ito, T., Mizutani, A. and Isoi, K. (1960), *J. Pharm. Soc. Jpn* **80**, 1488.
Taraskina, K.V., Chumbalov, T.K., Chekneneva, L.N. and Chukenova, T. (1973), Fenol'nye Soedin. Ikh. Fiziol. Svoista, Mater. Vises Simp. Fenol'nym Soedin. 141. from: C.A. 081-25-166323e.
Thivend, S., Lebreton, P., Ouabonzi, A. and Bouillant, M.L. (1979), *C.R. Acad. Sciences Paris, Ser. D* **289**, 465.
Timberlake, C.F. and Bridle, P. (1975). In *The Flavonoids* (eds. J.B. Harborne, T.J. Mabry and H. Mabry), Chapman and Hall, London, pp. 214–266.
Tissut, M. and Ravanel, P. (1980), *Phytochemistry* **19**, 2077.
Tronchet, J. (1968), *Ann. Sci. Univ. Besancon*, 3 Ser. Bot. **5**, 9.
van Genderen, H.H. and van Schaik, J. (1980), *Z. Naturforsch.* **35c**, 342.
van Genderen, H.H., Niemann, G.J. and van Brederode, J. (1983), *Protoplasma* **118**, 135.
Wallace, J.W. (1979), *Am. J. Bot.* **66**, 343.
Wallace, J.W., Porter, P.L., Besson, E. and Chopin, J. (1982), *Phytochemistry* **21**, 482.
Webby, R.F., Markham, K.R. and Molloy, B.P.J. (1986), *N. Z. J. Bot.* (in press).
Weissenböck, G. (1973), *Ber. Dtsch. Bot. Ges.* **86**, 351.
Weissenböck, G., Plesser, A. and Trinks, K. (1976), *Ber. Dtsch. Bot. Ges.* **89**, 457.
Weissenböck, G., Schnabl, H., Sachs, G., Elbert, C. and Heller, F.O. (1984), *Physiol. Plant.* **62**, 356.
Wollenweber, E. and Dietz, V.H. (1981), *Phytochemistry* **20**, 869.
Zapesochnaja, G.G., Ivanova, S.Z., Medvedeva, S.A. and Tjukavkina, N.A. (1978a), *Khim. Prir. Soedin.* **14**, 193.
Zapesochnaja, G.G., Ivanova, S.Z., Medvedeva, S.A. and Tjukavkina, N.A. (1978b), *Khim. Prir. Soedin.* **14**, 332.
Zapesochnaja, G.G., Ivanova, S.Z., Sheichenko, V.I. and Tjukavkina, N.A. (1978c), *Khim. Prir. Soedin.* **14**, 570.
Zapesochnaja, G.G., Ivanova, S.Z., Sheichenko, V.I. and Tjukavkina, N.A. (1980), *Khim. Prir. Soedin.* **16**, 186.

14

Flavonoids and evolution in the dicotyledons

DAVID E. GIANNASI

14.1 INTRODUCTION

The use of flavonoids in plant systematics and evolution, particularly in the angiosperms, has been reviewed on a regular basis (Harborne, 1967a, 1975, 1977; Swain, 1975; Young, 1981; Harborne and Mabry, 1982). Except for the pioneering survey work by Bate-Smith (1962, 1968), however, comprehensive discussion of flavonoid systematics of flowering plants within the context of a current classificatory scheme is rare (Harborne, 1967a; Gornall *et al.*, 1979). The study by Gornall *et al.* (1979) is particularly pertinent because the flavonoid data are arranged according to one of several contemporary angiosperm classifications (Dahlgren, 1977) and have provided the basis for other recent comparisons, both phytochemical (Young, 1981) and morphological/phytochemical (e.g. Rodman *et al.*, 1984).

The present summary also uses Gornall *et al.* (1979) as a starting point to update angiosperm flavonoid systematics. However, the flavonoid distributions in this paper are compared within the context of the classification by Cronquist (1981) and deal with only the Magnoliopsida (= Dicotyledonae; see Chapter 15 for monocotyledons).

The Flavonoids. Edited by J.B. Harborne
Published in 1988 by Chapman and Hall Ltd,
11 New Fetter Lane, London EC4P 4EE.

The use of six subclasses aggregates together larger numbers of families and orders than is observed in other classifications (Dahlgren, 1977, 1980, 1983; Thorne, 1976, 1981, 1983), which employ a larger number of ordinal and superordinal units. Needless to say families and orders in one classification do not necessarily occupy similar positions in another, nor is an order or family necessarily of the same generic composition or taxonomic level from one classification to another. Also, many of the flavonoid data have been used in, or compared with, the other classification systems of Dahlgren (1977) (see Gornall *et al.*, 1979) and Thorne (1976) (see Young, 1981). Thus, it is of considerable interest to see how the same data set fits into Cronquist's (1981) scheme (and with minor extrapolation, that of Takhtajan, 1980). The reader is obliged to do some shuffling between the various classifications, an accomplishment already completed for several other classes of metabolites (Gershenzon and Mabry, 1983). However, perhaps a more balanced consensus may be drawn in the future (Hegnauer, 1986).

Many of the same sources consulted earlier by Gornall *et al.* (1979) have been checked for subsequent new reports. Emphasis here, however, has been on larger surveys with systematic intent rather than numerous isolated reports of individual taxa or new compounds, except where the latter represent significant systematic distributions in a family. Note that flavonoid distinctions are based on various types of structural substitution patterns similar to those used in Gornall *et al.* (1979) as indicated in Table 14.1. Detailed distributions of individual compounds or classes of flavonoids are found in preceding chapters and additions may be inserted at the readers convenience in the general flavonoid distributions shown in Tables 14.2–14.7. Display of these data in tabular form should provide a basis for future additions and they are visually more manipulatable in comparing different angiosperm treatments.

The use of various structural features of flavonoids as representing primitive or advanced features in flavonoid evolution and their correlation with angiosperm evolution is based on earlier discussions (Harborne, 1967a, 1975, 1977; Swain, 1975) and more recently that by Gornall and Bohm (1978). Some of these character states are defined within known biosynthetic pathways. Thus 6-oxygenation and complex *O*-methylation represent advanced biosynthetic characters. Other character states are defined more by their distribution and association with various angiosperm groups. The presence of ellagic acid and myricetin, for example, are thought to be primitive characters since they are most often associated with angiosperm groups often considered to be primitive on the basis of other non-flavonoid characters, rather than any specific biosynthetic knowledge. Despite the possible circularity of such conclusions, correlations between chemical and non-chemical data are often quite high and do not radically defy each other (see Kubitzki and Gottlieb, 1984a, b for a discussion of secondary metabolite evolution in angiosperms). Finally, in any discussion of this type, there are few absolutes. A group of plants, such as the Hamamelidae, may be said to be characterized by the general presence of ellagic acid and myricetin, as opposed to the general absence of these characters in the Magnoliidae, for example (Giannasi, 1986). Thus with any character, chemical or non-chemical distinctions, especially at higher taxonomic levels, often are a matter of degrees or predominance of a character. And, of course, conclusions are only as strong as the data from available surveys. With these caveats in mind we can discuss flavonoids and evolution in the dicotyledons.

14.2 MAGNOLIIDAE

Earlier surveys (Bate-Smith, 1962; Kubitski and Reznik, 1966) clearly established the predominance of flavonols in the Magnoliidae, although flavones were occasionally reported in some taxa (Kubitzki and Reznik, 1966). For many years this has produced the impression that this taxon is characterized by a primitive flavonoid chemistry dominated by flavonols (Bate-Smith, 1962; Harborne, 1977). This certainly correlates with current angiosperm classifications (Takhtajan, 1980; Cronquist, 1981; Dahlgren, 1983; Thorne, 1983) which consider various elements of the Magnoliidae to show characteristics of the putative ancestral angiosperms. And yet two phenolic characters long associated with 'primitiveness' in flowering plants, ellagic acid and myricetin (Bate-Smith, 1962; Harborne, 1967a, 1977; Swain, 1975), are usually missing from the Magnoliidae (Table 14.2) with the exception of the Illiciales, Nymphaeaceae and Berberidaceae (Ranunculales). The presence of the primitive characters of ellagic acid and myricetin actually is more common in the Hamamelidae, Rosidae and Dilleniidae, which are presumably derived from the Magnoliidae! On the basis of these and other flavonoid characters, the Magnoliidae appear to be rather more advanced in biosynthetic capability than their supposed derivatives.

Leaf flavonoid patterns in the Winteraceae, for example, do show a common distribution of quercetin and kaempferol (but not myricetin) as well as simple flavones (apigenin and luteolin) and *C*-glycosylflavones in *Drimys* and *Tasmannia* (Williams and Harvey, 1982). Most

Table 14.1 Key to flavonoid characters used in flavonoid distributions in dicotyledons (Tables 14.2 to 14.7)[a]

No.	Character
1.	Ellagic acid
2.	Delphinidin; ⋆ = 3-deoxyanthocyanidins
3.	Cyanidin and/or pelargonidin; ⋆ = betalains
4.	*O*-Methylated anthocyanidins
5.	Acylated anthocyanidin glycosides
6.	Myricetin
7.	Quercetin and/or kaempferol
8.	7, 3′, 4′ or 5′-*O*-Methylflavonols
9.	3, 6, or 8-*O*-Methylflavonols
10.	6-Oxygenated flavonols
11.	8-Oxygenated flavonols
12.	2′-Oxygenated flavonols
13.	5-*O*-Methylflavonols
14.	Acylated flavonol glycosides
15.	Flavonol sulphates
16.	Dihydroflavonols
17.	Apigenin and/or luteolin; ⋆ = tricin
18.	7, 3′, 4′ or 5′-*O*-Methylflavones (excluding tricin)
19.	6 or 8-*O*-Methylflavones
20.	8-Oxygenated flavones
21.	6-Oxygenated flavones
22.	2′-Oxygenated flavones
23.	5-*O*-Methylflavones
24.	Acylated flavone glycosides
25.	Flavone sulphates
26.	Flavanones
27.	*C*-Glycosylflavonoids
28.	Isoflavones
29.	Biflavonyls
30.	5-Deoxyflavonoids; ⋆ = 7-deoxyflavonoids
31.	B-Ring deoxyflavonoids
32.	Proanthocyanidins
33.	Chalcones; ⋆ = aurones
34.	Dihydrochalcones
35.	Isoprenylated flavonoids
36.	Benzyl/allyl flavonoids
37.	*C*-Methylflavonoids

[a] +, presence of first or only character listed under each number;
⋆, presence of second or alternative character listed under some numbers;
+⋆, presence of both characters listed under some numbers;
(+), present in trace amount or only in some individuals of a taxon;
?, unconfirmed report;
–, no data available.

striking is the near universal occurrence of 7,3′-di-*O*-methyl-luteolin, certainly an advanced biosynthetic character state (Harborne, 1977; Gornall and Bohm, 1978). The scattered occurrence of 7,4′-di-*O*-methylluteolin and 7,4′-di-*O*-methylapigenin in various genera adds to this more advanced character state in this magnoliid taxon. Indeed, correlations of flavonoid distributions with morphological and cytological evolutionary trends in the Winteraceae led Williams and Harvey (1982) to state that flavonoid evolution in the family has proceeded by loss of highly methylated flavones back towards simpler flavones and *C*-glycosylflavones, hardly a primitive biosynthetic trend (Gornall and Bohm, 1978).

A survey of the Eupomatiaceae (Young, 1983), a taxon possibly related to the Winteraceae (among others), shows that they possess flavonols (kaempferol) and 7,3′-di-*O*-methylluteolin typical of the Winteraceae, as well as 7-*O*-methylapigenin, a compound apparently unique to the Eupomatiaceae. Suggested relationships of the Eupomatiaceae with the Degeneriaceae and Idiospermaceae seems unlikely due to: (1) the prominence of quercetin glycosides, (2) absence of flavones (methylated or otherwise) and *C*-glycosylflavones in the Degeneriaceae (Young and Sterner, 1981). Similary, the presence of only apigenin, luteolin and a single *C*-glycosylflavone in *Idiospermum* but lack of flavonols or methylated flavones distinguished this family from the Eupomatiaceae.

There are taxa that retain a rather simple flavonoid complement such as *Amborella* (Amborellaceae, Laurales) which produces only kaempferol 3-*O*-glycosides (Young, 1982), *Persea* (Lauraceae, Laurales) which produces flavonols and *C*-glycosylflavones (Wofford, 1974) and *Liriodendron* (Magnoliaceae, Magnoliales) which possesses only quercetin and kaempferol 3-*O*-glycosides along with traces of rhamnetin (Kubitzki and Reznik, 1966; Niklas *et al.*, 1985).

Examination of the flavonoid distributions in Table 14.2 shows a predominantly quercetin/kaempferol-based synthesis with considerable methylation of common hydroxyl positions (column 8). Apigenin and luteolin (column 17) are common to a number of orders and it is in some families of the Magnoliales and Laurales that exploitation of flavones and their variously substituted and methylated forms grows common, reaching their zenith in the Lauraceae (Table 14.2).

Certain elements within the Magnoliidae may actually be considered discordant. The Illiciales contain both ellagic acid and myricetin although they possess a conservative all-flavonol complement. Members of the Nymphaeales are heterogeneous in flavonoid composition. The Nymphaeaceae contain both primitive and advanced phenolic characters, such as ellagic acid, myricetin, common flavonols, flavones and *C*-glycosylflavones, while other families in the same order show a lesser combination of one or more of these phenolic characters. If the Nymphaeales are a natural phyletic group, the heterogeneity of their flavonoid distributions in microcosm seems to parallel those of the other of the Magnoliidae in macrocosm. It is perhaps possible that they are one of several parallel lines of evolution within the Magnoliidae as Cronquist (1981) suggests, although admittedly their flavonoid chemistry (at least the Nymphaeaceae) is not unlike that of a number of Hamamelidae as well (Giannasi, 1986).

The Ranunculales and Papaverales, included in the Magnoliidae by Cronquist (1981) but often treated as a distinct subclass (Tahktjan, 1980) or superorder (Dahlgren, 1983), also possess a conservative primarily flavonol content with some methylation and 8-hydroxylation (Table 14.2). Apigenin and luteolin (and their methyl derivatives) are found in several families, as are *C*-glycosylflavones. Ellagic acid occurs in at least one family, the Coriariaceae. Myricetin also occurs in the Berberidaceae along with biflavonyls (*Nandina*) and isoprenylated flavonoids. Earlier work also indicated the presence of myricetin in the Coriariaceae (Bate-Smith, 1962), but this was not confirmed by Bohm and Ornduff (1981), with only quercetin and kaempferol glycosides being observed. Attempts to use the flavonoid chemistry of the Coriariaceae at higher taxonomic levels to determine possible relationships to other subclasses were not successful due to its conservative flavonol complement. This conservative type of flavonol complement seems to limit the use of flavonoids to the generic level and below (e.g. Warnock, 1981) in orders, although occasionally unusual substitutions (e.g. benzyl/alkyl) are peculiar to a family (Menispermaceae). For other families, such as the Sabiaceae (doubtfully placed here by Cronquist), few flavonoid data are available for comment, and surveys are badly needed.

Concerning the origin of the Ranunculales, Cronquist (1981) suggests the Illiciales as an ancestral source. This certainly is compatible with flavonoid chemistry since both families of the Illiciales contain ellagic acid, and myricetin occurs in the Schizandraceae. The generally simple flavonol complement of these families includes no unusual compounds to dispute such a phyletic consideration.

In summary, the Magnoliidae do indeed possess a conservative flavonol pattern but generally lack the supposedly primitive traits of ellagic acid and myricetin. Sufficient taxa in the Magnoliidae have also been found to possess more advanced flavone and methylated flavones to suggest a less archaic flavonoid condition than was implied in previous discussions and is actually observed in other subclasses (e.g. Hamamelidae).

14.3 HAMAMELIDAE

The flavonoid (and other secondary metabolite) chemistry of the Hamamelidae has been reviewed by Giannasi (1986), in addition to earlier summaries (Egger and Reznik, 1961; Jay, 1968; Mears, 1973; Gornall *et al.*, 1979). For a group of plants which have seldom been

Table 14.2 Distribution of flavonoids in the Magnoliidae (sensu Cronquist, 1981)

Taxon	Compounds[a]																																				
Magnoliidae	*1*	*2*	*3*	*4*	*5*	*6*	*7*	*8*	*9*	*10*	*11*	*12*	*13*	*14*	*15*	*16*	*17*	*18*	*19*	*20*	*21*	*22*	*23*	*24*	*25*	*26*	*27*	*28*	*29*	*30*	*31*	*32*	*33*	*34*	*35*	*36*	*37*
MAGNOLIALES																																					
Winteraceae							+									+	+	+									+					+					
Degeneriaceae			+	+			+	+																			+					+					
Himantandraceae			+	+			+	+																								+					
Eupomatiaceae							+											+														+					
Austrobaileyaceae			+				+																														
Magnoliaceae			+	+			+	+																								+					
Lactoridaceae			+				+																														
Annonaceae							+							+					+							+					+	+	+	+		+	+
Myristacaceae							+									+	+									+		+				+	+	+			+
Canellaceae							+										+									+		+				+		+			
LAURALES																																					
Amborellaceae			+				+																									+					
Trimeniaceae				+			+	+					+			+			+	+	+		+			+	+				+	+	+	+			
Monimiaceae							+	+																								+					
Calycanthaceae			+				+	+																													
Idiospermaceae			+				+																														
Lauraceae			+	+	+	+	+	+	+				+			+	+		+	+	+		+			+	+				+	+	+	+		+	
Hernandiaceae[b]							+	+																													
Gomortegaceae			—	—	—	—																															
PIPERALES																																					
Chloranthaceae			+																																		
Saururaceae							+																														
Piperaceae							+											+								+	+				+		+				
ARISTOLOCHIALES																																					
Aristolochiaceae			+	+	+		+	+	+																												
ILLICIALES																																					
Illiciaceae	+						+																									+					
Schizandraceae	+		+			+	+																									+					
NYMPHAEALES																																					
Nelumbonaceae							+										+															+					
Nymphaeaceae	+	+	+			+	+										+										+					+					
Barclayaceae		+	+	+			+																									+					
Cabombaceae																											+										
Ceratophyllaceae		+	+	+			+																									+					

RANUNCULALES																		
Ranunculaceae[c]		+	+	+	+		+	+	+	+	+			+		+		
Circaeasteraceae[d]							+	+										
Berberidaceae[e]		+	+			+	+	+			+				+	+	+	
Sargentodoxaceae		+	+	+	+		+	+	+	+	+			+		+		
Lardizabalaceae			+		+		+	+			+					+		
Menispermaceae							+				+	+	+					+
Coriariaceae	+					+	+											
Sabiaceae[f]		+	+				+											
PAPAVERALES																		
Papaveraceae			+				+	+	+									
Fumariaceae[g]			+	+			+											

[a]See Table 14.1 for key to compound identities. Please note that in this and the following tables (14.3 to 14.7) some families are listed for which specific flavonoid data are unavailable.
[b]Includes Gyrocarpaceae.
[c]Includes Hydrastidiaceae, Glaucidiaceae, Podophyllaceae.
[d]Includes Kingdoniaceae.
[e]Includes Nandinaceae.
[f]Includes Meliosmaceae.
[g]Includes Hypecoaceae.

Table 14.3 Distribution of flavonoids in the Hamamelidae

Taxon	*Compounds*[a]																																				
Hamamelidae	*1*	*2*	*3*	*4*	*5*	*6*	*7*	*8*	*9*	*10*	*11*	*12*	*13*	*14*	*15*	*16*	*17*	*18*	*19*	*20*	*21*	*22*	*23*	*24*	*25*	*26*	*27*	*28*	*29*	*30*	*31*	*32*	*33*	*34*	*35*	*36*	*37*
TROCHODENDRALES																																					
Tetracentraceae							+	+								+																+					
Trochodendraceae	?					+	+																									+					
HAMAMELIDALES																																					
Cercidiphyllaceae	+						+									+																+					
Eupteliaceae						(+)	+																									+					
Platanaceae		+	+			+	+							+		+	+								+							+			+		
Hamamelidaceae[b]	+	+	+			+	+									+	+										+					+					
Myrothamnaceae	+					+	+																				?					+					
DAPHNIPHYLLALES																																					
Daphniphyllaceae		+					+										+																				
DIDYMELALES																																					
Didymelaceae			+				+																									+					
EUCOMMIALES																																					
Eucommiaceae							+	+																								+					
URTICALES																																					
Barbeyaceae	+						+																														
Ulmaceae	+					+	+									+	+										+					+					
Cannabaceae						+	+										+							+		+	+					+	+		+		
Moraceae		+	+	+		+	+					+				+	+	+				+				+	+	+				+	+		+		
Cecropiaceae							+																				+										
Urticaceae						+	+										+	+									?					+					
LEITNERIALES																																					
Leitneriaceae						+	+									+	+																				
JUGLANDALES																																					
Rhoipteleaceae	+					+	+																														
Juglandaceae	+					+	+	?	?				+			+										+	?					+					
MYRICALES																																					
Myricaceae	+		+			+	+									+															+	+	+	+			+
FAGALES																																					
Balanopaceae	unknown																																				
Fagaceae	+		+			+	+	+						+		+										+	+					+	+	+			
Betulaceae	+					+	+	+	+	+	+					+	+	+	+	+	+					+					+	+	+				
CASUARINALES																																					
Casuarinaceae	+					+	+									+													+			+					

[a]See Table 14.1 for key to compound identities.
[b]Includes Altingiaceae.

surveyed in any detail, it is surprising to see the differences of opinion as to where various orders of plants often included in this subclass should be placed (see Giannasi, 1986). Most authors at least support a Hamamelid concept which generally includes the Hamamelidales and Fagales in a separate subclass (Tahkajan, 1980; Cronquist 1981) or near (Thorne, 1983) or within (Dahlgren, 1980) the Rosidae. Beyond that, opinions differ wildly. Certainly, the *general* presence of ellagic acid and/or myricetin characterizes the Hamamelidales, Fagales, Myricales, Juglandales, Urticales and Casuarinales, but this is by predominance of these characters rather than an unequivocal distribution (Table 14.3). Several peripheral orders lack these characters (Daphniphyllales, Eucommiales, Leitneriales and Trochodendrales) and seem to represent a second group of orders in the Hamamelidae, most of which are placed in other subclasses or superorders by other authors (see Giannasi, 1986, for a review).

The Leitneriales, for example, lack ellagic acid and myricetin, but possess other common flavonols (quercetin and kaempferol) and basic flavones (apigenin, luteolin) suggesting a slightly more advanced status (Table 14.3). Serological studies indicate a closer relationship of the Leitneriales to the Simaroubaceae (Rutales – Rosidae; see Petersen and Fairbrothers, 1983, 1985). The primarily flavone-based pattern of the Eucommiaceae suggest a relatively advanced position for this family which also lack ellagic acid and myricetin. In neither case are the flavonoid data specific enough to indicate whether or not both taxa belong in the Hamamelidae.

While the presence of ellagic acid and myricetin and a predominance of flavonols characterize the Hamamelidales and Fagales as the 'core' of the Hamamelidae, taxa within these orders can vary considerably in their possession of these phenolic characters (Giannasi, 1986). In the Fagales, for example *Fagus* (Fagaceae) produces predominantly flavonols, but no ellagic acid or myricetin (Giannasi and Niklas, 1981), while *Castanea* produces all of these (Johnson and Beckmann, 1986). *Quercus*, in the same family, produces ellagic acid, myricetin and other flavonols, flavones and *C*-glycosylflavones (Niklas and Giannasi, 1978; Hunt, 1985; McDougal and Parks, 1986). In addition to earlier studies (Hänsel and Horhämmer, 1954) extensive *O*-methylation of flavonols and 6-substitution of flavones and flavonols has been observed subsequently in the Betulaceae (Wollenweber, 1975a). While it has been suggested that these highly modified flavonoids are peculiar to the bud exudates of Betulaceae (Gornall *et al.* 1979), many of them, in fact, occur in the leaves of *Betula* species (Giannasi and Niklas, unpublished). This high degree of methylation and extra substitution of flavonols and flavones makes the family a highly specialized, if not discordant, taxon within the Fagales at the least. This discordance is supported to some degree by serological data (Jensen and Greven, 1984).

The Juglandales are very conservative in their flavonoid patterns possessing ellagic acid, myricetin and other common flavonols but apparently no flavones or *C*-glycosylflavones, a pattern not unlike that of the Myricaceae (Giannasi, 1986) to which many feel they are related. The presence of 5-*O*-methylflavonols provides a distinctive marker for the Juglandales within the Hamamelidae, although the compound occurs in other orders and families in other subclasses. The suggested relationship of the Juglandaceae to the Anacardiaceae [Rosidae of Cronquist (1981) and Rutiflorae of Thorne (1983)] seems unlikely since 5-methoxy compounds are not found in the Anacardiaceae which in any case have a number of unusual 5-deoxy compounds and a series of anthochlor pigments not found in the Juglandaceae (Giannasi, 1986). Also, members of the Anacardiaceae possess biflavonyls, a flavonoid type not observed in the Juglandaceae. The hamamelid position of the Juglandales is also supported by two independent serological studies (Petersen and Fairbrothers, 1983, 1985; Polechko and Clarkston, 1986).

The Urticales have typically been placed in the Hamamelidae (Cronquist, 1981) although some authors currently place them in what is Cronquist's Dilleniidae [the Malviflorae of Dahlgren (1983) and Thorne (1976)]. Certainly, some elements of the Urticales could logically be placed in the latter subclass. The production of prenylated and/or sulphated flavonoids in the Moraceae (e.g. *Morus*) and Cannabaceae (e.g. *Humulus*) and occurrence of 2′-*O*-hydroxylation in these taxa is unusual in the Hamamelidae (but see prenylated flavanones in Platanaceae; Kaouadji *et al.*, 1986). These characters occur commonly elsewhere in the Dilleniidae in various combinations (prenylation, 2′-hydroxylation), the Rosidae (Rutales) and Berberidaceae (Ranunculales, Magnoliidae). Thus, removal of at least these two families (or parts) of the Urticales to other subclasses is not unreasonable (Thorne, 1983; Dahlgren, 1983), although some taxa (e.g. *Cannabis*) retain a simple flavonoid pattern (Clark and Bohm, 1979).

However, other taxa, such as the Ulmaceae, seem to be quite conservative in their flavonoids (common flavonols, *C*-glycosylflavones, *C*-glycosylflavonols and some flavones) although included in the Urticales (Bate-Smith and Richens, 1973). A dichotomy, flavonols versus *C*-glycosylflavones, exists between the two subfamilies (Ulmoideae, Celtidoideae, respectively; Giannasi, 1978; Giannasi and Niklas, 1977). This dichotomy supports the generic arrangements (ulmoid versus celtoid) of Grudzinskaya (1965) who actually feels that the two subfamilies are two distinct families. Regardless of the taxonomic categories, the flavonoids of the Ulmaceae fit comfortably within the hamamelid line, perhaps as the most primitive family of the Urticales. The Urticaceae may represent an advanced herbaceous terminus while the Moraceae and Cannabaceae (with their many unusual flavonoids) represent a highly derived/specialized line, an evolutionary scenario proposed in an earlier flavonoid study by Lebreton (1965). While the removal of the Urticales to the

Malviflorae (Thorne, 1983; Dahlgren, 1983) (along with the Euphorbiales from Cronquist's Rosidae) is plausible, the conservative flavonoid chemistry of the Ulmaceae sets it apart from the rest of the Urticales (and much of the Malviflorae). Its flavonoid complement as well as other non-chemical characters still suggests a stronger tie with the Hamamelidaceae (see Cronquist, 1981, p. 185 ff). In this situation dismemberment of the hamamelid Urticales might be considered.

With a few other orders in the Hamamelidae, flavonoid survey results confuse rather than confirm. Thus, myricetin is said to be present in (Gornall *et al.*, 1979) or absent from (along with ellagic acid) *Trochodendron* (Bate-Smith, 1962; Giannasi, 1978). Establishment of the presence of these two characters is required to decide between a magnoliid versus hamamelid affinity. Similarly, myricetin is described as being absent from the Casuarinales (Gornall *et al.*, 1979) and yet was found to be common (along with ellagic acid) in several *Casuarina* species (Saleh and El-Lakany, 1979; Giannasi, 1986) along with biflavonyls. Were it not for the presence of these biflavonyls, which occur sporadically in all other subclasses except the Caryophyllidae, the Casuarinales would reside without notice in the hamamelidalean core of orders. The Betulaceae, in contrast, are so 'bold' in their production of flavonols/flavone and methylation patterns as to stand out even within the Fagales, a distinction in agreement with serological data.

And yet the Hamamelidae can be circumscribed by their flavonoids as generally possessing the primitive characters of ellagic acid and myricetin, a real preponderance and development of flavonol chemistry, with some increased exploitation of the intermediate step of *C*-glycosylation and some development of flavone synthesis. In this respect, the Hamamelidae more closely approximate to a chemically primitive angiosperm group than do the Magnoliidae.

14.4 CARYOPHYLLIDAE

The Caryophyllales (Centrospermae) are unusual in the absence of anthocyanins in all but two families (Caryophyllaceae, Molluginaceae), their function apparently being assumed by the chromoalkaloid betalains (Mabry, 1977). However, flavonoid anthoxanthins (i.e. flavonols, flavones, glycoflavones, etc.) are commonly produced in leaves and flowers of Caryophyllales co-existing with the betalains (Table 14.4). Indeed, the majority of chemosystematic work below the family level utilizes flavonoid anthoxanthins (e.g. Crawford and Evans, 1978; Crawford and Mabry, 1978; Miller, 1981; Doyle, 1983; Mastenbroek *et al.*, 1983). Similarly, comprehensive studies at higher taxonomic levels include flavonoid surveys by Richardson (1978) and Burret *et al.* (1981).

An early survey by Bate-Smith (1962) established the presence of flavonols, absence of ellagic acid and limited occurrence of myricetin in the Caryophyllales (e.g. Burret *et al.*, 1982). The comprehensive ordinal survey by Richardson (1978) showed the presence of *C*-glycosylflavones as well as flavonols, results confirmed by Burret *et al.* (1981), in addition to simple flavones. In many respects (with the exception of betalains) the Caryophyllidae are not dissimilar from the Magnoliidae. Increased emphasis on *C*-glycosylflavone production, an intermediate evolutionary character (Harborne, 1977; Gornall and Bohm, 1978), and the presence of 6-hydroxylation suggest a more advanced and perhaps intermediate position from that of the Magnoliidae. On the bases of the latter character, Young (1981) suggests a close relationship of the Caryophyllidae with members of the Dilleniidae (or Thorne's equivalent). However, with the exception of the betalains, many of the flavonoids of the Caryophyllidae occur in several other subclasses as well (e.g. sulphated flavonoids).

Within the Caryophyllales, the various families contain flavonols or *C*-glycosylflavones or both in some cases (Table 14.3). On the basis of these data, Richardson (1978) suggests that the Didiereaceae originated near the Portulacaceae and the Cactaceae near the Aizoaceae. If flavonols are accepted as the most primitive state, with *C*-glycosylflavones intermediate toward advanced flavone production, a linear order of increasing advancement is suggested by Richardson (1978). A recent and comprehensive phonetic and cladistic analysis (including flavonoids) of the Caryophyllales (Rodman *et al.*, 1984) and several outgroups confirm several of the family pairs of Richardson, but suggest a much branched evolutionary sequence (Rodman *et al.*, 1984, Fig. 6) which is more like that from a later flavonoid survey by Burret *et al.* (1981).

Exclusive of the Caryophyllales, the Plumbaginales and Polygonales lack betalains, but, significantly, possess myricetin, a number of methylated flavonoids (including 5-*O*-methylation), 8-hydroxylation, chalcones/aurones, and sulphates not normally or only rarely found in the Caryophyllales (Table 14.4). If anything, the flavonoid chemistry of the Polygonales lies closer to members of Cronquist's Dilleniidae (see Thorne, 1983). The Plumbaginales similarly do not fit in the Caryophyllidae, but possess a number of 5-*O*-methylflavonoids (flavonols and anthocyanidins) not unlike these observed in the Ericaceae (Harborne, 1967b). Certainly they are not close to the Caryophyllales, as was confirmed by Rodman *et al.* (1984). At the least, the removal of these two orders from the Caryophyllidae (cf. Dahlgren, 1983) would certainly strengthen the taxonomic integrity of this latter taxon as Cronquist (1981) states. This certainly is true in the case of the exclusion of the Bataceae and Gyrostemonaceae from the Caryophyllales. The movement of these two families closer to the Capparales (Dilleniidae) is quite logical, based on several types of evidence including the presence of the glucosinolate biosynthetic apparatus.

Table 14.4 Distribution of flavonoids in the Caryophyllidae

Taxon	Compounds[a]																																				
Caryophyllidae	*1*	*2*	*3*	*4*	*5*	*6*	*7*	*8*	*9*	*10*	*11*	*12*	*13*	*14*	*15*	*16*	*17*	*18*	*19*	*20*	*21*	*22*	*23*	*24*	*25*	*26*	*27*	*28*	*29*	*30*	*31*	*32*	*33*	*34*	*35*	*36*	*37*
CARYOPHYLLALES																																					
Phytolaccaceae[b]			★				+	+																			+										
Achatocarpaceae		(See Portulacaceae)																																			
Nyctaginaceae			★				+																				+					+					
Aizoaceae			★			+	+	+	+	+																+						+					
Didiereaceae			★			+	+	+	+	(+)						+										+	+					+					+
Cactaceae			★				+	+	+							+	+									+											
Chenopodiaceae[c]			★			+	+	+	+	+					+			+									+	+								+	
Amaranthaceae			★				+												+		+					+	+	+		(+)	+					+	
Portulacaceae			★				+										+										+					+					
Basellaceae			★				+																				+										
Molluginaceae			+																					+			+										
Caryophyllaceae			+	+	+		+	+									+										+					+	+				
POLYGONALES																																					
Polygonaceae		+	+	+	+	+	+	+	+	+	+		+	+	+		+	+			+						+					+	+				
PLUMBAGINALES																																					
Plumbaginaceae		+★	+	+		+	+	+	+				+		+		+										+					+	+★				

[a]See Table 14.1 for key to compound identities.
[b]Includes Agdestidiaceae, Stegnospermataceae.
[c]Includes Dysplaniaceae, Hectorellaceae.

Table 14.5 Distribution of flavonoids in the Dilleniidae

Taxon	Compounds[a]																																				
Dilleniidae	1	2	3	4	5	6	7	8	9	10	11	12	13	14	15	16	17	18	19	20	21	22	23	24	25	26	27	28	29	30	31	32	33	34	35	36	37
DILLENIALES																																					
Dilleniaceae	+					+	+	+					+		+	+	+									+	+					+	+				
Paeoniaceae			+	+			+										+																				
THEALES																																					
Ochnaceae[b]		(*)	+			+	+	+			+						+										+		+			+					
Sphaerosepalaceae						+	+										+									+						+					
Sarcolaenaceae						+	+																														
Dipterocarpaceae	+					+	+																									+					
Caryocaraceae		+	+				+																														
Theaceae[c]	+	+*	+			+	+	+			+															+	+					+					
Actinidiaceae							+																									+					
Scytopetalaceae						+	+																														
Pentaphylacaceae						+	+																									+					
Tetrameristaceae	—		—		—		—																														
Pellicieraceae		+	+				+																														
Oncothecaceae		+*	+			+	+	+			+																+					+					
Marcgraviaceae						+	+																									+					
Quiinaceae						+	+																														
Elatinaceae							+																									+					
Paracryphiaceae	—		—		—		—																														
Medusagynaceae						+	+																														
Clusiaceae						+	+	+									+								+	+			+			+					
MALVALES																																					
Elaeocarpaceae	+	+	+	+		+	+	+																								+					+
Tiliaceae						+	+	+						+		+	+	+									+					+					
Sterculiaceae		+*	+				+	+			+						+				+						+					+					
Bombacaceae			+	+			+																									+					
Malvaceae		+	+	+		+	+	+	+		+		+		+	+		+		+					+	+	+					+	+				
LECYTHIDALES																																					
Lecythidaceae[d]	+	+	+				+																														
NEPENTHALES																																					
Sarraceniaceae							+	+																								+					
Nepenthaceae							+																									+					
Droseraceae	+		+				+																									+					
VIOLALES																																					
Flacourtiaceae							+										+	+														+					
Peridiscaceae	—		—		—		—																														
Bixaceae[e]	+																+			+					+	+						+					
Cistaceae						+	+	+	+						+		+	+														+					
Huaceae					+	+	+							+		+	+	+									+					+					
Lacistemataceae	—		—		—		—																														
Scyphostegiaceae		+	+	+	+	+	+	+									+	+									+					+					
Stachyuraceae	+						+																									+					
Violaceae		+	+	+	+	+	+	+									+	+									+					+					
Tamaricaceae	+		+				+	+						+	+																	+					
Frankeniaceae	+	+		+			+	+							+		+	+														+					
Dioncophyllaceae	—		—		—		—																														
Ancistrocladaceae	?					+	+																									+					

Turneraceae							+								+											+			
Malesherbiaceae							+								+	+						+							
Passifloraceae		+	+	+			+								+*	+						+				+			
Achariaceae							+								+	+						+							
Caricaceae							+																			+			
Fouquieriaceae		+	+	+	+		+																			+			
Hoplestigmataceae	—		—		—		—																						
Cucurbitaceae							+	+							+	+						+							
Datiscaceae							+	+			+														+				
Begoniaceae			+				+		+																	+			
Loasaceae							+																						
SALICALES																													
Salicaceae		+	+	+		+	+	+	+					+	+	+					+	+	+		+	+	+	+	
CAPPARALES																													
Tovariaceae							+																						
Capparaceae	+						+															+							
Brassicaceae[f]			+	+	+		+	+					+	+	+	+						+	+						
Moringaceae							+	+																					
Resedaceae							+	+							+														
BATALES																													
Gyrostemonaceae								+																					
Bataceae								+																					
ERICALES																													
Cyrillaceae	(+)	+	+	+		+	+																			+			
Clethraceae							+																			+			
Grubbiaceae	—		—		—		—																						
Empetraceae	(+)	+	+	+			+			+																+			
Epacridaceae	+	+	+	+		+	+																			+			
Ericaceae[g]	+	+	+	+	+	+	+	+	+	+		+	+	+	+*	+						+	+			+	+	+	+
Pyrolaceae	?						+							+															
Monotropaceae							+																						
DIAPENSIALES																													
Diapensiaceae	+						+			+			+													+			
EBENALES																													
Sapotaceae						+	+							+												+			
Ebenaceae	(+)					+	+	+						+												+			
Styracaceae							+																			+			
Lissocarpaceae							+																			+			
Symplocaceae		+					+	+																		+			
PRIMULALES																													
Theophrastaceae							+																						
Myrsinaceae		+	+	+		+	+	+																		+			
Primulaceae	+	+	+	+		+	+	+		+				+	+		+	+	+	+		+	+	+*	+	+			

[a]See Table 14.1 for key to compound identification.
[b]Includes Strasburgiaceae.
[c]Includes Bonnetiaceae.
[d]Includes Napoleonaceae, Foetidaceae.
[e]Includes Cochlospermaceae.
[f]Includes Pentadiplandraceae.
[g]Includes Vacciniaceae.

14.5 DILLENIIDAE

The Dilleniidae show a range of flavonoid types and other phenolics which parallel patterns in the Rosidae and the Hamamelidae. A survey of the Dilleniaceae (Gurni and Kubitzski, 1981) shows the presence of ellagic acid and myricetin (including *O*-methyl derivatives) in many of the species of the eight genera sampled. A number of flavonol sulphates of myricetin, quercetin and kaempferol also occur (Gurni *et al.*, 1981). Simple flavone types (apigenin and luteolin) are found but limited to one genus, *Tetracera*, and flavone *O*-methylation apparently is absent. Glycoflavones are present, but limited to *Doliocarpus*. 5-*O*-Methylflavonols also occur in the Dilleniaceae (as in some Hamamelidae, i.e. Juglandaceae) as well as other scattered types (dihydroflavonols, flavones, chalcones). The overall impression of the Dilleniaceae is that of a primitive taxon concentrating primarily on flavonol production with elaboration in *O*-methylmyricetin types and glycosylation patterns as well as some experimentation with simple flavones and *C*-glycosylflavones.

If, as Cronquist suggests, the Dilleniaceae trace their origin close to the Magnoliidae, possibly near the Illiciaceae, then the Dilleniaceae have evolved a more complex flavonoid chemistry in their methylated flavonols, *C*-glycosylflavones and flavones. These characters are not observed in the simple flavonol-producing Illiciaceae. Indeed, if one examines flavonoid distributions (Table 14.5), the Theales have a simple flavonoid chemistry (mainly flavonols) which comes close to that of the Illiciales. In this case, the Dilleniaceae (Dilleniales) would actually be a more derived taxon and placed to one side rather than directly in line with the Theales (see Cronquist, 1981, p. 289). The Theales share this rather simple flavonoid pattern with the Lecythidales, Nepenthales, Batales, Diapensiales and most of the Ericales and Ebenales. Orders such as the Malvales, Violales and Salicales have extensively exploited flavone and *C*-glycosylflavone production suggesting more advanced positions within the Dilleniidae. The Ericales and Primulales contain type families with rather complex flavonoid chemistry while other included families are actually quite simple in flavonoid complements. How much of this is due to patterns of sampling rather than actual distributions is unknown. The six or so orders with the more complex flavonoid patterns actually are presented by Cronquist (1981, p. 289) as derived terminal groups on several lines of evolution. The flavonoid trends certainly are compatible with the multiradiate pattern of evolution in the Dilleniidae pictured by Cronquist [but see John and Kolbe (1980) Kolbe and John (1980) for a more radical treatment of Theales and other orders in the Dilleniidae].

General characters, unusual or more common to the Dilleniidae (in contrast to the Magnoliidae), include flavone sulphates, 8-hydroxylation, biflavonyls (Theales), 2′-hydroxylation and isoprenylated flavonoids (Salicales). Many of these characters also occur in various taxa in the Rosidae and Hamamelidae. Thus in the Euphorbiaceae, 8-hydroxylation and biflavonyls are found, as they are in the Fabaceae which also produce isoprenylated flavonoids. These latter compounds also occur in the Rhamnaceae, while some combination of these three characters along with 2′-hydroxylation occur in the Anacardiaceae, Rutaceae and Meliaceae. Many of these same characters occur in the Hamamelidae (notably the Moraceae), biflavonyls, for example, being characteristic of a single family, the Casuarinaceae. A number of these characters (8-hydroxylation, sulphates, myricetin, ellagic acids and 5-*O*-methylation) clearly suggest realignment of the Polygonales and Plumbaginales closer to the Dilleniidae. The overwhelming presence of these chemical characters supports an evolutionarily intermediate position for the Dilleniidae comparable to that of the Rosidae and Hamamelidae.

At lower taxonomic levels, existing flavonoid reports tend to strengthen previous distributions of previous surveys (Gornall *et al.*, 1979). For example, *Galax* (Diapensiaceae) produces typical quercetin and kaempferol 3-*O*-glycosides (Soltis *et al.*, 1983) as does *Pyrola*, in the Pyrolaceae (Ericales) (Haber, 1983). The Ericaceae have been investigated intensively (Harborne and Williams, 1973) including the genus *Rhododendron* (Harborne, 1986; Harborne and Williams, 1971, 1973). The genus produces simple flavonols (myricetin, quercetin, kaempferol) and flavonol 5-*O*-methyl derivatives. Unusual flavonols include gossypetin (8-hydroxyquercetin), caryatin (3,5-di-*O*-methylquercetin) and 6,8-di-*C*-methylflavanones (farrerol, matteucinol). The genus seems to have retained primitive flavonoid characters, with *C*-glycosylflavones and flavones being absent. Detailed surveys of *Rhododendron*, section *Pentanthera*, show a similar array of aglycones along with extensive exploitation of glycosylation patterns and some aurone/chalcone glycosides, but again no flavones (King, 1977).

Other ericaceous genera such as *Menziesia* show a conservative flavonol chemistry similar to *Rhododendron* including the presence of gossypetin. Apparently unique to *Menziesia* is 7-*O*-methylation of flavonols (Bohm *et al.*, 1984). In *Cavendishia* (Luteyn *et al.*, 1980), quercetin and kaempferol are found but not myricetin or gossypetin. Instead, *C*-glycosylflavones occur in half of the *Cavendishia* species while flavone *O*-glycosides (luteolin, chrysoeriol and tricin) occur in several taxa. The lack of myricetin and 5-*O*-methylflavones (as in *Rhododendron*) and the presence of glycoflavones and flavones suggest that *Cavendishia* is more advanced in the family.

Other orders in the Dilleniidae vary radically in their flavonoid complement. The Malvales present one of the more complex flavonoid pictures containing many types of flavonols (including 8-hydroxylation/methylation), flavones and *C*-glycosylflavones as typified by *Gossypium* (Parks *et al.*, 1975). In the Salicaceae, a highly specialized group (Salicales), the survey of *Populus* bud exudates (Wollenweber, 1975b) shows a high degree of methylation of flavonols, flavones and chalcones, along with B-ring deoxy derivatives from each class of flavonoids. The considerable degree of methylation in *Populus* bud exudates may be characteristic of the buds only (Gornall *et al.*, 1979), as available studies of leaf flavonoids (Crawford, 1974) show the presence of much simpler flavonoids, a situation unlike that observed in the bud exudates of Betulaceae (Giannasi, 1986). Nevertheless, the occurrence of 2′-hydroxylation and isoprenylated flavonoids sets this order apart. The Cruciferae on the other hand seem to have more common flavonols (quercetin, kaempferol), flavones and glycoflavones with perhaps greater 'experimentation' in acylation/glycosylation at 3, 3′, 4′- and 3, 7 positions and sulphates (Aguinagalde *et al.*, 1982). In contrast to extensive surveys of glucosinolates in the Cruciferae, flavonoid surveys are limited and are badly needed, especially in the New World mustards. Also, as in other subclasses, biflavonyls are scattered in distribution, occurring in the Ochnaceae and Clusiaceae (Guttiferae), being recently reconfirmed in the latter family in *Garcinia* and extended to *Calophyllum* (Waterman and Hussain, 1983).

14.6 ROSIDAE

The Rosidae, in many ways, possess a flavonoid chemistry similar or parallel to that of the Dilleniidae (or vice versa). Both ellagic acid and myricetin are common in at least one or more families of the majority of orders (Table 14.6). There is considerable 'experimentation' in methylation of flavonoids in many orders, with some taxa (Fabales, Myrtales, Euphorbiales, Vitaceae, Rhamnales, Linales, Sapindales) showing considerable development of *C*-glycosylflavones. Flavone *O*-glycosides occur in some of the same orders but tend to be of the simpler types except for the Fabales, some of the Rosales and the Sapindales, where *O*-methylation, 2′-, 6- and/or 8-oxygenation (and other) substitution patterns are observed.

These unusual flavonoid substitutions find parallels in the Dilleniidae and Hamamelidae. For example, 5-methoxy substitution occurs in the Cunoniaceae (Rosales) and well as the Juglandaceae (Hamamelidae), Lauraceae (Magnoliidae), Plumbaginaceae (Caryophyllidae) and Ericaceae (Dilleniidae). Similarly, 2′-oxygenation (Saxifragaceae, Crassulaceae, Rutaceae–Rosidae) also occurs in the Urticales (Hamamelidae) and Salicaceae (Dilleniidae). Biflavonyls occur in the Anacardiaceae, Rhamnaceae and Euphorbiaceae (Rosidae) as they do in the Hamamelidae, Dilleniidae, and Magnoliidae (Berberidaceae). 8-Hydroxylation occurs in the flavonols (and some flavones) of several families in the Rosales, Fabales, Myrtales, Rutaceae, Zygophyllaceae, Geraniaceae and Euphorbiaceae, a character also found in other subclasses. Isoprenylated flavonoids occur in several Rosidae families (e.g. Fabaceae, Caesalpiniaceae, Rutaceae and Apiaceae). Flavonol sulphates occur in the Rosales and Myrtales as well as all other subclasses. One unusual character which seems to be more common in the Rosidae is the presence of 5-deoxyflavonoids (e.g. Anacardiaceae, Rutaceae, Julianiaceae, the Sapindales, some Celastrales, Geraniales, Myrtales, Fabales and Rosales). This character is rare in other groups (e.g. Primulales, Dilleniidae).

In summarizing the Rosidae, it is a taxon whose flavonoid chemistry parallels but is richer than that of the Dilleniidae and Hamamelidae. Emphasis on the primitive flavonols (especially myricetin) with various substitutions (*O*-methylation, acylation) with a gradual shift of synthesis toward *C*-glycosylation and flavone synthesis characterized the Rosidae (like the Dilleniidae) as an intermediate evolutionary group between the more primitive docots (Hamamelidae, Magnoliidae) and the advanced herbaceous dicots (i.e. the Asteridae). Indeed, the Rosidae, Dilleniidae and Hamamelidae probably share more in common with each other than they do with the other subclasses.

At lower taxonomic levels, earlier studies of the Rosaceae showed the general absence of both ellagic acid (except from subfamily Rosoideae) and myricetin, despite their common occurrence in most other families in the Rosidae (Bate-Smith, 1962). This was supported in more detailed surveys of many taxa in *Potentilla* and *Prunus* (Bate-Smith, 1961), although *Prunus* regularly produces ellagic acid. A survey of *Malus* (Maloideae) shows, not unexpectedly, glycosides of quercetin and kaempferol (Williams, 1982). In addition, however, isorhamnetin, galangin and azaleatin (5-methylquercetin) and toringin (chrysin 5-*O*-glucoside) are common, as are naringenin, phloridzin (a dihydrochalcone) and related sieboldin and trilobatin. Earlier work on the Pomoideae (Challice, 1972, 1973; Challice and Williams, 1968) shows the presence of flavone *O*-glycosides and *C*-glycosylflavones, including those of apigenin, luteolin and chrysin as well as typical flavonols. Thus it would seem that while the Rosaceae may be considered a basal group in the Rosidae, and presence of *C*-glycosylflavones and many specialized flavonols in the family suggests that several of the subfamilies, at least, are advanced. The Rosoideae are probably more primitive in their retention of myricetin as well as ellagic acid. The family then retains fewer generally distributed primitive flavonoid characters than the Fabaceae to which it is most often related.

Indeed, the Chrysobalanaceae, formerly a subfamily of the Rosaceae, retains more of these primitive characters. Early work (Bate-Smith, 1962) shows the presence of myricetin in *Chrysobalanus* and *Licania*. More recent work by Coradin *et al.* (1985) on the genus *Parinari* shows myricetin to be present in the African species but not in the Neotropical taxa. However, myricetin is common in other African genera (e.g. *Maranthes*) and neotropical taxa (*Exellodendron; Couepia*). The production of *C*-glycosylflavones and flavones in some neotropical taxa

Table 14.6 Distribution of flavonoids in the Rosidae

Taxon	Compounds[a]																																				
Rosidae	1	2	3	4	5	6	7	8	9	10	11	12	13	14	15	16	17	18	19	20	21	22	23	24	25	26	27	28	29	30	31	32	33	34	35	36	37
ROSALES																																					
Brunelliaceae							+	+	+				+			+										+						+					
Connaraceae						+	+																									+					
Eucryphiaceae							+	+	+				+			+										+						+					
Cunoniaceae	+	+	+			+	+	+					+		+		+															+					
Davidsoniaceae																																					
Dialypetalanthaceae			+	+		+	+	+	+	+	+																+			+		+					
Pihosporaceae							+	+									+																				
Byblidaceae[b]		+	+	+			+				+																					+					
Hydrangiaceae		+	+			+	+																									+					
Collumelliaceae							+																														
Grossulariaceae[c]	+	+	+		+	+	+	+								+																+					
Greyiaceae							+		+																												
Bruniaceae	—		—		—		—																														
Anisophyllaceae	—		—		—		—																														
Alseuosmiaceae			+				+										+									+	+		?			+		+			
Crassulaceae			+			+	+	+	+		+			+				+			+					+	+					+					
Cephalotaceae	+					+	+																			+											
Saxifragaceae[d]	+	+	+	+	+	+	+	+	+	+		+		+		+	+	+			+	+				+						+		+			
Rosaceae	+	+	+	+	+	+	+	+	+	+	+		+	+	+	+	+	+	+		+		+			+	+	+			+	+	+	+			
Neuradaceae	—		—		—		—																														
Crossosomataceae		+	+	+	+	+	+	+		+	+			+	+		+	+	+		+						+					+		+			
Chrysobalanaceae						+	+									+	+										+					+					
Surianaceae							+	+								+	+										+					+					
Rhabdodendraceae	—		—		—		—																														
FABALES																																					
Mimosaceae						+	+	+	+	+	+	+				+	+		+	+							+			+		+	+*		+		
Caesalpiniaceae			+	+		+	+	+	+	+		+	+			+	+*	+									+	+	+	+		+	+	+			
Fabaceae		+	+	+	+	+	+	+	+	+	+		+	+		+	+*	+	+		+	+	+				+	+	+	+*		+	+*	+	+	+	+
PROTEALES																																					
Elaeagnaceae	+				+		+	+																								+					
Proteales						+	+									+																+					
PODOSTEMALES																																					
Podostemaceae[e]																																					
HALORAGALES																																					
Haloragaceae	+		+				+																				+										
Gunneraceae							+																														
MYRTALES																																					
Sonneratiaceae	—		—		—		—																														
Lythraceae	+	+	+	+			+	+									+	+									+										
Penaeaceae		+	+	+		+	+	+								+										+	+					+					
Crypteroniaceae			+	+		+	+	+	+	+	+																+			+		+					
Thymelaeaceae			+				+										+	+								+	+					+					
Trapaceae			+				+																				+										
Myrtaceae	+	+	+	+		+	+	+								+	+									+	+				?	+					+
Punicaceae	+	+	+																																		
Onagraceae	+	+	+	+		+	+		+						+		+*	+							+		+					+	+				
Oliniaceae		+	+	+	+	+	+							+																		+					

Melastomataceae	+	+	+	+	+	+	+					+									+		
Combretaceae	+		+	+		+	+	+	+	+	+						+	+		+	+		
RHIZOPHORALES																							
Rhizophoraceae	(+)			+		+	+	+													+		
CORNALES																							
Alangiaceae		+	+				+														+		
Nyssaceae[f]	+						+														+		
Cornaceae[g]	+	+	+	+			+														+		
Garryaceae		+	+				+														+		
SANTALALES																							
Medusandraceae																							
Dipentodontaceae		+	+	+	+	+	+	+						+	+			+			+		
Olacaceae							+														+		
Opiliaceae							+														+		
Santalaceae							+	+					+										
Misodendraceae							+	+					+										
Loranthaceae							+	+						+				+			+		
Viscaceae	+					+	+								+							+	
Eremolepidaceae						+	+								+								
Balanophoraceae[h]							+						+				+						+
RAFFLESIALES																							
Hydnoraceae	—		—		—		—																
Mitrastemonaceae	—		—		—		—																
Rafflesiaceae	—		—		—		—																
CELASTRALES																							
Geissolomataceae			+				+														+		
Celastraceae[i]			+			+	+					+	+							+	+		
Hippocrateaceae																							
Stackhousiaceae																							
Salvadoraceae							+																
Aquifoliaceae[j]			+				+																
Icacinaceae							+																
Aextoxicaceae	—		—		—		—																
Cardiopteridaceae							+														+		
Corynocarpaceae	+						+																
Dichapetalaceae							+							+	+		+	+			+		
EUPHORBIALES																							
Buxaceae			+				+							?									
Simmondsiaceae	—		—		—		—																
Pandaceae	—		—		—		—																
Euphorbiaceae[k]	+	+	+			+	+	+	+		+	+		+			+	+	+		+	+	
RHAMNALES																							
Rhamnaceae				+			+	+					+		+	+	+		+		+	+*	
Leeaceae							+	+													+		
Vitaceae	+	+	+	+	+	+	+						+	+				+			+		
LINALES																							
Erythroxylaceae							+	+													+		
Humiriaceae		+	+	+	+		+	+										+					
Ixonanthaceae		+	+	+	+		+	+										+					
Hugoniaceae		+	+	+	+		+	+										+					
Linaceae		+	+	+	+		+	+										+					

(Contd.)

Table 14.6 (*Contd.*)

Taxon	*Compounds*[a]																																				
Rosidae	1	2	3	4	5	6	7	8	9	10	11	12	13	14	15	16	17	18	19	20	21	22	23	24	25	26	27	28	29	30	31	32	33	34	35	36	37
POLYGALALES																																					
Malpighiaceae							+																														
Vochysiaceae		+	+				+																														
Trigoniaceae		+	+				+																			+				+							
Tremandraceae	+			+			+	+																													
Polygalaceae[l]		+	+				+																														
Xanthophyllaceae		+	+				+																														
Krameriaceae	—		—		—		—																														
SAPINDALES																																					
Staphyleaceae			+				+																									+					
Melianthaceae	+						+																														
Bretschneideraceae							+																														
Akaniaceae							+																									+					
Sapindaceae			+			+	+	+	+	+																						+					
Hippocastanaceae						+	+	+	+					+																		+	+				
Aceraceae	+		+			+	+	+								+	+	+									+					+					
Burseraceae							+																						+			+					
Anacardiaceae[m]	+	+	+	+		+	+	+				+	+	+		+	+												+	+*		+	+*				
Julianiaceae						+	+									+										+	+			+			+*				
Simaroubaceae[n]	+					+	+																									+					
Cneoraceae							+																														
Meliaceae[o]							+	+	+			+				+		+	+		+	+				+						+			+		+
Rutaceae		+	+			+	+	+	+	+	+		+			+	+	+	+	+	+	+	+			+	+			+*	+	+	+		+	+	
Zygophyllaceae[p]							+	+	+		+			+			+*	+									+										
GERANIALES																																					
Oxalidaceae[q]							+																				+					+	*				
Geraniaceae[r]	+		+	+		+	+	+	+		+						+										+			+		+					
Limnanthaceae						+	+	+																								+					
Tropaeolaceae			+				+																									+					
Balsaminaceae		+	+	+	+	+	+																														
APIALES																																					
Araliaceae		+	+				+																			+											
Apiaceae			+	+	+	+	+	+						+	+		+	+						+	+	+	+	+				+	+		+		

[a]See Table 14.1 for key to compound identification
[b]Includes Roridulaceae.
[c]Includes Brexiaceae, Ribesiaceae, Escalloniaceae, Montiniaceae, lteaceae.
[d]Includes Lepuropetalaceae, Penthoraceae, Francoeaceae, Vahliaceae.
[e]Includes Trishchaceae.
[f]Includes Davidsoniaceae.
[g]Includes Torricelliaceae.
[h]Includes Cynothoriaceae.
[i]Includes Siphonodontaceae, Goupiaceae, Lophopyxidaceae.
[j]Includes Sphenostemonaceae.
[k]Includes Vapacaceae, Picrodendraceae.
[l]Includes Emblingiaceae.
[m]Includes Podoaceae.
[n]Includes Kirkiaceae.
[o]Includes Ailoniaceae.
[p]Includes Nitrariaceae, Balanitaceae, Peganiaceae, Tribulaceae.
[q]Includes Averrhoaceae, Hypseuchoridaceae, Lepidobotryaceae.
[r]Includes Dirachalaceae, Ledocaryaceae, Vivianaceae, Biebersteiniaceae.

(*Couepia*, *Hirtella*) parallels flavonoid evolution in the Rosaceae (sensu lato).

In the case of the Saxifragaceae quite extensive surveys have continued over the past 15 years by Bohm and colleagues (e.g. Bohm and Wilkins, 1978a, b; Bohm and Ornduff, 1978; Bohm *et al.*, 1977a, b; Wilkins and Bohm, 1976; Soltis and Soltis, 1986). The main emphasis is on flavonol glycosides (myricetin, quercetin and kaempferol) and various methylated derivatives in the majority of genera while *Heuchera* produces flavones (Bohm and Wilkins, 1978b; Soltis and Bohm, 1986; Bohm and Bhat, 1985). Flavones are also produced in *Sullivantia*, some of which undergo 6-oxygenation/glycosylation, along with quercetin glycosides (Soltis, 1980). Flavones also occur in *Tiarella* (Soltis and Bohm, 1984). A survey of the Hydrangiaceae (Bohm *et al.*, 1985) shows a series of glycosides of common flavonols typical of the Saxifragaceae with which it is often associated. Available data from the Crassulaceae also show an emphasis on flavonols (kaempferol and laricytrin) and a highly methylated flavone, a not unexpected result considering its usual association with the Saxifragaceae (Denton and Kerwin, 1980). The Cephalotaceae are thought to be related to the Crassulaceae/Saxifragaceae (Cronquist, 1981). Flavonoid analysis shows the presence of common flavonol glycosides (myricetin, quercetin, kaempferol), the flavone luteolin and the flavanone eriodictyol, a complement compatible with the Saxifragaceae/Crassulaceae (Nicholls *et al.*, 1985). *Ribes* (Bate-Smith, 1976) in the Grossulariaceae (at one time included within the Saxifragaceae) typically possesses common flavonols (myricetin, quercetin, kaempferol) with considerable similarity in glycosylation patterns to *Tellima* (Collins and Bohm, 1974) in the Saxifragaceae, indicative of the relationship between these two families. Several of these systematic associations are mirrored in serological work by Grund and Jensen (1981).

The Fabales is a very distinct group in terms of the variety of flavonoids produced. Myricetin occurs in many taxa along with other common flavonols and their methylated derivatives (Gornall *et al.*, 1979). Extra oxygenation patterns (6 and/or 8) are found in a number of taxa (Gomes *et al.*, 1981). *C*-Glycosylation is very common (Nicholls and Bohm, 1982). Most distinctive in the Fabales in the 'exploitation' of isoflavones/isoflavanones and related pterocarpans (Gomes *et al.*, 1981). 2′-Hydroxylation is found in some taxa (Williams *et al.*, 1983) while other taxa produce typical flavones and flavonols (Saleh *et al.*, 1982a). The 2′-hydroxylated flavonoids occur both in the Hamamelidae and Dilleniidae as well as some other members of the Rosidae (Rutaceae) and Asteridae, as do prenylated flavonoids (Pereira *et al.*, 1982) and biflavonyls (Castro and Valverde, 1985), characters again in common with the Rutaceae and/or Rhamnaceae.

The Myrtales are generally characterized by the presence of flavonols (myricetin, quercetin, kaempferol) and their methylated derivatives. *C*-Glycosides occur in the Lythraceae. Surveys of the Onagraceae show the presence of common flavonols, *C*-glycosylflavones, flavones and a few chalcones (Averett and Raven, 1984; Averett and Boufford, 1985). Common flavonols in *Rhynchocalyx* (variously treated by authors as Rhyncocalycaceae; or part of Lythraceae or Crypteroniaceae) allow it to rest comfortably within the Myrtales (Averett and Graham, 1984). *Alzatea* (variously treated by authors as Alzateaceae or included in Lythraceae or Crypteroniaceae) possesses the flavonol quercetin and the flavone apigenin 7-methyl ether, but lacks myricetin (Graham and Averett, 1984). It fits within the Myrtales, the absence of myricetin and the presence of the methylated flavone suggesting a derived origin.

The Santalales produce common flavonols with myricetin being present in the Viscaceae (e.g. Crawford and Hawksworth, 1979). The Euphorbiales also produce the three common flavonols as confirmed in recent studies (Stewart *et al.*, 1979, 1980; Kolterman and Breckon, 1982; Kolterman *et al.*, 1984) as well as some glycoflavones (Gornall *et al.*, 1979). Also unusual in the Euphorbiaceae is 8-oxygenation (cf Dilleniidae) and the presence of biflavonyls (see Rutaceae). The Rhamnaceae likewise possess biflavonyls, but myricetin is rare as in the Euphorbiaceae. Biflavonyls are unknown in the Vitaceae and Leeaceae. Recent surveys of the Vitaceae (Moore and Giannasi, 1986) show the presence of the three common flavonols, flavones (apigenin, luteolin) and *C*-glycosylflavones.

The Sapindales, sensu Cronquist (1981), is an order (ca. 15 families) of some considerable biosynthetic capacity. One of the 'simpler' families, the Aceraceae, has recently been surveyed by Delendick (1981). As with many of the Sapindales the flavonoid profile of *Acer* is based mainly on common flavonols and their *O*-methyl derivatives including myricetin (and ellagic and gallic acids in some taxa). *Dipteronia* is similar to several species of *Acer* and thus cannot be readily distinguished chemically as a distinct genus. *C*-Glycosylflavones are common in *Acer*, often delimiting whole sections. Flavones are also common in several series. The flavonoid data have been most helpful at the inter- and infra-serial level as well as in the systematics of the cultivated taxa (Delendick, 1984, 1985, 1987). Basically, however, its flavonoid pattern is similar to several other families in the Rosidae (Vitaceae, Onagraceae) with an intermediate position of evolution tending towards flavone exploitation.

The Anacardiaceae retain all these flavonoid characters and add synthesis of biflavonyls to their complement. Indeed *Blepharocarya*, placed in the Anacardiaceae (or sometimes in its own family or in the Sapindaceae), also produces biflavones and biflavanones, securely placing it in the Anacardiaceae (Wannan *et al.*, 1985). Surveys of Anacardiaceae heartwood flavonoids also show the characteristic presence of 5-deoxyflavonoids and anthochlor pigments, a capacity shared with the Fabales (Young, 1979). The same flavonoid characteristics have allowed

Young (1976) to clearly assign the Julianiaceae to a position close to the Anacardiaceae rather than the Juglandaceae (Hamamelidae) with which it has also been associated in the past.

The Rutaceae also possess some unusual biosynthetic capabilites. Like many families in the Sapindales it produces common flavones and methylated derivatives including myricetin as well as *C*-glycosylflavones and flavone *O*-glycosides. This seems to be a basic scheme in many Sapindales and Rutaceae in general (Waterman and Hussain, 1983). A recent comprehensive survey of the Aurantioideae (Rutaceae) shows just such a comprehensive complement (Grieve and Scora, 1980). In addition, 2′-hydroxylation and isoprenylated flavonoids also occur (cf. Urticales and Salicales), the latter terpene/flavone substitution also being observed in some Fabales.

The Zygophyllaceae possess common flavonols (myricetin, quercetin, kaempferol) as well as traces of herbacetin (8-hydroxykaempferol) and its 8-*O*-methyl ether (Saleh and El-Hadidi, 1977) in one taxon. Simple flavones and some *C*-glycosylflavones also occur (Gornall *et al.*, 1979). Recent examination of the subfamily Tribuloideae (sometimes treated as a separate family, Tribulaceae) of the Zygophyllaceae shows a flavonol pattern like the majority of taxa in the family, but herbacetin is absent (Saleh *et al.*, 1982b). The suggestion that the Tribuloideae be made a separate family on the basis of differences in a glycosylation pattern (gentiobioside versus rutinosides) does not seem as strong a character as would be differences in aglycone types.

The Geraniales are characterized by the three common flavonols (myricetin, quercetin, kaempferol), their methylated derivatives and ellagic acid. Flavones occur in the Geraniaceae and Oxalidaceae as do *C*-glycosylflavones (Gornal *et al.*, 1979; Bate-Smith, 1973; Saleh *et al.*, 1983a). 8-Hydroxylated flavones occur in the Geraniaceae while anthochlor pigments are known from *Oxalis* (*O. pescaprae*). 5-Deoxyflavonoids are also found in *Geranium*. Indeed these latter three flavonoid types occur in the Rutaceae, Anacardiaceae and Fabaceae, all together, or in combination, strongly supporting the placement of the Geraniales in the Rosidae. The Linales, sometimes included in the Geraniales, produce flavonols, flavones and *C*-glycosylflavones (Table 14.6). Foliar flavonoids of New World *Linum* are dominated by *C*-glycosylflavones (Giannasi and Rogers, 1971; Giannasi, unpublished). The Limnanthaceae are included by Cronquist (1981) in the Geraniales. The totally flavonol (with methylated deriva-

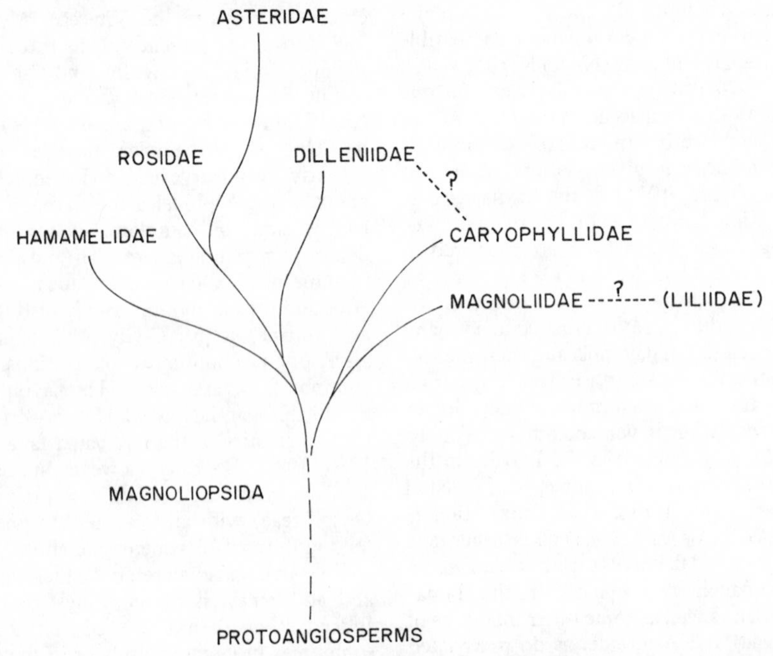

Fig. 14.1 Proposed systematic and evolutionary relationships of the subclasses of dicotyledons (Magnoliopsida) based on flavonoids and other data. The dotted line between the Dilleniidae and Caryophyllidae allows for consideration of the disputed positions of the Polygonales and Plumbaginales and possible alternative relationships of the Caryophyllales if the presence of betalains is not stressed. Relationships of the Magnoliidae to the Liliidae (Liliopsida, monocotyledons) based on flavonoids and alkaloids is indicated (dotted line) but specific phylogenetic relationships are not implied (see Chapter 15 for a discussion of the monocots).

tives) complement of the family (Parker and Bohm, 1979) certainly is primitive and fits within the Geraniales. However, its synthesis of glucosinolates argues for a placement within the Capparales (Dilleniidae), although in this position it is the only myricetin-producing taxon in an otherwise myricetin-free order (Gornall *et al.*, 1979).

The Apiales produce the common flavonols (quercetin, kaempferol with myricetin being rare) along with their methylated derivatives. Flavones, both *O*- and *C*-glycosyl types, are found (Saleh *et al.*, 1983b). Both isoprenylated and sulphated flavonoids occur in the Apiaceae (Umbelliferae), as do flavone 5-*O*-glycosides (Harborne, 1967a, 1977; Harborne and King, 1976).

Overall then, the Rosidae (like the Dilleniidae) seems to represent an intermediate group between primitive taxa (e.g. Magnoliidae) and more advanced taxa (e.g. Asteridae). Some members of the Rosidae (Table 14.6) only produce flavonol while others show the capacity to produce more advanced types (*C*-glycosyl- and ultimately *O*-glycosyl-flavones). Other taxa produce only advanced types (some *C*-glycosyl- and/or mostly *O*-glycosyl-flavones). 'Creative' biosynthetic modifications such as prenylation, 2′- and 8-hydroxylation (and some 6-hydroxylation), flavone coupling, 5-*O*-methylation, 5-deoxygenation and sulphation, find their counterparts in the Dilleniidae, Hamamelidae and Asteridae in multiple correlations. Indeed, these four subclasses seem to form an analogous (if not homologous) cluster of taxa distinct from those in the Magnoliidae or Caryophyllidae. Such a relationship is suggested by non-chemical characters and also by other secondary chemical characters (Giannasi, 1986). No doubt this has spurred the debate as to whether or not the Magnoliidae is the basal angiosperm group (at least conceptually).

14.7 ASTERIDAE

The Asteridae produce many of the same flavonols, *C*-glycosylflavones and *O*-glycosylflavones as seen in the Rosidae (or Dilleniidae and Hamamelidae for that matter) and hence its derivation from the Rosidae is quite logical on chemical grounds. One thing observed is the absence of ellagic acid and the gradual loss of myricetin on moving from the Solanales toward the advanced Asteraceae (Compositae). A gradual shift in emphasis also occurs towards *C*-glycosides and flavone *O*-glycosides, especially with 6-oxygenation in the latter. Flavonols in advanced groups tend to have considerable degrees of methylation as do flavones where they occur. This gives the impression of flavonoid advancement. It has also been noted that in the iridoid-producing Asteridae 6-hydroxyflavones are commonly found, whereas 6-hydroxyflavones are rare (mostly occurring as 6-methoxyflavones) in the non-iridoid producers (Gornall *et al.*, 1979). Anthochlor pigments are common in several groups (e.g. Asterales, Scrophulariales). Certainly, the presence of sesquiterpenes in the Asteraceae, Campanulaceae and Apiaceae (Umbelliferae) of the Rosidae provides yet another linkage between the two subclasses.

The Gentianales produce flavonols and their methylated derivatives and *C*-glycosyl- and *O*-glycosyl-flavones (including 6-hydroxy derivatives). The latter two flavonoid classes are characteristic of the Gentianaceae which also produce a number of xanthones (Hostettman-Kaldas *et al.*, 1981), but, in general, they lack flavonol diversity. In contrast, the Solanales seem to produce nearly all classes of flavonoids. *Solanum*, for example, is rich in flavonol- 3-*O*-glycosides but also produces a number of methylated kaempferol, quercetin and myricetin aglycones, several of which show 8-hydroxylation/glycosylation (Steinharter *et al.*, 1986). Also, both *C*-glycosylflavones and flavones (and their methyl derivatives) are found, including 8-hydroxylated chrysoeriol (Whalen, 1978). Certainly, this is a most accomplished biosynthetic repertoire and contrasts with *Capsicum* (also in Solanaceae) whose complement is based almost entirely on *C*- and *O*-glycosyl-flavones (Ballard *et al.*, 1970).

The Polemoniaceae produce common flavonols, 6-*O*-methylflavonols and *C*-glycosylflavones (Smith *et al.*, 1977). Recent work on *Collomia* shows similar types of flavonoids with the exception of *C*-glycosylflavones (Wilken *et al.*, 1982). In *Phlox*, however, *C*- and *O*-glycosylflavones are dominant (Levy, 1983).

The Menyanthaceae have been variously placed in the Gentianales or the Solanales. Production of flavonols only (although with considerable exploitation of glycosylation patterns) suggests that the family is not closely related to the Gentianaceae which in contrast contains *C*- and *O*-glycosylflavones as well as flavonols (Bohm *et al.*, 1986). Bohm *et al.* (1986) indicate that the Menyanthaceae flavonols (especially the 3-, 7- and 3′-*O*-methyl derivatives) do occur in other families in the Gentianales (but not Gentianaceae) and to a larger extent in the Solanales (including Solanaceae). While they emphasize that the flavonoid data are only suggestive of an alliance, the higher frequency of these methylated compounds in the Solanales (Polemoniaceae) just as easily tilts the Menyanthaceae toward a comfortable niche in the Solanales (sensu Cronquist, 1981).

The Lamiales show a shift towards flavones and *C*-glycosylflavones and less emphasis on flavonols; 6-hydroxylation is quite common (Adzet and Martinez, 1981). Similarly, surveys in the Verbenaceae show an emphasis on flavone (and methyl derivatives) *O*-glycosides, with flavone *C*-glycosides being absent (Umber, 1980).

The Scrophulariales are generally similar to the Lamiales in flavonoid chemistry, although the presence of anthochlor pigments in some taxa of the Scrophulariaceae sets the family apart. In general, however, flavones, including 6-hydroxy derivatives, predominate (Grayer-Barkmeijer, 1978). Most unusual is the recent finding of isoprenylated (6-geranyleriodictyol and 6-geranyldihydroquercetin) flavonoids which also occur in the Rosidae/Dilleniidae/Hamamelidae (Lincoln, 1980). The Oleaceae contain flavonol glycosides (quercetin, kaempferol) and flavone *O*-glycosides (apigenin, luteolin, chrysoeriol) but few if any *C*-glycosylflavones (Harborne and Green, 1980), but the flavonoid patterns do not indicate placement in or out of the Scrophulariales. Surveys of the Gesneriaceae and Bignoniaceae show a predominance of flavones, including 6-hydroxyflavones and anthochlors (Gesneriaceae) all of which are compatible with similar patterns in the Scrophulariaceae (Harborne, 1967c). The 3-deoxyanthocyanins are unique in angiosperms to the Gesneriaceae (but see Scogin, 1980, for occurrence in the Bignoniaceae).

The Campanulaceae show an equal mix of flavonols and flavones with associated methyl derivatives (Gornall *et al.*, 1979), a pattern not unlike the Solanales. The occurrence of flavone and common flavonol *O*-glycosides in the Goodeniaceae (Patterson, 1984) is compatible with the pattern in the Campanulaceae (as well as in other taxa).

6-Hydroxyflavones (along with flavonols to a greater or lesser degree) are common in the Dipsacales and Rubiales. Caprifoliaceae (Dipsacales), however, lack this character. Although common flavonols (quercetin, kaempferol) and flavone (apigenin, lutelin) *O*-glycosides are present they do not deny or confirm ordinal relationships between these taxa. The presence of biflavonyls in the Caprifoliaceae was unique to the Asteridae (Bohm and Glennie, 1971) until their recent discovery in *Strychnos* in the Loganiaceae (Gentianales; Nicoletti *et al.*, 1981). The recent report of the aurone, aureusidin, in *Mussaenda* (Rubiaceae) increases the number of families known to produce these compounds (Harborne *et al.*, 1983).

The Asterales (Asteraceae) produce a phenomenal array of flavonoids. The flavonols, quercetin and kaempferol, are common, especially polymethoxylated types including compounds with 6- and 8-hydroxyl/methoxyl substitutions. Similar patterns are observed among the flavones. Carbon–carbon-linked glycosides are rare. All three of these characters indicate an advanced state of flavonoid evolution for the Asteraceae (Compositae). The most recent compendium of the flavonoid chemistry of the Compositae is that of Heywood *et al.* (1977). Work since that time has added a tremendous number of new variations of the same themes, but little has changed in initial interpretations of flavonoid evolution in the Compositae. Also, besides the Fabaceae and Scrophulariaceae (and a few scattered families), the Compositae is the only family to 'experiment' extensively with anthochlors (see below). In this respect the Asteraceae have concentrated on the 'exploitation' of flavonoids (many with biological activity), sesquiterpene lactones (also with biological activity), polyacetylenes and in a few cases (*Senecio*) alkaloids.

Several salient points of flavonoid evolution are significant at lower taxonomic levels. Anthochlor pigments, chalcones and aurones have been long studied in the subtribe Coreopsidinae, tribe Heliantheae of the Compositae (Bohm, 1982). At the generic level they have been used with considerable facility in solving taxonomic problems (Giannasi, 1975a, b; Ballard, 1975, 1986; Hart, 1979; Smith and Crawford, 1981; Laduke, 1982; Stuessy and Crawford, 1983; Crawford and Smith, 1983a, b, 1984, 1985). These compounds were at one time thought to be restricted in the Asteraceae to the Coreopsidinae, thus providing a convenient marker (Cronquist, 1955). Surveys over more that twenty-five years showed them to be sporadically scattered through other tribes of the Compositae and other angiosperm families (Harborne, 1967a, 1977; Bohm, 1982; Harborne *et al.*, 1983). This led Cronquist (1977) to attribute less significance to anthochlors as a subtribal marker for the Coreopsidinae. However, in a survey of the Coreopsidinae, Crawford and Stuessy (1981) found anthochlors in 28 out of 30 genera sampled in the Coreopsidinae (32 total) and in only a few species of four other genera out of 69 sampled in the tribe Heliantheae. These latter four were *Helianthus, Simsia, Tithonia* and *Viguiera*, closely related genera of the subtribe Heliantheae. Despite some minor variation, anthochlors may be considered characteristic markers of the Coreopsidinae. Overall in the Asteraceae, the anthochlors produced fall into two groups: (1) A-ring phloroglucinol-based (Cardueae, Eupatorieae, Inuleae) and (2) A-ring resorcinol-based, (Heliantheae, Lactuceae). Neither of these groupings correspond to current subfamilial groupings in the Asteraceae, but do suggest two different basic pathways of anthochlor synthesis in the family. The seemingly random occurrence of anthochlors in a few Compositae other than the Coreopsidinae does not seem to be a matter of sampling procedures, since Crawford and Stuessy also present an extensive table of species sampled which lack anthochlors. Additional surveys for anthochlors in the Compositae may provide biosynthetic correlations in the future. Certainly, anthochlors in general are more widespread than formerly thought (see Chapter 9).

Earlier statements that myricetin is absent from the Asterales (Gornall *et al.*, 1979) are no longer valid. Myricetin and several methylated derivatives have recently been found, for example, in *Haplopappus* (Auanoglu *et al.*, 1981). Some taxa are particularly rich in these compounds and sulphated 6-methoxyflavonols (and 6-methoxyflavones) occur especially in *Brickellia* (Eupatorieae) (e.g. Timmerman *et al.*, 1981).

Another interesting flavonoid phenomenon has been observed in the Vernonieae. Several studies have shown a continuous range of typical flavonol and flavone *O*-glycosides (Mabry *et al.*, 1975; Stutts, 1981) as well as repeated series of highly methylated flavonols and flavones (King and Jones, 1982; King, 1986). In several genera it was noticed that foliar flavonoid glycosides were relatively common flavonol and flavone *O*-glycosides (often with *C*-glycosides), while large numbers of the free aglycones were highly methylated, suggesting the presence of two distinct biosynthetic systems (Smith, 1981). This does not appear to be a case of external versus internal leaf flavonoids (Wollenweber and Dietz, 1981).

In other studies the flavonoids of composite taxa seem to be common flavonols and flavones (e.g. Rees and Harborne, 1984; Pacheo *et al.*, 1985), or show an intermediate condition with the presence of *C*-glycosylflavones along with flavonols and/or flavones (Valant-Vetschera, 1982, 1985), while others have concentrated on flavones. While the Compositae still produce flavonols in many taxa, these are advanced biosynthetic types including polymethoxylation, 6- and/or 8-hydroxylation/methylation (Bohm and Stuessy, 1981a, b; Wollenweber and Dietz, 1981).

In microcosm, the Asteraceae, although putatively the most advanced, show a whole range of flavonoids found only in part in the other Asteridae. The Asteraceae have advanced biosynthetic character states for these compounds (methylation, 6-, 8-substitution, flavones) (Gornall and Bohm, 1978). Only the Rubiaceae, Scrophulariales and Lamiales approach the complexity of the Compositae.

14.8 CONCLUSIONS

In considering flavonoid evolution in the angiosperms, the presence of the 'primitive' characters, ellagic acid and myricetin, and derived characters, such as 2′-hydroxylation, 6- and/or 8-oxygenation, 5-methoxy substitution, isoprenylated flavonoids and biflavonyls, tie the Hamamelidae, Rosidae and Dilleniidae together as roughly comparable intermediate evolutionary groups between the primitive (Magnoliidae) and the advanced (Asteridae). These flavonoid and phenolic capabilities also suggest that they probably are closely related to each other beyond invoking the simple device of 'parallelism' (see Dahlgren, 1980).

The desire to recognize these and other chemical data is reflected in the taxonomic and philosophical conflict between the four or five major angiosperm treatments, where ordinal alliances vary with regularity and the movements are towards the more complex but perhaps more flexible concept of the superorder. Neither approach is optimal since the broad-scale subclass categories leave few options but parallelism, while the alternative fragmentation of the superorder enhances logical localized relationships but may obscure potential and reasonable broad-scale relationships.

If it is accepted that the Hamamelidae, Rosidae and Dilleniidae are very closely related to each other, from whence do they come? Origin of these three subclasses from a conceptual Magnoliid ancestral 'pool', as implied by Cronquist (1981, p. 21), is difficult to accept from a phytochemical standpoint. The list of flavonoid characters that hold these three subclasses together either do not occur, or are rare, in the Magnoliidae. As described above, the flavonoid chemistry of members of the Magnoliidae is either quite simple, so as to give no clue of advancement or primitiveness, or is derived in nature so as to suggest that it is not ancestral to the other three subclasses. The dominance of benzylisoquinoline alkaloids in the Magnoliidae (despite some occurrence in a few Rosidae taxa) also seems to suggest that it is isolated from the other three subclasses as a group, and may represent a distinct parallel line (Giannasi, 1986). Indeed, Ferreira *et al.* (1980) present some circumstantial evidence in *Aniba* (Lauraceae) for a biosynthetic reversal in this genus to phenylpropanoid synthesis in an otherwise tetrabenzylisoquinoline-producing family. Indeed, in terms of the alkaloid chemistry the Magnoliidae find their closest kinship in the Liliidae which also produce aromatic amino acid-derived isoquinoline alkaloids. Interestingly, Gornall *et al.* (1979) also point out that there is considerable similarity in flavonoid patterns between the Magnoliidae and Liliidae in the general lack of ellagic acid and myricetin and concentration on simple quercetin and kaempferol glycosides and some flavones.

This possibility that the magnoliid-like ancestors are not primal to the other subclasses is in conflict with previous fossil pollen records and seemingly logical patterns of pollen development which suggest that the magnoliid-like monosulcate pollen type is the earliest. Is this real or are we looking at an artificial relationship established more by the odds of discovery? For example, recent palynological finds of early Hamamelids by Crane *et al.* (1986) appear to push back the geological origin for this group for being less of a derivative of the Magnoliidae to perhaps a more co-equal status.

The Asteridae, while generally lacking ellagic acid, clearly are related to the Rosidae (and Hamamelidae/Dilleniidae by implication). Myricetin is probably more common than previously thought, and certainly its B-ring methyl ethers occur in the Asteraceae (and probably other taxa as surveys increase). In a host of other flavonoids and unusual substitutions the Asteridae, and especially the Asteraceae, mirror patterns in Rosidae, Hamamelidae and Dilleniidae. The more commonly invoked concept of the advanced nature of the Asteraceae is apparent in increased production of complex *O*-methyl and extra substituted flavonols as well as more emphasis on similarly substituted flavones (Table 14.7).

In fact, the Asteraceae share a number of compounds with the Rubiaceae, some Lamiales and Scrophulariales all of which have been suggested as possible ancestral sources. The Campanulales, in contrast, are so simple in flavonoids as to open to question their role as an ancestral source of the Asteraceae. Oddly enough, the Asteraceae produce nearly all possible flavonoids observed in the dicots, including some not observed in the other Asteridae but which occur in some combination in members of the more intermediate Rosidae. In this sense, a suggested earlier origin for the Compositae closer to the Rosidae rather than as a terminal family of the Asteridae arises

Table 14.7 Distribution of flavonoids in the Asteridae

Taxon Asteridae	Compounds[a] 1	2	3	4	5	6	7	8	9	10	11	12	13	14	15	16	17	18	19	20	21	22	23	24	25	26	27	28	29	30	31	32	33	34	35	36	37
GENTIANALES																																					
Loganiaceae	?		?				+	+													+														?		
Retziaceae							+																														
Gentianaceae		+	+				+										+				+					+	+										
Saccifoliaceae	—		—		—		—																														
Apocynaceae			+	+		+	+	+									+		+	+			+								+	+					
Asclepiadaceae			+	+			+	+																			+				+				+		
SOLANALES																																					
Duckeodendraceae	—		—		—		—																														
Nolanaceae							+																														
Solanaceae		+	+	+	+	+	+	+	+		+			+		+	+*	+								+	+				+		+			+	
Convolvulaceae			+	+	+		+																														
Cuscutaceae							+																									+					
Menyanthaceae							+	+																								+					
Polemoniaceae[b]		+	+		+	+	+	+	+	+							+	+									+										
Hydrophyllaceae			+				+											+								+											
LAMIALES																																					
Lennoaceae	—		—		—		—																														
Boraginaceae[c]		+	+	+			+	+	+							+	+									+											
Verbenaceae[d]		+	+	+			+	+	+	+	+						+	+	+		+					+	+				+	+					
Lamiaceae		+	+		+		+	+	+	+							+	+	+	+	+	+	+	+		+	+				+					+	
CALLITRICHALES																																					
Hippuridaceae			+				+																														
Callitrichaceae																	+																				
Hydrostachyaceae	—		—		—		—																														
PLANTAGINALES																																					
Plantaginaceae							+										+		+		+										+						
SCROPHULARIALES																																					
Buddlejaceae						+	+										+	+	+	+	+																
Oleaceae			+	+	+	+	+										+									+	(+)						+				
Scrophulariaceae		+	+	+	+		+	+	+	+	+						+*	+	+	+	+		+	+		+	+				+		+*		+		
Globulariaceae							+										+	+	+		+																
Myoporaceae							+	+		+						+	+*	+								+					+						
Orobanchaceae		+	+	+	+												+*																				
Gesneriaceae		+	+*	+			+										+	+								+					+				+*		+
Acanthaceae		+	+	+			+										+	+	+	+	+					+	+			*	+		+				
Pedaliaceae		+	+	+	+												+	+	+		+																
Bignoniaceae		+	+	+			+					+					+	+	+	+	+		+	+		+					+						
Mendonciaceae	—		—		—		—																														
Lentibulariaceae																	+	+		+	+											+					

CAMPANULALES																																			
Pentaphragmataceae		+	+		+																														
Sphenocleaceae	—		—		—		—																												
Campanulaceae		+	+	+	+		+	+									+		+																
Stylidiaceae							+																												
Donatiaceae	—		—		—		—																												
Brunoniaceae	—		—		—		—																												
Goodeniaceae		+		+			+		+																										
RUBIALES																																			
Rubiaceae		+	+			+	+	+	+	+	+				+		+	+	+	+	+			+	+	+			*	+	+	*	+		
Theligonaceae																																			
DIPSACALES																																			
Caprifoliaceae[e]			+				+										+	+								+		+			+		+		
Adoxaceae																																			
Valerianaceae[f]							+	+									+	+			+			+		+									
Dipsacaceae		+					+										+	+								+									
CALYCERALES																																			
Calyceraceae																																			
ASTERALES																																			
Asteraceae		+	+	+	+		+	+	+	+	+	+	+	+	+	+	+*	+	+	+	+	+	+	+	+	+	+		+*	+	+	+*		+	+

[a]See Table 14.1 for key to compound identity.
[b]Includes Coloaeceae.
[c]Includes Ehreliaceae, Wellstediaceae.
[d]Includes Avicenniaceae.
[e]Includes Sambucaceae.
[f]Includes Dipiostegiaceae.

again, although suggested ancestral types (Mabry and Bohlmann, 1977) seem unlikely. The lack of adequate surveys in many Asteridae provides few data to support this alternative conclusion.

Last but not least, the Caryophyllidae, at least the Caryophyllales, produce the unusual betalain alkaloid pigments. It is obvious that the Polygonales and Plumbaginales do not belong in this subclass and both Thorne (1983) and Dahlgren (1983) have isolated them or removed them to other taxonomic groups. Within the Caryophyllales, of course, the Caryophyllaceae and Molluginaceae lack these betalains, producing instead typical anthocyanins. All of the families produce anthoxanthins, however, suggesting that the 'loss' of anthocyanins and 'gain' of betalains may be coincidental terminal genetic changes during evolution of the groups (Giannasi, 1979; Giannasi and Crawford, 1986; see discussion on *Aniba*; Gottlieb and Kubitzki, 1981). This would be compatible with the scheme in Cronquist (1981). Indeed, the Caryophyllales are a natural group (Rodman *et al.*, 1984). The cladistic review of Rodman *et al.* (1984) using chemical and non-chemical characters shows no particular relationship between the Caryophyllidae and the Magnoliidae or Dilleniidae, although these workers did not include taxa from the Hamamelidae/Rosidae/Asteridae for effect. The presence of non-betalain Caryophyllaceae and Molluginaceae remain discordant elements within the Caryophyllales on chemical bases. Regrettably, there is no unequivocal relationship apparent for the Caryophyllidae. On the basis of several flavonoid types, Young (1981) supports a closer relationship of the Caryophyllales to the Dilleniidae (Theiflorae of Thorne, 1983). However, the absence of ellagic acid and rare occurrence of myricetin in this order suggest an equally possible relationship of the Caryophyllales to the Magnoliidae (Dahlgren, 1983). If the betalains are treated as tyrosine-derived alkaloids (i.e. chromoalkaloids; Mabry, 1977) rather than as an isolated pigment group, then a closer relationship to the aromatic amino acid-derived benzylisoquinoline-producing Magnoliidae is not unrealistic.

However, on the basis of the conclusions drawn here, an evolutionary tree for the dicots may be projected as shown in Fig. 14.1. The actual relationships suggested between the majority of Cronquist's (1981) subclasses are not unreasonable and are actually accepted to a greater or lesser degree in many of the current angiosperm treatments (Thorne, 1983; Dahlgren, 1983), which more actively integrate elements of Cronquist's subclasses into various superorders. The major conflict to a great extent remains the taxonomic categories used and, on the basis of the chemical data, the question of the temporal and primal status of the Magnoliidae.

ACKNOWLEDGEMENTS

I wish to thank Dr Samuel B. Jones and Mrs Nance Coile for their comments on this paper. My thanks also to Michael Moore and Jeff Rettig for their critical reading of the manuscript and Sally Pedigo and Ginger Vickery for preparation of the manuscript. This work was supported in part by NSF Grant BSR-8120515.

REFERENCES

Adzet, T. and Martinez, F. (1981), *Biochem. Syst. Ecol.* **9**, 293.
Aguinagalde, I. and del Pero Martinez, M.A. (1982), *Phytochemistry* **21**, 2875.
Averett, J.E. and Boufford, D.E. (1985), *Syst. Bot.* **10**, 363.
Averett, J.E. and Graham, S.A. (1984), *Ann. Missouri Bot. Gard.* **71**, 853.
Averett, J.E. and Raven, P.H. (1984), *Ann. Missouri Bot. Gard.* **71**, 30.
Auanoglu, E., Ulubelen, A., Clark, W.D., Brown, G.K., Kerr, R.R. and Mabry, T.J. (1981), *Phytochemistry* **21**, 1715.
Ballard, R. (1975), Ph.D. Dissertation, University of Iowa.
Ballard, R. (1986), *Am. J. Bot.* **73**, 1452.
Ballard, R.E., McClure, J.W., Eshbaugh, W.H. and Wilson, K.G. (1970), *Am. J. Bot.* **57**, 225.
Bate-Smith, E.C. (1961), *Bot. J. Linn. Soc.* **58**, 39.
Bate-Smith, E.C. (1962), *Bot. J. Linn. Soc.* **58**, 95.
Bate-Smith, E.C. (1968), *Bot. J. Linn. Soc.* **60**, 325.
Bate-Smith, E.C. (1973), *Bot. J. Linn. Soc.* **67**, 347.
Bate-Smith, E.C. (1976), *Biochem. Syst. Ecol.* **4**, 13.
Bate-Smith, E.C. and Richens, R.H. (1973), *Biochem. Syst. Ecol.* **1**, 141.
Bohm, B.A. (1982). In *The Flavonoids – Advances in Research* (eds J.B. Harborne and T.J. Mabry), Chapman and Hall, London, p. 313.
Bohm, B.A. and Bhat, U.G. (1985), *Biochem. Syst. Ecol.* **13**, 437.
Bohm, B.A. and Glennie, C.W. (1971), *Canad. J. Bot.* **49**, 1799.
Bohm, B.A. and Ornduff, R. (1978), *Madroño* **25**, 39.
Bohm, B.A. and Ornduff, R. (1981), *Syst. Bot.* **6**, 15.
Bohm, B.A. and Stuessy, T.F. (1981a), *Phytochemistry* **20**, 1053.
Bohm, B.A. and Stuessy, T.F. (1981b), *Phytochemistry* **20**, 1573.
Bohm, B.A. and Wilkins, C.K. (1978a), *Brittonia* **30**, 327.
Bohm, B.A. and Wilkins, C.K. (1978b), *Phytochemistry* **56**, 1174.
Bohm, B.A., Collins, F.E. and Bose, R. (1977a), *Phytochemistry* **16**, 1205.
Bohm, B.A., Collins, F.E. and Bose, R. (1977b), *Canad. J. Bot.* **49**, 1799.
Bohm, B.A., Banek, H.M. and Maze, J.R. (1984), *Syst. Bot.* **9**, 324.
Bohm, B.A., Nicholls, K.W. and Bhat, V.G. (1985), *Biochem. Syst. Ecol.* **13**, 441.
Bohm, B.A., Nicholls, K.W. and Ornduff, R. (1986), *Am. J. Bot.* **73**, 204.
Burret, F., Rabesa, Z., Zandonella, P. and Voirin, B. (1981), *Biochem. Syst. Ecol.* **9**, 257.
Burret, F., Lebreton, Ph. and Voirin, B. (1982), *J. Nat. Prod.* **45**, 687.
Castro, C.O. and Valverde, R.V. (1985), *Phytochemistry* **24**, 367.
Clark, M.N. and Bohm, B.A. (1979), *Bot. J. Linn. Soc.* **79**, 249.
Challice, J.S. (1972), *Phytochemistry* **11**, 3015.
Challice, J.S. (1973), *Phytochemistry* **12**, 1095.
Challice, J.S. and Williams, A.H. (1968), *Phytochemistry* **7**, 1781.
Collins, F.W. and Bohm, B.A. (1974), *Canad. J. Bot.* **52**, 307.
Coradin, L., Giannasi, D.E. and Prance, G.T. (1985), *Brittonia* **37**, 169.

Crawford, D.J. (1974), *Brittonia* **26**, 74.
Crawford, D.J. and Evans, K.A. (1978), *Brittonia* **30**, 313.
Crawford, D.J. and Hawksworth, F.G. (1979), *Brittonia* **31**, 212.
Crawford, D.J. and Mabry, T.J. (1978), *Biochem. Syst. Ecol.* **6**, 189.
Crawford, D.J. and Smith, E.B. (1983a), *Am. J. Bot.* **70**, 355.
Crawford, D.J. and Smith, E.B. (1983b), *Bot. Gaz.* **144**, 577.
Crawford, D.J. and Smith, E.B. (1984), *Madroño* **31**, 1.
Crawford, D.J. and Smith, E.B. (1985), *Biochem. Syst. Ecol.* **13**, 115.
Crawford, D.J. and Stuessy, T.F. (1981), *Am. J. Bot.* **68**, 107.
Crane, P.R., Friis, E.M. and Pedersen, K.R. (1986), *Science* **232**, 852.
Cronquist, A. (1955), *Am. Midl. Nat.* **53**, 478.
Cronquist, A. (1977), *Brittonia* **29**, 137.
Cronquist, A. (1981), *An Integrated System of Classification of Flowering Plants*. Columbia University Press, New York.
Dahlgren, R. (1977), *Plant Syst. Evol. Suppl.* **1**, 253.
Dahlgren, R. (1980), *Bot. J. Linn. Soc.* **80**, 91.
Dahlgren, R. (1983), *Nord. J. Bot.* **3**, 119.
Delendick, T.J. (1981), Ph.D. Dissertation, City University of New York, 693 pp.
Delendick, T.J. (1984), *Brittonia* **36**, 49.
Delendick, T.J. (1985), *Taxon* **34**, 96.
Delendick, T.J. (1987), *Mem. N.Y. Bot. Gard.*, (in press).
Denton, M.F. and Kerwin, J.L. (1980), *Canad. J. Bot.* **58**, 902.
Doyle, J.J. (1983), *Am. J. Bot.* **70**, 1085.
Egger, K. and Reznik, H. (1961), *Planta* **57**, 239.
Ferreira, Z.S. Gottlieb, O.R. and Rague, N.F. (1980), *Biochem. Syst. Ecol.* **8**, 51.
Gershenzon, J. and Mabry, T.J. (1983), *Nord. J. Bot.* **3**, 5.
Giannasi, D.E. (1975a), *Mem. N. Y. Bot. Gard.* **26**, 1.
Giannasi, D.E. (1975b), *Bull. Torrey Bot. Club* **102**, 404.
Giannasi, D.E. (1978), *Taxon* **27**, 331.
Giannasi, D.E. (1979), *Bot. Rev.* **44**, 399.
Giannasi, D.E. (1986), *Ann. Missouri Bot. Gard.* **73**, 417.
Giannasi, D.E. and Crawford, D.J. (1986). In *Evolutionary Biology*, (eds M.K. Hecht, B. Wallace and G.T. Prance), Plenum Press, New York, Vol. 20, p. 25.
Giannasi, D.E. and Niklas, K.J. (1977), *Science* **197**, 765.
Giannasi, D.E. and Niklas, K.J. (1981), *Am. J. Bot.* **68**, 762.
Giannasi, D.E. and Rogers, C.M. (1971), *Brittonia* **22**, 163.
Gomes, M.R.C., Gottlieb, O.R., Marini Bettolo, G.B., Delle Monache, F. and Polhill, R.M. (1981), *Biochem. Syst. Ecol.* **9**, 129.
Gornall, R.J. and Bohm, B.A. (1978), *Syst. Bot.* **3**, 353.
Gornall, R.J., Bohm, B.A. and Dahlgren, R. (1979), *Bot. Notiser* **132**, 1.
Gottlieb, O.R. and Kubitzki, K. (1981), *Biochem. Syst. Ecol.* **9**, 5.
Graham, S.A. and Averett, J.E. (1984), *Ann. Missouri. Bot. Gard.* **71**, 855.
Grayer-Barkmeijer, R.J. (1978), *Biochem. Syst. Ecol.* **6**, 131.
Grieve, C.M. and Scora, R.W. (1980), *Syst. Bot.* **5**, 39.
Grudzinskaya, I.A. (1965), *Bot. Zh.* **52**, 1723.
Grund, C. and Jensen, V. (1981), *Plant. Syst. Evol.* **137**, 1.
Gurni, A.A. and Kubitzki, K. (1981), *Biochem. Syst. Ecol.* **9**, 109.
Gurni, A.A. and König, W.A. and Kubitzki, K. (1981), *Phytochemistry* **20**, 1057.
Haber, E. (1983), *Syst. Bot.* **8**, 277.
Hänsel, R. and Hörhammer, I.L. (1954), *Arch. Pharm.* **287**, 117.
Harborne, J.B. (1967a), *Comparative Biochemistry of the Flavonoids*, Academic Press, London.
Harborne, J.B. (1967b), *Phytochemistry* **6**, 1415.
Harborne, J.B. (1967c), *Phytochemistry* **6**, 1643.
Harborne, J.B. (1975). In *The Flavonoids* (eds J.B. Harborne, T.J. Mabry and H. Mabry), Chapman and Hall, London, p. 1056.
Harborne, J.B. (1977), *Biochem. Syst. Ecol.* **5**, 7.
Harborne, J.B. (1986), *Phytochemistry* **25**, 1641.
Harborne, J. B. and Green, P.S. (1980), *Bot. J. Linn. Soc.* **81**, 155.
Harborne, J.B. and King, L. (1976), *Phytochemistry* **4**, 111.
Harborne, J.B. and Mabry, T.J. (eds) (1982). *The Flavonoids – Advances in Research*, Chapman and Hall, London.
Harborne, J.B. and Williams, C.A. (1971), *Phytochemistry* **10**, 2727.
Harborne, J.B. and Williams, C.A. (1973), *Bot. J. Linn. Soc.* **66**, 37.
Harborne, J.B. Girija, A.R., Devi, H.M. and Lakshmi, N.K.M. (1983), *Phytochemistry* **22**, 2741.
Hart, C.R. (1979). *Syst. Bot.* **4**, 130.
Hegnauer, R. (1986), *Phytochemistry* **25**, 1519.
Heywood, V.H., Harborne, J.B. and Turner, B.L. (eds) (1977). *The Biology and Chemistry of the Compositae*. 2 Vols. Academic Press, London.
Hostettmann-Kaldas, M., Hostettmann, K. and Sticher, O. (1981), *Phytochemistry* **20**, 443.
Hunt, D. (1985), *Am. J. Bot.* **72**, 932.
Jay, M. (1968), *Taxon* **17**, 136.
Jensen, U. and Greven, B. (1984), *Taxon* **33**, 563.
John, J. and Kolbe, K.P. (1980), *Biochem. Syst. Ecol.* **8**, 241.
Johnson, G.P. and Beckmann, R.L. (1986), *Am. J. Bot.* **73**, 769.
Kaouadji, M., Ravanel, P. and Mariotte, A.-M. (1986), *J. Nat. Prod.* **49**, 153.
King, B.L. (1977), *Am. J. Bot.* **64**, 350.
King, B.L. (1986), *Syst. Bot.* **11**, 403.
King, B.L. and Jones, S.B. (1982), *Bull. Torrey Bot. Club.* **109**, 279.
Kolbe, K.P. and John, J. (1980), *Biochem. Syst. Ecol.* **8**, 249.
Kolterman, D.A. and Breckon, G.J. (1982), *Syst. Bot.* **7**, 178.
Kolterman, D.A., Breckon, G.J. and Kowal, R.R. (1984), *Syst. Bot.* **9**, 22.
Kubitzki, K. and Gottlieb, O.R. (1984a), *Taxon* **33**, 375.
Kubitzki, K. and Gottlieb, O.R. (1984b), *Acta. Bot. Neerl.* **33**, 457.
Kubitzki, K. and Reznik, H. (1966), *Beitr. Biol. Pflanzen.* **42**, 445.
LaDuke, J. (1982), *Am. J. Bot.* **69**, 784.
Lebreton, P. (1965), *Bull. Soc. Bot. Fr.* **111**, 80.
Levy, M. (1983), *Syst. Bot.* **8**, 118.
Lincoln, D.E. (1980), *Biochem. Syst. Ecol.* **8**, 397.
Luteyn, J.L., Harborne, J.B. and Williams, C.A. (1980), *Brittonia* **32**, 1.
Mabry, T.J. (1977), *Ann. Missouri Bot. Gard.* **64**, 210.
Mabry, T.J. and Bohlmann, F. (1977). In *The Biology and Chemistry of the Compositae* (eds V.H. Heywood, J.R. Harborne and B.L. Turner). 2 Vols. Academic Press, London, p. 1098.
Mabry, T.J., Abdel-Baset, Z., Padolina, W.G. and Jones, S.B. (1975), *Biochem. Syst. Ecol.* **2**, 185.
Mastenbroek, O., Hogeweg, P., van Brederode, J. and van Nigtevecht, G. (1983), *Biochem. Syst. Ecol.* **11**, 91.
McDougal, K.M. and Parks, C.R. (1986), *Biochem. Syst. Ecol.* **14**, 291.
Mears, J.A. (1973), *Brittonia* **25**, 385.
Miller, J.M. (1981), *Syst. Bot.* **6**, 27.
Moore, M.O. and Giannasi, D.E. (1987), *Biochem. Syst. Ecol.* **15**, 79.

Nicholls, N.W. and Bohm, B.A. (1982), *Biochem. Syst. Ecol.* **10**, 225.
Nicholls, K.W., Bohm, B.A. and Ornduff, R. (1985), *Biochem. Syst. Ecol.* **13**, 261.
Nicoletti, M., Goulart, M.O.F., De Lima, R.A., Goulart, A.E., Della Monache, F. and Marini-Bettolo, G.B. (1984), *J. Nat. Prod.* **47**, 953.
Niklas, K.J., and Giannasi, D.E. (1978), *Am. J. Bot.* **65**, 943.
Niklas, K.J., Giannasi, D.E. and Baghai, N.L. (1985), *Biochem. Syst. Ecol.* **13**, 1.
Pacheo, P., Crawford, D.J., Stuessy, T.F. and Silva, M. (1985), *Am. J. Bot.* **72**, 989.
Parker, W.H. and Bohm, B.A. (1979), *Am. J. Bot.* **66**, 191.
Parks, C.R., Ezell, W.L., Williams, D.E. and Dreyer, D.L. (1975), *Bull. Torrey Bot. Club.* **102**, 350.
Patterson, R. (1984), *Syst. Bot.* **9**, 263.
Pereira, M.O. da S., Fantine, E.C. and de J.R. Sousa (1982), *Phytochemistry* **21**, 48.
Petersen, F.P. and Fairbrothers, D.E. (1983). *Syst. Bot.* **8**, 134.
Peterson, F.P. and Fairbrothers, D.E. (1985), *Bull. Torrey Bot. Club* **112**, 43.
Polechko, M.H. and Clarkson, R.B. (1986), *Biochem. Syst. Ecol.* **14**, 33.
Rees, S. and Harborne, J.B. (1984), *J. Linn. Soc. Bot.* **89**, 313.
Richardson, M. (1978), *Biochem. Syst. Ecol.* **6**, 283.
Rodman, J.E., Oliver, M.K. Nakamura, R.R., McClammer, Jr., J.V. and Bledsoe, A.H. (1984), *Syst. Bot.* **9**, 297.
Saleh, N.A.M. and El-Hadidi, M.N. (1977), *Biochem. Syst. Ecol.* **5**, 121.
Saleh, N.A.M. and El-Lakany, M.H. (1979), *Biochem. Syst. Ecol.* **7**, 13.
Saleh, N.A.M. and Boulos, L., El-Negoumy, S.I. and Abdalla, M.F. (1982a), *Biochem. Syst. Ecol.* **10**, 33.
Saleh, N.A.M., El-Hadidi, M.N. and Ahmad, A.A. (1982b), *Phytochemistry* **10**, 313.
Saleh, N.A.M., El-Karemy, Z.A.R., Mansour, R.M.A. and Fayed, A.A.A. (1983a), *Phytochemistry* **22**, 2501.
Saleh, N.A.M., El-Negoumy, S.I. and Hosni, H.A. (1983b), *Phytochemistry* **22**, 1417.
Scogin, R. (1980), *Biochem. Syst. Ecol.* **8**, 273.
Smith, D.M., Glennie, C.Wm., Harborne, J.B. and Williams, C.A. (1977), *Biochem. Syst. Ecol.* **5**, 107.
Smith, E.B. and Crawford, D.J. (1981), *Bull. Torrey Bot. Club* **108**, 7.
Smith, G.C. (1981), Ph.D. Dissertation, University of Georgia, Athens.
Soltis, D.E. (1980), *Biochem. Syst. Ecol.* **8**, 149.
Soltis, D.E. and Bohm, B.A. (1984), *Syst. Bot.* **9**, 441.
Soltis, D.E. and Bohm, B.A. (1986), *Syst. Bot.* **11**, 20.
Soltis, P.M. and Soltis, D.E. (1986), *Syst. Bot.* **11**, 32.
Soltis, D.E., Bohm, B.A. and Nesom, G.L. (1983), *Syst. Bot.* **8**, 15.
Steinharter, T.-P., Cooper-Driver, G.A. and Anderson, G.J. (1986), *Biochem. Syst. Ecol.* **14**, 299.
Stewart, R.N., Asen, S., Massie, D.R. and Norris, K.H. (1979), *Biochem. Syst. Ecol.* **7**, 281.
Stewart, R.N., Asen, S., Massie, D.R., Norris, K.H. (1980), *Biochem. Syst. Ecol.* **8**, 119.
Stuessy, T.F. and Crawford, D.J. (1983), *Plant. Syst. Evol.* **143**, 83.
Stutts, J.G. (1981), *Rhodora* **83**, 385.
Swain, T. (1975). In *The Flavonoids* (eds J.B. Harborne, T.J. Mabry and H. Mabry), Chapman and Hall, London, p. 1096.
Takhtajan, A.L. (1980), *Bot. Rev.* **46**, 225.
Thorne, R.F. (1976). In *Evolutionary Biology* (eds M.K. Hecht, W.C. Steere and J.W. Wallace), Plenum Press, New York, Vol. 9, p. 35.
Thorne, R.F. (1981). In *Phytochemistry and Angiosperm Phylogeny* (eds D.A. Young and D.S. Seigler), Praeger, New York, p. 233.
Thorne, R.F. (1983), *Nord. J. Bot.* **3**, 85.
Timmermann, B.N., Graham, S.A. and Mabry, T.J. (1981), *Phytochemistry* **20**, 1762.
Umber, R.E. (1980), *Am. J. Bot.* **67**, 935.
Valant-Vetschera, K. (1982), *Phytochemistry* **21**, 1067.
Valant-Vetschera, K. (1985), *Biochem. Syst. Ecol.* **13**, 15.
Wannan, B.S., Waterhouse, J.T., Gadek, P.A. and Quinn, C.J. (1985), *Biochem. Syst. Ecol.* **13**, 105.
Warnock, M.J. (1981), *Syst. Bot.* **6**, 15.
Waterman, P.G. and Hussain, R.A. (1983), *Bot. J. Linn. Soc.* **86**, 227.
Whalen, M.D. (1978), *Syst. Bot.* **3**, 257.
Wilken, D.H., Smith, D.M., Harborne, J.B. and Glennie, C.Wm. (1982), *Biochem. Syst. Ecol.* **10**, 239.
Wilkins, C.K. and Bohm, B.A. (1976), *Canad. J. Bot.* **54**, 2133.
Williams, A.H. (1982), *Bot. J. Linn. Soc.* **84**, 31.
Williams, C.A. and Harvey, W.J. (1982), *Phytochemistry* **21**, 329.
Williams, C.A., Demissie, A. and Harborne, J.B. (1983), *Biochem. Syst. Ecol.* **11**, 221.
Wofford, B.E. (1974), *Biochem. Syst. Ecol.* **3**, 35.
Wollenweber, E. (1975a), *Biochem. Syst. Ecol.* **3**, 47.
Wollenweber, E. (1975b), *Biochem. Syst. Ecol.* **3**, 35.
Wollenweber, E. and Dietz, V.H. (1981), *Phytochemistry* **20**, 869.
Young, D.A. (1976), *Syst. Bot.* **1**, 149.
Young, D.A. (1979), *Am. J. Bot.* **66**, 502.
Young, D.A. (1981). In *Phytochemistry and Angiosperm Phylogeny* (eds D.A. Young and D.S. Seigler), Praeger, New York, p. 205.
Young, D.A. (1982), *Biochem. Syst. Ecol.* **10**, 21.
Young, D.A. (1983), *Biochem. Syst. Ecol.* **11**, 209.
Young, D.A. and Sterner, R.W. (1981), *Biochem. Syst. Ecol.* **9**, 185.

15

Distribution and evolution of flavonoids in the monocotyledons

CHRISTINE A. WILLIAMS and JEFFREY B. HARBORNE

15.1 INTRODUCTION

The Monocotyledoneae comprise between 50 and 100 families, according to which system of classification one adopts. The treatment of Engler and Prantl (1930) recognizes 45 families but there has been a tendency in recent years to split off segregates from some of the larger more heterogeneous families, such as the Liliaceae. Dahlgren *et al.* (1985) in the most recent system regard the monocotyledons as containing 93 families. These range in size from two of the largest in the whole of the angiosperms, the Orchidaceae (20 000 spp.) and the Gramineae (10 000 spp.) to small groups like the Strelitziaceae (seven spp.) and the Joinvilleaceae (two spp.). Many families are economically important (e.g. the Gramineae, Palmae) while others are known for their ornamental value (e.g. Amaryllidaceae, Iridaceae). While anthocyanins are conspicuously present as the colouring matters of petals, fruits and leaves in many families, other flavonoids are universally found in leaf, pollen or inflorescence, although their presence is only revealed following extraction and detection.

The widely accepted separation of the monocots from the dicots, based on cotyledon number in the seedling, is to some extent a matter of convenience, since the two groups of plants are broadly similar in most other characteristics. There are a few other morphological differences, but these are not completely clearcut. For example, all monocots investigated have been shown to possess distinctive triangular protein bodies in their sieve-tube plastids, but these do occur exceptionally in the dicots in two genera of the Aristolochiaceae (Dahlgren *et al.*, 1985). The generally prevailing view that the monocots arose by evolution from the primitive dicots also emphasises similarities rather than differences and is based on the many morphological links that exist between monocots and the dicot families within the Ranalian complex.

It is not surprising, therefore, to find that when one compares the flavonoids of monocots with those of dicots there are only a few differences. Not only are the same widespread classes of flavonoid generally in both monocots and dicots, but also many of the same compounds can be isolated equally from plants of both types. There may be differences, however, in the frequencies of occurrence. Isoflavonoids, for example, are common in the Leguminosae and have been detected in nine other dicot families; in the monocots, they are restricted to two families, the Iridaceae and Liliaceae, and here they are confined to single genera (see Chapter 5). Again, biflavonoids have more family records in the dicots than in the monocots (see Section 15.15). On the other hand, the flavone tricin is much more widespread in monocot species than among dicotyledons.

The Flavonoids. Edited by J.B. Harborne
Published in 1988 by Chapman and Hall Ltd,
11 New Fetter Lane, London EC4P 4EE.

Flavonoid differences between the two major types of angiosperm thus appear to be rather subtle and complex in nature. One significant divergence is in the distribution and kind of tannins that are found. In the dicots, both hydrolysable (gallic acid based) and condensed (flavolan) tannins are correlated in their distribution with the woody habit and they are rare in or absent from the herbaceous groups (Bate-Smith, 1962). By contrast, in the monocots, condensed tannins are relatively widely distributed, in spite of the fact that the tree habit is rare among monocots and that most plants are herbs. This is underlined by the fact that even in the Gramineae, which is almost entirely herbaceous, some 3% of species contain tannin.

The type of tannin present is also different. Hydrolysable tannins are entirely absent from monocots, although they have been recorded in over 30 dicot families. There are also variations in the kind of condensed tannin found in the two plant types. Thus proanthocyanidins based on a mixture of epicatechin-4 and *ent*-epicatechin-4 units are characteristically found only among monocot groups (Ellis *et al.*, 1983).

While it is interesting to recognize that there are minor differences in flavonoid patterns between monocots and dicots, it is more rewarding to examine the variations that occur between the various families that make up the monocots. Such a study has been one of the objects of a long-term research programme carried out in the authors' laboratory. As a result, it is clear that most families do have a characteristic flavonoid profile. It is our purpose here to consider briefly the compounds that have been recorded family by family and then to compare the overall distributions that emerge among the monocots generally.

No-one writing about the flavonoids of the monocotyledons should fail to acknowledge the pioneering surveys of Bate-Smith. He first established that the common constituents previously recorded in the dicots were widely present in the monocots as well (Bate-Smith, 1968). Thus, flavonols, glycoflavones and proanthocyanidins were regularly recorded as leaf constituents in most families. In a later review, Bate-Smith (1969) discussed the taxonomic significance of his findings. Other workers, of course, have also contributed greatly to our present knowledge of flavonoids in these plants; for example Wiermann (1968) recorded flavonols in the pollen of 45 spp.

Biosynthetic studies in monocots have shown that an isozyme of dehydroquinate hydrolyase, which mediates in the shikimate pathway leading to flavonoids, is restricted to monocot plants (Boudet *et al.*, 1977). It occurs mainly in representatives of the Juncaceae, Cyperaceae and Poaceae. Other investigations of later stages in the pathway (e.g. of chalcone synthase in *Tulipa*) have not yet indicated any enzymological differences, but these may well appear when a larger number of enzymes have been compared (cf. Chapter 11).

For a detailed account of flavonoids in the monocotyledons, the reader is referred to Hegnauer (1963), and a later volume (Hegnauer, 1986), which updates the literature to 1984. Brief earlier reviews of flavonoid patterns in the families of the monocots can be found in Harborne (1967, 1982) and in Dahlgren and Clifford (1982).

15.2 ARALES

The order Arales of Dahlgren *et al.* (1985) contains two clearly related families, the Araceae and the Lemnaceae. The Araceae, a largely tropical/subtropical family of some 110 genera and over 2000 species is very diverse in its vegetative habit. In Bogner's (1978) modification of Engler's (1905–1920) classic system, eight subfamilies and 31 tribes are recognized.

Anthocyanins are largely responsible for the characteristic scarlet, red, purple and brown colours in the spathe and spadix of many flowers and in the leaves or petioles of some taxa but they are less frequent in the brightly coloured fruits, where carotenoids are the main pigments (Williams *et al.*, 1981). The family has not been extensively studied for its chemical constituents but there have been two major flavonoid surveys. Thus, Bate-Smith (1968) examined 24 members for their leaf flavonoids and concluded that the family showed a very diverse phenolic pattern. However, in a more recent survey of 59 species Williams *et al.* (1981) reported a simple flavonoid profile in terms of both leaf constituents and anthocyanin pigments. Flavone *C*-glycosides and proanthocyanidins are the most characteristic leaf components. However, in the subfamily Calloideae, subtribe Symplocarpeae, flavonols replace glycoflavones as major leaf components but are otherwise infrequent and usually co-occur with flavone *C*-glycosides. Simple flavones (luteolin and chrysoeriol) have been found only in the subtribes Arinae and Cryptocoryninae and subfamily Aroideae, and flavonoid sulphates were detected in only four taxa. Three unusual methylated glycoflavones: 6,8-di-*C*-arabinosylapigenin 7,4′-dimethyl ether, 2″-*O*-glucosyl-6-*C*-arabinosylapigenin 7,4′-dimethyl ether and 2″-*O*-(caffeyl)glucosyl-6-*C*-arabinosylapigenin 7,4′-dimethyl ether were identified in leaves of *Asterostigma riedelianum* (Schott) O. Kuntze (Markham and Williams, 1980). A more detailed examination of flavone *C*-glycosides in other Araceae could prove rewarding and possibly reveal more variation in what appears otherwise to be a uniform family profile.

The Lemnaceae is a much smaller family than the Araceae, containing only four genera and 30 species. The plants, which have undergone considerable morphological reduction, are quite small and consist mainly of fronds of 'duckweed', which float on water. The flavonoid chemistry has been exhaustively surveyed (McClure and Alston, 1966) and glycosylflavones emerge as the major constituents, occurring in 27 of the 30 taxa. Kaempferol and quercetin glycosides occur in all genera except *Lemna*, while flavone glycosides based on apigenin and luteolin are found in all genera except *Wolfiella*. Two anthocyanins, cyanidin 3-glucoside and petunidin 3, 5-diglucoside, have been characterized but these are restricted to a few species of *Spirodela* and *Lemna*. The overall pattern is a simple one. It has been suggested that the Lemnaceae originated from an ancestral species within the Araceae which is similar to the present day *Pistia stratiotes* L. *Pistia* is also of an aquatic habit and its leaf flavonoids, indeed, are identical to those found in the Lemnaceae (Zennie and McClure, 1977).

Overall, the Arales are unusual among the monocots in the simplicity and relative uniformity of the flavonoids. Methylated compounds are rare and 6- or 8-hydroxylated flavonoids are absent. The predominance of flavone *C*-glycosides, often considered to be retained primitive characters, and the frequency of proanthocyanidins suggest that these two families are chemically 'primitive' within the monocots.

15.3 BROMELIACEAE

The Bromeliaceae is a predominantly tropical family of some 45 genera and 1900 taxa, all New World except for one *Pitcairnia* species from West Africa. Three subfamilies are usually recognized, as in the system of Mez (1965): the Bromelioideae, Pitcairnioideae and Tillandsioideae. The one major leaf flavonoid survey (Williams, 1978), showed that the family is unique amongst the monocotyledons in the frequency and variety of flavonoids with extra hydroxylation or *O*-methylation at the 6 position. Simple flavonols and flavones occur throughout but the rarer flavonoids are more restricted. Thus, 6-hydroxyflavones were found in both the Pitcairnioideae and Tillandsioideae while patuletin, gossypetin and methylated 6-hydroxymyricetin derivatives were detected only in the Tillandsioideae. 6-Hydroxymyricetin 3, 6, 3′, 5′-tetramethyl ether, first characterized from *Tillandsia usneoides* (L.) L. (Lewis and Mabry, 1977), was found together with a new ether, 6-hydroxymyricetin 6, 3′, 5′-trimethyl ether, in the same species by Williams (1978). The latter ether was also found in *T. fasciculata* Swartz but 6-hydroxymyricetin derivatives were not detected in the remaining 15 *Tillandsia* species surveyed (Williams, 1978). However, another unusual flavonol, 6-hydroxykaempferol 3, 6, 7, 4′-tetramethyl ether, has more recently been identified from *T. urticulata* L. leaves (Ulubelen and Mabry, 1982) together with 6-hydroxyluteolin 6, 3′-dimethyl ether (jaceosidin) and 6, 7, 3′-trimethyl ether (cirsilineol). This is the first report of 6-hydroxykaempferol and methylated 6-hydroxyluteolin derivatives in the Tillandsioideae but Raffauf *et al.* (1981) previously identified 6-hydroxykaempferol 3, 6, 7-trimethyl ether (penduletin) together with quercetagetin 3, 6, 7, 4′-tetramethyl ether (casticin) and scutellarein 6, 7-dimethyl ether (cirsimaritin) in *Bromelia pinquin* L. (Bromelioideae). In *Aechmea glomerata* Hook. and *Billbergia vittata* Brongn. ex Morel, also in the Bromelioideae, apigenin and luteolin glycosides were found to be the major leaf constituents but tricin and quercetin glycosides were also detected (Ashtakala, 1975). This is the only record so far of tricin in the Bromeliaceae. Simple myricetin glycosides were found only in the subfamily Bromelioideae (Williams, 1978) and their presence there together with the near absence of 6-hydroxy flavonoids could indicate the more primitive condition of this subfamily.

However, the flavonoid results as a whole confirm the morphological view that the Bromeliaceae occupies an isolated position in the monocots. This is emphasized by the discovery of unique anthocyanins with a 3, 5, 3′-trisubstitution pattern (Saito and Harborne, 1983). In a survey of the leaf, bract, flower and fruit pigments of 34 taxa (16 genera) these workers showed that the dominant anthocyanin was cyanidin 3, 5, 3′-triglucoside. This pigment was found to be particularly abundant in the Bromelioideae, for example in the pineapple, *Ananas comosus* (L.) Merrill. Three other new cyanidin glycosides, the 3, 3′-diglucoside, the 3-rutinoside-3′-glucoside and the 3-rutinoside-5, 3′-diglucoside were also detected. All these novel anthocyanins are scarlet rather than crimson in colour because a glucose residue in the 3′ position causes a hypsochromic shift in the visible spectrum. The production of these pigments could be the result of natural selection for floral coloration preferred by bird pollinators. The presence of cyanidin 3′-glycosides also indicates some affinity with the Commelinaceae, where an acylated cyanidin 3, 7, 3′-triglucoside is widespread. The common 3, 5-diglucosides of cyanidin, peonidin and malvidin often accompany the new pigments but pelargonidin derivatives were found in only three taxa.

15.4 COMMELINACEAE

The tropical and subtropical Commelinaceae, comprising some 7000 species and 50 genera, is chiefly valued for its ornamental members (e.g. *Rhoeo*, *Tradescantia* and *Zebrina* species). The leaf flavonoids of 152 species have been analysed and the family appears to be generally uniform in pattern (Del Pero Martinez and Swain, 1985). *C*-Glycosylflavones are predominant (in 78% of species) with flavonol glycosides, mainly based on quercetin, in a minority (28%) of species. The glycosylflavones have been identified in *Gibasis schiedeana* (Kunth) D.R. Hunt, where their distribution is related to ploidy level, and mono-*C*-glycosides (e.g. isovitexin), mono-*C*-glycoside *O*-glycosides (iso-orientin rhamnoside) and di-*C*-glycosides (e.g. vicenin-3, schaftoside) have been characterized (Del Pero Martinez and Swain, 1977).

Two more distinctive characters present are 6-hydroxyluteolin (mainly in *Setcreasia* and *Tradescantia*) and tricin, but these are both quite rare in the family. The former flavone provides a link with the Cyperaceae, since 6-hydroxyluteolin is found in a few South American sedges, and the latter a link with the Cyperaceae and Poaceae, where tricin is common. A sulphated flavonoid has been reported in *Tradescantia hirta* D.R. Hunt but the character otherwise seems to be absent from the family.

The uniformity in leaf flavonoid pattern does not extend to leaf and petal anthocyanins, since several distinctive glycosidic patterns are present (Stirton and Harborne, 1980; Goto, 1984; Saito and Harborne, unpublished results). Four divisions are apparent based on anthocyanin chemistry. The genus *Commelina* is marked off, for example, by the presence of malonylawobanin, the 3-(*p*-coumarylglucoside)-5-(malonylglucoside) of delphinidin, which occurs *in vivo* as a magnesium complex with a flavone co-pigment. By contrast, most other genera analysed (e.g. *Gibasis, Zebrina*) contain triacylated cyanidin 3,7,3′-triglucosides, with either caffeic or caffeic and ferulic acids as acylating substituents. Yet other acylated pigments with more than three sugars occur in *Callisia, Rhoeo* and in some *Tradescantia* species. Finally, simple 3-glycosides of cyanidin and/or peonidin are found in *Polyspatha paniculata* Benth., *Bufforestia obovata* Brenan and *Stanfieldella oligantha* (Mildbr.) Brenan.

In Dahlgren *et al.* (1985), Commelinaceae are joined with the Mayaceae, Rapataceae, Xyridaceae and Eriocaulaceae in the order Commelinales. Unfortunately, too little is known about the flavonoids of most of these other families to be able to make comparisons. However, the Eriocaulaceae (six spp. of *Eriocaulon*) have been shown to contain quercetagetin and patuletin in the leaves (Bate-Smith and Harborne, 1969), so that it would seem to be quite different.

An attempt to link Commelinaceae on the basis of their leaf flavonoids with other families that they have sometimes been associated with by cladistic procedures (Del Pero Martinez, 1985) failed to show any close relative. However, the rich presence of *C*-glycosylflavones and the occasional occurrence of tricin, 6-hydroxyluteolin and sulphated flavones would appear to link Commelinaceae, at least distantly, with the Poaceae and Cyperaceae. Overall though the Commelinaceae are well separated from associated taxa by their flavonoids and this is reinforced by the distinctive nature of the anthocyanins found in leaf or petal.

15.5 CYPERACEAE

Although the sedges comprise a large family of 4000 species and 90 genera of worldwide distribution, they have been rather well surveyed for their flavonoids in leaf and inflorescence. In particular, the numerous Australian species have been extensively investigated in recent years. The results of the Australian surveys, together with those of earlier investigations, confirm the presence of a highly characteristic flavonoid profile. Aurones are probably the most distinctive components but the family is also unusual in the variety of flavonoid structures produced. Thus, 25 aglycones have been identified (Table 15.1) including all four possible monomethyl ethers of luteolin and six flavonol methyl ethers. Extra hydroxylation is rare but 6-hydroxyluteolin was found in the South American genus *Lagenocarpus* (Williams and Harborne, 1977a) and sudachitin (5,7,4′-trihydroxy-6,8,3′-trimethoxyflavone) in the northern temperate genus *Eriophorum* (Salmenkallio *et al.*, 1982); neither of these constituents was detected in any Australian taxa. 5-Desoxyflavonoids are also infrequent: the only reports are of sulphuretin (6,3,4′-trihydroxyaurone) from *Carex appressa* R.Br. (Harborne *et al.*, 1985b) and 7,3,4′-trihydroxyflavone mainly from *Cyperus* inflorescences (Harborne *et al.*, 1982). Flavonols are comparatively rare in the Cyperaceae especially in tropical African and South American members and were only detected in 15% of the European species surveyed (Williams and Harborne, 1977a). However, they are significantly more frequent in leaves of both temperate and tropical Australian Cyperaceae (in 25% of taxa) other than the genus *Cyperus* and are generally less common in the inflorescences (in *ca.* 10% of Australian species surveyed) (Harborne *et al.*, 1985b). Of the four common flavonols, quercetin is the most regular constituent followed by isorhamnetin. Kaempferol occurs infrequently and myricetin was identified in the leaf and inflorescence of only one species, *Baumea rubiginosa* (Spreng.) Boeck. Three other quercetin and two kaempferol methyl ethers (see Table 15.1) were reported from 15% of the Australian

Table 15.1 Flavonoid aglycones present in the Cyperaceae

Flavones	*Flavonols*
Apigenin*	Kaempferol
4′-methyl ether (acacetin)	3-methyl ether
Luteolin*†	3,7-dimethyl ether
5-methyl ether	Quercetin†
7-methyl ether	3-methyl ether
3′-methyl ether (chrysoerial)*	3′-methyl ether (isorhamnetin)
4′-methyl ether (diosmetin)	3,7-dimethyl ether
	3,6,3′-trimethyl ether
7,3′,4′-Trihydroxyflavone	Myricetin
Tricin†	*3-Desoxyanthocyanidin*
6-Hydroxyluteolin	Carexidin
Sudachitin	
Aurones	
Sulphuretin	
Aureusidin†	
Leptosidin	
Mariscetin	

*Present additionally in *C*-glycosidic combination.
†Widely occurring major constituent.

Table 15.2 A comparison of leaf flavonoid occurrences in European, African, South American and Australian species of the Cyperaceae

Flavonoid	*European species (65 spp.)*	*African and South American species (61 spp.)*	*Australian*		*Whole family (388 spp.)*
			Cyperus (170 spp.)	*Others (92 spp.)*	
Flavone *C*-glycosides	+++	++++	(+)	++	+++
Tricin	++	++	+++	++++	+++
Luteolin	++	+	++++	++	++
Apigenin	—	(+)	(+)	(+)	(+)
Chrysoeriol	—	—	—	(+)	(+)
Luteolin 5-methyl ether	—	++	+	(+)	+
6-Hydroxyluteolin	—	(+)	—	—	(+)
5, 7, 4′-Trihydroxy-6, 8, 3′-trimethoxyflavone	(+)	—	—	—	(+)
Flavonoid sulphates	—	(+)	+	+	(+)
Common flavonols	+	(+)	+	+	+
Methylated flavonols	—	—	+	+	(+)
3-Desoxyanthocyanidins	(+)	—	—	(+)	(+)
Proanthocyanidins	nd	++	+++	++	++
Aurones	++	—	++	++	+

(+), present in < 10% of species; +, in 10–25% of species; ++, in 25–49% species; +++, in 50–75% of species; ++++, in >75% of species; nd, not determined.

Cyperaceae surveyed (Harborne *et al.*, 1982, 1985b). Such constituents are usually found in bud exudates and in wax deposits on leaves (Wollenweber and Dietz, 1981) but their exact location in the Cyperaceae has not been determined. Apart from the Zingiberaceae (Williams and Harborne, 1977b) and a report of quercetin 3-methyl ether from the exudate of *Barbacenia purpurea* Hook. (Velloziaceae) (Wollenweber, 1984), methylated flavonols are not otherwise known in the monocots.

Leaf and inflorescence tissue in the sedges differ quite markedly in their flavonoids. Thus in leaf, flavone *C*-glycosides and tricin are the major constituents, luteolin is common and flavonols frequent, while in inflorescences, tricin and luteolin are the main components and glycoflavones are rare. In Australian taxa, aurones are more frequent in the inflorescence (in *ca.* 45% taxa) than in the leaf (in *ca.* 30% of species).

In most taxonomic treatments, the Cyperaceae are divided into two subfamilies: the Cyperoideae and Caricoideae. Analysis of the combined flavonoid data shows no marked differences between the two groups except that glycoflavones are more predominant in the Caricoideae and flavonol methyl ethers are restricted to the Cyperoideae. There is some variation in the frequency of flavonoids at tribal level but the results are more meaningful when plant distribution is considered. Thus a summary of the combined leaf flavonoid results (Table 15.2) shows a strong correlation between chemistry and geography. Glycoflavones are most frequent and luteolin less well represented in the tropical species from Africa and South America, while tricin and luteolin are characteristic constituents of most Australian taxa. Luteolin is also frequent in European species. Proanthocyanidins are common in tropical members of the family but were not determined in European taxa. Luteolin 5-methyl ether was found in some 37% of African and 25% of South American species and was detected in some 10% of Australian species but was absent from European taxa.

The Cyperaceae are chemically distinct from the related Juncaceae and the Gramineae, by the presence of aurones and flavonol methyl ethers, frequency of proanthocyanidins and the almost complete absence of flavonoid sulphates. But, the Gramineae and Cyperaceae do show strong chemical similarity in the frequency of tricin derivatives and glycoflavones in leaf tissue (see Table 15.7). However, the occurrence of luteolin 5-methyl ether, a characteristic leaf constituent of the Juncaceae, in some tropical members of the Cyperaceae, together with the presence of 3-desoxyanthocyanidins in flowers of some Cyperaceae (Clifford and Harborne, 1969; Harborne *et al.*, 1985b) and Juncaceae (Fredga *et al.*, 1974) also suggests that the Cyperaceae may be as near to the Juncaceae as to the Gramineae.

15.6 FLUVIALES

The Fluviales is a convenient collective term, coined by Emberger (1960), for a group of about 13 monocot families which share an aquatic habitat. They include semi-aquatics and marsh plants and some (e.g. *Zostera*) even grow in marine waters, a unique feature within the angiosperms. They are divided into two groupings by Emberger (1960) and Dahlgren *et al.* (1985) treat them as two orders within the superorder Alismatiflorae. The

Table 15.3 Flavonoid patterns in families of aquatic monocotyledons

Order and family★	*Number of species surveyed*	*Flavone sulphates*	*Phenolic sulphates*	*Flavone O-glycosides*	*Flavone*[†] *C-glycosides*
Alismatales					
Butomaceae	1	–	–	–	+
Limnocharitaceae	1	–	–	–	+
Alismataceae	14	(+)	++	–	++
Hydrocharitaceae	22	(+)	++	+	+
Najadales					
Juncaginaceae	2	–	+	+	+
Potamogetonaceae	18	–	–	++	++
Posidoniaceae	3	–	+++	–	–
Zosteraceae	17	+++	+	++	–
Zannichelliaceae	1	+++	–	++	–
Cymodoceaceae	12	–	+++	–	–
Najadaceae	4	–	+	–	++

Key: +++, in over 50% of species; ++, in 25–50%, +, in 10–25%; (+), in < 10% species.
★Classification according to Dahlgren *et al.* (1985); there are no data for two of the families, Aponogetonaceae and Scheuchzeriaceae
[†]Data from Boutard et *al.* (1973); Harborne and Williams (1976a); McMillan *et al.* (1980).

results of the three major flavonoid surveys of these families are combined in Table 15.3.

Boutard *et al.* (1973) and Harborne and Williams (1976a) showed that flavonols and proanthocyanidins are uncommon, flavonoids with B-ring trihydroxylation are absent and that flavone *O*- and *C*-glycosides are frequent constituents. In addition, flavone and/or phenolic acid sulphates are characteristically present in all the families surveyed except the Potamogetonaceae, Butomaceae and Limnocharitaceae. The results of Boutard *et al.* (1973) largely support Emberger's (1960) division of the Fluviales into two groups, the first being relatively rich in flavonols and proanthocyanidins, the second rich in flavones with flavonols absent and proanthocyanidins rare. However, Harborne and Williams (1976a) failed to find flavonols in any of the families they investigated, but their results clearly distinguish the families from each other, except the Butomaceae from the Limnocharitaceae. The Potamogetonaceae is most distinct in lacking both flavone and phenolic acid sulphates even though flavones and flavone *C*-glycosides are major constituents. The Hydrocharitaceae and Alismataceae, although both producing flavone and phenolic acid sulphates, may be separated by the occurrence of flavones in the former and their absence from the latter. The Zosteraceae and Zannichelliaceae both produce flavone sulphates and flavones while in the Cymodoceaceae and Posidoniaceae only phenolic acid sulphates were found.

In 1980 McMillan *et al.* confirmed the high frequency of flavone sulphates in *Zostera* species and also showed them to be characteristic of *Phyllospadix*. In the other genus of the Zosteraceae, *Heterozostera*, phenolic acid sulphates were present but flavone sulphates were absent. In the Posidoniaceae and Cymodoceaceae, phenolic acid sulphates were found as major components in agreement with Harborne and Williams (1976a).

More recently Roberts and Haynes (1986) have used stem/leaf flavonoid evidence to distinguish *Potamogeton praelongus* Wulfen and *P. perfoliatus* L. from each other and to show that their putative hybrid, *P. richardsonii* (A. Benn.) Rydberg is probably of different origin, since it lacked the 'parental' flavonoids.

Anthocyanins are also recorded in these water plants, but they appear to occur mainly in colourless pseudo-base form in otherwise green leaves, the colour only appearing after the tissue is treated with acid (Reznik and Neuhausel, 1959). True anthocyanins have been identified in a few species, e.g. delphinidin 3-gentiobioside in flowers of *Eichhornia crassipes* (Ishikura and Shibata, 1970) and 5-methylcyanidin 3-glucoside in leaves of *Egeria densa* (Momose *et al.*, 1977).

Although the available flavonoid evidence does not fully support any of the published taxonomic groupings within the Fluviales, it does suggest: (1) the separation of the Potamogetonaceae from other families, (2) a close alliance between the Zosteraceae and Zannichelliaceae, (3) a close alliance between the Posidoniaceae and Cymodoceaceae and, (4) a chemical similarity between the Alismataceae and Hydrocharitaceae. Of these suggestions, only the last is in line with the Dahlgren *et al.* (1985) treatment of the monocots.

15.7 GRAMINEAE

Despite its immense agricultural importance, the Gramineae has not been extensively surveyed for flavonoids. The sheer size of the family (10 000 species and 750 genera) and the difficulty of taxonomic verification are two of the main problems in screening this group. There has been only one major flavonoid survey, in which 274 species from 121 genera were examined (Harborne and Williams, 1976b), representing about one-fifth of the known genera. The results of this survey established a deceptively simple leaf flavonoid profile for the family, in which tricin *O*-glycosides and apigenin and luteolin *C*-glycosides are almost universal and other flavonoid types [sulphate conjugates (in 16% of species), apigenin and luteolin *O*-glycosides (9%), luteoforol and apigiforol (8%), flavonols (6%) and proanthocyanidins] are infrequent. Rarer constituents include 5, 7, 3′-trihydroxy-4′, 5′-dimethoxyflavone, an isomer of tricin from *Poa huecii* (Rofi and Pomilio, 1985) and tricin 4′-methyl ether, detected in four of 18 grass species by Kaneta and Sugiyama (1973). The only example of a 5-desoxyflavone is 7, 3′, 4′-trihydroxyflavone, which was found in *Bothriochloa bladhii* (Harborne and Williams, 1976b) but is also present in the inflorescences of a number of *Cyperus* species (Cyperaceae) (Harborne *et al.*, 1982). Similarly the only report of a 6-hydroxylated flavonoid is quercetagetin 3, 6-dimethyl ether, from flowers of *Saccharum officinarum* (Misra and Mishra, 1979).

Although the basic pattern in the family is a simple one, there is considerable complexity in the number and variety of glycosides present. In particular, a vast array of flavone *C*-glycosides have been identified, many with several sugar attachments, some with unusual sugars and others with aromatic or aliphatic acylation; the sinapyl ester of 6-arabinosyl-8-galactosylapigenin from leaves of *Triticum aestivum* (Wagner *et al.*, 1980) and 8-galactosylapigenin 6″-acetate from diploid *Briza media* L. (Chari *et al.*, 1980) are typical examples. Additionally, the silks of some *Zea mays* L. varieties contain a *C*-glycosylflavone with a unique carbon-linked sugar, namely 2″-rhamnosyl-6-*C*-(6-deoxyxylohexose-4-ulosyl)luteolin (Elliger *et al.*, 1980). Complex mixtures of *C*-glycosides are the rule rather than the exception. Thus, up to 22 flavone *C*-glycosides occur in leaves of *Hordeum* cultivars (Fröst *et al.*, 1977), while more than 30 are present in leaves of diploid *Triticum* species (Harborne *et al.*, 1986).

The other characteristic grass constituent, tricin, occurs most frequently as the 5-*O*-glucoside but the 7-glucoside, 7-glucuronide, 7-rutinoside, 7-neohesperidoside, 7-diglucoside, 7-glucoside sulphate and 7-(2″-rhamnosylgalacturonide) are all known.

Flavonol glycosides are uncommon as leaf constituents, but they may occur more widely in other tissues such as pollen. Wiermann (1968) reported flavonols in the pollen of all 13 species surveyed, and more recently Ceska and Styles (1984) identified a series of ten flavonol glycosides in maize pollen.

Anthocyanins, although regularly present in leaf and/or inflorescence, are usually of a simple type, e.g. cyanidin 3-glucoside, 3-galactoside or 3-arabinoside (Clifford and Harborne, 1967). Exceptionally, peonidin 3-arabinoside occurs in *Phalaris arundinacea* leaf, peonidin 3-galactoside in sugar cane leaf, delphinidin 3-rutinoside in *Secale cereale* seed and delphinidin 3, 5-diglucoside in *Nardus stricta* leaf. There is now good evidence that many of the grass anthocyanins occur acylated with malonic acid and two malonated cyanidin 3-glucosides have been isolated from maize leaf (Harborne and Self, 1987).

The distribution of flavonoid sulphates in the grasses shows some correlation with habitat. Thus, these constituents characterize the tropical and subtropical families, Panicoideae (in 18% of species), Chloridoideae (15%) and Arundinoideae (40%) and are generally absent from tribes of cool temperate regions (Harborne and Williams, 1976b). Proanthocyanidins based on both flavan-4-ols and catechins are similarly confined to two of these subfamilies, the Panicoideae and Chloridoideae (Harborne and Williams, 1976b).

At the lower levels of classification flavonoids have been useful in separating genera, species and cultivars within a species. One example is in *Briza media*, where doubling of the chromosome number has led to a change in flavonoid pattern. Thus, the diploid plants produce *inter alia* 8-galactosylapigenin and its 2″-acetate, while the tetraploid plants instead accumulate the corresponding luteolin derivatives (Williams and Murray, 1972; Chari *et al.*, 1980). The flavonoid patterns of artificially produced tetraploids were identical with those of the natural tetraploid (Murray and Williams, 1976). Thus, autotetraploidy in this species apparently upsets the regulatory genes controlling B-ring hydroxylation in these glycoflavones. These same workers in a survey of other *Briza* species also found sufficient divergence in the flavonoid chemistry between the mainly diploid Eurasian species and the tetraploid South American species to confirm a recent suggestion that these two groups should be separated at the generic level (Williams and Murray, 1972).

In Dahlgren *et al.* (1985), the grasses are allied with six other families, the largest being the Restionaceae. The universal occurrence of flavone *C*-glycosides with tricin throughout the Gramineae is a very distinctive feature, which is not yet recorded in any of these other groups. The Flagellariaceae and Anarthriaceae are different in having flavonols (see Williams *et al.*, 1971) and so is the Restionaceae (see Section 15.13). The reported occurrence of tricin in the latter family (Dahlgren *et al.*, 1985, p. 421) is incorrect and overall there is hardly any link with the grass pattern. The Joinvilleaceae (two spp.) is considered by some to be particularly close to the grasses, but unfortunately it is not yet possible to say whether there are any flavonoid links, since its constituents have not been determined.

15.8 IRIDACEAE

The Iridaceae is an ornamentally important family of 1500 species and 85 genera, which is most richly represented in

Table 15.4 The percentage occurrence of leaf flavonoids and other phenolics in tribes of the Iridaceae

*Subfamily tribe**	*No. of species surveyed*	*Pro-Cy*	*Flavone C-glycosides*	*Flavones*					*Flavonols*				*Biflavones*
				Lu	*Ap*	*Tr*	*Aca*	*6-OH flavones*	*Km*	*Qu*	*My*	*Isorh*	
Isophysidoideae	1	—	100 (trace only)	—	—	—	—	—	—	—	—	—	100
Iridoideae													
Aristeae	19	39	5	—	—	5	—	—	59	95	39	53	32[†]
Irideae	88	11	82	1	—	1	—	—	5	15	8	1	6[‡]
Sisyrinchieae	37	11	97	5	3	—	—	—	—	—	—	—	11[‡]
Tigrideae	26	19	100	—	—	4	—	—	4	—	—	4	4[‡]
Trimezieae	8	63	100	—	—	—	—	—	—	—	—	—	—
Ixioideae													
Watsonieae	6	17	—	—	—	—	—	—	50	100	67	50	—
Ixieae	62	11	44	8	10	15	2	5	40	47	19	15	—

Key: ProCy, proanthocyanidins; Lu, luteolin; Ap, apigenin; Tr, tricin; Aca, acacetin; 6-OH, 6-hydroxy; Km kaempferol; Qu, quercetin; My, myricetin; Isorh, isorhamnetin.

*Classification according to Harborne *et al.* (1986).

[†]One of these compounds was identified as amentoflavone but constituents from five other species with similar R_F values, UV colour reactions and spectral data were not fully characterized.

[‡]These compounds were not fully characterized but had similar R_F values, colour and UV spectral properties to biflavonoids.

the South African and tropical American flora; there are a number of species in Australia (*Isophysis*, *Patersonia*) while the genus *Iris* inhabits the Northern Hemisphere with a centre of origin in Asia. The flavonoid chemistry has been studied in some detail by Bate-Smith (1968) and more recently in these laboratories (Harborne and Williams, 1984; Williams and Harborne, 1985; Williams *et al.*, 1986).

The major flavonoids are flavone *C*-glycosides, flavonols and proanthocyanidins. The flavonol myricetin is a distinctive marker, occurring regularly in *Babiana*, *Iris*, *Klattia*, *Nivenia* and *Patersonia*. It occurs mainly in simple glycosidic forms: 3-glucoside, 3-arabinoside, 3-rutinoside, 3-rhamnosylgalactoside, etc. The two myricetin methyl ethers, larycitrin and syringetin have also been recorded, in flowers of *Gladiolus tristis* L.

Flavones such as apigenin and luteolin are known, as is tricin, which is present occasionally in such genera as *Crocus*, *Gladiolus*, *Homeria*, *Iris* and *Moraea*. Acacetin (4′-methylapigenin) is rare, recorded only in *Crocus laevigatus* Bory & Chaub. 6-Hydroxyflavones (scutellarein, 6-hydroxyluteolin and their 7-methyl ethers) are likewise rare and occur only in *Crocus*.

Iridaceae are unusual in synthesizing isoflavones, mostly in rhizomes, although the character is essentially restricted to a few species of *Iris* (Ingham, 1983). Another unusual flavonoid type, recently discovered in the family, is that of the biflavonoid (Williams and Harborne, 1985). Amentoflavone, in particular, has been characterized in *Isophysis*, where it occurs with the dihydro derivative, and in *Patersonia*. Biflavonoids may occur in a few other primitive members of the family. Other related phenolics which have been found in the Iridaceae include the xanthone *C*-glucoside, mangiferin, and the naphthoquinone, plumbagin.

All six common anthocyanidins have been recorded variously in flowers of the Iridaceae. They occur chiefly as 3-rutinoside, 3,5-diglucoside and 3-rutinoside-5-glucoside, but 3-gentiobioside is present in *Tritonia* and 3-sophoroside in *Watsonia*. Aromatic acylation is an added complication in the pigments of *Iris*, which include delphinidin and malvidin 3-(*p*-coumarylglucoside)-5-glucoside. Aliphatic acylation is rare, only being described in *Lapeirousia corymbosa* Ker-Gawl.

The distribution of these various flavonoids is mainly of interest at the tribal level (Table 15.4) and the data have been of some assistance in taxonomic revision within the family. The separation of *Isophysis tasmanica* (Hook, f.) T. Moore as the only taxon of the Isophysidoideae is confirmed by its distintive flavonoids. The Trimezieae and Sisyrincheae are likewise separated by the absence of flavonols, whereas the Aristeae and Nivenieae are distinguished from the other tribes by the rarity of glycoflavone occurrences. Again, members of the Watsonieae are separated by having only flavonols in the leaves, while the Ixieae have a number of distinctive flavones in both leaf and flower.

The Iridaceae are placed in the same order as the Liliaceae and Orchidaceae in Dahlgren *et al.* (1985) but comparison of the flavonoid data on these two families (see Sections 15.10 and 15.11) show only a few links. With the occurrence of rare flavonoid types (biflavonoids, isoflavones) and of rare derivatives (acacetin, 6-hydroxyflavones), the Iridaceae family is generally different from any other related family. It is also unique in the chemical diversity that is present within its members.

15.9 JUNCACEAE

The Juncaceae is a small family of 300 species and nine genera. There is an interesting geographical dichotomy in the family in that seven of the nine genera are restricted to the southern hemisphere: *Marsippospermum* and *Rostkovia* to Antarctica, *Distachia*, *Oxychloe*, *Patosia* and *Andesia* to the South American Andes and *Prionium* to South Africa. *Juncus* (with about 225 species) is cosmopolitan but most of its species are Northern in distribution. *Luzula* (with 80 species) is also more common in the northern hemisphere.

Early flavonoid studies on the Juncaceae were very limited. Bate-Smith (1968) detected caffeic acid in five taxa and quercetin in *Luzula sylvatica*. Luteolin and its 7-glucoside were identified in *Juncus effusus* flowers and were also detected in a further 10 *Juncus* species and in *Luzula sylvatica* (Hudson) Gaudin (Stabursvik, 1969). Several 3-desoxyanthocyanins based on luteolinidin were identified in two *Luzula* and three *Juncus* species (Fredga *et al.*, 1974). Since these findings there has been only one major flavonoid survey of the Juncaceae, in which 38 species from six genera were examined (Williams and Harborne, 1975). This survey showed that large quantities of free flavones (i.e. without *C*- or *O*-sugars attached) occur in the stem/leaf tissue (both fresh and herbarium).

Stabursvik (1969) previously noted this phenomenon in his survey of some Juncaceae species and free tricin has been found frequently in fresh leaves of the Cyperaceae (Harborne, 1971). In the Juncaceae luteolin was recorded in 95% of species, luteolin 5-methyl ether in 70% and chrysoeriol in 13% of the taxa surveyed. In all but three species luteolin and luteolin 5-methyl ether co-occurred with their 7-glucosides. A 5-desoxyflavone, 7,3′,4′-trihydroxyflavone, was tentatively identified in *Juncus trifidus* and *Luzula purpurea* and its *O*-glucoside in one sample of *Prionium seratum*. 7,3′,4′-Trihydroxyflavone is a rare plant constituent first identified in two species of the Leguminosae by Wong and Francis (1968) but more recently in some members of the Cyperaceae. This establishes a chemical link between the Juncaceae and the Cyperaceae.

However, the characteristic flavonoid profile of free luteolin, luteolin 5-methyl ether and their 7-glucosides serves to distinguish the Juncaceae from the Centrolepidaceae, Restionaceae and Thurniaceae, which lack these constituents but which are placed together by Hutchinson

(1959) in his order Juncales. Similarly luteolin 5-methyl ether and its 7-glucoside were not found in the Palmae or Gramineae, and flavone 5-glycosides, which are characteristic constituents of these families, were not detected in the Juncaceae. Also flavone *C*-glycosides, which are major leaf components in the Gramineae and Cyperaceae, were only found in *Prionium*, a genus which is anatomically anomalous in the Juncaceae.

Nevertheless, the Juncaceae, Cyperaceae and Gramineae all produce flavones in which the 5-hydroxyl is masked, although the masking is mainly 5-glycosylation in the Cyperaceae and Gramineae and 5-methylation in the case of the Juncaceae. In the latest treatment of the monocots, Juncaceae is included with the Cyperaceae in the order Cyperales (Dahlgren *et al.*, 1985), and the flavonoid data overall provide support for such an alignment.

15.10 LILIALES

There has been much dispute as to the delimitation and the number of families in the order Liliales: Cronquist (1968) recognizes 13 families, Takhtajan (1969) 20, Melchior (1964) 17 and Hutchinson (1959) only six. However, in the most recent classification, Dalgren *et al.* (1985) place only nine families in their order Liliales including the Iridaceae, Geosiridaceae and Orchidaceae. Many traditional members of the Liliales and of the Liliaceae have now been grouped into smaller families and placed into one of the new orders: Dioscoreales (seven families), Asparagales (30 families), Melanthiales (two families) and the Burmanniales (three families) in the superorder Liliiflorae. However, it is difficult to discuss the flavonoid data in relation to these new families since most of the published results have been related to the older classifications. In this summary, we shall consider alphabetically the families recognized by Hegnauer (1963) (based largely on Hutchinson's (1959) system) but excluding the two families, Petermanniaceae and Tecophilaeaceae, for which no data are available.

15.10.1 Agavaceae

There are only a few reports of flavonols: quercetin from leaves of *Yucca filamentosa* L. (Williams, 1975), kaempferol, its 3-xyloside and 3,4′-dixyloside from leaves of *Polyanthus tuberosa*, and kaempferol, its 3-glucoside and 3-rutinoside from *Agave americana* (El-Mohazy *et al.*, 1980). Three homoisoflavan-4-ones, two homoisoflavans, four flavans and a 5-deoxyflavonol were isolated from *Dracaena draco* (Camarda *et al.*, 1983) and in the leaves of *Phormium tenax* a flavanone, 6,8-di-*C*-methylnaringenin (farrerol), was discovered (Matsuo and Kubota, 1980). Another rare flavonoid, 3-*O*-methyl-8-*C*-methylquercetin, has been reported from leaves of both *Dasylirion acrotrichum* and *Xanthorrhoea hastilis* (Xanthorrhoeaceae) (Laracine *et al.*, 1982). In a current flavonoid investigation of the Liliales *sensu lato* (Williams, unpublished results) quercetin and kaempferol were detected in *Calibanus hookeri* and *Nolina recurvata*; kaempferol but no quercetin in *Yucca filamentosa*; quercetin, isorhamnetin and kaempferol in *Phormium tenax*; and kaempferol in four *Dianella* species. Proanthocyanidins were also detected in the latter five taxa.

Thus from the limited data flavonols appear to be the characteristic constituents, with *C*-methyl derivatives also being present occasionally.

15.10.2 Alstroemeriaceae

This is a small family of four genera and some 157 species. Only six taxa have been examined for their leaf flavonoids (Bate-Smith, 1968; Williams, 1975). Kaempferol was found in all four *Alstroemeria* species examined and in one of two *Bomarea* species; quercetin was detected in one of the *Alstroemeria* taxa. More interestingly, two novel anthocyanidin glycosides – the 3-rutinoside and 3-glucoside of 6-hydroxycyanidin – have been obtained from flowers of some *Alstroemeria* cultivars (Saito *et al.*, 1985). The only previous report of a 6-hydroxyanthocyanidin in nature is of 6-hydroxypelargonidin from flowers of *Impatiens* in the dicotyledons (Jurd and Harborne, 1968).

15.10.3 Amaryllidaceae

Here we follow Hegnauer (1963) who uses Hutchinson's (1959) system but excludes the Allioideae, which is considered under the Liliaceae. Only a small number of the 325 species from 45 genera have been surveyed for their leaf flavonoids. Thus Bate–Smith (1968) found quercetin in nine and kaempferol in seven of the 15 species he screened. Similarly, Williams (1975) found quercetin in 12 and kaempferol in 13 taxa in a survey of 17 Amaryllidaceae. No flavones have been recorded. However, a number of unusual flavonol glycosides have been characterized (Hegnauer, 1986) including kaempferol 3-leucoveroside-7-glucoside and kaempferol 3-leucoveroside from *Leucojum vernum* (Horhammer *et al.*, 1967). Leucoverose is the isomer of sambubiose, 2-*O*-α-D-xylosyl-D-glucose. The same workers also isolated kaempferol 3-sophoroside and 3-sophoroside-7-glucoside from flowers of *Galanthus nivalis*. Indeed, flavonol di- and tri-glycosides appear to be the characteristic components of the family.

15.10.4 Dioscoreaceae

Bate–Smith (1968) found the flavonols kaempferol and quercetin as the major leaf constituents in ten *Dioscorea* species but could find no flavonoids in *Tamus communis*. In a more recent examination (Williams, unpublished results), quercetin was detected in four and kaempferol in one of the four taxa examined which included two *Dioscorea* species, *Rajania hastata* and *T. comuunis*, confirming the predominance of flavonols in the family and their presence in the latter species. Hutchinson separated off the genus *Avetra* from the Dioscoreaceae and this is supported chemically by the finding of flavone *C*-glycosides in, and the absence of flavonols from, *Avetra sempervirens*. Flavone *C*-glycosides have also been tentatively identified in *Trichopus zeylanicus* which is another borderline taxon in the family. Dalgren *et al.* (1985) retain *Avetra* within their Dioscoreoideae, which is definitely not in agreement with the chemical evidence.

15.10.5 Haemodoraceae and Hypoxidaceae

There are only two reports of flavonoids from these small families: quercetin and a trace of kaempferol from

Table 15.5 Leaf flavonoid aglycones* in the Liliaceae[†] from Williams (1975) and unpublished data

Subfamily tribe[‡]	*Flavonols*			*Flavones*		
	Qu	*Km*	*Isorh*[§]	*Lu*	*Ap*	*Others*
Melanthioideae						
Tofieldieae (4)	3	3	—	—	—	
Melanthieae (4)	1	1	1	3	1	
Uvularieae (10)	2	2	—	3	2	Tricin (1) 6-OH Lu (1)
Tricyrteae (6)	6	4	—	—		
Asphodeloideae						
Dianelleae (4)	—	4	—	—	—	
Asphodeleae (17)	2	1	—	13	7	Flavone *C*-glycosides (1)
Aphyllantheae (1)	1	1	—	—	—	
Hemerocalleae (8)	4	7	1	—	—	
Aloeae (30)	9	7	—	—	—	
Wurmbaeoideae						
Colchiceae (6)	—	—	—	6	5	Diosmetin (3)
Glorioseae (1)	—	—	—	1	1	
Lilioideae						
Lloydieae (1)	—	1	—	—	—	
Tulipeae (8)	6	8	—	—	—	
Lilieae (15)	12	13	—	—	—	
Scilloideae						
Scilleae (40)	6	7	2	22	27	Tricetin (1) Diosmetin (1) Tricin (2) Flavone *C*-glycosides (5)
Allioideae						
Agapantheae (4)	4	2	—	—	—	
Allieae (10)	8	7	—	—	—	
Gillieseae (1)	1	1	—	—	—	
Asparagoideae						
Convallarieae (3)	3	1	2	—	—	
Milliganieae (4)	2	1	1	—	—	
Polygonatae (10)	8	6	2	—	—	Flavone *C*-glycosides (2)
Parideae (5)	4	5	1	—	—	
Asparageae (11)	11	7	—	—	—	
Ophiopogonoideae						
Ophiopogoneae (3)	—	—	—	—	—	
Luzuriagoideae (8)	1	5	—	—	—	Tricin (1) Flavone *C*-glycosides (1)
Smilacoideae (5)	5	5	—	—	—	

Key: Qu, quercetin; Km, kaempferol; Isorh, isorhamnetin; Lu, luteolin; Ap, apigenin; 6-OH Lu, 6-hydroxyluteolin.
*The figures in the table refer to the number of species in which each flavonoid aglycone was present out of the number of taxa surveyed in the relevant tribe.
[†]Classification according to Melchior (1964).
[‡]The value in brackets is the number of taxa surveyed for each tribe.
§Isorhamnetin was not determined in the survey of Williams (1975).

Haemodorum teretifolium R.Br. (Haemodoraceae) and quercetin from *Hypoxis krebsii*, Fisch. (Hypoxidaceae) (Bate–Smith, 1968).

15.10.6 Liliaceae

Bate–Smith (1968) in a leaf flavonoid survey of 121 Liliaceae species recorded flavonols in 70% of the taxa. Since then there have been two main leaf surveys – by Williams (1975) and Skrzypczak (1976) – of 168 and 172 species respectively. (The latter author also examined some floral tissues.) The results of the first survey (Williams, 1975), combined with unpublished data for a further 44 taxa by the same author, are summarized in Table 15.5. Flavonols are confirmed as the most frequent constituents in the family with both quercetin and kaempferol in 42% of taxa. Isorhamnetin was not determined in the 1975 survey but has since been identified in ten species (Williams, unpublished data) and is probably present in further taxa. The flavones, luteolin and apigenin, were identified in only 22% and 15% of species respectively, while methylated flavones, i.e. chrysoeriol (in one species), diosmetin (in three *Colchicum* species) and tricin (in four taxa), are all rare. Flavone *C*-glycosides were found only in five members of the Scilloideae and in two *Polygonatum* species. The only example of a 6-hydroxylated flavonoid is 6-hydroxyluteolin from *Schelhammera multiflora* R.Br. (Uvularieae) (Williams, unpublished results). Proanthocyanidins were found in only 17 species and flavonoid sulphates detected in *Bellevalia flexuosa* and in a later study (Williams *et al.*, 1976) in *Lachenalia unifolia*.

The flavonoid results are probably most meaningful at subfamily and tribal level (see Table 15.5). Thus the subfamilies may be grouped as those with: (1) only flavonols: Lilioideae, Allioideae, Asparagoideae and Smilacoideae; (2) only flavones. Wurmbaeoideae; (3) flavones plus flavonols: Melanthioideae, Asphodeloideae and Scilloideae; and (4) no apparent flavonoids: Ophiopogonoideae. At the tribal level, within the Melanthioideae both the Melantheae and Uvularieae appear to be heterogeneous with flavones and flavonols, whilst the Tricyrteae and Tofieldieae apparently produce only flavonols. Similarly, within the Asphodeloideae, the Asphodeleae differs from the other four tribes in producing mainly flavones. The Scilloideae is chemically the most variable subfamily, with all the major flavonoid classes represented. The flavonoid survey by Skrzypczak (1976) shows a similar distribution of flavonoids within the subfamilies but there is some disagreement between the two surveys on the flavonoids present in certain species.

Flavonoid glycosides, including some unusual di- and tri-glycosides, have been characterized from quite a number of Liliaceae species (see Hegnauer, 1986). For example, kaempferol and quercetin 3-rutinoside-7-glucuronide occur in leaves and flowers of *Tulipa* species (Harborne, 1965), kaempferol 3-cellobioside in *Paris quadrifolia* (Nohara *et al.*, 1982) and kaempferol 3-glucosyl(1 → 2) galactoside in *Lilium candidum* (Nagy *et al.*, 1984). The more unusual flavone glycosides include diosmetin 7-diglucoside from *Colchicum byzanthinum*, tricin 7-rutinoside-4′-glucoside from cultivated *Hyacinthus orientalis* (Williams, 1975) and the 3′-monosulphates and 7, 3′-disulphates of tricetin, diosmetin and luteolin from *Lachenalia unifolia* L. (Williams *et al.*, 1976). This is the only report of tricetin in the family or in the monocots. Similarly, homoisoflavonoids, which have been characterized in four *Eucomis* species, *Muscari comosum* and *Ophiopogon japonicus* (see Hegnauer, 1986), have not been recorded elsewhere in the monocotyledons apart from in *Dracaena draco* (Agavaceae) and in the Liliales.

Anthocyanins are mainly of a simple type with the 3-glucosides, 3-rutinosides and 3, 5-diglucosides of pelargonidin, cyanidin and delphinidin being very frequent (see Harborne, 1967; Timberlake and Bridle, 1975; Hrazdina, 1982). Acylation of the 3, 5-diglucosides with *p*-coumaric acid is also common but methylation is rare. Two most recently characterized pigments are delphinidin 3-(6″-*p*-coumarylglucoside)-5-(6″-malonylglucoside) from bluebell flowers (Takeda *et al.*, 1986) and petunidin 3-(2^G-glucosylrutinoside)-5′-glucoside in the seed coats of *Ophiopogon jaburan* (Ishikura and Yoshitama, 1984).

15.11 ORCHIDACEAE

The Orchidaceae is the largest monocot family with 20 000 species and 750 genera. Although the family is cosmopolitan, the majority of orchid species grow in the tropics. The size, morphological complexity and inaccessibility of many species make classification difficult. There are numerous taxonomic treatments but one of the most recent is that of Dressler (1974). Limited surveys of the anthocyanins (Arditti and Fisch, 1977) have suggested that only three anthocyanidins predominate: cyanidin, pelargonidin and petunidin and that complex mixtures of their glycosides and acylated derivatives frequently occur. More recently Strack *et al.* (1986) have identified a unique acylated pigment, cyanidin 3-*O*-β-(6″-*O*-oxalyl) glucoside from flowers of five *Nigritella*, six *Orchis*, one *Ophrys* and one *Anacamptis* species and as a trace constituent in a further eight orchid taxa. However, a report (Uphoff, 1979) of acylated cyanidin 3, 5-diglucosides in European species could not be verified by Strack *et al.* (1986).

The only major leaf flavonoid survey of orchids (Williams, 1979) has revealed such a strong correlation with plant geography that it is not possible to represent the family by a single flavonoid profile. For example, flavone *C*-glycosides are characteristic of tropical and subtropical species and flavonol glycosides of temperate taxa. 6-Hydroxyflavones, luteolin and tricin are all rare in the family. It is also surprising that more advanced characters, e.g. highly methylated or glycosylated derivatives, are apparently lacking in a group where morphological complexity has reached a climax. However, the almost complete absence of proanthocyanidins does suggest a high degree of advancement for the family as a whole and perhaps further flavonoid surveys of orchid flowers, morphologically the most derived tissue in these plants, could provide further evidence for this.

There have been few detailed glycosidic studies in orchids. Pagani (1976) identified quercetin 3-glucoside, 7-glucoside and 3,7-diglucoside in *Orchis sambucina* and quercetin 3-glucoside in *O. morio*. Williams (1979) identified luteolin 3,4′-diglucoside from *Listera ovata* (L.) R.Br., pectolinarigenin 7-glucoside, 7-rutinoside and scutellarein 6-methyl ether 7-rutinoside from two *Oncidium* species and pectolinarigenin 7-glucoside from *Eria javanica* Blume. The xanthones, mangiferin and isomangiferin, were detected in five taxa including a mangiferin sulphate in *Polystachya rizanzensis* Rendler (cf. the occurrence of these xanthones in Iridaceae).

The relationship of the Orchidaceae with other monocot families is uncertain in the absence of a fossil record but it has been suggested on morphological grounds that the orchids arose from Liliaceous ancestors. However, the flavonoid evidence apparently refutes this suggestion in that flavone *C*-glycosides, the characteristic leaf flavonoids of tropical orchids, are rarely found in leaves of the Liliaceae *sensu lato*. Instead, the Orchidaceae shows most similarity to the Commelinaceae, a family in which glycoflavones and flavonols are characteristic leaf components (see Section 15.4), the Iridaceae, where flavone *C*-glycosides, flavonols and mangiferin are common leaf constituents (see Section 15.8) and the Bromeliaceae, where glycoflavones, flavonols, flavones and 6-hydroxyflavones are all present (see Section 15.3). A possible common origin for the Bromeliales and Commelinales has been suggested by Takhtajan (1969) and from the flavonoid evidence, the orchids could have arisen from a similar stock or at an earlier stage in angiosperm evolution (see Garay, 1960).

15.12 PALMAE

The Palmae is an ancient family of 3400 mostly tropical species of varying habit from large trees to shrubs and lianas. Classification is based largely on morphological and anatomical characters of the leaf and fruit, and nine subfamilies are usually recognized. Moore (1961) produced a system which Dalgren *et al.* (1985) have followed.

Chemically the family has been rather neglected. In two early studies the unusual red pigments dracorubin (Hesse and Klingel, 1936; Brockmann and Haase, 1937) and dracorhodin (Freudenberg and Weinges, 1958) were identified in the 'dragon's blood' resin, which exudes from the floral parts of *Daemonorops draco*. More recently Cardillo *et al.* (1971) found two further related red pigments in this resin, nordracorhodin and nordracorubin, together with: (2*S*)-5-methoxy-β-methoxyflavan-7-ol, (2*S*)-5-methoxyflavan-7-ol, 2,4-dihydroxy-5-methyl-6-methoxychalcone and 2,4-dihydroxy-6-methoxy-chalcone. Also Delle Monache *et al.* (1971, 1972) isolated (+)-epicatechin and (+)-catechin from leaves of *Phoenix canariensis*, *Butia capitata*, *Howea fosteriana*, the leaves and fruits of *Chamaerops humilis* and the seeds of *Archontophoenix cunninghamiana* and both (+)-epicatechin and (+)-epiafzelechin from *Livistona chinensis* leaves. This is the first report of catechins with the (+)-epi configuration from natural sources. Delle Monache *et al.* (1971) also found two new procyanidins, formed from the C_4–C_6 bonding of two molecules of (+)-epicatechin, from *C. humilis*. Bate-Smith (1968) surveyed the leaves of 17 palm species and recorded proanthocyanidins in 13 and flavonols in two taxa respectively. Williams *et al.* (1971) later identified leaf flavonoids in a further five palm species, and characterized the flavone sulphate, luteolin 7-sulphate-3′-glucoside, from leaves of *Mascarena verschaffeltii*. This discovery led to a major leaf flavonoid survey in which 125 species from 72 genera were examined (Williams *et al.*, 1973). A complex leaf flavonoid profile was revealed in which flavone *C*-glycosides, proanthocyanidins and tricin glycosides were major constituents (in 84, 66 and 51% of species respectively) and luteolin and quercetin glycosides occurred frequently (in 30 and 24% of taxa respectively). Kaempferol and apigenin were detected in only a few species and isorhamnetin in only one taxon. Flavonoids identified include the 7-glucoside, 7-diglucoside and 7-rutinoside of tricin and luteolin, tricin 5-glucoside, apigenin 7-rutinoside, quercetin 3-rutinoside-7-galactoside, isorhamnetin 7-rutinoside, orientin, iso-orientin, vitexin and its 7-glucoside and isovitexin. Additionally many of the *O*- and *C*-flavonoid glycosides occurred as sulphates and negatively charged conjugates were detected in 50% of the species surveyed.

The flavonoid results are potentially useful in taxonomic revision at all levels of classification. Thus the subfamilies Phoenicoideae and Caryotoideae have distinctive flavonoid patterns and there is evidence to support the separation of the subfamilies Phytelephantoideae and Nypoideae. The flavone, tricin, is present in all subfamilies except the Caryotoideae, Borassoideae and Phytelephantoideae and may be a useful marker at tribal level. For example, within Potztal's (1964) Arecoideae, it was found to be frequent in only one tribe, the Areceae, and was almost completely absent from the other 13 tribes. The subfamilies Coryphideae, Borassoideae and Phoenicoideae all have relatively uniform flavonoid profiles but the Arecoideae, a morphologically diverse group, is also chemically variable. Within the Cocosoideae there is flavonoid evidence to support the separation of the tribe Cocoseae (Williams *et al.*, 1973) into several genera. Thus, all those taxa which have been split off from *Cocos* differ from this genus in having flavonoid sulphates, and the more recently created genus *Arecastrum* is also distinguished by the absence of proanthocyanidins. In further studies of the cocosoid palms (Williams *et al.*, 1983, 1985), *Attalea, Orbigyna, Scheelea* and *Syagrus* were found to be chemically heterogeneous groups with as much variation between species as between genera.

A survey of the flowers of nine palm species (Harborne *et al.*, 1974) revealed a very different flavonoid pattern from that found in leaf tissue in that both flavonoid sulphates and flavonols are more frequent and the phenol, 3-caffeylshikimic acid, is a major constituent. The latter substance is rare outside the Palmae, there being only a single report, in the gymnosperm *Tsuga canadensis* Carr. (Goldschmid and Hergert, 1961). It was found in all the palm flowers examined and may be a distinctive marker for the family. It is also of interest that male and female flowers of *Phoenix canariensis* have strikingly different flavonoid profiles, with flavonol *O*-glycoside and *O*-glycosylflavone sulphates in the former and luteolin 7-rutinoside and flavone and *C*-glycosylflavone sulphates in the latter.

In Dahlgren *et al.* (1985), the palms are separated from all other monocot families at both the ordinal and superordinal levels. The flavonoid data, discussed above, show quite clearly a relationship between the palms, the grasses and the Cyperaceae since all three families have tricin, *C*-glycosylflavones, sulphates and flavone 5-glycosides in common. There are also some differences (e.g. the characteristic aurones of sedges have not yet been recorded in palms) so that the leaf flavonoid similarities revealed here should not be overstressed.

15.13 RESTIONACEAE

The restionads are a rush-like group of xeromorphic usually dioecious plants of 40 genera and 400 species, which grow mainly in either South Africa or Australasia. In the absence of true leaves, the stems or aerial culms have been analysed for flavonoids (Harborne and Clifford, 1969; Harborne, 1979; Harborne *et al.*, 1985a). Although the often highly coloured inflorescences (various shades of yellow, red and brown) have been analysed as well, they have generally failed to yield recognizable flavonoid pigments, except in a few cases. The most distinctive flavonoids in the culms are 8-hydroxy substituted, a type otherwise uncommon in monocots. The first isolation of hypolaetin, the 8-hydroxy derivative of luteolin, from plants was from *Hypolaena fastigiata*. 8-Hydroxyflavonols are more regularly present, and gossypetin, its 7-methyl ether and herbacetin 4′-methyl ether have all been identified.

The flavonol myricetin is also present regularly in two anatomically related South African genera, *Chondropetalum* and *Elegia*, where it occurs with the two methyl ethers, larycitrin and syringetin. The 3′-methyl ether, larycitrin is also recorded in monocots in *Gladiolus* (Iridaceae) while syringetin, the 3′,5′-dimethyl ether of myricetin, has also been found in *Philydrum* (Philydraceae) (Bohm and Collins, 1975) and in *Hedychium* (Zingiberaceae) (Williams and Harborne, 1977b).

Surveys of flavonoids in 60 representative restionads have shown significant differences between the constituents of South African and of Australasian members (Table 15.6). Thus 8-hydroxyflavonoids are mainly confined to Australian taxa, whereas proanthocyanidins, flavones and glycosylflavones occur regularly only in the South African plants. These chemical differences are not unexpected in view of the geographical separation of the plants; there are also marked anatomical differences between the two groups of plants (Cutler, 1969).

Flavonoid studies have made a positive contribution to taxonomic revision in the Restionaceae, in the case of 11 South African species belonging to the genus *Chondropetalum* (Harborne *et al.*, 1985a). Here there is a striking dichotomy in flavonoid aglycone profiles, with seven species having myricetin, larycitrin and syringetin, and four having instead kaempferol, quercetin, gossypetin, gossypetin 7-methyl ether and herbacetin 4′-methyl ether. These strongly marked chemical data, combined with anatomical and morphological variations, have been

Table 15.6 Flavonoid differences in the Restionaceae according to phytogeography

Class	*Compounds in South African species*	*Compounds in Australasian species*
Anthocyanin	Rare, cyanidin glycoside in *Chondropetalum hookerianum*	Rare, cyanidin 3-glucoside in two *Restio* spp.
Proanthocyanidin	Widespread, in 29/34 spp.	Uncommon, in 3/13 spp.
Flavonol	Quercetin common; myricetin and derivatives in *Chondropetalum, Elegia*; gossypetin* rare, in 4 spp. *Chondropetalum* and in *Staberoah cernua*	Quercetin and myricetin only in *Lepyrodia scariosa*; gossypetin and derivatives common (in 7/17 spp.)
Flavone	Luteolin, apigenin and chrysoeriol in 8/34 spp.	Hypolaetin in 2 spp.
Glycosylflavone	In 11/34 spp.	Not recorded

*Gossypetin derivatives also occur in *Leptocarpus*, the only genus of the family which occurs in South America and Asia, as well as in Australasia.

employed by Linder (1984) to separate three of the four latter species into the genus *Askidiosperma* and the fourth species as the monotypic *Dovea macrocarpa*.

In the treatment of Dahlgren *et al.* (1985), the Restionaceae are included with the Gramineae in the order Poales. The flavonoid chemistry of the grasses, however (see Section 15.7) could not be more different. There are also no obvious flavonoid links between the restionads and the other five small families in the order Poales. The distinctive presence of 8-hydroxyflavonoids and of myricetin derivatives separates this family clearly from any of its apparent relatives.

15.14 ZINGIBERALES

A number of earlier taxonomists, e.g. Engler and Prantl (1930) and Engler (1900–1912), recognized only four families in the order Zingiberales: the Zingiberaceae, Marantaceae, Cannaceae and Musaceae. Hutchinson (1959) later separated the Lowioideae as the Lowiaceae and elevated the Strelitzioideae to family status leaving only *Musa* and *Erisete* in the Musaceae.

Many flavonoid investigations have concentrated on rhizome and seed constituents. In these organs most of the flavonoids identified are unusual in that there are no free hydroxyl groups on the B-ring, e.g. galangin (3,5,7-trihydroxyflavone) and its 3-methyl ether, which occur together with kaempferol 4′-methyl ether in the rhizome of *Alpinia officinarum* (Heap and Robinson, 1926). Galangin 7-methyl ether is present in seeds of both *A. chinensis* and *A. japonica* and the corresponding dihydroflavonol, alpinone, in seed of *A. japonica* (Gripenberg *et al.*, 1956). 5,7-Dimethoxyflavanone (alpinetin) occurs in seed of *A. chinensis* (Gripenberg *et al.*, 1956) and in rhizomes of *Kaempferia pandurata*, where it co-occurs with its isomer pinostrobin (5-hydroxy-7-methoxyflavanone) (Mongkolsuk and Dean, 1964). More recently Jaipetch *et al.* (1983) have characterized pinostrobin, alpinetin, five methylated flavones and four methylated flavonols with no B-ring hydroxyls from *Boesenbergia pandurata*.

In the Musaceae, Simmonds (1954) reported four anthocyanin patterns in bract tissue of various *Musa* species: (1) pelargonidin–cyanidin mixtures in *M. coccinea*, (2) cyanidin–delphinidin mixtures in *M. laterite*, *M. balbisiana* and *M. velutina*, (3) partly methylated cyanidin and delphinidin mixtures in *M. acuminata* and (4) peonidin and malvidin mixtures in *M. ornata* and *M. violascens*. In *M. velutina* Harborne (1967) later identified cyanidin and delphinidin 3-glucosides and 3-rutinosides. He also found an abundance of procyanidin and prodelphinidin in green banana and characterized delphinidin 3-rutinoside in *Strelitzia regina* petals.

Bate-Smith (1968) surveyed only five Zingiberaceae and four Marantaceae taxa, *Canna indica* and three *Musa* species for their leaf flavonoids. In a more extensive survey of leaf tissue of 81 species of the Zingiberales, Williams and Harborne (1977b) found that while most of the major classes of flavonoid are represented, only two families, the Zingiberaceae and Marantaceae, are at all rich in these constituents. Thus in the Cannaceae, Musaceae and Strelitziaceae flavonol glycosides were found in small amount and in *Orchidantha maxillarioides* (Lowiaceae) no flavonoids could be detected. Kaempferol, quercetin and proanthocyanidins are distributed throughout the Zingiberaceae but the two subfamilies may be distinguished by the occurrence of isorhamnetin, myricetin and syringetin in the Zingiberoideae and of glycoflavones in the Costoideae. The Marantaceae is chemically the most diverse group and is distinguished from other families in the order by the presence of both flavone *O*- and *C*-glycosides and flavonoid sulphates and the absence of isorhamnetin and kaempferol glycosides. The distribution of flavonoid components in the Marantaceae does not follow tribal classification (Potztal, 1964), since flavonols and flavone *C*-glycosides are found with similar frequency throughout.

Flavonoid glycosides have been characterized in leaves of only nine members of the Zingiberaceae and six Marantaceae taxa. Amongst the rarer constituents are syringetin and myricetin 3-rhamnosides from *Hedychium stenopetalum*, quercetin and isorhamnetin 3-glucuronides and 3-diglucosides from *Roscoea humeana* and quercetin 3-galactosylrhamnoside, schaftoside and isoschaftoside from *Costus sanguineus* (Zingiberaceae). In the Marantaceae, flavonoid sulphates are common leaf components: e.g., apigenin and luteolin 7-sulphates and luteolin 7,3′-disulphate from *Maranta bicolor* and orientin and vitexin 7-glucoside sulphate, orientin 7-glucoside *bis*-sulphate, isovitexin glucoside sulphate and vitexin and isovitexin sulphates from *Stromanthe sanguinea*.

In a limited survey (Williams and Harborne, 1977b) of leaf anthocyanins in 11 taxa of the Zingiberales, the 3-rutinosides of peonidin, petunidin and malvidin were reported for the first time. An anthocyanin pattern based on cyanidin and/or delphinidin, with occasional methylation and the 3-rutinoside as the main glycosidic class, appears to be the same throughout the Zingiberales. Delphinidin has only been found in the Marantaceae, and *Costus* may be distinguished from other Zingiberaceae in having only cyanidin 3-glucoside and no 3-rutinoside. This evidence together with the absence of myricetin and methylated flavonols and the occurrence of flavone *C*-glycosides only in leaves of *Costus* species lends chemical support for the separation of this genus at the family level, as suggested by Tomlinson (1969) from his anatomical evidence.

15.15 EVOLUTIONARY TRENDS IN FLAVONOID PATTERNS

The data base for flavonoids in the monocots is still relatively incomplete so that at present it is difficult to

Table 15.7 Leaf flavonoid patterns within the monocots

Order or family	*Flavone C-glycosides*	*Flavonols*	*Proantho-cyanidins*	*Flavones (Lu, Ap, Aca or Chrys)*	*Tricin*	*6-Hydroxy-flavonoids*	*Flavonoid sulphates*
Arales	+++	++	++	(+)	—	—	(+)
Amaryllidaceae	—	+++	(+)	—	—	—	—
Bromeliaceae	+	++	—	+	—	+	—
Commelinaceae	+++	++	nd	(+)	(+)	+	(+)
Cyperaceae	++	+	++	++	+++	(+)	+
Fluviales	++	(+)	(+)	+	—	—	+
Gramineae	+++	(+)	(+)	+	+++	—	+
Iridaceae	+++	++	+	(+)	(+)	(+)	(+)
Juncaceae	(+)	(+)	+	+++	—	—	+
Liliales	(+)	+	+	++	(+)	—	(+)
Orchidaceae	++	++	(+)	(+)	(+)	+	(+)
Palmae	+++	+	+++	+	++	—	+++
Restionaceae	+	+++	+++	+	—	+*	(+)
Zingiberales	+	++	+++	+	—	—	(+)

Key: (+), in <10% of species; +, in 10–25% of species; ++, in 25–50% of species; +++, in >50% of species; nd, not determined.
*8-Hydroxylation (not 6) in the case of Restionaceae; one 8-hydroxyflavonol, gossypetin, is also recorded in one bromeliad, but the pattern otherwise in bromeliads is 6-hydroxylation.

evaluate evolutionary trends within the group with much confidence. The available data, which have been described in previous sections, are summarized for the leaf flavonoids in Table 15.7 and for the anthocyanin pigments in Table 15.8. These refer to the widely occurring flavonoid classes. Other types of flavonoid are known in monocots but, generally speaking, occurrences are sporadic, often being restricted as far as is known to single species or genera.

Chalcones, for example, are known only in *Killinga* (Cyperaceae), *Polygonatum* and *Tulipa* (Liliaceae), *Xanthorrhoea* (Xanthorrhoaceae) and *Alpinia* (Zingiberaceae). Dihydrochalcones are described from *Smilax* (Liliaceae) and *Alpinia* (Zingiberaceae) and aurones mainly in Cyperaceae, where they are widespread, and in Zingiberaceae, where there is a single record in *Amomum subulatum*. Isoflavonoids are likewise rare, being restricted to *Iris* (Iridaceae) and *Hemerocallis* (Liliaceae). Flavanones and flavanonols have a few more records (see Bohm, 1982, and Chapter 9), having been found in Liliaceae (*Urginea*), Gramineae (*Sorghum*), Cyperaceae (*Cyperus*), Araceae (*Anthurium*) and Zingiberaceae (*Alpinia, Kaempferia*).

Biflavonoids, which are mainly considered to be gymnosperm constituents but do have a few dicot occurrences, have not been reported in monocots until very recently. A biflavanone was provisionally detected in *Lophiola americana* in 1980 by Xue and Edwards; this plant is of Lilialean affinities and is placed in the Melianthiaceae by Dahlgren *et al.* (1985). We have found biflavones for the first time in monocots in *Patersonia* and *Isophysis* of the Iridaceae (Williams *et al.*, 1986) and in *Xerophyta* of the Velloziaceae (Williams and Harborne, unpublished results). The main compound is amentoflavone, but its methyl ethers and a dihydro derivative, as well as hinokiflavone, have all been recorded. This finding in *Xerophyta* appears to be one of the first modern reports of flavonoids in Velloziaceae and suggests that the pattern in this family will be an interesting one to explore.

Phyletic trends are apparent in anthocyanin pigments found in the different families (Table 15.8), if it is accepted that structural complexity is associated with evolutionary advancement. The presence of *O*-methylation, increasing number of sugar substituents, increasing number of hydroxyl groups carrying sugars and presence of acylation may all be considered advanced features. On this basis, only a few families, i.e. Araceae, Lemnaceae, Restionaceae, might be considered to have a primitive pattern. By contrast, the Bromeliaceae and Commelinaceae with 3,7,3′- and 3,5,3′-trisubstitution patterns would be highly advanced families. The Orchidaceae also probably belongs in this group, but it is difficult to be sure at present since the anthocyanins have not been studied in enough species. Whether the presence of 3-desoxyanthocyanins in Cyperaceae, Gramineae and Juncaceae represents the retention of a primitive feature or is an example of advancement by reduction is not easy to decide; the latter is more likely, since these families are otherwise specialized groups. The complexity of anthocyanin structure would appear to be ecologically related to pollination mechanisms in that such pigments are generally absent from families which contain many wind- or fly-pollinated members.

Overall, the leaf flavonoid results (Table 15.7) would support the idea of reticulate evolution of families within

Table 15.8 Anthocyanin patterns in the monocots

	Anthocyanidins					Glycosidic patterns					Acylation	
Order or family	*Pg*	*Cy*	*Dp*	*Methyl ethers*	*3-Desoxy*	*3-Monoside*	*3-Diglycoside*	*5-Glycosylation*	*7-Glycosylation*	*3′-Glycosylation*	*Aromatic*	*Aliphatic*
Arales	+	+	(+)	(+)	—	+	+	(+)	—	—	—	(+)
Bromeliaceae	+	+	+	+	—	+	(+)	+	—	+	—	—
Commelinaceae	—	+	+	(+)	—	+	—	+	+	+	+	(+)
Cyperaceae	—	—	—	—	(+)	—	—	—	—	—	—	—
Fluviales	—	+	+	+	—	+	+	+	—	—	—	—
Gramineae	(+)	+	(+)	+	(+)	+	(+)	(+)	—	—	—	+
Iridaceae	+	+	+	+	—	+	+	+	—	—	(+)	(+)
Juncaceae	—	—	—	—	+	—	—	—	—	—	—	—
Liliales	+	+	+	+	—	+	+	+	—	—	—	—
Orchidaceae	+	+	+	—	—	+	(+)	+	—	—	+	+
Palmae	—	—	—	—	—	—	—	—	—	—	+	+
Restionaceae	—	+	—	—	—	+	—	—	—	—	—	—
Zingiberales	(+)	+	(+)	+	—	+	+	(+)	—	—	—	—

Key: (+), infrequent; +, frequent; —, not detected.

the monocots, since there are no sharp discontinuities in pattern. Nevertheless, it is possible to separate those families which have a fairly simple basic pattern – widespread presence of *C*-glycoside, flavonol and proanthocyanidin – and consider them less specialized than the remainder. Families in which flavones, especially tricin, 6-hydroxyflavonoids and sulphates are regularly found might be considered more advanced. Thus the families of the Arales, Fluviales and Zingiberales would make one group, with families of the Liliales being intermediate, and all other families surveyed as being more specialized.

Such a scenario is as yet incomplete, since many of the smaller families have yet to be investigated in any detail. Moreover, the Dioscoreales (including the Dioscoreaceae) is considered by some plant systematists to be ancestral within the monocots. Unfortunately, we know too little about its flavonoids (see Section 15.10) to be able to include it here. One discordant feature of the leaf flavonoid patterns is the apparent similarity between the palms, grasses and sedges. Since the palms are so well separated from the other two families by morphology, this would appear to represent an example of parallel evolution.

Dahlgren and Rasmussen (1983) have produced cladistic trees for the families of the main monocot orders, using biological characters. For example, the Typhaceae and Sparganiaceae appear as the climax families of the Bromeliiflorae, which otherwise include Pontederiaceae, Haemodoraceae, Philydraceae, Bromeliaceae and Velloziaceae in descending order. Insufficient flavonoid data are yet available on many of these families. It would, however, be a worthwhile goal for future investigation to see whether the flavonoid data are in agreement with such a phylogeny.

REFERENCES

Arditti, J. and Fisch, M.H. (1977). In *Orchid Biology Reviews and Perspectives* (ed. J. Arditti), Cornell University Press, New York.

Ashtakala, S.S. (1975), *J. Am. Soc. Hortic. Sci.* **100**, 546.

Bate-Smith, E.C. (1962), *J. Linn. Soc. (Bot.)* **58**, 95.

Bate-Smith, E.C. (1968), *J. Linn. Soc. (Bot.)* **60**, 325.

Bate-Smith, E.C. (1969). In *Perspectives in Phytochemistry* (eds J.B. Harborne and T. Swain), Academic Press, London, pp. 167–178.

Bate-Smith, E.C. and Harborne, J.B. (1969), *Phytochemistry* **8**, 1035.

Bogner, J. (1978), *Aroideana* **1**, 63.

Bohm, B.A. (1982). In *The Flavonoids – Advances in Research* (eds J.B. Harborne and T.J. Mabry), Chapman and Hall, London, pp. 313–416.

Bohm, B.A. and Collins, F.W. (1975), *Phytochemistry* **14**, 315.

Boudet, A.M., Boudet, A. and Bouysson, H. (1977), *Phytochemistry* **16**, 912.

Boutard, B., Bouillant, M.L., Chopin, J. and Lebreton, P. (1973), *Biochem. Syst.* **1**, 133.

Brockmann, H. and Haase, R. (1937), *Ber. Dtsch. Chem. Ges.* **70**, 1733.

Camarda, L., Merlini, L. and Nasini, G. (1983), *Heterocycles* **20**, 39.

Cardillo, G., Merlini, L. and Nasini, G. (1971), *J. Chem. Soc. C.*, 3967.

Ceska, O. and Styles, E.D. (1984), *Phytochemistry* **23**, 1822.

Chari, V.M., Harborne, J.B. and Williams, C.A. (1980), *Phytochemistry* **19**, 983.

Clifford, H.T. and Harborne, J.B. (1967), *Proc. Linn. Soc. London* **178**, 125.

Clifford, H.T. and Harborne, J.B. (1969), *Phytochemistry* **8**, 123.

Cronquist, A. (1968), *The Evolution and Classification of Flowering Plants*, Nelson, London.

Cutler, D.F. (1969). In *Anatomy of the Monocotyledons* (ed. C.R. Metcalfe), Vol. IV Juncales, Clarendon Press, Oxford.

Dalgren, R.M.T. and Clifford, H.T. (1982), *The Monocotyledons: A Comparative Study*, Academic Press, London.

Dahlgren, R.M.T. and Rasmussen, F.N. (1983), *Evol. Biol.* **16**, 255.

Dalgren, R.M.T., Clifford, H.T. and Yeo, P.F. (1985), *The Families of the Monocotyledons*, Springer-Verlag, Berlin.

Delle Monache, F., Ferrari, F. and Marini-Bettolo, G.B. (1971), *Gazz. Chem. Ital.* **101**, 387.

Delle Monache, F., Ferrari, F., Poce-Tucci, A. and Marini-Bettolo, G.B. (1972), *Phytochemistry* **11**, 2333.

Del Pero Martinez, M.A. (1985), *Biochem. Syst. Ecol.* **13**, 253.

Del Pero Martinez, M.A. and Swain, T. (1977), *Biochem. Syst. Ecol.* **5**, 37.

Del Pero Martinez, M.A. and Swain, T. (1985), *Biochem. Syst. Ecol.* **13**, 391.

Dressler, R.L. (1974). In *The 7th World Orchid Conference Proceedings*, Medellin, Colombia.

Elliger, C.A., Chan, B.G., Waiss, Jr., A.C., Lundin, R.E. and Haddon, W.F. (1980), *Phytochemistry* **19**, 293.

Ellis, C.J., Foo, J.Y. and Porter, L.J. (1983), *Phytochemistry* **22**, 483.

El-Moghazy, A.M., Ali, A.A., Ross, S.A. and El-Shanavany, M.A. (1980), *Fitoterapia* **LI**, 179.

Emberger, L. (1960). *Les Vegetaux Vasculaires*, Tome 111, Masson, Paris.

Engler, A. (1900–1912), *Das Pflanzenreich* **1**, **11**, **20**, **56** (IV45–IV48), Verlag von Engelmann (J. Cramer), Weinheim.

Engler, A. (in collaboration with Krause, K.) (1905–1920), *Das Pflanzenreich* **IV**, 23, A-F (Hefte 21, 37, 48, 55, 60, 71, 73 and 74), Engelmann, Leipzig.

Engler, A. and Prantl, K. (eds) (1930), *Die naturlichen Pflanzenfamilien* (2nd edn), Verlag von Wilhelm Engelmann, Leipzig.

Fredga, A., Bendz, G. and Apell, A. (1974). In *Chemistry in Botanical Classification* (eds G. Bendz and J. Santesson), Academic Press, New York, p. 121.

Freudenberg, K. and Weinges, K. (1958), *Liebigs Ann.* **613**, 61.

Fröst, S., Harborne, J.B. and King, L. (1977), *Hereditas* **85**, 163.

Garay, L.A. (1960), *Bot. Mus. Leafl. Harv. Univ.* **19**, 57.

Goldschmid, O. and Hergert, H.L. (1961), *Tappi* **44**, 858.

Goto, T. (1984), *Proc. 5th Asian Symp. Med. Plants and Spices* (Seoul, Korea), pp. 593–604.

Gripenberg, J., Honkanen, E. and Silander, K. (1956), *Acta Chem. Scand.* **10**, 393.

Harborne, J.B. (1965), *Phytochemistry* **4**, 107.

Harborne, J.B. (1967), *Comparative Biochemistry of the Flavonoids*, Academic Press, London.

Harborne, J.B. (1971), *Phytochemistry* **10**, 1569.
Harborne, J.B. (1979), *Phytochemistry* **18**, 1323.
Harborne, J.B. (1982), *Proceedings of the International Bioflavonoid Symposium* (eds L. Farkas, M. Gabor, F. Kallay and H. Wagner), Akademiai Kiado, Budapest, pp. 251–262.
Harborne, J.B. and Clifford, H.T. (1969), *Phytochemistry* **8**, 2071.
Harborne, J.B. and Self, R. (1987), *Phytochemistry* **26**, 2417.
Harborne, J.B. and Williams, C.A. (1976a), *Biochem. Syst. Ecol.* **4**, 37.
Harborne, J.B. and Williams, C.A. (1976b), *Biochem. Syst. Ecol.* **4**, 267.
Harborne, J.B and Williams, C.A. (1984), *Z. Naturforsch.* **39c**, 18.
Harborne, J.B., Williams, C.A. and Greenham, J. (1974), *Phytochemistry* **21**, 2491.
Harborne, J.B., Williams, C.A. and Wilson, K.L. (1982), *Phytochemistry* **21**, 2491.
Harborne, J.B., Boardley, M. and Linder, H.P. (1985a), *Phytochemistry* **24**, 273.
Harborne, J.B., Williams, C.A. and Wilson, K.L. (1985b), *Phytochemistry* **24**, 751.
Harborne, J.B., Boardley, M., Fröst, S. and Holm, G. (1986), *Plant Syst. Evol.* **154**, 251.
Heap, T. and Robinson, R. (1926), *J. Chem. Soc.*, 2336.
Hegnauer, R. (1963), *Chemotaxonomie der Pflanzen*, Vol. II. *Monocotyledonae*, Birkhauser, Verlag, Basel.
Hegnauer, R. (1986), *Chemotaxonomie der Pflanzen* Band 7, Birkhauser, Verlag, Basel.
Hesse, G. and Klingel, W. (1936), *Liebigs Ann.* **524**, 14.
Horhammer, L., Wagner, H. and Beck, K. (1967), *Z. Naturforsch.* **22b**, 896.
Hrazdina, G. (1982). In *The Flavonoids – Advances in Research* (eds J.B. Harborne and T.J. Mabry), Chapman and Hall, London, pp. 135–188.
Hutchinson, J. (1959). *The Families of Flowering Plants: Vol. II Monocotyledons*, Claredon Press, Oxford.
Ingham, J. (1983), *Fortschr. Chem. Org. Naturst.* **43**, 1.
Ishikura, N. and Shibata, M. (1970), *Bot. Mag. (Tokyo)* **83**, 179.
Ishikura N. and Yoshitama, K. (1984), *J. Plant Physiol.* **115**, 171.
Jaipetch, T., Reutrakul, V., Tuntiwachwuttikul, P. and Santisuk, T. (1983), *Phytochemistry* **22**, 625.
Jurd, L. and Harborne, J.B. (1968), *Phytochemistry* **7**, 1209.
Kaneta, M. and Sugiyama, N. (1973), *Agric. Biol. Chem.* **37**, 2663.
Laracine, C., Favre-Bonvin, J. and Lebreton, P. (1982), *Z. Naturforsch.* **37c**, 335.
Lewis, D.S. and Mabry, T.J. (1977), *Phytochemistry* **16**, 114.
Linder, H.P. (1984), *Bothalia* **15**, 11.
Markham, K.R. and Williams, C.A. (1980), *Phytochemistry* **19**, 2789.
Matsuo, K. and Kubota, T. (1980), *Agric. Biol. Chem.* **44A**, 18.
McClure, J.W. and Alston, R.E. (1966), *Am. J. Bot.* **53**, 849.
McMillan, C., Zapata, O. and Escobar, L. (1980), *Aq. Bot.* **8**, 267.
Melchior, H. (1964). In *Syllabus der Pflanzenfamilien* (ed. A. Engler), Gebruder Borntraeger, Berlin.
Mez, C. (1965). In *Das Pflanzenreich* (ed. A. Engler), **100** (IV.32), Verlag von Engelmann (J. Cramer), Weinheim.
Misra, K. and Mishra, C.S. (1979), *Indian J. Chem.* **18B**, 88.
Momose, J., Abe, K. and Yoshitama, K. (1977), *Phytochemistry* **16**, 1321.
Mongkolsuk, S. and Dean, F.M. (1964), *J. Chem. Soc.*, 4654.
Moore, H.E. (1961), *Am. Hortic. Mag.* **40**, 17.
Murray, B.G. and Williams, C.A. (1976), *Biochem. Genet.* **14**, 897.
Nagy, E., Seres, I., Verzar-Petri, G. and Neszmelyi, A. (1984), *Z. Naturforsch.* **39b**, 1813.
Nohara, T., Ito, Y., Seike, H., Komori, T., Moriyama, M., Gomita, Y. and Kawasaki, T. (1982), *Chem. Pharm. Bull.* **30**, 1851.
Pagani, F. (1976), *Boll. Chim. Farm.* **115**, 407.
Potztal, E. (1964). In *Syllabus der Pflanzenfamilien* (ed. A. Engler), Gebruder Borntraeger, Berlin.
Raffauf, R.F., Menachery, M.D., Le Quesne, P.W., Arnold, E.V. and Clardy, J. (1981), *J. Org. Chem.* **46**, 1094.
Řeznik, H. and Neuhäusel, R. (1959), *Z. Bot.* **47**, 471.
Roberts, M.L. and Haynes, R.R. (1986). *Nord. J. Bot.* **6**, 291.
Rofi, R.D. and Pomilio, A.B. (1985), *Phytochemistry* **24**, 2131.
Saito, N. and Harborne, J.B. (1983), *Phytochemistry* **22**, 1735.
Saito, N., Yokoi, M., Yamaji, M. and Honda, T. (1985), *Phytochemistry* **24**, 2125.
Salmenkallio, M., McCormick, S., Mabry, T.J., Dellamonica, G. and Chopin, J. (1982), *Phytochemistry* **21**, 2990.
Simmonds, N.W. (1954), *Nature (London)* **173**, 402.
Skrzypczak, L. (1976), *Herba Pol.* **22**, 336.
Stabursvik, A. (1969), *Acta Chem. Scand.* **22**, 2371.
Stirton, J.C. and Harborne, J.B. (1980), *Biochem. Syst. Ecol.* **8**, 285.
Strack, D., Busch, E., Wray, V., Grotjahn, L. and Klein, E. (1986), *Z. Naturforsch.* **41c**, 707.
Takeda, K., Harborne, J.B. and Self, R. (1986), *Phytochemistry* **25**, 2191.
Takhtajan, A. (1969), *Flowering Plants, Origin and Dispersal* (Translation by C. Jeffrey), Oliver and Boyd, Edinburgh.
Timberlake, C.F. and Bridle, P. (1975). In *The Flavonoids* (eds J.B. Harborne, T.J. Mabry and H. Mabry), Chapman and Hall, London, pp. 214–266.
Tomlinson, P.B. (1969). In *Anatomy of the Monocotyledons*, Vol. III (ed. C.R. Metcalfe), Clarendon Press, Oxford.
Ulubelen, A. and Mabry, T.J. (1982), *Rev. Latinoam. Quim.* **13**, 35.
Uphoff, W. (1979), *Experientia* **35**, 1013.
Wagner, H., Obermuer, G., Chari, V.M. and Galle, K. (1980), *J. Nat. Prod.* **43**, 583.
Wiermann, R. (1968), *Sonder. Ber. Dtsch. Bot. Gesellsch. Jg.* **81**, 3.
Williams, C.A. (1975), *Biochem. Syst. Ecol.* **3**, 229.
Williams, C.A. (1978), *Phytochemistry* **17**, 729.
Williams, C.A. (1979), *Phytochemistry* **18**, 803.
Williams, C.A. and Harborne, J.B. (1975), *Biochem. Syst. Ecol.* **3**, 181.
Williams, C.A. and Harborne, J.B. (1977a), *Biochem. Syst. Ecol.* **5**, 45.
Williams, C.A. and Harborne, J.B. (1977b), *Biochem. Syst. Ecol.* **5**, 221.
Williams, C.A. and Harborne, J.B. (1985), *Z. Naturforsch.* **40c**, 325.
Williams, C.A. and Murray, B.G. (1972), *Phytochemistry* **11**, 2507.
Williams, C.A., Harborne, J.B. and Clifford, H.T. (1971), *Phytochemistry* **10**, 1059.
Williams, C.A., Harborne, J.B. and Clifford, H.T. (1973), *Phytochemistry* **12**, 2417.
Williams, C.A., Harborne, J.B. and Crosby, T.S. (1976), *Phytochemistry* **15**, 349.
Williams, C.A., Harborne, J.B. and Mayo, S.J. (1981), *Phytochemistry* **20**, 217.

Williams, C.A., Harborne, J.B. and Glassman, S.F. (1983), *Plant Syst. Evol.* **142**, 157.
Williams, C.A., Harborne, J.B. and Glassman, S.F. (1985), *Plant Syst. Evol.* **149**, 233.
Williams, C.A., Harborne, J.B. and Goldblatt, P. (1986), *Phytochemistry* **25**, 2135.
Wollenweber, E. (1984). In *Biology and Chemistry of Plant Trichomes* (eds E. Rodriguez, P.L. Healey and I. Mehta), Plenum Press, New York and London.
Wollenweber, E. and Dietz, V.H. (1981), *Phytochemistry* **20**, 869.
Wong, E. and Francis, C.M. (1968), *Phytochemistry* 7, 2123.
Xue, L.-X. and Edwards, J.M. (1980), *Planta Med.* **39**, 220.
Zennie, T.M. and McClure, J.W. (1977), *Aq. Bot.* **3**, 49.

16

Flavonoids and flower colour

RAYMOND BROUILLARD

16.1 INTRODUCTION

Flower colour results from the preferential absorption of part of the visible light by one, or several, chemical compound(s) synthesized by higher plants. Since plants cannot 'see' their own colours, it is conceivable that the colour signals emerging from flowers are messages interpretable by mammals as well as by birds and insects (Harborne, 1976). Thus colour is most likely a link between animals and plants, as are flower shape, scent and taste. In particular, insects and birds attracted by colour pollinate the flowers and aid the survival of plant species. In contrast to man, some insects, especially bees, can perceive in the near ultraviolet (340–380 nm) as well as in the visible.

The pigments responsible for flower colour are essentially flavonoids and are located in the vacuoles (McClure, 1979). They are divided into several groups of which one represents anthocyanins. Anthocyanins provide most of the pink, orange, red, violet and blue colours of flowers (Harborne, 1967). Chalcones, aurones and yellow flavonols contribute sometimes to yellow flower colour. The remaining flavonoids are colourless to man, but may be visible to some insects. Flavones, flavonols and their glycosides are also important to flower colour because they act as co-pigments.

Harborne (1976) reviewed the contribution of flavonoids to flower colour. All the factors known at the time to influence flower pigmentation were discussed and it was concluded that, among the flavonoids, anthocyanins are the most important compounds. Since 1976, our knowledge of the chemistry and properties of these compounds has greatly increased. New very complex anthocyanins have been discovered (see Chapter 1). It is now clear what structural transformations anthocyanins undergo in aqueous solutions (Brouillard, 1982). Powerful analytical techniques, such as NMR, FAB-MS and Raman spectrometry have been successfully applied to the study of anthocyanins (Goto *et al.*, 1978; Bridle *et al.*, 1984; Merlin *et al.*, 1985). Taking advantage of the strong visible-light absorption by anthocyanins, microspectrometric techniques have been devised to investigate these pigments at the cellular and even subcellular level (Stewart *et al.*, 1975; Brouillard, 1983).

For further understanding of flower colour processes, however, it is important to study the interactions at the molecular level between anthocyanins and the other vacuolar constituents, e.g. other flavonoids, mineral ions, sugars, peptides and organic acids. Since anthocyanins are so widespread in flowers, and since they are the most important pigments, attention in this chapter will be focused primarily on these compounds and on the more significant research results obtained during the last decade. Nevertheless, since recent important work has been achieved on yellow flavonoids and UV patterning in flowers, the last section of this chapter will be devoted to these topics.

16.2 CHEMICAL STRUCTURE OF NATURAL ANTHOCYANINS

As has been pointed out, compounds capable of absorbing visible light are called pigments. Each pigment absorbs

The Flavonoids. Edited by J.B. Harborne
Published in 1988 by Chapman and Hall Ltd,
11 New Fetter Lane, London EC4P 4EE.

light in a specific manner which depends essentially on its chemical structure, and, for an appreciable part, on its environment. In epidermal flower cells, where anthocyanins are most abundantly found, the aqueous environment is always more or less acidic or neutral, and chemical structural changes are among the most important factors responsible for variations in flower colour. By structural changes we mean, not only pigments of different structures (e.g. pelargonin and cyanin), but also different structures of the same pigment. Effectively, what distinguishes anthocyanins from other flavonoids is not only the fact that they are coloured while most flavones and flavonols are colourless, but also that they give rise, especially in water, to many elementary reactions including proton transfer, isomerization and tautomerization. Under the same conditions, colourless flavonoids do not react and therefore appear only in one chemical state. Whether the first type of structural change (pelargonin to cyanin, for instance) is more important than the second type (proton transfer, tautomerization, for instance) is difficult to estimate. Both are probably closely linked, and once a given anthocyanin is formed (primary or basic structure), it could evolve towards numerous secondary structures derived from the primary structure, but possessing different colours, according to the conditions prevailing in a given cell. Relationships between flower colour and basic structure of anthocyanins have been reviewed (Harborne, 1976).

For a long time the structures of flavonoids could only be established by means of purely chemical analytical methods. The first instrumental technique to give good results in structural elucidation of flavonoids was electronic spectrometry (Harborne, 1957; Jurd, 1962; Swain, 1985). In the recent book by Markham (1982), the classical chemical methods as well as the more recent chromatographic and spectroscopic techniques for flavonoid identification are given. Of special interest are NMR spectroscopy and mass spectrometry. These two techniques, however, have only recently been applied with success to natural anthocyanins without derivatization. Much of our most recent knowledge of the new complex anthocyanin primary structures has been obtained by NMR spectroscopy. This technique was first applied to anthocyanins by Goto *et al.* (1978), and shown by these and other authors (Cornuz *et al.*, 1981; Bridle *et al.*, 1984) to be the most powerful technique currently available for structural elucidation. ^{1}H and ^{13}C NMR techniques were previously applied successfully to the identification of the colourless flavonoids (Markham and Mabry, 1975; Markham and Chari, 1982).

16.2.1 Primary structure

Natural anthocyanins are generally isolated in the form of flavylium salts. The flavylium cation bears hydroxyl as well as methoxyl and glycosyl groups. Natural anthocyanins always possess a sugar and hence are glycosides. An apparent exception to this rule, however, is carajurin, a 3-deoxyanthocyanidin present in both leaves and flowers of *Arrabidaea chica* (Scogin, 1980). The most frequently encountered sugars are β-D-glucose, β-D-galactose and α-L-rhamnose (Timberlake and Bridle, 1975). Many di- and tri-saccharides are also known to be formed from combinations of the previously mentioned monosaccharides. With a few exceptions (3-deoxyanthocyanins) there is always a glycosyl group at C-3. When more than one sugar is present in the molecule, the other sugar(s) can be attached to any one of the hydroxyls at C-5, C-7, C-3′, C-5′ and even at C-4′. Some very rare 6-hydroxyanthocyanins have been isolated from the red flowers of *Alstroemeria* (Saito *et al.*, 1985b). As previously demonstrated by Jurd and Harborne (1968), 6-hydroxylation causes a hypsochromic shift of the visible absorption band. A similar effect is produced by 3′-glucosylation (Saito and Harborne, 1983). Sugars may be acylated by acetic, malonic, succinic, benzoic or cinnamic acids giving rise to mono- or poly-acylated anthocyanins. For instance, the flavylium cation of platyconin, a diacylated pigment with a linear side chain, is shown in Fig. 16.1. Platyconin is found in the petals of the chinese bell-flower *Platycodon grandiflorum*. Its stereochemistry has been established by use of pulsed Fourier transform ^{1}H NMR spectrometry (Goto *et al.*, 1983a). Of the six delphinidin hydroxyl groups, four are free and two are substituted, the one at C-3 with a rutinosyl residue and the one at C-7 by a linear chain of alternating glucosyl and caffeyl groups. This unique substitution pattern gives this molecule a high degree of flexibility. A complete list of the new structures of the 1981–1985 period, especially the zwitterionic anthocyanins, is to be found in Chapter 1.

It should be noted, however, that whatever the extraction medium, it will undoubtedly be significantly different from the natural medium, and this may well destroy part of the *in vivo* structure of the pigment. In this regard, the structural changes caused by the processes of extraction, purification and analysis can cause the loss of a labile group, such as malonic acid, but it could also relate to a dilution effect or an ionic strength effect resulting in the dissociation of a complex in which the anthocyanin is involved. It is remarkable that, whenever mild extraction conditions have been used, the pigments identified have been frequently associated with other chemical compounds. In some cases, macromolecules have even been associated with natural pigment structures. For a summary on the so-called genuine anthocyanins, see Osawa (1982).

16.2.2 NMR spectroscopy

The first proton NMR spectra of anthocyanidins were reported by Nilsson (1973). Goto *et al.* (1978) reported the proton NMR spectra of awobanin, extracted from *Commelina communis*, of violanin found in *Viola tricolor* and of shisonin, the pigment of *Perilla ocimoides*. The spectra of

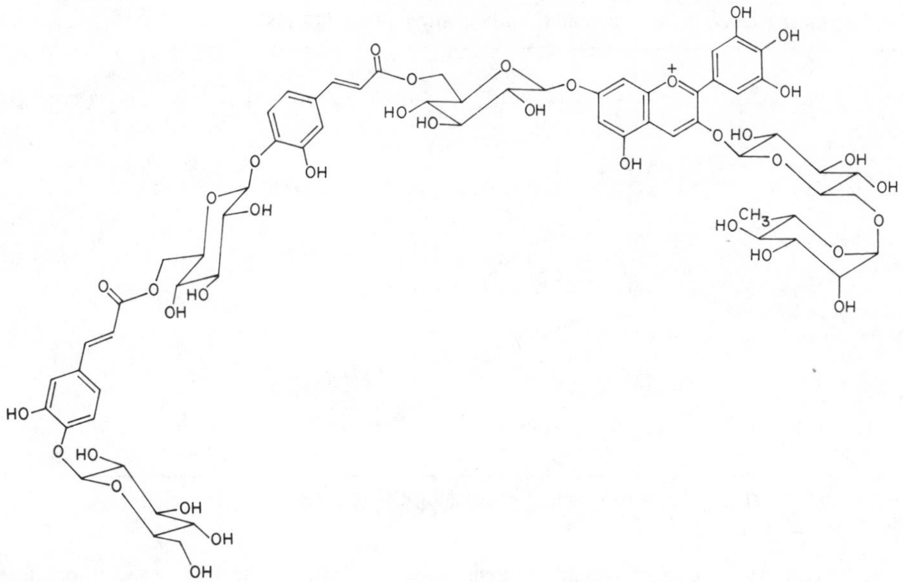

Fig. 16.1 Structure of platyconin, the anthocyanin of *Platycodon grandiflorum* flowers (Goto *et al.*, 1983a, redrawn by permission of the authors and Pergamon Press).

these three anthocyanins were recorded while the pigments were in the flavylium form, that is in acidified protic solvents (CD_3OD/DCl and D_2O/DCl). Since 1978, many new, sometimes very complex, anthocyanin stereostructures have been elucidated by NMR procedures. The following examples may be quoted: pelargonidin 3-malonylsophoroside from the flower of the red iceland poppy (Cornuz *et al.*, 1981); gentiodelphin from the petals of *Gentiana makinoi* (Goto *et al.*, 1982); platyconin (Goto *et al.*, 1983a); cinerarin from the blue garden cineraria *Senecio cruentus* (Goto *et al.*, 1984); cyanidin 3-malonylglucoside from the red leaves of *Cichorium intybus* (Bridle *et al.*, 1984); monardaein, which bears two malonyl residues, from the red petals of *Monarda didyma* (Kondo *et al.*, 1985) and ternatins from the blue flowers of *Clitoria ternatea* (Saito *et al.*, 1985a). NMR spectroscopy is very valuable in determining the stereochemistry of anthocyanins. For instance, the position of attachment of a glucosyl group can be established by the use of the nuclear Overhauser effect (n.O.e). The anomeric proton of glucose in the β-configuration is characterized by a coupling constant of 7–8 Hz. Attachment of an acyl group to the 6 position of a glucosyl moiety is apparent from the deshielding of the methylene protons as compared to the lines of the methylene protons of an unsubstituted glucosyl group. Characteristic coupling constants ($\simeq$ 16 Hz) are observed for the *trans* protons of the double bond of cinnamic esters. The anthocyanidin protons (protons at C-4, C-6, C-8, C-2′, C-6′, C-3′ and C-5′) usually appear at low field and exhibit signals with characteristic features. For instance, in the case of malvidin 3-glucoside, the flavylium nucleus protons have δ values, in D_2O/DCl, in the range 6.5–9.1 ppm, whereas the glucose protons resonate in the range 3.4 to 4 ppm with the exception of the anomeric proton whose signal is close to 5.4 ppm. The contribution of the sugar moiety to the spectrum is generally very complicated due to the fact that aliphatic protons have only slightly different chemical environments and also due to the strong coupling that occurs between vicinal protons (Goto *et al.*, 1982).

^{13}C NMR spectra of anthocyanins have also been recorded (Bridle *et al.*, 1984). The natural abundance of ^{13}C is 1.1%, and it is only recently with the advent of pulsed Fourier transform NMR apparatus that it has been used for the structural elucidation of organic molecules. For the application of this new technique to flavonoids in general, consult the review by Markham and Chari (1982) which clearly outlines its major drawbacks and advantages over proton NMR. In this regard, it should perhaps be mentioned that ^{13}C–^{13}C coupling does not exist and decoupling the ^{13}C atoms from the protons makes the spectrum relatively easy to interpret; in such a case, each resonance line corresponds to only one type of carbon atom, and the carbon skeleton of the molecule is readily obtainable. Here again, sugar carbon resonances are well separated from the aglycone carbon resonances (Bridle *et al.*, 1984).

16.2.3 Mass spectrometry

Field desorption was probably the first mass spectrometric technique applied to the study of anthocyanins without

Table 16.1 FAB mass spectrometry of malonated anthocyanins of Compositae

Pigment	$[M]^+$	Loss of malonic $[M-86]^+$	Loss of glucose $[M-162]^+$	Loss of malonylglucose $[M-248]^+$	Loss of 2 malonyl $[M-172]^+$	Aglycone $[M]^+$
Pg 3-(6″-malonylglucoside)	519	433	—	—	—	271
Pg 3-(6″-malonylglucoside)-5-5-glucoside	681	595	519	433	—	271
Pg 3-(6″-malonylglucoside)-5-malonylglucoside	767	681	—	519	595	271
Cy 3-dimalonylglucoside	621	535	—	—	—	287
Cy 3-malonylglucuronosylglucoside	711	625	535*	—	—	287
Cy 3-(6″-malonylglucoside)-5-glucoside	697	611	535	449	—	287
Cy 3-(6″-malonylglucoside)-5-malonylglucoside	783	697	—	535	611	287
Dp 3-(6″-malonylglucoside)-5-malonylglucoside	799	713	—	551	—	303

*Loss of glucuronic acid $[M-176]^+$; Pg, pelargonidin; Cy, cyanidin; Dp, delphinidin.
Data from Takeda *et al.* (1986).

derivatization (Goto *et al.*, 1981). Fast atom bombardment (FAB) mass spectrometry has also been used. FAB mass spectra were recorded by Saito *et al.* (1983) for cyanidin, malvidin 3-glucoside, violanin and platyconin. For violanin, the fragmentation pattern (losses of varied associations of glucosyl, rhamnosyl and coumaroyl groups), is in good agreement with the structure previously established by NMR techniques (Goto *et al.*, 1978). For platyconin, however, the structure proposed from the FAB spectra differs completely from the structure shown in Fig. 16.1. The reason for such a discrepancy is apparently due to the fact that glucosyl and caffeyl residues have the same molecular weight. Saito *et al.* (1983), however, clearly indicated the different theoretical combinations of glucosyl and caffeyl groups which could give rise to the measured mass peaks. Mass spectrometry alone is probably not sufficient to elucidate completely the structure of a complex pigment like platyconin. In the sinapyl ester of cyanidin 3-(xylosylglucosylgalactoside) present in carrot tissue culture, the detection by FAB-MS (Harborne and Self, unpublished results) of the loss of a sinapylglucose residue provided the means of locating the sinapic acid on the glucose moiety. Similarly, the separate loss of xylose and absence of other sugar fragments indicates that the 3-trisaccharide of the pigment is branched. Very recently, FAB-MS was used systematically in the structural elucidation of malonated anthocyanins from various members of the Compositae (Takeda *et al.*, 1986). Typical results are shown in Table 16.1.

16.3 ANTHOCYANINS IN AQUEOUS MEDIA

Among the factors involved in flower pigmentation, the pH of the vacuole is of utmost importance. It is well known that, for model solutions, many colour changes occur with changes in pH. Though the effect of pH has been studied for a long time, it was not until the late seventies that the structural transformations observed by increasing or decreasing the pH of an acidic or neutral anthocyanin solution were finally clarified (Brouillard and Dubois, 1977; Brouillard and Delaporte, 1977; Brouillard, 1982). For basic media there is still a gap in our knowledge of structural transformations, and no definitive work has yet been published. For slightly dilute anthocyanin acidic or neutral solutions, the main results are: (a) the finding of kinetic and thermodynamic competition between the hydration reaction of the pyrylium ring and the proton transfer reactions related to the acidic hydroxyl groups of the aglycone; (b) that no hydration of the quinonoidal bases takes place, and (c) the existence of an open form, the chalcone pseudobase (Brouillard, 1982). Very recently, Cheminat and Brouillard (1986), using proton NMR spectroscopy, were able to demonstrate the existence, in the case of malvidin 3-glucoside, of a water 4-adduct, as well as two chalcone stereoisomers *E* and *Z*. The mechanism associated with the structural transformations of anthocyanins in water is shown in Fig. 16.2. This is significantly different from the mechanism appearing in the previous edition of this book (Hrazdina, 1982), and is based on reliable kinetic, UV spectrometric, thermodynamic and NMR data (Brouillard and Dubois, 1977; Brouillard and Delaporte, 1977; Brouillard, 1982; Cheminat and Brouillard, 1986). The old mechanism was sustained only by visual and visible spectral measurements.

16.3.1 Prototropic tautomerism

The flavylium cation of natural anthocyanins behaves as a weak diacid, whereas a neutral quinonoidal base is, at the same time, a weak acid and a weak base. The coloured forms appearing on Fig. 16.2 (AH^+, A_5, $A_{4'}$, A_7, $A^-_{54'}$, A^-_{75}, $A^-_{4'7}$), are probably of utmost importance in flower and fruit pigmentation. The equilibrium position between the flavylium cation (AH^+) and the neutral quinonoidal bases (A_5, A_7 and $A_{4'}$) is pH-dependent and characterized by pK'_a values ranging from 3.50 [the *Zebrina pendula* anthocyanin (Brouillard, 1981)] to 4.85 [4′-methoxy-4-methyl-7-hydroxyflavylium chloride (Brouillard *et al.*, 1982)]. Loss of a proton can occur at any of the hydroxyl groups at C-4′, C-5 or C-7. These hydroxyls are much more acidic than the corresponding hydroxyls in flavones and flavonols, for instance. It is remarkable that all natural anthocyanins known to date always possess at least one free hydroxyl at the 4′, 5 or 7 positions. This fact clearly indicates that formation of a quinonoidal base is vital for flower pigmentation. If two acidic hydroxyls are present in the cation, an ionized quinonoidal base is formed at pH values higher than 6 (Brouillard, 1982). Ionized quinonoidal bases, shown on top of Fig. 16.2, give rise to large bathochromic and hyperchromic shifts as compared to the

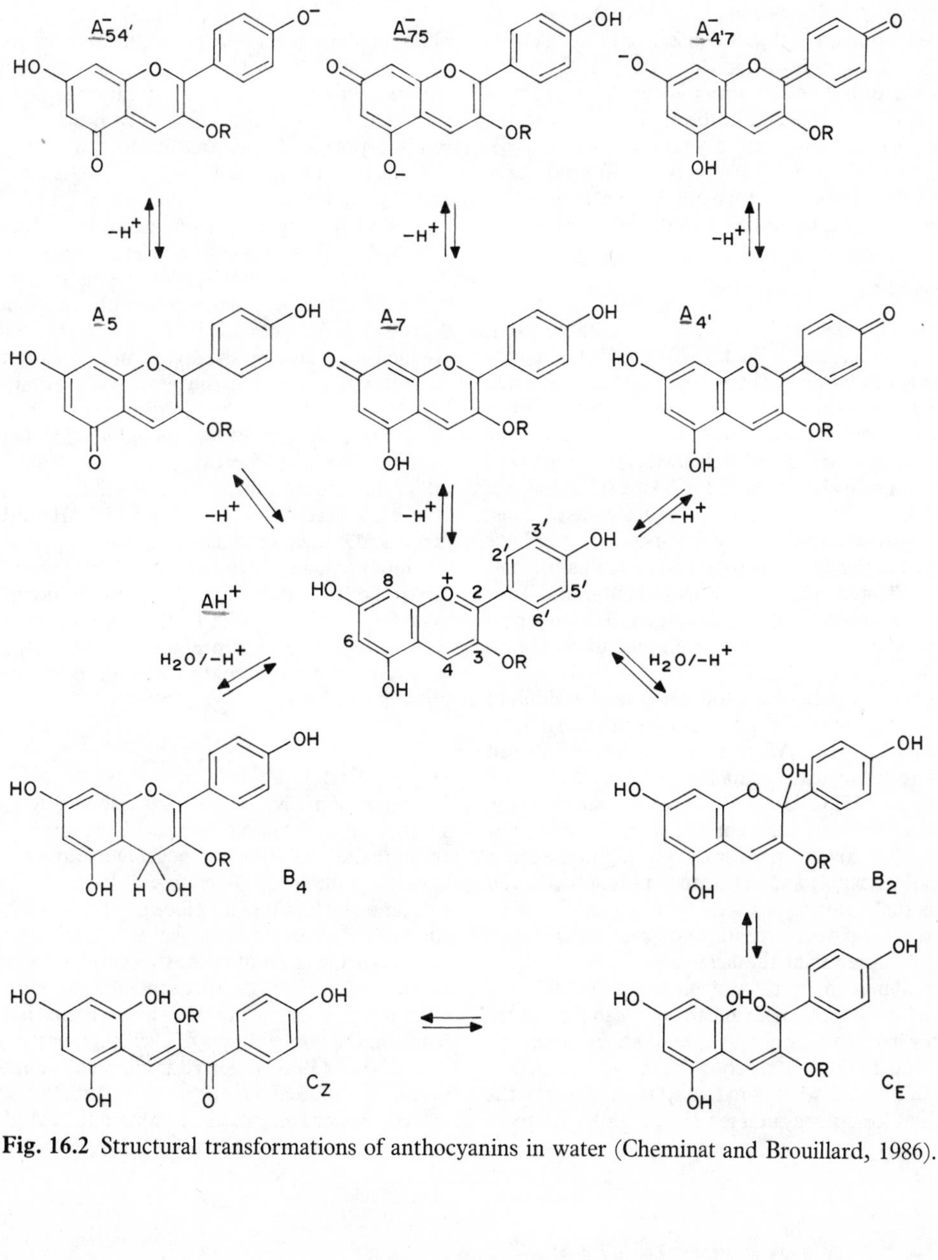

Fig. 16.2 Structural transformations of anthocyanins in water (Cheminat and Brouillard, 1986).

neutral bases. Since high pH values have been measured in some petal vacuoles, there is no doubt that anionic quinonoidal bases also contribute to flower coloration.

16.3.2 Covalent hydration of the flavylium cation

Natural anthocyanin flavylium cations are often rapidly and almost completely hydrated to colourless carbinol pseudobases at pH values ranging from 3 to 6. Water addition takes place at the 2 position (B_2), and only small amounts of the 4-adduct (B_4) can be formed due to unfavourable kinetic and thermodynamic conditions (Brouillard, 1982; Cheminat and Brouillard, 1986). In the absence of a 3-glycosyl substituent, the hydration process is less efficient and carbinol pseudobases only form at pH values higher than 4–5 (Brouillard *et al.*, 1982). The hydration–dehydration equilibrium is characterized by a pK'_h value which is expressed in the same manner as the pK'_a value. For ordinary anthocyanin mono- and di-glycosides, pK'_h is always lower than pK'_a and large quantities of the colourless carbinol B_2 are formed, at equilibrium, in slightly acidified water.

16.3.3 Ring–chain tautomerism

It was previously believed that natural anthocyanins do not form chalcone pseudobases (Jurd, 1972). By increasing the temperature, however, it has been possible to demonstrate the existence of a ring–chain prototropic tautomerism between the carbinol pseudobase B_2 and the chalcone pseudobase C_E (Brouillard and Delaporte, 1977). At room temperature, and in slightly acidic aqueous solutions, the interconversion of the carbinol to the chalcone is a fast reaction and only small amounts of the open tautomer have been observed in the case of natural anthocyanins (Brouillard, 1982). Preston and Timberlake (1981), and later Bronnum-Hansen and Hansen (1983), quantitatively prepared the chalcone forms of some common anthocyanins using HPLC.

It is possible to predict the production of the different secondary structures of a given anthocyanin according to the acidity of the solution. When dissolving a flavylium salt in a slightly acidic or neutral aqueous solution, the neutral and/or ionized quinonoidal bases appear immediately. However, for the more common 3-glycosides and 3,5-diglycosides, they change more or less rapidly to the much more stable carbinol and chalcone pseudobases. At equilibrium, such solutions are weakly coloured. The main reason for this is that the hydration constant K'_h is always 10–100 times larger than the acid–base constant K'_a. If such a phenomenon were preponderant in flowers, anthocyanins would not confer much colour to them. One can therefore assume that in vacuoles there are processes preventing the hydration reaction from occurring to a large extent. It is somewhat paradoxical that, although the natural solvent for anthocyanin is water, in order for these pigments to manifest their maximum colouring capacity they have to be protected against their own solvent. However, from time to time, reports of the occurrence *in vivo* of colourless pseudobases in plants have been made (Harborne, 1967; Crowden, 1982). These are probably exceptions to the general protection against hydration.

16.3.4 pH effect *in vivo*

Shibata *et al.* (1949) found that the pH values of crude extracts of numerous flower, fruit and leaf tissues ranged from 2.8 to 6.2. Stewart *et al.* (1975) were able to show that the pH of young epidermal flower cells is between 2.5 and 7.5. Interestingly, Asen *et al.* (1971a) found the vacuolar pH of epidermal cells of the 'Better Times' rose petals to change from 3.70–4.15 in fresh petals to 4.40–4.50 in 3-day-old cut petals. At the same time the colour changed from red to blue. For 'Heavenly Blue' flowers, as the reddish–purple buds transform into light-blue open flowers, the pH changes from 6.5 to 7.5 (Asen *et al.*, 1977). In general, it has been assumed that blueing of flower tissues, as they age, is accompanied by a decrease in the free acidity. A decrease in pH causes the opposite effect. For example, the blue–violet young petals of *Fuchsia* (pH 4.8), turn to purple–red with time and the pH decreases to 4.2 (Yazaki, 1976). All the pH effects *in vivo* are well explained by the appearance or disappearance of the strongly coloured species AH^+, A and A^- (see Fig. 16.2). Lowering the pH displaces the anionic quinonoidal bases (A^-) to the neutral quinonoidal bases (A) and, finally to the flavylium cation and *vice versa*. One should notice that the pK values governing the structural modifications fall exactly within the 2.5–7.5 pH range. For the more acidic media the flavylium cation predominates. In the intermediate range two situations can prevail: (a) a mixture of the cation and the neutral quinonoidal bases is to be found (pH from 3.5 to 4.5), and (b) only the neutral quinonoidal bases are present (pH from 4.5 to 6). At neutrality, there is a mixture of both neutral and anionic quinonoidal bases.

16.4 EFFECT OF CONCENTRATION

Concentration is a factor influencing colour of anthocyanin-containing media. Frequently large variations in the amount of anthocyanin in a given flower tissue have been observed. Asen *et al.* (1971a) found the concentration of cyanin in epidermal cells of a petal of the 'Better Times' rose to be as large as 2.4×10^{-2} M. A large increase in cyanin content was observed by Yasuda (1971), when comparing pale to deep-pink rose petals. When several pigments are found in a living tissue, it is not easy to measure the contribution of each pigment to the colour. Akavia *et al.* (1981) reported as many as 24 anthocyanins in petals of different *Gladiolus* cultivars; the six common aglycones were all present in four different glycosylation states. With the exception of the 3-glucosides, the 18 remaining anthocyanins make the major contribution to the pigmentation of a given *Gladiolus* flower. Saito *et al.* (1985a) observed as many as six blue acylated anthocyanins in the blue flowers of *Clitoria ternatea*. Two of them were shown to be highly acylated. It is difficult to explain at present why there should be so many different pigments in these blue flowers.

It has been assumed (Brouillard, 1983) that physical adsorption on a suitable surface of any of the visible light-absorbing chromophores will provide a good means of preventing colour loss, by taking the pigment out of the bulk of the solution. Adsorption is more likely to occur if the molecule is planar. It has been demonstrated that pure crystalline pelargonidin bromide monohydrate is a nearly planar molecule (Saito and Ueno, 1985). Similar results were reported 10 years ago for different 3-deoxyanthocyanidins (Busetta *et al.*, 1974; Ueno and Saito, 1977a, 1977b). A planar configuration for the flavylium nucleus of malvidin 3-glucoside probably also obtains in an aqueous solution. For instance, Cheminat and Brouillard (1986), observed formation, in equal amounts, of two epimers of malvidin 3-glucoside water 2-adduct. These epimers have opposite configurations at C-2. If the B-ring plane is to make a large dihedral angle with the benzopyrylium plane, one would expect different amounts of the two epimers to be formed, as a result of different steric effects when the water molecule approaches the reactive C-2 atom.

Another way of retaining anthocyanin colour is to remove as much water as possible. In this connection, Pecket and Small (1980) and Hemleben (1981) have observed the existence of red-coloured crystals *in vivo*. Previously, Asen *et al.* (1975) reported that blue crystals occur in vacuoles of aged flowers. An intriguing aspect of anthocyanin chemistry is their ability to self-associate. Self-association was first put forward by Asen *et al.* (1972) to explain the deviation from the Beer–Lambert law in connection with increasing concentrations of cyanin. A similar result was reported by Timberlake (1980) in the case of malvidin 3-glucoside. Self-association was only recently quoted among the factors modifying the colour of anthocyanin-containing media (Timberlake, 1980, 1981). It would be interesting to know what the solubility of anthocyanins is and how this is affected by pH or by the nature of the solvent system. It may well be that self-association and solubility are related.

Most of the work on anthocyanin self-association has been carried out by Hoshino *et al.* (1980a, 1981a, b, 1982). With one exception (Hoshino *et al.*, 1982), self-association has always been demonstrated to take place using UV-visible absorption spectroscopy and circular dichroism with aqueous solutions buffered to pH 7 with phosphate. Many natural anthocyanins have been investigated (e.g. cyanin, malvin, hirsutin), as well as structurally related derivatives. Cyanin and malvin exhibit opposite Cotton effects, association being shown to occur much faster for malvin than for cyanin. 4′-*O*-Methylmalvin does not associate. A 5-glucosyl is also necessary for self-association to take place. Structural arrangements of the monomeric forms present at pH 7 lead to chiral molecular aggregates with left- or right-handed helical stacking geometry depending on the anthocyanin structure. The proton NMR spectrum of malvin quinonoidal bases was demonstrated to be concentration-dependent. Flavylium cations also form aggregates (Hoshino, 1986). It is not yet fully established how much of the observed absorbance or ellipticity (circular dichroism measurements) can originate from optical effects other than absorption. The role played by the buffer, if any, remains to be established.

In the next sections, we focus attention on the molecular interactions between anthocyanins and compounds known to modify their initial colour.

16.5 INTERMOLECULAR CO-PIGMENTATION

A co-pigment has no colour in itself, but when present in sufficient amount, it enhances the stability of anthocyanin chromophores. As is frequently the case in chemistry, a given molecular interaction can be produced in two ways: intramolecularly or intermolecularly. This means that the co-pigment effect can be brought about by either a different molecule or by part of the pigment structure itself. If co-pigmentation is due to a different molecule, this is the classical co-pigment effect, now known as intermolecular co-pigmentation (Brouillard, 1982; Goto *et al.*, 1986). If the co-pigment is covalently bound to the chromophore, it is described as an intramolecular co-pigment effect (Brouillard, 1981, 1982, 1983; Iacobucci and Sweeny, 1983; Goto *et al.*, 1986). With pH, co-pigmentation is probably the most quoted of the many factors involved in flower pigmentation due to flavonoids. In fact, recent work, based on HPLC, has unambiguously demonstrated that co-pigments are often associated with anthocyanins in flowers (Asen and Griesbach, 1983; Asen, 1984; Van Sumere *et al.*, 1984, 1985). Since its discovery in 1931 (Robinson and Robinson), co-pigmentation has been extensively studied; we only discuss here the results emerging from the literature during the past decade or so. Although known for a long time, co-pigmentation has only recently been investigated at the molecular level (Goto *et al.*, 1979; Goto *et al.*, 1986; Mazza and Brouillard, 1987).

The intermolecular co-pigment effect results in both bathochromic and hyperchromic shifts in the visible maximum (Asen *et al.*, 1975; Harborne, 1976; Asen, 1976; Scheffeldt and Hrazdina, 1978; Timberlake, 1980, 1981; Timberlake and Bridle, 1980; Brouillard, 1982; Osawa, 1982; Takeda *et al.*, 1985a, b). Our understanding of intermolecular co-pigmentation has been advanced considerably by the elucidation of the structural transformations that anthocyanins undergo in solution (Brouillard and Delaporte, 1977). Since hydration only takes place with the flavylium cation (see Fig. 16.2), and since intermolecular co-pigmentation leads to increase in absorbance in the visible range (more chromophores are present), it appears (Brouillard, 1982) that the co-pigment molecule partly prevents the hydration reaction occurring. In other words, the flavylium cation–co-pigment complex does not hydrate and the co-pigment offers real protection to the pyrylium ring against water attack. Therefore formation of the complex competes with formation of the pseudobases. Complexation with quinonoidal bases, either neutral or ionized, reinforces the previous basic co-pigment effect, but in its absence, it would probably be insufficient to prevent hydration of the flavylium form occurring to a considerable extent. The stability constant of the cyanin flavylium–quercitrin complex is estimated to be close to 2 $\times 10^3\ \text{M}^{-1}$ which leads to an apparent reduction in the hydration constant from 10^{-2} to 7×10^{-4} M (Brouillard, 1982).

Most of the natural anthocyanins form co-pigment complexes when present in suitable concentrations (Asen *et al.*, 1972). Numerous substances are known to act as co-pigments, many of them being polyphenols (Asen *et al.*, 1972; Chen and Hrazdina, 1981). Hoshino *et al.* (1980b) quantitatively demonstrated that awobanin (delphinidin 3-*p*-coumarylglucoside 5-glucoside) forms a much stronger complex with flavocommelinin than delphinin (delphinidin 3,5-diglucoside). The conclusion was drawn that the *p*-coumaryl moiety strongly favours the stability of the complex. Since the true anthocyanin present in flower petals of *Commelina communis* has now been identified as malonylawobanin (Goto *et al.*, 1983b), it remains to be demonstrated whether or not the malonyl group is important to the stability of the natural commelinin complex. Stabilization of quinonoidal bases is readily obtained on addition of a co-pigment to an anthocyanin solution at a suitable pH (Hoshino *et al.*, 1980b; Takeda *et al.*, 1985a). Takeda *et al.* (1985a) reported that some blueing effect is observed when 3-caffeylquinic acid, extracted from *Hydrangea* blue petals, is mixed with delphinidin 3-glucoside. Surprisingly, chlorogenic acid (5-caffeylquinic acid), which is a good co-pigment (Asen *et al.*, 1972), and which also occurs in the blue petals, does not produce any blueing when mixed with delphinidin 3-glucoside.

Co-pigmentation is affected by several factors, among which pH is important. The co-pigment effect occurs from pH 1 to neutrality (Asen *et al.*, 1970; Yazaki, 1976; Williams and Hrazdina, 1979). Colour differences between young and old *Fuchsia* petals are readily explained by pH changes which affect the spiraeoside–malvin complex (Yazaki, 1976). Indirect information on the occurrence *in vivo* of co-pigmentation has been deduced from resonance Raman (RR) and fluorescence spectra (Merlin *et al.*, 1985). The RR spectrum of the malvin flavylium cation in aqueous 0.1% HCl was shown to be identical to the RR spectrum recorded from the upper epidermis of a *Malva sylvestris* petal. However, a strong fluorescence emission exists in the model solution, whereas the spectrum *in vivo* is characterized by a moderate fluorescence. Attenuation of the fluorescence intensity when going from the pure malvin solution to the malvin in the petals may well be explained by the formation in the petals of a loose malvin–co-pigment complex. This is in agreement with the demonstration that the fluorescence intensities of 7,4′-dihydroxy- and 4′-hydroxy-flavylium chlorides are strongly affected by the presence of co-pigments (Santhanam *et al.*, 1983).

Asen *et al.* (1972) were the first to propose that association takes place with both the flavylium cation and the neutral quinonoidal bases. Sweeny *et al.* (1981) demonstrated that pentamethylcyanidin chloride strongly associates with quercetin 5′-sulphonic acid in 0.01 M citric acid (pH 2.8). Hydrogen bonding has been suggested as the driving force for the pigment–co-pigment association (Somers and Evans, 1979; Chen and Hrazdina, 1981). However, hydrogen bonding can be ruled out on the basis that formation of an end-to-end complex does not prevent attack of water on the pyrylium ring (Brouillard, 1983). Intermolecular co-pigmentation probably follows a stacking process where the driving force is of a hydrophobic type (Goto *et al.*, 1979). A sandwich complex between pigment and co-pigment provides a good protection against nucleophilic addition by water (Goto *et al.*, 1986).

16.6 INTRAMOLECULAR CO-PIGMENTATION

Another way of protecting the flavylium cation of natural anthocyanins against water attack is by intramolecular co-pigmentation. In order to explain the extraordinary colour stability of the *Zebrina pendula* anthocyanin as a function of pH, Brouillard (1981) demonstrated that the hydration reaction does not occur with such a pigment, and he suggested that two acyl moieties stack above and below the pyrylium ring, thus protecting it from nucleophilic addition of water. A sandwich-type conformation is therefore adopted by the anthocyanin, while in the flavylium form the acyl groups are on either side of the pyrylium ring. One year later, this very efficient mode of preventing colour loss in water was called intramolecular co-pigmentation (Brouillard, 1982).

The first pigment exhibiting this interesting property is platyconin (Saito *et al.*, 1971) whose stereochemistry has been reinvestigated (Goto *et al.*, 1983a; Saito *et al.*, 1983). Also, Yoshitama and Hayashi (1974) isolated cinerarin from garden cineraria (*Senecio cruentus*) and its structure was further suggested to be dicaffeyldelphinidin 3,7,3′-triglucoside (Yoshitama *et al.*, 1975). More recently, Goto *et al.* (1984) using proton NMR and FAB–MS demonstrated cinerarin to be a monomalonated, tetra-acylated and tetraglucosylated delphinidin with a molecular weight of 1523. Another interesting pigment in this series was discovered by Asen *et al.* (1977). These authors showed that the main pigment of *Ipomoea tricolor* Cav. cv heavenly blue (HBA) possesses the same unusual colour stability in neutral aqueous solutions. In fact HBA is a peonidin-based aglycone with six glucosyl groups and three molecules of caffeic acid that make it the largest anthocyanin known today (Goto *et al.*, 1986). Very interestingly from nuclear Overhauser effect measurements, the HBA flavylium cation was shown to adopt a folded conformation in which two of its three caffeyl residues are close to the pyrylium ring (Goto *et al.*, 1986). Such a result deserves special attention since it is the first direct experimental proof for a sandwich-type intramolecular association among anthocyanins. Other polyacylated pigments have been discussed by Osawa (1982). To this group, one should now add the ternatins extracted from *Clitoria ternatea* flowers (Saito *et al.*, 1985a). Some ternatins appear to be even more stable in solution than platyconin, gentiodelphin, HBA, cinerarin and rubrocinerarin. Three of the ternatins, when kept for a month at room temperature in a pH 6.95 buffer, decreased in their absorbance at 580 nm by only 5–10% (Saito *et al.*, 1985a).

Although monoacylated anthocyanins are not so stable in their neutral and ionized quinonoidal forms, polyacylation does not seem to be sufficient for intramolecular copigmentation to occur (Asen *et al.*, 1979; Brouillard, 1983). It has been pointed out that the structure of the acyl group, its position of attachment to the sugar, as well as the structure of the sugar and its location are important factors for intramolecular co-pigmentation to take place. Malonation has been reported as a colour-stabilizing factor (Saito *et al.*, 1985a). It is noteworthy that intermolecular co-pigmentation is more efficient with monoacylated than with unacylated anthocyanins (Hoshino *et al.*, 1980b). In the former case, some weak intramolecular co-pigment effect may occur, which is impossible in the case of pigments lacking an acyl substituent.

On the basis of solvation–desolvation effects, it appears that intramolecular co-pigmentation should be more efficient than intermolecular co-pigmentation (Brouillard, 1983). For example, mixing caffeic acid with deacylated cinerarin does not regenerate the blue colour of cinerarin (Yoshitama and Hayashi, 1974). When the intramolecular effect occurs, little or no intermolecular co-pigmentation can take place. Thus Asen *et al.* (1977) and Ishikura and Yamamoto (1980) observed that adding rutin, which is an excellent co-pigment, to HBA does not significantly modify the visible absorption spectrum of HBA near neutrality. There is no doubt that intramolecular co-pigmentation occurs *in vivo*. For instance, it has been shown that the visible absorption spectrum of intact petals of *Ipomoea tricolor* matches the absorption spectrum of a pure aqueous solution of HBA (Asen *et al.*, 1977). Again, the visible absorption spectrum (400–700 nm) of a pH 5.4 aqueous solution of the *Zebrina* anthocyanin is identical to the spectrum taken from a purple leaf of this plant (Brouillard, unpublished results). Finally, it appears that many polyacylated anthocyanins can readily be characterized by a typical absorption pattern in the visible range (Saito, 1967; Ishikura, 1978). Usually, three bands and a shoulder are clearly seen at pH values ranging from 4 to neutrality. These spectroscopic features can be used to rapidly identify intramolecularly co-pigmented quinonoidal bases.

16.7 INTERACTIONS WITH METALS

A critical review of the importance of metal complexation by anthocyanins in blue flowers has been published by Harborne (1976). He pointed out that blue flowers in which interaction between anthocyanins and metals was thought to be the origin of blue colour always possess flavones in their vacuoles, and also that there are blue flowers which contain anthocyanidins which lack a free catechol group. It is therefore not surprising that for the *Hydrangea macrophylla* pigment, which was long believed to be an aluminium complex of delphinidin 3-glucoside, Takeda *et al.* (1985b) demonstrated that, in fact, it is a complex made of delphinidin 3-glucoside, aluminium and 3-caffeylquinic acid, aluminium alone being insufficient for producing a stable blue colour in model experiments. Of the metals known to form chelates with anthocyanin having a catechol nucleus, only aluminium and iron are found in appreciable amounts in plants (Asen, 1976). Calcium, magnesium and potassium, far more abundant in plant tissues, apparently do not form chelates with anthocyanins. Chelation in the B-ring is thus restricted to a few metals of biological interest, and to only half of the known anthocyanidins. It is therefore doubtful that this type of interaction is sufficient on its own to account for blue flower colour generally.

Does it mean that metals are unimportant for flower colour expression? My opinion is that there are probably other interactions between anthocyanins and metals which do not involve chelation with the B-ring catechol nucleus. A clue to such interactions is given by the controversy about the participation of Mg in the structure of commelinin. Commelinin is thought to consist of awobanin (delphinidin 3-*p*-coumarylglucoside-5-glucoside), a flavone flavocommelinin and magnesium (Takeda and Hayashi, 1977; Osawa, 1982; Takeda *et al.*, 1984). Goto *et al.* (1979) first questioned the participation of Mg in commelinin by demonstrating that this metal is not necessary for the production of the characteristic blue colour of the petals. However, Goto *et al.* (1986) and Tamura *et al.* (1986), in re-examining the structure of commelinin by use of electrophoresis and UV–visible spectrophotometry, concluded that the real anthocyanin in the petals is malonylawobanin which exhibits a negative charge at pH 4, and that on electrophoresis one spot is observed for freshly prepared natural commelinin whereas on ageing it gives several spots. Commelinin is now believed to be a complex composed of six molecules of malonylawobanin and six molecules of flavocommelinin and also two atoms of Mg. Commelinin-like blue pigments were also prepared using other metals: manganese, cobalt, nickel, zinc and cadmium (Takeda, 1977). Among several structurally related anthocyanins, only shisonin (cyanidin 3-*p*-coumarylglucoside-5-glucoside) gives stable blue metalloanthocyanins. It has been stressed that the nature of the 3-sugar substituent is a determining factor in the formation of metal–anthocyanin complexes (Takeda *et al.*, 1980). Goto *et al.* (1986) still retained the catechol structure in the B-ring as the major structural feature for metal complexation to take place, Mg being the co-ordinating atom. Osawa (1982), however, indicated that commelinin still possesses the cytochrome *c* reducing capacity of anthocyanins with a free catechol group. Probably, other groups within the malonylawobanin molecule interact with the magnesium atom.

The blue pigment of the cornflower (*Centaurea cyanus*) is thought to be another metalloanthocyanin. K, Al, Fe and Mg have all been considered, at some time, as being involved in its structure (Osawa, 1982). Goto *et al.* (1986) suggested the structure of this blue pigment to be similar to the structure of commelinin. The structures of both the anthocyanin and flavone in *Centaurea cyanus* petals have been revised (Tamura *et al.*, 1983; Takeda and Tominaga, 1983). Macromolecular compounds, which seemed to be important for metalloanthocyanin stability twenty years ago (Bayer *et al.*, 1966), apparently are no longer thought to play a role (Goto *et al.*, 1986).

16.8 INVESTIGATIONS *IN VIVO*

Investigations *in vivo* of anthocyanins have been limited because such investigations must meet two important requirements. Firstly, they must not be damaging to the natural medium in which the pigments exist, and secondly they must be based on a characteristic of anthocyanins which is not exhibited by other biochemicals present in the living cell. Visible light absorption and visible light scattering fulfil the above requirements. Therefore it is not surprising that the only techniques now available for the investigations *in vivo* of anthocyanins are based on visible spectrophotometry and Raman scattering of visible light (Brouillard, 1983). Measurements could be extended into the UV range as well. In a living tissue, however, there are so many UV light-absorbing substances that the characteristic spectral features of anthocyanins cannot be seen. Finally, Hoshino (1986) successfully applied circular dichroism (CD) absorption spectrophotometry to living flower petals.

Saito (1967), using the opal glass method, first applied visible absorption spectrometry to intact pigment-containing plant tissues. He, and later Ishikura (1978), classified the absorption spectra of intact pigmented tissues into four groups. It was concluded that colours of some flowers are due to the anthocyanin alone; in other cases, co-pigmentation was shown to exist. Finally, some pigments were considered to fall within the metalloanthocyanin group. Decisive progress was made by Asen and his co-workers in the 1970s when they adapted an optical microscope to their UV–visible spectrophotometer, thus transforming it into a microspectrophotometer capable of recording spectra from areas as small as a single cell. Asen *et al.* (1971b), for instance, were able to reproduce the absorption spectrum of a portion of a vacuole of the orange sport of 'Red Wing' azalea, by placing cyanin in an aqueous solution at pH 2.8. Owing to the small number of bands present in a visible spectrum, structural determination of the absorbing species is rather uncertain. Nevertheless, this microtechnique was applied to many flowers and the spectra of their intact cells were analysed (Asen *et al.*, 1975). It also permitted the measurement of the acidity of the vacuolar content of a few (5–10) epidermal cells (Stewart *et al.*, 1975).

CD, which is an extension of UV–visible absorption spectrophotometry to optically active material, indicates that the anthocyanins sometimes occur strongly associated in flower petals (Hoshino, 1986). For instance, characteristic Cotton effects, with negative and positive signs, were observed in the case of the blue *Centaurea cyanus* and *Commelina communis* petals, whereas petals of *Platycodon grandiflorum* and *Ipomoea tricolor*, which possess polyacylated pigments, do not show characteristic CD spectra. Since about one-third of the species investigated exhibit CD, it was concluded that the stacking of anthocyanin quinonoidal bases frequently takes place *in vivo*, and that it gives rise to optically active molecular aggregates.

Resonance Raman (RR) spectrometry has been shown to be a sensitive probe for analysis *in vivo* of pigments (Statoua *et al.*, 1982, 1983; Merlin, 1983; Brouillard, 1983). This technique, however, although proven very useful in the investigation of many biological systems (Carey, 1982), has not been extensively applied to the study of flavonoids. The information obtained is essentially the same as that provided by infrared spectrometry, but Raman spectrometry presents some advantages in biological investigation since water is a poor Raman scatterer. The resonance effect takes place when the excitation light falls within an electronic absorption band. Owing to the coupling of the electronic and vibronic transitions, the Raman lines relating to the chromophores are strongly enhanced. Dilute solutions (10^{-5} M or less) can be studied and strongly light-absorbing pigments in aqueous solutions are especially well suited for such investigations. When an optical microscope is attached to the spectrometer, the instrument functions as a Raman microspectrometer. Areas as small as a few μm^2 can be analysed. The Raman microanalysis technique was first described by Delhaye and Dhamelincourt (1975). Experimental details are to be found in Merlin *et al.* (1985).

Good RR spectra of some flower petals (Merlin, 1985;

Merlin *et al.*, 1985), and grape berry skins as well as leaves of the red cabbage have been reported (Merlin, 1983; Statoua *et al.*, 1983; Brouillard, 1983; Brouillard *et al.*, 1985; Merlin *et al.*, 1985). RR spectra of malvin chloride in water at pH 1 and at pH 6 were also obtained (Statoua *et al.*, 1983; Brouillard, 1983). These spectra are very different and are characteristic of the flavylium form (pH 1) and the quinonoidal form (pH 6). Moreover, malvidin, cyanidin and delphinidin 3-glucosides can be readily distinguished from the corresponding 3, 5-diglucosides. In effect whereas diglucosides show an intense line close to 630 cm^{-1}, the monoglucosides exhibit a strong band close to 540 cm^{-1}. The model experiments were compared to the spectra obtained from living tissues. For instance, *in vivo* RR spectra of a petal epidermal cell of *Malva sylvestris* with excitation light of different wavelengths have been recorded (Merlin *et al.*, 1985). The only existing malvin chromophore is the flavylium form, and no other pigments absorbing light in the 400–600 nm range could be found in these flower petals. In particular, the yellow or orange carotenoids are not present at all. For the flavylium cation to be the only existing coloured malvin structure, the pH should be lower than 4.

16.9 YELLOW TO COLOURLESS FLAVONOIDS

The contribution of flavonoids to yellow flower colour has been reviewed by Harborne (1967, 1976, 1977a). In the case of UV flower patterning, a recent survey is available (Bohm, 1982). Yellow to orange flower colours are usually due to the presence of carotenoids (Goodwin, 1976; Britton, 1983), whereas natural anthocyanins, in their natural environment (vacuoles), do not give yellow colour to plants. Flavonoids involved in this case are most frequently the anthochlor pigments (aurones and chalcones) as well as the yellow flavonols (Harborne, 1967, 1976, 1977a). These flavonoids, and the colourless flavone and flavonol glycosides, are frequently the origin of UV patterns in flowers visible, for instance, to honey bees (Harborne, 1976).

In *Lepidophorum repandum*, the 7-glucoside of the 6, 3′-dimethyl ether of quercetagetin (spinacetin 7-glucoside) has been reported for the first time as a yellow flower pigment (Harborne *et al.*, 1976). It was also demonstrated that among the Anthemideae, methylation in the 6 and 3′ positions is related to a highly developed species as in the case of *Lepidophorum*. Yellow flower colour in these plants originates from the presence of carotenoid as well as the flavonols gossypetin, quercetagetin and methyl ethers of quercetagetin.

Identification of the rare gossypetin methyl ethers acting as yellow flower pigments have been reported recently on several occasions. For instance, the 7-methyl ether occurs in four different glycosylated states, in *Eriogonum nudum* where it constitutes the major source of yellow colour (Harborne *et al.*, 1978). Small amounts of the 8-methyl ether of gossypetin were also detected in the same plant. Reinvestigation by these authors of *Lotus corniculatus* flowers showed that only the latter gossypetin-based flavonol is present. It also contributes, as the 3-galactoside, to flower colour in *Geraea canescens*. 3-Rutinosides of gossypetin 3′-methyl ether and gossypetin 8, 3′-dimethyl ether have been demonstrated to exist in the wings of the flowers *Coronilla valentina* where they contribute to yellow colour and provide UV signals visible to bees (Harborne, 1981). Gossypetin 7-methyl ether 3-rutinoside has been found in the Rutaceae and occurs, with carotenoid, as the yellow colouring matter of the flowers in *Ruta graveolens* and five other *Ruta* species, whereas in four *Haplophyllum* species gossypetin 3′-methyl ether 7-glucoside and a corresponding acylated flavonol are present (Harborne and Boardley, 1983a).

Remarkably, glycosides of the rare flavone isoetin (5, 7, 2′, 4′, 5′-pentahydroxyflavone) have been shown to be major contributors to yellow flower colour in three species of the Cichorieae (Harborne, 1978). It is suggested that isoetin can exist in a quinonoid structure made possible by the existence of two *p*-hydroxyl groups in the B-ring. Kaempferol 3-sophoroside and kaempferol 3-rutinoside-7-glucoside are both present in the petals of a few cultivated *Crocus* species (Harborne and Williams, 1984). As normally expected for common flavonols they appear to be colourless, petal colour being only related to the yellow crocetin-based carotenoids.

In flower buds of *Lotus corniculatus*, Jay *et al.* (1983) reported the specific accumulation of 8-methoxyflavonols as a result of 8-*O*-methyltransferase activity. Gossypetin and its 8-methyl ether were identified as the major flavonoids giving rise, in association with carotenoids, to the yellow flower colour in this Leguminosae (Jay and Ibrahim, 1986).

The number of families of the Compositae known to possess anthochlors (chalcones and aurones) as yellow flower pigments has increased continuously. For instance, among the Cichorieae, yellow ligules of *Pyrrhopappus* are essentially pigmented by the chalcone coreopsin (butein 4′-glucoside), whereas carotenoids are completely lacking in the genus (Harborne, 1977b). Aureusin (aureusidin 6-glucoside), first isolated from flowers of *Antirrhinum majus*, is also the main yellow pigment of the orange petals of *Mussaenda hirsutissima* (Harborne *et al.*, 1983). Cernuoside (aureusidin 4-glucoside) and aureusidin 4, 6-diglucoside, a new flavonoid, were identified as minor pigments in this species. Yellow flower colour of *Zinnia linearis* is largely due to a mixture of the aurone maritimein (6, 7, 3′, 4′-tetrahydroxyaurone 6-glucoside) with the structurally related chalcone marein (Harborne *et al.*, 1983).

An interesting field of floral chemistry is that of UV honey patterns which guide insects, especially bees, in their search for nectar. Yellow flavonoids, as well as colourless UV-absorbing flavonoids, can be involved in the production of UV honey guides. For instance, in a sample of yellow-flowered Compositae the usual flavonol glycosides, 6- or 8-substituted flavonol glycosides and anthochlor pigments were shown to be responsible for UV patterning in different species (Harborne and Smith, 1978). It was found by these authors that, although honey guides are frequently observed when yellow flavonoids are present, yellow colour does not seem to be a major factor in their elaboration. Confirmation of this fact is given by the discovery, in the wings of the flower of *Coronilla emerus*, of kaempferol and quercetin glycosides, which strongly absorb light in the 340–380 nm range, and therefore largely contribute to UV patterning without giving yellow colour which is due to carotenoids (Harborne and Boardley, 1983b). Another example is given by the genus *Potentilla* (Rosaceae) where UV-absorbing patterns are essentially given by the widespread chalcone isosalipurposide or colourless quercetin glycosides (Harborne and Nash, 1984). In ray flowers of different *Helianthus* species, UV nectar guides are related, in some cases to the presence of the chalcone coreopsin and its aurone analogue sulphurein, and in other cases to the presence of quercetin 7-glucoside, quercetin 3-glucoside being formed in all cases (Schilling, 1983). It is clear that yellow and colourless flavonoids are complementary in producing UV honey guides, a phenomenon previously pointed out by Harborne and Smith (1978). In some species of *Viguiera*, UV nectar guides were also demonstrated to occur in ligules (Scogin, 1978). Here again, existence of several different classes of flavonoids, i.e. flavone, flavonol, chalcone and aurone glycosides, was reported (Rieseberg and Schilling, 1985). Interestingly, these authors made the observation that chalcone cyclization to the corresponding aurone takes place easily. This might well explain why a given chalcone is frequently associated with its structurally related aurone.

ACKNOWLEDGEMENTS

I am very grateful to Dr A. Cheminat (Strasbourg, France) and Dr G. Mazza (Morden, Canada) for helpful comments during the preparation of this chapter.

REFERENCES

Akavia, N., Strack, D. and Cohen, A. (1981), *Z. Naturforsch.* **36c**, 378.

Asen, S. (1976), *Acta Hortic.* **63**, 217.

Asen, S. (1984), *Phytochemistry* **23**, 2523.

Asen, S. and Griesbach, R. (1983), *J. Am. Soc. Hortic. Sci.* **108**, 845.

Asen, S., Stewart, R.N., Norris, K.H. and Massie, D.R. (1970), *Phytochemistry* **9**, 619.

Asen, S., Norris, K.H. and Stewart, R.N. (1971a), *J. Am. Soc. Hortic. Sci.* **96**, 770.

Asen, S., Stewart, R.N. and Norris, K.H. (1971b), *Phytochemistry* **10**, 171.

Asen, S., Stewart, R.N. and Norris, K.H. (1972), *Phytochemistry* **11**, 1139.

Asen, S., Stewart, R.N. and Norris, K.H. (1975), *Phytochemistry* **14**, 2677.

Asen, S., Stewart, R.N. and Norris, K.H. (1977), *Phytochemistry* **16**, 1118.

Asen, S., Stewart, R.N. and Norris, K.H. (1979), *Phytochemistry* **18**, 1251.

Bayer, E., Egeter, H., Fink, A., Nether, K. and Wegmann, K. (1966), *Angew. Chem. Int. Ed. Engl.* **5**, 791.

Bohm, B.A. (1982). In *The Flavonoids – Advances in Research* (eds J.B. Harborne and T.J. Mabry), Chapman and Hall, London, p. 313.

Bridle, P., Loeffler, R.S.T., Timberlake, C.F. and Self, R. (1984), *Phytochemistry* **23**, 2968.

Britton, G. (1983), *The Biochemistry of Natural Pigments*, Cambridge University Press, Cambridge.

Bronnum-Hansen, K. and Hansen, S.H. (1983), *J. Chromatogr.* **262**, 385.

Brouillard, R. (1981), *Phytochemistry* **20**, 143.

Brouillard, R. (1982). In *Anthocyanins as Food Colors* (ed. P. Markakis), Academic Press, New York, p. 1.

Brouillard, R. (1983), *Phytochemistry* **22**, 1311.

Brouillard, R., and Delaporte, B. (1977), *J. Am. Chem. Soc.* **99**, 8461.

Brouillard, R. and Dubois, J.E. (1977), *J. Am. Chem. Soc.* **99**, 1359.

Brouillard, R., Iacobucci, G.A. and Sweeny, J.G. (1982), *J. Am. Chem. Soc.* **104**, 7585.

Brouillard, R., Statoua, A. and Merlin, J.C. (1985). In *Spectroscopy of Biological Molecules* (eds A.J.P. Alix, L. Bernard and M. Manfait), Wiley, Chichester, p. 445.

Busetta, P.B., Colleter, J.C. and Gadret, M. (1974), *Acta Crystallogr. Sect. B* **30**, 1448.

Carey, P.R. (1982), *Biochemical Applications of Raman and Resonance Raman Spectroscopies*, Academic Press, New York.

Cheminat, A. and Brouillard, R. (1986), *Tetrahedron Lett.*, 4457.

Chen, L.J. and Hrazdina, G. (1981), *Phytochemistry* **20**, 297.

Cornuz, G., Wyler, H. and Lauterwein, J. (1981), *Phytochemistry* **20**, 1461.

Crowden, R.K. (1982), *Phytochemistry* **21**, 2989.

Delhaye, M. and Dhamelincourt, P. (1975), *J. Raman Spectrosc.* **3**, 33.

Goodwin, T.W. (1976). In *Chemistry and Biochemistry of Plant Pigments* (ed. T.W. Goodwin), 2nd edn, Vol. 1, Academic Press, New York, p. 225.

Goto, T., Takase, S. and Kondo, T. (1978), *Tetrahedron Lett.*, 2413.

Goto, T., Hoshino, T. and Takase, S. (1979), *Tetrahedron Lett.*, 2905.

Goto, T., Kondo, T., Imagawa, H., Takase, S., Atobe, M. and Miura, I. (1981), *Chem. Lett.*, 883.

Goto, T., Kondo, T., Tamura, H., Imagawa, H., Iino, A. and Takeda, K. (1982), *Tetrahedron Lett.*, 3695.

Goto, T., Kondo, T., Tamura, H., Kawahori, K. and Hattori, H. (1983a), *Tetrahedron Lett.*, 2181.

Goto, T., Kondo, T., Tamura, H. and Takase, S. (1983b), *Tetrahedron Lett.*, 4863.

Goto, T., Kondo, T., Kawai, T. and Tamura, H. (1984), *Tetrahedron Lett.*, 6021.

Goto, T., Tamura, H., Kawai, T., Hoshino, T., Harada, N. and Kondo, T. (1986), *Ann. N. Y. Acad. Sci.* **471**, 155.
Harborne, J.B. (1957), *Biochem. J.* **70**, 22.
Harborne, J.B. (1967), *Comparative Biochemistry of the Flavonoids*, Academic Press, New York.
Harborne, J.B. (1976). In *Chemistry and Biochemistry of Plant Pigments* (ed. T.W. Goodwin), 2nd edn, Vol. 1, Academic Press, New York, p. 736.
Harborne, J.B. (1977a), In *The Biology and Chemistry of The Compositae* (eds V.H. Heywood, J.B. Harborne and B.L. Turner), Academic Press, London, p. 359.
Harborne, J.B. (1977b), *Phytochemistry* **16**, 927.
Harborne, J.B. (1978), *Phytochemistry* **17**, 915.
Harborne, J.B. (1981), *Phytochemistry* **20**, 1117.
Harborne, J.B. and Boardley, M. (1983a), *Z. Naturforsch.* **38c**, 148.
Harborne, J.B. and Boardley, M. (1983b), *Phytochemistry* **22**, 622.
Harborne, J.B. and Nash, R.J. (1984), *Biochem. Syst. Ecol.* **12**, 315.
Harborne, J.B. and Smith, D.M. (1978), *Biochem. Syst. Ecol.* **6**, 287.
Harborne J.B. and Williams, C.A. (1984), *Z. Naturforsch.* **39c**, 18.
Harborne, J.B., Heywood, V.H. and King, L. (1976), *Biochem. Syst. Ecol.* **4**, 1.
Harborne, J.B., Saleh, N.A.M. and Smith, D.M. (1978), *Phytochemistry* **17**, 589.
Harborne, J.B., Girija, A.R., Devi, H.M. and Lakshmi, N.K.M. (1983), *Phytochemistry* **22**, 2741.
Hemleben, V. (1981), *Z. Naturforsch.* **36c**, 925.
Hoshino, T. (1986), *Phytochemistry* **25**, 829.
Hoshino, T., Matsumoto, U. and Goto, T. (1980a), *Tetrahedron Lett.*, 1751.
Hoshino, T., Matsumoto, U. and Goto, T. (1980b), *Phytochemistry* **19**, 663.
Hoshino, T., Matsumoto, U. and Goto, T. (1981a), *Phytochemistry* **20**, 1971.
Hoshino, T., Matsumoto, U., Harada, N. and Goto, T. (1981b), *Tetrahedron Lett.*, 3621.
Hoshino, T., Matsumoto, U., Goto, T. and Harada, N. (1982), *Tetrahedron Lett.*, 433.
Hrazdina, G. (1982). In *The Flavonoids – Advances in Research* (eds J.B. Harborne and T.J. Mabry), Chapman and Hall, London, p. 135.
Iacobucci, G.A. and Sweeny, J.G. (1983), *Tetrahedron* **39**, 3005.
Ishikura, N. (1978), *Plant Cell Physiol.* **19**, 887.
Ishikura, N. and Yamamoto, E. (1980), *Nippon Nogei Kagaku Kaishi* **54**, 637.
Jay, M. and Ibrahim, R. (1986), *Biochem. Physiol. Pflanz.* **181**, 199.
Jay, M. De Luca, V. and Ibrahim, R. (1983), *Z. Naturforsch.* **38c**, 413.
Jurd, L. (1962). In *Chemistry of the Flavonoid Compounds* (ed. T.A. Geissman), Pergamon, Oxford, p. 107.
Jurd, L. (1972), *Adv. Food. Res.* (suppl.) **3**, 123.
Jurd, L. and Harborne, J.B. (1968), *Phytochemistry* **7**, 1209.
Kondo, T., Nakane, Y., Tamura, H., Goto, T. and Eugster, C.H. (1985), *Tetrahedron Lett.*, 5879.
Markham, K.R. (1982), *Techniques of Flavonoid Identification*, Academic Press, London.
Markham, K.R. and Chari, V.M. (1982) In *The Flavonoids – Advances in Research* (eds J.B. Harborne and T.J. Mabry), Chapman and Hall, London, p. 19.
Markham, K.R. and Mabry, T.J. (1975), In *The Flavonoids* (eds J.B. Harborne, T.J. Mabry and H. Mabry), Chapman and Hall, London, p. 45.
Mazza, G. and Brouillard, R. (1987), *Food Chem.* **25**, 207.
McClure, J.W. (1979), In *Biochemistry of Plant Phenolics* (eds T. Swain, J.B. Harborne and C.F. Van Sumere), Plenum Press, New York, p. 525.
Merlin, J.C. (1983), *Spectrosc. Int. J.* **2**, 52.
Merlin J.C. (1985). In *Spectroscopy of Biological Molecules* (eds A.J.P. Alix, L. Bernard and M. Manfait), Wiley, Chichester, p. 427.
Merlin, J.C., Statoua, A. and Brouillard, R. (1985), *Phytochemistry* **24**, 1575.
Nilsson, E. (1973), *Chem. Script.* **4**, 49.
Osawa, Y. (1982). In *Anthocyanins as Food Colors* (ed P. Markakis), Academic Press, New York, p. 41.
Pecket, R.C. and Small, C.J. (1980), *Phytochemistry* **19**, 2571.
Preston, N.W. and Timberlake, C.F. (1981), *J. Chromatogr.* **214**, 222.
Rieseberg, L.H. and Schilling, E.E. (1985), *Am. J. Bot.* **72**, 999.
Robinson, G.M. and Robinson, R. (1931), *Biochem. J.* **25**, 1687.
Saito, N. (1967), *Phytochemistry* **6**, 1013.
Saito, N. and Harborne, J.B. (1983), *Phytochemistry* **22**, 1735.
Saito, N. and Ueno, K. (1985), *Heterocycles* **23**, 2709.
Saito, N., Osawa, Y. and Hayashi, K. (1971), *Phytochemistry* **10**, 445.
Saito, N., Timberlake, C.F., Tucknott, O. and Lewis, I.A.S. (1983), *Phytochemistry* **22**, 1007.
Saito, N., Abe, K. Honda, T., Timberlake, C.F. and Bridle, P. (1985a), *Phytochemistry* **24**, 1583.
Saito, N., Yokoi, M., Yamaji, M. and Honda, T. (1985b), *Phytochemistry* **24**, 2125.
Santhanam, M., Hautala, R.R., Sweeny, J.G. and Iacobucci, G.A. (1983), *Photochem. Photobiol.* **38**, 477.
Scheffeldt, P. and Hrazdina, G. (1978), *J. Food Sci.* **43**, 517.
Schilling, E.E. (1983), *Biochem. Syst. Ecol.* **11**, 341.
Scogin, R. (1978), *Southwest. Nat.* **23**, 371.
Scogin, R. (1980), *Biochem. Syst. Ecol.* **8**, 273.
Shibata, K., Hayashi, K. and Isaka, T. (1949), *Acta Phytochem. Jpn* **15**, 17.
Somers, T.C. and Evans, M.E. (1979), *J. Sci. Food Agric.* **30**, 623.
Statoua, A., Merlin, J.C., Delhaye, M. and Brouillard, R. (1982). In *Raman Spectroscopy Linear and Nonlinear* (eds J. Lascombe and P.V. Huong), John Wiley, Chichester, p. 629.
Statoua, A., Merlin, J.C., Brouillard, R. and Delhaye, M. (1983), *C.R. Acad. Sci. Paris* **296**, 1397.
Stewart, R.N., Norris, K.H. and Asen, S. (1975), *Phytochemistry* **14**, 937.
Swain, T. (1985). In *The Biochemistry of Plant Phenolics* (eds C.F. Van Sumere and P.J. Lea), Clarendon Press, Oxford, p. 453.
Sweeny, J.G., Wilkinson, M.M. and Iacobucci, G.A. (1981), *J. Agric. Food Chem.* **29**, 563.
Takeda, K. (1977), *Proc. Jpn Acad.* **53**, 257.
Takeda, K. and Hayashi, K. (1977), *Proc. Jpn Acad.* **53**, 1.
Takeda, K. and Tominaga, S. (1983), *Bot. Mag. Tokyo* **96**, 359.
Takeda, K., Narashima, F. and Nonaka, S. (1980), *Phytochemistry* **19**, 2175.
Takeda, K., Fujii, T. and Iida, M. (1984), *Phytochemistry* **23**, 879.
Takeda, K., Kubota, R. and Yagioka, C. (1985a), *Phytochemistry* **24**, 1207.

Takeda, K. Kariuda, M. and Itoi, H. (1985b), *Phytochemistry* **24**, 2251.
Takeda, K., Harborne, J.B. and Self, R. (1986), *Phytochemistry* **25**, 1337.
Tamura, H., Kondo, T., Kato, Y. and Goto, T. (1983), *Tetrahedron Lett.*, 5749.
Tamura, H. Kondo, T. and Goto, T. (1986), *Tetrahedron Lett.* 1801.
Timberlake, C.F. (1980), *Food. Chem.* **5**, 69.
Timberlake, C.F. (1981). In *Recent Advances in the Biochemistry of Fruits and Vegetables* (eds J. Friend and M.J.C. Rhodes), Academic Press, New York, p. 221.
Timberlake, C.F. and Bridle, P. (1975). In *The Flavonoids* (eds J.B. Harborne, T.J. Mabry and H. Mabry), Chapman and Hall, London, p. 214.
Timberlake, C.F. and Bridle, P. (1980). In *Developments in Food Colours – 1* (ed J. Walford), Applied Science, London, p. 115.
Ueno, K. and Saito, N. (1977a), *Acta Crystallogr. Sect. B* **33**, 111.
Ueno, K. and Saito, N. (1977b), *Acta Crystallogr. Sect. B* **33**, 114.
Van Sumere, C.F., Vande Casteele, K., De Loose, R. and Heursel, J. (1984), *Bull. Liaison Groupe Polyphénols* **12**, 53.
Van Sumere, C.F., Vande Casteele, K., De Loose, R. and Heursel, J. (1985). In *The Biochemistry of Plant Phenolics* (eds C.F. Van Sumere and P.J. Lea), Clarendon Press, Oxford, p. 17.
Williams, M. and Hrazdina, G. (1979), *J. Food Sci.* **44**, 66.
Yasuda, H. (1971), *Bot. Mag. Tokyo* **84**, 256.
Yazaki, Y. (1976), *Bot. Mag. Tokyo* **89**, 45.
Yoshitama, K. and Hayashi, K. (1974), *Bot. Mag. Tokyo* **87**, 33.
Yoshitama, K., Hayashi, K., Abe, K. and Kakisawa, H. (1975), *Bot. Mag. Tokyo* **88**, 213.

Appendix

FLAVONOID CHECKLISTS

Chapter references	*Class*	*No. of structures*
1	Anthocyanidins and Anthocyanins prepared by J.B. Harborne and R.J. Grayer	256
2	Flavans, Leucoanthocyanidins and Proanthocyanidins prepared by L.J. Porter	309
3	*C*-Glycosylflavonoids prepared by J. Chopin and G. Dellamonica	303
4	Biflavonoids prepared by H. Geiger	134
5	Isoflavonoids prepared by P.M. Dewick	630
6	Neoflavonoids prepared by J.B. Harborne	70
7	Flavones and Flavonols prepared by E. Wollenweber	668
8	Flavone and Flavonol Glycosides prepared by J.B. Harborne and C.A. Williams	992
9	Chalcones, Aurones, Dihydrochalcones, Flavanones and Dihydroflavonols prepared by B.A. Bohm	723
10	Miscellaneous flavonoids prepared by H. Geiger	100
11	Alphabetical key to trivial names of flavones and flavonols prepared by E. Wollenweber	

Notes

These checklists are numbered 1–10 corresponding to Chapters 1–10 and some structures referred to in the checklist with bold numbers can be found in the appropriate chapter. There are considerable difficulties in meaningfully listing some of the more complex structures, and in the case of neoflavonoids, only the simpler derivatives are shown.

Since a few structures are listed in more than one chapter, the total number of known flavonoids does not necessarily correspond to 4185, as indicated above. However, because of pressure on space, some 130 isoflavonoid glycosides are not included here (but see Ingham, J.L. (1983) in reference list of Chapter 5). One can therefore predict that the total number of natural flavonoids known up to the end of 1985 is well over four thousand.

Compounds newly described in the period 1980–1985 are asterisked in these lists.

(1) CHECKLIST OF ALL KNOWN ANTHOCYANIDINS AND ANTHOCYANINS

1. Apigeninidin (5, 7, 4′-triOH flavylium)
2. 5-Glucoside
3. Luteolinidin (5, 7, 3′, 4′-tetraOH flavylium)
4. 5-Glucoside
5. 5-Diglucoside
6. Tricetinidin (5, 7, 3′, 4′, 5′-pentaOH flavylium)
7. 5-Glucoside
8. Pelargonidin (3, 5, 7, 4′-tetraOH flavylium)
9. 3-Galactoside
10. 3-Glucoside
11. 3-Rhamnoside
12. 5-Glucoside
13. 7-Glucoside
14. 3-Gentiobioside
15. 3-Lathyroside
16. 3-Sambubioside
17. 3-Sophoroside
18. 3-Glucosylxyloside
19. 3-Rhamnosylgalactoside
20. 3-Robinobioside
21. 3-Rutinoside
22. 3-Galactoside-5-glucoside
23. 3, 5-Diglucoside
24. 3-Rhamnoside-5-glucoside

25. 3, 7-Diglucoside
26. 3-Gentiotrioside
27. 3-(2^G-Glucosylrutinoside)
28. 3-Rutinoside-5-glucoside
29. 3-Sambubioside-5-glucoside
30. 3-Sophoroside-5-glucoside
31. 3-Sophoroside-7-glucoside
32. 3-(6″-Malonylglucoside)*
33. 3-(6″-Malylglucoside)*
34. 3-Malonylsophoroside*
35. 3-(4‴-*p*-Coumarylrhamnoglucoside)
36. 3-(*p*-Coumarylglucoside)-5-glucoside
37. 3-(Caffeylglucoside)-5-glucoside
38. 3-(6″-Malonylglucoside)-5-glucoside*
39. 3-(*p*-Coumarylrutinoside)-5-glucoside (pelanin)
40. 3-(*p*-Coumarylsophoroside)-5-glucoside
41. 3-(Ferulylsophoroside)-5-glucoside
42. 3-(*p*-Coumarylcaffeyldiglucoside)-5-glucoside
43. 3-(*p*-Coumarylferulylsambubioside)-5-glucoside
44. 3-(*p*-Coumarylferulylxylosylglucoside)-5-glucoside (matthiolanin)
45. 3-(Dicaffeylglucoside)-5-glucoside
46. 3-(Dicaffeyldiglucoside)-5-glucoside
47. 3-(*p*-Coumaryldicaffeyldiglucoside)-5-glucoside
48. 3-(Di-*p*-hydroxybenzylrutinoside)-7-glucoside
49. 3, 5-Di(malonylglucoside)*
50. 3-(6″-*p*-Coumarylglucoside)-5-(4‴, 6‴-dimalonylglucoside)* (monardaein, 1.6)
51. Aurantinidin (3, 5, 6, 7, 4′-pentaOH flavylium)
52. 3-Sophoroside
53. 3, 5-Diglucoside
54. Cyanidin (3, 5, 7, 3′, 4′-pentaOH flavylium)
55. 3-Arabinoside
56. 3-Galactoside
57. 3-Glucoside
58. 3-Rhamnoside
59. 5-Glucoside
60. 4′-Glucoside
61. 3-Arabinosylgalactoside
62. 3-Arabinosylglucoside
63. 3-Gentiobioside
64. 3-Lathyroside
65. 3-Rhamnosylarabinoside*
66. 3-Robinobioside
67. 3-Rutinoside
68. 3-Sambubioside
69. 3-Sophoroside
70. 3-Xylosylarabinoside
71. 3-Gentiotrioside
72. 3-(2^G-Glucosylrutinoside)
73. 3-Rhamnosyldiglucoside
74. 3-Xylosylglucosylgalactoside
75. 3-(2^G-Xylosylrutinoside)
76. 3-Arabinoside-5-glucoside
77. 3-Galactoside-5-glucoside
78. 3-Rhamnoside-5-glucoside
79. 3, 5-Diglucoside
80. 3, 7-Diglucoside
81. 3-Glucoside-7-rhamnoside
82. 3, 3′-Diglucoside*
83. 3, 4′-Diglucoside
84. 3-Rutinoside-5-glucoside
85. 3-Sambubioside-5-glucoside
86. 3-Sophoroside-5-glucoside
87. 3-Rhamnosylglucoside-7-xyloside
88. 3, 5, 3′-Triglucoside*
89. 3-Rutinoside-3′-glucoside*
90. 3-Rutinoside-5, 3′-diglucoside*
91. 3-(Acetylglucoside)
92. 3-(*p*-Coumarylglucoside) (hyacinthin)
93. 3-(Caffeylglucoside)
94. 3-(Ferulylglucoside)*
95. 3-(Sinapylglucoside)*
96. 3-(6″-Oxalylglucoside)*
97. 3-(6″-Malonylglucoside)*
98. 3-(6″-Malylglucoside)*
99. 3-(*p*-Coumarylgentiobioside)
100. 3-(6″-*p*-Coumarylsophoroside)*
101. 3-(*p*-Coumarylxylosylglucoside)
102. 3-(Caffeylgentiobioside)
103. 3-(Caffeylsophoroside)*
104. 3-(Caffeylglucosylarabinoside)
105. 3-(Caffeylrhamnosylglucoside)
106. 3-(Ferulylglucosylgalactoside)
107. 3-(Sinapylglucosylgalactoside)
108. 3-(Malonylglucuronosylglucoside)*
109. 3-(Caffeylrhamnosyldiglucoside)
110. 3-(Ferulylxylosylglucosylgalactoside)
111. 3-(Synapylxylosylglucosylgalactoside)* (**1.4**)
112. 3-(*p*-Coumarylferulylglucoside)*
113. 3-(Dicaffeylsophoroside)*
114. 3-(Dimalonylglucoside)*
115. 3-(6″-*p*-Coumarylglucoside)-5-glucoside (perillanin)
116. 3-(Caffeylglucoside)-5-glucoside
117. 3-(6″-Malonylglucoside)-5-glucoside*
118. 3-(6″-Succinylglucoside)-5-glucoside*
119. 3-(*p*-Coumarylrutinoside)-5-glucoside (cyananin)
120. 3-(*p*-Coumarylsophoroside)-5-glucoside
121. 3-(Caffeylsophoroside)-5-glucoside
122. 3-(Ferulylsophoroside)-5-glucoside
123. 3-(Sinapylsophoroside)-5-glucoside
124. 3-(Malonylsophoroside)-5-glucoside
125. 3, 5-Di(malonylglucoside)*
126. 3-(*p*-Coumarylcaffeyl)sophoroside-5-glucoside
127. 3-(Di-*p*-coumaryl)sophoroside-5-glucoside
128. 3-(Diferulyl)sophoroside-5-glucoside
129. 3-(Disinapyl)sophoroside-5-glucoside
130. 3-(*p*-Coumarylglucoside)-5-(malonylglucoside)*
131. 3, 7, 3′-Tri(caffeylglucoside)*
132. Rubrocinerarin*

133. 5-Methylcyanidin (3, 7, 3′, 4′-tetraOH-5-OMe flavylium)
134. 3-Glucoside
135. Peonidin (3, 5, 7, 4′-tetraOH-3′-OMe flavylium)
136. 3-Arabinoside
137. 3-Galactoside
138. 3-Glucoside
139. 3-Rhamnoside
140. 5-Glucoside
141. 3-Gentiobioside
142. 3-Lathyroside
143. 3-Rutinoside
144. 3-Arabinoside-5-glucoside*
145. 3-Galactoside-5-glucoside
146. 3, 5-Diglucoside
147. 3-Rhamnoside-5-glucoside
148. 3-Rutinoside-5-glucoside
149. 3-Gentiotrioside
150. 3-Glucosylrhamnosylglucoside
151. 3-(Acetylglucoside)
152. 3-(*p*-Coumarylglucoside)
153. 3-(Caffeylglucoside)
154. 3-(*p*-Coumarylgentiobioside)
155. 3-(Caffeylgentiobioside)
156. 3-(*p*-Coumarylglucoside)-5-glucoside
157. 3-(*p*-Coumarylrutinoside)-5-glucoside (peonanin)
158. 3-(*p*-Coumarylsophoroside)-5-glucoside
159. 3-(Di-*p*-coumaryl)glucoside
160. 3-Dicaffeylsophoroside)-5-glucoside
161. 3-(*p*-Coumarylcaffeylsophoroside)-5-glucoside
162. 3-Sophoroside-5-glucoside tri(caffeylglucose) ester* (heavenly blue anthocyanin, **1.1**)
163. Rosinidin (3, 5, 4′-triOH-7, 3′-di-OMe flavylium)
164. 3, 5-Diglucoside
165. 6-Hydroxycyanidin (3, 5, 6, 7, 3′, 4′-hexaOH flavylium)*
166. 3-Glucoside*
167. 3-Rutinoside*
168. Delphinidin (3, 5, 7, 3′, 4′, 5′-hexaOH flavylium)
169. 3-Arabinoside
170. 3-Galactoside
171. 3-Glucoside
172. 3-Rhamnoside
173. 7-Galactoside
174. 7-Glucoside
175. 3-Glucosylglucoside
176. 3-Lathyroside
177. 3-Rhamnosylgalactoside
178. 3-Rutinoside
179. 3-Sambubioside
180. 3, 5-Diglucoside
181. 3-Rhamnoside-5-glucoside
182. 3-Rutinoside-5-glucoside
183. 3, 7-Diglucoside*
184. 3-Rhamnosylglucoside-7-xyloside*
185. 3, 7, 3′-Triglucoside*
186. 3-(Acetyglucoside)
187. 3-(*p*-Coumarylglucoside)
188. 3-(Caffeylglucoside)
189. 3-(Di-*p*-coumaryl)glucoside
190. 3-(*p*-Coumarylrutinoside)-5-glucoside (delphanin)
191. 3-(*p*-Coumarylsophoroside)-5-glucoside (cayratinin)
192. 3-(Caffeylglucoside)-5-glucoside
193. 3-(Dicaffeylrutinoside)-5-glucoside
194. 3, 5-Di(malonylglucoside)*
195. 3-(6″-*p*-Coumarylglucoside)-5-(6‴-malonylglucoside)* (malonylawobanin, **1.5**)
196. 3-Glucoside-5, 3′-di(6″caffeylglucoside)* (gentiodelphin, **1.2**)
197. 3, 5, 3′-Triglucosides acylated with *p*-coumaric and malonic acids* (ternatins A–F)
198. 3-Rutinoside-5, 3′, 5′-triglucoside acylated with caffeic and *p*-coumaric acids
199. Cinerarin* (**1.7**)
200. Petunidin (3, 5, 7, 3′, 4′-pentaOH-5′-OMe flavylium)
201. 3-Arabinoside
202. 3-Galactoside
203. 3-Glucoside
204. 3-Rhamnoside
205. 5-Glucoside
206. 3-Gentiobioside
207. 3-Rutinoside
208. 3-Sophoroside
209. 3, 5-Diglucoside
210. 3-Rhamnoside-5-glucoside
211. 3-Rutinoside-5-glucoside
212. 3-Gentiotrioside
213. 3-(2^G-Glucosylrutinoside)-5′ glucoside*
214. 3-(Acetylglucoside)
215. 3-(*p*-Coumarylglucoside)
216. 3-(Caffeylglucoside)
217. 3-(Dicaffeyl)glucoside
218. 3-(*p*-Coumarylglucoside)-5-glucoside
219. 3-(*p*-Coumarylrutinoside)-5-glucoside (petanin)
220. 3-(Di-*p*-coumaryl)rutinoside-5-glucoside (guineesin)
221. Malvidin (3, 5, 7, 4′, tetraOH-3′, 5′-diOMe flavylium)
222. 3-Arabinoside
223. 3-Galactoside
224. 3-Glucoside
225. 3-Rhamnoside
226. 5-Glucoside
227. 3-Gentiobioside

228. 3-Laminaribioside
229. 3-Rutinoside
230. 3,5-Diglucoside
231. 3-Rhamnoside-5-glucoside
232. 3,7-Diglucoside*
233. 3-Gentiotrioside
234. 3-Arabinosylglucoside-5-glucoside
235. 3-Rutinoside-5-glucoside
236. 3-Sophoroside-5-glucoside
237. 3-(Acetylglucoside)
238. 3-(*p*-Coumarylglucoside)
239. 3-(Caffeylglucoside)
240. 3-(Di-*p*-coumaryl)glucoside
241. 3-(Acetylglucoside)-5-glucoside
242. 3-(*p*-Coumarylglucoside)-5-glucoside
243. 3-(*p*-Coumarylxyloside)-5-glucoside*
244. 3-(Caffeylglucoside)-5-glucoside
245. 3-(Di-*p*-coumarylxyloside)-5-glucoside*
246. 3-(*p*-Coumarylrutinoside)-5-glucoside (negretein)
247. 3-(*p*-Coumarylsambubioside)-5-glucoside*
248. Pulchellidin (3,7,3′,4′,5′-pentaOH-S-OMe flavylium)
249. 3-Glucoside
250. Europinidin (3,7,4′,5′-tetraOH-5,3′-diOMe flavylium)
251. 3-Glucoside
252. 3-Galactoside
253. Capensinidin (3,7,4′-triOH-5,3′,5′-triOMe flavylium)
254. 3-Rhamnoside
255. Hirsutidin (3,5,4′-triOH-7,3′,5′-triOMe flavylium)
256. 3,5-Diglucoside

(2) CHECKLIST OF FLAVANS, LEUCOANTHOCYANIDINS, AND PROANTHOCYANIDINS

Flavans (all with 2*S* absolute stereochemistry, unless stated otherwise).

Hydroxylated

1. 7-OH
2. 4′,7-diOH

O-Methylated and/or methylenated

3. 4′-OH-7-OMe
4. 7-OH-5-OMe
5. 7-OH-4′-OMe (broussin)
6. 5,7-diOMe (tephrowatsin E)
7. 4′,5,7-triOMe[b]
8. 4′,7-diOH-3′-OMe
9. 7-OH-3′,4′-methylenedioxy
10. 3′,4′-diOH-5,7-diOH[b]
11. 2′-OH-7-OMe-4′,5′-methylenedioxy[b]

C-Methylated

12. 7-OH-5-OMe-6-Me
13. 4′-OH-7-OMe-8-Me
14. 4′,5-diOH-7-OMe-8-Me
15. 4′,7-diOH-8-Me
16. 3′,7-diOH-4′-OMe-8-Me
17. 4′,7-diOH-3′-OMe-8-Me
18. 2′,7-diOH-4′, 5′-methylenedioxy-6, 8-diMe[b]
19. 2′,7-diOH-4′, 5′-methylenedioxy-5, 8-diMe[b]
20. 2′,5-diOH-4′, 5′-methylenedioxy-6, 8-diMe[b]

Prenylated

21. 5,7-diOMe-8-pr[a]
22. 3′,4′,7-triOH-2′,5′-di-pr[a] (kazinol A)
23. Kazinol B (**2.15**)
24. Nitenin (**2.16**)

O-Glycosides[c]

25. 5-OH-7-O-Glc (koaburanin)
26. 7,4′-diOMe-5-O-Glc (dichotosin)
27. 4′,7-diOH-5-O-Xyl
28. 3′,7-diOH-4′-OMe-5′-O-Glc (auriculoside)
29. 3′,4′,7-triOMe-5-O-Glc (dichotosinin)[b]
30. 7-OH-3′,4′-diOMe-5′-O-Glc (diffutin)

Biflavonoids

31. Daphnodorin A (**2.17**)
32. Daphnodorin C (**2.18**)
33. Compound (**2.19**)
34. Dracorubin (**2.20**)
35. Nordracorubin (**2.21**)
36. Compound (**2.22**)[b]
37. Compound (**2.23**)[b]
38. Compound (**2.24**)[b]
39. Compound (**2.25**)[b]
40. Compound (**2.26**)[b]
41. Compound (**2.27**)[b]
42. Compound (**2.28**)[b]

Key: [a]pr = 3,3-dimethylallyl; [b]absolute stereochemistry unknown; [c]Sugars are abbreviated: Glc = glucoside, Ara = arabinoside, Xyl = Xyloside, All = alloside, f = furano, p = pyrano.

Flavan-3-ols and their natural derivatives

Hydroxylated A- and B-rings

1. Afzelechin
2. Epiafzelechin

3. *ent*-Epiafzelechin
4. Catechin
5. *ent*-Catechin
6. Epicatechin
7. *ent*-Epicatechin
8. Gallocatechin
9. Epigallocatechin
10. Fisetinidol
11. *ent*-Fisetinidol
12. *ent*-Epifisetinidol
13. Robinetinidol
14. (2*R*, 3*R*)-3, 3′, 5, 5′, 7-Pentahydroxyflavan[a]
15. Prosopin

O-Methylated

16. 4′, 5-diOMe-Afzelechin[b]
17. 4′-OMe-Catechin
18. 4′, 7-diOMe-Catechin
19. 3′, 5, 7-triOMe-Catechin[b]
20. 4′, 5, 7-triOMe-Catechin
21. 3′-OMe-Epicatechin
22. 3′, 4′-diOMe-Epicatechin
23. 3′, 5-diOMe-Epicatechin
24. 3′, 5, 7-triOMe-Epicatechin
25. 3′, 4′-Methylenedioxy-5, 7-diOMe-epicatechin[b]
26. 4′-OMe-Epigallocatechin (ourateacatechin)

Simple esters

27. *ent*-Epicatechin 3-O-δ-(3, 4-diOH-phenyl)-β-OH-pentanoate (phylloflavan)
28. Catechin 3-O-(1-OH-6-oxo-2-cyclohexene-1-carboxylate)
29. Catechin 3-O-(1, 6-diOH-2-cyclohexene-1-carboxylate)
30. Epigallocatechin 3-O-*p*-coumaroate
31. Catechin 3-O-Ga[c]
32. Catechin 7-O-Ga
33. *ent*-Catechin 3-O-Ga
34. Epicatechin 3-O-Ga
35. Epicatechin 3-O-(3-O-Me)-Ga
36. Epicatechin 3, 5-di-O-Ga
37. Epigallocatechin 3-O-Ga
38. Epigallocatechin 3-O-(3-O-Me)-Ga
39. Epigallocatechin 7-O-Ga
40. Epigallocatechin 3, 5-di-O-Ga
41. Epigallocatechin 5, 7-di-O-Ga

C- and O-glycosides

42. Epiafzelechin 5-O-β-D-Glcp
43. Catechin 5-O-β-D-Glcp
44. Catechin 7-O-α-L-Araf
45. Catechin 7-O-Apioside
46. Catechin 7-O-β-D-Xylp
47. Catechin 5-O-β-D-(2″-O-ferulyl-6″-O-*p*-coumaryl)-Glcp
48. *ent*-Catechin 7-O-β-D-Glcp
49. Epicatechin 3-O-β-D-Glcp
50. Epicatechin 3-O-β-D-Allp
51. Epicatechin 3-O-β-D-(2″-*trans*-cinnamyl)-Allp
52. Epicatechin 3-O-β-D-(3″-*trans*-cinnamyl)-Allp
53. 3′-OMe-Epicatechin 7-O-β-D-Glcp (symplocoside)
54. Epicatechin 6-C-β-D-Glcp
55. Epicatechin 8-C-β-D-Glcp

A- or B-ring C-*substituted flavan-3-ols*

56. Broussinol (**2.29**)
57. Daphnodorin B (**2.30**)
58. Larixinol (**2.31**)
59. Catechin 6-carboxylic acid
60. Gambiriin A1 (**2.32**)
61. Gambiriin A2 (**2.33**)
62. Gambiriin B1 (**2.34**)
63. Gambiriin B2 (**2.35**)
64. Gambiriin B3 (**2.36**)
65. Kopsirachin (**2.37**)[b]
66. Stenophyllanin A (**2.38**)
67. Stenophyllanin B (**2.39**)
68. Stenophyllanin C (**2.40**)
69. Cinchonain 1a (**2.41**)
70. Cinchonain 1b (**2.42**)
71. Cinchonain 1c (**2.43**)
72. Cinchonain 1d (**2.44**)

Biflavan-3-ols

73. Prosopin-(5 → 5)-prosopin
74. Prosopin-(5 → 6)-prosopin
75. Epigallocatechin-(2′ → 2′)-Epigallocatechin (**2.45**)
76. Theasinensin A (**2.46**)
77. Theasinensin B (**2.47**)

Key for flavan-3-ols: [a]structure uncertain; [b]absolute stereochemistry unknown; [c]Ga = gallate

Leucoanthocyanidins

Leucopelargonidin
1. Afzelechin-4β-ol

Leucocyanidin
2. Catechin-4β-ol

Leucoguibourtinidins
3. Epiguibourtinidol-4β-ol
4. Guibourtinidol-4α-ol

Leucofisetinidins
5. Fisetinidol-4α-ol (mollisacacidin)
6. Fisetinidol-4β-ol
7. *ent*-Fisetinidol-4β-ol
8. Epifisetinidol-4α-ol
9. Epifisetinidol-4β-ol
10. *ent*-Epifisetinidol-4β-ol

Leucorobinetinidin
11. Robinetinidol-4α-ol

Leucoteracacinidin
12. Oritin-4β-ol
13. Epioritin-4α-ol (teracacidin)
14. Epioritin-4β-ol (isoteracacidin)

Leucomelacacinidins
15. Prosopin-4β-ol
16. Epiprosopin-4α-ol (melacacidin)
17. Epiprosopin-4α-ol (isomelacacidin)

Miscellaneous
18. Cyanomaclurin (**2.48**)[a]

Alkylated flavanoids
19. 4α-OEt-Epiprosopin
20. 4β-OEt-Epiprosopin
21. 8-OMe-Oritin-4α-ol
22. 8-OMe-Prosopin-4β-ol
23. 8-OMe-Prosopin-4α-ol
24. 3, 8-DiOMe-prosopin-4β-ol
25. 3′, 4′, 7-TriOMe-fisetinidol-4β-ol
26. 3, 3′, 4, 4′, 7-PentaOMe-fisetinidol-4β-ol
27. 5-OMe-8-pr-afzelechin-4β-ol[a,b]
28. 4, 5-DiOMe-8-pr-afzelechin-4β-ol[a,b]
29. 3, 5-DiOMe-8-pr-afzelechin-4β-ol[a,b]
30. compound (**2.49**)[a]
31. compound (**2.50**)[a]

Key for leucoanthocyanidins: [a]absolute stereochemistry unknown; [b]pr = 3, 3-dimethylallyl

Flavan-4-ols

1. 4′, 5, 7-TriOMe-2, 4-*trans*-flavan-4-ol
2. Tephrowatsin A (**2.51**)
3. 4′, 5-DiOH-7-OMe-8-pr-2, 4-*trans*-flavan-4-ol[a]
4. Erubin A (**2.52**)
5. Erubin B (**2.53**)
6. Triphyllin A (**2.54**)
7. Triphyllin B (**2.55**)

Key: [a]pr = 3, 3-dimethylallyl

Peltogynoids

1. Pubeschin (**2.56**)
2. Peltogynol (**2.57**)
3. Peltogynol B (**2.58**)
4. Mopanol (**2.59**)
5. Mopanol B (**2.60**)
6. *ent*-12a, 6a-*cis*-Peltogynol B (**2.61**)
7. 10-OMe-Peltogynol
8. 7-OMe-2, 3-O, O-Methylenemopanol
9. 3-OMe-4-OH-5, 5-DiMe-peltogynol

Proanthocyanidins

Unsubstituted flavanoid units

Dimers

Propelargonidins
1. Afzelechin-(4α → 8)-afzelechin
2. Afzelechin-(4α → 8)-catechin
3. Afzelechin-(4α → 8)-epicatechin
4. Epiafzelechin-(4β → 8)-catechin (gambiriin C)
5. Epiafzelechin-(4β → 8)-4′-OMe-epigallocatechin

Procyanidins
6. Catechin-(4α → 6)-catechin (procyanidin B6)
7. Catechin-(4α → 8)-catechin (procyanidin B3)
8. Catechin-(4α → 6)-epicatechin (procyanidin B8)
9. Catechin-(4α → 8)-epicatechin (procyanidin B4)
10. Epicatechin-(4β → 6)-catechin (procyanidin B7)
11. Epicatechin-(4β → 8)-catechin (procyanidin B1)
12. Epicatechin-(4β → 6)-epicatechin (procyanidin B5)
13. Epicatechin-(4β → 8)-epicatechin (procyanidin B2)
14. *ent*-Epicatechin-(4α → 8)-*ent*-epicatechin
15. Catechin-(4α → 8)-epiafzelechin
16. Epicatechin-(4β → 8)-epiafzelechin
17. Epicatechin-(4β → 8)-*ent*-epicatechin

Prodelphinidins
18. Gallocatechin-(4α → 8)-epigallocatechin
19. Gallocatechin-(4α → 8)-catechin
20. Epigallocatechin-(4β → 8)-catechin

Proguibourtinidins
21. Guibourtinidol-(4α → 8)-epiafzelechin
22. Guibourtinidol-(4α → 6)-catechin
23. Guibourtinidol-(4α → 8)-catechin
24. Guibourtinidol-(4α → 6)-epicatechin
25. Guibourtinidol-(4β → 8)-epicatechin

Profisetinidins
26. Fisetinidol-(4α → 6)-fisetinidol
27. Fisetinidol-(4β → 6)-fisetinidol
28. Fisetinidol-(4α → 6)-fisetinidol-4α-ol
29. Fisetinidol-(4α → 6)-fisetinidol-4β-ol
30. Fisetinidol-(4β → 6)-fisetinidol-4α-ol
31. Fisetinidol-(4β → 6)-fisetinidol-4β-ol
32. Fisetinidol-(4α → 8)-catechin
33. Fisetinidol-(4β → 8)-catechin
34. *ent*-Fisetinidol-(4β → 6)-catechin
35. *ent*-Fisetinidol-(4β → 8)-catechin
36. *ent*-Fisetinidol-(4α → 6)-catechin
37. *ent*-Fisetinidol-(4α → 8)-catechin
38. *ent*-Fisetinidol-(4β → 8)-epicatechin

Prorobinetinidins
39. Robinetinidol-(4α → 8)-catechin
40. Robinetinidol-(4α → 8)-gallocatechin

Promelacacinidins

41. Prosopin-(4α → 6)-prosopin
42. Epiprosopin-(4α → 6)-epiprosopin-4α-ol

Trimers

Procyanidins

43. [Catechin-(4α → 8)]$_2$-catechin
44. [Catechin-(4α → 8)]$_2$-epicatechin
45. [Epicatechin-(4β → 8)]$_2$-epicatechin
46. Epicatechin-(4β → 6)-epicatechin-(4β → 8)-epicatechin
47. [Epicatechin-(4β → 6)]$_2$-epicatechin
48. [Epicatechin-(4β → 8)]$_2$-catechin
49. Epicatechin-(4β → 8)-epicatechin-(4β → 6)-catechin
50. Epicatechin-(4β → 6)-epicatechin-(4β → 8)-catechin
51. Catechin-(4α → 6)-epicatechin-(4β → 8)-catechin
52. Epicatechin-(4β → 8)-catechin-(4α → 8)-catechin
53. Epicatechin-(4β → 8)-catechin-(4α → 8)-epicatechin

Prodelphinidins

54. [Gallocatechin-(4α → 8)]$_2$ catechin

Procyanidin and prodelphinidin

55. Catechin-(4α → 8)-gallocatechin-(4α → 8)-catechin
56. Gallocatechin-(4α → 8)-catechin-(4α → 8)-catechin

Profisetinidins

57. Fisetinidol-(4α → 8)-catechin-(6 → 4α)-fisetinidol
58. Fisetinidol-(4α → 8)-catechin-(6 → 4β)-fisetinidol
59. Fisetinidol-(4β → 8)-catechin-(6 → 4α)-fisetinidol
60. Fisetinidol-(4β → 8)-catechin-(6 → 4β)-fisetinidol
61. *ent*-Fisetinidol-(4β → 8)-catechin (6 → 4β)-*ent*-fisetinidol
62. *ent*-Fisetinidol-(4β → 8)-catechin (6 → 4α)-*ent*-fisetinidol
63. *ent*-Fisetinidol-(4α → 8)-catechin (6 → 4β)-*ent*-fisetinidol
64. *ent*-Fisetinidol-(4α → 8)-catechin (6 → 4α)-*ent*-fisetinidol

Prorobinetinidins

65. Robinetinidol-(4α → 8)-catechin-(6 → 4β)-robinetinidol
66. Robinetinidol-(4β → 8)-catechin-(6 → 4β)-robinetinidol
67. Robinetinidol-(4α → 8)-gallocatechin-(6 → 4α)-robinetinidol
68. Robinetinidol-(4α → 8)-gallocatechin-(6 → 4β)-robinetinidol

Tetramers

Procyanidins

69. [Epicatechin-(4β → 8)]$_3$-epicatechin
70. [Epicatechin-(4β → 8)]$_3$-catechin
71. [Epicatechin-(4β → 8)]$_2$-epicatechin (4β → 6)-catechin
72. Catechin-(4α → 6)-[epicatechin-(4β → 8)]$_2$-epicatechin

Profisetinidins

73. Fisetinidol-(4β → 6)-fisetinidol-(4β → 8)-catechin-(6 → 4β)-fisetinidol
74. Fisetinidol-(4β → 6)-fisetinidol-(4β → 8)-catechin-(6 → 4α)-fisetinidol
75. Fisetinidol-(4α → 6)-fisetinidol-(4α → 8)-catechin-(6 → 4α)-fisetinidol
76. Fisetinidol-(4α → 6)-fisetinidol-(4α → 8)-catechin-(6 → 4β)-fisetinidol
77. *ent*-Fisetinidol-(4β → 6)-*ent*-fisetinidol-(4β → 8)-catechin-(6 → 4β)-*ent*-fisetinidol
78. *ent*-Fisetinidol-(4β → 6)-*ent*-fisetinidol-(4β → 8)-catechin-(6 → 4α)-*ent*-fisetinidol
79. *ent*-Fisetinidol-(4α → 6)-*ent*-fisetinidol-(4α → 8)-catechin-(6 → 4β)-*ent*-fisetinidol
80. *ent*-Fisetinidol-(4α → 6)-*ent*-fisetinidol-(4α → 8)-catechin-(6 → 4α)-*ent*-fisetinidol

Pentamer

Procyanidin

81. [Epicatechin-(4β → 8)]$_4$-epicatechin

Hexamer

Procyanidin

82. [Epicatechin-(4β → 8)]$_5$-epicatechin

Simple esters, Gallate esters

Dimers

Procyanidins

83. 3-O-Ga-Catechin-(4α → 8)-catechin
84. Catechin-(4α → 8)-epicatechin-3-O-Ga
85. Epicatechin-(4β → 8)-epicatechin-3-O-Ga
86. 3-O-Ga-Epicatechin-(4β → 8)-epicatechin-3-O-Ga
87. 3-O-Ga-Epicatechin-(4β → 8)-catechin

Prodelphinidins

88. Epigallocatechin-(4β → 8)-epicatechin-3-O-Ga
89. Epigallocatechin-(4β → 8)-epigallocatechin-3-O-Ga
90. 3-O-Ga-Epigallocatechin-(4β → 6)-epigallocatechin-3-O-Ga
91. 3-O-Ga-Epigallocatechin-(4β → 8)-epigallocatechin-3-O-Ga
92. 3-O-Ga-Epigallocatechin-(4β → 8)-gallocatechin-3-O-Ga

Cyclohexene carboxylates

Dimers

Procyanidins
93. Epicatechin-(4$\beta \rightarrow$8)-catechin-3-O-(1-OH-6-oxo-2-cyclohexene-1-carboxylate)
94. Catechin-(4$\alpha \rightarrow$8)-catechin-3-O-(1-OH-6-oxo-2-cyclohexene-1-carboxylate)

Trimer

Procyanidin
95. Epicatechin-(4$\beta \rightarrow$8)-catechin-(4$\alpha \rightarrow$8)-catechin-3-O-(1-OH-6-oxo-2-cyclohexene-1-carboxylate)

C- *and* O-*Glycosides*

Proluteolinidins
96. 5-O-Glc-luteoliflavan-(4$\beta \rightarrow$8)-eriodictyol
97. 5-O-Glc-luteoliflavan-(4$\beta \rightarrow$8)-eriodictyol-3-O-Glc

Procyanidins
98. 6-*C*-β-D-Glcp-epicatechin-(4$\beta \rightarrow$8)-epicatechin
99. 8-*C*-β-D-Glcp-epicatechin-(4$\beta \rightarrow$8)-epicatechin

Trimers

Proluteolinidin

100. [5-O-Glc-luteoliflavan-(4$\beta \rightarrow$8)]$_2$-eriodictyol
101. [5-O-Glc-luteoliflavan-(4$\beta \rightarrow$8)]$_2$-eriodictyol-5-O-Glc

C-*Substituted flavanoid units*

Dimers

Proguibourtinidins
102. Guibourtinidol-(4$\alpha \rightarrow$8)-epicatechin-6-carboxylic acid
103. Guibourtinidol-(4$\alpha \rightarrow$6)-epicatechin-8-carboxylic acid
104. Guibourtinidol-(4$\alpha \rightarrow$8)-catechin-6-carboxylic acid
105. Guibourtinidol-(4$\alpha \rightarrow$6)-catechin-8-carboxylic acid

Procyanidins
106. Cinchonain-1a-(4$\beta \rightarrow$8)-epicatechin (cinchonain IIa)
107. Cinchonain-1b-(4$\beta \rightarrow$8)-epicatechin (cinchonain IIb)
108. Cinchonain-1a-(4$\beta \rightarrow$8)-catechin (kandelin A-1)
109. Cinchonain-1b-(4$\beta \rightarrow$8)-catechin (kandelin A-2)

Trimers

Procyanidins
110. Cinchonain-1a-(4$\beta \rightarrow$8)-epicatechin-(4$\beta \rightarrow$8)-epicatechin (kandelin B-1)
111. Cinchonain-1b-(4$\beta \rightarrow$8)-epicatechin-(4$\beta \rightarrow$8)-epicatechin (kandelin B-2)
112. Cinchonian-1b-(4$\beta \rightarrow$8)-epicatechin-(4$\beta \rightarrow$8)-catechin (kandelin B-4)
113. Epicatechin-(4$\beta \rightarrow$6)-cinchonain-1a-(4$\beta \rightarrow$8)-epicatechin (kandelin B-3)

Miscellaneous

Monomers

Flavan
114. 4′-OH-7-OMe-4-(4-OH-styryl)-flavan
115. 4′-OH-5, 7-diOMe-4-(4-OH-styryl)-flavan

Procyanidin
116. Dryopterin (**2.62**)
117. Epicatechin-(4$\beta \rightarrow$2)-phloroglucinol

Proguibourtinidin
118. Guibourtinidol-(4$\alpha \rightarrow$2)-3, 5, 3′, 4′-tetra-hydroxystilbene
119. Epiguibourtinidol-(4$\beta \rightarrow$2)-3, 5, 3′, 4′-tetra-hydroxystilbene (**2.63**)

Dimers

Proguibourtinidins
120. Guibourtinidol-(4$\alpha \rightarrow$2)-3, 5, 3′, 4′-tetra-hydroxystilbene-(6$\rightarrow$4β)-epiguibourtinidol
121. Epiguibourtinidol-(4$\beta \rightarrow$2)-3, 5, 3′, 4′-tetra-hydroxystilbene-(6$\rightarrow$4β)-epiguibourtinidol

Isoflavanoid proanthocyanidins

122. Compound (**2.64**)
123. Compound (**2.65**)
124. Compound (**2.66**)
125. Compound (**2.67**)
126. Compound (**2.68**)

A-Type Proanthocyanidins

Dimers

Proapigeninidins
1. *ent*-Apigeniflavan-(2$\alpha \rightarrow$7, 4$\alpha \rightarrow$8)-epiafzelechin (mahuannin D)

Propelargonidins
2. *ent*-Epiafzelechin-(2$\alpha \rightarrow$7, 4$\alpha \rightarrow$8)-afzelechin
3. *ent*-Epiafzelechin-(2$\alpha \rightarrow$7, 4$\alpha \rightarrow$8)-catechin
4. Epiafzelechin-(2$\beta \rightarrow$7, 4$\beta \rightarrow$8)-epiafzelechin (mahuannin B)
5. Epiafzelechin-(2$\beta \rightarrow$7, 4$\beta \rightarrow$6)-epiafzelechin (mahuannin C)

6. *ent*-Epiafzelechin-(2α→7, 4α→8)-epiafzelechin (mahuannin A)
7. *ent*-Epiafzelechin-(2α→7, 4α→8)-kaempferol (ephedrannin A)

Procyanidins
8. Epicatechin-(2β→7, 4β→8)-catechin (proanthocyanidin A1)
9. Epicatechin-(2β→7, 4β→8)-epicatechin (proanthocyanidin A2)

Trimers

Procyanidins
10. Epicatechin-(2β→7, 4β→8)-epicatechin-(4β→8)-epicatechin
11. Epicatechin-(2β→7, 4β→8)-epicatechin-(4β→8)-catechin

Tetramers

Procyanidins
12. Epicatechin-(4β→8)-epicatechin-(2β→7, 4β→8)-epicatechin-(4β→8)-epicatechin
13. Epicatechin-(4β→6)-epicatechin-(2β→7, 4β→8)-epicatechin-(4β→8)-epicatechin
14. Epicatechin-(4β→8)-epicatechin-(2β→7, 4β→8)-epicatechin-(4β→8)-catechin
15. Epicatechin-(4β→6)-epicatechin-(2β→7, 4β→8)-epicatechin-(4β→8)-catechin

Pentamers

Procyanidins
16. [Epicatechin-(4β→8)]$_2$-epicatechin-(2β→7, 4β→8)-epicatechin-(4β→8)-epicatechin
17. [Epicatechin-(4β→8)]$_2$-epicatechin-(2β→7, 4β→8)-epicatechin-(4β→8)-catechin

(3) *C*-GLYCOSYLFLAVONOIDS

Mono-*C*-glycosylflavones

1. Bayin (7, 4′-diOH 8-Glc)
2. Isovitexin (5, 7, 4′-triOH 6-Glc)
3. 7-*O*-Glucoside
4. 7-*O*-Galactoside
5. 7-*O*-Xyloside
6. 7-*O*-Rhamnoside*
7. 4′-*O*-Glucoside
8. 4′-*O*-Arabinoside*
9. 2″-*O*-Glucoside
10. 2″-*O*-Galactoside*
11. 2″-*O*-Xyloside
12. 2″-*O*-Arabinoside
13. 2″-*O*-Rhamnoside
14. 6″-*O*-Arabinoside
15. 7-*O*-Rhamnosylglucoside*
16. 7, 2″-Di-*O*-glucoside
17. 7, 2″-Di-*O*-galactoside*
18. 7-*O*-Glucoside-2″-*O*-arabinoside
19. 7-*O*-Glucoside-2″-*O*-rhamnoside
20. 7-*O*-Galactoside-2″-*O*-glucoside
21. 7-*O*-Galactoside-2″-*O*-arabinoside*
22. 7-*O*-Galactoside-2″-*O*-rhamnoside
23. 7-*O*-Xyloside-2″-*O*-glucoside
24. 7-*O*-Xyloside-2″-*O*-arabinoside
25. 7-*O*-Xyloside-2″-*O*-rhamnoside
26. 7-*O*-Arabinoside-2″-*O*-glucoside*
27. 4′, 2″-Di-*O*-glucoside
28. 4′-*O*-Glucoside-2″-*O*-arabinoside
29. X″-*O*-Arabinosylglucoside*
30. X″-*O*-Diglucoside*
31. 7-*O*-Ferulylglucoside*
32. 2″-*O*-(*E*)-Ferulyl
33. 2″-*O*-(*E*)-Ferulyl-4′-*O*-glucoside
34. X″-*O*-(*E*)-Caffeyl-2″-*O*-glucoside
35. 7-Sulfate
36. Vitexin (5, 7, 4′-triOH 8-Glc)
37. 7-*O*-Glucoside
38. 4′-*O*-Glucoside
39. 4′-*O*-Galactoside*
40. 2″-*O*-Glucoside
41. X″-*O*-Glucoside*
42. 2″-*O*-Xyloside
43. 6″-*O*-Xyloside
44. X″-*O*-Arabinoside
45. 2″-*O*-Rhamnoside
46. 6″-*O*-Rhamnoside
47. 7-*O*-Rutinoside
48. 2″-*O*-Sophoroside
49. 6″-*O*-Gentiobioside (marginatoside)
50. 4′-*O*-Glucoside-2″-*O*-rhamnoside*
51. 2″-*O*-Acetyl*
52. 2″-*O*-*p*-Hydroxybenzoyl
53. 2″-*O*-*p*-Coumaryl
54. 2″-*O*-*p*-Coumaryl-7-*O*-glucoside*
55. 4‴-*O*-Acetyl-2″-*O*-rhamnoside
56. 7-Sulfate
57. 7-*O*-Rutinosidesulfate
58. 8-*C*-β-D-Galactopyranosylapigenin (5, 7, 4′-triOH 8-Gal)
59. 6″-*O*-Acetyl
60. Cerarvensin (5, 7, 4′-triOH 6-Xyl)*
61. 7-*O*-Glucoside*
62. 2″-*O*-Rhamnoside
63. Isomollupentin (5, 7, 4′-triOH 6-Ara)
64. 7-*O*-Glucoside*
65. 4′-*O*-Glucoside*
66. 2″-*O*-Glucoside*
67. 7-*O*-Rhamnosylglucoside*
68. 7, 2″-Di-*O*-glucoside
69. 7-*O*-Glucoside-2″-*O*-xyloside*

70. 7-*O*-Glucoside-2″-*O*-arabinoside*
71. Mollupentin (5, 7, 4′-triOH 8-Ara)
72. Isofurcatain (5, 7, 4′-triOH 6-Rha)*
73. 7-*O*-Glucoside*
74. 3′-Deoxyderhamnosylmaysin [5,7,4′-triOH 6-(6-deoxyxylohexos-4-ulosyl)]
75. 2″-*O*-Rhamnoside (3′-deoxymaysin)
76. Swertisin (5,4′-diOH 7-OMe 6-Glc)
77. 5-*O*-Glucoside*
78. 4′-*O*-Glucoside
79. 4′-*O*-Rhamnoside*
80. 2″-*O*-Glucoside (spinosin, flavoayamenin)
81. 2″-*O*-Rhamnoside*
82. X″-*O*-Xyloside
83. 6‴-*O*-*p*-Coumaryl-2″-*O*-glucoside*
84. 6‴-*O*-Ferulyl-2″-*O*-glucoside*
85. 6‴-*O*-Sinapyl-2″-*O*-glucoside*
86. Isoswertisin (5,4′-diOH 7-OMe 8-Glc)
87. 5-*O*-Glucoside*
88. 4′-*O*-Glucoside
89. 2″-*O*-Glucoside*
90. 2″-*O*-Xyloside*
91. 2″-*O*-Rhamnoside
92. 2″-*O*-Acetyl
93. Isomolludistin (5,4′-diOH 7-OMe 6-Ara)
94. 2″-*O*-Glucoside
95. Molludistin (5,4′-diOH 7-OMe 8-Ara)
96. 2″-*O*-Glucoside
97. 2″-*O*-Xyloside*
98. 2″-*O*-Rhamnoside
99. Isocytisoside (5,7-diOH 4′-OMe 6-Glc)
100. 7-*O*-Glucoside
101. 2″-*O*-Glucoside
102. 2″-*O*-Rhamnoside
103. Cytisoside (5,7-diOH 4′-OMe 8-Glc)
104. 7-*O*-Glucoside
105. 2″-*O*-Rhamnoside
106. *O*-Acetyl-7-*O*-glucoside (tremasperin)
107. Embigenin (5-OH 7,4′-diOMe 6-Glc)
108. 2″-*O*-Glucoside* (embinoidin)
109. X″-*O*-Glucoside*
110. 2″-*O*-Rhamnoside (embinin)
111. 2‴-*O*-Acetyl-2″-*O*-rhamnoside*
112. X″,2‴-Di-*O*-acetyl-2″-*O*-rhamnoside*
113. Isoembigenin (5-OH 7,4′-diOMe 8-Glc)*
114. 7,4′-Di-*O*-methylisomollupentin (5-OH 7,4′-diOMe 6-Ara)
115. 2″-*O*-Glucoside
116. 2″-*O*-Caffeylglucoside
117. Isoorientin (5,7,3′,4′-tetraOH 6-Glc)
118. 7-*O*-Glucoside
119. 3′-*O*-Glucoside
120. 3′-*O*-Glucuronide
121. 4′-*O*-Glucoside
122. 2″-*O*-Glucoside
123. 2″-*O*-Mannoside*
124. 2″-*O*-Xyloside
125. 2″-*O*-Arabinoside
126. 2″-*O*-β-L-Arabinofuranoside
127. 2″-*O*-Rhamnoside
128. 6″-*O*-Glucoside*
129. 6″-*O*-Arabinoside
130. 7-*O*-Rutinoside
131. 3′-*O*-Sophoroside*
132. 3′-*O*-Neohesperidoside*
133. 3′,6″-Di-*O*-glucoside*
134. 4′,2″-Di-*O*-glucoside
135. X″-*O*-Acetyl
136. 2″-*O*-*p*-Hydroxybenzoyl
137. 2″-*O*-(*E*)-*p*-Coumaryl
138. 2″-*O*-(*E*)-Caffeyl
139. 2″-*O*-(*E*)-Ferulyl
140. 2″-*O*-*p*-Hydroxybenzoyl-4′-*O*-glucoside
141. 2″-*O*-(*E*)-Caffeyl-4′-*O*-glucoside
142. 2″-(*p*-*O*-Glucosyl)-(*E*)-caffeyl-4′-*O*-glucoside
143. 2″-*O*-(*E*)-Ferulyl-4′-*O*-glucoside
144. 2″-*O*-(*E*)-Caffeylglucoside
145. 2″-*O*-[2-*O*-Glucosyl-2,4,5-trihydroxy-(*E*)-cinnamyl]-4′-*O*-glucoside*
146. 7-Sulfate
147. Orientin (5,7,3′,4′-tetraOH 8-Glc)
148. 7-*O*-Rhamnoside
149. 4′-*O*-Glucoside
150. 2″-*O*-Glucoside
151. X″-*O*-Glucoside*
152. 2″-*O*-Xyloside (adonivernith)
153. 2″-*O*-β-L-Arabinofuranoside
154. 2″-*O*-Rhamnoside
155. 4′-*O*-Glucoside-2″-*O*-rhamnoside*
156. X″-*O*-Rutinoside*
157. 2″-*O*-Acetyl*
158. 7-Sulfate
159. 7-*O*-Glucosidesulfate
160. 6-*C*-β-D-Xylopyranosylluteolin (5,7,3′,4′-tetraOH 6-Xyl)
161. 2″-*O*-Rhamnoside
162. Derhamnosylmaysin [5, 7, 3′, 4′-tetraOH 6-(6-deoxyxylohexos-4-ulosyl)]
163. 2″-*O*-Rhamnoside (maysin)
164. Swertiajaponin (5,3′,4′-triOH 7-OMe 6-Glc)
165. 3′-*O*-Glucoside
166. 4′-*O*-Rhamnoside*
167. 2″-*O*-Glucoside* (luteoayamenin)
168. 2″-*O*-Rhamnoside
169. 3′-*O*-Gentiobioside
170. Isoswertiajaponin (5,3′,4′-triOH 7-OMe 8-Glc)
171. Isoscoparin (5,7,4′-triOH 3′-OMe 6-Glc)
172. 7-*O*-Glucoside
173. 2″-*O*-Glucoside
174. 2″-*O*-Rhamnoside*
175. 6‴-*O*-*p*-Coumaryl-2″-*O*-glucoside*
176. 6‴-*O*-Ferulyl-2″-*O*-glucoside*

177. Scoparin (5, 7, 4′-triOH 3′-OMe 8-Glc)
178. 2″-O-Rhamnoside*
179. X″-O-Rhamnosylglucoside
180. Episcoparin 7-O-glucoside (knautoside)
181. 3′-O-Methylderhamnosylmaysin [5, 7, 4′-triOH 3′-OMe 6-(6-deoxyxylohexos-4-ulosyl)]
182. 2″-O-Rhamnoside (3′-O-methylmaysin)
183. 6-C-β-D-Glucopyranosyldiosmetin (5, 7, 3′-triOH 4′-OMe 6-Glc)
184. 8-C-β-D-Glucopyranosyldiosmetin (5, 7, 3′-triOH 4′-OMe 8-Glc)
185. 2″-O-Rhamnoside*
186. 7, 3′-Di-O-methylisoorientin (5, 4′-diOH 7, 3′-diOMe 6-Glc)
187. 7, 3′-Di-O-methylorientin (5, 4′-diOH 7, 3′-diOMe 8-Glc)*
188. 7, 3′, 4′-Tri-O-methylisoorientin (5-OH 7, 3′, 4′-triOMe 6-Glc)
189. 2″-O-Rhamnoside (linoside B)
190. 6″-O-Acetyl-2″-O-rhamnoside (linoside A)
191. 6-C-Galactosylisoscutellarein (5, 7, 8, 4′-tetraOH 6-Gal)*
192. Isoaffinetin (5, 7, 3′, 4′, 5′-pentaOH 6-Glc)
193. Affinetin (5, 7, 3′, 4′, 5′-pentaOH 8-Glc)
194. Isopyrenin (5, 7, 4′-triOH 3′, 5′-diOMe 6-Glc)
195. 7-O-Glucoside
196. 6-C-Glucosyl-5, 7-dihydroxy-8, 3′, 4′, 5′-tetramethoxyflavone
197. 5-O-Rhamnoside

Di-C-Glycosylflavones

1. 6-C-Glucosyl-8-C-arabinosylchrysin (5, 7-diOH 6-Glc-8-Ara)*
2. 6-C-Arabinosyl-8-C-glucosylchrysin (5, 7-diOH 6-Ara-8-Glc)*
3. Vicenin-2 (5, 7, 4′-triOH 6, 8-di-Glc)
4. 6″-O-Glucoside*
5. X‴-O-Diferulylglucoside*
6. 3, 6-Di-C-glucosylapigenin (5, 7, 4′-triOH 3, 6-diGlc)*
7. 3, 8-Di-C-glucosylapigenin (5, 7, 4′-triOH 3, 8-di-Glc)*
8. 6, 8-Di-C-galactosylapigenin (5, 7, 4′-triOH 6, 8-diGal)*
9. 6-C-Glucosyl-8-C-galactosylapigenin (5, 7, 4′-triOH 6-Glc-8-Gal)*
10. 6, 8-Di-C-hexosylapigenin (5, 7, 4′-triOH 6, 8-diHex)
11. Vicenin-3 (5, 7, 4′-triOH 6-Glc-8-Xyl)
12. Vicenin-1 (5, 7, 4′-triOH 6-Xyl-8-Glc)
13. Sinapyl
14. Violanthin (5, 7, 4′-triOH 6-Glc-8-Rha)
15. Isoviolanthin (5, 7, 4′-triOH 6-Rha-8-Glc)
16. Schaftoside (5, 7, 4′-triOH 6-Glc-8-Ara)
17. 6″-O-Glucoside*
18. Isoschaftoside (5, 7, 4′-triOH 6-Ara-8-Glc)
19. 2‴-O-Ferulyl*
20. Sinapyl
21. Neoschaftoside (5, 7, 4′-triOH 6-Glc-8-Ara)
22. Neoisoschaftoside (5, 7, 4′-triOH 6-Ara-8-Glc)
23. Isocorymboside (5, 7, 4′-triOH 6-Gal-8-Ara)
24. Corymboside (5, 7, 4′-triOH 6-Ara-8-Gal)
25. X‴-O-Ferulyl*
26. X‴-O-Sinapyl*
27. 6, 8-Di-C-arabinosylapigenin (5, 7, 4′-triOH 6, 8-diAra)
28. 6-C-Xylosyl-8-C-arabinosylapigenin (5, 7, 4′-triOH 6-Xyl-8-Ara)*
29. 6-C-Arabinosyl-8-C-xylosylapigenin (5, 7, 4′-triOH 6-Ara-8-Xyl)*
30. 6-C-Glucosyl-8-C-galactosylgenkwanin (5, 4′-diOH 7-OMe 6-Glc-8-Gal)*
31. 6-C-Glucosyl-8-C-arabinosylgenkwanin (5, 4′-diOH 7-OMe 6-Glc-8-Ara)*
32. Almeidein (5, 4′-diOH 7-OMe 6, 8-diAra)
33. 3, 6-Di-C-glucosylacacetin (5, 7-diOH 4′-OMe 3, 6-diGlc)*
34. 6-C-Pentosyl-8-C-hexosylacacetin (5, 7-diOH 4′-OMe 6-Pen-8-Hex)
35. 7, 4′-Di-O-methyl-6, 8-di-C-arabinosylapigenin (5-OH 7, 4′-diOMe 6, 8-diAra)
36. Lucenin-2 (5, 7, 3′, 4′-tetraOH 6, 8-di-Glc)
37. Lucenin-3 (5, 7, 3′, 4′-tetraOH 6-Glc-8-Xyl)
38. Lucenin-1 (5, 7, 3′, 4′-tetraOH 6-Xyl-8-Glc)
39. Carlinoside (5, 7, 3′, 4′-tetraOH 6-Glc-8-Ara)
40. Isocarlinoside (5, 7, 3′, 4′-tetraOH 6-Ara-8-Glc)*
41. Neocarlinoside (5, 7, 3′, 4′-tetraOH 6-Glc-8-Ara)
42. 6, 8-Di-C-pentosylluteolin (5, 7, 3′, 4′-tetraOH 6, 8-diPen)
43. 6, 8-Di-C-Glucosylchrysoeriol (5, 7, 4′-triOH 3′-OMe 6, 8-diGlc)
44. 6-C-Glucosyl-8-C-arabinosylchrysoeriol (5, 7, 4′-triOH 3′-OMe 6-Glc-8-Ara)*
45. 6-C-Arabinosyl-8-C-glucosylchrysoeriol (5, 7, 4′-triOH 3′-OMe 6-Ara-8-Glc)*
46. 6-C-Arabinosyl-8-C-hexosylchrysoeriol (5, 7, 4′-triOH 3′-OMe 6-Ara-8-Hex)
47. 6, 8-Di-C-glucosyldiosmetin (5, 7, 3′-triOH 4′-OMe 6, 8-diGlc)*
48. 3, 8-Di-C-glucosyldiosmetin 5, 7, 3′-triOH 4′-OMe 3, 8-diGlc)*
49. 6, 8-Di-C-glucosyltricetin (5, 7, 3′, 4′, 5′-pentaOH 6, 8-diGlc)
50. X‴-O-Rhamnoside*
51. 6, 8-Di-C-hexosyltricetin (5, 7, 3′, 4′, 5′-pentaOH 6, 8-diHex)
52. 6-C-Glucosyl-8-C-arabinosyltricetin*
53. 6-C-Hexosyl-8-C-pentosyltricetin
54. 6-C-Arabinosyl-8-C-glucosyltricetin
55. 6, 8-Di-C-pentosyltricetin
56. 6-C-Hexosyl-8-C-pentosyl-3′-O-methyltricetin

57. 6, 8-Di-*C*-glucosyltricin (5, 7, 4′-triOH 3′, 5-diOMe 6, 8-di-Glc)
58. 6-*C*-Glucosyl-8-*C*-arabinosyltricin
59. 6-*C*-Arabinosyl-8-*C*-glucosyltricin
60. 6-*C*-Xylosyl-8-*C*-hexosyltricin
61. 6, 8-Di-*C*-arabinosyltricin
62. 6-*C*-Hexosyl-8-*C*-pentosylapometzgerin (5, 7, 5′-triOH 3′, 4′-diOMe 6-Hex-8-Pen)
63. 6, 8-Di-*C*-arabinosylapometzgerin

C-Glycosylflavonols

1. 8-*C*-Glucosyl-5-deoxykaempferol (3, 7, 4′-triOH 8-Glc)*
2. 8-*C*-Glucosylfisetin (3, 7, 3′, 4′-tetraOH 8-Glc)*
3. 6-*C*-Glucosylkaempferol (3, 5, 7, 4′-tetraOH 6-Glc)
4. Keyakinin (3, 5, 4′-triOH 7-OMe 6-Glc)
5. 6-*C*-Glucosylquercetin (3, 5, 7, 3′, 4′-pentaOH 6-Glc)
6. Keyakinin B (3, 5, 3′, 4′-tetraOH 7-OMe 6-Glc)
7. 8-*C*-Rhamnosyleuropetin (3, 5, 3′, 4′, 5′-pentaOH 7-OMe 8-Rha)*

C-Glycosylflavanones

1. Aervanone (7, 4′-diOH 8-Gal)
2. Hemiphloin (5, 7, 4′-triOH 6-Glc)
3. Isohemiphloin (5, 7, 4′-triOH 8-Glc)

Di-C-glycosylflavanones

1. 6, 8-Di-*C*-glucosylnaringenin (5, 7, 4′-triOH 6, 8-diGlc)*

C-Glycosylflavanonols

1. 6-*C*-Glucosyldihydrokaempferol (3, 5, 7, 4′-tetraOH 6-Glc)
2. Keyakinol (3, 5, 4′-triOH 7-OMe 6-Glc)
3. 6-*C*-Glucosyldihydroquercetin (3, 5, 7, 3′, 4′-pentaOH 6-Glc)

C-Glycosylchalcone

1. 3′-*C*-Glucosylisoliquiritigenin (2′, 4′, 4-triOH 3′-Glc)

C-Glycosyldihydrochalcones

1. Nothofagin (2′, 4′, 6′, 4-tetraOH *C*-Gly)
2. Konnanin (2′, 4′, 6′, 3, 4-pentaOH *C*-Gly)
3. Aspalathin (2′, 4′, 6′, 3, 4-pentaOH 3′-Glc)

Di-C-glycosyldihydrochalcone

1. 3′, 5′-Di-*C*-glycosylphloretin (2′, 4′, 6′, 4-tetraOH 3′, 5′-diGly)

C-Glycosyl-α-hydroxydihydrochalcones

1. Coatline A (α, 2′, 4′, 4-tetraOH 3′-Glc)* (**3.1**)
2. Coatline B (α, 2′, 4′, 3, 4-pentaOH 3′-Glc)* (**3.2**)

C-Glycosyl-β-hydroxydihydrochalcone

1. Pterosupin (β, 2′, 4′, 4-tetraOH 3′-Glc)* (**3.3**)

C-Glycosylquinochalcones

1. Safflor yellow A* (**3.8**)
2. Safflor yellow B* (**3.9**)
3. Carthamin* (**3.6**)

C-Glycosylisoflavones

1. Puerarin (7, 4′-diOH 8-Glc)
2. *O*-Xyloside
3. 6″-*O*-β-Apiofuranoside (mirificin)*
4. 4′, 6″-Di-*O*-acetyl
5. 8-*C*-Glucosylgenistein (5, 7, 4′-triOH 8-Glc)
6. 8-*C*-Glucosylprunetin (5, 4′-diOH 7-OMe 8-Glc)
7. Isovolubilin (5-OH 7, 4′-diOMe 6-Rha)
8. Volubilin (5-OH 7, 4′-diOMe 8-Rha)
9. 8-*C*-Glucosylorobol (5, 7, 3′, 4′-tetraOH 8-Glc)
10. Dalpanitin (5, 7, 4′-triOH 3′-OMe 8-Glc)
11. Volubilinin (5, 7-diOH 6, 4′-diOMe 8-Glc)

Di-C-glycosylisoflavones

1. Paniculatin (5, 7, 4′-triOH 6, 8-diGlc)
2. 6, 8-Di-*C*-glucosylorobol (5, 7, 3′, 4′-tetraOH 6, 8-diGlc)*

C-Glycosylisoflavanones

1. Dalpanin (cf Table 3.2)

C-Glycosylflavanols

1. 6-*C*-Glucosylepicatechin (3, 5, 7, 3′, 4′-pentaOH 6-Glc)*
2. 8-*C*-Glucosylepicatechin (3, 5, 7, 3′, 4′-pentaOH 8-Glc)*

C-Glycosylprocyanidins

1. 6-*C*-Glucosylprocyanidin B-2* (**3.4**)
2. 8-*C*-Glucosylprocyanidin B-2* (**3.5**)

(4) BIFLAVONOIDS

1. 3, 3″-Biapigenin
2. Brackenin*
3. Chamaejasmin
4. Isochamaejasmin*

5. Neochamaejasmin B*
6. Neochamaejasmin A*
7. 7-Methylchamaejasmin*
8. Chamaejasmenin A*
9. Chamaejasmenin B*
10. Chamaejasmenin C*
11. 7,7′-Dimethylchamaejasmenin A*
12. 7,4′,7″,4‴-Tetramethylisochamaejasmin*
13. 7,4′,7″,4‴-Tetramethylneochamaejasmin B*
14. 7,4′,7″,4‴-Tetramethylneochamaejasmin A*
15. Bisdehydro Garcinia biflavonoid 1a
16. Sahranflavone
17. Volkensiflavone (Talbotiflavone)
18. Spicatiside
19. Fukugetin (Morelloflavone)
20. 3-Methyl ether
21. Fukugiside
22. Garcinia biflavonoid 1a
23. 7″-Glucoside
24. Garcinia biflavonoid 1
25. Garcinia biflavonoid 2
26. Garcinia biflavonoid 2a
27. Kolaflavanone
28. Xanthochymusside
29. Manniflavanone
30. Zeyherin
31. Compound 5*
32. I, II-3′-Linked dihydrobichalkone*
33. Taiwaniaflavone
34. 7″-Methyl ether
35. 4′,7″-Dimethyl ether
36. Hexamethyl ether
37. Didehydrosuccedaneaflavone*
38. Hexamethyl ether*
39. Succedaneaflavone
40. Hexamethyl ether
41. Agathisflavone
42. 7-Methyl ether
43. 7,7″-Dimethyl ether
44. 7,4‴-Dimethyl ether
45. 7,7″,4‴-Trimethyl ether
46. Tetramethyl ether
47. Hexamethyl ether*
48. Rhusflavone
49. Rhusflavanone
50. Cupressuflavone
51. 7-Methyl ether
52. 4′-Methyl ether
53. 5,5″-Dimethyl ether
54. 7,7″-Dimethyl ether
55. 7,4′ or 4‴-Dimethyl ether
56. 7,4′,7″-Trimethyl ether
57. Tetramethyl ether
58. Pentamethyl ether
59. Hexamethyl ether
60. Mesuaferrone B
61. Hexamethyl ether*
62. Neorhusflavanone
63. Hexamethyl ether*
64. Dehydrohegoflavone A*
65. Heptamethyl ether*
66. Dehydrohegoflavone B*
67. Octamethyl ether*
68. Hegoflavone A*
69. Hegoflavone B*
70. Robustaflavone
71. Hexamethyl ether
72. Abiesin
73. 5′,3‴-Dihydroxyrobustaflavone*
74. Octamethyl ether*
75. Strychnobiflavone*
76. Hexamethyl ether*
77. Octamethyl ether*
78. Amentoflavone
79. Glucoside Pn II*
80. Glucoside Pn IV*
81. Glucoside Pn V*
82. 7-Methyl ether (sequoiflavone)
83. 4′-Methyl ether (bilobetin)
84. 7″-Methyl ether (sotetsuflavone)
85. 4‴-Methyl ether (podocarpus flavone A)
86. 7,4′-Dimethyl ether (ginkgetin)
87. 7,7″-Dimethyl ether
88. 7,4‴-Dimethyl ether (podocarpus flavone B)
89. 4′,7″-Dimethyl ether
90. 4′,4‴-Dimethyl ether (isoginkgetin)
91. 7″,4‴-Dimethyl ether
92. 7,4′,7″-Trimethyl ether
93. 7,4′,4‴-Trimethyl ether (sciadopitysin)
94. 7,7″,4‴-Trimethyl ether (heveaflavone)
95. 4′,7″,4‴-Trimethyl ether (kayaflavone)
96. 7,7″,4′,4‴-Tetramethyl ether
97. Hexamethyl ether (dioonflavone)
98. 5′-Methoxybilobetin
99. 5‴-Hydroxyamentoflavone*
100. 5′,8″-Biluteolin
101. 7-*O*-Methyl-6-methylamentoflavone
102. 2,3-Dihydroamentoflavone
103. 7″,4‴-Dimethyl ether
104. 2,3-Dihydrosciadipitysin
105. 2,3-Dihydroamentoflavone hexamethyl ether
106. Biflavanone C*
107. Tetrahydroamentoflavone*
108. Biflavanone A*
109. Semecarpuflavanone*
110. Jeediflavanone*
111. Galluflavanone*
112. II-7-*O*-Methyltetrahydroamentoflavone*
113. 3′,3‴-Biapigenin 5,5″,7,7″,4′,4‴-hexamethyl ether
114. Hinokiflavone
115. 7-Methyl ether (neocryptomerin)
116. 7″-Methyl ether (isocryptomerin)
117. 4‴-Methyl ether (cryptomerin A)

118. 7,7″-Dimethyl ether (chamaecyparin)
119. 7,4‴-Dimethyl ether
120. 7″,4‴-Dimethyl ether (cryptomerin B)
121. 7,7″,4‴-Trimethyl ether
122. Pentamethyl ether*
123. 2,3-Dihydrohinokiflavone
124. Pentamethyl ether
125. Occidentoside*
126. 4‴,5,5″,7,7″-pentamethoxy-4′,8″-biflavonyl ether*
127. Ochnaflavone
128. 4′-Methyl ether
129. 7,4′-Dimethyl ether
130. 4′,7,7″-Trimethyl ether
131. Pentamethyl ether
132. Heterobryoflavone*
133. Bryoflavone*

ISOFLAVONOIDS

Isoflavones

Isoflavones

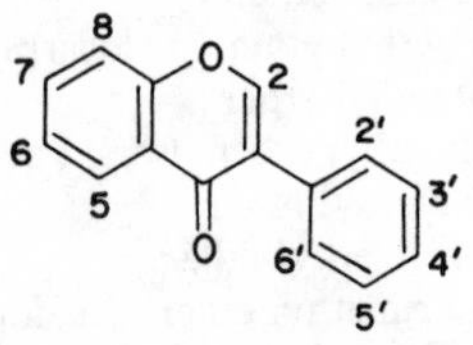

1. Daidzein (7,4′-diOH)
2. Formononetin (7-OH, 4′-OMe)
3. Isoformononetin (4′-OH, 7-OMe)
4. Dimethyldaidzein (7,4′-diOMe)
5. Durlettone (7-OMe, 4′-OCH$_2$CH=CMe$_2$)
6. Maximaisoflavone J (4′-OCH$_2$CH=CMe$_2$, 7-OMe)*
7. 2′-Hydroxydaidzein (7,2′,4′-triOH)
8. Theralin (7,4′-diOH, 2′-OMe)
9. 2′-Hydroxyformononetin (xenognosin B, 7,2′-diOH, 4′-OMe)
10. 3′-Hydroxydaidzein (7,3′,4′-triOH)
11. 3′-Methoxydaidzein (7,4′-diOH, 3′-OMe)
12. 3′-Hydroxyformononetin (calycosin, 7,3′-diOH, 4′-OMe)
13. Sayanedin (4′-OH, 7,3′-diOH)
14. Cladrin (7-OH, 3′,4′-diOMe)
15. 3′-OH, 4′-OMe, 7-OCH$_2$CH=CMe$_2$*
16. Cabreuvin (7,3′,4′-triOMe)
17. 3′,4′-DiOMe, 7-OCH$_2$CH=CMe$_2$*
18. Pseudobaptigenin (Ψ-baptigenin, 7-OH, 3′,4′-OCH$_2$O-)
19. 7-OMe, 3′,4′-OCH$_2$O-
20. Maximaisoflavone B (7-OCH$_2$CH=CMe$_2$, 3′,4′-OCH$_2$O-)
21. 7,2′,4′-TriOH, 3′-OMe*
22. Koparin (7,2′,3′-triOH, 4′-OMe)
23. 2′-OH, 7,3′,4′-triOMe
24. Glyzaglabrin (7,2′-diOH, 3′,4′-OCH$_2$O-)
25. 7-OH, 2′,4′,5′-triOMe*
26. 7,2′,4′,5′-TetraOMe
27. Cuneatin (maximaisoflavone G 7-OH, 2′-OMe, 4′,5′-OCH$_2$O-)*
28. 7,2′-DiOMe, 4′,5′-OCH$_2$O-
29. Maximaisoflavone C (2′-OMe, 7-OCH$_2$CH=CMe$_2$, 4′,5′-OCH$_2$O-)
30. Baptigenin (7,3′,4′,5′-tetraOH)
31. Gliricidin (7,3′,5′-triOH, 4′-OMe)
32. 5,7-DiOH*
33. 5-OH, 7-OMe
34. 5,7-DiOMe
35. 5,7,2′-TriOH
36. 5,7,3′-TriOH*
37. Genistein (5,7,4′-triOH)
38. Biochanin A (5,7-diOH, 4′-OMe)
39. Isoprunetin (5-methylgenistein, 7,4′-diOH, 5-OMe)
40. Prunetin (5,4′-diOH, 7-OMe)
41. 5-Methylbiochanin A (7-OH, 5,4′-diOMe)
42. 7-Methylbiochanin A (5-OH, 7,4′-diOMe)
43. 2′-Hydroxygenistein (5,7,2′,4′-tetraOH)
44. 2′-Hydroxybiochanin A (5,7,2′-triOH, 4′-OMe)
45. 2′-Hydroxyisoprunetin (7,2′,4′-triOH, 5-OMe)*
46. Cajanin (5,2′,4′-triOH, 7-OMe)
47. 2′-Methoxybiochanin A (5,7-diOH, 2′,4′-diOMe)
48. 5,2′-DiOH, 7,4′-triOMe (?)
49. Orobol (5,7,3′,4′-tetraOH)
50. 3′-Methylorobol (5,7,4′-triOH, 3′-OMe)
51. Pratensein (5,7,3′-triOH, 4′-OMe)
52. Santal (5,3′,4′-triOH, 7-OMe)
53. 3′-Methylpratensein (5,7-diOH, 3′,4′-diOMe)*
54. 5,3′-DiOH, 7,4′-diOMe*
55. Glabrescione B (5,7-diOMe, 3′,4′-di-(OCH$_2$CH=CMe$_2$)
56. Methylenedioxyorobol (5,7-diOH, 3′,4′-OCH$_2$O-)
57. Derrustone (5,7-diOMe, 3′,4′-OCH$_2$O-)
58. Derrugenin (5,5′-diOH, 7,2′,4′-triOMe)
59. Robustigenin (5-OH, 7,2′,4′,5′-tetraOMe)
60. Junipegenin A (5,7,3′,5′-tetraOH, 4′-OMe)
61. 7,2′-DiOH, 6-OMe*
62. 2′-OH, 6,7-OCH$_2$O-*
63. 6,7,4′-triOH
64. Texasin (6,7-diOH, 4′-OMe)
65. Glycitein (7,4′-diOH, 6-OMe)
66. Kakkatin (6,4′-diOH, 7-OMe)
67. Afrormosin (7-OH, 6,4′-diOMe)
68. Hemerocallone (2′,5′-diOMe, 6,7-OCH$_2$O-)*
69. Odoratin (7,3′-diOH, 6,4′-diOMe)
70. Cladrastin (7-OH, 6,3′,4′-triOMe)

71. 6, 7, 3′, 4′-TetraOMe
72. Fujikinetin (7-OH, 6-OMe, 3′, 4′-OCH_2O-)
73. 6, 7-DiOMe, 3′, 4′-OCH_2O-
74. 3′, 4′-DiOMe, 6, 7-OCH_2O-
75. 6, 7, 2′, 3′, 4′-PentaOMe
76. 7-OH, 6, 2′, 4′, 5′-TetraOMe
77. 6, 7, 2′, 4′, 5′-pentaOMe
78. Dalpatein (7-OH, 6, 2′-diOMe, 4′, 5′-OCH_2O-)
79. 6-OH, 7, 2′-diOMe, 4′, 5′-OCH_2O-
80. Milldurone (6, 7, 2′-triOMe, 4′, 5′-OCH_2O-)
81. 6, 7, 3′, 4′, 5′-PentaOMe
82. 6, 7, 3′-TriOMe, 4′, 5′-OCH_2O-
83. Retusin (7, 8-diOH, 4′-OMe)
84. 8-Methylretusin (7-OH, 8, 4′-diOMe)
85. Maximaisoflavone H (4′-OMe, 7, 8-OCH_2O-)*
86. 8-Methyl-3′-hydroxyretusin (7, 3′-diOH, 8, 4′-diOMe)
87. 7-OH, 8, 3′, 4′-triOMe
88. Maximaisoflavone E (7-OH, 8-OMe, 3′, 4′-OCH_2O-)*
89. 8-OMe, 7-$OCH_2CH{=}CMe_2$, 3′, 4′-OCH_2O-*
90. Maximaisoflavone D (3′, 4′-diOMe, 7, 8-OCH_2O-)*
91. Maximaisoflavone A (7, 8; 3′, 4′-(-OCH_2O-$)_2$)
92. Maximaisoflavone F (7-OH, 8, 2′-diOMe, 4′, 5′-OCH_2O)*
93. 7, 8, 2′-TriOMe, 4′, 5′-OCH_2O-
94. 5, 7-DiOH, 6-OMe*
95. 5, 7-DiOH, 6, 2′-diOMe*
96. Betavulgarin (2′-OH, 5-OMe, 6, 7-OCH_2O-)
97. Tlatlancuayin (5, 2′-diOMe, 6, 7-OCH_2O-)
98. 6-Hydroxygenistein (5, 6, 7, 4′-tetraOH)
99. Tectorigenin (5, 7, 4′-triOH, 6-OMe)
100. Muningin (6, 4′-diOH, 5, 7-diOMe)
101. 7-Methyltectorigenin (5, 4′-diOH, 6, 7-diOMe)
102. Irisolidone (5, 7-diOH, 6, 4′-diOMe)
103. Isoaurmillone (5, 7-diOH, 6-OMe, 4′-$OCH_2CH{=}CMe_2$)*
104. 7, 4′-Dimethyltectorigenin (5-OH, 6, 7, 4′-triOMe)
105. 5-Methoxyafrormosin (7-OH, 5, 6, 4′-triOMe)
106. Irilone (5, 4′-diOH, 6, 7-OCH_2O-)
107. Irisolone (4′-OH, 5-OMe, 6, 7-OCH_2O-)
108. Methylirisolone (5, 4′-diOMe, 6, 7-OCH_2O-)
109. Podospicatin (5, 7, 2′-triOH, 6, 5′-diOMe)
110. 5, 6, 7, 4′-TetraOH, 3′-OMe*
111. Iristectorigenin B (5, 7, 4′-triOH, 6, 3′-diOMe)
112. Iristectorigenin A (5, 7, 3′-triOH, 6, 4′-diOMe)
113. Junipegenin B (dalspinosin, 5, 7-diOH, 6, 3′, 4′-triOMe)*
114. Dalspinin (5, 7-diOH, 6-OMe, 3′, 4′-OCH_2O-)*
115. 7-OH, 5, 6-DiOMe, 3′, 4′-OCH_2O-
116. 5, 6, 7-TriOMe, 3′, 4′-OCH_2O-
117. Iriflogenin (5, 4′-diOH, 3′-OMe, 6, 7-OCH_2O-)
118. Iriskumaonin (3′-OH, 5, 4′-diOMe, 6, 7-OCH_2O-)
119. Methyliriskumaonin (5, 3′, 4′-triOMe, 6, 7-OCH_2O-)
120. Caviunin (5, 7-diOH, 6, 2′, 4′, 5′-tetraOMe)
121. 5, 7, 4′-TriOH, 6, 3′, 5′-triOMe*
122. Irigenin (5, 7, 3′-triOH, 6, 4′, 5′-triOMe)
123. Junipegenin C (5, 7-diOH, 6, 3′, 4′, 5′-tetraOMe)*
124. 5, 3′-DiOH, 4′, 5′-diOMe, 6, 7-OCH_2O-*
125. Irisflorentin (5, 3′, 4′, 5′-tetraOMe, 6, 7-OCH_2O-)
126. 8-Hydroxygenistein (5, 7, 8, 4′-tetraOH)
127. Isotectorigenin (pseudotectorigenin, 5, 7, 4′-triOH, 8-OMe)
128. Aurmillone (5, 7-diOH, 8-OMe, 4′-$OCH_2CH{=}CMe_2$)
129. 5, 7, 3′, 4′-TetraOH, 8-OMe
130. 5, 7, 4′-TriOH, 8, 3′-diOMe
131. 5, 7, 3′-TriOH, 8, 4′-diOMe
132. 5, 7-DiOH, 8, 3′, 4′-triOMe*
133. Platycarpanetin (7-OH, 5, 8-diOMe, 3′, 4′-OCH_2O-) (?)
134. Isocaviunin (5, 7-diOH, 8, 2′, 4′, 5′-tetraOMe)
135. Dipteryxin (7, 8-diOH, 6, 4′-diOMe)
136. 8, 3′-DiOH, 6, 7, 4′-triOMe
137. Petalostetin (6, 7, 8-triOMe, 3′, 4′-OCH_2O-)
138. 8, 3′, 4′-TriOMe, 6, 7-OCH_2O-
139. 5, 6, 7, 4′-TetraOH, 8-OMe*
140. 5, 6, 7, 8-TetraOMe 3′, 4′-OCH_2O-
141. Corylinal (7, 4′-diOH, 3′-CHO)
142. Neobavaisoflavone (7, 4′-diOH, 3′-$CH_2CH{=}CMe_2$)
143. Corylin (7-OH, 3′, 4′-$CH{=}CHCMe_2O$—)
144. Psoralenol (7-OH, 3′, 4′-$CH_2CH(OH)CMe_2O$—)
145. Glabrone (7, 2′-diOH, 3′, 4′-$CH{=}CHCMe_2O$—)
146. Licoricone (7, 6′-diOH, 2′, 4′-diOMe, 3′-$CH_2CH{=}CMe_2$)
147. 3′-Dimethylallylgenistein (5, 7, 4′-triOH, 3′-$CH_2CH{=}CMe_2$)*
148. Lupinisoflavone C (5, 7-diOH, 3′, 4′-$CH_2CH(CMe_2OH)O$-)*
149. Licoisoflavone A (phaseoluteone, 5, 7, 2′, 4′-tetraOH, 3′-$CH_2CH{=}CMe_2$)
150. 5, 7, 2′, 4′-TetraOH, 3′-$CH_2CH(OH)CMe_2OH$ Metabolite*
151. 5, 7, 4′-TriOH, 2′, 3′-$OCMe_2CH(OH)CH_2$- Metabolite*
152. 5, 7, 4′-TriOH, 2′, 3′-$OCH(CMe_2OH)CH_2$- Metabolite*
153. Licoisoflavone B (5, 7, 2′-triOH, 3′, 4′-$CH{=}CHCMe_2O$-)
154. 5, 7, 2′-TriOH, 3′, 4′-$CH_2CH(OH)CMe_2O$- Metabolite*
155. Lupinisoflavone D (5, 7, 2′-triOH, 3′, 4′-$CH_2CH(CMe_2OH)O$-)*
156. 2′-Deoxypiscerythrone (5, 7, 4′-triOH, 5′-OMe, 3′-$CH_2CH{=}CMe_2$)*
157. Piscerythrone (5, 7, 2′, 4′-tetraOH, 5′-OMe, 3′-$CH_2CH{=}CMe_2$)
158. Piscidone, (5, 7, 3′, 4′-tetraOH,

6′-OMe, 2′-CH_2CH=CMe_2 *or* 5,7,3′,6′-tetraOH, 4′-OMe, 2′-CH_2CH=CMe_2) (?)
159. 5,7,2′,4′-TetraOH, 5′-OMe, 3′,6′-di(CH_2CH=CMe_2)
160. Erythrinin A (4′-OH, 6,7-CH=$CHCMe_2O$-)*
161. Dehydroneotenone (2′-OMe, 4′,5′-OCH_2O-, 6,7-CH=CHO-)
162. Calopogonium isoflavone A (4′-OMe, 7,8-$OCMe_2CH$=CH-)
163. Calopogonium isoflavone B (3′,4′-OCH_2O-, 7,8-$OCMe_2CH$=CH-)
164. Barbigerone (2′,4′,5′-triOMe, 7,8-$OCMe_2CH$=CH-)
165. Jamaicin (2′-OMe, 4′,5′-OCH_2O-, 7,8-$OCMe_2CH$=CH-)
166. Ferrugone (2′,5′-diOMe, 3′,4′-OCH_2O-, 7,8-$OCMe_2CH$=CH-)
167. 6 (or 8)-(1,1-Dimethylallyl)-genistein (5,7,4′-triOH, 6 (or 8)-CMe_2CH=CH_2)*
168. Wighteone (erythrinin B, 5,7,4′-triOH, 6-CH_2CH=CMe_2)
169. Luteone (5,7,2′,4′-tetraOH, 6-CH_2CH=CMe_2)
170. Lupisoflavone (5,7,4′-triOH, 3′-OMe, 6-CH_2CH=CMe_2)*
171. Derrubone (5,7-diOH, 3′,4′-OCH_2O-, 6-CH_2CH=CMe_2)
172. Viridiflorin (5,7,4′-triOH, 2′,5′-diOMe, 6-CH_2CH=CMe_2)*
173. Wighteone hydrate (5,7,4′-triOH, 6-$CH_2CH_2CMe_2OH$) Metabolite*
174. Luteone hydrate (5,7,2′,4′-tetraOH, 6-$CH_2CH_2CMe_2OH$) Metabolite*
175. 5,7,4′-TriOH, 6-$CH_2CH(OH)CMe_2OH$ Metabolite*
176. 5,7,2′,4′-TetraOH, 6-$CH_2CH(OH)CMe_2OH$ Metabolite*
177. Alpinumisoflavone (5,4′-diOH, 6,7-CH=$CHCMe_2O$-)
178. 4′-Methylalpinumisoflavone (5-OH, 4′-OMe, 6,7-CH=$CHCMe_2O$-)
179. 4′-Dimethylallylalpinumisoflavone (5-OH, 4′-OCH_2CH=CMe_2, 6,7-CH=$CHCMe_2O$-)*
180. Dimethylalpinumisoflavone (5,4′-diOMe, 6,7-CH=$CHCMe_2O$-)
181. Parvisoflavone B (5,2′,4′-triOH, 6,7-CH=$CHCMe_2O$-)
182. Isoauriculatin (5,2′-diOH, 4′-OCH_2CH=CMe_2, 6,7-CH=$CHCMe_2O$-)
183. 5,3′-DiOH, 4′-OMe, 6,7-CH=$CHCMe_2O$-*
184. Isoauriculasin (5,3′-diOH, 4′-OCH_2CH=CMe_2, 6,7-CH=$CHCMe_2O$-)
185. Robustone (5-OH, 3′,4′-OCH_2O-, 6,7-CH=$CHCMe_2O$-)
186. Methylrobustone (5-OMe, 3′,4′-OCH_2O-, 6,7-CH=$CHCMe_2O$-)
187. Elongatin (5,4′-diOH, 2′,5′-diOMe, 6,7-CH=$CHCMe_2O$-)
188. Dihydroalpinumisoflavone (5,4′-diOH, 6,7-$CH_2CH_2CMe_2O$-)*
189. 5,4′-DiOH, 6,7-$CH_2CH(OH)CMe_2O$- Metabolite*
190. 5,2′,4′-TriOH, 6,7-$CH_2CH(OH)CMe_2O$- Metabolite*
191. Lupinisoflavone A (5,2′,4′-triOH, 6,7-CH_2CH(CMe=CH_2)O-)*
192. Glabrescione A (5-OMe, 3′,4′-OCH_2O-, 6,7-CH_2CH(CMe=CH_2)O-)
193. Erythrinin C (5,4′-diOH, 6,7-$CH_2CH(CMe_2OH)O$-)
194. Lupinisoflavone B (5,2′,4′-triOH, 6,7-$CH_2CH(CMe_2OH)O$-)*
195. 7-OH, 5,4′-diOMe, 8-Me*
196. 5-OH, 7,3′,4′-triOMe, 8-Me*
197. 2,3-Dehydrokievitone (5,7,2′,4′-tetraOH, 8-CH_2CH=CMe_2)
198. Derrone (5,4′-diOH, 7,8-$OCMe_2CH$=CH-)
199. 4′-Methylderrone (5-OH, 4′-OMe, 7,8-$OCMe_2CH$=CH-)
200. Parvisoflavone A (5,2′,4′-triOH, 7,8-$OCMe_2CH$=CH-)
201. Toxicarol isoflavone (5-OH, 2′,4′,5′-triOMe 7,8-$OCMe_2CH$=CH-)
202. Durmillone (6-OMe, 3′,4′-OCH_2O-, 7,8-$OCMe_2CH$=CH-)
203. Ichthynone (6,2′-diOMe, 4′,5′-OCH_2O-, 7,8-$OCMe_2CH$=CH-)
204. Munetone (2′-OMe, 6,7; 3′,4′-di(-CH=$CHCMe_2O$-))
205. Mundulone(2′-OMe, 6,7-$CH_2CH(OH)CMe_2O$-, 3′,4′-CH=$CHCMe_2O$-)
206. Cajaisoflavone (2′,4′-diOH, 6′-OMe, 3′-CH_2CH=CMe_2 6,7-CH=$CHCMe_2O$-)
207. Lupalbigenin (5,7,4′-triOH, 6,3′-di(CH_2CH=CMe_2))
208. 2′-Hydroxylupalbigenin (5,7,2′,4′-tetraOH6,3′-di(CH_2CH=CMe_2))*
209. 2′-Methoxylupalbigenin (5,7,4′-triOH, 2′-OMe, 6,3′-di(CH_2CH=CMe_2))*
210. Chandalone (5,4′-diOH, 3′-CH_2CH=CMe_2, 6,7-CH=$CHCMe_2O$-)
211. Lupinisoflavone E (5-OH, 6,7; 3′,4′-di[-$CH_2CH(CMe_2OH)O$-])
212. Lupinisoflavone F (5,2′-diOH, 6,7; 3′,4′-di[-$CH_2CH(CMe_2OH)O$-])

213. Flemiphyllin
(5, 7, 4′-triOH, 8, 3′, 5′-tri($CH_2CH{=}CMe_2$))*
214. 6, 8-Di (dimethylallyl) genistein
(5, 7, 4′-triOH, 6, 8-di($CH_2CH{=}CMe_2$))
215. 6, 8-Di (dimethylallyl) orobol
(5, 7, 3′, 4′-tetraOH, 6, 8-di($CH_2CH{=}CMe_2$))
216. 6, 8-Di (dimethylallyl) pratensein
(5, 7, 3′-triOH, 4′-OMe,
6, 8-di($CH_2CH{=}CMe_2$))*
217. Warangalone (scandenone, 5, 4′-diOH,
8-$CH_2CH{=}CMe_2$, 6, 7-$CH{=}CHCMe_2O$-)
218. Auriculatin (5, 2′, 4′-triOH, 8-$CH_2CH{=}CMe_2$,
6, 7-$CH{=}CHCMe_2O$-)
219. Auriculin
(5, 2′-diOH, 4′-OMe, 8-$CH_2CH{=}CMe_2$,
6, 7-$CH{=}CHCMe_2O$-)
220. Auriculasin (5, 3′, 4′-triOH, 8-$CH_2CH{=}CMe_2$,
6, 7-$CH{=}CHCMe_2O$-)
221. 5-OH, 2′, 4′-diOMe, 8-$CH_2CH_2CMe_2OH$,
6, 7-$CH_2CH_2CMe_2O$-*
222. 5-OH, 3′, 4′-diOMe, 8-$CH_2CH_2CMe_2OH$,
6, 7-$CH_2CH_2CMe_2O$-*
223. Osajin (5, 4′-diOH, 6-$CH_2CH{=}CMe_2$
7, 8-$OCMe_2CH{=}CH$-)
224. Scandinone
(nallanin, 4′-OH, 5-OMe, 6-$CH_2CH{=}CMe_2$,
7, 8-$OCMe_2CH{=}CH$-)
225. Pomiferin (5, 3′, 4′-triOH, 6-$CH_2CH{=}CMe_2$
7, 8-$OCMe_2CH{=}CH$-)
226. 2′-OH, 4′-OMe, 5, 6; 7, 8-di(-$CH_2CH_2CMe_2O$-)*
227. 4′-OH, 3′-OMe, (or 3′-OH, 4′-OMe),
5, 6; 7, 8-di(-$CH_2CH_2CMe_2O$-)*
228. 7-OH, 2-Me
229. 7-OMe, 2-Me
230. 7-OCOMe, 2-Me
231. Glyzarin (7-OH, 8-COMe, 2-Me)
232. 6-Chlorogenistein (5, 7, 4′-triOH, 6-Cl)
233. 6, 3′-Dichlorogenistein (5, 7, 4′-triOH, 6, 3′-diCl)

Isoflavonequinone

1. Bowdichione (7-OH, 4′-OMe)

Isoflavanones

1. Dihydrodaidzein (7, 4′-diOH)
2. Dihydroformononetin (7-OH, 4′-OMe)
3. 2′-Hydroxydihydrodaidzein (7, 2′, 4′-triOH)
4. Vestitone (7, 2′-diOH, 4′-OMe)
5. Sativanone (7-OH, 2′, 4′-diOMe)
6. Isosativanone (2′-OH, 7, 4′-diOMe)*
7. 7, 3′-DiOH, 4′-OMe
8. Lespedeol C(7, 4′-diOH, 2′, 3′-diOMe)*
9. Violanone (7, 3′-diOH, 2′, 4′-diOMe)
10. 3′-Methylviolanone (7-OH, 2′, 3′, 4′-triOMe)
11. 2′-OH, 7, 3′, 4′-triOMe
12. Sophorol (7, 2′-diOH, 4′, 5′-OCH_2O-)
13. Onogenin (7-OH, 2′-OMe, 4′, 5′-OCH_2O-)
14. Dihydrogenistein (5, 7, 4′-triOH) Metabolite*
15. Dihydrobiochanin A (5, 7-diOH, 4′-OMe)
16. Padmakastein (5, 4′-diOH, 7-OMe)
17. Dalbergioidin (5, 7, 2′, 4′-tetraOH)
18. Isoferreirin (5, 7, 4′-triOH, 2′-OMe)*
19. Ferreirin (5, 7, 2′-triOH, 4′-OMe)
20. Homoferreirin (5, 7-diOH, 2′, 4′-diOMe)
21. 7, 4′-DiOH, 5, 2′-diOMe*
22. Cajanol (5, 4′-diOH, 7, 2′-diOMe)
23. Parvisoflavanone (5, 7, 4′-triOH, 2′, 3′-diOMe)
24. 6, 7-DiOMe, 3′, 4′-OCH_2O-
25. Bolusanthin (5, 7, 3, 3′-tetraOH, 4′-OMe)*
26. Sophoraisoflavanone A
(5, 7, 4′-triOH, 2′-OMe, 3′-$CH_2CH{=}CMe_2$)
27. Licoisoflavanone
(5, 7, 2′-triOH, 3′, 4′-$CH{=}CHCMe_2O$-)
28. Isosophoronol
(5, 7-diOH, 2′-OMe, 3′, 4′-$CH{=}CHCMe_2O$-)
29. 5, 7, 2′-TriOH, 4′, 5′-OCH (CMe_2OH)CH_2-
30. Neoraunone (2′, 4′-diOMe, 6, 7-$CH{=}CHO$-)
31. Nepseudin (2′, 3′, 4′-triOMe, 6, 7-$CH{=}CHO$-)
32. Ambonone (2′, 4′, 5′-triOMe, 6, 7-$CH{=}CHO$-)
33. Neotenone
(2′-OMe, 4′, 5′-OCH_2O-, 6, 7-$CH{=}CHO$-)
34. Erosenone
(2′-OH, 6′-OMe, 3′, 4′-OCH_2O-,
6, 7-$CH{=}CHO$-)(?)
35. 5-Deoxykievitone
(7, 2′, 4′-triOH, 8-$CH_2CH{=}CMe_2$)
36. 5-Deoxykievitone hydrate
(7, 2′, 4′-triOH, 8-$CH_2CH_2CMe_2OH$)*
37. Ougenin (5, 2′, 4′-triOH, 7, 3′-diOMe, 6-Me)
38. Diphysolone
(5, 7, 2′, 4′-tetraOH, 6-$CH_2CH{=}CMe_2$)*
39. Diphysolidone (5, 7, 4′-triOH, 2′-OMe
[or 5, 7, 2′-triOH, 4′-OMe], 6-$CH_2CH{=}CMe_2$)*
40. Lespedeol A (5, 7, 2′, 4′-tetraOH,
6-$CH_2CH{=}CMeCH_2CH_2CH{=}CMe_2$)
41. Lespedeol B (5, 2′, 4′-triOH,
6, 7-$CH{=}CHCMe(CH_2CH_2CH{=}CMe_2)O$-)
42. Kievitone (5, 7, 2′, 4′-tetraOH, 8-$CH_2CH{=}CMe_2$)
43. 4′-Methylkievitone
(5, 7, 2′-triOH, 4′-OMe, 8-$CH_2CH{=}CMe_2$)*

44. Kievitone hydrate
(5, 7, 2′, 4′-tetraOH, 8-$CH_2CH_2CMe_2OH$)
45. Cyclokievitone
(5, 2′, 4′-triOH, 7, 8-$OCMe_2CH{=}CH$-)
46. Cyclokievitone hydrate
(5, 2′, 4′-triOH, 7, 8-$OCMe_2CH(OH)CH_2$-
[or 7, 8-$OCMe_2CH_2CH(OH)$-])*
47. Isosophoranone
(5, 7, 4′-triOH, 2′-OMe, 6, 3′-di($CH_2CH{=}CMe_2$))
48. Sophoraisoflavone B
(5, 7, 2′-triOH, 4′-OMe, 6, 5′-di($CH_2CH{=}CMe_2$))*
49. Cajanone
(5, 2′, 4′-triOH, 5′-$CH_2CH{=}CMe_2$, 6, 7-
$CH{=}CHCMe_2O$-)
50. 2′-Methylcajanone
(5, 4′-diOH, 2′-OMe, 5′
-$CH_2CH{=}CMe_2$ 6, 7-$CH{=}CHCMe_2O$-)
51. Secondifloran
(7, 3, 2′, 3′-tetraOH, 4′-OMe, 5′-$CMe_2CH{=}CH_2$)

Rotenoids

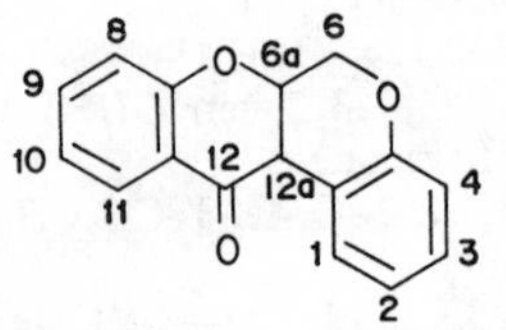

Simple Rotenoids

1. Munduserone (2, 3, 9-triOMe)
2. Sermundone (11-OH, 2, 3, 9-triOMe)
3. Erosone (2, 3-diOMe, 9, 10-$OCH{=}CH$-)
4. Dolineone (2, 3-OCH_2O-, 9, 10-$OCH{=}CH$-)
5. Rot-2′-enonic acid
(9-OH, 2, 3-diOMe, 8-$CH_2CH{=}CMe_2$)*
6. Deguelin (2, 3-diOMe, 8, 9-$CH{=}CHCMe_2O$-)
7. Millettone (2, 3-OCH_2O-, 8, 9-$CH{=}CHCMe_2O$-)
8. Elliptone (2, 3-diOMe, 8, 9-$CH{=}CHO$-)
9. Rotenone
(2, 3-diOMe, 8, 9-$CH_2CH(CMe{=}CH_2)O$-)
10. Isomillettone
(2, 3-OCH_2O-, 8, 9-$CH_2CH(CMe{=}CH_2)O$-)
11. 3-Demethylamorphigenin
(3-OH, 2-OMe, 8, 9-$CH_2CH[C(CH_2OH)$
$=CH_2]O$-)*
12. Amorphigenin
(2, 3-diOMe, 8, 9-$CH_2CH[C(CH_2OH){=}CH_2]O$-)
13. Dalpanol (2, 3-diOMe, 8, 9-$CH_2CH(CMe_2OH)O$-)
14. Dihydroamorphigenin
(2, 3-diOMe, 8, 9-$CH_2CH(CHMeCH_2OH)O$-)
15. Amorphigenol
(2, 3-diOMe, 8, 9-$CH_2CH[CMe(OH)CH_2OH]O$-)
16. Toxicarol
(11-OH, 2, 3-diOMe, 8, 9-$CH{=}CHCMe_2O$-)
17. Malaccol
(11-OH, 2, 3-diOMe, 8, 9-$CH{=}CHO$-)
18. Sumatrol
(11-OH, 2, 3-diOMe, 8, 9-$CH_2CH(CMe{=}CH_2)O$-)
19. Pachyrrhizone
(8-OMe, 2, 3-OCH_2O-, 9, 10-$OCH{=}CH$-)
20. Villosin
(6, 11-diOH, 2, 3-(diOMe, 8, 9-CH_2CH
$(CMe{=}CH_2)O$-)

12a-Hydroxyrotenoids

21. 12a-Hydroxymunduserone
(12a-OH, 2, 3, 9-triOMe)
22. 12a-Hydroxyerosene
(12a-OH, 2, 3-diOMe 9, 10-$OCH{=}CH$-)
23. Neobanone (2, 3, 12a-triOMe, 9, 10-$OCH{=}CH$-)
24. 12a-Hydroxydolineone
(12a-OH, 2, 3-OCH_2O-, 9, 10-$OCH{=}CH$-)
25. 12a-Methoxydolineone
(12a-OMe, 2, 3-OCH_2O-, 9, 10-$OCH{=}CH$-)
26. 12a-Hydroxyrot-2′-enonic acid
(9, 12a-diOH, 2, 3-diOMe, 8-$CH_2CH{=}CMe_2$)*
27. Tephrosin
(12a-OH, 2, 3-diOMe, 8, 9-$CH{=}CHCMe_2O$-)
28. Millettosin
(12a-OH, 2, 3-OCH_2O-, 8, 9-$CH{=}CHCMe_2O$-)
29. 12a-Hydroxyrotenone
(12a-OH, 2, 3-diOMe), 8, 9-CH_2CH
$(CMe{=}CH_2)O$-)
30. 12a-Methoxyrotenone
(2, 3, 12a-triOMe, 8, 9-$CH_2CH(CMe{=}CH_2)O$-)
31. 12a-Hydroxyisomillettone
(12a-OH, 2, 3-diOMe, 8, 9-CH_2CH
$(CMe{=}CH_2)O$-)
32. 12a-Hydroxydihydrorotenone
(12a-OH, 2, 3-diOMe, 8, 9-$CH_2CH(CHMe_2)O$-)
Metabolite*
33. Dalbinol (12a-hydroxyamorphigenin,
12a-OH, 2, 3-diOMe, 8, 9-CH_2CH
$[C(CH_2OH){=}CH_2]O$-)
34. 12a-Hydroxydalpanol
(12a-OH, 2, 3-diOMe, 8, 9-$CH_2CH(CMe_2OH)O$-)
Metabolite
35. Volubinol (3, 12a-diOH, 2-OMe,
8, 9-$CH_2CH(CHMeCH_2OH)O$-)*
36. 12a-Hydroxydihydroamorphigen
(12a-OH, 2, 3-diOMe, 8, 9-
$CH_2CH(CHMeCH_2OH)O$-) Metabolite*
37. 11-Hydroxytephrosin
(11, 12a-diOH, 2, 3-diOMe, 8, 9-$CH{=}CHCMe_2O$-)
38. Villosinol (12a-hydroxysumatrol,
11, 12a-diOH, 2, 3-diOMe, 8, 9-$CH_2CH(CMe{=}$
$CH_2)O$-)
39. 12a-Hydroxypachyrrhizone
(12a-OH, 8-OMe, 2, 3-OCH_2O-, 9, 10-$OCH{=}CH$-)

40. Clitoriacetal (6, 11, 12a-triOH, 2, 3, 9-triOMe)
41. Villol (6, 11, 12a-triOH, 2, 3-diOMe, 8, 9-$CH_2CH(CMe{=}CH_2)O$-)
42. 12-Dihydrodalbinol (12-dihydro, 12a-OH, 2, 3-diOMe, 8, 9-$CH_2CH[C(CH_2OH){=}CH_2]O$-)*

Dehydrorotenoids

43. Dehydrodolineone (6a, 12a-dehydro, 2, 3-OCH_2O-9, 10-OCH=CH-)
44. Dehydrodeguelin (6a, 12a-dehydro, 2, 3-diOMe, 8, 9-$CH{=}CHCMe_2O$-)
45. Dehydromillettone (6a, 12a-dehydro, 2, 3-OCH_2O- 8, 9-$CH{=}CHCMe_2O$-)
46. Dehydrorotenone (6a, 12a-dehydro, 2, 3-diOMe, 8, 9-$CH_2CH(CMe{=}CH_2)O$-)
47. Dehydroamorphigenin (6a, 12a-dehydro, 2, 3-diOMe, 8, 9-$CH_2CH[C(CH_2OH){=}CH_2]O$-)*
48. Dehydrodalpanol (6a, 12a-dehydro, 2, 3-diOMe, 8, 9-$CH_2CH(CMe_2OH)O$-)
49. Dehydrotoxicarol (6a, 12a-dehydro, 11-OH, 2, 3-diOMe, 8, 9-$CH{=}CHCMe_2O$-)
50. Villosol (dehydrosumatrol, 6a, 12a-dehydro, 11-OH, 2, 3-diOMe, 8, 9-$CH_2CH(CMe{=}CH_2)O$-)
51. Dehydropachyrrhizone (6a,12a-dehydro, 8-OMe, 2, 3-OCH_2O-9, 10-OCH=CH-)
52. Stemonal (6a, 12a-dehydro, 6, 11-diOH, 2, 3, 9-triOMe)
53. Stemonacetal (6a, 12a-dehydro, 11-OH, 2, 3, 9-triOMe, 6-OEt)
54. Amorpholone (6a, 12a-dehydro, 6-OH, 2, 3-diOMe, 8, 9-$CH_2CH(CMe{=}CH_2)O$-)
55. 6-Hydroxydehydrotoxicarol (6a, 12a-dehydro, 6, 11-diOH, 2, 3-diOMe, 8, 9-$CH{=}CHCMe_2O$-)*
56. Villinol (6a, 12a-dehydro, 11-OH, 2, 3, 6-triOMe 8, 9-$CH_2CH(CMe{=}CH_2)O$-)
57. Stemonone 6a, 12a-dehydro, 6-oxo, 11-OH, 2, 3, 9-triOMe)
58. Rotenonone (6a, 12a-dehydro, 6-oxo, 2, 3-diOMe, 8, 9-$CH_2CH(CMe{=}CH_2)O$-)
59. Villosone (6a, 12a-dehydro, 6-oxo 11-OH, 2, 3-diOMe, 8, 9-$CH_2CH(CMe{=}CH_2)O$-)

Pterocarpans

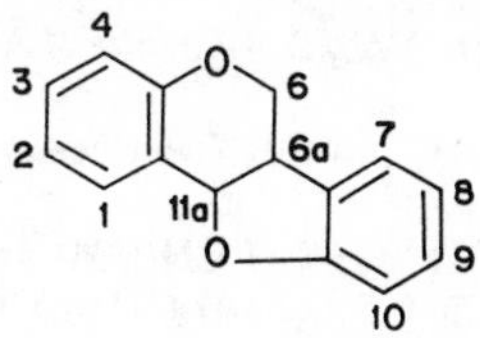

Simple pterocarpans

1. Demethylmedicarpin (3, 9-diOH)
2. Medicarpin (demethylhomopterocarpin, 3-OH, 9-OMe)
3. Isomedicarpin (9-OH, 3-OMe)
4. Homopterocarpin (3, 9-diOMe)
5. Nissicarpin (3, 7-diOH, 9-OMe)*
6. Fruticarpin (7-OH, 3, 9-diOMe)*
7. Kushenin (3, 9-diOH, 8-OMe)*
8. Maackiain (3-OH, 8, 9-OCH_2O-)
9. Pterocarpin (3-OMe, 8, 9-OCH_2O-)
10. Vesticarpin (3, 10-diOH, 9-OMe)
11. Nissolin (3, 9-diOH, 10-OMe)
12. Methylnissolin (3-OH, 9, 10-diOMe)
13. Philenopteran (3, 9-diOH, 7, 10-diOMe)
14. 9-Methylphilenopteran (3-OH, 7, 9, 10-triOMe)
15. Desmocarpin (1, 9-diOH, 3-OMe)*
16. 3-OH, 2, 9-DiOMe
17. Sparticarpin (9-OH, 2, 3-diOMe)
18. 2, 3, 9-TriOMe
19. Nissolicarpin (3, 7-diOH, 2, 9-diOMe)*
20. 2, 8-DiOH, 3, 9-diOMe
21. 2-Hydroxypterocarpin (2-OH, 3-OMe, 8, 9-OCH_2O-)
22. 2-Methoxypterocarpin (2, 3-diOMe, 8, 9-OCH_2O-)
23. Mucronucarpan (2, 10-diOH, 3, 9-diOMe)
24. 2, 8-(DiOH, 3, 9, 10-triOMe)
25. 3, 4, 9-TriOH
26. 3, 4-DiOH, 9-OMe
27. Melilotocarpan B (4, 9-diOH, 3-OMe)*
28. 4-Methoxymedicarpin (3-OH, 4, 9-diOMe)
29. Melilotocarpan A (4-OH, 3, 9-diOMe)
30. 3, 4, 9-TriOMe
31. 3, 4-DiOH, 8, 9-OCH_2O-
32. 4-Methoxymaackiain (3-OH, 4-OMe, 8, 9-OCH_2O-)
33. 4-OH, 3-OMe, 8, 9-OCH_2O-
34. 3, 4-DiOMe, 8, 9-OCH_2O-
35. Melilotocarpan D (4,10-diOH, 3, 9-diOMe)*
36. Melilotocarpan E (4, 9-diOH, 3, 10-diOMe)*
37. Melilotocarpan C (4-OH, 3, 9, 10-triOMe)*
38. Odoricarpan (10-OH, 3, 4, 9-triOMe)*
39. 8-OH, 3, 4, 9, 10-tetraOMe
40. Trifolian (1, 3-diOH, 2-OMe, 8, 9-OCH_2O-)
41. 4-OH, 2, 3, 9-triOMe
42. 2, 8-DiOH, 3, 4, 9, 10-tetraOMe
43. Sophorapterocarpan A (3, 9-diOH, 8-$CH_2CH{=}CMe_2$)*
44. Isoneorautenol (3-OH, 8, 9-$CH{=}CHCMe_2O$-)*
45. Phaseollidin (3, 9-diOH, 10-$CH_2CH{=}CMe_2$)
46. Sandwicensin (9-Methylphaseollidin, 3-OH, 9-OMe, 10-$CH_2CH{=}CMe_2$)
47. Dolichin A (3, 9-diOH, 10-$CH_2CH(OH)CMe{=}CH_2$)*
48. Dolichin B (3, 9-diOH, 10-$CH_2CH(OH)CMe{=}CH_2$)*

49. Phaseollidin hydrate ($3,9$-diOH, 10-$CH_2CH_2CMe_2OH$) Metabolite
50. Lespedezin ($3,9$-diOH, 10-$CH_2CH{=}CMeCH_2CH_2CH{=}CMe_2$)
51. Phaseollin (3-OH, $9,10$-$OCMe_2CH{=}CH$-)
52. 3-OH, $9,10$-$OCMe_2CH(OH)CH(OH)$- Metabolite
53. 1-Methoxyphaseollidin ($3,9$-diOH, 1-OMe, 10-$CH_2CH{=}CMe_2$)
54. Calopocarpin ($3,9$-diOH, 2-$CH_2CH{=}CMe_2$)
55. Edunol (3-OH, $8,9$-OCH_2O-, 2-$CH_2CH{=}CMe_2$)
56. Cabenegrin A-II (3-OH, $8,9$-OCH_2O-, 2-$CH_2CH_2CHMeCH_2OH$)*
57. Neorautenol (9-OH, $2,3$-$CH{=}CHCMe_2O$-)
58. Neorautenane ($8,9$-OCH_2O-, $2,3$-$CH{=}CHCMe_2O$-)
59. Neorautane ($8,9$-OCH_2O-, $2,3$-$CH_2CH_2CMe_2O$-)
60. Neorautanol ($8,9$-OCH_2O-, $2,3$-$CH_2CH(OH)CMe_2O$-)
61. Neodunol (9-OH, $2,3$-$CH{=}CHO$-)
62. Methylneodunol (9-OMe, $2,3$-$CH{=}CHO$-)*
63. Neodulin ($8,9$-OCH_2O-, $2,3$-$CH{=}CHO$-)
64. Ambonane ($9,10$-diOMe, $2,3$-$CH{=}CHO$-)
65. Cabenegrin A-I (3-OH, $8,9$-OCH_2O-, 4-$CH_2CH{=}CMeCH_2OH$)*
66. Nitiducol (3-OH, $8,9$-OCH_2O-, 4-$CH_2CH{=}CMeCH_2CH_2CH{=}CMe_2$)
67. Hemileiocarpin (9-OMe, $3,4$-$OCMe_2CH{=}CH$-)
68. Leiocarpin ($8,9$-OCH_2O-, $3,4$-$OCMe_2CH{=}CH$-)
69. Nitiducarpin ($8,9$-OCH_2O-, $3,4$-$OCMe(CH_2CH_2CH{=}CMe_2)CH{=}CH$-)
70. Edudiol ($3,9$-diOH, 1-OMe, 2-$CH_2CH{=}CMe_2$)
71. Edulenol (3-OH, $1,9$-diOMe, 2-$CH_2CH{=}CMe_2$)
72. Edulenanol (9-OH, 1-OMe, $2,3$-$CH{=}CHCMe_2O$-)
73. Edulenane ($1,9$-diOMe, $2,3$-$CH{=}CHCMe_2O$-)
74. Desmodin (9-OH, $1,8$-diOMe $2,3$-$CH{=}CHCMe_2O$-)
75. Neorautenanol (1-OH, $8,9$-OCH_2O-, $2,3$-$CH{=}CHCMe_2O$-)
76. Edulane (1, 9-diOMe, 2, 3-$CH_2CH_2CMe_2O$-)
77. Neorautanin (1-OMe, $8,9$-OCH_2O-, $2,3$-$CH_2CH_2CMe_2O$-)
78. Apiocarpin ($1,9$-diOH, $3,4$-$OCH(CMe{=}CH_2)CH_2$-)
79. Neoraucarpanol (3-OH, 4-OMe, $8,9$-OCH_2O-, 2-$CH_2CH{=}CMe_2$)
80. Neuraucarpan ($3,4$-diOMe, $8,9$-OCH_2O-, 2-$CH_2CH{=}CMe_2$)
81. Ficinin (4-OMe, $8,9$-OCH_2O-, $2,3$-$CH{=}CHO$-)
82. Ficifolinol ($3,9$-diOH, $2,8$-di($CH_2CH{=}CMe_2$))
83. Erythrabyssin II ($3,9$-diOH, $2,10$-di($CH_2CH{=}CMe_2$))*
84. Folitenol (3-OH, 2-$CH_2CH{=}CMe_2$, 9, 10-$OCMe_2CH{=}$ CH-)
85. Folinin ($2,3$-$CH_2CH_2CMe_2O$-, $9,10$-$OCMe_2CH{=}CH$-)
86. Gangetin (9-OH, 1-OMe, 10-$CH_2CH{=}CMe_2$, $2,3$-$CH{=}CHCMe_2O$-)
87. Gangetinin (1-OMe, 2, 3; 10, 9-di(-$CH{=}CHCMe_2O$-))
88. Lespein ($3,9$-diOH, 6a, 10-di($CH_2CH{=}CMe_2$))
89. 6-Methoxyhomopterocarpin ($3,6,9$-triOMe)*
90. 6-Methoxypterocarpin ($3,6$-diOMe, $8,9$-OCH_2O-)*

6a-Hydroxypterocarpans

91. Glycinol ($3,9,6a$-triOH)
92. 6a-Hydroxymedicarpin ($3,6a$-diOH, 9-OMe)
93. 6a-Hydroxyisomedicarpin ($9,6a$-diOH, 3-OMe)
94. Variabilin (6a-OH, $3,9$-diOMe)
95. 6a-Hydroxymaackiain ($3,6a$-diOH, $8,9$-OCH_2O-)
96. Pisatin (6a-OH, 3-OMe, $8,9$-OCH_2O-)
97. $3,7,6a$-TriOH, 9-OMe
98. $3,7,6a$-TriOH, $8,9$-OCH_2O-
99. Hildecarpin ($3,6a$-diOH, 2-OMe, $8,9$-OCH_2O-)*
100. Lathycarpin (6a-OH, $2,3$-diOMe, $8,9$-OCH_2O-)*
101. Tephrocarpin ($3,6a$-diOH, 4-OMe, $8,9$-OCH_2O-)*
102. Acanthocarpan (6a-OH, 3, 4; 8, 9-di(-OCH_2O-))
103. Tuberosin ($3,6a$-diOH, $8,9$-$CH{=}CHCMe_2O$-)
104. Sandwicarpin (6a-hydroxyphaseollidin, $3,9,6a$-triOH, 10-$CH_2CH{=}CMe_2$)
105. Cristacarpin (erythrabyssin I, $3,6a$-diOH, 9-OMe, 10-$CH_2CH{=}CMe_2$)
106. 6a-Hydroxyphaseollin ($3,6a$-diOH, $9,10$-$OCMe_2CH{=}CH$-)
107. $3,7,6a$-TriOH, $9,10$-$OCMe_2CH{=}CH$- Metabolite
108. Glyceocarpin (glyceollidin II, $3,9,6a$-triOH, 2-$CH_2CH{=}CMe_2$)*
109. Glyceollin IV ($9,6a$-diOH, 3-OMe, 2-$CH_2CH{=}CMe_2$)
110. Glyceollin II ($9,6a$-diOH, $2,3$-$CH{=}CHCMe_2O$-)
111. Neobanol (6a-OH, $8,9$-OCH_2O-, $2,3$-$CH{=}CHO$-)
112. Glyceofuran ($9,6a$-diOH, $2,3$-$CH{=}C(CMe_2OH)O$-)*
113. 9-Methylglyceofuran (6a-OH, 9-OMe, $2,3$-$CH{=}C(CMe_2OH)O$-)
114. Canescacarpin ($9,6a$-diOH, $2,3$-$CH_2CH(CMe{=}CH_2)O$-)*
115. Glyceollin III ($9,6a$-diOH, $2,3$-$CH_2CH(CMe{=}CH_2)O$-)
116. Glyceollidin I ($3,9,6a$-triOH, 4-$CH_2CH{=}CMe_2$)*
117. Glyceollin I ($9,6a$-diOH, $3,4$-$OCMe_2CH{=}CH$-)
118. Glandestacarpin ($9,6a$-diOH, $3,4$-$OC(CMe{=}CH_2){=}C$-)*

Pterocarpenes

119. 6a, 11a-Dehydro, 3, 9-diOH
120. 6a, 11a-Dehydro, 3, 9-diOMe
121. Anhydropisatin (flemichapparin B, 6a, 11a-dehydro, 3-OMe, 8, 9-OCH_2O-)
122. Bryacarpene-5 (6a, 11a-dehydro, 3, 9, 10-triOMe)
123. Bryacarpene-2 (6a, 11a-dehydro, 10-OH, 3, 8, 9-triOMe)
124. Bryacarpene-3 (6a, 11a-dehydro, 3, 8, 9, 10-tetraOMe)
125. 6a, 11a-Dehydro 3-OH, 4-OMe, 8, 9-OCH_2O-
126. Bryacarpene-4 (6a, 11a-dehydro, 4-OH, 3, 9, 10-triOMe)
127. Bryacarpene-1 (6a, 11a-dehydro, 4, 10-diOH, 3, 8, 9-triOMe)
128. Leiocalycin (6a, 11a-dehydro, 2-OH, 1, 3-diOMe, 8, 9-OCH_2O-)
129. Anhydrotuberosin (6a, 11a-dehydro, 3-OH, 8, 9-CH=$CHCMe_2O$-)*
130. Methylanhydrotuberosin (6a, 11a-dehydro, 3-OMe, 9, 9-CH=$CHCMe_2O$-)*
131. Neorauteen (6a, 11a-dehydro, 9-OH, 2, 3-CH=CHO-)
132. Neoduleen (6a, 11a-dehydro, 8, 9-OCH_2O-, 2, 3-CH=CHO-)
133. Erycristagallin (6a, 11a-dehydro, 3, 9-diOH, 2, 10-di(CH_2CH=CMe_2))*

Pterocarpanones

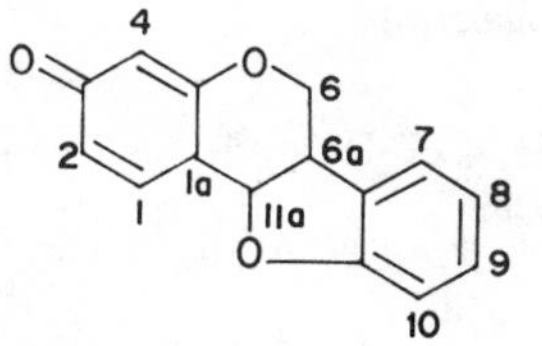

134. 1a-OH, 9-OMe — Metabolite*
135. 1a-OH, 8, 9-OCH_2O- — Metabolite*
136. 1a-Hydroxyphaseollone (1a-OH, 9, 10-$OCMe_2CH$=CH-) Metabolite
137. Hydroxytuberosone (1a, 6a-diOH, 8, 9-CH=$CHCMe_2O$-)*

Pterocarpenequinones

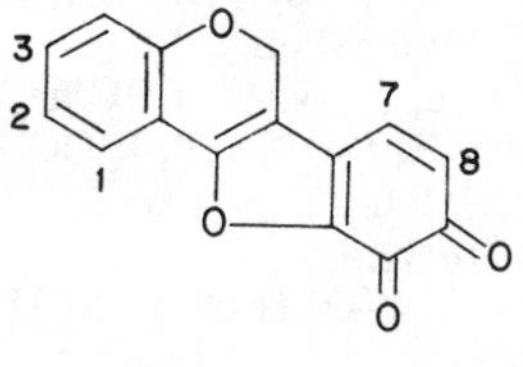

138. 4-Deoxybryaquinone (3, 7-diOMe) (?)
139. Bryaquinone (4-OH, 3, 7-diOMe) (?)

Isoflavans

Isoflavans

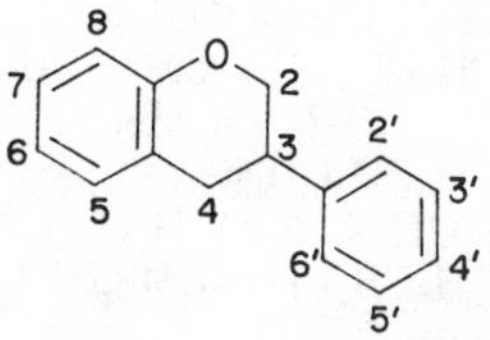

1. 7, 3′-DiOH — Metabolite*
2. Equol (7, 4′-diOH) — Metabolite
3. 4′-OH, 7-OMe — Metabolite*
4. Demethylvestitol (7, 2′, 4′-triOH)
5. Isovestitol (7, 4′-diOH, 2′-OMe)
6. Vestitol (7, 2′-diOH, 4′-OMe)
7. Neovestitol (2′, 4′-diOH, 7-OMe)
8. Sativan (7-OH, 2′, 4′-diOMe)
9. Isosativan (2′-OH, 7, 4′-diOMe)
10. Arvensan (4′-OH, 7, 2′-diOMe)
11. Laxifloran (spherosin, 7, 4′-diOH, 2′, 3′-diOMe)
12. Mucronulatol (7, 3′-diOH, 2′, 4′-diOMe)
13. Isomucronulatol (7, 2′-diOH, 3′, 4′-diOMe)
14. 5′-Methoxyvestitol (7, 2′-diOH, 4′, 5′-diOMe)*
15. Maackiainisoflavan (7, 2′-diOH, 4′, 5′-OCH_2O-) Metabolite
16. Astraciceran (7-OH, 2′-OMe, 4′, 5′-OCH_2O-)
17. Lonchocarpan (7, 4′-diOH, 2′, 3′, 6′triOMe)
18. 5-Methoxyvestitol (7, 2′-diOH, 5, 4′-diOMe)
19. Lotisoflavan (2′, 4′-diOH, 5, 7-diOMe)*
20. 7, 2′-DiOH, 6, 4′-diOMe — Metabolite
21. Bryaflavan (6, 7, 3′-triOH, 2′, 4′-diOMe)
22. 8-Demethylduartin (7, 8, 3′-triOH, 2′, 4′-diOMe)
23. Duartin (7, 3′-diOH, 8, 2′, 4′-triOMe)
24. Isoduartin (7, 2′-diOH, 8, 3′, 4′-triOMe)*
25. 5, 8, 2′-TriOH, 7, 3′, 4′-triOMe*
26. 5, 2′-DiOH, 7, 8, 3′, 4′-tetraOMe*
27. 2′-Methylphaseollidinisoflavan (7, 4′-diOH, 2′-OMe, 3′-CH_2CH=CMe_2)
28. Phaseollinisoflavan (7, 2′-diOH, 3′, 4′-CH=$CHCMe_2O$-)
29. 2′-Methylphaseollinisoflavan (7-OH, 2′-OMe, 3′, 4′-CH=$CHCMe_2O$-)
30. Crotmarine (7, 2′-diOH, 4′, 5′-OCH(CMe=CH_2)CH_2-)*
31. α, α-Dimethylallylcyclolobin (7, 3′, 4′-triOH, 2′-OMe, 5′-CMe_2CH=CH_2)
32. Unanisoflavan (7, 3′-diOH, 2′, 4′-diOMe, 5′-CMe_2CH=CH_2)
33. Spherosinin (4′-OH, 2′, 3′-diOMe, 6, 7-CH=$CHCMe_2O$-)

34. Glabridin (2′,4′-diOH, 7,8-$OCMe_2CH{=}CH$-)
35. 4′-Methylglabridin
(2′-OH, 4′-OMe, 7,8-$OCMe_2CH{=}CH$-)
36. 3′-Methoxyglabridin
(2′,4′-diOH, 3′-OMe, 7,8-$OCMe_2CH{=}CH$-)
37. Leiocin
(2′-OH, 4′,5′-OCH_2O-, 7,8-$OCMe_2CH{=}CH$-)
38. Heminitidulin (2′-OH, 4′-OMe,
7,8-$OCMe(CH_2CH_2CH{=}CMe_2)CH{=}CH$-)
39. Nitidulin (2′,3′-diOH, 4′-OMe,
7,8-$OCMe(CH_2CH_2CH{=}CMe_2)CH{=}CH$-)
40. Nitidulan (2′-OH, 4′,5′-OCH_2O-,
7,8-$OCMe(CH_2CH_2CH{=}CMe_2)CH{=}CH$-)
41. Neorauflavane
(2′,4′-diOH, 5-OMe, 6,7-$CH{=}CHCMe_2O$-)
42. Leiocinol
(6,2′-diOH, 4′,5′-OCH_2O-, 7,8-$OCMe_2CH{=}CH$-)
43. Hispaglabridin A (2′,4′-diOH, 3′-$CH_2CH{=}CMe_2$,
7,8-$OCMe_2CH{=}CH$-)
44. Hispaglabridin B
(2′-OH, 7,8; 4′,3′-di(-$OCMe_2CH{=}CH$-))
45. Licoricidin
(5,2′,4′-triOH, 7-OMe, 6,3′-di($CH_2CH{=}CMe_2$))
46. 3-Hydroxymaackiainisoflavan
(7,3,2′-triOH, 4′,5′-OCH_2O-) Metabolite
47. Biscyclolobin [7,3′-diOH, 2′-OMe, 4′-]$_2$O (?)

Isoflavanquinones

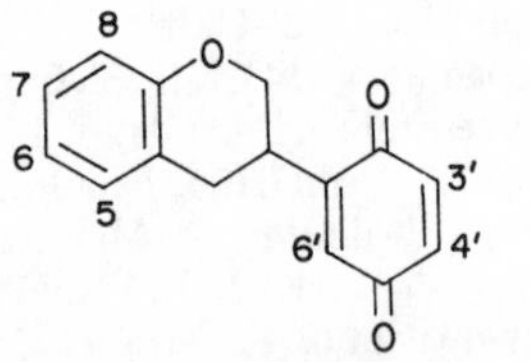

48. Claussequinone (7-OH, 4′-OMe)
49. Pendulone (7-OH, 3′,4′-diOMe)
50. Abruquinone A (6,7,3′,4′-tetraOMe)
51. Mucroquinone (7-OH, 8,4′-diOMe)
52. Amorphaquinone (7-OH, 8,3′,4′-triOMe)
53. Abruquinone C (6-OH, 7,8,3′,4′-tetraOMe)
54. Abruquinone B (6,7,8,3′,4′-tetraOMe)

Isoflavanols

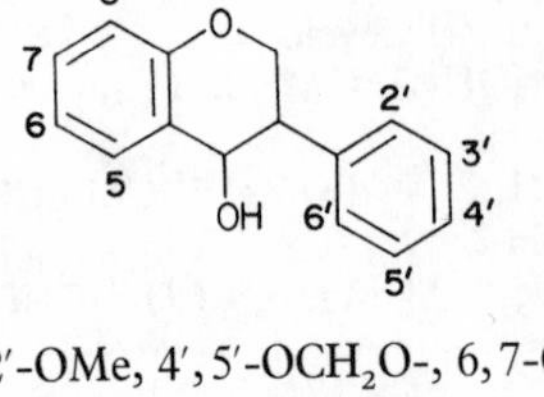

1. Ambanol (2′-OMe, 4′,5′-OCH_2O-, 6,7-$CH{=}CHO$-)

Isoflav-3-enes

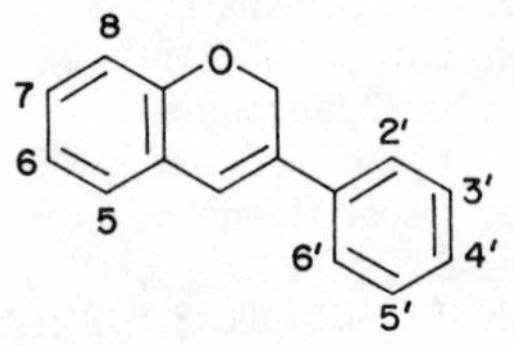

1. Haginin D (7,2′,4′-triOH)*
2. Haginin B (7,4′-diOH, 2′-OMe)
3. 7,2′-DiOH, 4′-OMe Metabolite*
4. Haginin C (7,2′,4′-triOH, 3′-OMe)*
5. Sepiol (7,2′,3′-triOH, 4′-OMe)
6. Haginin A (7,4′-diOH, 2′,3′-diOMe)
7. 2′-Methylsepiol (7,3′-diOH, 2′,4′-diOMe)
8. Odoriflavene (7,2′-diOH, 3′,4′-diOMe)*
9. 6,7,3′-TriOH, 2′, 4′-diOMe*
10. Glabrene (7,2′-diOH, 3′,4′-$CH{=}CHCMe_2O$-)
11. Neorauflavene
(2′,4′-diOH, 5-OMe, 6,7-$CH{=}CHCMe_2O$-)

3-Arylcoumarins

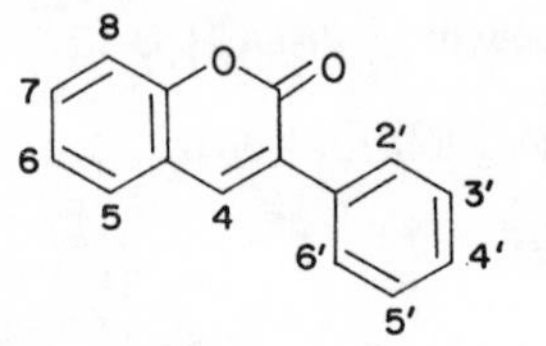

Simple 3-arylcoumarins

1. 7, 2′-DiOH, 4′-OMe
2. 7, 2′-DiOH, 4′-5′-OCH_2O-
3. Pachyrrhizin
(2′-OMe, 4′,5′-OCH_2O-, 6,7-$CH{=}CHO$-)
4. Glycycoumarin
(7,2′,4′-triOH, 5-OMe, 6-$CH_2CH{=}CMe_2$)*
5. Glycyrin
(2′,4′-diOH, 5,7-diOMe, 6-$CH_2CH{=}CMe_2$)
6. Neofolin
(8,2′-diOMe, 4′,5′-OCH_2O-, 6,7-$CH{=}CHO$-)

3-Aryl-4-hydroxycoumarins

7. Derrusnin (4,5,7-triOMe, 3′,4′-OCH_2O-)
8. Robustic acid
(4-OH, 5,4′-diOMe, 6,7-$CH{=}CHCMe_2O$-)
9. Robustic acid methyl ether
(4,5,4′-triOMe, 6,7-$CH{=}CHCMe_2O$-)
10. Robustin (4-OH, 5-OMe, 3′,4′-OCH_2O-,
6,7-$CH{=}CHCMe_2O$-)
11. Robustin methyl ether
(4,5-diOMe, 3′,4′-OCH_2O-, 6,7-$CH{=}$
$CHCMe_2O$-)

12. Glabrescin (4, 5-diOMe, 3′, 4′-OCH_2O-, 6, 7-$CH{=}CHCMe_2O$-)
13. Thonningine-B (4-OH, 5, 8, 4′-triOMe, 6, 7-$CH{=}CHCMe_2O$-)*
14. Thonningine-A (4-OH, 5, 8-diOMe, 3′, 4′-OCH_2O-, 6, 7-$CH{=}CHCMe_2O$-)*
15. Lonchocarpic acid (4, 4′-diOH, 5-OMe, 8-$CH_2CH{=}CMe_2$ 6, 7-$CH{=}CHCMe_2O$-)
16. Lonchocarpenin (4-OH, 5, 4′-diOMe, 8-$CH_2CH{=}CMe_2$ 6, 7-$CH{=}CHCMe_2O$-)
17. Scandenin (4, 4′-diOH, 5-OMe, 6-$CH_2CH{=}CMe_2$, 7, 8-$OCMe_2CH{=}CH$-)

Coumestans

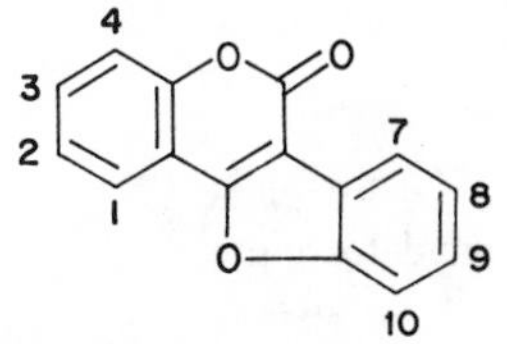

1. Coumestrol (3, 9-diOH)
2. 9-Methylcoumestrol (3-OH, 9-OMe)
3. Dimethylcoumestrol (3, 9-diOMe)
4. Repensol (3, 7, 9-triOH)
5. Trifoliol (3, 7-diOH, 9-OMe)
6. Wairol (3-OH, 7, 9-diOMe)*
7. 3, 9-DiOH, 8-OMe
8. 3-OH, 8, 9-diOMe
9. Medicagol (3-OH, 8, 9-OCH_2O-)
10. Flemichapparin C (3-OMe, 8, 9-OCH_2O-)
11. Aureol (1, 3, 9-triOH)*
12. Demethylwedelolactone (norwedelolactone, 1, 3, 8, 9-tetraOH)
13. Wedelolactone (1, 8, 9-triOH, 3-OMe)
14. Lucernol (2, 3, 9-triOH)
15. Tephrosol (3-OH, 2-OMe, 8, 9-OCH_2O-)
16. 2-OH, 3-OMe, 8, 9-OCH_2O-
17. Sativol (4, 9-diOH, 3-OMe)
18. Sophoracoumestan B (3-OH, 4-OMe, 8, 9-OCH_2O-)*
19. 2-OH, 1, 3-diOMe, 8, 9-OCH_2O-
20. Sophoracoumestan A (3-OH, 8, 9-$CH{=}CHCMe_2O$-)*
21. Tuberostan (3-OMe, 8, 9-$CH{=}CHCMe_2O$-)*
22. Isosojagol (3, 9-diOH, 10-$CH_2CH{=}CMe_2$)
23. Sojagol (3-OH, 9, 10-$OCMe_2CH_2CH_2$-)
24. Psoralidin (3, 9-diOH, 2-$CH_2CH{=}CMe_2$)
25. Psoralidin oxide (3, 9-diOH, 2-CH_2CH—CMe_2 [O bridge])
26. Corylidin (9-OH, 2, 3-CH(OH)CH(OH)CMe_2O-)
27. Erosnin (8, 9-OCH_2O-, 2, 3-CH=CHO-)
28. Phaseol (3, 9-diOH, 4-$CH_2CH{=}CMe_2$)*
29. Glycyrol (1, 9-diOH, 3-OMe, 2-$CH_2CH{=}CMe_2$)
30. 1-Methylglycyrol (9-OH, 1, 3-diOMe, 2-$CH_2CH{=}CMe_2$)
31. Isoglycyrol (9-OH, 3-OMe, 1, 2-$OCMe_2CH_2CH_2$-)
32. Norwedelic acid (2-(2, 4, 6-trihydroxyphenyl)-5, 6-dihydroxy-3-carboxybenzofuran)*

Coumaronochromones

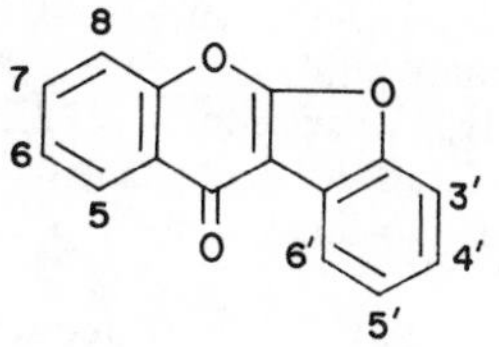

1. Lupinalbin A (5, 7, 4′-triOH)*
2. Lupinalbin D (5, 7, 4′-triOH, 3′-$CH_2CH{=}CMe_2$)*
3. Lupinalbin E (5, 7-diOH, 3′, 4′-$CH_2CH(CMe_2OH)O$-)*
4. Lupinalbin B (5, 7, 4′-triOH, 6-$CH_2CH{=}CMe_2$)*
5. Lisetin (5, 7, 4′-triOH, 5′-OMe, 3′-$CH_2CH{=}CMe_2$)
6. Lupinalbin C (5, 4′-diOH, 6, 7-$CH_2CH(CMe_2OH)O$-)*
7. Millettin (5, 4′-diOH, 8-$CH_2CH{=}CMe_2$, 6, 7-$CH{=}CHCMe_2O$-)*

α-Methyldeoxybenzoins

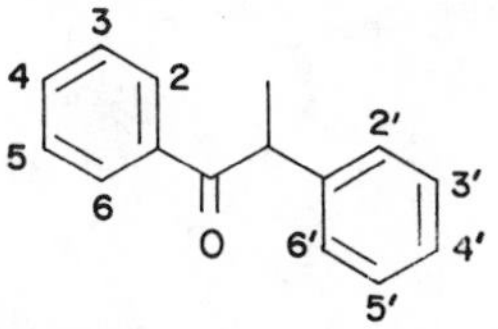

1. Angolensin (2, 4-diOH, 4′-OMe)
2. 2-Methylangolensin (4-OH, 2, 4′-diOMe)
3. 4-Methylangolensin (2-OH, 4, 4′-diOMe)
4. 4-Cadinylangolensin (2-OH, 4′-OMe, 4-O-$C_{15}H_{25}$)

2-Arylbenzofurans

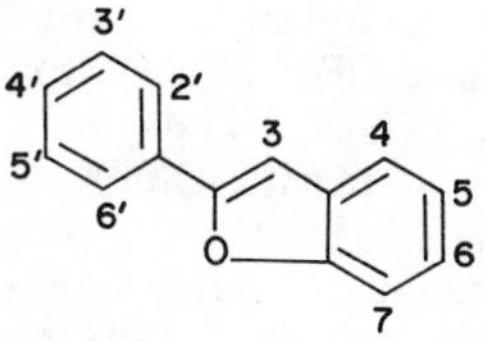

1. 2′, 4′, 6 -TriOH*
2. 6-Demethylvignafuran (4′, 6-diOH, 2′-OMe)
3. Centrolobofuran (2′, 6-diOH, 4′-OMe)*
4. Vignafuran (4′-OH, 2′, 6-diOMe)
5. Sainfuran (2′, 5-diOH, 4′, 6-diOMe)*

6. 2′, 4′-DiOH, 5, 6-diOMe (?)
7. 2′, 4′-DiOH, 5, 6-OCH_2O-
8. Methylsainfuran (5-OH, 2′, 4′, 6-triOMe)*
9. 2′-OH, 4′-OMe, 5, 6-OCH_2O-
10. Isopterofuran (4′, 6-diOH, 2′, 3′-diOMe)
11. Pterofuran (3′, 6-diOH, 2′, 4′-diOMe)
12. Sophorafuran A (2′, 4′-diOH, 3′-OMe, 6, 7-OCH_2O-)*
13. Neoraufurane (2′, 4′-diOH, 4-OMe, 5, 6-$CH{=}CHCMe_2O$-)
14. Ambofuranol (2′, 4-diOH, 4′, 3-diOMe, 6-$CH_2CH{=}CMe_2$)
15. Licobenzofuran (liconeolignan, 4′, 3-diOH, 5, 6-diOMe, 3′-$CH_2CH{=}CMe_2$)*
16. Bryebinal (3′, 4′-diOH, 2′, 4′, 7, 5 (or 6)-tetraOMe, 3-CHO) (?)

Bi-isoflavans

1. Vestitol (4 → 5′) vestitol*
2. 3′-Hydroxyvestitol (4 → 5′) vestitol*
3. Mucronulatol (4 → 5′) vestitol*
4. 5′-Methoxyvestitol (4 → 5′) vestitol*
5. Claussequinone (4 → 5′) vestitol*

*Newly reported during 1980–1985

(6) NEOFLAVONOIDS

4-Arylcoumarins

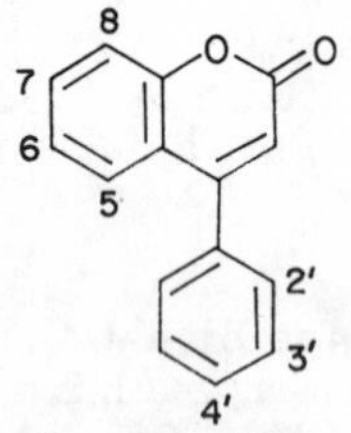

1. Dalbergin (6-OH, 7-OMe)
2. Methyldalbergin (6, 7-diOMe)
3. Isodalbergin (6-OMe, 7-OH)
4. Nordalbergin (6, 7-diOH)
5. 5, 7-Dihydroxy-4-phenylcoumarin 7-glucoside
6. Kuhlmannin (6-OH, 7, 8-diOMe)
7. Stevenin (6, 3′-dioH, 7-OMe)
8. Melannin (6, 4′-diOH, 7-OMe)
9. 4′-Hydroxy-5, 7-dimethoxy-4-phenylcoumarin
10. 5, 7, 4′-Trimethoxy-4-phenylcoumarin
11. Melannein (6, 3′-diOH, 7, 4′-diOMe)
12. 5, 2′, 5′-Trihydroxy-7-methoxy-4-phenylcoumarin
13. Exostemin (5, 7, 4′-triOMe, 8-OH)
14. 3′, 4′-Dihydroxy-5, 7-dimethoxy-4-phenylcoumarin
15. 3′-Hydroxy-5, 7, 4′-trimethoxy-4-phenylcoumarin
16. 5, 7-Dimethoxy-3′, 4′-methylenedioxy-4-phenylcoumarin
17. Sisafolin (5, 4′-diOH, 7, 2′-diOMe, 6-CHO)
18–38. Isopentenyl substituted compounds (see (Donnelly, 1975 and Chapter 5 for structures).

Dalbergiones

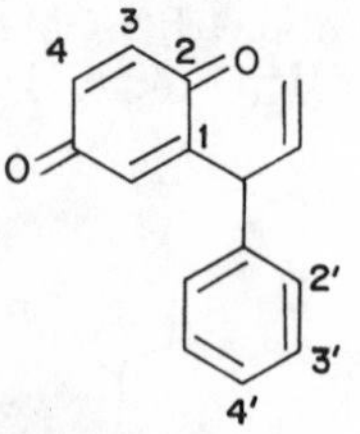

1. (*R*)-4-Methoxydalbergione
2. (*S*)-4-Methoxydalbergione
3. (*R*, *S*)-4-Methoxydalbergione
4. (*S*)-4′-Hydroxy-4-methoxydalbergione
5. (*S*)-4, 4′-Dimethoxydalbergione
6. (*R*)-3, 4-Dimethoxydalbergione
7. (*R*)-4′-Hydroxy-3, 4-dimethoxydalbergione
8. (*S*)-3′-Hydroxy-4, 4′-dimethoxydalbergione

Neoflavenes

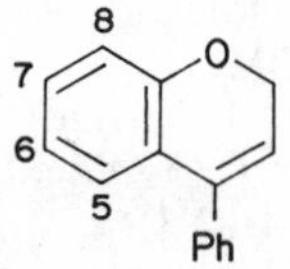

1. 6-Hydroxy-7-methoxyneoflavene
2. 6-Hydroxy-7, 8-dimethoxyneoflavene

Dalbergiquinols

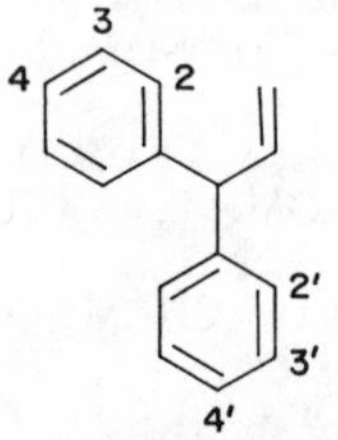

1. (*R*, *S*)-Obtusaquinol (2, 5-diOH, 4-OMe)
2. (*R*)-5-Hydroxy-2, 4-dimethoxydalbergiquinol
3. (*R*)-2, 4, 5-Trimethoxydalbergiquinol
4. (*R*)-Latifolin (5, 2′-diOH, 2, 4-diOMe)
5. (*R*)-5-*O*-Methyllatifolin (2′-OH, 2, 4, 5-triOMe)
6. 5, 3′-Dihydroxy-2, 4-dimethoxydalbergiquinol
7. (*R*-2, 5-Dihydroxy-3, 4-dimethoxydalbergiquinol
8. (*R*)-Kuhlmanniquinol (5, 4′-diOH, 2, 3, 4-triOMe)

4-Arylchromans

1. Brazilin
2. Haematoxylin

Coumarinic acids

1. Calophyllic acid
2. Chapelieric acid methyl ester

(7) FLAVONES AND FLAVONOLS

Hydroxyflavones and their methyl ethers

1. Flavone
2. Primuletin (5-OH)
3. 6-Methoxyflavone*
4. 7-Methoxyflavone*
5. 2′-Hydroxyflavone
6. 2′-Methoxyflavone*
7. 3′-Methoxyflavone*
8. 4′-Methoxyflavone*
9. 5-Hydroxy-6-methoxyflavone
10. 5,6-Dimethoxyflavone
11. Chrysin (5,7-diOH)
12. 7-Methyl ether (tectochrysin)
13. 5,7-Dimethyl ether
14. Primetin (5,8-diOH)
15. 5,2′-Dihydroxyflavone
16. 2′-Methyl ether*
17. 6,3′-Dimethoxyflavone*
18. 6-Hydroxy-4′-methoxyflavone*
19. 7,4′-Dihydroxyflavone
20. 7-Methyl ether
21. 4′-Methyl ether (pratol)
22. 3′,4′-Dihydroxyflavone
23. Baicalein (5,6,7-triOH)
24. 6-Methyl ether (oroxylin A)
25. 7-Methyl ether (negletein)*
26. 5,6-Dimethyl ether*
27. 6,7-Dimethyl ether
28. 5,6,7-Trimethyl ether
29. Norwogonin (5,7,8-triOH)
30. 8-Methyl ether (wogonin)
31. 5,8-Dimethyl ether*
32. 7,8-Dimethyl ether
33. 5,7,8-Trimethyl ether
34. 5,6,2′-Trimethoxyflavone
35. 5,6,3′-Trimethoxyflavone
36. 5,7,2′-Trihydroxyflavone*
37. 7-Methyl ether (echioidinin)
38. Apigenin (5,7,4′-triOH)
39. 5-Methyl ether (thevetiaflavone)
40. 7-Methyl ether (genkwanin)
41. 4′-Methyl ether (acacetin)
42. 5,7-Dimethyl ether
43. 7,4′-Dimethyl ether
44. 5,7,4′-Trimethyl ether
45. 6-Chloroapigenin*
46. 5,8,2′-Trihydroxyflavone
47. 6,2′,3′-Trimethoxyflavone*
48. 7,3′,4′-Trihydroxyflavone
49. 3′-Methyl ether (geraldone)
50. 7,4′-Dimethyl ether (tithonine)
51. 5,8-Dihydroxy-6,7-dimethoxyflavone
52. 8-Hydroxy-5,6,7-trimethoxyflavone
53. Alnetin (5-OH 6,7,8-triOMe)
54. 5,6,7,8-Tetramethoxyflavone
55. 5,7,2′-Trihydroxy-6-methoxyflavone*
55. Scutellarein (5,6,7,4′-tetraOH)
57. 6-Methyl ether (hispidulin)
58. 7-Methyl ether (sorbifolin)
59. 4′-Methyl ether
60. 6,7-Dimethyl ether (cirsimaritin)
61. 6,4′-Dimethyl ether (pectolinarigenin)
62. 7,4′-Dimethyl ether (ladanein)
63. 5,6,4′-Trimethyl ether*
64. 6,7,4′-Trimethyl ether (salvigenin)
65. 5,6,7,4′-Tetramethyl ether
66. 5,8,2′-Trihydroxy-7-methoxyflavone*
67. Scutevulin (5,7,2′-triOH, 8-OMe)*
68. Skullcapflavone I (5,2′-diOH, 7,8-diOMe)
69. 5,7-Dihydroxy-8,2′-dimethoxyflavone*
70. 7-Hydroxy-5,8,2′-trimethoxyflavone*
71. 5-Hydroxy-7,8,2′-trimethoxyflavone
72. Isoscutellarein (5,7,8,4′-tetraOH)
73. 7-Methyl ether
74. 8-Methyl ether
75. 4′-Methyl ether (takakin)
76. 7,8-Dimethyl ether
77. 8,4′-Dimethyl ether (bucegin)
78. 7,8,4′-Trimethyl ether
79. 5,7,8,4′-Tetramethyl ether
80. Zapotinin (5-OH,6,2′,6′-triOMe)
81. Zapotin (5,6,2′,6′-tetraOMe)
82. Cerosillin (5,6,3′,5′-tetraOMe)
83. 5,7,2′,3′-Tetrahydroxyflavone*
84. Norartocarpetin (5,7,2′,4′-tetraOH)
85. 7-Methyl ether (artocarpetin)
86. 5,7,2′,4′-Tetramethyl ether
87. 5,7,2′-Trihydroxy-6′-methoxyflavone*
88. Luteolin (5,7,3′,4′-tetraOH)
89. 5-Methyl ether
90. 7-Methyl ether
91. 3′-Methyl ether (chrysoeriol)
92. 4′-Methyl ether (diosmetin)
93. 5,3′-Dimethyl ether
94. 7,3′-Dimethyl ether (velutin)
95. 7,4′-Dimethyl ether (pilloin)
96. 3′,4′-Dimethyl ether

*compound newly reported during 1980–1986

97. 7, 3′, 4′-Trimethyl ether
98. 5, 7, 3′, 4′-Tetramethyl ether
99. Abrectorin (7, 3′-di-OH, 6, 4′-di-OMe)
100. 7, 3′, 4′, 5′-Tetrahydroxyflavone★
101. 5, 8, 2′-Trihydroxy-6, 7-dimethoxyflavone★
102. 5, 2′-Dihydroxy-6, 7, 8-trimethoxyflavone★
103. Desmethoxysudachitin (5, 7, 4′-triOH, 6, 8-diOMe)
104. Thymusin (5, 6, 4′-triOH, 7, 8-diOMe)★
105. Xanthomicrol (5, 4′-diOH, 6, 7, 8-triOMe)
106. Pedunculin (5, 8-diOH, 6, 7, 4′-triOMe)★
107. Nevadensin (5, 7-diOH, 6, 8, 4′-triOMe)
108. 5, 6-Dihydroxy-7, 8, 4′-trimethoxyflavone★
109. 7-Hydroxy-5, 6, 8, 4′-tetramethoxyflavone
110. Gardenin B (5-OH, 6, 7, 8, 4′-tetraOMe)
111. Tangeretin (5, 6, 7, 8, 4′-pentaOMe)
112. 6-Hydroxyluteolin (5, 6, 7, 3′, 4′-pentaOH)
113. 6-Methyl ether (nepetin)
114. 7-Methyl ether (pedalitin)
115. 3′-Methyl ether (nodifloretin)
116. 4′-Methyl ether★
117. 5, 6-Dimethyl ether★
118. 6, 7-Dimethyl ether (cirsiliol)
119. 6, 3′-Dimethyl ether (jaceosidin)
120. 6, 4′-Dimethyl ether (desmethoxycentaureidin)
121. 7, 3′-Dimethyl ether
122. 7, 4′-Dimethyl ether
123. 6, 7, 3′-Trimethyl ether (cirsilineol)
124. 6, 7, 4′-Trimethyl ether (eupatorin)
125. 6, 3′, 4′-Trimethyl ether (eupalitin)
126. 7, 3′, 4′-Trimethyl ether★
127. 5, 6, 7, 4′-Tetramethyl ether
128. 6, 7, 3′, 4′-Tetramethyl ether
129. 5, 6, 7, 3′, 4′-Pentamethyl ether (sinensetin)
130. Wightin (5, 3′-diOH, 7, 8, 2′-triOMe)
131. Rehderianin I (5, 2′, 5′-triOH, 7, 8-diOMe)
132. 5, 2′, 6′-Trihydroxy-7, 8-dimethoxyflavone★
133. 5, 7, 2′-Trihydroxy-8, 6′-dimethoxyflavone★
134. Rivularin (5, 2′-diOH, 7, 8, 6′-triOMe)★
135. 5, 7-Dihydroxy-8, 2′, 6′-trimethoxyflavone★
136. Hypolaetin (5, 7, 8, 3′, 4′-pentaOH)
137. 8-Methyl ether (onopordin)
138. 3′-Methyl ether
139. 8, 3′-Dimethyl ether
140. 7, 8, 3′, 4′-Tetramethyl ether
141. 5, 7, 8, 3′, 4′-Pentamethyl ether (isosinensetin)
142. Cerosillin B (5, 6, 3′, 4′, 5′-pentaOMe)★
143. Isoetin (5, 7, 2′, 4′, 5′-pentaOH)
144. Tricetin (5, 7, 3′, 4′, 5′-pentaOH)
145. 4′-Methyl ether★
146. 3′-Methyl ether (selgin)
147. 3′, 5′-Dimethyl ether (tricin)
148. 3′, 4′-Dimethyl ether★
149. 3′, 4′, 5′-Trimethyl ether
150. 7, 3′, 4′, 5′-Tetramethyl ether (corymbosin)
151. 5, 7, 3′, 4′, 5′-Pentamethyl ether
152. Prosogerin D (7-OH, 6, 3′, 4′, 5′-tetraOMe)
153. Prosogerin C (6, 7, 3′, 4′, 5′-pentaOMe)
154. 5, 2′, 5′-Trihydroxy-6, 7, 8-trimethoxyflavone★
155. Skullcapflavone II (5, 2′-diOH, 6, 7, 8, 6′-tetraOMe)
156. 5, 7, 8, 3′, 4′-Pentahydroxy-6-methoxyflavone★
157. Leucanthoflavone (5, 8, 3′, 4′-tetraOH, 6, 7-diOMe)★
158. 5, 7, 3′, 4′-Tetrahydroxy-6, 8-dimethoxyflavone
159. 5, 6, 7, 8-Tetrahydroxy-3′, 4′-dimethoxyflavone
160. Sideritiflavone (5, 3′, 4′-triOH, 6, 7, 8-triOMe)
161. Sudachitin (5, 7, 4′-triOH, 6, 8, 3′-triOMe)
162. Acerosin (5, 7, 3′-triOH, 6, 8, 4′-triOMe)
163. Thymonin (5, 6, 4′-triOH, 7, 8, 3′-triOMe)★
164. 5, 4′-Dihydroxy-6, 7, 8, 3′-tetramethoxyflavone
165. Gardenin D (5, 3′-diOH, 6, 7, 8, 4′-tetraOMe)
166. Hymenoxin (5, 7-diOH, 6, 8, 3′, 4′-tetraOMe)
167. 5, 6-Dihydroxy-7, 8, 3′, 4′-tetramethoxyflavone★
168. 4′-Hydroxy-5, 6, 7, 8, 3′-pentamethoxyflavone
169. 5-Desmethoxynobiletin (5-OH, 6, 7, 8, 3′, 4′-pentaOMe)
170. Nobiletin (5, 6, 7, 8, 3′, 4′-hexaOMe)
171. 5, 6, 7, 2′, 4′, 5′-Hexahydroxyflavone
172. 6, 5′-Dimethyl ether★
173. 6, 7, 5′-Trimethyl ether (arcapillin)★
174. 6, 2′, 4′, 5′-Tetramethyl ether (tabularin)
175. 5, 7, 3′, 4′-Tetrahydroxy-6, 5′-dimethoxyflavone★
166. 5, 7, 4′-Trihydroxy-6, 3′, 5′-trimethoxyflavone
177. 5, 7, 3′-Trihydroxy-6, 4′, 5′-trimethoxyflavone
178. 5, 4′-Dihydroxy-6, 7, 3′, 5′-tetramethoxyflavone★
179. 5, 3′-Dihydroxy-6, 7, 4′, 5′-tetramethoxyflavone★
180. 5-Hydroxy-6, 7, 3′, 4′, 5′-pentamethoxyflavone
181. 5, 6, 7, 3′, 4′, 5′-Hexamethoxyflavone★
182. Serpyllin (5-OH, 7, 8, 2′, 3′, 4′-pentaOMe)
183. Ganhuangenin (5, 7, 3′, 6′-tetraOH, 8, 2′-diOMe)★
184. 5, 7, 2′, 4′-Tetrahydroxy-8, 5′-dimethoxyflavone★
185. 5, 7, 2′, 5′-Tetrahydroxy-8, 6′-dimethoxyflavone★
186. 5, 7, 3′, 4′, 5′-Pentahydroxy-8-methoxyflavone
187. 5, 7, 3′, 5′-Tetrahydroxy-8, 4′-dimethoxyflavone
188. 5, 7, 5′-Trihydroxy-8, 3′, 4′-trimethoxyflavone
189. 5, 3′-Dihydroxy-7, 8, 4′, 5′-tetramethoxyflavone
190. 5, 7-Dihydroxy-8, 3′, 4′, 5′-tetramethoxyflavone
191. 5, 7, 8, 3′, 4′, 5′-Hexamethoxyflavone★
192. 5, 6, 2′, 3′, 4′, 6′-Hexamethoxyflavone★
193. 5, 7, 2′, 4′-Tetrahydroxy-6, 8, 5′-trimethoxyflavone★
194. Agecorynin-D (5, 2′, 4′-triOH, 6, 7, 8, 5′-tetraOMe)
195. Agecorynin-C (5, 6, 7, 8, 2′, 4′, 5′-heptaOMe)
196. Gardenin E (5, 3′, 5′-triOH, 6, 7, 8, 4′-tetraOMe)
197. Scaposin (5, 7, 5′-triOH, 6, 8, 3′, 4′-tetraOMe)
198. Trimethoxywogonin (5, 6, 7-triOH, 8, 3′, 4′, 5′-tetraOMe)
199. Gardenin C (5, 3′-diOH, 6, 7, 8, 4′, 5′-pentaOMe)

200. 4′-Hydroxy-5, 6, 7, 8, 3′, 5′-hexamethoxyflavone*
201. 3′-Hydroxy-5, 6, 7, 8, 4′, 5′-hexamethoxyflavone*
202. Gardenin A (5-OH, 6, 7, 8, 3′, 4′, 5′-hexaOMe)
203. 5, 6, 7, 8, 3′, 4′, 5′-Heptamethoxyflavone
204. 5, 3′-Dihydroxy-6, 7, 2′, 4′, 5′-pentamethoxyflavone*
205. Agehoustin B (5, 6, 7, 2′, 3′, 4′, 5′-heptaOMe)*
206. Agehoustin D (5, 3′-diOH, 6, 7, 8, 2′, 4′, 5′-hexaOMe)*
207. Agehoustin C (3′-OH, 5, 6, 7, 8, 2′, 4′, 5′-heptaOMe)*
208. Agehoustin A (5, 6, 7, 8, 2′, 3′, 4′, 5′-octaOMe)*

Hydroxyflavonols and their methyl ethers

1. 3, 7-Dihydroxyflavone*
2. Galangin (3, 5, 7-triOH)
3. 3-Methyl ether
4. 5-Methyl ether*
5. 7-Methyl ether (izalpinin)
6. 3, 7-Dimethyl ether
7. 5, 7-Dimethyl ether*
8. 3, 5, 7-Trimethyl ether
9. 3, 7-Dihydroxy-8-methoxyflavone*
10. 3, 7, 4′-Trihydroxyflavone
11. 6-Hydroxygalangin 6-methyl ether (alnusin)
12. 3, 6-Dimethyl ether
13. 3, 7-Dimethyl ether
14. 5, 6-Dimethyl ether*
15. 5, 7-Dimethyl ether
16. 3, 6, 7-Trimethyl ether (alnustin)
17. 3, 5, 6, 7-Tetramethyl ether
18. 8-Hydroxygalangin 3-methyl ether*
19. 7-Methyl ether
20. 8-Methyl ether*
21. 3, 7-Dimethyl ether (isognaphalin)
22. 3, 8-Dimethyl ether (gnaphalin)
23. 7, 8-Dimethyl ether*
24. 3, 5, 8-Trimethyl ether*
25. 3, 7, 8-Trimethyl ether (methylgnaphalin)
26. 3, 5, 7, 8-Tetramethyl ether*
27. Datiscetin (3, 5, 7, 2′-tetraOH)
28. 7-Methyl ether (datin)
29. 2′-Methyl ether (ptaeroxylol)
30. Kaempferol (3, 5, 7, 4′-tetraOH)
31. 3-Methyl ether (isokaempferide)
32. 5-Methyl ether
33. 7-Methyl ether (rhamnocitrin)
34. 4′-Methyl ether (kaempferide)
35. 3, 5-Dimethyl ether*
36. 3, 7-Dimethyl ether (kumatakenin)
37. 3, 4′-Dimethyl ether (ermanin)
38. 7, 4′-Dimethyl ether
39. 3, 5, 7-Trimethyl ether
40. 3, 7, 4′-Trimethyl ether
41. Pratoletin (3, 5, 8, 4′-tetraOH)
42. 3, 6, 7, 4′-Tetrahydroxyflavone
43. 3, 6-Dimethyl ether*
44. 3, 7, 8, 4′-Tetrahydroxyflavone
45. 3-Methyl ether
46. Fisetin (3, 7, 3′, 4′-tetraOH)
47. 3-Methyl ether
48. 3′-Methyl ether (geraldol)
49. 4′-Methyl ether
50. 3, 7, 3′, 4′-Tetramethyl ether
51. 3, 5, 7-Trihydroxy-6, 8-dimethoxyflavone
52. 5, 8-Dihydroxy-3, 6, 7-trimethoxyflavone*
53. Araneol (5, 7-diOH, 3, 6, 8-triOMe)
54. 3, 5-Dihydroxy-6, 7, 8-trimethoxyflavone
55. 5-Hydroxy-3, 6, 7, 8-tetramethoxyflavone*
56. 6-Hydroxykaempferol (3, 5, 6, 7, 4′-pentaOH)
57. 3-Methyl ether
58. 5-Methyl ether (vogeletin)
59. 6-Methyl ether
60. 4′-Methyl ether
61. 3, 6-Dimethyl ether
62. 3, 7-Dimethyl ether
63. 5, 6-Dimethyl ether*
64. 6, 7-Dimethyl ether (eupalitin)
65. 6, 4′-Dimethyl ether (betuletol)
66. 3, 6, 7-Trimethyl ether (penduletin)
67. 3, 6, 4′-Trimethyl ether (santin)
68. 3, 7, 4′-Trimethyl ether
69. 5, 6, 7-Trimethyl ether (candidol)*
70. 6, 7, 4′-Trimethyl ether (mikanin)
71. 3, 6, 7, 4′-Tetramethyl ether
72. Dechlorochlorflavonin (5, 2′-diOH, 3, 7, 8-triOMe)*
73. Chlorflavonin (5, 2′-diOH, 3, 7, 8-triOMe, 3′-chloro)*
74. 5-Hydroxy-3, 7, 8, 2′-tetramethoxyflavone*
75. Herbacetin (3, 5, 7, 8, 4′-pentaOH)
76. 3-Methyl ether*
77. 7-Methyl ether (pollenitin)
78. 8-Methyl ether (sexangularetin)
79. 4′-Methyl ether*
80. 3, 7-Dimethyl ether
81. 3, 8-Dimethyl ether
82. 7, 8-Dimethyl ether
83. 7, 4′-Dimethyl ether
84. 8, 4′-Dimethyl ether (prudomestin)
85. 3, 7, 8-Trimethyl ether
86. 3, 8, 4′-Trimethyl ether
87. 7, 8, 4′-Trimethyl ether (tambulin)
88. 3, 7, 8, 4′-Tetramethyl ether (flindulatin)
89. Auranetin (3, 6, 7, 8, 4′-pentaOMe)
90. Morin (3, 5, 7, 2′, 4′-pentaOH)
91. 3, 5, 2′-Trihydroxy-7, 5′-dimethoxyflavone
92. 3, 5, 7, 2′, 6′-Pentahydroxyflavone*
93. 3, 5, 7, 2′, 6′-Pentamethyl ether
94. Quercetin (3, 5, 7, 3′, 4′-pentaOH)
95. 3-Methyl ether

96. 5-Methyl ether (azaleatin)
97. 7-Methyl ether (rhamnetin)
98. 3′-Methyl ether (isorhamnetin)
99. 4′-Methyl ether (tamarixetin)
100. 3, 5-Dimethyl ether (caryatin)
101. 3, 7-Dimethyl ether
102. 3, 3′-Dimethyl ether
103. 3, 4′-Dimethyl ether
104. 5, 3′-Dimethyl ether
105. 7, 3′-Dimethyl ether (rhamnazin)
106. 7, 4′-Dimethyl ether (ombuin)
107. 3′, 4′-Dimethyl ether (dillenetin)
108. 3, 7, 3′-Trimethyl ether (pachypodol)
109. 3, 7, 4′-Trimethyl ether (ayanin)
110. 3, 3′, 4′-Trimethyl ether
111. 7, 3′, 4′-Trimethyl ether
112. 3, 5, 7, 3′-Tetramethyl ether*
113. 3, 5, 7, 4′-Tetramethyl ether
114. 3, 5, 3′, 4′-Tetramethyl ether
115. 3, 7, 3′, 4′-Tetramethyl ether (retusin)
116. 5, 7, 3′, 4′-Tetramethyl ether*
117. 3, 5, 8, 3′, 4′-Pentahydroxyflavone
118. Rhynchosin (3, 6, 7, 3′, 4′-pentaOH)
119. Melanoxetin (3, 7, 8, 3′, 4′-pentaOH)
120. 3-Methyl ether (transilitin)
121. 8-Methyl ether
122. 3, 8-Dimethyl ether
123. 3, 7, 2′, 3′, 4′-Pentahydroxyflavone*
124. Robinetin (3, 7, 3′, 4′, 5′-pentaOH)
125. Emmaosunin (5-OH, 3, 6, 7, 8, 3′-pentaOMe)
126. 5, 7, 8, 4′-Tetrahydroxy-3, 6-dimethoxyflavone*
127. Sarothrin (5, 7, 4′-triOH, 3, 6, 8-triOMe)*
128. Calycopterin (5, 4′-diOH, 3, 6, 7, 8-tetraOMe)*
129. Araneosol (5, 7-diOH, 3, 6, 8, 4′-tetraOMe)
130. Eriostemin (3, 8-diOH, 5, 6, 7, 4′-tetraOMe)
131. 5-Hydroxyauranetin (5-OH, 3, 6, 7, 8, 4′-pentaOMe)
132. Chrysosplin (5, 4′-diOH, 3, 6, 7, 2′-tetraOMe)
133. Quercetagetin (3, 5, 6, 7, 3′, 4′-hexaOH)
134. 3-Methyl ether
135. 5-Methyl ether
136. 6-Methyl ether (patuletin)
137. 3′-Methyl ether
138. 4′-Methyl ether*
139. 3, 6-Dimethyl ether (axillarin)
140. 3, 7-Dimethyl ether (tomentin)
141. 3, 3′-Dimethyl ether*
142. 6, 7-Dimethyl ether (eupatolitin)
143. 6, 3′-Dimethyl ether (spinacetin)
144. 6, 4′-Dimethyl ether (laciniatin)
145. 3, 5, 7-Trimethyl ether*
146. 3, 6, 7-Trimethyl ether (chrysosplenol-D)
147. 3, 6, 3′-Trimethyl ether (jaceidin)
148. 3, 6, 4′-Trimethyl ether (centaureidin)
149. 3, 7, 3′-Trimethyl ether (chrysosplenol-C)
150. 3, 7, 4′-Trimethyl ether (oxyayanin-B)
151. 6, 7, 3′-Trimethyl ether (veronicafolin)
152. 6, 7, 4′-Trimethyl ether (eupatin)
153. 6, 3′, 4′-Trimethyl ether*
154. 3, 5, 6, 3′-Tetramethyl ether
155. 3, 5, 7, 3′-Tetramethyl ether*
156. 3, 6, 7, 3′-Tetramethyl ether (chrysosplenetin)
157. 3, 6, 7, 4′-Tetramethyl ether (casticin)
158. 3, 6, 3′, 4′-Tetramethyl ether (bonanzin)
159. 3, 7, 3′, 4′-Tetramethyl ether
160. 5, 6, 7, 4′-Tetramethyl ether (eupatoretin)
161. 6, 7, 3′, 4′-Tetramethyl ether*
162. 3, 6, 7, 3′, 4′-Pentamethyl ether (artemetin)
163. 3, 5, 6, 7, 3′, 4′-Hexamethyl ether
164. 5, 4′-Dihydroxy-3, 7, 8, 2′-tetramethoxyflavone
165. 5-Hydroxy-3, 7, 8, 2′, 4′-pentamethoxyflavone
166. 3, 5, 8-Trihydroxy-7, 2′, 5′-trimethoxyflavone*
167. Gossypetin (3, 5, 7, 8, 3′, 4′-hexaOH)
168. 3-Methyl ether*
169. 7-Methyl ether (ranupenin)
170. 8-Methyl ether (corniculatusin)
171. 3′-Methyl ether
172. 3, 7-Dimethyl ether
173. 3, 8-Dimethyl ether
174. 3, 3′-Dimethyl ether*
175. 7, 4′-Dimethyl ether (tambuletin)
176. 8, 3′-Dimethyl ether (limocitrin)
177. 8, 4′-Dimethyl ether
178. 3, 7, 8-Trimethyl ether
179. 3, 7, 3′-Trimethyl ether
180. 3, 8, 3′-Trimethyl ether
181. 7, 8, 3′-Trimethyl ether
182. 3, 7, 8, 3′-Tetramethyl ether (ternatin)
183. 3, 7, 8, 4′-Tetramethyl ether
184. 3, 7, 3′, 4′-Tetramethyl ether
185. 3, 8, 3′, 4′-Tetramethyl ether
186. 7, 8, 3′, 4′-Tetramethyl ether*
187. 3, 7, 8, 3′, 4′-Pentamethyl ether
188. 3, 5, 7, 8, 3′, 4′-Hexamethyl ether
189. 3, 5, 7, 3′-Tetrahydroxy-2′, 4′-dimethoxyflavone
190. Apuleidin (5, 2′, 3′-triOH, 3, 7, 4′-triOMe)
191. 5′-Hydroxymorin (3, 5, 7, 2′, 4′, 5′-hexaOH)
192. 3, 7, 4′-Trimethyl ether (oxyayanin-A)
193. 3, 5, 7, 4′-Tetramethyl ether
194. 3, 7, 4′, 5′-Tetramethyl ether
195. 3, 5, 7, 4′, 5′-Pentamethyl ether
196. 3, 7, 2′, 4′, 5′-Pentamethyl ether
197. 3, 5, 7, 2′, 4′, 5′-Hexamethyl ether
198. 5-*O*-Demethylapulein (2′, 5′-diOH, 3, 6, 7, 4′-tetraOMe)
199. Myricetin (3, 5, 7, 3′, 4′, 5′-hexaOH)
200. 3-Methyl ether (annulatin)
201. 5-Methyl ether
202. 7-Methyl ether (europetin)
203. 3′-Methyl ether (larycitrin)
204. 4′-Methyl ether (mearnsetin)
205. 3, 4′-Dimethyl ether

206. 3, 5′-Dimethyl ether
207. 7, 4′-Dimethyl ether*
208. 3′, 4′-Dimethyl ether*
209. 3′, 5′-Dimethyl ether (syringetin)
210. 3, 7, 3′-Trimethyl ether*
211. 3, 7, 4′-Trimethyl ether
212. 3, 3′, 4′-Trimethyl ether*
213. 7, 3′, 4′-Trimethyl ether
214. 3, 7, 3′, 4′-Tetramethyl ether
215. 3, 7, 3′, 5′-Tetramethyl ether*
216. 3, 3′, 4′, 5′-Tetramethyl ether*
217. 3, 7, 3′, 4′, 5′-Pentamethyl ether (combretol)
218. 5, 2′, 4′-Trihydroxy-3, 6, 7, 8-tetramethoxyflavone
219. 5, 4′-Dihydroxy-3, 6, 7, 8, 2′-pentamethoxyflavone
220. 5, 2′-Dihydroxy-3, 6, 7, 8, 4′-pentamethoxyflavone
221. 5-Hydroxy-3, 6, 7, 8, 2′, 4′-hexamethoxyflavone
222. 5, 7, 8, 3′, 4′-Pentahydroxy-3, 6-dimethoxyflavone
223. 3, 5, 7, 3′, 4′-Pentahydroxy-6, 8-dimethoxyflavone*
224. 5, 7, 3′, 4′-Tetrahydroxy-3, 6, 8-trimethoxyflavone*
225. 5, 6, 3′, 4′-Tetrahydroxy-3, 7, 8-trimethoxyflavone*
226. 3, 5, 3′, 4′-Tetrahydroxy-6, 7, 8-trimethoxyflavone*
227. Limocitrol (3, 5, 7, 4′-tetraOH, 6, 8, 3′-triOMe)
228. Isolimocitrol (3, 5, 7, 3′-tetraOH, 6, 8, 4′-triOMe)
229. 5, 3′, 4′-Trihydroxy-3, 6, 7, 8-tetramethoxyflavone*
230. 5, 7, 4′-Trihydroxy-3, 6, 8, 3′-tetramethoxyflavone
231. 5, 7, 3′-Trihydroxy-3, 6, 8, 4′-tetramethoxyflavone*
232. 5, 4′-Dihydroxy-3, 6, 7, 8, 3′-pentamethoxyflavone
223. 5, 3′-Dihydroxy-3, 6, 7, 8, 4′-pentamethoxyflavone
234. 5-Hydroxy-3, 6, 7, 8, 3′, 4′-hexamethoxyflavone
235. Natsudaidain (3-OH, 5, 6, 7, 8, 3′, 4′-hexaOMe)
236. 3, 5, 6, 7, 8, 3′, 4′-Heptamethoxyflavone
237. Apuleisin (5, 6, 2′, 3′-tetraOH, 3, 7, 4′-triOMe)
238. 3, 5, 6, 7, 2′, 3′, 4′-Heptamethoxyflavone
239. 5, 7, 2′, 5′-Tetrahydroxy-3, 6, 4′-trimethoxyflavone
240. 5, 7, 2′, 4′-Tetrahydroxy-3, 6, 5′-trimethoxyflavone*
241. 5, 2′, 5′-Trihydroxy-3, 6, 7, 4′-tetramethoxyflavone*
242. 5, 7, 2′-Trihydroxy-3, 6, 4′, 5′-tetramethoxyflavone*
243. 5, 6, 5′-Trihydroxy-3, 7, 2′, 4′-tetramethoxyflavone*
244. Apulein (2′, 5′-diOH, 3, 5, 6, 7, 4′-pentaOMe)
245. 6, 5′-Dihydroxy-3, 5, 7, 2′, 4′-pentamethoxyflavone*
246. 6, 2′-Dihydroxy-3, 5, 7, 4′, 5′-pentamethoxyflavone
247. 5, 5′-Dihydroxy-3, 6, 7, 2′, 4′-pentamethoxyflavone*
248. Brickellin (5, 2′-diOH, 3, 6, 7, 4′, 5′-pentaOMe)*
249. 5, 6-Dihydroxy-3, 7, 2′, 4′, 5′-pentamethoxyflavone
250. 3, 5, 6, 7, 2′, 4′, 5′-Heptamethoxyflavone
251. 6-Hydroxymyricetin 3, 6-dimethyl ether
252. 3, 4′-Dimethyl ether
253. 6, 4′-Dimethyl ether
254. 3, 6, 3′-Trimethyl ether
255. 3, 6, 4′-Trimethyl ether
256. 3, 6, 5′-Trimethyl ether*
257. 3, 3′, 5′-Trimethyl ether
258. 6, 3′, 5′-Trimethyl ether*
259. 3, 6, 7, 4′-Tetramethyl ether
260. 3, 6, 3′, 4′-Tetramethyl ether
261. 3, 6, 3′, 5′-Tetramethyl ether
262. 3, 7, 3′, 4′-Tetramethyl ether (apuleitrin)
263. 3, 5, 7, 3′-4′-Pentamethyl ether (apuleirin)
264. 3, 6, 7, 3′, 4′-Pentamethyl ether
265. 3, 6, 7, 3′, 5′-Pentamethyl ether (murrayanol)
266. 3, 5, 6, 7, 3′, 5′-Hexamethyl ether
267. 3, 5, 6, 7, 3′, 4′, 5′-Heptamethyl ether
268. 5, 8-Dihydroxy-3, 7, 2′, 3′, 4′-pentamethoxyflavone
269. 5, 7, 2′, 4′-Tetrahydroxy-3, 8, 5′-trimethoxyflavone*
270. 5, 2′, 4′-Trihydroxy-3, 7, 8, 5′-tetramethoxyflavone*
271. Hibiscetin (3, 5, 7, 8, 3′, 4′, 5′-heptaOH)
272. 3, 7, 4′-Trimethyl ether
273. 3, 8, 4′-Trimethyl ether*
274. 3, 7, 8, 4′-Tetramethyl ether
275. 3, 8, 4′, 5′-Tetramethyl ether
276. 3, 8, 3′, 4′, 5′-Pentamethyl ether (conyzatin)
277. 7, 8, 3′, 4′, 5′-Pentamethyl ether
278. 3, 5, 7, 8, 3′, 4′, 5′-Heptamethyl ether
279. 5, 7, 2′, 5′-Tetrahydroxy-3, 6, 8, 4′-tetramethoxyflavone*
280. 5, 7, 2′, 4′-Tetrahydroxy-3, 6, 8, 5′-tetramethoxyflavone*
281. 5, 6.2′, 5′-Tetrahydroxy-3, 7, 8, 4′-tetramethoxyflavone*
282. 5, 7, 2′-Trihydroxy-3, 6, 8, 4′, 5′-pentamethoxyflavone*
283. 5, 2′-Dihydroxy-3, 6, 7, 8, 4′, 5′-hexamethoxyflavone*
284. Purpurascenin (3, 5, 6, 7, 8, 2′, 4′, 5′-octaOMe)*
285. 5, 7, 3′, 4′, 5′-Pentahydroxy-3, 6, 8-trimethoxyflavone*
286. 5, 7, 3′, 5′-Tetrahydroxy-3, 6, 8, 4′-tetramethoxyflavone*
287. 5, 7, 3′, 4′-Tetrahydroxy-3, 6, 8, 5′-tetramethoxyflavone*
288. 5, 6, 3′, 5′-Tetrahydroxy-3, 7, 8, 4′-tetramethoxyflavone*
289. 5, 3′, 5′-Trihydroxy-3, 6, 7, 8, 4′-

pentamethoxyflavone*
290. 5, 7, 4′-Trihydroxy-3, 6, 8, 3′, 5′-pentamethoxyflavone*
291. 5, 7, 3′-Trihydroxy-3, 6, 8, 4′, 5′-pentamethoxyflavone*
292. 3′, 5′-Dihydroxy-3, 5, 6, 7, 8, 4′-hexamethoxyflavone*
293. Digicitrin (5, 3′-diOH, 3, 6, 7, 8, 4′, 5′-hexaOMe)
294. 5, 7-Dihydroxy-3, 6, 8, 3′, 4′, 5′-hexamethoxyflavone*
295. 3′-Hydroxy-3, 5, 6, 7, 8, 4′, 5′-heptamethoxyflavone*
296. 5-Hydroxy-3, 6, 7, 8, 3′, 4′, 5′-heptamethoxyflavone*
297. Exoticin (3, 5, 6, 7, 8, 3′, 4′, 5′-octaOMe)

_C_-Methylflavones

1. 6-Methylchrysin (strobochrysin, 5, 7-diOH, 6-Me)
2. 5-Hydroxy-7-methoxy-6-methylflavone*
3. Saltillin (5-OH, 4′-OMe, 7-Me)*
4. 5, 7, 4′-Trihydroxy-3-methylflavone*
5. 8-Desmethylsideroxylin (5, 4′-diOH, 7-OMe, 6-Me)
6. Sideroxylin (5, 4′-diOH, 7-OMe, 6, 8-diMe)
7. 8-Desmethyleucalyptin (5-OH, 7, 4′-diOMe, 6-Me)
8. Eucalyptin (5, OH, 7, 4′-diOMe, 6, 8-diMe)
9. 5, 7, 3′, 4′-Tetrahydroxy-3-methoxyflavone*
10. Unonal (5, 7-diOH, 8-Me, 6-CHO)
11. Isounonal (5, 7-diOH, 6-Me,8-CHO)
12. Unonal 7-methyl ether (5-OH, 7-OMe, 8-Me, 6-CHO)
13. 5, 7-Dihydroxy-3′, 4′-dimethoxy-6, 8-dimethyl flavone
14. 5, 7, 3′-Trihydroxy-4′, 5′-dimethoxy-6, 8-dimethylflavone

_C_-Methylflavonols

1. 8-_C_-Methylgalangin (3, 5, 7-triOH, 8-Me)*
2. 3, 5, 7-Trihydroxy-6, 8-dimethylflavone*
3. 5, 7-Dihydroxy-3-methoxy-8-methylflavone*
4. 5, 7-Dihydroxy-3-methoxy-6, 8-dimethylflavone*
5. 3, 5-Dihydroxy-7-methoxy-8-methylflavone*
6. 3, 5-Dihydroxy-7-methoxy-6, 8-dimethylflavone*
7. Pityrogrammin (3, 5, 7-triOH, 8-OMe, 6-Me)
8. Sylpin (5, 6, 4′-triOH, 3-OMe, 8-Me)
9. 6-Methylkaempferol (3, 5, 7, 4′-tetraOH, 6-Me)
10. 5, 7, 4′-Trihydroxy-3-methoxy-6-methylflavone
11. 5, 7, 4′-Trihydroxy-3-methoxy-6, 8-dimethylflavone*
12. 5, 4′-Dihydroxy-3, 7-dimethoxy-6-methylflavone*
13. 5, 4′-Dihydroxy-3, 7-dimethoxy-6, 8-dimethylflavone*
14. 8-Desmethylkalmiatin (5-OH, 3, 7, 4′-triOMe, 6-Me)*
15. Kalmiatin (5-OH, 3, 7, 4′-triOMe, 6, 8-diMe)*
16. 3, 5, 8, 4′-Tetrahydroxy-7-methoxy-6-methylflavone*
17. Pinoquercetin (3, 5, 7, 3′, 4′-pentaOH, 6-Me)
18. 3, 5, 7, 3′, 4′-Pentahydroxy-8-methylflavone*
19. 3, 5, 7, 3′, 4′-Pentahydroxy-6, 8-dimethylflavone
20. 5, 7, 3′, 4′-Tetrahydroxy-3-methoxy-6-methylflavone*
21. 5, 7, 4′-Trihydroxy-3, 3′-dimethoxy-6-methylflavone
22. 5, 7, 4′-Trihydroxy-3, 3′-dimethoxy-6, 8-dimethylflavone*
23. 5, 4′-Dihydroxy-3, 7, 3′-trimethoxy-6-methylflavone*
24. 5, 4′-Dihydroxy-3, 7, 3′-trimethoxy-6, 8-dimethylflavone*
25. 5, 8, 4′-Trihydroxy-3, 7-dimethoxy-6-methylflavone*
26. 3, 5, 7, 3′, 4′, 5′-Hexahydroxy-6-methylflavone
27. Alluaudiol (5, 7, 3′, 4′, 5′-pentaOH, 3-OMe, 6-Me)
28. 5, 7, 3′, 4′, 5′-Pentahydroxy-3-methoxy-6, 8-dimethylflavone
29. Dumosol (3, 5, 7, 3′, 5′-pentaOH, 4′-OMe, 6-Me)
30. 3, 5, 7, 3′, 5′-Pentahydroxy-4′-methoxy-6, 8-dimethylflavone
31. 5, 7, 3′, 5′-Tetrahydroxy-3, 4′-dimethoxy-6-methylflavone
32. 5, 7, 3′, 5′-Tetrahydroxy-3, 4′-dimethoxy-6, 8-dimethoxyflavone

Methylenedioxyflavones

1. 5-Methoxy-7, 8-methylenedioxyflavone
2. 7-Methoxy-3′, 4′-methylenedioxyflavone*
3. 7-Methoxy-5, 6-methylenedioxy-3′, 4′-methylenedioxyflavone*
4. 5, 6-Dimethoxy-3′, 4′-methylenedioxyflavone
5. Prosogerin-A ((7-OH, 6-OMe, 3′, 4′-OCH_2O-)
6. Kanzakiflavone-2 (5, 4′-diOH, 6, 7-OCH_2O-)
7. 5, 7, 5′-Trimethoxy-3′, 4′-methylenedioxyflavone*
8. Kanzakiflavone-1 (5, 8-diOH, 4′-OMe, 6, 7-OCH_2O-)
9. Linderoflavone A (5, 7-diOH, 6, 8-diOMe, 3′, 4′-OCH_2O-)
10. Linderoflavone B (5, 6, 7, 8-tetraOMe, 3′, 4′-OCH_2O-)
11. 5, 6, 7, 5′-Tetramethoxy-3′, 4′-methylenedioxyflavone*
12. Eupalestin (5, 6, 7, 8, 5′-pentaOMe, 3′, 4′-OCH_2O-)
13. 5, 6, 7, 3′, 4′, 5′-Hexamethoxy-3′, 4′-methylenedioxyflavone*

Methylenedioxyflavonols

1. 3, 5-Dimethoxy-6, 7-methylenedioxyflavone
2. Meliternatin (3, 5-diOMe, 6, 7-OCH_2O-, 3′, 4′-OCH_2O-)
3. Demethoxykanungin (3, 7-diOMe, 3′, 4′-OCH_2O-)
4. 3, 5, 8-Trimethoxy-6, 7-methylenedioxyflavone*
5. 3, 5, 8-Trimethoxy-3′, 4′-methylenedioxyflavone*
6. Gomphrenol (3, 5, 4′-triOH, 6, 7-OCH_2O-)
7. Kanugin (3, 7, 3′-triOMe, 3′, 4′-OCH_2O-)
8. Melisimplin (5-OH, 3, 6, 7-triOMe, 3′, 4′-OCH_2O-)
9. Melisimplexin (3, 5, 6, 7-tetraOMe, 3′, 4′-OCH_2O-)
10. 5-Hydroxy-3, 7, 8-trimethoxy-3′, 4′-methylenedioxyflavone
11. Meliternin (3, 5, 7, 8-tetraOMe, 3′, 4′-OCH_2O-)
12. 3, 5, 8, 3′-Tetramethoxy-6, 7-methylenedioxy-3′, 4-methylenedioxyflavone*
13. 5, 3′, 4′-Trihydroxy-3-methoxy-6, 7-methylenedioxyflavone
14. Wharangin (5, 3′, 4′-triOH, 3-OMe, 7, 8-OCH_2O-)
15. 5, 4′-Dihydroxy-3, 3′-dimethoxy-6, 7-methylenedioxyflavone*
16. Melinervin (3, 5, 7-triOH, 6, 8-diOMe, 3′, 4′-OCH_2O-)
17. 5-Hydroxy-3, 6, 7, 8-tetramethoxy-3′, 4′-methylenedioxyflavone
18. Melibentin (3, 5, 6, 7, 8-pentaOMe, 3′, 4′-OCH_2O-)

Isopentenyl flavones

1. *Trans*-Lanceolatin (7-OMe, 8-C_5-OH)
2. Tephrinone (5-OH, 7-OMe, 8-C_5)*
3. *Cis*-Tephrostachin
4. *Trans*-Tephrostachin (5, 7-diOMe, 8-C_5-OH)*
5. *Trans*-Anhydrotephrostachin (5, 7-diOMe, 8-C_5) *
6. 5-Methoxy-7, 8-di-*O*-isopentenylflavone*
7. 4′-Hydroxy-5-methoxy-7-*O*-C_5-flavone*
8. Kuwanone S (5, 7, 4′-triOH, 3′-C_{10})*
9. 5, 7-Dihydroxy-4′-methoxy-8, 3′-di-C_5-flavone*
10. 6, 7, 4′-Trimethoxy, 5-*O*-C_3-flavone
11. Rubraflavone A (7, 4′, 6′-triOH, 3-C_{10})
12. Rubraflavone B (7, 4′, 6′-triOH, 3-C_{10}, 8-C_5)
13. Artocarpesin (5, 7, 2′, 4′-tetraOH, 6-C_5)
14. Oxidihydroartocarpesin (5, 7, 2′, 4′-tetraOH, 6-C_5-OH)
15. Norartocarpin (mulberrin, 5, 7, 2′, 4′-tetraOH, 3, 6-di-C_5)
16. Kuwanone T (5, 7, 2′, 4′-tetraOH, 3, 3′-di-C_5)*
17. Integrin (5, 2′, 4′-triOH, 7-OMe, 3-C_5)
18. Artocarpin (5, 2′, 4′-triOH, 7-OMe, 3, 6-di-C_5)
19. Kuwanone C (5, 7, 2′, 4′-tetraOH, 3, 8-di-C_5)
20. 5, 7, 3′, 4′-Tetrahydroxy-8-C_5-flavone
21. Rubraflavone C (5, 7, 4′, 6′-tetraOH, 3-C_{10}, 6-C_5)
22. Asplenetin (5, 7, 3′, 4′, 5′-pentaOH, 3-C_5)

Isopentenylflavonols

1. Glepidotin A (3, 5, 7-triOH, 8-C_5)*
2. 3, 7, 4′-Trihydroxy-8-(C_5-OH)-flavone
3. 3, 5, 7, 8-Tetrahydroxy-6-C_5-flavone
4. 5-Hydroxy-3, 8-dimethoxy, 7-*O*-C_5-flavone*
5. Noranhydroicaritin (3, 5, 7, 4′-tetraOH, 8-C_5)
6. Noricaritin (3, 5, 7, 4′-tetraOH, 8-C_5-OH)
7. Isoanhydroicaritin (3, 5, 4′-triOH, 7-OMe, 8-C_5)
8. Icaritin (3, 5, 7-triOH, 4′-OMe, 8-C_5-OH)
9. Aliarin (5, 7, 4′-triOH, 3, 6-diOMe, 3′-C_5-OH)*
10. Viscosol (5, 7-diOH, 3, 6, 4′-triOMe, 3′-C_5)*
11. 5, 7-Dihydroxy-3, 6, 4′-trimethoxy-3′-(C_5-OH)-flavone*
12. 3, 5, 7, 3′, 4′-Pentahydroxy-6-(C_5-OH)-flavone
13. 3, 5, 7, 4′-Tetrahydroxy-3′-isopentenylflavone*
14. Broussoflavonol B (5, 7, 3′, 4′-tetraOH, 3-OMe, 6, 8-di-C_5)*
15. 5, 4′-Dihydroxy-3, 3′-dimethoxy-7-*O*-C_5-flavone
16. Rhynchospermin (3, 5, 3′-triOH, 7, 4′-diOMe, 8-C_5)
17. Broussoflavonol C (3, 5, 7, 3′, 4′-OH, 8, 2′, 6′-C_5)
18. 5, 4′-Dihydroxy-3, 7, 3′-trimethoxy-8-C_5-flavone

*C_5, isopentenyl; C_5-OH, hydroxyisopentanyl; C_{10}, geranyl.

Pyranoflavones

1. 5-Hydroxy-7, 8-ODmp-flavone
2. Isopongaflavone (5-OMe, 7, 8-ODmp)
3. Fulvinervin B (5-OH, 6-C_5, 7, 8-ODmp)
4. Fulvinervin C (5-OH, 6-C_5-OH, 7, 8-ODmp)
5. Isopongachromene (6-OMe, 7, 8-ODmp, 3′, 4′-OCH_2O-)*
6. 5, 4′-Dihydroxy-7, 6-ODmp-flavone
7. Cyclomulberrochromene (5, 4′-diOH, 7, 6-ODmp, 6′, 3-ODmp)
8. Cyclomorusin (5, 4′-diOH, 7, 8-ODmp, 6′, 3′-ODmp)
9. Cycloartocarpesin (5, 2′, 4′-triOH, 7, 6-ODmp)
10. Mulberrochromene (5, 2′, 4′-triOH, 3-C_5, 7, 6-ODmp)
11. Kuwanone B (5, 7, 2′-triOH, 3-C_5, 3′, 4′-ODmp)
12. Kuwanone A (5, 7, 4′-triOH, C-C_5, 2′, 3′-ODmp)
13. Morusin (5, 2′, 4′-triOH, 3-C_5, 7, 8-ODmp)
14. Oxydihydromorusin (5, 2′, 4′-triOH, 3-C_5-OH, 7, 8-ODmp)
15. Rubraflavone D (5, 2′, 4′-triOH, 3-C_{10}, 7, 6-ODmp)
16. Cyclomulberrin (5, 7, 4′-triOH, 6-C_5, 6′, 3-ODmp)

17. Cycloartocarpin
    (5, 4′-diOH, 7-OMe, 6-C_5, 6′, 3-ODmp)
18. Desmodol (5, 3′, 4′-triOH, 6-Me, 7, 8-ODmp)
19. Cycloheterophyllin
    (5, 3′, 4′-triOH, 8-C_5, 7, 6-ODmp, 6′, 3-ODmp)
20. Artobilichromene
    (5, 3′, 4′, 6′-tetraOH, 5′-C_5, 7, 6-ODmp)

Pyranoflavonols

1. Karanjachromene
   (Pongaflavone 3-OMe, 7, 8-ODmp)
2. Sericetin (3, 5-diOH, 8-C_5, 7, 6-ODmp)
3. Pongachromene
   (3, 5-diOMe, 3′, 4′-OCH_2O-, 7, 8-ODmp)
4. 3, 6-Dimethoxy-7, 8-ODmp-flavone
5. Macaflavone I (3, 3′, 4′-triOH, 7, 6-ODmp, 8-C_5)*
6. Macaflavone II
   (3′, 4′-diOH, 3-OMe, 7, 6-ODmp, 8-C_5)*
7. 3, 5, 3′, 4′-Tetrahydroxy-7, 8-ODmp-flavone
8. Broussoflavonol A
   (5, 3′, 4′-triOH, 3-OMe, 8-C_5, 7, 6-ODmp)*
9. Broussoflavonol D
   (3, 5, 7, 4′-tetraOH, 8, 6′-C_5, 3′, 2′-ODmp)

*ODmp, O-linked dimethylallyl.

Furanoflavones

1. Lanceolatin B (7, 8-fur)
2. Pongaglabrone (3′, 4′-OCH_2O-, 7, 8-fur)
3. Pongaglabol (5-OH, 7, 8-fur)
4. Pinnatin (5-OMe, 7, 8-fur)
5. Gamatin (5-OMe, 3′, 4′-OCH_2O-, 7, 8-fur)
6. 6-Methoxy-7, 8-fur-flavone
7. 8-OMe, 7, 6-fur-flavone
8. 2′-Methoxy-7, 8-fur-flavone*
9. 3′-Hydroxy-7, 8-fur-flavone
10. Isopongaglabol (4′-OH, 7, 8-fur)
11. 6-Methoxyisopongaglabol
    (4′-OH, 6-OMe, 7, 8-fur)*
12. Dihydrofurano-artobilichromene b_1 and b_2
    (5, 3′, 6′-triOH, 4′, 5′-OIPf)
13. Dihydrofurano-artobilichromene a
    (5, 3′, 4′-triOH, 6′, 5′-OIPf)
14. Mulberranol (5, 4′, 6′-triOH, 3-C_5, 7, 6-OIPf-OH)

*fur, furanyl: OIPf, isopentenylfuranyl.

Furanoflavonols

1. Karanjin (3-OMe, 7, 8-fur)
2. Pongapin (3-OMe, 3′, 4′-OCH_2O-, 7, 8-fur)
2. 3′-Methoxypongapin
   (3, 3′-diOMe, 4′, 5′-OCH_2O-, 7, 8-fur)

(8) FLAVONE AND FLAVONOL GLYCOSIDES

Flavone glycosides

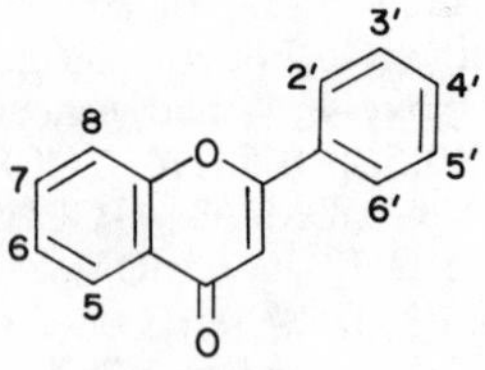

5, 7-Dihydroxyflavone (Chrysin)
1. 5-Xyloside*
2. 5-Glucoside (toringin)
3. 7-Glucoside
4. 7-Galactoside
5. 7-Glucuronide
6. 7-Rutinoside
7. 7-Gentiobioside*

6-Hydroxy-4′-methoxyflavone
8. 6-Arabinoside*

7, 4′-Dihydroxyflavone
9. 7-Glucoside
10. 7-Rutinoside

3′, 4′-Dihydroxyflavone
11. 4′-Glucoside

5, 6, 7-Trihydroxyflavone (Baicalein)
12. 6-Glucoside
13. 6-Glucuronide
14. 7-Rhamnoside
15. 7-Glucuronide

Baicalein 6-methyl ether
16. 7-Glucoside
17. 7-Glucuronide

Baicalein 7-methyl ether
18. 5-Glucuronide
19. 5-Glucuronosylglucoside

Baicalein 5, 6-dimethyl ether
20. 7-Glucoside*

5, 7-Dihydroxy-8-methoxyflavone (Wogonin)
21. 5-Glucoside*
22. 7-Glucoside*
23. 7-Glucuronide

7, 3′, 4′-Trihydroxyflavone
24. 7-Glucoside
25. 7-Galactoside*
26. 7-Rutinoside

7, 4′-Dihydroxy-3′-methoxyflavone
27. 7-Glucoside

5, 2′-Dihydroxy-7-methoxyflavone
28. 2′-Glucoside (echioidin)

5, 7, 4′-Trihydroxyflavone (Apigenin)
29. 5-Glucoside

30. 5-Galactoside*
31. 7-Arabinoside
32. 7-Xyloside
33. 7-Rhamnoside
34. 7-Glucoside (cosmosiin)
35. 7-Galactoside
36. 7-Glucuronide
37. 7-Galacturonide
38. 7-Methylglucuronide*
39. 7-Methylgalacturonide
40. 7-(6″-Ethylglucuronide)*
41. 4′-Arabinoside*
42. 4′-Glucoside
43. 4′-Glucuronide
44. 7-Arabinosylglucoside
45. 7-Xylosyl (1 → 2) glucoside
46. 7-Apiosylglucoside (apiin)
47. 7-Rutinoside
48. 7-Neohesperidoside
49. 7-Rhamnosylglucuronide
50. 7-Diglucoside
51. 7-Galactosyl (1 → 4) mannoside*
52. 7-Xylosylglucuronide
53. 7-Rhamnosylglucuronide
54. 7-Glucuronosyl (1 → 2) glucuronide*
55. 7, 4′-Diglucoside
56. 7, 4′-Dialloside*
57. 7, 4′-Diglucuronide
58. 7-Glucuronide-4′-rhamnoside
59. 4′-Diglucoside
60. 7-Rutinoside-4′-glucoside
61. 7-Neohesperidoside-4′-glucoside
62. 7-Diglucuronide-4′-glucuronide
63. 7-Neohesperidoside-4′-sophoroside
64. 7-(4″-*p*-Coumarylglucoside)
65. 7-(6″-*p*-Coumarylglucoside)*
66. 7-Caffeylglucoside*
67. 7-(Malonylglucoside)
68. 7-(2″-Acetylglucoside)
69. 7-(6″-Acetylglucoside)
70. 7-Glucoside-4′-*p*-coumarate
71. 7-Glucoside-4′-caffeate
72. 7-(2″, 6″-Di-*p*-coumarylglucoside)*
73. 7-(3″, 6″-Di-*p*-coumarylglucoside)*
74. 7-(4″, 6″-Di-*p*-coumarylglucoside)*
75. 7-(2″, 3″-Diacetylglucoside)*
76. 7-(3″, 4″-Diacetylglucoside)*
77. 7-(Malonylapiosyl) glucoside
78. 7-Rutinoside-4′-caffeate
79. 7-(6″-Acetylalloside)-4′-alloside*
80. 7-(4″, 6″-Diacetylalloside)-4′-alloside*
81. 7-Glucuronide-4′-(6″-malonylglucoside)*
82. 7-Sulphatoglucoside
83. 7-Sulphatogalactoside*
84. 7-Sulphate

Apigenin 7-methyl ether (Genkwanin)
85. 5-Glucoside
86. 4′-Glucoside
87. 5-Xylosylglucoside

Apigenin 4′-methyl ether (Acacetin)
88. 7-Glucoside (tilianine)
89. 7-Galactoside*
90. 7-Glucuronide
91. 7-(6″-Methylglucuronide)*
92. 7-Arabinosylrhamnoside*
93. 7-Rutinoside (linarin)
94. 7-Neohesperidoside (fortunellin)
95. 7-Diglucoside
96. 7-Rhamnosylgalacturonide
97. 7-Glucuronosyl (1 → 2)glucuronide
98. 7-(2″-Acetylglucoside)*
99. 7-(6″-Acetylglucoside)*
100. 7-(4″-Acetylrutinoside)
101. 7-[2‴-(2-Methylbutyryl)rutinoside]
102. 7-[3‴-(2-Methylbutyryl)rutinoside]
103. Di-6″-(acacetin-7-glucosyl)malonate*

Apigenin 5, 7-Dimethyl ether
104. 4′-Galactoside

Apigenin 7, 4′-Dimethyl ether
105. 5-Xylosylglucoside

6-Hydroxyapigenin (Scutellarein)
106. 5-Glucuronide
107. 6-Xyloside
108. 6-Glucoside*
109. 7-Rhamnoside
110. 7-Glucuronide
111. 4′-Arabinoside
112. 7-Xylosyl(1 → 4)rhamnoside
113. 7-Rutinoside
114. 7-Neohesperidoside*
115. 7-Diglucoside
116. 6-Xyloside-7-rhamnoside*
117. 7, 4′-Dirhamnoside
118. 7-(6″-Ferulylglucuronide)*
119. 7-(Sinapylglucuronide)

Scutellarein 6-methyl ether (Hispidulin)
120. 7-Glucoside (homoplantaginin)
121. 7-Glucuronide
122. 7-Rutinoside

Scutellarein 7-methyl ether
123. 6-Rhamnosylxyloside

Scutellarein 4′-methyl ether
124. 6-Glucoside
125. 7-Glucoside*
126. 7-Glucuronide
127. 7-Sophoroside
128. 7-[*p*-Coumarylglucosyl(1 → 2)mannoside]

Scutellarein 6, 7-dimethyl ether
129. 4′-Glucoside
130. 4′-Rutinoside

Scutellarein 6,4′-dimethyl ether (Pectolinarigenin)
131. 7-Glucoside
132. 7-Glucuronide
133. 7-Glucuronic acid methyl ester
134. 7-Rutinoside
135. 7-Acetylrutinoside
Scutellarein 7,4′-dimethyl ether
136. 6-Glucoside
Scutellarein 6,7,4′-trimethyl ether
137. 5-Glucoside
8-Hydroxyapigenin (isoscutellarein)
138. 8-Glucuronide
139. 7-Allosyl(1→2)glucoside*
140. 7-(6‴-Acetylallosyl) (1→2)glucoside*
8-Hydroxyapigenin 4′-methyl ether
141. 8-Glucuronide
142. 8-Xylosylglucoside
143. 7-(6‴-Acetylallosyl) (1→2)glucoside
8-Hydroxyapigenin 8,4′-dimethyl ether
144. 7-Glucuronide*
5,7,4′-Trihydroxy-6,8-dimethoxyflavone
145. 7-Glucoside*
7,3′,4′,5′-Tetrahydroxyflavone
146. 7-Rhamnoside*
147. 7-Glucoside*
5,7,3′,4′-Tetrahydroxyflavone (Luteolin)
148. 5-Glucoside (galuteolin)
149. 5-Galactoside
150. 5-Glucuronide*
151. 7-Xyloside*
152. 7-Rhamnoside*
153. 7-Glucoside
154. 7-Galactoside
155. 7-Glucuronide
156. 7-Galacturonide
157. 7-Methylglucuronide*
158. 3′-Xyloside*
159. 3′-Rhamnoside*
160. 3′-Glucoside
161. 3′-Glucuronide
162. 3′-Galacturonide
163. 4′-Arabinoside
164. 4′-Glucoside
165. 4′-Glucuronide
166. 7-Arabinosylglucoside
167. 7-Glucosylarabinoside
168. 7-Sambubioside*
169. 7-Apiosylglucoside
170. 7-Rutinoside
171. 7-Neohesperidoside (veronicastroside)
172. 7-Glucosylrhamnoside*
173. 7-Gentiobioside
174. 7-Laminaribioside
175. 7-Digalactoside
176. 7-Glucosylgalactoside
177. 7-Galactosylglucoside
178. 7-Glucosylglucuronide
179. 7-Glucuronosyl(1→2)glucuronide*
180. 7-Glucoside-3′-xyloside*
181. 7,3′-Diglucoside
182. 7-Glucuronide-3′-glucoside
183. 7,3′-Diglucuronide
184. 7,3′-Digalacturonide
185. 7,4′-Diglucoside*
186. 7-Galactoside-4′-glucoside
187. 7-Glucuronide-4′-rhamnoside
188. 7,4′-Diglucuronide
189. 3′,4′-Diglucoside
190. 3′,4′-Diglucuronide
191. 3′,4′-Digalacturonide
192. 7-Rhamnosyldiglucoside*
193. 7-Glucosylarabinoside-4′-glucoside
194. 7-Rutinoside-4′-glucoside
195. 7-Neohesperidoside-4′-glucoside
196. 7-Glucoside-4′-neohesperidoside
197. 7-Gentiobioside-4′-glucoside*
198. 7-Glucuronide-3′,4′-dirhamnoside
199. 7,4′-Diglucuronide-3′-glucoside
200. 7-Glucuronosyl(1→2)glucuronide-4′-glucuronide*
201. 7,3′,4′-Triglucuronide
202. 7-Neohesperidoside-4′-sophoroside
203. 7-(2″-*p*-Coumarylglucoside)*
204. 7-Caffeylglucoside*
205. 7-(6″-Malonylglucoside)*
206. 7-(3″-Acetylapiosyl)(1→2)xyloside*
207. 7-Glucuronide-3′-ferulylglucoside
208. 7,4′-Diglucuronide-3′-ferulylglucoside
209. 7-Sulphatoglucoside
210. 7-Sulphatoglucuronide*
211. 7-Sulphate-3′-glucoside
212. 7-Sulphatorutinoside
213. 7-Sulphate-3′-rutinoside
214. 7-Sulphate
215. 3′-Sulphate
216. 4′-Sulphate
217. 7,3′-Disulphate
Luteolin 5-methyl ether
218. 7-Glucoside
Luteolin 7-methyl ether
219. 5-Glucoside*
220. 4′-Rhamnoside
221. 5-Xylosylglucoside
222. 4′-Gentiobioside
Luteolin 3′-methyl ether (chrysoeriol)
223. 5-Glucoside
224. 7-Rhamnoside
225. 7-Glucoside
226. 7-Glucuronide
227. 4′-Glucoside
228. 5-Diglucoside

229. 7-Arabinofuranosyl(1 → 2)galactoside
230. 7-Apiosylglucoside
231. 7-Rutinoside
232. 7-Rhamnosylgalactoside*
233. 7-Rhamnosylglucuronide
234. 7-Digalactoside
235. 7-Allosyl(1 → 2)glucoside
236. 7-Glucuronosyl(1 → 2)glucuronide*
237. 5, 4′-Diglucoside
238. 7, 4′-Diglucoside
239. 7′-Glucuronide-4′-rhamnoside
240. 7-Sophorotrioside
241. 7-(Malonylglucoside)
242. 7-*p*-Coumarylglucosylglucuronide
243. 7-(6‴-Acetylglucosyl)(1 → 2)mannoside
244. 7-Sulphatoglucoside
245. 7-Sulphate
Luteolin 4′-methyl ether (Diosmetin)
246. 7-Glucoside
247. 7-Glucuronide
248. 7-α-[Glucosyl(1 → 6)arabinoside]
249. 7-β-[Glucosyl(1 → 6)arabinoside]
250. 7-Xylosylglucoside
251. 7-Rutinoside (diosmin)
252. 7-Diglucoside
253. 7-(6″-Malonylglucoside)*
254. 7-Sulphate
255. 3′-Sulphate
256. 7, 3′-Disulphate
Luteolin 7, 3′-dimethyl ether
257. 5-Rhamnoside*
258. 5-Glucoside*
259. 4′-Glucoside
260. 4′-Apiosylglucoside
Luteolin 7, 4′-dimethyl ether
261. 5-Xylosylglucoside
Luteolin 3′, 4′-dimethyl ether
262. 7-Rhamnoside
263. 7-Glucuronide
6-Hydroxyluteolin
264. 5-Glucoside*
265. 6-Xyloside
266. 6-Glucoside
267. 6-Glucuronide
268. 7-Arabinoside
269. 7-Xyloside
270. 7-Apioside
271. 7-Glucoside
272. 7-Galactoside
273. 7-Glucuronide
274. 7-Rhamnosyl(1 → 4)xyloside
275. 7-Gentiobioside*
276. 7-Arabinoside-4′-rhamnoside
277. 7-(6″-Malonylglucoside)*
6-Methoxyluteolin
278. 7-Glucoside
279. 7-Rutinoside
6-Hydroxyluteolin 7-methyl ether (Pedalitin)
280. 6-Glucoside (pedaliin)
281. 6-Galactoside
282. 7-Glucuronide*
283. 7-Methylglucuronide*
284. 6-Galactosylglucoside
6-Hydroxyluteolin 3′-methyl ether
285. 7-Diglucoside
6-Hydroxyluteolin 4′-methyl ether
286. 7-Allosyl(1 → 2)glucoside*
6-Hydroxyluteolin 6, 7-dimethyl ether (Cirsiliol)
287. 4′-Glucoside
6-Hydroxyluteolin 6, 3′-dimethyl ether
288. 7-Rhamnoside
289. 7-Glucoside
6-Hydroxyluteolin 6, 4′-dimethyl ether
290. 7-Glucoside
291. 7-Rutinoside
6-Hydroxyluteolin 7, 3′-dimethyl ether
292. 6-Glucoside
6-Hydroxyluteolin 6, 7, 3′-trimethyl ether (Cirsilineol)
293. 4′-Glucoside
8-Hydroxyluteolin (Hypolaetin)
294. 7-Glucoside
295. 8-Glucoside*
296. 8-Glucuronide
297. 7-Allosyl(1 → 2)glucoside*
298. 8, 4′-Diglucuronide
299. 8-Glucoside-3′-sulphate
300. 8-Sulphate
8-Hydroxyluteolin 3′-methyl ether
301. 7-Glucoside
302. 7-Allosyl(1 → 2)glucoside*
303. 7-Mannosyl(1 → 2)glucoside
8-Hydroxyluteolin 4′-methyl ether
304. 7-(6‴-Acetylallosyl)(1 → 2)glucoside*
305. 8-Glucoside*
306. 8-Glucoside-3′-sulphate*
5, 7, 3′, 4′-Tetrahydroxy-6, 8-dimethoxyflavone
307. 7-Glucoside
5, 8, 3′, 4′-Tetrahydroxy-6, 7-dimethoxyflavone
308. 8-Glucoside*
5, 7, 3′-Trihydroxy-6, 8, 4′-trimethoxyflavone (Acerosin)
309. 5-(6″-Acetylglucoside)
5, 7, 4′-Trihydroxy-6, 8, 3′-trimethoxyflavone (Sudachitin)
310. 7-Glucoside
311. 4′-Glucoside*
312. 7-(3-Hydroxy-3-methylglutarate)-4′-glucoside
5, 7, 3′, 4′, 5′-Pentahydroxyflavone (Tricetin)
313. 7-Glucoside
314. 3′-Glucoside
315. 3′-Sulphate

316. 7-Diglucoside*
317. 3′,5′-Diglucoside*
318. 3′-Disulphate

Tricetin 3′-methyl ether

319. 7-Glucoside

Tricetin 3′,5′-dimethyl ether (Tricin)

320. 5-Glucoside
321. 7-Glucoside
322. 7-Glucuronide
323. 4′-Glucoside
324. 5-Diglucoside
325. 7-Rutinoside
326. 7-Neohesperidoside
327. 7-Diglucoside
328. 7-Fructosylglucoside
329. 7-Rhamnosylglucuronide
330. 7-Rhamnosyl(1 → 2)galacturonide*
331. 7-Diglucuronide
332. 5,7-Diglucoside
333. 7-Rutinoside-4′-glucoside
334. 7-Sulphatoglucoside
335. 7-Sulphatoglucuronide*
336. 7-Disulphatoglucuronide*

3-(3-Methylbutyl)tricetin

337. 5-Neohesperidoside*

5,7,2′,4′,5′-Pentahydroxyflavone (Isoetin)

338. 7-Glucoside
339. 2′-Arabinoside
340. 2′-Glucoside
341. 4′ or 5′-Glucoside
342. 2′-Xylosylarabinosylglucoside

3-*C*-Methylapigenin

343. 5-Rhamnoside*

5,7,4′-Trihydroxy-3′-*C*-methylflavone

344. 4′-Rhamnoside

3-*C*-Methylluteolin

345. 5-Rhamnoside*

Flavonol Glycosides

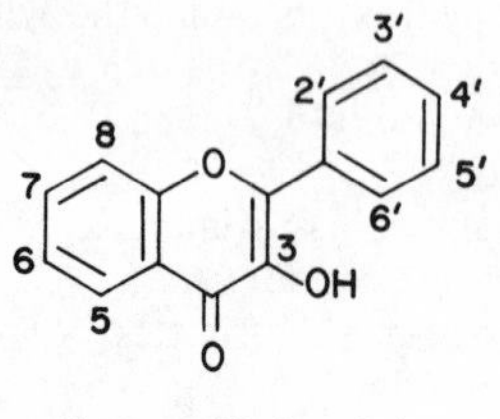

3,5,7-Trihydroxyflavone (Galangin)

1. 3-Glucoside
2. 7-Glucoside
3. 3-Rutinoside
4. 3-Galactosyl(1 → 4)rhamnoside

3,7,4′-Trihydroxyflavone

5. 3-Glucoside
6. 7-Glucoside
7. 4′-Glucoside
8. 7-Rutinoside

8-Hydroxygalangin 7-methyl ether

9. 8-Acetate
10. 8-Butyrate

3,7,3′,4′-Tetrahydroxyflavone (Fisetin)

11. 3-Glucoside
12. 7-Glucoside
13. 4′-Glucoside*
14. 7-Rutinoside

3,7,4′-Trihydroxy-3′-methoxyflavone (Geraldol)

15. 4′-Glucoside

3,5,7,4′-Tetrahydroxyflavone (Kaempferol)

16. 3-Arabinofuranoside (juglanin)
17. 3-Xyloside
18. 3-Rhamnoside (afzelin)
19. 3-Glucoside (astragalin)
20. 3-Galactoside (trifolin)
21. 3-Alloside (asiaticalin)
22. 3-Glucuronide
23. 3-(6″-Ethylglucuronide)*
24. 5-Rhamnoside*
25. 5-Glucoside
26. 7-Arabinoside
27. 7-Xyloside
28. 7-Rhamnoside
29. 7-Glucoside (populnin)
30. 4′-Rhamnoside
31. 4′-Glucoside
32. 3-Xylosylrhamnoside
33. 3-Rhamnosylxyloside
34. 3-Arabinosyl(1 → 6)galactoside
35. 3-Xylosylglucoside
36. 3-Xylosyl(1 → 2)galactoside*
37. 3-Apiosyl(1 → 2)glucoside*
38. 3-Glucosyl(1 → 4)rhamnoside
39. 3-Rutinoside
40. 3-Neohesperidoside
41. 3-Rhamnosyl(1 → 3)glucoside (rungioside)
42. 3-Sambubioside
43. 3-Gentiobioside
44. 3-Sophoroside
45. 3-Robinobioside
46. 3-Rhamnosyl(1 → 2)galactoside
47. 3-Glucosyl(1 → 6)galactoside
48. 3-Galactosylglucoside
49. 3-Glucosyl(1 → 2)galactoside*
50. 3-Digalactoside
51. 3,5-Digalactoside*
52. 3-Arabinoside-7-rhamnoside
53. 3-α-L-Arabinofuranoside-7-α-L-rhamnopyranoside*
54. 3-Rhamnoside-7-arabinoside
55. 3-Glucoside-7-arabinoside

56. 3-Xyloside-7-rhamnoside
57. 3-Xyloside-7-glucoside*
58. 3-Glucoside-7-xyloside
59. 3, 7-Dirhamnoside
60. 3-Rhamnoside-7-glucoside*
61. 3-Glucoside-7-rhamnoside
62. 3-Galactoside-7-rhamnoside*
63. 3-Glucoside-7-galactoside*
64. 3, 7-Diglucoside
65. 3-Glucuronide-7-glucoside*
66. 3, 4′-Dixyloside*
67. 3-Rhamnoside-4′-arabinoside
68. 7, 4′-Dirhamnoside*
69. 3-Glucosyl-β-(1 → 4)arabinofuranosyl-α-(1 → 2)-arabinopyranoside (primflasine)
70. 3-Glucosylxylosylarabinoside
71. 3-Rhamnosyl(1 → 2)rhamnosyl(1 → 6)glucoside
72. 3-(2^G-Rhamnosylrutinoside)
73. 3-(2^G-Rhamnosylgentiobioside)
74. 3-(2^G-Glucosylrutinoside)
75. 3-(2′-Rhamnosyllaminaribioside)
76. 3-Rhamnosylglucosylgalactoside
77. 3-(2^G-Glucosylgentiobioside)
78. 3-Glucosyl(1 → 2)gentiobioside
79. 3-Sophorotrioside
80. 3-β-Maltosyl(1 → 6)glucoside
81. 4′-Rhamnosyl(1 → 2)-[rhamnosyl(1 → 6)galactoside]*
82. 3-Rhamnosylarabinoside-7-rhamnoside
83. 3-Rhamnosylgalactoside-7-arabinoside
84. 3-Glucosylxyloside-7-xyloside
85. 3-β-D-Apiofuranosyl(1 → 2)-α-L-arabinofuranosyl-7-α-L-rhamnopyranoside*
86. 3-Rhamnosylxyloside-7-glucoside
87. 3-Glucosylrhamnoside-7-rhamnoside
88. 3-Rutinoside-7-rhamnoside
89. 3-Glucosyl(1 → 3)rhamnoside-7-rhamnoside*
90. 3-Robinobioside-7-rhamnoside (robinin)
91. 3-Robinobioside-7-glucoside
92. 3-Lathyroside-7-rhamnoside
93. 3-Rutinoside-7-glucoside
94. 3-Neohesperidoside-7-glucoside*
95. 3-Rutinoside-7-galactoside
96. 3-Rutinoside-7-glucuronide
97. 3-Sophoroside-7-rhamnoside
98. 3-Sophoroside-7-glucoside
99. 3-Gentiobioside-7-glucoside*
100. 3-Glucoside-7-gentiobioside
101. 3-Sambubioside-7-glucoside
102. 3-Rutinoside-4′-glucoside*
103. 3-Glucoside-7, 4′-dirhamnoside*
104. 3, 7, 4′-Triglucoside
105. 3-Sophorotrioside-7-rhamnoside
106. 3-Galactosyl(1 → 6)glucoside-7-dirhamnoside (malvitin)*
107. 3-Sophorotrioside-7-glucoside*
108. 3-Rutinoside-4′-diglucoside
109. 3-(p-Coumarylarabinoside)
110. 3-(X″-p-Coumarylglucoside) (tiliroside)
111. 3-(3″-p-Coumarylglucoside)
112. 3-(6″-p-Coumarylglucoside) (tribuloside)
113. 3-(2″-Galloylglucoside)
114. 3-(6″-Galloylglucoside)
115. 3-p-Hydroxybenzoylglucoside
116. 3-Benzoylglucoside
117. 3-(6″-Succinylglucoside)
118. 3-(6″-Malonylglucoside)*
119. 3-(6″-Malonylgalactoside)*
120. 7-Galloylglucoside*
121. 7-(6″-Succinylglucoside)
122. 3-(2″, 4″-Di-p-coumarylglucoside)*
123. 3-(2″, 6″-Di-p-coumarylglucoside)*
124. 3-(3″-p-Coumaryl-6″-ferulylglucoside)
125. 3-(6″-p-Coumarylacetylglucoside)
126. 3-(2^G-p-Coumarylrutinoside)*
127. 3-(Ferulylsophoroside) (petunoside)
128. 3-[2‴-Acetylarabinosyl(1 → 6)galactoside]*
129. 3-(3″-Acetylarabinofuranoside)-7-rhamnoside*
130. 3-Glucoside-7-(p-coumarylglucoside)*
131. 3-(Caffeylglucoside)-7-glucoside*
132. 3-Ferulylglucoside-7-glucoside*
133. 3-Benzoylglucoside-7-glucoside
134. 3-(p-Hydroxybenzoylglucoside)-7-glucoside
135. 3-(6″-Acetylglucoside)-7-rhamnoside
136. 3-(6″-Acetylglucoside)-7-glucoside
137. 3-(p-Coumarylglucoside)-4′-glucoside*
138. 3-(2-Hydroxypropionylglucoside)-4′-glucoside*
139. 3-[Glucosyl(1 → 3)-4‴-acetylrhamnosyl(1 → 6)-galactoside]
140. 3-(p-Coumarylsophorotrioside)
141. 3-(Ferulylsophorotrioside)
142. 3-(p-Coumarylrutinoside)-7-glucoside*
143. 3-(4″-Caffeyllaminaribioside)-7-rhamnoside
144. 3-Sinapylsophoroside-7-glucoside
145. 3-(p-Coumarylglucoside)-7, 4′-diglucoside*
146. 3-(p-Coumarylferulyldiglucoside)-7-rhamnoside*
147. 3-Gentiobioside-7-(caffeylarabinosylrhamnoside)*
148. 3-Sulphatorhamnoside
149. 3-α-(6″-Sulphatoglucoside)
150. 3-β-(3″-Sulphatoglucoside)*
151. 3-β-(6″-Sulphatoglucoside)*
152. 3-Glucuronide-7-sulphate
153. 3-Sulphatorutinoside*
154. 3-(6″-Sulphatogentiobioside)
155. 3-Sulphate*
156. 7-Sulphate
157. 3, 7-Disulphate
158. 3, 7, 4′-Trisulphate*

Kaempferol 3-methyl ether

159. 7-Rhamnoside*

160. 7-Glucoside*
161. 7-Rutinoside
Kaempferol 5-methyl ether
162. 3-Galactoside
Kaempferol 7-methyl ether (Rhamnocitrin)
163. 3-Rhamnoside*
164. 3-Glucoside
165. 3-Galactoside*
166. 3-Glucuronide
167. 5-Glucoside*
168. 3-Rutinoside
169. 3-[Rhamnosyl(1→3)-rhamnosyl(1→6)-galactoside] (alaternin = catharticin)
170. 3-Sulphate
Kaempferol 4′-methyl ether (Kaempferide)
171. 3-Galactoside*
172. 3-Glucuronide
173. 3-Diglucoside
174. 3-Rhamnoside-7-glucoside*
175. 3-Glucoside-7-rhamnoside*
176. 3,7-Diglucoside*
Kaempferol 3,4′-dimethyl ether
177. 7-Glucoside
Kaempferol 7,4′-dimethyl ether
178. 3-[6″-(*E*)-*p*-Coumarylglucoside]*
179. 3-Sulphate
6-*C*-Methylkaempferol
180. 3-Glucoside
6-Hydroxykaempferol
181. 7-Glucoside
6-Hydroxykaempferol 3-methyl ether
182. 6-Glucoside
183. 7-Glucoside
184. 7-Sulphate*
6-Hydroxykaempferol 5-methyl ether
185. 4′-Rhamnoside
186. 3-Arabinosylrhamnoside
6-Hydroxykaempferol 6-methyl ether (Eupafolin)
187. 3-Glucoside
188. 3-Galactoside
189. 7-Glucoside
190. 3-Rutinoside
191. 3-Robinobioside*
6-Hydroxykaempferol 7-methyl ether
192. 6-Rhamnosyl(1→4)xyloside
6-Hydroxykaempferol 4′-methyl ether
193. 3,7-Dirhamnoside
6-Hydroxykaempferol 3,6-dimethyl ether
194. 7-Glucoside
6-Hydroxykaempferol 6,7-dimethyl ether (Eupalitin)
195. 3-Rhamnoside (eupalin)
196. 3-Galactoside
197. 3-Galactosylrhamnoside*
198. 3-Diglucoside
199. 3-Glucosylgalactoside*
6-Hydroxykaempferol 3,6,7-trimethyl ether
200. 4′-Glucoside
6-Hydroxykaempferol 6,7,4′-trimethyl ether (Mikanin)
201. 3-Glucoside*
202. 3-Galactoside
8-Hydroxykaempferol (Herbacetin)
203. 3-Glucoside
204. 7-Arabinoside
205. 7-Rhamnoside*
206. 7-Glucoside
207. 8-α-L-Arabinopyranoside*
208. 8-Xyloside
209. 8-Rhamnoside
210. 8-Glucoside
211. 4′-Glucoside
212. 7-Glucosyl(1→3)rhamnoside*
213. 8-Rutinoside
214. 8-Gentiobioside
215. 3-Glucoside-8-xyloside*
216. 7-Rhamnoside-8-glucoside*
217. 8,4′-Dixyloside*
218. 8-Arabinoside-4′-xyloside
219. 3-Sophoroside-8-glucoside*
220. 7-(6″-Quinylglucoside)*
221. 8-(3″-Acetyl-α-L-arabinopyranoside)*
222. 8-(3″-Acetylxyloside)
223. 8-(2″, 3″-Diacetylxyloside)*
224. 8-(Diacetylglucoside)
225. 8-(2″, 3″, 4″-Triacetylxyloside)*
226. 8-Acetate
227. 8-Butyrate
Herbacetin 7-methyl ether
228. 8-Acetate
229. 8-Butyrate
Herbacetin 8-methyl ether (Sexangularetin)
230. 3-Glucoside
231. 3-Galactoside*
232. 3-Rutinoside
233. 3-Glucoside-7-rhamnoside*
234. 3-Rhamnosylglucoside-7-rhamnoside
235. 3-Glucoside-7-rutinoside*
Herbacetin 7,8-dimethyl ether
236. 3-Rhamnoside
Herbacetin 7,4′-dimethyl ether
237. 8-Acetate
238. 8-Butyrate
3,5,7,3′,4′-Pentahydroxyflavone (Quercetin)
239. 3-α-L-Arabinofuranoside (avicularin)
240. 3-α-L-Arabinopyranoside (guaijaverin, foeniculin)
241. 3-β-L-Arabinoside (polystachioside)
242. 3-Xyloside (reynoutrin)
243. 3-Rhamnoside (quercitrin)
244. 3-Glucoside (isoquercitrin)
245. 3-Galactoside (hyperin)

246. 3-Alloside*
247. 3-Glucuronide (miquelianin)
248. 3-Galacturonide
249. 3-(6″-Methylglucuronide)*
250. 3-(6″-Ethylglucuronide)*
251. 5-Glucoside
252. 7-Arabinoside
253. 7-Xyloside*
254. 7-Rhamnoside
255. 7-Glucoside (quercimeritrin)
256. 7-α-Galactoside
257. 3′-Xyloside
258. 3′-Glucoside
259. 4′-Glucoside (spiraeoside)
260. 3-Diarabinoside
261. 3-Arabinosylxyloside*
262. 3-Rhamnosyl(1→2)arabinoside*
263. 3-Arabinosyl(1→6)glucoside (vicianoside)
264. 3-Arabinosyl(1→6)galactoside
265. 3-Galactosylarabinoside
266. 3-Dixyloside
267. 3-Xylosyl(1→2)galactoside*
268. 3-Apiofuranosyl(1→2)arabinoside*
269. 3-Apiofuranosyl(1→2)xyloside*
270. 3-Rhamnosylxyloside
271. 3-Xylosylglucoside
272. 3-Apiofuranosyl(1→2)glucoside*
273. 3-Apiofuranosyl(1→2)galactoside*
274. 3-Rutinoside (rutin)
275. 3-Neohesperidoside
276. 3-Galactosyl(1→4)rhamnoside*
277. 3-Rhamnosyl(1→6)galactoside
278. 3-Gentiobioside
279. 3-Sophoroside
280. 3-Sambubioside*
281. 3-Galactosylglucoside
282. 3-Glucosyl(1→2)galactoside*
283. 3-Glucosyl(1→6)galactoside*
284. 3-Digalactoside
285. 3-Glucosylmannoside
286. 3-Glucosylglucuronide
287. 3-Galactosylglucuronide
288. 7-Rutinoside
289. 7-Glucosylrhamnoside*
290. 3,5-Digalactoside
291. 3-Arabinoside-7-glucoside
292. 3-Glucoside-7-arabinoside
293. 3-Xyloside-7-glucoside
294. 3-Galactoside-7-xyloside
295. 3,7-Dirhamnoside
296. 3-Rhamnoside-7-glucoside
297. 3-Glucoside-7-rhamnoside
298. 3-Galactoside-7-rhamnoside
299. 3,7-Diglucoside
300. 3-Galactoside-7-glucoside
301. 3-Glucuronide-7-glucoside*
302. 3,7-Diglucuronide
303. 3,3′-Diglucoside
304. 3,4′-Diglucoside
305. 7,4′-Diglucoside
306. 3-(2^G-Apiosylrutinoside)
307. 3-Rhamnosyl(1→2)rutinoside
308. 3-(2^G-Rhamnosylrutinoside)
309. 3-(2^G-Glucosylrutinoside)
310. 3-(2^G-Rhamnosylgentiobioside)
311. 3-Glucosyl(1→4)galactosylrhamnoside*
312. 3-Sophorotrioside
313. 3-(2^G-Glucosylgentiobioside)
314. 3-Dixyloside-7-glucoside
315. 3-Xylosylglucoside-7-glucoside
316. 3-Rutinoside-7-rhamnoside*
317. 3-Robinobioside-7-rhamnoside
318. 3-Rutinoside-7-glucoside
319. 3-Glucoside-7-rutinoside
320. 3-Rutinoside-7-galactoside
321. 3-Robinobioside-7-glucoside
322. 3-Rutinoside-7-glucuronide
323. 3-Gentiobioside-7-glucoside
324. 3-Sophoroside-7-glucoside
325. 3-Sambubioside-7-glucoside
326. 3-Sambubioside-3′-glucoside
327. 3-Xylosyl(1→2)rhamnoside-4′-rhamnoside*
328. 3-Rutinoside-4′-glucoside*
329. 3-(2^G-Rhamnosylrutinoside)-7-glucoside
330. 3-Rhamnosyldiglucoside-7-glucoside
331. 3-Rutinoside-7,3′-bisglucoside
332. 3-Rutinoside-4′-diglucoside
333. 3-(3″-*p*-Coumarylglucoside)
334. 3-(6″-*p*-Coumarylglucoside) (helichrysoside)
335. 3-Isoferulylglucuronide
336. 3-(6″-Caffeylgalactoside)*
337. 3-(2″-Galloylrhamnoside)*
338. 3-(2″-Galloylglucoside)
339. 3-(6″-Galloylglucoside)
340. 3-(2″-Galloylgalactoside)
341. 3-(6″-Galloylgalactoside)
342. 3-(6″-*p*-Hydroxybenzoylgalactoside)*
343. 3-(6″-Malonylglucoside)*
344. 3-(6″-Malonylgalactoside)*
345. 3-(3″-Acetylrhamnoside)
346. 3-(4″-Acetylrhamnoside)
347. 3-(6″-Acetylglucoside)
348. 3-(3″-Acetylgalactoside)*
349. 3-(6″-Acetylgalactoside)*
350. 7-Acetyl-3′-glucoside
351. 3-(3″,6″-Di-*p*-coumarylglucoside)
352. 3-(X″-Benzoyl-X″-xylosylglucoside)*
353. 3-(X″ or X‴-Benzoyl-X″-glucosylglucoside)*
354. 3-Ferulylglucoside-7-glucoside*
355. 3-Ferulylglucoside-4′-glucoside*

356. 3-Malonylglucoside-4′-glucoside*
357. 3-(*p*-Coumarylsophorotrioside)
358. 3-(Ferulylsophorotrioside)
359. 3-(3‴-Benzoylsophoroside)-7-rhamnoside*
360. 3-Glucosyl(1 → 3)-4‴-acetylrhamnosyl(1 → 6)-galactoside
361. 3-*p*-Coumarylsophoroside-7-rhamnoside
362. 3-Caffeylsophoroside-7-glucoside
363. 3-Acetylsophoroside-7-rhamnoside
364. 3-Rhamnoside-7-glucoside-4′-(caffeylgalactoside)
365. 3-Ferulylglucoside-7, 4′-diglucoside*
366. 3-Sulphatorhamnoside
367. 3-(3″-Sulphatoglucoside)*
368. 3-Glucuronide-7-sulphate
369. 3-Acetyl-7, 3′, 4′-trisulphate
370. 3-Sulphate
371. 3′, 4′-Disulphate
372. 3, 7, 3′-Trisulphate*
373. 3, 7, 4′-Trisulphate
374. 3, 7, 3′, 4′-Tetrasulphate

Quercetin 3-methyl ether
375. 7-Rhamnoside*
376. 7-Glucoside
377. 3′-Xyloside*
378. 4′-Glucoside*
379. 7-Rhamnosyl-3′-xyloside*
380. 7-Diglucoside-4′-glucoside

Quercetin 5-methyl ether (Azaleatin)
381. 3-Arabinoside
382. 3-Rhamnoside (azalein)
383. 3-Galactoside
384. 3-Glucuronide
385. 3-Xylosylarabinoside
386. 3-Rhamnosylarabinoside
387. 3-Arabinosylgalactoside
388. 3-Rutinoside
389. 3-Diglucoside

Quercetin 7-methyl ether (Rhamnetin)
390. 3-α-L-Arabinofuranoside
391. 3-α-L-Arabinopyranoside
392. 3-Glucoside
393. 3-Galactoside
394. 5-Glucoside*
395. 3′-Glucuronide
396. 3-α-Diarabinoside
397. 3-β-Diarabinoside
398. 3-Rutinoside
399. 3-Neohesperidoside*
400. 3-Galactosyl(1 → 4)galactoside*
401. 3-Galactosyl(1 → 6)galactoside*
402. 3-Galactoside-3′-rhamnoside*
403. 3-Rhamnoside
404. 3-Rhamnosyl(1 → 2)rhamnosyl(1 → 6)galactoside (xanthorhamnin A and B)
405. 3-Rhamnosyl(1 → 3)-4″-acetylrhamnosyl(1 → 6)-galactoside*
406. 3-Sulphate
407. 3, 5, 4′-Trisulphate-3′-glucuronide

Quercetin 3′-methyl ether (Isorhamnetin)
408. 3-α-L-Arabinofuranoside
409. 3-α-L-Arabinopyranoside (distichin)
410. 3-Xyloside
411. 3-Glucoside
412. 3-Galactoside
413. 3-Glucuronide
414. 5-Glucoside
415. 7-Rhamnoside*
416. 7-Glucoside*
417. 3-Arabinosyl(1 → 2)rhamnoside
418. 3-Arabinosyl(1 → 6)glucoside
419. 3-Xylosylglucoside
420. 3-Dirhamnoside
421. 3-Rutinoside (narcissin)
422. 3-Neohesperidoside*
423. 3-Robinobioside*
424. 3-Sophoroside
425. 3-Gentiobioside
426. 3-Lactoside
427. 3-Arabinoside-7-glucoside
428. 3-Glucoside-7-arabinoside
429. 3-Glucoside-7-xyloside
430. 3-Glucoside-7-rhamnoside
431. 3, 7-Diglucoside
432. 3-Galactoside-7-glucoside
433. 3-Glucoside-4′-rhamnoside
434. 3, 4′-Diglucoside (dactylin)
435. 3-Galactoside-4′-glucoside
436. 3-Xylosylrutinoside
437. 3-Xylosylrhamnosylgalactoside
438. 3-Rhamnosylrutinoside
439. 3-Rutinosylglucoside
440. 3-Glucosylrhamnosylgalactoside
441. 3-Galactosylrhamnosylgalactoside
442. 3-Arabinoside-7-rhamnoside
443. 3-Rutinoside-7-glucoside
444. 3-Sophoroside-7-rhamnoside
445. 3-Sophoroside-7-glucoside
446. 3-Glucoside-7-gentiobioside
447. 3-Gentiobioside-7-glucoside*
448. 3-Rutinoside-4′-rhamnoside*
449. 3-Gentiotrioside-7-glucoside*
450. 3-*p*-Coumarylglucoside*
451. 3-(6″-Acetylglucoside)
452. 3-(6″-Acetylgalactoside)
453. 3-Sulphatorutinoside
454. 3-Glucuronide-7-sulphate
455. 3-Sulphate
456. 7-Sulphate
457. 3, 7-Disulphate
458. 3, 7, 4′-Trisulphate*

Quercetin 4′-methyl ether (Tamarixetin)
459. 3-Rhamnoside*

460. 3-Glucoside
461. 3-Digalactoside*
462. 3-Sulphate
Quercetin 3, 5-dimethyl ether (Caryatin)
463. 3′-(or 4′-)Glucoside
Quercetin 3, 3′-dimethyl ether
464. 7-Glucoside*
465. 4′-Glucoside*
Quercetin 5, 3′-dimethyl ether
466. 3-Glucoside
Quercetin 7, 3′-dimethyl ether (Rhamnazin)
467. 3-Glucoside
468. 3-Rhamnoside
469. 4′-Glucoside*
470. 3-Rutinoside*
471. 3-Neohesperidoside
472. 3-Rhamnosyl(1 → 4)rhamnosyl(1 → 6)galactoside (xanthorhamnin)
473. 3-Sulphate
Quercetin 7, 4′-dimethyl ether (Ombuin)
474. 3-Galactoside
475. 3-Rutinoside (ombuoside)
476. 3, 5-Diglucoside
477. 3-Rutinoside-5-glucoside*
Quercetin 3′, 4′-dimethyl ether
478. 3-Rutinoside
Quercetin 7, 3′, 4′-trimethyl ether
479. 3-Arabinoside*
480. 3-Digalactoside
Quercetin 5, 7, 3′, 4′-tetramethyl ether
481. 3-Arabinoside*
6-Hydroxyquercetin (Quercetagetin)
482. 3-Rhamnoside
483. 3-Glucoside (tagetiin)
484. 7-Glucoside
485. 3, 7-Diglucoside
Quercetagetin 3-methyl ether
486. 7-Glucoside
487. 7-Sulphate
Quercetagetin 6-methyl ether (Patuletin)
488. 3-Xyloside
489. 3-Rhamnoside
490. 3-Glucoside
491. 3-Galactoside
492. 3-Glucuronide
493. 5-Glucoside*
494. 7-Glucoside
495. 3-Rutinoside
496. 3-Robinobioside*
497. 3-Galactosylrhamnoside*
498. 3-Diglucoside
499. 3-Digalactoside*
500. 3, 7-Dirhamnoside
501. 3-Digalactosylrhamnoside*
502. 3-Acetylglucoside
503. 3-Glucoside-7-sulphate
504. 3-Sulphate
505. 7-Sulphate
Quercetagetin 7-methyl ether
506. 3-Glucoside
Quercetagetin 3′-methyl ether
507. 7-Glucoside
Quercetagetin 4′-methyl ether
508. 3-Arabinoside*
Quercetagetin 3, 6-dimethyl ether
509. 7-Glucoside (axillaroside)
510. 4′-Glucuronide*
Quercetagetin 3, 7-dimethyl ether
511. 6-Glucoside
512. 6-Galactoside
513. 4′-Glucoside
Quercetagetin 6, 7-dimethyl ether
514. 3-Rhamnoside (eupatolin)
515. 3-Glucoside
516. 3-Galactoside
517. 3-Glucosylgalactoside*
518. 3-Sulphate
Quercetagetin 3, 3′-dimethyl ether
519. 7-Glucoside*
Quercetagetin 6, 3′-dimethyl ether (Spinacetin)
520. 7-Glucoside
521. 3-Rutinoside
Quercetagetin 3, 6, 7-trimethyl ether
522. 4′-Glucoside
Quercetagetin 3, 6, 3′-trimethyl ether (Jaceidin)
523. 7-Glucoside (jacein)
524. 7-Neohesperidoside*
525. 4′-Sulphate*
Quercetagetin 3, 6, 4′-trimethyl ether
526. 7-Glucoside (centaurein)*
Quercetagetin 3, 7, 4′-trimethyl ether
527. 6-Glucoside
528. 3′-Glucoside
Quercetagetin 6, 7, 3′-trimethyl ether (Veronicafolin)
529. 3-Rutinoside
530. 3-Digalactoside
531. 3-Sulphate
Quercetagetin 6, 7, 4′-trimethyl ether (Eupatin)
532. 3-Sulphate
Quercetagetin 6, 3′, 4′-trimethyl ether
533. 3-Sulphate*
Quercetagetin 7, 3′, 4′-trimethyl ether
534. 3-Rhamnoside
Quercetagetin 3, 6, 7, 3′-tetramethyl ether
535. 4′-Glucoside (chrysosplenin)
Quercetagetin 3, 6, 7, 3′, 4′-pentamethyl ether (Artemetin)
536. 5-Glucosylrhamnoside
8-Hydroxyquercetin (Gossypetin)
537. 3-Glucoside
538. 3-Galactoside
539. 3-Glucuronide

540. 7-Rhamnoside*
541. 7-Glucoside
542. 8-Rhamnoside*
543. 8-Glucoside
544. 8-Glucuronide*
545. 7-Rhamnoside-8-glucoside*
546. 3-Gentiotrioside
547. 3-Sophoroside-8-glucoside*
548. 8-Glucuronide-3-sulphate*
549. 3-Sulphate
Gossypetin 7-methyl ether
550. 3-Arabinoside
551. 3-Rhamnoside
552. 3-Galactoside
553. 8-Glucoside
554. 3-Rutinoside*
555. 3-Galactoside-8-glucoside
Gossypetin 8-methyl ether (Corniculatusin)
556. 3-α-L-Arabinofuranoside*
557. 3-Glucoside*
558. 3-Galactoside
559. 7-Glucoside *
Gossypetin 3′-methyl ether
560. 7-Glucoside*
561. 3-Rutinoside
562. 7-Neohesperidoside (Haploside F)*
563. 7-(6″-Acetylglucoside)
564. 7-[6″-Acetylrhamnosyl (1 → 2) glucoside]*
Gossypetin 7, 4′-dimethyl ether
565. 8-Glucoside
566. 8-Acetate
567. 8-Butyrate
Gossypetin 8, 3′-dimethyl ether (Limocitrin)
568. 3-Rhamnoside*
569. 3-Glucoside*
570. 3-Rutinoside
571. 7-Neohesperidoside*
572. 7-(6″-Acetylglucoside)*
3, 5, 7, 3′, 4′, 5′-Hexahydroxyflavone (Myricetin)
573. 3-Arabinoside
574. 3-Xyloside
575. 3-Rhamnoside (myricitrin)
576. 3-Glucoside
577. 3-Galactoside
578. 7-Glucoside
579. 3′-Arabinoside
580. 3′-Xyloside*
581. 3′-Glucoside
582. 3-Dixyloside
583. 3-Dirhamnoside*
584. 3-Xylosylglucoside
585. 3-Rutinoside
586. 3-Diglucoside
587. 3-Galactosylglucoside
588. 3-Digalactoside
589. 3-Rhamnoside-7-glucoside*
590. 3, 3′-Digalactoside
591. 3-Rhamnosylrutinoside
592. 3-Glucosylrutinoside*
593. 3-Triglucoside
594. 3-Rutinoside-7-rhamnoside
595. 3-Robinobioside-7-rhamnoside
596. 3-Rutinoside-7-glucoside
597. 3-(2″-Galloylrhamnoside)*
598. 3-(4″-Galloylrhamnoside)*
599. 3-(6″-Galloylglucoside)
600. 3-(6″-Galloylgalactoside)
601. 3-Sulphatorhamnoside
Myricetin 3-methyl ether
602. 7-Rhamnoside*
603. 3′-Xyloside*
604. 3′-Glucoside
605. 7-Rhamnoside-3′-xyloside
Myricetin 5-methyl ether
606. 3-Rhamnoside
607. 3-Galactoside
Myricetin 7-methyl ether (Europetin)
608. 3-Rhamnoside
Myricetin 3′-methyl ether (Larycitrin)
609. 3-Rhamnoside
610. 3-Glucoside
611. 3-Galactoside
612. 7-Glucoside*
613. 3-Rutinoside
514. 3, 5′-Diglucoside*
615. 3-Rhamnosylrutinoside
616. 3-Rutinoside-7-glucoside
617. 3, 7, 5′-Triglucoside*
618. 3-*p*-Coumarylglucoside
Myricetin 4′-methyl ether
619. 3-Rhamnoside (mearnsitrin)
620. 3-Galactosyl (1 → 4) galactoside
Myricetin 3, 4′-dimethyl ether
621. 3′-Xyloside*
622. 7-Rhamnoside-3′-xyloside*
Myricetin 7, 4′-dimethyl ether
623. 3-Galactoside*
Myricetin 3′, 5′-dimethyl ether (Syringetin)
624. 3-Arabinoside*
625. 3-Rhamnoside
626. 3-Glucoside
627. 3-Galactoside*
628. 3-Rutinoside
629. 3-Rhamnosylrutinoside
630. 3-Rutinoside-7-glucoside
631. 3-*p*-Coumarylglucoside
6-Hydroxymyricetin 6, 3′, 5′-trimethyl ether
672. 3-Glucoside
6-Hydroxymyricetin 3, 6, 3′, 5′-tetramethyl ether
633. 7-Glucoside
8-Hydroxymyricetin (Hibiscetin)
634. 3-Glucoside

635. 8-Glucosylxyloside
3, 5, 7, 2′-Tetrahydroxyflavone (Datiscetin)
636. 3-Glucoside*
637. 3-Rutinoside
3, 7, 2′, 3′, 4′-Pentahydroxyflavone
638. 3-Neohesperidoside*
5, 2′-Dihydroxy-3, 7, 4′-trimethoxyflavone
639. 2′-Glucoside
5, 2′-5′-Trihydroxy-3, 7, 4′-trimethoxyflavone
640. 2′-Glucoside
5, 6′, 5′-Trihydroxy-3, 7, 4′-trimethoxyflavone
641. 5′-Glucoside*
5, 2′-Dihydroxy-3, 7, 4′, 5′-tetramethoxyflavone
642. 2′-Glucoside
5, 5′-Dihydroxy-3, 6, 7, 4′-tetramethoxyflavone
643. 5′-Glucoside*
5, 8, 4′-Trihydroxyflavone-3, 7, 3′-trimethoxyflavone
644. 8-Acetate*
5, 8-Dihydroxy-3, 7, 2′, 3′, 4′-pentamethoxyflavone
645. 8-Acetate*
5, 2′, 5′-Trihydroxy-3, 6, 7, 4′-tetramethoxyflavone
646. 5′-Glucoside*
5, 5′-Dihydroxy-3, 6, 7, 2′, 4′-pentamethoxyflavone
647. 5′-Glucoside*

(9) CHALCONES, AURONES, DIHYDROCHALCONES, FLAVANONES AND DIHYDROFLAVONOLS

Chalcones

1. 4′-Methoxychalcone
2. 4′-Hydroxychalcone glucoside*
3. 2′, 4′-Dihydroxychalcone
4. 4′-Hydroxy-2′-methoxychalcone*
5. 2′-Hydroxy-4′-prenyloxychalcone
6. 2′-Methoxy-4′-*O*-prenyloxychalcone*
7. 2′, 4′-Dihydroxy-3′-*C*-prenylchalcone
8. 2′-Hydroxy-3′-*C*-prenyl-4′-methoxychalcone
9. 2′, 4′-Dihydroxy-3′-(αα-dimethylallyl)chalcone
10. 2′-Methoxyfurano(2″, 3″, 4′, 3′)chalcone
11. 2′-Methoxy[furano(2″, 3″, 4′, 3′)]-4-*C*-methylchalcone*
12. β, 2′-Dimethoxy[furano(2″, 3″, 4′, 3′)]-chalcone*
13. 2′-Hydroxy-6″, 6″-dimethylpyrano(2″, 3″, 4′, 3′)chalcone
14. β-Hydroxy-2′-methoxy[6″, 6″-dimethoxyl-pyrano(2″, 3″, 4′, 3′)]chalcone*
15. 2′, 4′-Dihydroxy-3′-methoxychalcone
16. 2′, 4′-Dihydroxy-5′-methoxychalcone
17. 2′, 4′, 6′-Trihydroxychalcone
18. 2′, 4′-Dihydroxy-6-methoxychalcone
19. 2′, 6′-Dihydroxy-4′-methoxychalcone
20. 2′-Hydroxy-4′, 6′-dimethoxychalcone
21. 2′, 4′, 6′-Trihydroxy-3′-*C*-prenylchalcone
22. 2′, 6′-Dihydroxy-4′-prenyloxychalcone*
23. 2′, 6′-Dihydroxy-4′-methoxy-3′-*C*-prenylchalcone
24. 2′-Hydroxy-3′-*C*-prenyl-4′, 6′-dimethoxychalcone
25. 2′, 5′-Dihydroxy-6″, 6″-dimethylpyrano(2″, 3″, 4′, 3′)chalcone
26. 2′, 4′-Dihydroxy-3′-*C*-methyl-6″, 6″-dimethylpyrano(2″, 3″, 6′, 5′)chalcone
27. 2′-Hydroxy-6′-methoxy-6″, 6″-dimethylpyrano(2″, 3″, 4′, 3′)chalcone
28. 2′-Hydroxy-6′-methoxy[6″, 6″-dimethylpyrano(2″, 3″, 4′, 5′)]chalcone*
29. β, 2′-Dihydroxy-6′-methoxy[6″, 6″-dimethylpyrano(2″, 3″, 4′, 3′)]chalcone*
30. 2′-Hydroxy-[6″, 6″-dimethylpyrano-(2″, 3″, 4′, 3′)]-[6‴, 6‴-dimethylpyrano-2‴, 3‴, 6′, 5′)]chalcone*
31. 2′, 6′, β-Trimethoxy-6″, 6″-dimethylpyrano(2″, 3″, 4′, 3′)chalcone
32. β-Hydroxy-2′, 6′-dimethoxy-6″, 6″-dimethylpyrano(2″, 3″, 4′, 3′)chalcone
33. 2′, 4′, 6′-Trihydroxy-3′-*C*-geranylchalcone
34. 2′, 6′-Dimethoxy-3′, 4′-methylenedioxychalcone
35. 2′-Hydroxy-4′, 5′, 6′-trimethoxychalcone
36. 2′, 3′, 4′, 5′, 6′-Pentamethoxychalcone
37. 2′-Hydroxy-3′, 6′-diketo-4′, 5′-dimethoxychalcone
38. 2′, 5′-Dihydroxy-3′, 6′-diketo-4′-methoxychalcone
39. 2′, 5′-Dihydroxy-3′, 4′, 6′-trimethoxychalcone
40. 2′, 4′-Dihydroxy-5′-*C*-methyl-6′-methoxychalcone
41. 2′, 4′-Dihydroxy-3′-*C*-methyl-6′-methoxychalcone*
42. 2′, 6′-Dihydroxy-3′-*C*-methyl-4′-methoxychalcone
43. 2′-Hydroxy-3′-*C*-methyl-4′, 6′-dimethoxychalcone
44. 2′, 4′-Dihydroxy-3′, 5′-di-*C*-methyl-6′-methoxychalcone
45. 2′, 6′-Dihydroxy-3, 5′-di-*C*-methyl-4′-methoxychalcone*
46. 2′, 3′-Dihydroxy-4′, 6′-dimethoxychalcone*
47. 2′-Hydroxy-3′, 4′, 6′-trimethoxychalcone
48. 2′-Hydroxy-4′, 5, 6′-trimethoxychalcone*
49. 2′, 3′, 4′, 6′-Tetrahydroxychalcone
50. 2′, 4′-Dihydroxy-3′, 6′-dimethoxychalcone
51. 2′, 6′-Dihydroxy-3′, 4′-dimethoxychalcone
52. 2′, 4′-Dihydroxy-3′, 5′, 6′-trimethoxychalcone
53. 2′-Hydroxy-3′, 4′, 5′, 6′-tetramethoxychalcone
54. 3′-Hydroxy-2′, 4′, 5′, 6′-tetramethoxychalcone*
55. 2′, 4-Dihydroxychalcone-4-*O*-glucoside*
56. 2′, 4′, 4-Trihydroxychalcone
57. 4′-*O*-Glucoside
58. 4′-*O*-Glucosylglucoside
59. 4′, 4-Di-*O*-glucoside
60. 4′-*O*-Glucosylglucoside-4-*O*-glucoside
61. 4-*O*-Glucoside
62. 4-*O*-Apiosylglucoside

63. 4-*O*-Rhamnosylglucoside
64. 2′-*O*-Rhamnosylglucoside*
65. 2′,4′,4-Trihydroxy-3′-*C*-glucosylchalcone
66. 2′,4-Dihydroxy-4′-methoxychalcone
67. 4,4′-Dihydroxy-2′-methoxychalcone*
68. 2′,4-Dihydroxy-4′-methoxy-5′-formylchalcone
69. 2′-Methoxy-4′,4-dihydroxy-5′-formylchalcone
70. 2′,4′,4-Trihydroxy-3′-*C*-prenylchalcone
71. 2′,4-Dihydroxy-3′-*C*-prenyl-4′-methoxychalcone
72. 2′,4-Dihydroxy-4′-methoxy-5′-*C*-prenylchalcone
73. 2′,4,4′-Trihydroxy-3,5-di-*C*-prenylchalcone*
74. 2,2′,4′-Trihydroxy-3′-*C*-lavandulylchalcone*
75. 2′,4′,4-Trihydroxy-3′*C*-geranylchalcone
76. 2′,4′,4-Trihydroxy-3,3′,5-tri-*C*-prenylchalcone
77. 2′,4-Dihydroxy-6″,6″-dimethylpyrano(2″,3″,4′,5′)chalcone
78. 2′-Hydroxy-4-methoxy[6″,6″-dimethylpyrano(2″,3″,4′,3′)]chalcone*
79. 4,4′-Dihydroxy[6″,6″-dimethyl-4″,5″-dihydropyrano(2″,3″,2′,3′)]chalcone*
80. 2′,4′,6′,4-Tetrahydroxychalcone
81. 4-*O*-Glucoside
82. 2′-*O*-Glucoside
83. 2′-*O*-Xyloside*
84. 4′-*O*-Glucoside*
85. 2′-*O*-Rhamnosylglucoside
86. 2′,4′,4-Trihydroxy-3′-*C*-prenyl-6′-methoxychalcone
87. 2′,4′,4-Trihydroxy-2-methoxy-5′-*C*-lavandulylchalcone*
88. 2′,4′,4-Trihydroxy-6′-methoxychalcone
89. 4′-*O*-Glucoside
90. 6′,6′,4-Trihydroxy-4′-methoxychalcone
91. 2′-*O*-Glucoside
92. 2′,4-Dihydroxy-4′,6′-dimethoxychalcone
93. 4-*O*-Glucoside
94. 2′,6′-Dihydroxy-4′,4-dimethoxychalcone
95. 2′-Hydroxy-4′,6′,4-trimethoxychalcone
96. 2′,4,6′-Trihydroxy-4′-prenyloxychalcone*
97. 2′,4-Dihydroxy-4′,5′,6′-trimethoxychalcone
98. 2′-Hydroxy-4′,5′,6′,4-tetramethoxychalcone
99. 2′,4-Dihydroxy-3′,4′-methylenedioxy-6′-methoxychalcone
100. 4′,4-Dihydroxy-2-methoxychalcone
101. 4′,3,4-Trihydroxy-2-methoxychalcone
102. 4′,4-Dihydroxy-2-methoxy-5-(α,α-dimethylallyl)chalcone
103. 2′,4′,3,4,α-Pentahydroxychalcone
104. 2′,3′,4′,4-Tetrahydroxychalcone
105. 2′,3′,4′,6′,4-Pentahydroxychalcone 6′-*O*-glucoside
106. 2′,4′,4-Trihydroxy-3′,6′-diketochalcone
107. 2′-*O*-Glucoside
108. 2′,5′-Dihydroxy-3′,4,4′,6′-tetramethoxychalcone
109. 2′,3,4-Trihydroxy-6″,6″-dimethylpyrano(2″,3″,4′,3′)chalcone
110. 2′,4-Dihydroxy-3-methoxy-6″,6″-dimethylpyrano(2″,3″,4′,3′)chalcone
111. 2′-Hydroxy-3,4-methylenedioxy-6″,6″-dimethylpyrano(2″,3″,4′,3′)chalcone
112. 2′-Hydroxy-3,4-methylenedioxy-6′-methoxy-6″,6″-dimethylpyrano-(2″,3″,4′,3′)chalcone
113. 2′-Methoxy-3,4-methylene-dioxyfurano(2″,3″,4′,3′)chalcone
114. 2′,4′,3,4-Tetrahydroxychalcone
115. 3-*O*-Glucoside
116. 4′-*O*-Glucoside
117. 3,4′-Di-*O*-glucoside
118. 2′,3-Di-*O*-glucoside
119. 4′-*O*-Glucosylglucoside
120. 4′-*O*-Arabinosylgalactoside
121. 2′,4′,3-Trihydroxy-4-methoxychalcone
122. 2′,4′,4-Trihydroxy-3-methoxychalcone
123. 4-*O*-Glucoside
124. 2′,4′-Dihydroxy-3,4-dimethoxychalcone
125. 2′,3′,4′,3,4-Pentahydroxychalcone
126. 4′-*O*-Glucoside
127. 4′-*O*-Arabinosylglucoside*
128. 2′,4′,3,4,α-Pentahydroxychalcone
129. 2′,4′,3,4-Tetrahydroxy-3′-methoxychalcone
130. 4′-*O*-Glucoside
131. 2′,3′,4′,3-Tetrahydroxy-4-methoxychalcone 4′-*O*-glucoside*
132. 2′,4′,5′,3,4-Pentahydroxychalcone
133. 4′-*O*-Glucoside
134. 2′,3′,4′,4-Tetrahydroxy-3-methoxychalcone
135. 2′,3′,4′-Trihydroxy-3,4-dimethoxychalcone
136. 4′-*O*-Glucoside*
137. 2′,4′-Dihydroxy-3,3′,4-trimethoxychalcone 4′-*O*-glucoside*
138. 2′-Hydroxy-3,3′,4,4′-tetramethoxychalcone*
139. 2′,4′-Dihydroxy-3,4-methylenedioxy-5′-methoxychalcone
140. 2′,4′,6′,3,4-Pentahydroxychalcone
141. 2′-*O*-Glucoside
142. 2′,3-Dihydroxy-4′,6′,4-trimethoxychalcone
143. 2′,4′-Dihydroxy-2,3,6′-trimethoxychalcone*
144. 2′-Hydroxy-4′,6′,3,4-tetramethoxychalcone
145. 2′-Hydroxy-3,4-methylenedioxy-3′-*C*-prenyl-4′,6′-tetramethoxychalcone
146. 2′,4′-Dihydroxy-6′,2,4-trimethoxychalcone
147. 2′-Hydroxy-4′,6′,2,4-tetramethoxychalcone
148. 2′-Hydroxy-4′,5′,6′,3,4-pentamethoxychalcone
149. 2′,4′,4,2-Tetrahydroxy-3′-(2-isopropylidene-5-methylhex-4-enyl)-5′-methoxychalcone
150. 2′,4′,2-Trihydroxy-3′-geranyl-5′-methoxychalcone
151. 2′,4′,2,5-Tetrahydroxy-3′-geranyl-5′-methoxychalcone

152. 2′, 4′, 3, 4, 5-Pentahydroxychalcone
153. 2′, 4′, 6′, 3, 4, 5-Hexahydroxychalcone
154. 2′-*O*-Glucoside
155. 2′, 3′, 4′, 5′, 3, 4, α-Heptahydroxychalcone
156. 2′-*O*-Glucoside
157. 2′, 3-Dihydroxy-4′, 6′, 4, 5-tetramethoxychalcone
158. 2′-Hydroxy-4′, 6′, 3, 4, 5-pentamethoxychalcone
159. 2′-Hydroxy-4′, 6′, 2, 4, 5-pentamethoxychalcone
160. 2′, 5′-Dihydroxy-4′, 6′-dimethoxy-3, 4-methylenedioxychalcone*
161. 2′-Hydroxy-2, 4, 5-trimethoxy[6″, 6″-dimethylpyrano(2″, 3″, 4′, 3′)]chalcone*
162. 3, 4-Dihydroxychalcone 4-*O*-arabinosylgalactoside*
163. 3, 4, 2′-Trihydroxychalcone*
164. 2′, 4-Dihydroxy-2-methoxychalcone*
165. 5, 7-Dihydroxy-8-cinnamoyl-4-phenyl-2*H*-1-benzopyran-2-one
166. 5, 7-Dihydroxy-8-(4-hydroxycinnamoyl)-4-phenyl-2*H*-1-benzopyran-2-one
167. 5, 7-Dihydroxy-8-(3, 4-dihydroxycinnamoyl)-4-phenyl-2*H*-1-benzopyran-2-one
168. 2-Cinnamoyl-3-hydroxy-4, 4-dimethyl-5-methoxycyclohexadienone (ceroptin)
169. 2′, 4′-Dihydroxy-3′-*C*-(2, 4, 6-trihydroxy-3-acetyl-5-methylbenzyl)-6″, 6″-dimethylpyrano-(2″, 3″, 4′, 5′)chalcone (rottlerin)
170. 2′, 4′, 4-Trihydroxy-3′-*C*-(2, 4, 6-trihydroxy-3-acetyl-5-methylbenzyl)-6″, 6″-dimethylpyrano(2″, 3″, 4′, 5′)chalcone-(× 4-hydroxyrottlerin)
171. 4′, 2, 3, 4-Tetrahydroxy-3′-*C*-(2, 4, 6-trihydroxy-3-acetyl-5-methylbenzyl)-6″, 6″-dimethylpyrano(2″, 3″, 4′, 5′) chalcone (3, 4-dihydroxyrottlerin)
172. 2′, 4′-Dihydroxy-3′, 3-di-*C*-prenyl-6″, 6″-dimethylpyrano(2″, 3″, 4, 5)chalcone
173. Cryptocaryone
174. Flemingin-A
175. Flemingin-B
176. Flemingin-C
177. Peltogynoid chalcone
178. (−)-Rubranine
179. Bavachromanol
180. 4′-*O*-Methyl-4-deoxybavachromanol
181. Flemistrictin-B
182. Flemistrictin-C
183. Lespeol
184. Flemistrictin-E
185. Flemistrictin-F
186. Flemiwallichin-F
187. Flemiwallichin-D
188. Flemiwallichin-E
189. Boesenbergin-A
190. Boesenbergin-B
191. 2-Hydroxy-7, 8-dehydrograndiflorone
192. Kuwanon-J
193. Kuwanon-Q
194. Kuwanon-R
195. Kuwanon-V
196. Bakuchalcone
197. 3′, 4′-Dihydrooxepino-6-hydroxybutein

Aurones

1. Furano(2″, 3″, 6, 7)aurone
2. 4′-Hydroxyfurano(2″, 3″, 6, 7)aurone
3. 4′-Methoxyfurano(2″, 3″, 6, 7)aurone
4. 6, 4′-Dihydroxyaurone
5. 6-*O*-Glucoside
6. 4, 6, 4′-Trihydroxyaurone 4-*O*-glucoside
7. 6, 3′, 4′-Trihydroxyaurone
8. 6-*O*-Glucoside
9. 6, 3′-Di-*O*-glucoside
10. 6-*O*-Glucosylglucoside
11. 6, 7, 3′, 4′-Tetrahydroxyaurone
12. 6-*O*-Glucoside
13. 6, 3′, 4′-Trihydroxy-7-methoxyaurone
14. 6-*O*-Glucoside
15. 6-*O*-Glucosylrhamnoside*
16. 6-*O*-Xylosylarabinoside*
17. 4, 3-Methylenedioxyfurano(2″, 3″, 6, 7)aurone
18. 4, 6, 3′, 4′-Tetrahydroxyaurone
19. 6-*O*-Glucoside
20. 6-*O*-Glucuronide
21. 4, 6-Di-*O*-glucoside*
22. 2, 6, 4′-Trihydroxy-4-methoxy aurone*
23. 6, 3′, 4′-Trihydroxy-4-methoxyaurone
24. 4, 6, 3′, 4′, 5′-Pentahydroxyaurone
25. 4-*O*-Glucoside
26. 6-*O*-Glucoside
27. 6, 3′, 4′, 5′-Tetrahydroxy-4-methoxyaurone
28. 6-*O*-Rhamnosylglucoside
29. Derriobtusone

Dihydrochalcones

1. Dihydrochalcone
2. 2′-Hydroxy-4′-prenyloxydihydrochalcone
3. 2′, β-Dimethoxyfurano-(2″, 3″, 4′, 3′)dihydrochalcone
2. 2′-Methoxyfurano(2″, 3″, 4′, 3′)-β-ketodihydrochalcone
5. 2′, 4′, 6′-Trihydroxydihydrochalcone
6. 2′-*O*-Glucoside
7. 4′-*O*-Glucoside
8. 2′, 6′-Dihydroxy-4′-prenyloxydihydrochalcone
9. 2′, 4′, 6′-Trihydroxy-β-ketodihydrochalcone 2′-*O*-glucoside
10. 2′, 6′-Dihydroxy-4′-methoxydihydrochalcone

11. 2′,4′-Dihydroxy-6′-methoxydihydrochalcone
12. 2′-Hydroxy-3′-*C*-methyl-4′,6′-dimethoxydihydrochalcone
13. 2′,6′-Dihydroxy-3′,5′-di-*C*-methyl-4′-methoxydihydrochalcone
14. 2′,4′-Dihydroxy-3′,5′-di-*C*-methyl-6′-methoxydihydrochalcone
15. 2′,6′,*β*-Trimethoxy-3′,4′-methylenedioxydihydrochalcone
16. 1-Phenyl-3-(2,6-dimethoxy-3,4-methylenedioxyphenyl)propan-1,3-diol
17. 2′,4′,6′-Trihydroxy-3′-*C*-methyl-5′-formyl-*β*-ketodihydrochalcone
18. 2′,6′-Dihydroxy-3′-*C*-methyl-4′-methoxy-5′-formyl-*β*-ketodihydrochalcone
19. 2′,4′-Dihydroxy-3′-*C*-(2-hydroxybenzyl)-6′-methoxydihydrochalcone
20. 2′,4′-Dihydroxy-5′-*C*-(2-hydroxybenzyl)-6′-methoxydihydrochalcone
21. 2′,4′,6′-Trihydroxy-3′-methoxy-5′-*C*-prenyldihydrochalcone
22. 2′,4′,6′-Trihydroxy-3′,5′-di-*C*-prenyldihydrochalcone
23. 4′,6′-Dihydroxy-3′,3′,5′-tri-*C*-prenyl-*β*-ketodihydrochalcone
24. 4′,6′-Dihydroxy-3′-methoxy-3′,5′-di-*C*-prenyl-*β*-ketodihydrochalcone
25. 4′,6′-Dihydroxy-2′-methoxydihydrochalcone*
26. 6′-Hydroxy-2′,4′-dimethoxydihydrochalcone*
27. 2′,4′,6′-Trihydroxy-3′-*C*-prenyldihydrochalcone*
28. 2′,6′-Dihydroxy-4′-methoxy-5′-*C*-prenyldihydrochalcone*
29. 2′,4′-Dihydroxy-6′-methoxy-3′-*C*-(2-hydroxybenzyl)-5′-*C*-methyldihydrochalcone*
30. 2′,4′-Dihydroxy-6′-methoxy-3′,5′-di-*C*-(2-hydroxybenzyl)dihydrochalcone*
31. *Uvaria angolensis* compound*
32. 2′,5′-Dihydroxy-6′-methoxy [6″,6″-dimethylpyrano(2″,3″,4′,3′)]-dihydrochalcone*
33. 2′,4′,6′-Trihydroxy-3′-*C*-(3-methyl-6-isopropylcyclohex-2-enyl)-dihydrochalcone*
34. 2′,4′,4-Trihydroxydihydrochalcone
35. 2′-*O*-Glucoside
36. 4′-*O*-Glucoside*
37. 2′,4-Dihydroxy-4′-methoxydihydrochalcone
38. 2′,4′,4-Trihydroxy-*β*-ketodihydrochalcone
39. 2′,4′,6′,4-Tetrahydroxydihydrochalcone
40. 2′-*O*-Glucoside
41. 2′-*O*-Rhamnoside
42. 4′-*O*-Glucoside
43. 2′-*O*-Xylosylglucoside
44. 2′,6′,4-Trihydroxy-4′-methoxydihydrochalcone 2′-*O*-glucoside
45. 2′,6′-Dihydroxy-4,4′-dimethoxydihydrochalcone
46. 2′,4′-Dihydroxy-4,6′-dimethoxydihydrochalcone
47. 2′,6′,4-Trihydroxy-3′-*C*-methyl-4′-methoxydihydrochalcone
48. 2′,4′,6′,4-Tetrahydroxy-3′-*C*-glycosyldihydrochalcone
49. 2′,3′,4′,6′,4-Pentahydroxy-5′-geranyldihydrochalcone
50. 2′,3′,4′,6′,4-Pentahydroxy-5′-neryldihydrochalcone
51. 2′,4′,4-Trihydroxy-3′-methoxydihydrochalcone
52. 2′,4′,6′,3,4-Pentahydroxydihydrochalcone
53. 4′-*O*-Glucoside
54. 2′-Methoxy-3,4-methylenedioxyfurano-(2″,3″,4′,3′)-*β*-ketodihydrochalcone
55. 2′,4′,6′,3,4-Pentahydroxy-3′-*C*-glucosyldihydrochalcone
56. 2′,4′,3,5,*β*-Pentahydroxy-4-methoxydihydrochalcone
57. 2′,6′-Dihydroxy-3′,4-dimethoxy-4′,5′-methylenedioxydihydrochalcone*
58. 2′,4,6′-Trihydroxy-4′-methoxy-3′,5′-di-*C*-methyldihydrochalcone*
59. 2′,3,4,4′,6′-Pentahydroxydihydrochalcone 2′-*O*-glucoside*
60. 2′-*O*-Galactoside*
61. 2′-Hydroxy-5′,6′-dimethoxy-[furano(2″,3″,4′,3′)]-3,4-methylenedioxy-dihydrochalcone*
62. *β*,2′,4′-Trimethoxy-3,4-methylenedioxydihydrochalcone*
63. *β*,2,4-Trimethoxy-3′,4′-methylenedioxydihydrochalcone*
64. α,2′,4,4′-Tetrahydroxydihydrochalcone*
65. α,2′,4,4′-Tetrahydroxy-3′-*C*-glucosyldihydrochalcone*
66. α,2′3,4,4′-Pentahydroxy-3′-*C*-glucosyldihydrochalcone*
67. α,2′-Dihydroxy-4,4′-dimethoxydihydrochalcone*
68. Grenoblone*
69. 4-Hydroxygrenoblone*
70. α-Hydroxy-2′,4,4′-trimethoxydihydrochalcone*
71. α,2′,4,4′,6′-Pentahydroxydihydrochalcone*

Flavanones

1. 7-Hydroxyflavanone
2. 7-Methoxyflavanone
3. 7-Hydroxy-8-methoxyflavanone
4. 7-Methoxy-8-*C*-(3-hydroxy-3-methyl-*t*-buten-1-yl)flavanone*
5. 7,8-Dimethoxyflavanone
6. 7-Prenyloxyflavanone
7. 6″,6″-Dimethylpyrano(2″,3″,7,8)flavanone
8. [6″,6″-Dimethylpyrano(2″,3″,7,6)]-8-*C*-prenylflavanone*

9. 6-Methoxy-6″, 6″-dimethylpyrano-(2″, 3″, 7, 8)flavanone
10. 7-Hydroxy-6, 8-di-*C*-prenylflavanone
11. 7-Hydroxy-8-*C*-prenylflavanone
12. 5, 7-Dihydroxyflavanone
13. 5-*O*-Glucoside
14. 7-*O*-Rhamnoside
15. 7-*O*-Neohesperidoside
16. 7-*O*-Neohesperidoside 2″-*O*-acetate*
17. 3″-*O*-Acetate*
18. 4″-*O*-Acetate*
19. 6″-*O*-Acetate*
20. 5-Methoxy-7-hydroxyflavanone
21. 5-Hydroxy-7-methoxyflavanone
22. 5, 7-Dimethoxyflavanone
23. 5, 6, 7-Trihydroxyflavanone*
24. 7-*O*-Glucuronide*
25. 5, 7-Dihydroxy-6-methoxyflavanone
26. 5-Hydroxy-6, 7-dimethoxyflavanone
27. 6-Hydroxy-5, 7-dimethoxy flavanone*
28. 5, 7, 8-Trihydroxyflavanone*
29. 7-*O*-Glucuronide*
30. 5, 8-Dihydroxy-7-methoxyflavanone 8-*O*-acetate*
31. 5-Hydroxy-7, 8-dimethoxyflavanone*
32. 8-Hydroxy-5, 7-dimethoxyflavanone*
33. 5, 7-Dihydroxy-8-methoxyflavanone
34. 7-*O*-Glucoside
35. 5, 7, 8-Trimethoxyflavanone
36. 7-Hydroxy-5, 6, 8-trimethoxyflavanone
37. 6-Hydroxy-5, 7, 8-trimethoxyflavanone
38. 5-Hydroxy-6, 7, 8-trimethoxyflavanone*
39. 5, 6, 7, 8-Tetramethoxyflavanone
40. 5, 7-Dihydroxy-6-*C*-methylflavanone
41. 5-Hydroxy-7-*O*-(3-methyl-2, 3-epoxybutyl)flavanone*
42. 7-Hydroxy-5-methoxy-6-*C*-methylflavanone*
43. 5-Hydroxy-6-*C*-methyl-7-methoxyflavanone
44. 5, 7-Dihydroxy-8-*C*-methylflavanone
45. 5-Hydroxy-7-methoxy-8-*C*-methylflavanone*
46. 5, 7-Dihydroxy-6, 8-di-*C*-methylflavanone
47. 7-*O*-Glucoside
48. 5-Hydroxy-6, 8-di-*C*-methyl-7-methoxyflavanone
49. 2, 5, 7-Trihydroxy-6-(or 8)-*C*-methyl-8-(or 6)-formylflavanone
50. 5, 7-Dihydroxy-6-formyl-8-*C*-methylflavanone
51. 5, 7-Dihydroxy-6-*C*-prenylflavanone
52. 5, 7-Dihydroxy-8-*C*-prenylflavanone
53. 5-Hydroxy-6-*C*-prenyl-7-methoxyflavanone
54. 5-Methoxy-6″, 6″-dimethylpyrano-(2″, 3″, 7, 8)flavanone
55. 5, 7-Dihydroxy-8-*C*-(β-methyl-β-hydroxy-methylallyl)flavanone
56. 5, 7-Dihydroxy-8-*C*-(β-methyl-β-formylallyl)flavanone
57. 5, 7-Dihydroxy-8-*C*-geranylflavanone
58. 5-Hydroxy-7-methoxy-8-*C*-prenylflavanone
59. 5, 7-Dihydroxy-6-*C*-(2-hydroxybenzyl)flavanone
60. 5, 7-Dihydroxy-8-*C*-(2-hydroxybenzyl)flavanone
61. 6, 8-Dihydroxy-7-*C*-(5-hydroxygeranyl)flavanone
62. 7-Hydroxy-5-methoxy-6, 8-di-*C*-methylflavanone*
63. 7-Hydroxy-5-methoxy-8-*C*-(2-hydroxybenzyl)flavanone*
64. 5-Hydroxy-7-*O*-prenyloxy-8-*C*-prenylflavanone*
65. 7-Hydroxy-5-methoxy-8-*C*-prenylflavanone*
66. 5-Methoxy-7-*O*-prenyloxy-8-*C*-prenylflavanone*
67. 5-Hydroxy-7-*O*-prenyloxy-8-*C*-(3-hydroxy-3-methyl-*trans*-buten-l-yl)flavanone*
68. 5-Hydroxy-7-methoxy-8-*C*-(3-hydroxy-3-methylbutyl)flavanone*
69. 5, 7-Dihydroxy-6-*C*-prenyl-8-*C*-methylflavanone*
70. 5, 7-Dihydroxy-6, 8-di-*C*-prenylflavanone*
71. 5-Hydroxy-7-*O*-prenyloxyflavanone*
72. 5-Hydroxy-7-*O*-prenyloxy-8-*C*-methyflavanone*
73. 5-Methoxy-[6″, 6″-dimethylpyrano(2″, 3″, 7, 8)]-flavanone*
74. 5-Hydroxy-6-*C*-prenyl-[6″, 6″-dimethylpyrano(2″, 3″, 7, 8)]flavanone*
75. 7, 8, 4′-Trihydroxyflavanone
76. 7, 4′-Dihydroxyflavanone
77. 4′-*O*-Glucoside
78. 7-*O*-Glucoside
79. 4′-*O*-Apiosylglucoside*
80. 4′-*O*-Rhamnosylglucoside
81. 7-*O*-Glucosylglucoside
82. 7, 4′-Dihydroxy-8-*C*-galactosylflavanone
83. 7, 4′-Dihydroxy-6, 8-di-*C*-prenylflavanone
84. 7, 4′-Dihydroxy-8-*C*-prenylflavanone*
85. 7, 4′-Dihydroxy-8, 3′, 5′-tri-*C*-prenylflavanone*
86. 4′-Hydroxy-7-methoxyflavanone*
87. 7, 4′-Dimethoxy-6-*C*-methylflavanone*
88. 7, 3′, 4′-Trihydroxyflavanone
89. 3′-Hydroxy-7, 4′-dimethoxyflavanone*
90. 7, 4′-Dihydroxy-3′-*C*-prenylflavanone*
91. 7, 4′-Dihydroxy-3′, 5′-di-*C*-prenylflavanone*
92. 7-Hydroxy-[6″, 6″-dimethylpyrano(2″, 3″, 4′, 3′)]-flavanone*
93. 7-Hydroxy-5′-*C*-prenyl-[6″, 6″-dimethylpyrano(2″, 3″, 4′, 3′)]flavanone*
94. 5, 7, 4′-Trihydroxyflavanone
95. 5-*O*-Glucoside
96. 5-*O*-Rhamnoside*
97. 5-*O*-Neohesperidoside*
98. 7-*O*-Glucoside
99. 7-*O*-Rhamnoside
100. 4′-*O*-Glucoside
101. 4′-*O*-Galactoside
102. ?-*O*-Fructoside
103. 7-*O*-Glucoside-6″-*O*-gallate
104. 7-*O*-Glucoside-6″-*p*-coumarate
105. 7-*O*-Glucoside-3″-*p*-coumarate*
106. 7-*O*-Glucoside-3″, 6″-di-*p*-coumarate*

107. 5,7-Di-*O*-glucoside
108. 5-*O*-Glucosylglucoside
109. 7-*O*-Rutinoside
110. 7-*O*-Neohesperidoside
111. 7-*O*-Arabinosylglucoside*
112. 7-*O*-Galactosylglucoside*
113. 4′-*O*-Rutinoside
114. 4′-*O*-Xylosylglucoside
115. 5,7,4′-Trihydroxy-6-*C*-glucosylflavanone
116. 5,4′-Dihydroxy-7-methoxyflavanone
117. 5-*O*-Glucoside
118. 5,7-Dihydroxy-4′-methoxyflavanone
119. 5-*O*-Glucoside
120. 7-*O*-Xyloside*
121. 7-*O*-Glucoside
122. 7-*O*-Rutinoside
123. 7-*O*-Neohesperidoside
124. 7-*O*-4′-(Rhamnosyl)glucoside
125. 5-Methoxy-7,4′-dihydroxyflavanone
126. 5-Hydroxy-7,4′-dimethoxyflavanone
127. 4′-Hydroxy-5,7-dimethoxyflavanone
128. 4′-*O*-Xylosylarabinoside
129. 4′-*O*-Rhamnosylglucoside
130. 5,7,4′-Trimethoxyflavanone
131. 5,6,7,4′-Tetrahydroxyflavanone
132. 5-*O*-Glucoside
133. 5,7,8,4′-Tetrahydroxyflavanone*
134. 5,7,4′-Trihydroxy-6-methoxyflavanone*
135. 5,6,7,4′-Tetramethoxyflavanone*
136. 4′-Prenyloxy-5,7-dihydroxyflavanone
137. 5,7,4′-Trihydroxy-6-*C*-glucosylflavanone
138. 5,7,4′-Trihydroxy-8-*C*-glucosylflavanone
139. 5,2′-Dihydroxy-7-methoxyflavanone
140. 2′-*O*-Glucoside
141. 5,7,4′-Trihydroxy-6-*C*-methylflavanone
142. 7-*O*-Glucoside
143. 5,7,4′-Trihydroxy-6,8-di-*C*-methylflavanone
144. 7-*O*-Glucoside
145. 5,7-Di-*O*-glucoside
146. 5,4′-Dihydroxy-6,8-di-*C*-methyl-7-methoxyflavanone
147. 4′-Methoxy-5,7-dihydroxy-6,8-di-*C*-methylflavanone
148. 7-*O*-Glucoside
149. 7-*O*-Glucosylglucoside
150. 5,7,4′-Trihydroxy-6-*C*-prenylflavanone*
151. 5,7,4′-Trihydroxy-8-*C*-prenylflavanone
152. 7,4′-Dihydroxy-5-methoxy-8-*C*-prenylflavanone
153. 7,4′-Dihydroxy-8-*C*-prenylflavanone
154. 7,4′-Dihydroxy-6-*C*-prenylflavanone
155. 4′-Hydroxy-6-*C*-prenyl-7-methoxyflavanone
156. 5,4′-Dihydroxy-6-*C*-prenyl-6″,6″-dimethylpyrano(2″,3″,7,8)flavanone
157. 5,7,4′-Trihydroxy-6-(or 8)-*C*-prenylflavanone 7,4′-di-*O*-glucoside
158. 7,4′-Dihydroxy-8,3′,5′-tri-*C*-prenylflavanone
159. 7-Hydroxy-8,5′-di-*C*-prenyl-6″,6″-dimethylpyrano(2″,3″,4′,3′)flavanone
160. 5,4′-Dihydroxy-6,7-dimethoxyflavanone
161. 5-Hydroxy-6,7,4′-trimethoxyflavanone
162. 5,4′-Dihydroxy-7,8-dimethoxyflavanone
163. 5-Hydroxy-7,8,4′-trimethoxyflavanone
164. 5,6-Dihydroxy-7,8,4′-trimethoxyflavanone
165. 5,4′-Dihydroxy-6,7,8-trihydroxyflavanone
166. 5,4′-Dimethoxy-6,7-methylenedioxyflavanone*
167. 5,7-Dihydroxy-8,2′-dimethoxyflavanone*
168. 7-Hydroxy-5,8,2′-trimethoxyflavanone*
169. 5,2′-Dihydroxy-8-*C*-prenyl-[6″,6″-dimethylpyrano(2″,3″,7,6)]flavanone*
170. 5,4′-Dihydroxy-8-*C*-prenyl-[6″,6″-dimethylpyrano(2″,3″,7,6)]flavanone*
171. 5,4′-Dihydroxy-7-methoxy-6-*C*-methylflavanone*
172. 5,7-Dihydroxy-8,4′-dimethoxy-6-*C*-methylflavanone*
173. 5-Hydroxy-7,4′-dimethoxy-6,8-di-*C*-methylflavanone 5-*O*-galactoside*
174. 5,4′-Dihydroxy-7-*O*-prenyloxyflavanone*
175. 5,7,4′-Trihydroxy-6,3′-di-*C*-prenylflavanone*
176. 5,4′-Dihydroxy-6-*C*-prenyl-[6″,6″-dimethylpyrano(2″,3″,7,8)]flavanone*
177. 5,4′-Dihydroxy-8-*C*-prenyl-[6″,6″-dimethylpyrano(2″,3″,7,6)]flavanone*
178. 5,7-Dihydroxy-4′-methoxy-8,3′-di-*C*-prenylflavanone*
179. 5,7,4′-Trihydroxy-8,3′,5′-tri-*C*-prenylflavanone*
180. 5,7,4′-Trihydroxy-6,8,3′-tri-*C*-prenylflavanone*
181. 5,7,2′-Trihydroxy-8-*C*-lavandulylflavanone*
182. 5,6,7,4′-Tetrahydroxy-8-*C*-prenylflavanone 5-*O*-rutinoside*
183. 5,7,4′-Trihydroxy-6-*C*-prenyl-3′-(3-methyl-2,3-epoxybutyl)flavanone*
184. 5,7,4′-Trihydroxy-6-*C*-geranylflavanone*
185. 5,7,4′-Trihydroxy-8-*C*-geranylflavanone*
186. 5,7,4′-Trihydroxy-3′-*C*-geranylflavanone*
187. 6,3′,4′-Trimethoxyflavanone*
188. 5-Hydroxy-6,7,8,4′-tetramethoxyflavanone
189. 4′-Hydroxy-5,6,7-trimethoxyflavanone
190. 5,4′-Dimethoxy-6,7-methylenedioxyflavanone
191. 7,3′,4′-Trihydroxyflavanone
192. 3′-*O*-Glucoside
193. 7-*O*-Glucoside
194. 7,3′-Di-*O*-glucoside
195. 6,7,3′,4′-Tetrahydroxyflavanone
196. 7,8,3′,4′-Tetrahydroxyflavanone
197. 7-*O*-Glucoside
198. 7-*O*-Rhamnoside*
199. 7,3′,4′-Trihydroxy-8-methoxyflavanone

200. 3′,4′-Methylenedioxy-7-hydroxy-8-*C*-prenylflavanone
201. 3′,4′-Methylenedioxy-6,8-di-*C*-prenyl-7-hydroxyflavanone
202. 3′,4′-Methylenedioxy-6″,6″-dimethylpyrano(2″,3″,7,8)flavanone
203. 3′,4′-Methylenedioxy-6-methoxy-6″,6″-dimethylpyrano(2″,3″,7,8)flavanone
204. 5,7,3′,4′-Tetrahydroxyflavanone
205. 5-*O*-Rhamnoside
206. 5-*O*-Glucoside*
207. 7-*O*-Glucoside
208. 7-*O*-Rhamnoside
209. 3′-*O*-Glucoside*
210. 5,3′-Di-*O*-glucoside
211. 7-*O*-Rhamnosylglucoside
212. 7-*O*-Rutinoside
213. 7-*O*-Neohesperidoside
214. 5,7,4′-Trihydroxy-3′-methoxyflavanone
215. 5,7,3′-Trihydroxy-4′-methoxyflavanone
216. 5-*O*-Glucoside
217. 7-*O*-Glucoside*
218. 7-*O*-Rhamnoside
219. 7-*O*-Rutinoside
220. 7-*O*-Neohesperidoside
221. 5,4′-Dihydroxy-7,3′-dimethoxyflavanone
222. 5,3′-Dihydroxy-7,4′-dimethoxyflavanone
223. 5-*O*-Glucoside
224. 5,7-Dihydroxy-3′,4′-methylenedioxyflavanone 5-*O*-glucoside
225. 7,3′,4′-Trihydroxy-5-methoxyflavanone 7-*O*-xylosylgalactoside*
226. 7,3′,4′-Trihydroxy-5-methoxyflavanone 7-*O*-xylosylarabinoside *
227. 5,3′,4′-Trihydroxy-7-methoxyflavanone*
228. 7,3′-Dihydroxy-5,4′-dimethoxyflavanone 7-*O*-glucoside*
229. 5,7,4′-Trihydroxy-8,3′-dimethoxyflavanone*
230. 5,3′-Dihydroxy-6,7,4′-trimethoxyflavanone*
231. 5,3′-Dihydroxy-7,8,4′-trimethoxyflavanone*
232. 6,4′-Dihydroxy-5,7,3′-trimethoxyflavanone*
233. 6-Hydroxy-5,7,3′,4′-tetramethoxyflavanone*
234. 6-Hydroxy-5,7-dimethoxy-3′,4′-methylenedioxyflavanone*
235. 5-Hydroxy-6-methoxy-3′,4′-methylenedioxy-[furano(2″,3″,7,8)]flavanone*
236. 7,3′,4′-Trihydroxy-5-methoxy-6-*C*-methylflavanone 7-*O*-glucoside*
237. 3′-Hydroxy-5,7,4′-trimethoxy-8-*C*-methylflavanone*
238. 5,4′-Dihydroxy-3′-methoxy-6,8-di-*C*-methylflavanone*
239. 5,7,3′,4′-Tetrahydroxy-6-*C*-prenylflavanone*
240. 5,7,3′,4′-Tetrahydroxy-8-*C*-prenylflavanone*
241. 5,7,3′,4′-Tetrahydroxy-5′-*C*-prenylflavanone*
242. 5,7,3′,4′-Tetrahydroxy-2′,5′-di-*C*-prenylflavanone*
243. 5,7,3′-Trihydroxy 6″,6″-dimethylpyrano-(2″,3″,4′,5′)flavanone*
244. 5,3′,4′-Trihydroxy-7-methoxy-6,8-di-*C*-prenylflavanone*
245. 5,4′-Dihydroxy-7,3′-dimethoxy-6,8-di-*C*-prenylflavanone*
246. 5,7,3′,4′-Tetrahydroxy-6,8,5′-tri-*C*-prenylflavanone*
247. 5,7,4′-Trihydroxy-3′-methoxy-6,8,5′-tri-*C*-prenylflavanone*
248. 5,7,3′-Trihydroxy-6,8-di-*C*-prenyl-6″,6″-dimethylpyrano(2″,3″,4′,5′)flavanone*
249. 5,7,3′,4′-Tetrahydroxy-6-*C*-geranylflavanone*
250. 5,7,3′-Trihydroxy-4′-methoxy-6-*C*-geranylflavanone*
251. 5,7,2′,4′-Tetrahydroxy-5′-*C*-geranylflavanone*
252. 5,7-Dihydroxy-2′-methoxy-6″,6″-dimethylpyrano(2″,3″,4′,3′)flavanone*
253. 5,7,2′-Trihydroxy-6″,6″-dimethylpyrano (2″,3″,4′,3′)flavanone*
254. 5,7,4′-Trihydroxy-6″,6″-dimethylpyrano-(2″,3″,2′,3′)flavanone*
255. 5,7,2′,4′,4′-Tetrahydroxy-6,5′-di-*C*-prenylflavanone*
256. 5,7,2′-Trihydroxy-8-*C*-prenyl-6″,6″-dimethylpyrano(2″,3″,4′,5′)flavanone*
257. 5,7,2′-Trihydroxy-6-*C*-prenyl-6″,6″-dimethylpyrano(2″,3″,4′,5′)flavanone*
258. 5,2′-Dihydroxy-4′-methoxy-6-*C*-prenyl-6″,6″-dimethylpyrano(2″,3″,7,8)flavanone*
259. 5,7,2′,4′-Tetrahydroxy-6-*C*-lavandulylflavanone*
260. 5,7,4′-Trihydroxy-2′-methoxy-6-*C*-lavandulylflavanone*
261. 5,7,2′,4′-Tetrahydroxy-6-*C*-prenyl-8-*C*-lavandulylflavanone*
262. 5,7,2′,5′-Tetrahydroxy-6-methoxyflavanone*
263. 5,7,2′,6′-Tetrahydroxyflavanone*
264. 7,2′,6′-Trihydroxy-5-methoxyflavanone*
265. 5,2′-Dihydroxy-6,7,6′-trimethoxyflavanone*
266. 5,2′-Dihydroxy-7,8,6′-trimethoxyflavanone*
267. 5,7,3′,4′,5′-Pentahydroxyflavanone*
268. 3′-*O*-Glucoside*
269. 5,3′,4′-Trihydroxy-7,5′-dimethoxyflavanone*
270. 5,7,3′-Trihydroxy-4′,5′-dimethoxy-6,8-di-*C*-methylflavanone*
271. 5,6,7,3′-Tetramethoxy-4,5-methylenedioxyflavanone*
272. 5,6,7,8,3′-Pentamethoxy-4′,5′-methylenedioxyflavanone*
273. 5,6,7,2′,3′,4′,5′-Heptamethoxyflavanone*
274. 5-Hydroxy-7,3′,4′-trimethoxyflavanone
275. 5-Hydroxy-7,3′-dimethoxy-4′-prenyloxyflavanone
276. 5,7,3′,4′-Tetrahydroxy-6-*C*-methylflavanone 7-*O*-glucoside

277. 5, 7, 3′, 4′-Tetrahydroxy-6, 8-di-*C*-methylflavanone
278. 2, 5, 7, 3′, 4′-Pentahydroxyflavanone 5-*O*-glucoside
279. 5, 7, 2′, 5′-Tetrahydroxyflavanone
280. 7-Hydroxy-5, 2′, 4′-trimethoxyflavanone
281. 5, 7, 2′, 4′-Tetramethoxyflavanone
282. 5, 3′, 4′-Trihydroxy-7-methoxy-8-*C*-prenylflavanone
283. 5, 7, 2′-Trihydroxy-4′-methoxy-6, 8-di-*C*-methylflavanone
284. 5, 2′, 4′-Trihydroxy-8-*C*-prenyl-6″, 6″-dimethylpyrano(2″, 3″, 7, 6)flavanone
285. 5-Hydroxy-6, 7, 8, 3′, 4′-pentamethoxyflavanone
286. 5, 6, 7, 8, 3′, 4′-Hexamethoxyflavanone
287. 5-Hydroxy-7, 3′, 4′, 5′-tetramethoxyflavanone
288. 5-Hydroxy-6, 7, 3′, 4′, 5′-pentamethoxyflavanone 5-*O*-rhamnoside
289. Artocarpanone
290. Breverin
291. Kurarinone
292. Norkurarinone
293. Remerin
294. Scaberin
295. Steppogenin
296. Stepposide
297. Flemichin-A
298. Flemichin-E
299. Louisfieserone-A
300. Louisfieserone-B
301. Purpurin
302. Uvarinol
303. (−)-Silyandrin
304. (+)-Silymonin
305. Silyherman
306. Neosilyhermin-A
307. Neosilyhermin-B
308. Kuwanon-F
309. Kuwanon-D
310. Sangganon-I
311. Sangganon-N
312. Sangganon-A
313. Sangganon-M
314. Sangganon-L
315. Sangganon-D
316. Sangganon-B
317. Sangganon-O
318. Sangganon-G
319. Kuwanon-L

Dihydroflavonols

1. 7-Hydroxydihydroflavonol
2. 6-Methoxy-7-hydroxydihydroflavonol
3. 5, 7-Dihydroxydihydroflavonol
4. 3-Acetate*
5. 5-*O*-Glucosylgalactoside*
6. 7-Hydroxy-5-methoxydihydroflavonol*
7. 5, 7-Dihydroxy-6-*C*-prenyldihydroflavonol*
8. 5, 7-Dihydroxy-6-*C*-geranyldihydroflavonol*
9. 5, 7-Dihydroxy-8-*C*-prenyldihydroflavonol*
10. 5, 7-Dihydroxy-6-methoxydihydroflavonol*
11. 5-Methoxy-6, 7-methylenedioxydihydroflavonol*
12. 5-Hydroxy-7-methoxydihydroflavonol
13. 3-Acetyl-5-hydroxy-7-methoxydihydroflavonol
14. 5, 7-Dimethoxydihydroflavonol
15. 5, 7-Dihydroxy-6-methoxydihydroflavonol
16. 5, 7-Dihydroxy-6-*C*-methyldihydroflavonol
17. 6″, 6″-Dimethylpyrano(2″, 3″, 7, 8)-dihydroflavonol
18. 7, 4′-Dihydroxydihydroflavonol
19. 3-*O*-Glucoside
20. 7, 8, 4′-Trihydroxydihydroflavonol
21. 4′-Hydroxy-7-methoxydihydroflavonol
22. 7-Hydroxy-4′-methoxydihydroflavonol 7-*O*-glucosylxyloside*
23. 7, 4′-Dihydroxy-8, 3′, 5′-tri-*C*-prenyldihydroflavonol*
24. 5, 7, 4′-Trihydroxydihydroflavonol
25. 3-*O*-Galactoside*
26. 3-*O*-Glucoside
27. 3-*O*-Rhamnoside
28. 7-*O*-Glucoside
29. 3-*O*-Acetate*
30. 4′-*O*-Xyloside
31. 5, 7-Dihydroxy-4′-methoxydihydroflavonol
32. 5, 4′-Dihydroxy-7-methoxydihydroflavonol*
33. 3-*O*-Acetate*
34. 4′-Hydroxy-5, 7-dimethoxydihydroflavonol
35. 5-Hydroxy-7, 4′-dimethoxydihydroflavonol
36. 3-*O*-Rhamnoside*
37. 5-*O*-Glucoside
38. 5, 4′-Dihydroxy-7-*O*-prenyldihydroflavonol*
39. 5, 7, 8, 4′-Tetrahydroxydihydroflavonol 8-*O*-glucoside*
40. 5, 7, 4′-Trihydroxy-6-*C*-methyldihydroflavonol 7-*O*-glucoside*
41. 7, 4′-Dihydroxy-5-methoxy-6-*C*-methyldihydroflavonol*
42. 5, 7, 4′-Trihydroxy-6-*C*-geranyldihydroflavonol*
43. 5, 4′-Dihydroxy-6, 7-dimethoxydihydroflavonol
49. 5, 7-Dihydroxy-8, 4′-dimethoxydihydroflavonol 7-*O*-glucoside
45. 5, 4′-Dihydroxy-6-*C*-glucosyl-7-methoxydihydroflavonol
46. 5, 7, 4′-Trihydroxy-6-(3-hydroxy-3-methylbutyl)dihydroflavonol 7-*O*-glucoside
47. 5, 7, 4′-Trihydroxy-8-(3-hydroxy-3-methylbutyl)dihydroflavonol 7-*O*-glucoside

48. 5, 7, 4′-Trihydroxy-8-(3-glucosyloxy-3-methylbutyl)dihydroflavonol 7-*O*-glucoside
49. 5, 7, 4′-Trihydroxy-8-*C*-prenyldihydroflavonol
50. 5, 4′-Dihydroxy-6″, 6″-dimethylpyrano (2″, 3″, 7, 8)dihydroflavonol
51. 5, 4′-Dihydroxy-8-*C*-prenyl-6″, 6″-dimethylpyrano(2″, 3″, 7, 6)dihydroflavonol
52. 7, 3′, 4′-Trihydroxydihydroflavonol
53. 3-*O*-Glucoside
54. 3, 7-Di-*O*-Glucoside
55. 7, 4′-Dihydroxy-5-methoxy-6-*C*-geranyldihydroflavonol*
56. 7, 3′, 4′-Trihydroxydihydroflavonol
57. 7, 8, 3′, 4′-Tetrahydroxydihydroflavonol
58. 7, 3′, 4′-Trihydroxy-3-*O*-methyldihydroflavonol
59. 5, 7, 3′, 4′-Tetrahydroxydihydroflavonol
60. 7-*O*-Glucoside*
61. 5-*O*-Glucoside*
62. 5-*O*-Galactoside*
63. 3-*O*-Rhamnoside
64. 3-*O*-Glucoside
65. 3-*O*-Glucoside-6″-*O*-phenylacetate*
66. 3-*O*-Glucoside-6″-gallate*
67. 3-*O*-Glucosylrhamnoside*
68. 3-*O*-Acetate*
69. 2-*O*-Galactoside
70. 3-*O*-Xyloside
71. 4′-*O*-Glucoside
72. 7-*O*-Galactoside
73. 3, 5-Di-*O*-Rhamnoside
74. 5, 7, 4′-Trihydroxy-3′-methoxydihydroflavonol*
75. 5, 3′, 4′-Trihydroxy-7-methoxydihydroflavonol*
76. 3-*O*-Glucoside
77. 3-*O*-Acetate*
78. 5, 7, 3′-Trihydroxy-4′-methoxydihydroflavonol*
79. 3-*O*-Rhamnosylglucoside*
80. 5, 3′-Dihydroxy-7, 4′-dimethoxydihydroflavonol*
81. 5, 4′-Dihydroxy-7, 3′-dimethoxydihydroflavonol*
82. 3, 5, 6-Trimethoxy[furano(2″, 3″, 7, 8)]3′, 4′-methylenedioxydihydroflavonol*
83. 5, 7, 3′, 4′-Tetrahydroxy-7-*O*-prenyldihydroflavonol*
84. 5, 7, 4′-Trihydroxy-3′-methoxy-8-*C*-prenyldihydroflavonol*
85. 5, 7, 3′, 4′-Tetrahydroxy-6-*C*-geranyldihydroflavonol*
86. 5, 7, 3′-Trihydroxy-4′-methoxy-6-*C*-geranyldihydroflavonol*
87. 7, 8-Dihydrooxepinodihydroquercetin*
88. 5, 7, 2′, 4′-Tetrahydroxydihydroflavonol*
89. 5, 7, 3′, 4′-Tetrahydroxy-3-*O*-methyldihydroflavonol*
90. 5, 7, 3′, 4′-Tetrahydroxy-6-*C*-methyldihydroflavonol
91. 5, 3′-Dihydroxy-7, 4′-dimethoxy-8-*C*-prenyldihydroflavonol
92. 5, 7, 2′-Trihydroxy-5′-methoxydihydroflavonol
93. 5, 7-Dihydroxy-2′-methoxy-6″, 6″-dimethylpyrano(2″, 3″, 4′, 3′)dihydroflavanol
94. 7, 3′, 4′, 5′-Tetrahydroxydihydroflavonol
95. 7, 3′, 5′-Trihydroxy-4′-methoxydihydroflavonol
96. 5, 7, 3′, 4′, 5′-Pentahydroxy-6-*C*-methyldihydroflavonol
97. 3′-*O*-Glucoside
98. 5, 7, 3′, 4′-Tetrahydroxy-5′-methoxydihydroflavonol 4′-*O*-rhamnoside
99. 5, 7, 4′-Trihydroxy-3′-5′-dimethoxydihydroflavonol
100. 5, 7, 2′, 4′-Tetrahydroxy-6, 8-di-*C*-prenyldihydroflavonol*
101. 7, 2′, 4′-Trihydroxy-5-methoxy-8-*C*-lavandulyldihydroflavonol (2*R*, 3*R*)*
102. 7, 2′, 4′-Trihydroxy-5-methoxy-8-*C*-lavandulyldihydroflavonol (2*R*, 3*S*)*
103. 5, 7, 2′, 4′-Tetrahydroxy-6-*C*-prenyl-8-*C*-lavandulyldihydroflavonol*
104. 7, 2′, 4′-Trihydroxy-5-methoxy-8-*C*-(2-isopropenyl-5-methyl-5-hydroxyhexyl)dihydroflavonol*
105. 5, 7, 3′, 4′, 5′-Pentahydroxydihydroflavonol*
106. 3′-*O*-Glucoside*
107. 3-*O*-Rhamnoside*
108. 5, 7, 3′, 5′-Tetrahydroxy-4′-methoxydihydroflavonol*
109. 5, 7, 3′, 4′, 5′-Pentahydroxy-6, 8-di-*C*-methyldihydroflavonol*
110. 5, 7, 2′, 4′, 5′-Pentahydroxy-3-*O*-gallate-2′-sulfate*

(10) MISCELLANEOUS FLAVONOIDS

Diarylpropanes

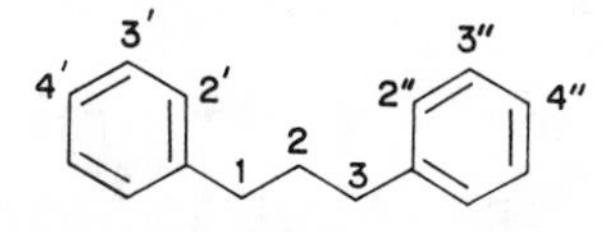

1. Propterol (2, 4′, 4″-triOH)
2. Broussonin C (2′, 4′, 4″-triOH, 3″-pren)
3. Broussonin A (2′, 4″-diOH, 4′-OMe)
4. Broussonin B (4′, 4″-diOH, 2′-OMe)
5. Broussonin D (2, 2′, 4″-triOH, 4′-OMe)
6. Kazinol F (2′, 4′, 4″-triOH, 2″, 3″-dipren)
7. Kazinol C (2′, 4′, 4″, 5″-tetraOH, 2″, 3″, 5′-tripren)
8. Kazinol J (4′, 4″, 5″-triOH, 2′-OMe, 2″, 3″-dipren)
9. Kazinol G (2′, 4′, 4″-triOH, 2″, 3″, 5′-tripren, 5″-O-pren)

10. 3″,4′-Dihydroxy-2′,4″-dimethoxydiphenylpropane
11. Broussonin F (4′,4″-diOH, 2′,3″-diOMe)
12. Broussonin E (2′,3″-diOH, 4′,4″-diOMe)
13. 2′,4″-Dihydroxy-3″,4′-dimethoxydiphenylpropane
14. 2′,4′-Dihydroxy-3″,4″-methylenedioxydiphenylpropane
15. Virolane (2′-OH, 4′-OMe, 3″,4″-OCH_2O-)
16. 2″-Hydroxy-4′-methoxy-4″,5″-methylenedioxydiphenylpropane
17. Virolanol B (2,2′,4′,4″-tetraOH, 3″-OMe)
18. Virolanol C (2,2′,4″-triOH, 3″,4′-diOMe)
19. Virolanol (2,2′-diOH, 4′-OMe, 3″,4″-OCH_2O-)
20. 4″,6′-Dihydroxy-2′,3″,4′-trimethoxydiphenylpropane
21. 2′,4′-Dihydroxy-6′-methoxy-3″,4″-methylenedioxydiphenylpropane
22. 2′,2″,4′-Trihydroxy-3′,5′-dimethyl-3″,4″-methylenedioxydiphenylpropane
23. 2″,4′-Dihydroxy-2′-methoxy-5′-methyl-3″,4″-methylenedioxydiphenylpropane
24. 2′,4′-Dihydroxy-2″-methoxy-3′-methyl-3″,4″-methylenedioxydiphenylpropane
25. 6′-Hydroxy-2′,4′-dimethoxy-3″,4″-methylenedioxydiphenylpropane
26. 4′,4″,6′-Trihydroxy-2′,3″-dimethoxydiphenyl-2-propanol
27. Kazinol D
28. Kazinol M
29. Kazinol N
30. Kazinol K
31. Kazinol K
32. Viorolaflorin
33. Spiroelliptin

(pren = CH:$CHCMe_2$ or CMe_2CH:CH_2, structures 27–33 are given in Chapter 10).

Cinnamylphenols

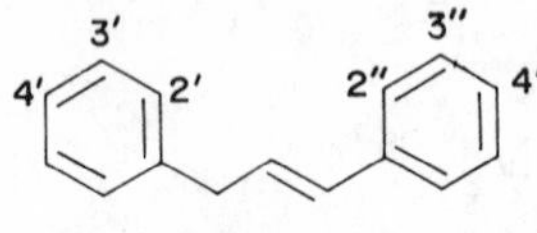

1. Obtusastyrene (4′-OH)
2. Obtustyrene (4′-OH, 2′-OMe)
3. Mucronustyrene (4′-OH, 2′,3′-diOMe)
4. 4′,5′-Dihydroxy-2′-methoxycinnamylbenzene
5. Violastyrene (4′-OH, 2′,5′-diOMe)
6. Isoviolastyrene (5′-OH, 2′,4′-diOMe)
7. Xenognosin (4′,4″-diOH, 2′-OMe)
8. Mucronulastyrene (2″,4′-diOH, 2′,3′-diOMe)
9. Villostyrene (4′-OH, 2′,2″,3′-triOMe)
10. Petrostyrene (2″,5′-diOH, 2′,3′,4′-triOMe, *trans*)
11. Kuhlmannistyrene (2″,5′-diOH, 2′,3′,4′-triOMe, *cis*)
12. Obtusaquinone (5′-OH, 2′-OMe, 4″-quinone)

Homoisoflavonoids

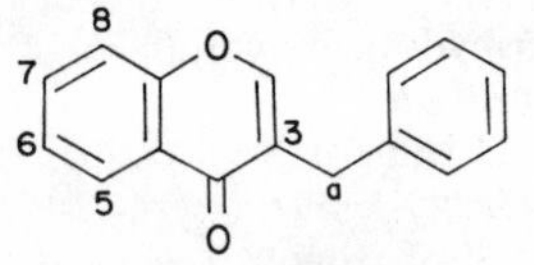

1. 7-Hydroxy-3-(4-hydroxybenzyl)chroman-4-one
2. 4′-Demethyl-3,9-dihydroeucomin
3. 5-Methyl-4′-demethyl-3,9-dihydroeucomin
4. 3,9-Dihydroeucomin
5. 7-Methyl-3,9-dihydroeucomin
6. 3′-Hydroxy-3,9-dihydroeucomin
7. 5,7-Dihydroxy-6-methyl-3-(4-hydroxybenzyl)chroman-4-one
8. Ophiopogonanone B
9. Ophiopogonanone A
10. Methylophiopogonanone B
11. Methylophiopogonanone A
12. 3,9-Dihydroeucomnalin
13. 8-Demethyl-7-methyl-3,9-dihydropunctatin
14. 8-Demethyl-8-acetyl-7-methyl-3,9-dihydropunctatin
15. 3,9-Dihydropunctatin
16. 7-Methyl-3,9-dihydropunctatin
17. 4′-Methyl-3,9-dihydropunctatin
18. Muscomin
19. 6-Formylisoophiopogonanone B
20. 7-Methyl-6-formylisoophiopogonanone B
21. 6-Formylisoophiopogonanone A
22. 7-Methyl-6-formylisoophiopogonanone A

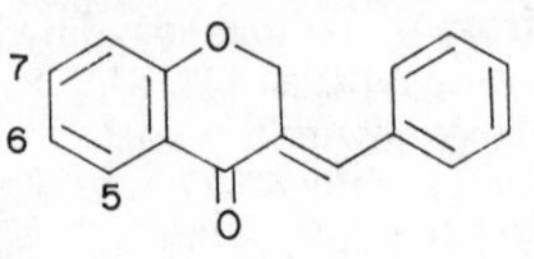

23. Bonducellin
24. 8-Methoxybonducellin
25. (*E*)-4′-Dimethyleucomin
26. (*E*)-Eucomin (5,7-diOH, 4′-OMe)
27. (*E*)-7-Methyleucomin
28. (*E*)-Eucomnalin (5,7,4′-triOH, 6-OMe)
29. (*E*)-Punctatin (5,7,4′-triOH, 8-OMe)
30. (*E*)-4′-Methylpunctatin

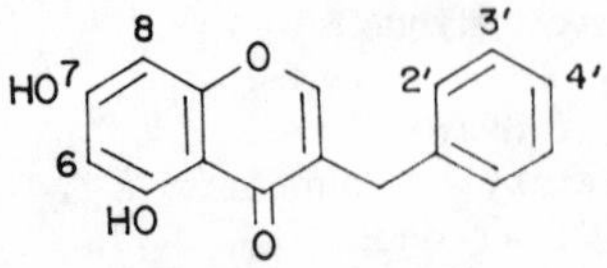

31. Ophiopogonone B (6-Me, 4′-OMe)
32. Ophiopogonone A (6-Me, 3′4′-OCH_2O-)
33. Desmethylisoophiopogonone B
34. Isoophiopogonone A
35. Methylophiopogonone B
36. Methylophiopogonone A

Miscellaneous

37. 7-Hydroxy-3-(4-hydroxybenzyl)chroman
38. 7-Hydroxy-8-methoxy-3-(4-hydroxybenzyl)chroman
39. Demethyleucomol
40. Eucomol
41. 7-Methyleucomol
42. (*Z*)-Eucomin
43. (*Z*)-4′-Methylpunctatin
44. Muscomosin
45. Scillascillin
46. 2-Hydroxy-7-methylscillascillin
47. Comosin

(11) ALPHABETICAL KEY TO TRIVIAL NAMES OF FLAVONES AND FLAVONOLS

Abrectorin	7, 3′-OH, 6, 4′-OMe-flavone
Acacetin	5, 7-OH, 4′-OMe-flavone
Acerosin	5, 7, 3′-OH, 6, 8, 4′-OMe-flavone
Agecorynin-C	5, 6, 7, 8, 2′, 4′, 5′-OMe-flavone
Agecorynin-D	5, 2′, 4′-OH, 6, 7, 8, 5′-OMe-flavone
Agehoustin A	5, 6, 7, 8, 2′, 3′, 4′, 5′-OMe-flavone
Agehoustin B	5, 6, 7, 2′, 3′, 4′, 5′-OMe-flavone
Agehoustin C	3′-OH, 5, 6, 7, 8, 2′, 4′, 5′-OMe-flavone
Agehoustin D	5, 3′-OH, 6, 7, 8, 2′, 4′, 5′-OMe-flavone
Aliarin	5, 7, 4′-OH, 3, 6-OMe, 3′-C_5-OH-flavone
Alluaudiol	5, 7, 3′, 4′, 5′-OH, 3-OMe, 6-Me-flavone
Alnetin	5-OH, 6, 7, 8-OMe-flavone
Alnusin	3, 5, 7-OH, 6-OMe-flavone
Alnustin	5-OH, 3, 6, 7-OMe-flavone
Anhydrotephrostachin	(**7.18**)
Annulatin	5, 7, 3′, 4′, 5′-OH, 3-OMe-flavone
Apigenin	5, 7, 4′-OH-flavone
Apuleidin	5, 2′, 3′-OH, 3, 7, 4′-OMe-flavone
Apulein	2′, 5′OH, 3, 5, 6, 7, 4′-OMe-flavone
Apuleirin	6, 5′-OH, 3, 5, 7, 3′, 4′-OMe-flavone
Apuleisin	5, 6, 2′, 3′-OH, 3, 7, 4′-OMe-flavone
Apuleitrin	5, 6, 5′-OH, 3, 7, 3′, 4′-OMe-flavone
Araneol	5, 7-OH, 3, 6, 8-OMe-flavone
Araneosol	5, 7-OH, 3, 6, 8, 4′-OMe-flavone
Arcapillin	5, 2′, 4′-OH, 6, 7, 5′-OMe-flavone
Artemetin	5-OH, 3, 6, 7, 3′, 4′-OMe-flavone
Artemisetin = artemetin	
Artobilichromene	5, 3′, 4′, 6′-OH, 5′-C_5, 7, 6-ODmp-flavone
Artocarpesin	5, 7, 2′, 4′-OH, 6-C_5-flavone
Artocarpetin	5, 2′, 4′-OH, 7-OMe-flavone
Artocarpin	5, 2′, 4′-OH, 7-OMe, 3, 6-C_5-flavone
Asplenetin	5, 7, 3′, 4′, 5′-OH, 3-C_5-flavone
Auranetin	3, 6, 7, 8, 4′-OMe-flavone
Axillarin	5, 7, 3′, 4′-OH, 3, 6-OMe-flavone
Ayanin	5, 3′-OH, 3, 7, 4′-OMe-flavone
Azaleatin	3, 7, 3′, 4′-OH, 5-OMe-flavone
Baicalein	5, 6, 7-OH-flavone
Batatifolin = nodifloretin	
Betuletol	3, 5, 7-OH, 6, 4′-OMe-flavone
Bonanzin	5, 7-OH, 3, 6, 3′, 4′-OMe-flavone
Brickellin	5, 2′-OH, 3, 6, 7, 4′, 5′-OMe-flavone
Broussoflavonol A	5, 3′, 4′-OH, 3-OMe, 8-C_5, 7, 6-ODmp-flavone
Broussoflavonol B	5, 7, 3′, 4′-OH, 3-OMe, 6, 8-C_5-flavone
Broussoflavonol C	3, 5, 7, 3′, 4′-OH, 8, 2′, 6′-C_5-flavone
Broussoflavonol D	3, 5, 7, 4′-OH, 8, 6′-C_5, 3′, 2′-ODmp-flavone (**7.20**)
Bucegin	5, 7-OH, 8, 4′-OMe-flavone
Calycopterin	5, 4′-OH, 3, 6, 7, 8-OMe-flavone

Candidol	3, 4′-OH, 5, 6, 7-OMe-flavone
Caryatin	7, 3′, 4′-OH, 3, 5-OMe-flavone
Casticin	5, 3′-OH, 3, 6, 7, 4′-OMe-flavone
Centaureidin	5, 7, 3′-OH, 3, 6, 4′-OMe-flavone
Cerosillin	5, 6, 3′, 5′-OMe-flavone
Cerosillin B	5,6, 3′, 4′, 5′-OMe-flavone
Chlorflavonin	5, 2′-OH, 3, 7, 8-OMe, 3′-chloroflavone
6-Chloroapigenin	5, 7, 4′-OH, 6-chloroflavone
Chrysin	5, 7-OH-flavone
Chrysoeriol	5, 7, 4′-OH, 3′-OMe-flavone
Chrysosplenetin	5, 4′-OH, 3, 6, 7, 3′-OMe-flavone
Chrysosplenol C	5, 6, 4′-OH, 3, 7, 3′-OMe-flavone
Chrysosplenol D	5, 3′, 4′-OH, 3, 6, 7-OMe-flavone
Chrysosplin	5, 4′-OH, 3, 6, 7, 2′-OMe-flavone
Cirsilineol	5, 4′-OH, 6, 7, 3′-OMe-flavone
Cirsiliol	5, 3′, 4′-OH, 6, 7-OMe-flavone
Cirsimaritin	5, 4′-OH, 6, 7-OMe-flavone
Combretol	5-OH, 3, 7, 3′, 4′, 5′-OMe-flavone
Conyzatin	5, 7-OH, 3, 8, 3′, 4′, 5′-OMe-flavone
Corniculatusin	3, 5, 7, 3′, 4′-OH, 8-OMe-flavone
Corymbosin	5-OH, 7, 3′, 4′, 5′-OMe-flavone
Cycloartocarpesin	5, 2′, 4′-OH, 7, 6-ODmp-flavone
Cycloartocarpin	5, 4′-OH, 7-OMe, 6-C_5, 6′, 3-ODmp-flavone
Cycloheterophyllin	5, 3′, 4′-OH, 8-C_5, 7, 6-ODmp, 6′, 3-ODmp-flavone
Cyclomorusin	5, 4′-OH, 7, 8-ODmp, 6′, 3-ODmp-flavone
Cyclomulberrin	5, 7, 4′-OH, 6-C_5, 6′, 3-ODmp-flavone
Cyclomulberrochromene	5, 4′-OH, 7, 6-ODmp, 6′, 3-ODmp-flavone
Datin	3, 5, 2′-OH, 7-OMe-flavone
Datiscetin	3, 5, 7, 2′-OH-flavone
Demethoxykanungin	3, 7-OMe, 3′, 4′-O_2CH_2-flavone
5-O-Demethylapulein	2′, 5′-OH, 3, 6, 7, 4′-OMe-flavone
Desmethoxycentaureidin	5, 7, 3′-OH, 6, 4′-OMe-flavone
5-Desmethoxynobiletin	5-OH, 6, 7, 8, 3′, 4′-OMe-flavone
Desmethoxysudachitin	5, 7, 4′-OH, 6, 8-OMe-flavone
8-Desmethyleucalyptin	5-OH, 7, 4′-OMe, 6-Me-flavone
Desmodol	5, 3′, 4′-OH, 6-Me, 7, 8-ODmp-flavone
Digicitrin	5, 3′-OH, 3, 6, 7, 8, 4′, 5′-OMe-flavone
Dillenetin	3, 5, 7-OH, 3′, 4′-OMe-flavone
Dinatin = hispidulin	
Diosmetin	5, 7, 3′-OH, 4′-OMe-flavone
Dumosol	3, 5, 7, 3′, 5′-OH, 4′-OMe, 6-Me-flavone
Echioidinin	5, 2′-OH, 7-OMe-flavone
Emmaosunin	5-OH, 3, 6, 7, 8, 3′-OMe-flavone
Erianthin = artemetin	
Eriostemin	3, 8-OH, 5, 6, 7, 4′-OMe-flavone
Ermanin	5, 7-OH, 3, 4′-OMe-flavone
Eucalyptin	5-OH, 7, 4′-OMe, 6, 8-Me-flavone
Eupafolin = nepetin	
Eupalestin	5, 6, 7, 8, 5′-OMe, 3′, 4′-O_2CH_2-flavone
Eupalitin	3, 5, 4′-OH, 6, 7-OMe-flavone
Eupatilin	5, 7-OH, 6, 3′, 4′-OMe-flavone
Eupatin	3, 5, 3′-OH, 6, 7, 4′-OMe-flavone
Eupatolitin	3, 5, 3′, 4′-OH, 6, 7-OMe-flavone
Eupatoretin	3, 3′-OH, 5, 6, 7, 4′-OMe-flavone
Eupatorin	5, 3′-OH, 6, 7, 4′-OMe-flavone
Europetin	3, 5, 3′, 4′, 5′-OH, 7-OMe-flavone
Exoticin	3, 5, 6, 7, 8, 3′, 4′, 5′-OMe-flavone
Fastigenin = cirsilineol	
Fisetin	3, 7, 3′, 4′-OH-flavone
Flindulatin	5-OH, 3, 7, 8, 4′-OMe-flavone
Fulvinervin B	5-OH, 6-C_5, 7, 8-ODmp-flavone

Fulvinervin C	5-OH, 6-C_5-OH, 7, 8-ODmp-flavone
Galangin	3, 5, 7-OH-flavone
Gamatin	5-OMe, 3′, 4′-O_2CH_2, 7, 8-fur-flavone
Ganhuangenin	5, 7, 3′, 6′-OH, 8, 2′-OMe-flavone
Gardenin A	5-OH, 6, 7, 8, 3′, 4′, 5′-OMe-flavone
Gardenin B	5-OH, 6, 7, 8, 4′-OMe-flavone
Gardenin C	5, 3′-OH, 6, 7, 8, 4′, 5′-OMe-flavone
Gardenin D	5, 3′-OH, 6, 7, 8, 4′-OMe-flavone
Gardenin E	5, 3′, 5′-OH, 6, 7, 8, 4′-OMe-flavone
Genkwanin	5, 4′-OH, 7-OMe-flavone
Geraldol	3, 7, 4′-OH, 3′-OMe-flavone
Geraldone	7, 4′-OH, 3′-OMe-flavone
Glepidotin A	3, 5, 7-OH, 8-C_5-flavone
Gnaphalin	5, 7-OH, 3, 8-OMe-flavone
Gomphrenol	3, 5, 4′-OH, 6, 7-O_2CH_2-flavone
Gossypetin	3, 5, 7, 8, 3′, 4′-OH-flavone
Herbacetin	3, 5, 7, 8, 4′-OH-flavone
Hibiscetin	3, 5, 7, 8, 3′, 4′, 5′-OH-flavone
Hispidulin	5, 7, 4′-OH, 6-OMe-flavone
6-Hydroxygalangin	(3, 5, 6, 7-OH-flavone)
8-Hydroxygalangin	(3, 5, 7, 8-OH-flavone)
6-Hydroxykaempferol	3, 5, 6, 7, 4′-OH-flavone
6-Hydroxyluteolin	5, 6, 7, 3′, 4′-OH-flavone
6-Hydroxymyricetin	(3, 5, 6, 7, 3′, 4′, 5′-OH-flavone)
Hymenoxin	5, 7-OH, 6, 8, 3′, 4′-OMe-flavone
Hypolaetin	5, 7, 8, 3′, 4′-OH- flavone
Icaritin	3, 5, 7-OH, 4′-OMe, 8-C_5-OH-flavone
Integrin	5, 2′, 4′-OH, 7-OMe, 3-C_5-flavone
Isoanhydroicaritin	3, 5, 4′-OH, 7-OMe, 8-C_5-flavone
Isoëtin	5, 7, 2′, 4′, 5′-OH-flavone
Isognaphalin	5, 8-OH, 3, 7-OMe-flavone
Isokaempferide	5, 7, 4′-OH, 3-OMe-flavone
Isolimocitrol	3, 5, 7, 3′-OH, 6, 8, 4′-OMe-flavone
Isopongachromene	6-OMe, 7, 8-ODmp, 3′, 4′-O_2CH_2-flavone (**7.21**).
Isopongaflavon	5-OMe, 7, 8-ODmp-flavone
Isopongaglabol	4′-OH, 7, 8-fur-flavone
Isorhamnetin	3, 5, 7, 4′-OH, 3′-OMe-flavone
Isoscutellarein	5, 7, 8, 4′-OH-flavone
Isosinensetin	5, 7, 8, 3′, 4′-OMe-flavone
Isounonal	5, 7-OH, 6-Me, 8-CHO-flavone
Izalpinin	3, 5-OH, 7-OMe-flavone
Jaceidin	5, 7, 4′-OH, 3, 6, 3′-OMe-flavone
Jaceosidin	5, 7, 4′-OH, 6, 3′-OMe-flavone
Jaranol = kumatakenin	
Kaempferide	3, 5, 7-OH, 4′-OMe-flavone
Kaempferol	3, 5, 7, 4′-OH-flavone
Kalmiatin	5-OH, 3, 7, 4′-OMe, 6, 8-Me-flavone
Kanugin	3, 7, 3′-OMe, 3′, 4′-O_2CH_2-flavone
Kanzakiflavon-1	5, 8-OH, 4′-OMe, 6, 7-O_2CH_2-flavone
Kanzakiflavon-2	5, 4′-OH, 6, 7-O_2CH_2-flavone
Karanjachromene	3-OMe, 7, 8-ODmp-flavone
Karanjin	3-OMe, 7, 8-fur-flavone
Kumatakenin	5, 4′-OH, 3, 7-OMe-flavone
Kuwanone A	5, 7, 4′-OH, 3-C_5, 2′, 3′-ODmp-flavone
Kuwanone B	5, 7, 2′-OH, 3-C_5-3′, 4′-ODmp-flavone
Kuwanone C	5, 7, 2′, 4′-OH, 3, 8-C_5-flavone
Kuwanone G	(**7.26**)
Kuwanone H	(**7.27**)
Kuwanone K	(**7.31**)
Kuwanone M	(**7.30**)
Kuwanone N	(**7.32**)
Kuwanone S	5, 7, 4′-OH, 3′-C_{10}-flavone
Kuwanone T	5, 7, 2′, 4′-OH, 3, 3′-C_5-flavone
	(**7.19**)
Kuwanone W	(**7.28**)
Laciniatin	3, 5, 7, 3′-OH, 6, 4′-OMe-flavone
Ladanein	5, 6-OH, 7, 4′-OMe-flavone
Lanceolatin B	7, 8-fur-flavone
Laricitrin	3, 5, 7, 4′, 5′-OH, 3′-OMe-flavone
Leucanthoflavon	5, 8, 3′, 4′-OH, 6, 7-OMe-flavone
Limocitrin	3, 5, 7, 4′-OH, 8, 3′-OMe-flavone
Limocitrol	3, 5, 7, 4′-OH, 6, 8, 3′-OMe-flavone
Linderoflavone A	5, 7-OH, 6, 8-OMe, 3′, 4′-O_2CH_2-flavone
Linderoflavone B	5, 6, 7, 8-OMe, 3′, 4′-O_2CH_2-flavone

Lucidin = linderoflavone A

Luteolin	5, 7, 3′, 4′-OH-flavone
Macaflavon I	3, 3′, 4′-OH, 7, 6-ODmp, 8-C_5-flavone
Macaflavon II	3′, 4′-OH, 3-OMe, 7, 6-OMe, 7, 6-ODmp, 8-C_5-flavone
Mearnsetin	3, 5, 7, 3′, 5′-OH, 4′-OMe-flavone
Melanoxetin	3, 7, 8, 3′, 4′-OH-flavone
Melibentin	3, 5, 6, 7, 8-OMe, 3′, 4′-O_2CH_2-flavone
Melinervin	3, 5, 7-OH, 6, 8-OMe, 3′, 4′-O_2CH_2-flavone
Melisimplexin	3, 5, 6, 7-OMe, 3′, 4′-O_2CH_2-flavone
Melisimplin	5-OH, 3, 6, 7-OMe, 3′, 4′-O_2CH_2-flavone
Meliternatin	3, 5-OMe, 6, 7-O_2CH_2, 3′, 4′-O_2CH_2-flavone
Meliternin	3, 5, 7, 8-OMe, 3′, 4′-O_2CH_2-flavone
Methylgnaphalin	5-OH, 3, 7, 8-OMe-flavone
Mikanin	3, 5-OH, 6, 7, 4′-OMe-flavone
Moracein D	(**7.29**)
Morin	3, 5, 7, 2′, 4′-OH-flavone
Morusin	5, 2′, 4′-OH, 3-C_5, 7, 8-ODmp-flavone
Mulberranol	5, 4′, 6′-OH, 3-C_5, 7, 6-OIPf-OH-flavone

Mulberrin = norartocarpin

Mulberrochromene	5, 2′, 4′-OH, 3-C_5, 7, 6-ODmp-flavone
Murrayanol	5, 4′-OH, 3, 6, 7, 3′, 5′-OMe-flavone
Myricetin	3, 5, 7, 3′, 4′, 5′-OH-flavone
Natsudaidain	3-OH, 5, 6, 7, 8, 3′, 4′-OMe-flavone
Negletein	5, 6-OH, 7-OMe-flavone
Nepetin	5, 7, 3′, 4′-OH, 6-OMe-flavone
Nevadensin	5, 7-OH, 6, 8, 4′-OMe-flavone
Nobiletin	5, 6, 7, 8, 3′, 4′-OMe-flavone
Nodifloretin	5, 6, 7, 4′-OH, 3′-OMe-flavone
Noranhydroicaritin	3, 5, 7, 4′-OH, 8-C_5-flavone
Norartocarpetin	5, 7, 2′, 4′-OH-flavone
Norartocarpin	5, 7, 2′, 4′-OH, 3, 6-C_5-flavone
Noricaritin	3, 5, 7, 4′-OH, 8-C_5-OH-flavone
Norwightin	(5, 7, 8, 2′, 3′-OH-flavone)
Norwogonin	5, 7, 8-OH-flavone
Ombuin	3, 5, 3′-OH, 7, 4′-OMe-flavone
Onopordin	5, 7, 3′, 4′-OH, 8-OMe-flavone
Oroxylin-A	5, 7-OH, 6-OMe-flavone
Oxidihydroartocarpesin	5, 7, 2′, 4′-OH, 6-C_5-OH-flavone
Oxyayanin A	5, 2′, 5′-OH, 3, 7, 4′-OMe-flavone
Oxyayanin B	5, 6, 3′-OH, 3, 7, 4′-OMe-flavone
Oxydihydromorusin	5, 2′, 4′-OH, 3-C_5-OH, 7, 8-ODmp-flavone
Pachypodol	5, 4′-OH, 3, 7, 3′-OMe-flavone
Patuletin	3, 5, 7, 3′, 4′-OH, 6-OMe-flavone
Pectolinarigenin	5, 7-OH, 6, 4′-OMe-flavone
Pedalitin	5, 6, 3′, 4′-OH, 7-OMe-flavone
Pedunculin	5, 8-OH, 6, 7, 4′-OMe-flavone
Penduletin	5, 4′-OH, 3, 6, 7-OMe-flavone
Pilloin	5, 3′-OH, 7, 4′-OMe-flavone
Pinnatin	5-OMe, 7, 8-fur-flavone
Pinoquercetin	3, 5, 7, 3′, 4′-OH, 6-Me-flavone
Pityrogrammin	3, 5, 7-OH, 8-OMe, 6-Me-flavone
Pollenitin	3, 5, 8, 4′-OH, 7-OMe-flavone

Polycladin = chrysosplenetin

Pongachromene	3, 5-OMe, 3′, 4′-O_2CH_2-7, 8-ODmp-flavone

Pongaflavon = karanjachromene

Pongaglabol	5-OH, 7, 8-fur-flavone
Pongaglabrone	3′, 4′-O_2CH_2, 7, 8-fur-flavone
Pongapin	3-OMe, 3′, 4′-O_2CH_2, 7, 8-fur-flavone

Ponkanetin = tangeretin

Pratol	7-OH, 4′-OMe-flavone
Pratoletin	3, 5, 8, 4′-OH-flavone
Primetin	5, 8-OH-flavone
Primuletin	5-OH-flavone
Prosogerin A	7-OH, 6-OMe, 3′, 4′-O_2CH_2-flavone
Prosogerin C	6, 7, 3′, 4′, 5′-OMe-flavone
Prosogerin D	7-OH, 6, 3′, 4′, 5′-OMe-flavone
Prudomestin	3, 5, 7-OH, 8, 4′-OMe-flavone
Pseudosemiglabrinol	(**7.23**)
Ptaeroxylol	3, 5, 7-OH, 2′-OMe-flavone

Purpurascenin	3, 5, 6, 7, 8, 2′, 4′, 5′-OMe-flavone
Quercetagetin	3, 5, 6, 7, 3′, 4′-OH-flavone
Quercetin	3, 5, 7, 3′, 4′-OH-flavone
Ranupenin	3, 5, 8, 3′, 4′-OH, 7-OMe-flavone
Rehderianin I	5, 2′, 5′-OH, 7, 8-OMe-flavone
Retusin	5-OH, 3, 7, 3′, 4′-OMe-flavone
Rhamnazin	3, 5, 4′-OH, 7, 3′-OMe-flavone
Rhamnetin	3, 5, 3′, 4′-OH, 7-OMe-flavone
Rhamnocitrin	3, 5, 4′-OH, 7-OMe-flavone
Rhynchosin	3, 6, 7, 3′, 4′-OH-flavone
Rhynchospermin	3, 5, 3′-OH, 7, 4′-OMe, 8-C_5-flavone
Rivularin	5, 2′-OH, 7, 8, 6′-OMe-flavone
Robinetin	3, 7, 3′, 4′, 5′-OH-flavone
Rubraflavone A	7, 4′, 6′-OH, 3-C_{10}-flavone
Rubraflavone B	7, 4′, 6′-OH, 3-C_{10}, 8-C_5-flavone
Rubraflavone C	5, 7, 4′, 6′-OH, 3-C_{10}, 6-C_5-flavone
Rubraflavone D	5, 2′, 4′-OH, 3-C_{10}, 7, 6-ODmp-flavone
Saltillin	5-OH, 4′-OMe, 7-Me-flavone
Salvigenin	5-OH, 6, 7, 4′-OMe-flavone
Santin	5, 7-OH, 3, 6, 4′-OMe-flavone
Sarothrin	5, 7, 4′-OH, 3, 6, 8-OMe-flavone
Scaposin	5, 7, 5′-OH, 6, 8, 3′, 4′-OMe-flavone
Scutellarein	5, 6, 7, 4′-OH-flavone
Scutevulin	5, 7, 2′-OH, 8-OMe-flavone
Selgin	5, 7, 3′, 4′-OH, 5′-OMe-flavone
Sericetin	3, 5-OH, 8-C_5, 7, 6-ODmp-flavone
Serpyllin	5-OH, 7, 8, 2′, 3′, 4′-OMe-flavone
Sexangularetin	3, 5, 7, 4′-OH, 8-OMe-flavone
Sideritiflavon	5, 3′, 4′-OH, 6, 7, 8-OMe-flavone
Sideroxylin	5, 4′-OH, 7-OMe, 6, 8-Me-flavone
Sinensetin	5, 6, 7, 3′, 4′-OMe-flavone
Skullcapflavone I	5, 2′-OH, 7, 8-OMe-flavone
Skullcapflavone II	5, 2′-OH, 6, 7, 8, 6′-OMe-flavone
Sorbifolin	5, 6, 4′-OH, 7-OMe-flavone
Spinacetin	3, 5, 7, 4′-OH, 6, 3′-OMe-flavone
Strobochrysin	5, 7-OH, 6-Me-flavone
Sudachitin	5, 7, 4′-OH, 6, 8, 3′-OMe-flavone
Sylpin	5, 6, 4′-OH, 3-OMe, 8-Me-flavone
Syringetin	3, 5, 7, 4′-OH, 3′, 5′-OMe-flavone
Tabularin	5, 7-OH, 6, 2′, 4′, 5′-OMe-flavone
Takakin	5, 7, 8-OH, 4′-OMe-flavone
Tamarixetin	3, 5, 7, 3′-OH, 4′-OMe-flavone
Tambuletin	3, 5, 8, 3′-OH, 7, 4′-OMe-flavone
Tambulin	3, 5-OH, 7, 8, 4′-OMe-flavone
Tangeretin	5, 6, 7, 8, 4′-OMe-flavone
Tectochrysin	5-OH, 7-OMe-flavone
Tephrinone	5-OH, 7-OMe, 8-C_5-flavone
Tephroglabrin	(**7.24**)
Tephrostachin	(**7.16**/**7.17**)
Tepurindiol	(**7.25**)
Ternatin	5, 4′-OH, 3, 7, 8, 3′-OMe-flavone
Thapsin = calycopterin	
Thevetiaflavon	7, 4′-OH, 5-OMe-flavone
Thymonin	5, 6, 4′-OH, 7, 8, 3′-OMe-flavone
Thymusin	5, 6, 4′-OH, 7, 8-OMe-flavone
Tithonine	3-OH, 7, 4′-OMe-flavone
Tomentin	5, 6, 3′, 4′-OH, 3, 7-OMe-flavone
Transilitin	7, 8, 3′, 4′-OH, 3-OMe-flavone
Tricetin	5, 7, 3′, 4′, 5′-OH-flavone
Tricin	5, 7, 4′-OH, 3′, 5′-OMe-flavone
Trimethoxywogonin	5, 6, 7-OH, 8, 3′, 4′, 5′-OMe-flavone
Unonal	5, 7-OH, 8-Me, 6-CHO-flavone
Velutin	5, 4′-OH, 7, 3′-OMe-flavone
Veronicafolin	3, 5, 4′-OH, 6, 7, 3′-OMe-flavone
Viscosol	5, 7-OH, 3, 6, 4′-OMe, 3′-C_5-flavone
Vitexicarpin = casticin	
Vogeletin	3, 6, 7, 4′-OH, 5-OMe-flavone

Wharangin	5, 3′, 4′-OH, 3-OMe, 7, 8-O_2CH_2-flavone
Wightin	5, 3′-OH, 7, 8, 2′-OMe-flavone
Wogonin	5, 7-OH, 8-OMe-flavone
Xanthomicrol	5, 4′-OH, 6, 7, 8-OMe-flavone
Zapotin	5, 6, 2′, 6′-OMe-flavone
Zapotinin	5-OH, 6, 2′, 6′-OMe-flavone

Plant species index

Subject index

Individual flavonoids are listed in the various check lists of known compounds in the Appendix. In this Appendix an alphabetical key to the trivial names of flavones and flavonols can also be found.